$115.95 per copy, (In United States).
Price subject to change without prior notice.

0025

RSMeans

Mechanical Cost Data

28th Annual Edition

2005

RSMeans
Construction Publishers & Consultants
63 Smiths Lane
Kingston, MA 02364-0800
(781) 585-7880

Printed in the United States of America
ISSN 0748-2698
ISBN 0-87629-753-X

Senior Editor
Melville J. Mossman, PE

Contributing Editors
Barbara Balboni
Robert A. Bastoni
John H. Chiang, PE
Robert J. Kuchta
Robert C. McNichols
Robert W. Mewis, CCC
John J. Moylan
Jeannene D. Murphy
Stephen C. Plotner
Michael J. Regan
Eugene R. Spencer
Marshall J. Stetson
Phillip R. Waier, PE

Senior Engineering Operations Manager
John H. Ferguson, PE

Senior Vice President & General Manager
John Ware

Product Manager
Jane Crudden

Vice President of Sales
John M. Shea

Production Manager
Michael Kokernak

Production Coordinator
Marion E. Schofield

Technical Support
Thomas J. Dion
Jonathan Forgit
Mary Lou Geary
Gary L. Hoitt
Alice McSharry
Paula Reale-Camelio
Robin Richardson
Kathryn S. Rodriguez
Sheryl A. Rose

Book & Cover Design
Norman R. Forgit

This book is recyclable.

The this book is printed on recycled stock.

First Printing

Foreword

RSMeans is a product line of Reed Construction Data, Inc., a leading provider of construction information, products, and services in North America and globally. Reed Construction Data's project information products include more than 100 regional editions, national construction data, sales leads, and local plan rooms in major business centers. Reed Construction Data's PlansDirect provides surveys, plans and specifications. The First Source suite of products consists of *First Source for Products,* SPEC-DATA™, MANU-SPEC™, CADBlocks, Manufacturer Catalogs and First Source Exchange (www.firstsourceexchange.com) for the selection of nationally available building products. Reed Construction Data also publishes ProFile, a database of more than 20,000 U.S. architectural firms. RSMeans provides construction cost data, training, and consulting services in print, CD-ROM and online. Reed Construction Data, headquartered in Atlanta, is owned by Reed Business Information (www.reedconstructiondata.com), a leading provider of critical information and marketing solutions to business professionals in the media, manufacturing, electronics, construction and retail industries. Its market-leading properties include more than 135 business-to-business publications, over 125 Webzines and Web portals, as well as online services, custom publishing, directories, research and direct-marketing lists. Reed Business Information is a member of the Reed Elsevier plc group (NYSE: RUK and ENL)—a world-leading publisher and information provider operating in the science and medical, legal, education and business-to-business industry sectors.

Our Mission

Since 1942, RSMeans has been actively engaged in construction cost publishing and consulting throughout North America.

Today, over 60 years after RSMeans began, our primary objective remains the same: to provide you, the construction and facilities professional, with the most current and comprehensive construction cost data possible.

Whether you are a contractor, an owner, an architect, an engineer, a facilities manager, or anyone else who needs a fast and reliable construction cost estimate, you'll find this publication to be a highly useful and necessary tool.

Today, with the constant flow of new construction methods and materials, it's difficult to find the time to look at and evaluate all the different construction cost possibilities. In addition, because labor and material costs keep changing, last year's cost information is not a reliable basis for today's estimate or budget.

That's why so many construction professionals turn to RSMeans. We keep track of the costs for you, along with a wide range of other key information, from city cost indexes . . . to productivity rates . . . to crew composition . . . to contractor's overhead and profit rates.

RSMeans performs these functions by collecting data from all facets of the industry, and organizing it in a format that is instantly accessible to you. From the preliminary budget to the detailed unit price estimate, you'll find the data in this book useful for all phases of construction cost determination.

The Staff, the Organization, and Our Services

When you purchase one of RSMeans' publications, you are in effect hiring the services of a full-time staff of construction and engineering professionals.

Our thoroughly experienced and highly qualified staff works daily at collecting, analyzing, and disseminating comprehensive cost information for your needs. These staff members have years of practical construction experience and engineering training prior to joining the firm. As a result, you can count on them not only for the cost figures, but also for additional background reference information that will help you create a realistic estimate.

The RSMeans organization is always prepared to help you solve construction problems through its five major divisions: Construction and Cost Data Publishing, Electronic Products and Services, Consulting Services, Insurance Services, and Educational Services.

Besides a full array of construction cost estimating books, Means also publishes a number of other reference works for the construction industry. Subjects include construction estimating and project and business management; special topics such as HVAC, roofing, plumbing, and hazardous waste remediation; and a library of facility management references.

In addition, you can access all of our construction cost data through your computer with RSMeans *CostWorks 2005* CD-ROM, an electronic tool that offers over 50,000 lines of Means detailed construction cost data, along with assembly and whole building cost data. You can also access Means cost information from our Web site at www.rsmeans.com

What's more, you can increase your knowledge and improve your construction estimating and management performance with a Means Construction Seminar or In-House Training Program. These two-day seminar programs offer unparalleled opportunities for everyone in your organization to get updated on a wide variety of construction-related issues.

RSMeans also is a worldwide provider of construction cost management and analysis services for commercial and government owners and of claims and valuation services for insurers.

In short, RSMeans can provide you with the tools and expertise for constructing accurate and dependable construction estimates and budgets in a variety of ways.

Robert Snow Means Established a Tradition of Quality That Continues Today

Robert Snow Means spent years building RSMeans, making certain he always delivered a quality product.

Today, at RSMeans, we do more than talk about the quality of our data and the usefulness of our books. We stand behind all of our data, from historical cost indexes to construction materials and techniques to current costs.

If you have any questions about our products or services, please call us toll-free at 1-800-334-3509. Our customer service representatives will be happy to assist you or visit our Web site at www.rsmeans.com

Table of Contents

Related Means Products and Services

Melville J. Mossman, P.E., Senior Editor of this cost data book, suggests the following Means products and services as **companion** information resources to *Mechanical Cost Data*:

Construction Cost Data Books

Electrical Cost Data 2005
Plumbing Cost Data 2005

Reference Books

Green Building: Project Planning & Cost Estimating
Mechanical Estimating, 3rd Edition
Plumbing Estimating, 2nd Edition
Cyberplaces 2nd Ed.: The Internet Guide for Architects, Engineers, Contractors & Facility Managers
Electrical Estimating, 3rd Edition
HVAC: Design Criteria, Options, Selection, 2nd Edition
HVAC Systems Evaluation
Understanding Building Automation Systems

Seminars and In-House Training

Mechanical & Electrical Estimating

Means on the Internet

Visit Means at **http://www.rsmeans.com.** The site contains useful ***interactive*** cost and reference material. Request or download ***FREE*** estimating software demos. Visit our bookstore for convenient ordering, plus learn more about new publications and companion products.

Means Data on CD-ROM

Get the information found in Means traditional cost books on Means new, easy to use CD-ROM, Means *CostWorks 2005*. Means *CostWorks* users can now enhance their Costlist with the Means *CostWorks Estimator*. For more information see the special full color brochure inserted in this book.

Means Consulting Services

Means Consulting Group provides a number of construction cost-related services including: Cost Estimates, Customized Databases, Benchmark Studies, Estimating Audits, Feasibility Studies, Litigation Support, Staff Assessments and Customized Training.

Means Data for Computer Estimating

Construction costs for software applications.
Over 25 unit price and assemblies cost databases are available through a number of leading estimating and facilities management software providers (listed below). For more information see the "yellow" pages at the back of this publication.

- 3D International
- 4Clicks-Solutions, LLC
- Aepco, Inc.
- Applied Flow Technology
- ArenaSoft Estimating
- ARES Corporation
- Barchan Associates
- BSD – Building Systems Design, Inc.
- CMS – Computerized Micro Solutions
- Corecon Technologies, Inc.
- CorVet Systems
- Discover Software
- Estimating Systems, Inc.
- Entech Engineering
- FAMIS Software, Inc.
- ISES Corporation
- Magellan K-12
- Maximus Asset Solutions
- MC^2 – Management Computer Controls
- Neddam Software Technologies
- Prime Time
- Quest Solutions, Inc.
- RIB Software (Americas), Inc.
- Shaw Beneco Enterprises, Inc.
- Schwab Technology Solutions, Inc.
- Timberline Software Corp.
- TMA Systems, LLC
- US Cost, Inc.
- Vanderweil Facility Advisors
- Vertigraph, Inc.

How the Book Is Built: An Overview

A Powerful Construction Tool

You have in your hands one of the most powerful construction tools available today. A successful project is built on the foundation of an accurate and dependable estimate. This book will enable you to construct just such an estimate.

For the casual user the book is designed to be:

- quickly and easily understood so you can get right to your estimate
- filled with valuable information so you can understand the necessary factors that go into the cost estimate

For the regular user, the book is designed to be:

- a handy desk reference that can be quickly referred to for key costs
- a comprehensive, fully reliable source of current construction costs and productivity rates, so you'll be prepared to estimate any project
- a source book for preliminary project cost, product selections, and alternate materials and methods

To meet all of these requirements we have organized the book into the following clearly defined sections.

How To Use the Book: The Details

This section contains an in-depth explanation of how the book is arranged . . . and how you can use it to determine a reliable construction cost estimate. It includes information about how we develop our cost figures and how to completely prepare your estimate.

Unit Price Section

All cost data has been divided into the 16 divisions according to the MasterFormat system of classification and numbering as developed by the Construction Specifications Institute (CSI) and Construction Specifications Canada (CSC). For a listing of these divisions and an outline of their subdivisions, see the Unit Price Section Table of Contents.

Estimating tips are included at the beginning of each division.

Division 17: Quick Project Estimates
In addition to the 16 Unit Price Divisions there is a S.F. (Square Foot) and C.F. (Cubic Foot) Cost Division, Division 17. It contains costs for 59 different building types that allow you to quickly make a rough estimate for the overall cost of a project or its major components.

Assemblies Section

The cost data in this section has been organized in an "Assemblies" format. These assemblies are the functional elements of a building and are arranged according to the 7 divisions of the UNIFORMAT II classification system. For a complete explanation of a typical "Assemblies" page, see "How To Use the Assemblies Cost Tables."

Reference Section

This section includes information on Reference Numbers, Change Orders, Crew Listings, Historical Cost Indexes, City Cost Indexes, Location Factors and a listing of Abbreviations. It is visually identified by a vertical gray bar on the edge of pages.

Reference Numbers: Following selected major classifications throughout the book are "reference numbers" shown in bold squares. These numbers refer you to related information in other sections.

In these other sections, you'll find related tables, explanations, and estimating information. Also included are alternate pricing methods and technical data, along with information on design and economy in construction. You'll also find helpful tips on estimating and construction.

Change Orders: This section includes information on the factors that influence the pricing of change orders.

Crew Listings: This section lists all the crews referenced. For the purposes of this book, a crew is composed of more than one trade classification and/or the addition of equipment to any trade classification. Power equipment is included in the cost of the crew. Costs are shown both with bare labor rates and with the installing contractor's overhead and profit added. For each, the total crew cost per eight-hour day and the composite cost per labor-hour are listed.

Historical Cost Indexes: These indexes provide you with data to adjust construction costs over time. If you know costs for a past project, you can use these indexes to estimate the cost to construct the same project today.

City Cost Indexes: Obviously, costs vary depending on the regional economy. You can adjust the "national average" costs in this book to 316 major cities throughout the U.S. and Canada by using the data in this section.

Location Factors, to quickly adjust the data to over 930 zip code areas, are included .

Abbreviations: A listing of the abbreviations and the terms they represent, is included.

Index

A comprehensive listing of all terms and subjects in this book to help you find what you need quickly.

The Scope of This Book

This book is designed to be comprehensive and easy to use. To that end we have made certain assumptions:

1. We have established material prices based on a "national average."
2. We have computed labor costs based on a 30-city "national average" of union wage rates.
3. We have targeted the data for projects of a certain size range.

For a more detailed explanation of how the cost data is developed, see "How To Use the Book: The Details."

Project Size

This book is aimed primarily at commercial and industrial projects costing $1,000,000 and up, or large multi-family housing projects. Costs are primarily for new construction or major renovation of buildings rather than repairs or minor alterations.

With reasonable exercise of judgment the figures can be used for any building work. *However, for civil engineering structures such as bridges, dams, highways, or the like, please refer to* ***Means Heavy Construction Cost Data.***

How to Use the Book: The Details

What's Behind the Numbers? The Development of Cost Data

The staff at RSMeans continuously monitors developments in the construction industry in order to ensure reliable, thorough and up-to-date cost information.

While *overall* construction costs may vary relative to general economic conditions, price fluctuations within the industry are dependent upon many factors. Individual price variations may, in fact, be opposite to overall economic trends. Therefore, costs are continually monitored and complete updates are published yearly. Also, new items are frequently added in response to changes in materials and methods.

Costs—$ (U.S.)

All costs represent U.S. national averages and are given in U.S. dollars. The Means City Cost Indexes can be used to adjust costs to a particular location. The City Cost Indexes for Canada can be used to adjust U.S. national averages to local costs in Canadian dollars.

Material Costs

The RSMeans staff contacts manufacturers, dealers, distributors, and contractors all across the U.S. and Canada to determine national average material costs. If you have access to current material costs for your specific location, you may wish to make adjustments to reflect differences from the national average. Included within material costs are fasteners for a normal installation. RSMeans engineers use manufacturers' recommendations, written specifications and/or standard construction practice for size and spacing of fasteners. Adjustments to material costs may be required for your specific application or location. Material costs do not include sales tax.

Labor Costs

Labor costs are based on the average of wage rates from 30 major U.S. cities. Rates are determined from labor union agreements or prevailing wages for construction trades for the current year. Rates along with overhead and profit markups are listed on the inside back cover of this book.

- If wage rates in your area vary from those used in this book, or if rate increases are expected within a given year, labor costs should be adjusted accordingly.

Labor costs reflect productivity based on actual working conditions. These figures include time spent during a normal workday on tasks other than actual installation, such as material receiving and handling, mobilization at site, site movement, breaks, and cleanup.

Productivity data is developed over an extended period so as not to be influenced by abnormal variations and reflects a typical average.

Equipment Costs

Equipment costs include not only rental, but also operating costs for equipment under normal use. The operating costs include parts and labor for routine servicing such as repair and replacement of pumps, filters and worn lines. Normal operating expendables such as fuel, lubricants, tires and electricity (where applicable) are also included. Extraordinary operating expendables with highly variable wear patterns such as diamond bits and blades are excluded. These costs are included under materials. Equipment rental rates are obtained from industry sources throughout North America—contractors, suppliers, dealers, manufacturers, and distributors.

Crew Equipment Cost/Day—The power equipment required for each crew is included in the crew cost. The daily cost for crew equipment is based on dividing the weekly bare rental rate by 5 (number of working days per week), and then adding the hourly operating cost times 8 (hours per day). This "Crew Equipment Cost/Day" is listed in Subdivision 01590.

Mobilization/Demobolization—The cost to move construction equipment from an equipment yard or rental company to the job site and back again is not included in equipment costs. Mobilization (to the site) and demolization (from the site) costs can be found in Section 02305-250. If a piece of equipment is already at the job site, it is not appropriate to utlize mob/demob costs again in an estimate.

General Conditions

Cost data in this book is presented in two ways: Bare Costs and Total Cost including O&P (Overhead and Profit). General Conditions, when applicable, should also be added to the Total Cost including O&P. The costs for General Conditions are listed in Division 1 of the Unit Price Section and the Reference Section of this book. General Conditions for the *Installing Contractor* may range from 0% to 10% of the Total Cost including O&P. For the *General* or *Prime Contractor*, costs for General Conditions may range from 5% to 15% of the Total Cost including O&P, with a figure of 10% as the most typical allowance.

Overhead and Profit

Total Cost including O&P for the *Installing Contractor* is shown in the last column on both the Unit Price and the Assemblies pages of this book. This figure is the sum of the bare material cost plus 10% for profit, the base labor cost plus total overhead and profit, and the bare equipment cost plus 10% for profit. Details for the calculation of Overhead and Profit on labor are shown on the inside back cover and in the Reference Section of this book. (See the "How to Use the Unit Price Pages" for an example of this calculation.)

Factors Affecting Costs

Costs can vary depending upon a number of variables. Here's how we have handled the main factors affecting costs.

Quality—The prices for materials and the workmanship upon which productivity is based represent sound construction work. They are also in line with U.S. government specifications.

Overtime—We have made no allowance for overtime. If you anticipate premium time or work beyond normal working hours, be sure to make an appropriate adjustment to your labor costs.

Productivity—The productivity, daily output, and labor-hour figures for each line item are based on working an eight-hour day in daylight hours in moderate temperatures. For work that extends beyond normal work hours or is performed under adverse conditions, productivity may decrease. (See the section in "How To Use the Unit Price Pages" for more on productivity.)

Size of Project—The size, scope of work, and type of construction project will have a significant impact on cost. Economies of scale can reduce costs for large projects. Unit costs can often run higher for small projects. Costs in this book are intended for the size and type of project as previously described in "How the Book Is Built: An Overview." Costs for projects of a significantly different size or type should be adjusted accordingly.

Location—Material prices in this book are for metropolitan areas. However, in dense urban areas, traffic and site storage limitations may increase costs. Beyond a 20-mile radius of large cities, extra trucking or transportation charges may also increase the material costs slightly. On the other hand, lower wage rates may be in effect. Be sure to consider both these factors when preparing an estimate, particularly if the job site is located in a central city or remote rural location.

In addition, highly specialized subcontract items may require travel and per diem expenses for mechanics.

Other factors —

- season of year
- contractor management
- weather conditions
- local union restrictions
- building code requirements
- availability of:
    - adequate energy
    - skilled labor
    - building materials
- owner's special requirements/restrictions
- safety requirements
- environmental considerations

Unpredictable Factors—General business conditions influence "in-place" costs of all items. Substitute materials and construction methods may have to be employed. These may affect the installed cost and/or life cycle costs. Such factors may be difficult to evaluate and cannot necessarily be predicted on the basis of the job's location in a particular section of the country. Thus, where these factors apply, you may find significant, but unavoidable cost variations for which you will have to apply a measure of judgment to your estimate.

Rounding of Costs

In general, all unit prices in excess of $5.00 have been rounded to make them easier to use and still maintain adequate precision of the results. The rounding rules we have chosen are in the following table.

Prices from . . .	Rounded to the nearest . . .
$.01 to $5.00	$.01
$5.01 to $20.00	$.05
$20.01 to $100.00	$.50
$100.01 to $300.00	$1.00
$300.01 to $1,000.00	$5.00
$1,000.01 to $10,000.00	$25.00
$10,000.01 to $50,000.00	$100.00
$50,000.01 and above	$500.00

Contingencies

Contingencies: The allowance for contingencies generally is to provide for unforeseen construction difficulties. On alterations or repair jobs 20% is none too much. If drawings are final and only field contingencies are being considered, 2% or 3% is probably sufficient and often nothing need be added. As far as the contract is concerned, future changes in plans will be covered by extras. The contractor should consider inflationary price trends and possible material shortages during the course of the job. If drawings are not complete or approved or a budget cost wanted, it is wise to add 5% to 10%. Contingencies, then, are a matter of judgment. Additional allowances are shown in Division 01250-200 for contingencies and job conditions and Division 01250-400 for factors to convert prices for repair and remodeling jobs.

Note: For an explanation of "Selective Demolition" versus "Removal for Replacement" see Reference number R15050-720.

Important Estimating Considerations

One reason for listing a job size "minimum" is to ensure that the construction craftsmen are productive for 8 hours a day on a continuous basis. Otherwise, tasks that only require 6 to 7 hours could be billed as an 8-hour day, thus providing an erroneous productivity estimate. The "productivity," or daily output of each craftsman, includes mobilization and cleanup time, break time, plan layout time, as well as an allowance to carry stock from the storage trailer or location on the job site up to 200′ into the building and on the first or second floor. If material has to be transported over greater distances or to higher floors, an additional allowance should be considered by the estimator. An allowance has also been included in the piping and fittings installation time for a leak check and minor tightening.

Equipment installation time includes all applicable items following: positioning, leveling and securing the unit in place, connecting all associated piping, ducts, vents, etc., which shall have been estimated separately, connecting to an adjacent power source, filling/bleeding, startup, adjusting the controls up and down to assure proper response, setting the integral controls/valves/regulators/thermostats for proper operation (does not include external building type control systems, DDC systems, etc.), explaining/training owner's operator, and warranty. A reasonable breakdown of the labor costs would be as follows:

1. Movement into building, installation/setting of equipment 35%
2. Connecting to piping/duct/power, etc. 25%
3. Filling/flushing/cleaning/touchup, etc. 15%
4. Startup/running adjustments 5%
5. Training owner's rep. 5%
6. Warranty/call back/service 15%

Note that cranes or other lifting equipment are not included on any lines in Division 15. For example, if a crane is required to lift a heavy piece of pipe into place high above a gym floor, or to put a rooftop unit on the roof of a four-story building, etc., it must be added. Due to the potential for extreme variation—from nothing additional required, to a major crane or helicopter—we feel that including a nominal amount for "lifting contingency" would be useless and detract from the accuracy of the estimate. When using equipment rental from Means do not forget to include the cost of the operator(s).

Estimating Labor-Hours

The labor-hours expressed in this publication are based on Average Installation time, using an efficiency level of approximately 60-65% (see item 7) which has been found reasonable and acceptable by many contractors.

The book uses this national efficiency average to establish a consistent benchmark.

For bid situations, adjustments to this efficiency level should be the responsibility of the contractor bidding the project.

The unit labor-hour is divided in the following manner. A typical day for a journeyman might be:

1.	Study Plans	3%	14.4 min.
2.	Material Procurement	3%	14.4 min.
3.	Receiving and Storing	3%	14.4 min
4.	Mobilization	5%	24.0 min.
5.	Site Movement	5%	24.0 min.
6.	Layout and Marking	8%	38.4 min.
7.	Actual Installation	64%	307.2 min.
8.	Cleanup	3%	14.4 min.
9.	Breaks, Non-Productive	6%	28.8 min.
		100%	480.0 min.

If any of the percentages expressed in this breakdown do not apply to the particular work or project situation, then that percentage or a portion of it may be deducted from or added to labor hours.

Final Checklist

Estimating can be a straightforward process provided you remember the basics. Here's a checklist of some of the items you should remember to do before completing your estimate.

Did you remember to . . .

- factor in the City Cost Index for your locale
- take into consideration which items have been marked up and by how much
- mark up the entire estimate sufficiently for your purposes
- read the background information on techniques and technical matters that could impact your project time span and cost
- include all components of your project in the final estimate
- double check your figures to be sure of your accuracy
- for more information, please see "Tips for Accurate Estimating," R01100-005 in the Reference Section
- call RSMeans if you have any questions about your estimate or the data you've found in our publications

Remember, RSMeans stands behind its publications. If you have any questions about your estimate . . . about the costs you've used from our books . . . or even about the technical aspects of the job that may affect your estimate, feel free to call the RSMeans editors at 1-800-334-3509.

Unit Price Section

Table of Contents

How to Use the Unit Price Pages

The following is a detailed explanation of a sample entry in the Unit Price Section. Next to each bold number below is the item being described with appropriate component of the sample entry following in parenthesis. Some prices are listed as bare costs, others as costs that include overhead and profit of the installing contractor. In most cases, if the work is to be subcontracted, the general contractor will need to add an additional markup (RSMeans suggests using 10%) to the figures in the column "Total Incl. O&P."

1 *Division Number/Title (03300/Cast-In-Place Concrete)*

Use the Unit Price Section Table of Contents to locate specific items. The sections are classified according to the CSI MasterFormat (1995 Edition).

2 *Line Numbers (03310 240 3900)*

Each unit price line item has been assigned a unique 12-digit code based on the CSI MasterFormat classification.

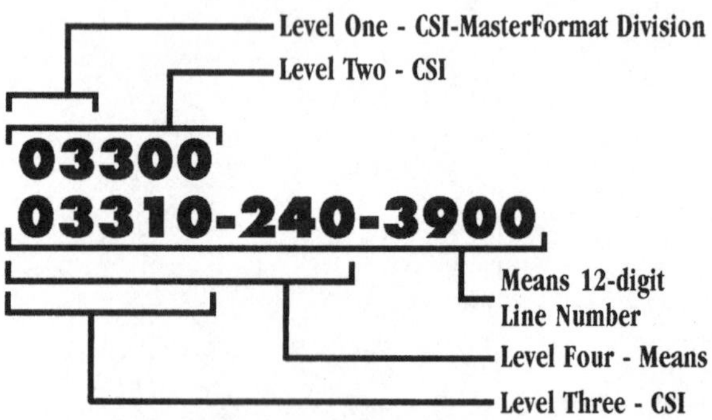

3 *Description (Concrete-In-Place, etc.)*

Each line item is described in detail. Sub-items and additional sizes are indented beneath the appropriate line items. The first line or two after the main item (in boldface) may contain descriptive information that pertains to all line items beneath this boldface listing.

4 *Reference Number Information*

R03310 -010 You'll see reference numbers shown in bold rectangles at the beginning of some sections. These refer to related items in the Reference Section, visually identified by a vertical gray bar on the edge of pages.

The relation may be: (1) an estimating procedure that should be read before estimating, (2) an alternate pricing method, or (3) technical information.

The "R" designates the Reference Section. The numbers refer to the MasterFormat classification system.

It is strongly recommended that you review all reference numbers that appear within the section in which you are working.

Note: **Not all reference numbers appear in all Means publications.**

03300 | Cast-In-Place Concrete

03310 | Structural Concrete			CREW	DAILY OUTPUT	LABOR-HOURS	UNIT	2005 BARE COSTS MAT.	LABOR	EQUIP.	TOTAL	TOTAL INCL O&P	
240	0010	**CONCRETE IN PLACE** Including forms (4 uses), reinforcing R03310 -010										240
	0050	steel and finishing unless ot[illegible] indi[illegible]ated										
	0300	Beams, 5 kip per L.F., 10′ span R03310 -100	C-14A	[illegible]	[illegible]	C.Y.	284	[illegible]	46.50	770.50	1,050	
	0350	25′ span	"	18.55	10.782		214	370	39.50	623.50	860	
	3850	Over 5 C.Y.		81.04	1.382		226	45	.29	271.29	320	
	3900	Footings, strip, 18″ x 9″, unreinforced		40	2.800		101	91.50	.59	193.09	255	
	3920	18″ x 9″, reinforced		35	3.200		120	105	.67	225.67	297	
	3925	20″ x 10″, unreinforced		45	2.489		98	81.50	.52	180.02	236	
	3930	20″ x 10″, reinforced		40	2.800		114	91.50	.59	206.09	[illegible]	
	3935	24″ x 12″, unreinforced		55	2.036		97	66.50	.43	163.93	[illegible]	
	3940	24″ x 12″, reinforced		48	2.333		113	76	.49	189.49	[illegible]	
	3945	36″ x 12″, unreinforced		70	1.600		93	52.50	.34	145.84	184	
	3950	36″ x 12″, reinforced		60	1.867		108	61	.39	169.39	214	
	4000	Foundation mat, under 10 C.Y.		38.67	2.896		164	94.50	.61	259.11	330	

Crew (C-14C)

The "Crew" column designates the typical trade or crew used to install the item. If an installation can be accomplished by one trade and requires no power equipment, that trade and the number of workers are listed (for example, "2 Carpenters"). If an installation requires a composite crew, a crew code designation is listed (for example, "C-14C"). You'll find full details on all composite crews in the Crew Listings.

- For a complete list of all trades utilized in this book and their abbreviations, see the inside back cover.

Crews

Crew No.	Bare Costs		Incl. Subs O & P		Cost Per Labor-Hour	
Crew C-14C	Hr.	Daily	Hr.	Daily	Bare Costs	Incl. O&P
1 Carpenter Foreman (out)	$36.25	$290.00	$56.45	$451.60	$32.66	$51.14
6 Carpenters	34.25	1644.00	53.35	2560.80		
2 Rodmen (reinf.)	37.95	607.20	62.60	1001.60		
4 Laborers	26.70	854.40	41.55	1329.60		
1 Cement Finisher	32.85	262.80	48.35	386.80		
1 Gas Engine Vibrator		24.00		26.40	.21	.24
112 L.H., Daily Totals		$3682.40		$5756.80	$32.87	$51.38

Productivity: Daily Output (40.0)/Labor-Hours (2.80)

The "Daily Output" represents the typical number of units the designated crew will install in a normal 8-hour day. To find out the number of days the given crew would require to complete the installation, divide your quantity by the daily output. For example:

Quantity	÷	Daily Output	=	Duration
100 C.Y.	÷	40.0/ Crew Day	=	2.50 Crew Days

The "Labor-Hours" figure represents the number of labor-hours required to install one unit of work. To find out the number of labor-hours required for your particular task, multiply the quantity of the item times the number of labor-hours shown. For example:

Quantity	x	Productivity Rate	=	Duration
100 C.Y.	x	2.80 Labor-Hours/ C.Y.	=	280 Labor-Hours

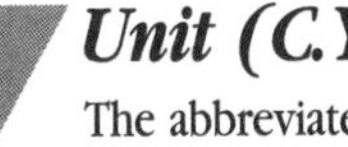

Unit (C.Y.)

The abbreviated designation indicates the unit of measure upon which the price, production, and crew are based (C.Y. = Cubic Yard). For a complete listing of abbreviations refer to the Abbreviations Listing in the Reference Section of this book.

Bare Costs:

Mat. (Bare Material Cost) (101)

The unit material cost is the "bare" material cost with no overhead and profit included. *Costs shown reflect national average material prices for January of the current year and include delivery to the job site. No sales taxes are included.*

Labor (91.50)

The unit labor cost is derived by multiplying bare labor-hour costs for Crew C-14C by labor-hour units. The bare labor-hour cost is found in the Crew Section under C-14C. (If a trade is listed, the hourly labor cost—the wage rate—is found on the inside back cover.)

Labor-Hour Cost Crew C-14C	x	Labor-Hour Units	=	Labor
$32.66	x	2.80	=	$91.00

Equip. (Equipment) (.59)

Equipment costs for each crew are listed in the description of each crew. Tools or equipment whose value justifies purchase or ownership by a contractor are considered overhead as shown on the inside back cover. The unit equipment cost is derived by multiplying the bare equipment hourly cost by the labor-hour units.

Equipment Cost Crew C-14C	x	Labor-Hour Units	=	Equip.
.21	x	2.80	=	$.59

Total (193.09)

The total of the bare costs is the arithmetic total of the three previous columns: mat., labor, and equip.

Material	+	Labor	+	Equip.	=	Total
$101	+	$91.50	+	$.59	=	$193.09

Total Costs Including O&P

This figure is the sum of the bare material cost plus 10% for profit; the bare labor cost plus total overhead and profit (per the inside back cover or, if a crew is listed, from the crew listings); and the bare equipment cost plus 10% for profit.

Material is Bare Material Cost + 10% = 101 + 10.10	=	$111.10
Labor for Crew C-14C = Labor-Hour Cost (51.14) x Labor-Hour Units (2.80)	=	$143.19
Equip. is Bare Equip. Cost + 10% = .59 + .06	=	$ .65
Total (Rounded)	=	$255

Division 1
General Requirements

Estimating Tips

The General Requirements of any contract are very important to both the bidder and the owner. These lay the ground rules under which the contract will be executed and have a significant influence on the cost of operations. Therefore, it is extremely important to thoroughly read and understand the General Requirements both before preparing an estimate and when the estimate is complete, to ascertain that nothing in the contract is overlooked. Caution should be exercised when applying items listed in Division 1 to an estimate. Many of the items are included in the unit prices listed in the other divisions such as mark-ups on labor and company overhead.

01200 Price & Payment Procedures

- When estimating historic preservation projects (depending on the condition of the existing structure and the owner's requirements), a 15-20% contingency or allowance is recommended, regardless of the stage of the drawings.

01300 Administrative Requirements

- Before determining a final cost estimate, it is a good practice to review all the items listed in subdivision 01300 to make final adjustments for items that may need customizing to specific job conditions.
- Historic preservation projects may require specialty labor and methods, as well as extra time to protect existing materials that must be preserved and/or restored. Some additional expenses may be incurred in architectural fees for facility surveys and other special inspections and analyses.
- Requirements for initial and periodic submittals can represent a significant cost to the General Requirements of a job. Thoroughly check the submittal specifications when estimating a project to determine any costs that should be included.

01400 Quality Requirements

- All projects will require some degree of Quality Control. This cost is not included in the unit cost of construction listed in each division. Depending upon the terms of the contract, the various costs of inspection and testing can be the responsibility of either the owner or the contractor. Be sure to include the required costs in your estimate.

01500 Temporary Facilities & Controls

- Barricades, access roads, safety nets, scaffolding, security and many more requirements for the execution of a safe project are elements of direct cost. These costs can easily be overlooked when preparing an estimate. When looking through the major classifications of this subdivision, determine which items apply to each division in your estimate.

01590 Equipment Rental

- This subdivision contains transportation, handling, storage, protection and product options and substitutions. Listed in this cost manual are average equipment rental rates for all types of equipment. This is useful information when estimating the time and materials requirement of any particular operation in order to establish a unit or total cost.
- A good rule of thumb is that weekly rental is 3 times daily rental and that monthly rental is 3 times weekly rental.
- The figures in the column for Crew Equipment Cost represent the rental rate used in determining the daily cost of equipment in a crew. It is calculated by dividing the weekly rate by 5 days and adding the hourly operating cost times 8 hours.

01740 Execution Requirements

- When preparing an estimate, read the specifications to determine the requirements for Contract Closeout thoroughly. Final cleaning, record documentation, operation and maintenance data, warranties and bonds, and spare parts and maintenance materials can all be elements of cost for the completion of a contract. Do not overlook these in your estimate.

01830 Operations & Maintenance

- If maintenance and repair are included in your contract, they require special attention. To estimate the cost to remove and replace any unit usually requires a site visit to determine the accessibility and the specific difficulty at that location. Obstructions, dust control, safety, and often overtime hours must be considered when preparing your estimate.

Reference Numbers

Reference numbers are shown in bold squares at the beginning of some major classifications. These numbers refer to related items in the Reference Section. The reference information may be an estimating procedure, an alternate pricing method or technical information.

Note: Not all subdivisions listed here necessarily appear in this publication.

01100 | Summary

	01107 | Professional Consultant		Crew	Daily Output	Labor-Hours	Unit	2005 Bare Costs Mat.	Labor	Equip.	Total	Total Incl O&P	
300	0010	**ENGINEERING FEES** R01107-030										300
	0020	Educational planning consultant, minimum				Project					.50%	
	0100	Maximum				″					2.50%	
	0200	Electrical, minimum				Contrct					4.10%	
	0300	Maximum									10.10%	
	0400	Elevator & conveying systems, minimum									2.50%	
	0500	Maximum									5%	
	0600	Food service & kitchen equipment, minimum									8%	
	0700	Maximum									12%	
	1000	Mechanical (plumbing & HVAC), minimum									4.10%	
	1100	Maximum									10.10%	

01200 | Price & Payment Procedures

	01250 | Contract Modification Procedures		Crew	Daily Output	Labor-Hours	Unit	2005 Bare Costs Mat.	Labor	Equip.	Total	Total Incl O&P	
200	0010	**CONTINGENCIES** for estimate at conceptual stage				Project					20%	200
	0050	Schematic stage									15%	
	0100	Preliminary working drawing stage (Design Dev.)									10%	
	0150	Final working drawing stage									3%	
300	0010	**CREWS** For building construction, see How To Use This Book										300
400	0010	**FACTORS** Cost adjustments R01250-010										400
	0100	Add to construction costs for particular job requirements										
	0500	Cut & patch to match existing construction, add, minimum				Costs	2%	3%				
	0550	Maximum					5%	9%				
	0800	Dust protection, add, minimum					1%	2%				
	0850	Maximum					4%	11%				
	1100	Equipment usage curtailment, add, minimum					1%	1%				
	1150	Maximum					3%	10%				
	1400	Material handling & storage limitation, add, minimum					1%	1%				
	1450	Maximum					6%	7%				
	1700	Protection of existing work, add, minimum					2%	2%				
	1750	Maximum					5%	7%				
	2000	Shift work requirements, add, minimum						5%				
	2050	Maximum						30%				
	2300	Temporary shoring and bracing, add, minimum					2%	5%				
	2350	Maximum					5%	12%				
	2400	Work inside prisons and high security areas, add, minimum						30%				
	2450	Maximum						50%				
500	0010	**JOB CONDITIONS** Modifications to total										500
	0020	project cost summaries										
	0100	Economic conditions, favorable, deduct				Project					2%	
	0200	Unfavorable, add									5%	
	0300	Hoisting conditions, favorable, deduct									2%	
	0400	Unfavorable, add									5%	
	0700	Labor availability, surplus, deduct									1%	
	0800	Shortage, add									10%	
	0900	Material storage area, available, deduct									1%	
	1000	Not available, add									2%	

01200 | Price & Payment Procedures

	01250 | Contract Modification Procedures		CREW	DAILY OUTPUT	LABOR-HOURS	UNIT	2005 BARE COSTS MAT.	LABOR	EQUIP.	TOTAL	TOTAL INCL O&P	
500	1100	Subcontractor availability, surplus, deduct				Project					5%	500
	1200	Shortage, add									12%	
	1300	Work space, available, deduct									2%	
	1400	Not available, add				↓					5%	
600	0010	**OVERTIME** For early completion of projects or where (R01100-110)										600
	0020	labor shortages exist, add to usual labor, up to				Costs		100%				

	01255 | Cost Indexes		CREW	DAILY OUTPUT	LABOR-HOURS	UNIT	MAT.	LABOR	EQUIP.	TOTAL	TOTAL INCL O&P	
200	0010	**CONSTRUCTION COST INDEX** (Reference) over 930 zip code locations in										200
	0020	The U.S. and Canada, total bldg cost, min. (Clarksdale, MS)				%					66.80%	
	0050	Average									100%	
	0100	Maximum (New York, NY)				↓					132.40%	
400	0010	**HISTORICAL COST INDEXES** (Reference) Back to 1955										400
500	0010	**LABOR INDEX** (Reference) For over 930 zip code locations in										500
	0020	the U.S. and Canada, minimum (Clarksdale, MS)				%		31.70%				
	0050	Average						100%				
	0100	Maximum (New York, NY)				↓		162.30%				
600	0011	**MATERIAL INDEX** For over 930 zip code locations in										600
	0020	the U.S. and Canada, minimum (Elizabethtown, KY)				%	91.10%					
	0040	Average					100%					
	0060	Maximum (Ketchikan, AK)				↓	143.10%					

	01290 | Payment Procedures		CREW	DAILY OUTPUT	LABOR-HOURS	UNIT	MAT.	LABOR	EQUIP.	TOTAL	TOTAL INCL O&P	
800	0010	**TAXES** Sales tax, State, average (R01100-090)				%	4.85%					800
	0050	Maximum					7.25%					
	0200	Social Security, on first $87,900 of wages (R01100-100)						7.65%				
	0300	Unemployment, MA, combined Federal and State, minimum						1.92%				
	0350	Average						6.20%				
	0400	Maximum				↓		11.76%				

01300 | Administrative Requirements

	01310 | Project Management/Coordination		CREW	DAILY OUTPUT	LABOR-HOURS	UNIT	2005 BARE COSTS MAT.	LABOR	EQUIP.	TOTAL	TOTAL INCL O&P	
150	0010	**PERMITS** Rule of thumb, most cities, minimum				Job					.50%	150
	0100	Maximum				"					2%	
200	0010	**PERFORMANCE BOND** For buildings, minimum (R01100-080)				Job					.60%	200
	0100	Maximum				"					2.50%	
350	0010	**INSURANCE** Builders risk, standard, minimum (R01100-040)				Job					.24%	350
	0050	Maximum									.64%	
	0200	All-risk type, minimum (R01100-060)									.25%	
	0250	Maximum				↓					.62%	
	0400	Contractor's equipment floater, minimum				Value					.50%	
	0450	Maximum				"					1.50%	
	0600	Public liability, average				Job					2.02%	
	0800	Workers' compensation & employer's liability, average										
	0850	by trade, carpentry, general				Payroll		18.42%				
	1000	Electrical						6.46%				
	1150	Insulation						15.53%				
	1450	Plumbing				↓		7.96%				

01300 | Administrative Requirements

		01310 | Project Management/Coordination		CREW	DAILY OUTPUT	LABOR-HOURS	UNIT	2005 BARE COSTS MAT.	LABOR	EQUIP.	TOTAL	TOTAL INCL O&P	
350	1550	Sheet metal work (HVAC)	R01100-040				Payroll		11.46%				350
500	0010	**MARK-UP** For General Contractors for change	R01100-070										500
	0100	of scope of job as bid											
	0200	Extra work, by subcontractors, add					%					10%	
	0250	By General Contractor, add										15%	
	0400	Omitted work, by subcontractors, deduct all but										5%	
	0450	By General Contractor, deduct all but										7.50%	
	0600	Overtime work, by subcontractors, add										15%	
	0650	By General Contractor, add										10%	
	1000	Installing contractors, on his own labor, minimum							46.90%				
	1100	Maximum							87.30%				
600	0010	**OVERHEAD** As percent of direct costs, minimum	R01100-070				%				5%		600
	0050	Average									13%		
	0100	Maximum									30%		
620	0010	**OVERHEAD & PROFIT** Allowance to add to items in this	R01100-070										620
	0020	book that do not include Subs O&P, average					%				25%		
	0100	Allowance to add to items in this book that											
	0110	do include Subs O&P, minimum					%					5%	
	0150	Average										10%	
	0200	Maximum										15%	
	0300	Typical, by size of project, under $100,000									30%		
	0350	$500,000 project									25%		
	0400	$2,000,000 project									20%		
	0450	Over $10,000,000 project									15%		

		01321 | Construction Photos		CREW	DAILY OUTPUT	LABOR-HOURS	UNIT	MAT.	LABOR	EQUIP.	TOTAL	TOTAL INCL O&P	
500	0010	**PHOTOGRAPHS** 8″ x 10″, 4 shots, 2 prints ea., std. mounting					Set	292			292	320	500
	0100	Hinged linen mounts						315			315	350	
	0200	8″ x 10″, 4 shots, 2 prints each, in color						340			340	370	
	0300	For I.D. slugs, add to all above						4.09			4.09	4.50	
	1500	Time lapse equipment, camera and projector, buy						3,675			3,675	4,050	
	1550	Rent per month						545			545	600	
	1700	Cameraman and film, including processing, B.&W.					Day	1,350			1,350	1,475	
	1720	Color					″	1,350			1,350	1,475	

01500 | Temporary Facilities & Controls

		01510 | Temporary Utilities	CREW	DAILY OUTPUT	LABOR-HOURS	UNIT	2005 BARE COSTS MAT.	LABOR	EQUIP.	TOTAL	TOTAL INCL O&P	
800	0010	**TEMPORARY UTILITIES**										800
	0100	Heat, incl. fuel and operation, per week, 12 hrs. per day	1 Skwk	100	.080	CSF Flr	6.25	2.79		9.04	11.25	
	0200	24 hrs. per day	″	60	.133		9.40	4.65		14.05	17.60	
	0350	Lighting, incl. service lamps, wiring & outlets, minimum	1 Elec	34	.235		2.32	9.60		11.92	16.80	
	0360	Maximum	″	17	.471		5.05	19.20		24.25	34	
	0400	Power for temp lighting only, per month, min/month 6.6 KWH								.75	1.18	
	0450	Maximum/month 23.6 KWH								2.85	3.14	
	0600	Power for job duration incl. elevator, etc., minimum								47	51.70	
	0650	Maximum								110	121	
	1000	Toilet, portable, see division 01590-400										

		01520 | Construction Facilities	CREW	DAILY OUTPUT	LABOR-HOURS	UNIT	2005 BARE COSTS MAT.	LABOR	EQUIP.	TOTAL	TOTAL INCL O&P	
500	0010	**OFFICE** Trailer, furnished, no hookups, 20′ x 8′, buy	2 Skwk	1	16	Ea.	6,200	560		6,760	7,700	500
	0250	Rent per month					154			154	169	
	0300	32′ x 8′, buy	2 Skwk	.70	22.857		9,250	795		10,045	11,500	
	0350	Rent per month					163			163	180	
	0400	50′ x 10′, buy	2 Skwk	.60	26.667		17,000	930		17,930	20,200	
	0450	Rent per month					254			254	280	
	0500	50′ x 12′, buy	2 Skwk	.50	32		20,300	1,125		21,425	24,000	
	0550	Rent per month					286			286	315	
	0700	For air conditioning, rent per month, add					39.50			39.50	43.50	
	0800	For delivery, add per mile				Mile	1.53			1.53	1.68	
	1000	Portable buildings, prefab, on skids, economy, 8′ x 8′	2 Carp	265	.060	S.F.	81.50	2.07		83.57	93	
	1100	Deluxe, 8′ x 12′	″	150	.107	″	88.50	3.65		92.15	103	
	1200	Storage boxes, 20′ x 8′, buy	2 Skwk	1.80	8.889	Ea.	3,175	310		3,485	3,975	
	1250	Rent per month					74.50			74.50	82	
	1300	40′ x 8′, buy	2 Skwk	1.40	11.429		4,250	400		4,650	5,300	
	1350	Rent per month					105			105	116	
550	0010	**FIELD OFFICE EXPENSE**										550
	0100	Field office expense, office equipment rental average				Month	143			143	157	
	0120	Office supplies, average				″	85			85	93.50	
	0125	Office trailer rental, see division 01520-500										
	0140	Telephone bill; avg. bill/month incl. long dist.				Month	204			204	224	
	0160	Field office lights & HVAC				″	98			98	108	
		01530 | Temporary Construction										
700	0010	**PROTECTION** Stair tread, 2″ x 12″ planks, 1 use	1 Carp	75	.107	Tread	4.80	3.65		8.45	11	700
	0100	Exterior plywood, 1/2″ thick, 1 use		65	.123		1.86	4.22		6.08	8.60	
	0200	3/4″ thick, 1 use		60	.133		2.82	4.57		7.39	10.20	
900	0010	**WINTER PROTECTION** Reinforced plastic on wood										900
	0100	framing to close openings	2 Clab	750	.021	S.F.	.38	.57		.95	1.31	
	0200	Tarpaulins hung over scaffolding, 8 uses, not incl. scaffolding		1,500	.011		.18	.28		.46	.64	
	0250	Tarpaulin polyester reinf. w/ integral fastening system 11 mils thick		1,600	.010		.78	.27		1.05	1.28	
	0300	Prefab fiberglass panels, steel frame, 8 uses		1,200	.013		.74	.36		1.10	1.36	
		01540 | Construction Aids										
750	0010	**SCAFFOLDING** R01540-100										750
	0015	Steel tubular, reg, rent/mo, no plank, incl erect or dismantle										
	0090	Building exterior, wall face, 1 to 5 stories, 6′-4″ x 5′ frames	3 Carp	24	1	C.S.F.	24.50	34.50		59	80	
	0200	6 to 12 stories	4 Carp	21.20	1.509		24.50	51.50		76	107	
	0310	13 to 20 stories	5 Carp	20	2		24.50	68.50		93	134	
	0460	Building interior, wall face area, up to 16′ high	3 Carp	25	.960		24.50	33		57.50	77.50	
	0560	16′ to 40′ high		23	1.043		24.50	35.50		60	82	
	0800	Building interior floor area, up to 30′ high		312	.077	C.C.F.	2.57	2.63		5.20	6.95	
	0900	Over 30′ high	4 Carp	275	.116	″	2.57	3.99		6.56	9.05	
	0910	Steel tubular, heavy duty shoring, buy										
	0920	Frames 5′ high 2′ wide				Ea.	82.50			82.50	91	
	0925	5′ high 4′ wide					93.50			93.50	103	
	0930	6′ high 2′ wide					94.50			94.50	104	
	0935	6′ high 4′ wide					111			111	122	
	0940	Accessories										
	0945	Cross braces				Ea.	17.60			17.60	19.35	
	0950	U-head, 8″ x 8″					19.25			19.25	21	
	0955	J-head, 4″ x 8″					14.10			14.10	15.50	
	0960	Base plate, 8″ x 8″					15.60			15.60	17.20	
	0965	Leveling jack					33.50			33.50	37	

01500 | Temporary Facilities & Controls

	01540 | Construction Aids		CREW	DAILY OUTPUT	LABOR-HOURS	UNIT	2005 BARE COSTS MAT.	LABOR	EQUIP.	TOTAL	TOTAL INCL O&P	
750	1000	Steel tubular, regular, buy (R01540 -100)										750
	1100	Frames 3′ high 5′ wide				Ea.	64			64	70	
	1150	5′ high 5′ wide					73.50			73.50	81	
	1200	6′-4″ high 5′ wide					92.50			92.50	102	
	1350	7′-6″ high 6′ wide					160			160	175	
	1500	Accessories cross braces					16.50			16.50	18.15	
	1550	Guardrail post					16.50			16.50	18.15	
	1600	Guardrail 7′ section					8			8	8.80	
	1650	Screw jacks & plates					26.50			26.50	29	
	1700	Sidearm brackets					31			31	34	
	1750	8″ casters					36.50			36.50	40	
	1800	Plank 2″ x 10″ x 16′-0″					47			47	51.50	
	1900	Stairway section					270			270	296	
	1910	Stairway starter bar					32			32	35	
	1920	Stairway inside handrail					58.50			58.50	64	
	1930	Stairway outside handrail					80.50			80.50	88.50	
	1940	Walk-thru frame guardrail					40.50			40.50	44.50	
	2000	Steel tubular, regular, rent/mo.										
	2100	Frames 3′ high 5′ wide				Ea.	3.75			3.75	4.13	
	2150	5′ high 5′ wide					3.75			3.75	4.13	
	2200	6′-4″ high 5′ wide					3.75			3.75	4.13	
	2250	7′-6″ high 6′ wide					7			7	7.70	
	2500	Accessories, cross braces					.60			.60	.66	
	2550	Guardrail post					1			1	1.10	
	2600	Guardrail 7′ section					.75			.75	.83	
	2650	Screw jacks & plates					1.50			1.50	1.65	
	2700	Sidearm brackets					1.50			1.50	1.65	
	2750	8″ casters					6			6	6.60	
	2800	Outrigger for rolling tower					3			3	3.30	
	2850	Plank 2″ x 10″ x 16′-0″					5			5	5.50	
	2900	Stairway section					10			10	11	
	2910	Stairway starter bar					.10			.10	.11	
	2920	Stairway inside handrail					5			5	5.50	
	2930	Stairway outside handrail					5			5	5.50	
	2940	Walk-thru frame guardrail					2			2	2.20	
	3000	Steel tubular, heavy duty shoring, rent/mo.										
	3250	5′ high 2′ & 4′ wide				Ea.	5			5	5.50	
	3300	6′ high 2′ & 4′ wide					5			5	5.50	
	3500	Accessories, cross braces					1			1	1.10	
	3600	U - head, 8″ x 8″					1			1	1.10	
	3650	J - head, 4″ x 8″					1			1	1.10	
	3700	Base plate, 8″ x 8″					1			1	1.10	
	3750	Leveling jack					2			2	2.20	
	5700	Planks, 2x10x16′-0″, labor only, erect or remove to 50′ H	3 Carp	144	.167			5.70		5.70	8.90	
	5800	Over 50′ high	4 Carp	160	.200			6.85		6.85	10.65	
	6000	Heavy duty shoring for elevated slab forms to 8′-2″ high, floor area										
	6010	incl. erection or dismantle labor										
	6100	1 use/month	4 Carp	36	.889	C.S.F.	29.50	30.50		60	80	
	6150	2 uses/month	″	36	.889	″	14.80	30.50		45.30	64	
	6500	To 14′-8″ high										
	6600	1 use/month	4 Carp	18	1.778	C.S.F.	43	61		104	143	
	6650	2 uses/month	″	18	1.778	″	21.50	61		82.50	119	
755	0010	**SCAFFOLDING SPECIALTIES**										755
	1200	Sidewalk bridge, heavy duty steel posts & beams, including										
	1210	parapet protection & waterproofing										
	1220	8′ to 10′ wide, 2 posts	3 Carp	15	1.600	L.F.	34.50	55		89.50	124	

		01540 | Construction Aids	CREW	DAILY OUTPUT	LABOR-HOURS	UNIT	2005 BARE COSTS MAT.	LABOR	EQUIP.	TOTAL	TOTAL INCL O&P	
755	1230	3 posts	3 Carp	10	2.400	L.F.	53.50	82		135.50	187	755
	1500	Sidewalk bridge using tubular steel										
	1510	scaffold frames, including planking	3 Carp	45	.533	L.F.	4.72	18.25		22.97	33.50	
	1600	For 2 uses per month, deduct from all above					50%					
	1700	For 1 use every 2 months, add to all above					100%					
	1900	Catwalks, 20" wide, no guardrails, 7' span, buy				Ea.	138			138	151	
	2000	10' span, buy				"	176			176	194	
	2800	Hand winch-operated masons scaffolding, no plank										
	2810	plank moving not required										
	2900	98' long, 10'-6" high, buy				Ea.	28,900			28,900	31,700	
	3000	Rent per month					1,150			1,150	1,275	
	3100	28'-6" high, buy					35,400			35,400	38,900	
	3200	Rent per month					1,425			1,425	1,550	
	3400	196' long, 28'-6" high, buy					68,500			68,500	75,000	
	3500	Rent per month					2,725			2,725	3,000	
	3600	64'-6" high, buy					93,500			93,500	103,000	
	3700	Rent per month					3,750			3,750	4,125	
	3720	Putlog, standard, 8' span, with hangers, buy					67			67	74	
	3730	Rent per month					10			10	11	
	3750	12' span, buy					101			101	111	
	3755	Rent per month					15			15	16.50	
	3760	Trussed type, 16' span, buy					231			231	254	
	3770	Rent per month					20			20	22	
	3790	22' span, buy					277			277	305	
	3795	Rent per month					30			30	33	
	3800	Rolling ladders with handrails, 30" wide, buy, 2 step					173			173	191	
	4000	7 step					585			585	645	
	4050	10 step					765			765	845	
	4100	Rolling towers, buy, 5' wide, 7' long, 10' high					1,250			1,250	1,375	
	4200	For 5' high added sections, to buy, add					209			209	230	
	4300	Complete incl. wheels, railings, outriggers,										
	4350	21' high, to buy				Ea.	2,125			2,125	2,325	
	4400	Rent/month				"	159			159	175	
800	0010	**TARPAULINS** Cotton duck, 10 oz. to 13.13 oz. per S.Y., minimum				S.F.	.49			.49	.54	800
	0050	Maximum					.58			.58	.64	
	0100	Polyvinyl coated nylon, 14 oz. to 18 oz., minimum					.48			.48	.53	
	0150	Maximum					.68			.68	.75	
	0200	Reinforced polyethylene 3 mils thick, white					.11			.11	.12	
	0300	4 mils thick, white, clear or black					.14			.14	.15	
	0400	5.5 mils thick, clear					.19			.19	.21	
	0500	White, fire retardant					.18			.18	.20	
	0600	7.5 mils, oil resistant, fire retardant					.19			.19	.21	
	0700	8.5 mils, black					.24			.24	.26	
	0710	Woven polyethylene, 6 mils thick					.48			.48	.53	
	0730	Polyester reinforced w/ integral fastening system 11 mils thick					1.07			1.07	1.18	
	0740	Mylar polyester, non-reinforced, 7 mils thick					1.17			1.17	1.29	
820	0010	**SMALL TOOLS** As % of contractor's work, minimum				Total					.50%	820
	0100	Maximum				"					2%	

		01550 | Vehicular Access & Parking	CREW	DAILY OUTPUT	LABOR-HOURS	UNIT	MAT.	LABOR	EQUIP.	TOTAL	TOTAL INCL O&P	
700	0010	**ROADS AND SIDEWALKS** Temporary										700
	0050	Roads, gravel fill, no surfacing, 4" gravel depth	B-14	715	.067	S.Y.	3.23	1.89	.30	5.42	6.80	
	0100	8" gravel depth	"	615	.078	"	6.45	2.19	.35	8.99	10.90	
	1000	Ramp, 3/4" plywood on 2" x 6" joists, 16" O.C.	2 Carp	300	.053	S.F.	1.48	1.83		3.31	4.47	

01500 | Temporary Facilities & Controls

		01550 | Vehicular Access & Parking	Crew	Daily Output	Labor-Hours	Unit	2005 Bare Costs Mat.	Labor	Equip.	Total	Total Incl O&P	
700	1100	On 2" x 10" joists, 16" O.C.	2 Carp	275	.058	S.F.	2.11	1.99		4.10	5.40	700
	2200	Sidewalks, 2" x 12" planks, 2 uses	1 Carp	350	.023		.80	.78		1.58	2.10	
	2300	Exterior plywood, 2 uses, 1/2" thick		750	.011		.31	.37		.68	.91	
	2400	5/8" thick		650	.012		.39	.42		.81	1.09	
	2500	3/4" thick	↓	600	.013	↓	.47	.46		.93	1.23	
		01560 | Barriers & Enclosures										
100	0010	**BARRICADES** 5' high, 3 rail @ 2" x 8", fixed	2 Carp	30	.533	L.F.	11	18.25		29.25	40.50	100
	0150	Movable		20	.800		11	27.50		38.50	54.50	
	1000	Guardrail, wooden, 3' high, 1" x 6", on 2" x 4" posts		200	.080		1.02	2.74		3.76	5.40	
	1100	2" x 6", on 4" x 4" posts	↓	165	.097		1.85	3.32		5.17	7.20	
	1200	Portable metal with base pads, buy					15.50			15.50	17.05	
	1250	Typical installation, assume 10 reuses	2 Carp	600	.027	↓	1.60	.91		2.51	3.18	
	1300	Barricade tape, polyethelyne, 7 mil, 3" wide x 500' long roll				Ea.	25			25	27.50	
250	0010	**TEMPORARY FENCING** Chain link, 11 ga, 5' high	2 Clab	400	.040	L.F.	4.65	1.07		5.72	6.75	250
	0100	6' high		300	.053		4.48	1.42		5.90	7.15	
	0200	Rented chain link, 6' high, to 1000' (up to 12 mo.)		400	.040		2.94	1.07		4.01	4.89	
	0250	Over 1000' (up to 12 mo.)	↓	300	.053		2.13	1.42		3.55	4.56	
	0350	Plywood, painted, 2" x 4" frame, 4' high	A-4	135	.178		5.30	5.85		11.15	14.85	
	0400	4" x 4" frame, 8' high	"	110	.218		9.90	7.20		17.10	22	
	0500	Wire mesh on 4" x 4" posts, 4' high	2 Carp	100	.160		7.90	5.50		13.40	17.25	
	0550	8' high	"	80	.200	↓	12.10	6.85		18.95	24	
400	0010	**TEMPORARY CONSTRUCTION** See also division 01530										400
		01580 | Project Signs										
700	0010	**SIGNS** Hi-intensity reflectorized, no posts, buy				S.F.	16.40			16.40	18	700

01590 | Equipment Rental

			UNIT	HOURLY OPER. COST	RENT PER DAY	RENT PER WEEK	RENT PER MONTH	CREW EQUIPMENT COST/DAY
100	0010	**CONCRETE EQUIPMENT RENTAL**						
	0100	without operators						
	0150	For batch plant, see div. 01590-500						
	0200	Bucket, concrete lightweight, 1/2 C.Y.	Ea.	.55	15.35	46	138	13.60
	0300	1 C.Y.		.60	18.65	56	168	16
	0400	1-1/2 C.Y.		.75	25.50	76	228	21.20
	0500	2 C.Y.		.80	30.50	91	273	24.60
	0580	8 C.Y.		4.40	200	600	1,800	155.20
	0600	Cart, concrete, self propelled, operator walking, 10 C.F.		2.10	55	165	495	49.80
	0700	Operator riding, 18 C.F.		3.35	83.50	250	750	76.80
	0800	Conveyer for concrete, portable, gas, 16" wide, 26' long		7.35	117	350	1,050	128.80
	0900	46' long		7.70	142	425	1,275	146.60
	1000	56' long		7.85	150	450	1,350	152.80
	1100	Core drill, electric, 2-1/2 H.P., 1" to 8" bit diameter		1.61	62.50	187	560	50.30
	1150	11 HP, 8" to 18" cores		6.61	83.50	250.80	750	103.05
	1200	Finisher, concrete floor, gas, riding trowel, 48" diameter		4.95	86.50	260	780	91.60
	1300	Gas, manual, 3 blade, 36" trowel		1.05	16.65	50	150	18.40
	1400	4 blade, 48" trowel		1.50	21	63	189	24.60
	1500	Float, hand-operated (Bull float) 48" wide		.08	13.35	40	120	8.65
	1570	Curb builder, 14 H.P., gas, single screw		9.50	195	585	1,750	193
	1590	Double screw		9.90	228	685	2,050	216.20
	1600	Grinder, concrete and terrazzo, electric, floor		2.10	88.50	266	800	70
	1700	Wall grinder		1.05	44.50	133	400	35
	1800	Mixer, powered, mortar and concrete, gas, 6 C.F., 18 H.P.		5.25	102	305	915	103
	1900	10 C.F., 25 H.P.		6.40	120	360	1,075	123.20
	2000	16 C.F.		6.70	143	430	1,300	139.60
	2100	Concrete, stationary, tilt drum, 2 C.Y.		5.25	200	600	1,800	162
	2120	Pump, concrete, truck mounted 4" line 80' boom		21.65	885	2,655	7,975	704.20
	2140	5" line, 110' boom		28.05	1,175	3,510	10,500	926.40
	2160	Mud jack, 50 C.F. per hr.		4.82	115	346	1,050	107.75
	2180	225 C.F. per hr.		7.56	154	461.60	1,375	152.80
	2190	Shotcrete pump rig, 12 CY/hr		11.10	225	675	2,025	223.80
	2600	Saw, concrete, manual, gas, 18 H.P.		3.60	35	105	315	49.80
	2650	Self-propelled, gas, 30 H.P.		7.05	96.50	290	870	114.40
	2700	Vibrators, concrete, electric, 60 cycle, 2 H.P.		.37	10.65	32	96	9.35
	2800	3 H.P.		.56	15.65	47	141	13.90
	2900	Gas engine, 5 H.P.		.95	18.35	55	165	18.60
	3000	8 H.P.		1.35	22	66	198	24
	3050	Vibrating screed, gas engine, 8 H.P.		1.68	49.50	148	445	43.05
	3100	Concrete transit mixer, hydraulic drive						
	3120	6 x 4, 250 H.P., 8 C.Y., rear discharge		34.95	550	1,650	4,950	609.60
	3200	Front discharge		40.65	670	2,015	6,050	728.20
	3300	6 x 6, 285 H.P., 12 C.Y., rear discharge		40.60	670	2,010	6,025	726.80
	3400	Front discharge		41.55	680	2,040	6,125	740.40
200	0010	**EARTHWORK EQUIPMENT RENTAL** Without operators (R01590-100)						
	0040	Aggregate spreader, push type 8' to 12' wide	Ea.	1.80	34.50	103	310	35
	0045	Tailgate type, 8' wide	"	1.75	31.50	95	285	33
	0050	Augers for vertical drilling						
	0055	Earth auger, truck-mounted, for fence & sign posts	Ea.	8.30	485	1,455	4,375	357.40
	0060	For borings and monitoring wells		29.20	620	1,855	5,575	604.60
	0070	Earth auger, portable, trailer mounted		1.55	22.50	67	201	25.80
	0075	Earth auger, truck-mounted, for caissons, water wells, utility poles		153.10	3,250	9,770	29,300	3,179
	0080	Auger, horizontal boring machine, 12" to 36" diameter, 45 H.P.		15.85	182	545	1,625	235.80
	0090	12" to 48" diameter, 65 H.P.		22.30	325	970	2,900	372.40
	0095	Auger, for fence posts, gas engine, hand held		.30	4.33	13	39	5
	0100	Excavator, diesel hydraulic, crawler mounted, 1/2 C.Y. cap.		15.80	345	1,030	3,100	332.40
	0120	5/8 C.Y. capacity		19.30	465	1,390	4,175	432.40
	0140	3/4 C.Y. capacity		22.50	490	1,470	4,400	474

01590 | Equipment Rental

			UNIT	HOURLY OPER. COST	RENT PER DAY	RENT PER WEEK	RENT PER MONTH	CREW EQUIPMENT COST/DAY	
200	0150	1 C.Y. capacity R01590 -100	Ea.	27.10	570	1,705	5,125	557.80	200
	0200	1-1/2 C.Y. capacity		33.55	755	2,260	6,775	720.40	
	0300	2 C.Y. capacity		42.35	950	2,855	8,575	909.80	
	0320	2-1/2 C.Y. capacity		55.75	1,275	3,845	11,500	1,215	
	0340	3-1/2 C.Y. capacity		94.45	2,100	6,320	19,000	2,020	
	0341	Attachments							
	0342	Bucket thumbs		2.50	208	625	1,875	145	
	0345	Grapples		1	200	600.40	1,800	128.10	
	0350	Gradall type, truck mounted, 3 ton @ 15' radius, 5/8 C.Y.		38.35	885	2,655	7,975	837.80	
	0370	1 C.Y. capacity		44.25	1,025	3,085	9,250	971	
	0400	Backhoe-loader, 40 to 45 H.P., 5/8 C.Y. capacity		8.40	177	530	1,600	173.20	
	0450	45 H.P. to 60 H.P., 3/4 C.Y. capacity		10.60	218	655	1,975	215.80	
	0460	80 H.P., 1-1/4 C.Y. capacity		13.30	250	750	2,250	256.40	
	0470	112 H.P., 1-1/2 C.Y. capacity		18.10	390	1,165	3,500	377.80	
	0480	Attachments							
	0482	Compactor, 20,000 lb		4.25	117	350	1,050	104	
	0485	Hydraulic hammer, 750 ft-lbs		1.95	68.50	205	615	56.60	
	0486	Hydraulic hammer, 1200 ft-lbs		4	133	400	1,200	112	
	0500	Brush chipper, gas engine, 6" cutter head, 35 H.P.		5.90	93.50	280	840	103.20	
	0550	12" cutter head, 130 H.P.		9.35	150	450	1,350	164.80	
	0600	15" cutter head, 165 H.P.		13.40	158	475	1,425	202.20	
	0750	Bucket, clamshell, general purpose, 3/8 C.Y.		1	35	105	315	29	
	0800	1/2 C.Y.		1.10	41.50	125	375	33.80	
	0850	3/4 C.Y.		1.25	51.50	155	465	41	
	0900	1 C.Y.		1.30	55	165	495	43.40	
	0950	1-1/2 C.Y.		2.05	75	225	675	61.40	
	1000	2 C.Y.		2.15	85	255	765	68.20	
	1010	Bucket, dragline, medium duty, 1/2 C.Y.		.60	22.50	67	201	18.20	
	1020	3/4 C.Y.		.60	23.50	70	210	18.80	
	1030	1 C.Y.		.65	25.50	76	228	20.40	
	1040	1-1/2 C.Y.		.95	38.50	115	345	30.60	
	1050	2 C.Y.		1.05	43.50	130	390	34.40	
	1070	3 C.Y.		1.55	58.50	175	525	47.40	
	1200	Compactor, manually guided 2-drum vibratory smooth roller, 7.5 H.P.		4.60	143	430	1,300	122.80	
	1250	Rammer compactor, gas, 1000 lb. blow		1.60	36.50	110	330	34.80	
	1300	Vibratory plate, gas, 18" plate, 3000 lb. blow		1.50	22	66	198	25.20	
	1350	21" plate, 5000 lb. blow		1.80	37.50	112	335	36.80	
	1370	Curb builder/extruder, 14 H.P., gas, single screw		9.50	197	590	1,775	194	
	1390	Double screw		9.90	228	685	2,050	216.20	
	1500	Disc harrow attachment, for tractor		.36	60.50	181	545	39.10	
	1750	Extractor, piling, see lines 2500 to 2750							
	1810	Feller buncher, shearing & accumulating trees, 100 H.P.	Ea.	19.80	455	1,370	4,100	432.40	
	1860	Grader, self-propelled, 25,000 lb.		18.20	400	1,205	3,625	386.60	
	1910	30,000 lb.		20.70	485	1,455	4,375	456.60	
	1920	40,000 lb.		30.50	730	2,190	6,575	682	
	1930	55,000 lb.		40.25	1,025	3,080	9,250	938	
	1950	Hammer, pavement demo., hyd., gas, self-prop., 1000 to 1250 lb.		20.15	390	1,170	3,500	395.20	
	2000	Diesel 1300 to 1500 lb.		27.80	590	1,765	5,300	575.40	
	2050	Pile driving hammer, steam or air, 4150 ft.-lb. @ 225 BPM		6.35	275	825	2,475	215.80	
	2100	8750 ft.-lb. @ 145 BPM		8.25	450	1,350	4,050	336	
	2150	15,000 ft.-lb. @ 60 BPM		8.60	485	1,455	4,375	359.80	
	2200	24,450 ft.-lb. @ 111 BPM		11.35	535	1,605	4,825	411.80	
	2250	Leads, 15,000 ft.-lb. hammers	L.F.	.03	1.69	5.06	15.20	1.25	
	2300	24,450 ft.-lb. hammers and heavier	"	.05	2.67	8	24	2	
	2350	Diesel type hammer, 22,400 ft.-lb.	Ea.	24	620	1,860	5,575	564	
	2400	41,300 ft.-lb.		32.20	675	2,025	6,075	662.60	
	2450	141,000 ft.-lb.		60.90	1,475	4,430	13,300	1,373	
	2500	Vib. elec. hammer/extractor, 200 KW diesel generator, 34 H.P.		26.25	655	1,965	5,900	603	

01590 | Equipment Rental

			UNIT	HOURLY OPER. COST	RENT PER DAY	RENT PER WEEK	RENT PER MONTH	CREW EQUIPMENT COST/DAY	
200	2550	80 H.P. R01590-100	Ea.	44.80	960	2,885	8,650	935.40	200
	2600	150 H.P.		64	1,475	4,455	13,400	1,403	
	2700	Extractor, steam or air, 700 ft.-lb.		14.50	410	1,235	3,700	363	
	2750	1000 ft.-lb.		16.60	515	1,545	4,625	441.80	
	2800	Log chipper, up to 22" diam, 600 H.P.		38.53	1,250	3,770	11,300	1,062	
	2850	Logger, for skidding & stacking logs, 150 H.P.		33.80	790	2,375	7,125	745.40	
	2900	Rake, spring tooth, with tractor		7.99	213	638	1,925	191.50	
	3000	Roller, vibratory, tandem, smooth drum, 20 H.P.		5.05	112	335	1,000	107.40	
	3050	35 H.P.		7.15	203	610	1,825	179.20	
	3100	Towed type vibratory compactor, smooth drum, 50 H.P.		31.95	560	1,680	5,050	591.60	
	3150	Sheepsfoot, 50 H.P.		32.65	580	1,745	5,225	610.20	
	3170	Landfill compactor, 220 HP		46.75	1,150	3,470	10,400	1,068	
	3200	Pneumatic tire roller, 80 H.P.		8.35	292	875	2,625	241.80	
	3250	120 H.P.		13.60	505	1,515	4,550	411.80	
	3300	Sheepsfoot vibratory roller, 200 H.P.		35.35	855	2,560	7,675	794.80	
	3320	340 H.P.		48.90	1,225	3,670	11,000	1,125	
	3350	Smooth drum vibratory roller, 75 H.P.		13.60	395	1,180	3,550	344.80	
	3400	125 H.P.		17.15	495	1,480	4,450	433.20	
	3410	Rotary mower, brush, 60", with tractor		11.35	230	690	2,075	228.80	
	3450	Scrapers, towed type, 9 to 12 C.Y. capacity		4.12	196	589	1,775	150.75	
	3500	12 to 17 C.Y. capacity		1.51	262	785.60	2,350	169.20	
	3550	Scrapers, self-propelled, 4 x 4 drive, 2 engine, 14 C.Y. capacity		82.40	1,500	4,465	13,400	1,552	
	3600	2 engine, 24 C.Y. capacity		119.45	2,275	6,805	20,400	2,317	
	3640	32 - 44 C.Y. capacity		142.65	2,650	7,935	23,800	2,728	
	3650	Self-loading, 11 C.Y. capacity		40.55	825	2,475	7,425	819.40	
	3700	22 C.Y. capacity		76.40	1,725	5,185	15,600	1,648	
	3710	Screening plant 110 H.P. w/ 5' x 10' screen		21.75	370	1,105	3,325	395	
	3720	5' x 16' screen		23.80	465	1,390	4,175	468.40	
	3850	Shovels, see Cranes division 01590-600							
	3860	Shovel/backhoe bucket, 1/2 C.Y.	Ea.	1.80	55	165	495	47.40	
	3870	3/4 C.Y.		1.85	61.50	185	555	51.80	
	3880	1 C.Y.		1.95	71.50	215	645	58.60	
	3890	1-1/2 C.Y.		2.10	85	255	765	67.80	
	3910	3 C.Y.		2.40	118	355	1,075	90.20	
	3950	Stump chipper, 18" deep, 30 H.P.		4.56	57	171	515	70.70	
	4110	Tractor, crawler, with bulldozer, torque converter, diesel 80 H.P.		16.10	310	930	2,800	314.80	
	4150	105 H.P.		22.10	460	1,385	4,150	453.80	
	4200	140 H.P.		27.25	590	1,770	5,300	572	
	4260	200 H.P.		40.95	985	2,960	8,875	919.60	
	4310	300 H.P.		53.10	1,275	3,850	11,600	1,195	
	4360	410 H.P.		71.95	1,575	4,740	14,200	1,524	
	4370	500 H.P.		95.25	2,125	6,365	19,100	2,035	
	4380	700 H.P.		142.20	3,325	10,005	30,000	3,139	
	4400	Loader, crawler, torque conv., diesel, 1-1/2 C.Y., 80 H.P.		14.95	320	965	2,900	312.60	
	4450	1-1/2 to 1-3/4 C.Y., 95 H.P.		17.35	385	1,150	3,450	368.80	
	4510	1-3/4 to 2-1/4 C.Y., 130 H.P.		23.60	605	1,815	5,450	551.80	
	4530	2-1/2 to 3-1/4 C.Y., 190 H.P.		35.50	840	2,520	7,550	788	
	4560	3-1/2 to 5 C.Y., 275 H.P.		46.95	1,200	3,590	10,800	1,094	
	4610	Tractor loader, wheel, torque conv., 4 x 4, 1 to 1-1/4 C.Y., 65 H.P.		9.85	192	575	1,725	193.80	
	4620	1-1/2 to 1-3/4 C.Y., 80 H.P.		12.50	235	705	2,125	241	
	4650	1-3/4 to 2 C.Y., 100 H.P.		14.20	278	835	2,500	280.60	
	4710	2-1/2 to 3-1/2 C.Y., 130 H.P.		15.25	305	915	2,750	305	
	4730	3 to 4-1/2 C.Y., 170 H.P.		20.95	465	1,390	4,175	445.60	
	4760	5-1/4 to 5-3/4 C.Y., 270 H.P.		34.55	705	2,120	6,350	700.40	
	4810	7 to 8 C.Y., 375 H.P.		58.95	1,250	3,755	11,300	1,223	
	4870	12-1/2 C.Y., 690 H.P.		82.30	1,950	5,860	17,600	1,830	
	4880	Wheeled, skid steer, 10 C.F., 30 H.P. gas		8.80	138	415	1,250	153.40	
	4890	1 C.Y., 78 H.P., diesel		10.35	192	575	1,725	197.80	

01590 | Equipment Rental

		01590 Equipment Rental	UNIT	HOURLY OPER. COST	RENT PER DAY	RENT PER WEEK	RENT PER MONTH	CREW EQUIPMENT COST/DAY	
200	4891	Attachments for all skid steer loaders R01590 -100							200
	4892	Auger	Ea.	.40	66.50	200	600	43.20	
	4893	Backhoe		.65	109	326	980	70.40	
	4894	Broom		.63	105	314	940	67.85	
	4895	Forks		.22	37.50	112	335	24.15	
	4896	Grapple		.50	83	249	745	53.80	
	4897	Concrete hammer		.98	163	489	1,475	105.65	
	4898	Tree spade		1	167	500	1,500	108	
	4899	Trencher		.68	114	342	1,025	73.85	
	4900	Trencher, chain, boom type, gas, operator walking, 12 H.P.		2.75	43.50	130	390	48	
	4910	Operator riding, 40 H.P.		8.55	242	725	2,175	213.40	
	5000	Wheel type, diesel, 4' deep, 12" wide		46.90	735	2,205	6,625	816.20	
	5100	Diesel, 6' deep, 20" wide		63.75	1,575	4,750	14,300	1,460	
	5150	Ladder type, diesel, 5' deep, 8" wide		23.10	705	2,115	6,350	607.80	
	5200	Diesel, 8' deep, 16" wide		56.80	1,600	4,830	14,500	1,420	
	5210	Tree spade, self-propelled		9.60	267	800	2,400	236.80	
	5250	Truck, dump, tandem, 12 ton payload		20.45	272	815	2,450	326.60	
	5300	Three axle dump, 16 ton payload		27.95	420	1,265	3,800	476.60	
	5350	Dump trailer only, rear dump, 16-1/2 C.Y.		4.25	115	345	1,025	103	
	5400	20 C.Y.		4.60	132	395	1,175	115.80	
	5450	Flatbed, single axle, 1-1/2 ton rating		11.85	56.50	170	510	128.80	
	5500	3 ton rating		15.15	91.50	275	825	176.20	
	5550	Off highway rear dump, 25 ton capacity		40.40	980	2,945	8,825	912.20	
	5600	35 ton capacity		41.30	1,000	3,015	9,050	933.40	
	5610	50 ton capacity		53.15	1,250	3,780	11,300	1,181	
	5620	65 ton capacity		56.85	1,375	4,105	12,300	1,276	
	5630	100 ton capacity		73	1,775	5,325	16,000	1,649	
	6000	Vibratory plow, 25 H.P., walking	↓	3.90	58.50	175	525	66.20	
400	0010	**GENERAL EQUIPMENT RENTAL** Without operators							400
	0150	Aerial lift, scissor type, to 15' high, 1000 lb. cap., electric	Ea.	2.30	41.50	125	375	43.40	
	0160	To 25' high, 2000 lb. capacity		2.70	60	180	540	57.60	
	0170	Telescoping boom to 40' high, 500 lb. capacity, gas		11.45	268	805	2,425	252.60	
	0180	To 45' high, 500 lb. capacity		12.30	310	925	2,775	283.40	
	0190	To 60' high, 600 lb. capacity		14.25	410	1,230	3,700	360	
	0195	Air compressor, portable, 6.5 CFM, electric		.42	14.65	44	132	12.15	
	0196	Gasoline		.51	22	66	198	17.30	
	0200	Air compressor, portable, gas engine, 60 C.F.M.		4.50	32.50	97	291	55.40	
	0300	160 C.F.M.		7.80	41.50	125	375	87.40	
	0400	Diesel engine, rotary screw, 250 C.F.M.		7.80	83.50	250	750	112.40	
	0500	365 C.F.M.		10.35	107	320	960	146.80	
	0550	450 C.F.M.		12.55	127	380	1,150	176.40	
	0600	600 C.F.M.		21.35	212	635	1,900	297.80	
	0700	750 C.F.M.		23.05	222	665	2,000	317.40	
	0800	For silenced models, small sizes, add		3%	5%	5%	5%		
	0900	Large sizes, add	↓	5%	7%	7%	7%		
	0920	Air tools and accessories							
	0930	Breaker, pavement, 60 lb.	Ea.	.40	11	33	99	9.80	
	0940	80 lb.		.40	12.65	38	114	10.80	
	0950	Drills, hand (jackhammer) 65 lb.		.45	15	45	135	12.60	
	0960	Track or wagon, swing boom, 4" drifter		35.05	640	1,920	5,750	664.40	
	0970	5" drifter		47.70	735	2,205	6,625	822.60	
	0975	Track mounted quarry drill, 6" diameter drill		50.45	795	2,380	7,150	879.60	
	0980	Dust control per drill		.78	12	36	108	13.45	
	0990	Hammer, chipping, 12 lb.		.40	21.50	64	192	16	
	1000	Hose, air with couplings, 50' long, 3/4" diameter		.03	5.65	17	51	3.65	
	1100	1" diameter		.03	5.65	17	51	3.65	
	1200	1-1/2" diameter		.04	7.35	22	66	4.70	
	1300	2" diameter	↓	.10	17.35	52	156	11.20	

01590 | Equipment Rental

			UNIT	HOURLY OPER. COST	RENT PER DAY	RENT PER WEEK	RENT PER MONTH	CREW EQUIPMENT COST/DAY	
400	1400	2-1/2" diameter	Ea.	.12	19.65	59	177	12.75	400
	1410	3" diameter		.16	27.50	82	246	17.70	
	1450	Drill, steel, 7/8" x 2'		.05	8	24	72	5.20	
	1460	7/8" x 6'		.06	9.35	28	84	6.10	
	1520	Moil points		.03	4.33	13	39	2.85	
	1525	Pneumatic nailer w/accessories		.41	27	81	243	19.50	
	1530	Sheeting driver for 60 lb. breaker		.10	6.95	20.80	62.50	4.95	
	1540	For 90 lb. breaker		.15	10	30	90	7.20	
	1550	Spade, 25 lb.		.35	6.35	19	57	6.60	
	1560	Tamper, single, 35 lb.		.58	38.50	115	345	27.65	
	1570	Triple, 140 lb.		.87	57.50	173	520	41.55	
	1580	Wrenches, impact, air powered, up to 3/4" bolt		.25	7.35	22	66	6.40	
	1590	Up to 1-1/4" bolt		.35	15.35	46	138	12	
	1600	Barricades, barrels, reflectorized, 1 to 50 barrels		.02	2.60	7.80	23.50	1.70	
	1610	100 to 200 barrels		.01	1.93	5.80	17.40	1.25	
	1620	Barrels with flashers, 1 to 50 barrels		.02	3.27	9.80	29.50	2.10	
	1630	100 to 200 barrels		.02	2.60	7.80	23.50	1.70	
	1640	Barrels with steady burn type C lights		.03	4.33	13	39	2.85	
	1650	Illuminated board, trailer mounted, with generator		.65	117	350	1,050	75.20	
	1670	Portable barricade, stock, with flashers, 1 to 6 units		.02	3.27	9.80	29.50	2.10	
	1680	25 to 50 units		.02	3.03	9.10	27.50	2	
	1690	Butt fusion machine, electric		22.35	425	1,270	3,800	432.80	
	1695	Electro fusion machine		8.85	168	505	1,525	171.80	
	1700	Carts, brick, hand powered, 1000 lb. capacity		.30	49.50	148	445	32	
	1800	Gas engine, 1500 lb., 7-1/2' lift		2.77	94	282	845	78.55	
	1822	Dehumidifier, medium, 6 lb/hr, 150 CFM		.68	41.50	124	370	30.25	
	1824	Large, 18 lb/hr, 600 CFM		1.36	82.50	248	745	60.50	
	1830	Distributor, asphalt, trailer mtd, 2000 gal., 38 H.P. diesel		7.35	243	730	2,200	204.80	
	1840	3000 gal., 38 H.P. diesel		8.35	278	835	2,500	233.80	
	1850	Drill, rotary hammer, electric, 1-1/2" diameter		.40	24.50	73	219	17.80	
	1860	Carbide bit for above		.03	5.35	16	48	3.45	
	1865	Rotary, crawler, 250 H.P.		87.35	1,600	4,835	14,500	1,666	
	1870	Emulsion sprayer, 65 gal., 5 H.P. gas engine		1.80	72	216	650	57.60	
	1880	200 gal., 5 H.P. engine		4.85	120	360	1,075	110.80	
	1920	Floodlight, mercury vapor, or quartz, on tripod							
	1930	1000 watt	Ea.	.30	12	36	108	9.60	
	1940	2000 watt		.52	22	66	198	17.35	
	1950	Floodlights, trailer mounted with generator, 1 - 300 watt light		2.45	65	195	585	58.60	
	1960	2 - 1000 watt lights		3.30	108	325	975	91.40	
	2000	4 - 300 watt lights		2.75	76.50	230	690	68	
	2020	Forklift, wheeled, for brick, 18', 3000 lb., 2 wheel drive, gas		13.70	177	530	1,600	215.60	
	2040	28', 4000 lb., 4 wheel drive, diesel		11.75	232	695	2,075	233	
	2050	For rough terrain, 8000 lb., 16' lift, 68 H.P.		15.70	360	1,075	3,225	340.60	
	2060	For plant, 4 T. capacity, 80 H.P., 2 wheel drive, gas		7.80	83.50	250	750	112.40	
	2080	10 T. capacity, 120 H.P., 2 wheel drive, diesel		11.95	157	470	1,400	189.60	
	2100	Generator, electric, gas engine, 1.5 KW to 3 KW		1.75	15	45	135	23	
	2200	5 KW		2.40	21.50	65	195	32.20	
	2300	10 KW		3.95	46.50	140	420	59.60	
	2400	25 KW		7.65	71.50	215	645	104.20	
	2500	Diesel engine, 20 KW		5.95	61.50	185	555	84.60	
	2600	50 KW		11.05	68.50	205	615	129.40	
	2700	100 KW		16.20	78.50	235	705	176.60	
	2800	250 KW		45.65	142	425	1,275	450.20	
	2850	Hammer, hydraulic, for mounting on boom, to 500 ft.-lb.		1.75	63.50	190	570	52	
	2860	1000 ft.-lb.		3.10	102	305	915	85.80	
	2900	Heaters, space, oil or electric, 50 MBH		.92	9.65	29	87	13.15	
	3000	100 MBH		1.66	13.65	41	123	21.50	
	3100	300 MBH		5.33	35	105	315	63.65	

01590 | Equipment Rental

			UNIT	HOURLY OPER. COST	RENT PER DAY	RENT PER WEEK	RENT PER MONTH	CREW EQUIPMENT COST/DAY	
400	3150	500 MBH	Ea.	10.70	50	150	450	115.60	400
	3200	Hose, water, suction with coupling, 20' long, 2" diameter		.02	4	12	36	2.55	
	3210	3" diameter		.03	6.65	20	60	4.25	
	3220	4" diameter		.03	8	24	72	5.05	
	3230	6" diameter		.10	20	60	180	12.80	
	3240	8" diameter		.25	41.50	125	375	27	
	3250	Discharge hose with coupling, 50' long, 2" diameter		.01	3	9	27	1.90	
	3260	3" diameter		.01	4	12	36	2.50	
	3270	4" diameter		.02	5.65	17	51	3.55	
	3280	6" diameter		.06	14	42	126	8.90	
	3290	8" diameter		.31	52	156	470	33.70	
	3295	Insulation blower		.11	7	21	63	5.10	
	3300	Ladders, extension type, 16' to 36' long		.17	28.50	86	258	18.55	
	3400	40' to 60' long		.26	36.50	110	330	24.10	
	3405	Lance for cutting concrete		2.78	105	315	945	85.25	
	3407	Lawn mower, rotary, 22", 5HP		.97	27	81	243	23.95	
	3408	48" self propelled		3.06	100	300	900	84.50	
	3410	Level, laser type, for pipe and sewer leveling		1.27	84.50	253	760	60.75	
	3430	Electronic		.76	50.50	151	455	36.30	
	3440	Laser type, rotating beam for grade control		1.04	69	207	620	49.70	
	3460	Builders level with tripod and rod		.07	12.35	37	111	7.95	
	3500	Light towers, towable, with diesel generator, 2000 watt		2.75	76.50	230	690	68	
	3600	4000 watt		3.30	108	325	975	91.40	
	3700	Mixer, powered, plaster and mortar, 6 C.F., 7 H.P.		1.30	26.50	80	240	26.40	
	3800	10 C.F., 9 H.P.		1.55	41.50	125	375	37.40	
	3850	Nailer, pneumatic		.41	27	81	243	19.50	
	3900	Paint sprayers complete, 8 CFM		.69	45.50	137	410	32.90	
	4000	17 CFM		1.10	73.50	220	660	52.80	
	4020	Pavers, bituminous, rubber tires, 8' wide, 50 H.P., diesel		26.80	775	2,320	6,950	678.40	
	4030	10' wide, 150 H.P.		59.45	1,300	3,895	11,700	1,255	
	4050	Crawler, 8' wide, 100 H.P., diesel		55.60	1,450	4,335	13,000	1,312	
	4060	10' wide, 150 H.P.		66.95	1,875	5,590	16,800	1,654	
	4070	Concrete paver, 12' to 24' wide, 250 H.P.		57.90	1,325	3,995	12,000	1,262	
	4080	Placer-spreader-trimmer, 24' wide, 300 H.P.		87.60	2,025	6,045	18,100	1,910	
	4100	Pump, centrifugal gas pump, 1-1/2", 4 MGPH		2.60	36.50	110	330	42.80	
	4200	2", 8 MGPH		3.25	41.50	125	375	51	
	4300	3", 15 MGPH		3.45	43.50	130	390	53.60	
	4400	6", 90 MGPH		15.75	152	455	1,375	217	
	4500	Submersible electric pump, 1-1/4", 55 GPM		.35	17	51	153	13	
	4600	1-1/2", 83 GPM		.41	19.65	59	177	15.10	
	4700	2", 120 GPM		.43	24.50	73	219	18.05	
	4800	3", 300 GPM		.90	33.50	100	300	27.20	
	4900	4", 560 GPM		6.16	135	405	1,225	130.30	
	5000	6", 1590 GPM		9.27	200	600	1,800	194.15	
	5100	Diaphragm pump, gas, single, 1-1/2" diameter		.78	38.50	116	350	29.45	
	5200	2" diameter		2.70	48.50	145	435	50.60	
	5300	3" diameter		2.75	48.50	145	435	51	
	5400	Double, 4" diameter		3.55	76.50	230	690	74.40	
	5500	Trash pump, self-priming, gas, 2" diameter		2.55	26.50	80	240	36.40	
	5600	Diesel, 4" diameter		6.05	53.50	160	480	80.40	
	5650	Diesel, 6" diameter		18.35	112	335	1,000	213.80	
	5655	Grout Pump	↓	8.30	58.50	175	525	101.40	
	5660	Rollers, see division 01590-200							
	5700	Salamanders, L.P. gas fired, 100,000 BTU	Ea.	1.66	10.65	32	96	19.70	
	5705	50,000 BTU		1.25	7.65	23	69	14.60	
	5720	Sandblaster, portable, open top, 3 C.F. capacity		.40	20	60	180	15.20	
	5730	6 C.F. capacity		.65	29	87	261	22.60	
	5740	Accessories for above	↓	.11	18.35	55	165	11.90	

01590 | Equipment Rental

			UNIT	HOURLY OPER. COST	RENT PER DAY	RENT PER WEEK	RENT PER MONTH	CREW EQUIPMENT COST/DAY	
400	5750	Sander, floor	Ea.	.72	17	51	153	15.95	400
	5760	Edger		.71	25	75	225	20.70	
	5800	Saw, chain, gas engine, 18" long		1.20	16	48	144	19.20	
	5900	36" long		.55	48.50	145	435	33.40	
	5950	60" long		.55	50	150	450	34.40	
	6000	Masonry, table mounted, 14" diameter, 5 H.P.		1.26	54.50	163	490	42.70	
	6050	Portable cut-off, 8 H.P.		1.30	28.50	85	255	27.40	
	6100	Circular, hand held, electric, 7-1/4" diameter		.20	7.35	22	66	6	
	6200	12" diameter		.27	11	33	99	8.75	
	6250	Wall saw, w/hydraulic power, 10 H.P		2.04	95	284.80	855	73.30	
	6275	Shot blaster, walk behind, 20" wide		1.42	545	1,635	4,900	338.35	
	6300	Steam cleaner, 100 gallons per hour		2.25	61.50	185	555	55	
	6310	200 gallons per hour		3	75	225	675	69	
	6340	Tar Kettle/Pot, 400 gallon		2.69	51.50	155	465	52.50	
	6350	Torch, cutting, acetylene-oxygen, 150' hose		1.50	10	30	90	18	
	6360	Hourly operating cost includes tips and gas		8.10				64.80	
	6410	Toilet, portable chemical		.11	17.65	53	159	11.50	
	6420	Recycle flush type		.13	21.50	65	195	14.05	
	6430	Toilet, fresh water flush, garden hose,		.15	24.50	73	219	15.80	
	6440	Hoisted, non-flush, for high rise		.13	21.50	64	192	13.85	
	6450	Toilet, trailers, minimum		.22	36.50	110	330	23.75	
	6460	Maximum		.66	110	330	990	71.30	
	6465	Tractor, farm with attachment		10.25	222	665	2,000	215	
	6470	Trailer, office, see division 01520-500							
	6500	Trailers, platform, flush deck, 2 axle, 25 ton capacity	Ea.	4.25	90	270	810	88	
	6600	40 ton capacity		5.55	125	375	1,125	119.40	
	6700	3 axle, 50 ton capacity		6	138	415	1,250	131	
	6800	75 ton capacity		7.45	182	545	1,625	168.60	
	6810	Trailer mounted cable reel for H.V. line work		4.47	213	639	1,925	163.55	
	6820	Trailer mounted cable tensioning rig		8.75	415	1,250	3,750	320	
	6830	Cable pulling rig		55.44	2,375	7,120	21,400	1,868	
	6850	Trailer, storage, see division 01520-500							
	6900	Water tank, engine driven discharge, 5000 gallons	Ea.	5.60	120	360	1,075	116.80	
	6925	10,000 gallons		7.70	168	505	1,525	162.60	
	6950	Water truck, off highway, 6000 gallons		49.25	700	2,095	6,275	813	
	7010	Tram car for H.V. line work, powered, 2 conductor		5.73	116	347	1,050	115.25	
	7020	Transit (builder's level) with tripod		.07	12.35	37	111	7.95	
	7030	Trench box, 3000 lbs. 6'x8'		.67	90.50	272	815	59.75	
	7040	7200 lbs. 6'x20'		.82	136	409	1,225	88.35	
	7050	8000 lbs., 8' x 16'		.88	147	442	1,325	95.45	
	7060	9500 lbs., 8'x20'		1.19	199	597	1,800	128.90	
	7065	11,000 lbs., 8'x24'		1.34	224	671	2,025	144.90	
	7070	12,000 lbs., 10' x 20'		1.77	294	883	2,650	190.75	
	7100	Truck, pickup, 3/4 ton, 2 wheel drive		5.75	53.50	160	480	78	
	7200	4 wheel drive		5.90	63.50	190	570	85.20	
	7250	Crew carrier, 9 passenger		5.44	92	276.40	830	98.80	
	7290	Tool van, 24,000 G.V.W.		8.40	91.50	274.80	825	122.15	
	7300	Tractor, 4 x 2, 30 ton capacity, 195 H.P.		13.65	165	495	1,475	208.20	
	7410	250 H.P.		18.20	243	730	2,200	291.60	
	7500	6 x 2, 40 ton capacity, 240 H.P.		17.45	270	810	2,425	301.60	
	7600	6 x 4, 45 ton capacity, 240 H.P.		22.20	293	880	2,650	353.60	
	7620	Vacuum truck, hazardous material, 2500 gallon		6.51	300	902	2,700	232.50	
	7625	5,000 gallon		11.37	400	1,202.80	3,600	331.50	
	7640	Tractor, with A frame, boom and winch, 225 H.P.		15.75	223	670	2,000	260	
	7650	Vacuum, H.E.P.A., 16 gal., wet/dry		.27	24	72	216	16.55	
	7655	55 gal, wet/dry		.60	36	108	325	26.40	
	7660	Water tank, portable		1	9.35	28	84	13.60	
	7690	Large production vacuum loader, 3150 CFM		15	600	1,800	5,400	480	

01590 | Equipment Rental

			UNIT	HOURLY OPER. COST	RENT PER DAY	RENT PER WEEK	RENT PER MONTH	CREW EQUIPMENT COST/DAY	
400	7700	Welder, electric, 200 amp	Ea.	3.79	60.50	181	545	66.50	400
	7800	300 amp		5.28	64	192	575	80.65	
	7900	Gas engine, 200 amp		5.70	28	84	252	62.40	
	8000	300 amp		7.70	32.50	98	294	81.20	
	8100	Wheelbarrow, any size		.06	10.65	32	96	6.90	
	8200	Wrecking ball, 4000 lb.	↓	1.85	68.50	205	615	55.80	
500	0010	**HIGHWAY EQUIPMENT RENTAL**							500
	0050	Asphalt batch plant, portable drum mixer, 100 ton/hr.	Ea.	55.75	1,275	3,850	11,600	1,216	
	0060	200 ton/hr.		61.85	1,350	4,045	12,100	1,304	
	0070	300 ton/hr.		72.05	1,600	4,785	14,400	1,533	
	0100	Backhoe attachment, long stick, up to 185 HP, 10.5' long		.30	20	60	180	14.40	
	0140	Up to 250 HP, 12' long		.33	21.50	65	195	15.65	
	0180	Over 250 HP, 15' long		.43	28.50	85	255	20.45	
	0200	Special dipper arm, up to 100 HP, 32' long		.88	58.50	175	525	42.05	
	0240	Over 100 HP, 33' long		1.10	73.50	220	660	52.80	
	0300	Concrete batch plant, portable, electric, 200 CY/Hr	↓	10.40	570	1,715	5,150	426.20	
	0500	Grader attachment, ripper/scarifier, rear mounted							
	0520	Up to 135 HP	Ea.	2.75	58.50	175	525	57	
	0540	Up to 180 HP		3.30	75	225	675	71.40	
	0580	Up to 250 HP		3.65	85	255	765	80.20	
	0700	Pvmt. removal bucket, for hyd. excavator, up to 90 HP		1.40	43.50	130	390	37.20	
	0740	Up to 200 HP		1.60	65	195	585	51.80	
	0780	Over 200 HP		1.70	76.50	230	690	59.60	
	0900	Aggregate spreader, self-propelled, 187 HP		37.15	790	2,370	7,100	771.20	
	1000	Chemical spreader, 3 C.Y.		2.20	61.50	185	555	54.60	
	1900	Hammermill, traveling, 250 HP		39.98	1,700	5,120	15,400	1,344	
	2000	Horizontal borer, 3" diam, 13 HP gas driven		3.95	51.50	155	465	62.60	
	2200	Hydromulchers, gas power, 3000 gal., for truck mounting		10.50	187	560	1,675	196	
	2400	Joint & crack cleaner, walk behind, 25 HP		2.05	46.50	140	420	44.40	
	2500	Filler, trailer mounted, 400 gal., 20 HP		6	182	545	1,625	157	
	3000	Paint striper, self propelled, double line, 30 HP		5.15	155	465	1,400	134.20	
	3200	Post drivers, 6" I-Beam frame, for truck mounting		8.15	415	1,250	3,750	315.20	
	3400	Road sweeper, self propelled, 8' wide, 90 HP		23.50	430	1,285	3,850	445	
	4000	Road mixer, self-propelled, 130 HP		29.05	570	1,715	5,150	575.40	
	4100	310 HP	↓	54	1,850	5,570	16,700	1,546	
	4200	Cold mix paver, incl pug mill and bitumen tank,							
	4220	165 HP	Ea.	66.95	1,875	5,600	16,800	1,656	
	4250	Paver, asphalt, wheel or crawler, 130 H.P., diesel		65.65	1,775	5,325	16,000	1,590	
	4300	Paver, road widener, gas 1' to 6', 67 HP		29.65	630	1,890	5,675	615.20	
	4400	Diesel, 2' to 14', 88 HP		39.95	955	2,870	8,600	893.60	
	4600	Slipform pavers, curb and gutter, 2 track, 75 HP		24.85	640	1,920	5,750	582.80	
	4700	4 track, 165 HP		34.75	745	2,235	6,700	725	
	4800	Median barrier, 215 HP		35.35	775	2,325	6,975	747.80	
	4901	Trailer, low bed, 75 ton capacity		7.90	178	535	1,600	170.20	
	5000	Road planer, walk behind, 10" cutting width, 10 HP		2	26	78	234	31.60	
	5100	Self propelled, 12" cutting width, 64 HP		5.30	195	585	1,750	159.40	
	5200	Pavement profiler, 4' to 6' wide, 450 HP		146.05	2,675	8,040	24,100	2,776	
	5300	8' to 10' wide, 750 HP		233.70	4,000	11,980	35,900	4,266	
	5400	Roadway plate, steel, 1"x8'x20'		.06	10	30	90	6.50	
	5600	Stabilizer,self-propelled, 150 HP		27.40	540	1,625	4,875	544.20	
	5700	310 HP		44.55	1,150	3,445	10,300	1,045	
	5800	Striper, thermal, truck mounted 120 gal. paint, 150 H.P.		32.40	485	1,450	4,350	549.20	
	6000	Tar kettle, 330 gal., trailer mounted		2.37	36.50	110	330	40.95	
	7000	Tunnel locomotive, diesel, 8 to 12 ton		20.95	540	1,615	4,850	490.60	
	7005	Electric, 10 ton		20.65	610	1,830	5,500	531.20	
	7010	Muck cars, 1/2 C.Y. capacity		1.50	20	60	180	24	
	7020	1 C.Y. capacity		1.70	28	84	252	30.40	
	7030	2 C.Y. capacity	↓	1.85	32.50	98	294	34.40	

01590 | Equipment Rental

			Unit	Hourly Oper. Cost	Rent Per Day	Rent Per Week	Rent Per Month	Crew Equipment Cost/Day	
500	7040	Side dump, 2 C.Y. capacity	Ea.	2	40	120	360	40	500
	7050	3 C.Y. capacity		2.70	45	135	405	48.60	
	7060	5 C.Y. capacity		3.85	58.50	175	525	65.80	
	7100	Ventilating blower for tunnel, 7-1/2 H.P.		1.22	44	132	395	36.15	
	7110	10 H.P.		1.41	45.50	137	410	38.70	
	7120	20 H.P.		2.30	56.50	170	510	52.40	
	7140	40 H.P.		4.02	83.50	250	750	82.15	
	7160	60 H.P.		6.09	127	380	1,150	124.70	
	7175	75 H.P.		7.79	168	505	1,525	163.30	
	7180	200 H.P.		17.45	245	735	2,200	286.60	
	7800	Windrow loader, elevating		33.40	875	2,620	7,850	791.20	
600	0010	**LIFTING AND HOISTING EQUIPMENT RENTAL** R01590-100							600
	0100	without operators							
	0120	Aerial lift truck, 2 person, to 80'	Ea.	18.60	580	1,740	5,225	496.80	
	0140	Boom work platform, 40' snorkel		9.80	203	610	1,825	200.40	
	0150	Crane, flatbed mntd, 3 ton cap.		12.45	183	550	1,650	209.60	
	0200	Crane, climbing, 106' jib, 6000 lb. capacity, 410 FPM		43.35	1,350	4,070	12,200	1,161	
	0300	101' jib, 10,250 lb. capacity, 270 FPM		48.75	1,725	5,150	15,500	1,420	
	0400	Tower, static, 130' high, 106' jib,							
	0500	6200 lb. capacity at 400 FPM	Ea.	46.50	1,575	4,700	14,100	1,312	
	0600	Crawler mounted, lattice boom, 1/2 C.Y., 15 tons at 12' radius		20.76	570	1,710	5,125	508.10	
	0700	3/4 C.Y., 20 tons at 12' radius		27.68	680	2,045	6,125	630.45	
	0800	1 C.Y., 25 tons at 12' radius		36.90	895	2,690	8,075	833.20	
	0900	1-1/2 C.Y., 40 tons at 12' radius		41.15	930	2,795	8,375	888.20	
	1000	2 C.Y., 50 tons at 12' radius		51.35	1,325	3,975	11,900	1,206	
	1100	3 C.Y., 75 tons at 12' radius		49.10	1,300	3,885	11,700	1,170	
	1200	100 ton capacity, 60' boom		62.80	1,700	5,085	15,300	1,519	
	1300	165 ton capacity, 60' boom		92.20	2,175	6,560	19,700	2,050	
	1400	200 ton capacity, 70' boom		98.10	2,350	7,015	21,000	2,188	
	1500	350 ton capacity, 80' boom		141.90	3,325	9,945	29,800	3,124	
	1600	Truck mounted, lattice boom, 6 x 4, 20 tons at 10' radius		27.76	960	2,875	8,625	797.10	
	1700	25 tons at 10' radius		29.66	1,025	3,070	9,200	851.30	
	1800	8 x 4, 30 tons at 10' radius		36.66	1,075	3,260	9,775	945.30	
	1900	40 tons at 12' radius		32.74	1,150	3,450	10,400	951.90	
	2000	8 x 4, 60 tons at 15' radius		39.81	1,250	3,770	11,300	1,072	
	2050	82 tons at 15' radius		46.50	1,500	4,470	13,400	1,266	
	2100	90 tons at 15' radius		44.08	1,625	4,860	14,600	1,325	
	2200	115 tons at 15' radius		50.55	1,800	5,400	16,200	1,484	
	2300	150 tons at 18' radius		43.10	1,825	5,500	16,500	1,445	
	2350	165 tons at 18' radius		77.25	2,150	6,470	19,400	1,912	
	2400	Truck mounted, hydraulic, 12 ton capacity		32.45	570	1,715	5,150	602.60	
	2500	25 ton capacity		32.60	595	1,780	5,350	616.80	
	2550	33 ton capacity		33.45	625	1,870	5,600	641.60	
	2560	40 ton capacity		31.85	595	1,790	5,375	612.80	
	2600	55 ton capacity		47.05	865	2,595	7,775	895.40	
	2700	80 ton capacity		56	910	2,735	8,200	995	
	2720	100 ton capacity		80.75	2,325	6,945	20,800	2,035	
	2740	120 ton capacity		85.55	2,525	7,550	22,700	2,194	
	2760	150 ton capacity		104.65	3,175	9,535	28,600	2,744	
	2800	Self-propelled, 4 x 4, with telescoping boom, 5 ton		15.50	320	965	2,900	317	
	2900	12-1/2 ton capacity		25.10	510	1,525	4,575	505.80	
	3000	15 ton capacity		25.70	585	1,755	5,275	556.60	
	3050	20 ton capacity		26.80	610	1,835	5,500	581.40	
	3100	25 ton capacity		27.25	595	1,785	5,350	575	
	3150	40 ton capacity		45.20	845	2,540	7,625	869.60	
	3200	Derricks, guy, 20 ton capacity, 60' boom, 75' mast		12.72	330	996	3,000	300.95	
	3300	100' boom, 115' mast		20.60	570	1,710	5,125	506.80	

01590 | Equipment Rental

				UNIT	HOURLY OPER. COST	RENT PER DAY	RENT PER WEEK	RENT PER MONTH	CREW EQUIPMENT COST/DAY	
600	3400	Stiffleg, 20 ton capacity, 70' boom, 37' mast	R01590 -100	Ea.	14.71	425	1,280	3,850	373.70	600
	3500	100' boom, 47' mast			23.19	695	2,080	6,250	601.50	
	3550	Helicopter, small, lift to 1250 lbs. maximum, w/pilot			67.85	2,675	8,050	24,200	2,153	
	3600	Hoists, chain type, overhead, manual, 3/4 ton			.10	2	6	18	2	
	3900	10 ton			.60	13	39	117	12.60	
	4000	Hoist and tower, 5000 lb. cap., portable electric, 40' high			4.19	191	574	1,725	148.30	
	4100	For each added 10' section, add			.09	15	45	135	9.70	
	4200	Hoist and single tubular tower, 5000 lb. electric, 100' high			5.66	267	801	2,400	205.50	
	4300	For each added 6'-6" section, add			.15	25	75	225	16.20	
	4400	Hoist and double tubular tower, 5000 lb., 100' high			6.06	294	882	2,650	224.90	
	4500	For each added 6'-6" section, add			.17	28.50	85	255	18.35	
	4550	Hoist and tower, mast type, 6000 lb., 100' high			6.56	305	915	2,750	235.50	
	4570	For each added 10' section, add			.11	18.35	55	165	11.90	
	4600	Hoist and tower, personnel, electric, 2000 lb., 100' @ 125 FPM			13.52	815	2,440	7,325	596.15	
	4700	3000 lb., 100' @ 200 FPM			15.40	915	2,750	8,250	673.20	
	4800	3000 lb., 150' @ 300 FPM			17.10	1,025	3,090	9,275	754.80	
	4900	4000 lb., 100' @ 300 FPM			17.73	1,050	3,150	9,450	771.85	
	5000	6000 lb., 100' @ 275 FPM			19.14	1,100	3,300	9,900	813.10	
	5100	For added heights up to 500', add		L.F.	.01	1.67	5	15	1.10	
	5200	Jacks, hydraulic, 20 ton		Ea.	.05	6.35	19	57	4.20	
	5500	100 ton		"	.30	18.65	56	168	13.60	
	6000	Jacks, hydraulic, climbing with 50' jackrods								
	6010	and control consoles, minimum 3 mo. rental								
	6100	30 ton capacity		Ea.	1.65	110	329	985	79	
	6150	For each added 10' jackrod section, add			.05	3.33	10	30	2.40	
	6300	50 ton capacity			2.65	177	530	1,600	127.20	
	6350	For each added 10' jackrod section, add			.06	4	12	36	2.90	
	6500	125 ton capacity			6.95	465	1,390	4,175	333.60	
	6550	For each added 10' jackrod section, add			.48	31.50	95	285	22.85	
	6600	Cable jack, 10 ton capacity with 200' cable			1.38	91.50	275	825	66.05	
	6650	For each added 50' of cable, add			.15	10	30	90	7.20	
700	0010	**WELLPOINT EQUIPMENT RENTAL** See also division 02240								700
	0020	Based on 2 months rental								
	0100	Combination jetting & wellpoint pump, 60 H.P. diesel		Ea.	9.17	272	817	2,450	236.75	
	0200	High pressure gas jet pump, 200 H.P., 300 psi		"	16.39	233	698	2,100	270.70	
	0300	Discharge pipe, 8" diameter		L.F.	.01	.44	1.32	3.96	.35	
	0350	12" diameter			.01	.65	1.96	5.90	.45	
	0400	Header pipe, flows up to 150 G.P.M., 4" diameter			.01	.40	1.20	3.60	.30	
	0500	400 G.P.M., 6" diameter			.01	.47	1.42	4.26	.35	
	0600	800 G.P.M., 8" diameter			.01	.65	1.96	5.90	.45	
	0700	1500 G.P.M., 10" diameter			.01	.69	2.06	6.20	.50	
	0800	2500 G.P.M., 12" diameter			.02	1.30	3.89	11.65	.95	
	0900	4500 G.P.M., 16" diameter			.02	1.66	4.98	14.95	1.15	
	0950	For quick coupling aluminum and plastic pipe, add			.03	1.72	5.15	15.45	1.25	
	1100	Wellpoint, 25' long, with fittings & riser pipe, 1-1/2" or 2" diameter		Ea.	.05	3.43	10.28	31	2.45	
	1200	Wellpoint pump, diesel powered, 4" diameter, 20 H.P.			4.45	157	471	1,425	129.80	
	1300	6" diameter, 30 H.P.			5.81	195	584	1,750	163.30	
	1400	8" suction, 40 H.P.			7.91	267	801	2,400	223.50	
	1500	10" suction, 75 H.P.			10.86	310	936	2,800	274.10	
	1600	12" suction, 100 H.P.			16.25	500	1,500	4,500	430	
	1700	12" suction, 175 H.P.			21.61	550	1,650	4,950	502.90	
800	0010	**MARINE EQUIPMENT RENTAL**								800
	0200	Barge, 400 Ton, 30' wide x 90' long		Ea.	16.15	240	720	2,150	273.20	
	0240	800 Ton, 45' wide x 90' long			26.65	345	1,030	3,100	419.20	
	2000	Tugboat, diesel, 100 HP			16.95	168	505	1,525	236.60	
	2040	250 HP			31.45	315	945	2,825	440.60	
	2080	380 HP			73.10	925	2,780	8,350	1,141	

Important: See the Reference Section for critical supporting data - Reference Nos., Crews, & City Cost Indexes

01700 | Execution Requirements

	01740 | Cleaning		CREW	DAILY OUTPUT	LABOR-HOURS	UNIT	2005 BARE COSTS MAT.	LABOR	EQUIP.	TOTAL	TOTAL INCL O&P	
500	0010	**CLEANING UP** After job completion, allow, minimum				Job					.30%	500
	0040	Maximum				"					1%	
	0050	Cleanup of floor area, continuous, per day, during const.	A-5	24	.750	M.S.F.	1.70	20	1.34	23.04	34.50	
	0100	Final by GC at end of job	"	11.50	1.565	"	2.65	42	2.80	47.45	71	
	1000	Mechanical demolition, see division 02220										

01800 | Facility Operation

	01810 | Commissioning		CREW	DAILY OUTPUT	LABOR-HOURS	UNIT	2005 BARE COSTS MAT.	LABOR	EQUIP.	TOTAL	TOTAL INCL O&P	
100	0010	**COMMISSIONING** Including documentation of design intent										100
	0100	performance verification, O&M, training, min				Project					.50%	
	0150	Maximum				"					.75%	

For information about Means Estimating Seminars, see yellow pages 12 and 13 in back of book

Division Notes

		CREW	DAILY OUTPUT	LABOR-HOURS	UNIT	2005 BARE COSTS MAT.	LABOR	EQUIP.	TOTAL	TOTAL INCL O&P

Division 2 Site Construction

Estimating Tips

02200 Site Preparation

- If possible visit the site and take an inventory of the type, quantity and size of the trees. Certain trees may have a landscape resale value or firewood value. Stump disposal can be very expensive, particularly if they cannot be buried at the site. Consider using a bulldozer in lieu of hand cutting trees.
- Estimators should visit the site to determine the need for haul road, access, storage of materials, and security considerations. When estimating for access roads on unstable soil, consider using a geotextile stabilization fabric. It can greatly reduce the quantity of crushed stone or gravel. Sites of limited size and access can cause cost overruns due to lost productivity. Theft and damage is another consideration if the location is isolated. A temporary fence or security guards may be required. Investigate the site thoroughly.

02210 Subsurface Investigation

In preparing estimates on structures involving earthwork or foundations, all information concerning soil characteristics should be obtained. Look particularly for hazardous waste, evidence of prior dumping of debris, and previous stream beds.

02220 Selective Demolition

The costs shown for selective demolition do not include rubbish handling or disposal. These items should be estimated separately using Means data or other sources.

- Historic preservation often requires that the contractor remove materials from the existing structure, rehab them and replace them. The estimator must be aware of any related measures and precautions that must be taken when doing selective demolition, and cutting and patching. Requirements may include special handling and storage, as well as security.
- In addition to Section 02220, you can find selective demolition items in each division. Example: Roofing demolition is in division 7.

02300 Earthwork

- Estimating the actual cost of performing earthwork requires careful consideration of the variables involved. This includes items such as type of soil, whether or not water will be encountered, dewatering, whether or not banks need bracing, disposal of excavated earth, length of haul to fill or spoil sites, etc. If the project has large quantities of cut or fill, consider raising or lowering the site to reduce costs while paying close attention to the effect on site drainage and utilities if doing this.
- If the project has large quantities of fill, creating a borrow pit on the site can significantly lower the costs.
- It is very important to consider what time of year the project is scheduled for completion. Bad weather can create large cost overruns from dewatering, site repair and lost productivity from cold weather.

02500 Utility Services
02600 Drainage & Containment

- Never assume that the water, sewer and drainage lines will go in at the early stages of the project. Consider the site access needs before dividing the site in half with open trenches, loose pipe, and machinery obstructions. Always inspect the site to establish that the site drawings are complete. Check off all existing utilities on your drawing as you locate them. If you find any discrepancies, mark up the site plan for further research. Differing site conditions can be very costly if discovered later in the project.
- See also Section 02955 for restoration of pipe where removal/replacement may be undesirable. Use of new types of piping materials can reduce the overall project cost. Owners/design engineers should consider the installing construction as a valuable source of current information on piping products that could lead to significant utility cost savings.

02700 Bases, Ballasts, Pavements/Appurtenances

- When estimating paving, keep in mind the project schedule. If an asphaltic paving project is in a colder climate and runs through to the spring, consider placing the base course in the autumn, then topping it in the spring just prior to completion. This could save considerable costs in spring repair. Keep in mind that prices for asphalt and concrete are generally higher in the cold seasons.
- See also Sections 02960/02965.

02900 Planting

- The timing of planting and guarantee specifications often dictate the costs for establishing tree and shrub growth and a stand of grass or ground cover. Establish the work performance schedule to coincide with the local planting season. Maintenance and growth guarantees can add from 20% to 100% to the total landscaping cost. The cost to replace trees and shrubs can be as high as 5% of the total cost depending on the planting zone, soil conditions and time of year.

02960 & 02965 Flexible Pavement Surfacing Recovery

- Recycling of asphalt pavement is becoming very popular and is an alternative to removal and replacement of asphalt pavement. It can be a good value engineering proposal if removed pavement can be recycled either at the site or another site that is reasonably close to the project site.

Reference Numbers

Reference numbers are shown in bold squares at the beginning of some major classifications. These numbers refer to related items in the Reference Section. The reference information may be an estimating procedure, an alternate pricing method or technical information.

Note: Not all subdivisions listed here necessarily appear in this publication.

02065 | Cement & Concrete

		02065 Cement & Concrete	CREW	DAILY OUTPUT	LABOR-HOURS	UNIT	2005 BARE COSTS MAT.	LABOR	EQUIP.	TOTAL	TOTAL INCL O&P	
300	0010	**PLANT MIXED BITUMINOUS CONCRETE**										300
	0020	Asphaltic concrete plant mix (145 LB per c.f.)				Ton	34.50			34.50	38	
	0040	Asphaltic concrete less than 300 tons add trucking costs										
	0050	See 02315-490-0010 for hauling costs										
	0200	All weather patching mix, hot					33.50			33.50	37	
	0250	Cold patch					40			40	44	
	0300	Berm mix					33.50			33.50	37	

02080 | Utility Materials

		02080 Utility Materials	CREW	DAILY OUTPUT	LABOR-HOURS	UNIT	MAT.	LABOR	EQUIP.	TOTAL	TOTAL INCL O&P	
500	0010	**VALVES** Water distribution, see also div. 15110										500
	3000	Butterfly valves with boxes, cast iron, mech. jt.										
	3100	4″ diameter	B-20	6	4	Ea.	214	120		334	420	
	3140	6″ diameter	″	5	4.800		246	144		390	495	
	3180	8″ diameter	B-21	4	7		298	216	39.50	553.50	705	
	3300	10″ diameter		3.50	8		440	247	45.50	732.50	915	
	3340	12″ diameter		3	9.333		570	288	53	911	1,125	
	3400	14″ diameter		2	14		980	435	79	1,494	1,825	
	3440	16″ diameter		2	14		1,325	435	79	1,839	2,200	
	3460	18″ diameter		1.50	18.667		1,650	575	106	2,331	2,800	
	3480	20″ diameter		1	28		2,075	865	158	3,098	3,800	
	3500	24″ diameter		.50	56		2,875	1,725	315	4,915	6,200	
	3600	With lever operator										
	3610	4″ diameter	B-20	6	4	Ea.	116	120		236	315	
	3614	6″ diameter	″	5	4.800		162	144		306	405	
	3616	8″ diameter	B-21	4	7		235	216	39.50	490.50	640	
	3618	10″ diameter		3.50	8		375	247	45.50	667.50	850	
	3620	12″ diameter		3	9.333		700	288	53	1,041	1,275	
	3622	14″ diameter		2	14		1,125	435	79	1,639	1,975	
	3624	16″ diameter		2	14		1,525	435	79	2,039	2,425	
	3626	18″ diameter		1.50	18.667		1,750	575	106	2,431	2,925	
	3628	20″ diameter		1	28		2,325	865	158	3,348	4,075	
	3630	24″ diameter		.50	56		2,500	1,725	315	4,540	5,775	
	3700	Check valves, flanged										
	3710	4″ diameter	B-20	6	4	Ea.	425	120		545	650	
	3714	6″ diameter	″	5	4.800		720	144		864	1,025	
	3716	8″ diameter	B-21	4	7		1,475	216	39.50	1,730.50	1,975	
	3718	10″ diameter		3.50	8		2,325	247	45.50	2,617.50	3,000	
	3720	12″ diameter		3	9.333		3,575	288	53	3,916	4,450	
	3722	14″ diameter		2	14		5,725	435	79	6,239	7,050	
	3724	16″ diameter		2	14		7,900	435	79	8,414	9,425	
	3726	18″ diameter		1.50	18.667		11,700	575	106	12,381	13,800	
	3728	20″ diameter		1	28		13,600	865	158	14,623	16,400	
	3730	24″ diameter		.50	56		19,500	1,725	315	21,540	24,500	
	3800	Gate valves, C.I., 250 PSI, mechanical joint, w/boxes										
	3810	4″ diameter	B-21A	8	5	Ea.	166	165	63.50	394.50	505	
	3814	6″ diameter		6.80	5.882		214	194	74.50	482.50	615	
	3816	8″ diameter		5.60	7.143		340	236	90.50	666.50	830	
	3818	10″ diameter		4.80	8.333		530	275	105	910	1,125	
	3820	12″ diameter		4.08	9.804		730	325	124	1,179	1,425	
	3822	14″ diameter	B-21	2	14		760	435	79	1,274	1,600	
	3824	16″ diameter		1	28		900	865	158	1,923	2,500	
	3826	18″ diameter		.80	35		1,100	1,075	198	2,373	3,125	
	3828	20″ diameter		.80	35		1,400	1,075	198	2,673	3,450	
	3830	24″ diameter		.50	56		1,725	1,725	315	3,765	4,925	
	3831	30″ diameter		.35	80		2,075	2,475	455	5,005	6,600	

02050 | Basic Site Materials & Methods

	02080 | Utility Materials		Crew	Daily Output	Labor-Hours	Unit	2005 Bare Costs Mat.	Labor	Equip.	Total	Total Incl O&P	
500	3832	36" diameter	B-21	.30	93.333	Ea.	2,475	2,875	530	5,880	7,775	500
	3900	Globe valves, flanged, iron body, class 125										
	3910	4" diameter	B-20	10	2.400	Ea.	1,125	72		1,197	1,350	
	3914	6" diameter	"	9	2.667		2,175	80		2,255	2,525	
	3916	8" diameter	B-21	6	4.667		4,050	144	26.50	4,220.50	4,700	
	3918	10" diameter		5	5.600		6,375	173	31.50	6,579.50	7,300	
	3920	12" diameter		4	7		8,900	216	39.50	9,155.50	10,200	
	3922	14" diameter		3	9.333		9,550	288	53	9,891	11,000	
	3924	16" diameter		2	14		8,025	435	79	8,539	9,575	
	3926	18" diameter		.80	35		8,825	1,075	198	10,098	11,600	
	3928	20" diameter		.60	46.667		8,700	1,450	264	10,414	12,100	
	3930	24" diameter		.50	56		10,300	1,725	315	12,340	14,400	
600	0010	UTILITY ACCESSORIES G1030 805										600
	0400	Underground tape, detectable, reinforced, alum. foil core, 2"	1 Clab	150	.053	C.L.F.	1.43	1.42		2.85	3.79	
	0500	6"	"	140	.057	"	3.58	1.53		5.11	6.30	

02100 | Site Remediation

	02115 | Underground Storage Tank Removal		Crew	Daily Output	Labor-Hours	Unit	2005 Bare Costs Mat.	Labor	Equip.	Total	Total Incl O&P	
200	0010	REMOVAL OF UNDERGROUND STORAGE TANKS R02115 -200										200
	0011	Petroleum storage tanks, non-leaking										
	0100	Excavate & load onto trailer										
	0110	3000 gal. to 5000 gal. tank	B-14	4	12	Ea.		335	54	389	580	
	0120	6000 gal to 8000 gal tank	B-3A	3	13.333			375	240	615	845	
	0130	9000 gal to 12000 gal tank	"	2	20			565	360	925	1,275	
	0190	Known leaking tank add				%				100%	100%	
	0200	Remove sludge, water and remaining product from tank bottom										
	0201	of tank with vacuum truck										
	0300	3000 gal to 5000 gal tank	A-13	5	1.600	Ea.		53	96	149	186	
	0310	6000 gal to 8000 gal tank		4	2			66	120	186	232	
	0320	9000 gal to 12000 gal tank		3	2.667			88	160	248	310	
	0390	Dispose of sludge off-site, average				Gal.					4.40	
	0400	Insert inert solid CO2 "dry ice" into tank										
	0401	For cleaning/transporting tanks (1.5 lbs./100 gal. cap)	1 Clab	500	.016	Lb.	1.34	.43		1.77	2.13	
	1020	Haul tank to certified salvage dump, 100 miles round trip										
	1023	3000 gal. to 5000 gal. tank				Ea.				550	690	
	1026	6000 gal. to 8000 gal. tank								650	825	
	1029	9,000 gal. to 12,000 gal. tank								875	1,100	
	1100	Disposal of contaminated soil to landfill										
	1110	Minimum				C.Y.					110	
	1111	Maximum				"					330	
	1120	Disposal of contaminated soil to										
	1121	bituminous concrete batch plant										
	1130	Minimum				C.Y.					55	
	1131	Maximum				"					110	
	2010	Decontamination of soil on site incl poly tarp on top/bottom										
	2011	Soil containment berm, and chemical treatment										
	2020	Minimum	B-11C	100	.160	C.Y.	5.60	4.91	2.16	12.67	16	
	2021	Maximum	"	100	.160		7.25	4.91	2.16	14.32	17.85	

02100 | Site Remediation

	02115 | Underground Storage Tank Removal	CREW	DAILY OUTPUT	LABOR-HOURS	UNIT	2005 BARE COSTS MAT.	LABOR	EQUIP.	TOTAL	TOTAL INCL O&P	
200	2050 Disposal of decontaminated soil, minimum R02115-200				C.Y.					66	200
	2055 Maximum				↓					135	

02200 | Site Preparation

	02210 | Subsurface Investigation	CREW	DAILY OUTPUT	LABOR-HOURS	UNIT	2005 BARE COSTS MAT.	LABOR	EQUIP.	TOTAL	TOTAL INCL O&P	
200	0010 CORE DRILLING Reinforced concrete slab, up to 6" thick slab										200
	0020 Including bit, layout and set up										
	0100 1" diameter core	B-89A	28	.571	Ea.	2.40	17.60	3.68	23.68	34	
	0150 Each added inch thick, add		300	.053		.43	1.64	.34	2.41	3.40	
	0300 3" diameter core		23	.696		5.35	21.50	4.48	31.33	44.50	
	0350 Each added inch thick, add		186	.086		.96	2.65	.55	4.16	5.80	
	0500 4" diameter core		19	.842		5.35	26	5.40	36.75	52.50	
	0550 Each added inch thick, add		170	.094		1.21	2.90	.61	4.72	6.50	
	0700 6" diameter core		14	1.143		8.80	35	7.35	51.15	72.50	
	0750 Each added inch thick, add		140	.114		1.49	3.52	.74	5.75	7.90	
	0900 8" diameter core		11	1.455		12	45	9.35	66.35	93	
	0950 Each added inch thick, add		95	.168		2.02	5.20	1.08	8.30	11.45	
	1100 10" diameter core		10	1.600		16	49.50	10.30	75.80	105	
	1150 Each added inch thick, add		80	.200		2.65	6.15	1.29	10.09	13.95	
	1300 12" diameter core		9	1.778		19.20	54.50	11.45	85.15	119	
	1350 Each added inch thick, add		68	.235		3.18	7.25	1.52	11.95	16.40	
	1500 14" diameter core		7	2.286		23.50	70.50	14.70	108.70	151	
	1550 Each added inch thick, add		55	.291		4.04	8.95	1.87	14.86	20.50	
	1700 18" diameter core		4	4		30	123	26	179	254	
	1750 Each added inch thick, add	↓	28	.571		5.30	17.60	3.68	26.58	37.50	
	1760 For horizontal holes, add to above				↓				30%	30%	
	1770 Prestressed hollow core plank, 6" thick										
	1780 1" diameter core	B-89A	52	.308	Ea.	1.59	9.45	1.98	13.02	18.70	
	1790 Each added inch thick, add		350	.046		.28	1.41	.29	1.98	2.82	
	1800 3" diameter core		50	.320		3.51	9.85	2.06	15.42	21.50	
	1810 Each added inch thick, add		240	.067		.58	2.05	.43	3.06	4.30	
	1820 4" diameter core		48	.333		4.67	10.25	2.15	17.07	23.50	
	1830 Each added inch thick, add		216	.074		.81	2.28	.48	3.57	4.96	
	1840 6" diameter core		44	.364		5.80	11.20	2.34	19.34	26.50	
	1850 Each added inch thick, add		175	.091		.96	2.81	.59	4.36	6.10	
	1860 8" diameter core		32	.500		7.75	15.40	3.22	26.37	36	
	1870 Each added inch thick, add		118	.136		1.34	4.17	.87	6.38	8.95	
	1880 10" diameter core		28	.571		10.45	17.60	3.68	31.73	43	
	1890 Each added inch thick, add		99	.162		1.44	4.97	1.04	7.45	10.45	
	1900 12" diameter core		22	.727		12.75	22.50	4.68	39.93	54	
	1910 Each added inch thick, add		85	.188	↓	2.12	5.80	1.21	9.13	12.65	
	1950 Minimum charge for above, 3" diameter core		7	2.286	Total		70.50	14.70	85.20	125	
	2000 4" diameter core		6.80	2.353			72.50	15.15	87.65	130	
	2050 6" diameter core		6	2.667			82	17.15	99.15	147	
	2100 8" diameter core		5.50	2.909			89.50	18.75	108.25	160	
	2150 10" diameter core		4.75	3.368			104	21.50	125.50	185	
	2200 12" diameter core		3.90	4.103			126	26.50	152.50	225	
	2250 14" diameter core		3.38	4.734			146	30.50	176.50	261	
	2300 18" diameter core	↓	3.15	5.079	↓		156	32.50	188.50	279	

02220 | Site Demolition

		02220 Site Demolition	Crew	Daily Output	Labor-Hours	Unit	2005 Bare Costs Mat.	Labor	Equip.	Total	Total Incl O&P	
240	0010	**MINOR SITE DEMOLITION** R02220-510										240
	0015	No hauling, abandon catch basin or manhole	B-6	7	3.429	Ea.		99	31	130	186	
	0020	Remove existing catch basin or manhole, masonry		4	6			173	54	227	325	
	0030	Catch basin or manhole frames and covers, stored		13	1.846			53	16.60	69.60	100	
	0040	Remove and reset		7	3.429			99	31	130	186	
	0900	Hydrants, fire, remove only	B-21A	5	8			264	101	365	510	
	0950	Remove and reset	"	2	20			660	253	913	1,275	
	4000	Sidewalk removal, bituminous, 2-1/2" thick	B-6	325	.074	S.Y.		2.13	.66	2.79	4	
	4050	Brick, set in mortar		185	.130			3.74	1.17	4.91	7.05	
	4100	Concrete, plain, 4"		160	.150			4.32	1.35	5.67	8.15	
	4200	Mesh reinforced		150	.160			4.61	1.44	6.05	8.70	
250	0010	**DEMOLISH, REMOVE PAVEMENT AND CURB** R02220-510										250
	5010	Pavement removal, bituminous roads, 3" thick	B-38	690	.058	S.Y.		1.74	1.20	2.94	3.98	
	5050	4" to 6" thick		420	.095			2.85	1.96	4.81	6.55	
	5100	Bituminous driveways		640	.063			1.87	1.29	3.16	4.29	
	5200	Concrete to 6" thick, hydraulic hammer, mesh reinforced		255	.157			4.70	3.24	7.94	10.75	
	5300	Rod reinforced		200	.200			6	4.13	10.13	13.75	
	5400	Concrete, 7" to 24" thick, plain		33	1.212	C.Y.		36.50	25	61.50	83	
	5500	Reinforced		24	1.667	"		50	34.50	84.50	115	
	5600	With hand held air equipment, bituminous, to 6" thick	B-39	1,900	.025	S.F.		.71	.07	.78	1.18	
	5700	Concrete to 6" thick, no reinforcing		1,600	.030			.84	.09	.93	1.40	
	5800	Mesh reinforced		1,400	.034			.96	.10	1.06	1.60	
	5900	Rod reinforced		765	.063			1.76	.19	1.95	2.93	
330	0010	**SELECTIVE DEMOLITION, DUMP CHARGES** R02220-510										330
	0020	Dump charges, typical urban city, tipping fees only										
	0100	Building construction materials				Ton					70	
	0200	Trees, brush, lumber									50	
	0300	Rubbish only									60	
	0500	Reclamation station, usual charge									85	
350	0010	**SELECTIVE DEMOLITION, RUBBISH HANDLING** R02220-510										350
	0020	The following are to be added to the demolition prices										
	0400	Chute, circular, prefabricated steel, 18" diameter	B-1	40	.600	L.F.	28	16.40		44.40	56.50	
	0440	30" diameter	"	30	.800	"	37.50	22		59.50	75.50	
	0725	Dumpster, weekly rental, 1 dump/week, 20 C.Y. capacity (8 Tons)				Week					440	
	0800	30 C.Y. capacity (10 Tons)									665	
	0840	40 C.Y. capacity (13 Tons)									805	
	1000	Dust partition, 6 mil polyethylene, 1" x 3" frame	2 Carp	2,000	.008	S.F.	.17	.27		.44	.62	
	1080	2" x 4" frame	"	2,000	.008	"	.28	.27		.55	.74	
	2000	Load, haul, and dump, 50' haul	2 Clab	24	.667	C.Y.		17.80		17.80	27.50	
	2040	100' haul		16.50	.970			26		26	40.50	
	2080	Over 100' haul, add per 100 L.F.		35.50	.451			12.05		12.05	18.75	
	2120	In elevators, per 10 floors, add		140	.114			3.05		3.05	4.75	
	3000	Loading & trucking, including 2 mile haul, chute loaded	B-16	45	.711			19.50	10.60	30.10	41.50	
	3040	Hand loading truck, 50' haul	"	48	.667			18.25	9.95	28.20	39.50	
	3080	Machine loading truck	B-17	120	.267			7.60	4.52	12.12	16.60	
	5000	Haul, per mile, up to 8 C.Y. truck	B-34B	1,165	.007			.19	.41	.60	.74	
	5100	Over 8 C.Y. truck	"	1,550	.005			.14	.31	.45	.56	
370	0010	**SELECTIVE DEMOLITION, TORCH CUTTING** R02220-510										370
	0020	Steel, 1" thick plate	1 Clab	360	.022	L.F.	.18	.59		.77	1.12	
	0040	1" diameter bar	"	210	.038	Ea.		1.02		1.02	1.58	
	1000	Oxygen lance cutting, reinforced concrete walls										
	1040	12" to 16" thick walls	1 Clab	10	.800	L.F.		21.50		21.50	33	
	1080	24" thick walls	"	6	1.333	"		35.50		35.50	55.50	

02200 | Site Preparation

02240 | Dewatering

		02240 Dewatering	CREW	DAILY OUTPUT	LABOR-HOURS	UNIT	2005 BARE COSTS MAT.	LABOR	EQUIP.	TOTAL	TOTAL INCL O&P	
500	0010	**DEWATERING**										500
	0020	Excavate drainage trench, 2′ wide, 2′ deep	B-11C	90	.178	C.Y.		5.45	2.40	7.85	11	
	0100	2′ wide, 3′ deep, with backhoe loader	″	135	.119			3.64	1.60	5.24	7.30	
	0200	Excavate sump pits by hand, light soil	1 Clab	7.10	1.127			30		30	47	
	0300	Heavy soil	″	3.50	2.286			61		61	95	
	0500	Pumping 8 hr., attended 2 hrs. per day, including 20 L.F.										
	0550	of suction hose & 100 L.F. discharge hose										
	0600	2″ diaphragm pump used for 8 hours	B-10H	4	3	Day		96	14.25	110.25	162	
	0650	4″ diaphragm pump used for 8 hours	B-10I	4	3			96	21.50	117.50	170	
	0800	8 hrs. attended, 2″ diaphragm pump	B-10H	1	12			385	57	442	650	
	0900	3″ centrifugal pump	B-10J	1	12			385	63	448	655	
	1000	4″ diaphragm pump	B-10I	1	12			385	86.50	471.50	680	
	1100	6″ centrifugal pump	B-10K	1	12			385	248	633	855	
	1300	CMP, incl. excavation 3′ deep, 12″ diameter	B-6	115	.209	L.F.	9.65	6	1.88	17.53	22	
	1400	18″ diameter		100	.240	″	12	6.90	2.16	21.06	26	
	1600	Sump hole construction, incl. excavation and gravel, pit		1,250	.019	C.F.	.70	.55	.17	1.42	1.81	
	1700	With 12″ gravel collar, 12″ pipe, corrugated, 16 ga.		70	.343	L.F.	15.10	9.90	3.08	28.08	35	
	1800	15″ pipe, corrugated, 16 ga.		55	.436		19.30	12.60	3.92	35.82	44.50	
	1900	18″ pipe, corrugated, 16 ga.		50	.480		22.50	13.85	4.32	40.67	51.50	
	2000	24″ pipe, corrugated, 14 ga.		40	.600		27	17.30	5.40	49.70	62.50	
	2200	Wood lining, up to 4′ x 4′, add		300	.080	SFCA	12.80	2.31	.72	15.83	18.40	
	9950	See div. 02240-900 for wellpoints										
	9960	See div. 02240-700 for deep well systems										
700	0010	**WELLS** For dewatering 10′ to 20′ deep, 2′ diameter										700
	0020	with steel casing, minimum	B-6	165	.145	V.L.F.	2.35	4.19	1.31	7.85	10.50	
	0050	Average		98	.245		4.68	7.05	2.20	13.93	18.40	
	0100	Maximum		49	.490		12.50	14.10	4.40	31	40	
	0300	For pumps for dewatering, see division 01590-400										
	0500	For domestic water wells, see division 02520-510										
900	0010	**WELLPOINTS** For wellpoint equipment rental, see div. 01590-700 R02240-900										900
	0100	Installation and removal of single stage system										
	0110	Labor only, .75 labor-hours per L.F., minimum	1 Clab	10.70	.748	LF Hdr		19.95		19.95	31	
	0200	2.0 labor-hours per L.F., maximum	″	4	2	″		53.50		53.50	83	
	0400	Pump operation, 4 @ 6 hr. shifts										
	0410	Per 24 hour day	4 Eqlt	1.27	25.197	Day		835		835	1,250	
	0500	Per 168 hour week, 160 hr. straight, 8 hr. double time		.18	177	Week		5,875		5,875	8,850	
	0550	Per 4.3 week month		.04	800	Month		26,400		26,400	39,800	
	0600	Complete installation, operation, equipment rental, fuel &										
	0610	removal of system with 2″ wellpoints 5′ O.C.										
	0700	100′ long header, 6″ diameter, first month	4 Eqlt	3.23	9.907	LF Hdr	125	325		450	630	
	0800	Thereafter, per month		4.13	7.748		99.50	256		355.50	495	
	1000	200′ long header, 8″ diameter, first month		6	5.333		110	176		286	385	
	1100	Thereafter, per month		8.39	3.814		56	126		182	252	
	1300	500′ long header, 8″ diameter, first month		10.63	3.010		43.50	99.50		143	198	
	1400	Thereafter, per month		20.91	1.530		31	50.50		81.50	111	
	1600	1,000′ long header, 10″ diameter, first month		11.62	2.754		37.50	91		128.50	178	
	1700	Thereafter, per month		41.81	.765		18.70	25.50		44.20	58.50	
	1900	Note: above figures include pumping 168 hrs. per week										
	1910	and include the pump operator and one stand-by pump.										

02250 | Shoring & Underpinning

		02250 Shoring & Underpinning	CREW	DAILY OUTPUT	LABOR-HOURS	UNIT	MAT.	LABOR	EQUIP.	TOTAL	TOTAL INCL O&P	
400	0010	**SHEET PILING**										400
	0020	Sheet piling steel, not incl. wales, 22 psf, 15′ excav., left in place	B-40	10.81	5.920	Ton	865	200	212	1,277	1,500	
	0100	Drive, extract & salvage R02250-450		6	10.667		385	360	380	1,125	1,425	
	0300	20′ deep excavation, 27 psf, left in place		12.95	4.942		865	167	177	1,209	1,400	

02200 | Site Preparation

02250 | Shoring & Underpinning

		02250 Shoring & Underpinning	CREW	DAILY OUTPUT	LABOR-HOURS	UNIT	MAT.	LABOR	EQUIP.	TOTAL	TOTAL INCL O&P	
							2005 BARE COSTS					
400	0400	Drive, extract & salvage (R02250-450)	B-40	6.55	9.771	Ton	385	330	350	1,065	1,325	400
	1200	15′ deep excavation, 22 psf, left in place		983	.065	S.F.	10.05	2.20	2.33	14.58	17.10	
	1300	Drive, extract & salvage		545	.117		4.31	3.97	4.20	12.48	15.70	
	1500	20′ deep excavation, 27 psf, left in place		960	.067		12.65	2.25	2.39	17.29	20	
	1600	Drive, extract & salvage		485	.132		5.60	4.46	4.72	14.78	18.45	
	2100	Rent steel sheet piling and wales, first month				Ton	206			206	226	
	2200	Per added month					20.50			20.50	22.50	
	2300	Rental piling left in place, add to rental					685			685	755	
	2500	Wales, connections & struts, 2/3 salvage					210			210	231	
	3900	Wood, solid sheeting, incl. wales, braces and spacers,										
	3910	drive, extract & salvage, 8′ deep excavation	B-31	330	.121	S.F.	1.72	3.47	.39	5.58	7.70	
	4000	10′ deep, 50 S.F./hr. in & 150 S.F./hr. out		300	.133		1.77	3.81	.43	6.01	8.35	
	4100	12′ deep, 45 S.F./hr. in & 135 S.F./hr. out		270	.148		1.82	4.24	.48	6.54	9.15	
	4200	14′ deep, 42 S.F./hr. in & 126 S.F./hr. out		250	.160		1.88	4.58	.52	6.98	9.80	
	4300	16′ deep, 40 S.F./hr. in & 120 S.F./hr. out		240	.167		1.94	4.77	.54	7.25	10.10	
	4400	18′ deep, 38 S.F./hr. in & 114 S.F./hr. out		230	.174		2	4.98	.56	7.54	10.55	
	4520	Left in place, 8′ deep, 55 S.F./hr.		440	.091		3.10	2.60	.29	5.99	7.80	
	4540	10′ deep, 50 S.F./hr.		400	.100		3.26	2.86	.32	6.44	8.40	
	4560	12′ deep, 45 S.F./hr.		360	.111		3.44	3.18	.36	6.98	9.15	
	4565	14′ deep, 42 S.F./hr.		335	.119		3.65	3.42	.39	7.46	9.75	
	4570	16′ deep, 40 S.F./hr.		320	.125		3.88	3.58	.40	7.86	10.25	
	4700	Alternate pricing, left in place, 8′ deep		1.76	22.727	M.B.F.	695	650	73.50	1,418.50	1,850	
	4800	Drive, extract and salvage, 8′ deep		1.32	30.303	"	620	865	98	1,583	2,150	

02260 | Excavation Support/Protection

		02260 Excavation Support/Protection	CREW	DAILY OUTPUT	LABOR-HOURS	UNIT	MAT.	LABOR	EQUIP.	TOTAL	TOTAL INCL O&P	
200	0010	**COFFERDAMS**, incl. mobilization and temporary sheeting										200
	0080	Soldier beams & lagging H piles with 3″ wood sheeting										
	0090	horizontal between piles, including removal of wales & braces										
	0100	No hydrostatic head, 15′ deep, 1 line of braces, minimum	B-50	545	.205	S.F.	8	6.65	3.11	17.76	23	
	0200	Maximum		495	.226		8.90	7.30	3.42	19.62	25.50	
	1300	No hydrostatic head, left in place, 15′ dp., 1 line of braces, min.		635	.176		10.65	5.70	2.67	19.02	24	
	1400	Maximum		575	.195		11.45	6.30	2.94	20.69	26	
	2350	Lagging only, 3″ thick wood between piles 8′ O.C., minimum	B-46	400	.120		1.78	3.64	.08	5.50	7.90	
	2370	Maximum		250	.192		2.67	5.80	.13	8.60	12.45	
	2400	Open sheeting no bracing, for trenches to 10′ deep, min.		1,736	.028		.80	.84	.02	1.66	2.25	
	2450	Maximum		1,510	.032		.89	.96	.02	1.87	2.55	
	2500	Tie-back method, add to open sheeting, add, minimum								20%	20%	
	2550	Maximum								60%	60%	
	2700	Tie-backs only, based on tie-backs total length, minimum	B-46	86.80	.553	L.F.	10.70	16.75	.39	27.84	39	
	2750	Maximum		38.50	1.247	"	18.80	38	.87	57.67	82.50	
	3500	Tie-backs only, typical average, 25′ long		2	24	Ea.	470	730	16.80	1,216.80	1,700	
	3600	35′ long		1.58	30.380	"	625	920	21.50	1,566.50	2,200	

02300 | Earthwork

02305 | Equipment

		02305 Equipment	CREW	DAILY OUTPUT	LABOR-HOURS	UNIT	MAT.	LABOR	EQUIP.	TOTAL	TOTAL INCL O&P	
							2005 BARE COSTS					
250	0010	**MOBILIZATION OR DEMOB.** (One or the other, unless noted)										250
	0015	Up to 25 mi haul dist (50 mi round trip for mob/demob crew)										

02300 | Earthwork

		02305 | Equipment	CREW	DAILY OUTPUT	LABOR-HOURS	UNIT	2005 BARE COSTS MAT.	LABOR	EQUIP.	TOTAL	TOTAL INCL O&P	
250	0020	Dozer, loader, backhoe, excav., grader, paver, roller, 70 to 150 H.P.	B-34N	4	2	Ea.		55	112	167	207	250
	0100	Above 150 HP	B-34K	3	2.667			73.50	175	248.50	305	
	0900	Shovel or dragline, 3/4 C.Y.	"	3.60	2.222			61	146	207	254	
	1100	Small equipment, placed in rear of, or towed by pickup truck	A-3A	8	1			27	10.65	37.65	52.50	
	1150	Equip up to 70 HP, on flatbed trailer behind pickup truck	A-3D	4	2			53.50	43.50	97	129	
	2000	Mob & demob truck-mounted crane up to 75 ton, driver only	1 EQHV	3.60	2.222			80		80	120	
	2100	Crane, truck-mounted, over 75 ton	A-3E	2.50	6.400			203	34	237	350	
	2200	Crawler-mounted, up to 75 ton	A-3F	2	8			254	278	532	690	
	2300	Over 75 ton	A-3G	1.50	10.667			340	405	745	960	
	2500	For each additional 5 miles haul distance, add						10%	10%			
	3000	For large pieces of equipment, allow for assembly/knockdown										
	3100	For mob/demob of micro-tunneling equip, see section 02441-400										
		02315 | Excavation and Fill										
110	0010	**BACKFILL, GENERAL** G1030 805										110
	0015	By hand, no compaction, light soil	1 Clab	14	.571	L.C.Y.		15.25		15.25	23.50	
	0100	Heavy soil		11	.727	"		19.40		19.40	30	
	0300	Compaction in 6" layers, hand tamp, add to above		20.60	.388	E.C.Y.		10.35		10.35	16.15	
	0400	Roller compaction operator walking, add	B-10A	100	.120			3.84	1.23	5.07	7.20	
	0500	Air tamp, add	B-9D	190	.211			5.70	.93	6.63	9.95	
	0600	Vibrating plate, add	A-1D	60	.133			3.56	.42	3.98	6	
	0800	Compaction in 12" layers, hand tamp, add to above	1 Clab	34	.235			6.30		6.30	9.80	
	0900	Roller compaction operator walking, add	B-10A	150	.080			2.56	.82	3.38	4.79	
	1000	Air tamp, add	B-9	285	.140			3.80	.50	4.30	6.45	
	1100	Vibrating plate, add	A-1E	90	.089			2.37	.41	2.78	4.14	
	1300	Dozer backfilling, bulk, up to 300' haul, no compaction	B-10B	1,200	.010	L.C.Y.		.32	.77	1.09	1.33	
	1400	Air tamped, add	B-11B	80	.200	E.C.Y.		6	2.30	8.30	11.70	
	1600	Compacting backfill, 6" to 12" lifts, vibrating roller	B-10C	800	.015			.48	1.89	2.37	2.81	
	1700	Sheepsfoot roller	B-10D	750	.016			.51	2.04	2.55	3.02	
	1900	Dozer backfilling, trench, up to 300' haul, no compaction	B-10B	900	.013	L.C.Y.		.43	1.02	1.45	1.77	
	2000	Air tamped, add	B-11B	80	.200	E.C.Y.		6	2.30	8.30	11.70	
	2200	Compacting backfill, 6" to 12" lifts, vibrating roller	B-10C	700	.017			.55	2.16	2.71	3.20	
	2300	Sheepsfoot roller	B-10D	650	.018			.59	2.35	2.94	3.49	
416	0010	**DRILLING AND BLASTING ROCK**										416
	0020	Rock, open face, under 1500 CY	B-47	225	.107	B.C.Y.	1.89	3.14	4.43	9.46	11.80	
	0100	Over 1500 C.Y.		300	.080		1.89	2.36	3.33	7.58	9.35	
	0200	Areas where blasting mats are required, under 1500 C.Y.		175	.137		1.89	4.04	5.70	11.63	14.55	
	0250	Over 1500 C.Y.		250	.096		1.89	2.83	3.99	8.71	10.80	
	2200	Trenches, up to 1500 C.Y.		22	1.091		5.50	32	45.50	83	106	
	2300	Over 1500 C.Y.		26	.923		5.50	27	38.50	71	90	
462	0010	**EXCAVATION, STRUCTURAL**										462
	0015	Hand, pits to 6' deep, sandy soil	1 Clab	8	1	B.C.Y.		26.50		26.50	41.50	
	0100	Heavy soil or clay		4	2			53.50		53.50	83	
	0300	Pits 6' to 12' deep, sandy soil		5	1.600			42.50		42.50	66.50	
	0500	Heavy soil or clay		3	2.667			71		71	111	
	0700	Pits 12' to 18' deep, sandy soil		4	2			53.50		53.50	83	
	0900	Heavy soil or clay		2	4			107		107	166	
	1500	For wet or muck hand excavation, add to above				%				50%	50%	
490	0010	**HAULING**, excavated or borrow, loose cubic yards										490
	0012	no loading included, highway haulers										
	0020	6 C.Y. dump truck, 1/4 mile round trip, 5.0 loads/hr.	B-34A	195	.041	L.C.Y.		1.13	1.68	2.81	3.56	
	0030	1/2 mile round trip, 4.1 loads/hr.		160	.050			1.38	2.04	3.42	4.35	
	0040	1 mile round trip, 3.3 loads/hr. R01590 -100		130	.062			1.70	2.51	4.21	5.35	
	0100	2 mile round trip, 2.6 loads/hr.		100	.080			2.20	3.27	5.47	6.95	

02315 | Excavation and Fill

		02315 Excavation and Fill	CREW	DAILY OUTPUT	LABOR-HOURS	UNIT	MAT.	LABOR	EQUIP.	TOTAL	TOTAL INCL O&P	
							2005 BARE COSTS					
490	0150	3 mile round trip, 2.1 loads/hr. R01590-100	B-34A	80	.100	L.C.Y.		2.76	4.08	6.84	8.70	490
	0200	4 mile round trip, 1.8 loads/hr.		70	.114			3.15	4.67	7.82	9.95	
	0310	12 C.Y. dump truck, 1/4 mile round trip 3.7 loads/hr.	B-34B	288	.028			.77	1.66	2.43	2.99	
	0400	2 mile round trip, 2.2 loads/hr.		180	.044			1.22	2.65	3.87	4.78	
	0450	3 mile round trip, 1.9 loads/hr.		170	.047			1.30	2.80	4.10	5.05	
	0500	4 mile round trip, 1.6 loads/hr.		125	.064			1.76	3.81	5.57	6.90	
	1300	Hauling in medium traffic, add								20%	20%	
	1400	Heavy traffic, add								30%	30%	
	1600	Grading at dump, or embankment if required, by dozer	B-10B	1,000	.012			.38	.92	1.30	1.59	
	1800	Spotter at fill or cut, if required	1 Clab	8	1	Hr.		26.50		26.50	41.50	
520	0010	**FILL**, spread dumped material, no compaction										520
	0020	By dozer, no compaction	B-10B	1,000	.012	L.C.Y.		.38	.92	1.30	1.59	
	0100	By hand	1 Clab	12	.667	"		17.80		17.80	27.50	
	0500	Gravel fill, compacted, under floor slabs, 4" deep	B-37	10,000	.005	S.F.	.16	.13	.01	.30	.39	
	0600	6" deep		8,600	.006		.24	.16	.01	.41	.51	
	0700	9" deep		7,200	.007		.39	.19	.01	.59	.74	
	0800	12" deep		6,000	.008		.55	.22	.02	.79	.98	
	1000	Alternate pricing method, 4" deep		120	.400	E.C.Y.	11.80	11.25	.90	23.95	31.50	
	1100	6" deep		160	.300		11.80	8.45	.67	20.92	27	
	1200	9" deep		200	.240		11.80	6.75	.54	19.09	24	
	1300	12" deep		220	.218		11.80	6.15	.49	18.44	23	
	1400	Granular fill				L.C.Y.	7.90			7.90	8.65	
610	0010	**EXCAVATING, TRENCH** or continuous footing, common earth G1030 805										610
	0020	No sheeting or dewatering included										
	0050	1' to 4' deep, 3/8 C.Y. tractor loader/backhoe	B-11C	150	.107	B.C.Y.		3.27	1.44	4.71	6.60	
	0060	1/2 C.Y. tractor loader/backhoe	B-11M	200	.080			2.45	1.28	3.73	5.15	
	0090	4' to 6' deep, 1/2 C.Y. tractor loader/backhoe	"	200	.080			2.45	1.28	3.73	5.15	
	0100	5/8 C.Y. hydraulic backhoe	B-12Q	250	.064			2	1.73	3.73	4.96	
	0110	3/4 C.Y. hydraulic backhoe	B-12F	300	.053			1.67	1.58	3.25	4.29	
	0300	1/2 C.Y. hydraulic excavator, truck mounted	B-12J	200	.080			2.50	4.19	6.69	8.45	
	0500	6' to 10' deep, 3/4 C.Y. hydraulic backhoe, 6' to 10' deep	B-12F	225	.071			2.23	2.11	4.34	5.70	
	0600	1 C.Y. hydraulic excavator, truck mounted	B-12K	400	.040			1.25	2.43	3.68	4.58	
	0900	10' to 14' deep, 3/4 C.Y. hydraulic backhoe	B-12F	200	.080			2.50	2.37	4.87	6.45	
	1000	1-1/2 C.Y. hydraulic backhoe	B-12B	540	.030			.93	1.33	2.26	2.89	
	1300	14' to 20' deep, 1 C.Y. hydraulic backhoe	B-12A	320	.050			1.57	1.74	3.31	4.31	
	1400	By hand with pick and shovel 2' to 6' deep, light soil	1 Clab	8	1			26.50		26.50	41.50	
	1500	Heavy soil	"	4	2			53.50		53.50	83	
	1700	For tamping backfilled trenches, air tamp, add	A-1G	100	.080	E.C.Y.		2.14	.35	2.49	3.70	
	1900	Vibrating plate, add	B-18	180	.133	"		3.65	.20	3.85	5.95	
	2100	Trim sides and bottom for concrete pours, common earth		1,500	.016	S.F.		.44	.02	.46	.71	
	2300	Hardpan		600	.040	"		1.09	.06	1.15	1.77	
620	0010	**EXCAVATING, UTILITY TRENCH** Common earth										620
	0050	Trenching with chain trencher, 12 H.P., operator walking										
	0100	4" wide trench, 12" deep	B-53	800	.010	L.F.		.33	.06	.39	.57	
	0150	18" deep		750	.011			.35	.06	.41	.60	
	0200	24" deep		700	.011			.38	.07	.45	.65	
	0300	6" wide trench, 12" deep		650	.012			.41	.07	.48	.69	
	0350	18" deep		600	.013			.44	.08	.52	.75	
	0400	24" deep		550	.015			.48	.09	.57	.82	
	0450	36" deep		450	.018			.59	.11	.70	1.01	
	0600	8" wide trench, 12" deep		475	.017			.56	.10	.66	.95	
	0650	18" deep		400	.020			.66	.12	.78	1.13	
	0700	24" deep		350	.023			.76	.14	.90	1.29	
	0750	36" deep		300	.027			.88	.16	1.04	1.51	
	1000	Backfill by hand including compaction, add										

For expanded coverage of these items see *Means Heavy Construction Cost Data 2005*

02315 | Excavation and Fill

		02315 Excavation and Fill	Crew	Daily Output	Labor-Hours	Unit	2005 Bare Costs Mat.	Labor	Equip.	Total	Total Incl O&P	
620	1050	4″ wide trench, 12″ deep	A-1G	800	.010	L.F.		.27	.04	.31	.47	620
	1100	18″ deep		530	.015			.40	.07	.47	.70	
	1150	24″ deep		400	.020			.53	.09	.62	.93	
	1300	6″ wide trench, 12″ deep		540	.015			.40	.06	.46	.69	
	1350	18″ deep		405	.020			.53	.09	.62	.91	
	1400	24″ deep		270	.030			.79	.13	.92	1.37	
	1450	36″ deep		180	.044			1.19	.19	1.38	2.06	
	1600	8″ wide trench, 12″ deep		400	.020			.53	.09	.62	.93	
	1650	18″ deep		265	.030			.81	.13	.94	1.39	
	1700	24″ deep		200	.040			1.07	.17	1.24	1.85	
	1750	36″ deep	▼	135	.059	▼		1.58	.26	1.84	2.74	
	2000	Chain trencher, 40 H.P. operator riding										
	2050	6″ wide trench and backfill, 12″ deep	B-54	1,200	.007	L.F.		.22	.18	.40	.53	
	2100	18″ deep		1,000	.008			.26	.21	.47	.63	
	2150	24″ deep		975	.008			.27	.22	.49	.65	
	2200	36″ deep		900	.009			.29	.24	.53	.70	
	2250	48″ deep		750	.011			.35	.28	.63	.84	
	2300	60″ deep		650	.012			.41	.33	.74	.97	
	2400	8″ wide trench and backfill, 12″ deep		1,000	.008			.26	.21	.47	.63	
	2450	18″ deep		950	.008			.28	.22	.50	.67	
	2500	24″ deep		900	.009			.29	.24	.53	.70	
	2550	36″ deep		800	.010			.33	.27	.60	.79	
	2600	48″ deep		650	.012			.41	.33	.74	.97	
	2700	12″ wide trench and backfill, 12″ deep		975	.008			.27	.22	.49	.65	
	2750	18″ deep		860	.009			.31	.25	.56	.73	
	2800	24″ deep		800	.010			.33	.27	.60	.79	
	2850	36″ deep		725	.011			.36	.29	.65	.87	
	3000	16″ wide trench and backfill, 12″ deep		835	.010			.32	.26	.58	.76	
	3050	18″ deep		750	.011			.35	.28	.63	.84	
	3100	24″ deep	▼	700	.011	▼		.38	.31	.69	.91	
	3200	Compaction with vibratory plate, add								50%	50%	
	5100	Hand excavate and trim for pipe bells after trench excavation										
	5200	8″ pipe	1 Clab	155	.052	L.F.		1.38		1.38	2.14	
	5300	18″ pipe	″	130	.062	″		1.64		1.64	2.56	
640	0010	**UTILITY BEDDING** For pipe and conduit, not incl. compaction										640
	0050	Crushed or screened bank run gravel	B-6	150	.160	L.C.Y.	20	4.61	1.44	26.05	30.50	
	0100	Crushed stone 3/4″ to 1/2″		150	.160		27.50	4.61	1.44	33.55	38.50	
	0200	Sand, dead or bank	▼	150	.160	▼	4.13	4.61	1.44	10.18	13.20	
	0500	Compacting bedding in trench	A-1D	90	.089	E.C.Y.		2.37	.28	2.65	4	
	0600	If material source exceeds 2 miles, add for extra mileage.				L.C.Y.						
	0610	See 02315-490-0010 for hauling mileage add.				″						

02400 | Tunneling, Boring & Jacking

		02441 Microtunneling	Crew	Daily Output	Labor-Hours	Unit	2005 Bare Costs Mat.	Labor	Equip.	Total	Total Incl O&P	
400	0010	**MICROTUNNELING** Not including excavation, backfill, shoring,										400
	0020	or dewatering, average 50′/day, slurry method										
	0100	24″ to 48″ outside diameter, minimum				L.F.					640	
	0110	Adverse conditions, add				%					50%	

02400 | Tunneling, Boring & Jacking

	02441 | Microtunneling		CREW	DAILY OUTPUT	LABOR-HOURS	UNIT	2005 BARE COSTS MAT.	LABOR	EQUIP.	TOTAL	TOTAL INCL O&P	
400	1000	Rent microtunneling machine, average monthly lease				Month					85,500	400
	1010	Operating technician				Day					640	
	1100	Mobilization and demobilization, minimum				Job					42,800	
	1110	Maximum				"					430,000	

	02445 | Boring or Jacking Conduits		CREW	DAILY OUTPUT	LABOR-HOURS	UNIT	MAT.	LABOR	EQUIP.	TOTAL	TOTAL INCL O&P	
300	0010	**HORIZONTAL BORING** Casing only, 100′ minimum,										300
	0020	not incl. jacking pits or dewatering										
	0100	Roadwork, 1/2″ thick wall, 24″ diameter casing	B-42	20	3.200	L.F.	67	96	53.50	216.50	284	
	0200	36″ diameter		16	4		107	120	67	294	380	
	0300	48″ diameter		15	4.267		157	128	71.50	356.50	455	
	0500	Railroad work, 24″ diameter		15	4.267		67	128	71.50	266.50	355	
	0600	36″ diameter		14	4.571		107	137	76.50	320.50	415	
	0700	48″ diameter		12	5.333		157	160	89	406	520	
	0900	For ledge, add								155	190	
	1000	Small diameter boring, 3″, sandy soil	B-82	900	.018		18.45	.53	.07	19.05	21.50	
	1040	Rocky soil	"	500	.032		18.45	.96	.13	19.54	22	
	1100	Prepare jacking pits, incl. mobilization & demobilization, minimum				Ea.				2,850	3,375	
	1101	Maximum				"				16,100	19,300	

02500 | Utility Services

	02510 | Water Distribution		CREW	DAILY OUTPUT	LABOR-HOURS	UNIT	2005 BARE COSTS MAT.	LABOR	EQUIP.	TOTAL	TOTAL INCL O&P	
600	0010	**VALVES**										600
	9000	Valves, gate valve, N.R.S. post type, 4″ diameter	B-21	32	.875	Ea.	365	27	4.95	396.95	450	
	9020	6″ diameter		20	1.400		460	43.50	7.90	511.40	585	
	9040	8″ diameter		16	1.750		705	54	9.90	768.90	870	
	9060	10″ diameter		16	1.750		1,150	54	9.90	1,213.90	1,350	
	9080	12″ diameter		13	2.154		1,425	66.50	12.20	1,503.70	1,675	
	9100	14″ diameter		11	2.545		1,650	78.50	14.40	1,742.90	1,975	
	9120	O.S.&Y., 4″ diameter		32	.875		291	27	4.95	322.95	365	
	9140	6″ diameter		20	1.400		425	43.50	7.90	476.40	540	
	9160	8″ diameter		16	1.750		695	54	9.90	758.90	860	
	9180	10″ diameter		16	1.750		1,125	54	9.90	1,188.90	1,350	
	9200	12″ diameter		13	2.154		1,475	66.50	12.20	1,553.70	1,750	
	9220	14″ diameter		11	2.545		4,150	78.50	14.40	4,242.90	4,725	
	9400	Check valves, rubber disc, 2-1/2″ diameter	B-20	44	.545		350	16.40		366.40	410	
	9420	3″ diameter	"	38	.632		350	19		369	415	
	9440	4″ diameter	B-21	32	.875		430	27	4.95	461.95	520	
	9480	6″ diameter		20	1.400		650	43.50	7.90	701.40	790	
	9500	8″ diameter		16	1.750		1,050	54	9.90	1,113.90	1,250	
	9520	10″ diameter		16	1.750		1,700	54	9.90	1,763.90	1,975	
	9540	12″ diameter		13	2.154		2,675	66.50	12.20	2,753.70	3,075	
	9542	14″ diameter		11	2.545		3,775	78.50	14.40	3,867.90	4,300	
	9700	Detector check valves, reducing, 4″ diameter		32	.875		705	27	4.95	736.95	820	
	9720	6″ diameter		20	1.400		1,075	43.50	7.90	1,126.40	1,275	
	9740	8″ diameter		16	1.750		1,775	54	9.90	1,838.90	2,050	
	9760	10″ diameter		16	1.750		3,500	54	9.90	3,563.90	3,950	
	9800	Galvanized, 4″ diameter		32	.875		920	27	4.95	951.95	1,075	

	02510 | Water Distribution		CREW	DAILY OUTPUT	LABOR-HOURS	UNIT	2005 BARE COSTS MAT.	LABOR	EQUIP.	TOTAL	TOTAL INCL O&P	
600	9820	6″ diameter	B-21	20	1.400	Ea.	1,350	43.50	7.90	1,401.40	1,550	600
	9840	8″ diameter		16	1.750		2,150	54	9.90	2,213.90	2,450	
	9860	10″ diameter	↓	16	1.750	↓	4,250	54	9.90	4,313.90	4,775	
710	0010	TAPPING, CROSSES AND SLEEVES										710
	4000	Drill and tap pressurized main (labor only)										
	4100	6″ main, 1″ to 2″ service	Q-1	3	5.333	Ea.		196		196	295	
	4150	8″ main, 1″ to 2″ service	″	2.75	5.818	″		214		214	320	
	4500	Tap and insert gate valve										
	4600	8″ main, 4″ branch	B-21	3.20	8.750	Ea.		270	49.50	319.50	475	
	4650	6″ branch		2.70	10.370			320	58.50	378.50	560	
	4700	10″ Main, 4″ branch		2.70	10.370			320	58.50	378.50	560	
	4750	6″ branch		2.35	11.915			370	67.50	437.50	645	
	4800	12″ main, 6″ branch		2.35	11.915			370	67.50	437.50	645	
	4850	8″ branch		2.35	11.915			370	67.50	437.50	645	
	7020	Crosses, 4″ x 4″		37	.757		455	23.50	4.28	482.78	540	
	7030	6″ x 4″		25	1.120		540	34.50	6.35	580.85	650	
	7040	6″ x 6″		25	1.120		540	34.50	6.35	580.85	650	
	7060	8″ x 6″		21	1.333		660	41	7.55	708.55	795	
	7080	8″ x 8″		21	1.333		715	41	7.55	763.55	855	
	7100	10″ x 6″		21	1.333		1,300	41	7.55	1,348.55	1,525	
	7120	10″ x 10″		21	1.333		1,425	41	7.55	1,473.55	1,625	
	7140	12″ x 6″		18	1.556		1,300	48	8.80	1,356.80	1,525	
	7160	12″ x 12″		18	1.556		1,650	48	8.80	1,706.80	1,900	
	7180	14″ x 6″		16	1.750		3,275	54	9.90	3,338.90	3,700	
	7200	14″ x 14″		16	1.750		3,450	54	9.90	3,513.90	3,900	
	7220	16″ x 6″		14	2		3,500	62	11.30	3,573.30	3,950	
	7240	16″ x 10″		14	2		3,550	62	11.30	3,623.30	4,000	
	7260	16″ x 16″		14	2		3,725	62	11.30	3,798.30	4,200	
	7280	18″ x 6″		10	2.800		5,225	86.50	15.85	5,327.35	5,900	
	7300	18″ x 12″		10	2.800		5,275	86.50	15.85	5,377.35	5,950	
	7320	18″ x 18″		10	2.800		4,825	86.50	15.85	4,927.35	5,450	
	7340	20″ x 6″		8	3.500		4,225	108	19.80	4,352.80	4,850	
	7360	20″ x 12″		8	3.500		4,525	108	19.80	4,652.80	5,200	
	7380	20″ x 20″		8	3.500		6,625	108	19.80	6,752.80	7,500	
	7400	24″ x 6″		6	4.667		5,475	144	26.50	5,645.50	6,275	
	7420	24″ x 12″		6	4.667		5,550	144	26.50	5,720.50	6,350	
	7440	24″ x 18″		6	4.667		8,275	144	26.50	8,445.50	9,350	
	7460	24″ x 24″		6	4.667		8,625	144	26.50	8,795.50	9,725	
	7600	Cut-in sleeves with rubber gaskets, 4″		18	1.556		150	48	8.80	206.80	249	
	7620	6″		12	2.333		191	72	13.20	276.20	335	
	7640	8″		10	2.800		260	86.50	15.85	362.35	435	
	7660	10″		10	2.800		360	86.50	15.85	462.35	545	
	7680	12″		9	3.111		430	96	17.60	543.60	640	
	7800	Cut-in valves with rubber gaskets, 4″		18	1.556		380	48	8.80	436.80	505	
	7820	6″		12	2.333		515	72	13.20	600.20	690	
	7840	8″		10	2.800		795	86.50	15.85	897.35	1,025	
	7860	10″		10	2.800		995	86.50	15.85	1,097.35	1,250	
	7880	12″		9	3.111		1,200	96	17.60	1,313.60	1,500	
	7900	Tapping Valve 4 inch, MJ, ductile iron		18	1.556		380	48	8.80	436.80	505	
	7920	Tapping Valve 6 inch, MJ, ductile iron		12	2.333		455	72	13.20	540.20	625	
	8000	Sleeves with rubber gaskets, 4″ x 4″		37	.757		775	23.50	4.28	802.78	895	
	8010	6″ x 4″		25	1.120		435	34.50	6.35	475.85	535	
	8020	6″ x 6″		25	1.120		470	34.50	6.35	510.85	575	
	8030	8″ x 4″		21	1.333		470	41	7.55	518.55	590	
	8040	8″ x 6″	↓	21	1.333	↓	520	41	7.55	568.55	645	

02510 | Water Distribution

			Crew	Daily Output	Labor-Hours	Unit	2005 Bare Costs Mat.	Labor	Equip.	Total	Total Incl O&P	
710	8060	8″ x 8″	B-21	21	1.333	Ea.	610	41	7.55	658.55	740	710
	8070	10″ x 4″		21	1.333		475	41	7.55	523.55	595	
	8080	10″ x 6″		21	1.333		530	41	7.55	578.55	655	
	8090	10″ x 8″		21	1.333		890	41	7.55	938.55	1,050	
	8100	10″ x 10″		21	1.333		1,000	41	7.55	1,048.55	1,175	
	8110	12″ x 4″		18	1.556		515	48	8.80	571.80	650	
	8120	12″ x 6″		18	1.556		805	48	8.80	861.80	970	
	8130	12″ x 8″		18	1.556		715	48	8.80	771.80	870	
	8135	12″ x 10″		18	1.556		985	48	8.80	1,041.80	1,150	
	8140	12″ x 12″		18	1.556		1,200	48	8.80	1,256.80	1,375	
	8160	14″ x 6″		16	1.750		2,600	54	9.90	2,663.90	2,975	
	8180	14″ x 14″		16	1.750		2,775	54	9.90	2,838.90	3,150	
	8200	16″ x 6″		14	2		2,800	62	11.30	2,873.30	3,175	
	8220	16″ x 10″		14	2		2,850	62	11.30	2,923.30	3,225	
	8240	16″ x 16″		14	2		2,975	62	11.30	3,048.30	3,375	
	8260	18″ x 6″		10	2.800		4,175	86.50	15.85	4,277.35	4,750	
	8280	18″ x 12″		10	2.800		4,225	86.50	15.85	4,327.35	4,775	
	8300	18″ x 18″		10	2.800		4,325	86.50	15.85	4,427.35	4,925	
	8320	20″ x 6″		8	3.500		3,375	108	19.80	3,502.80	3,925	
	8340	20″ x 12″		8	3.500		3,625	108	19.80	3,752.80	4,200	
	8360	20″ x 20″		8	3.500		5,300	108	19.80	5,427.80	6,025	
	8380	24″ x 6″		6	4.667		4,375	144	26.50	4,545.50	5,050	
	8400	24″ x 12″		6	4.667		4,450	144	26.50	4,620.50	5,125	
	8420	24″ x 18″		6	4.667		6,625	144	26.50	6,795.50	7,525	
	8440	24″ x 24″	↓	6	4.667		6,900	144	26.50	7,070.50	7,825	
	8800	Curb box, 6′ long	B-20	20	1.200		69.50	36		105.50	133	
	8820	8′ long	″	18	1.333	↓	81	40		121	152	
730	0010	**WATER SUPPLY, DUCTILE IRON PIPE** cement lined R02510-800										730
	0020	Not including excavation or backfill										
	2000	Pipe, class 50 water piping, 18′ lengths										
	2020	Mechanical joint, 4″ diameter	B-21A	200	.200	L.F.	12.50	6.60	2.53	21.63	26.50	
	2040	6″ diameter		160	.250		14.50	8.25	3.16	25.91	32	
	2060	8″ diameter		133.33	.300		16.05	9.90	3.80	29.75	37	
	2080	10″ diameter		114.29	.350		21.50	11.55	4.43	37.48	46.50	
	2100	12″ diameter		105.26	.380		26.50	12.55	4.81	43.86	54	
	2120	14″ diameter		100	.400		34	13.20	5.05	52.25	63	
	2140	16″ diameter		72.73	.550		37	18.15	6.95	62.10	76	
	2160	18″ diameter		68.97	.580		46.50	19.10	7.35	72.95	88.50	
	2170	20″ diameter		57.14	.700		54.50	23	8.85	86.35	105	
	2180	24″ diameter		47.06	.850		70	28	10.75	108.75	131	
	3000	Tyton, Push-on joint, 4″ diameter		400	.100		7.35	3.30	1.27	11.92	14.50	
	3020	6″ diameter		333.33	.120		8.40	3.96	1.52	13.88	16.85	
	3040	8″ diameter		200	.200		11.50	6.60	2.53	20.63	25.50	
	3060	10″ diameter		181.82	.220		18.15	7.25	2.78	28.18	34	
	3080	12″ diameter		160	.250		19.15	8.25	3.16	30.56	37	
	3100	14″ diameter		133.33	.300		21	9.90	3.80	34.70	42	
	3120	16″ diameter		114.29	.350		29.50	11.55	4.43	45.48	55	
	3140	18″ diameter		100	.400		32.50	13.20	5.05	50.75	61.50	
	3160	20″ diameter		88.89	.450		36	14.85	5.70	56.55	68.50	
	3180	24″ diameter	↓	76.92	.520	↓	47	17.15	6.60	70.75	85	
	8000	Fittings, mechanical joint										
	8006	90° bend, 4″ diameter	B-20A	16	2	Ea.	126	64.50		190.50	237	
	8020	6″ diameter		12.80	2.500		169	80.50		249.50	310	
	8040	8″ diameter	↓	10.67	2.999		245	96.50		341.50	420	
	8060	10″ diameter	B-21A	11.43	3.500	↓	480	115	44.50	639.50	755	

2 SITE CONSTRUCTION

02510 | Water Distribution

		02510 | Water Distribution	CREW	DAILY OUTPUT	LABOR-HOURS	UNIT	2005 BARE COSTS MAT.	LABOR	EQUIP.	TOTAL	TOTAL INCL O&P	
730	8080	12″ diameter (R02510-800)	B-21A	10.53	3.799	Ea.	495	125	48	668	790	730
	8100	14″ diameter		10	4		685	132	50.50	867.50	1,000	
	8120	16″ diameter		7.27	5.502		735	181	69.50	985.50	1,175	
	8140	18″ diameter		6.90	5.797		1,000	191	73.50	1,264.50	1,475	
	8160	20″ diameter		5.71	7.005		2,550	231	88.50	2,869.50	3,250	
	8180	24″ diameter		4.70	8.511		3,150	281	108	3,539	4,000	
	8200	Wye or tee, 4″ diameter	B-20A	10.67	2.999		153	96.50		249.50	315	
	8220	6″ diameter		8.53	3.751		192	121		313	395	
	8240	8″ diameter		7.11	4.501		284	145		429	535	
	8260	10″ diameter	B-21A	7.62	5.249		490	173	66.50	729.50	870	
	8280	12″ diameter		7.02	5.698		610	188	72	870	1,050	
	8300	14″ diameter		6.67	5.997		1,250	198	76	1,524	1,750	
	8320	16″ diameter		4.85	8.247		1,300	272	104	1,676	1,950	
	8340	18″ diameter		4.60	8.696		2,650	287	110	3,047	3,450	
	8360	20″ diameter		3.81	10.499		3,375	345	133	3,853	4,375	
	8380	24″ diameter .		3.14	12.739		3,850	420	161	4,431	5,075	
	8398	45° bends, 4″ diameter	B-20A	16	2		133	64.50		197.50	245	
	8400	6″ diameter	″	12.80	2.500		146	80.50		226.50	284	
	8410	12″ diameter	B-21A	10.53	3.799		355	125	48	528	635	
	8420	16″ diameter		7.27	5.502		460	181	69.50	710.50	860	
	8430	20″ diameter		5.71	7.005		2,125	231	88.50	2,444.50	2,800	
	8440	24″ diameter		4.70	8.511		2,575	281	108	2,964	3,375	
	8450	Decreaser, 6″ x 4″ diameter	B-20A	14.22	2.250		104	72.50		176.50	226	
	8460	8″ x 6″ diameter	″	11.64	2.749		213	88.50		301.50	370	
	8470	10″ x 6 ″ diameter	B-21A	13.33	3.001		217	99	38	354	430	
	8480	12″ x 6″ diameter		12.70	3.150		245	104	40	389	470	
	8490	16″ x 6″ diameter		10	4		885	132	50.50	1,067.50	1,225	
	8500	20″ x 6″ diameter		8.42	4.751		1,400	157	60	1,617	1,850	
	8550	Butterfly valves with boxes, cast iron										
	8560	4″ diameter	B-20	6	4	Ea.	760	120		880	1,025	
	8570	6″ diameter	″	5	4.800		460	144		604	730	
	8580	8″ diameter	B-21	4	7		760	216	39.50	1,015.50	1,225	
	8590	10″ diameter		3.50	8		1,050	247	45.50	1,342.50	1,575	
	8600	12″ diameter		3	9.333		1,350	288	53	1,691	1,975	
	8610	14″ diameter		2	14		1,875	435	79	2,389	2,800	
	8620	16″ diameter		2	14		2,400	435	79	2,914	3,400	
	9600	Steel sleeve and tap, 4″ diameter	B-20	3	8		430	241		671	845	
	9620	6″ diameter		2	12		505	360		865	1,125	
	9630	8″ diameter		2	12		680	360		1,040	1,300	
750	0010	**WATER SUPPLY, POLYVINYL CHLORIDE PIPE**										750
	0020	Not including excavation or backfill, unless specified										
	2100	AWWA Class 160, S.D.R. 26, 1-1/2″ diameter	B-20	750	.032	L.F.	.45	.96		1.41	2	
	2120	2″ diameter		686	.035		1.09	1.05		2.14	2.84	
	2140	2-1/2″ diameter		500	.048		1.62	1.44		3.06	4.03	
	2160	3″ diameter		430	.056		2.32	1.68		4	5.15	
	2180	4″ diameter		375	.064		3.78	1.93		5.71	7.15	
	2200	6″ diameter		316	.076		8.15	2.28		10.43	12.50	
	2210	8″ diameter		260	.092		13.80	2.78		16.58	19.50	
	3010	AWWA C905, PR 100, DR 41										
	3030	14″ diameter	B-20A	213	.150	L.F.	12.60	4.84		17.44	21.50	
	3040	16″ diameter		200	.160		16.70	5.15		21.85	26	
	3050	18″ diameter		160	.200		21	6.45		27.45	33	
	3060	20″ diameter		133	.241		26	7.75		33.75	41	
	3070	24″ diameter		107	.299		36.50	9.65		46.15	55	
	3080	30″ diameter		80	.400		56.50	12.90		69.40	82	

02510 | Water Distribution

			CREW	DAILY OUTPUT	LABOR-HOURS	UNIT	2005 BARE COSTS MAT.	LABOR	EQUIP.	TOTAL	TOTAL INCL O&P	
750	3090	36″ diameter	B-20A	80	.400	L.F.	80	12.90		92.90	108	750
	3100	42″ diameter		60	.533		109	17.20		126.20	146	
	3200	48″ diameter		60	.533		143	17.20		160.20	183	
	4520	Class 150, SDR 18, AWWA C900, 4″		380	.084		2.38	2.71		5.09	6.75	
	4530	6″		316	.101		4.70	3.26		7.96	10.15	
	4540	8″		264	.121		8.15	3.91		12.06	14.90	
	4550	10″		220	.145		12.30	4.69		16.99	20.50	
	4560	12″	↓	186	.172	↓	17.40	5.55		22.95	27.50	
	8000	Fittings with rubber gasket										
	8003	Class 150, D.R. 18										
	8006	90° Bend , 4″ diameter	B-20	100	.240	Ea.	39.50	7.20		46.70	55	
	8020	6″ diameter		90	.267		70	8		78	89.50	
	8040	8″ diameter		80	.300		136	9		145	163	
	8060	10″ diameter		50	.480		235	14.45		249.45	282	
	8080	12″ diameter		30	.800		300	24		324	370	
	8100	Tee, 4″ diameter		90	.267		54.50	8		62.50	72	
	8120	6″ diameter		80	.300		121	9		130	147	
	8140	8″ diameter		70	.343		173	10.30		183.30	206	
	8160	10″ diameter		40	.600		196	18.05		214.05	243	
	8180	12″ diameter		20	1.200		420	36		456	515	
	8200	45° Bend, 4″ diameter		100	.240		38.50	7.20		45.70	54	
	8220	6″ diameter		90	.267		68.50	8		76.50	87.50	
	8240	8″ diameter		50	.480		130	14.45		144.45	166	
	8260	10″ diameter		50	.480		199	14.45		213.45	241	
	8280	12″ diameter		30	.800		258	24		282	320	
	8300	Reducing tee 6″x4″		100	.240		118	7.20		125.20	141	
	8320	8″ x 6″		90	.267		210	8		218	244	
	8330	10″ x 6″		90	.267		270	8		278	310	
	8340	10″x8″		90	.267		281	8		289	325	
	8350	12″ x 6″		90	.267		320	8		328	365	
	8360	12″ x 8″		90	.267		335	8		343	385	
	8400	Tapped service tee (threaded type) 6″ x 6″ x 3/4		100	.240		42	7.20		49.20	57.50	
	8420	6″ x 6″ x 3/4″		90	.267		42	8		50	58.50	
	8430	6″ x 6″ x 1″		90	.267		42	8		50	58.50	
	8440	6″ x 6″ x 1 1/2″		90	.267		42	8		50	58.50	
	8450	6″ x 6″ x 2″		90	.267		42	8		50	58.50	
	8460	8″ x 8″ x 3/4″		90	.267		177	8		185	208	
	8470	8″ x 8″ x 1″		90	.267		177	8		185	208	
	8480	8″ x 8″ x 1 1/2″		90	.267		177	8		185	208	
	8490	8″ x 8″ x 2″		90	.267		177	8		185	208	
	8500	Repair coupling 4″		100	.240		25	7.20		32.20	39	
	8520	6″ diameter		90	.267		38.50	8		46.50	54.50	
	8540	8″ diameter		50	.480		84.50	14.45		98.95	116	
	8560	10″ diameter		50	.480		143	14.45		157.45	180	
	8580	12″ diameter		50	.480		209	14.45		223.45	253	
	8600	Plug end 4″		100	.240		21	7.20		28.20	35	
	8620	6″ diameter		90	.267		38	8		46	54.50	
	8640	8″ diameter		50	.480		65	14.45		79.45	94	
	8660	10″ diameter		50	.480		68.50	14.45		82.95	97.50	
	8680	12″ diameter	↓	50	.480	↓	84.50	14.45		98.95	116	
760	0010	**WATER SUPPLY, HDPE,** butt fusion joints, SDR 21, 40′ lengths										760
	0100	4″ diameter	B-22A	400	.095	L.F.	1.84	2.88	1.76	6.48	8.40	
	0200	6″ diameter		380	.100		2.65	3.03	1.85	7.53	9.65	
	0300	8″ diameter		320	.119		4.48	3.60	2.20	10.28	12.90	
	0400	10″ diameter		300	.127		6.95	3.84	2.34	13.13	16.20	
	0500	12″ diameter	↓	260	.146	↓	9.80	4.43	2.70	16.93	20.50	

02510 | Water Distribution

		02510 Water Distribution	CREW	DAILY OUTPUT	LABOR-HOURS	UNIT	2005 BARE COSTS MAT.	LABOR	EQUIP.	TOTAL	TOTAL INCL O&P	
760	0600	14″ diameter	B-22A	220	.173	L.F.	11.80	5.25	3.19	20.24	24.50	760
	0700	16″ diameter		180	.211		15.45	6.40	3.90	25.75	31	
	0800	18″ diameter		140	.271		19.50	8.20	5	32.70	39.50	
	0900	24″ diameter		100	.380		35	11.50	7.05	53.55	63.50	
	1000	Fittings										
	1100	Elbows, 90 degrees										
	1200	4″ diameter	B-22B	32	.500	Ea.	25.50	15.40	5.35	46.25	58	
	1300	6″ diameter		28	.571		67	17.60	6.15	90.75	108	
	1400	8″ diameter		24	.667		168	20.50	7.15	195.65	225	
	1500	10″ diameter		18	.889		290	27.50	9.55	327.05	375	
	1600	12″ diameter		12	1.333		490	41	14.30	545.30	620	
	1700	14″ diameter		9	1.778		490	54.50	19.10	563.60	640	
	1800	16″ diameter		6	2.667		590	82	28.50	700.50	805	
	1900	18″ diameter		4	4		795	123	43	961	1,125	
	2000	24″ diameter		3	5.333		1,625	164	57.50	1,846.50	2,100	
	2100	Tees										
	2200	4″ diameter	B-22B	30	.533	Ea.	35.50	16.40	5.75	57.65	71	
	2300	6″ diameter		26	.615		93	18.95	6.60	118.55	140	
	2400	8″ diameter		22	.727		236	22.50	7.80	266.30	305	
	2500	10″ diameter		15	1.067		310	33	11.45	354.45	410	
	2600	12″ diameter		10	1.600		425	49.50	17.20	491.70	565	
	2700	14″ diameter		8	2		505	61.50	21.50	588	675	
	2800	16″ diameter		6	2.667		590	82	28.50	700.50	810	
	2900	18″ diameter		4	4		800	123	43	966	1,125	
	3000	24″ diameter		2	8		1,650	246	86	1,982	2,300	

02520 | Wells

		02520 Wells	CREW	DAILY OUTPUT	LABOR-HOURS	UNIT	MAT.	LABOR	EQUIP.	TOTAL	TOTAL INCL O&P	
510	0010	**WELLS & ACCESSORIES**, domestic										510
	0100	Drilled, 4″ to 6″ diameter	B-23	120	.333	L.F.		9.05	28	37.05	45	
	0200	8″ diameter	″	95.20	.420	″		11.40	35	46.40	56.50	
	0400	Gravel pack well, 40′ deep, incl. gravel & casing, complete										
	0500	24″ diameter casing x 18″ diameter screen	B-23	.13	307	Total	22,900	8,350	25,800	57,050	66,500	
	0600	36″ diameter casing x 18″ diameter screen		.12	333	″	24,700	9,025	28,000	61,725	72,000	
	0800	Observation wells, 1-1/4″ riser pipe		163	.245	V.L.F.	12.60	6.65	20.50	39.75	47	
	0900	For flush Buffalo roadway box, add	1 Skwk	16.60	.482	Ea.	34.50	16.80		51.30	64	
	1200	Test well, 2-1/2″ diameter, up to 50′ deep (15 to 50 GPM)	B-23	1.51	26.490	″	515	720	2,225	3,460	4,150	
	1300	Over 50′ deep, add	″	121.80	.328	L.F.	13.75	8.90	27.50	50.15	59.50	
	1500	Pumps, installed in wells to 100′ deep, 4″ submersible										
	1510	1/2 H.P.	Q-1	3.22	4.969	Ea.	330	183		513	640	
	1520	3/4 H.P.		2.66	6.015		390	221		611	765	
	1600	1 H.P.		2.29	6.987		415	257		672	840	
	1700	1-1/2 H.P.	Q-22	1.60	10		1,075	370	375	1,820	2,150	
	1800	2 H.P.		1.33	12.030		1,250	440	455	2,145	2,550	
	1900	3 H.P.		1.14	14.035		1,425	515	530	2,470	2,925	
	2000	5 H.P.		1.14	14.035		1,950	515	530	2,995	3,500	
	2050	Remove and install motor only, 4 H.P.		1.14	14.035		700	515	530	1,745	2,125	
	3000	Pump, 6″ submersible, 25′ to 150′ deep, 25 H.P., 249 to 297 GPM		.89	17.978		4,250	660	675	5,585	6,425	
	3100	25′ to 500′ deep, 30 H.P., 100 to 300 GPM		.73	21.918		4,850	805	825	6,480	7,425	
	8110	Well screen assembly, stainless steel, 2″ diameter	B-23A	273	.088	L.F.	47.50	2.64	11.95	62.09	69	
	8120	3″ diameter		253	.095		65.50	2.85	12.85	81.20	90.50	
	8130	4″ diameter		200	.120		75	3.60	16.30	94.90	106	
	8140	5″ diameter		168	.143		87.50	4.29	19.40	111.19	125	
	8150	6″ diameter		126	.190		105	5.70	26	136.70	152	
	8160	8″ diameter		98.50	.244		137	7.30	33	177.30	198	
	8170	10″ diameter		73	.329		172	9.85	44.50	226.35	253	

Important: See the Reference Section for critical supporting data - Reference Nos., Crews, & City Cost Indexes

		02520 | Wells	CREW	DAILY OUTPUT	LABOR-HOURS	UNIT	2005 BARE COSTS MAT.	LABOR	EQUIP.	TOTAL	TOTAL INCL O&P	
510	8180	12″ diameter	B-23A	62.50	.384	L.F.	202	11.55	52	265.55	297	510
	8190	14″ diameter		54.30	.442		229	13.25	60	302.25	340	
	8200	16″ diameter		48.30	.497		253	14.90	67.50	335.40	375	
	8210	18″ diameter		39.20	.612		315	18.40	83	416.40	465	
	8220	20″ diameter		31.20	.769		360	23	104	487	545	
	8230	24″ diameter		23.80	1.008		445	30.50	137	612.50	690	
	8240	26″ diameter		21	1.143		500	34.50	155	689.50	775	
	8300	Slotted PVC, 1-1/4″ diameter		521	.046		1.69	1.38	6.25	9.32	10.90	
	8310	1-1/2″ diameter		488	.049		2.46	1.48	6.65	10.59	12.35	
	8320	2″ diameter		273	.088		3.41	2.64	11.95	18	21	
	8330	3″ diameter		253	.095		3.62	2.85	12.85	19.32	22.50	
	8340	4″ diameter		200	.120		4.17	3.60	16.30	24.07	28	
	8350	5″ diameter		168	.143		4.38	4.29	19.40	28.07	33	
	8360	6″ diameter		126	.190		6.10	5.70	26	37.80	44	
	8370	8″ diameter		98.50	.244		9.20	7.30	33	49.50	58	
	8400	Artificial gravel pack, 2″ screen, 6″ casing	B-23B	174	.138		2.77	4.14	19.95	26.86	31.50	
	8405	8″ casing		111	.216		3.78	6.50	31.50	41.78	48.50	
	8410	10″ casing		74.50	.322		4.75	9.65	46.50	60.90	71.50	
	8415	12″ casing		60	.400		6.40	12	58	76.40	89	
	8420	14″ casing		50.20	.478		8.30	14.35	69	91.65	107	
	8425	16″ casing		40.70	.590		11.45	17.70	85.50	114.65	134	
	8430	18″ casing		36	.667		13.30	20	96.50	129.80	152	
	8435	20″ casing		29.50	.814		15.30	24.50	118	157.80	184	
	8440	24″ casing		25.70	.934		17	28	135	180	211	
	8445	26″ casing		24.60	.976		19	29.50	141	189.50	221	
	8450	30″ casing		20	1.200		22	36	174	232	271	
	8455	36″ casing		16.40	1.463		23.50	44	212	279.50	325	
	8500	Develop well		8	3	Hr.	215	90	435	740	855	
	8550	Pump test well		8	3		57	90	435	582	680	
	8560	Standby well	B-23A	8	3		55.50	90	405	550.50	650	
	8570	Standby, drill rig		8	3			90	405	495	590	
	8580	Surface seal well, concrete filled		1	24	Ea.	540	720	3,250	4,510	5,275	
	8590	Well test pump, install & remove	B-23	1	40			1,075	3,350	4,425	5,375	
	8600	Well sterilization, chlorine	2 Clab	1	16		400	425		825	1,100	
	9950	See div. 02240-900 for wellpoints										
	9960	See div. 02240-700 for drainage wells										
520	0010	**WATER SUPPLY WELLS, PUMPS** with pressure control										520
	1000	Deep well, jet, 42 gal. galvanized tank										
	1040	3/4 HP	1 Plum	.80	10	Ea.	585	410		995	1,250	
	3000	Shallow well, jet, 30 gal. galvanized tank										
	3040	1/2 HP	1 Plum	2	4	Ea.	370	163		533	655	

		02530 | Sanitary Sewerage	CREW	DAILY OUTPUT	LABOR-HOURS	UNIT	MAT.	LABOR	EQUIP.	TOTAL	TOTAL INCL O&P	
730	0010	**SEWAGE COLLECTION, CONCRETE PIPE**										730
	0020	See 02630-530 for sewage/drainage collection, concrete pipe										
780	0010	**SEWAGE COLLECTION, POLYVINYL CHLORIDE PIPE**										780
	0020	Not including excavation or backfill										
	2000	10′ lengths, S.D.R. 35, B&S, 4″ diameter	B-20	375	.064	L.F.	1.92	1.93		3.85	5.10	
	2040	6″ diameter		350	.069		3.44	2.06		5.50	7	
	2080	8″ diameter		335	.072		5.80	2.15		7.95	9.75	
	2120	10″ diameter	B-21	330	.085		8.75	2.62	.48	11.85	14.25	
	2160	12″ diameter		320	.087		9.75	2.70	.50	12.95	15.50	
	2200	15″ diameter		190	.147		14.75	4.56	.83	20.14	24	
	4000	Piping, DWV PVC,no exc/bkfill, 10′ L, Sch 40, 4″ dia	B-20	375	.064		1.88	1.93		3.81	5.05	
	4010	6″ dia		350	.069		4.06	2.06		6.12	7.70	

	02530 | Sanitary Sewerage		Crew	Daily Output	Labor-Hours	Unit	2005 Bare Costs Mat.	Labor	Equip.	Total	Total Incl O&P	
780	4020	8" dia	B-20	335	.072	L.F.	12.50	2.15		14.65	17.10	780

	02550 | Piped Energy Distribution		Crew	Daily Output	Labor-Hours	Unit	Mat.	Labor	Equip.	Total	Total Incl O&P	
200	0010	PIPE CONDUIT, PREFABRICATED / PREINSULATED										200
	0020	Does not include trenching, fittings or crane.										
	0300	For cathodic protection, add 12 to 14%										
	0310	of total built-up price (casing plus service pipe)										
	0580	Polyurethane insulated system, 250°F. max. temp.										
	0620	Black steel service pipe, standard wt., 1/2" insulation										
	0660	3/4" diam. pipe size	Q-17	54	.296	L.F.	24.50	10.90	1.49	36.89	45	
	0670	1" diam. pipe size		50	.320		27	11.80	1.61	40.41	49	
	0680	1-1/4" diam. pipe size		47	.340		30	12.55	1.72	44.27	53.50	
	0690	1-1/2" diam. pipe size		45	.356		32.50	13.10	1.79	47.39	57.50	
	0700	2" diam. pipe size		42	.381		34	14.05	1.92	49.97	60	
	0710	2-1/2" diam. pipe size		34	.471		34.50	17.35	2.37	54.22	66	
	0720	3" diam. pipe size		28	.571		39.50	21	2.88	63.38	78	
	0730	4" diam. pipe size		22	.727		50	27	3.67	80.67	99.50	
	0740	5" diam. pipe size		18	.889		64	33	4.48	101.48	124	
	0750	6" diam. pipe size	Q-18	23	1.043		74	40	3.51	117.51	145	
	0760	8" diam. pipe size		19	1.263		109	48.50	4.24	161.74	197	
	0770	10" diam. pipe size		16	1.500		143	57.50	5.05	205.55	250	
	0780	12" diam. pipe size		13	1.846		178	70.50	6.20	254.70	310	
	0790	14" diam. pipe size		11	2.182		199	83.50	7.35	289.85	350	
	0800	16" diam. pipe size		10	2.400		228	91.50	8.05	327.55	400	
	0810	18" diam. pipe size		8	3		262	115	10.10	387.10	470	
	0820	20" diam. pipe size		7	3.429		292	131	11.50	434.50	530	
	0830	24" diam. pipe size		6	4		355	153	13.45	521.45	640	
	0900	For 1" thick insulation, add					10%					
	0940	For 1-1/2" thick insulation, add					13%					
	0980	For 2" thick insulation, add					20%					
	1500	Gland seal for system, 3/4" diam. pipe size	Q-17	32	.500	Ea.	395	18.45	2.52	415.97	465	
	1510	1" diam. pipe size		32	.500		395	18.45	2.52	415.97	465	
	1540	1-1/4" diam. pipe size		30	.533		425	19.65	2.69	447.34	495	
	1550	1-1/2" diam. pipe size		30	.533		425	19.65	2.69	447.34	495	
	1560	2" diam. pipe size		28	.571		505	21	2.88	528.88	590	
	1570	2-1/2" diam. pipe size		26	.615		545	22.50	3.10	570.60	635	
	1580	3" diam. pipe size		24	.667		585	24.50	3.36	612.86	680	
	1590	4" diam. pipe size		22	.727		690	27	3.67	720.67	805	
	1600	5" diam. pipe size		19	.842		850	31	4.24	885.24	985	
	1610	6" diam. pipe size	Q-18	26	.923		905	35.50	3.10	943.60	1,050	
	1620	8" diam. pipe size		25	.960		1,050	36.50	3.23	1,089.73	1,200	
	1630	10" diam. pipe size		23	1.043		1,275	40	3.51	1,318.51	1,475	
	1640	12" diam. pipe size		21	1.143		1,400	43.50	3.84	1,447.34	1,625	
	1650	14" diam. pipe size		19	1.263		1,575	48.50	4.24	1,627.74	1,825	
	1660	16" diam. pipe size		18	1.333		1,850	51	4.48	1,905.48	2,100	
	1670	18" diam. pipe size		16	1.500		1,975	57.50	5.05	2,037.55	2,275	
	1680	20" diam. pipe size		14	1.714		2,250	65.50	5.75	2,321.25	2,575	
	1690	24" diam. pipe size		12	2		2,500	76.50	6.70	2,583.20	2,875	
	2000	Elbow, 45° for system										
	2020	3/4" diam. pipe size	Q-17	14	1.143	Ea.	250	42	5.75	297.75	345	
	2040	1" diam. pipe size		13	1.231		257	45.50	6.20	308.70	360	
	2050	1-1/4" diam. pipe size		11	1.455		292	53.50	7.35	352.85	410	
	2060	1-1/2" diam. pipe size		9	1.778		305	65.50	8.95	379.45	445	
	2070	2" diam. pipe size		6	2.667		320	98.50	13.45	431.95	515	
	2080	2-1/2" diam. pipe size		4	4		345	147	20	512	625	

02550 | Piped Energy Distribution

			CREW	DAILY OUTPUT	LABOR-HOURS	UNIT	2005 BARE COSTS MAT.	LABOR	EQUIP.	TOTAL	TOTAL INCL O&P
200	2090	3″ diam. pipe size	Q-17	3.50	4.571	Ea.	400	168	23	591	720
	2100	4″ diam. pipe size		3	5.333		465	197	27	689	835
	2110	5″ diam. pipe size		2.80	5.714		595	211	29	835	1,000
	2120	6″ diam. pipe size	Q-18	4	6		675	229	20	924	1,100
	2130	8″ diam. pipe size		3	8		980	305	27	1,312	1,575
	2140	10″ diam. pipe size		2.40	10		1,250	380	33.50	1,663.50	1,975
	2150	12″ diam. pipe size		2	12		1,650	460	40.50	2,150.50	2,550
	2160	14″ diam. pipe size		1.80	13.333		2,050	510	45	2,605	3,075
	2170	16″ diam. pipe size		1.60	15		2,425	575	50.50	3,050.50	3,600
	2180	18″ diam. pipe size		1.30	18.462		3,050	705	62	3,817	4,475
	2190	20″ diam. pipe size		1	24		3,850	915	80.50	4,845.50	5,700
	2200	24″ diam. pipe size		.70	34.286		4,825	1,300	115	6,240	7,425
	2260	For elbow, 90°, add					25%				
	2300	For tee, straight, add					85%	30%			
	2340	For tee, reducing, add					170%	30%			
	2380	For weldolet, straight, add					50%				
	2400	Polyurethane insulation, 1″									
	2410	FRP carrier and casing									
	2420	4″	Q-5	18	.889	L.F.	37.50	33		70.50	90.50
	2422	6″	Q-6	23	1.043		65	40		105	132
	2424	8″		19	1.263		106	48.50		154.50	190
	2426	10″		16	1.500		143	57.50		200.50	244
	2428	12″		13	1.846		186	70.50		256.50	310
	2430	FRP carrier and PVC casing									
	2440	4″	Q-5	18	.889	L.F.	14.80	33		47.80	65.50
	2444	8″	Q-6	19	1.263		31	48.50		79.50	107
	2446	10″		16	1.500		44.50	57.50		102	135
	2448	12″		13	1.846		57	70.50		127.50	169
	2450	PVC carrier and casing									
	2460	4″	Q-1	36	.444	L.F.	7.30	16.35		23.65	32.50
	2462	6″	″	29	.552		10.25	20.50		30.75	42
	2464	8″	Q-2	36	.667		14.25	25.50		39.75	53.50
	2466	10″		32	.750		19.50	28.50		48	64.50
	2468	12″		31	.774		22.50	29.50		52	69.50
	2800	Calcium silicate insulated system, high temp. (1200°F)									
	2840	Steel casing with protective exterior coating									
	2850	6-5/8″ diameter	Q-18	52	.462	L.F.	45.50	17.65	1.55	64.70	78.50
	2860	8-5/8″ diameter		50	.480		50	18.35	1.61	69.96	84.50
	2870	10-3/4″ diameter		47	.511		58.50	19.50	1.72	79.72	96
	2880	12-3/4″ diameter		44	.545		63.50	21	1.83	86.33	104
	2890	14″ diameter		41	.585		71.50	22.50	1.97	95.97	115
	2900	16″ diameter		39	.615		77	23.50	2.07	102.57	122
	2910	18″ diameter		36	.667		85	25.50	2.24	112.74	134
	2920	20″ diameter		34	.706		95.50	27	2.37	124.87	148
	2930	22″ diameter		32	.750		133	28.50	2.52	164.02	192
	2940	24″ diameter		29	.828		151	31.50	2.78	185.28	217
	2950	26″ diameter		26	.923		173	35.50	3.10	211.60	246
	2960	28″ diameter		23	1.043		215	40	3.51	258.51	300
	2970	30″ diameter		21	1.143		228	43.50	3.84	275.34	320
	2980	32″ diameter		19	1.263		257	48.50	4.24	309.74	360
	2990	34″ diameter		18	1.333		260	51	4.48	315.48	365
	3000	36″ diameter		16	1.500		279	57.50	5.05	341.55	395
	3040	For multi-pipe casings, add					10%				
	3060	For oversize casings, add					2%				
	3400	Steel casing gland seal, single pipe									
	3420	6-5/8″ diameter	Q-18	25	.960	Ea.	635	36.50	3.23	674.73	760

02550 | Piped Energy Distribution

			CREW	DAILY OUTPUT	LABOR-HOURS	UNIT	2005 BARE COSTS MAT.	LABOR	EQUIP.	TOTAL	TOTAL INCL O&P	
200	3440	8-5/8" diameter	Q-18	23	1.043	Ea.	745	40	3.51	788.51	885	200
	3450	10-3/4" diameter		21	1.143		835	43.50	3.84	882.34	990	
	3460	12-3/4" diameter		19	1.263		995	48.50	4.24	1,047.74	1,175	
	3470	14" diameter		17	1.412		1,100	54	4.74	1,158.74	1,275	
	3480	16" diameter		16	1.500		1,275	57.50	5.05	1,337.55	1,500	
	3490	18" diameter		15	1.600		1,425	61	5.40	1,491.40	1,650	
	3500	20" diameter		13	1.846		1,575	70.50	6.20	1,651.70	1,850	
	3510	22" diameter		12	2		1,775	76.50	6.70	1,858.20	2,075	
	3520	24" diameter		11	2.182		1,975	83.50	7.35	2,065.85	2,300	
	3530	26" diameter		10	2.400		2,250	91.50	8.05	2,349.55	2,625	
	3540	28" diameter		9.50	2.526		2,550	96.50	8.50	2,655	2,950	
	3550	30" diameter		9	2.667		2,600	102	8.95	2,710.95	3,025	
	3560	32" diameter		8.50	2.824		2,925	108	9.50	3,042.50	3,375	
	3570	34" diameter		8	3		3,175	115	10.10	3,300.10	3,675	
	3580	36" diameter	↓	7	3.429		3,375	131	11.50	3,517.50	3,925	
	3620	For multi-pipe casings, add				↓	5%					
	4000	Steel casing anchors, single pipe										
	4020	6-5/8" diameter	Q-18	8	3	Ea.	570	115	10.10	695.10	815	
	4040	8-5/8" diameter		7.50	3.200		595	122	10.75	727.75	850	
	4050	10-3/4" diameter		7	3.429		785	131	11.50	927.50	1,075	
	4060	12-3/4" diameter		6.50	3.692		835	141	12.40	988.40	1,150	
	4070	14" diameter		6	4		980	153	13.45	1,146.45	1,325	
	4080	16" diameter		5.50	4.364		1,150	167	14.65	1,331.65	1,525	
	4090	18" diameter		5	4.800		1,275	183	16.15	1,474.15	1,725	
	4100	20" diameter		4.50	5.333		1,425	204	17.90	1,646.90	1,875	
	4110	22" diameter		4	6		1,575	229	20	1,824	2,100	
	4120	24" diameter		3.50	6.857		1,725	262	23	2,010	2,325	
	4130	26" diameter		3	8		1,950	305	27	2,282	2,650	
	4140	28" diameter		2.50	9.600		2,125	365	32.50	2,522.50	2,925	
	4150	30" diameter		2	12		2,275	460	40.50	2,775.50	3,225	
	4160	32" diameter		1.50	16		2,725	610	54	3,389	3,975	
	4170	34" diameter		1	24		3,050	915	80.50	4,045.50	4,825	
	4180	36" diameter	↓	1	24		3,325	915	80.50	4,320.50	5,125	
	4220	For multi-pipe, add				↓	5%	20%				
	4800	Steel casing elbow										
	4820	6-5/8" diameter	Q-18	15	1.600	Ea.	785	61	5.40	851.40	960	
	4830	8-5/8" diameter		15	1.600		835	61	5.40	901.40	1,025	
	4850	10-3/4" diameter		14	1.714		1,000	65.50	5.75	1,071.25	1,200	
	4860	12-3/4" diameter		13	1.846		1,200	70.50	6.20	1,276.70	1,450	
	4870	14" diameter		12	2		1,275	76.50	6.70	1,358.20	1,525	
	4880	16" diameter		11	2.182		1,400	83.50	7.35	1,490.85	1,650	
	4890	18" diameter		10	2.400		1,600	91.50	8.05	1,699.55	1,900	
	4900	20" diameter		9	2.667		1,700	102	8.95	1,810.95	2,050	
	4910	22" diameter		8	3		1,825	115	10.10	1,950.10	2,200	
	4920	24" diameter		7	3.429		2,025	131	11.50	2,167.50	2,425	
	4930	26" diameter		6	4		2,200	153	13.45	2,366.45	2,675	
	4940	28" diameter		5	4.800		2,375	183	16.15	2,574.15	2,925	
	4950	30" diameter		4	6		2,400	229	20	2,649	3,000	
	4960	32" diameter		3	8		2,725	305	27	3,057	3,500	
	4970	34" diameter		2	12		2,975	460	40.50	3,475.50	4,000	
	4980	36" diameter	↓	2	12	↓	3,175	460	40.50	3,675.50	4,225	
	5500	Black steel service pipe, std. wt., 1" thick insulation										
	5510	3/4" diameter pipe size	Q-17	54	.296	L.F.	19.90	10.90	1.49	32.29	40	
	5540	1" diameter pipe size		50	.320		20.50	11.80	1.61	33.91	42.50	
	5550	1-1/4" diameter pipe size		47	.340		23	12.55	1.72	37.27	46	
	5560	1-1/2" diameter pipe size	↓	45	.356	↓	25.50	13.10	1.79	40.39	49.50	

02550 | Piped Energy Distribution		CREW	DAILY OUTPUT	LABOR-HOURS	UNIT	2005 BARE COSTS MAT.	LABOR	EQUIP.	TOTAL	TOTAL INCL O&P
200 5570	2" diameter pipe size	Q-17	42	.381	L.F.	28	14.05	1.92	43.97	54
5580	2-1/2" diameter pipe size		34	.471		29.50	17.35	2.37	49.22	61
5590	3" diameter pipe size		28	.571		33.50	21	2.88	57.38	71.50
5600	4" diameter pipe size		22	.727		43	27	3.67	73.67	92
5610	5" diameter pipe size		18	.889		58.50	33	4.48	95.98	118
5620	6" diameter pipe size	Q-18	23	1.043		63.50	40	3.51	107.01	134
6000	Black steel service pipe, std. wt., 1-1/2" thick insul.									
6010	3/4" diameter pipe size	Q-17	54	.296	L.F.	20	10.90	1.49	32.39	40
6040	1" diameter pipe size		50	.320		22.50	11.80	1.61	35.91	44.50
6050	1-1/4" diameter pipe size		47	.340		25.50	12.55	1.72	39.77	48.50
6060	1-1/2" diameter pipe size		45	.356		27.50	13.10	1.79	42.39	52
6070	2" diameter pipe size		42	.381		30	14.05	1.92	45.97	56
6080	2-1/2" diameter pipe size		34	.471		32	17.35	2.37	51.72	63.50
6090	3" diameter pipe size		28	.571		36	21	2.88	59.88	74
6100	4" diameter pipe size		22	.727		46	27	3.67	76.67	95
6110	5" diameter pipe size		18	.889		58.50	33	4.48	95.98	118
6120	6" diameter pipe size	Q-18	23	1.043		66.50	40	3.51	110.01	137
6130	8" diameter pipe size		19	1.263		95.50	48.50	4.24	148.24	182
6140	10" diameter pipe size		16	1.500		125	57.50	5.05	187.55	229
6150	12" diameter pipe size		13	1.846		149	70.50	6.20	225.70	277
6190	For 2" thick insulation, add					15%				
6220	For 2-1/2" thick insulation, add					25%				
6260	For 3" thick insulation, add					30%				
6800	Black steel service pipe, ex. hvy. wt., 1" thick insul.									
6820	3/4" diameter pipe size	Q-17	50	.320	L.F.	21	11.80	1.61	34.41	42.50
6840	1" diameter pipe size		47	.340		22.50	12.55	1.72	36.77	45
6850	1-1/4" diameter pipe size		44	.364		25.50	13.40	1.83	40.73	50.50
6860	1-1/2" diameter pipe size		42	.381		27.50	14.05	1.92	43.47	53
6870	2" diameter pipe size		40	.400		29	14.75	2.02	45.77	56
6880	2-1/2" diameter pipe size		31	.516		36	19	2.60	57.60	71
6890	3" diameter pipe size		27	.593		40.50	22	2.99	65.49	81
6900	4" diameter pipe size		21	.762		53	28	3.84	84.84	105
6910	5" diameter pipe size		17	.941		74.50	34.50	4.74	113.74	139
6920	6" diameter pipe size	Q-18	22	1.091		82.50	41.50	3.67	127.67	157
7400	Black steel service pipe, ex. hvy. wt., 1-1/2" thick insul.									
7420	3/4" diameter pipe size	Q-17	50	.320	L.F.	21	11.80	1.61	34.41	42.50
7440	1" diameter pipe size		47	.340		24	12.55	1.72	38.27	47
7450	1-1/4" diameter pipe size		44	.364		27.50	13.40	1.83	42.73	52.50
7460	1-1/2" diameter pipe size		42	.381		30.50	14.05	1.92	46.47	56.50
7470	2" diameter pipe size		40	.400		30.50	14.75	2.02	47.27	57.50
7480	2-1/2" diameter pipe size		31	.516		36.50	19	2.60	58.10	71.50
7490	3" diameter pipe size		27	.593		43	22	2.99	67.99	83.50
7500	4" diameter pipe size		21	.762		56	28	3.84	87.84	108
7510	5" diameter pipe size		17	.941		77	34.50	4.74	116.24	142
7520	6" diameter pipe size	Q-18	22	1.091		85	41.50	3.67	130.17	160
7530	8" diameter pipe size		18	1.333		127	51	4.48	182.48	221
7540	10" diameter pipe size		15	1.600		151	61	5.40	217.40	264
7550	12" diameter pipe size		13	1.846		186	70.50	6.20	262.70	315
7590	For 2" thick insulation, add					13%				
7640	For 2-1/2" thick insulation, add					18%				
7680	For 3" thick insulation, add					24%				
9000	Combined steam pipe with condensate return									
9010	8" and 4" in 24" case	Q-18	100	.240	L.F.	268	9.15	.81	277.96	310
9020	6" and 3" in 20" case		104	.231		183	8.80	.78	192.58	216
9030	3" and 1-1/2" in 16" case		110	.218		137	8.35	.73	146.08	164
9040	2" and 1-1/4" in 12-3/4" case		114	.211		119	8.05	.71	127.76	144 200

02550 | Piped Energy Distribution

		02550 Piped Energy Distribution	CREW	DAILY OUTPUT	LABOR-HOURS	UNIT	2005 BARE COSTS MAT.	LABOR	EQUIP.	TOTAL	TOTAL INCL O&P	
200	9050	1-1/2" and 1-1/4" in 10-3/4" case	Q-18	116	.207	L.F.	104	7.90	.70	112.60	128	200
	9100	Steam pipe only (no return)										
	9110	6" in 18" case	Q-18	108	.222	L.F.	151	8.50	.75	160.25	180	
	9120	4" in 14" case		112	.214		119	8.20	.72	127.92	144	
	9130	3" in 14" case		114	.211		110	8.05	.71	118.76	134	
	9140	2-1/2" in 12-3/4" case		118	.203		103	7.75	.68	111.43	125	
	9150	2" in 12-3/4" case		122	.197		98.50	7.50	.66	106.66	120	
464	0010	**PIPING, GAS SERVICE & DISTRIBUTION, POLYETHYLENE**										464
	0020	not including excavation or backfill										
	1000	60 psi coils, comp cplg @ 100', 1/2" diameter, SDR 9.3	B-20A	608	.053	L.F.	.47	1.70		2.17	3.11	
	1040	1-1/4" diameter, SDR 11		544	.059		.87	1.90		2.77	3.85	
	1100	2" diameter, SDR 11		488	.066		1.08	2.11		3.19	4.42	
	1160	3" diameter, SDR 11		408	.078		2.25	2.53		4.78	6.35	
	1500	60 PSI 40' joints with coupling, 3" diameter, SDR 11	B-21A	408	.098		2.25	3.23	1.24	6.72	8.75	
	1540	4" diameter, SDR 11		352	.114		5.15	3.75	1.44	10.34	13	
	1600	6" diameter, SDR 11		328	.122		16.10	4.02	1.54	21.66	25.50	
	1640	8" diameter, SDR 11		272	.147		22	4.85	1.86	28.71	33.50	
550	0010	**GAS STATION PRODUCT LINE**										550
	0020	Primary containment pipe, fiberglass-reinforced										
	0030	Plastic pipe 15' & 30' lengths										
	0040	2" diameter	Q-6	425	.056	L.F.	3.23	2.16		5.39	6.80	
	0050	3" diameter		400	.060		4.23	2.29		6.52	8.10	
	0060	4" diameter		375	.064		5.45	2.45		7.90	9.70	
	0100	Fittings										
	0110	Elbows, 90° & 45°, bell-ends, 2"	Q-6	24	1	Ea.	32.50	38		70.50	93.50	
	0120	3" diameter		22	1.091		34.50	41.50		76	100	
	0130	4" diameter		20	1.200		45.50	46		91.50	119	
	0200	Tees, bell ends, 2"		21	1.143		39.50	43.50		83	109	
	0210	3" diameter		18	1.333		40	51		91	121	
	0220	4" diameter		15	1.600		55	61		116	153	
	0230	Flanges bell ends, 2"		24	1		13.10	38		51.10	72	
	0240	3" diameter		22	1.091		16.50	41.50		58	80.50	
	0250	4" diameter		20	1.200		22.50	46		68.50	93.50	
	0260	Sleeve couplings, 2"		21	1.143		8.35	43.50		51.85	74.50	
	0270	3" diameter		18	1.333		11.85	51		62.85	89.50	
	0280	4" diameter		15	1.600		16.35	61		77.35	110	
	0290	Threaded adapters 2"		21	1.143		10.95	43.50		54.45	77.50	
	0300	3" diameter		18	1.333		19.25	51		70.25	97.50	
	0310	4" diameter		15	1.600		26	61		87	121	
	0320	Reducers, 2"		27	.889		14.60	34		48.60	67	
	0330	3" diameter		22	1.091		16.90	41.50		58.40	81	
	0340	4" diameter		20	1.200		22	46		68	93	
	1010	Gas station product line for secondary containment (double wall)										
	1100	Fiberglass reinforced plastic pipe 25' lengths										
	1120	Pipe, plain end, 3"	Q-6	375	.064	L.F.	5.55	2.45		8	9.80	
	1130	4" diameter		350	.069		9.10	2.62		11.72	13.95	
	1140	5" diameter		325	.074		11.85	2.82		14.67	17.25	
	1150	6" diameter		300	.080		12.20	3.06		15.26	18	
	1200	Fittings										
	1230	Elbows, 90° & 45°, 3"	Q-6	18	1.333	Ea.	40	51		91	121	
	1240	4" diameter		16	1.500		69.50	57.50		127	163	
	1250	5" diameter		14	1.714		162	65.50		227.50	277	
	1260	6" diameter		12	2		164	76.50		240.50	295	
	1270	Tees, 3"		15	1.600		59	61		120	157	
	1280	4" diameter		12	2		87	76.50		163.50	211	

02500 | Utility Services

02550 | Piped Energy Distribution

			CREW	DAILY OUTPUT	LABOR-HOURS	UNIT	2005 BARE COSTS MAT.	LABOR	EQUIP.	TOTAL	TOTAL INCL O&P	
550	1290	5″ diameter	Q-6	9	2.667	Ea.	176	102		278	345	550
	1300	6″ diameter		6	4		184	153		337	430	
	1310	Couplings, 3″		18	1.333		28	51		79	108	
	1320	4″ diameter		16	1.500		72.50	57.50		130	166	
	1330	5″ diameter		14	1.714		150	65.50		215.50	265	
	1340	6″ diameter		12	2		156	76.50		232.50	287	
	1350	Cross-over nipples, 3″		18	1.333		6.40	51		57.40	83.50	
	1360	4″ diameter		16	1.500		7.50	57.50		65	94.50	
	1370	5″ diameter		14	1.714		11.15	65.50		76.65	111	
	1380	6″ diameter		12	2		11.70	76.50		88.20	128	
	1400	Telescoping, reducers, concentric 4″ x 3″		18	1.333		21.50	51		72.50	100	
	1410	5″ x 4″		17	1.412		55.50	54		109.50	143	
	1420	6″ x 5″		16	1.500		134	57.50		191.50	233	

02600 | Drainage & Containment

02620 | Subdrainage

			CREW	DAILY OUTPUT	LABOR-HOURS	UNIT	2005 BARE COSTS MAT.	LABOR	EQUIP.	TOTAL	TOTAL INCL O&P	
610	0010	**PIPING, SUBDRAINAGE, CONCRETE**										610
	0021	Not including excavation and backfill										
	3000	Porous wall concrete underdrain, std. strength, 4″ diameter	B-20	335	.072	L.F.	1.95	2.15		4.10	5.50	
	3020	6″ diameter	″	315	.076		2.54	2.29		4.83	6.35	
	3040	8″ diameter	B-21	310	.090		3.13	2.79	.51	6.43	8.30	
	3100	18″ diameter	″	165	.170		10.05	5.25	.96	16.26	20	
	4000	Extra strength, 6″ diameter	B-20	315	.076		2.57	2.29		4.86	6.40	
	4020	8″ diameter	B-21	310	.090		3.86	2.79	.51	7.16	9.10	
	4040	10″ diameter		285	.098		7.70	3.04	.56	11.30	13.80	
	4060	12″ diameter		230	.122		8.35	3.76	.69	12.80	15.80	
	4080	15″ diameter		200	.140		9.25	4.33	.79	14.37	17.75	
	4100	18″ diameter		165	.170		13.50	5.25	.96	19.71	24	
620	0010	**PIPING, SUBDRAINAGE, CORRUGATED METAL**										620
	0021	Not including excavation and backfill										
	2010	Aluminum, perforated										
	2020	6″ diameter, 18 ga.	B-14	380	.126	L.F.	2.76	3.55	.57	6.88	9.15	
	2200	8″ diameter, 16 ga.		370	.130		3.94	3.64	.58	8.16	10.60	
	2220	10″ diameter, 16 ga.		360	.133		4.93	3.75	.60	9.28	11.85	
	2240	12″ diameter, 16 ga.		285	.168		5.50	4.73	.76	10.99	14.20	
	2260	18″ diameter, 16 ga.		205	.234		8.25	6.60	1.05	15.90	20.50	
	3000	Uncoated galvanized, perforated										
	3020	6″ diameter, 18 ga.	B-20	380	.063	L.F.	4.37	1.90		6.27	7.75	
	3200	8″ diameter, 16 ga.	″	370	.065		6	1.95		7.95	9.65	
	3220	10″ diameter, 16 ga.	B-21	360	.078		9	2.40	.44	11.84	14.10	
	3240	12″ diameter, 16 ga.		285	.098		9.45	3.04	.56	13.05	15.70	
	3260	18″ diameter, 16 ga.		205	.137		14.40	4.22	.77	19.39	23.50	
	4000	Steel, perforated, asphalt coated										
	4020	6″ diameter 18 ga.	B-20	380	.063	L.F.	3.49	1.90		5.39	6.80	
	4030	8″ diameter 18 ga	″	370	.065		5.45	1.95		7.40	9.05	
	4040	10″ diameter 16 ga	B-21	360	.078		6.30	2.40	.44	9.14	11.10	
	4050	12″ diameter 16 ga		285	.098		7.20	3.04	.56	10.80	13.25	
	4060	18″ diameter 16 ga		205	.137		9.85	4.22	.77	14.84	18.20	

02620 | Subdrainage

		02620 Subdrainage	CREW	DAILY OUTPUT	LABOR-HOURS	UNIT	2005 BARE COSTS MAT.	LABOR	EQUIP.	TOTAL	TOTAL INCL O&P	
630	0010	**PIPING, SUBDRAINAGE, PLASTIC**										630
	0020	Not including excavation and backfill										
	1110	10″				Ea.	3.01			3.01	3.31	
	2100	Perforated PVC, 4″ diameter	B-14	314	.153	L.F.	.71	4.29	.69	5.69	8.20	
	2110	6″ diameter		300	.160		1.31	4.49	.72	6.52	9.20	
	2120	8″ diameter		290	.166		1.49	4.65	.74	6.88	9.65	
	2130	10″ diameter		280	.171		2.31	4.82	.77	7.90	10.85	
	2140	12″ diameter		270	.178		3.20	4.99	.80	8.99	12.10	

02630 | Storm Drainage

		02630 Storm Drainage	CREW	DAILY OUTPUT	LABOR-HOURS	UNIT	MAT.	LABOR	EQUIP.	TOTAL	TOTAL INCL O&P	
110	0010	**CATCH BASIN GRATES AND FRAMES** not including footing, excavation										110
	1600	Frames & covers, C.I., 24″ square, 500 lb.	B-6	7.80	3.077	Ea.	239	88.50	27.50	355	430	
	1700	26″ D shape, 600 lb.		7	3.429		380	99	31	510	605	
	1800	Light traffic, 18″ diameter, 100 lb.		10	2.400		128	69	21.50	218.50	271	
	1900	24″ diameter, 300 lb.		8.70	2.759		196	79.50	25	300.50	365	
	2000	36″ diameter, 900 lb.		5.80	4.138		425	119	37	581	690	
	2100	Heavy traffic, 24″ diameter, 400 lb.		7.80	3.077		193	88.50	27.50	309	380	
	2200	36″ diameter, 1150 lb.		3	8		610	231	72	913	1,100	
	2300	Mass. State standard, 26″ diameter, 475 lb.		7	3.429		465	99	31	595	695	
	2400	30″ diameter, 620 lb.		7	3.429		325	99	31	455	540	
	2500	Watertight, 24″ diameter, 350 lb.		7.80	3.077		325	88.50	27.50	441	520	
	2600	26″ diameter, 500 lb.		7	3.429		310	99	31	440	525	
	2700	32″ diameter, 575 lb.		6	4		665	115	36	816	945	
	2800	3 piece cover & frame, 10″ deep,										
	2900	1200 lbs., for heavy equipment	B-6	3	8	Ea.	990	231	72	1,293	1,525	
	3000	Raised for paving 1-1/4″ to 2″ high,										
	3100	4 piece expansion ring										
	3200	20″ to 26″ diameter	1 Clab	3	2.667	Ea.	114	71		185	237	
	3300	30″ to 36″ diameter	″	3	2.667	″	159	71		230	286	
	3320	Frames and covers, existing, raised for paving, 2″, including										
	3340	row of brick, concrete collar, up to 12″ wide frame	B-6	18	1.333	Ea.	35.50	38.50	12	86	112	
	3360	20″ to 26″ wide frame		11	2.182		56.50	63	19.60	139.10	180	
	3380	30″ to 36″ wide frame		9	2.667		70	77	24	171	222	
	3400	Inverts, single channel brick	D-1	3	5.333		78.50	166		244.50	340	
	3500	Concrete		5	3.200		61	99.50		160.50	220	
	3600	Triple channel, brick		2	8		119	249		368	510	
	3700	Concrete		3	5.333		105	166		271	370	
400	0010	**STORM DRAINAGE MANHOLES, FRAMES & COVERS** not including										400
	0020	footing, excavation, backfill (See line items for frame & cover)										
	0050	Brick, 4′ inside diameter, 4′ deep	D-1	1	16	Ea.	320	495		815	1,100	
	0100	6′ deep		.70	22.857		445	710		1,155	1,575	
	0150	8′ deep		.50	32		565	995		1,560	2,150	
	0200	For depths over 8′, add		4	4	V.L.F.	174	124		298	380	
	0400	Concrete blocks (radial), 4′ I.D., 4′ deep		1.50	10.667	Ea.	255	330		585	785	
	0500	6′ deep		1	16		335	495		830	1,125	
	0600	8′ deep		.70	22.857		415	710		1,125	1,525	
	0700	For depths over 8′, add		5.50	2.909	V.L.F.	41.50	90.50		132	184	
	0800	Concrete, cast in place, 4′ x 4′, 8″ thick, 4′ deep	C-14H	2	24	Ea.	445	810	12	1,267	1,750	
	0900	6′ deep		1.50	32		650	1,075	16	1,741	2,400	
	1000	8′ deep		1	48		920	1,625	24	2,569	3,575	
	1100	For depths over 8′, add		8	6	V.L.F.	107	202	3	312	435	
	1110	Precast, 4′ I.D., 4′ deep	B-22	4.10	7.317	Ea.	675	229	58	962	1,175	
	1120	6′ deep		3	10		870	315	79.50	1,264.50	1,525	
	1130	8′ deep		2	15		1,025	470	119	1,614	1,975	
	1140	For depths over 8′, add		16	1.875	V.L.F.	143	58.50	14.85	216.35	264	

02630 | Storm Drainage

		02630 Storm Drainage	CREW	DAILY OUTPUT	LABOR-HOURS	UNIT	2005 BARE COSTS MAT.	LABOR	EQUIP.	TOTAL	TOTAL INCL O&P	
400	1150	5′ I.D., 4′ deep	B-6	3	8	Ea.	720	231	72	1,023	1,225	400
	1160	6′ deep		2	12		970	345	108	1,423	1,725	
	1170	8′ deep		1.50	16		1,225	460	144	1,829	2,225	
	1180	For depths over 8′, add		12	2	V.L.F.	159	57.50	18	234.50	283	
	1190	6′ I.D., 4′ deep		2	12	Ea.	1,175	345	108	1,628	1,950	
	1200	6′ deep		1.50	16		1,525	460	144	2,129	2,550	
	1210	8′ deep		1	24		1,875	690	216	2,781	3,375	
	1220	For depths over 8′, add		8	3	V.L.F.	246	86.50	27	359.50	435	
	1250	Slab tops, precast, 8″ thick										
	1300	4′ diameter manhole	B-6	8	3	Ea.	165	86.50	27	278.50	345	
	1400	5′ diameter manhole		7.50	3.200		325	92	29	446	535	
	1500	6′ diameter manhole		7	3.429		405	99	31	535	630	
	3800	Steps, heavyweight cast iron, 7″ x 9″	1 Bric	40	.200		12.20	7.05		19.25	24	
	3900	8″ x 9″		40	.200		18.35	7.05		25.40	31	
	3928	12″ x 10-1/2″		40	.200		16.40	7.05		23.45	29	
	4000	Standard sizes, galvanized steel		40	.200		15.30	7.05		22.35	27.50	
	4100	Aluminum		40	.200		18.90	7.05		25.95	32	
510	0010	**PIPING, STORM DRAINAGE, CORRUGATED METAL**										510
	0020	Not including excavation or backfill										
	2000	Corrugated metal pipe, galvanized and coated										
	2020	Bituminous coated with paved invert, 20′ lengths										
	2040	8″ diameter, 16 ga.	B-14	330	.145	L.F.	8.95	4.09	.65	13.69	16.85	
	2060	10″ diameter, 16 ga.		260	.185		10.75	5.20	.83	16.78	20.50	
	2080	12″ diameter, 16 ga.		210	.229		12.85	6.40	1.03	20.28	25	
	2100	15″ diameter, 16 ga.		200	.240		15.65	6.75	1.08	23.48	29	
	2120	18″ diameter, 16 ga.		190	.253		20	7.10	1.14	28.24	34.50	
	2140	24″ diameter, 14 ga.		160	.300		24.50	8.45	1.35	34.30	41.50	
	2160	30″ diameter, 14 ga.	B-13	120	.467		32.50	13.45	5.15	51.10	61.50	
	2180	36″ diameter, 12 ga.		120	.467		47.50	13.45	5.15	66.10	78.50	
	2200	48″ diameter, 12 ga.		100	.560		72.50	16.10	6.15	94.75	111	
	2220	60″ diameter, 10 ga.	B-13B	75	.747		93.50	21.50	11.95	126.95	149	
	2240	72″ diameter, 8 ga.	″	45	1.244		140	36	19.90	195.90	231	
	2500	Galvanized, uncoated, 20′ lengths										
	2520	8″ diameter, 16 ga.	B-14	355	.135	L.F.	6.90	3.80	.61	11.31	14.10	
	2540	10″ diameter, 16 ga.		280	.171		7.60	4.82	.77	13.19	16.65	
	2560	12″ diameter, 16 ga.		220	.218		8.65	6.15	.98	15.78	20	
	2580	15″ diameter, 16 ga.		220	.218		11	6.15	.98	18.13	22.50	
	2600	18″ diameter, 16 ga.		205	.234		14.55	6.60	1.05	22.20	27.50	
	2620	24″ diameter, 14 ga.		175	.274		21	7.70	1.23	29.93	36.50	
	2640	30″ diameter, 14 ga.	B-13	130	.431		27	12.40	4.74	44.14	54	
	2660	36″ diameter, 12 ga.		130	.431		44	12.40	4.74	61.14	73	
	2680	48″ diameter, 12 ga.		110	.509		59	14.65	5.60	79.25	93.50	
	2690	60″ diameter, 10 ga.	B-13B	78	.718		93	20.50	11.50	125	147	
	2780	End sections, 8″ diameter	B-14	24	2	Ea.	71.50	56	9	136.50	175	
	2785	10″ diameter		22	2.182		73	61.50	9.80	144.30	186	
	2790	12″ diameter		35	1.371		82.50	38.50	6.15	127.15	157	
	2800	18″ diameter		30	1.600		98	45	7.20	150.20	185	
	2810	24″ diameter	B-13	25	2.240		143	64.50	24.50	232	284	
	2820	30″ diameter		25	2.240		272	64.50	24.50	361	425	
	2825	36″ diameter		20	2.800		360	80.50	31	471.50	555	
	2830	48″ diameter		10	5.600		745	161	61.50	967.50	1,125	
	2835	60″ diameter	B-13B	5	11.200		970	320	179	1,469	1,775	
	2840	72″ diameter	″	4	14		1,750	405	224	2,379	2,800	
530	0010	**SEWAGE/DRAINAGE COLLECTION, CONCRETE PIPE**										530
	0020	Not including excavation or backfill										

For expanded coverage of these items see *Means Heavy Construction Cost Data 2005*

	02630 | Storm Drainage		CREW	DAILY OUTPUT	LABOR-HOURS	UNIT	2005 BARE COSTS MAT.	LABOR	EQUIP.	TOTAL	TOTAL INCL O&P	
530	1000	Non-reinforced pipe, extra strength, B&S or T&G joints										530
	1010	6″ diameter	B-14	265.04	.181	L.F.	4.13	5.10	.82	10.05	13.30	
	1020	8″ diameter		224	.214		4.54	6	.96	11.50	15.35	
	1030	10″ diameter		216	.222		5.05	6.25	1	12.30	16.30	
	1040	12″ diameter		200	.240		6.20	6.75	1.08	14.03	18.45	
	1050	15″ diameter		180	.267		7.25	7.50	1.20	15.95	21	
	1060	18″ diameter		144	.333		8.90	9.35	1.50	19.75	26	
	1070	21″ diameter		112	.429		10.95	12.05	1.93	24.93	33	
	1080	24″ diameter		100	.480		13.40	13.50	2.16	29.06	38	
	2000	Reinforced culvert, class 3, no gaskets										
	2010	12″ diameter	B-14	150	.320	L.F.	11.20	9	1.44	21.64	28	
	2020	15″ diameter		150	.320		14.35	9	1.44	24.79	31.50	
	2030	18″ diameter		132	.364		15.10	10.20	1.64	26.94	34	
	2035	21″ diameter		120	.400		19.60	11.25	1.80	32.65	41	
	2040	24″ diameter		100	.480		22.50	13.50	2.16	38.16	48	
	2045	27″ diameter	B-13	92	.609		28	17.50	6.70	52.20	65.50	
	2050	30″ diameter		88	.636		30.50	18.30	7	55.80	69	
	2060	36″ diameter		72	.778		43.50	22.50	8.55	74.55	92	
	2070	42″ diameter	B-13B	72	.778		59.50	22.50	12.45	94.45	114	
	2080	48″ diameter		64	.875		73.50	25	14	112.50	135	
	2090	60″ diameter		48	1.167		118	33.50	18.65	170.15	201	
	2100	72″ diameter		40	1.400		166	40.50	22.50	229	270	
	2120	84″ diameter		32	1.750		281	50.50	28	359.50	420	
	2140	96″ diameter		24	2.333		335	67	37.50	439.50	515	
	2200	With gaskets, class 3, 12″ diameter	B-21	168	.167		12.25	5.15	.94	18.34	22.50	
	2220	15″ diameter		160	.175		14.70	5.40	.99	21.09	25.50	
	2230	18″ diameter		152	.184		18.40	5.70	1.04	25.14	30.50	
	2240	24″ diameter		136	.206		27.50	6.35	1.17	35.02	41.50	
	2260	30″ diameter	B-13	88	.636		37	18.30	7	62.30	76	
	2270	36″ diameter	″	72	.778		55	22.50	8.55	86.05	104	
	2290	48″ diameter	B-13B	64	.875		90	25	14	129	153	
	2310	72″ diameter	″	40	1.400		233	40.50	22.50	296	345	
	2330	Flared ends, 6′-1″ long, 12″ diameter	B-21	190	.147		34	4.56	.83	39.39	45	
	2340	15″ diameter		155	.181		38.50	5.60	1.02	45.12	52.50	
	2400	6′-2″ long, 18″ diameter		122	.230		40	7.10	1.30	48.40	56.50	
	2420	24″ diameter		88	.318		46.50	9.85	1.80	58.15	68	
	2440	36″ diameter	B-13	60	.933		84.50	27	10.30	121.80	145	
	3080	Radius pipe, add to pipe prices, 12″ to 60″ diameter					50%					
	3090	Over 60″ diameter, add					20%					
	3500	Reinforced elliptical, 8′ lengths, C507 class 3										
	3520	14″ x 23″ inside, round equivalent 18″ diameter	B-21	82	.341	L.F.	26.50	10.55	1.93	38.98	47.50	
	3530	24″ x 38″ inside, round equivalent 30″ diameter	B-13	58	.966		47	28	10.65	85.65	106	
	3540	29″ x 45″ inside, round equivalent 36″ diameter		52	1.077		60.50	31	11.85	103.35	128	
	3550	38″ x 60″ inside, round equivalent 48″ diameter		38	1.474		79	42.50	16.25	137.75	170	
	3560	48″ x 76″ inside, round equivalent 60″ diameter		26	2.154		120	62	23.50	205.50	254	
	3570	58″ x 91″ inside, round equivalent 72″ diameter		22	2.545		171	73.50	28	272.50	335	
	3780	Concrete slotted pipe, class 4 mortar joint										
	3800	12″ diameter	B-21	168	.167	L.F.	13.15	5.15	.94	19.24	23.50	
	3840	18″ diameter	″	152	.184	″	20.50	5.70	1.04	27.24	32.50	
	3900	Class 4 O-ring										
	3940	12″ diameter	B-21	168	.167	L.F.	13.75	5.15	.94	19.84	24	
	3960	18″ diameter	″	152	.184	″	18.45	5.70	1.04	25.19	30.50	
810	0010	**OVAL ARCH CULVERTS**										810
	3000	Corrugated galvanized or aluminum, coated & paved										
	3020	17″ x 13″, 16 ga., 15″ equivalent	B-14	200	.240	L.F.	26.50	6.75	1.08	34.33	40.50	
	3040	21″ x 15″, 16 ga., 18″ equivalent		150	.320		34	9	1.44	44.44	53	

02600 | Drainage & Containment

	02630 | Storm Drainage		CREW	DAILY OUTPUT	LABOR-HOURS	UNIT	2005 BARE COSTS MAT.	LABOR	EQUIP.	TOTAL	TOTAL INCL O&P	
810	3060	28″ x 20″, 14 ga., 24″ equivalent	B-14	125	.384	L.F.	49	10.80	1.73	61.53	72	810
	3080	35″ x 24″, 14 ga., 30″ equivalent	↓	100	.480		59.50	13.50	2.16	75.16	89	
	3100	42″ x 29″, 12 ga., 36″ equivalent	B-13	100	.560		88.50	16.10	6.15	110.75	129	
	3120	49″ x 33″, 12 ga., 42″ equivalent		90	.622		107	17.90	6.85	131.75	152	
	3140	57″ x 38″, 12 ga., 48″ equivalent	↓	75	.747	↓	119	21.50	8.20	148.70	173	
	3160	Steel, plain oval arch culverts, plain										
	3180	17″ x 13″, 16 ga., 15″ equivalent	B-14	225	.213	L.F.	14.30	6	.96	21.26	26	
	3200	21″ x 15″, 16 ga., 18″ equivalent		175	.274		16.95	7.70	1.23	25.88	32	
	3220	28″ x 20″, 14 ga., 24″ equivalent	↓	150	.320		27	9	1.44	37.44	45.50	
	3240	35″ x 24″, 14 ga., 30″ equivalent	B-13	108	.519		34	14.95	5.70	54.65	67	
	3260	42″ x 29″, 12 ga., 36″ equivalent		108	.519		56.50	14.95	5.70	77.15	91.50	
	3280	49″ x 33″, 12 ga., 42″ equivalent		92	.609		66.50	17.50	6.70	90.70	108	
	3300	57″ x 38″, 12 ga., 48″ equivalent		75	.747	↓	58.50	21.50	8.20	88.20	107	
	3320	End sections, 17″ x 13″		22	2.545	Ea.	78.50	73.50	28	180	230	
	3340	42″ x 29″	↓	17	3.294	″	320	95	36.50	451.50	540	
	3360	Multi-plate arch, steel	B-20	1,690	.014	Lb.	.90	.43		1.33	1.65	

02700 | Bases, Ballasts, Pavements & Appurtenances

	02740 | Flexible Pavement		CREW	DAILY OUTPUT	LABOR-HOURS	UNIT	2005 BARE COSTS MAT.	LABOR	EQUIP.	TOTAL	TOTAL INCL O&P	
310	0010	ASPHALTIC CONCRETE PAVEMENT, HIGHWAYS										310
	0020	and large paved areas										
	0080	Binder course, 1-1/2″ thick	B-25	7,725	.011	S.Y.	2.63	.33	.26	3.22	3.69	
	0120	2″ thick		6,345	.014		3.50	.40	.32	4.22	4.82	
	0160	3″ thick		4,905	.018		5.20	.52	.41	6.13	6.95	
	0200	4″ thick	↓	4,140	.021		6.95	.62	.49	8.06	9.15	
	0300	Wearing course, 1″ thick	B-25B	10,575	.009		1.86	.27	.21	2.34	2.68	
	0340	1-1/2″ thick		7,725	.012		2.83	.37	.28	3.48	3.98	
	0380	2″ thick		6,345	.015		3.80	.45	.35	4.60	5.25	
	0420	2-1/2″ thick		5,480	.018		4.69	.52	.40	5.61	6.40	
	0460	3″ thick	↓	4,900	.020		5.60	.58	.45	6.63	7.55	
	0500	Open graded friction course	B-25C	5,000	.010	↓	1.66	.28	.35	2.29	2.66	
	0800	Alternate method of figuring paving costs										
	0810	Binder course, 1-1/2″ thick	B-25	630	.140	Ton	34.50	4.06	3.19	41.75	48	
	0811	2″ thick		690	.128		34.50	3.71	2.91	41.12	47	
	0812	3″ thick		800	.110		34.50	3.20	2.51	40.21	45.50	
	0813	4″ thick	↓	850	.104		34.50	3.01	2.37	39.88	45	
	0850	Wearing course, 1″ thick	B-25B	575	.167		34.50	4.93	3.81	43.24	49.50	
	0851	1-1/2″ thick		630	.152		34.50	4.50	3.48	42.48	48.50	
	0852	2″ thick		690	.139		34.50	4.11	3.17	41.78	48	
	0853	2-1/2″ thick		745	.129		34.50	3.80	2.94	41.24	47	
	0854	3″ thick	↓	800	.120	↓	34.50	3.54	2.74	40.78	46.50	
	1000	Pavement replacement over trench, 2″ thick	B-37	90	.533	S.Y.	3.87	15	1.19	20.06	28.50	
	1050	4″ thick		70	.686		7.65	19.25	1.54	28.44	40	
	1080	6″ thick	↓	55	.873	↓	12.20	24.50	1.95	38.65	53.50	
	1200	For paving projects 300 tons or less add for trucking										
	1300	See 02315-490-0010 for hauling costs										
315	0011	ASPHALTIC CONCRETE PAVEMENT, LOTS & DRIVEWAYS										315
	0020	6″ stone base, 2″ binder course, 1″ topping	B-25C	9,000	.005	S.F.	1.38	.16	.20	1.74	1.98	

02700 | Bases, Ballasts, Pavements & Appurtenances

02740 | Flexible Pavement

		02740 Flexible Pavement	CREW	DAILY OUTPUT	LABOR-HOURS	UNIT	2005 BARE COSTS MAT.	LABOR	EQUIP.	TOTAL	TOTAL INCL O&P	
315	0300	Binder course, 1-1/2" thick	B-25C	35,000	.001	S.F.	.31	.04	.05	.40	.46	315
	0400	2" thick		25,000	.002		.41	.06	.07	.54	.62	
	0500	3" thick		15,000	.003		.63	.10	.12	.85	.97	
	0600	4" thick		10,800	.004		.82	.13	.16	1.11	1.28	
	0800	Sand finish course, 3/4" thick		41,000	.001		.19	.03	.04	.26	.31	
	0900	1" thick	↓	34,000	.001		.23	.04	.05	.32	.38	
	1000	Fill pot holes, hot mix, 2" thick	B-16	4,200	.008		.42	.21	.11	.74	.90	
	1100	4" thick		3,500	.009		.61	.25	.14	1	1.21	
	1120	6" thick	↓	3,100	.010		.82	.28	.15	1.25	1.51	
	1140	Cold patch, 2" thick	B-51	3,000	.016		.50	.43	.04	.97	1.27	
	1160	4" thick		2,700	.018		.95	.48	.05	1.48	1.85	
	1180	6" thick	↓	1,900	.025	↓	1.48	.68	.07	2.23	2.76	

02800 | Site Improvements and Amenities

02810 | Irrigation System

		02810 Irrigation System	CREW	DAILY OUTPUT	LABOR-HOURS	UNIT	2005 BARE COSTS MAT.	LABOR	EQUIP.	TOTAL	TOTAL INCL O&P	
300	0010	**SPRINKLER IRRIGATION SYSTEM** For lawns										300
	0100	Golf course with fully automatic system	C-17	.05	1,600	9 holes	82,500	56,500		139,000	178,500	
	0200	24' diam. head at 15' O.C incl. piping, auto oper., minimum	B-20	70	.343	Head	18.20	10.30		28.50	36	
	0300	Maximum		40	.600		42	18.05		60.05	74	
	0500	60' diam. head at 40' O.C. incl. piping, auto oper., minimum		28	.857		55	26		81	101	
	0600	Maximum		23	1.043	↓	154	31.50		185.50	219	
	0800	Residential system, custom, 1" supply		2,000	.012	S.F.	.27	.36		.63	.86	
	0900	1-1/2" supply	↓	1,800	.013	"	.32	.40		.72	.97	

02950 | Site Restoration & Rehabilitation

02955 | Restoration of Underground Piping

		02955 Restoration of Underground Piping	CREW	DAILY OUTPUT	LABOR-HOURS	UNIT	2005 BARE COSTS MAT.	LABOR	EQUIP.	TOTAL	TOTAL INCL O&P	
210	0010	**PIPE INTERNAL CLEANING & INSPECTION**										210
	0100	Cleaning, pressure pipe systems										
	0120	Pig method, lengths 1000' to 10,000'										
	0140	4" diameter thru 24" diameter, minimum				L.F.				2.14	2.46	
	0160	Maximum				"				5.35	6.40	
	6000	Sewage/sanitary systems										
	6100	Power rodder with header & cutters										
	6110	Mobilization charge, minimum				Total				320	375	
	6120	Mobilization charge, maximum				"				750	850	
	6140	4" diameter				L.F.				1.07	1.23	
	6150	6" diameter								1.34	1.55	
	6160	8" diameter								1.60	1.87	
	6170	10" diameter								1.87	2.14	
	6180	12" diameter								2.03	2.35	
	6190	14" diameter								2.14	2.46	
	6200	16" diameter				↓				2.41	2.78	

Important: See the Reference Section for critical supporting data - Reference Nos., Crews, & City Cost Indexes

	02955 | Restoration of Underground Piping		Crew	Daily Output	Labor-Hours	Unit	2005 Bare Costs Mat.	Labor	Equip.	Total	Total Incl O&P	
210	6210	18″ diameter				L.F.				2.68	3.05	210
	6220	20″ diameter								2.78	3.21	
	6230	24″ diameter								3.10	3.58	
	6240	30″ diameter								3.75	4.28	
	6250	36″ diameter								4.28	4.92	
	6260	48″ diameter								4.60	5.30	
	6270	60″ diameter								6	6.90	
	6280	72″ diameter								6.65	7.65	
	9000	Inspection, television camera with film										
	9060	500 linear feet				Total				1,075	1,230	
220	0010	**PIPEBURSTING**, 300′ runs, replace with HDPE pipe										220
	0020	Not including excavation, backfill, shoring, or dewatering										
	0100	6″ to 15″ diameter, minimum				L.F.					80	
	0200	Maximum									160	
	0300	18″ to 36″ diameter, minimum									187	
	0400	Maximum									320	
	0500	Mobilize and demobilize, minimum				Job					2,675	
	0600	Maximum				″					26,800	
240	0010	**HDPE PIPE LINING**, excludes cleaning and video inspection										240
	0020	Pipe relined with one pipe size smaller than original (4″ for 6 ″)										
	0100	6″ diameter, original size	B-6B	600	.080	L.F.	1.84	2.19	1.46	5.49	7.05	
	0150	8″ diameter, original size		600	.080		2.65	2.19	1.46	6.30	7.95	
	0200	10″ diameter, original size		600	.080		4.48	2.19	1.46	8.13	9.95	
	0250	12″ diameter, original size		400	.120		6.95	3.28	2.20	12.43	15.15	
	0300	14″ diameter, original size		400	.120		9.80	3.28	2.20	15.28	18.30	
	0350	16″ diameter, original size		300	.160		11.80	4.38	2.93	19.11	23	
	0400	18″ diameter, original size		300	.160		15.45	4.38	2.93	22.76	27	
	1000	Pipe HDPE lining, make service line taps	B-6	4	6	Ea.	61	173	54	288	395	
310	0010	**CORROSION RESISTANCE** Wrap & coat, add to pipe, 4″ dia.				L.F.	1.46			1.46	1.61	310
	0020	5″ diameter				″	1.53			1.53	1.68	
	0040	6″ diameter				L.F.	1.53			1.53	1.68	
	0060	8″ diameter					2.37			2.37	2.61	
	0080	10″ diameter					3.54			3.54	3.89	
	0100	12″ diameter					3.54			3.54	3.89	
	0120	14″ diameter					5.20			5.20	5.70	
	0140	16″ diameter					5.20			5.20	5.70	
	0160	18″ diameter					5.25			5.25	5.75	
	0180	20″ diameter					5.50			5.50	6.05	
	0200	24″ diameter					6.70			6.70	7.35	
	0220	Small diameter pipe, 1″ diameter, add					1.10			1.10	1.21	
	0240	2″ diameter					1.21			1.21	1.33	
	0260	2-1/2″ diameter					1.33			1.33	1.46	
	0280	3″ diameter					1.33			1.33	1.46	
	0300	Fittings, field covered, add				S.F.	7			7	7.70	
	0500	Coating, bituminous, per diameter inch, 1 coat, add				L.F.	.27			.27	.30	
	0540	3 coat					.42			.42	.46	
	0560	Coal tar epoxy, per diameter inch, 1 coat, add					.16			.16	.18	
	0600	3 coat					.34			.34	.37	
	1000	Polyethylene H.D. extruded, .025″ thk., 1/2″ diameter add					.24			.24	.26	
	1020	3/4″ diameter					.25			.25	.28	
	1040	1″ diameter					.28			.28	.31	
	1060	1-1/4″ diameter					.34			.34	.37	
	1080	1-1/2″ diameter					.38			.38	.42	
	1100	.030″ thk., 2″ diameter					.41			.41	.45	
	1120	2-1/2″ diameter					.48			.48	.53	
	1140	.035″ thk., 3″ diameter					.60			.60	.66	

	02955 Restoration of Underground Piping		CREW	DAILY OUTPUT	LABOR-HOURS	UNIT	MAT.	LABOR	EQUIP.	TOTAL	TOTAL INCL O&P	
							2005 BARE COSTS					
310	1160	3-1/2" diameter				L.F.	.68			.68	.75	310
	1180	4" diameter					.76			.76	.84	
	1200	.040" thk, 5" diameter					1.05			1.05	1.16	
	1220	6" diameter					1.09			1.09	1.20	
	1240	8" diameter					1.43			1.43	1.57	
	1260	10" diameter					1.74			1.74	1.91	
	1280	12" diameter					2.08			2.08	2.29	
	1300	.060" thk., 14" diameter					2.87			2.87	3.16	
	1320	16" diameter					3.36			3.36	3.70	
	1340	18" diameter					3.97			3.97	4.37	
	1360	20" diameter				↓	4.46			4.46	4.91	
	1380	Fittings, field wrapped, add				S.F.	4.61			4.61	5.05	
400	0010	PIPE REPAIR										400
	0020	Not including excavation or backfill										
	0100	Clamp, stainless steel, lightweight, for steel pipe										
	0110	3" long, 1/2" diameter pipe	1 Plum	34	.235	Ea.	10.45	9.60		20.05	26	
	0120	3/4" diameter pipe		32	.250		10.80	10.20		21	27	
	0130	1" diameter pipe		30	.267		11.60	10.90		22.50	29	
	0140	1-1/4" diameter pipe		28	.286		12.30	11.65		23.95	31	
	0150	1-1/2" diameter pipe		26	.308		12.70	12.55		25.25	33	
	0160	2" diameter pipe		24	.333		13.85	13.60		27.45	36	
	0170	2-1/2" diameter pipe		23	.348		15.10	14.20		29.30	38	
	0180	3" diameter pipe		22	.364		16.65	14.85		31.50	41	
	0190	3-1/2" diameter pipe	↓	21	.381		17.85	15.55		33.40	43	
	0200	4" diameter pipe	B-20	56	.429		18.65	12.90		31.55	40.50	
	0210	5" diameter pipe		53	.453		21.50	13.60		35.10	44.50	
	0220	6" diameter pipe		48	.500		24	15.05		39.05	49.50	
	0230	8" diameter pipe		30	.800		28	24		52	68	
	0240	10" diameter pipe		28	.857		71.50	26		97.50	119	
	0250	12" diameter pipe		24	1		76	30		106	131	
	0260	14" diameter pipe		22	1.091		79.50	33		112.50	138	
	0270	16" diameter pipe		20	1.200		85.50	36		121.50	150	
	0280	18" diameter pipe		18	1.333		90.50	40		130.50	162	
	0290	20" diameter pipe		16	1.500		100	45		145	180	
	0300	24" diameter pipe	↓	14	1.714		111	51.50		162.50	203	
	0360	For 6" long, add					100%	40%				
	0370	For 9" long, add					200%	100%				
	0380	For 12" long, add					300%	150%				
	0390	For 18" long, add				↓	500%	200%				
	0400	Pipe Freezing for live repairs of systems 3/8 inch to 6 inch										
	0410	Note: Pipe Freezing can also used to install a valve into a live system										
	0420	Pipe Freezing each side 3/8 inch	2 Skwk	8	2	Ea.	500	69.50		569.50	660	
	0425	Pipe Freezing each side 3/8 inch , second location same kit		8	2		19	69.50		88.50	129	
	0430	Pipe Freezing each side 3/4 inch		8	2		430	69.50		499.50	585	
	0435	Pipe Freezing each side 3/4 inch , second location same kit		8	2		19	69.50		88.50	129	
	0440	Pipe Freezing each side 1 1/2 inch		6	2.667		430	93		523	620	
	0445	Pipe Freezing each side 1 1/2 inch , second location same kit		6	2.667		19	93		112	166	
	0450	Pipe Freezing each side 2 inch		6	2.667		635	93		728	845	
	0455	Pipe Freezing each side 2 inch , second location same kit		6	2.667		19	93		112	166	
	0460	Pipe Freezing each side 2 1/2 inch-3 inch		6	2.667		890	93		983	1,125	
	0465	Pipe Freezing each side 2 1/2-3 inch , second loc. same kit		6	2.667		38	93		131	187	
	0470	Pipe Freezing each side 4 inch		4	4		1,450	139		1,589	1,800	
	0475	Pipe Freezing each side 4 inch , second location same kit		4	4		60	139		199	283	
310	0480	Pipe Freezing each side 5-6 inch	↓	4	4	↓	3,475	139		3,614	4,050	310

02950 | Site Restoration & Rehabilitation

	02955 | Restoration of Underground Piping	CREW	DAILY OUTPUT	LABOR-HOURS	UNIT	2005 BARE COSTS MAT.	LABOR	EQUIP.	TOTAL	TOTAL INCL O&P	
400	0485 Pipe Freezing each side 5-6 inch , second location same kit	2 Skwk	4	4	Ea.	180	139		319	415	400
	0490 Pipe Freez extra 20 lb CO2 cylin. (3/8 to 2 " - 1 ea, 3 " -2 ea)					189			189	208	
	0500 Pipe Freezing extra 50 lb CO2 cylinders (4 " - 2 ea, 5-6 " -6 ea)					425			425	465	
	1000 Clamp, stainless steel, with threaded service tap										
	1040 Full seal for iron, steel, PVC pipe										
	1100 6" long, 2" diameter pipe	1 Plum	17	.471	Ea.	135	19.20		154.20	177	
	1110 2-1/2" diameter pipe		16	.500		143	20.50		163.50	188	
	1120 3" diameter pipe		15.60	.513		152	21		173	199	
	1130 3-1/2" diameter pipe		15	.533		159	22		181	207	
	1140 4" diameter pipe	B-20	40	.600		174	18.05		192.05	220	
	1150 6" diameter pipe		34	.706		201	21		222	255	
	1160 8" diameter pipe		21	1.143		251	34.50		285.50	330	
	1170 10" diameter pipe		20	1.200		278	36		314	360	
	1180 12" diameter pipe		17	1.412		315	42.50		357.50	415	
	1205 For 9" long, add					20%	45%				
	1210 For 12" long, add					40%	80%				
	1220 For 18" long, add					70%	110%				
	1600 Clamp, stainless steel, single section										
	1640 Full seal for iron, steel, PVC pipe										
	1700 6" long, 2" diameter pipe	1 Plum	17	.471	Ea.	70	19.20		89.20	106	
	1710 2-1/2" diameter pipe		16	.500		76	20.50		96.50	115	
	1720 3" diameter pipe		15.60	.513		79.50	21		100.50	119	
	1730 3-1/2" diameter pipe		15	.533		87	22		109	129	
	1740 4" diameter pipe	B-20	40	.600		95	18.05		113.05	133	
	1750 6" diameter pipe		34	.706		119	21		140	164	
	1760 8" diameter pipe		21	1.143		159	34.50		193.50	228	
	1770 10" diameter pipe		20	1.200		182	36		218	257	
	1780 12" diameter pipe		17	1.412		198	42.50		240.50	284	
	1800 For 9" long, add					40%	45%				
	1810 For 12" long, add					60%	80%				
	1820 For 18" long, add					120%	110%				
	2000 Clamp, stainless steel, two section										
	2040 Full seal, for iron, steel, PVC pipe										
	2100 6" long, 4" diameter pipe	B-20	24	1	Ea.	135	30		165	195	
	2110 6" diameter pipe		20	1.200		159	36		195	230	
	2120 8" diameter pipe		13	1.846		174	55.50		229.50	279	
	2130 10" diameter pipe		12	2		254	60		314	375	
	2140 12" diameter pipe		10	2.400		278	72		350	415	
	2200 9" long, 4" diameter pipe		16	1.500		174	45		219	262	
	2210 6" diameter pipe		13	1.846		198	55.50		253.50	305	
	2220 8" diameter pipe		9	2.667		206	80		286	350	
	2230 10" diameter pipe		8	3		285	90		375	455	
	2240 12" diameter pipe		7	3.429		315	103		418	510	
	2250 14" diameter pipe		6.40	3.750		475	113		588	700	
	2260 16" diameter pipe		6	4		500	120		620	735	
	2270 18" diameter pipe		5	4.800		515	144		659	790	
	2280 20" diameter pipe		4.60	5.217		555	157		712	855	
	2290 24" diameter pipe		4	6		950	180		1,130	1,325	
	2320 For 12" long, add to 9"					15%	25%				
	2330 For 18" long, add to 9"					70%	55%				
	8000 For internal cleaning and inspection, see Div. 02955-210										
	8100 For pipe testing, see Div. 15955-700										

Division Notes

		CREW	DAILY OUTPUT	LABOR-HOURS	UNIT	2005 BARE COSTS MAT.	LABOR	EQUIP.	TOTAL	TOTAL INCL O&P

Division 3 Concrete

Estimating Tips

General

- Carefully check all the plans and specifications. Concrete often appears on drawings other than structural drawings, including mechanical and electrical drawings for equipment pads. The cost of cutting and patching is often difficult to estimate. See Subdivisions 02220 and 03055 for demolition costs.
- Always obtain concrete prices from suppliers near the job site. A volume discount can often be negotiated depending upon competition in the area. Remember to add for waste, particularly for slabs and footings on grade.

03100 Concrete Forms & Accessories

- A primary cost for concrete construction is forming. Most jobs today are constructed with prefabricated forms. The selection of the forms best suited for the job and the total square feet of forms required for efficient concrete forming and placing are key elements in estimating concrete construction. Enough forms must be available for erection to make efficient use of the concrete placing equipment and crew.
- Concrete accessories for forming and placing depend upon the systems used. Study the plans and specifications to assure that all special accessory requirements have been included in the cost estimate such as anchor bolts, inserts and hangers.

03200 Concrete Reinforcement

- Ascertain that the reinforcing steel supplier has included all accessories, cutting, bending and an allowance for lapping, splicing and waste. A good rule of thumb is 10% for lapping, splicing and waste. Also, 10% waste should be allowed for welded wire fabric.

03300 Cast-in-Place Concrete

- When estimating structural concrete, pay particular attention to requirements for concrete additives, curing methods and surface treatments. Special consideration for climate, hot or cold, must be included in your estimate. Be sure to include requirements for concrete placing equipment and concrete finishing.

03400 Precast Concrete
03500 Cementitious Decks & Toppings

- The cost of hauling precast concrete structural members is often an important factor. For this reason, it is important to get a quote from the nearest supplier. It may become economically feasible to set up precasting beds on the site if the hauling costs are prohibitive.

Reference Numbers

Reference numbers are shown in bold squares at the beginning of some major classifications. These numbers refer to related items in the Reference Section. The reference information may be an estimating procedure, an alternate pricing method or technical information.

Note: Not all subdivisions listed here necessarily appear in this publication.

03100 | Concrete Forms & Accessories

		03110 | Structural C.I.P. Forms	CREW	DAILY OUTPUT	LABOR-HOURS	UNIT	2005 BARE COSTS MAT.	LABOR	EQUIP.	TOTAL	TOTAL INCL O&P	
425	0010	**FORMS IN PLACE, EQUIPMENT FOUNDATIONS** job built										425
	0020	1 use	C-2	160	.300	SFCA	2.67	10		12.67	18.50	
	0050	2 use		190	.253		1.47	8.40		9.87	14.70	
	0100	3 use		200	.240		1.07	8		9.07	13.60	
	0150	4 use		205	.234		.87	7.80		8.67	13.10	
430	0010	**FORMS IN PLACE, FOOTINGS** Continuous wall, plywood, 1 use	C-1	375	.085	SFCA	2.31	2.76		5.07	6.85	430
	0050	2 use		440	.073		1.27	2.35		3.62	5.05	
	0100	3 use		470	.068		.92	2.20		3.12	4.45	
	0150	4 use		485	.066		.75	2.14		2.89	4.16	
	5000	Spread footings, job-built lumber, 1 use		305	.105		1.66	3.40		5.06	7.15	
	5050	2 use		371	.086		.92	2.79		3.71	5.35	
	5100	3 use		401	.080		.67	2.58		3.25	4.75	
	5150	4 use		414	.077		.54	2.50		3.04	4.49	
445	0010	**FORMS IN PLACE, SLAB ON GRADE**										445
	3000	Edge forms, wood, 4 use, on grade, to 6″ high	C-1	600	.053	L.F.	.27	1.73		2	2.98	
	6000	Trench forms in floor, wood, 1 use		160	.200	SFCA	2.20	6.45		8.65	12.50	
	6050	2 use		175	.183		1.22	5.90		7.12	10.55	
	6100	3 use		180	.178		.89	5.75		6.64	9.95	
	6150	4 use		185	.173		.72	5.60		6.32	9.50	
	8760	Void form, corrugated fiberboard, 6″ x 12″, 10′ long		240	.133	S.F.	9.20	4.31		13.51	16.80	
620	0010	**ACCESSORIES, SLEEVES AND CHASES**										620
	0100	Plastic, 1 use, 9″ long, 2″ diameter	1 Carp	100	.080	Ea.	.53	2.74		3.27	4.85	
	0150	4″ diameter		90	.089		1.56	3.04		4.60	6.45	
	0200	6″ diameter		75	.107		2.75	3.65		6.40	8.75	
	0250	12″ diameter		60	.133		18.05	4.57		22.62	27	
	5000	Sheet metal, 2″ diameter		100	.080		.68	2.74		3.42	5	
	5100	4″ diameter		90	.089		.85	3.04		3.89	5.70	
	5150	6″ diameter		75	.107		1.23	3.65		4.88	7.05	
	5200	12″ diameter		60	.133		2.46	4.57		7.03	9.80	
	6000	Steel pipe, 2″ diameter		100	.080		3.49	2.74		6.23	8.10	
	6100	4″ diameter		90	.089		11.90	3.04		14.94	17.80	
	6150	6″ diameter		75	.107		25.50	3.65		29.15	34	
	6200	12″ diameter		60	.133		64	4.57		68.57	77.50	

03200 | Concrete Reinforcement

		03210 | Reinforcing Steel	CREW	DAILY OUTPUT	LABOR-HOURS	UNIT	2005 BARE COSTS MAT.	LABOR	EQUIP.	TOTAL	TOTAL INCL O&P	
600	0010	**REINFORCING IN PLACE** A615 Grade 60, incl. access. labor										600
	0502	Footings, #4 to #7	4 Rodm	4,200	.008	Lb.	.42	.29		.71	.94	
	0552	#8 to #18		7,200	.004		.42	.17		.59	.74	
	0602	Slab on grade, #3 to #7		4,200	.008		.40	.29		.69	.92	
	3000	For epoxy dowel anchoring, see Div 05090-300										
		03220 | Welded Wire Fabric										
200	0010	**WELDED WIRE FABRIC** ASTM A185										200
	0040	Reinforcing sheets, 6x6-W1.4xW1.4				S.F.	.44			.44	.48	

		03310 | Structural Concrete	CREW	DAILY OUTPUT	LABOR-HOURS	UNIT	2005 BARE COSTS MAT.	LABOR	EQUIP.	TOTAL	TOTAL INCL O&P	
220	0010	**CONCRETE, READY MIX** Normal weight										220
	0020	2000 psi				C.Y.	77.50			77.50	85.50	
	0100	2500 psi					79.50			79.50	87.50	
	0150	3000 psi					81			81	89	
	0200	3500 psi					82			82	90	
	0300	4000 psi					84			84	92.50	
	0350	4500 psi					86			86	94.50	
	0400	5000 psi					90			90	99	
	0411	6000 psi					103			103	113	
	0412	8000 psi					167			167	184	
	0413	10,000 psi					238			238	261	
	0414	12,000 psi					287			287	315	
	1000	For high early strength cement, add					10%					
	2000	For all lightweight aggregate, add				↓	45%					
240	0010	**CONCRETE IN PLACE** Including forms (4 uses), reinforcing										240
	0050	steel and finishing unless otherwise indicated										
	0500	Chimney foundations, industrial, minimum	C-14C	32.22	3.476	C.Y.	112	114	.73	226.73	300	
	0510	Maximum		23.71	4.724		119	154	.99	273.99	375	
	3800	Footings, spread under 1 C.Y.		38.07	2.942		162	96	.62	258.62	330	
	3850	Over 5 C.Y.		81.04	1.382		226	45	.29	271.29	320	
	3900	Footings, strip, 18″ x 9″, unreinforced		40	2.800		101	91.50	.59	193.09	255	
	3920	18″ x 9″, reinforced		35	3.200		120	105	.67	225.67	297	
	3925	20″ x 10″, unreinforced		45	2.489		98	81.50	.52	180.02	236	
	3930	20″ x 10″, reinforced		40	2.800		114	91.50	.59	206.09	269	
	3935	24″ x 12″, unreinforced		55	2.036		97	66.50	.43	163.93	211	
	3940	24″ x 12″, reinforced		48	2.333		113	76	.49	189.49	244	
	3945	36″ x 12″, unreinforced		70	1.600		93	52.50	.34	145.84	184	
	3950	36″ x 12″, reinforced		60	1.867		108	61	.39	169.39	214	
	4000	Foundation mat, under 10 C.Y.		38.67	2.896		164	94.50	.61	259.11	330	
	4050	Over 20 C.Y.	↓	56.40	1.986		144	65	.42	209.42	260	
	4650	Slab on grade, not including finish, 4″ thick	C-14E	60.75	1.449		104	48.50	.39	152.89	192	
	4700	6″ thick	″	92	.957	↓	98	32	.26	130.26	159	
	4751	Slab on grade, incl. troweled finish, not incl. forms										
	4760	or reinforcing, over 10,000 S.F., 4″ thick	C-14F	3,425	.021	S.F.	1.07	.65	.01	1.73	2.15	
	4820	6″ thick		3,350	.021		1.56	.67	.01	2.24	2.71	
	4840	8″ thick		3,184	.023		2.13	.70	.01	2.84	3.39	
	4900	12″ thick		2,734	.026		3.20	.82	.01	4.03	4.74	
	4950	15″ thick	↓	2,505	.029	↓	4.01	.89	.01	4.91	5.75	
700	0010	**PLACING CONCRETE** and vibrating, including labor & equipment										700
	1900	Footings, continuous, shallow, direct chute	C-6	120	.400	C.Y.		11.20	.40	11.60	17.70	
	1950	Pumped	C-20	150	.427			12.25	5	17.25	24.50	
	2000	With crane and bucket	C-7	90	.800			23	10.65	33.65	47.50	
	2100	Footings, continuous, deep, direct chute	C-6	140	.343			9.60	.34	9.94	15.20	
	2150	Pumped	C-20	160	.400			11.50	4.70	16.20	23	
	2200	With crane and bucket	C-7	110	.655			18.90	8.75	27.65	38.50	
	2400	Footings, spread, under 1 C.Y., direct chute	C-6	55	.873			24.50	.87	25.37	38.50	
	2450	Pumped	C-20	65	.985			28.50	11.55	40.05	56.50	
	2500	With crane and bucket	C-7	45	1.600			46	21.50	67.50	94.50	
	2600	Over 5 C.Y., direct chute	C-6	120	.400			11.20	.40	11.60	17.70	
	2650	Pumped	C-20	150	.427			12.25	5	17.25	24.50	
	2700	With crane and bucket	C-7	100	.720			21	9.60	30.60	42.50	
	2900	Foundation mats, over 20 C.Y., direct chute	C-6	350	.137			3.85	.14	3.99	6.05	
	2950	Pumped	C-20	400	.160			4.59	1.88	6.47	9.10	
	3000	With crane and bucket	C-7	300	.240	↓		6.95	3.20	10.15	14.10	

For expanded coverage of these items see *Means Concrete & Masonry Cost Data 2005*

03350 | Concrete Finishing

		03350 Concrete Finishing	CREW	DAILY OUTPUT	LABOR-HOURS	UNIT	2005 BARE COSTS MAT.	LABOR	EQUIP.	TOTAL	TOTAL INCL O&P	
300	0010	**FINISHING FLOORS** Monolithic, screed finish	1 Cefi	900	.009	S.F.		.29		.29	.43	300
	0100	Screed and bull float (darby) finish		725	.011			.36		.36	.53	
	0150	Screed, float, and broom finish		630	.013			.42		.42	.61	
	0200	Screed, float, and hand trowel		600	.013			.44		.44	.64	
	0250	Machine trowel		550	.015			.48		.48	.70	
325	0010	**CONTROL JOINT**, concrete floor slab										325
	0100	Sawcut in green concrete										
	0120	1" depth	C-27	2,000	.008	L.F.		.26	.06	.32	.44	
	0140	1-1/2" depth		1,800	.009			.29	.06	.35	.50	
	0160	2" depth		1,600	.010			.33	.07	.40	.56	
	0200	Clean out control joint of debris	C-28	6,000	.001			.04		.04	.06	
	0300	Joint sealant										
	0320	Backer rod, polyethylene, 1/4" diameter	1 Cefi	460	.017	L.F.	.02	.57		.59	.86	
	0340	Sealant, polyurethane										
	0360	1/4" x 1/4" (308 LF/Gal)	1 Cefi	270	.030	L.F.	.15	.97		1.12	1.60	
	0380	1/4" x 1/2" (154 LF/Gal)	"	255	.031	"	.31	1.03		1.34	1.86	

03920 | Concrete Resurfacing

		03920 Concrete Resurfacing	CREW	DAILY OUTPUT	LABOR-HOURS	UNIT	MAT.	LABOR	EQUIP.	TOTAL	TOTAL INCL O&P	
600	0010	**PATCHING CONCRETE**										600
	0100	Floors, 1/4" thick, small areas, regular grout	1 Cefi	170	.047	S.F.	.07	1.55		1.62	2.36	
	0150	Epoxy grout	"	100	.080	"	4.11	2.63		6.74	8.40	
	0300	Slab on Grade, cut outs, up to 50 C.F.	2 Cefi	50	.320	C.F.	6	10.50		16.50	22	
	2000	Walls, including chipping, cleaning and epoxy grout										
	2100	Minimum	1 Cefi	65	.123	S.F.	3.31	4.04		7.35	9.60	
	2150	Average		50	.160		6.60	5.25		11.85	15.05	
	2200	Maximum		40	.200		13.25	6.55		19.80	24	
	2510	Underlayment, P.C based self-leveling, 4100 psi, pumped, 1/4"	C-8	20,000	.003		1.42	.08	.04	1.54	1.73	
	2520	1/2"		19,000	.003		2.56	.09	.04	2.69	2.98	
	2530	3/4"		18,000	.003		3.98	.09	.04	4.11	4.55	
	2540	1"		17,000	.003		5.40	.10	.04	5.54	6.15	
	2550	1-1/2"		15,000	.004		8.25	.11	.05	8.41	9.25	
	2560	Hand mix, 1/2"	C-18	4,000	.002		2.56	.06	.01	2.63	2.91	
	2610	Topping, P.C. based self-level/dry 6100 psi, pumped, 1/4"	C-8	20,000	.003		2	.08	.04	2.12	2.37	
	2620	1/2"		19,000	.003		3.60	.09	.04	3.73	4.13	
	2630	3/4"		18,000	.003		5.60	.09	.04	5.73	6.35	
	2660	1"		17,000	.003		7.60	.10	.04	7.74	8.55	
	2670	1-1/2"		15,000	.004		11.60	.11	.05	11.76	12.95	
	2680	Hand mix, 1/2"	C-18	4,000	.002		3.60	.06	.01	3.67	4.06	

For information about Means Estimating Seminars, see yellow pages 12 and 13 in back of book

Division 4 Masonry

Estimating Tips

04050 Basic Masonry Materials & Methods

- The terms *mortar* and *grout* are often used interchangeably, and incorrectly. Mortar is used to bed masonry units, seal the entry of air and moisture, provide architectural appearance, and allow for size variations in the units. Grout is used primarily in reinforced masonry construction and is used to bond the masonry to the reinforcing steel. Common mortar types are M(2500 psi), S(1800 psi), N(750 psi), and O(350 psi), and conform to ASTM C270. Grout is either fine or coarse, conforms to ASTM C476, and in-place strengths generally exceed 2500 psi. Mortar and grout are different components of masonry construction and are placed by entirely different methods. An estimator should be aware of their unique uses and costs.
- Waste, specifically the loss/droppings of mortar and the breakage of brick and block, is included in all masonry assemblies in this division. A factor of 25% is added for mortar and 3% for brick and concrete masonry units.
- Scaffolding or staging is not included in any of the Division 4 costs. Refer to section 01540 for scaffolding and staging costs.

04800 Masonry Assemblies

- The most common types of unit masonry are brick and concrete masonry. The major classifications of brick are building brick (ASTM C62), facing brick (ASTM C216) and glazed brick, fire brick and pavers. Many varieties of texture and appearance can exist within these classifications, and the estimator would be wise to check local custom and availability within the project area. On repair and remodeling jobs, matching the existing brick may be the most important criteria.
- Brick and concrete block are priced by the piece and then converted into a price per square foot of wall. Openings less than two square feet are generally ignored by the estimator because any savings in units used is offset by the cutting and trimming required.
- It is often difficult and expensive to find and purchase small lots of historic brick. Costs can vary widely. Many design issues affect costs, selection of mortar mix, and repairs or replacement of masonry materials. Cleaning techniques must be reflected in the estimate.
- All masonry walls, whether interior or exterior, require bracing. The cost of bracing walls during construction should be included by the estimator and this bracing must remain in place until permanent bracing is complete. Permanent bracing of masonry walls is accomplished by masonry itself, in the form of pilasters or abutting wall corners, or by anchoring the walls to the structural frame. Accessories in the form of anchors, anchor slots and ties are used, but their supply and installation can be by different trades. For instance, anchor slots on spandrel beams and columns are supplied and welded in place by the steel fabricator, but the ties from the slots into the masonry are installed by the bricklayer. Regardless of the installation method the estimator must be certain that these accessories are accounted for in pricing.

Reference Numbers

Reference numbers are shown in bold squares at the beginning of some major classifications. These numbers refer to related items in the Reference Section. The reference information may be an estimating procedure, an alternate pricing method or technical information.

Note: Not all subdivisions listed here necessarily appear in this publication.

04050 | Basic Masonry Materials and Methods

		04090 | Masonry Accessories	CREW	DAILY OUTPUT	LABOR-HOURS	UNIT	2005 BARE COSTS MAT.	LABOR	EQUIP.	TOTAL	TOTAL INCL O&P	
860	0010	**VENT BOX** Extruded aluminum, 4" deep, 2-3/8" x 8-1/8"	1 Bric	30	.267	Ea.	20	9.40		29.40	36.50	860
	0050	5" x 8-1/8"		25	.320		26.50	11.30		37.80	46	
	0100	2-1/4" x 25"		25	.320		54.50	11.30		65.80	77	
	0150	5" x 16-1/2"		22	.364		47.50	12.80		60.30	72	
	0200	6" x 16-1/2"		22	.364		64	12.80		76.80	90	
	0250	7-3/4" x 16-1/2"	↓	20	.400		50	14.10		64.10	76.50	
	0400	For baked enamel finish, add					35%					
	0500	For cast aluminum, painted, add					60%					
	1000	Stainless steel ventilators, 6" x 6"	1 Bric	25	.320		92	11.30		103.30	118	
	1050	8" x 8"		24	.333		97	11.75		108.75	125	
	1100	12" x 12"		23	.348		112	12.25		124.25	142	
	1150	12" x 6"		24	.333		98	11.75		109.75	126	
	1200	Foundation block vent, galv., 1-1/4" thk, 8" high, 16" long, no damper	↓	30	.267		18	9.40		27.40	34	
	1250	For damper, add				↓	6			6	6.60	

04800 | Masonry Assemblies

		04810 | Unit Masonry Assemblies	CREW	DAILY OUTPUT	LABOR-HOURS	UNIT	2005 BARE COSTS MAT.	LABOR	EQUIP.	TOTAL	TOTAL INCL O&P	
160	0010	**CHIMNEY** See Div. 03310-240 for foundation, add to prices below										160
	1800	Metal, high temp. steel jacket, factory lining, 24" diam.	E-2	65	.862	V.L.F.	186	32	20.50	238.50	281	
	1900	60" diameter	"	30	1.867		675	69	44	788	910	
	2100	Poured concrete, brick lining, 200' high x 10' diam.					5,600			5,600	6,150	
	2800	500' x 20' diameter				↓	9,800			9,800	10,800	

For information about Means Estimating Seminars, see yellow pages 12 and 13 in back of book

Division 5 Metals

Estimating Tips

05050 Basic Metal Materials & Methods

- Nuts, bolts, washers, connection angles and plates can add a significant amount to both the tonnage of a structural steel job as well as the estimated cost. As a rule of thumb add 10% to the total weight to account for these accessories.
- Type 2 steel construction, commonly referred to as "simple construction," consists generally of field bolted connections with lateral bracing supplied by other elements of the building, such as masonry walls or x-bracing. The estimator should be aware, however, that shop connections may be accomplished by welding or bolting. The method may be particular to the fabrication shop and may have an impact on the estimated cost.

05200 Metal Joists

- In any given project the total weight of open web steel joists is determined by the loads to be supported and the design. However, economies can be realized in minimizing the amount of labor used to place the joists. This is done by maximizing the joist spacing and therefore minimizing the number of joists required to be installed on the job. Certain spacings and locations may be required by the design, but in other cases maximizing the spacing and keeping it as uniform as possible will keep the costs down.

05300 Metal Deck

- The takeoff and estimating of metal deck involves more than simply the area of the floor or roof and the type of deck specified or shown on the drawings. Many different sizes and types of openings may exist. Small openings for individual pipes or conduits may be drilled after the floor/roof is installed, but larger openings may require special deck lengths as well as reinforcing or structural support. The estimator should determine who will be supplying this reinforcing. Additionally, some deck terminations are part of the deck package, such as screed angles and pour stops, and others will be part of the steel contract, such as angles attached to structural members and cast-in-place angles and plates. The estimator must ensure that all pieces are accounted for in the complete estimate.

05500 Metal Fabrications

- The most economical steel stairs are those that use common materials, standard details and most importantly, a uniform and relatively simple method of field assembly. Commonly available A36 channels and plates are very good choices for the main stringers of the stairs, as are angles and tees for the carrier members. Risers and treads are usually made by specialty shops, and it is most economical to use a typical detail in as many places as possible. The stairs should be pre-assembled and shipped directly to the site. The field connections should be simple and straightforward to be accomplished efficiently and with a minimum of equipment and labor.

Reference Numbers

Reference numbers are shown in bold squares at the beginning of some major classifications. These numbers refer to related items in the Reference Section. The reference information may be an estimating procedure, an alternate pricing method or technical information.

Note: Not all subdivisions listed here necessarily appear in this publication.

	05090 | Metal Fastenings		CREW	DAILY OUTPUT	LABOR-HOURS	UNIT	2005 BARE COSTS MAT.	LABOR	EQUIP.	TOTAL	TOTAL INCL O&P	
080	0010	**ANCHOR BOLTS**										080
	0100	J-type, incl. hex nut & washer, 1/2" diameter x 6" long	2 Carp	70	.229	Ea.	.94	7.85		8.79	13.25	
	0110	12" long		65	.246		1.17	8.45		9.62	14.45	
	0120	18" long		60	.267		1.52	9.15		10.67	15.90	
	0130	3/4" diameter x 8" long		50	.320		1.39	10.95		12.34	18.60	
	0140	12" long		45	.356		1.74	12.20		13.94	21	
	0150	18" long		40	.400		2.26	13.70		15.96	24	
	0160	1" diameter x 12" long		35	.457		3.32	15.65		18.97	28	
	0170	18" long		30	.533		4	18.25		22.25	33	
	0180	24" long		25	.640		4.88	22		26.88	39.50	
	0190	36" long		20	.800		6.70	27.50		34.20	50	
	0200	1-1/2" diameter x 18" long		22	.727		11.85	25		36.85	52	
	0210	24" long		16	1		14.10	34.50		48.60	69	
	0300	L-type, incl. hex nut & washer, 3/4" diameter x 12" long		45	.356		1.31	12.20		13.51	20.50	
	0310	18" long		40	.400		1.67	13.70		15.37	23.50	
	0320	24" long		35	.457		2.04	15.65		17.69	26.50	
	0330	30" long		30	.533		2.59	18.25		20.84	31.50	
	0340	36" long		25	.640		2.96	22		24.96	37.50	
	0350	1" diameter x 12" long		35	.457		2.28	15.65		17.93	27	
	0360	18" long		30	.533		2.84	18.25		21.09	31.50	
	0370	24" long		25	.640		3.50	22		25.50	38	
	0380	30" long		23	.696		4.13	24		28.13	41.50	
	0390	36" long		20	.800		4.73	27.50		32.23	47.50	
	0400	42" long		18	.889		5.75	30.50		36.25	54	
	0410	48" long		15	1.067		6.45	36.50		42.95	64	
	0420	1-1/4" diameter x 18" long		25	.640		5.10	22		27.10	39.50	
	0430	24" long		22	.727		6.05	25		31.05	45.50	
	0440	30" long		20	.800		7	27.50		34.50	50	
	0450	36" long		18	.889		8	30.50		38.50	56.50	
	0460	42" long		16	1		9	34.50		43.50	63.50	
	0470	48" long		14	1.143		10.30	39		49.30	72.50	
	0480	54" long		12	1.333		12.15	45.50		57.65	84.50	
	0490	60" long		10	1.600		13.35	55		68.35	100	
	0500	1-1/2" diameter x 18" long		22	.727		7.85	25		32.85	47.50	
	0510	24" long		19	.842		9.15	29		38.15	55	
	0520	30" long		17	.941		10.35	32		42.35	61.50	
	0540	42" long		15	1.067		13.55	36.50		50.05	72	
	0550	48" long		13	1.231		15.25	42		57.25	82.50	
	0560	54" long		11	1.455		18.55	50		68.55	98	
	0570	60" long		9	1.778		20.50	61		81.50	118	
	0580	1-3/4" diameter x 18" long		20	.800		11.85	27.50		39.35	55.50	
	0590	24" long		18	.889		13.90	30.50		44.40	63	
	0600	30" long		17	.941		16.20	32		48.20	68	
	0610	36" long		16	1		18.45	34.50		52.95	74	
	0620	42" long		14	1.143		20.50	39		59.50	84	
	0630	48" long		12	1.333		23	45.50		68.50	96	
	0640	54" long		10	1.600		28.50	55		83.50	117	
	0650	60" long		8	2		30.50	68.50		99	141	
	0660	2" diameter x 24" long		17	.941		17.70	32		49.70	69.50	
	0670	30" long		15	1.067		19.95	36.50		56.45	79	
	0680	36" long		13	1.231		22	42		64	89.50	
	0690	42" long		11	1.455		24.50	50		74.50	105	
	0700	48" long		10	1.600		28	55		83	117	
	0710	54" long		9	1.778		33.50	61		94.50	132	
	0720	60" long		8	2		36	68.50		104.50	147	
	0730	66" long	↓	7	2.286	↓	38.50	78.50		117	165	

	05090 | Metal Fastenings		Crew	Daily Output	Labor-Hours	Unit	2005 Bare Costs Mat.	Labor	Equip.	Total	Total Incl O&P	
080	0740	72″ long	2 Carp	6	2.667	Ea.	42	91.50		133.50	189	080
	0990	For galvanized, add					75%					
300	0010	**CHEMICAL ANCHORS**, Includes layout & drilling										300
	1430	Chemical anchor, w/rod & epoxy cartridge, 3/4″ diam. x 9-1/2″ long	B-89A	27	.593	Ea.	11.65	18.25	3.82	33.72	45.50	
	1435	1″ diameter x 11-3/4″ long		24	.667		22.50	20.50	4.29	47.29	61.50	
	1440	1-1/4″ diameter x 14″ long		21	.762		43	23.50	4.91	71.41	89	
	1445	1-3/4″ diameter x 15″ long		20	.800		81	24.50	5.15	110.65	134	
	1450	18″ long		17	.941		97.50	29	6.05	132.55	159	
	1455	2″ diameter x 18″ long		16	1		124	31	6.45	161.45	192	
	1460	24″ long		15	1.067		162	33	6.85	201.85	238	
	1500	Chemical anchoring, epoxy cartridge, excludes layout, drilling, fastener										
	1530	For fastener 3/4″ dia x 6″ embedment	B-89A	27	.593	Ea.	4.58	18.25	3.82	26.65	38	
	1535	1″ dia x 8″ embedment		24	.667		6.85	20.50	4.29	31.64	44.50	
	1540	1-1/4″ dia x 10″ embedment		21	.762		13.75	23.50	4.91	42.16	57	
	1545	1-3/4″ dia x 12″ embedment		20	.800		23	24.50	5.15	52.65	69	
	1550	14″ embedment		17	.941		27.50	29	6.05	62.55	81.50	
	1555	2″ dia x 12″ embedment		16	1		36.50	31	6.45	73.95	95.50	
	1560	18″ embedment		15	1.067		46	33	6.85	85.85	109	
340	0010	**DRILLING** For anchors, up to 4″ deep, incl. bit and layout										340
	0050	in concrete or brick walls and floors, no anchor										
	0100	Holes, 1/4″ diameter	1 Carp	75	.107	Ea.	.08	3.65		3.73	5.80	
	0150	For each additional inch of depth, add		430	.019		.02	.64		.66	1.01	
	0200	3/8″ diameter		63	.127		.08	4.35		4.43	6.85	
	0250	For each additional inch of depth, add		340	.024		.02	.81		.83	1.28	
	0300	1/2″ diameter		50	.160		.08	5.50		5.58	8.65	
	0350	For each additional inch of depth, add		250	.032		.02	1.10		1.12	1.73	
	0400	5/8″ diameter		48	.167		.15	5.70		5.85	9.05	
	0450	For each additional inch of depth, add		240	.033		.04	1.14		1.18	1.82	
	0500	3/4″ diameter		45	.178		.18	6.10		6.28	9.70	
	0550	For each additional inch of depth, add		220	.036		.04	1.25		1.29	1.99	
	0600	7/8″ diameter		43	.186		.21	6.35		6.56	10.20	
	0650	For each additional inch of depth, add		210	.038		.05	1.30		1.35	2.09	
	0700	1″ diameter		40	.200		.24	6.85		7.09	10.90	
	0750	For each additional inch of depth, add		190	.042		.06	1.44		1.50	2.32	
	0800	1-1/4″ diameter		38	.211		.35	7.20		7.55	11.65	
	0850	For each additional inch of depth, add		180	.044		.09	1.52		1.61	2.47	
	0900	1-1/2″ diameter		35	.229		.53	7.85		8.38	12.80	
	0950	For each additional inch of depth, add		165	.048		.13	1.66		1.79	2.73	
	1000	For ceiling installations, add						40%				
	1100	Drilling & layout for drywall/plaster walls, up to 1″ deep, no anchor										
	1200	Holes, 1/4″ diameter	1 Carp	150	.053	Ea.	.01	1.83		1.84	2.86	
	1300	3/8″ diameter		140	.057		.01	1.96		1.97	3.06	
	1400	1/2″ diameter		130	.062		.01	2.11		2.12	3.29	
	1500	3/4″ diameter		120	.067		.02	2.28		2.30	3.58	
	1600	1″ diameter		110	.073		.03	2.49		2.52	3.91	
	1700	1-1/4″ diameter		100	.080		.04	2.74		2.78	4.32	
	1800	1-1/2″ diameter		90	.089		.07	3.04		3.11	4.81	
	1900	For ceiling installations, add						40%				
	1910	Drilling & layout for steel, up to 1/4″ deep, no anchor										
	1920	Holes, 1/4″ diameter	1 Sswk	112	.071	Ea.	.11	2.73		2.84	5	
	1925	For each additional 1/4″ depth, add		336	.024		.11	.91		1.02	1.75	
	1930	3/8″ diameter		104	.077		.13	2.93		3.06	5.40	
	1935	For each additional 1/4″ depth, add		312	.026		.13	.98		1.11	1.89	
	1940	1/2″ diameter		96	.083		.14	3.18		3.32	5.85	
	1945	For each additional 1/4″ depth, add		288	.028		.14	1.06		1.20	2.06	
	1950	5/8″ diameter		88	.091		.24	3.47		3.71	6.45	

	05090 | Metal Fastenings		CREW	DAILY OUTPUT	LABOR-HOURS	UNIT	2005 BARE COSTS MAT.	LABOR	EQUIP.	TOTAL	TOTAL INCL O&P	
340	1955	For each additional 1/4" depth, add	1 Sswk	264	.030	Ea.	.24	1.16		1.40	2.33	340
	1960	3/4" diameter		80	.100		.26	3.82		4.08	7.15	
	1965	For each additional 1/4" depth, add		240	.033		.26	1.27		1.53	2.56	
	1970	7/8" diameter		72	.111		.31	4.24		4.55	7.95	
	1975	For each additional 1/4" depth, add		216	.037		.31	1.41		1.72	2.87	
	1980	1" diameter		64	.125		.35	4.77		5.12	8.95	
	1985	For each additional 1/4" depth, add	↓	192	.042		.35	1.59		1.94	3.23	
	1990	For drilling up, add				↓		40%				
380	0010	**EXPANSION ANCHORS** & shields										380
	0100	Bolt anchors for concrete, brick or stone, no layout and drilling										
	0200	Expansion shields, zinc, 1/4" diameter, 1-5/16" long, single	1 Carp	90	.089	Ea.	1.01	3.04		4.05	5.85	
	0300	1-3/8" long, double		85	.094		1.11	3.22		4.33	6.20	
	0400	3/8" diameter, 1-1/2" long, single		85	.094		1.66	3.22		4.88	6.85	
	0500	2" long, double		80	.100		2.05	3.43		5.48	7.60	
	0600	1/2" diameter, 2-1/16" long, single		80	.100		2.75	3.43		6.18	8.40	
	0700	2-1/2" long, double		75	.107		2.65	3.65		6.30	8.60	
	0800	5/8" diameter, 2-5/8" long, single		75	.107		3.93	3.65		7.58	10	
	0900	2-3/4" long, double		70	.114		3.93	3.91		7.84	10.40	
	1000	3/4" diameter, 2-3/4" long, single		70	.114		5.85	3.91		9.76	12.50	
	1100	3-15/16" long, double		65	.123		7.80	4.22		12.02	15.10	
	1500	Self drilling anchor, snap-off, for 1/4" diameter bolt		26	.308		.83	10.55		11.38	17.30	
	1600	3/8" diameter bolt		23	.348		1.20	11.90		13.10	19.85	
	1700	1/2" diameter bolt		20	.400		1.84	13.70		15.54	23.50	
	1800	5/8" diameter bolt		18	.444		3.08	15.20		18.28	27	
	1900	3/4" diameter bolt	↓	16	.500	↓	5.20	17.15		22.35	32	
	2100	Hollow wall anchors for gypsum wall board, plaster or tile										
	2300	1/8" diameter, short	1 Carp	160	.050	Ea.	.26	1.71		1.97	2.96	
	2400	Long		150	.053		.31	1.83		2.14	3.19	
	2500	3/16" diameter, short		150	.053		.55	1.83		2.38	3.46	
	2600	Long		140	.057		.59	1.96		2.55	3.70	
	2700	1/4" diameter, short		140	.057		.68	1.96		2.64	3.80	
	2800	Long		130	.062		.76	2.11		2.87	4.12	
	3000	Toggle bolts, bright steel, 1/8" diameter, 2" long		85	.094		.25	3.22		3.47	5.30	
	3100	4" long		80	.100		.37	3.43		3.80	5.75	
	3200	3/16" diameter, 3" long		80	.100		.42	3.43		3.85	5.80	
	3300	6" long		75	.107		.58	3.65		4.23	6.35	
	3400	1/4" diameter, 3" long		75	.107		.47	3.65		4.12	6.20	
	3500	6" long		70	.114		.66	3.91		4.57	6.85	
	3600	3/8" diameter, 3" long		70	.114		.90	3.91		4.81	7.10	
	3700	6" long		60	.133		1.58	4.57		6.15	8.85	
	3800	1/2" diameter, 4" long		60	.133		2.33	4.57		6.90	9.65	
	3900	6" long	↓	50	.160	↓	3.85	5.50		9.35	12.80	
	5000	Screw anchors for concrete, masonry,										
	5100	stone & tile, no layout or drilling included										
	5200	Jute fiber, #6, #8, & #10, 1" long	1 Carp	240	.033	Ea.	.24	1.14		1.38	2.04	
	5300	#12, 1-1/2" long		200	.040		.35	1.37		1.72	2.52	
	5400	#14, 2" long		160	.050		.55	1.71		2.26	3.28	
	5500	#16, 2" long		150	.053		.57	1.83		2.40	3.48	
	5600	#20, 2" long		140	.057		.91	1.96		2.87	4.05	
	5700	Lag screw shields, 1/4" diameter, short		90	.089		.44	3.04		3.48	5.20	
	5800	Long		85	.094		.51	3.22		3.73	5.55	
	5900	3/8" diameter, short		85	.094		.80	3.22		4.02	5.90	
	6000	Long		80	.100		.94	3.43		4.37	6.40	
	6100	1/2" diameter, short		80	.100		1.11	3.43		4.54	6.55	
	6200	Long		75	.107		1.39	3.65		5.04	7.25	
	6300	3/4" diameter, short	↓	70	.114	↓	3.12	3.91		7.03	9.55	

	05090 | Metal Fastenings		CREW	DAILY OUTPUT	LABOR-HOURS	UNIT	2005 BARE COSTS MAT.	LABOR	EQUIP.	TOTAL	TOTAL INCL O&P	
380	6400	Long	1 Carp	65	.123	Ea.	3.78	4.22		8	10.70	380
	6600	Lead, #6 & #8, 3/4″ long		260	.031		.16	1.05		1.21	1.82	
	6700	#10 - #14, 1-1/2″ long		200	.040		.23	1.37		1.60	2.38	
	6800	#16 & #18, 1-1/2″ long		160	.050		.31	1.71		2.02	3.01	
	6900	Plastic, #6 & #8, 3/4″ long		260	.031		.10	1.05		1.15	1.75	
	7000	#8 & #10, 7/8″ long		240	.033		.04	1.14		1.18	1.82	
	7100	#10 & #12, 1″ long		220	.036		.13	1.25		1.38	2.08	
	7200	#14 & #16, 1-1/2″ long		160	.050		.07	1.71		1.78	2.75	
	8000	Wedge anchors, not including layout or drilling										
	8050	Carbon steel, 1/4″ diameter, 1-3/4″ long	1 Carp	150	.053	Ea.	.42	1.83		2.25	3.31	
	8100	3 1/4″ long		140	.057		.55	1.96		2.51	3.66	
	8150	3/8″ diameter, 2-1/4″ long		145	.055		.63	1.89		2.52	3.63	
	8200	5″ long		140	.057		1.10	1.96		3.06	4.26	
	8250	1/2″ diameter, 2-3/4″ long		140	.057		.96	1.96		2.92	4.11	
	8300	7″ long		125	.064		1.65	2.19		3.84	5.20	
	8350	5/8″ diameter, 3-1/2″ long		130	.062		1.90	2.11		4.01	5.35	
	8400	8-1/2″ long		115	.070		4.05	2.38		6.43	8.15	
	8450	3/4″ diameter, 4-1/4″ long		115	.070		2.29	2.38		4.67	6.25	
	8500	10″ long		95	.084		5.20	2.88		8.08	10.25	
	8550	1″ diameter, 6″ long		100	.080		7.70	2.74		10.44	12.70	
	8575	9″ long		85	.094		10	3.22		13.22	16	
	8600	12″ long		75	.107		10.80	3.65		14.45	17.55	
	8650	1-1/4″ diameter, 9″ long		70	.114		14	3.91		17.91	21.50	
	8700	12″ long		60	.133		17.90	4.57		22.47	27	
	8750	For type 303 stainless steel, add					350%					
	8800	For type 316 stainless steel, add					450%					
	8950	Self-drilling concrete screw, hex washer head, 3/16″ dia x 1-3/4″ long	1 Carp	300	.027	Ea.	.32	.91		1.23	1.77	
	8960	2-1/4″ long		250	.032		.49	1.10		1.59	2.25	
	8970	Phillips flat head, 3/16″ dia x 1-3/4″ long		300	.027		.33	.91		1.24	1.78	
	8980	2-1/4″ long		250	.032		.48	1.10		1.58	2.24	
460	0010	**LAG SCREWS**										460
	0020	Steel, 1/4″ diameter, 2″ long	1 Carp	200	.040	Ea.	.08	1.37		1.45	2.22	
	0100	3/8″ diameter, 3″ long		150	.053		.23	1.83		2.06	3.10	
	0200	1/2″ diameter, 3″ long		130	.062		.38	2.11		2.49	3.70	
	0300	5/8″ diameter, 3″ long		120	.067		.75	2.28		3.03	4.39	
500	0010	**MACHINE SCREWS**										500
	0020	Steel, round head, #8 x 1″ long				C	2.28			2.28	2.51	
	0110	#8 x 2″ long					4.96			4.96	5.45	
	0200	#10 x 1″ long					3.25			3.25	3.58	
	0300	#10 x 2″ long					6.05			6.05	6.70	
540	0010	**MACHINERY ANCHORS**, heavy duty, incl. sleeve, floating base nut,										540
	0020	lower stud & coupling nut, fiber plug, connecting stud, washer & nut.										
	0030	For flush mounted embedment in poured concrete heavy equip. pads.										
	0200	Material only, 1/2″ diameter stud & bolt				Ea.	52.50			52.50	58	
	0300	5/8″ diameter					58.50			58.50	64	
	0500	3/4″ diameter					67.50			67.50	74	
	0600	7/8″ diameter					73.50			73.50	81	
	0800	1″ diameter					77.50			77.50	85	
	0900	1-1/4″ diameter					103			103	113	
580	0010	**POWDER ACTUATED** Tools & fasteners										580
	0020	Stud driver, .22 caliber, buy, minimum				Ea.	310			310	340	
	0100	Maximum				″	495			495	545	
	0300	Powder charges for above, low velocity				C	16.60			16.60	18.25	

05050 | Basic Metal Materials & Methods

	05090 | Metal Fastenings		CREW	DAILY OUTPUT	LABOR-HOURS	UNIT	2005 BARE COSTS MAT.	LABOR	EQUIP.	TOTAL	TOTAL INCL O&P
580	0400	Standard velocity				C	23.50			23.50	26
	0600	Drive pins & studs, 1/4" & 3/8" diam., to 3" long, minimum					11.85			11.85	13.05
	0700	Maximum					46.50			46.50	51
	0800	Pneumatic stud driver for 1/8" diameter studs				Ea.	2,150			2,150	2,350
	0900	Drive pins for above, 1/2" to 3/4" long				M	490			490	540
600	0010	RIVETS									
	0100	Aluminum rivet & mandrel, 1/2" grip length x 1/8" diameter				C	5			5	5.50
	0200	3/16" diameter					7.70			7.70	8.45
	0300	Aluminum rivet, steel mandrel, 1/8" diameter					7.30			7.30	8.05
	0400	3/16" diameter					6.70			6.70	7.35
	0500	Copper rivet, steel mandrel, 1/8" diameter					6.65			6.65	7.30
	1200	Steel rivet and mandrel, 1/8" diameter					5.75			5.75	6.30
	1300	3/16" diameter					8.60			8.60	9.50
	1400	Hand riveting tool, minimum				Ea.	111			111	122
	1500	Maximum					212			212	233
	1600	Power riveting tool, minimum					800			800	880
	1700	Maximum					2,025			2,025	2,225

5 METALS

05100 | Structural Metal Framing

	05120 | Structural Steel		CREW	DAILY OUTPUT	LABOR-HOURS	UNIT	2005 BARE COSTS MAT.	LABOR	EQUIP.	TOTAL	TOTAL INCL O&P
440	0010	LIGHTWEIGHT FRAMING									
	0400	Angle framing, field fabricated, 4" and larger	E-3	440	.055	Lb.	.51	2.12	.18	2.81	4.55
	0450	Less than 4" angles		265	.091		.53	3.52	.31	4.36	7.20
	0600	Channel framing, field fabricated, 8" and larger		500	.048		.53	1.86	.16	2.55	4.09
	0650	Less than 8" channels		335	.072		.53	2.78	.24	3.55	5.85
	1000	Continuous slotted channel framing system, shop fab, min	2 Sswk	2,400	.007		2.71	.25		2.96	3.44
	1200	Maximum	"	1,600	.010		3.06	.38		3.44	4.05
	1250	Plate & bar stock for reinforcing beams and trusses					.96			.96	1.06
	1300	Cross bracing, rods, shop fabricated, 3/4" diameter	E-3	700	.034		1.05	1.33	.12	2.50	3.67
	1310	7/8" diameter		850	.028		1.05	1.10	.10	2.25	3.23
	1320	1" diameter		1,000	.024		1.05	.93	.08	2.06	2.92
	1330	Angle, 5" x 5" x 3/8"		2,800	.009		1.05	.33	.03	1.41	1.79
	1350	Hanging lintels, shop fabricated, average		850	.028		1.05	1.10	.10	2.25	3.23
	1380	Roof frames, shop fabricated, 3'-0" square, 5' span	E-2	4,200	.013		1.05	.49	.32	1.86	2.36
	1400	Tie rod, not upset, 1-1/2" to 4" diameter, with turnbuckle	2 Sswk	800	.020		1.14	.76		1.90	2.62
	1420	No turnbuckle		700	.023		1.09	.87		1.96	2.76
	1500	Upset, 1-3/4" to 4" diameter, with turnbuckle		800	.020		1.14	.76		1.90	2.62
	1520	No turnbuckle		700	.023		1.09	.87		1.96	2.76
520	0010	PIPE SUPPORT FRAMING									
	0020	Under 10#/L.F.	E-4	3,900	.008	Lb.	1.17	.32	.02	1.51	1.88
	0200	10.1 to 15#/L.F.		4,300	.007		1.16	.29	.02	1.47	1.80
	0400	15.1 to 20#/L.F.		4,800	.007		1.14	.26	.02	1.42	1.73
	0600	Over 20#/L.F.		5,400	.006		1.12	.23	.02	1.37	1.66

Division 6
Wood & Plastics

Estimating Tips

06050 Basic Wood & Plastic Materials & Methods

- Common to any wood framed structure are the accessory connector items such as screws, nails, adhesives, hangers, connector plates, straps, angles and holdowns. For typical wood framed buildings, such as residential projects, the aggregate total for these items can be significant, especially in areas where seismic loading is a concern. For floor and wall framing, the material cost is based on 10 to 25 lbs. per MBF. Holdowns, hangers and other connectors should be taken off by the piece.

06100 Rough Carpentry

- Lumber is a traded commodity and therefore sensitive to supply and demand in the marketplace. Even in "budgetary" estimating of wood framed projects, it is advisable to call local suppliers for the latest market pricing.
- Common quantity units for wood framed projects are "thousand board feet" (MBF). A board foot is a volume of wood, 1″ x 1′ x 1′, or 144 cubic inches. Board foot quantities are generally calculated using nominal material dimensions—dressed sizes are ignored. Board foot per lineal foot of any stick of lumber can be calculated by dividing the nominal cross sectional area by 12. As an example, 2,000 lineal feet of 2 x 12 equates to 4 MBF by dividing the nominal area, 2 x 12, by 12, which equals 2, and multiplying by 2,000 to give 4,000 board feet. This simple rule applies to all nominal dimensioned lumber.
- Waste is an issue of concern at the quantity takeoff for any area of construction. Framing lumber is sold in even foot lengths, i.e., 10′, 12′, 14′, 16′, and depending on spans, wall heights and the grade of lumber, waste is inevitable. A rule of thumb for lumber waste is 5% to 10% depending on material quality and the complexity of the framing.
- Wood in various forms and shapes is used in many projects, even where the main structural framing is steel, concrete or masonry. Plywood as a back-up partition material and 2x boards used as blocking and cant strips around roof edges are two common examples. The estimator should ensure that the costs of all wood materials are included in the final estimate.

06200 Finish Carpentry

- It is necessary to consider the grade of workmanship when estimating labor costs for erecting millwork and interior finish. In practice, there are three grades: premium, custom and economy. The Means daily output for base and case moldings is in the range of 200 to 250 L.F. per carpenter per day. This is appropriate for most average custom grade projects. For premium projects an adjustment to productivity of 25% to 50% should be made depending on the complexity of the job.

Reference Numbers

Reference numbers are shown in bold squares at the beginning of some major classifications. These numbers refer to related items in the Reference Section. The reference information may be an estimating procedure, an alternate pricing method or technical information.

Note: Not all subdivisions listed here necessarily appear in this publication.

06052 Selective Demolition			CREW	DAILY OUTPUT	LABOR-HOURS	UNIT	2005 BARE COSTS MAT.	LABOR	EQUIP.	TOTAL	TOTAL INCL O&P	
600	0010	**NAILS** Prices of material only, based on 50# box purchase, copper, plain				Lb.	4.99			4.99	5.50	600
	0400	Stainless steel, plain					5.75			5.75	6.30	
	0500	Box, 3d to 20d, bright					1.32			1.32	1.45	
	0520	Galvanized					1.44			1.44	1.58	
	0600	Common, 3d to 60d, plain					1.02			1.02	1.12	
	0700	Galvanized					1.39			1.39	1.53	
	0800	Aluminum					3.38			3.38	3.72	
	1000	Annular or spiral thread, 4d to 60d, plain					1.60			1.60	1.76	
	1200	Galvanized					1.80			1.80	1.98	
	1400	Drywall nails, plain					.79			.79	.87	
	1600	Galvanized					1.54			1.54	1.69	
	1800	Finish nails, 4d to 10d, plain					.94			.94	1.03	
	2000	Galvanized					1.32			1.32	1.45	
	2100	Aluminum					4.10			4.10	4.51	
	2300	Flooring nails, hardened steel, 2d to 10d, plain					1.40			1.40	1.54	
	2400	Galvanized					2.25			2.25	2.48	
	2500	Gypsum lath nails, 1-1/8", 13 ga. flathead, blued					1.54			1.54	1.69	
	2600	Masonry nails, hardened steel, 3/4" to 3" long, plain					1.48			1.48	1.63	
	2700	Galvanized					1.87			1.87	2.06	
	5000	Add to prices above for cement coating					.10			.10	.11	
	5200	Zinc or tin plating					.13			.13	.14	
	5500	Vinyl coated sinkers, 8d to 16d				↓	.57			.57	.63	
650	0010	**NAILS** mat. only, for pneumatic tools, framing, per carton of 5000, 2"				Ea.	37			37	41	650
	0100	2-3/8"					42.50			42.50	46.50	
	0200	Per carton of 4000, 3"					37.50			37.50	41.50	
	0300	3-1/4"					40			40	44	
	0400	Per carton of 5000, 2-3/8", galv.					57.50			57.50	63.50	
	0500	Per carton of 4000, 3", galv.					65			65	71.50	
	0600	3-1/4", galv.					80.50			80.50	88.50	
	0700	Roofing, per carton of 7200, 1"					35			35	38.50	
	0800	1-1/4"					32.50			32.50	36	
	0900	1-1/2"					37.50			37.50	41.50	
	1000	1-3/4"				↓	45.50			45.50	50	
700	0010	**SHEET METAL SCREWS** Steel, standard, #8 x 3/4", plain				C	2.81			2.81	3.09	700
	0100	Galvanized					3.44			3.44	3.78	
	0300	#10 x 1", plain					3.76			3.76	4.14	
	0400	Galvanized					4.35			4.35	4.79	
	0600	With washers, #14 x 1", plain					10.20			10.20	11.20	
	0700	Galvanized					10.20			10.20	11.20	
	0900	#14 x 2", plain					10.20			10.20	11.20	
	1000	Galvanized					18.15			18.15	19.95	
	1500	Self-drilling, with washers, (pinch point) #8 x 3/4", plain					6.10			6.10	6.70	
	1600	Galvanized					6.10			6.10	6.70	
	1800	#10 x 3/4", plain					6.10			6.10	6.70	
	1900	Galvanized					6.10			6.10	6.70	
	3000	Stainless steel w/aluminum or neoprene washers, #14 x 1", plain					18.35			18.35	20	
	3100	#14 x 2", plain				↓	25			25	27.50	
750	0010	**WOOD SCREWS** #8, 1" long, steel				C	3.36			3.36	3.70	750
	0100	Brass					11.20			11.20	12.35	
	0200	#8, 2" long, steel					3.76			3.76	4.14	
	0300	Brass					11.70			11.70	12.90	
	0400	#10, 1" long, steel					4.30			4.30	4.73	
	0500	Brass					23			23	25	
	0600	#10, 2" long, steel					7.65			7.65	8.45	
	0700	Brass				↓	40.50			40.50	44.50	

06050 | Basic Wood / Plastic Materials / Methods

		06090 | Wood & Plastic Fastenings	CREW	DAILY OUTPUT	LABOR-HOURS	UNIT	2005 BARE COSTS MAT.	LABOR	EQUIP.	TOTAL	TOTAL INCL O&P	
750	0800	#10, 3" long, steel				C	11.95			11.95	13.15	750
	1000	#12, 2" long, steel					4.89			4.89	5.40	
	1100	Brass					16.30			16.30	17.95	
	1500	#12, 3" long, steel					16.20			16.20	17.85	
	2000	#12, 4" long, steel				▼	29			29	32	

06100 | Rough Carpentry

		06160 | Sheathing	CREW	DAILY OUTPUT	LABOR-HOURS	UNIT	2005 BARE COSTS MAT.	LABOR	EQUIP.	TOTAL	TOTAL INCL O&P	
800	0010	**SHEATHING** Plywood on roof, CDX										800
	0030	5/16" thick	2 Carp	1,600	.010	S.F.	.63	.34		.97	1.22	
	0035	Pneumatic nailed		1,952	.008		.63	.28		.91	1.13	
	0050	3/8" thick		1,525	.010		.65	.36		1.01	1.28	
	0055	Pneumatic nailed		1,860	.009		.65	.29		.94	1.18	
	0100	1/2" thick		1,400	.011		.62	.39		1.01	1.29	
	0105	Pneumatic nailed		1,708	.009		.62	.32		.94	1.18	
	0200	5/8" thick		1,300	.012		.78	.42		1.20	1.52	
	0205	Pneumatic nailed		1,586	.010		.78	.35		1.13	1.40	
	0300	3/4" thick		1,200	.013		.94	.46		1.40	1.74	
	0305	Pneumatic nailed		1,464	.011		.94	.37		1.31	1.61	
	0500	Plywood on walls with exterior CDX, 3/8" thick		1,200	.013		.65	.46		1.11	1.43	
	0505	Pneumatic nailed		1,488	.011		.65	.37		1.02	1.29	
	0600	1/2" thick		1,125	.014		.62	.49		1.11	1.44	
	0605	Pneumatic nailed		1,395	.011		.62	.39		1.01	1.29	
	0700	5/8" thick		1,050	.015		.78	.52		1.30	1.67	
	0705	Pneumatic nailed		1,302	.012		.78	.42		1.20	1.52	
	0800	3/4" thick		975	.016		.94	.56		1.50	1.91	
	0805	Pneumatic nailed	▼	1,209	.013	▼	.94	.45		1.39	1.74	

06400 | Architectural Woodwork

		06440 | Wood Ornaments	CREW	DAILY OUTPUT	LABOR-HOURS	UNIT	2005 BARE COSTS MAT.	LABOR	EQUIP.	TOTAL	TOTAL INCL O&P	
400	0010	**LOUVERS** Redwood, 2'-0" diameter, full circle	1 Carp	16	.500	Ea.	136	17.15		153.15	177	400
	0100	Half circle		16	.500		130	17.15		147.15	170	
	0200	Octagonal		16	.500		104	17.15		121.15	141	
	0300	Triangular, 5/12 pitch, 5'-0" at base	▼	16	.500	▼	220	17.15		237.15	269	

For information about Means Estimating Seminars, see yellow pages 12 and 13 in back of book

Division Notes

			DAILY	LABOR-		2005 BARE COSTS				TOTAL
		CREW	OUTPUT	HOURS	UNIT	MAT.	LABOR	EQUIP.	TOTAL	INCL O&P

Division 7
Thermal & Moisture Protection

Estimating Tips

07100 Dampproofing & Waterproofing

- Be sure of the job specifications before pricing this subdivision. The difference in cost between waterproofing and dampproofing can be great. Waterproofing will hold back standing water. Dampproofing prevents the transmission of water vapor. Also included in this section are vapor retarding membranes.

07200 Thermal Protection

- Insulation and fireproofing products are measured by area, thickness, volume or R value. Specifications may only give what the specific R value should be in a certain situation. The estimator may need to choose the type of insulation to meet that R value.

07300 Shingles, Roof Tiles & Roof Coverings
07400 Roofing & Siding Panels

- Many roofing and siding products are bought and sold by the square. One square is equal to an area that measures 100 square feet. This simple change in unit of measure could create a large error if the estimator is not observant. Accessories necessary for a complete installation must be figured into any calculations for both material and labor.

07500 Membrane Roofing
07600 Flashing & Sheet Metal
07700 Roof Specialties & Accessories

- The items in these subdivisions compose a roofing system. No one component completes the installation and all must be estimated. Built-up or single ply membrane roofing systems are made up of many products and installation trades. Wood blocking at roof perimeters or penetrations, parapet coverings, reglets, roof drains, gutters, downspouts, sheet metal flashing, skylights, smoke vents or roof hatches all need to be considered along with the roofing material. Several different installation trades will need to work together on the roofing system. Inherent difficulties in the scheduling and coordination of various trades must be accounted for when estimating labor costs.

07900 Joint Sealers

- To complete the weather-tight shell the sealants and caulkings must be estimated. Where different materials meet—at expansion joints, at flashing penetrations, and at hundreds of other locations throughout a construction project—they provide another line of defense against water penetration. Often, an entire system is based on the proper location and placement of caulking or sealants. The detail drawings that are included as part of a set of architectural plans, show typical locations for these materials. When caulking or sealants are shown at typical locations, this means the estimator must include them for all the locations where this detail is applicable. Be careful to keep different types of sealants separate, and remember to consider backer rods and primers if necessary.

Reference Numbers

Reference numbers are shown in bold squares at the beginning of some major classifications. These numbers refer to related items in the Reference Section. The reference information may be an estimating procedure, an alternate pricing method or technical information.

Note: Not all subdivisions listed here necessarily appear in this publication.

07600 | Flashing and Sheet Metal

07650 | Flexible Flashing

		07650 Flexible Flashing	CREW	DAILY OUTPUT	LABOR-HOURS	UNIT	2005 BARE COSTS MAT.	LABOR	EQUIP.	TOTAL	TOTAL INCL O&P	
600	0010	**FLASHING** Aluminum, mill finish, .013″ thick	1 Rofc	145	.055	S.F.	.35	1.62		1.97	3.16	600
	0030	.016″ thick		145	.055		.51	1.62		2.13	3.33	
	0060	.019″ thick		145	.055		.65	1.62		2.27	3.49	
	0100	.032″ thick		145	.055		1.06	1.62		2.68	3.94	
	0200	.040″ thick		145	.055		1.45	1.62		3.07	4.37	
	0300	.050″ thick		145	.055		1.84	1.62		3.46	4.79	
	0400	Painted finish, add					.24			.24	.26	
	1300	Asphalt flashing cement, 5 gallon				Gal.	3.98			3.98	4.38	
	1600	Copper, 16 oz, sheets, under 1000 lbs.	1 Rofc	115	.070	S.F.	2.62	2.05		4.67	6.35	
	1700	Over 4000 lbs.		155	.052		2.91	1.52		4.43	5.80	
	1900	20 oz sheets, under 1000 lbs.		110	.073		3.90	2.14		6.04	7.95	
	2000	Over 4000 lbs.		145	.055		3.62	1.62		5.24	6.75	
	2200	24 oz sheets, under 1000 lbs.		105	.076		4.69	2.24		6.93	8.95	
	2300	Over 4000 lbs.		135	.059		4.34	1.75		6.09	7.75	
	2500	32 oz sheets, under 1000 lbs.		100	.080		6.20	2.36		8.56	10.85	
	2600	Over 4000 lbs.		130	.062		5.80	1.81		7.61	9.50	
	4300	Copper-clad stainless steel, .015″ thick, under 500 lbs.		115	.070		3.16	2.05		5.21	6.95	
	4400	Over 2000 lbs.		155	.052		3.05	1.52		4.57	5.95	
	4600	.018″ thick, under 500 lbs.		100	.080		4.18	2.36		6.54	8.60	
	4700	Over 2000 lbs.		145	.055		3.06	1.62		4.68	6.15	
	5800	Lead, 2.5 lb. per SF, up to 12″ wide		135	.059		2.94	1.75		4.69	6.20	
	5900	Over 12″ wide		135	.059		3.25	1.75		5	6.55	
	8500	Shower pan, bituminous membrane, 7 oz		155	.052		1.08	1.52		2.60	3.78	
	8550	3 ply copper and fabric, 3 oz		155	.052		1.63	1.52		3.15	4.38	
	8600	7 oz		155	.052		3.37	1.52		4.89	6.30	
	8650	Copper, 16 oz		100	.080		3.11	2.36		5.47	7.45	
	8700	Lead on copper and fabric, 5 oz		155	.052		1.62	1.52		3.14	4.37	
	8800	7 oz		155	.052		2.93	1.52		4.45	5.80	
	8850	Polyvinyl chloride, .030″ thick		160	.050		.30	1.47		1.77	2.84	
	8900	Stainless steel sheets, 32 ga, .010″ thick		155	.052		2.16	1.52		3.68	4.97	
	9000	28 ga, .015″ thick		155	.052		2.68	1.52		4.20	5.55	
	9100	26 ga, .018″ thick		155	.052		3.25	1.52		4.77	6.15	
	9200	24 ga, .025″ thick		155	.052		4.22	1.52		5.74	7.25	
	9400	Terne coated stainless steel, .015″ thick, 28 ga		155	.052		4.05	1.52		5.57	7.05	
	9500	.018″ thick, 26 ga		155	.052		4.57	1.52		6.09	7.65	
	9600	Zinc and copper alloy (brass), .020″ thick		155	.052		3.26	1.52		4.78	6.20	
	9700	.027″ thick		155	.052		4.37	1.52		5.89	7.40	
	9800	.032″ thick		155	.052		5.10	1.52		6.62	8.20	
	9900	.040″ thick		155	.052		6.20	1.52		7.72	9.45	

07700 | Roof Specialties and Accessories

07710 | Manufactured Roof Specialties

		07710 Manufactured Roof Specialties	CREW	DAILY OUTPUT	LABOR-HOURS	UNIT	2005 BARE COSTS MAT.	LABOR	EQUIP.	TOTAL	TOTAL INCL O&P	
400	0010	**DOWNSPOUTS** Aluminum 2″ x 3″, .020″ thick, embossed	1 Shee	190	.042	L.F.	.68	1.69		2.37	3.35	400
	0100	Enameled		190	.042		.97	1.69		2.66	3.67	
	4800	Steel, galvanized, round, corrugated, 2″ or 3″ diam, 28 ga		190	.042		.75	1.69		2.44	3.43	
	4900	4″ diameter, 28 gauge		145	.055		1.02	2.22		3.24	4.53	
	5100	5″ diameter, 28 gauge		130	.062		1.54	2.47		4.01	5.50	
	5200	26 gauge		130	.062		1.36	2.47		3.83	5.30	

Important: See the Reference Section for critical supporting data - Reference Nos., Crews, & City Cost Indexes

07700 | Roof Specialties and Accessories

	07710 | Manufactured Roof Specialties		Crew	Daily Output	Labor-Hours	Unit	2005 Bare Costs Mat.	Labor	Equip.	Total	Total Incl O&P	
400	5400	6″ diameter, 28 gauge	1 Shee	105	.076	L.F.	2.15	3.06		5.21	7.10	400
	5500	26 gauge		105	.076		1.42	3.06		4.48	6.25	
	7500	Steel pipe, black, extra heavy, 4″ diameter		20	.400		11.90	16.10		28	37.50	
	7600	6″ diameter		18	.444		25.50	17.85		43.35	56	
560	0010	SNOW GUARDS										560
	0100	Slate & asphalt shingle roofs	1 Rofc	160	.050	Ea.	8	1.47		9.47	11.30	
	0200	Standing seam metal roofs		48	.167		12.25	4.91		17.16	22	
	0300	Surface mount for metal roofs		48	.167		6.75	4.91		11.66	15.80	

07800 | Fire and Smoke Protection

	07840 | Firestopping		Crew	Daily Output	Labor-Hours	Unit	2005 Bare Costs Mat.	Labor	Equip.	Total	Total Incl O&P	
100	0010	FIRESTOPPING R07800-030										100
	0100	Metallic piping, non insulated										
	0110	Through walls, 2″ diameter	1 Carp	16	.500	Ea.	9.95	17.15		27.10	37.50	
	0120	4″ diameter		14	.571		15.20	19.55		34.75	47.50	
	0130	6″ diameter		12	.667		20.50	23		43.50	58	
	0140	12″ diameter		10	.800		36.50	27.50		64	82.50	
	0150	Through floors, 2″ diameter		32	.250		6.05	8.55		14.60	20	
	0160	4″ diameter		28	.286		8.70	9.80		18.50	25	
	0170	6″ diameter		24	.333		11.40	11.40		22.80	30.50	
	0180	12″ diameter		20	.400		19.25	13.70		32.95	42.50	
	0190	Metallic piping, insulated										
	0200	Through walls, 2″ diameter	1 Carp	16	.500	Ea.	14.15	17.15		31.30	42	
	0210	4″ diameter		14	.571		19.40	19.55		38.95	52	
	0220	6″ diameter		12	.667		24.50	23		47.50	62.50	
	0230	12″ diameter		10	.800		40.50	27.50		68	87	
	0240	Through floors, 2″ diameter		32	.250		10.20	8.55		18.75	24.50	
	0250	4″ diameter		28	.286		12.85	9.80		22.65	29.50	
	0260	6″ diameter		24	.333		15.60	11.40		27	35	
	0270	12″ diameter		20	.400		19.25	13.70		32.95	42.50	
	0280	Non metallic piping, non insulated										
	0290	Through walls, 2″ diameter	1 Carp	12	.667	Ea.	41	23		64	81	
	0300	4″ diameter		10	.800		51.50	27.50		79	99.50	
	0310	6″ diameter		8	1		72	34.50		106.50	133	
	0330	Through floors, 2″ diameter		16	.500		32	17.15		49.15	62	
	0340	4″ diameter		6	1.333		40	45.50		85.50	115	
	0350	6″ diameter		6	1.333		48	45.50		93.50	124	
	0370	Ductwork, insulated & non insulated, round										
	0380	Through walls, 6″ diameter	1 Carp	12	.667	Ea.	21	23		44	58.50	
	0390	12″ diameter		10	.800		41.50	27.50		69	88.50	
	0400	18″ diameter		8	1		67.50	34.50		102	128	
	0410	Through floors, 6″ diameter		16	.500		11.45	17.15		28.60	39	
	0420	12″ diameter		14	.571		21	19.55		40.55	53.50	
	0430	18″ diameter		12	.667		36.50	23		59.50	75.50	
	0440	Ductwork, insulated & non insulated, rectangular										
	0450	With stiffener/closure angle, through walls, 6″ x 12″	1 Carp	8	1	Ea.	17.35	34.50		51.85	72.50	
	0460	12″ x 24″		6	1.333		23	45.50		68.50	96.50	
	0470	24″ x 48″		4	2		65.50	68.50		134	179	
	0480	With stiffener/closure angle, through floors, 6″ x 12″		10	.800		9.35	27.50		36.85	53	

07840 | Firestopping

			CREW	DAILY OUTPUT	LABOR-HOURS	UNIT	2005 BARE COSTS MAT.	LABOR	EQUIP.	TOTAL	TOTAL INCL O&P	
100	0490	12" x 24" R07800-030	1 Carp	8	1	Ea.	16.85	34.50		51.35	72	100
	0500	24" x 48"	↓	6	1.333	↓	33	45.50		78.50	108	
	0510	Multi trade openings										
	0520	Through walls, 6" x 12"	1 Carp	2	4	Ea.	36.50	137		173.50	253	
	0530	12" x 24"	"	1	8		147	274		421	585	
	0540	24" x 48"	2 Carp	1	16		585	550		1,135	1,500	
	0550	48" x 96"	"	.75	21.333		2,350	730		3,080	3,750	
	0560	Through floors, 6" x 12"	1 Carp	2	4		36.50	137		173.50	253	
	0570	12" x 24"	"	1	8		147	274		421	585	
	0580	24" x 48"	2 Carp	.75	21.333		585	730		1,315	1,800	
	0590	48" x 96"	"	.50	32	↓	2,350	1,100		3,450	4,300	
	0600	Structural penetrations, through walls										
	0610	Steel beams, W8 x 10	1 Carp	8	1	Ea.	23	34.50		57.50	78.50	
	0620	W12 x 14		6	1.333		36.50	45.50		82	111	
	0630	W21 x 44		5	1.600		73	55		128	166	
	0640	W36 x 135		3	2.667		177	91.50		268.50	335	
	0650	Bar joists, 18" deep		6	1.333		33.50	45.50		79	108	
	0660	24" deep		6	1.333		41.50	45.50		87	117	
	0670	36" deep		5	1.600		62.50	55		117.50	154	
	0680	48" deep	↓	4	2	↓	73	68.50		141.50	187	
	0690	Construction joints, floor slab at exterior wall										
	0700	Precast, brick, block or drywall exterior										
	0710	2" wide joint	1 Carp	125	.064	L.F.	5.20	2.19		7.39	9.10	
	0720	4" wide joint	"	75	.107	"	10.40	3.65		14.05	17.15	
	0730	Metal panel, glass or curtain wall exterior										
	0740	2" wide joint	1 Carp	40	.200	L.F.	12.30	6.85		19.15	24	
	0750	4" wide joint	"	25	.320	"	16.80	10.95		27.75	35.50	
	0760	Floor slab to drywall partition										
	0770	Flat joint	1 Carp	100	.080	L.F.	5.10	2.74		7.84	9.85	
	0780	Fluted joint		50	.160		10.40	5.50		15.90	20	
	0790	Etched fluted joint	↓	75	.107	↓	6.75	3.65		10.40	13.15	
	0800	Floor slab to concrete/masonry partition										
	0810	Flat joint	1 Carp	75	.107	L.F.	11.45	3.65		15.10	18.30	
	0820	Fluted joint	"	50	.160	"	13.55	5.50		19.05	23.50	
	0830	Concrete/CMU wall joints										
	0840	1" wide	1 Carp	100	.080	L.F.	6.25	2.74		8.99	11.10	
	0850	2" wide		75	.107		11.45	3.65		15.10	18.30	
	0860	4" wide	↓	50	.160	↓	22	5.50		27.50	32.50	
	0870	Concrete/CMU floor joints										
	0880	1" wide	1 Carp	200	.040	L.F.	3.12	1.37		4.49	5.55	
	0890	2" wide		150	.053		5.70	1.83		7.53	9.15	
	0900	4" wide	↓	100	.080	↓	10.90	2.74		13.64	16.25	

07900 | Joint Sealers

07920 | Joint Sealants

			CREW	DAILY OUTPUT	LABOR-HOURS	UNIT	2005 BARE COSTS MAT.	LABOR	EQUIP.	TOTAL	TOTAL INCL O&P	
800	0010	CAULKING AND SEALANTS										800
	0020	Acoustical sealant, elastomeric, cartridges				Ea.	2.21			2.21	2.43	
	0030	Backer rod, polyethylene, 1/4" diameter	1 Bric	4.60	1.739	C.L.F.	2.12	61.50		63.62	96	
	0050	1/2" diameter	↓	4.60	1.739	↓	3.39	61.50		64.89	97	

07920 | Joint Sealants

		07920 Joint Sealants	CREW	DAILY OUTPUT	LABOR-HOURS	UNIT	2005 BARE COSTS MAT.	LABOR	EQUIP.	TOTAL	TOTAL INCL O&P	
800	0070	3/4" diameter	1 Bric	4.60	1.739	C.L.F.	5.80	61.50		67.30	100	800
	0090	1" diameter		4.60	1.739		9.35	61.50		70.85	104	
	0100	Acrylic latex caulk, white										
	0200	11 fl. oz cartridge				Ea.	1.88			1.88	2.07	
	0500	1/4" x 1/2"	1 Bric	248	.032	L.F.	.15	1.14		1.29	1.90	
	0600	1/2" x 1/2"		250	.032		.31	1.13		1.44	2.06	
	0800	3/4" x 3/4"		230	.035		.69	1.23		1.92	2.63	
	0900	3/4" x 1"		200	.040		.92	1.41		2.33	3.16	
	1000	1" x 1"		180	.044		1.15	1.57		2.72	3.66	
	1400	Butyl based, bulk				Gal.	22			22	24.50	
	1500	Cartridges				"	27			27	29.50	
	1700	Bulk, in place 1/4" x 1/2", 154 L.F./gal.	1 Bric	230	.035	L.F.	.14	1.23		1.37	2.03	
	1800	1/2" x 1/2", 77 L.F./gal.	"	180	.044	"	.29	1.57		1.86	2.71	
	2000	Latex acrylic based, bulk				Gal.	23			23	25.50	
	2100	Cartridges					26			26	28.50	
	2300	Polysulfide compounds, 1 component, bulk					44			44	48	
	2600	1 or 2 component, in place, 1/4" x 1/4", 308 L.F./gal.	1 Bric	145	.055	L.F.	.14	1.94		2.08	3.12	
	2700	1/2" x 1/4", 154 L.F./gal.		135	.059		.28	2.09		2.37	3.49	
	2900	3/4" x 3/8", 68 L.F./gal.		130	.062		.64	2.17		2.81	4.01	
	3000	1" x 1/2", 38 L.F./gal.		130	.062		1.15	2.17		3.32	4.57	
	3200	Polyurethane, 1 or 2 component				Gal.	47.50			47.50	52.50	
	3300	Cartridges				"	46.50			46.50	51.50	
	3500	Bulk, in place, 1/4" x 1/4"	1 Bric	150	.053	L.F.	.15	1.88		2.03	3.03	
	3600	1/2" x 1/4"		145	.055		.31	1.94		2.25	3.30	
	3800	3/4" x 3/8", 68 L.F./gal.		130	.062		.70	2.17		2.87	4.07	
	3900	1" x 1/2"		110	.073		1.24	2.56		3.80	5.25	
	4100	Silicone rubber, bulk				Gal.	34.50			34.50	38	
	4200	Cartridges				"	40.50			40.50	44.50	
	4400	Neoprene gaskets, closed cell, adhesive, 1/8" x 3/8"	1 Bric	240	.033	L.F.	.20	1.17		1.37	2.01	
	4500	1/4" x 3/4"		215	.037		.48	1.31		1.79	2.53	
	4700	1/2" x 1"		200	.040		1.40	1.41		2.81	3.69	
	4800	3/4" x 1-1/2"		165	.048		2.91	1.71		4.62	5.80	
	5500	Resin epoxy coating, 2 component, heavy duty				Gal.	26			26	28.50	
	5800	Tapes, sealant, P.V.C. foam adhesive, 1/16" x 1/4"				C.L.F.	4.60			4.60	5.05	
	5900	1/16" x 1/2"					6.80			6.80	7.50	
	5950	1/16" x 1"					11.30			11.30	12.45	
	6000	1/8" x 1/2"					7.65			7.65	8.40	
	6200	Urethane foam, 2 component, handy pack, 1 C.F.				Ea.	27.50			27.50	30.50	
	6300	50.0 C.F. pack				C.F.	14.05			14.05	15.45	

Division Notes

		CREW	DAILY OUTPUT	LABOR-HOURS	UNIT	2005 BARE COSTS MAT.	LABOR	EQUIP.	TOTAL	TOTAL INCL O&P

Division 8 Doors & Windows

Estimating Tips

08100 Metal Doors & Frames

- Most metal doors and frames look alike, but there may be significant differences among them. When estimating these items be sure to choose the line item that most closely compares to the specification or door schedule requirements regarding:
  - type of metal
  - metal gauge
  - door core material
  - fire rating
  - finish

08200 Wood & Plastic Doors

- Wood and plastic doors vary considerably in price. The primary determinant is the veneer material. Lauan, birch and oak are the most common veneers. Other variables include the following:
  - hollow or solid core
  - fire rating
  - flush or raised panel
  - finish
- If the specifications require compliance with AWI (Architectural Woodwork Institute) standards or acoustical standards, the cost of the door may increase substantially. All wood doors are priced pre-mortised for hinges and predrilled for cylindrical locksets.
- Frequently doors, frames, and windows are unique in old buildings. Specified replacement units could be stock, custom (similar to the original) or exact reproduction. The estimator should work closely with a window consultant to determine any extra costs that may be associated with the unusual installation requirements.

08300 Specialty Doors

- There are many varieties of special doors, and they are usually priced per each. Add frames, hardware or operators required for a complete installation.

08510 Steel Windows

- Most metal windows are delivered preglazed. However, some metal windows are priced without glass. Refer to 08800 Glazing for glass pricing. The grade C indicates commercial grade windows, usually ASTM C-35.

08550 Wood Windows

- All wood windows are priced preglazed. The two glazing options priced are single pane float glass and insulating glass 1/2″ thick. Add the cost of screens and grills if required.

08700 Hardware

- Hardware costs add considerably to the cost of a door. The most efficient method to determine the hardware requirements for a project is to review the door schedule. This schedule, in conjunction with the specifications, is all you should need to take off the door hardware.
- Door hinges are priced by the pair, with most doors requiring 1-1/2 pairs per door. The hinge prices do not include installation labor because it is included in door installation. Hinges are classified according to the frequency of use.

08800 Glazing

- Different openings require different types of glass. The three most common types are:
  - float
  - tempered
  - insulating
- Most exterior windows are glazed with insulating glass. Entrance doors and window walls, where the glass is less than 18″ from the floor, are generally glazed with tempered glass. Interior windows and some residential windows are glazed with float glass.
- Energy efficient coatings are also available

08900 Glazed Curtain Wall

- Glazed curtain walls consist of the metal tube framing and the glazing material. The cost data in this subdivision is presented for the metal tube framing alone or the composite wall. If your estimate requires a detailed takeoff of the framing, be sure to add the glazing cost.

Reference Numbers

Reference numbers are shown in bold squares at the beginning of some major classifications. These numbers refer to related items in the Reference Section. The reference information may be an estimating procedure, an alternate pricing method or technical information.

Note: Not all subdivisions listed here necessarily appear in this publication.

08340 | Special Function Doors

			CREW	DAILY OUTPUT	LABOR-HOURS	UNIT	2005 BARE COSTS MAT.	LABOR	EQUIP.	TOTAL	TOTAL INCL O&P	
100	0010	COLD STORAGE										100
	0020	Single, 20 ga. galvanized steel										
	0300	Horizontal sliding, 5′ x 7′, manual operation, 3.5″ thick	2 Carp	2	8	Ea.	2,400	274		2,674	3,050	
	0400	4″ thick		2	8		2,900	274		3,174	3,600	
	0500	6″ thick		2	8		2,700	274		2,974	3,400	
	0800	5′ x 7′, power operation, 2″ thick		1.90	8.421		4,500	288		4,788	5,400	
	0900	4″ thick		1.90	8.421		4,575	288		4,863	5,475	
	1000	6″ thick		1.90	8.421		5,200	288		5,488	6,175	
	1300	9′ x 10′, manual operation, 2″ insulation		1.70	9.412		3,625	320		3,945	4,500	
	1400	4″ insulation		1.70	9.412		3,750	320		4,070	4,625	
	1500	6″ insulation		1.70	9.412		4,525	320		4,845	5,475	
	1800	Power operation, 2″ insulation		1.60	10		6,250	345		6,595	7,400	
	1900	4″ insulation		1.60	10		6,400	345		6,745	7,575	
	2000	6″ insulation		1.70	9.412		7,250	320		7,570	8,475	
	2300	For stainless steel face, add					20%					
	3000	Hinged, lightweight, 3′ x 7′-0″, galvanized 1 face, 2″ thick	2 Carp	2	8	Ea.	1,125	274		1,399	1,675	
	3050	4″ thick		1.90	8.421		1,325	288		1,613	1,900	
	3300	Aluminum doors, 3′ x 7′-0″, 4″ thick		1.90	8.421		1,075	288		1,363	1,625	
	3350	6″ thick		1.40	11.429		1,900	390		2,290	2,700	
	3600	Stainless steel, 3′ x 7′-0″, 4″ thick		1.90	8.421		1,375	288		1,663	1,975	
	3650	6″ thick		1.40	11.429		2,300	390		2,690	3,150	
	3900	Painted, 3′ x 7′-0″, 4″ thick		1.90	8.421		985	288		1,273	1,525	
	3950	6″ thick		1.40	11.429		1,875	390		2,265	2,675	
	5000	Bi-parting, electric operated										
	5010	6′ x 8′ opening, galv. faces, 4″ thick for cooler	2 Carp	.80	20	Opng.	5,875	685		6,560	7,550	
	5050	For freezer, 4″ thick		.80	20		6,475	685		7,160	8,200	
	5300	For door buck framing and door protection, add		2.50	6.400		480	219		699	870	
	6000	Galvanized batten door, galvanized hinges, 4′ x 7′		2	8		1,450	274		1,724	2,025	
	6050	6′ x 8′		1.80	8.889		1,975	305		2,280	2,650	
	6500	Fire door, 3 hr., 6′ x 8′, single slide		.80	20		6,800	685		7,485	8,550	
	6550	Double, bi-parting		.70	22.857		9,825	785		10,610	12,000	

For information about Means Estimating Seminars, see yellow pages 12 and 13 in back of book

Division 9
Finishes

Estimating Tips

General

- Room Finish Schedule: A complete set of plans should contain a room finish schedule. If one is not available, it would be well worth the time and effort to put one together. A room finish schedule should contain the room number, room name (for clarity), floor materials, base materials, wainscot materials, wainscot height, wall materials (for each wall), ceiling materials, ceiling height and special instructions.
- Surplus Finishes: Review the specifications to determine if there is any requirement to provide certain amounts of extra materials for the owner's maintenance department. In some cases the owner may require a substantial amount of materials, especially when it is a special order item or long lead time item.

09200 Plaster & Gypsum Board

- Lath is estimated by the square yard for both gypsum and metal lath, plus usually 5% allowance for waste. Furring, channels and accessories are measured by the linear foot. An extra foot should be allowed for each accessory miter or stop.
- Plaster is also estimated by the square yard. Deductions for openings vary by preference, from zero deduction to 50% of all openings over 2 feet in width. Some estimators deduct a percentage of the total yardage for openings. The estimator should allow one extra square foot for each linear foot of horizontal interior or exterior angle located below the ceiling level. Also, double the areas of small radius work.
- Each room should be measured, perimeter times maximum wall height. Floors and ceiling areas are equal to length times width.
- Drywall accessories, studs, track, and acoustical caulking are all measured by the linear foot. Drywall taping is figured by the square foot. Gypsum wallboard is estimated by the square foot. No material deductions should be made for door or window openings under 32 S.F. Coreboard can be obtained in a 1″ thickness for solid wall and shaft work. Additions should be made to price out the inside or outside corners.
- Different types of partition construction should be listed separately on the quantity sheets. There may be walls with studs of various widths, double studded, and similar or dissimilar surface materials. Shaft work is usually different construction from surrounding partitions requiring separate quantities and pricing of the work.

09300 Tile
09400 Terrazzo

- Tile and terrazzo areas are taken off on a square foot basis. Trim and base materials are measured by the linear foot. Accent tiles are listed per each. Two basic methods of installation are used. Mud set is approximately 30% more expensive than the thin set. In terrazzo work, be sure to include the linear footage of embedded decorative strips, grounds, machine rubbing and power cleanup.

09600 Flooring

- Wood flooring is available in strip, parquet, or block configuration. The latter two types are set in adhesives with quantities estimated by the square foot. The laying pattern will influence labor costs and material waste. In addition to the material and labor for laying wood floors, the estimator must make allowances for sanding and finishing these areas unless the flooring is prefinished.
- Most of the various types of flooring are all measured on a square foot basis. Base is measured by the linear foot. If adhesive materials are to be quantified, they are estimated at a specified coverage rate by the gallon depending upon the specified type and the manufacturer's recommendations.
- Sheet flooring is measured by the square yard. Roll widths vary, so consideration should be given to use the most economical width, as waste must be figured into the total quantity. Consider also the installation methods available, direct glue down or stretched.

09700 Wall Finishes

- Wall coverings are estimated by the square foot. The area to be covered is measured, length by height of wall above baseboards, to calculate the square footage of each wall. This figure is divided by the number of square feet in the single roll which is being used. Deduct, in full, the areas of openings such as doors and windows. Where a pattern match is required allow 25%-30% waste. One gallon of paste should be sufficient to hang 12 single rolls of light to medium weight paper.

09800 Acoustical Treatment

- Acoustical systems fall into several categories. The takeoff of these materials should be by the square foot of area with a 5% allowance for waste. Do not forget about scaffolding, if applicable, when estimating these systems.

09900 Paints & Coatings

- A major portion of the work in painting involves surface preparation. Be sure to include cleaning, sanding, filling and masking costs in the estimate.
- Painting is one area where bids vary to a greater extent than almost any other section of a project. This arises from the many methods of measuring surfaces to be painted. The estimator should check the plans and specifications carefully to be sure of the required number of coats.
- Protection of adjacent surfaces is not included in painting costs. When considering the method of paint application, an important factor is the amount of protection and masking required. These must be estimated separately and may be the determining factor in choosing the method of application.

Reference Numbers

Reference numbers are shown in bold squares at the beginning of some major classifications. These numbers refer to related items in the Reference Section. The reference information may be an estimating procedure, an alternate pricing method or technical information.

Note: Not all subdivisions listed here necessarily appear in this publication.

09910 | Paints		CREW	DAILY OUTPUT	LABOR-HOURS	UNIT	2005 BARE COSTS MAT.	LABOR	EQUIP.	TOTAL	TOTAL INCL O&P
630 0010	**MISCELLANEOUS, INTERIOR**									
3800	Grilles, per side, oil base, primer coat, brushwork	1 Pord	520	.015	S.F.	.10	.47		.57	.82
3850	Spray		1,140	.007		.10	.21		.31	.43
3880	Paint 1 coat, brushwork		520	.015		.11	.47		.58	.83
3900	Spray		1,140	.007		.12	.21		.33	.45
3920	Paint 2 coats, brushwork		325	.025		.21	.75		.96	1.36
3940	Spray		650	.012		.24	.38		.62	.83
3950	Prime & paint 1 coat		325	.025		.20	.75		.95	1.35
3960	Prime & paint 2 coats		270	.030		.20	.91		1.11	1.58
4250	Paint 1 coat, brushwork	2 Pord	1,300	.012	L.F.	.11	.38		.49	.69
4500	Louvers, one side, primer, brushwork	1 Pord	524	.015	S.F.	.06	.47		.53	.76
4520	Paint one coat, brushwork		520	.015		.06	.47		.53	.77
4530	Spray		1,140	.007		.06	.21		.27	.39
4540	Paint two coats, brushwork		325	.025		.11	.75		.86	1.25
4550	Spray		650	.012		.12	.38		.50	.70
4560	Paint three coats, brushwork		270	.030		.16	.91		1.07	1.54
4570	Spray		500	.016		.18	.49		.67	.94
5000	Pipe, to 4" diameter, primer or sealer coat, oil base, brushwork	2 Pord	1,250	.013	L.F.	.06	.39		.45	.66
5100	Spray		2,165	.007		.06	.23		.29	.40
5200	Paint 1 coat, brushwork		1,250	.013		.06	.39		.45	.66
5300	Spray		2,165	.007		.06	.23		.29	.40
5350	Paint 2 coats, brushwork		775	.021		.11	.63		.74	1.07
5400	Spray		1,240	.013		.13	.39		.52	.73
5420	Paint 3 coats, brushwork		775	.021		.17	.63		.80	1.14
5450	To 8" diameter, primer or sealer coat, brushwork		620	.026		.12	.79		.91	1.32
5500	Spray		1,085	.015		.20	.45		.65	.90
5550	Paint 1 coat, brushwork		620	.026		.17	.79		.96	1.38
5600	Spray		1,085	.015		.19	.45		.64	.89
5650	Paint 2 coats, brushwork		385	.042		.23	1.27		1.50	2.16
5700	Spray		620	.026		.25	.79		1.04	1.47
5720	Paint 3 coats, brushwork		385	.042		.34	1.27		1.61	2.28
5750	To 12" diameter, primer or sealer coat, brushwork		415	.039		.18	1.18		1.36	1.97
5800	Spray		725	.022		.21	.68		.89	1.25
5850	Paint 1 coat, brushwork		415	.039		.17	1.18		1.35	1.96
6000	Spray		725	.022		.19	.68		.87	1.23
6200	Paint 2 coats, brushwork		260	.062		.34	1.88		2.22	3.20
6250	Spray		415	.039		.38	1.18		1.56	2.18
6270	Paint 3 coats, brushwork		260	.062		.50	1.88		2.38	3.38
6300	To 16" diameter, primer or sealer coat, brushwork		310	.052		.24	1.58		1.82	2.64
6350	Spray		540	.030		.27	.91		1.18	1.66
6400	Paint 1 coat, brushwork		310	.052		.23	1.58		1.81	2.63
6450	Spray		540	.030		.26	.91		1.17	1.65
6500	Paint 2 coats, brushwork		195	.082		.45	2.51		2.96	4.27
6550	Spray		310	.052		.50	1.58		2.08	2.92
6600	Radiators, per side, primer, brushwork	1 Pord	520	.015	S.F.	.06	.47		.53	.77
6620	Paint one coat, brushwork		520	.015		.05	.47		.52	.77
6640	Paint two coats, brushwork		340	.024		.11	.72		.83	1.20
6660	Paint three coats, brushwork		283	.028		.16	.87		1.03	1.48

For information about Means Estimating Seminars, see yellow pages 12 and 13 in back of book

Division 10 Specialties

Estimating Tips

General

- The items in this division are usually priced per square foot or each.
- Many items in Division 10 require some type of support system or special anchors that are not usually furnished with the item. The required anchors must be added to the estimate in the appropriate division.
- Some items in Division 10, such as lockers, may require assembly before installation. Verify the amount of assembly required. Assembly can often exceed installation time.

10150 Compartments & Cubicles

- Support angles and blocking are not included in the installation of toilet compartments, shower/dressing compartments or cubicles. Appropriate line items from Divisions 5 or 6 may need to be added to support the installations.
- Toilet partitions are priced by the stall. A stall consists of a side wall, pilaster and door with hardware. Toilet tissue holders and grab bars are extra.

10600 Partitions

- The required acoustical rating of a folding partition can have a significant impact on costs. Verify the sound transmission coefficient rating of the panel priced to the specification requirements.

10800 Toilet/Bath/Laundry Accessories

- Grab bar installation does not include supplemental blocking or backing to support the required load. When grab bars are installed at an existing facility provisions must be made to attach the grab bars to solid structure.

Reference Numbers

Reference numbers are shown in bold squares at the beginning of some major classifications. These numbers refer to related items in the Reference Section. The reference information may be an estimating procedure, an alternate pricing method or technical information.

Note: Not all subdivisions listed here necessarily appear in this publication.

10200 | Louvers & Vents

		10210 | Wall Louvers	CREW	DAILY OUTPUT	LABOR-HOURS	UNIT	2005 BARE COSTS MAT.	LABOR	EQUIP.	TOTAL	TOTAL INCL O&P	
800	0010	**LOUVERS** Aluminum with screen, residential, 8" x 8"	1 Carp	38	.211	Ea.	9.35	7.20		16.55	21.50	800
	0100	12" x 12"		38	.211		10.20	7.20		17.40	22.50	
	0200	12" x 18"		35	.229		13.30	7.85		21.15	27	
	0250	14" x 24"		30	.267		17.50	9.15		26.65	33.50	
	0300	18" x 24"		27	.296		21	10.15		31.15	39.50	
	0500	24" x 30"		24	.333		25.50	11.40		36.90	46	
	0700	Triangle, adjustable, small		20	.400		25.50	13.70		39.20	49.50	
	0800	Large	↓	15	.533	↓	42	18.25		60.25	74.50	
	1200	Extruded aluminum, see division 15850-600										
	2100	Midget, aluminum, 3/4" deep, 1" diameter	1 Carp	85	.094	Ea.	.87	3.22		4.09	5.95	
	2150	3" diameter		60	.133		1.81	4.57		6.38	9.10	
	2200	4" diameter		50	.160		2.81	5.50		8.31	11.65	
	2250	6" diameter	↓	30	.267	↓	3.35	9.15		12.50	17.95	
	2300	Ridge vent strip, mill finish	1 Shee	155	.052	L.F.	2.53	2.07		4.60	5.95	
	2400	Under eaves vent, aluminum, mill finish, 16" x 4"	1 Carp	48	.167	Ea.	1.78	5.70		7.48	10.85	
	2500	16" x 8"	"	48	.167	"	1.97	5.70		7.67	11.05	

10 SPECIALTIES

10270 | Access Flooring

		10275 | Access Flooring	CREW	DAILY OUTPUT	LABOR-HOURS	UNIT	2005 BARE COSTS MAT.	LABOR	EQUIP.	TOTAL	TOTAL INCL O&P	
150	0010	**PEDESTAL ACCESS FLOORS** Computer room application, metal										150
	0020	Particle board or steel panels, no covering, under 6,000 S.F.	2 Carp	400	.040	S.F.	13.80	1.37		15.17	17.30	
	0300	Metal covered, over 6,000 S.F.		450	.036		8	1.22		9.22	10.70	
	0400	Aluminum, 24" panels	↓	500	.032		29.50	1.10		30.60	34	
	0600	For carpet covering, add					5.70			5.70	6.25	
	0700	For vinyl floor covering, add					5.80			5.80	6.35	
	0900	For high pressure laminate covering, add					3.85			3.85	4.24	
	0910	For snap on stringer system, add	2 Carp	1,000	.016	↓	1.32	.55		1.87	2.30	
	0950	Office applications, to 8" high, steel panels,										
	0960	no covering, over 6,000 S.F.	2 Carp	500	.032	S.F.	9.20	1.10		10.30	11.85	
	1000	Machine cutouts after initial installation	1 Carp	10	.800	Ea.	4.21	27.50		31.71	47	
	1050	Pedestals, 6" to 12"	2 Carp	85	.188		8.10	6.45		14.55	18.95	
	1100	Air conditioning grilles, 4" x 12"	1 Carp	17	.471		58	16.10		74.10	89	
	1150	4" x 18"	"	14	.571	↓	79.50	19.55		99.05	118	
	1200	Approach ramps, minimum	2 Carp	85	.188	S.F.	22	6.45		28.45	34	
	1300	Maximum	"	60	.267	"	29.50	9.15		38.65	46.50	
	1500	Handrail, 2 rail, aluminum	1 Carp	15	.533	L.F.	85	18.25		103.25	122	

10300 | Fireplaces & Stoves

		10305 | Manufactured Fireplaces	CREW	DAILY OUTPUT	LABOR-HOURS	UNIT	2005 BARE COSTS MAT.	LABOR	EQUIP.	TOTAL	TOTAL INCL O&P	
100	0010	**FIREPLACE, PREFABRICATED** Free standing or wall hung										100
	0100	with hood & screen, minimum	1 Carp	1.30	6.154	Ea.	1,150	211		1,361	1,575	

10300 | Fireplaces & Stoves

10305 | Manufactured Fireplaces

			CREW	DAILY OUTPUT	LABOR-HOURS	UNIT	2005 BARE COSTS MAT.	LABOR	EQUIP.	TOTAL	TOTAL INCL O&P	
100	0150	Average	1 Carp	1	8	Ea.	1,375	274		1,649	1,925	100
	0200	Maximum		.90	8.889	↓	3,350	305		3,655	4,175	
	0500	Chimney dbl. wall, all stainless, over 8′-6″, 7″ diam., add		33	.242	V.L.F.	52	8.30		60.30	70	
	0600	10″ diameter, add		32	.250		55	8.55		63.55	74	
	0700	12″ diameter, add		31	.258		72	8.85		80.85	93	
	0800	14″ diameter, add		30	.267	↓	91	9.15		100.15	114	
	1000	Simulated brick chimney top, 4′ high, 16″ x 16″		10	.800	Ea.	196	27.50		223.50	259	
	1100	24″ x 24″		7	1.143	″	365	39		404	460	
	1500	Simulated logs, gas fired, 40,000 BTU, 2′ long, minimum		7	1.143	Set	480	39		519	585	
	1600	Maximum		6	1.333		670	45.50		715.50	805	
	1700	Electric, 1,500 BTU, 1′-6″ long, minimum		7	1.143		136	39		175	210	
	1800	11,500 BTU, maximum	↓	6	1.333	↓	293	45.50		338.50	390	

10320 | Stoves

			CREW	DAILY OUTPUT	LABOR-HOURS	UNIT	MAT.	LABOR	EQUIP.	TOTAL	TOTAL INCL O&P	
100	0010	**WOODBURNING STOVES** Cast iron, minimum	2 Carp	1.30	12.308	Ea.	705	420		1,125	1,425	100
	0020	Average		1	16		1,300	550		1,850	2,275	
	0030	Maximum	↓	.80	20		2,175	685		2,860	3,450	
	0050	For gas log lighter, add				↓	39			39	43	

For information about Means Estimating Seminars, see yellow pages 12 and 13 in back of book

Division Notes

		CREW	DAILY OUTPUT	LABOR-HOURS	UNIT	2005 BARE COSTS MAT.	LABOR	EQUIP.	TOTAL	TOTAL INCL O&P

Division 11
Equipment

Estimating Tips

General

- The items in this division are usually priced per square foot or each. Many of these items are purchased by the owner for installation by the contractor. Check the specifications for responsibilities, and include time for receiving, storage, installation and mechanical and electrical hook-ups in the appropriate divisions.
- Many items in Division 11 require some type of support system that is not usually furnished with the item. Examples of these systems include blocking for the attachment of casework and support angles for ceiling hung projection screens. The required blocking or supports must be added to the estimate in the appropriate division.
- Some items in Division 11 may require assembly or electrical hook-ups. Verify the amount of assembly required or the need for a hard electrical connection and add the appropriate costs.

Reference Numbers

Reference numbers are shown in bold squares at the beginning of some major classifications. These numbers refer to related items in the Reference Section. The reference information may be an estimating procedure, an alternate pricing method or technical information.

Note: Not all subdivisions listed here necessarily appear in this publication.

11010 | Maintenance Equipment

	11013 | Floor/Wall Cleaning Equipment	CREW	DAILY OUTPUT	LABOR-HOURS	UNIT	2005 BARE COSTS MAT.	LABOR	EQUIP.	TOTAL	TOTAL INCL O&P	
800	0010 VACUUM CLEANING										800
	0020 Central, 3 inlet, residential	1 Skwk	.90	8.889	Total	610	310		920	1,150	
	0200 Commercial		.70	11.429		1,150	400		1,550	1,900	
	0400 5 inlet system, residential		.50	16		925	560		1,485	1,900	
	0600 7 inlet system, commercial		.40	20		1,025	695		1,720	2,225	
	0800 9 inlet system, residential		.30	26.667		1,300	930		2,230	2,875	
	4010 Rule of thumb: First 1200 S.F., installed									1,125	
	4020 For each additional S.F., add				S.F.					.18	

11100 | Mercantile Equipment

	11104 | Display Cases & Systems	CREW	DAILY OUTPUT	LABOR-HOURS	UNIT	2005 BARE COSTS MAT.	LABOR	EQUIP.	TOTAL	TOTAL INCL O&P	
700	0010 REFRIGERATED FOOD CASES										700
	0030 Dairy, multi-deck, 12′ long	Q-5	3	5.333	Ea.	8,200	197		8,397	9,300	
	0100 For rear sliding doors, add					1,125			1,125	1,250	
	0200 Delicatessen case, service deli, 12′ long, single deck	Q-5	3.90	4.103		5,500	151		5,651	6,275	
	0300 Multi-deck, 18 S.F. shelf display		3	5.333		6,750	197		6,947	7,725	
	0400 Freezer, self-contained, chest-type, 30 C.F.		3.90	4.103		4,025	151		4,176	4,650	
	0500 Glass door, upright, 78 C.F.		3.30	4.848		7,650	179		7,829	8,700	
	0600 Frozen food, chest type, 12′ long		3.30	4.848		5,525	179		5,704	6,375	
	0700 Glass door, reach-in, 5 door		3	5.333		10,600	197		10,797	12,000	
	0800 Island case, 12′ long, single deck		3.30	4.848		6,275	179		6,454	7,175	
	0900 Multi-deck		3	5.333		13,200	197		13,397	14,900	
	1000 Meat case, 12′ long, single deck		3.30	4.848		4,550	179		4,729	5,275	
	1050 Multi-deck		3.10	5.161		7,825	190		8,015	8,875	
	1100 Produce, 12′ long, single deck		3.30	4.848		6,025	179		6,204	6,900	
	1200 Multi-deck		3.10	5.161		6,700	190		6,890	7,650	

11110 | Commercial Laundry & Dry Cleaning Equipment

	11119 | Laundry Cleaning	CREW	DAILY OUTPUT	LABOR-HOURS	UNIT	2005 BARE COSTS MAT.	LABOR	EQUIP.	TOTAL	TOTAL INCL O&P	
450	0010 **LAUNDRY EQUIPMENT** Not incl. rough-in										450
	0500 Dryers, gas fired residential, 16 lb. capacity, average	1 Plum	3	2.667	Ea.	565	109		674	785	
	1000 Commercial, 30 lb. capacity, coin operated, single		3	2.667		2,575	109		2,684	3,000	
	1100 Double stacked		2	4		5,525	163		5,688	6,325	
	1500 Industrial, 30 lb. capacity		2	4		2,200	163		2,363	2,650	
	1600 50 lb. capacity		1.70	4.706		2,875	192		3,067	3,450	
	5000 Washers, residential, 4 cycle, average		3	2.667		645	109		754	875	
	5300 Commercial, coin operated, average		3	2.667		1,050	109		1,159	1,325	
	6000 Combination washer/extractor, 20 lb. capacity	L-6	1.50	8		3,725	325		4,050	4,600	
	6100 30 lb. capacity		.80	15		7,200	610		7,810	8,850	
	6200 50 lb. capacity		.68	17.647		8,950	720		9,670	10,900	
	6300 75 lb. capacity		.30	40		18,500	1,625		20,125	22,800	

11110 | Commercial Laundry & Dry Cleaning Equipment

	11119 | Laundry Cleaning		Crew	Daily Output	Labor-Hours	Unit	2005 Bare Costs Mat.	Labor	Equip.	Total	Total Incl O&P
450	6350	125 lb. capacity	L-6	.16	75	Ea.	22,400	3,050		25,450	29,300

11140 | Vehicle Service Equipment

	11141 | Service Station Equipment		Crew	Daily Output	Labor-Hours	Unit	2005 Bare Costs Mat.	Labor	Equip.	Total	Total Incl O&P
100	0010	COMPRESSED AIR EQUIPMENT									
	0030	Compressors, electric, 1-1/2 H.P., standard controls	L-4	1.50	16	Ea.	296	510		806	1,125
	0550	Dual controls		1.50	16		475	510		985	1,325
	0600	5 H.P., 115/230 volt, standard controls		1	24		1,700	760		2,460	3,050
	0650	Dual controls	↓	1	24	↓	1,800	760		2,560	3,175
200	0010	FUEL DISPENSING EQUIPMENT									
	1100	Product dispenser with vapor recovery for 6 nozzles, installed, not									
	1110	including piping to storage tanks				Ea.	18,000			18,000	19,800
300	0010	LUBRICATION EQUIPMENT									
	3000	Lube equipment, 3 reel type, with pumps, not including piping	L-4	.50	48	Set	6,825	1,525		8,350	9,875
	3700	Pump lubrication, pneumatic, not incl. air compressor									
	3710	Oil/gear lube	Q-1	9.60	1.667	Ea.	615	61.50		676.50	770
	3720	Grease	"	9.60	1.667	"	640	61.50		701.50	795
400	0010	SPRAY PAINTING EQUIPMENT									
	4000	Spray painting booth, 26′ long, complete	L-4	.40	60	Ea.	12,700	1,900		14,600	16,900

11170 | Solid Waste Handling Equipment

	11179 | Waste Handling Equipment		Crew	Daily Output	Labor-Hours	Unit	2005 Bare Costs Mat.	Labor	Equip.	Total	Total Incl O&P
150	0010	WASTE HANDLING									
	0020	Compactors, 115 volt, 250#/hr., chute fed	L-4	1	24	Ea.	9,475	760		10,235	11,600
	0100	Hand fed		2.40	10		6,825	320		7,145	8,000
	0300	Multi-bag, 230 volt, 600#/hr, chute fed		1	24		8,550	760		9,310	10,600
	0400	Hand fed		1	24		7,325	760		8,085	9,250
	0500	Containerized, hand fed, 2 to 6 C.Y. containers, 250#/hr.		1	24		8,975	760		9,735	11,100
	0550	For chute fed, add per floor		1	24		1,025	760		1,785	2,300
	1000	Heavy duty industrial compactor, 0.5 C.Y. capacity		1	24		6,075	760		6,835	7,850
	1050	1.0 C.Y. capacity		1	24		9,350	760		10,110	11,500
	1100	3 C.Y. capacity		.50	48		12,700	1,525		14,225	16,400
	1150	5.0 C.Y. capacity		.50	48		15,800	1,525		17,325	19,700
	1200	Combination shredder/compactor (5,000 lbs./hr.)	↓	.50	48		30,900	1,525		32,425	36,400
	1400	For handling hazardous waste materials, 55 gallon drum packer, std.					14,300			14,300	15,800
	1410	55 gallon drum packer w/HEPA filter					17,900			17,900	19,700
	1420	55 gallon drum packer w/charcoal & HEPA filter					23,900			23,900	26,300
	1430	All of the above made explosion proof, add				↓	10,900			10,900	12,000

11170 | Solid Waste Handling Equipment

	11179 | Waste Handling Equipment		Crew	Daily Output	Labor-Hours	Unit	2005 Bare Costs Mat.	Labor	Equip.	Total	Total Incl O&P	
150	1500	Crematory, not including building, 1 place	Q-3	.20	160	Ea.	51,000	6,225		57,225	65,500	150
	1750	2 place		.10	320		72,500	12,500		85,000	98,500	
	4400	Incinerator, gas, not incl. chimney, elec. or pipe, 50#/hr., minimum		.80	40		19,900	1,550		21,450	24,200	
	4420	Maximum		.70	45.714		25,900	1,775		27,675	31,100	
	4440	200 lb. per hr., minimum (batch type)		.60	53.333		25,900	2,075		27,975	31,500	
	4460	Maximum (with feeder)		.50	64		50,500	2,500		53,000	59,500	
	4480	400 lb. per hr., minimum (batch type)		.30	106		30,600	4,150		34,750	40,000	
	4500	Maximum (with feeder)		.25	128		58,000	4,975		62,975	71,000	
	4520	800 lb. per hr., with feeder, minimum		.20	160		75,500	6,225		81,725	92,500	
	4540	Maximum		.17	188		103,000	7,325		110,325	124,500	
	4560	1,200 lb. per hr., with feeder, minimum		.15	213		109,000	8,300		117,300	132,500	
	4580	Maximum		.11	290		131,000	11,300		142,300	161,500	
	4600	2,000 lb. per hr., with feeder, minimum		.10	320		190,500	12,500		203,000	228,000	
	4620	Maximum		.05	640		320,500	24,900		345,400	390,000	
	4700	For heat recovery system, add, minimum		.25	128		62,000	4,975		66,975	75,500	
	4710	Add, maximum		.11	290		198,000	11,300		209,300	234,500	
	4720	For automatic ash conveyer, add		.50	64		26,000	2,500		28,500	32,400	
	4750	Large municipal incinerators, incl. stack, minimum		.25	128	Ton/day	15,800	4,975		20,775	24,900	
	4850	Maximum		.10	320	"	42,000	12,500		54,500	65,000	
	5800	Shredder, industrial, minimum				Ea.	18,500			18,500	20,400	
	5850	Maximum					99,500			99,500	109,000	
	5900	Baler, industrial, minimum					7,425			7,425	8,150	
	5950	Maximum					433,500			433,500	476,500	

11300 | Fluid Waste Treatment & Disposal Equipment

	11310 | Sewage & Sludge Pumps		Crew	Daily Output	Labor-Hours	Unit	2005 Bare Costs Mat.	Labor	Equip.	Total	Total Incl O&P	
350	0010	**PUMPS, PNEUMATIC EJECTOR**										350
	0020	With steel receiver, level controls, inlet/outlet gate/check valves										
	0030	Cross connect. not incl. compressor, fittings or piping										
	0040	Duplex										
	0050	30 GPM	Q-2	1.70	14.118	Ea.	14,400	540		14,940	16,600	
	0060	50 GPM		1.56	15.385		20,300	585		20,885	23,200	
	0070	100 GPM		1.30	18.462		24,500	705		25,205	28,100	
	0080	150 GPM		1.10	21.818		47,300	830		48,130	53,500	
	0090	200 GPM		.85	28.235		60,000	1,075		61,075	67,500	
	0100	250 GPM	Q-3	.91	35.165		77,500	1,375		78,875	87,000	
	0110	300 GPM	"	.57	56.140		94,500	2,175		96,675	107,500	

Important: See the Reference Section for critical supporting data - Reference Nos., Crews, & City Cost Indexes

11400 | Food Service Equipment

		11405 | Food Storage Equipment	Crew	Daily Output	Labor-Hours	Unit	2005 Bare Costs Mat.	Labor	Equip.	Total	Total Incl O&P	
110	0010	**FOOD STORAGE EQUIPMENT**										110
	2350	Cooler, reach-in, beverage, 6′ long	Q-1	6	2.667	Ea.	3,550	98		3,648	4,075	
	4300	Freezers, reach-in, 44 C.F.		4	4		8,125	147		8,272	9,175	
	4500	68 C.F.		3	5.333		9,150	196		9,346	10,400	
	8300	Refrigerators, reach-in type, 44 C.F.		5	3.200		4,950	118		5,068	5,625	
	8310	With glass doors, 68 C.F.		4	4		7,150	147		7,297	8,075	

		11420 | Food Cooking Equipment	Crew	Daily Output	Labor-Hours	Unit	Mat.	Labor	Equip.	Total	Total Incl O&P	
110	0010	**COOKING EQUIPMENT**										110
	0020	Bake oven, gas, one section	Q-1	8	2	Ea.	4,225	73.50		4,298.50	4,750	
	0300	Two sections		7	2.286		8,625	84		8,709	9,625	
	0600	Three sections		6	2.667		12,200	98		12,298	13,600	
	0900	Electric convection, single deck	L-7	4	7		4,525	231		4,756	5,325	
	6350	Kettle, w/steam jacket, tilting, w/positive lock, SS, 20 gallons		7	4		5,350	132		5,482	6,100	
	6600	60 gallons		6	4.667		6,825	154		6,979	7,775	

		11425 | Hood and Ventilation Equipment	Crew	Daily Output	Labor-Hours	Unit	Mat.	Labor	Equip.	Total	Total Incl O&P	
110	0010	**HOOD & VENTILATION EQUIPMENT**										110
	7950	Hood fire protection system, minimum	Q-1	3	5.333	Ea.	3,225	196		3,421	3,850	
	8050	Maximum	"	1	16	"	24,100	590		24,690	27,500	
210	0010	**KITCHEN VENTILATION,** Commercial										210
	1010	Heat reclaim unit, air to air										
	1100	Heat pipe exchanger										
	1110	Combined supply/exhaust air volume										
	1120	2.5 to 6.0 MCFM	Q-10	2.80	8.571	MCFM	5,600	320		5,920	6,650	
	1130	6 to 16 MCFM		5	4.800		3,425	180		3,605	4,050	
	1140	16 to 22 MCFM		6	4		2,675	150		2,825	3,175	
	2010	Packaged ventilation										
	2100	Combined supply/exhaust air volume										
	2110	With ambient supply										
	2120	3 to 7 MCFM	Q-10	5	4.800	MCFM	1,000	180		1,180	1,375	
	2130	7 to 12 MCFM		7.10	3.380		675	127		802	935	
	2140	12 to 22 MCFM		10.70	2.243		475	84		559	650	
	2400	With tempered supply										
	2410	3 to 7 MCFM	Q-10	5	4.800	MCFM	1,750	180		1,930	2,200	
	2420	7 to 12 MCFM		7.10	3.380		1,125	127		1,252	1,450	
	2430	12 to 22 MCFM		10:70	2.243		825	84		909	1,050	
	4000	Electrostatic precipitators										
	4100	Smoke pollution control										
	4110	Exhaust air volume										
	4120	1 to 2.4 MCFM	Q-10	2.15	11.163	MCFM	6,375	420		6,795	7,650	
	4130	2.4 to 4.8 MCFM		2.70	8.889		7,250	335		7,585	8,500	
	4140	4.8 to 7.2 MCFM		3.60	6.667		7,375	250		7,625	8,500	
	4150	7.2 to 12 MCFM		5.40	4.444		7,725	167		7,892	8,750	
	4400	Electrostatic precipitators with built-in exhaust fans										
	4410	Smoke pollution control										
	4420	Exhaust air volume										
	4440	1 to 2.4 MCFM	Q-10	2.15	11.163	MCFM	7,675	420		8,095	9,100	
	4450	2.4 to 4.8 MCFM		2.70	8.889		6,075	335		6,410	7,225	
	4460	4.8 to 7.2 MCFM		3.60	6.667		5,200	250		5,450	6,100	
	4470	7.2 to 12 MCFM		5.40	4.444		4,500	167		4,667	5,175	
	6000	Exhaust hoods										
	6100	Centrifugal grease extraction										
	6110	Water wash type										

11400 | Food Service Equipment

11425 | Hood and Ventilation Equipment

			CREW	DAILY OUTPUT	LABOR-HOURS	UNIT	2005 BARE COSTS MAT.	LABOR	EQUIP.	TOTAL	TOTAL INCL O&P	
210	6120	Per foot of length, minimum	Q-10	7.10	3.380	L.F.	1,300	127		1,427	1,625	210
	6130	Maximum	"	7.10	3.380	"	2,225	127		2,352	2,650	
	6200	Non water wash type										
	6210	Per foot of length	Q-10	13.40	1.791	L.F.	950	67		1,017	1,150	
	6400	Island style ventilator										
	6410	Water wash type										
	6420	Per foot of length, minimum	Q-10	7.10	3.380	L.F.	1,725	127		1,852	2,100	
	6430	Maximum	"	7.10	3.380	"	2,875	127		3,002	3,375	
	6460	Non water wash type										
	6470	Per foot of length, minimum	Q-10	13.40	1.791	L.F.	750	67		817	930	
	6480	Maximum	"	13.40	1.791	"	1,525	67		1,592	1,775	

11435 | Ice Machines

			CREW	DAILY OUTPUT	LABOR-HOURS	UNIT	MAT.	LABOR	EQUIP.	TOTAL	TOTAL INCL O&P	
110	0010	ICE MACHINES										110
	5800	Ice cube maker, 50 pounds per day	Q-1	6	2.667	Ea.	1,325	98		1,423	1,600	
	6050	500 pounds per day	"	4	4	"	2,875	147		3,022	3,400	

11440 | Cleaning and Disposal Equipment

			CREW	DAILY OUTPUT	LABOR-HOURS	UNIT	MAT.	LABOR	EQUIP.	TOTAL	TOTAL INCL O&P	
110	0010	CLEANING & DISPOSAL EQUIPMENT										110
	2700	Dishwasher, commercial, rack type										
	2720	10 to 12 racks per hour	Q-1	3.20	5	Ea.	3,650	184		3,834	4,300	
	2750	Semi-automatic 38 to 50 racks per hour	"	1.30	12.308	↓	6,550	455		7,005	7,875	
	2800	Automatic, 190 to 230 racks per hour	L-6	.35	34.286	↓	8,200	1,400		9,600	11,100	
	2820	235 to 275 racks per hour	↓	.25	48	↓	19,000	1,950		20,950	23,800	
	2840	8,750 to 12,500 dishes per hour	↓	.10	120	↓	41,200	4,900		46,100	52,500	

11450 | Residential Equipment

11454 | Residential Appliances

			CREW	DAILY OUTPUT	LABOR-HOURS	UNIT	2005 BARE COSTS MAT.	LABOR	EQUIP.	TOTAL	TOTAL INCL O&P	
500	0010	RESIDENTIAL APPLIANCES										500
	0020	Cooking range, 30" free standing, 1 oven, minimum	2 Clab	10	1.600	Ea.	241	42.50		283.50	330	
	0050	Maximum	↓	4	4	↓	1,475	107		1,582	1,800	
	0150	2 oven, minimum	↓	10	1.600	↓	1,475	42.50		1,517.50	1,700	
	0200	Maximum	↓	10	1.600	↓	1,550	42.50		1,592.50	1,775	
	2450	Dehumidifier, portable, automatic, 15 pint				↓	149			149	164	
	2550	40 pint				↓	166			166	183	
	2750	Dishwasher, built-in, 2 cycles, minimum	L-1	4	4	↓	248	163		411	515	
	2800	Maximum	↓	2	8	↓	289	325		614	810	
	2950	4 or more cycles, minimum	↓	4	4	↓	262	163		425	530	
	2960	Average	↓	4	4	↓	350	163		513	630	
	3000	Maximum	↓	2	8	↓	540	325		865	1,075	
	3200	Dryer, automatic, minimum	L-2	3	5.333	↓	282	159		441	560	
	3250	Maximum	"	2	8	↓	790	239		1,029	1,250	
	3300	Garbage disposal, sink type, minimum	L-1	10	1.600	↓	42.50	65.50		108	144	
	3350	Maximum	"	10	1.600	↓	145	65.50		210.50	258	
	3550	Heater, electric, built-in, 1250 watt, ceiling type, minimum	1 Elec	4	2	↓	72	81.50		153.50	201	
	3600	Maximum	↓	3	2.667	↓	118	109		227	292	

11450 | Residential Equipment

		11454 | Residential Appliances	CREW	DAILY OUTPUT	LABOR-HOURS	UNIT	2005 BARE COSTS MAT.	LABOR	EQUIP.	TOTAL	TOTAL INCL O&P	
500	3700	Wall type, minimum	1 Elec	4	2	Ea.	102	81.50		183.50	234	500
	3750	Maximum		3	2.667		136	109		245	310	
	3900	1500 watt wall type, with blower		4	2		127	81.50		208.50	260	
	3950	3000 watt	↓	3	2.667		258	109		367	445	
	4150	Hood for range, 2 speed, vented, 30" wide, minimum	L-3	5	3.200		37.50	120		157.50	224	
	4200	Maximum		3	5.333		595	199		794	960	
	4300	42" wide, minimum		5	3.200		225	120		345	430	
	4330	Custom		5	3.200		625	120		745	870	
	4350	Maximum	↓	3	5.333		760	199		959	1,150	
	4500	For ventless hood, 2 speed, add					15			15	16.50	
	4650	For vented 1 speed, deduct from maximum					39			39	43	
	4850	Humidifier, portable, 8 gallons per day					149			149	164	
	5000	15 gallons per day					179			179	197	
	5200	Icemaker, automatic, 20 lb. per day	1 Plum	7	1.143		360	46.50		406.50	465	
	5350	51 lb. per day		2	4		1,000	163		1,163	1,350	
	6400	Sump pump cellar drainer, pedestal, 1/3 H.P., molded PVC base		3	2.667		87	109		196	260	
	6450	Solid brass	↓	2	4	↓	179	163		342	445	
	6460	Sump pump, see also division 15440-940										
	6650	Washing machine, automatic, minimum	1 Plum	3	2.667	Ea.	279	109		388	470	
	6700	Maximum	"	1	8		1,025	325		1,350	1,625	
	6900	Water heater, electric, glass lined, 30 gallon, minimum	L-1	5	3.200		252	131		383	470	
	6950	Maximum		3	5.333		350	218		568	710	
	7100	80 gallon, minimum		2	8		485	325		810	1,025	
	7150	Maximum	↓	1	16		670	655		1,325	1,725	
	7180	Water heater, gas, glass lined, 30 gallon, minimum	2 Plum	5	3.200		345	131		476	575	
	7220	Maximum		3	5.333		480	218		698	855	
	7260	50 gallon, minimum		2.50	6.400		480	261		741	925	
	7300	Maximum	↓	1.50	10.667	↓	665	435		1,100	1,400	
	7310	Water heater, see also division 15480-200										
	7350	Water softener, automatic, to 30 grains per gallon	2 Plum	5	3.200	Ea.	430	131		561	670	
	7400	To 100 grains per gallon	"	4	4		600	163		763	905	
	7450	Vent kits for dryers	1 Carp	10	.800	↓	12.65	27.50		40.15	56.50	

		11460 | Unit Kitchens	CREW	DAILY OUTPUT	LABOR-HOURS	UNIT	MAT.	LABOR	EQUIP.	TOTAL	TOTAL INCL O&P	
100	0010	UNIT KITCHENS										100
	1500	Combination range, refrigerator and sink, 30" wide, minimum	L-1	2	8	Ea.	730	325		1,055	1,300	
	1550	Maximum		1	16		1,450	655		2,105	2,575	
	1570	60" wide, average		1.40	11.429		2,400	465		2,865	3,350	
	1590	72" wide, average		1.20	13.333		2,725	545		3,270	3,825	
	1600	Office model, 48" wide		2	8		2,050	325		2,375	2,775	
	1620	Refrigerator and sink only	↓	2.40	6.667	↓	2,050	272		2,322	2,675	
	1640	Combination range, refrigerator, sink, microwave										
	1660	oven and ice maker	L-1	.80	20	Ea.	4,025	815		4,840	5,650	

11470 | Darkroom Equipment

		11471 | Darkroom Processing	CREW	DAILY OUTPUT	LABOR-HOURS	UNIT	2005 BARE COSTS MAT.	LABOR	EQUIP.	TOTAL	TOTAL INCL O&P	
700	0010	DARKROOM EQUIPMENT										700
	0020	Developing sink, 5" deep, 24" x 48"	Q-1	2	8	Ea.	4,225	294		4,519	5,100	

11470 | Darkroom Equipment

11471 | Darkroom Processing

			CREW	DAILY OUTPUT	LABOR-HOURS	UNIT	2005 BARE COSTS MAT.	LABOR	EQUIP.	TOTAL	TOTAL INCL O&P	
700	0050	48″ x 52″	Q-1	1.70	9.412	Ea.	4,300	345		4,645	5,250	700
	0200	10″ deep, 24″ x 48″		1.70	9.412		5,275	345		5,620	6,325	
	0250	24″ x 108″		1.50	10.667		1,900	390		2,290	2,700	
	3500	Washers, round, minimum sheet 11″ x 14″		2	8		2,600	294		2,894	3,325	
	3550	Maximum sheet 20″ x 24″		1	16		3,050	590		3,640	4,250	
	3800	Square, minimum sheet 20″ x 24″		1	16		2,750	590		3,340	3,900	
	3900	Maximum sheet 50″ x 56″	↓	.80	20	↓	4,275	735		5,010	5,800	
	4500	Combination tank sink, tray sink, washers, with										
	4510	dry side tables, average	Q-1	.45	35.556	Ea.	8,100	1,300		9,400	10,900	

11500 | Industrial & Process Equipment

11520 | Industrial Equipment

			CREW	DAILY OUTPUT	LABOR-HOURS	UNIT	2005 BARE COSTS MAT.	LABOR	EQUIP.	TOTAL	TOTAL INCL O&P	
250	0010	**DUST COLLECTION SYSTEMS** Commercial / industrial										250
	0120	Central vacuum units										
	0130	Includes stand, filters and motorized shaker										
	0200	500 CFM, 10″ inlet, 2 HP	Q-20	2.40	8.333	Ea.	3,475	310		3,785	4,300	
	0220	1000 CFM, 10″ inlet, 3 HP		2.20	9.091		3,625	335		3,960	4,525	
	0240	1500 CFM, 10″ inlet, 5 HP		2	10		3,700	370		4,070	4,650	
	0260	3000 CFM, 13″ inlet, 10 HP		1.50	13.333		9,700	495		10,195	11,500	
	0280	5000 CFM, 16″ inlet, 2 @ 10 HP	↓	1	20	↓	10,200	740		10,940	12,400	
	1000	Vacuum tubing, galvanized										
	1100	2-1/8″ OD, 16 ga.	Q-9	440	.036	L.F.	1.84	1.32		3.16	4.04	
	1110	2-1/2″ OD, 16 ga.		420	.038		2.13	1.38		3.51	4.46	
	1120	3″ OD, 16 ga.		400	.040		2.71	1.45		4.16	5.20	
	1130	3-1/2″ OD, 16 ga.		380	.042		3.84	1.52		5.36	6.55	
	1140	4″ OD, 16 ga.		360	.044		3.93	1.61		5.54	6.80	
	1150	5″ OD, 14 ga.		320	.050		7.25	1.81		9.06	10.80	
	1160	6″ OD, 14 ga.		280	.057		8.15	2.07		10.22	12.15	
	1170	8″ OD, 14 ga.		200	.080		12.35	2.89		15.24	18	
	1180	10″ OD, 12 ga.		160	.100		12.35	3.62		15.97	19.10	
	1190	12″ OD, 12 ga.		120	.133		35.50	4.82		40.32	46.50	
	1200	14″ OD, 12 ga.	↓	80	.200	↓	39	7.25		46.25	54	
	1940	Hose, flexible wire reinforced rubber										
	1956	3″ dia.	Q-9	400	.040	L.F.	5.85	1.45		7.30	8.70	
	1960	4″ dia.		360	.044		7.90	1.61		9.51	11.15	
	1970	5″ dia.		320	.050		9.10	1.81		10.91	12.80	
	1980	6″ dia.	↓	280	.057	↓	10.50	2.07		12.57	14.75	
	2000	90° Elbow, slip fit										
	2110	2-1/8″ dia.	Q-9	70	.229	Ea.	7.70	8.25		15.95	21	
	2120	2-1/2″ dia.		65	.246		10.75	8.90		19.65	25.50	
	2130	3″ dia.		60	.267		14.55	9.65		24.20	31	
	2140	3-1/2″ dia.		55	.291		18	10.55		28.55	36	
	2150	4″ dia.		50	.320		22.50	11.60		34.10	42.50	
	2160	5″ dia.		45	.356		42.50	12.85		55.35	66.50	
	2170	6″ dia.		40	.400		58.50	14.45		72.95	87	
	2180	8″ dia.	↓	30	.533	↓	111	19.30		130.30	152	
	2400	45° Elbow, slip fit										
	2410	2-1/8″ dia.	Q-9	70	.229	Ea.	6.75	8.25		15	20	

Important: See the Reference Section for critical supporting data - Reference Nos., Crews, & City Cost Indexes

	11520 | Industrial Equipment		CREW	DAILY OUTPUT	LABOR-HOURS	UNIT	MAT.	LABOR	EQUIP.	TOTAL	TOTAL INCL O&P	
250	2420	2-1/2" dia.	Q-9	65	.246	Ea.	9.90	8.90		18.80	24.50	250
	2430	3" dia.		60	.267		11.95	9.65		21.60	28	
	2440	3-1/2" dia.		55	.291		15.05	10.55		25.60	33	
	2450	4" dia.		50	.320		19.50	11.60		31.10	39.50	
	2460	5" dia.		45	.356		33	12.85		45.85	56	
	2470	6" dia.		40	.400		44.50	14.45		58.95	71.50	
	2480	8" dia.		35	.457		87.50	16.55		104.05	122	
	2800	90° TY, slip fit thru 6" dia										
	2810	2-1/8" dia.	Q-9	42	.381	Ea.	15.05	13.80		28.85	37.50	
	2820	2-1/2" dia.		39	.410		19.55	14.85		34.40	44.50	
	2830	3" dia.		36	.444		26	16.10		42.10	53	
	2840	3-1/2" dia.		33	.485		33	17.55		50.55	63.50	
	2850	4" dia.		30	.533		47.50	19.30		66.80	81.50	
	2860	5" dia.		27	.593		90	21.50		111.50	132	
	2870	6" dia.		24	.667		123	24		147	172	
	2880	8" dia., butt end					260			260	286	
	2890	10" dia., butt end					470			470	515	
	2900	12" dia., butt end					470			470	515	
	2910	14" dia., butt end					855			855	940	
	2920	6" x 4" dia., butt end					74.50			74.50	82	
	2930	8" x 4" dia., butt end					141			141	156	
	2940	10" x 4" dia., butt end					185			185	204	
	2950	12" x 4" dia., butt end					213			213	234	
	3100	90° Elbow, butt end segmented										
	3110	8" dia., butt end, segmented				Ea.	168			168	185	
	3120	10" dia., butt end, segmented					276			276	305	
	3130	12" dia., butt end, segmented					350			350	385	
	3140	14" dia., butt end, segmented					440			440	480	
	3200	45° Elbow, butt end segmented										
	3210	8" dia., butt end, segmented				Ea.	143			143	157	
	3220	10" dia., butt end, segmented					199			199	219	
	3230	12" dia., butt end, segmented					204			204	225	
	3240	14" dia., butt end, segmented					251			251	276	
	3400	All butt end fittings require one coupling per joint.										
	3410	Labor for fitting included with couplings.										
	3460	Compression coupling, galvanized, neoprene gasket										
	3470	2-1/8" dia.	Q-9	44	.364	Ea.	11.70	13.15		24.85	33	
	3480	2-1/2" dia.		44	.364		11.70	13.15		24.85	33	
	3490	3" dia.		38	.421		15.35	15.25		30.60	40.50	
	3500	3-1/2" dia.		35	.457		17.20	16.55		33.75	44.50	
	3510	4" dia.		33	.485		18.70	17.55		36.25	47.50	
	3520	5" dia.		29	.552		21.50	19.95		41.45	54	
	3530	6" dia.		26	.615		25	22.50		47.50	62	
	3540	8" dia.		22	.727		45.50	26.50		72	90.50	
	3550	10" dia.		20	.800		66.50	29		95.50	118	
	3560	12" dia.		18	.889		82.50	32		114.50	141	
	3570	14" dia.		16	1		117	36		153	185	
	3800	Air gate valves, galvanized										
	3810	2-1/8" dia.	Q-9	30	.533	Ea.	89	19.30		108.30	128	
	3820	2-1/2" dia.		28	.571		93.50	20.50		114	135	
	3830	3" dia.		26	.615		109	22.50		131.50	154	
	3840	4" dia.		23	.696		142	25		167	195	
	3850	6" dia.		18	.889		196	32		228	266	

(2005 BARE COSTS: MAT., LABOR, EQUIP., TOTAL)

11600 | Laboratory Equipment

		11620 | Laboratory Equipment	CREW	DAILY OUTPUT	LABOR-HOURS	UNIT	2005 BARE COSTS MAT.	LABOR	EQUIP.	TOTAL	TOTAL INCL O&P	
110	0010	**LABORATORY EQUIPMENT**										110
	0600	Fume hood, with countertop & base, not including HVAC										
	0610	Simple, minimum	2 Carp	5.40	2.963	L.F.	520	101		621	730	
	0620	Complex, including fixtures		2.40	6.667		1,225	228		1,453	1,700	
	0630	Special, maximum		1.70	9.412		1,275	320		1,595	1,900	
	0650	Ductwork, minimum	2 Shee	1	16	Hood	860	645		1,505	1,950	
	0660	Maximum	″	.50	32	″	4,350	1,275		5,625	6,775	
	0670	Service fixtures, average				Ea.	44			44	48.50	
	0680	For sink assembly with hot and cold water, add	1 Plum	1.40	5.714		590	233		823	1,000	
	0700	Glassware washer, undercounter, minimum	L-1	1.80	8.889		6,200	365		6,565	7,375	
	0710	Maximum	″	1	16		8,175	655		8,830	9,975	
	1400	Safety equipment, eye wash, hand held					294			294	325	
	1450	Deluge shower					565			565	625	
	1600	Sink, one piece plastic, flask wash, hose, free standing	1 Plum	1.60	5		1,500	204		1,704	1,950	
	1610	Epoxy resin sink, 25″ x 16″ x 10″		2	4		166	163		329	430	
	1850	Utensil washer-sanitizer		2	4		8,975	163		9,138	10,100	
	8000	Alternate pricing method: as percent of lab furniture										
	8050	Installation, not incl. plumbing & duct work				% Furn.					22%	
	8100	Plumbing, final connections, simple system									10%	
	8110	Moderately complex system									15%	
	8120	Complex system									20%	
	8150	Electrical, simple system									10%	
	8160	Moderately complex system									20%	
	8170	Complex system									35%	

11700 | Medical Equipment

		11710 | Medical Sterilizing Equipment	CREW	DAILY OUTPUT	LABOR-HOURS	UNIT	2005 BARE COSTS MAT.	LABOR	EQUIP.	TOTAL	TOTAL INCL O&P	
110	0010	**MEDICAL STERILIZING EQUIPMENT**										110
	0700	Distiller, water, steam heated, 50 gal. capacity	1 Plum	1.40	5.714	Ea.	15,200	233		15,433	17,100	
	5600	Sterilizers, floor loading, 26″ x 62″ x 42″, single door, steam					130,500			130,500	143,500	
	5650	Double door, steam					167,500			167,500	184,500	
	5800	General purpose, 20″ x 20″ x 38″, single door					12,600			12,600	13,800	
	6000	Portable, counter top, steam, minimum					3,125			3,125	3,450	
	6020	Maximum					4,900			4,900	5,375	
	6050	Portable, counter top, gas, 17″x15″x32-1/2″					32,300			32,300	35,500	
	6150	Manual washer/sterilizer, 16″x16″x26″	1 Plum	2	4		44,300	163		44,463	48,900	
	6200	Steam generators, electric 10 KW to 180 KW, freestanding										
	6250	Minimum	1 Elec	3	2.667	Ea.	6,575	109		6,684	7,400	
	6300	Maximum	″	.70	11.429		23,700	465		24,165	26,800	
	8200	Bed pan washer-sanitizer	1 Plum	2	4		5,875	163		6,038	6,700	

		11730 | Patient Care Equipment	CREW	DAILY OUTPUT	LABOR-HOURS	UNIT	MAT.	LABOR	EQUIP.	TOTAL	TOTAL INCL O&P	
110	0010	**PATIENT CARE EQUIPMENT**										110
	1800	Heat therapy unit, humidified, 26″ x 78″ x 28″				Ea.	2,800			2,800	3,075	
	8400	Whirlpool bath, mobile, sst, 18″ x 24″ x 60″					3,775			3,775	4,150	
	8450	Fixed, incl. mixing valves	1 Plum	2	4		7,575	163		7,738	8,575	

11740 | Dental Equipment

			CREW	DAILY OUTPUT	LABOR-HOURS	UNIT	2005 BARE COSTS MAT.	LABOR	EQUIP.	TOTAL	TOTAL INCL O&P	
200	0010	DENTAL EQUIPMENT										200
	0020	Central suction system, minimum	1 Plum	1.20	6.667	Ea.	2,900	272		3,172	3,575	
	0100	Maximum	"	.90	8.889		22,300	365		22,665	25,000	
	0600	Chair, electric or hydraulic, minimum	1 Skwk	.50	16		3,625	560		4,185	4,850	
	0700	Maximum		.25	32		7,675	1,125		8,800	10,200	
	2000	Light, ceiling mounted, minimum		8	1		1,875	35		1,910	2,125	
	2100	Maximum		8	1		3,800	35		3,835	4,225	
	2200	Unit light, minimum	2 Skwk	5.33	3.002		1,400	105		1,505	1,725	
	2210	Maximum		5.33	3.002		4,600	105		4,705	5,250	
	2220	Track light, minimum		3.20	5		2,500	174		2,674	3,025	
	2230	Maximum		3.20	5		5,475	174		5,649	6,300	
	2300	Sterilizers, steam portable, minimum					3,175			3,175	3,500	
	2350	Maximum					11,000			11,000	12,100	
	2600	Steam, institutional					13,700			13,700	15,000	
	2650	Dry heat, electric, portable, 3 trays					1,075			1,075	1,175	
		11781 | Mortuary Equipment										
110	0010	MORTUARY EQUIPMENT										110
	0015	Autopsy table, standard	1 Plum	1	8	Ea.	7,275	325		7,600	8,525	
	0020	Deluxe	"	.60	13.333		13,400	545		13,945	15,600	
	3200	Mortuary refrigerator, end operated, 2 capacity					11,500			11,500	12,700	
	3300	6 capacity					21,400			21,400	23,500	

Division Notes

	CREW	DAILY OUTPUT	LABOR-HOURS	UNIT	2005 BARE COSTS MAT.	LABOR	EQUIP.	TOTAL	TOTAL INCL O&P

Division 13 Special Construction

Estimating Tips

General

- The items and systems in this division are usually estimated, purchased, supplied and installed as a unit by one or more subcontractors. The estimator must ensure that all parties are operating from the same set of specifications and assumptions and that all necessary items are estimated and will be provided. Many times the complex items and systems are covered but the more common ones such as excavation or a crane are overlooked for the very reason that everyone assumes nobody could miss them. The estimator should be the central focus and be able to ensure that all systems are complete.
- Another area where problems can develop in this division is at the interface between systems. The estimator must ensure, for instance, that anchor bolts, nuts and washers are estimated and included for the air-supported structures and pre-engineered buildings to be bolted to their foundations. Utility supply is a common area where essential items or pieces of equipment can be missed or overlooked due to the fact that each subcontractor may feel it is the others' responsibility. The estimator should also be aware of certain items which may be supplied as part of a package but installed by others, and ensure that the installing contractor's estimate includes the cost of installation. Conversely, the estimator must also ensure that items are not costed by two different subcontractors, resulting in an inflated overall estimate.

13120 Pre-Engineered Structures

- The foundations and floor slab, as well as rough mechanical and electrical, should be estimated, as this work is required for the assembly and erection of the structure. Generally, as noted in the book, the pre-engineered building comes as a shell and additional features, such as windows and doors, must be included by the estimator. Here again, the estimator must have a clear understanding of the scope of each portion of the work and all the necessary interfaces.

13200 Storage Tanks

- The prices in this subdivision for above and below ground storage tanks do not include foundations or hold-down slabs. The estimator should refer to Divisions 2 and 3 for foundation system pricing. In addition to the foundations, required tank accessories such as tank gauges, leak detection devices, and additional manholes and piping must be added to the tank prices.

Reference Numbers

Reference numbers are shown in bold squares at the beginning of some major classifications. These numbers refer to related items in the Reference Section. The reference information may be an estimating procedure, an alternate pricing method or technical information.

Note: Not all subdivisions listed here necessarily appear in this publication.

		13035 | Special Purpose Rooms	CREW	DAILY OUTPUT	LABOR-HOURS	UNIT	2005 BARE COSTS MAT.	LABOR	EQUIP.	TOTAL	TOTAL INCL O&P	
200	0010	CLEAN ROOMS										200
	1100	Clean room, soft wall, 12′ x 12′, Class 100	1 Carp	.18	44.444	Ea.	12,600	1,525		14,125	16,300	
	1110	Class 1,000		.18	44.444		9,775	1,525		11,300	13,200	
	1120	Class 10,000		.21	38.095		8,325	1,300		9,625	11,200	
	1130	Class 100,000		.21	38.095		7,725	1,300		9,025	10,500	
	2800	Ceiling grid support, slotted channel struts 4′-0″ O.C., ea. way				S.F.					6.50	
	3000	Ceiling panel, vinyl coated foil on mineral substrate										
	3020	Sealed, non-perforated				S.F.					1.40	
	4000	Ceiling panel seal, silicone sealant, 150 L.F./gal.	1 Carp	150	.053	L.F.	.22	1.83		2.05	3.09	
	4100	Two sided adhesive tape	"	240	.033	"	.10	1.14		1.24	1.89	
	4200	Clips, one per panel				Ea.	.88			.88	.97	
	6000	HEPA filter,2′x4′,99.97% eff.,3″ dp beveled frame (silicone seal)					269			269	296	
	6040	6″ deep skirted frame (channel seal)					310			310	340	
	6100	99.99% efficient, 3″ deep beveled frame (silicone seal)					290			290	320	
	6140	6″ deep skirted frame (channel seal)					330			330	365	
	6200	99.999% efficient, 3″ deep beveled frame (silicone seal)					330			330	365	
	6240	6″ deep skirted frame (channel seal)					375			375	410	
	7000	Wall panel systems, including channel strut framing										
	7020	Polyester coated aluminum, particle board				S.F.					20	
	7100	Porcelain coated aluminum, particle board									35	
	7400	Wall panel support, slotted channel struts, to 12′ high									18	
700	0010	REFRIGERATION Curbs, 12″ high, 4″ thick, concrete	2 Carp	58	.276	L.F.	3.20	9.45		12.65	18.20	700
	1000	Doors, see division 08341-200										
	2400	Finishes, 2 coat portland cement plaster, 1/2″ thick	1 Plas	48	.167	S.F.	.88	5.25		6.13	8.90	
	2500	For galvanized reinforcing mesh, add	1 Lath	335	.024		.59	.75		1.34	1.76	
	2700	3/16″ thick latex cement	1 Plas	88	.091		1.51	2.85		4.36	6	
	2900	For glass cloth reinforced ceilings, add	"	450	.018		.36	.56		.92	1.25	
	3100	Fiberglass panels, 1/8″ thick	1 Carp	149.45	.054		1.99	1.83		3.82	5.05	
	3200	Polystyrene, plastic finish ceiling, 1″ thick		274	.029		1.84	1		2.84	3.58	
	3400	2″ thick		274	.029		2.09	1		3.09	3.86	
	3500	4″ thick		219	.037		2.33	1.25		3.58	4.51	
	3800	Floors, concrete, 4″ thick	1 Cefi	93	.086		.87	2.83		3.70	5.10	
	3900	6″ thick	"	85	.094		1.31	3.09		4.40	6	
	4000	Insulation, 1″ to 6″ thick, cork				B.F.	.85			.85	.94	
	4100	Urethane					.84			.84	.92	
	4300	Polystyrene, regular					.57			.57	.63	
	4400	Bead board					.43			.43	.47	
	4600	Installation of above, add per layer	2 Carp	657.60	.024	S.F.	.27	.83		1.10	1.60	
	4700	Wall and ceiling juncture		298.90	.054	L.F.	1.36	1.83		3.19	4.36	
	4900	Partitions, galvanized sandwich panels, 4″ thick, stock		219.20	.073	S.F.	5.75	2.50		8.25	10.25	
	5000	Aluminum or fiberglass		219.20	.073	"	6.30	2.50		8.80	10.80	
	5200	Prefab walk-in, 7′-6″ high, aluminum, incl. door & floors,										
	5210	not incl. partitions or refrigeration, 6′ x 6′ O.D. nominal	2 Carp	54.80	.292	SF Flr.	103	10		113	129	
	5500	10′ x 10′ O.D. nominal		82.20	.195		82.50	6.65		89.15	101	
	5700	12′ x 14′ O.D. nominal		109.60	.146		74	5		79	89.50	
	5800	12′ x 20′ O.D. nominal		109.60	.146		64.50	5		69.50	79	
	6100	For 8′-6″ high, add					5%					
	6300	Rule of thumb for complete units, w/o doors & refrigeration, cooler	2 Carp	146	.110		93.50	3.75		97.25	109	
	6400	Freezer		109.60	.146		110	5		115	129	
	6600	Shelving, plated or galvanized, steel wire type		360	.044	SF Hor.	8.15	1.52		9.67	11.35	
	6700	Slat shelf type		375	.043		10.05	1.46		11.51	13.40	
	6900	For stainless steel shelving, add					300%					
	7000	Vapor barrier, on wood walls	2 Carp	1,644	.010	S.F.	.12	.33		.45	.65	
	7200	On masonry walls	"	1,315	.012	"	.31	.42		.73	.99	
	7500	For air curtain doors, see division 15840-200										

13030 | Special Purpose Rooms

		13035 | Special Purpose Rooms	CREW	DAILY OUTPUT	LABOR-HOURS	UNIT	2005 BARE COSTS MAT.	LABOR	EQUIP.	TOTAL	TOTAL INCL O&P	
800	0010	**SAUNA** Prefabricated, incl. heater & controls, 7′ high, 6′ x 4′, C/C	L-7	2.20	12.727	Ea.	3,525	420		3,945	4,525	800
	0050	6′ x 4′, C/P		2	14		3,275	460		3,735	4,325	
	0400	6′ x 5′, C/C		2	14		3,950	460		4,410	5,075	
	0450	6′ x 5′, C/P		2	14		3,700	460		4,160	4,800	
	0600	6′ x 6′, C/C		1.80	15.556		4,200	515		4,715	5,425	
	0650	6′ x 6′, C/P		1.80	15.556		3,925	515		4,440	5,125	
	0800	6′ x 9′, C/C		1.60	17.500		5,275	580		5,855	6,700	
	0850	6′ x 9′, C/P		1.60	17.500		5,000	580		5,580	6,400	
	1000	8′ x 12′, C/C		1.10	25.455		8,175	840		9,015	10,300	
	1050	8′ x 12′, C/P		1.10	25.455		7,500	840		8,340	9,550	
	1200	8′ x 8′, C/C		1.40	20		6,225	660		6,885	7,875	
	1250	8′ x 8′, C/P		1.40	20		5,850	660		6,510	7,475	
	1400	8′ x 10′, C/C		1.20	23.333		6,925	770		7,695	8,800	
	1450	8′ x 10′, C/P		1.20	23.333		6,425	770		7,195	8,275	
	1600	10′ x 12′, C/C		1	28		8,650	925		9,575	11,000	
	1650	10′ x 12′, C/P	↓	1	28		7,825	925		8,750	10,100	
	2500	Heaters only (incl. above), wall mounted, to 200 C.F.					460			460	505	
	2750	To 300 C.F.					555			555	615	
	3000	Floor standing, to 720 C.F., 10,000 watts, w/controls	1 Elec	3	2.667		1,450	109		1,559	1,725	
	3250	To 1,000 C.F., 16,000 watts	"	3	2.667	↓	1,500	109		1,609	1,800	
940	0010	**STEAM BATH** Heater, timer & head, single, to 140 C.F.	1 Plum	1.20	6.667	Ea.	975	272		1,247	1,475	940
	0500	To 300 C.F.		1.10	7.273		1,100	297		1,397	1,650	
	1000	Commercial size, with blow-down assembly, to 800 C.F.		.90	8.889		3,800	365		4,165	4,725	
	1500	To 2500 C.F.	↓	.80	10		5,975	410		6,385	7,200	
	2000	Multiple, motels, apts., 2 baths, w/ blow-down assm., 500 C.F.	Q-1	1.30	12.308		3,750	455		4,205	4,800	
	2500	4 baths	"	.70	22.857		4,075	840		4,915	5,775	
	2700	Conversion unit for residential tub, including door				↓	3,075			3,075	3,375	

13120 | Pre-Engineered Structures

		13128 | Pre-Engineered Structures	CREW	DAILY OUTPUT	LABOR-HOURS	UNIT	2005 BARE COSTS MAT.	LABOR	EQUIP.	TOTAL	TOTAL INCL O&P	
200	0010	**COMFORT STATIONS** Prefab., stock, w/doors, windows & fixt.										200
	0100	Not incl. interior finish or electrical										
	0300	Mobile, on steel frame, minimum				S.F.	36.50			36.50	40	
	0350	Maximum					57.50			57.50	63.50	
	0400	Permanent, including concrete slab, minimum	B-12J	50	.320		142	10	16.75	168.75	190	
	0500	Maximum	"	43	.372	↓	206	11.65	19.50	237.15	266	
	0600	Alternate pricing method, mobile, minimum				Fixture	1,775			1,775	1,950	
	0650	Maximum					2,650			2,650	2,925	
	0700	Permanent, minimum	B-12J	.70	22.857		10,200	715	1,200	12,115	13,600	
	0750	Maximum	"	.50	32	↓	17,000	1,000	1,675	19,675	22,100	

13150 | Swimming Pools

	13151 | Swimming Pools		CREW	DAILY OUTPUT	LABOR-HOURS	UNIT	2005 BARE COSTS MAT.	LABOR	EQUIP.	TOTAL	TOTAL INCL O&P	
200	0010	**SWIMMING POOLS** Residential in-ground, vinyl lined, concrete sides										200
	0020	Sides including equipment, sand bottom	B-52	300	.187	SF Surf	10.60	5.85	1.31	17.76	22	
	0100	Metal or polystyrene sides	B-14	410	.117	↓	8.85	3.29	.53	12.67	15.45	
	0200	Add for vermiculite bottom				↓	.68			.68	.75	
	0500	Gunite bottom and sides, white plaster finish										
	0600	12′ x 30′ pool	B-52	145	.386	SF Surf	17.55	12.10	2.72	32.37	41	
	0720	16′ x 32′ pool	↓	155	.361	↓	15.85	11.35	2.54	29.74	38	
	0750	20′ x 40′ pool	↓	250	.224	↓	14.15	7.05	1.58	22.78	28	
	0810	Concrete bottom and sides, tile finish										
	0820	12′ x 30′ pool	B-52	80	.700	SF Surf	17.75	22	4.93	44.68	59	
	0830	16′ x 32′ pool	↓	95	.589	↓	14.65	18.50	4.15	37.30	49	
	0840	20′ x 40′ pool	↓	130	.431	↓	11.65	13.50	3.03	28.18	37	
	1100	Motel, gunite with plaster finish, incl. medium										
	1150	capacity filtration & chlorination	B-52	115	.487	SF Surf	21.50	15.30	3.43	40.23	51.50	
	1200	Municipal, gunite with plaster finish, incl. high										
	1250	capacity filtration & chlorination	B-52	100	.560	SF Surf	28	17.60	3.94	49.54	62.50	
	1350	Add for formed gutters				L.F.	48.50			48.50	53.50	
	1360	Add for stainless steel gutters				"	144			144	158	
	1700	Filtration and deck equipment only, as % of total				Total				20%	20%	
	1800	Deck equipment, rule of thumb, 20′ x 40′ pool				SF Pool					1.30	
	1900	5000 S.F. pool				"					1.90	
	3000	Painting pools, preparation + 3 coats, 20′ x 40′ pool, epoxy	2 Pord	.33	48.485	Total	605	1,475		2,080	2,900	
	3100	Rubber base paint, 18 gallons	"	.33	48.485	↓	460	1,475		1,935	2,725	
	3500	42′ x 82′ pool, 75 gallons, epoxy paint	3 Pord	.14	171	↓	2,550	5,250		7,800	10,700	
	3600	Rubber base paint	"	.14	171	↓	1,975	5,250		7,225	10,100	
700	0010	**SWIMMING POOL EQUIPMENT** Diving stand, stainless steel, 3 meter	2 Carp	.40	40	Ea.	4,925	1,375		6,300	7,550	700
	0300	1 meter	"	2.70	5.926	"	3,650	203		3,853	4,325	
	0900	Filter system, sand or diatomite type, incl. pump, 6,000 gal./hr.	2 Plum	1.80	8.889	Total	1,125	365		1,490	1,775	
	1020	Add for chlorination system, 800 S.F. pool	↓	3	5.333	Ea.	203	218		421	550	
	1040	5,000 S.F. pool	↓	3	5.333	"	1,150	218		1,368	1,600	
	1100	Gutter system, stainless steel, with grating, stock,										
	1110	contains supply and drainage system	E-1	20	1.200	L.F.	170	44.50	4.06	218.56	267	
	1120	Integral gutter and 5′ high wall system, stainless steel	"	10	2.400	"	254	89	8.10	351.10	440	
	2100	Lights, underwater, 12 volt, with transformer, 300 watt	1 Elec	.40	20	Ea.	145	815		960	1,375	
	2200	110 volt, 500 watt, standard	↓	.40	20	↓	130	815		945	1,375	
	2400	Low water cutoff type	↓	.40	20	↓	135	815		950	1,375	
	2800	Heaters, see division 15510-880										

13175 | Ice Rinks

	13176 | Ice Rinks		CREW	DAILY OUTPUT	LABOR-HOURS	UNIT	2005 BARE COSTS MAT.	LABOR	EQUIP.	TOTAL	TOTAL INCL O&P	
500	0010	**ICE SKATING** Equipment incl. refrigeration, plumbing & cooling										500
	0020	coils & concrete slab, 85′ x 200′ rink										
	0300	55° system, 5 mos., 100 ton				Total					545,500	
	0700	90° system, 12 mos., 135 ton				"					605,500	
	1000	Dasher boards, 1/2″ H.D. polyethylene faced steel frame, 3′ acrylic										
	1020	screen at sides, 5′ acrylic ends, 85′ x 200′	F-5	.06	533	Ea.	115,000	18,500		133,500	155,500	
	1100	Fiberglass & aluminum construction, same sides and ends	"	.06	533	↓	160,000	18,500		178,500	205,000	
	1200	Subsoil heating system (recycled from compressor), 85′ x 200′	Q-7	.27	118	↓	25,000	4,625		29,625	34,500	

13175 | Ice Rinks

		13176 | Ice Rinks	CREW	DAILY OUTPUT	LABOR-HOURS	UNIT	2005 BARE COSTS MAT.	LABOR	EQUIP.	TOTAL	TOTAL INCL O&P	
500	1300	Subsoil insulation, 2 lb. polystyrene with vapor barrier, 85′ x 200′	2 Carp	.14	114	Ea.	30,000	3,925		33,925	39,100	500

13200 | Storage Tanks

		13201 | Storage Tanks	CREW	DAILY OUTPUT	LABOR-HOURS	UNIT	2005 BARE COSTS MAT.	LABOR	EQUIP.	TOTAL	TOTAL INCL O&P	
200	0010	**ELEVATED STORAGE TANKS**, not incl pipe, pumps or foundation										200
	3000	Elevated water tanks, 100′ to bottom capacity line, incl painting										
	3010	50,000 gallons				Ea.					193,764	
	3300	100,000 gallons									267,300	
	3400	250,000 gallons									368,851	
	3600	500,000 gallons									595,298	
	3700	750,000 gallons									820,577	
	3900	1,000,000 gallons				↓					824,180	
300	0010	**GROUND TANKS** Not incl. pipe or pumps, prestress conc., 250,000 gal.				Ea.					284,200	300
	0100	500,000 gallons									385,700	
	0300	1,000,000 gallons									548,100	
	0400	2,000,000 gallons									828,240	
	0600	4,000,000 gallons									1,309,350	
	0700	6,000,000 gallons									1,786,400	
	0750	8,000,000 gallons									2,273,600	
	0800	10,000,000 gallons									2,740,500	
	0900	Steel, ground level, ht/dia less than 1, not incl. fdn, 100,000 gallons									141,500	
	1000	250,000 gallons									159,910	
	1200	500,000 gallons									243,955	
	1250	750,000 gallons									290,645	
	1300	1,000,000 gallons									444,360	
	1500	2,000,000 gallons									717,858	
	1600	4,000,000 gallons									1,114,723	
	1800	6,000,000 gallons									1,546,606	
	1850	8,000,000 gallons									2,025,179	
	1900	10,000,000 gallons				↓					2,784,500	
	2100	Steel standpipes, hgt/dia more than 1, 100′ to overflow, no fdn										
	2200	500,000 gallons				Ea.					320,993	
	2400	750,000 gallons									391,029	
	2500	1,000,000 gallons									476,238	
	2700	1,500,000 gallons									665,333	
	2800	2,000,000 gallons				↓					817,075	
	3000	Steel, storage, above ground, including cradles, coating,										
	3020	fittings, not including fdn, pumps or piping										
	3040	Single wall, interior, 275 gallon	Q-5	5	3.200	Ea.	305	118		423	515	
	3060	550 gallon	"	2.70	5.926		1,400	218		1,618	1,850	
	3080	1,000 gallon	Q-7	5	6.400		2,200	250		2,450	2,800	
	3100	1,500 gallon		4.75	6.737		3,325	263		3,588	4,050	
	3120	2,000 gallon		4.60	6.957		3,800	272		4,072	4,600	
	3140	5,000 gallon		3.20	10		5,175	390		5,565	6,275	
	3150	10,000 gallon		2	16		15,500	625		16,125	17,900	
	3160	15,000 gallon		1.70	18.824		16,400	735		17,135	19,100	
	3170	20,000 gallon		1.45	22.069		20,400	860		21,260	23,800	
	3180	25,000 gallon capacity	↓	1.30	24.615	↓	23,500	960		24,460	27,300	

13201 | Storage Tanks

		13201 Storage Tanks	CREW	DAILY OUTPUT	LABOR-HOURS	UNIT	2005 BARE COSTS MAT.	LABOR	EQUIP.	TOTAL	TOTAL INCL O&P	
300	3190	30,000 gallon	Q-7	1.10	29.091	Ea.	28,100	1,125		29,225	32,600	300
	3320	Double wall, 500 gallon capacity	Q-5	2.40	6.667		2,425	246		2,671	3,050	
	3330	2000 gallon capacity	Q-7	4.15	7.711		5,575	300		5,875	6,575	
	3340	4000 gallon capacity		3.60	8.889		9,925	345		10,270	11,400	
	3350	6000 gallon capacity		2.40	13.333		11,700	520		12,220	13,700	
	3360	8000 gallon capacity		2	16		15,000	625		15,625	17,400	
	3370	10000 gallon capacity		1.80	17.778		16,600	695		17,295	19,300	
	3380	15000 gallon capacity		1.50	21.333		25,200	835		26,035	29,000	
	3390	20000 gallon capacity		1.30	24.615		28,600	960		29,560	33,000	
	3400	25000 gallon capacity		1.15	27.826		34,800	1,075		35,875	39,900	
	3410	30000 gallon capacity		1	32		38,200	1,250		39,450	43,900	
	4000	Fixed roof oil storage tanks, steel, (1 BBL=42 GAL w/ fdn 3'D x 1'W)										
	4200	5,000 barrels				Ea.					165,600	
	4300	25,000 barrels									232,300	
	4500	55,000 barrels									393,300	
	4600	100,000 barrels									632,500	
	4800	150,000 barrels									891,250	
	4900	225,000 barrels									1,343,200	
	5100	Floating roof gasoline tanks, steel, 5,000 barrels									140,300	
	5200	25,000 barrels									313,950	
	5400	55,000 barrels									474,950	
	5500	100,000 barrels									718,750	
	5700	150,000 barrels									972,900	
	5800	225,000 barrels									1,348,950	
	6000	Wood tanks, ground level, 2" cypress, 3,000 gallons	C-1	.19	168		6,800	5,450		12,250	16,000	
	6100	2-1/2" cypress, 10,000 gallons		.12	266		18,000	8,625		26,625	33,200	
	6300	3" redwood or 3" fir, 20,000 gallons		.10	320		27,500	10,400		37,900	46,400	
	6400	30,000 gallons		.08	400		34,000	12,900		46,900	57,500	
	6600	45,000 gallons		.07	457		51,500	14,800		66,300	79,500	
	6700	Larger sizes, minimum				Gal.					.66	
	6900	Maximum				"					.81	
	7000	Vinyl coated fabric pillow tanks, freestanding, 5,000 gallons	4 Clab	4	8	Ea.	3,625	214		3,839	4,325	
	7100	Supporting embankment not included, 25,000 gallons	6 Clab	2	24		6,900	640		7,540	8,575	
	7200	50,000 gallons	8 Clab	1.50	42.667		13,600	1,150		14,750	16,700	
	7300	100,000 gallons	9 Clab	.90	80		28,200	2,125		30,325	34,300	
	7400	150,000 gallons		.50	144		34,500	3,850		38,350	44,000	
	7500	200,000 gallons		.40	180		40,400	4,800		45,200	52,000	
	7600	250,000 gallons		.30	240		45,500	6,400		51,900	60,000	
800	0010	**UNDERGROUND STORAGE TANKS**										800
	0210	Fiberglass, underground, single wall, U.L. listed, not including										
	0220	manway or hold-down strap										
	0225	550 gallon capacity	Q-5	2.67	5.993	Ea.	1,500	221		1,721	1,975	
	0230	1,000 gallon capacity	"	2.46	6.504		1,875	240		2,115	2,425	
	0240	2,000 gallon capacity	Q-7	4.57	7.002		2,525	273		2,798	3,175	
	0250	4,000 gallon capacity		3.55	9.014		3,650	350		4,000	4,550	
	0260	6,000 gallon capacity		2.67	11.985		4,075	470		4,545	5,175	
	0270	8,000 gallon capacity		2.29	13.974		4,950	545		5,495	6,275	
	0280	10,000 gallon capacity		2	16		5,550	625		6,175	7,050	
	0282	12,000 gallon capacity		1.88	17.021		7,150	665		7,815	8,875	
	0284	15,000 gallon capacity		1.68	19.048		8,075	745		8,820	10,000	
	0290	20,000 gallon capacity		1.45	22.069		11,200	860		12,060	13,600	
	0300	25,000 gallon capacity		1.28	25		17,500	975		18,475	20,800	
	0320	30,000 gallon capacity		1.14	28.070		21,900	1,100		23,000	25,800	
	0340	40,000 gallon capacity		.89	35.955		30,800	1,400		32,200	35,900	
	0360	48,000 gallon capacity		.81	39.506		50,500	1,550		52,050	58,000	
	0500	For manway, fittings and hold-downs, add					20%	15%				

	13201 | Storage Tanks		CREW	DAILY OUTPUT	LABOR-HOURS	UNIT	MAT.	LABOR	EQUIP.	TOTAL	TOTAL INCL O&P	
							2005 BARE COSTS					
800	0600	For manways, add				Ea.	850			850	935	800
	1000	For helical heating coil, add	Q-5	2.50	6.400	↓	2,225	236		2,461	2,800	
	1020	Fiberglass, underground, double wall, U.L. listed										
	1030	includes manways, not incl. hold-down straps										
	1040	600 gallon capacity	Q-5	2.42	6.612	Ea.	3,325	244		3,569	4,025	
	1050	1,000 gallon capacity	"	2.25	7.111		4,650	262		4,912	5,525	
	1060	2,500 gallon capacity	Q-7	4.16	7.692		6,800	300		7,100	7,950	
	1070	3,000 gallon capacity		3.90	8.205		7,325	320		7,645	8,550	
	1080	4,000 gallon capacity		3.64	8.791		8,700	345		9,045	10,100	
	1090	6,000 gallon capacity		2.42	13.223		9,725	515		10,240	11,500	
	1100	8,000 gallon capacity		2.08	15.385		10,800	600		11,400	12,800	
	1110	10,000 gallon capacity		1.82	17.582		12,000	685		12,685	14,200	
	1120	12,000 gallon capacity		1.70	18.824		14,800	735		15,535	17,400	
	1122	15,000 gallon capacity		1.52	21.053		20,500	820		21,320	23,700	
	1124	20,000 gallon capacity		1.33	24.060		24,700	940		25,640	28,600	
	1126	25,000 gallon capacity		1.16	27.586		32,500	1,075		33,575	37,400	
	1128	30,000 gallon capacity	↓	1.03	31.068		39,000	1,225		40,225	44,600	
	1140	For hold-down straps, add				↓	2%	10%				
	1150	For hold-downs 500-4000 gal, add	Q-7	16	2	Set	237	78		315	375	
	1160	For hold-downs 5000-15000 gal, add		8	4		475	156		631	755	
	1170	For hold-downs 20,000 gal, add		5.33	6.004		710	234		944	1,125	
	1180	For hold-downs 25,000 gal, add		4	8		945	310		1,255	1,525	
	1190	For hold-downs 30,000 gal, add	↓	2.60	12.308	↓	1,425	480		1,905	2,275	
	2210	Fiberglass, underground, single wall, U.L. listed, including										
	2220	hold-down straps, no manways										
	2225	550 gallon capacity	Q-5	2	8	Ea.	1,725	295		2,020	2,350	
	2230	1,000 gallon capacity	"	1.88	8.511		2,125	315		2,440	2,800	
	2240	2,000 gallon capacity	Q-7	3.55	9.014		2,750	350		3,100	3,550	
	2250	4,000 gallon capacity		2.90	11.034		3,900	430		4,330	4,925	
	2260	6,000 gallon capacity		2	16		4,550	625		5,175	5,950	
	2270	8,000 gallon capacity		1.78	17.978		5,425	700		6,125	7,025	
	2280	10,000 gallon capacity		1.60	20		6,025	780		6,805	7,800	
	2282	12,000 gallon capacity		1.52	21.053		7,625	820		8,445	9,625	
	2284	15,000 gallon capacity		1.39	23.022		8,550	900		9,450	10,800	
	2290	20,000 gallon capacity		1.14	28.070		11,900	1,100		13,000	14,800	
	2300	25,000 gallon capacity		.96	33.333		18,500	1,300		19,800	22,300	
	2320	30,000 gallon capacity	↓	.80	40	↓	23,400	1,550		24,950	28,100	
	3020	Fiberglass, underground, double wall, U.L. listed										
	3030	includes manways and hold-down straps										
	3040	600 gallon capacity	Q-5	1.86	8.602	Ea.	3,550	315		3,865	4,400	
	3050	1,000 gallon capacity	"	1.70	9.412		4,875	345		5,220	5,900	
	3060	2,500 gallon capacity	Q-7	3.29	9.726		7,050	380		7,430	8,325	
	3070	3,000 gallon capacity		3.13	10.224		7,575	400		7,975	8,925	
	3080	4,000 gallon capacity		2.93	10.921		8,950	425		9,375	10,500	
	3090	6,000 gallon capacity		1.86	17.204		10,200	670		10,870	12,200	
	3100	8,000 gallon capacity		1.65	19.394		11,300	755		12,055	13,500	
	3110	10,000 gallon capacity		1.48	21.622		12,500	845		13,345	15,100	
	3120	12,000 gallon capacity		1.40	22.857		15,300	890		16,190	18,200	
	3122	15,000 gallon capacity		1.28	25		20,900	975		21,875	24,500	
	3124	20,000 gallon capacity		1.06	30.189		25,400	1,175		26,575	29,800	
	3126	25,000 gallon capacity		.90	35.556		33,500	1,400		34,900	38,900	
	3128	30,000 gallon capacity	↓	.74	43.243	↓	40,400	1,700		42,100	46,900	
	5000	Steel underground, sti-P3, set in place, not incl. hold-down bars.										
	5500	Excavation, pad, pumps and piping not included										
	5510	Single wall, 500 gallon capacity, 7 gauge shell	Q-5	2.70	5.926	Ea.	1,000	218		1,218	1,425	
	5520	1,000 gallon capacity, 7 gauge shell	"	2.50	6.400	↓	1,650	236		1,886	2,150	

13200 | Storage Tanks

13201 | Storage Tanks

			CREW	DAILY OUTPUT	LABOR-HOURS	UNIT	2005 BARE COSTS MAT.	LABOR	EQUIP.	TOTAL	TOTAL INCL O&P	
800	5530	2,000 gallon capacity, 1/4" thick shell	Q-7	4.60	6.957	Ea.	3,025	272		3,297	3,725	800
	5535	2,500 gallon capacity, 7 gauge shell	Q-5	3	5.333		3,675	197		3,872	4,350	
	5540	5,000 gallon capacity, 1/4" thick shell	Q-7	3.20	10		6,850	390		7,240	8,100	
	5560	10,000 gallon capacity, 1/4" thick shell		2	16		11,100	625		11,725	13,200	
	5580	15,000 gallon capacity, 5/16" thick shell		1.70	18.824		13,200	735		13,935	15,700	
	5600	20,000 gallon capacity, 5/16" thick shell		1.50	21.333		17,800	835		18,635	20,800	
	5610	25,000 gallon capacity, 3/8" thick shell		1.30	24.615		21,500	960		22,460	25,200	
	5620	30,000 gallon capacity, 3/8" thick shell		1.10	29.091		28,600	1,125		29,725	33,200	
	5630	40,000 gallon capacity, 3/8" thick shell		.90	35.556		37,400	1,400		38,800	43,200	
	5640	50,000 gallon capacity, 3/8" thick shell		.80	40		46,800	1,550		48,350	54,000	
	6200	Steel, underground, 360°, double wall, U.L. listed,										
	6210	with sti-P3 corrosion protection,										
	6220	(dielectric coating, cathodic protection, electrical										
	6230	isolation) 30 year warranty,										
	6240	not incl. manholes or hold-downs.										
	6250	500 gallon capacity	Q-5	2.40	6.667	Ea.	3,025	246		3,271	3,700	
	6260	1,000 gallon capactiy	"	2.25	7.111		4,400	262		4,662	5,250	
	6270	2,000 gallon capacity	Q-7	4.16	7.692		5,800	300		6,100	6,825	
	6280	3,000 gallon capacity		3.90	8.205		6,400	320		6,720	7,525	
	6290	4,000 gallon capacity		3.64	8.791		8,150	345		8,495	9,500	
	6300	5,000 gallon capacity		2.91	10.997		10,700	430		11,130	12,300	
	6310	6,000 gallon capacity		2.42	13.223		12,700	515		13,215	14,700	
	6320	8,000 gallon capacity		2.08	15.385		13,200	600		13,800	15,400	
	6330	10,000 gallon capacity		1.82	17.582		14,900	685		15,585	17,400	
	6340	12,000 gallon capacity		1.70	18.824		18,000	735		18,735	20,800	
	6350	15,000 gallon capacity		1.33	24.060		19,900	940		20,840	23,200	
	6360	20,000 gallon capacity		1.33	24.060		23,900	940		24,840	27,700	
	6370	25,000 gallon capacity		1.16	27.586		40,100	1,075		41,175	45,700	
	6380	30,000 gallon capacity		1.03	31.068		48,300	1,225		49,525	55,000	
	6390	40,000 gallon capacity		.80	40		62,000	1,550		63,550	70,500	
	6395	50,000 gallon capacity		.73	43.836		79,500	1,700		81,200	90,000	
	6400	For hold-downs 500-2000 gal, add		16	2	Set	370	78		448	525	
	6410	For hold-downs 3000-6000 gal, add		12	2.667		460	104		564	660	
	6420	For hold-downs 8000-12,000 gal, add		11	2.909		1,175	114		1,289	1,475	
	6430	For hold-downs 15,000 gal, add		9	3.556		1,475	139		1,614	1,825	
	6440	For hold-downs 20,000 gal, add		8	4		1,650	156		1,806	2,025	
	6450	For hold-downs 20,000 gal plus, add		6	5.333		1,650	208		1,858	2,125	
	6500	For manways, add				Ea.	640			640	705	
	6600	In place with hold-downs										
	6652	550 gallon capacity	Q-5	1.84	8.696	Ea.	3,400	320		3,720	4,200	

13280 | Hazardous Material Remediation

13281 | Hazardous Material Remediation

			CREW	DAILY OUTPUT	LABOR-HOURS	UNIT	2005 BARE COSTS MAT.	LABOR	EQUIP.	TOTAL	TOTAL INCL O&P	
120	0010	BULK ASBESTOS REMOVAL										120
	0020	Includes disposable tools and 2 suits and 1respirator filter/day/worker										
	0200	Boiler insulation	A-9	480	.133	S.F.	.44	4.95		5.39	8.30	
	0210	With metal lath add				%				50%		
	0300	Boiler breeching or flue insulation	A-9	520	.123	S.F.	.35	4.57		4.92	7.60	
	0310	For active boiler, add				%				100%		

13 SPECIAL CONSTRUCTION

Important: See the Reference Section for critical supporting data - Reference Nos., Crews, & City Cost Indexes

13281 | Hazardous Material Remediation

			CREW	DAILY OUTPUT	LABOR-HOURS	UNIT	2005 BARE COSTS MAT.	LABOR	EQUIP.	TOTAL	TOTAL INCL O&P
120	0400	Duct or AHU insulation	A-10B	440	.073	S.F.	.21	2.71		2.92	4.50
	0500	Duct vibration isolation joints, up to 24 Sq. In. duct	A-9	56	1.143	Ea.	3.24	42.50		45.74	70.50
	0520	25 Sq. In. to 48 Sq. In. duct		48	1.333		3.78	49.50		53.28	82
	0530	49 Sq. In. to 76 Sq. In. duct		40	1.600		4.54	59.50		64.04	99
	0600	Pipe insulation, air cell type, up to 4″ diameter pipe		900	.071	L.F.	.20	2.64		2.84	4.39
	0610	4″ to 8″ diameter pipe		800	.080		.23	2.97		3.20	4.94
	0620	10″ to 12″ diameter pipe		700	.091		.26	3.40		3.66	5.65
	0630	14″ to 16″ diameter pipe		550	.116		.33	4.32		4.65	7.15
	0650	Over 16″ diameter pipe		650	.098	S.F.	.28	3.66		3.94	6.05
	0700	With glove bag up to 3″ diameter pipe		300	.213	L.F.	3.15	7.95		11.10	15.95
	1000	Pipe fitting insulation up to 4″ diameter pipe		320	.200	Ea.	.57	7.45		8.02	12.35
	1100	6″ to 8″ diameter pipe		304	.211		.60	7.80		8.40	13
	1110	10″ to 12″ diameter pipe		192	.333		.95	12.40		13.35	20.50
	1120	14″ to 16″ diameter pipe		128	.500		1.42	18.60		20.02	31
	1130	Over 16″ diameter pipe		176	.364	S.F.	1.03	13.50		14.53	22.50
	1200	With glove bag, up to 8″ diameter pipe		100	.640	L.F.	6.55	24		30.55	44.50
	2000	Scrape foam fireproofing from flat surface		2,400	.027	S.F.	.08	.99		1.07	1.64
	2100	Irregular surfaces		1,200	.053		.15	1.98		2.13	3.30
	3000	Remove cementitious material from flat surface		1,800	.036		.10	1.32		1.42	2.20
	3100	Irregular surface		1,400	.046		.13	1.70		1.83	2.82
	6000	Remove contaminated soil from crawl space by hand		400	.160	C.F.	.45	5.95		6.40	9.90
	6100	With large production vacuum loader	A-12	700	.091	″	.26	3.40	.69	4.35	6.40
	7000	Radiator backing, not including radiator removal	A-9	1,200	.053	S.F.	.15	1.98		2.13	3.30
	9000	For type C (supplied air) respirator equipment, add				%					10%
125	0010	**ASBESTOS ABATEMENT WORK AREA** Containment and preparation.									
	0100	Pre-cleaning, HEPA vacuum and wet wipe, flat surfaces	A-10	12,000	.005	S.F.	.02	.20		.22	.33
	0200	Protect carpeted area, 2 layers 6 mil poly on 3/4″ plywood	″	1,000	.064		1.50	2.38		3.88	5.40
	0300	Separation barrier, 2″ x 4″ @ 16″, 1/2″ plywood ea. side, 8′ high	2 Carp	400	.040		1.25	1.37		2.62	3.51
	0310	12′ high R13281 -120		320	.050		1.40	1.71		3.11	4.21
	0320	16′ high		200	.080		1.50	2.74		4.24	5.90
	0400	Personnel decontam. chamber, 2″ x 4″ @ 16″, 3/4″ ply ea. side		280	.057		2.50	1.96		4.46	5.80
	0450	Waste decontam. chamber, 2″ x 4″ studs @ 16″, 3/4″ ply ea. side		360	.044		3	1.52		4.52	5.65
	0500	Cover surfaces with polyethelene sheeting									
	0501	Including glue and tape									
	0550	Floors, each layer, 6 mil	A-10	8,000	.008	S.F.	2.85	.30		3.15	3.61
	0551	4 mil		9,000	.007		1.94	.26		2.20	2.55
	0560	Walls, each layer, 6 mil		6,000	.011		2.84	.40		3.24	3.75
	0561	4 mil		7,000	.009		1.95	.34		2.29	2.69
	0570	For heights above 12′, add						20%			
	0575	For heights above 20′, add						30%			
	0580	For fire retardant poly, add					100%				
	0590	For large open areas, deduct					10%	20%			
	0600	Seal floor penetrations with foam firestop to 36 Sq. In.	2 Carp	200	.080	Ea.	6.25	2.74		8.99	11.15
	0610	36 Sq. In. to 72 Sq. In.		125	.128		12.50	4.38		16.88	20.50
	0615	72 Sq. In. to 144 Sq. In.		80	.200		25	6.85		31.85	38
	0620	Wall penetrations, to 36 square inches		180	.089		6.25	3.04		9.29	11.65
	0630	36 Sq. In. to 72 Sq. In.		100	.160		12.50	5.50		18	22.50
	0640	72 Sq. In. to 144 Sq. In.		60	.267		25	9.15		34.15	42
	0800	Caulk seams with latex	1 Carp	230	.035	L.F.	.15	1.19		1.34	2.03
	0900	Set up neg. air machine, 1-2k C.F.M. /25 M.C.F. volume	1 Asbe	4.30	1.860	Ea.		69		69	109
130	0010	**DEMOLITION IN ASBESTOS CONTAMINATED AREA**									
	0200	Ceiling, including suspension system, plaster and lath	A-9	2,100	.030	S.F.	.09	1.13		1.22	1.89
	0210	Finished plaster, leaving wire lath		585	.109		.31	4.07		4.38	6.75
	0220	Suspended acoustical tile		3,500	.018		.05	.68		.73	1.13
	0230	Concealed tile grid system		3,000	.021		.06	.79		.85	1.32
	0240	Metal pan grid system		1,500	.043		.12	1.59		1.71	2.63

	13281 | Hazardous Material Remediation	CREW	DAILY OUTPUT	LABOR-HOURS	UNIT	2005 BARE COSTS MAT.	LABOR	EQUIP.	TOTAL	TOTAL INCL O&P	
130	0250 Gypsum board	A-9	2,500	.026	S.F.	.07	.95		1.02	1.58	130
	0260 Lighting fixtures up to 2′ x 4′	↓	72	.889	Ea.	2.52	33		35.52	55	
	0400 Partitions, non load bearing										
	0410 Plaster, lath, and studs	A-9	690	.093	S.F.	.81	3.45		4.26	6.35	
	0450 Gypsum board and studs	″	1,390	.046	″	.13	1.71		1.84	2.84	
	9000 For type C (supplied air) respirator equipment, add				%					10%	
135	0010 **ASBESTOS ABATEMENT EQUIPMENT** and supplies, buy R13281 -120										135
	0200 Air filtration device, 2000 C.F.M.				Ea.	2,500			2,500	2,750	
	0250 Large volume air sampling pump, minimum					365			365	400	
	0260 Maximum					680			680	745	
	0300 Airless sprayer unit, 2 gun					2,000			2,000	2,200	
	0350 Light stand, 500 watt				↓	250			250	275	
	0400 Personal respirators										
	0410 Negative pressure, 1/2 face, dual operation, min.				Ea.	22.50			22.50	25	
	0420 Maximum					24			24	26.50	
	0450 P.A.P.R., full face, minimum					400			400	440	
	0460 Maximum					700			700	770	
	0470 Supplied air, full face, incl. air line, minimum					450			450	495	
	0480 Maximum					600			600	660	
	0500 Personnel sampling pump, minimum					450			450	495	
	0510 Maximum					750			750	825	
	1500 Power panel, 20 unit, incl. G.F.I.					1,800			1,800	1,975	
	1600 Shower unit, including pump and filters					1,125			1,125	1,250	
	1700 Supplied air system (type C)					10,000			10,000	11,000	
	1750 Vacuum cleaner, HEPA, 16 gal., stainless steel, wet/dry					1,000			1,000	1,100	
	1760 55 gallon					2,200			2,200	2,425	
	1800 Vacuum loader, 9-18 ton/hr					90,000			90,000	99,000	
	1900 Water atomizer unit, including 55 gal. drum					230			230	253	
	2000 Worker protection, whole body, foot, head cover & gloves, plastic					5.50			5.50	6.05	
	2500 Respirator, single use					10.50			10.50	11.55	
	2550 Cartridge for respirator					11.70			11.70	12.85	
	2570 Glove bag, 7 mil, 50″ x 64″					8.50			8.50	9.35	
	2580 10 mil, 44″ x 60″					8.40			8.40	9.25	
	3000 HEPA vacuum for work area, minimun					1,050			1,050	1,150	
	3050 Maximum					2,775			2,775	3,050	
	6000 Disposable polyethelene bags, 6 mil, 3 C.F.					1.15			1.15	1.27	
	6300 Disposable fiber drums, 3 C.F.					6.50			6.50	7.15	
	6400 Pressure sensitive caution lables, 3″ x 5″					.88			.88	.97	
	6450 11″ x 17″					6.50			6.50	7.15	
	6500 Negative air machine, 1800 C.F.M.				↓	775			775	855	
140	0010 **DECONTAMINATION CONTAINMENT AREA DEMOLITION** and clean-up										140
	0100 Spray exposed substrate with surfactant (bridging)										
	0200 Flat surfaces	A-9	6,000	.011	S.F.	.36	.40		.76	1.03	
	0250 Irregular surfaces		4,000	.016	″	.31	.59		.90	1.28	
	0300 Pipes, beams, and columns		2,000	.032	L.F.	.56	1.19		1.75	2.50	
	1000 Spray encapsulate polyethelene sheeting		8,000	.008	S.F.	.31	.30		.61	.81	
	1100 Roll down polyethelene sheeting		8,000	.008	″		.30		.30	.47	
	1500 Bag polyethelene sheeting		400	.160	Ea.	.77	5.95		6.72	10.25	
	2000 Fine clean exposed substrate, with nylon brush		2,400	.027	S.F.		.99		.99	1.56	
	2500 Wet wipe substrate		4,800	.013			.50		.50	.78	
	2600 Vacuum surfaces, fine brush	↓	6,400	.010	↓		.37		.37	.59	
	3000 Structural demolition										
	3100 Wood stud walls	A-9	2,800	.023	S.F.		.85		.85	1.34	
	3500 Window manifolds, not incl. window replacement		4,200	.015			.57		.57	.89	
	3600 Plywood carpet protection	↓	2,000	.032	↓		1.19		1.19	1.88	
	4000 Remove custom decontamination facility	A-10A	8	3	Ea.	15	112		127	193	

13281 | Hazardous Material Remediation

			CREW	DAILY OUTPUT	LABOR-HOURS	UNIT	2005 BARE COSTS MAT.	LABOR	EQUIP.	TOTAL	TOTAL INCL O&P	
140	4100	Remove portable decontamination facility	3 Asbe	12	2	Ea.	12.50	74		86.50	131	140
	5000	HEPA vacuum, shampoo carpeting	A-9	4,800	.013	S.F.	.05	.50		.55	.84	
	9000	Final cleaning of protected surfaces	A-10A	8,000	.003	″		.11		.11	.18	
145	0010	**OSHA TESTING**										145
	0100	Certified technician, minimum				Day					300	
	0110	Maximum				″					500	
	0200	Personal sampling, PCM analysis, NIOSH 7400, minimum	1 Asbe	8	1	Ea.	2.75	37		39.75	61.50	
	0210	Maximum	″	4	2	″	3	74		77	120	
	0300	Industrial hygenist, minimum				Day					400	
	0310	Maximum				″					550	
	1000	Cleaned area samples	1 Asbe	8	1	Ea.	2.63	37		39.63	61.50	
	1100	PCM air sample analysis, NIOSH 7400, minimum		8	1		30	37		67	91.50	
	1110	Maximum		4	2		3.09	74		77.09	120	
	1200	TEM air sample analysis, NIOSH 7402, minimum								100	125	
	1210	Maximum								400	500	
150	0010	**ENCAPSULATION WITH SEALANTS**										150
	0100	Ceilings and walls, minimum	A-9	21,000	.003	S.F.	.27	.11		.38	.48	
	0110	Maximum		10,600	.006	″	.42	.22		.64	.81	
	0300	Pipes to 12″ diameter including minor repairs, minimum		800	.080	L.F.	.37	2.97		3.34	5.10	
	0310	Maximum		400	.160	″	1.04	5.95		6.99	10.55	
155	0010	**WASTE PACKAGING, HANDLING, & DISPOSAL**										155
	0100	Collect and bag bulk material, 3 C.F. bags, by hand	A-9	400	.160	Ea.	1.15	5.95		7.10	10.65	
	0200	Large production vacuum loader	A-12	880	.073		.80	2.70	.55	4.05	5.75	
	1000	Double bag and decontaminate	A-9	960	.067		2.30	2.48		4.78	6.45	
	2000	Containerize bagged material in drums, per 3 C.F. drum	″	800	.080		6.50	2.97		9.47	11.85	
	3000	Cart bags 50′ to dumpster	2 Asbe	400	.040			1.48		1.48	2.34	
	5000	Disposal charges, not including haul, minimum				C.Y.					50	
	5020	Maximum				″					175	
	5100	Remove refrigerant from system	1 Plum	40	.200	Lb.		8.15		8.15	12.30	
	9000	For type C (supplied air) respirator equipment, add				%					10%	
600	0010	**MOLD ABATEMENT WORK AREA** Containment and preparation R13281-120										600
	0100	Pre-cleaning, HEPA vacuum and wet wipe, flat surfaces	A-10	12,000	.005	S.F.	.02	.20		.22	.33	
	0300	Separation barrier, 2″ x 4″ @ 16″, 1/2″ plywood ea. side, 8′ high	2 Carp	400	.040		1.25	1.37		2.62	3.51	
	0310	12′ high		320	.050		1.40	1.71		3.11	4.21	
	0320	16′ high		200	.080		1.50	2.74		4.24	5.90	
	0400	Personnel decontam. chamber, 2″ x 4″ @ 16″, 3/4″ ply ea. side		280	.057		2.50	1.96		4.46	5.80	
	0450	Waste decontam. chamber, 2″ x 4″ studs @ 16″, 3/4″ ply each side		360	.044		3	1.52		4.52	5.65	
	0500	Cover surfaces with polyethelene sheeting										
	0501	Including glue and tape										
	0550	Floors, each layer, 6 mil	A-10	8,000	.008	S.F.	2.85	.30		3.15	3.61	
	0551	4 mil		9,000	.007		1.94	.26		2.20	2.55	
	0560	Walls, each layer, 6 mil		6,000	.011		2.84	.40		3.24	3.75	
	0561	4 mil		7,000	.009		1.95	.34		2.29	2.69	
	0570	For heights above 12′, add						20%				
	0575	For heights above 20′, add						30%				
	0580	For fire retardant poly, add					100%					
	0590	For large open areas, deduct					10%	20%				
	0600	Seal floor penetrations with foam firestop to 36 sq in	2 Carp	200	.080	Ea.	6.25	2.74		8.99	11.15	
	0610	36 sq in to 72 sq in		125	.128		12.50	4.38		16.88	20.50	
	0615	72 sq in to 144 sq in		80	.200		25	6.85		31.85	38	
	0620	Wall penetrations, to 36 square inches		180	.089		6.25	3.04		9.29	11.65	
	0630	36 Sq. in. to 72 sq. in.		100	.160		12.50	5.50		18	22.50	
	0640	72 Sq. in. to 144 sq. in.		60	.267		25	9.15		34.15	42	
	0800	Caulk seams with latex	1 Carp	230	.035	L.F.	.15	1.19		1.34	2.03	

13280 | Hazardous Material Remediation

	13281 | Hazardous Material Remediation		Crew	Daily Output	Labor-Hours	Unit	2005 Bare Costs Mat.	Labor	Equip.	Total	Total Incl O&P	
600	0900	Set up neg. air machine, 1-2k C.F.M. /25 M.C.F. volume (R13281-120)	1 Asbe	4.30	1.860	Ea.		69		69	109	600
610	0010	DEMOLITION IN MOLD CONTAMINATED AREA										610
	0200	Ceiling, including suspension system, plaster and lath	A-9	2,100	.030	S.F.	.09	1.13		1.22	1.89	
	0210	Finished plaster, leaving wire lath		585	.109		.31	4.07		4.38	6.75	
	0220	Suspended acoustical tile		3,500	.018		.05	.68		.73	1.13	
	0230	Concealed tile grid system		3,000	.021		.06	.79		.85	1.32	
	0240	Metal pan grid system		1,500	.043		.12	1.59		1.71	2.63	
	0250	Gypsum board		2,500	.026		.07	.95		1.02	1.58	
	0255	Plywood		2,500	.026		.07	.95		1.02	1.58	
	0260	Lighting fixtures up to 2′ x 4′		72	.889	Ea.	2.52	33		35.52	55	
	0400	Partitions, non load bearing										
	0410	Plaster, lath, and studs	A-9	690	.093	S.F.	.81	3.45		4.26	6.35	
	0450	Gypsum board and studs		1,390	.046		.13	1.71		1.84	2.84	
	0465	Carpet & pad		1,390	.046		.13	1.71		1.84	2.84	
	0600	Pipe insulation, air cell type, up to 4″ diameter pipe		900	.071	L.F.	.20	2.64		2.84	4.39	
	0610	4″ to 8″ diameter pipe		800	.080		.23	2.97		3.20	4.94	
	0620	10″ to 12″ diameter pipe		700	.091		.26	3.40		3.66	5.65	
	0630	14″ to 16″ diameter pipe		550	.116		.33	4.32		4.65	7.15	
	0650	Over 16″ diameter pipe		650	.098	S.F.	.28	3.66		3.94	6.05	
	9000	For type C (supplied air) respirator equipment, add				%					10%	

13600 | Solar and Wind Energy Equipment

	13630 | Solar Collector Components		Crew	Daily Output	Labor-Hours	Unit	2005 Bare Costs Mat.	Labor	Equip.	Total	Total Incl O&P	
200	0010	SOLAR ENERGY (R13600-610)										200
	0020	System/Package prices, not including connecting										
	0030	pipe, insulation, or special heating/plumbing fixtures										
	0150	For solar ultraviolet pipe insulation see Div. 15086										
	0500	Hot water, standard package, low temperature										
	0540	1 collector, circulator, fittings, 65 gal. tank	Q-1	.50	32	Ea.	1,450	1,175		2,625	3,350	
	0580	2 collectors, circulator, fittings, 120 gal. tank		.40	40		2,050	1,475		3,525	4,450	
	0620	3 collectors, circulator, fittings, 120 gal. tank		.34	47.059		2,475	1,725		4,200	5,325	
	0700	Medium temperature package										
	0720	1 collector, circulator, fittings, 80 gal. tank	Q-1	.50	32	Ea.	1,550	1,175		2,725	3,475	
	0740	2 collectors, circulator, fittings, 120 gal. tank		.40	40		2,200	1,475		3,675	4,625	
	0780	3 collectors, circulator, fittings, 120 gal. tank		.30	53.333		4,000	1,950		5,950	7,350	
	0980	For each additional 120 gal. tank, add					700			700	770	
	2250	Controller, liquid temperature	1 Plum	5	1.600		87.50	65.50		153	194	
	2300	Circulators, air										
	2310	Blowers										
	2330	100-300 S.F. system, 1/10 HP	Q-9	16	1	Ea.	136	36		172	206	
	2340	300-500 S.F. system, 1/5 HP		15	1.067		188	38.50		226.50	266	
	2350	Two speed, 100-300 S.F., 1/10 HP		14	1.143		136	41.50		177.50	214	
	2400	Reversible fan, 20″ diameter, 2 speed		18	.889		108	32		140	169	
	2520	Space & DHW system, less duct work		.50	32		1,475	1,150		2,625	3,400	
	2550	Booster fan 6″ diameter, 120 CFM		16	1		35	36		71	94	
	2570	6″ diameter, 225 CFM		16	1		43	36		79	103	
	2580	8″ diameter, 150 CFM		16	1		39.50	36		75.50	99	

13630 | Solar Collector Components

		Description	Crew	Daily Output	Labor-Hours	Unit	2005 Bare Costs Mat.	Labor	Equip.	Total	Total Incl O&P	
200	2590	8″ diameter, 310 CFM (R13600-610)	Q-9	14	1.143	Ea.	60.50	41.50		102	130	200
	2600	8″ diameter, 425 CFM		14	1.143		68	41.50		109.50	139	
	2650	Rheostat		32	.500		14.80	18.10		32.90	44.50	
	2660	Shutter/damper		12	1.333		39	48		87	117	
	2670	Shutter motor		16	1		92.50	36		128.50	158	
	2800	Circulators, liquid, 1/25 HP, 5.3 GPM	Q-1	14	1.143		147	42		189	225	
	2820	1/20 HP, 17 GPM		12	1.333		154	49		203	244	
	2850	1/20 HP, 17 GPM, stainless steel		12	1.333		243	49		292	340	
	2870	1/12 HP, 30 GPM		10	1.600		330	59		389	455	
	3000	Collector panels, air with aluminum absorber plate										
	3010	Wall or roof mount										
	3040	Flat black, plastic glazing										
	3080	4′ x 8′	Q-9	6	2.667	Ea.	635	96.50		731.50	845	
	3100	4′ x 10′		5	3.200	"	780	116		896	1,050	
	3200	Flush roof mount, 10′ to 16′ x 22″ wide		96	.167	L.F.	52	6.05		58.05	66.50	
	3210	Manifold, by L.F. width of collectors		160	.100	"	27	3.62		30.62	35	
	3300	Collector panels, liquid with copper absorber plate										
	3320	Black chrome, tempered glass glazing										
	3330	Alum. frame, 4′ x 8′, 5/32″ single glazing	Q-1	9.50	1.684	Ea.	545	62		607	695	
	3390	Alum. frame, 4′ x 10′, 5/32″ single glazing		6	2.667		650	98		748	855	
	3450	Flat black, alum. frame, 3′ x 8′		9	1.778		430	65.50		495.50	575	
	3500	4′ x 8.5′		5.50	2.909		490	107		597	700	
	3520	4′ x 10.5′		10	1.600		595	59		654	745	
	3540	4′ x 12.5′		5	3.200		735	118		853	985	
	3550	Liquid with fin tube absorber plate										
	3560	Alum. frame 4′ x 8′ tempered glass	Q-1	10	1.600	Ea.	495	59		554	635	
	3580	Liquid with vacuum tubes, 4′ x 6′-10″		9	1.778		880	65.50		945.50	1,075	
	3600	Liquid, full wetted, plastic, alum. frame, 3′ x 10′		5	3.200		195	118		313	390	
	3650	Collector panel mounting, flat roof or ground rack		7	2.286		45	84		129	176	
	3670	Roof clamps		70	.229	Set	1.61	8.40		10.01	14.40	
	3700	Roof strap, teflon	1 Plum	205	.039	L.F.	9.50	1.59		11.09	12.85	
	3900	Differential controller with two sensors										
	3930	Thermostat, hard wired	1 Plum	8	1	Ea.	87.50	41		128.50	158	
	3950	Line cord and receptacle		12	.667		83.50	27		110.50	133	
	4050	Pool valve system		2.50	3.200		215	131		346	430	
	4070	With 12 VAC actuator		2	4		276	163		439	550	
	4080	Pool pump system, 2″ pipe size		6	1.333		172	54.50		226.50	272	
	4100	Five station with digital read-out		3	2.667		218	109		327	405	
	4150	Sensors										
	4200	Brass plug, 1/2″ MPT	1 Plum	32	.250	Ea.	17.95	10.20		28.15	35	
	4210	Brass plug, reversed		32	.250		24.50	10.20		34.70	42.50	
	4220	Freeze prevention		32	.250		22	10.20		32.20	39.50	
	4240	Screw attached		32	.250		10.95	10.20		21.15	27.50	
	4250	Brass, immersion		32	.250		26.50	10.20		36.70	44.50	
	4300	Heat exchanger										
	4330	Fluid to air coil, up flow, 45 MBH	Q-1	4	4	Ea.	295	147		442	545	
	4380	70 MBH		3.50	4.571		335	168		503	620	
	4400	80 MBH		3	5.333		445	196		641	785	
	4580	Fluid to fluid package includes two circulating pumps										
	4590	expansion tank, check valve, relief valve										
	4600	controller, high temperature cutoff and sensors	Q-1	2.50	6.400	Ea.	695	235		930	1,125	
	4650	Heat transfer fluid										
	4700	Propylene glycol, inhibited anti-freeze	1 Plum	28	.286	Gal.	8.80	11.65		20.45	27.50	
	4800	Solar storage tanks, knocked down										
	4810	Air, galvanized steel clad, double wall, 4″ fiberglass										
	4820	insulation, 20 Mil PVC lining										

13630 | Solar Collector Components

	Line	Description	Crew	Daily Output	Labor-Hours	Unit	Mat. (2005 Bare Costs)	Labor (2005 Bare Costs)	Equip. (2005 Bare Costs)	Total (2005 Bare Costs)	Total Incl O&P	
200	4870	4′ high, 4′ x 4′, = 64 C.F./450 gallons (R13600-610)	Q-9	2	8	Ea.	2,000	289		2,289	2,675	200
	4880	4′ x 8′ = 128 C.F./900 gallons		1.50	10.667		2,600	385		2,985	3,450	
	4890	4′ x 12′ = 190 C.F./1300 gallons		1.30	12.308		3,475	445		3,920	4,500	
	4900	8′ x 8′ = 250 C.F./1700 gallons		1	16		3,875	580		4,455	5,175	
	5010	6′-3″ high, 7′ x 7′ = 306 C.F./2000 gallons	Q-10	1.20	20		9,450	750		10,200	11,600	
	5020	7′ x 10′-6″ = 459 C.F./3000 gallons		.80	30		11,900	1,125		13,025	14,800	
	5030	7′ x 14′ = 613 C.F./4000 gallons		.60	40		13,900	1,500		15,400	17,600	
	5040	10′-6″ x 10′-6″ = 689 C.F./4500 gallons		.50	48		14,400	1,800		16,200	18,600	
	5050	10′-6″ x 14′ = 919 C.F./6000 gallons		.40	60		16,900	2,250		19,150	22,100	
	5060	14′ x 14′ = 1225 C.F./8000 gallons	Q-11	.40	80		19,800	3,075		22,875	26,500	
	5070	14′ x 17′-6″ = 1531 C.F./10,000 gallons		.30	106		22,300	4,075		26,375	30,800	
	5080	17′-6″ x 17′-6″ = 1914 C.F./12,500 gallons		.25	128		26,200	4,900		31,100	36,500	
	5090	17′-6″ x 21′ = 2297 C.F./15,000 gallons		.20	160		29,300	6,125		35,425	41,600	
	5100	21′ x 21′ = 2756 C.F./18,000 gallons		.18	177		33,500	6,800		40,300	47,400	
	5120	30 Mil reinforced Chemflex lining,										
	5140	4′ high, 4′ x 4′ = 64 C.F./450 gallons	Q-9	2	8	Ea.	2,050	289		2,339	2,700	
	5150	4′ x 8′ = 128 C.F./900 gallons		1.50	10.667		3,075	385		3,460	4,000	
	5160	4′ x 12′ = 190 C.F./1300 gallons		1.30	12.308		4,075	445		4,520	5,175	
	5170	8′ x 8′ = 250 C.F./1700 gallons		1	16		4,550	580		5,130	5,925	
	5190	6′-3″ high, 7′ x 7′ = 306 C.F./2000 gallons	Q-10	1.20	20		10,100	750		10,850	12,300	
	5200	7′ x 10′-6″ = 459 C.F./3000 gallons		.80	30		12,300	1,125		13,425	15,200	
	5210	7′ x 14′ = 613 C.F./4000 gallons		.60	40		14,300	1,500		15,800	18,000	
	5220	10′-6″ x 10′-6″ = 689 C.F./4500 gallons		.50	48		14,800	1,800		16,600	19,000	
	5230	10′-6″ x 14′ = 919 C.F./6000 gallons		.40	60		17,300	2,250		19,550	22,600	
	5240	14′ x 14′ = 1225 C.F./8000 gallons	Q-11	.40	80		20,300	3,075		23,375	27,000	
	5250	14′ x 17′-6″ = 1531 C.F./10,000 gallons		.30	106		22,700	4,075		26,775	31,300	
	5260	17′-6″ x 17′-6″ = 1914 C.F./12,500 gallons		.25	128		26,800	4,900		31,700	37,000	
	5270	17′-6″ x 21′ = 2297 C.F./15,000 gallons		.20	160		30,100	6,125		36,225	42,500	
	5280	21′ x 21′ = 2756 C.F./18,000 gallons		.18	177		34,500	6,800		41,300	48,500	
	5290	30 Mil reinforced Hypalon lining, add					.02%					
	7000	Solar control valves and vents										
	7050	Air purger, 1″ pipe size	1 Plum	12	.667	Ea.	43.50	27		70.50	88.50	
	7070	Air eliminator, automatic 3/4″ size		32	.250		28.50	10.20		38.70	47	
	7090	Air vent, automatic, 1/8″ fitting		32	.250		10.20	10.20		20.40	26.50	
	7100	Manual, 1/8″ NPT		32	.250		2.24	10.20		12.44	17.80	
	7120	Backflow preventer, 1/2″ pipe size		16	.500		58.50	20.50		79	94.50	
	7130	3/4″ pipe size		16	.500		62.50	20.50		83	99	
	7150	Balancing valve, 3/4″ pipe size		20	.400		26	16.35		42.35	53	
	7180	Draindown valve, 1/2″ copper tube		9	.889		202	36.50		238.50	277	
	7200	Flow control valve, 1/2″ pipe size		22	.364		58.50	14.85		73.35	87	
	7220	Expansion tank, up to 5 gal.		32	.250		61.50	10.20		71.70	83	
	7250	Hydronic controller (aquastat)		8	1		43.50	41		84.50	110	
	7400	Pressure gauge, 2″ dial		32	.250		22	10.20		32.20	40	
	7450	Relief valve, temp. and pressure 3/4″ pipe size		30	.267		9.10	10.90		20	26.50	
	7500	Solenoid valve, normally closed										
	7520	Brass, 3/4″ NPT, 24V	1 Plum	9	.889	Ea.	212	36.50		248.50	288	
	7530	1″ NPT, 24V		9	.889		212	36.50		248.50	288	
	7750	Vacuum relief valve, 3/4″ pipe size		32	.250		26.50	10.20		36.70	44.50	
	7800	Thermometers										
	7820	Digital temperature monitoring, 4 locations	1 Plum	2.50	3.200	Ea.	128	131		259	335	
	7900	Upright, 1/2″ NPT		8	1		24	41		65	88	
	7970	Remote probe, 2″ dial		8	1		30.50	41		71.50	95	
	7990	Stem, 2″ dial, 9″ stem		16	.500		20	20.50		40.50	52.50	
	8250	Water storage tank with heat exchanger and electric element										
	8270	66 gal. with 2″ x 2 lb. density insulation	1 Plum	1.60	5	Ea.	730	204		934	1,100	
	8300	80 gal. with 2″ x 2 lb. density insulation		1.60	5		835	204		1,039	1,225	

13 SPECIAL CONSTRUCTION

13600 | Solar and Wind Energy Equipment

13630 | Solar Collector Components

			CREW	DAILY OUTPUT	LABOR-HOURS	UNIT	2005 BARE COSTS MAT.	LABOR	EQUIP.	TOTAL	TOTAL INCL O&P	
200	8380	120 gal. with 2" x 2 lb. density insulation R13600-610	1 Plum	1.40	5.714	Ea.	940	233		1,173	1,375	200
	8400	120 gal. with 2" x 2 lb. density insul., 40 S.F. heat coil	↓	1.40	5.714	↓	1,075	233		1,308	1,550	
	8500	Water storage module, plastic										
	8600	Tubular, 12" diameter, 4' high	1 Carp	48	.167	Ea.	80	5.70		85.70	97.50	
	8610	12" diameter, 8' high		40	.200		122	6.85		128.85	146	
	8620	18" diameter, 5' high		38	.211		134	7.20		141.20	159	
	8630	18" diameter, 10' high	↓	32	.250		176	8.55		184.55	207	
	8640	58" diameter, 5' high	2 Carp	32	.500		490	17.15		507.15	560	
	8650	Cap, 12" diameter					12.65			12.65	13.95	
	8660	18" diameter				↓	15.85			15.85	17.40	

13700 | Security Access and Surveillance

13720 | Detection & Alarm

			CREW	DAILY OUTPUT	LABOR-HOURS	UNIT	2005 BARE COSTS MAT.	LABOR	EQUIP.	TOTAL	TOTAL INCL O&P	
065	0010	**DETECTION SYSTEMS**, not including wires & conduits										065
	0100	Burglar alarm, battery operated, mechanical trigger	1 Elec	4	2	Ea.	254	81.50		335.50	400	
	0200	Electrical trigger		4	2		305	81.50		386.50	455	
	0400	For outside key control, add		8	1		72	41		113	140	
	0600	For remote signaling circuitry, add		8	1		114	41		155	187	
	0800	Card reader, flush type, standard		2.70	2.963		850	121		971	1,125	
	1000	Multi-code	↓	2.70	2.963	↓	1,100	121		1,221	1,375	
	3594	Fire, alarm control panel										
	3600	4 zone	2 Elec	2	8	Ea.	940	325		1,265	1,500	
	3800	8 zone		1	16		1,400	650		2,050	2,525	
	4000	12 zone	↓	.67	23.988		1,825	980		2,805	3,450	
	4020	Alarm device	1 Elec	8	1		122	41		163	195	
	4050	Actuating device		8	1		292	41		333	380	
	4200	Battery and rack		4	2		690	81.50		771.50	880	
	4400	Automatic charger		8	1		445	41		486	550	
	4600	Signal bell		8	1		49.50	41		90.50	115	
	4800	Trouble buzzer or manual station		8	1		37	41		78	101	
	5000	Detector, rate of rise		8	1		33.50	41		74.50	97.50	
	5100	Fixed temperature		8	1		28	41		69	91.50	
	5200	Smoke detector, ceiling type		6.20	1.290		75	52.50		127.50	161	
	5400	Duct type		3.20	2.500		250	102		352	425	
	5600	Strobe and horn		5.30	1.509		95	61.50		156.50	197	
	5800	Fire alarm horn		6.70	1.194		36.50	48.50		85	113	
	6000	Door holder, electro-magnetic		4	2		77.50	81.50		159	207	
	6200	Combination holder and closer		3.20	2.500		430	102		532	625	
	6400	Code transmitter		4	2		690	81.50		771.50	880	
	6600	Drill switch		8	1		86.50	41		127.50	156	
	6800	Master box		2.70	2.963		3,100	121		3,221	3,575	
	7000	Break glass station		8	1		50	41		91	116	
	7800	Remote annunciator, 8 zone lamp	↓	1.80	4.444		175	181		356	465	
	8000	12 zone lamp	2 Elec	2.60	6.154		300	251		551	705	
	8200	16 zone lamp	"	2.20	7.273	↓	300	296		596	770	

13832 | Direct Digital Controls

			CREW	DAILY OUTPUT	LABOR-HOURS	UNIT	2005 BARE COSTS MAT.	LABOR	EQUIP.	TOTAL	TOTAL INCL O&P	
200	0010	**CONTROL COMPONENTS / DDC SYSTEMS** (Sub's quote incl. M & L)										200
	0100	Analog inputs										
	0110	Sensors (avg. 50' run in 1/2" EMT)										
	0120	Duct temperature				Ea.					353.73	
	0130	Space temperature									250	
	0140	Duct humidity, +/- 3%									665.85	
	0150	Space humidity, +/- 2%									1,019.59	
	0160	Duct static pressure									541	
	0170	C.F.M./Transducer									728.28	
	0172	Water temp. (see 15120 for well tap add)									624.24	
	0174	Water flow (see 15120 for circuit sensor add)									2,288.88	
	0176	Water pressure differential (see 15120 for tap add)									936.36	
	0177	Steam flow (see 15120 for circuit sensor add)									2,288.88	
	0178	Steam pressure (see 15120 for tap add)									977.97	
	0180	K.W./Transducer									1,300.50	
	0182	K.W,H. totalization (not incl. elec. meter pulse xmtr.)									597.72	
	0190	Space static pressure				↓					1,020	
	1000	Analog outputs (avg. 50' run in 1/2" EMT)										
	1010	P/I Transducer				Ea.					603.43	
	1020	Analog output, matl. in MUX									291.31	
	1030	Pneumatic (not incl. control device)									613.83	
	1040	Electric (not incl control device)				↓					364.14	
	2000	Status (Alarms)										
	2100	Digital inputs (avg. 50' run in 1/2" EMT)										
	2110	Freeze				Ea.					416.16	
	2120	Fire									374.54	
	2130	Differential pressure, (air)									572.22	
	2140	Differential pressure, (water)									811.51	
	2150	Current sensor									416.16	
	2160	Duct high temperature thermostat									546.21	
	2170	Duct smoke detector				↓					676.26	
	2200	Digital output (avg. 50' run in 1/2" EMT)										
	2210	Start/stop				Ea.					326.40	
	2220	On/off (maintained contact)				"					561	
	3000	Controller M.U.X. panel, incl. function boards										
	3100	48 point				Ea.					5,045.94	
	3110	128 point				"					6,918.66	
	3200	D.D.C. controller (avg. 50' run in conduit)										
	3210	Mechanical room										
	3214	16 point controller (incl. 120v/1ph power supply)				Ea.					3,121.20	
	3229	32 point controller (incl. 120v/1ph power supply)				"					5,202	
	3230	Includes software programming and checkout										
	3260	Space										
	3266	V.A.V. terminal box (incl. space temp. sensor)				Ea.					806.31	
	3280	Host computer (avg. 50' run in conduit)										
	3281	Package complete with PC, keyboard,										
	3282	printer, color CRT, modem, basic software				Ea.					9,363.60	
	4000	Front end costs										
	4100	Computer (P.C.)/software program				Ea.					6,242.40	
	4200	Color graphics software									3,745.44	
	4300	Color graphics slides									468.18	
	4350	Additional dot matrix printer				↓					936.36	
	4400	Communications trunk cable				L.F.					3.64	
	4500	Engineering labor, (not incl. dftg.)				Point					79.07	
	4600	Calibration labor									79.07	
	4700	Start-up, checkout labor				↓					119.64	

Important: See the Reference Section for critical supporting data - Reference Nos., Crews, & City Cost Indexes

		13832 | Direct Digital Controls	Crew	Daily Output	Labor-Hours	Unit	2005 Bare Costs				Total Incl O&P	
							Mat.	Labor	Equip.	Total		
200	4800	Drafting labor, as req'd										200
	5000	Communications bus (data transmission cable)										
	5010	#18 twisted shielded pair in 1/2" EMT conduit				C.L.F.					364.14	
	8000	Applications software										
	8050	Basic maintenance manager software (not incl. data base entry)				Ea.					1,872.72	
	8100	Time program				Point					6.55	
	8120	Duty cycle									13.05	
	8140	Optimum start/stop									39.53	
	8160	Demand limiting									19.56	
	8180	Enthalpy program				↓					39.53	
	8200	Boiler optimization				Ea.					1,170.45	
	8220	Chiller optimization				"					1,560.60	
	8240	Custom applications										
	8260	Cost varies with complexity										
		13834 | Electric/Electronic Control										
200	0010	CONTROL SYSTEMS, ELECTRONIC R15050-710										200
	0020	For electronic costs, add to division 13836-200				Ea.					15%	
		13836 | Pneumatic Controls										
200	0010	CONTROL SYSTEMS, PNEUMATIC (Sub's quote incl. mat. & labor)										200
	0011	and nominal 50 Ft. of tubing. Add control panelboard if req'd.										
	0100	Heating and Ventilating, split system R15050-710										
	0200	Mixed air control, economizer cycle,panel readout,tubing										
	0220	Up to 10 tons	Q-19	.68	35.294	Ea.	2,975	1,350		4,325	5,300	
	0240	For 10 to 20 tons		.63	37.915		3,175	1,450		4,625	5,675	
	0260	For over 20 tons		.58	41.096		3,450	1,575		5,025	6,150	
	0270	Enthalpy cycle, up to 10 tons		.50	48.387		3,275	1,850		5,125	6,400	
	0280	For 10 to 20 tons		.46	52.174		3,550	2,000		5,550	6,875	
	0290	For over 20 tons	↓	.42	56.604	↓	3,850	2,150		6,000	7,450	
	0300	Heating coil, hot water, 3 way valve,										
	0320	Freezestat, limit control on discharge, readout	Q-5	.69	23.088	Ea.	2,200	850		3,050	3,700	
	0500	Cooling coil, chilled water, room										
	0520	Thermostat, 3 way valve	Q-5	2	8	Ea.	985	295		1,280	1,525	
	0600	Cooling tower, fan cycle, damper control,										
	0620	Control system including water readout in/out at panel	Q-19	.67	35.821	Ea.	3,900	1,375		5,275	6,350	
	1000	Unit ventilator, day/night operation,										
	1100	freezestat, ASHRAE, cycle 2	Q-19	.91	26.374	Ea.	2,175	1,000		3,175	3,875	
	2000	Compensated hot water from boiler, valve control,										
	2100	readout and reset at panel, up to 60 GPM	Q-19	.55	43.956	Ea.	4,050	1,675		5,725	6,950	
	2120	For 120 GPM		.51	47.059		4,325	1,800		6,125	7,475	
	2140	For 240 GPM		.49	49.180		4,525	1,875		6,400	7,800	
	3000	Boiler room combustion air, damper to 5 SF, controls		1.36	17.582		1,950	670		2,620	3,150	
	3500	Fan coil, heating and cooling valves, 4 pipe control system		3	8		880	305		1,185	1,425	
	3600	Heat exchanger system controls	↓	.86	27.907	↓	1,900	1,075		2,975	3,700	
	3900	Multizone control (one per zone), includes thermostat, damper										
	3910	motor and reset of discharge temperature	Q-5	.51	31.373	Ea.	1,950	1,150		3,100	3,875	
	4000	Pneumatic thermostat, including controlling room radiator valve	"	2.43	6.593		590	243		833	1,000	
	4040	Program energy saving optimizer	Q-19	1.21	19.786		4,875	755		5,630	6,500	
	4060	Pump control system	"	3	8		905	305		1,210	1,450	
	4080	Reheat coil control system, not incl coil	Q-5	2.43	6.593	↓	765	243		1,008	1,200	
	4500	Air supply for pneumatic control system										
	4600	Tank mounted duplex compressor, starter, alternator,										
	4620	piping, dryer, PRV station and filter										
	4630	1/2 HP	Q-19	.68	35.139	Ea.	7,275	1,350		8,625	10,000	
	4640	3/4 HP	↓	.64	37.383	↓	7,625	1,425		9,050	10,500	

13836 | Pneumatic Controls

				CREW	DAILY OUTPUT	LABOR-HOURS	UNIT	2005 BARE COSTS MAT.	LABOR	EQUIP.	TOTAL	TOTAL INCL O&P	
200	4650	1 HP	R15050-710	Q-19	.61	39.539	Ea.	8,325	1,500		9,825	11,400	200
	4660	1-1/2 HP			.57	41.739		8,875	1,600		10,475	12,100	
	4680	3 HP			.55	43.956		12,100	1,675		13,775	15,800	
	4690	5 HP			.42	57.143		21,100	2,175		23,275	26,500	
	4800	Main air supply, includes 3/8" copper main and labor		Q-5	1.82	8.791	C.L.F.	244	325		569	755	
	4810	If poly tubing used, deduct										30%	
	7000	Static pressure control for air handling unit, includes pressure											
	7010	sensor, receiver controller, readout and damper motors		Q-19	.64	37.383	Ea.	5,775	1,425		7,200	8,475	
	7020	If return air fan requires control, add										70%	
	8600	VAV boxes, incl. thermostat, damper motor, reheat coil & tubing		Q-5	1.46	10.989		905	405		1,310	1,600	
	8610	If no reheat coil, deduct										204	
	9400	Sub assemblies for assembly systems											

13838 | Pneumatic/Electric Controls

			CREW	DAILY OUTPUT	LABOR-HOURS	UNIT	2005 BARE COSTS MAT.	LABOR	EQUIP.	TOTAL	TOTAL INCL O&P	
200	0010	CONTROL COMPONENTS										200
	0500	Aquastats										
	0508	Immersion type										
	0510	High/low limit, breaks contact w/temp rise	1 Stpi	4	2	Ea.	110	82		192	244	
	0514	Sequencing, break 2 switches w/temp rise		4	2		89.50	82		171.50	222	
	0518	Circulating, makes contact w/temp rise		4	2		81	82		163	213	
	0600	Carbon monoxide detector system										
	0606	Panel	1 Stpi	4	2	Ea.	1,875	82		1,957	2,200	
	0610	Sensor	"	7.30	1.096	"	385	45		430	495	
	0700	Controller, receiver										
	0730	Pneumatic, panel mount, single input	1 Plum	8	1	Ea.	280	41		321	370	
	0740	With conversion mounting bracket		8	1		290	41		331	380	
	0750	Dual input, with control point adjustment		7	1.143		385	46.50		431.50	495	
	0850	Electric, single snap switch	1 Elec	4	2		360	81.50		441.50	515	
	0860	Dual snap switches		3	2.667		450	109		559	655	
	0870	Humidity controller		8	1		210	41		251	292	
	0880	Load limiting controller		8	1		460	41		501	570	
	0890	Temperature controller		8	1		263	41		304	350	
	0900	Control panel readout										
	0910	Panel up to 12 indicators	1 Stpi	1.20	6.667	Ea.	140	273		413	565	
	0914	Panel up to 24 indicators		.86	9.302		174	380		554	765	
	0918	Panel up to 48 indicators		.50	16		415	655		1,070	1,450	
	1000	Enthalpy control, boiler water temperature control										
	1010	governed by outdoor temperature, with timer	1 Elec	3	2.667	Ea.	208	109		317	390	
	1300	Energy control/monitor										
	1320	BTU computer meter and controller	1 Stpi	1	8	Ea.	1,400	330		1,730	2,050	
	1600	Flow meters										
	1610	Gas	1 Stpi	4	2	Ea.	82	82		164	213	
	1620	Liquid	"	4	2	"	82	82		164	213	
	1640	Freezestat										
	1644	20' sensing element, adjustable	1 Stpi	3.50	2.286	Ea.	85	93.50		178.50	235	
	2000	Gauges, pressure or vacuum										
	2100	2" diameter dial	1 Stpi	32	.250	Ea.	8.35	10.25		18.60	24.50	
	2200	2-1/2" diameter dial		32	.250		9.10	10.25		19.35	25.50	
	2300	3-1/2" diameter dial		32	.250		20	10.25		30.25	37.50	
	2400	4-1/2" diameter dial		32	.250		30	10.25		40.25	48.50	
	2700	Flanged iron case, black ring										
	2800	3-1/2" diameter dial	1 Stpi	32	.250	Ea.	63.50	10.25		73.75	85	
	2900	4-1/2" diameter dial		32	.250		86.50	10.25		96.75	110	
	3000	6" diameter dial		32	.250		68.50	10.25		78.75	91	
	3010	Steel case, 0 - 300 psi										
	3012	2" diam. dial	1 Stpi	16	.500	Ea.	18.15	20.50		38.65	51	

Important: See the Reference Section for critical supporting data - Reference Nos., Crews, & City Cost Indexes

	13838 | Pneumatic/Electric Controls	CREW	DAILY OUTPUT	LABOR-HOURS	UNIT	2005 BARE COSTS MAT.	LABOR	EQUIP.	TOTAL	TOTAL INCL O&P	
200	3014 4″ diam. dial	1 Stpi	16	.500	Ea.	42.50	20.50		63	77.50	200
	3020 Aluminum case, 0 - 300 psi										
	3022 3-1/2″ diam. dial	1 Stpi	16	.500	Ea.	56	20.50		76.50	92.50	
	3024 4-1/2″ diam. dial		16	.500		71	20.50		91.50	110	
	3026 6″ diam. dial		16	.500		118	20.50		138.50	161	
	3028 8-1/2″ diam. dial		16	.500		281	20.50		301.50	340	
	3030 Brass case, 0 - 300 psi										
	3032 2″ diam. dial	1 Stpi	16	.500	Ea.	23.50	20.50		44	57	
	3034 4-1/2″ diam. dial	″	16	.500	″	53.50	20.50		74	90	
	3040 Steel case, high pressure, 0 -10,000 psi										
	3042 4-1/2″ diam. dial	1 Stpi	16	.500	Ea.	134	20.50		154.50	178	
	3044 6-1/2″ diam. dial		16	.500		215	20.50		235.50	267	
	3046 8-1/2″ diam. dial		16	.500		390	20.50		410.50	460	
	3080 Pressure gauge, differential, magnehelic										
	3084 0 - 2″ W.C., with air filter kit	1 Stpi	6	1.333	Ea.	79.50	54.50		134	169	
	3300 For compound pressure-vacuum, add					18%					
	3350 Humidistat										
	3360 Pneumatic operation										
	3361 Room humidistat, direct acting	1 Stpi	12	.667	Ea.	165	27.50		192.50	222	
	3362 Room humidistat, reverse acting		12	.667		165	27.50		192.50	222	
	3363 Room humidity transmitter		17	.471		178	19.25		197.25	225	
	3364 Duct mounted controller		12	.667		217	27.50		244.50	280	
	3365 Duct mounted transmitter		12	.667		196	27.50		223.50	256	
	3366 Humidity indicator, 3-1/2″		28	.286		79	11.70		90.70	105	
	3390 Electric operated	1 Shee	8	1		31	40		71	96.50	
	3400 Relays										
	3430 Pneumatic/electric	1 Plum	16	.500	Ea.	192	20.50		212.50	242	
	3440 Pneumatic proportioning		8	1		104	41		145	177	
	3450 Pneumatic switching		12	.667		92.50	27		119.50	143	
	3460 Selector, 3 point		6	1.333		66.50	54.50		121	155	
	3470 Pneumatic time delay		8	1		246	41		287	330	
	3500 Sensor, air operated										
	3520 Humidity	1 Plum	16	.500	Ea.	217	20.50		237.50	270	
	3540 Pressure		16	.500		48.50	20.50		69	83.50	
	3560 Temperature		12	.667		104	27		131	155	
	3600 Electric operated										
	3620 Humidity	1 Elec	8	1	Ea.	27	41		68	90.50	
	3650 Pressure		8	1		870	41		911	1,025	
	3680 Temperature		10	.800		100	32.50		132.50	159	
	3700 Switches										
	3710 Minimum position										
	3720 Electrical	1 Stpi	4	2	Ea.	25.50	82		107.50	151	
	3730 Pneumatic	″	4	2	″	39.50	82		121.50	166	
	4000 Thermometers										
	4100 Dial type, 3-1/2″ diameter, vapor type, union connection	1 Stpi	32	.250	Ea.	96.50	10.25		106.75	121	
	4120 Liquid type, union connection		32	.250		135	10.25		145.25	163	
	4130 Remote reading, 15′ capillary		32	.250		111	10.25		121.25	138	
	4500 Stem type, 6-1/2″ case, 2″ stem, 1/2″ NPT		32	.250		55	10.25		65.25	76	
	4520 4″ stem, 1/2″ NPT		32	.250		75.50	10.25		85.75	98.50	
	4600 9″ case, 3-1/2″ stem, 3/4″ NPT		28	.286		90	11.70		101.70	117	
	4620 6″ stem, 3/4″ NPT		28	.286		96	11.70		107.70	123	
	4640 8″ stem, 3/4″ NPT		28	.286		118	11.70		129.70	148	
	4660 12″ stem, 1″ NPT		26	.308		130	12.60		142.60	162	
	4670 Bi-metal, dial type, steel case brass stem										
	4672 2″ dial, 4″ - 9″ stem	1 Stpi	16	.500	Ea.	29.50	20.50		50	63.50	
	4673 2-1/2″ dial, 4″ - 9″ stem		16	.500		34	20.50		54.50	68	

	13838 | Pneumatic/Electric Controls		CREW	DAILY OUTPUT	LABOR-HOURS	UNIT	2005 BARE COSTS MAT.	LABOR	EQUIP.	TOTAL	TOTAL INCL O&P
200	4674	3-1/2" dial, 4" - 9" stem	1 Stpi	16	.500	Ea.	33	20.50		53.50	67
	4680	Mercury filled, industrial, union connection type									
	4682	Angle stem, 7" scale	1 Stpi	16	.500	Ea.	74.50	20.50		95	113
	4683	9" scale		16	.500		109	20.50		129.50	151
	4684	12" scale		16	.500		121	20.50		141.50	164
	4686	Straight stem, 7" scale		16	.500		96	20.50		116.50	136
	4687	9" scale		16	.500		90	20.50		110.50	130
	4688	12" scale		16	.500		110	20.50		130.50	152
	4690	Mercury filled, industrial, separable socket type									
	4692	Angle stem, with socket, 7" scale	1 Stpi	16	.500	Ea.	64.50	20.50		85	102
	4693	9" scale		16	.500		77	20.50		97.50	116
	4694	12" scale		16	.500		156	20.50		176.50	202
	4696	Straight stem, with socket, 7" scale		16	.500		64.50	20.50		85	102
	4697	9" scale		16	.500		77	20.50		97.50	116
	4698	12" scale		16	.500		90.50	20.50		111	131
	5000	Thermostats									
	5030	Manual	1 Shee	8	1	Ea.	27	40		67	92
	5040	1 set back, electric, timed		8	1		81	40		121	151
	5050	2 set back, electric, timed		8	1		178	40		218	258
	5100	Locking cover					24			24	26
	5200	24 hour, automatic, clock	1 Shee	8	1		124	40		164	198
	5220	Electric, low voltage, 2 wire	1 Elec	13	.615		18	25		43	57.50
	5230	3 wire		10	.800		22	32.50		54.50	73
	5236	Heating/cooling, low voltage, with clock		8	1		102	41		143	173
	5240	Pneumatic									
	5250	Single temp., single pressure	1 Stpi	8	1	Ea.	120	41		161	194
	5251	Dual pressure		8	1		177	41		218	257
	5252	Dual temp., dual pressure		8	1		168	41		209	247
	5253	Reverse acting w/averaging element		8	1		154	41		195	231
	5254	Heating-cooling w/deadband		8	1		167	41		208	246
	5255	Integral w/piston top valve actuator		8	1		165	41		206	243
	5256	Dual temp., dual pressure		8	1		168	41		209	247
	5257	Low limit, 8' averaging element		8	1		117	41		158	190
	5258	Room single temp. proportional		8	1		48	41		89	115
	5260	Dual temp, direct acting for VAV fan		8	1		158	41		199	236
	5262	Capillary tube type, 20', -30 to +100°F		8	1		103	41		144	176
	5300	Transmitter, pneumatic									
	5320	Temperature averaging element	Q-1	8	2	Ea.	99.50	73.50		173	221
	5350	Pressure differential	1 Plum	7	1.143		585	46.50		631.50	710
	5370	Humidity, duct		8	1		196	41		237	277
	5380	Room		12	.667		178	27		205	237
	5390	Temperature, with averaging element		6	1.333		112	54.50		166.50	205
	5420	Electric operated, humidity	1 Elec	8	1		50.50	41		91.50	116
	5430	DPST	"	8	1		71	41		112	139
	5500	Timer/time clocks									
	5510	7 day, 12 hr. battery	1 Stpi	2.60	3.077	Ea.	248	126		374	460
	5600	Recorders									
	5610	Hydrograph humidity, wall mtd., aluminum case, grad. chart									
	5612	8"	1 Stpi	1.60	5	Ea.	530	205		735	895
	5614	10"	"	1.60	5	"	980	205		1,185	1,375
	5620	Thermo-hydrograph, wall mtd., grad. chart, 5' SS probe									
	5622	8"	1 Stpi	1.60	5	Ea.	480	205		685	840
	5624	10"	"	1.60	5	"	830	205		1,035	1,225
	5630	Time of operation, wall mtd., 1 pen cap, 24 hr. clock									
	5632	6"	1 Stpi	1.60	5	Ea.	390	205		595	740
	5634	8"	"	1.60	5	"	390	205		595	740

Important: See the Reference Section for critical supporting data - Reference Nos., Crews, & City Cost Indexes

	13838 | Pneumatic/Electric Controls	CREW	DAILY OUTPUT	LABOR-HOURS	UNIT	2005 BARE COSTS MAT.	LABOR	EQUIP.	TOTAL	TOTAL INCL O&P	
200	5640 Pressure & vacuum, bourdon tube type										200
	5642 Flush mount, 1 pen										
	5644 6"	1 Stpi	1.60	5	Ea.	370	205		575	720	
	5646 8"	"	1.60	5	"	700	205		905	1,075	
	6000 Valves, motorized zone										
	6100 Sweat connections, 1/2" C x C	1 Stpi	20	.400	Ea.	67.50	16.40		83.90	99	
	6110 3/4" C x C		20	.400		72.50	16.40		88.90	104	
	6120 1" C x C		19	.421		84.50	17.25		101.75	119	
	6140 1/2" C x C, with end switch, 2 wire		20	.400		71.50	16.40		87.90	103	
	6150 3/4" C x C, with end switch, 2 wire		20	.400		76	16.40		92.40	108	
	6160 1" C x C, with end switch, 2 wire		19	.421		86	17.25		103.25	121	
	6180 1-1/4" C x C, w/end switch, 2 wire	↓	15	.533	↓	103	22		125	147	
	7090 Valves, motor controlled, including actuator										
	7100 Electric motor actuated										
	7200 Brass, two way, screwed										
	7210 1/2" pipe size	L-6	36	.333	Ea.	206	13.60		219.60	247	
	7220 3/4" pipe size		30	.400		299	16.35		315.35	355	
	7230 1" pipe size		28	.429		335	17.50		352.50	395	
	7240 1-1/2" pipe size		19	.632		375	26		401	455	
	7250 2" pipe size	↓	16	.750	↓	585	30.50		615.50	690	
	7350 Brass, three way, screwed										
	7360 1/2" pipe size	L-6	33	.364	Ea.	222	14.85		236.85	266	
	7370 3/4" pipe size		27	.444		248	18.15		266.15	300	
	7380 1" pipe size		25.50	.471		219	19.20		238.20	270	
	7384 1-1/4" pipe size		21	.571		278	23.50		301.50	340	
	7390 1-1/2" pipe size		17	.706		315	29		344	390	
	7400 2" pipe size	↓	14	.857	↓	490	35		525	590	
	7550 Iron body, two way, flanged										
	7560 2-1/2" pipe size	L-6	4	3	Ea.	360	122		482	585	
	7570 3" pipe size		3	4		480	163		643	770	
	7580 4" pipe size	↓	2	6	↓	685	245		930	1,125	
	7850 Iron body, three way, flanged										
	7860 2-1/2" pipe size	L-6	3	4	Ea.	805	163		968	1,125	
	7870 3" pipe size		2.50	4.800		975	196		1,171	1,375	
	7880 4" pipe size	↓	2	6	↓	1,250	245		1,495	1,750	
	8000 Pneumatic, air operated										
	8050 Brass, two way, screwed										
	8060 1/2" pipe size, class 250	1 Plum	24	.333	Ea.	105	13.60		118.60	136	
	8070 3/4" pipe size, class 250		20	.400		139	16.35		155.35	177	
	8080 1" pipe size, class 250		19	.421		158	17.20		175.20	200	
	8090 1-1/4" pipe size, class 125		15	.533		226	22		248	282	
	8100 1-1/2" pipe size, class 125		13	.615		262	25		287	325	
	8110 2" pipe size, class 125	↓	11	.727	↓	297	29.50		326.50	370	
	8180 Brass, three way, screwed										
	8190 1/2" pipe size, class 250	1 Plum	22	.364	Ea.	115	14.85		129.85	149	
	8200 3/4" pipe size, class 250		18	.444		140	18.15		158.15	182	
	8210 1" pipe size, class 250		17	.471		160	19.20		179.20	205	
	8214 1-1/4" pipe size, class 250		14	.571		220	23.50		243.50	277	
	8220 1-1/2" pipe size, class 125		11	.727		257	29.50		286.50	325	
	8230 2" pipe size, class 125	↓	9	.889	↓	325	36.50		361.50	410	
	8250 Iron body, two way, screwed										
	8260 1/2" pipe size, class 300	1 Plum	24	.333	Ea.	289	13.60		302.60	340	
	8270 3/4" pipe size, class 300		20	.400		320	16.35		336.35	375	
	8280 1" pipe size, class 300		19	.421		520	17.20		537.20	595	
	8290 1-1/4" pipe size, class 300		15	.533		550	22		572	640	
	8300 1-1/2" pipe size, class 300	↓	13	.615	↓	725	25		750	840	

13838 | Pneumatic/Electric Controls		CREW	DAILY OUTPUT	LABOR-HOURS	UNIT	2005 BARE COSTS MAT.	LABOR	EQUIP.	TOTAL	TOTAL INCL O&P	
200	8310 2" pipe size, class 300	1 Plum	11	.727	Ea.	800	29.50		829.50	925	200
	8320 1-1/4" pipe size, class 125		15	.533		435	22		457	515	
	8330 1-1/2" pipe size, class 125		13	.615		390	25		415	470	
	8340 2" pipe size, class 125	↓	11	.727	↓	455	29.50		484.50	545	
	8450 Iron body, two way, flanged										
	8460 2-1/2" pipe size, 250 lb. flanges	Q-1	5	3.200	Ea.	2,900	118		3,018	3,375	
	8470 3" pipe size, 250 lb. flanges		4.50	3.556		3,100	131		3,231	3,600	
	8480 4" pipe size, 250 lb. flanges		3	5.333		3,750	196		3,946	4,450	
	8510 2-1/2" pipe size, class 125		5	3.200		625	118		743	860	
	8520 3" pipe size, class 125		4.50	3.556		745	131		876	1,025	
	8530 4" pipe size, class 125	↓	3	5.333		955	196		1,151	1,350	
	8540 5" pipe size, class 125	Q-2	3.40	7.059		1,975	269		2,244	2,575	
	8550 6" pipe size, class 125	"	3	8	↓	2,350	305		2,655	3,025	
	8560 Iron body, three way, flanged										
	8570 2-1/2" pipe size, class 125	Q-1	4.50	3.556	Ea.	710	131		841	975	
	8580 3" pipe size, class 125		4	4		885	147		1,032	1,200	
	8590 4" pipe size, class 125	↓	2.50	6.400		1,775	235		2,010	2,300	
	8600 6" pipe size, class 125	Q-2	3	8	↓	2,425	305		2,730	3,100	
	9005 Pneumatic system misc. components										
	9010 Adding/subtracting repeater	1 Stpi	20	.400	Ea.	103	16.40		119.40	138	
	9020 Adjustable ratio network		16	.500		96	20.50		116.50	137	
	9030 Comparator	↓	20	.400	↓	197	16.40		213.40	241	
	9040 Cumulator										
	9041 Air switching	1 Stpi	20	.400	Ea.	92.50	16.40		108.90	127	
	9042 Averaging		20	.400		96	16.40		112.40	131	
	9043 2:1 ratio		20	.400		86.50	16.40		102.90	120	
	9044 Sequencing		20	.400		59	16.40		75.40	89.50	
	9045 Two-position		16	.500		106	20.50		126.50	148	
	9046 Two-position pilot	↓	16	.500	↓	100	20.50		120.50	141	
	9050 Damper actuator										
	9051 For smoke control	1 Stpi	8	1	Ea.	134	41		175	209	
	9052 For ventilation	"	8	1	"	146	41		187	223	
	9053 Series duplex pedestal mounted										
	9054 For inlet vanes on fans, compressors	1 Stpi	6	1.333	Ea.	670	54.50		724.50	815	
	9060 Enthalpy logic center		16	.500		208	20.50		228.50	260	
	9070 High/low pressure selector		22	.364		37.50	14.90		52.40	63.50	
	9080 Signal limiter		20	.400		59	16.40		75.40	89.50	
	9081 Transmitter	↓	28	.286	↓	22.50	11.70		34.20	42	
	9090 Optimal start										
	9091 Warmup or cooldown programmer	1 Stpi	10	.800	Ea.	223	33		256	294	
	9092 Mass temperature transmitter	"	18	.444	"	97	18.20		115.20	135	
	9100 Step controller with positioner, time delay restrictor										
	9110 recycler air valve, switches and cam settings										
	9112 With cabinet										
	9113 6 points	1 Stpi	4	2	Ea.	455	82		537	625	
	9114 6 points, 6 manual sequences		2	4		480	164		644	775	
	9115 8 points	↓	3	2.667	↓	580	109		689	805	
	9200 Pressure controller and switches										
	9210 High static pressure limit	1 Stpi	8	1	Ea.	134	41		175	210	
	9220 Pressure transmitter		8	1		145	41		186	222	
	9230 Differential pressure transmitter		8	1		585	41		626	700	
	9240 Static pressure transmitter		8	1		340	41		381	435	
	9250 Proportional-only control, single input		6	1.333		280	54.50		334.50	390	
	9260 Proportional plus integral, single input		6	1.333		415	54.50		469.50	540	
	9270 Proportional-only, dual input		6	1.333		385	54.50		439.50	505	
	9280 Proportional plus integral, dual input	↓	6	1.333	↓	510	54.50		564.50	640	

	13838 Pneumatic/Electric Controls	CREW	DAILY OUTPUT	LABOR-HOURS	UNIT	MAT.	LABOR	EQUIP.	TOTAL	TOTAL INCL O&P	
200	9281 Time delay for above proportional units	1 Stpi	16	.500	Ea.	110	20.50		130.50	152	200
	9290 Proportional/2 position controller units		6	1.333		295	54.50		349.50	405	
	9300 Differential press. control dir./rev.	↓	6	1.333	↓	199	54.50		253.50	300	
	9310 Air pressure reducing valve										
	9311 1/8" size	1 Stpi	18	.444	Ea.	21.50	18.20		39.70	51	
	9315 Precision valve, 1/4" size		17	.471		114	19.25		133.25	154	
	9320 Air flow controller		12	.667		115	27.50		142.50	167	
	9330 Booster relay volume amplifier		18	.444		63	18.20		81.20	97	
	9340 Series restrictor, straight		60	.133		3.38	5.45		8.83	11.90	
	9341 T-fitting		40	.200		4.96	8.20		13.16	17.75	
	9342 In-line adjustable		48	.167		21.50	6.85		28.35	34	
	9350 Diode tee		40	.200		11.40	8.20		19.60	25	
	9351 Restrictor tee		40	.200		11.40	8.20		19.60	25	
	9360 Pneumatic gradual switch		8	1		151	41		192	228	
	9361 Selector switch		8	1		66.50	41		107.50	135	
	9370 Electro-pneumatic motor driven servo		10	.800		715	33		748	835	
	9380 Regulated power supply		10	.800		495	33		528	595	
	9390 Fan control switch and mounting base		16	.500		67	20.50		87.50	105	
	9400 Circulating pump sequencer	↓	14	.571	↓	690	23.50		713.50	795	
	9410 Pneumatic tubing, fittings and accessories										
	9414 Tubing, urethane										
	9415 1/8" OD x 1/16" ID	1 Stpi	120	.067	L.F.	.45	2.73		3.18	4.60	
	9416 1/4" OD x 1/8" ID		115	.070		.89	2.85		3.74	5.25	
	9417 5/32" OD x 3/32" ID	↓	110	.073	↓	.50	2.98		3.48	5.05	
	9420 Coupling, straight										
	9422 Barb x barb										
	9423 1/4" x 5/32"	1 Stpi	160	.050	Ea.	.50	2.05		2.55	3.63	
	9424 1/4" x 1/4"		158	.051		.26	2.07		2.33	3.41	
	9425 3/8" x 3/8"		154	.052		.34	2.13		2.47	3.57	
	9426 3/8" x 1/4"		150	.053		.34	2.18		2.52	3.65	
	9427 1/2" x 1/4"		148	.054		.74	2.21		2.95	4.14	
	9428 1/2" x 3/8"		144	.056		.68	2.28		2.96	4.17	
	9429 1/2" x 1/2"	↓	140	.057	↓	.50	2.34		2.84	4.07	
	9440 Tube x tube										
	9441 1/4" x 1/4"	1 Stpi	100	.080	Ea.	1.40	3.28		4.68	6.45	
	9442 3/8" x 1/4"		96	.083		1.72	3.41		5.13	7.05	
	9443 3/8" x 3/8"		92	.087		1.72	3.56		5.28	7.25	
	9444 1/2" x 3/8"		88	.091		2.43	3.72		6.15	8.25	
	9445 1/2" x 1/2"	↓	84	.095	↓	3.16	3.90		7.06	9.35	
	9450 Elbow coupling										
	9452 Barb x barb										
	9454 1/4" x 1/4"	1 Stpi	158	.051	Ea.	.56	2.07		2.63	3.74	
	9455 3/8" x 3/8"		154	.052		.74	2.13		2.87	4.01	
	9456 1/2" x 1/2"	↓	140	.057	↓	1	2.34		3.34	4.62	
	9460 Tube x tube										
	9462 1/2" x 1/2"	1 Stpi	84	.095	Ea.	3.87	3.90		7.77	10.10	
	9470 Tee coupling										
	9472 Barb x barb x barb										
	9474 1/4" x 1/4" x 5/32"	1 Stpi	108	.074	Ea.	.90	3.03		3.93	5.55	
	9475 1/4" x 1/4" x 1/4"		104	.077		.60	3.15		3.75	5.40	
	9476 3/8" x 3/8" x 5/32"		102	.078		1.94	3.21		5.15	6.95	
	9477 3/8" x 3/8" x 1/4"		99	.081		.90	3.31		4.21	5.95	
	9478 3/8" x 3/8" x 3/8"		98	.082		.96	3.34		4.30	6.05	
	9479 1/2" x 1/2" x 1/4"		96	.083		1.86	3.41		5.27	7.20	
	9480 1/2" x 1/2" x 3/8"		95	.084		1.04	3.45		4.49	6.35	
	9481 1/2" x 1/2" x 1/2"	↓	92	.087	↓	1.08	3.56		4.64	6.55	

(2005 BARE COSTS: MAT., LABOR, EQUIP., TOTAL)

13800 | Building Automation & Control

13838 | Pneumatic/Electric Controls			CREW	DAILY OUTPUT	LABOR-HOURS	UNIT	2005 BARE COSTS MAT.	LABOR	EQUIP.	TOTAL	TOTAL INCL O&P	
200	9484	Tube x tube x tube										200
	9485	1/4" x 1/4" x 1/4"	1 Stpi	66	.121	Ea.	2.69	4.96		7.65	10.40	
	9486	3/8" x 1/4" x 1/4"		64	.125		5.10	5.10		10.20	13.30	
	9487	3/8" x 3/8" x 1/4"		61	.131		3.99	5.35		9.34	12.45	
	9488	3/8" x 3/8" x 3/8"		60	.133		4.19	5.45		9.64	12.80	
	9489	1/2" x 1/2" x 3/8"		58	.138		6.50	5.65		12.15	15.65	
	9490	1/2" x 1/2" x 1/2"	↓	56	.143	↓	5.75	5.85		11.60	15.15	
	9492	Needle valve										
	9494	Tube x tube										
	9495	1/4" x 1/4"	1 Stpi	86	.093	Ea.	4.89	3.81		8.70	11.15	
	9496	3/8" x 3/8"	"	82	.098	"	5.45	4		9.45	12	
	9600	Electronic system misc. components										
	9610	Electric motor damper actuator	1 Elec	8	1	Ea.	254	41		295	340	
	9620	Damper position indicator		10	.800		80.50	32.50		113	137	
	9630	Pneumatic-electronic transducer		12	.667		296	27		323	365	
	9640	Electro-pneumatic transducer		12	.667		296	27		323	365	
	9650	Remote reset control with motor drive		6	1.333		340	54.50		394.50	450	
	9660	Manual remote set point control		16	.500		65	20.50		85.50	102	
	9670	Alarm unit with two adjustable points	↓	12	.667	↓	310	27		337	385	
	9700	Modulating step controller										
	9701	2-5 steps	1 Elec	5.30	1.509	Ea.	1,525	61.50		1,586.50	1,775	
	9702	6-10 steps		3.20	2.500		1,900	102		2,002	2,250	
	9703	11-15 steps		2.70	2.963		2,350	121		2,471	2,750	
	9704	16-20 steps	↓	1.60	5	↓	2,775	204		2,979	3,350	
	9710	Staging network										
	9711	Master one stage	1 Elec	10	.800	Ea.	410	32.50		442.50	500	
	9712	Two stage		9	.889		535	36		571	645	
	9720	Sequencing network		8	1		615	41		656	735	
	9730	Regulated DC power supply	↓	16	.500	↓	375	20.50		395.50	440	
	9740	Accessory unit regulator										
	9741	Three function	1 Elec	9	.889	Ea.	203	36		239	277	
	9742	Five function		8	1		390	41		431	490	
	9750	Accessory power supply		18	.444		78	18.10		96.10	113	
	9760	Step down transformer	↓	16	.500	↓	131	20.50		151.50	176	

13850 | Detection & Alarm

13851 | Detection & Alarm			CREW	DAILY OUTPUT	LABOR-HOURS	UNIT	2005 BARE COSTS MAT.	LABOR	EQUIP.	TOTAL	TOTAL INCL O&P	
350	0010	**TANK LEAK DETECTION SYSTEMS** Liquid and vapor										350
	0100	For hydrocarbons and hazardous liquids/vapors										
	0120	Controller, data acquisition, incl. printer, modem, RS232 port										
	0140	24 channel, for use with all probes				Ea.	2,100			2,100	2,325	
	0160	9 channel, for external monitoring				"	1,625			1,625	1,800	
	0200	Probes										
	0210	Well monitoring										
	0220	Liquid phase detection				Ea.	735			735	805	
	0230	Hydrocarbon vapor, fixed position					675			675	745	
	0240	Hydrocarbon vapor, float mounted					600			600	660	
	0250	Both liquid and vapor hydrocarbon				↓	910			910	1,000	
	0300	Secondary containment, liquid phase										

	13851 | Detection & Alarm		CREW	DAILY OUTPUT	LABOR-HOURS	UNIT	2005 BARE COSTS MAT.	LABOR	EQUIP.	TOTAL	TOTAL INCL O&P	
350	0310	Pipe trench/manway sump				Ea.	325			325	360	350
	0320	Double wall pipe and manual sump				↓	320			320	350	
	0330	Double wall fiberglass annular space				↓	295			295	325	
	0340	Double wall steel tank annular space				↓	295			295	325	
	0500	Accessories										
	0510	Modem, non-dedicated phone line				Ea.	264			264	290	
	0600	Monitoring, internal										
	0610	Automatic tank gauge, incl. overfill				Ea.	1,050			1,050	1,150	
	0620	Product line				"	1,125			1,125	1,250	
	0700	Monitoring, special										
	0710	Cathodic protection				Ea.	630			630	690	
	0720	Annular space chemical monitor				"	855			855	940	

For information about Means Estimating Seminars, see yellow pages 12 and 13 in back of book

Division Notes

		CREW	DAILY OUTPUT	LABOR-HOURS	UNIT	2005 BARE COSTS MAT.	LABOR	EQUIP.	TOTAL	TOTAL INCL O&P

Division 14 Conveying Systems

Estimating Tips

General

- Many products in Division 14 will require some type of support or blocking for installation not included with the item itself. Examples are supports for conveyors or tube systems, attachment points for lifts, and footings for hoists or cranes. Add these supports in the appropriate division.

14100 Dumbwaiters
14200 Elevators

- Dumbwaiters and elevators are estimated and purchased in a method similar to buying a car. The manufacturer has a base unit with standard features. Added to this base unit price will be whatever options the owner or specifications require. Increased load capacity, additional vertical travel, additional stops, higher speed, and cab finish options are items to be considered. When developing an estimate for dumbwaiters and elevators, remember that some items needed by the installers may have to be included as part of the general contract.

  Examples are:
  - shaftway
  - rail support brackets
  - machine room
  - electrical supply
  - sill angles
  - electrical connections
  - pits
  - roof penthouses
  - pit ladders

  Check the job specifications and drawings before pricing.
- Installation of elevators and handicapped lifts in historic structures can require significant additional costs. The associated structural requirements may involve cutting into and repairing finishes, mouldings, flooring, etc. The estimator must account for these special conditions.

14300 Escalators & Moving Walks

- Escalators and moving walks are specialty items installed by specialty contractors. There are numerous options associated with these items. For specific options contact a manufacturer or contractor. In a method similar to estimating dumbwaiters and elevators, you should verify the extent of general contract work and add items as necessary.

14400 Lifts
14500 Material Handling
14600 Hoists & Cranes

- Products such as correspondence lifts, conveyors, chutes, pneumatic tube systems, material handling cranes and hoists as well as other items specified in this subdivision may require trained installers. The general contractor might not have any choice as to who will perform the installation or when it will be performed. Long lead times are often required for these products, making early decisions in scheduling necessary.

Reference Numbers

Reference numbers are shown in bold squares at the beginning of some major classifications. These numbers refer to related items in the Reference Section. The reference information may be an estimating procedure, an alternate pricing method or technical information.

Note: Not all subdivisions listed here necessarily appear in this publication.

14560 | Chutes

			CREW	DAILY OUTPUT	LABOR-HOURS	UNIT	2005 BARE COSTS MAT.	LABOR	EQUIP.	TOTAL	TOTAL INCL O&P	
250	0010	**CHUTES** Linen or refuse, incl. sprinklers, 12′ floor height										250
	0050	Aluminized steel, 16 ga., 18″ diameter	2 Shee	3.50	4.571	Floor	860	184		1,044	1,225	
	0100	24″ diameter		3.20	5		935	201		1,136	1,325	
	0200	30″ diameter		3	5.333		1,025	214		1,239	1,450	
	0300	36″ diameter		2.80	5.714		1,175	230		1,405	1,625	
	0400	Galvanized steel, 16 ga., 18″ diameter		3.50	4.571		765	184		949	1,125	
	0500	24″ diameter		3.20	5		860	201		1,061	1,250	
	0600	30″ diameter		3	5.333		965	214		1,179	1,400	
	0700	36″ diameter		2.80	5.714		1,150	230		1,380	1,600	
	0800	Stainless steel, 18″ diameter		3.50	4.571		1,775	184		1,959	2,225	
	0900	24″ diameter		3.20	5		1,875	201		2,076	2,375	
	1000	30″ diameter		3	5.333		2,225	214		2,439	2,775	
	1005	36″ diameter		2.80	5.714		2,475	230		2,705	3,075	
	1200	Linen chute bottom collector, aluminized steel		4	4	Ea.	1,075	161		1,236	1,425	
	1300	Stainless steel		4	4		1,375	161		1,536	1,750	
	1500	Refuse, bottom hopper, aluminized steel, 18″ diameter		3	5.333		695	214		909	1,100	
	1600	24″ diameter		3	5.333		765	214		979	1,175	
	1800	36″ diameter		3	5.333		1,300	214		1,514	1,775	
	2900	Package chutes, spiral type, minimum		4.50	3.556	Floor	1,850	143		1,993	2,250	
	3000	Maximum		1.50	10.667	″	4,825	430		5,255	5,975	

14580 | Pneumatic Tube Systems

			CREW	DAILY OUTPUT	LABOR-HOURS	UNIT	MAT.	LABOR	EQUIP.	TOTAL	TOTAL INCL O&P	
800	0010	**PNEUMATIC TUBE SYSTEM** Single tube, 2 stations, blower										800
	0020	100′ long, stock										
	0100	3″ diameter	2 Stpi	.12	133	Total	5,025	5,450		10,475	13,700	
	0300	4″ diameter	″	.09	177	″	5,675	7,275		12,950	17,200	
	0400	Twin tube, two stations or more, conventional system										
	0600	2-1/2″ round	2 Stpi	62.50	.256	L.F.	10.15	10.50		20.65	27	
	0700	3″ round		46	.348		11.65	14.25		25.90	34.50	
	0900	4″ round		49.60	.323		12.90	13.20		26.10	34	
	1000	4″ x 7″ oval		37.60	.426		18.80	17.45		36.25	46.50	
	1050	Add for blower		2	8	System	3,975	330		4,305	4,875	
	1110	Plus for each round station, add		7.50	2.133	Ea.	450	87.50		537.50	625	
	1150	Plus for each oval station, add		7.50	2.133	″	450	87.50		537.50	625	
	1200	Alternate pricing method: base cost, minimum		.75	21.333	Total	4,875	875		5,750	6,675	
	1300	Maximum		.25	64	″	9,675	2,625		12,300	14,600	
	1500	Plus total system length, add, minimum		93.40	.171	L.F.	6.25	7		13.25	17.45	
	1600	Maximum		37.60	.426	″	18.50	17.45		35.95	46.50	
	1800	Completely automatic system, 4″ round, 15 to 50 stations		.29	55.172	Station	15,300	2,250		17,550	20,300	
	2200	51 to 144 stations		.32	50		11,900	2,050		13,950	16,200	
	2400	6″ round or 4″ x 7″ oval, 15 to 50 stations		.24	66.667		19,200	2,725		21,925	25,200	
	2800	51 to 144 stations		.23	69.565		16,100	2,850		18,950	22,000	

For information about Means Estimating Seminars, see yellow pages 12 and 13 in back of book

Division 15 Mechanical

Estimating Tips

15100 Building Services Piping

This subdivision is primarily basic pipe and related materials. The pipe may be used by any of the mechanical disciplines, i.e., plumbing, fire protection, heating, and air conditioning.

- The piping section lists the add to labor for elevated pipe installation. These adds apply to all elevated pipe, fittings, valves, insulation, etc., that are placed above 10′ high. CAUTION: the correct percentage may vary for the same pipe. For example, the percentage add for the basic pipe installation should be based on the maximum height that the craftsman must install for that particular section. If the pipe is to be located 14′ above the floor but it is suspended on threaded rod from beams, the bottom flange of which is 18′ high (4′ rods), then the height is actually 18′ and the add is 20%. The pipe coverer, however, does not have to go above the 14′ and so his add should be 10%.
- Most pipe is priced first as straight pipe with a joint (coupling, weld, etc.) every 10′ and a hanger usually every 10′. There are exceptions with hanger spacing such as: for cast iron pipe (5′) and plastic pipe (3 per 10′). Following each type of pipe there are several lines listing sizes and the amount to be subtracted to delete couplings and hangers. This is for pipe that is to be buried or supported together on trapeze hangers. The reason that the couplings are deleted is that these runs are usually long and frequently longer lengths of pipe are used. By deleting the couplings the estimator is expected to look up and add back the correct reduced number of couplings.
- When preparing an estimate it may be necessary to approximate the fittings. Fittings usually run between 25% and 50% of the cost of the pipe. The lower percentage is for simpler runs, and the higher number is for complex areas like mechanical rooms.
- For historic restoration projects, the systems must be as invisible as possible, and pathways must be sought for pipes, conduits, and ductwork. While installations in accessible spaces (such as basements and attics) are relatively straightforward to estimate, labor costs may be more difficult to determine when delivery systems must be concealed.

15400 Plumbing Fixtures & Equipment

- Plumbing fixture costs usually require two lines, the fixture itself and its "rough-in, supply and waste".
- In the Assemblies Section (Plumbing D2010) for the desired fixture, the System Components Group in the center of the page shows the fixture itself on the first line while the rest of the list (fittings, pipe, tubing, etc.) will total up to what we refer to in the Unit Price section as "Rough-in, supply, waste and vent". Note that for most fixtures we allow a nominal 5′ of tubing to reach from the fixture to a main or riser.
- Remember that gas and oil fired units need venting.

15500 Heat Generation Equipment

- When estimating the cost of an HVAC system, check to see who is responsible for providing and installing the temperature control system. It is possible to overlook controls, assuming that they would be included in the electrical estimate.
- When looking up a boiler be careful on specified capacity. Some manufacturers rate their products on output while others use input.
- Include HVAC insulation for pipe, boiler and duct (wrap and liner).
- Be careful when looking up mechanical items to get the correct pressure rating and connection type (thread, weld, flange).

15700 Heating/Ventilation/ Air Conditioning Equipment

- Combination heating and cooling units are sized by the air conditioning requirements. (See Reference No. R15710-020 for preliminary sizing guide.)
- A ton of air conditioning is nominally 400 CFM.
- Rectangular duct is taken off by the linear foot for each size, but its cost is usually estimated by the pound. Remember that SMACNA standards now base duct on internal pressure.
- Prefabricated duct is estimated and purchased like pipe: straight sections and fittings.
- Note that cranes or other lifting equipment are not included on any lines in Division 15. For example, if a crane is required to lift a heavy piece of pipe into place high above a gym floor, or to put a rooftop unit on the roof of a four-story building, etc., it must be added. Due to the potential for extreme variation—from nothing additional required, to a major crane or helicopter—we feel that including a nominal amount for "lifting contingency" would be useless and detract from the accuracy of the estimate. When using equipment rental from Means do not forget to include the cost of the operator(s).

Reference Numbers

Reference numbers are shown in bold squares at the beginning of some major classifications. These numbers refer to related items in the Reference Section. The reference information may be an estimating procedure, an alternate pricing method or technical information.

Note: Not all subdivisions listed here necessarily appear in this publication.

Note: **i2 Trade Service,** *in part, has been used as a reference source for some of the material prices used in Division 15.*

		15051 | Mechanical General	CREW	DAILY OUTPUT	LABOR-HOURS	UNIT	2005 BARE COSTS MAT.	LABOR	EQUIP.	TOTAL	TOTAL INCL O&P	
100	0010	**BOILERS, GENERAL** Prices do not include flue piping, elec. wiring,										100
	0020	gas or oil piping, boiler base, pad, or tankless unless noted										
	0100	Boiler H.P.: 10 KW = 34 lbs/steam/hr = 33,475 BTU/hr. R15500 -020										
	0120											
	0150	To convert SFR to BTU rating: Hot water, 150 x SFR;										
	0160	Forced hot water, 180 x SFR; steam, 240 x SFR										
700	0010	**PIPING** See also divisions 02500 & 02600 for site work R15100 -070										700
	1000	Add to labor for elevated installation										
	1080	10′ to 15′ high						10%				
	1100	15′ to 20′ high						20%				
	1120	20′ to 25′ high						25%				
	1140	25′ to 30′ high						35%				
	1160	30′ to 35′ high						40%				
	1180	35′ to 40′ high						50%				
	1200	Over 40′ high						55%				

		15055 | Selective Mech Demolition	CREW	DAILY OUTPUT	LABOR-HOURS	UNIT	MAT.	LABOR	EQUIP.	TOTAL	TOTAL INCL O&P	
300	0010	**HVAC DEMOLITION** R15050 -720										300
	0100	Air conditioner, split unit, 3 ton	Q-5	2	8	Ea.		295		295	445	
	0150	Package unit, 3 ton	Q-6	3	8			305		305	460	
	0190	Rooftop, self contained, up to 5 ton	1 Plum	1.20	6.667			272		272	410	
	0250	Air curtain	Q-9	20	.800	L.F.		29		29	44.50	
	0254	Air filters, up thru 16,000 CFM		20	.800	Ea.		29		29	44.50	
	0256	20,000 thru 60,000 CFM		16	1			36		36	55.50	
	0297	Boiler blowdown	Q-5	8	2			73.50		73.50	111	
	0298	Boilers										
	0300	Electric, up thru 148 kW	Q-19	2	12	Ea.		460		460	685	
	0310	150 thru 518 kW	"	1	24			915		915	1,375	
	0320	550 thru 2000 kW	Q-21	.40	80			3,100		3,100	4,650	
	0330	2070 kW and up	"	.30	106			4,150		4,150	6,200	
	0340	Gas and/or oil, up thru 150 MBH	Q-7	2.20	14.545			570		570	855	
	0350	160 thru 2000 MBH		.80	40			1,550		1,550	2,350	
	0360	2100 thru 4500 MBH		.50	64			2,500		2,500	3,750	
	0370	4600 thru 7000 MBH		.30	106			4,175		4,175	6,250	
	0380	7100 thru 12,000 MBH		.16	200			7,800		7,800	11,700	
	0390	12,200 thru 25,000 MBH		.12	266			10,400		10,400	15,600	
	0400	Central station air handler unit, up thru 15 ton	Q-5	1.60	10			370		370	555	
	0410	17.5 thru 30 ton	"	.80	20			735		735	1,100	
	0430	Computer room unit										
	0434	Air cooled split, up thru 10 ton	Q-5	.67	23.881	Ea.		880		880	1,325	
	0436	12 thru 23 ton		.53	30.189			1,100		1,100	1,675	
	0440	Chilled water, up thru 10 ton		1.30	12.308			455		455	680	
	0444	12 thru 23 ton		1	16			590		590	885	
	0450	Glycol system, up thru 10 ton		.53	30.189			1,100		1,100	1,675	
	0454	12 thru 23 ton		.40	40			1,475		1,475	2,225	
	0460	Water cooled, not including condenser, up thru 10 ton		.80	20			735		735	1,100	
	0464	12 thru 23 ton		.60	26.667			985		985	1,475	
	0600	Condenser, up thru 50 ton		1	16			590		590	885	
	0610	51 thru 100 ton	Q-6	.80	30			1,150		1,150	1,725	
	0620	150 thru 1000 ton	"	.16	150			5,725		5,725	8,625	
	0660	Condensing unit, up thru 10 ton	Q-5	1.25	12.800			470		470	710	
	0670	11 thru 50 ton	"	.40	40			1,475		1,475	2,225	
	0680	60 thru 100 ton	Q-6	.30	80			3,050		3,050	4,600	
	0700	Cooling tower, up thru 400 ton		.80	30			1,150		1,150	1,725	
	0710	450 thru 600 ton		.53	45.283			1,725		1,725	2,600	

15055 | Selective Mech Demolition			DAILY OUTPUT	LABOR-HOURS	UNIT	2005 BARE COSTS MAT.	LABOR	EQUIP.	TOTAL	TOTAL INCL O&P
		CREW								
300 0720	700 thru 1300 ton (R15050-720)	Q-6	.40	60	Ea.		2,300		2,300	3,450
0780	Dehumidifier, up thru 155 lb./hr.	Q-1	8	2			73.50		73.50	111
0790	240 lb./hr. and up	″	2	8			294		294	440
1560	Ductwork									
1570	Metal, steel, sst, fabricated	Q-9	1,000	.016	Lb.		.58		.58	.89
1580	Aluminum, fabricated		485	.033	″		1.19		1.19	1.84
1590	Spiral, prefabricated		400	.040	L.F.		1.45		1.45	2.23
1600	Fiberglass, prefabricated		400	.040			1.45		1.45	2.23
1610	Flex, prefabricated		500	.032			1.16		1.16	1.78
1620	Glass fiber reinforced plastic, prefabricated		280	.057			2.07		2.07	3.18
1630	Diffusers, registers or grills, up to 20″ max dimension	1 Shee	50	.160	Ea.		6.45		6.45	9.90
1640	21 thru 36″ max dimension		36	.222			8.95		8.95	13.75
1650	Above 36″ max dimension		30	.267			10.70		10.70	16.50
1700	Evaporator, up thru 12,000 BTUH	Q-5	5.30	3.019			111		111	167
1710	12,500 thru 30,000 BTUH	″	2.70	5.926			218		218	330
1720	31,000 BTUH and up	Q-6	1.50	16			610		610	920
1730	Evaporative cooler, up thru 5 H.P.	Q-9	2.70	5.926			214		214	330
1740	10 thru 30 H.P.	″	.67	23.881			865		865	1,325
1750	Exhaust systems									
1760	Exhaust components	1 Shee	8	1	System		40		40	62
1770	Weld fume hoods	″	20	.400	Ea.		16.10		16.10	24.50
2120	Fans, up thru 1 H.P. or 2000 CFM	Q-9	8	2			72.50		72.50	111
2124	2 thru 10 H.P. or 20,000 CFM		5.30	3.019			109		109	168
2128	15 thru 30 H.P. or above 20,000 CFM		4	4			145		145	223
2150	Fan coil air conditioner, chilled water, up thru 7.5 ton	Q-5	14	1.143			42		42	63.50
2154	Direct expansion, up thru 10 ton		8	2			73.50		73.50	111
2158	11 thru 30 ton		2	8			295		295	445
2170	Flue shutter damper	Q-9	8	2			72.50		72.50	111
2200	Furnace, electric	Q-20	2	10			370		370	565
2300	Gas or oil, under 120 MBH	Q-9	4	4			145		145	223
2340	Over 120 MBH	″	3	5.333			193		193	297
2730	Heating and ventilating unit	Q-5	2.70	5.926			218		218	330
2740	Heater, electric, wall, baseboard and quartz	1 Elec	10	.800			32.50		32.50	48.50
2750	Heater, electric, unit, cabinet, fan and convector	″	8	1			41		41	60.50
2760	Heat exchanger, shell and tube type	Q-5	1.60	10			370		370	555
2770	Plate type	Q-6	.60	40			1,525		1,525	2,300
2810	Heat pump									
2820	Air source, split, up thru 10 ton	Q-5	.90	17.778	Ea.		655		655	985
2830	15 thru 25 ton	Q-6	.80	30			1,150		1,150	1,725
2850	Single package, up thru 12 ton	Q-5	1	16			590		590	885
2860	Water source, split, up thru 15 ton	″	.90	17.778			655		655	985
2870	20 thru 50 ton	Q-6	.80	30			1,150		1,150	1,725
2910	Heat recovery package, up thru 20,000 CFM	Q-5	2	8			295		295	445
2920	25,000 CFM and up		1.20	13.333			490		490	740
2930	Heat transfer package, up thru 130 GPM		.80	20			735		735	1,100
2934	255 thru 800 GPM		.42	38.095			1,400		1,400	2,100
2940	Humidifier		10.60	1.509			55.50		55.50	83.50
2961	Hydronic unit heaters, up thru 200 MBH		14	1.143			42		42	63.50
2962	Above 200 MBH		8	2			73.50		73.50	111
2964	Valance units		32	.500			18.45		18.45	27.50
2966	Radiant floor heating									
2967	System valves, controls, manifolds	Q-5	16	1	Ea.		37		37	55.50
2968	Per room distribution		8	2			73.50		73.50	111
2970	Hydronic heating, baseboard radiation		16	1			37		37	55.50
2976	Convectors and free standing radiators		18	.889			33		33	49
2980	Induced draft fan, up thru 1 H.P.	Q-9	4.60	3.478			126		126	194

		15055 | Selective Mech Demolition	CREW	DAILY OUTPUT	LABOR-HOURS	UNIT	MAT.	LABOR	EQUIP.	TOTAL	TOTAL INCL O&P
							2005 BARE COSTS				
300	2984	1-1/2 H.P. thru 7-1/2 H.P. R15050-720	Q-9	2.20	7.273	Ea.		263		263	405
	2988	Infra-red unit	Q-5	16	1			37		37	55.50
	2992	Louvers	1 Shee	46	.174	S.F.		7		7	10.75
	3000	Mechanical equipment, light items. Unit is weight, not cooling.	Q-5	.90	17.778	Ton		655		655	985
	3600	Heavy items	"	1.10	14.545	"		535		535	805
	3700	Deduct for salvage (when applicable), minimum				Job					55
	3710	Maximum				"					420
	3720	Make up air unit, up thru 6000 CFM	Q-5	3	5.333	Ea.		197		197	295
	3730	8000 thru 30,000 CFM	"	1.60	10			370		370	555
	3740	35,000 thru 75,000 CFM	Q-6	1	24			915		915	1,375
	3800	Mixing boxes, constant and VAV	Q-9	18	.889			32		32	49.50
	4000	Packaged terminal air conditioner, up thru 18,000 BTUH	Q-5	8	2			73.50		73.50	111
	4010	24,000 thru 48,000 BTUH	"	2.80	5.714			211		211	315
	5000	Refrigerant compressor, reciprocating or scroll									
	5010	Up thru 5 ton	1 Stpi	6	1.333	Ea.		54.50		54.50	82
	5020	5.08 thru 10 ton	Q-5	6	2.667			98.50		98.50	148
	5030	15 thru 50 ton	"	.40	40			1,475		1,475	2,225
	5040	60 thru 130 ton	Q-6	.48	50			1,900		1,900	2,875
	5100	Roof top air conditioner, up thru 10 ton	Q-5	1.40	11.429			420		420	635
	5110	12-1/2 thru 40 ton	Q-6	1	24			915		915	1,375
	5120	50 thru 140 ton		.50	48			1,825		1,825	2,750
	5130	150 thru 300 ton		.30	80			3,050		3,050	4,600
	6000	Self contained single package air conditioner, up thru 10 ton	Q-5	1.60	10			370		370	555
	6010	15 thru 60 ton	Q-6	1.20	20			765		765	1,150
	6100	Space heaters, up thru 200 MBH	Q-5	10	1.600			59		59	88.50
	6110	Over 200 MBH		5	3.200			118		118	177
	6200	Split ductless, both sections		8	2			73.50		73.50	111
	6300	Steam condensate meter	1 Stpi	11	.727			30		30	45
	6600	Thru-the-wall air conditioner	L-2	8	2			60		60	93
	7000	Vent chimney, prefabricated, up thru 12" diameter	Q-9	94	.170	V.L.F.		6.15		6.15	9.45
	7010	14" thru 36" diameter		40	.400			14.45		14.45	22.50
	7020	38" thru 48" diameter		32	.500			18.10		18.10	28
	7030	54" thru 60" diameter	Q-10	14	1.714			64.50		64.50	99
	7400	Ventilators, up thru 14" neck diameter	Q-9	58	.276	Ea.		10		10	15.35
	7410	16" thru 50" neck diameter		40	.400			14.45		14.45	22.50
	7450	Relief vent, up thru 24" x 96"		22	.727			26.50		26.50	40.50
	7460	48" x 60" thru 96" x 144"		10	1.600			58		58	89
	8000	Water chiller up thru 10 ton	Q-5	2.50	6.400			236		236	355
	8010	15 thru 100 ton	Q-6	.48	50			1,900		1,900	2,875
	8020	110 thru 500 ton	Q-7	.29	110			4,300		4,300	6,475
	8030	600 thru 1000 ton		.23	139			5,425		5,425	8,150
	8040	1100 ton and up		.20	160			6,250		6,250	9,375
	8400	Window air conditioner	1 Carp	16	.500			17.15		17.15	26.50
600	0010	**PLUMBING DEMOLITION** R15050-720									
	0400	Air compressor, up thru 2 H.P.	Q-1	10	1.600	Ea.		59		59	88.50
	0410	3 H.P. thru 7-1/2 H.P.		5.60	2.857			105		105	158
	0420	10 H.P. thru 15 H.P.		1.40	11.429			420		420	630
	0430	20 H.P. thru 30 H.P.	Q-2	1.30	18.462			705		705	1,050
	0500	Backflow preventer, up thru 2" diameter	1 Plum	17	.471			19.20		19.20	29
	0510	2-1/2" thru 3" diameter	Q-1	10	1.600			59		59	88.50
	0520	4" thru 6" diameter	"	5	3.200			118		118	177
	0530	8" thru 10" diameter	Q-2	3	8			305		305	460
	0700	Carriers and supports									
	0710	Fountains, sinks, lavatories and urinals	1 Plum	14	.571	Ea.		23.50		23.50	35
	0720	Water closets	"	12	.667	"		27		27	41

15 MECHANICAL

15055 | Selective Mech Demolition

			CREW	DAILY OUTPUT	LABOR-HOURS	UNIT	2005 BARE COSTS MAT.	LABOR	EQUIP.	TOTAL	TOTAL INCL O&P
600	0730	Grinder pump or sewage ejector system R15050-720									
	0732	Simplex	Q-1	7	2.286	Ea.		84		84	126
	0734	Duplex	"	2.80	5.714			210		210	315
	0738	Hot water dispenser	1 Plum	36	.222			9.10		9.10	13.65
	0740	Hydrant, wall		26	.308			12.55		12.55	18.90
	0744	Ground		12	.667			27		27	41
	0750	Hot water dispenser		36	.222			9.10		9.10	13.65
	0760	Cleanouts and drains, up thru 4" pipe diameter		10	.800			32.50		32.50	49
	0764	5" thru 8" pipe diameter	Q-1	10	1.600			59		59	88.50
	0780	Industrial safety fixtures	1 Plum	8	1			41		41	61.50
	1020	Fixtures, including 10' piping									
	1100	Bath tubs, cast iron	1 Plum	4	2	Ea.		81.50		81.50	123
	1120	Fiberglass		6	1.333			54.50		54.50	82
	1140	Steel		5	1.600			65.50		65.50	98
	1150	Bidet	Q-1	7	2.286			84		84	126
	1200	Lavatory, wall hung	1 Plum	10	.800			32.50		32.50	49
	1220	Counter top		8	1			41		41	61.50
	1300	Sink, steel or cast iron, single		8	1			41		41	61.50
	1320	Double		7	1.143			46.50		46.50	70
	1340	Shower, stall and receptor	Q-1	6	2.667			98		98	147
	1350	Group	"	7	2.286			84		84	126
	1400	Water closet, floor mounted	1 Plum	8	1			41		41	61.50
	1420	Wall mounted	"	7	1.143			46.50		46.50	70
	1440	Wash fountain, 36" diameter	Q-2	8	3			114		114	172
	1442	54" diameter	"	7	3.429			131		131	197
	1500	Urinal, floor mounted	1 Plum	4	2			81.50		81.50	123
	1520	Wall mounted	"	7	1.143			46.50		46.50	70
	1590	Whirl pool or hot tub	Q-1	2.60	6.154			226		226	340
	1600	Water fountains, free standing	1 Plum	8	1			41		41	61.50
	1620	Recessed		6	1.333			54.50		54.50	82
	1800	Medical gas specialties		8	1			41		41	61.50
	1900	Piping fittings, single connection, up thru 1-1/2" diameter		30	.267			10.90		10.90	16.35
	1910	2" thru 4" diameter		14	.571			23.50		23.50	35
	1980	Pipe hanger/support removal		80	.100			4.09		4.09	6.15
	1990	Glass pipe with fittings, 1" thru 3" diameter		200	.040	L.F.		1.63		1.63	2.46
	1992	4" thru 6" diameter		150	.053			2.18		2.18	3.27
	2000	Piping, metal, up thru 1-1/2" diameter		200	.040			1.63		1.63	2.46
	2050	2" thru 3-1/2" diameter		150	.053			2.18		2.18	3.27
	2100	4" thru 6" diameter	2 Plum	100	.160			6.55		6.55	9.80
	2150	8" thru 14" diameter	"	60	.267			10.90		10.90	16.35
	2153	16" thru 20 " diameter	Q-18	70	.343			13.10	1.15	14.25	21
	2155	24" thru 26 " diameter		55	.436			16.70	1.47	18.17	26.50
	2156	30" thru 36 " diameter		40	.600			23	2.02	25.02	36.50
	2160	Glass pipe with fittings, up thru 1-1/2" diameter	1 Plum	250	.032			1.31		1.31	1.96
	2162	2" thru 3" diameter	"	200	.040			1.63		1.63	2.46
	2164	4" thru 6" diameter	Q-1	200	.080			2.94		2.94	4.42
	2166	8" thru 14" diameter		150	.107			3.92		3.92	5.90
	2168	16" diameter		100	.160			5.90		5.90	8.85
	2170	Prison fixtures, lavatory or sink		18	.889	Ea.		32.50		32.50	49
	2172	Shower		5.60	2.857			105		105	158
	2174	Urinal or water closet		13	1.231			45.50		45.50	68
	2180	Pumps, all fractional horse-power		12	1.333			49		49	73.50
	2184	1 H.P. thru 5 H.P.		6	2.667			98		98	147
	2186	7-1/2 H.P. thru 15 H.P.		2.50	6.400			235		235	355
	2188	20 H.P. thru 25 H.P.	Q-2	4	6			229		229	345
	2190	30 H.P. thru 60 H.P.		.80	30			1,150		1,150	1,725

	15055 | Selective Mech Demolition		CREW	DAILY OUTPUT	LABOR-HOURS	UNIT	2005 BARE COSTS MAT.	LABOR	EQUIP.	TOTAL	TOTAL INCL O&P	
600	2192	75 H.P. thru 100 H.P. (R15050-720)	Q-2	.60	40	Ea.		1,525		1,525	2,300	600
	2194	150 H.P.		.50	48			1,825		1,825	2,750	
	2198	Pump, sump or submersible	1 Plum	12	.667			27		27	41	
	2200	Receptors and interceptors, up thru 20 GPM	"	8	1			41		41	61.50	
	2204	25 thru 100 GPM	Q-1	6	2.667			98		98	147	
	2208	125 thru 300 GPM	"	2.40	6.667			245		245	370	
	2211	325 thru 500 GPM	Q-2	2.60	9.231			350		350	530	
	2212	Deduct for salvage, aluminum scrap				Ton					525	
	2214	Brass scrap									420	
	2216	Copper scrap									1,025	
	2218	Lead scrap									235	
	2220	Steel scrap									65	
	2230	Temperature maintenance cable	1 Plum	1,200	.007	L.F.		.27		.27	.41	
	2250	Water heater, 40 gal.	"	6	1.333	Ea.		54.50		54.50	82	
	3100	Tanks, water heaters and liquid containers										
	3110	Up thru 40 gallons	Q-1	22	.727	Ea.		27		27	40	
	3120	50 thru 120 gallons		14	1.143			42		42	63	
	3130	130 thru 240 gallons		7.60	2.105			77.50		77.50	116	
	3140	250 thru 500 gallons		5.40	2.963			109		109	164	
	3150	600 thru 1000 gallons	Q-2	1.60	15			570		570	860	
	3160	1100 thru 2000 gallons		.70	34.286			1,300		1,300	1,975	
	3170	2100 thru 4000 gallons		.50	48			1,825		1,825	2,750	
	6000	Remove and reset fixtures, minimum	1 Plum	6	1.333			54.50		54.50	82	
	6100	Maximum	"	4	2			81.50		81.50	123	
	9100	Valve, metal, up thru 1-1/2" diameter	1 Stpi	28	.286			11.70		11.70	17.60	
	9110	2" thru 3" diameter	Q-1	11	1.455			53.50		53.50	80.50	
	9120	4" thru 6" diameter	"	8	2			73.50		73.50	111	
	9130	8" thru 14" diameter	Q-2	8	3			114		114	172	
	9140	16" thru 20" diameter		2	12			460		460	690	
	9150	24" diameter		1.20	20			765		765	1,150	
	9200	Valve, plastic, up thru 1-1/2" diameter	1 Plum	42	.190			7.80		7.80	11.70	
	9210	2" thru 3" diameter		15	.533			22		22	33	
	9220	4" thru 6" diameter		12	.667			27		27	41	
	9300	Vent flashing and caps		55	.145			5.95		5.95	8.95	
	9350	Water filter, commercial, 1" thru 1-1/2"	Q-1	2	8			294		294	440	
	9360	2" thru 2-1/2"	"	1.60	10			370		370	555	
	9400	Water heaters (when tank size given see line 15055 600 3000)										
	9410	Up thru 245 GPH	Q-1	2.40	6.667	Ea.		245		245	370	
	9420	250 thru 756 GPH		1.60	10			370		370	555	
	9430	775 thru 1640 GPH		.80	20			735		735	1,100	
	9440	1650 thru 4000 GPH	Q-2	.50	48			1,825		1,825	2,750	
	9470	Water softener	Q-1	2	8			294		294	440	
	15060 | Hangers & Supports											
300	0010	PIPE HANGERS AND SUPPORTS, TYPE numbers per MSS-SP58										300
	0050	Brackets										
	0060	Beam side or wall, malleable iron, TYPE 34										
	0070	3/8" threaded rod size	1 Plum	48	.167	Ea.	3.39	6.80		10.19	14	
	0080	1/2" threaded rod size		48	.167		6.55	6.80		13.35	17.45	
	0090	5/8" threaded rod size		48	.167		8.70	6.80		15.50	19.80	
	0100	3/4" threaded rod size		48	.167		10.45	6.80		17.25	22	
	0110	7/8" threaded rod size		48	.167		12.55	6.80		19.35	24	
	0120	For concrete installation, add						30%				
	0150	Wall, welded steel, medium, TYPE 32										
	0160	0 size, 12" wide, 18" deep	1 Plum	34	.235	Ea.	255	9.60		264.60	294	
	0170	1 size, 18" wide 24" deep		34	.235		305	9.60		314.60	350	

15060 | Hangers & Supports		CREW	DAILY OUTPUT	LABOR-HOURS	UNIT	MAT.	LABOR	EQUIP.	TOTAL	TOTAL INCL O&P	
300	0180 2 size, 24" wide, 30" deep	1 Plum	34	.235	Ea.	375	9.60		384.60	425	300
	0200 Beam attachment, welded, TYPE 22										
	0202 3/8"	Q-15	80	.200	Ea.	15.35	7.35	1.01	23.71	29	
	0203 1/2"		76	.211		16.35	7.75	1.06	25.16	31	
	0204 5/8"		72	.222		17.35	8.15	1.12	26.62	32.50	
	0205 3/4"		68	.235		23.50	8.65	1.19	33.34	40.50	
	0206 7/8"		64	.250		30	9.20	1.26	40.46	48	
	0207 1"	↓	56	.286	↓	45.50	10.50	1.44	57.44	67.50	
	0300 Clamps										
	0310 C-clamp, for mounting on steel beam flange, w/locknut, TYPE 23										
	0320 3/8" threaded rod size	1 Plum	160	.050	Ea.	1.81	2.04		3.85	5.05	
	0330 1/2" threaded rod size		160	.050		2.26	2.04		4.30	5.55	
	0340 5/8" threaded rod size		160	.050		3.46	2.04		5.50	6.90	
	0350 3/4" threaded rod size		160	.050		4.68	2.04		6.72	8.20	
	0352 7/8" threaded rod size	↓	140	.057	↓	16.40	2.33		18.73	21.50	
	0400 High temperature to 1050°F, alloy steel										
	0410 4" pipe size	Q-1	106	.151	Ea.	27.50	5.55		33.05	38.50	
	0420 6" pipe size		106	.151		46	5.55		51.55	59	
	0430 8" pipe size		97	.165		52	6.05		58.05	66	
	0440 10" pipe size		84	.190		70.50	7		77.50	88	
	0450 12" pipe size		72	.222		81.50	8.15		89.65	102	
	0460 14" pipe size		64	.250		171	9.20		180.20	202	
	0470 16" pipe size	↓	56	.286	↓	195	10.50		205.50	230	
	0480 Beam clamp, flange type, TYPE 25										
	0482 For 3/8" bolt	1 Plum	48	.167	Ea.	4.09	6.80		10.89	14.75	
	0483 For 1/2" bolt		44	.182		9.45	7.45		16.90	21.50	
	0484 For 5/8" bolt		40	.200		18	8.15		26.15	32	
	0485 For 3/4" bolt		36	.222		18.35	9.10		27.45	33.50	
	0486 For 1" bolt	↓	32	.250	↓	88	10.20		98.20	112	
	0500 I-beam, for mounting on bottom flange, strap iron, TYPE 21										
	0510 2" flange size	1 Plum	96	.083	Ea.	4.42	3.40		7.82	9.95	
	0520 3" flange size		95	.084		4.56	3.44		8	10.15	
	0530 4" flange size		93	.086		4.67	3.51		8.18	10.45	
	0540 5" flange size		92	.087		4.84	3.55		8.39	10.65	
	0550 6" flange size		90	.089		4.95	3.63		8.58	10.90	
	0560 7" flange size		88	.091		5.10	3.71		8.81	11.25	
	0570 8" flange size	↓	86	.093	↓	5.30	3.80		9.10	11.50	
	0600 One hole, vertical mounting, malleable iron										
	0610 1/2" pipe size	1 Plum	160	.050	Ea.	.76	2.04		2.80	3.91	
	0620 3/4" pipe size		145	.055		.91	2.25		3.16	4.39	
	0630 1" pipe size		136	.059		1.35	2.40		3.75	5.10	
	0640 1-1/4" pipe size		128	.063		1.51	2.55		4.06	5.50	
	0650 1-1/2" pipe size		120	.067		1.69	2.72		4.41	5.95	
	0660 2" pipe size		112	.071		2.45	2.92		5.37	7.10	
	0670 2-1/2" pipe size		104	.077		4.77	3.14		7.91	9.95	
	0680 3" pipe size		96	.083		6.80	3.40		10.20	12.55	
	0690 3-1/2" pipe size		90	.089		6.80	3.63		10.43	12.90	
	0700 4" pipe size	↓	84	.095	↓	10.45	3.89		14.34	17.30	
	0750 Riser or extension pipe, carbon steel, TYPE 8										
	0756 1/2" pipe size	1 Plum	52	.154	Ea.	3.33	6.30		9.63	13.10	
	0760 3/4" pipe size		48	.167		3.33	6.80		10.13	13.90	
	0770 1" pipe size		47	.170		3.35	6.95		10.30	14.15	
	0780 1-1/4" pipe size		46	.174		4.09	7.10		11.19	15.20	
	0790 1-1/2" pipe size		45	.178		4.30	7.25		11.55	15.65	
	0800 2" pipe size		43	.186		4.52	7.60		12.12	16.35	
	0810 2-1/2" pipe size	↓	41	.195	↓	4.77	7.95		12.72	17.25	

Column group "2005 BARE COSTS" covers MAT., LABOR, EQUIP., TOTAL.

15060 | Hangers & Supports		CREW	DAILY OUTPUT	LABOR-HOURS	UNIT	2005 BARE COSTS MAT.	LABOR	EQUIP.	TOTAL	TOTAL INCL O&P	
300 0820	3″ pipe size	1 Plum	40	.200	Ea.	5.20	8.15		13.35	18.05	300
0830	3-1/2″ pipe size		39	.205		6.35	8.40		14.75	19.60	
0840	4″ pipe size		38	.211		6.55	8.60		15.15	20	
0850	5″ pipe size		37	.216		9.05	8.85		17.90	23.50	
0860	6″ pipe size		36	.222		10.90	9.10		20	25.50	
0870	8″ pipe size		34	.235		17.65	9.60		27.25	34	
0880	10″ pipe size		32	.250		26.50	10.20		36.70	44.50	
0890	12″ pipe size	↓	28	.286	↓	32	11.65		43.65	53	
0900	For plastic coating 3/4″ to 4″, add					190%					
0910	For copper plating 3/4″ to 4″, add					58%					
0950	Two piece, complete, carbon steel, medium weight, TYPE 4										
0960	1/2″ pipe size	Q-1	137	.117	Ea.	2.32	4.30		6.62	9	
0970	3/4″ pipe size		134	.119		2.32	4.39		6.71	9.15	
0980	1″ pipe size		132	.121		2.45	4.46		6.91	9.40	
0990	1-1/4″ pipe size		130	.123		3.04	4.53		7.57	10.15	
1000	1-1/2″ pipe size		126	.127		3.04	4.67		7.71	10.35	
1010	2″ pipe size		124	.129		3.76	4.75		8.51	11.30	
1020	2-1/2″ pipe size		120	.133		4.05	4.90		8.95	11.80	
1030	3″ pipe size		117	.137		4.52	5.05		9.57	12.50	
1040	3-1/2″ pipe size		114	.140		6.40	5.15		11.55	14.80	
1050	4″ pipe size		110	.145		6.40	5.35		11.75	15.10	
1060	5″ pipe size		106	.151		9.25	5.55		14.80	18.50	
1070	6″ pipe size		104	.154		15.60	5.65		21.25	25.50	
1080	8″ pipe size		100	.160		22.50	5.90		28.40	33.50	
1090	10″ pipe size		96	.167		33.50	6.15		39.65	46	
1100	12″ pipe size		89	.180		39	6.60		45.60	53	
1110	14″ pipe size		82	.195		68	7.20		75.20	86	
1120	16″ pipe size	↓	68	.235		73.50	8.65		82.15	94	
1130	For galvanized, add				↓	45%					
1150	Insert, concrete										
1160	Wedge type, carbon steel body, malleable iron nut										
1170	1/4″ threaded rod size	1 Plum	96	.083	Ea.	1.77	3.40		5.17	7.05	
1180	3/8″ threaded rod size		96	.083		1.81	3.40		5.21	7.10	
1190	1/2″ threaded rod size		96	.083		1.91	3.40		5.31	7.20	
1200	5/8″ threaded rod size		96	.083		1.95	3.40		5.35	7.25	
1210	3/4″ threaded rod size		96	.083		2.30	3.40		5.70	7.65	
1220	7/8″ threaded rod size	↓	96	.083		2.89	3.40		6.29	8.30	
1230	For galvanized, add				↓	1.99			1.99	2.19	
1250	Pipe guide sized for insulation										
1260	No. 1, 1″ pipe size, 1″ thick insulation	1 Stpi	26	.308	Ea.	114	12.60		126.60	144	
1270	No. 2, 1-1/4″-2″ pipe size, 1″ thick insulation		23	.348		134	14.25		148.25	170	
1280	No. 3, 1-1/4″-2″ pipe size, 1-1/2″ thick insulation		21	.381		134	15.60		149.60	172	
1290	No. 4, 2-1/2″-3-1/2″ pipe size, 1-1/2″ thick insulation		18	.444		134	18.20		152.20	176	
1300	No. 5, 4″-5″ pipe size, 1-1/2″ thick insulation	↓	16	.500		149	20.50		169.50	195	
1310	No. 6, 5″-6″ pipe size, 2″ thick insulation	Q-5	21	.762		166	28		194	224	
1320	No. 7, 8″ pipe size, 2″ thick insulation		16	1		290	37		327	375	
1330	No. 8, 10″ pipe size, 2″ thick insulation	↓	12	1.333		550	49		599	680	
1340	No. 9, 12″ pipe size, 2″ thick insulation	Q-6	17	1.412		550	54		604	685	
1350	No. 10, 12″-14″ pipe size, 2-1/2″ thick insulation		16	1.500		715	57.50		772.50	870	
1360	No. 11, 16″ pipe size, 2-1/2″ thick insulation		10.50	2.286		715	87.50		802.50	915	
1370	No. 12, 16″-18″ pipe size, 3″ thick insulation		9	2.667		755	102		857	985	
1380	No. 13, 20″ pipe size, 3″ thick insulation		7.50	3.200		755	122		877	1,025	
1390	No. 14, 24″ pipe size, 3″ thick insulation	↓	7	3.429	↓	965	131		1,096	1,250	
1400	Bands										
1410	Adjustable band, carbon steel, for non-insulated pipe, TYPE 7										
1420	1/2″ pipe size	Q-1	142	.113	Ea.	.19	4.14		4.33	6.45	

	15060 | Hangers & Supports	CREW	DAILY OUTPUT	LABOR-HOURS	UNIT	MAT.	LABOR	EQUIP.	TOTAL	TOTAL INCL O&P	
300	1430 3/4" pipe size	Q-1	140	.114	Ea.	.85	4.20		5.05	7.25	300
	1440 1" pipe size		137	.117		.99	4.30		5.29	7.55	
	1450 1-1/4" pipe size		134	.119		1.03	4.39		5.42	7.75	
	1460 1-1/2" pipe size		131	.122		1.14	4.49		5.63	8	
	1470 2" pipe size		129	.124		1.58	4.56		6.14	8.60	
	1480 2-1/2" pipe size		125	.128		1.65	4.71		6.36	8.90	
	1490 3" pipe size		122	.131		1.81	4.82		6.63	9.25	
	1500 3-1/2" pipe size		119	.134		1.96	4.95		6.91	9.60	
	1510 4" pipe size		114	.140		2.82	5.15		7.97	10.85	
	1520 5" pipe size		110	.145		2.98	5.35		8.33	11.35	
	1530 6" pipe size		108	.148		4.13	5.45		9.58	12.75	
	1540 8" pipe size	↓	104	.154	↓	6.45	5.65		12.10	15.60	
	1550 For copper plated, add					50%					
	1560 For galvanized, add					30%					
	1570 For plastic coating, add					30%					
	1600 Adjusting nut malleable iron, steel band, TYPE 9										
	1610 1/2" pipe size, galvanized band	Q-1	137	.117	Ea.	2.73	4.30		7.03	9.45	
	1620 3/4" pipe size, galvanized band		135	.119		2.73	4.36		7.09	9.55	
	1630 1" pipe size, gavanized band		132	.121		2.73	4.46		7.19	9.70	
	1640 1-1/4" pipe size, galvanized band		129	.124		3.30	4.56		7.86	10.50	
	1650 1-1/2" pipe size, galvanized band		126	.127		3.30	4.67		7.97	10.65	
	1660 2" pipe size, galvanized band		124	.129		3.58	4.75		8.33	11.10	
	1670 2-1/2" pipe size, galvanized band		120	.133		6	4.90		10.90	13.95	
	1680 3" pipe size, galvanized band		117	.137		6.55	5.05		11.60	14.75	
	1690 3-1/2" pipe size, galvanized band		114	.140		8.20	5.15		13.35	16.80	
	1700 4" pipe size, cadmium plated band	↓	110	.145		10.50	5.35		15.85	19.60	
	1740 For plastic coated band, add					35%					
	1750 For completely copper coated, add				↓	45%					
	1800 Clevis, adjustable, carbon steel, for non-insulated pipe, TYPE 1										
	1810 1/2" pipe size	Q-1	137	.117	Ea.	.34	4.30		4.64	6.80	
	1820 3/4" pipe size		135	.119		.34	4.36		4.70	6.90	
	1830 1" pipe size		132	.121		.36	4.46		4.82	7.10	
	1840 1-1/4" pipe size		129	.124		.41	4.56		4.97	7.30	
	1850 1-1/2" pipe size		126	.127		.45	4.67		5.12	7.50	
	1860 2" pipe size		124	.129		.50	4.75		5.25	7.70	
	1870 2-1/2" pipe size		120	.133		.82	4.90		5.72	8.25	
	1880 3" pipe size		117	.137		1.03	5.05		6.08	8.70	
	1890 3-1/2" pipe size		114	.140		1.08	5.15		6.23	8.95	
	1900 4" pipe size		110	.145		1.24	5.35		6.59	9.40	
	1910 5" pipe size		106	.151		1.72	5.55		7.27	10.25	
	1920 6" pipe size		104	.154		1.98	5.65		7.63	10.70	
	1930 8" pipe size		100	.160		3.26	5.90		9.16	12.45	
	1940 10" pipe size		96	.167		27	6.15		33.15	39	
	1950 12" pipe size		89	.180		34	6.60		40.60	47.50	
	1960 14" pipe size		82	.195		51	7.20		58.20	67.50	
	1970 16" pipe size		68	.235		53	8.65		61.65	71.50	
	1971 18" pipe size		54	.296		60.50	10.90		71.40	83	
	1972 20" pipe size	↓	38	.421	↓	127	15.50		142.50	163	
	1980 For galvanized, add					66%					
	1990 For copper plated 1/2" to 4", add					77%					
	2000 For light weight 1/2" to 4", deduct					13%					
	2010 Insulated pipe type, 3/4" to 12" pipe, add					180%					
	2020 Insulated pipe type, chrome-moly U-strap, add					530%					
	2250 Split ring, malleable iron, for non-insulated pipe, TYPE 11										
	2260 1/2" pipe size	Q-1	137	.117	Ea.	3.89	4.30		8.19	10.75	
	2270 3/4" pipe size	↓	135	.119	↓	3.92	4.36		8.28	10.85	

15060 | Hangers & Supports		CREW	DAILY OUTPUT	LABOR-HOURS	UNIT	2005 BARE COSTS MAT.	LABOR	EQUIP.	TOTAL	TOTAL INCL O&P
300	2280 1" pipe size	Q-1	132	.121	Ea.	4.14	4.46		8.60	11.25
	2290 1-1/4" pipe size		129	.124		5.15	4.56		9.71	12.50
	2300 1-1/2" pipe size		126	.127		6.25	4.67		10.92	13.85
	2310 2" pipe size		124	.129		7.10	4.75		11.85	14.95
	2320 2-1/2" pipe size		120	.133		10.25	4.90		15.15	18.60
	2330 3" pipe size		117	.137		12.75	5.05		17.80	21.50
	2340 3-1/2" pipe size		114	.140		13.05	5.15		18.20	22
	2350 4" pipe size		110	.145		13.30	5.35		18.65	22.50
	2360 5" pipe size		106	.151		17.70	5.55		23.25	28
	2370 6" pipe size		104	.154		39.50	5.65		45.15	52
	2380 8" pipe size	↓	100	.160	↓	70	5.90		75.90	86
	2390 For copper plated, add					8%				
	2500 Washer, flat steel									
	2502 3/8"	1 Plum	240	.033	Ea.	.05	1.36		1.41	2.11
	2503 1/2"		220	.036		.07	1.49		1.56	2.31
	2504 5/8"		200	.040		.15	1.63		1.78	2.63
	2505 3/4"		180	.044		.11	1.82		1.93	2.85
	2506 7/8"		160	.050		.16	2.04		2.20	3.25
	2507 1"		140	.057		.16	2.33		2.49	3.69
	2508 1-1/8"	↓	120	.067	↓	.48	2.72		3.20	4.62
	2520 Nut, steel, hex									
	2522 3/8"	1 Plum	200	.040	Ea.	.15	1.63		1.78	2.63
	2523 1/2"		180	.044		.27	1.82		2.09	3.03
	2524 5/8"		160	.050		.56	2.04		2.60	3.69
	2525 3/4"		140	.057		.76	2.33		3.09	4.35
	2526 7/8"		120	.067		1.27	2.72		3.99	5.50
	2527 1"		100	.080		2.83	3.27		6.10	8
	2528 1-1/8"	↓	80	.100	↓	5.40	4.09		9.49	12.10
	2532 Turnbuckle, TYPE 13									
	2534 3/8"	1 Plum	80	.100	Ea.	12.30	4.09		16.39	19.65
	2535 1/2"		72	.111		15.25	4.54		19.79	23.50
	2536 5/8"		64	.125		15.95	5.10		21.05	25
	2538 7/8"		48	.167		29.50	6.80		36.30	43
	2539 1"		40	.200		37	8.15		45.15	53
	2540 1-1/8"	↓	32	.250	↓	67	10.20		77.20	89
	2544 Eye rod, welded									
	2546 3/8"	1 Plum	56	.143	Ea.	16.30	5.85		22.15	26.50
	2547 1/2"		48	.167		24	6.80		30.80	37
	2548 5/8"		40	.200		36.50	8.15		44.65	52.50
	2549 3/4"		32	.250		46.50	10.20		56.70	66.50
	2550 7/8"		24	.333		57	13.60		70.60	83
	2551 1"	↓	16	.500	↓	72	20.50		92.50	110
	2556 Eye rod, welded, linked									
	2558 3/8"	1 Plum	56	.143	Ea.	49.50	5.85		55.35	63.50
	2559 1/2"		48	.167		67.50	6.80		74.30	84.50
	2560 5/8"		40	.200		97	8.15		105.15	119
	2561 3/4"		32	.250		126	10.20		136.20	153
	2562 7/8"		24	.333		153	13.60		166.60	190
	2563 1"	↓	16	.500	↓	197	20.50		217.50	247
	2650 Rods, carbon steel									
	2660 Continuous thread									
	2670 1/4" thread size	1 Plum	144	.056	L.F.	.25	2.27		2.52	3.69
	2680 3/8" thread size		144	.056		.27	2.27		2.54	3.71
	2690 1/2" thread size		144	.056		.33	2.27		2.60	3.77
	2700 5/8" thread size		144	.056		.50	2.27		2.77	3.96
	2710 3/4" thread size	↓	144	.056	↓	1.03	2.27		3.30	4.54

15 MECHANICAL

15060 | Hangers & Supports		CREW	DAILY OUTPUT	LABOR-HOURS	UNIT	2005 BARE COSTS MAT.	LABOR	EQUIP.	TOTAL	TOTAL INCL O&P
300 2720	7/8" thread size	1 Plum	144	.056	L.F.	1.55	2.27		3.82	5.10 300
2721	1" thread size	Q-1	160	.100		2.59	3.68		6.27	8.40
2722	1-1/8" thread size	"	120	.133		3.96	4.90		8.86	11.70
2725	1/4" thread size, bright finish	1 Plum	144	.056		.79	2.27		3.06	4.28
2726	1/2" thread size, bright finish	"	144	.056	↓	.59	2.27		2.86	4.06
2730	For galvanized, add					40%				
2750	Both ends machine threaded 18" length									
2760	3/8" thread size	1 Plum	240	.033	Ea.	2.06	1.36		3.42	4.32
2770	1/2" thread size		240	.033		3.30	1.36		4.66	5.70
2780	5/8" thread size		240	.033		4.79	1.36		6.15	7.30
2790	3/4" thread size		240	.033		7.30	1.36		8.66	10.05
2800	7/8" thread size		240	.033		16.35	1.36		17.71	20
2810	1" thread size	↓	240	.033	↓	22.50	1.36		23.86	26.50
2820	Rod couplings									
2821	3/8"	1 Plum	60	.133	Ea.	2.23	5.45		7.68	10.65
2822	1/2"		54	.148		2.59	6.05		8.64	11.95
2823	5/8"		48	.167		2.94	6.80		9.74	13.50
2824	3/4"		44	.182		4.04	7.45		11.49	15.60
2825	7/8"		40	.200		4.52	8.15		12.67	17.25
2826	1"		34	.235		6.10	9.60		15.70	21
2827	1-1/8"	↓	30	.267	↓	17.25	10.90		28.15	35.50
2860	Pipe hanger assy, adj. clevis, saddle, rod, clamp, insul. allowance									
2864	1/2" pipe size	Q-5	35	.457	Ea.	17.05	16.85		33.90	44.50
2866	3/4" pipe size		34.80	.460		17.65	16.95		34.60	45
2868	1" pipe size		34.60	.462		31.50	17.05		48.55	60
2869	1-1/4" pipe size		34.30	.466		31.50	17.20		48.70	60.50
2870	1-1/2" pipe size		33.90	.472		32	17.40		49.40	61.50
2872	2" pipe size		33.30	.480		34	17.70		51.70	63.50
2874	2-1/2" pipe size		32.30	.495		39	18.25		57.25	70.50
2876	3" pipe size		31.20	.513		42	18.90		60.90	74.50
2880	4" pipe size		30.70	.521		44	19.20		63.20	77.50
2884	6" pipe size		29.80	.537		53.50	19.80		73.30	88.50
2888	8" pipe size		28	.571		98.50	21		119.50	140
2892	10" pipe size		25.20	.635		182	23.50		205.50	235
2896	12" pipe size	↓	23.20	.690	↓	224	25.50		249.50	285
2900	Rolls									
2910	Adjustable yoke, carbon steel with CI roll, TYPE 43									
2918	2" pipe size	Q-1	140	.114	Ea.	10.70	4.20		14.90	18.05
2920	2-1/2" pipe size		137	.117		10.70	4.30		15	18.20
2930	3" pipe size		131	.122		11.90	4.49		16.39	19.85
2940	3-1/2" pipe size		124	.129		15.80	4.75		20.55	24.50
2950	4" pipe size		117	.137		16.05	5.05		21.10	25
2960	5" pipe size		110	.145		18.75	5.35		24.10	28.50
2970	6" pipe size		104	.154		23.50	5.65		29.15	34
2980	8" pipe size		96	.167		35	6.15		41.15	47.50
2990	10" pipe size		80	.200		41.50	7.35		48.85	56.50
3000	12" pipe size		68	.235		61	8.65		69.65	80
3010	14" pipe size		56	.286		163	10.50		173.50	196
3020	16" pipe size	↓	48	.333	↓	209	12.25		221.25	248
3050	Chair, carbon steel with CI roll									
3060	2" pipe size	1 Plum	68	.118	Ea.	14.40	4.81		19.21	23
3070	2-1/2" pipe size		65	.123		15.35	5.05		20.40	24.50
3080	3" pipe size		62	.129		16.90	5.25		22.15	26.50
3090	3-1/2" pipe size		60	.133		20.50	5.45		25.95	30.50
3100	4" pipe size		58	.138		23	5.65		28.65	33.50
3110	5" pipe size	↓	56	.143	↓	25	5.85		30.85	36.50

	15060 Hangers & Supports		Crew	Daily Output	Labor-Hours	Unit	2005 Bare Costs Mat.	Labor	Equip.	Total	Total Incl O&P	
300	3120	6" pipe size	1 Plum	53	.151	Ea.	34.50	6.15		40.65	47.50	300
	3130	8" pipe size		50	.160		46.50	6.55		53.05	61.50	
	3140	10" pipe size		48	.167		58.50	6.80		65.30	75	
	3150	12" pipe size		46	.174		85	7.10		92.10	105	
	3170	Single pipe roll, (see line 2650 for rods), TYPE 41, 1" pipe size	Q-1	137	.117		7.75	4.30		12.05	15	
	3180	1-1/4" pipe size		131	.122		7.90	4.49		12.39	15.45	
	3190	1-1/2" pipe size		129	.124		8.05	4.56		12.61	15.70	
	3200	2" pipe size		124	.129		8.20	4.75		12.95	16.15	
	3210	2-1/2" pipe size		118	.136		9.65	4.99		14.64	18.15	
	3220	3" pipe size		115	.139		11.30	5.10		16.40	20	
	3230	3-1/2" pipe size		113	.142		12.75	5.20		17.95	22	
	3240	4" pipe size		112	.143		14.35	5.25		19.60	23.50	
	3250	5" pipe size		110	.145		15.90	5.35		21.25	25.50	
	3260	6" pipe size		101	.158		20.50	5.85		26.35	31.50	
	3270	8" pipe size		90	.178		32	6.55		38.55	45	
	3280	10" pipe size		80	.200		38	7.35		45.35	52.50	
	3290	12" pipe size		68	.235		42	8.65		50.65	59	
	3291	14" pipe size		56	.286		66.50	10.50		77	89.50	
	3292	16" pipe size		48	.333		81	12.25		93.25	107	
	3293	18" pipe size		40	.400		88	14.70		102.70	119	
	3294	20" pipe size		35	.457		128	16.80		144.80	166	
	3296	24" pipe size		30	.533		202	19.60		221.60	252	
	3297	30" pipe size		25	.640		485	23.50		508.50	570	
	3298	36" pipe size		20	.800		495	29.50		524.50	590	
	3300	Saddles (add vertical pipe riser, usually 3" diameter)										
	3310	Pipe support, complete, adjust., CI saddle, TYPE 36										
	3320	2-1/2" pipe size	1 Plum	96	.083	Ea.	183	3.40		186.40	206	
	3330	3" pipe size		88	.091		185	3.71		188.71	210	
	3340	3-1/2" pipe size		79	.101		186	4.14		190.14	211	
	3350	4" pipe size		68	.118		243	4.81		247.81	274	
	3360	5" pipe size		64	.125		245	5.10		250.10	277	
	3370	6" pipe size		59	.136		249	5.55		254.55	282	
	3380	8" pipe size		53	.151		264	6.15		270.15	299	
	3390	10" pipe size		50	.160		291	6.55		297.55	330	
	3400	12" pipe size		48	.167		300	6.80		306.80	345	
	3450	For standard pipe support, one piece, CI deduct					34%					
	3460	For stanchion support, CI with steel yoke, deduct					60%					
	3500	Multiple pipe support 2' long strut	1 Plum	30	.267	Ea.	4.46	10.90		15.36	21.50	
	3510	3' long strut		20	.400		6.70	16.35		23.05	32	
	3520	4' long strut		17	.471		8.90	19.20		28.10	39	
	3550	Insulation shield 1" thick, 1/2" pipe size, TYPE 40	1 Asbe	100	.080		4.93	2.97		7.90	10.10	
	3560	3/4" pipe size		100	.080		6.10	2.97		9.07	11.45	
	3570	1" pipe size		98	.082		6.10	3.03		9.13	11.55	
	3580	1-1/4" pipe size		98	.082		6.45	3.03		9.48	11.90	
	3590	1-1/2" pipe size		96	.083		6.45	3.09		9.54	12	
	3600	2" pipe size		96	.083		6.75	3.09		9.84	12.35	
	3610	2-1/2" pipe size		94	.085		7	3.16		10.16	12.70	
	3620	3" pipe size		94	.085		7.45	3.16		10.61	13.15	
	3630	2" thick, 3-1/2" pipe size		92	.087		8.25	3.23		11.48	14.20	
	3640	4" pipe size		92	.087		9.70	3.23		12.93	15.80	
	3650	5" pipe size		90	.089		14.40	3.30		17.70	21	
	3660	6" pipe size		90	.089		16.45	3.30		19.75	23.50	
	3670	8" pipe size		88	.091		24.50	3.37		27.87	32.50	
	3680	10" pipe size		88	.091		36.50	3.37		39.87	45.50	
	3690	12" pipe size		86	.093		43	3.45		46.45	53	
	3700	14" pipe size		86	.093		68.50	3.45		71.95	81	

		15060 | Hangers & Supports	CREW	DAILY OUTPUT	LABOR-HOURS	UNIT	2005 BARE COSTS MAT.	LABOR	EQUIP.	TOTAL	TOTAL INCL O&P	
300	3710	16" pipe size	1 Asbe	84	.095	Ea.	72	3.53		75.53	84.50	300
	3720	18" pipe size		84	.095		83.50	3.53		87.03	97.50	
	3730	20" pipe size		82	.098		91	3.62		94.62	106	
	3732	24" pipe size		80	.100		160	3.71		163.71	183	
	3733	30" pipe size		78	.103		207	3.81		210.81	233	
	3735	36" pipe size	↓	75	.107	↓	220	3.96		223.96	248	
	3750	Covering protection saddle, TYPE 39										
	3760	1" covering size										
	3770	3/4" pipe size	1 Plum	68	.118	Ea.	3.65	4.81		8.46	11.20	
	3780	1" pipe size		68	.118		3.65	4.81		8.46	11.20	
	3790	1-1/4" pipe size		68	.118		3.65	4.81		8.46	11.20	
	3800	1-1/2" pipe size		66	.121		3.96	4.95		8.91	11.80	
	3810	2" pipe size		66	.121		4.04	4.95		8.99	11.90	
	3820	2-1/2" pipe size		64	.125		5.70	5.10		10.80	13.95	
	3830	3" pipe size		64	.125		7.30	5.10		12.40	15.75	
	3840	3-1/2" pipe size		62	.129		7.30	5.25		12.55	15.95	
	3850	4" pipe size		62	.129		9.20	5.25		14.45	18	
	3860	5" pipe size		60	.133		9.20	5.45		14.65	18.30	
	3870	6" pipe size	↓	60	.133	↓	10.95	5.45		16.40	20	
	3900	1-1/2" covering size										
	3910	3/4" pipe size	1 Plum	68	.118	Ea.	11.05	4.81		15.86	19.35	
	3920	1" pipe size		68	.118		11.05	4.81		15.86	19.35	
	3930	1-1/4" pipe size		68	.118		11.05	4.81		15.86	19.35	
	3940	1-1/2" pipe size		66	.121		11.05	4.95		16	19.60	
	3950	2" pipe size		66	.121		11.80	4.95		16.75	20.50	
	3960	2-1/2" pipe size		64	.125		14.15	5.10		19.25	23.50	
	3970	3" pipe size		64	.125		14.15	5.10		19.25	23.50	
	3980	3-1/2" pipe size		62	.129		14.35	5.25		19.60	23.50	
	3990	4" pipe size		62	.129		14.35	5.25		19.60	23.50	
	4000	5" pipe size		60	.133		14.35	5.45		19.80	24	
	4010	6" pipe size		60	.133		20.50	5.45		25.95	30.50	
	4020	8" pipe size		58	.138		27	5.65		32.65	38	
	4022	10" pipe size		56	.143		23.50	5.85		29.35	35	
	4024	12" pipe size	↓	54	.148	↓	38	6.05		44.05	50.50	
	4028	2" covering size										
	4029	2-1/2" pipe size	1 Plum	62	.129	Ea.	12	5.25		17.25	21	
	4032	3" pipe size		60	.133		15.95	5.45		21.40	26	
	4033	4" pipe size		58	.138		15.95	5.65		21.60	26	
	4034	6" pipe size		56	.143		23	5.85		28.85	34	
	4035	8" pipe size		54	.148		27	6.05		33.05	38.50	
	4080	10" pipe size		58	.138		28.50	5.65		34.15	40	
	4090	12" pipe size		56	.143		42.50	5.85		48.35	55.50	
	4100	14" pipe size		56	.143		42.50	5.85		48.35	55.50	
	4110	16" pipe size		54	.148		59.50	6.05		65.55	74.50	
	4120	18" pipe size		54	.148		59.50	6.05		65.55	74.50	
	4130	20" pipe size		52	.154		66	6.30		72.30	82	
	4150	24" pipe size		50	.160		77	6.55		83.55	95	
	4160	30" pipe size		48	.167		85	6.80		91.80	104	
	4180	36" pipe size	↓	45	.178	↓	93.50	7.25		100.75	114	
	4186	2-1/2" covering size										
	4187	3" pipe size	1 Plum	58	.138	Ea.	15.55	5.65		21.20	25.50	
	4188	4" pipe size		56	.143		16.90	5.85		22.75	27.50	
	4189	6" pipe size		52	.154		26	6.30		32.30	38	
	4190	8" pipe size		48	.167		31.50	6.80		38.30	45	
	4191	10" pipe size		44	.182		34.50	7.45		41.95	49	
	4192	12" pipe size	↓	40	.200	↓	46	8.15		54.15	63	

15060 Hangers & Supports		CREW	DAILY OUTPUT	LABOR-HOURS	UNIT	2005 BARE COSTS MAT.	LABOR	EQUIP.	TOTAL	TOTAL INCL O&P
300 4193	14" pipe size	1 Plum	36	.222	Ea.	46	9.10		55.10	64
4194	16" pipe size		32	.250		68	10.20		78.20	90
4195	18" pipe size	↓	28	.286	↓	72.50	11.65		84.15	97.50
4200	Sockets									
4210	Rod end, malleable iron, TYPE 16									
4220	1/4" thread size	1 Plum	240	.033	Ea.	1.22	1.36		2.58	3.39
4230	3/8" thread size		240	.033		1.31	1.36		2.67	3.49
4240	1/2" thread size		230	.035		1.57	1.42		2.99	3.87
4250	5/8" thread size		225	.036		2.84	1.45		4.29	5.30
4260	3/4" thread size		220	.036		7.55	1.49		9.04	10.60
4270	7/8" thread size	↓	210	.038		9.90	1.56		11.46	13.25
4290	Strap, 1/2" pipe size, TYPE 26	Q-1	142	.113		.47	4.14		4.61	6.75
4300	3/4" pipe size		140	.114		.51	4.20		4.71	6.85
4310	1" pipe size		137	.117		.53	4.30		4.83	7.05
4320	1-1/4" pipe size		134	.119		.60	4.39		4.99	7.25
4330	1-1/2" pipe size		131	.122		.67	4.49		5.16	7.50
4340	2" pipe size		129	.124		.73	4.56		5.29	7.65
4350	2-1/2" pipe size		125	.128		1.14	4.71		5.85	8.35
4360	3" pipe size		122	.131		1.31	4.82		6.13	8.70
4370	3-1/2" pipe size		119	.134		14.30	4.95		19.25	23
4380	4" pipe size	↓	114	.140	↓	16	5.15		21.15	25.50
4400	U-bolt, carbon steel									
4410	Standard, with nuts, TYPE 42									
4420	1/2" pipe size	1 Plum	160	.050	Ea.	.55	2.04		2.59	3.68
4430	3/4" pipe size		158	.051		.55	2.07		2.62	3.72
4450	1" pipe size		152	.053		.62	2.15		2.77	3.91
4460	1-1/4" pipe size		148	.054		.71	2.21		2.92	4.10
4470	1-1/2" pipe size		143	.056		.79	2.29		3.08	4.30
4480	2" pipe size		139	.058		.86	2.35		3.21	4.48
4490	2-1/2" pipe size		134	.060		1.42	2.44		3.86	5.25
4500	3" pipe size		128	.063		1.50	2.55		4.05	5.50
4510	3-1/2" pipe size		122	.066		1.53	2.68		4.21	5.70
4520	4" pipe size		117	.068		1.60	2.79		4.39	5.95
4530	5" pipe size		114	.070		1.70	2.87		4.57	6.20
4540	6" pipe size		111	.072		2.99	2.94		5.93	7.70
4550	8" pipe size		109	.073		3.42	3		6.42	8.25
4560	10" pipe size		107	.075		5.50	3.05		8.55	10.65
4570	12" pipe size	↓	104	.077	↓	7.35	3.14		10.49	12.75
4580	For plastic coating on 1/2" thru 6" size, add					150%				
4700	U-hook, carbon steel, requires mounting screws or bolts									
4710	3/4" thru 2" pipe size									
4720	6" long	1 Plum	96	.083	Ea.	4.09	3.40		7.49	9.60
4730	8" long		96	.083		4.36	3.40		7.76	9.90
4740	10" long		96	.083		5.50	3.40		8.90	11.15
4750	12" long	↓	96	.083		6.10	3.40		9.50	11.80
4760	For copper plated, add					50%				
8000	Pipe clamp, plastic, 1/2" CTS	1 Plum	80	.100		.20	4.09		4.29	6.35
8010	3/4" CTS		73	.110		.21	4.48		4.69	7
8020	1" CTS		68	.118		.50	4.81		5.31	7.75
8080	Economy clamp, 1/4" CTS		175	.046		.05	1.87		1.92	2.87
8090	3/8" CTS		168	.048		.06	1.95		2.01	2.99
8100	1/2" CTS		160	.050		.06	2.04		2.10	3.14
8110	3/4" CTS		145	.055		.12	2.25		2.37	3.52
8200	Half clamp, 1/2" CTS		80	.100		.07	4.09		4.16	6.25
8210	3/4" CTS		73	.110		.10	4.48		4.58	6.85
8300	Suspension clamp, 1/2" CTS	↓	80	.100	↓	.20	4.09		4.29	6.35

15060 | Hangers & Supports

			Crew	Daily Output	Labor-Hours	Unit	2005 Bare Costs Mat.	Labor	Equip.	Total	Total Incl O&P	
300	8310	3/4" CTS	1 Plum	73	.110	Ea.	.21	4.48		4.69	7	300
	8320	1" CTS		68	.118		.50	4.81		5.31	7.75	
	8400	Insulator, 1/2" CTS		80	.100		.34	4.09		4.43	6.50	
	8410	3/4" CTS		73	.110		.35	4.48		4.83	7.15	
	8420	1" CTS		68	.118		.36	4.81		5.17	7.60	
	8500	J hook clamp with nail 1/2" CTS		240	.033		.11	1.36		1.47	2.17	
	8501	3/4" CTS		240	.033		.12	1.36		1.48	2.18	
	8800	Wire cable support system										
	8810	Cable with hook terminal and locking device										
	8830	2 mm, (.079 ") dia cable, (100 lb. cap.)										
	8840	1 m, (3.3') length, with hook	1 Shee	96	.083	Ea.	3.74	3.35		7.09	9.25	
	8850	2 m, (6.6') length, with hook		84	.095		4.32	3.83		8.15	10.65	
	8860	3 m, (9.9') length, with hook		72	.111		4.90	4.47		9.37	12.25	
	8870	5 m, (16.4') length, with hook	Q-9	60	.267		6.15	9.65		15.80	21.50	
	8880	10 m, (32.8') length, with hook	"	30	.533		8.90	19.30		28.20	39.50	
	8900	3mm, (.118 ") dia cable, (200 lb. cap.)										
	8910	1 m, (3.3') length, with hook	1 Shee	96	.083	Ea.	4.80	3.35		8.15	10.45	
	8920	2 m, (6.6') length, with hook		84	.095		5.40	3.83		9.23	11.85	
	8930	3 m, (9.9') length, with hook		72	.111		5.95	4.47		10.42	13.40	
	8940	5 m, (16.4') length, with hook	Q-9	60	.267		7.35	9.65		17	23	
	8950	10 m, (32.8') length, with hook	"	30	.533		10.60	19.30		29.90	41	
	9000	Cable system accessories										
	9010	Anchor bolt, 3/8", with nut	1 Shee	140	.057	Ea.	1.47	2.30		3.77	5.15	
	9020	Air duct corner protector		160	.050		.64	2.01		2.65	3.79	
	9030	Air duct support attachment		140	.057		.24	2.30		2.54	3.79	
	9040	Flange clip, hammer-on style										
	9044	For flange thickness 3/32" - 9/64", 160 lb. cap.	1 Shee	180	.044	Ea.	.29	1.79		2.08	3.07	
	9048	For flange thickness 1/8" - 1/4", 200 lb. cap.		160	.050		.33	2.01		2.34	3.45	
	9052	For flange thickness 5/16" - 1/2", 200 lb. cap.		150	.053		.36	2.14		2.50	3.70	
	9056	For flange thickness 9/16" - 3/4", 200 lb. cap.		140	.057		.53	2.30		2.83	4.11	
	9060	Wire insulation protection tube		180	.044	L.F.	.41	1.79		2.20	3.20	
	9070	Wire cutter				Ea.	40			40	44	

15070 | Mech Sound/Vibration/Seismic Ctrl

			Crew	Daily Output	Labor-Hours	Unit	Mat.	Labor	Equip.	Total	Total Incl O&P	
800	0010	**VIBRATION ABSORBERS**										800
	0100	Hangers, neoprene flex										
	0200	10 - 120 lb. capacity				Ea.	13.55			13.55	14.90	
	0220	75 - 550 lb. capacity					17.65			17.65	19.40	
	0240	250 - 1100 lb. capacity					37			37	40.50	
	0260	1000 - 4000 lb. capacity					59.50			59.50	65.50	
	0500	Spring flex, 60 lb. capacity					19			19	21	
	0520	450 lb. capacity					23.50			23.50	26	
	0540	900 lb. capacity					31			31	34.50	
	0560	1100-1300 lb. capacity					31			31	34.50	
	0600	Rubber in shear										
	0610	45 - 340 lb., up to 1/2" rod size	1 Stpi	22	.364	Ea.	12	14.90		26.90	35.50	
	0620	130 - 700 lb., up to 3/4" rod size		20	.400		27	16.40		43.40	54	
	0630	50 - 1000 lb., up to 3/4" rod size		18	.444		31	18.20		49.20	61.50	
	1000	Mounts, neoprene, 45 - 380 lb. capacity					10.25			10.25	11.25	
	1020	250 - 1100 lb. capacity					26.50			26.50	29.50	
	1040	1000 - 4000 lb. capacity					60			60	66	
	1100	Spring flex, 60 lb. capacity					37			37	40.50	
	1120	165 lb. capacity					37			37	40.50	
	1140	260 lb. capacity					37			37	40.50	
	1160	450 lb. capacity					48.50			48.50	53.50	
	1180	600 lb. capacity					48.50			48.50	53.50	

15070 | Mech Sound/Vibration/Seismic Ctrl

				DAILY	LABOR-		2005 BARE COSTS				TOTAL	
			CREW	OUTPUT	HOURS	UNIT	MAT.	LABOR	EQUIP.	TOTAL	INCL O&P	
800	1200	750 lb. capacity				Ea.	48.50			48.50	53.50	800
	1220	900 lb. capacity					49			49	54	
	1240	1100 lb. capacity					50.50			50.50	55.50	
	1260	1300 lb. capacity					53.50			53.50	59	
	1280	1500 lb. capacity					70.50			70.50	77.50	
	1300	1800 lb. capacity					74			74	81.50	
	1320	2200 lb. capacity					82.50			82.50	90.50	
	1340	2600 lb. capacity					84			84	92.50	
	1399	Spring type										
	1400	2 Piece										
	1410	50 - 1000 lb.	1 Stpi	12	.667	Ea.	102	27.50		129.50	153	
	1420	1100 - 1600 lb.	"	12	.667	"	110	27.50		137.50	162	
	1500	Double spring open										
	1510	150 - 450 lb.	1 Stpi	24	.333	Ea.	74	13.65		87.65	102	
	1520	500 - 1000 lb.		24	.333		74	13.65		87.65	102	
	1530	1100 - 1600 lb.		24	.333		74	13.65		87.65	102	
	1540	1700 - 2400 lb.		24	.333		80	13.65		93.65	109	
	1550	2500 - 3400 lb.		24	.333		136	13.65		149.65	171	
	2000	Pads, cork rib, 18" x 18" x 1", 10-50 psi					100			100	110	
	2020	18" x 36" x 1", 10-50 psi					201			201	221	
	2100	Shear flexible pads, 18" x 18" x 3/8", 20-70 psi					47.50			47.50	52.50	
	2120	18" x 36" x 3/8", 20-70 psi					88.50			88.50	97.50	
	2150	Laminated neoprene and cork										
	2160	1" thick	1 Stpi	16	.500	S.F.	47	20.50		67.50	82.50	
	3000	Note overlap in capacities due to deflections										

15075 | Mechanical Identification

			CREW	DAILY OUTPUT	LABOR-HOURS	UNIT	MAT.	LABOR	EQUIP.	TOTAL	TOTAL INCL O&P	
400	0010	**PIPING SYSTEM IDENTIFICATION LABELS**										400
	0100	Indicate contents and flow direction										
	0106	Pipe markers										
	0110	Plastic snap around										
	0114	1/2" pipe	1 Plum	80	.100	Ea.	3.40	4.09		7.49	9.90	
	0116	3/4" pipe		80	.100		3.40	4.09		7.49	9.90	
	0118	1" pipe		80	.100		3.40	4.09		7.49	9.90	
	0120	2" pipe		75	.107		4.40	4.36		8.76	11.40	
	0122	3" pipe		70	.114		8.20	4.67		12.87	16	
	0124	4" pipe		60	.133		8.20	5.45		13.65	17.20	
	0126	6" pipe		60	.133		9	5.45		14.45	18.10	
	0128	8" pipe		56	.143		12.30	5.85		18.15	22.50	
	0130	10" pipe		56	.143		12.30	5.85		18.15	22.50	
	0200	Over 10" pipe size		50	.160		15	6.55		21.55	26.50	
	1110	Self adhesive										
	1114	1" pipe	1 Plum	80	.100	Ea.	2.40	4.09		6.49	8.80	
	1116	2" pipe		75	.107		2.40	4.36		6.76	9.20	
	1118	3" pipe		70	.114		3.80	4.67		8.47	11.20	
	1120	4" pipe		60	.133		3.80	5.45		9.25	12.40	
	1122	6" pipe		60	.133		3.80	5.45		9.25	12.40	
	1124	8" pipe		56	.143		3.80	5.85		9.65	12.95	
	1126	10" pipe		56	.143		5.30	5.85		11.15	14.60	
	1200	Over 10" pipe size		50	.160		7.30	6.55		13.85	17.85	
	2000	Valve tags										
	2010	Numbered plus identifying legend										
	2100	Brass, 2" diameter	1 Plum	40	.200	Ea.	2.10	8.15		10.25	14.60	
	2200	Plastic, 1-1/2" Dia.	"	40	.200	"	1.80	8.15		9.95	14.30	

15080 | Mechanical Insulation

			CREW	DAILY OUTPUT	LABOR-HOURS	UNIT	2005 BARE COSTS MAT.	LABOR	EQUIP.	TOTAL	TOTAL INCL O&P	
200	0010	**DUCT INSULATION**										200
	0100	Rule of thumb, as a percentage of total mechanical costs				Job				10%		
	0110	Insulation req'd is based on the surface size/area to be covered										
	3000	Ductwork										
	3020	Blanket type, fiberglass, flexible										
	3030	Fire resistant liner, black coating one side										
	3050	1/2" thick, 2 lb. density	Q-14	380	.042	S.F.	.38	1.41		1.79	2.64	
	3060	1" thick, 1-1/2 lb. density		350	.046		.50	1.53		2.03	2.96	
	3070	1-1/2" thick, 1-1/2 lb. density		320	.050		.88	1.67		2.55	3.61	
	3080	2" thick, 1-1/2 lb. density	↓	300	.053	↓	1.04	1.78		2.82	3.95	
	3140	FRK vapor barrier wrap, .75 lb. density										
	3160	1" thick	Q-14	350	.046	S.F.	.25	1.53		1.78	2.69	
	3170	1-1/2" thick		320	.050		.27	1.67		1.94	2.94	
	3180	2" thick		300	.053		.38	1.78		2.16	3.23	
	3190	3" thick		260	.062		.52	2.06		2.58	3.81	
	3200	4" thick	↓	242	.066	↓	.74	2.21		2.95	4.29	
	3210	Vinyl jacket, same as FRK										
	3280	Unfaced, 1 lb. density										
	3310	1" thick	Q-14	360	.044	S.F.	.33	1.48		1.81	2.70	
	3320	1-1/2" thick		330	.048		.45	1.62		2.07	3.05	
	3330	2" thick	↓	310	.052	↓	.50	1.72		2.22	3.27	
	3340	Board type fiberglass liner, 1-1/2 lb. density										
	3344	1" thick	Q-14	150	.107	S.F.	.30	3.56		3.86	5.95	
	3345	1-1/2" thick		130	.123		.38	4.11		4.49	6.90	
	3346	2" thick	↓	120	.133	↓	.44	4.45		4.89	7.55	
	3490	Board type, fiberglass liner, 3 lb. density										
	3500	Fire resistant, black pigmented, 1 side										
	3520	1" thick	Q-14	150	.107	S.F.	1.44	3.56		5	7.20	
	3540	1-1/2" thick		130	.123		1.76	4.11		5.87	8.45	
	3560	2" thick	↓	120	.133	↓	2.12	4.45		6.57	9.40	
	3600	FRK vapor barrier										
	3620	1" thick	Q-14	150	.107	S.F.	1.57	3.56		5.13	7.35	
	3630	1-1/2" thick		130	.123		2.03	4.11		6.14	8.75	
	3640	2" thick	↓	120	.133	↓	2.46	4.45		6.91	9.75	
	3680	No finish										
	3700	1" thick	Q-14	170	.094	S.F.	.84	3.14		3.98	5.90	
	3710	1-1/2" thick		140	.114		1.25	3.82		5.07	7.40	
	3720	2" thick	↓	130	.123	↓	1.57	4.11		5.68	8.25	
	3730	Sheet insulation										
	3760	Polyethylene foam, closed cell, UV resistant										
	3770	Standard temperature (-90° to +212° F)										
	3771	1/4" thick	Q-14	450	.036	S.F.	1.27	1.19		2.46	3.27	
	3772	3/8" thick		440	.036		1.78	1.21		2.99	3.88	
	3773	1/2" thick		420	.038		2.21	1.27		3.48	4.44	
	3774	3/4" thick		400	.040		3.23	1.34		4.57	5.65	
	3775	1" thick	↓	380	.042	↓	4.06	1.41		5.47	6.70	
	3779	Adhesive (see line 7878)										
	3780	Foam, rubber										
	3782	1" thick	1 Stpi	50	.160	S.F.	2.66	6.55		9.21	12.80	
	3795	Finishes										
	3800	1/2" cement over 1" wire mesh, incl. corner bead	Q-14	100	.160	S.F.	1.50	5.35		6.85	10.10	
	3820	Glass cloth, pasted on		170	.094		2.43	3.14		5.57	7.65	
	3900	Weatherproof, non-metallic, 2 lb. per S.F.	↓	100	.160	↓	3.47	5.35		8.82	12.25	
	7878	Contact cement, quart can				Ea.	22			22	24	
	9600	Minimum labor/equipment charge	1 Stpi	4	2	Job		82		82	123	

MECHANICAL 15

	15080 | Mechanical Insulation		CREW	DAILY OUTPUT	LABOR-HOURS	UNIT	2005 BARE COSTS MAT.	LABOR	EQUIP.	TOTAL	TOTAL INCL O&P	
400	0010	**EQUIPMENT INSULATION**										400
	0100	Rule of thumb, as a percentage of total mechanical costs				Job				10%		
	0110	Insulation req'd is based on the surface size/area to be covered										
	1000	Boiler, 1-1/2" calcium silicate, 1/2" cement finish	Q-14	50	.320	S.F.	3.30	10.70		14	20.50	
	1020	2" fiberglass	"	80	.200	"	2.03	6.70		8.73	12.80	
	2000	Breeching, 2" calcium silicate with 1/2" cement finish, no lath										
	2020	Rectangular	Q-14	42	.381	S.F.	3.87	12.70		16.57	24.50	
	2040	Round	"	38.70	.413	"	4.33	13.80		18.13	27	
	2300	Calcium silicate block, + 200°F to + 1200°F										
	2310	On irregular surfaces, valves and fittings										
	2340	1" thick	Q-14	30	.533	S.F.	2.70	17.80		20.50	31	
	2360	1-1/2" thick		25	.640		3.15	21.50		24.65	37	
	2380	2" thick		22	.727		3.46	24.50		27.96	42.50	
	2400	3" thick		18	.889		5.40	29.50		34.90	53	
	2410	On plane surfaces										
	2420	1" thick	Q-14	126	.127	S.F.	2.70	4.24		6.94	9.65	
	2430	1-1/2" thick		120	.133		3.15	4.45		7.60	10.50	
	2440	2" thick		100	.160		3.46	5.35		8.81	12.25	
	2450	3" thick		70	.229		5.40	7.65		13.05	18	
	2900	Domestic water heater wrap kit										
	2920	1-1/2" with vinyl jacket, 20-60 gal.	1 Plum	8	1	Ea.	16.60	41		57.60	80	
600	0010	**PIPING INSULATION**										600
	0100	Rule of thumb, as a percentage of total mechanical costs				Job				10%		
	0110	Insulation req'd is based on the surface size/area to be covered										
	2930	Insulated protectors, (ADA)										
	2935	For exposed piping under sinks or lavatories.										
	2940	Vinyl coated foam, velcro tabs										
	2945	P Trap, 1-1/4" or 1-1/2"	1 Plum	32	.250	Ea.	17.15	10.20		27.35	34	
	2960	Valve and supply cover										
	2965	1/2", 3/8", and 7/16" pipe size	1 Plum	32	.250	Ea.	17.15	10.20		27.35	34	
	2970	Extension drain cover										
	2975	1-1/4", or 1-1/2" pipe size	1 Plum	32	.250	Ea.	17.60	10.20		27.80	34.50	
	2980	Tailpiece offset (wheelchair)										
	2985	1-1/4" pipe size	1 Plum	32	.250	Ea.	19.60	10.20		29.80	37	
	4000	Pipe covering (price copper tube one size less than IPS)										
	4040	Air cell, asbestos free corrugated felt, with cover										
	4100	3 ply, 1/2" iron pipe size	Q-14	145	.110	L.F.	.75	3.69		4.44	6.65	
	4130	3/4" iron pipe size		145	.110		.85	3.69		4.54	6.75	
	4140	1" iron pipe size		145	.110		.90	3.69		4.59	6.80	
	4150	1-1/4" iron pipe size		140	.114		1	3.82		4.82	7.10	
	4160	1-1/2" iron pipe size		140	.114		1.10	3.82		4.92	7.20	
	4170	2" iron pipe size		135	.119		1.20	3.96		5.16	7.55	
	4180	2-1/2" iron pipe size		135	.119		1.35	3.96		5.31	7.75	
	4190	3" iron pipe size		130	.123		1.50	4.11		5.61	8.15	
	4200	3-1/2" iron pipe size		130	.123		1.76	4.11		5.87	8.45	
	4210	4" iron pipe size		125	.128		1.95	4.28		6.23	8.90	
	4220	5" iron pipe size		120	.133		2.25	4.45		6.70	9.55	
	4230	6" iron pipe size		115	.139		2.40	4.65		7.05	10	
	4240	8" iron pipe size		105	.152		3.30	5.10		8.40	11.70	
	4250	10" iron pipe size		95	.168		3.90	5.65		9.55	13.20	
	4260	12" iron pipe size		90	.178		4.50	5.95		10.45	14.30	
	4280	Cellular glass, closed cell foam, all service jacket, sealant,										
	4281	working temp. (-450°F to +900°F), 0 water vapor xmission										
	4284	1" wall,										
	4286	1/2" iron pipe size	Q-14	120	.133	L.F.	1.65	4.45		6.10	8.85	

15080 | Mechanical Insulation

			CREW	DAILY OUTPUT	LABOR-HOURS	UNIT	2005 BARE COSTS MAT.	LABOR	EQUIP.	TOTAL	TOTAL INCL O&P
600	4300	1-1/2″ wall,									
	4301	1″ iron pipe size	Q-14	105	.152	L.F.	2.65	5.10		7.75	10.95
	4304	2-1/2″ iron pipe size		90	.178		4.46	5.95		10.41	14.25
	4306	3″ iron pipe size		85	.188		4.51	6.30		10.81	14.85
	4308	4″ iron pipe size		70	.229		6	7.65		13.65	18.65
	4310	5″ iron pipe size		65	.246		6.90	8.20		15.10	20.50
	4320	2″ wall,									
	4322	1″ iron pipe size	Q-14	100	.160	L.F.	3.94	5.35		9.29	12.80
	4324	2-1/2″ iron pipe size		85	.188		5.85	6.30		12.15	16.30
	4326	3″ iron pipe size		80	.200		6	6.70		12.70	17.15
	4328	4″ iron pipe size		65	.246		6.70	8.20		14.90	20.50
	4330	5″ iron pipe size		60	.267		8.05	8.90		16.95	23
	4332	6″ iron pipe size		50	.320		9.55	10.70		20.25	27.50
	4336	8″ iron pipe size		40	.400		12.30	13.35		25.65	34.50
	4338	10″ iron pipe size		35	.457		12.75	15.25		28	38
	4350	2-1/2″ wall,									
	4360	12″ iron pipe size	Q-14	32	.500	L.F.	20	16.70		36.70	48.50
	4362	14″ iron pipe size	"	28	.571	"	21.50	19.10		40.60	54
	4370	3″ wall,									
	4378	6″ iron pipe size	Q-14	48	.333	L.F.	12.75	11.15		23.90	31.50
	4380	8″ iron pipe size		38	.421		14.85	14.05		28.90	38.50
	4382	10″ iron pipe size		33	.485		19.65	16.20		35.85	47
	4384	16″ iron pipe size		25	.640		25.50	21.50		47	61.50
	4386	18″ iron pipe size		22	.727		28	24.50		52.50	69
	4388	20″ iron pipe size		20	.800		32.50	26.50		59	78
	4400	3-1/2″ wall,									
	4412	12″ iron pipe size	Q-14	27	.593	L.F.	28.50	19.80		48.30	62.50
	4414	14″ iron pipe size	"	25	.640	"	30.50	21.50		52	67
	4430	4″ wall,									
	4446	16″ iron pipe size	Q-14	22	.727	L.F.	31	24.50		55.50	72.50
	4448	18″ iron pipe size		20	.800		40	26.50		66.50	86
	4450	20″ iron pipe size		18	.889		41.50	29.50		71	92.50
	4480	Fittings, average with fabric and mastic									
	4484	1″ wall,									
	4486	1/2″ iron pipe size	1 Asbe	40	.200	Ea.	2.93	7.40		10.33	14.90
	4500	1-1/2″ wall,									
	4502	1″ iron pipe size	1 Asbe	38	.211	Ea.	5.05	7.80		12.85	17.90
	4504	2-1/2″ iron pipe size		32	.250		6.25	9.30		15.55	21.50
	4506	3″ iron pipe size		30	.267		6.90	9.90		16.80	23
	4508	4″ iron pipe size		28	.286		10.60	10.60		21.20	28.50
	4510	5″ iron pipe size		24	.333		11.70	12.35		24.05	32.50
	4520	2″ wall,									
	4522	1″ iron pipe size	1 Asbe	36	.222	Ea.	5.85	8.25		14.10	19.40
	4524	2-1/2″ iron pipe size		30	.267		6.90	9.90		16.80	23
	4526	3″ iron pipe size		28	.286		8.30	10.60		18.90	26
	4528	4″ iron pipe size		24	.333		11.15	12.35		23.50	32
	4530	5″ iron pipe size		22	.364		14.05	13.50		27.55	37
	4532	6″ iron pipe size		20	.400		16.45	14.85		31.30	41.50
	4536	8″ iron pipe size		12	.667		25	24.50		49.50	66.50
	4538	10″ iron pipe size		8	1		31	37		68	92.50
	4550	2-1/2″ wall,									
	4560	12″ iron pipe size	1 Asbe	6	1.333	Ea.	53	49.50		102.50	137
	4562	14″ iron pipe size	"	4	2	"	61.50	74		135.50	185
	4570	3″ wall,									
	4578	6″ iron pipe size	1 Asbe	16	.500	Ea.	19.10	18.55		37.65	50.50
	4580	8″ iron pipe size		10	.800		28	29.50		57.50	78

15080 | Mechanical Insulation

			CREW	DAILY OUTPUT	LABOR-HOURS	UNIT	2005 BARE COSTS MAT.	LABOR	EQUIP.	TOTAL	TOTAL INCL O&P	
600	4582	10" iron pipe size	1 Asbe	6	1.333	Ea.	40	49.50		89.50	122	600
	4584	16" iron pipe size		2	4		95.50	148		243.50	340	
	4586	18" iron pipe size		2	4		123	148		271	370	
	4588	20" iron pipe size	↓	2	4	↓	145	148		293	395	
	4600	3-1/2" wall,										
	4612	12" iron pipe size	1 Asbe	4	2	Ea.	69	74		143	193	
	4614	14" iron pipe size	"	2	4	"	76.50	148		224.50	320	
	4630	4" wall,										
	4646	16" iron pipe size	1 Asbe	2	4	Ea.	115	148		263	360	
	4648	18" iron pipe size		2	4		140	148		288	390	
	4650	20" iron pipe size	↓	2	4	↓	163	148		311	415	
	4900	Calcium silicate, with cover										
	5100	1" wall, 1/2" iron pipe size	Q-14	170	.094	L.F.	2.18	3.14		5.32	7.35	
	5130	3/4" iron pipe size		170	.094		2.19	3.14		5.33	7.35	
	5140	1" iron pipe size		170	.094		2.22	3.14		5.36	7.40	
	5150	1-1/4" iron pipe size		165	.097		2.37	3.24		5.61	7.70	
	5160	1-1/2" iron pipe size		165	.097		2.49	3.24		5.73	7.85	
	5170	2" iron pipe size		160	.100		2.53	3.34		5.87	8.05	
	5180	2-1/2" iron pipe size		160	.100		2.69	3.34		6.03	8.20	
	5190	3" iron pipe size		150	.107		2.89	3.56		6.45	8.80	
	5200	4" iron pipe size		140	.114		3.56	3.82		7.38	9.90	
	5210	5" iron pipe size		135	.119		3.79	3.96		7.75	10.40	
	5220	6" iron pipe size		130	.123		4.01	4.11		8.12	10.90	
	5280	1-1/2" wall, 1/2" iron pipe size		150	.107		2.35	3.56		5.91	8.20	
	5310	3/4" iron pipe size		150	.107		2.36	3.56		5.92	8.20	
	5320	1" iron pipe size		150	.107		2.59	3.56		6.15	8.45	
	5330	1-1/4" iron pipe size		145	.110		2.80	3.69		6.49	8.90	
	5340	1-1/2" iron pipe size		145	.110		3	3.69		6.69	9.10	
	5350	2" iron pipe size		140	.114		3.31	3.82		7.13	9.65	
	5360	2-1/2" iron pipe size		140	.114		3.60	3.82		7.42	9.95	
	5370	3" iron pipe size		135	.119		3.79	3.96		7.75	10.40	
	5380	4" iron pipe size		125	.128		4.39	4.28		8.67	11.60	
	5390	5" iron pipe size		120	.133		4.94	4.45		9.39	12.50	
	5400	6" iron pipe size		110	.145		5.10	4.86		9.96	13.25	
	5402	8" iron pipe size		95	.168		6.70	5.65		12.35	16.25	
	5404	10" iron pipe size		85	.188		8.75	6.30		15.05	19.55	
	5406	12" iron pipe size		80	.200		10.35	6.70		17.05	22	
	5408	14" iron pipe size		75	.213		11.65	7.15		18.80	24	
	5410	16" iron pipe size		70	.229		13.05	7.65		20.70	26.50	
	5412	18" iron pipe size		65	.246		16.40	8.20		24.60	31	
	5460	2" wall, 1/2" iron pipe size		135	.119		3.63	3.96		7.59	10.25	
	5490	3/4" iron pipe size		135	.119		3.82	3.96		7.78	10.45	
	5500	1" iron pipe size		135	.119		4.01	3.96		7.97	10.65	
	5510	1-1/4" iron pipe size		130	.123		4.31	4.11		8.42	11.25	
	5520	1-1/2" iron pipe size		130	.123		4.49	4.11		8.60	11.45	
	5530	2" iron pipe size		125	.128		4.73	4.28		9.01	11.95	
	5540	2-1/2" iron pipe size		125	.128		5.65	4.28		9.93	12.95	
	5550	3" iron pipe size		120	.133		5.70	4.45		10.15	13.30	
	5560	4" iron pipe size		115	.139		6.55	4.65		11.20	14.55	
	5570	5" iron pipe size		110	.145		7.45	4.86		12.31	15.80	
	5580	6" iron pipe size		105	.152		8.10	5.10		13.20	16.95	
	5581	8" iron pipe size		95	.168		9.65	5.65		15.30	19.50	
	5582	10" iron pipe size		85	.188		12.10	6.30		18.40	23	
	5583	12" iron pipe size		80	.200		13.45	6.70		20.15	25.50	
	5584	14" iron pipe size		75	.213		14.90	7.15		22.05	27.50	
	5585	16" iron pipe size	↓	70	.229	↓	16.40	7.65		24.05	30	

		15080 | Mechanical Insulation	CREW	DAILY OUTPUT	LABOR-HOURS	UNIT	2005 BARE COSTS MAT.	LABOR	EQUIP.	TOTAL	TOTAL INCL O&P	
600	5586	18" iron pipe size	Q-14	65	.246	L.F.	18.40	8.20		26.60	33.50	600
	5587	3" wall, 1-1/4" iron pipe size		100	.160		6.70	5.35		12.05	15.80	
	5588	1-1/2" iron pipe size		100	.160		6.75	5.35		12.10	15.85	
	5589	2" iron pipe size		95	.168		7.15	5.65		12.80	16.80	
	5590	2-1/2" iron pipe size		95	.168		8.15	5.65		13.80	17.85	
	5591	3" iron pipe size		90	.178		8.25	5.95		14.20	18.40	
	5592	4" iron pipe size		85	.188		10.40	6.30		16.70	21.50	
	5593	6" iron pipe size		80	.200		12.70	6.70		19.40	24.50	
	5594	8" iron pipe size		75	.213		15.10	7.15		22.25	28	
	5595	10" iron pipe size		65	.246		18.15	8.20		26.35	33	
	5596	12" iron pipe size		60	.267		20	8.90		28.90	36	
	5597	14" iron pipe size		55	.291		22.50	9.70		32.20	40.50	
	5598	16" iron pipe size		50	.320		25	10.70		35.70	44.50	
	5599	18" iron pipe size		45	.356		28	11.90		39.90	50	
	5600	Calcium silicate, no cover										
	5720	1" wall, 1/2" iron pipe size	Q-14	180	.089	L.F.	2	2.97		4.97	6.90	
	5740	3/4" iron pipe size		180	.089		2	2.97		4.97	6.90	
	5750	1" iron pipe size		180	.089		1.92	2.97		4.89	6.80	
	5760	1-1/4" iron pipe size		175	.091		1.96	3.05		5.01	7	
	5770	1-1/2" iron pipe size		175	.091		1.97	3.05		5.02	7	
	5780	2" iron pipe size		170	.094		2.27	3.14		5.41	7.45	
	5790	2-1/2" iron pipe size		170	.094		2.40	3.14		5.54	7.60	
	5800	3" iron pipe size		160	.100		2.58	3.34		5.92	8.10	
	5810	4" iron pipe size		150	.107		3.17	3.56		6.73	9.10	
	5820	5" iron pipe size		145	.110		3.36	3.69		7.05	9.50	
	5830	6" iron pipe size		140	.114		3.53	3.82		7.35	9.90	
	5900	1-1/2" wall, 1/2" iron pipe size		160	.100		2.11	3.34		5.45	7.55	
	5920	3/4" iron pipe size		160	.100		2.16	3.34		5.50	7.65	
	5930	1" iron pipe size		160	.100		2.35	3.34		5.69	7.85	
	5940	1-1/4" iron pipe size		155	.103		2.53	3.45		5.98	8.25	
	5950	1-1/2" iron pipe size		155	.103		2.71	3.45		6.16	8.45	
	5960	2" iron pipe size		150	.107		3	3.56		6.56	8.90	
	5970	2-1/2" iron pipe size		150	.107		3.27	3.56		6.83	9.20	
	5980	3" iron pipe size		145	.110		3.42	3.69		7.11	9.55	
	5990	4" iron pipe size		135	.119		3.78	3.96		7.74	10.40	
	6000	5" iron pipe size		130	.123		4.46	4.11		8.57	11.40	
	6010	6" iron pipe size		120	.133		4.55	4.45		9	12.05	
	6020	7" iron pipe size		115	.139		5.35	4.65		10	13.20	
	6030	8" iron pipe size		105	.152		6.05	5.10		11.15	14.70	
	6040	9" iron pipe size		100	.160		7.25	5.35		12.60	16.40	
	6050	10" iron pipe size		95	.168		8	5.65		13.65	17.70	
	6060	12" iron pipe size		90	.178		9.50	5.95		15.45	19.80	
	6070	14" iron pipe size		85	.188		10.70	6.30		17	21.50	
	6080	16" iron pipe size		80	.200		12	6.70		18.70	24	
	6090	18" iron pipe size		75	.213		15.25	7.15		22.40	28	
	6120	2" wall, 1/2" iron pipe size		145	.110		3.35	3.69		7.04	9.50	
	6140	3/4" iron pipe size		145	.110		3.53	3.69		7.22	9.70	
	6150	1" iron pipe size		145	.110		3.71	3.69		7.40	9.90	
	6160	1-1/4" iron pipe size		140	.114		3.99	3.82		7.81	10.40	
	6170	1-1/2" iron pipe size		140	.114		4.15	3.82		7.97	10.55	
	6180	2" iron pipe size		135	.119		4.37	3.96		8.33	11.05	
	6190	2-1/2" iron pipe size		135	.119		5.25	3.96		9.21	12.05	
	6200	3" iron pipe size		130	.123		5.25	4.11		9.36	12.30	
	6210	4" iron pipe size		125	.128		6.10	4.28		10.38	13.45	
	6220	5" iron pipe size		120	.133		6.95	4.45		11.40	14.70	
	6230	6" iron pipe size		115	.139		7.60	4.65		12.25	15.70	

15080 | Mechanical Insulation		CREW	DAILY OUTPUT	LABOR-HOURS	UNIT	2005 BARE COSTS MAT.	LABOR	EQUIP.	TOTAL	TOTAL INCL O&P	
600 6240	7" iron pipe size	Q-14	110	.145	L.F.	8.25	4.86		13.11	16.70	600
6250	8" iron pipe size		105	.152		9.05	5.10		14.15	18	
6260	9" iron pipe size		100	.160		10.25	5.35		15.60	19.75	
6270	10" iron pipe size		95	.168		11.35	5.65		17	21.50	
6280	12" iron pipe size		90	.178		12.65	5.95		18.60	23.50	
6290	14" iron pipe size		85	.188		13.90	6.30		20.20	25	
6300	16" iron pipe size		80	.200		15.30	6.70		22	27.50	
6310	18" iron pipe size		75	.213		17.20	7.15		24.35	30	
6320	20" iron pipe size		65	.246		21	8.20		29.20	36.50	
6330	22" iron pipe size		60	.267		23.50	8.90		32.40	40	
6340	24" iron pipe size		55	.291		24	9.70		33.70	42	
6360	3" wall, 1/2" iron pipe size		115	.139		6.05	4.65		10.70	14.05	
6380	3/4" iron pipe size		115	.139		6.10	4.65		10.75	14.05	
6390	1" iron pipe size		115	.139		6.15	4.65		10.80	14.10	
6400	1-1/4" iron pipe size		110	.145		6.25	4.86		11.11	14.50	
6410	1-1/2" iron pipe size		110	.145		6.30	4.86		11.16	14.55	
6420	2" iron pipe size		105	.152		6.70	5.10		11.80	15.40	
6430	2-1/2" iron pipe size		105	.152		7.65	5.10		12.75	16.45	
6440	3" iron pipe size		100	.160		7.70	5.35		13.05	16.90	
6450	4" iron pipe size		95	.168		9.80	5.65		15.45	19.70	
6460	5" iron pipe size		90	.178		10.65	5.95		16.60	21	
6470	6" iron pipe size		90	.178		12	5.95		17.95	22.50	
6480	7" iron pipe size		85	.188		13.35	6.30		19.65	24.50	
6490	8" iron pipe size		85	.188		14.25	6.30		20.55	25.50	
6500	9" iron pipe size		80	.200		16.20	6.70		22.90	28.50	
6510	10" iron pipe size		75	.213		17.25	7.15		24.40	30	
6520	12" iron pipe size		70	.229		19.05	7.65		26.70	33	
6530	14" iron pipe size		65	.246		19.75	8.20		27.95	35	
6540	16" iron pipe size		60	.267		20.50	8.90		29.40	36.50	
6550	18" iron pipe size		55	.291		26.50	9.70		36.20	45	
6560	20" iron pipe size		50	.320		32	10.70		42.70	52	
6570	22" iron pipe size		45	.356		34.50	11.90		46.40	57	
6580	24" iron pipe size	↓	40	.400	↓	37	13.35		50.35	62	
6600	Fiberglass, with all service jacket										
6640	1/2" wall, 1/2" iron pipe size	Q-14	250	.064	L.F.	.60	2.14		2.74	4.03	
6660	3/4" iron pipe size		240	.067		.67	2.23		2.90	4.25	
6670	1" iron pipe size		230	.070		.71	2.32		3.03	4.45	
6680	1-1/4" iron pipe size		220	.073		.79	2.43		3.22	4.70	
6690	1-1/2" iron pipe size		220	.073		.83	2.43		3.26	4.74	
6700	2" iron pipe size		210	.076		.91	2.54		3.45	5	
6710	2-1/2" iron pipe size		200	.080		1.04	2.67		3.71	5.35	
6720	3" iron pipe size		190	.084		1.15	2.81		3.96	5.70	
6730	3-1/2" iron pipe size		180	.089		1.24	2.97		4.21	6.05	
6740	4" iron pipe size		160	.100		1.47	3.34		4.81	6.85	
6750	5" iron pipe size		150	.107		1.70	3.56		5.26	7.45	
6760	6" iron pipe size		120	.133		1.80	4.45		6.25	9.05	
6840	1" wall, 1/2" iron pipe size		240	.067		.71	2.23		2.94	4.29	
6860	3/4" iron pipe size		230	.070		.83	2.32		3.15	4.58	
6870	1" iron pipe size		220	.073		.83	2.43		3.26	4.74	
6880	1-1/4" iron pipe size		210	.076		.96	2.54		3.50	5.10	
6890	1-1/2" iron pipe size		210	.076		1.02	2.54		3.56	5.15	
6900	2" iron pipe size		200	.080		1.12	2.67		3.79	5.45	
6910	2-1/2" iron pipe size		190	.084		1.23	2.81		4.04	5.80	
6920	3" iron pipe size		180	.089		1.39	2.97		4.36	6.20	
6930	3-1/2" iron pipe size		170	.094		1.45	3.14		4.59	6.55	
6940	4" iron pipe size	↓	150	.107	↓	1.80	3.56		5.36	7.60	

15 MECHANICAL

	15080 | Mechanical Insulation		CREW	DAILY OUTPUT	LABOR-HOURS	UNIT	2005 BARE COSTS MAT.	LABOR	EQUIP.	TOTAL	TOTAL INCL O&P	
600	6950	5" iron pipe size	Q-14	140	.114	L.F.	2.07	3.82		5.89	8.30	600
	6960	6" iron pipe size		120	.133		2.20	4.45		6.65	9.45	
	6970	7" iron pipe size		110	.145		2.71	4.86		7.57	10.65	
	6980	8" iron pipe size		100	.160		3.14	5.35		8.49	11.90	
	6990	9" iron pipe size		90	.178		3.41	5.95		9.36	13.10	
	7000	10" iron pipe size		90	.178		3.67	5.95		9.62	13.40	
	7010	12" iron pipe size		80	.200		4.17	6.70		10.87	15.15	
	7020	14" iron pipe size		80	.200		4.84	6.70		11.54	15.85	
	7030	16" iron pipe size		70	.229		5.75	7.65		13.40	18.35	
	7040	18" iron pipe size		70	.229		7	7.65		14.65	19.75	
	7050	20" iron pipe size		60	.267		7.80	8.90		16.70	22.50	
	7060	24" iron pipe size		60	.267		9.55	8.90		18.45	24.50	
	7080	1-1/2" wall, 1/2" iron pipe size		230	.070		1.47	2.32		3.79	5.30	
	7100	3/4" iron pipe size		220	.073		1.53	2.43		3.96	5.50	
	7110	1" iron pipe size		210	.076		1.61	2.54		4.15	5.80	
	7120	1-1/4" iron pipe size		200	.080		1.73	2.67		4.40	6.10	
	7130	1-1/2" iron pipe size		200	.080		1.82	2.67		4.49	6.20	
	7140	2" iron pipe size		190	.084		1.98	2.81		4.79	6.60	
	7150	2-1/2" iron pipe size		180	.089		2.18	2.97		5.15	7.10	
	7160	3" iron pipe size		170	.094		2.25	3.14		5.39	7.45	
	7170	3-1/2" iron pipe size		160	.100		2.47	3.34		5.81	7.95	
	7180	4" iron pipe size		140	.114		2.58	3.82		6.40	8.85	
	7190	5" iron pipe size		130	.123		2.88	4.11		6.99	9.65	
	7200	6" iron pipe size		110	.145		2.96	4.86		7.82	10.90	
	7210	7" iron pipe size		100	.160		3.41	5.35		8.76	12.20	
	7220	8" iron pipe size		90	.178		3.70	5.95		9.65	13.40	
	7230	9" iron pipe size		85	.188		4.22	6.30		10.52	14.55	
	7240	10" iron pipe size		80	.200		4.66	6.70		11.36	15.70	
	7250	12" iron pipe size		75	.213		5.10	7.15		12.25	16.85	
	7260	14" iron pipe size		70	.229		5.85	7.65		13.50	18.50	
	7270	16" iron pipe size		65	.246		6.60	8.20		14.80	20	
	7280	18" iron pipe size		60	.267		8.55	8.90		17.45	23.50	
	7290	20" iron pipe size		55	.291		9.05	9.70		18.75	25.50	
	7300	24" iron pipe size		50	.320		10.60	10.70		21.30	28.50	
	7320	2" wall, 1/2" iron pipe size		220	.073		2.29	2.43		4.72	6.35	
	7340	3/4" iron pipe size		210	.076		2.37	2.54		4.91	6.65	
	7350	1" iron pipe size		200	.080		2.54	2.67		5.21	7	
	7360	1-1/4" iron pipe size		190	.084		2.69	2.81		5.50	7.40	
	7370	1-1/2" iron pipe size		190	.084		2.84	2.81		5.65	7.55	
	7380	2" iron pipe size		180	.089		2.95	2.97		5.92	7.95	
	7390	2-1/2" iron pipe size		170	.094		3.16	3.14		6.30	8.45	
	7400	3" iron pipe size		160	.100		3.41	3.34		6.75	9	
	7410	3-1/2" iron pipe size		150	.107		3.68	3.56		7.24	9.65	
	7420	4" iron pipe size		130	.123		3.91	4.11		8.02	10.80	
	7430	5" iron pipe size		120	.133		4.42	4.45		8.87	11.90	
	7440	6" iron pipe size		100	.160		4.54	5.35		9.89	13.45	
	7450	7" iron pipe size		90	.178		5.20	5.95		11.15	15.05	
	7460	8" iron pipe size		80	.200		5.60	6.70		12.30	16.70	
	7470	9" iron pipe size		75	.213		6.20	7.15		13.35	18.05	
	7480	10" iron pipe size		70	.229		6.70	7.65		14.35	19.45	
	7490	12" iron pipe size		65	.246		7.40	8.20		15.60	21	
	7500	14" iron pipe size		60	.267		8.35	8.90		17.25	23.50	
	7510	16" iron pipe size		55	.291		9.50	9.70		19.20	26	
	7520	18" iron pipe size		50	.320		11.35	10.70		22.05	29.50	
	7530	20" iron pipe size		45	.356		12.60	11.90		24.50	32.50	
	7540	24" iron pipe size	↓	40	.400	↓	14.10	13.35		27.45	36.50	

15080 | Mechanical Insulation			CREW	DAILY OUTPUT	LABOR-HOURS	UNIT	2005 BARE COSTS MAT.	LABOR	EQUIP.	TOTAL	TOTAL INCL O&P	
600	7600	For fiberglass with standard canvas jacket, deduct					5%					600
	7660	For fittings, add 3 L.F. for each fitting										
	7680	plus 4 L.F. for each flange of the fitting										
	7700	For equipment or congested areas, add						20%				
	7720	Finishes, for .010" aluminum jacket, add	Q-14	200	.080	S.F.	.53	2.67		3.20	4.80	
	7740	For .016" aluminum jacket, add		200	.080		.68	2.67		3.35	4.97	
	7760	For .010" stainless steel, add	↓	160	.100	↓	1.99	3.34		5.33	7.45	
	7780	For single layer of felt, add					10%	10%				
	7782	For roofing paper, 45 lb to 55 lb, add					25%	10%				
	7784	Polyethylene tubing flexible closed cell foam, UV resistant										
	7828	Standard temperature (-90° to +212° F)										
	7830	3/8" wall, 1/8" iron pipe size	1 Asbe	130	.062	L.F.	.41	2.28		2.69	4.05	
	7831	1/4" iron pipe size		130	.062		.43	2.28		2.71	4.07	
	7832	3/8" iron pipe size		130	.062		.48	2.28		2.76	4.13	
	7833	1/2" iron pipe size		126	.063		.53	2.36		2.89	4.30	
	7834	3/4" iron pipe size		122	.066		.61	2.43		3.04	4.51	
	7835	1" iron pipe size		120	.067		.66	2.47		3.13	4.63	
	7836	1-1/4" iron pipe size		118	.068		.87	2.52		3.39	4.93	
	7837	1-1/2" iron pipe size		118	.068		1.06	2.52		3.58	5.15	
	7838	2" iron pipe size		116	.069		1.19	2.56		3.75	5.35	
	7839	2-1/2" iron pipe size		114	.070		1.72	2.60		4.32	6	
	7840	3" iron pipe size		112	.071		2.26	2.65		4.91	6.65	
	7842	1/2" wall, 1/8" iron pipe size		120	.067		.60	2.47		3.07	4.56	
	7843	1/4" iron pipe size		120	.067		.65	2.47		3.12	4.62	
	7844	3/8" iron pipe size		120	.067		.70	2.47		3.17	4.67	
	7845	1/2" iron pipe size		118	.068		.75	2.52		3.27	4.80	
	7846	3/4" iron pipe size		116	.069		.88	2.56		3.44	5	
	7847	1" iron pipe size		114	.070		.95	2.60		3.55	5.15	
	7848	1-1/4" iron pipe size		112	.071		1.19	2.65		3.84	5.50	
	7849	1-1/2" iron pipe size		110	.073		1.43	2.70		4.13	5.85	
	7850	2" iron pipe size		108	.074		1.67	2.75		4.42	6.20	
	7851	2-1/2" iron pipe size		106	.075		2.26	2.80		5.06	6.90	
	7852	3" iron pipe size		104	.077		2.83	2.85		5.68	7.60	
	7853	3-1/2" iron pipe size		102	.078		3.16	2.91		6.07	8.05	
	7854	4" iron pipe size		100	.080		3.92	2.97		6.89	9	
	7855	3/4" wall, 1/8" iron pipe size		110	.073		.92	2.70		3.62	5.25	
	7856	1/4" iron pipe size		110	.073		.97	2.70		3.67	5.35	
	7857	3/8" iron pipe size		108	.074		1.11	2.75		3.86	5.55	
	7858	1/2" iron pipe size		106	.075		1.24	2.80		4.04	5.80	
	7859	3/4" iron pipe size		104	.077		1.61	2.85		4.46	6.25	
	7860	1" iron pipe size		102	.078		1.85	2.91		4.76	6.65	
	7861	1-1/4" iron pipe size		100	.080		2.51	2.97		5.48	7.45	
	7862	1-1/2" iron pipe size		100	.080		2.94	2.97		5.91	7.90	
	7863	2" iron pipe size		98	.082		3.44	3.03		6.47	8.55	
	7864	2-1/2" iron pipe size		96	.083		4.33	3.09		7.42	9.65	
	7865	3" iron pipe size		94	.085		5.20	3.16		8.36	10.75	
	7866	3-1/2" iron pipe size		92	.087		5.60	3.23		8.83	11.25	
	7867	4" iron pipe size		90	.089		6.85	3.30		10.15	12.75	
	7868	1" wall, 1/4" iron pipe size		100	.080		1.67	2.97		4.64	6.50	
	7869	3/8" iron pipe size		98	.082		1.74	3.03		4.77	6.70	
	7870	1/2" iron pipe size		96	.083		1.85	3.09		4.94	6.90	
	7871	3/4" iron pipe size		94	.085		2.42	3.16		5.58	7.65	
	7872	1" iron pipe size		92	.087		2.79	3.23		6.02	8.15	
	7873	1-1/4" iron pipe size		90	.089		3.18	3.30		6.48	8.70	
	7874	1-1/2" iron pipe size		90	.089		3.63	3.30		6.93	9.20	
	7875	2" iron pipe size	↓	88	.091	↓	4.77	3.37		8.14	10.55	

	15080 | Mechanical Insulation		CREW	DAILY OUTPUT	LABOR-HOURS	UNIT	2005 BARE COSTS MAT.	LABOR	EQUIP.	TOTAL	TOTAL INCL O&P	
600	7876	2-1/2" iron pipe size	1 Asbe	86	.093	L.F.	6.20	3.45		9.65	12.25	600
	7877	3" iron pipe size	↓	84	.095	↓	7.30	3.53		10.83	13.60	
	7878	Contact cement, quart can				Ea.	22			22	24	
	7879	Rubber tubing, flexible closed cell foam										
	7880	3/8" wall, 1/4" iron pipe size	1 Asbe	120	.067	L.F.	.28	2.47		2.75	4.21	
	7900	3/8" iron pipe size		120	.067		.30	2.47		2.77	4.23	
	7910	1/2" iron pipe size		115	.070		.35	2.58		2.93	4.46	
	7920	3/4" iron pipe size		115	.070		.40	2.58		2.98	4.51	
	7930	1" iron pipe size		110	.073		.49	2.70		3.19	4.80	
	7940	1-1/4" iron pipe size		110	.073		.60	2.70		3.30	4.92	
	7950	1-1/2" iron pipe size		110	.073		.72	2.70		3.42	5.05	
	8100	1/2" wall, 1/4" iron pipe size		90	.089		.37	3.30		3.67	5.60	
	8120	3/8" iron pipe size		90	.089		.49	3.30		3.79	5.75	
	8130	1/2" iron pipe size		89	.090		.57	3.33		3.90	5.90	
	8140	3/4" iron pipe size		89	.090		.63	3.33		3.96	5.95	
	8150	1" iron pipe size		88	.091		.68	3.37		4.05	6.05	
	8160	1-1/4" iron pipe size		87	.092		.86	3.41		4.27	6.35	
	8170	1-1/2" iron pipe size		87	.092		1.03	3.41		4.44	6.55	
	8180	2" iron pipe size		86	.093		1.20	3.45		4.65	6.75	
	8190	2-1/2" iron pipe size		86	.093		1.65	3.45		5.10	7.25	
	8200	3" iron pipe size		85	.094		1.72	3.49		5.21	7.40	
	8210	3-1/2" iron pipe size		85	.094		2.30	3.49		5.79	8.05	
	8220	4" iron pipe size		80	.100		2.73	3.71		6.44	8.85	
	8230	5" iron pipe size		80	.100		3.75	3.71		7.46	10	
	8300	3/4" wall, 1/4" iron pipe size		90	.089		.67	3.30		3.97	5.95	
	8320	3/8" iron pipe size		90	.089		.74	3.30		4.04	6	
	8330	1/2" iron pipe size		89	.090		.81	3.33		4.14	6.15	
	8340	3/4" iron pipe size		89	.090		.97	3.33		4.30	6.30	
	8350	1" iron pipe size		88	.091		1.18	3.37		4.55	6.60	
	8360	1-1/4" iron pipe size		87	.092		1.36	3.41		4.77	6.90	
	8370	1-1/2" iron pipe size		87	.092		2.07	3.41		5.48	7.70	
	8380	2" iron pipe size		86	.093		2.43	3.45		5.88	8.10	
	8390	2-1/2" iron pipe size		86	.093		2.85	3.45		6.30	8.60	
	8400	3" iron pipe size		85	.094		3.32	3.49		6.81	9.15	
	8410	3-1/2" iron pipe size		85	.094		4.06	3.49		7.55	9.95	
	8420	4" iron pipe size		80	.100		4.64	3.71		8.35	10.95	
	8430	5" iron pipe size		80	.100		5.65	3.71		9.36	12.10	
	8440	6" iron pipe size		80	.100		6.85	3.71		10.56	13.40	
	8444	1" wall, 1/2" iron pipe size		86	.093		1.53	3.45		4.98	7.15	
	8445	3/4" iron pipe size		84	.095		1.90	3.53		5.43	7.70	
	8446	1" iron pipe size		84	.095		2.30	3.53		5.83	8.15	
	8447	1-1/4" iron pipe size		82	.098		2.68	3.62		6.30	8.65	
	8448	1-1/2" iron pipe size		82	.098		3.02	3.62		6.64	9	
	8449	2" iron pipe size		80	.100		4.70	3.71		8.41	11	
	8450	2-1/2" iron pipe size	↓	80	.100	↓	6.15	3.71		9.86	12.60	
	8456	Rubber insulation tape, 1/8" x 2" x 30'				Ea.	10.80			10.80	11.90	
	8460	Polyolefin tubing, flexible closed cell foam, UV stabilized, work										
	8462	temp.-165° F to +210° F, 0 water vapor transmission.										
	8464	3/8" wall, 1/8" iron pipe size	1 Asbe	140	.057	L.F.	.30	2.12		2.42	3.68	
	8466	1/4" iron pipe size		140	.057		.31	2.12		2.43	3.69	
	8468	3/8" iron pipe size		140	.057		.35	2.12		2.47	3.74	
	8470	1/2" iron pipe size		136	.059		.38	2.18		2.56	3.86	
	8472	3/4" iron pipe size		132	.061		.44	2.25		2.69	4.03	
	8474	1" iron pipe size		130	.062		.48	2.28		2.76	4.13	
	8476	1-1/4" iron pipe size		128	.063		.63	2.32		2.95	4.35	
	8478	1-1/2" iron pipe size	↓	128	.063	↓	.76	2.32		3.08	4.50	

	15080 | Mechanical Insulation	CREW	DAILY OUTPUT	LABOR-HOURS	UNIT	2005 BARE COSTS MAT.	LABOR	EQUIP.	TOTAL	TOTAL INCL O&P	
600	8480 2" iron pipe size	1 Asbe	126	.063	L.F.	.86	2.36		3.22	4.67	600
	8482 2-1/2" iron pipe size		123	.065		1.25	2.41		3.66	5.20	
	8484 3" iron pipe size		121	.066		1.63	2.45		4.08	5.65	
	8486 4" iron pipe size		118	.068		2.69	2.52		5.21	6.95	
	8500 1/2" wall, 1/8" iron pipe size		130	.062		.43	2.28		2.71	4.07	
	8502 1/4" iron pipe size		130	.062		.47	2.28		2.75	4.12	
	8504 3/8" iron pipe size		130	.062		.50	2.28		2.78	4.15	
	8506 1/2" iron pipe size		128	.063		.54	2.32		2.86	4.25	
	8508 3/4" iron pipe size		126	.063		.63	2.36		2.99	4.41	
	8510 1" iron pipe size		123	.065		.68	2.41		3.09	4.56	
	8512 1-1/4" iron pipe size		121	.066		.86	2.45		3.31	4.82	
	8514 1-1/2" iron pipe size		119	.067		1.03	2.49		3.52	5.05	
	8516 2" iron pipe size		117	.068		1.20	2.54		3.74	5.30	
	8518 2-1/2" iron pipe size		114	.070		1.63	2.60		4.23	5.90	
	8520 3" iron pipe size		112	.071		2.04	2.65		4.69	6.40	
	8522 4" iron pipe size		110	.073		2.82	2.70		5.52	7.35	
	8534 3/4" wall, 1/8" iron pipe size		120	.067		.66	2.47		3.13	4.63	
	8536 1/4" iron pipe size		120	.067		.70	2.47		3.17	4.67	
	8538 3/8" iron pipe size		117	.068		.82	2.54		3.36	4.90	
	8540 1/2" iron pipe size		114	.070		.89	2.60		3.49	5.10	
	8542 3/4" iron pipe size		112	.071		1.16	2.65		3.81	5.45	
	8544 1" iron pipe size		110	.073		1.33	2.70		4.03	5.70	
	8546 1-1/4" iron pipe size		108	.074		1.80	2.75		4.55	6.30	
	8548 1-1/2" iron pipe size		108	.074		2.12	2.75		4.87	6.65	
	8550 2" iron pipe size		106	.075		2.48	2.80		5.28	7.15	
	8552 2-1/2" iron pipe size		104	.077		3.12	2.85		5.97	7.95	
	8554 3" iron pipe size		102	.078		3.75	2.91		6.66	8.70	
	8556 4" iron pipe size		100	.080		4.92	2.97		7.89	10.10	
	8570 1" wall, 1/8" iron pipe size		110	.073		1.20	2.70		3.90	5.60	
	8572 1/4" iron pipe size		108	.074		1.23	2.75		3.98	5.70	
	8574 3/8" iron pipe size		106	.075		1.25	2.80		4.05	5.80	
	8576 1/2" iron pipe size		104	.077		1.33	2.85		4.18	5.95	
	8578 3/4" iron pipe size		102	.078		1.74	2.91		4.65	6.50	
	8580 1" iron pipe size		100	.080		2.01	2.97		4.98	6.90	
	8582 1-1/4" iron pipe size		97	.082		2.29	3.06		5.35	7.35	
	8584 1-1/2" iron pipe size		97	.082		2.61	3.06		5.67	7.70	
	8586 2" iron pipe size		95	.084		3.44	3.12		6.56	8.70	
	8588 2-1/2" iron pipe size		93	.086		4.33	3.19		7.52	9.80	
	8590 3" iron pipe size	↓	91	.088	↓	5.25	3.26		8.51	10.90	
	8600 Seam Roller				Ea.	7.50			7.50	8.25	
	8602 Fuse Seal applicator (Butane)					200			200	220	
	8604 Fuse Seal 1/2" sticks				↓	.91			.91	1	
	8606 Leaktite adhesive				Qt.	15.45			15.45	16.95	
	8608 Leaktite adhesive				Gal.	53.50			53.50	58.50	
	8610 NOTE: Preslit/preglued vs unslit, same price										
900	0010 **PIPE INSULATION PROTECTIVE JACKETING**										900
	0100 PVC, white, 48" lengths										
	0120 20 mil thick										
	0140 Size based on OD of insulation										
	0150 1-1/2" ID	Q-14	270	.059	L.F.	1.23	1.98		3.21	4.47	
	0152 2" ID		260	.062		1.22	2.06		3.28	4.58	
	0154 2-1/2" ID		250	.064		1.28	2.14		3.42	4.78	
	0156 3" ID		240	.067		1.32	2.23		3.55	4.96	
	0158 3-1/2" ID		230	.070		1.43	2.32		3.75	5.25	
	0160 4" ID	↓	220	.073	↓	1.48	2.43		3.91	5.45	

15080 | Mechanical Insulation		CREW	DAILY OUTPUT	LABOR-HOURS	UNIT	2005 BARE COSTS MAT.	LABOR	EQUIP.	TOTAL	TOTAL INCL O&P
900	0162 4-1/2" ID	Q-14	210	.076	L.F.	1.56	2.54		4.10	5.75
	0164 5" ID		200	.080		1.69	2.67		4.36	6.10
	0166 5-1/2" ID		190	.084		1.77	2.81		4.58	6.40
	0168 6" ID		180	.089		1.90	2.97		4.87	6.75
	0170 6-1/2" ID		175	.091		2	3.05		5.05	7
	0172 7" ID		170	.094		2.05	3.14		5.19	7.20
	0174 7-1/2" ID		164	.098		2.15	3.26		5.41	7.50
	0176 8" ID		161	.099		2.25	3.32		5.57	7.75
	0178 8-1/2" ID		158	.101		2.39	3.38		5.77	8
	0180 9" ID		155	.103		2.49	3.45		5.94	8.20
	0182 9-1/2" ID		152	.105		2.57	3.52		6.09	8.40
	0184 10" ID		149	.107		2.62	3.59		6.21	8.55
	0186 10-1/2" ID		146	.110		2.83	3.66		6.49	8.90
	0188 11" ID		143	.112		2.93	3.74		6.67	9.10
	0190 11-1/2" ID		140	.114		3.06	3.82		6.88	9.35
	0192 12" ID		137	.117		3.03	3.90		6.93	9.50
	0194 12-1/2" ID		134	.119		3.21	3.99		7.20	9.85
	0196 13-1/2" ID		132	.121		3.27	4.05		7.32	10
	0198 14" ID		130	.123		3.71	4.11		7.82	10.60
	0200 15" ID		128	.125		3.84	4.18		8.02	10.80
	0202 16" ID		126	.127		3.97	4.24		8.21	11.05
	0204 17" ID		124	.129		4.23	4.31		8.54	11.45
	0206 18" ID		122	.131		4.43	4.38		8.81	11.75
	0208 19" ID		120	.133		4.64	4.45		9.09	12.15
	0210 20" ID		118	.136		5.05	4.53		9.58	12.70
	0212 21" ID		116	.138		5.75	4.61		10.36	13.60
	0214 22" ID		114	.140		5.95	4.69		10.64	13.95
	0216 23" ID		112	.143		6.30	4.77		11.07	14.50
	0218 24" ID		110	.145		6.45	4.86		11.31	14.75
	0220 25" ID		108	.148		6.80	4.95		11.75	15.30
	0222 26" ID		106	.151		7	5.05		12.05	15.65
	0224 27" ID		104	.154		7.25	5.15		12.40	16.05
	0226 28" ID		102	.157		7.55	5.25		12.80	16.55
	0228 29" ID		100	.160		7.75	5.35		13.10	17
	0230 30" ID	↓	98	.163	↓	8.05	5.45		13.50	17.50
	0300 For colors, add				Ea.	10%				
	1000 30 mil thick									
	1010 Size based on OD of insulation									
	1020 2" ID	Q-14	260	.062	L.F.	1.34	2.06		3.40	4.71
	1022 2-1/2" ID		250	.064		1.57	2.14		3.71	5.10
	1024 3" ID		240	.067		1.71	2.23		3.94	5.40
	1026 3-1/2" ID		230	.070		1.91	2.32		4.23	5.75
	1028 4" ID		220	.073		2	2.43		4.43	6.05
	1030 4-1/2" ID		210	.076		2.17	2.54		4.71	6.40
	1032 5" ID		200	.080		2.43	2.67		5.10	6.90
	1034 5-1/2" ID		190	.084		2.62	2.81		5.43	7.30
	1036 6" ID		180	.089		2.82	2.97		5.79	7.80
	1038 6-1/2" ID		175	.091		3.02	3.05		6.07	8.15
	1040 7" ID		170	.094		3.08	3.14		6.22	8.35
	1042 7-1/2" ID		164	.098		3.19	3.26		6.45	8.65
	1044 8" ID		161	.099		3.28	3.32		6.60	8.85
	1046 8-1/2" ID		158	.101		3.48	3.38		6.86	9.20
	1048 9" ID		155	.103		3.79	3.45		7.24	9.60
	1050 9-1/2" ID		152	.105		3.88	3.52		7.40	9.80
	1052 10" ID		149	.107		4.07	3.59		7.66	10.15
	1054 10-1/2" ID	↓	146	.110	↓	4.31	3.66		7.97	10.55

15080 | Mechanical Insulation

			CREW	DAILY OUTPUT	LABOR-HOURS	UNIT	2005 BARE COSTS MAT.	LABOR	EQUIP.	TOTAL	TOTAL INCL O&P	
900	1056	11" ID	Q-14	143	.112	L.F.	4.45	3.74		8.19	10.80	900
	1058	11-1/2" ID		140	.114		4.59	3.82		8.41	11.05	
	1060	12" ID		137	.117		4.67	3.90		8.57	11.30	
	1062	12-1/2" ID		134	.119		4.81	3.99		8.80	11.60	
	1064	13-1/2" ID		132	.121		5.20	4.05		9.25	12.15	
	1066	14" ID		130	.123		5.65	4.11		9.76	12.70	
	1068	15" ID		128	.125		5.85	4.18		10.03	13.05	
	1070	16" ID		126	.127		6.20	4.24		10.44	13.50	
	1072	17" ID		124	.129		6.40	4.31		10.71	13.85	
	1074	18" ID		122	.131		6.70	4.38		11.08	14.25	
	1076	19" ID		120	.133		6.95	4.45		11.40	14.70	
	1078	20" ID		118	.136		7.25	4.53		11.78	15.10	
	1080	21" ID		116	.138		7.45	4.61		12.06	15.45	
	1082	22" ID		114	.140		7.75	4.69		12.44	15.95	
	1084	23" ID		112	.143		8.05	4.77		12.82	16.40	
	1086	24" ID		110	.145		8.30	4.86		13.16	16.80	
	1088	25" ID		108	.148		8.60	4.95		13.55	17.25	
	1090	26" ID		106	.151		8.85	5.05		13.90	17.70	
	1092	27" ID		104	.154		8.95	5.15		14.10	17.95	
	1094	28" ID		102	.157		9.35	5.25		14.60	18.55	
	1096	29" ID		100	.160		9.70	5.35		15.05	19.15	
	1098	30" ID	▼	98	.163	▼	9.90	5.45		15.35	19.50	
	1300	For colors, add				Ea.	10%					
	2000	PVC, white, fitting covers										
	2020	Fiberglass insulation inserts included with sizes 1-3/4" thru 9-3/4".										
	2030	Size is based on OD of insulation										
	2040	90° Elbow fitting										
	2060	1-3/4"	Q-14	135	.119	Ea.	.59	3.96		4.55	6.90	
	2062	2"		130	.123		.66	4.11		4.77	7.25	
	2064	2-1/4"		128	.125		.81	4.18		4.99	7.50	
	2068	2-1/2"		126	.127		.81	4.24		5.05	7.60	
	2070	2-3/4"		123	.130		1.05	4.34		5.39	8	
	2072	3"		120	.133		.93	4.45		5.38	8.05	
	2074	3-3/8"		116	.138		1.21	4.61		5.82	8.60	
	2076	3-3/4"		113	.142		1.17	4.73		5.90	8.75	
	2078	4-1/8"		110	.145		1.47	4.86		6.33	9.25	
	2080	4-3/4""		105	.152		1.79	5.10		6.89	10	
	2082	5-1/4""		100	.160		2.08	5.35		7.43	10.75	
	2084	5-3/4""		95	.168		2.91	5.65		8.56	12.10	
	2086	6-1/4""		90	.178		5.95	5.95		11.90	15.90	
	2088	6-3/4""		87	.184		5.15	6.15		11.30	15.40	
	2090	7-1/4""		85	.188		7.95	6.30		14.25	18.60	
	2092	7-3/4""		83	.193		6.75	6.45		13.20	17.55	
	2094	8-3/4""		80	.200		8.80	6.70		15.50	20	
	2096	9-3/4""		77	.208		12.25	6.95		19.20	24.50	
	2098	10-7/8"		74	.216		16.65	7.20		23.85	29.50	
	2100	11-7/8"		71	.225		18.95	7.55		26.50	33	
	2102	12-7/8"		68	.235		23.50	7.85		31.35	38.50	
	2104	14-1/8"		66	.242		28	8.10		36.10	44	
	2106	15-1/8"		64	.250		31	8.35		39.35	47	
	2108	16-1/8"		63	.254		34	8.50		42.50	51	
	2110	17-1/8"		62	.258		38.50	8.60		47.10	55.50	
	2112	18-1/8"		61	.262		47.50	8.75		56.25	66	
	2114	19-1/8"		60	.267		65.50	8.90		74.40	86	
	2116	20-1/8"	▼	59	.271	▼	84.50	9.05		93.55	107	
	2200	45° Elbow fitting										

	15080 | Mechanical Insulation		CREW	DAILY OUTPUT	LABOR-HOURS	UNIT	2005 BARE COSTS MAT.	LABOR	EQUIP.	TOTAL	TOTAL INCL O&P	
900	2220	1-3/4″ thru 9-3/4″ same price as 90° Elbow fitting										900
	2320	10-7/8″	Q-14	74	.216	Ea.	14.50	7.20		21.70	27.50	
	2322	11-7/8″		71	.225		16	7.55		23.55	29.50	
	2324	12-7/8″		68	.235		21	7.85		28.85	35.50	
	2326	14-1/8″		66	.242		24.50	8.10		32.60	40	
	2328	15-1/8″		64	.250		26.50	8.35		34.85	42.50	
	2330	16-1/8″		63	.254		29.50	8.50		38	46	
	2332	17-1/8″		62	.258		32.50	8.60		41.10	49.50	
	2334	18-1/8″		61	.262		40	8.75		48.75	58	
	2336	19-1/8″		60	.267		58	8.90		66.90	78	
	2338	20-1/8″		59	.271		84.50	9.05		93.55	107	
	2400	Tee fitting										
	2410	1-3/4″	Q-14	96	.167	Ea.	1.02	5.55		6.57	9.90	
	2412	2″		94	.170		1.12	5.70		6.82	10.20	
	2414	2-1/4″		91	.176		1.27	5.85		7.12	10.65	
	2416	2-1/2″		88	.182		1.38	6.05		7.43	11.10	
	2418	2-3/4″		85	.188		1.52	6.30		7.82	11.55	
	2420	3″		82	.195		1.63	6.50		8.13	12.10	
	2422	3-3/8″		79	.203		1.88	6.75		8.63	12.70	
	2424	3-3/4″		76	.211		2.17	7.05		9.22	13.50	
	2426	4-1/8″		73	.219		2.54	7.30		9.84	14.35	
	2428	4-3/4″″		70	.229		3.14	7.65		10.79	15.50	
	2430	5-1/4″″		67	.239		3.75	8		11.75	16.75	
	2432	5-3/4″″		63	.254		5	8.50		13.50	18.90	
	2434	6-1/4″″		60	.267		6.60	8.90		15.50	21.50	
	2436	6-3/4″″		59	.271		8.15	9.05		17.20	23.50	
	2438	7-1/4″″		57	.281		13.15	9.40		22.55	29.50	
	2440	7-3/4″″		54	.296		14.45	9.90		24.35	31.50	
	2442	8-3/4″″		52	.308		17.50	10.30		27.80	35.50	
	2444	9-3/4″″		50	.320		20.50	10.70		31.20	39.50	
	2446	10-7/8″		48	.333		23.50	11.15		34.65	43	
	2448	11-7/8″		47	.340		24	11.35		35.35	44	
	2450	12-7/8″		46	.348		28.50	11.60		40.10	49.50	
	2452	14-1/8″		45	.356		31	11.90		42.90	53	
	2454	15-1/8″		44	.364		33.50	12.15		45.65	56	
	2456	16-1/8″		43	.372		36	12.45		48.45	59.50	
	2458	17-1/8″		42	.381		38.50	12.70		51.20	62.50	
	2460	18-1/8″		41	.390		42.50	13.05		55.55	67.50	
	2462	19-1/8″		40	.400		46.50	13.35		59.85	72	
	2464	20-1/8″		39	.410		99	13.70		112.70	131	
	4000	Mechanical grooved fitting cover										
	4020	90° Elbow fitting										
	4030	3/4″ & 1″	Q-14	140	.114	Ea.	4.84	3.82		8.66	11.30	
	4040	1-1/4″ & 1-1/2″		135	.119		6.20	3.96		10.16	13.10	
	4042	2″		130	.123		9.85	4.11		13.96	17.35	
	4044	2-1/2″		125	.128		10.50	4.28		14.78	18.30	
	4046	3″		120	.133		11.30	4.45		15.75	19.45	
	4048	3-1/2″		115	.139		15.90	4.65		20.55	25	
	4050	4″		110	.145		15.20	4.86		20.06	24.50	
	4052	5″		100	.160		19.70	5.35		25.05	30	
	4054	6″		90	.178		26.50	5.95		32.45	39	
	4056	8″		80	.200		31.50	6.70		38.20	45.50	
	4058	10″		75	.213		40.50	7.15		47.65	56	
	4060	12″		68	.235		58.50	7.85		66.35	76.50	
	4062	14″		65	.246		81	8.20		89.20	102	
	4064	16″		63	.254		110	8.50		118.50	134	

15050 | Basic Materials & Methods

15080 Mechanical Insulation		CREW	DAILY OUTPUT	LABOR-HOURS	UNIT	2005 BARE COSTS MAT.	LABOR	EQUIP.	TOTAL	TOTAL INCL O&P	
900	4066 18"	Q-14	61	.262	Ea.	141	8.75		149.75	169	900
	4100 45° Elbow fitting										
	4120 3/4" & 1"	Q-14	140	.114	Ea.	4.13	3.82		7.95	10.55	
	4130 1-1/4" & 1-1/2"		135	.119		5.50	3.96		9.46	12.30	
	4140 2"		130	.123		9	4.11		13.11	16.40	
	4142 2-1/2"		125	.128		10.05	4.28		14.33	17.80	
	4144 3"		120	.133		11	4.45		15.45	19.15	
	4146 3-1/2"		115	.139		14.40	4.65		19.05	23	
	4148 4"		110	.145		13.30	4.86		18.16	22.50	
	4150 5"		100	.160		17.80	5.35		23.15	28	
	4152 6"		90	.178		24.50	5.95		30.45	36.50	
	4154 8"		80	.200		28	6.70		34.70	41.50	
	4156 10"		75	.213		35.50	7.15		42.65	50.50	
	4158 12"		68	.235		50.50	7.85		58.35	68.50	
	4160 14"		65	.246		57	8.20		65.20	76	
	4162 16"		63	.254		93.50	8.50		102	116	
	4164 18"	↓	61	.262	↓	121	8.75		129.75	147	
	4200 Tee fitting										
	4220 3/4" & 1"	Q-14	93	.172	Ea.	6.60	5.75		12.35	16.30	
	4230 1-1/4" & 1-1/2"		90	.178		8.50	5.95		14.45	18.70	
	4240 2"		87	.184		15.55	6.15		21.70	27	
	4242 2-1/2"		84	.190		15.55	6.35		21.90	27	
	4244 3"		80	.200		16.60	6.70		23.30	29	
	4246 3-1/2"		77	.208		21.50	6.95		28.45	35	
	4248 4"		73	.219		22	7.30		29.30	35.50	
	4250 5"		67	.239		26.50	8		34.50	42	
	4252 6"		60	.267		31.50	8.90		40.40	49	
	4254 8"		54	.296		44	9.90		53.90	64	
	4256 10"		50	.320		50.50	10.70		61.20	73	
	4258 12"		46	.348		67	11.60		78.60	92.50	
	4260 14"		43	.372		95	12.45		107.45	124	
	4262 16"		42	.381		126	12.70		138.70	159	
	4264 18"	↓	41	.390	↓	161	13.05		174.05	198	

15100 | Building Services Piping

15107 Metal Pipe & Fittings		CREW	DAILY OUTPUT	LABOR-HOURS	UNIT	2005 BARE COSTS MAT.	LABOR	EQUIP.	TOTAL	TOTAL INCL O&P	
220	0010 **PIPE, BRASS** Plain end,										220
	0900 Field threaded, coupling & clevis hanger 10' O.C.										
	0920 Regular weight										
	0980 1/8" diameter	1 Plum	62	.129	L.F.	2.23	5.25		7.48	10.35	
	1000 1/4" diameter		57	.140		2.86	5.75		8.61	11.75	
	1100 3/8" diameter		52	.154		3.47	6.30		9.77	13.25	
	1120 1/2" diameter		48	.167		4.20	6.80		11	14.85	
	1140 3/4" diameter		46	.174		5.75	7.10		12.85	17	
	1160 1" diameter	↓	43	.186		8.30	7.60		15.90	20.50	
	1180 1-1/4" diameter	Q-1	72	.222		12.55	8.15		20.70	26	
	1200 1-1/2" diameter		65	.246		14.90	9.05		23.95	30	
	1220 2" diameter	↓	53	.302	↓	20.50	11.10		31.60	39	

15 MECHANICAL

15107 | Metal Pipe & Fittings

			CREW	DAILY OUTPUT	LABOR-HOURS	UNIT	2005 BARE COSTS MAT.	LABOR	EQUIP.	TOTAL	TOTAL INCL O&P	
220	1240	2-1/2" diameter	Q-1	41	.390	L.F.	31	14.35		45.35	56	220
	1260	3" diameter		31	.516		44	19		63	77	
	1280	3-1/2" diameter	Q-2	39	.615		73.50	23.50		97	117	
	1300	4" diameter	"	37	.649		71.50	24.50		96	116	
	1930	To delete coupling & hanger, subtract										
	1940	1/8" diam. to 1/2" diam.					14%	46%				
	1950	3/4" diam. to 1-1/2" diam.					8%	47%				
	1960	2" diam. to 4" diam.					10%	37%				
260	0010	**PIPE, BRASS, FITTINGS** Rough bronze, threaded										260
	1000	Standard wt., 90° Elbow										
	1040	1/8"	1 Plum	13	.615	Ea.	7.80	25		32.80	46.50	
	1060	1/4"		13	.615		7.80	25		32.80	46.50	
	1080	3/8"		13	.615		7.80	25		32.80	46.50	
	1100	1/2"		12	.667		7.80	27		34.80	49.50	
	1120	3/4"		11	.727		10.40	29.50		39.90	56	
	1140	1"		10	.800		16.90	32.50		49.40	67.50	
	1160	1-1/4"	Q-1	17	.941		27.50	34.50		62	82	
	1180	1-1/2"		16	1		34	37		71	92.50	
	1200	2"		14	1.143		54.50	42		96.50	123	
	1220	2-1/2"		11	1.455		132	53.50		185.50	226	
	1240	3"		8	2		202	73.50		275.50	335	
	1260	4"	Q-2	11	2.182		410	83		493	575	
	1280	5"		8	3		1,125	114		1,239	1,425	
	1300	6"		7	3.429		1,675	131		1,806	2,025	
	1500	45° Elbow, 1/8"	1 Plum	13	.615		9.55	25		34.55	48.50	
	1540	1/4"		13	.615		9.55	25		34.55	48.50	
	1560	3/8"		13	.615		9.55	25		34.55	48.50	
	1580	1/2"		12	.667		9.55	27		36.55	51.50	
	1600	3/4"		11	.727		13.55	29.50		43.05	59.50	
	1620	1"		10	.800		23	32.50		55.50	74.50	
	1640	1-1/4"	Q-1	17	.941		37	34.50		71.50	92.50	
	1660	1-1/2"		16	1		46.50	37		83.50	107	
	1680	2"		14	1.143		75	42		117	146	
	1700	2-1/2"		11	1.455		143	53.50		196.50	238	
	1720	3"		8	2		220	73.50		293.50	355	
	1740	4"	Q-2	11	2.182		505	83		588	680	
	1760	5"		8	3		930	114		1,044	1,200	
	1780	6"		7	3.429		1,300	131		1,431	1,625	
	2000	Tee, 1/8"	1 Plum	9	.889		9.15	36.50		45.65	64.50	
	2040	1/4"		9	.889		9.15	36.50		45.65	64.50	
	2060	3/8"		9	.889		9.15	36.50		45.65	64.50	
	2080	1/2"		8	1		9.15	41		50.15	71.50	
	2100	3/4"		7	1.143		13	46.50		59.50	84.50	
	2120	1"		6	1.333		23.50	54.50		78	108	
	2140	1-1/4"	Q-1	10	1.600		40.50	59		99.50	133	
	2160	1-1/2"		9	1.778		45.50	65.50		111	149	
	2180	2"		8	2		75.50	73.50		149	194	
	2200	2-1/2"		7	2.286		180	84		264	325	
	2220	3"		5	3.200		275	118		393	475	
	2240	4"	Q-2	7	3.429		680	131		811	945	
	2260	5"		5	4.800		1,425	183		1,608	1,850	
	2280	6"		4	6		2,225	229		2,454	2,800	
	2500	Coupling, 1/8"	1 Plum	26	.308		6.50	12.55		19.05	26	
	2540	1/4"		22	.364		6.50	14.85		21.35	29.50	
	2560	3/8"		18	.444		6.50	18.15		24.65	34.50	
	2580	1/2"		15	.533		6.50	22		28.50	40	

	15107 | Metal Pipe & Fittings		Crew	Daily Output	Labor-Hours	Unit	2005 Bare Costs Mat.	Labor	Equip.	Total	Total Incl O&P	
260	2600	3/4"	1 Plum	14	.571	Ea.	9.15	23.50		32.65	45	260
	2620	1"		13	.615		15.60	25		40.60	55	
	2640	1-1/4"	Q-1	22	.727		26	27		53	68.50	
	2660	1-1/2"		20	.800		34	29.50		63.50	81	
	2680	2"		18	.889		56	32.50		88.50	111	
	2700	2-1/2"		14	1.143		95.50	42		137.50	168	
	2720	3"		10	1.600		132	59		191	234	
	2740	4"	Q-2	12	2		272	76.50		348.50	415	
	2760	5"		10	2.400		510	91.50		601.50	700	
	2780	6"		9	2.667		725	102		827	950	
	3000	Union, 125 lb										
	3020	1/8"	1 Plum	12	.667	Ea.	22	27		49	65.50	
	3040	1/4"		12	.667		22	27		49	65.50	
	3060	3/8"		12	.667		22	27		49	65.50	
	3080	1/2"		11	.727		22	29.50		51.50	69	
	3100	3/4"		10	.800		30.50	32.50		63	82.50	
	3120	1"		9	.889		40	36.50		76.50	98.50	
	3140	1-1/4"	Q-1	16	1		58	37		95	120	
	3160	1-1/2"		15	1.067		69	39		108	135	
	3180	2"		13	1.231		107	45.50		152.50	185	
	3200	2-1/2"		10	1.600		279	59		338	395	
	3220	3"		7	2.286		435	84		519	605	
	3240	4"	Q-2	10	2.400		1,075	91.50		1,166.50	1,350	
	3320	For 250 lb. (navy pattern), add					100%					
420	0010	**PIPE, COPPER** Solder joints R15100-050										420
	0100	Solder										
	0120	Solder, lead free, roll				Lb.	9.30			9.30	10.25	
	1000	Type K tubing, couplings & clevis hangers 10′ O.C.										
	1100	1/4" diameter	1 Plum	84	.095	L.F.	1.37	3.89		5.26	7.35	
	1120	3/8" diameter		82	.098		1.57	3.99		5.56	7.70	
	1140	1/2" diameter		78	.103		1.79	4.19		5.98	8.25	
	1160	5/8" diameter		77	.104		3.05	4.24		7.29	9.75	
	1180	3/4" diameter		74	.108		3.02	4.42		7.44	9.95	
	1200	1" diameter		66	.121		3.97	4.95		8.92	11.80	
	1220	1-1/4" diameter		56	.143		4.91	5.85		10.76	14.15	
	1240	1-1/2" diameter		50	.160		6.50	6.55		13.05	16.95	
	1260	2" diameter		40	.200		9.95	8.15		18.10	23.50	
	1280	2-1/2" diameter	Q-1	60	.267		14.85	9.80		24.65	31	
	1300	3" diameter		54	.296		20.50	10.90		31.40	39.50	
	1320	3-1/2" diameter		42	.381		28	14		42	52	
	1330	4" diameter		38	.421		35	15.50		50.50	62	
	1340	5" diameter		32	.500		91	18.40		109.40	128	
	1360	6" diameter	Q-2	38	.632		133	24		157	182	
	1380	8" diameter	"	34	.706		234	27		261	299	
	1390	For other than full hard temper, add					13%					
	1440	For silver solder, add						15%				
	1800	For medical clean, (oxygen class), add					12%					
	1950	To delete cplgs. & hngrs., 1/4"-1" pipe, subtract					27%	60%				
	1960	1-1/4" -3" pipe, subtract					14%	52%				
	1970	3-1/2"-5" pipe, subtract					10%	60%				
	1980	6"-8" pipe, subtract					19%	53%				
	2000	Type L tubing, couplings & hangers 10′ O.C.										
	2100	1/4" diameter	1 Plum	88	.091	L.F.	.94	3.71		4.65	6.65	
	2120	3/8" diameter		84	.095		1.28	3.89		5.17	7.25	
	2140	1/2" diameter		81	.099		1.50	4.03		5.53	7.70	
	2160	5/8" diameter		79	.101		2.23	4.14		6.37	8.65	

15107 | Metal Pipe & Fittings

			CREW	DAILY OUTPUT	LABOR-HOURS	UNIT	2005 BARE COSTS MAT.	LABOR	EQUIP.	TOTAL	TOTAL INCL O&P
420	2180	3/4" diameter R15100-050	1 Plum	76	.105	L.F.	2.20	4.30		6.50	8.85
	2200	1" diameter		68	.118		3.10	4.81		7.91	10.60
	2220	1-1/4" diameter		58	.138		4.25	5.65		9.90	13.15
	2240	1-1/2" diameter		52	.154		5.40	6.30		11.70	15.40
	2260	2" diameter	↓	42	.190		8.40	7.80		16.20	21
	2280	2-1/2" diameter	Q-1	62	.258		12.75	9.50		22.25	28.50
	2300	3" diameter		56	.286		17.25	10.50		27.75	35
	2320	3-1/2" diameter		43	.372		23.50	13.70		37.20	46.50
	2340	4" diameter		39	.410		29.50	15.10		44.60	55
	2360	5" diameter	↓	34	.471		82.50	17.30		99.80	117
	2380	6" diameter	Q-2	40	.600		103	23		126	148
	2400	8" diameter	"	36	.667		180	25.50		205.50	236
	2410	For other than full hard temper, add				↓	21%				
	2590	For silver solder, add						15%			
	2900	For medical clean, (oxygen class), add					12%				
	2940	To delete cplgs. & hngrs., 1/4"-1" pipe, subtract					37%	63%			
	2960	1-1/4"-3" pipe, subtract					12%	53%			
	2970	3-1/2"-5" pipe, subtract					12%	63%			
	2980	6"-8" pipe, subtract					24%	55%			
	3000	Type M tubing, couplings & hangers 10' O.C.									
	3100	1/4" diameter	1 Plum	90	.089	L.F.	1.04	3.63		4.67	6.60
	3120	3/8" diameter		87	.092		1.05	3.76		4.81	6.80
	3140	1/2" diameter		84	.095		1.18	3.89		5.07	7.15
	3160	5/8" diameter		81	.099		1.71	4.03		5.74	7.95
	3180	3/4" diameter		78	.103		1.73	4.19		5.92	8.20
	3200	1" diameter		70	.114		2.38	4.67		7.05	9.60
	3220	1-1/4" diameter		60	.133		3.48	5.45		8.93	12.05
	3240	1-1/2" diameter		54	.148		4.68	6.05		10.73	14.25
	3260	2" diameter	↓	44	.182		7.30	7.45		14.75	19.20
	3280	2-1/2" diameter	Q-1	64	.250		10.95	9.20		20.15	26
	3300	3" diameter		58	.276		14.65	10.15		24.80	31.50
	3320	3-1/2" diameter		45	.356		21	13.10		34.10	42.50
	3340	4" diameter		40	.400		26.50	14.70		41.20	51
	3360	5" diameter	↓	36	.444		79.50	16.35		95.85	112
	3370	6" diameter	Q-2	42	.571		103	22		125	146
	3380	8" diameter	"	38	.632	↓	170	24		194	222
	3440	For silver solder, add						15%			
	3960	To delete cplgs. & hngrs., 1/4"-1" pipe, subtract					35%	65%			
	3970	1-1/4"-3" pipe, subtract					19%	56%			
	3980	3-1/2"-5" pipe, subtract					13%	65%			
	3990	6"-8" pipe, subtract					28%	58%			
	4000	Type DWV tubing, couplings & hangers 10' O.C.									
	4100	1-1/4" diameter	1 Plum	60	.133	L.F.	3.67	5.45		9.12	12.25
	4120	1-1/2" diameter		54	.148		4.56	6.05		10.61	14.10
	4140	2" diameter	↓	44	.182		6.10	7.45		13.55	17.85
	4160	3" diameter	Q-1	58	.276		10.90	10.15		21.05	27.50
	4180	4" diameter		40	.400		19.15	14.70		33.85	43
	4200	5" diameter	↓	36	.444		57	16.35		73.35	87
	4220	6" diameter	Q-2	42	.571		81.50	22		103.50	123
	4240	8" diameter	"	38	.632	↓	201	24		225	257
	4730	To delete cplgs. & hngrs., 1-1/4"-2" pipe, subtract					16%	53%			
	4740	2"-4" pipe, subtract					13%	60%			
	4750	5"-8" pipe, subtract					23%	58%			
	5200	ACR tubing, type L, hard temper, cleaned and									
	5220	capped, no couplings or hangers									
	5240	3/8" OD				L.F.	.77			.77	.85

15107 | Metal Pipe & Fittings

			CREW	DAILY OUTPUT	LABOR-HOURS	UNIT	2005 BARE COSTS MAT.	LABOR	EQUIP.	TOTAL	TOTAL INCL O&P	
420	5250	1/2″ OD (R15100-050)				L.F.	1.22			1.22	1.34	420
	5260	5/8″ OD					1.49			1.49	1.64	
	5270	3/4″ OD					2.10			2.10	2.31	
	5280	7/8″ OD					2.34			2.34	2.57	
	5290	1-1/8″ OD					3.37			3.37	3.71	
	5300	1-3/8″ OD					4.54			4.54	4.99	
	5310	1-5/8″ OD					5.80			5.80	6.40	
	5320	2-1/8″ OD					9.15			9.15	10.05	
	5330	2-5/8″ OD					13.60			13.60	15	
	5340	3-1/8″ OD					18.30			18.30	20	
	5350	3-5/8″ OD					24			24	26.50	
	5360	4-1/8″ OD					30.50			30.50	33.50	
	5380	ACR tubing, type L, annealed, cleaned and capped										
	5381	No couplings or hangers										
	5384	3/8″	1 Stpi	160	.050	L.F.	.77	2.05		2.82	3.93	
	5385	1/2″		160	.050		1.22	2.05		3.27	4.42	
	5386	5/8″		160	.050		1.49	2.05		3.54	4.72	
	5387	3/4″		130	.062		2.10	2.52		4.62	6.10	
	5388	7/8″		130	.062		2.34	2.52		4.86	6.35	
	5389	1-1/8″		115	.070		3.37	2.85		6.22	8	
	5390	1-3/8″		100	.080		4.54	3.28		7.82	9.90	
	5391	1-5/8″		90	.089		5.80	3.64		9.44	11.85	
	5392	2-1/8″		80	.100		9.15	4.10		13.25	16.20	
	5393	2-5/8″	Q-5	125	.128		13.60	4.72		18.32	22	
	5394	3-1/8″		105	.152		18.30	5.60		23.90	28.50	
	5395	4-1/8″		95	.168		30.50	6.20		36.70	43	
	5800	Refrigeration tubing, dryseal, 50′ coils										
	5840	1/8″ OD				Coil	20			20	22	
	5850	3/16″ OD					22			22	24.50	
	5860	1/4″ OD					25			25	27.50	
	5870	5/16″ OD					32.50			32.50	35.50	
	5880	3/8″ OD					36.50			36.50	40	
	5890	1/2″ OD					46.50			46.50	51.50	
	5900	5/8″ OD					63			63	69.50	
	5910	3/4″ OD					73.50			73.50	81	
	5920	7/8″ OD					109			109	120	
	5930	1-1/8″ OD					160			160	176	
	5940	1-3/8″ OD					246			246	271	
	5950	1-5/8″ OD					315			315	350	
	9400	Sub assemblies used in assembly systems										
	9410	Chilled water unit, coil connections per unit under 10 ton	Q-5	.80	20	System	555	735		1,290	1,700	
	9420	Chilled water unit, coil connections per unit 10 ton and up		1	16		1,000	590		1,590	1,975	
	9430	Chilled water dist. piping per ton, less than 61 ton systems		26	.615		10.20	22.50		32.70	45	
	9440	Chilled water dist. piping per ton, 61 through 120 ton systems	Q-6	31	.774		31	29.50		60.50	78.50	
	9450	Chilled water dist. piping/ton, 135 ton systems and up	Q-8	25.40	1.260		42	49	3.17	94.17	123	
	9510	Refrigerant piping/ton of cooling for remote condensers	Q-5	2	8		168	295		463	630	
	9520	Refrigerant piping per ton up to 10 ton w/remote condensing unit		2.40	6.667		75	246		321	455	
	9530	Refrigerant piping per ton, 20 ton w/remote condensing unit		2	8		110	295		405	565	
	9540	Refrigerant piping per ton, 40 ton w/remote condensing unit		1.90	8.421		146	310		456	625	
	9550	Refrigerant piping per ton, 75-80 ton w/remote condensing unit	Q-6	2.40	10		212	380		592	810	
	9560	Refrigerant piping per ton, 100 ton w/remote condensing unit	″	2.20	10.909		282	415		697	935	
460	0010	**PIPE, COPPER, FITTINGS** Wrought unless otherwise noted										460
	0040	Solder joints, copper x copper										
	0070	90° elbow, 1/4″	1 Plum	22	.364	Ea.	1.57	14.85		16.42	24	
	0090	3/8″		22	.364		1.49	14.85		16.34	24	

15107 | Metal Pipe & Fittings

			CREW	DAILY OUTPUT	LABOR-HOURS	UNIT	2005 BARE COSTS MAT.	LABOR	EQUIP.	TOTAL	TOTAL INCL O&P
460	0100	1/2"	1 Plum	20	.400	Ea.	.48	16.35		16.83	25
	0110	5/8"		19	.421		2.99	17.20		20.19	29.50
	0120	3/4"		19	.421		1.06	17.20		18.26	27
	0130	1"		16	.500		2.61	20.50		23.11	33.50
	0140	1-1/4"		15	.533		3.94	22		25.94	37.50
	0150	1-1/2"		13	.615		5.55	25		30.55	44
	0160	2"		11	.727		11.20	29.50		40.70	57
	0170	2-1/2"	Q-1	13	1.231		23.50	45.50		69	94
	0180	3"		11	1.455		31.50	53.50		85	115
	0190	3-1/2"		10	1.600		110	59		169	210
	0200	4"		9	1.778		72.50	65.50		138	179
	0210	5"		6	2.667		405	98		503	590
	0220	6"	Q-2	9	2.667		450	102		552	650
	0230	8"	"	8	3		1,675	114		1,789	2,000
	0250	45° elbow, 1/4"	1 Plum	22	.364		2.65	14.85		17.50	25.50
	0270	3/8"		22	.364		2.16	14.85		17.01	25
	0280	1/2"		20	.400		.86	16.35		17.21	25.50
	0290	5/8"		19	.421		4.56	17.20		21.76	31
	0300	3/4"		19	.421		1.52	17.20		18.72	27.50
	0310	1"		16	.500		3.82	20.50		24.32	34.50
	0320	1-1/4"		15	.533		5.45	22		27.45	39
	0330	1-1/2"		13	.615		6.55	25		31.55	45
	0340	2"		11	.727		10.95	29.50		40.45	56.50
	0350	2-1/2"	Q-1	13	1.231		21	45.50		66.50	91
	0360	3"		13	1.231		34.50	45.50		80	106
	0370	3-1/2"		10	1.600		55	59		114	149
	0380	4"		9	1.778		73.50	65.50		139	180
	0390	5"		6	2.667		285	98		383	460
	0400	6"	Q-2	9	2.667		450	102		552	650
	0410	8"	"	8	3		1,525	114		1,639	1,850
	0450	Tee, 1/4"	1 Plum	14	.571		3.30	23.50		26.80	38.50
	0470	3/8"		14	.571		2.53	23.50		26.03	38
	0480	1/2"		13	.615		.81	25		25.81	39
	0490	5/8"		12	.667		5.50	27		32.50	47
	0500	3/4"		12	.667		1.95	27		28.95	43
	0510	1"		10	.800		6	32.50		38.50	55.50
	0520	1-1/4"		9	.889		8.65	36.50		45.15	64
	0530	1-1/2"		8	1		12	41		53	74.50
	0540	2"		7	1.143		18.75	46.50		65.25	90.50
	0550	2-1/2"	Q-1	8	2		42	73.50		115.50	158
	0560	3"		7	2.286		64	84		148	197
	0570	3-1/2"		6	2.667		186	98		284	350
	0580	4"		5	3.200		155	118		273	350
	0590	5"		4	4		510	147		657	780
	0600	6"	Q-2	6	4		695	153		848	995
	0610	8"	"	5	4.800		2,675	183		2,858	3,225
	0612	Tee, reducing on the outlet, 1/4"	1 Plum	15	.533		5.70	22		27.70	39.50
	0613	3/8"		15	.533		4.99	22		26.99	38.50
	0614	1/2"		14	.571		4.38	23.50		27.88	40
	0615	5/8"		13	.615		8.85	25		33.85	47.50
	0616	3/4"		12	.667		2.82	27		29.82	44
	0617	1"		11	.727		5.75	29.50		35.25	51
	0618	1-1/4"		10	.800		9.35	32.50		41.85	59.50
	0619	1-1/2"		9	.889		9.95	36.50		46.45	65.50
	0620	2"		8	1		16.25	41		57.25	79.50
	0621	2-1/2"	Q-1	9	1.778		44	65.50		109.50	147

15107 | Metal Pipe & Fittings

			CREW	DAILY OUTPUT	LABOR-HOURS	UNIT	2005 BARE COSTS MAT.	LABOR	EQUIP.	TOTAL	TOTAL INCL O&P
460	0622	3"	Q-1	8	2	Ea.	54	73.50		127.50	171
	0623	4"		6	2.667		110	98		208	268
	0624	5"		5	3.200		510	118		628	735
	0625	6"	Q-2	7	3.429		695	131		826	960
	0626	8"	"	6	4		2,675	153		2,828	3,175
	0630	Tee, reducing on the run, 1/4"	1 Plum	15	.533		6.65	22		28.65	40.50
	0631	3/8"		15	.533		7.55	22		29.55	41.50
	0632	1/2"		14	.571		8.90	23.50		32.40	45
	0633	5/8"		13	.615		9.40	25		34.40	48.50
	0634	3/4"		12	.667		2.30	27		29.30	43.50
	0635	1"		11	.727		7.55	29.50		37.05	53
	0636	1-1/4"		10	.800		12	32.50		44.50	62
	0637	1-1/2"		9	.889		21	36.50		57.50	78
	0638	2"		8	1		27	41		68	91.50
	0639	2-1/2"	Q-1	9	1.778		66	65.50		131.50	171
	0640	3"		8	2		93.50	73.50		167	214
	0641	4"		6	2.667		207	98		305	375
	0642	5"		5	3.200		480	118		598	705
	0643	6"	Q-2	7	3.429		735	131		866	1,000
	0644	8"	"	6	4		2,675	153		2,828	3,175
	0650	Coupling, 1/4"	1 Plum	24	.333		.37	13.60		13.97	21
	0670	3/8"		24	.333		.48	13.60		14.08	21
	0680	1/2"		22	.364		.33	14.85		15.18	23
	0690	5/8"		21	.381		1.02	15.55		16.57	24.50
	0700	3/4"		21	.381		.71	15.55		16.26	24.50
	0710	1"		18	.444		1.44	18.15		19.59	29
	0715	1-1/4"		17	.471		2.69	19.20		21.89	32
	0716	1-1/2"		15	.533		3.55	22		25.55	37
	0718	2"		13	.615		5.95	25		30.95	44.50
	0721	2-1/2"	Q-1	15	1.067		12.60	39		51.60	73
	0722	3"		13	1.231		18.90	45.50		64.40	89
	0724	3-1/2"		8	2		36.50	73.50		110	151
	0726	4"		7	2.286		40	84		124	170
	0728	5"		6	2.667		98	98		196	255
	0731	6"	Q-2	8	3		162	114		276	350
	0732	8"	"	7	3.429		525	131		656	770
	0741	Coupling, reducing, concentric									
	0743	1/2"	1 Plum	23	.348	Ea.	.96	14.20		15.16	22.50
	0745	3/4"		21.50	.372		1.37	15.20		16.57	24.50
	0747	1"		19.50	.410		3.05	16.75		19.80	28.50
	0748	1-1/4"		18	.444		3.50	18.15		21.65	31.50
	0749	1-1/2"		16	.500		6.15	20.50		26.65	37.50
	0751	2"		14	.571		9.60	23.50		33.10	45.50
	0752	2-1/2"		13	.615		23	25		48	63.50
	0753	3"	Q-1	14	1.143		28.50	42		70.50	94.50
	0755	4"	"	8	2		61	73.50		134.50	178
	0757	5"	Q-2	7.50	3.200		210	122		332	415
	0759	6"		7	3.429		345	131		476	575
	0761	8"		6.50	3.692		570	141		711	835
	0771	Cap, sweat									
	0773	1/2"	1 Plum	40	.200	Ea.	.34	8.15		8.49	12.65
	0775	3/4"		38	.211		.63	8.60		9.23	13.65
	0777	1"		32	.250		1.44	10.20		11.64	16.95
	0778	1-1/4"		29	.276		2.09	11.25		13.34	19.25
	0779	1-1/2"		26	.308		3.14	12.55		15.69	22.50
	0781	2"		22	.364		5.75	14.85		20.60	29

Important: See the Reference Section for critical supporting data - Reference Nos., Crews, & City Cost Indexes

15107 | Metal Pipe & Fittings

		Crew	Daily Output	Labor-Hours	Unit	2005 Bare Costs Mat.	Labor	Equip.	Total	Total Incl O&P
460 0791	Flange, sweat									460
0793	3"	Q-1	22	.727	Ea.	46.50	27		73.50	91
0795	4"		18	.889		65	32.50		97.50	121
0797	5"		12	1.333		121	49		170	207
0799	6"	Q-2	18	1.333		127	51		178	217
0801	8"	"	16	1.500		525	57		582	660
0850	Unions, 1/4"	1 Plum	21	.381		9.50	15.55		25.05	34
0870	3/8"		21	.381		9.60	15.55		25.15	34
0880	1/2"		19	.421		4.89	17.20		22.09	31.50
0890	5/8"		18	.444		22	18.15		40.15	51.50
0900	3/4"		18	.444		6.40	18.15		24.55	34.50
0910	1"		15	.533		11	22		33	45
0920	1-1/4"		14	.571		19.05	23.50		42.55	56
0930	1-1/2"		12	.667		25	27		52	68.50
0940	2"		10	.800		43	32.50		75.50	96
0950	2-1/2"	Q-1	12	1.333		93.50	49		142.50	177
0960	3"	"	10	1.600		242	59		301	355
0980	Adapter, copper x male IPS, 1/4"	1 Plum	20	.400		5.20	16.35		21.55	30
0990	3/8"		20	.400		2.58	16.35		18.93	27.50
1000	1/2"		18	.444		.99	18.15		19.14	28.50
1010	3/4"		17	.471		1.66	19.20		20.86	31
1020	1"		15	.533		4.30	22		26.30	37.50
1030	1-1/4"		13	.615		6.70	25		31.70	45.50
1040	1-1/2"		12	.667		7.70	27		34.70	49.50
1050	2"		11	.727		13	29.50		42.50	59
1060	2-1/2"	Q-1	10.50	1.524		50	56		106	139
1070	3"		10	1.600		66	59		125	161
1080	3-1/2"		9	1.778		86.50	65.50		152	194
1090	4"		8	2		86.50	73.50		160	206
1200	5"		6	2.667		500	98		598	695
1210	6"	Q-2	8.50	2.824		500	108		608	710
1214	Adapter, copper x female IPS									
1216	1/2"	1 Plum	18	.444	Ea.	1.58	18.15		19.73	29
1218	3/4"		17	.471		2.15	19.20		21.35	31.50
1220	1"		15	.533		5.25	22		27.25	39
1221	1/1/4"		13	.615		7.70	25		32.70	46.50
1222	1-1/2"		12	.667		12	27		39	54
1224	2"		11	.727		16.35	29.50		45.85	62.50
1250	Cross, 1/2"		10	.800		6.90	32.50		39.40	56.50
1260	3/4"		9.50	.842		13.40	34.50		47.90	66.50
1270	1"		8	1		23	41		64	86.50
1280	1-1/4"		7.50	1.067		32.50	43.50		76	102
1290	1-1/2"		6.50	1.231		46.50	50.50		97	127
1300	2"		5.50	1.455		88	59.50		147.50	187
1310	2-1/2"	Q-1	6.50	2.462		203	90.50		293.50	360
1320	3"	"	5.50	2.909		256	107		363	440
1500	Tee fitting, mechanically formed, (Type 1).									
1520	1/2" run size, 3/8" to 1/2" branch size	1 Plum	80	.100	Ea.		4.09		4.09	6.15
1530	3/4" run size, 3/8" to 3/4" branch size		60	.133			5.45		5.45	8.20
1540	1" run size, 3/8" to 1" branch size		54	.148			6.05		6.05	9.10
1550	1-1/4" run size, 3/8" to 1-1/4" branch size		48	.167			6.80		6.80	10.25
1560	1-1/2" run size, 3/8" to 1-1/2" branch size		40	.200			8.15		8.15	12.30
1570	2" run size, 3/8" to 2" branch size		35	.229			9.35		9.35	14.05
1580	2-1/2" run size, 1/2" to 2" branch size		32	.250			10.20		10.20	15.35
1590	3" run size, 1" to 2" branch size		26	.308			12.55		12.55	18.90
1600	4" run size, 1" to 2" branch size		24	.333			13.60		13.60	20.50

15107 | Metal Pipe & Fittings

		CREW	DAILY OUTPUT	LABOR-HOURS	UNIT	2005 BARE COSTS MAT.	LABOR	EQUIP.	TOTAL	TOTAL INCL O&P
460	1640	Tee fitting, mechanically formed, (Type 2)								
	1650	2-1/2" run size, 2-1/2" branch size	1 Plum	12.50	.640	Ea.		26		26	39.50
	1660	3" run size, 2-1/2" to 3" branch size		12	.667			27		27	41
	1670	3-1/2" run size, 2-1/2" to 3-1/2" branch size		11	.727			29.50		29.50	44.50
	1680	4" run size, 2-1/2" to 4" branch size		10.50	.762			31		31	47
	1698	5" run size, 2" to 4" branch size		9.50	.842			34.50		34.50	51.50
	1700	6" run size, 2" to 4" branch size		8.50	.941			38.50		38.50	58
	1710	8" run size, 2" to 4" branch size		7	1.143			46.50		46.50	70
	1800	ACR fittings, OD size									
	1802	Tee, straight									
	1808	5/8"	1 Stpi	12	.667	Ea.	.81	27.50		28.31	42
	1810	3/4"		12	.667		5.50	27.50		33	47
	1812	7/8"		10	.800		2.07	33		35.07	51.50
	1813	1"		10	.800		14.90	33		47.90	65.50
	1814	1-1/8"		10	.800		6	33		39	55.50
	1816	1-3/8"		9	.889		8.65	36.50		45.15	64
	1818	1-5/8"		8	1		13.35	41		54.35	76
	1820	2-1/8"		7	1.143		21	47		68	93.50
	1822	2-5/8"	Q-5	8	2		42	73.50		115.50	158
	1824	3-1/8"		7	2.286		64	84		148	198
	1826	4-1/8"		5	3.200		155	118		273	350
	1830	90° elbow									
	1836	5/8"	1 Stpi	19	.421	Ea.	1.66	17.25		18.91	28
	1838	3/4"		19	.421		3.03	17.25		20.28	29.50
	1840	7/8"		16	.500		3.01	20.50		23.51	34.50
	1842	1-1/8"		16	.500		4.03	20.50		24.53	35.50
	1844	1-3/8"		15	.533		3.94	22		25.94	37.50
	1846	1-5/8"		13	.615		6.15	25		31.15	45
	1848	2-1/8"		11	.727		11.20	30		41.20	57.50
	1850	2-5/8"	Q-5	13	1.231		23.50	45.50		69	94
	1852	3-1/8"		11	1.455		31.50	53.50		85	115
	1854	4-1/8"		9	1.778		80.50	65.50		146	188
	1860	Coupling									
	1866	5/8"	1 Stpi	21	.381	Ea.	.37	15.60		15.97	24
	1868	3/4"		21	.381		1.13	15.60		16.73	24.50
	1870	7/8"		18	.444		.71	18.20		18.91	28.50
	1871	1"		18	.444		2.58	18.20		20.78	30.50
	1872	1-1/8"		18	.444		1.53	18.20		19.73	29
	1874	1-3/8"		17	.471		2.69	19.25		21.94	32
	1876	1-5/8"		15	.533		3.55	22		25.55	37
	1878	2-1/8"		13	.615		5.95	25		30.95	44.50
	1880	2-5/8"	Q-5	15	1.067		13.35	39.50		52.85	73.50
	1882	3-1/8"		13	1.231		19.30	45.50		64.80	89.50
	1884	4-1/8"		7	2.286		44	84		128	176
	2000	DWV, solder joints, copper x copper									
	2030	90° Elbow, 1-1/4"	1 Plum	13	.615	Ea.	4.82	25		29.82	43.50
	2050	1-1/2"		12	.667		7.20	27		34.20	49
	2070	2"		10	.800		9.40	32.50		41.90	59.50
	2090	3"	Q-1	10	1.600		23.50	59		82.50	114
	2100	4"	"	9	1.778		100	65.50		165.50	209
	2150	45° Elbow, 1-1/4"	1 Plum	13	.615		3.99	25		28.99	42.50
	2170	1-1/2"		12	.667		3.30	27		30.30	44.50
	2180	2"		10	.800		7.60	32.50		40.10	57.50
	2190	3"	Q-1	10	1.600		16.10	59		75.10	106
	2200	4"	"	9	1.778		76.50	65.50		142	183
	2250	Tee, Sanitary, 1-1/4"	1 Plum	9	.889		8.55	36.50		45.05	64

Important: See the Reference Section for critical supporting data - Reference Nos., Crews, & City Cost Indexes

15107 Metal Pipe & Fittings		CREW	DAILY OUTPUT	LABOR-HOURS	UNIT	2005 BARE COSTS MAT.	LABOR	EQUIP.	TOTAL	TOTAL INCL O&P	
460	2270 1-1/2"	1 Plum	8	1	Ea.	10.65	41		51.65	73.50	460
	2290 2"	↓	7	1.143		12.45	46.50		58.95	83.50	
	2310 3"	Q-1	7	2.286		50	84		134	181	
	2330 4"	"	6	2.667		127	98		225	287	
	2400 Coupling, 1-1/4"	1 Plum	14	.571		2.03	23.50		25.53	37	
	2420 1-1/2"		13	.615		2.53	25		27.53	41	
	2440 2"	↓	11	.727		3.50	29.50		33	48.50	
	2460 3"	Q-1	11	1.455		6.80	53.50		60.30	88	
	2480 4"	"	10	1.600	↓	21.50	59		80.50	113	
	2600 Traps, see division 15155										
	3500 Compression joint fittings										
	3510 As used for plumbing and oil burner work										
	3520 Fitting price includes nuts and sleeves										
	3540 Sleeve, 1/8"				Ea.	.11			.11	.12	
	3550 3/16"					.11			.11	.12	
	3560 1/4"					.05			.05	.06	
	3570 5/16"					.11			.11	.12	
	3580 3/8"					.23			.23	.25	
	3600 1/2"					.31			.31	.34	
	3620 Nut, 1/8"					.17			.17	.19	
	3630 3/16"					.18			.18	.20	
	3640 1/4"					.17			.17	.19	
	3650 5/16"					.25			.25	.28	
	3660 3/8"					.27			.27	.30	
	3670 1/2"					.48			.48	.53	
	3710 Union, 1/8"	1 Plum	26	.308		1.01	12.55		13.56	20	
	3720 3/16"		24	.333		.97	13.60		14.57	21.50	
	3730 1/4"		24	.333		1.02	13.60		14.62	21.50	
	3740 5/16"		23	.348		1.30	14.20		15.50	23	
	3750 3/8"		22	.364		1.25	14.85		16.10	24	
	3760 1/2"		22	.364		1.91	14.85		16.76	24.50	
	3780 5/8"		21	.381		2.12	15.55		17.67	26	
	3820 Union tee, 1/8"		17	.471		3.65	19.20		22.85	33	
	3830 3/16"		16	.500		2.84	20.50		23.34	33.50	
	3840 1/4"		15	.533		2.38	22		24.38	35.50	
	3850 5/16"		15	.533		4.80	22		26.80	38.50	
	3860 3/8"		15	.533		2.75	22		24.75	36	
	3870 1/2"		15	.533		4.62	22		26.62	38	
	3910 Union elbow, 1/4"		24	.333		1.88	13.60		15.48	22.50	
	3920 5/16"		23	.348		3.10	14.20		17.30	25	
	3930 3/8"		22	.364		2.24	14.85		17.09	25	
	3940 1/2"		22	.364		3.65	14.85		18.50	26.50	
	3980 Female connector, 1/8"		26	.308		.83	12.55		13.38	19.80	
	4000 3/16" x 1/8"		24	.333		1.30	13.60		14.90	22	
	4010 1/4" x 1/8"		24	.333		.88	13.60		14.48	21.50	
	4020 1/4"		24	.333		1.11	13.60		14.71	21.50	
	4030 3/8" x 1/4"		22	.364		1.23	14.85		16.08	24	
	4040 1/2" x 3/8"		22	.364		1.96	14.85		16.81	24.50	
	4050 5/8" x 1/2"		21	.381		2.46	15.55		18.01	26	
	4090 Male connector, 1/8"		26	.308		.60	12.55		13.15	19.55	
	4100 3/16" x 1/8"		24	.333		.62	13.60		14.22	21	
	4110 1/4" x 1/8"		24	.333		.64	13.60		14.24	21	
	4120 1/4"		24	.333		.64	13.60		14.24	21	
	4130 5/16" x 1/8"		23	.348		.75	14.20		14.95	22.50	
	4140 5/16" x 1/4"		23	.348		.86	14.20		15.06	22.50	
	4150 3/8" x 1/8"	↓	22	.364	↓	.89	14.85		15.74	23.50	

	15107 | Metal Pipe & Fittings	CREW	DAILY OUTPUT	LABOR-HOURS	UNIT	2005 BARE COSTS MAT.	LABOR	EQUIP.	TOTAL	TOTAL INCL O&P
460 4160	3/8" x 1/4"	1 Plum	22	.364	Ea.	1.03	14.85		15.88	23.50
4170	3/8"		22	.364		1.26	14.85		16.11	24
4180	3/8" x 1/2"		22	.364		1.41	14.85		16.26	24
4190	1/2" x 3/8"		22	.364		1.37	14.85		16.22	24
4200	1/2"		22	.364		1.91	14.85		16.76	24.50
4210	5/8" x 1/2"		21	.381		1.78	15.55		17.33	25.50
4240	Male elbow, 1/8"		26	.308		1.30	12.55		13.85	20.50
4250	3/16" x 1/8"		24	.333		1.43	13.60		15.03	22
4260	1/4" x 1/8"		24	.333		1.08	13.60		14.68	21.50
4270	1/4"		24	.333		1.23	13.60		14.83	22
4280	3/8" x 1/4"		22	.364		1.91	14.85		16.76	24.50
4290	3/8"		22	.364		2.69	14.85		17.54	25.50
4300	1/2" x 1/4"		22	.364		2.60	14.85		17.45	25.50
4310	1/2" x 3/8"		22	.364		4.07	14.85		18.92	27
4340	Female elbow, 1/8"		26	.308		2.78	12.55		15.33	22
4350	1/4" x 1/8"		24	.333		1.58	13.60		15.18	22
4360	1/4"		24	.333		2.08	13.60		15.68	23
4370	3/8" x 1/4"		22	.364		2.06	14.85		16.91	25
4380	1/2" x 3/8"		22	.364		3.50	14.85		18.35	26.50
4390	1/2"		22	.364		4.71	14.85		19.56	27.50
4420	Male run tee, 1/4" x 1/8"		15	.533		1.94	22		23.94	35
4430	5/16" x 1/8"		15	.533		3.80	22		25.80	37
4440	3/8" x 1/4"		15	.533		3	22		25	36.50
4480	1/4" x 1/8"		15	.533		2.25	22		24.25	35.50
4490	1/4"		15	.533		2.42	22		24.42	35.50
4500	3/8" x 1/4"		15	.533		2.82	22		24.82	36
4510	1/2" x 3/8"		15	.533		4.38	22		26.38	38
4520	1/2"	↓	15	.533	↓	7	22		29	40.50
4800	Flare joint fittings									
4810	Refrigeration fittings									
4820	Flare joint nuts and labor not incl. in price. Add 1 nut per jnt.									
4830	90° Elbow, 1/4"				Ea.	1.86			1.86	2.05
4840	3/8"					2.16			2.16	2.38
4850	1/2"					3.13			3.13	3.44
4860	5/8"					4.58			4.58	5.05
4870	3/4"					6.90			6.90	7.60
5030	Tee, 1/4"					1.61			1.61	1.77
5040	5/16"					2.28			2.28	2.51
5050	3/8"					2.60			2.60	2.86
5060	1/2"					3.96			3.96	4.36
5070	5/8"					6.15			6.15	6.75
5080	3/4"					8.70			8.70	9.55
5140	Union, 3/16"					1.86			1.86	2.05
5150	1/4"					.63			.63	.69
5160	5/16"					1.18			1.18	1.30
5170	3/8"					1.26			1.26	1.39
5180	1/2"					1.32			1.32	1.45
5190	5/8"					2.15			2.15	2.37
5200	3/4"					3.21			3.21	3.53
5260	Long flare nut, 3/16"	1 Stpi	42	.190		.78	7.80		8.58	12.55
5270	1/4"		41	.195		.86	8		8.86	12.95
5280	5/16"		40	.200		2.11	8.20		10.31	14.60
5290	3/8"		39	.205		1.18	8.40		9.58	13.95
5300	1/2"		38	.211		1.35	8.60		9.95	14.45
5310	5/8"		37	.216		2.87	8.85		11.72	16.45
5320	3/4"	↓	34	.235	↓	2.45	9.65		12.10	17.20

	15107 | Metal Pipe & Fittings	CREW	DAILY OUTPUT	LABOR-HOURS	UNIT	2005 BARE COSTS MAT.	LABOR	EQUIP.	TOTAL	TOTAL INCL O&P	
460	5380 Short flare nut, 3/16"	1 Stpi	42	.190	Ea.	1.15	7.80		8.95	12.95	460
	5390 1/4"		41	.195		.39	8		8.39	12.45	
	5400 5/16"		40	.200		.63	8.20		8.83	13	
	5410 3/8"		39	.205		.54	8.40		8.94	13.25	
	5420 1/2"		38	.211		.79	8.60		9.39	13.80	
	5430 5/8"		36	.222		1.11	9.10		10.21	14.90	
	5440 3/4"	↓	34	.235		1.52	9.65		11.17	16.15	
	5500 90° Elbow flare by MIPS, 1/4"					1.35			1.35	1.49	
	5510 3/8"					1.69			1.69	1.86	
	5520 1/2"					2.70			2.70	2.97	
	5530 5/8"					3.13			3.13	3.44	
	5540 3/4"					5.25			5.25	5.80	
	5600 Flare by FIPS, 1/4"					4.05			4.05	4.46	
	5610 3/8"					2.87			2.87	3.16	
	5620 1/2"					4.31			4.31	4.74	
	5670 Flare by sweat, 1/4"					1.35			1.35	1.49	
	5680 3/8"					1.52			1.52	1.67	
	5690 1/2"					2.70			2.70	2.97	
	5700 5/8"					3.04			3.04	3.34	
	5760 Tee flare by IPS, 1/4"					2.34			2.34	2.57	
	5770 3/8"					3.19			3.19	3.51	
	5780 1/2"					4.89			4.89	5.40	
	5790 5/8"					6.25			6.25	6.90	
	5850 Connector, 1/4"					.80			.80	.88	
	5860 3/8"					1			1	1.10	
	5870 1/2"					1.27			1.27	1.40	
	5880 5/8"					1.40			1.40	1.54	
	5890 3/4"					4.64			4.64	5.10	
	5950 Seal cap, 1/4"					.30			.30	.33	
	5960 3/8"					.59			.59	.65	
	5970 1/2"					.68			.68	.75	
	5980 5/8"					1.01			1.01	1.11	
	5990 3/4"				↓	2.11			2.11	2.32	
	6000 Water service fittings										
	6010 Flare joints nut and labor are included in the fitting price.										
	6020 90° Elbow, cxc, 3/8"	1 Plum	19	.421	Ea.	13.70	17.20		30.90	41	
	6030 1/2"		18	.444		13.70	18.15		31.85	42.50	
	6040 3/4"		16	.500		17.45	20.50		37.95	49.50	
	6050 1"		15	.533		33	22		55	69.50	
	6060 1-1/4"		12	.667		64	27		91	112	
	6070 1-1/2"		12	.667		64	27		91	112	
	6080 2"		10	.800		114	32.50		146.50	174	
	6090 C x MPT, 3/8"		19	.421		9.85	17.20		27.05	37	
	6100 1/2"		18	.444		9.85	18.15		28	38.50	
	6110 3/4"		16	.500		11.40	20.50		31.90	43	
	6120 1"		15	.533		31	22		53	67	
	6130 1-1/4"		13	.615		64	25		89	109	
	6140 1-1/2"		12	.667		64	27		91	112	
	6150 2"		10	.800		89.50	32.50		122	147	
	6160 C x FPT, 3/8"		19	.421		10.60	17.20		27.80	37.50	
	6170 1/2"		18	.444		10.60	18.15		28.75	39	
	6180 3/4"		16	.500		13.10	20.50		33.60	45	
	6190 1"		15	.533		30	22		52	66	
	6200 1-1/4"		13	.615		30	25		55	71	
	6210 1-1/2"		12	.667		138	27		165	193	
	6220 2"	↓	10	.800	↓	211	32.50		243.50	282	

15107 | Metal Pipe & Fittings

			CREW	DAILY OUTPUT	LABOR-HOURS	UNIT	2005 BARE COSTS MAT.	LABOR	EQUIP.	TOTAL	TOTAL INCL O&P	
460	6230	Tee, cxcxc, 3/8"	1 Plum	13	.615	Ea.	20	25		45	60	460
	6240	1/2"		12	.667		20	27		47	63	
	6250	3/4"		11	.727		23.50	29.50		53	70.50	
	6260	1"		10	.800		41.50	32.50		74	94.50	
	6270	1-1/4"		8	1		109	41		150	181	
	6280	1-1/2"		8	1		130	41		171	205	
	6290	2"		7	1.143		233	46.50		279.50	325	
	6300	Tee, cxcxFPT, 3/8"		13	.615		12.25	25		37.25	51.50	
	6310	1/2"		12	.667		18.15	27		45.15	61	
	6320	3/4"		11	.727		25.50	29.50		55	73	
	6330	Tube nut, cxnut seat, 3/8"		40	.200		4.05	8.15		12.20	16.75	
	6340	1/2"		38	.211		4.05	8.60		12.65	17.40	
	6350	3/4"		34	.235		4.05	9.60		13.65	18.90	
	6360	1"		32	.250		7.35	10.20		17.55	23.50	
	6370	1-1/4"		25	.320		32.50	13.05		45.55	55	
	6380	Coupling, cxc, 3/8"		19	.421		12.25	17.20		29.45	39.50	
	6390	1/2"		18	.444		12.25	18.15		30.40	41	
	6400	3/4"		16	.500		15.50	20.50		36	47.50	
	6410	1"		15	.533		28.50	22		50.50	64.50	
	6420	1-1/4"		12	.667		46.50	27		73.50	92	
	6430	1-1/2"		12	.667		70.50	27		97.50	119	
	6440	2"		10	.800		106	32.50		138.50	165	
	6450	Adapter, cxFPT, 3/8"		19	.421		9.25	17.20		26.45	36	
	6460	1/2"		18	.444		9.25	18.15		27.40	37.50	
	6470	3/4"		16	.500		11.20	20.50		31.70	43	
	6480	1"		15	.533		24	22		46	59.50	
	6490	1-1/4"		13	.615		51	25		76	94	
	6500	1-1/2"		12	.667		51	27		78	97	
	6510	2"		10	.800		66	32.50		98.50	122	
	6520	Adapter, cxMPT, 3/8"		19	.421		8.30	17.20		25.50	35	
	6530	1/2"		18	.444		8.30	18.15		26.45	36.50	
	6540	3/4"		16	.500		11.45	20.50		31.95	43	
	6550	1"		15	.533		20	22		42	55	
	6560	1-1/4"		13	.615		50.50	25		75.50	93.50	
	6570	1-1/2"		12	.667		50.50	27		77.50	96.50	
	6580	2"	↓	10	.800	↓	68.50	32.50		101	124	
	6990	Polybutylene/polyethylene pipe, See 15108 for plastic ftng.										
	7000	Insert type Brass/copper, 100 psi @ 180°F, CTS										
	7010	Adapter MPT 3/8" x 3/8" CTS	1 Plum	29	.276	Ea.	.97	11.25		12.22	18	
	7020	1/2" x 1/2"		26	.308		.97	12.55		13.52	19.95	
	7030	3/4" x 1/2"		26	.308		1.44	12.55		13.99	20.50	
	7040	3/4" x 3/4"		25	.320		1.42	13.05		14.47	21	
	7050	Adapter CTS 1/2" x 1/2" sweat		24	.333		.38	13.60		13.98	21	
	7060	3/4" x 3/4" sweat		22	.364		.47	14.85		15.32	23	
	7070	Coupler center set 3/8" CTS		25	.320		.49	13.05		13.54	20	
	7080	1/2" CTS		23	.348		.38	14.20		14.58	22	
	7090	3/4" CTS		22	.364		.47	14.85		15.32	23	
	7100	Elbow 90°, copper 3/8"		25	.320		1.08	13.05		14.13	21	
	7110	1/2" CTS		23	.348		.69	14.20		14.89	22.50	
	7120	3/4" CTS		22	.364		.84	14.85		15.69	23.50	
	7130	Tee copper 3/8" CTS		17	.471		1.28	19.20		20.48	30.50	
	7140	1/2" CTS		15	.533		.88	22		22.88	34	
	7150	3/4" CTS		14	.571		1.35	23.50		24.85	36.50	
	7160	3/8" x 3/8" x 1/2"		16	.500		.88	20.50		21.38	31.50	
	7170	1/2" x 3/8" x 1/2"		15	.533		.88	22		22.88	34	
	7180	3/4" x 1/2" x 3/4"	↓	14	.571	↓	1.35	23.50		24.85	36.50	

15107 | Metal Pipe & Fittings

			CREW	DAILY OUTPUT	LABOR-HOURS	UNIT	2005 BARE COSTS MAT.	LABOR	EQUIP.	TOTAL	TOTAL INCL O&P
480	0010	**PIPE/TUBE, GROOVED-JOINT FOR COPPER**									
	4000	Fittings: coupling and labor required at joints not incl.									
	4001	in fitting price. add 1 per joint for installed price.									
	4010	Coupling, rigid style									
	4018	2" diameter	1 Plum	50	.160	Ea.	8.10	6.55		14.65	18.70
	4020	2-1/2" diameter	Q-1	80	.200		9.20	7.35		16.55	21
	4022	3" diameter		67	.239		10.30	8.80		19.10	24.50
	4024	4" diameter		50	.320		15.80	11.75		27.55	35
	4026	5" diameter		40	.400		27	14.70		41.70	52
	4028	6" diameter	Q-2	50	.480		36	18.30		54.30	67
	4100	Elbow, 90° or 45°									
	4108	2" diameter	1 Plum	25	.320	Ea.	12.35	13.05		25.40	33.50
	4110	2-1/2" diameter	Q-1	40	.400		13.50	14.70		28.20	37
	4112	3" diameter		33	.485		19.10	17.85		36.95	48
	4114	4" diameter		25	.640		44.50	23.50		68	84.50
	4116	5" diameter		20	.800		127	29.50		156.50	184
	4118	6" diameter	Q-2	25	.960		204	36.50		240.50	279
	4200	Tee									
	4208	2" diameter	1 Plum	17	.471	Ea.	20.50	19.20		39.70	52
	4210	2-1/2" diameter	Q-1	27	.593		22	22		44	57
	4212	3" diameter		22	.727		33	27		60	76.50
	4214	4" diameter		17	.941		73	34.50		107.50	132
	4216	5" diameter		13	1.231		205	45.50		250.50	294
	4218	6" diameter	Q-2	17	1.412		254	54		308	360
	4300	Reducer, concentric									
	4310	3" x 2-1/2" diameter	Q-1	35	.457	Ea.	18.05	16.80		34.85	45.50
	4312	4" x 2-1/2" diameter		32	.500		37.50	18.40		55.90	68.50
	4314	4" x 3" diameter		29	.552		37.50	20.50		58	71.50
	4316	5" x 3" diameter		25	.640		106	23.50		129.50	153
	4318	5" x 4" diameter		22	.727		106	27		133	157
	4320	6" x 3" diameter	Q-2	28	.857		115	32.50		147.50	176
	4322	6" x 4" diameter		26	.923		115	35		150	180
	4324	6" x 5" diameter		24	1		115	38		153	185
	4350	Flange, w/groove gasket									
	4351	ANSI class 125 and 150									
	4355	2" diameter	1 Plum	23	.348	Ea.	65.50	14.20		79.70	94
	4356	2-1/2" diameter	Q-1	37	.432		68.50	15.90		84.40	99.50
	4358	3" diameter		31	.516		71.50	19		90.50	108
	4360	4" diameter		23	.696		78.50	25.50		104	125
	4362	5" diameter		19	.842		106	31		137	163
	4364	6" diameter	Q-2	23	1.043		115	40		155	187
620	0010	**PIPE, STEEL** R15100-050									
	0020	All pipe sizes are to Spec. A-53 unless noted otherwise									
	0030	Schedule 10, see 15107-690-0500									
	0050	Schedule 40, threaded, with couplings, and clevis type									
	0060	hangers sized for covering, 10' O.C.									
	0540	Black, 1/4" diameter	1 Plum	66	.121	L.F.	2.09	4.95		7.04	9.75
	0550	3/8" diameter		65	.123		2.02	5.05		7.07	9.80
	0560	1/2" diameter		63	.127		1.28	5.20		6.48	9.20
	0570	3/4" diameter		61	.131		1.54	5.35		6.89	9.75
	0580	1" diameter		53	.151		2.16	6.15		8.31	11.65
	0590	1-1/4" diameter	Q-1	89	.180		2.75	6.60		9.35	12.95
	0600	1-1/2" diameter		80	.200		3.16	7.35		10.51	14.55
	0610	2" diameter		64	.250		4.13	9.20		13.33	18.35
	0620	2-1/2" diameter		50	.320		7	11.75		18.75	25.50

15107 | Metal Pipe & Fittings

		Description	Crew	Daily Output	Labor-Hours	Unit	2005 Bare Costs Mat.	2005 Bare Costs Labor	2005 Bare Costs Equip.	2005 Bare Costs Total	Total Incl O&P	
620	0630	3" diameter (R15100-050)	Q-1	43	.372	L.F.	9.05	13.70		22.75	30.50	620
	0640	3-1/2" diameter		40	.400		11.95	14.70		26.65	35	
	0650	4" diameter		36	.444		14.25	16.35		30.60	40	
	0660	5" diameter		26	.615		26	22.50		48.50	62.50	
	0670	6" diameter	Q-2	31	.774		32.50	29.50		62	80.50	
	0680	8" diameter		27	.889		51	34		85	107	
	0690	10" diameter		23	1.043		78	40		118	146	
	0700	12" diameter		18	1.333		86	51		137	171	
	1220	To delete coupling & hanger, subtract										
	1230	1/4" diam. to 3/4" diam.					31%	56%				
	1240	1" diam. to 1-1/2" diam.					23%	51%				
	1250	2" diam. to 4" diam.					23%	41%				
	1260	5" diam. to 12" diam.					21%	45%				
	1290	Galvanized, 1/4" diameter	1 Plum	66	.121	L.F.	2.36	4.95		7.31	10.05	
	1300	3/8" diameter		65	.123		2.69	5.05		7.74	10.50	
	1310	1/2" diameter		63	.127		1.52	5.20		6.72	9.45	
	1320	3/4" diameter		61	.131		1.76	5.35		7.11	10	
	1330	1" diameter		53	.151		2.46	6.15		8.61	11.95	
	1340	1-1/4" diameter	Q-1	89	.180		3.20	6.60		9.80	13.45	
	1350	1-1/2" diameter		80	.200		3.71	7.35		11.06	15.15	
	1360	2" diameter		64	.250		4.85	9.20		14.05	19.15	
	1370	2-1/2" diameter		50	.320		7.90	11.75		19.65	26.50	
	1380	3" diameter		43	.372		10.20	13.70		23.90	32	
	1390	3-1/2" diameter		40	.400		16.50	14.70		31.20	40	
	1400	4" diameter		36	.444		16.65	16.35		33	43	
	1410	5" diameter		26	.615		27.50	22.50		50	64.50	
	1420	6" diameter	Q-2	31	.774		38	29.50		67.50	86	
	1430	8" diameter		27	.889		57	34		91	114	
	1440	10" diameter		23	1.043		84	40		124	153	
	1450	12" diameter		18	1.333		115	51		166	204	
	1750	To delete coupling & hanger, subtract										
	1760	1/4" diam. to 3/4" diam.					31%	56%				
	1770	1" diam. to 1-1/2" diam.					23%	51%				
	1780	2" diam. to 4" diam.					23%	41%				
	1790	5" diam. to 12" diam.					21%	45%				
	2000	Welded, sch. 40, on yoke & roll hangers, sized for covering,										
	2010	10' O.C. (no hangers incl. for 14" diam. and up)										
	2040	Black, 1" diameter	Q-15	93	.172	L.F.	2.48	6.35	.87	9.70	13.20	
	2050	1-1/4" diameter		84	.190		3.21	7	.96	11.17	15.15	
	2060	1-1/2" diameter		76	.211		2.98	7.75	1.06	11.79	16.10	
	2070	2" diameter		61	.262		4.07	9.65	1.32	15.04	20.50	
	2080	2-1/2" diameter		47	.340		5.35	12.50	1.72	19.57	26.50	
	2090	3" diameter		43	.372		6.45	13.70	1.88	22.03	29.50	
	2100	3-1/2" diameter		39	.410		10.70	15.10	2.07	27.87	36.50	
	2110	4" diameter		37	.432		12	15.90	2.18	30.08	39.50	
	2120	5" diameter		32	.500		21	18.40	2.52	41.92	53.50	
	2130	6" diameter	Q-16	36	.667		26.50	25.50	2.24	54.24	70	
	2140	8" diameter		29	.828		39	31.50	2.78	73.28	93.50	
	2150	10" diameter		24	1		64	38	3.36	105.36	132	
	2160	12" diameter		19	1.263		83.50	48	4.24	135.74	169	
	2170	14" diameter		15	1.600		36	61	5.40	102.40	137	
	2180	16" diameter		13	1.846		41.50	70.50	6.20	118.20	158	
	2190	18" diameter		11	2.182		43.50	83	7.35	133.85	181	
	2200	20" diameter		9	2.667		61	102	8.95	171.95	230	
	2220	24" diameter		8	3		63	114	10.10	187.10	252	
	2230	26" diameter		7.50	3.200		145	122	10.75	277.75	355	

15107 | Metal Pipe & Fittings

			CREW	DAILY OUTPUT	LABOR-HOURS	UNIT	2005 BARE COSTS				TOTAL INCL O&P	
							MAT.	LABOR	EQUIP.	TOTAL		
620	2240	30″ diameter R15100 -050	Q-16	6	4	L.F.	158	153	13.45	324.45	420	620
	2260	36″ diameter		4.50	5.333		181	203	17.90	401.90	525	
	2334	.50″ wall, 8″ diameter		25	.960		33	36.50	3.23	72.73	95	
	2335	10″ diameter		20	1.200		50.50	46	4.03	100.53	129	
	2336	12″ diameter		16	1.500		57	57	5.05	119.05	154	
	2337	14″ diameter		13	1.846		64	70.50	6.20	140.70	183	
	2345	Sch. 40, A-53, gr. A/B, ERW, welded w/hngrs.										
	2346	2″ diameter	Q-15	61	.262	L.F.	6.55	9.65	1.32	17.52	23	
	2347	2-1/2″ diameter		47	.340		10.35	12.50	1.72	24.57	32	
	2348	3″ diameter		43	.372		13.50	13.70	1.88	29.08	37.50	
	2349	4″ diameter		38	.421		18.95	15.50	2.12	36.57	47	
	2350	6″ diameter	Q-16	37	.649		15.65	24.50	2.18	42.33	56.50	
	2351	8″ diameter		29	.828		22	31.50	2.78	56.28	74.50	
	2352	10″ diameter		24	1		38	38	3.36	79.36	103	
	2363	.375″ wall, A-53, gr. A/B, ERW, welded w/hngrs.										
	2364	12″ diameter	Q-16	20	1.200	L.F.	46.50	46	4.03	96.53	124	
	2365	14″ diameter		17	1.412		53.50	54	4.74	112.24	145	
	2366	16″ diameter		14	1.714		61	65.50	5.75	132.25	172	
	2560	To delete hanger, subtract										
	2570	1″ diam. to 1-1/2″ diam.					15%	34%				
	2580	2″ diam. to 3-1/2″ diam.					9%	21%				
	2590	4″ diam. to 12″ diam.					5%	12%				
	2640	Galvanized, 1″ diameter	Q-15	93	.172	L.F.	2.76	6.35	.87	9.98	13.50	
	2650	1-1/4″ diameter		84	.190		3.60	7	.96	11.56	15.55	
	2660	1-1/2″ diameter		76	.211		3.95	7.75	1.06	12.76	17.15	
	2670	2″ diameter		61	.262		4.68	9.65	1.32	15.65	21	
	2680	2-1/2″ diameter		47	.340		6.30	12.50	1.72	20.52	27.50	
	2690	3″ diameter		43	.372		7.70	13.70	1.88	23.28	31	
	2700	3-1/2″ diameter		39	.410		9.60	15.10	2.07	26.77	35.50	
	2710	4″ diameter		37	.432		11	15.90	2.18	29.08	38.50	
	2720	5″ diameter		32	.500		27	18.40	2.52	47.92	60	
	2730	6″ diameter	Q-16	36	.667		34.50	25.50	2.24	62.24	78.50	
	2740	8″ diameter		29	.828		50.50	31.50	2.78	84.78	106	
	2750	10″ diameter		24	1		80.50	38	3.36	121.86	150	
	2760	12″ diameter		19	1.263		107	48	4.24	159.24	195	
	3160	To delete hanger, subtract										
	3170	1″ diam. to 1-1/2″ diam.					30%	34%				
	3180	2″ diam. to 3-1/2″ diam.					19%	21%				
	3190	4″ diam. to 12″ diam.					10%	12%				
	3250	Flanged, 150 lb. weld neck, on yoke & roll hangers										
	3260	sized for covering, 10′ O.C.										
	3290	Black, 1″ diameter	Q-15	70	.229	L.F.	5.80	8.40	1.15	15.35	20.50	
	3300	1-1/4″ diameter		64	.250		6.50	9.20	1.26	16.96	22.50	
	3310	1-1/2″ diameter		58	.276		6.30	10.15	1.39	17.84	23.50	
	3320	2″ diameter		45	.356		7.95	13.10	1.79	22.84	30.50	
	3330	2-1/2″ diameter		36	.444		9.95	16.35	2.24	28.54	38	
	3340	3″ diameter		32	.500		11.80	18.40	2.52	32.72	43.50	
	3350	3-1/2″ diameter		29	.552		18.50	20.50	2.78	41.78	54	
	3360	4″ diameter		26	.615		19.70	22.50	3.10	45.30	59	
	3370	5″ diameter		21	.762		30.50	28	3.84	62.34	79.50	
	3380	6″ diameter	Q-16	25	.960		38	36.50	3.23	77.73	101	
	3390	8″ diameter		19	1.263		57	48	4.24	109.24	140	
	3400	10″ diameter		16	1.500		86.50	57	5.05	148.55	187	
	3410	12″ diameter		14	1.714		127	65.50	5.75	198.25	244	
	3470	For 300 lb. flanges, add					63%					
	3480	For 600 lb. flanges, add					310%					

15107 | Metal Pipe & Fittings

620		CREW	DAILY OUTPUT	LABOR-HOURS	UNIT	2005 BARE COSTS MAT.	LABOR	EQUIP.	TOTAL	TOTAL INCL O&P
3960	To delete flanges & hanger, subtract (R15100-050)									
3970	1" diam. to 2" diam.					76%	65%			
3980	2-1/2" diam. to 4" diam.					62%	59%			
3990	5" diam. to 12" diam.					60%	46%			
4040	Galvanized, 1" diameter	Q-15	70	.229	L.F.	6.05	8.40	1.15	15.60	20.50
4050	1-1/4" diameter		64	.250		6.90	9.20	1.26	17.36	23
4060	1-1/2" diameter		58	.276		7.25	10.15	1.39	18.79	25
4070	2" diameter		45	.356		8.15	13.10	1.79	23.04	30.50
4080	2-1/2" diameter		36	.444		10.50	16.35	2.24	29.09	38.50
4090	3" diameter		32	.500		12.30	18.40	2.52	33.22	44
4100	3-1/2" diameter		29	.552		15.10	20.50	2.78	38.38	50
4110	4" diameter		26	.615		18.75	22.50	3.10	44.35	58
4120	5" diameter		21	.762		36.50	28	3.84	68.34	86
4130	6" diameter	Q-16	25	.960		44	36.50	3.23	83.73	107
4140	8" diameter		19	1.263		62	48	4.24	114.24	145
4150	10" diameter		16	1.500		100	57	5.05	162.05	202
4160	12" diameter		14	1.714		131	65.50	5.75	202.25	250
4220	For 300 lb. flanges, add					65%				
4240	For 600 lb. flanges, add					350%				
4660	To delete flanges & hanger, subtract									
4670	1" diam. to 2" diam.					73%	65%			
4680	2-1/2" diam. to 4" diam.					58%	59%			
4690	5" diam. to 12" diam.					43%	46%			
4750	Schedule 80, threaded, with couplings, and clevis type hangers									
4760	sized for covering, 10' O.C.									
4790	Black, 1/4" diameter	1 Plum	54	.148	L.F.	2.59	6.05		8.64	11.95
4800	3/8" diameter		53	.151		2.38	6.15		8.53	11.85
4810	1/2" diameter		52	.154		2.84	6.30		9.14	12.55
4820	3/4" diameter		50	.160		3.75	6.55		10.30	13.90
4830	1" diameter		45	.178		3.83	7.25		11.08	15.10
4840	1-1/4" diameter	Q-1	75	.213		5.25	7.85		13.10	17.55
4850	1-1/2" diameter		69	.232		6.30	8.55		14.85	19.70
4860	2" diameter		56	.286		7.55	10.50		18.05	24
4870	2-1/2" diameter		44	.364		9.85	13.35		23.20	31
4880	3" diameter		38	.421		12.85	15.50		28.35	37.50
4890	3-1/2" diameter		35	.457		18.45	16.80		35.25	46
4900	4" diameter		32	.500		25	18.40		43.40	55
4910	5" diameter		23	.696		28	25.50		53.50	69.50
4920	6" diameter	Q-2	28	.857		48.50	32.50		81	103
4930	8" diameter		23	1.043		92.50	40		132.50	162
4940	10" diameter		20	1.200		119	46		165	200
4950	12" diameter		15	1.600		160	61		221	268
5061	A-106, gr. A/B seamless with cplgs. & hangers, 1/4" dia.	1 Plum	63	.127		4.47	5.20		9.67	12.70
5062	3/8" diameter		62	.129		4.57	5.25		9.82	12.95
5063	1/2" diameter		61	.131		4.90	5.35		10.25	13.45
5064	3/4" diameter		57	.140		5.20	5.75		10.95	14.30
5065	1" diameter		51	.157		5.45	6.40		11.85	15.65
5066	1-1/4" diameter	Q-1	85	.188		6.30	6.90		13.20	17.35
5067	1-1/2" diameter		77	.208		7.45	7.65		15.10	19.70
5071	A-53 2" diameter		61	.262		8.45	9.65		18.10	24
5072	2-1/2" diameter		48	.333		13	12.25		25.25	33
5073	3" diameter		41	.390		17.30	14.35		31.65	40.50
5074	4" diameter		35	.457		12.05	16.80		28.85	39
5430	To delete coupling & hanger, subtract									
5440	1/4" diam. to 1/2" diam.					31%	54%			
5450	3/4" diam. to 1-1/2" diam.					28%	49%			

15107 | Metal Pipe & Fittings

	Line	Description	Crew	Daily Output	Labor-Hours	Unit	2005 Bare Costs Mat.	2005 Bare Costs Labor	2005 Bare Costs Equip.	2005 Bare Costs Total	Total Incl O&P	
620	5460	2″ diam. to 4″ diam.					21%	40%				620
	5470	5″ diam. to 12″ diam.					17%	43%				
		R15100-050										
	5510	Galvanized, 1/4″ diameter	1 Plum	54	.148	L.F.	4.81	6.05		10.86	14.40	
	5520	3/8″ diameter		53	.151		5.50	6.15		11.65	15.30	
	5530	1/2″ diameter		52	.154		4.90	6.30		11.20	14.85	
	5540	3/4″ diameter		50	.160		4.95	6.55		11.50	15.25	
	5550	1″ diameter		45	.178		6.85	7.25		14.10	18.45	
	5560	1-1/4″ diameter	Q-1	75	.213		8.20	7.85		16.05	21	
	5570	1-1/2″ diameter		69	.232		12.15	8.55		20.70	26	
	5580	2″ diameter		56	.286		13.35	10.50		23.85	30.50	
	5590	2-1/2″ diameter		44	.364		15.45	13.35		28.80	37	
	5600	3″ diameter		38	.421		19.50	15.50		35	45	
	5610	3-1/2″ diameter		35	.457		21.50	16.80		38.30	49	
	5620	4″ diameter		32	.500		30.50	18.40		48.90	61	
	5630	5″ diameter		23	.696		38	25.50		63.50	80.50	
	5640	6″ diameter	Q-2	28	.857		60	32.50		92.50	115	
	5650	8″ diameter		23	1.043		100	40		140	170	
	5660	10″ diameter		20	1.200		117	46		163	198	
	5670	12″ diameter		15	1.600		157	61		218	265	
	5930	To delete coupling & hanger, subtract										
	5940	1/4″ diam. to 1/2″ diam.					31%	54%				
	5950	3/4″ diam. to 1-1/2″ diam.					28%	49%				
	5960	2″ diam. to 4″ diam.					21%	40%				
	5970	5″ diam. to 12″ diam.					17%	43%				
	6000	Welded, on yoke & roller hangers										
	6010	sized for covering, 10′ O.C.										
	6040	Black, 1″ diameter	Q-15	85	.188	L.F.	4.56	6.90	.95	12.41	16.45	
	6050	1-1/4″ diameter		79	.203		6.15	7.45	1.02	14.62	19.05	
	6060	1-1/2″ diameter		72	.222		7.20	8.15	1.12	16.47	21.50	
	6070	2″ diameter		57	.281		8.15	10.30	1.41	19.86	26	
	6080	2-1/2″ diameter		44	.364		10.30	13.35	1.83	25.48	33.50	
	6090	3″ diameter		40	.400		13.05	14.70	2.02	29.77	38.50	
	6100	3-1/2″ diameter		34	.471		18.55	17.30	2.37	38.22	49	
	6110	4″ diameter		33	.485		24.50	17.85	2.44	44.79	56.50	
	6120	5″ diameter		26	.615		28	22.50	3.10	53.60	68	
	6130	6″ diameter	Q-16	30	.800		46.50	30.50	2.69	79.69	100	
	6140	8″ diameter		25	.960		88.50	36.50	3.23	128.23	156	
	6150	10″ diameter		20	1.200		119	46	4.03	169.03	203	
	6160	12″ diameter		15	1.600		160	61	5.40	226.40	273	
	6540	To delete hanger, subtract										
	6550	1″ diam. to 1-1/2″ diam.					30%	14%				
	6560	2″ diam. to 3″ diam.					23%	9%				
	6570	3-1/2″ diam. to 5″ diam.					12%	6%				
	6580	6″ diam. to 12″ diam.					10%	4%				
	6610	Galvanized, 1″ diameter	Q-15	85	.188	L.F.	7.60	6.90	.95	15.45	19.80	
	6650	1-1/4″ diameter		79	.203		9.10	7.45	1.02	17.57	22.50	
	6660	1-1/2″ diameter		72	.222		13.05	8.15	1.12	22.32	28	
	6670	2″ diameter		57	.281		13.95	10.30	1.41	25.66	32.50	
	6680	2-1/2″ diameter		44	.364		15.85	13.35	1.83	31.03	39.50	
	6690	3″ diameter		40	.400		19.70	14.70	2.02	36.42	45.50	
	6700	3-1/2″ diameter		34	.471		21.50	17.30	2.37	41.17	52.50	
	6710	4″ diameter		33	.485		30	17.85	2.44	50.29	62.50	
	6720	5″ diameter		26	.615		38	22.50	3.10	63.60	79	
	6730	6″ diameter	Q-16	30	.800		58	30.50	2.69	91.19	113	
	6740	8″ diameter		25	.960		96.50	36.50	3.23	136.23	165	
	6750	10″ diameter		20	1.200		116	46	4.03	166.03	201	

15107	Metal Pipe & Fittings	CREW	DAILY OUTPUT	LABOR-HOURS	UNIT	2005 BARE COSTS MAT.	LABOR	EQUIP.	TOTAL	TOTAL INCL O&P
620 6760	12" diameter (R15100-050)	Q-16	15	1.600	L.F.	157	61	5.40	223.40	270
7150	To delete hanger, subtract									
7160	1" diam. to 1-1/2" diam.					26%	14%			
7170	2" diam. to 3" diam.					16%	9%			
7180	3-1/2" diam. to 5" diam.					10%	6%			
7190	6" diam. to 12" diam.					7%	4%			
7250	Flanged, 300 lb. weld neck, on yoke & roll hangers									
7260	sized for covering, 10' O.C.									
7290	Black, 1" diameter	Q-15	66	.242	L.F.	8.60	8.90	1.22	18.72	24
7300	1-1/4" diameter		61	.262		10.20	9.65	1.32	21.17	27
7310	1-1/2" diameter		54	.296		11.20	10.90	1.49	23.59	30.50
7320	2" diameter		42	.381		12.70	14	1.92	28.62	37
7330	2-1/2" diameter		33	.485		16.20	17.85	2.44	36.49	47.50
7340	3" diameter		29	.552		20.50	20.50	2.78	43.78	56
7350	3-1/2" diameter		24	.667		30	24.50	3.36	57.86	73.50
7360	4" diameter		23	.696		36.50	25.50	3.51	65.51	82.50
7370	5" diameter		19	.842		42	31	4.24	77.24	97
7380	6" diameter	Q-16	23	1.043		66.50	40	3.51	110.01	137
7390	8" diameter		17	1.412		118	54	4.74	176.74	215
7400	10" diameter		14	1.714		164	65.50	5.75	235.25	286
7410	12" diameter		12	2		223	76.50	6.70	306.20	365
7470	For 600 lb. flanges, add					100%				
7940	To delete flanges & hanger, subtract									
7950	1" diam. to 1-1/2" diam.					75%	66%			
7960	2" diam. to 3" diam.					62%	60%			
7970	3-1/2" diam. to 5" diam.					54%	66%			
7980	6" diam. to 12" diam.					55%	62%			
8040	Galvanized, 1" diameter	Q-15	66	.242	L.F.	11.60	8.90	1.22	21.72	27.50
8050	1-1/4" diameter		61	.262		13.10	9.65	1.32	24.07	30.50
8060	1-1/2" diameter		54	.296		17.10	10.90	1.49	29.49	37
8070	2" diameter		42	.381		18.50	14	1.92	34.42	43.50
8080	2-1/2" diameter		33	.485		22	17.85	2.44	42.29	53.50
8090	3" diameter		29	.552		27	20.50	2.78	50.28	63.50
8100	3-1/2" diameter		24	.667		33	24.50	3.36	60.86	77
8110	4" diameter		23	.696		41.50	25.50	3.51	70.51	88.50
8120	5" diameter		19	.842		52	31	4.24	87.24	109
8130	6" diameter	Q-16	23	1.043		78	40	3.51	121.51	150
8140	8" diameter		17	1.412		125	54	4.74	183.74	224
8150	10" diameter		14	1.714		162	65.50	5.75	233.25	283
8160	12" diameter		12	2		220	76.50	6.70	303.20	365
8240	For 600 lb. flanges, add					100%				
8900	To delete flanges & hangers, subtract									
8910	1" diam. to 1-1/2" diam.					72%	66%			
8920	2" diam. to 3" diam.					59%	60%			
8930	3-1/2" diam. to 5" diam.					51%	66%			
8940	6" diam. to 12" diam.					49%	62%			
9000	Threading pipe labor, one end, all schedules through 80									
9010	1/4" through 3/4" pipe size	1 Plum	80	.100	Ea.		4.09		4.09	6.15
9020	1" through 2" pipe size		73	.110			4.48		4.48	6.75
9030	2-1/2" pipe size		53	.151			6.15		6.15	9.25
9040	3" pipe size		50	.160			6.55		6.55	9.80
9050	3-1/2" pipe size	Q-1	89	.180			6.60		6.60	9.95
9060	4" pipe size		73	.219			8.05		8.05	12.10
9070	5" pipe size		53	.302			11.10		11.10	16.70
9080	6" pipe size		46	.348			12.80		12.80	19.25
9090	8" pipe size		29	.552			20.50		20.50	30.50

Important: See the Reference Section for critical supporting data - Reference Nos., Crews, & City Cost Indexes

15107 | Metal Pipe & Fittings

		15107 Metal Pipe & Fittings	CREW	DAILY OUTPUT	LABOR-HOURS	UNIT	2005 BARE COSTS MAT.	LABOR	EQUIP.	TOTAL	TOTAL INCL O&P	
620	9100	10" pipe size R15100-050	Q-1	21	.762	Ea.		28		28	42	620
	9110	12" pipe size		13	1.231			45.50		45.50	68	
	9200	Welding labor per joint										
	9210	Schedule 40,										
	9230	1/2" pipe size	Q-15	32	.500	Ea.		18.40	2.52	20.92	30.50	
	9240	3/4" pipe size		27	.593			22	2.99	24.99	36.50	
	9250	1" pipe size		23	.696			25.50	3.51	29.01	42.50	
	9260	1-1/4" pipe size		20	.800			29.50	4.03	33.53	48.50	
	9270	1-1/2" pipe size		19	.842			31	4.24	35.24	51	
	9280	2" pipe size		16	1			37	5.05	42.05	61	
	9290	2-1/2" pipe size		13	1.231			45.50	6.20	51.70	75	
	9300	3" pipe size		12	1.333			49	6.70	55.70	81	
	9310	4" pipe size		10	1.600			59	8.05	67.05	97.50	
	9320	5" pipe size		9	1.778			65.50	8.95	74.45	108	
	9330	6" pipe size		8	2			73.50	10.10	83.60	122	
	9340	8" pipe size		5	3.200			118	16.15	134.15	195	
	9350	10" pipe size		4	4			147	20	167	243	
	9360	12" pipe size		3	5.333			196	27	223	325	
	9370	14" pipe size		2.60	6.154			226	31	257	375	
	9380	16" pipe size		2.20	7.273			267	36.50	303.50	440	
	9390	18" pipe size		2	8			294	40.50	334.50	485	
	9400	20" pipe size		1.80	8.889			325	45	370	540	
	9410	22" pipe size		1.70	9.412			345	47.50	392.50	570	
	9420	24" pipe size		1.50	10.667			390	54	444	650	
	9450	Schedule 80,										
	9460	1/2" pipe size	Q-15	27	.593	Ea.		22	2.99	24.99	36.50	
	9470	3/4" pipe size		23	.696			25.50	3.51	29.01	42.50	
	9480	1" pipe size		20	.800			29.50	4.03	33.53	48.50	
	9490	1-1/4" pipe size		19	.842			31	4.24	35.24	51	
	9500	1-1/2" pipe size		18	.889			32.50	4.48	36.98	54	
	9510	2" pipe size		15	1.067			39	5.40	44.40	65	
	9520	2-1/2" pipe size		12	1.333			49	6.70	55.70	81	
	9530	3" pipe size		11	1.455			53.50	7.35	60.85	88.50	
	9540	4" pipe size		8	2			73.50	10.10	83.60	122	
	9550	5" pipe size		6	2.667			98	13.45	111.45	162	
	9560	6" pipe size		5	3.200			118	16.15	134.15	195	
	9570	8" pipe size		4	4			147	20	167	243	
	9580	10" pipe size		3	5.333			196	27	223	325	
	9590	12" pipe size		2	8			294	40.50	334.50	485	
	9600	14" pipe size	Q-16	2.60	9.231			350	31	381	565	
	9610	16" pipe size		2.30	10.435			400	35	435	640	
	9620	18" pipe size		2	12			460	40.50	500.50	735	
	9630	20" pipe size		1.80	13.333			510	45	555	815	
	9640	22" pipe size		1.60	15			570	50.50	620.50	915	
	9650	24" pipe size		1.50	16			610	54	664	975	
640	0010	**PIPE, STEEL, FITTINGS** Threaded										640
	0020	Cast Iron										
	0040	Standard weight, black										
	0060	90° Elbow, straight										
	0070	1/4"	1 Plum	16	.500	Ea.	4.01	20.50		24.51	35	
	0080	3/8"		16	.500		5.80	20.50		26.30	37	
	0090	1/2"		15	.533		2.54	22		24.54	36	
	0100	3/4"		14	.571		2.65	23.50		26.15	38	
	0110	1"		13	.615		3.14	25		28.14	41.50	
	0120	1-1/4"	Q-1	22	.727		4.46	27		31.46	45	

15107 | Metal Pipe & Fittings		CREW	DAILY OUTPUT	LABOR-HOURS	UNIT	2005 BARE COSTS MAT.	LABOR	EQUIP.	TOTAL	TOTAL INCL O&P
640	0130 1-1/2"	Q-1	20	.800	Ea.	6.15	29.50		35.65	51
	0140 2"		18	.889		9.65	32.50		42.15	59.50
	0150 2-1/2"		14	1.143		23	42		65	88.50
	0160 3"		10	1.600		38	59		97	130
	0170 3-1/2"		8	2		103	73.50		176.50	224
	0180 4"		6	2.667		70	98		168	224
	0190 5"		5	3.200		149	118		267	340
	0200 6"	Q-2	7	3.429		193	131		324	410
	0210 8"	"	6	4		475	153		628	750
	0250 45° Elbow, straight									
	0260 1/4"	1 Plum	16	.500	Ea.	4.81	20.50		25.31	36
	0270 3/8"		16	.500		5.15	20.50		25.65	36
	0280 1/2"		15	.533		3.89	22		25.89	37.50
	0300 3/4"		14	.571		3.92	23.50		27.42	39.50
	0320 1"		13	.615		4.59	25		29.59	43
	0330 1-1/4"	Q-1	22	.727		6.20	27		33.20	47
	0340 1-1/2"		20	.800		10.25	29.50		39.75	55.50
	0350 2"		18	.889		11.80	32.50		44.30	62
	0360 2-1/2"		14	1.143		30.50	42		72.50	97
	0370 3"		10	1.600		48.50	59		107.50	142
	0380 3-1/2"		8	2		108	73.50		181.50	230
	0400 4"		6	2.667		101	98		199	258
	0420 5"		5	3.200		222	118		340	420
	0440 6"	Q-2	7	3.429		285	131		416	510
	0460 8"	"	6	4		570	153		723	855
	0500 Tee, straight									
	0510 1/4"	1 Plum	10	.800	Ea.	6.30	32.50		38.80	56
	0520 3/8"		10	.800		6.10	32.50		38.60	55.50
	0530 1/2"		9	.889		3.97	36.50		40.47	59
	0540 3/4"		9	.889		4.63	36.50		41.13	59.50
	0550 1"		8	1		4.11	41		45.11	66
	0560 1-1/4"	Q-1	14	1.143		7.50	42		49.50	71.50
	0570 1-1/2"		13	1.231		9.75	45.50		55.25	79
	0580 2"		11	1.455		13.60	53.50		67.10	95.50
	0590 2-1/2"		9	1.778		35.50	65.50		101	138
	0600 3"		6	2.667		54	98		152	207
	0610 3-1/2"		5	3.200		109	118		227	297
	0620 4"		4	4		106	147		253	335
	0630 5"		3	5.333		281	196		477	605
	0640 6"	Q-2	4	6		320	229		549	695
	0650 8"	"	3	8		720	305		1,025	1,250
	0660 Tee, reducing, run or outlet									
	0661 1/2"	1 Plum	9	.889	Ea.	8.25	36.50		44.75	63.50
	0662 3/4"		9	.889		7.45	36.50		43.95	62.50
	0663 1"		8	1		8.15	41		49.15	70.50
	0664 1-1/4"	Q-1	14	1.143		15.70	42		57.70	80.50
	0665 1-1/2"		13	1.231		17.45	45.50		62.95	87
	0666 2"		11	1.455		25	53.50		78.50	108
	0667 2-1/2"		9	1.778		46.50	65.50		112	150
	0668 3"		6	2.667		70	98		168	224
	0669 3-1/2"		5	3.200		126	118		244	315
	0670 4"		4	4		143	147		290	380
	0671 5"		3	5.333		249	196		445	570
	0672 6"	Q-2	4	6		320	229		549	695
	0673 8"	"	3	8		420	305		725	925
	0674 Reducer, concentric									

Important: See the Reference Section for critical supporting data - Reference Nos., Crews, & City Cost Indexes

15107 | Metal Pipe & Fittings

			CREW	DAILY OUTPUT	LABOR-HOURS	UNIT	2005 BARE COSTS MAT.	LABOR	EQUIP.	TOTAL	TOTAL INCL O&P	
640	0675	3/4″	1 Plum	18	.444	Ea.	7.10	18.15		25.25	35.50	640
	0676	1″	″	15	.533		4.81	22		26.81	38.50	
	0677	1-1/4″	Q-1	26	.615		13.95	22.50		36.45	49.50	
	0678	1-1/2″		24	.667		22	24.50		46.50	61	
	0679	2″		21	.762		25.50	28		53.50	70.50	
	0680	2-1/2″		18	.889		36.50	32.50		69	89	
	0681	3″		14	1.143		55.50	42		97.50	124	
	0682	3-1/2″		12	1.333		96.50	49		145.50	180	
	0683	4″		10	1.600		103	59		162	203	
	0684	5″		6	2.667		143	98		241	305	
	0685	6″	Q-2	8	3		188	114		302	380	
	0686	8″	″	7	3.429		820	131		951	1,100	
	0687	Reducer, eccentric										
	0688	3/4″	1 Plum	16	.500	Ea.	14.95	20.50		35.45	47	
	0689	1″	″	14	.571		15.70	23.50		39.20	52.50	
	0690	1-1/4″	Q-1	25	.640		26	23.50		49.50	64	
	0691	1-1/2″		22	.727		34	27		61	77.50	
	0692	2″		20	.800		48	29.50		77.50	96.50	
	0693	2-1/2″		16	1		66	37		103	128	
	0694	3″		12	1.333		104	49		153	188	
	0695	3-1/2″		10	1.600		144	59		203	248	
	0696	4″		9	1.778		188	65.50		253.50	305	
	0697	5″		5	3.200		229	118		347	430	
	0698	6″	Q-2	8	3		291	114		405	490	
	0699	8″	″	7	3.429		1,125	131		1,256	1,450	
	0700	Standard weight, galvanized cast iron										
	0720	90° Elbow, straight										
	0730	1/4″	1 Plum	16	.500	Ea.	9.35	20.50		29.85	41	
	0740	3/8″		16	.500		9.35	20.50		29.85	41	
	0750	1/2″		15	.533		7.55	22		29.55	41.50	
	0760	3/4″		14	.571		10.50	23.50		34	46.50	
	0770	1″		13	.615		12.10	25		37.10	51.50	
	0780	1-1/4″	Q-1	22	.727		18.85	27		45.85	60.50	
	0790	1-1/2″		20	.800		26	29.50		55.50	72.50	
	0800	2″		18	.889		38	32.50		70.50	91	
	0810	2-1/2″		14	1.143		78.50	42		120.50	149	
	0820	3″		10	1.600		119	59		178	220	
	0830	3-1/2″		8	2		190	73.50		263.50	320	
	0840	4″		6	2.667		218	98		316	385	
	0850	5″		5	3.200		440	118		558	660	
	0860	6″	Q-2	7	3.429		575	131		706	825	
	0870	8″	″	6	4		1,400	153		1,553	1,775	
	0900	45° Elbow, straight										
	0910	1/4″	1 Plum	16	.500	Ea.	14.25	20.50		34.75	46	
	0920	3/8″		16	.500		15.30	20.50		35.80	47.50	
	0930	1/2″		15	.533		10.65	22		32.65	45	
	0940	3/4″		14	.571		12.25	23.50		35.75	48.50	
	0950	1″		13	.615		13.60	25		38.60	53	
	0960	1-1/4″	Q-1	22	.727		18.30	27		45.30	60	
	0970	1-1/2″		20	.800		31.50	29.50		61	79	
	0980	2″		18	.889		44.50	32.50		77	98	
	0990	2-1/2″		14	1.143		87.50	42		129.50	159	
	1000	3″		10	1.600		139	59		198	242	
	1010	3-1/2″		8	2		257	73.50		330.50	395	
	1020	4″		6	2.667		257	98		355	430	
	1030	5″		5	3.200		655	118		773	900	

15107 | Metal Pipe & Fittings		CREW	DAILY OUTPUT	LABOR-HOURS	UNIT	2005 BARE COSTS MAT.	LABOR	EQUIP.	TOTAL	TOTAL INCL O&P
640 1040	6"	Q-2	7	3.429	Ea.	845	131		976	1,125
1050	8"	"	6	4	↓	1,675	153		1,828	2,075
1100	Tee, straight									
1110	1/4"	1 Plum	10	.800	Ea.	18.60	32.50		51.10	69.50
1120	3/8"		10	.800		11.95	32.50		44.45	62
1130	1/2"		9	.889		11.75	36.50		48.25	67.50
1140	3/4"		9	.889		15.05	36.50		51.55	71
1150	1"	↓	8	1		16.35	41		57.35	79.50
1160	1-1/4"	Q-1	14	1.143		28.50	42		70.50	94.50
1170	1-1/2"		13	1.231		38	45.50		83.50	110
1180	2"		11	1.455		47	53.50		100.50	133
1190	2-1/2"		9	1.778		102	65.50		167.50	211
1200	3"		6	2.667		263	98		361	435
1210	3-1/2"		5	3.200		270	118		388	475
1220	4"		4	4		305	147		452	555
1230	5"	↓	3	5.333		835	196		1,031	1,200
1240	6"	Q-2	4	6		945	229		1,174	1,400
1250	8"	"	3	8	↓	2,150	305		2,455	2,800
1300	Extra heavy weight, black									
1310	Couplings, steel straight									
1320	1/4"	1 Plum	19	.421	Ea.	1.76	17.20		18.96	28
1330	3/8"		19	.421		1.92	17.20		19.12	28
1340	1/2"		19	.421		2.59	17.20		19.79	29
1350	3/4"		18	.444		2.77	18.15		20.92	30.50
1360	1"	↓	15	.533		3.53	22		25.53	37
1370	1-1/4"	Q-1	26	.615		5.65	22.50		28.15	40.50
1380	1-1/2"		24	.667		5.65	24.50		30.15	43.50
1390	2"		21	.762		8.65	28		36.65	51.50
1400	2-1/2"		18	.889		12.80	32.50		45.30	63
1410	3"		14	1.143		15.20	42		57.20	80
1420	3-1/2"		12	1.333		20.50	49		69.50	96
1430	4"		10	1.600		24	59		83	115
1440	5"	↓	6	2.667		34.50	98		132.50	185
1450	6"	Q-2	8	3		42	114		156	219
1460	8"		7	3.429		76.50	131		207.50	281
1470	10"		6	4		120	153		273	360
1480	12"	↓	4	6	↓	161	229		390	520
1510	90° Elbow, straight									
1520	1/2"	1 Plum	15	.533	Ea.	12.60	22		34.60	47
1530	3/4"		14	.571		13.35	23.50		36.85	49.50
1540	1"	↓	13	.615		16.15	25		41.15	56
1550	1-1/4"	Q-1	22	.727		24	27		51	66.50
1560	1-1/2"		20	.800		30	29.50		59.50	77
1580	2"		18	.889		37	32.50		69.50	89.50
1590	2-1/2"		14	1.143		86	42		128	158
1600	3"		10	1.600		114	59		173	214
1610	4"	↓	6	2.667		238	98		336	410
1620	6"	Q-2	7	3.429	↓	620	131		751	875
1650	45° Elbow, straight									
1660	1/2"	1 Plum	15	.533	Ea.	18	22		40	53
1670	3/4"		14	.571		17.35	23.50		40.85	54
1680	1"	↓	13	.615		21	25		46	61
1690	1-1/4"	Q-1	22	.727		34.50	27		61.50	78
1700	1-1/2"		20	.800		38	29.50		67.50	85.50
1710	2"		18	.889		54	32.50		86.50	108
1720	2-1/2"	↓	14	1.143	↓	93	42		135	165

15107 | Metal Pipe & Fittings

			CREW	DAILY OUTPUT	LABOR-HOURS	UNIT	2005 BARE COSTS MAT.	LABOR	EQUIP.	TOTAL	TOTAL INCL O&P	
640	1800	Tee, straight										640
	1810	1/2"	1 Plum	9	.889	Ea.	19.90	36.50		56.40	76.50	
	1820	3/4"		9	.889		19.90	36.50		56.40	76.50	
	1830	1"		8	1		24	41		65	88	
	1840	1-1/4"	Q-1	14	1.143		36	42		78	103	
	1850	1-1/2"		13	1.231		46	45.50		91.50	119	
	1860	2"		11	1.455		57	53.50		110.50	144	
	1870	2-1/2"		9	1.778		122	65.50		187.50	233	
	1880	3"		6	2.667		166	98		264	330	
	1890	4"		4	4		325	147		472	575	
	1900	6"	Q-2	4	6		715	229		944	1,125	
	4000	Standard weight, black										
	4010	Couplings, steel straight, merchants										
	4030	1/4"	1 Plum	19	.421	Ea.	.47	17.20		17.67	26.50	
	4040	3/8"		19	.421		.57	17.20		17.77	26.50	
	4050	1/2"		19	.421		.61	17.20		17.81	26.50	
	4060	3/4"		18	.444		.77	18.15		18.92	28.50	
	4070	1"		15	.533		1.08	22		23.08	34	
	4080	1-1/4"	Q-1	26	.615		1.38	22.50		23.88	35.50	
	4090	1-1/2"		24	.667		1.75	24.50		26.25	39	
	4100	2"		21	.762		2.50	28		30.50	45	
	4110	2-1/2"		18	.889		7.10	32.50		39.60	57	
	4120	3"		14	1.143		10	42		52	74	
	4130	3-1/2"		12	1.333		17.70	49		66.70	93	
	4140	4"		10	1.600		17.70	59		76.70	108	
	4150	5"		6	2.667		32.50	98		130.50	183	
	4160	6"	Q-2	8	3		39	114		153	215	
	4166	Plug, 1/4"	1 Plum	38	.211		1.31	8.60		9.91	14.40	
	4167	3/8"		38	.211		1.31	8.60		9.91	14.40	
	4168	1/2"		38	.211		1.13	8.60		9.73	14.20	
	4169	3/4"		32	.250		3.42	10.20		13.62	19.10	
	4170	1"		30	.267		3.65	10.90		14.55	20.50	
	4171	1-1/4"	Q-1	52	.308		4.19	11.30		15.49	21.50	
	4172	1-1/2"		48	.333		5.95	12.25		18.20	25	
	4173	2"		42	.381		7.65	14		21.65	29.50	
	4176	2-1/2"		36	.444		11.35	16.35		27.70	37	
	4200	Standard weight, galvanized										
	4210	Couplings, steel straight, merchants										
	4230	1/4"	1 Plum	19	.421	Ea.	.55	17.20		17.75	26.50	
	4240	3/8"		19	.421		.70	17.20		17.90	27	
	4250	1/2"		19	.421		.74	17.20		17.94	27	
	4260	3/4"		18	.444		.93	18.15		19.08	28.50	
	4270	1"		15	.533		1.30	22		23.30	34.50	
	4280	1-1/4"	Q-1	26	.615		1.66	22.50		24.16	36	
	4290	1-1/2"		24	.667		2.07	24.50		26.57	39.50	
	4300	2"		21	.762		3.08	28		31.08	45.50	
	4310	2-1/2"		18	.889		8.75	32.50		41.25	58.50	
	4320	3"		14	1.143		11.60	42		53.60	76	
	4330	3-1/2"		12	1.333		20.50	49		69.50	96	
	4340	4"		10	1.600		20.50	59		79.50	111	
	4350	5"		6	2.667		38	98		136	189	
	4360	6"	Q-2	8	3		47	114		161	224	
	4370	Plug, galvanized, square head										
	4374	1/2"	1 Plum	38	.211	Ea.	2.21	8.60		10.81	15.40	
	4375	3/4"		32	.250		2.40	10.20		12.60	18	
	4376	1"		30	.267		3.99	10.90		14.89	20.50	

MECHANICAL 15

15107 Metal Pipe & Fittings		CREW	DAILY OUTPUT	LABOR-HOURS	UNIT	2005 BARE COSTS MAT.	LABOR	EQUIP.	TOTAL	TOTAL INCL O&P
640	4377 1-1/4"	Q-1	52	.308	Ea.	4.31	11.30		15.61	21.50
	4378 1-1/2"		48	.333		5.80	12.25		18.05	25
	4379 2"		42	.381		7.20	14		21.20	29
	4380 2-1/2"		36	.444		15.30	16.35		31.65	41.50
	4381 3"		28	.571		19.55	21		40.55	53
	4382 4"		20	.800		52	29.50		81.50	102
	4384 6"	↓	8	2	↓	86	73.50		159.50	206
	5000 Malleable iron, 150 lb.									
	5020 Black									
	5040 90° elbow, straight									
	5060 1/4"	1 Plum	16	.500	Ea.	1.99	20.50		22.49	32.50
	5070 3/8"		16	.500		1.99	20.50		22.49	32.50
	5080 1/2"		15	.533		1.38	22		23.38	34.50
	5090 3/4"		14	.571		1.67	23.50		25.17	37
	5100 1"	↓	13	.615		2.90	25		27.90	41
	5110 1-1/4"	Q-1	22	.727		4.77	27		31.77	45.50
	5120 1-1/2"		20	.800		6.25	29.50		35.75	51
	5130 2"		18	.889		10.80	32.50		43.30	61
	5140 2-1/2"		14	1.143		24	42		66	89.50
	5150 3"		10	1.600		35.50	59		94.50	128
	5160 3-1/2"		8	2		94	73.50		167.50	215
	5170 4"		6	2.667		75.50	98		173.50	231
	5180 5"	↓	5	3.200		215	118		333	415
	5190 6"	Q-2	7	3.429	↓	255	131		386	480
	5250 45° elbow, straight									
	5270 1/4"	1 Plum	16	.500	Ea.	3	20.50		23.50	34
	5280 3/8"		16	.500		3	20.50		23.50	34
	5290 1/2"		15	.533		2.27	22		24.27	35.50
	5300 3/4"		14	.571		2.81	23.50		26.31	38
	5310 1"	↓	13	.615		3.55	25		28.55	42
	5320 1-1/4"	Q-1	22	.727		6.30	27		33.30	47
	5330 1-1/2"		20	.800		7.75	29.50		37.25	52.50
	5340 2"		18	.889		11.75	32.50		44.25	62
	5350 2-1/2"		14	1.143		34	42		76	101
	5360 3"		10	1.600		44	59		103	137
	5370 3-1/2"		8	2		86.50	73.50		160	206
	5380 4"		6	2.667		86.50	98		184.50	242
	5390 5"	↓	5	3.200		340	118		458	550
	5400 6"	Q-2	7	3.429	↓	340	131		471	570
	5450 Tee, straight									
	5470 1/4"	1 Plum	10	.800	Ea.	3.36	32.50		35.86	52.50
	5480 3/8"		10	.800		3.36	32.50		35.86	52.50
	5490 1/2"		9	.889		1.84	36.50		38.34	56.50
	5500 3/4"		9	.889		2.65	36.50		39.15	57.50
	5510 1"	↓	8	1		4.52	41		45.52	66.50
	5520 1-1/4"	Q-1	14	1.143		7.35	42		49.35	71
	5530 1-1/2"		13	1.231		9.15	45.50		54.65	78
	5540 2"		11	1.455		15.55	53.50		69.05	97.50
	5550 2-1/2"		9	1.778		33.50	65.50		99	136
	5560 3"		6	2.667		49.50	98		147.50	202
	5570 3-1/2"		5	3.200		119	118		237	310
	5580 4"		4	4		119	147		266	350
	5590 5"	↓	3	5.333		340	196		536	670
	5600 6"	Q-2	4	6	↓	380	229		609	760
	5601 Tee, reducing, on outlet									
	5602 1/2"	1 Plum	9	.889	Ea.	4.03	36.50		40.53	59

15 MECHANICAL

	15107 | Metal Pipe & Fittings		CREW	DAILY OUTPUT	LABOR-HOURS	UNIT	2005 BARE COSTS MAT.	LABOR	EQUIP.	TOTAL	TOTAL INCL O&P	
640	5603	3/4"	1 Plum	9	.889	Ea.	4.09	36.50		40.59	59	640
	5604	1"		8	1		5.15	41		46.15	67	
	5605	1-1/4"	Q-1	14	1.143		8.95	42		50.95	73	
	5606	1-1/2"		13	1.231		12.90	45.50		58.40	82	
	5607	2"		11	1.455		16.60	53.50		70.10	99	
	5608	2-1/2"		9	1.778		43.50	65.50		109	147	
	5609	3"		6	2.667		69	98		167	223	
	5610	3-1/2"		5	3.200		154	118		272	345	
	5611	4"		4	4		154	147		301	390	
	5612	5"		3	5.333		415	196		611	750	
	5650	Coupling										
	5670	1/4"	1 Plum	19	.421	Ea.	2.46	17.20		19.66	28.50	
	5680	3/8"		19	.421		2.46	17.20		19.66	28.50	
	5690	1/2"		19	.421		1.90	17.20		19.10	28	
	5700	3/4"		18	.444		2.23	18.15		20.38	30	
	5710	1"		15	.533		3.34	22		25.34	36.50	
	5720	1-1/4"	Q-1	26	.615		4.33	22.50		26.83	39	
	5730	1-1/2"		24	.667		5.85	24.50		30.35	43.50	
	5740	2"		21	.762		8.65	28		36.65	51.50	
	5750	2-1/2"		18	.889		24	32.50		56.50	75.50	
	5760	3"		14	1.143		32.50	42		74.50	98.50	
	5770	3-1/2"		12	1.333		65	49		114	145	
	5780	4"		10	1.600		65	59		124	160	
	5790	5"		6	2.667		186	98		284	350	
	5800	6"	Q-2	8	3		186	114		300	375	
	5840	Reducer, concentric, 1/4"	1 Plum	19	.421		2.61	17.20		19.81	29	
	5850	3/8"		19	.421		3.33	17.20		20.53	29.50	
	5860	1/2"		19	.421		2.64	17.20		19.84	29	
	5870	3/4"		16	.500		3.12	20.50		23.62	34	
	5880	1"		15	.533		5.25	22		27.25	39	
	5890	1-1/4"	Q-1	26	.615		5.90	22.50		28.40	40.50	
	5900	1-1/2"		24	.667		8.45	24.50		32.95	46.50	
	5910	2"		21	.762		12.20	28		40.20	55.50	
	5911	2-1/2"		18	.889		27	32.50		59.50	79	
	5912	3"		14	1.143		33	42		75	99.50	
	5913	3-1/2"		12	1.333		77	49		126	159	
	5914	4"		10	1.600		77	59		136	174	
	5915	5"		6	2.667		145	98		243	305	
	5916	6"	Q-2	8	3		186	114		300	375	
	5981	Bushing, 1/4"	1 Plum	19	.421		2.52	17.20		19.72	29	
	5982	3/8"		19	.421		2.52	17.20		19.72	29	
	5983	1/2"		19	.421		2.52	17.20		19.72	29	
	5984	3/4"		16	.500		2.48	20.50		22.98	33	
	5985	1"		15	.533		3.63	22		25.63	37	
	5986	1-1/4"	Q-1	26	.615		4.51	22.50		27.01	39	
	5987	1-1/2"		24	.667		5.20	24.50		29.70	43	
	5988	2"		21	.762		5.60	28		33.60	48	
	5989	Cap, 1/4"	1 Plum	38	.211		1.82	8.60		10.42	14.95	
	5991	3/8"		38	.211		1.54	8.60		10.14	14.65	
	5992	1/2"		38	.211		1.39	8.60		9.99	14.50	
	5993	3/4"		32	.250		1.88	10.20		12.08	17.40	
	5994	1"		30	.267		2.28	10.90		13.18	18.85	
	5995	1-1/4"	Q-1	52	.308		3.01	11.30		14.31	20.50	
	5996	1-1/2"		48	.333		4.14	12.25		16.39	23	
	5997	2"		42	.381		6.05	14		20.05	27.50	
	6000	For galvanized elbows, tees, and couplings add					20%					

15107 | Metal Pipe & Fittings

640		CREW	DAILY OUTPUT	LABOR-HOURS	UNIT	2005 BARE COSTS MAT.	LABOR	EQUIP.	TOTAL	TOTAL INCL O&P
6058	For galvanized reducers, caps and plugs add					20%				
6100	90° elbow, galvanized, 150 lb., reducing									
6110	3/4" x 1/2"	1 Plum	15.40	.519	Ea.	3.75	21		24.75	36
6112	1" x 3/4"		14	.571		4.81	23.50		28.31	40.50
6114	1" x 1/2"		14.50	.552		5.10	22.50		27.60	39.50
6116	1-1/4" x 1"	Q-1	24.20	.661		8.05	24.50		32.55	45.50
6118	1-1/4" x 3/4"		25.40	.630		9.70	23		32.70	45.50
6120	1-1/4" x 1/2"		26.20	.611		7.60	22.50		30.10	42.50
6122	1-1/2" x 1-1/4"		21.60	.741		12.90	27		39.90	55
6124	1-1/2" x 1"		23.50	.681		12.90	25		37.90	51.50
6126	1-1/2" x 3/4"		24.60	.650		12.90	24		36.90	50
6128	2" x 1-1/2"		20.50	.780		15.05	28.50		43.55	59.50
6130	2" x 1-1/4"		21	.762		17.25	28		45.25	61
6132	2" x 1"		22.80	.702		17.90	26		43.90	58.50
6134	2" x 3/4"		23.90	.669		18.25	24.50		42.75	57
6136	2-1/2" x 2"		12.30	1.301		51	48		99	128
6138	2-1/2" x 1-1/2"		12.50	1.280		51	47		98	127
6140	3" x 2-1/2"		8.60	1.860		90	68.50		158.50	202
6142	3" x 2"		11.80	1.356		80.50	50		130.50	164
6144	4" x 3"		8.20	1.951		204	72		276	330
6160	90° elbow, black, 150 lb., reducing									
6170	1" x 3/4"	1 Plum	14	.571	Ea.	3.47	23.50		26.97	39
6174	1-1/2" x 1"	Q-1	23.50	.681		8.25	25		33.25	46.50
6178	1-1/2" x 3/4"		24.60	.650		9.40	24		33.40	46.50
6182	2" x 1-1/2"		20.50	.780		11.90	28.50		40.40	56
6186	2" x 1"		22.80	.702		13.55	26		39.55	54
6190	2" x 3/4"		23.90	.669		14.40	24.50		38.90	53
6194	2-1/2" x 2"		12.30	1.301		32.50	48		80.50	108
7000	Union, with brass seat									
7010	1/4"	1 Plum	15	.533	Ea.	10.70	22		32.70	45
7020	3/8"		15	.533		7.40	22		29.40	41
7030	1/2"		14	.571		6.70	23.50		30.20	42.50
7040	3/4"		13	.615		7	25		32	45.50
7050	1"		12	.667		10	27		37	52
7060	1-1/4"	Q-1	21	.762		13.75	28		41.75	57
7070	1-1/2"		19	.842		17.90	31		48.90	66
7080	2"		17	.941		21	34.50		55.50	75
7090	2-1/2"		13	1.231		62	45.50		107.50	136
7100	3"		9	1.778		74.50	65.50		140	181
7120	Union, galvanized									
7124	1/2"	1 Plum	14	.571	Ea.	7.50	23.50		31	43.50
7125	3/4"		13	.615		8.80	25		33.80	47.50
7126	1"		12	.667		11.50	27		38.50	53.50
7127	1-1/4"	Q-1	21	.762		16.65	28		44.65	60.50
7128	1-1/2"		19	.842		20	31		51	68.50
7129	2"		17	.941		23	34.50		57.50	77.50
7130	2-1/2"		13	1.231		80.50	45.50		126	157
7131	3"		9	1.778		113	65.50		178.50	223
7500	Malleable iron, 300 lb									
7520	Black									
7540	90° Elbow, straight, 1/4"	1 Plum	16	.500	Ea.	7.65	20.50		28.15	39
7560	3/8"		16	.500		6.80	20.50		27.30	38
7570	1/2"		15	.533		8.80	22		30.80	42.50
7580	3/4"		14	.571		9.95	23.50		33.45	46
7590	1"		13	.615		12.75	25		37.75	52
7600	1-1/4"	Q-1	22	.727		18.60	27		45.60	60.50

15107 | Metal Pipe & Fittings			CREW	DAILY OUTPUT	LABOR-HOURS	UNIT	2005 BARE COSTS MAT.	LABOR	EQUIP.	TOTAL	TOTAL INCL O&P	
640	7610	1-1/2"	Q-1	20	.800	Ea.	22	29.50		51.50	68	640
	7620	2"		18	.889		32	32.50		64.50	84	
	7630	2-1/2"		14	1.143		81.50	42		123.50	153	
	7640	3"		10	1.600		94	59		153	192	
	7650	4"	↓	6	2.667		214	98		312	385	
	7700	45° Elbow, straight, 1/4"	1 Plum	16	.500		11.40	20.50		31.90	43	
	7720	3/8"		16	.500		11.40	20.50		31.90	43	
	7730	1/2"		15	.533		12.70	22		34.70	47	
	7740	3/4"		14	.571		14.10	23.50		37.60	50.50	
	7750	1"	↓	13	.615		15.60	25		40.60	55	
	7760	1-1/4"	Q-1	22	.727		25	27		52	67.50	
	7770	1-1/2"		20	.800		32.50	29.50		62	80	
	7780	2"		18	.889		49.50	32.50		82	103	
	7790	2-1/2"		14	1.143		107	42		149	181	
	7800	3"		10	1.600		111	59		170	211	
	7810	4"	↓	6	2.667		320	98		418	495	
	7850	Tee, straight, 1/4"	1 Plum	10	.800		9.65	32.50		42.15	59.50	
	7870	3/8"		10	.800		10.20	32.50		42.70	60.50	
	7880	1/2"		9	.889		13	36.50		49.50	69	
	7890	3/4"		9	.889		14.05	36.50		50.55	70	
	7900	1"	↓	8	1		16.90	41		57.90	80	
	7910	1-1/4"	Q-1	14	1.143		24	42		66	89.50	
	7920	1-1/2"		13	1.231		28	45.50		73.50	99	
	7930	2"		11	1.455		41.50	53.50		95	127	
	7940	2-1/2"		9	1.778		98.50	65.50		164	207	
	7950	3"		6	2.667		144	98		242	305	
	7960	4"	↓	4	4		400	147		547	660	
	7980	6"	Q-2	4	6		425	229		654	810	
	7990	8"	"	3	8		665	305		970	1,200	
	8050	Couplings, straight, 1/4"	1 Plum	19	.421		6.90	17.20		24.10	33.50	
	8070	3/8"		19	.421		6.90	17.20		24.10	33.50	
	8080	1/2"		19	.421		7.70	17.20		24.90	34.50	
	8090	3/4"		18	.444		8.85	18.15		27	37.50	
	8100	1"	↓	15	.533		10.15	22		32.15	44	
	8110	1-1/4"	Q-1	26	.615		12.05	22.50		34.55	47.50	
	8120	1-1/2"		24	.667		18.05	24.50		42.55	57	
	8130	2"		21	.762		26	28		54	70.50	
	8140	2-1/2"		18	.889		46.50	32.50		79	100	
	8150	3"		14	1.143		67	42		109	137	
	8160	4"	↓	10	1.600	↓	84	59		143	181	
	8200	Galvanized										
	8220	90° Elbow, straight, 1/4"	1 Plum	16	.500	Ea.	12.95	20.50		33.45	45	
	8222	3/8"		16	.500		13.20	20.50		33.70	45	
	8224	1/2"		15	.533		15.75	22		37.75	50.50	
	8226	3/4"		14	.571		17.65	23.50		41.15	54.50	
	8228	1"	↓	13	.615		22.50	25		47.50	63	
	8230	1-1/4"	Q-1	22	.727		35.50	27		62.50	79	
	8232	1-1/2"		20	.800		38.50	29.50		68	86.50	
	8234	2"		18	.889		65	32.50		97.50	121	
	8236	2-1/2"		14	1.143		139	42		181	216	
	8238	3"		10	1.600		170	59		229	276	
	8240	4"	↓	6	2.667	↓	495	98		593	690	
	8280	45° Elbow, straight										
	8282	1/2"	1 Plum	15	.533	Ea.	23.50	22		45.50	58.50	
	8284	3/4"		14	.571		27	23.50		50.50	64.50	
	8286	1"	↓	13	.615	↓	29.50	25		54.50	70.50	

15107 | Metal Pipe & Fittings

			CREW	DAILY OUTPUT	LABOR-HOURS	UNIT	2005 BARE COSTS MAT.	LABOR	EQUIP.	TOTAL	TOTAL INCL O&P	
640	8288	1-1/4"	Q-1	22	.727	Ea.	46	27		73	90.50	640
	8290	1-1/2"		20	.800		59.50	29.50		89	109	
	8292	2"	↓	18	.889		82	32.50		114.50	140	
	8310	Tee, straight, 1/4"	1 Plum	10	.800		18.45	32.50		50.95	69.50	
	8312	3/8"		10	.800		18.80	32.50		51.30	69.50	
	8314	1/2"		9	.889		23.50	36.50		60	80.50	
	8316	3/4"		9	.889		26	36.50		62.50	83	
	8318	1"	↓	8	1		32	41		73	96.50	
	8320	1-1/4"	Q-1	14	1.143		45.50	42		87.50	113	
	8322	1-1/2"		13	1.231		47.50	45.50		93	120	
	8324	2"		11	1.455		79.50	53.50		133	168	
	8326	2-1/2"		9	1.778		233	65.50		298.50	355	
	8328	3"		6	2.667		272	98		370	445	
	8330	4"	↓	4	4		680	147		827	965	
	8380	Couplings, straight, 1/4"	1 Plum	19	.421		6.90	17.20		24.10	33.50	
	8382	3/8"		19	.421		9.90	17.20		27.10	37	
	8384	1/2"		19	.421		10.20	17.20		27.40	37	
	8386	3/4"		18	.444		11.50	18.15		29.65	40	
	8388	1"	↓	15	.533		15.20	22		37.20	49.50	
	8390	1-1/4"	Q-1	26	.615		19.80	22.50		42.30	56	
	8392	1-1/2"		24	.667		31.50	24.50		56	72	
	8394	2"		21	.762		38.50	28		66.50	84	
	8396	2-1/2"		18	.889		98	32.50		130.50	157	
	8398	3"		14	1.143		117	42		159	192	
	8399	4"	↓	10	1.600	↓	124	59		183	226	
	8529	Black										
	8530	Reducer, concentric, 1/4"	1 Plum	19	.421	Ea.	8.40	17.20		25.60	35.50	
	8531	3/8"		19	.421		8.40	17.20		25.60	35.50	
	8532	1/2"		17	.471		10.10	19.20		29.30	40	
	8533	3/4"		16	.500		14	20.50		34.50	46	
	8534	1"	↓	15	.533		16.80	22		38.80	51.50	
	8535	1-1/4"	Q-1	26	.615		25	22.50		47.50	61.50	
	8536	1-1/2"		24	.667		26	24.50		50.50	65.50	
	8537	2"	↓	21	.762		38	28		66	84	
	8550	Cap, 1/4"	1 Plum	38	.211		6.50	8.60		15.10	20	
	8551	3/8"		38	.211		6.50	8.60		15.10	20	
	8552	1/2"		34	.235		6.45	9.60		16.05	21.50	
	8553	3/4"		32	.250		8.15	10.20		18.35	24.50	
	8554	1"	↓	30	.267		10.70	10.90		21.60	28	
	8555	1-1/4"	Q-1	52	.308		11.85	11.30		23.15	30	
	8556	1-1/2"		48	.333		17.30	12.25		29.55	37.50	
	8557	2"	↓	42	.381		24	14		38	47.50	
	8570	Plug, 1/4"	1 Plum	38	.211		1.87	8.60		10.47	15	
	8571	3/8"		38	.211		2.30	8.60		10.90	15.50	
	8572	1/2"		34	.235		2.30	9.60		11.90	17	
	8573	3/4"		32	.250		2.99	10.20		13.19	18.65	
	8574	1"	↓	30	.267		4.58	10.90		15.48	21.50	
	8575	1-1/4"	Q-1	52	.308		8.70	11.30		20	26.50	
	8576	1-1/2"		48	.333		10	12.25		22.25	29.50	
	8577	2"	↓	42	.381		16.70	14		30.70	39.50	
	9500	Union with brass seat, 1/4"	1 Plum	15	.533		14.25	22		36.25	48.50	
	9530	3/8"		15	.533		14.45	22		36.45	49	
	9540	1/2"		14	.571		11.70	23.50		35.20	48	
	9550	3/4"		13	.615		12.95	25		37.95	52	
	9560	1"	↓	12	.667		16.90	27		43.90	59.50	
	9570	1-1/4"	Q-1	21	.762	↓	27.50	28		55.50	72.50	

15107 Metal Pipe & Fittings		CREW	DAILY OUTPUT	LABOR-HOURS	UNIT	MAT.	LABOR	EQUIP.	TOTAL	TOTAL INCL O&P
640	9580 1-1/2"	Q-1	19	.842	Ea.	29	31		60	78
	9590 2"		17	.941		35.50	34.50		70	91.50
	9600 2-1/2"		13	1.231		115	45.50		160.50	195
	9610 3"		9	1.778		149	65.50		214.50	263
	9620 4"	↓	5	3.200		475	118		593	695
	9630 Union, all iron, 1/4"	1 Plum	15	.533		24	22		46	59.50
	9650 3/8"		15	.533		24	22		46	59.50
	9660 1/2"		14	.571		25.50	23.50		49	63
	9670 3/4"		13	.615		27.50	25		52.50	68
	9680 1"	↓	12	.667		33.50	27		60.50	78
	9690 1-1/4"	Q-1	21	.762		51.50	28		79.50	98.50
	9700 1-1/2"		19	.842		63	31		94	116
	9710 2"		17	.941		79.50	34.50		114	140
	9720 2-1/2"		13	1.231		141	45.50		186.50	223
	9730 3"	↓	9	1.778		268	65.50		333.50	395
	9750 For galvanized unions, add				↓	15%				
	9757 Forged steel, 3000 lb.									
	9758 Black									
	9760 90° Elbow, 1/4"	1 Plum	16	.500	Ea.	9.50	20.50		30	41
	9761 3/8"		16	.500		13.05	20.50		33.55	45
	9762 1/2"		15	.533		9.70	22		31.70	43.50
	9763 3/4"		14	.571		12.35	23.50		35.85	48.50
	9764 1"	↓	13	.615		18.70	25		43.70	58.50
	9765 1-1/4"	Q-1	22	.727		34.50	27		61.50	78
	9766 1-1/2"		20	.800		44	29.50		73.50	92
	9767 2"	↓	18	.889		54.50	32.50		87	109
	9780 45° Elbow 1/4"	1 Plum	16	.500		16.70	20.50		37.20	49
	9781 3/8"		16	.500		16.70	20.50		37.20	49
	9782 1/2"		15	.533		16.70	22		38.70	51.50
	9783 3/4"		14	.571		19.30	23.50		42.80	56
	9784 1"	↓	13	.615		26.50	25		51.50	67
	9785 1-1/4"	Q-1	22	.727		38	27		65	81.50
	9786 1-1/2"		20	.800		49	29.50		78.50	97.50
	9787 2"	↓	18	.889		67.50	32.50		100	123
	9800 Tee, 1/4"	1 Plum	10	.800		14.25	32.50		46.75	64.50
	9801 3/8"		10	.800		14.75	32.50		47.25	65.50
	9802 1/2"		9	.889		13.35	36.50		49.85	69
	9803 3/4"		9	.889		18.85	36.50		55.35	75
	9804 1"	↓	8	1		24.50	41		65.50	88.50
	9805 1-1/4"	Q-1	14	1.143		49.50	42		91.50	118
	9806 1-1/2"		13	1.231		55	45.50		100.50	129
	9807 2"	↓	11	1.455		74.50	53.50		128	163
	9820 Reducer, concentric, 1/4"	1 Plum	19	.421		7.20	17.20		24.40	34
	9821 3/8"		19	.421		7.20	17.20		24.40	34
	9822 1/2"		17	.471		7.20	19.20		26.40	37
	9823 3/4"		16	.500		8.15	20.50		28.65	39.50
	9824 1"	↓	15	.533		11.15	22		33.15	45.50
	9825 1-1/4"	Q-1	26	.615		17.55	22.50		40.05	53.50
	9826 1-1/2"		24	.667		20	24.50		44.50	59
	9827 2"	↓	21	.762		29	28		57	74
	9840 Cap, 1/4"	1 Plum	38	.211		4.71	8.60		13.31	18.15
	9841 3/8"		38	.211		4.71	8.60		13.31	18.15
	9842 1/2"		34	.235		4.71	9.60		14.31	19.65
	9843 3/4"		32	.250		6.55	10.20		16.75	22.50
	9844 1"	↓	30	.267		10.05	10.90		20.95	27.50
	9845 1-1/4"	Q-1	52	.308	↓	15.10	11.30		26.40	33.50

Note: MAT., LABOR, EQUIP., TOTAL are 2005 BARE COSTS.

	15107 | Metal Pipe & Fittings		CREW	DAILY OUTPUT	LABOR-HOURS	UNIT	2005 BARE COSTS MAT.	LABOR	EQUIP.	TOTAL	TOTAL INCL O&P	
640	9846	1-1/2"	Q-1	48	.333	Ea.	17.40	12.25		29.65	37.50	640
	9847	2"		42	.381		26.50	14		40.50	50.50	
	9860	Plug, 1/4"	1 Plum	38	.211		2	8.60		10.60	15.15	
	9861	3/8"		38	.211		2.30	8.60		10.90	15.50	
	9862	1/2"		34	.235		2.30	9.60		11.90	17	
	9863	3/4"		32	.250		2.99	10.20		13.19	18.65	
	9864	1"		30	.267		4.58	10.90		15.48	21.50	
	9865	1-1/4"	Q-1	52	.308		8.70	11.30		20	26.50	
	9866	1-1/2"		48	.333		10	12.25		22.25	29.50	
	9867	2"		42	.381		16.70	14		30.70	39.50	
	9880	Union, bronze seat, 1/4"	1 Plum	15	.533		32	22		54	68.50	
	9881	3/8"		15	.533		32	22		54	68.50	
	9882	1/2"		14	.571		31	23.50		54.50	69	
	9883	3/4"		13	.615		41.50	25		66.50	83.50	
	9884	1"		12	.667		46.50	27		73.50	92.50	
	9885	1-1/4"	Q-1	21	.762		90.50	28		118.50	142	
	9886	1-1/2"		19	.842		90	31		121	146	
	9887	2"		17	.941		112	34.50		146.50	175	
	9900	Coupling, 1/4"	1 Plum	19	.421		4.64	17.20		21.84	31	
	9901	3/8"		19	.421		5.75	17.20		22.95	32.50	
	9902	1/2"		17	.471		4.51	19.20		23.71	34	
	9903	3/4"		16	.500		5.95	20.50		26.45	37	
	9904	1"		15	.533		8.45	22		30.45	42.50	
	9905	1-1/4"	Q-1	26	.615		14.50	22.50		37	50	
	9906	1-1/2"		24	.667		20.50	24.50		45	59.50	
	9907	2"		21	.762		23	28		51	67	
660	0010	**PIPE, STEEL, FITTINGS** Flanged, welded and special type										660
	0020	flanged joints, C.I., standard weight, black. One gasket & bolt										
	0040	set, mat'l only, required at each joint, not included (see line 0620)										
	0060	90° Elbow, straight, 1-1/2" pipe size	Q-1	14	1.143	Ea.	220	42		262	305	
	0080	2" pipe size		13	1.231		134	45.50		179.50	215	
	0090	2-1/2" pipe size		12	1.333		145	49		194	233	
	0100	3" pipe size		11	1.455		147	53.50		200.50	243	
	0110	4" pipe size		8	2		149	73.50		222.50	275	
	0120	5" pipe size		7	2.286		355	84		439	515	
	0130	6" pipe size	Q-2	9	2.667		234	102		336	410	
	0140	8" pipe size		8	3		405	114		519	615	
	0150	10" pipe size		7	3.429		885	131		1,016	1,175	
	0160	12" pipe size		6	4		1,775	153		1,928	2,175	
	0171	90° Elbow, reducing										
	0172	2-1/2" by 2" pipe size	Q-1	12	1.333	Ea.	425	49		474	545	
	0173	3" by 2-1/2" pipe size		11	1.455		450	53.50		503.50	570	
	0174	4" by 3" pipe size		8	2		330	73.50		403.50	475	
	0175	5" by 3" pipe size		7	2.286		685	84		769	880	
	0176	6" by 4" pipe size	Q-2	9	2.667		410	102		512	605	
	0177	8" by 6" pipe size		8	3		600	114		714	830	
	0178	10" by 8" pipe size		7	3.429		1,150	131		1,281	1,450	
	0179	12" by 10" pipe size		6	4		2,200	153		2,353	2,625	
	0200	45° Elbow, straight, 1-1/2" pipe size	Q-1	14	1.143		265	42		307	355	
	0220	2" pipe size		13	1.231		194	45.50		239.50	281	
	0230	2-1/2" pipe size		12	1.333		145	49		194	233	
	0240	3" pipe size		11	1.455		121	53.50		174.50	214	
	0250	4" pipe size		8	2		149	73.50		222.50	275	
	0260	5" pipe size		7	2.286		355	84		439	515	
	0270	6" pipe size	Q-2	9	2.667		234	102		336	410	
	0280	8" pipe size		8	3		405	114		519	615	

15107 | Metal Pipe & Fittings

		15107 Metal Pipe & Fittings	CREW	DAILY OUTPUT	LABOR-HOURS	UNIT	2005 BARE COSTS MAT.	LABOR	EQUIP.	TOTAL	TOTAL INCL O&P	
660	0290	10″ pipe size	Q-2	7	3.429	Ea.	885	131		1,016	1,175	660
	0300	12″ pipe size		6	4		1,775	153		1,928	2,175	
	0310	Cross, straight										
	0311	2-1/2″ pipe size	Q-1	6	2.667	Ea.	425	98		523	615	
	0312	3″ pipe size		5	3.200		445	118		563	665	
	0313	4″ pipe size		4	4		585	147		732	860	
	0314	5″ pipe size		3	5.333		950	196		1,146	1,350	
	0315	6″ pipe size	Q-2	5	4.800		1,275	183		1,458	1,675	
	0316	8″ pipe size		4	6		1,975	229		2,204	2,525	
	0317	10″ pipe size		3	8		2,325	305		2,630	3,025	
	0318	12″ pipe size		2	12		3,550	460		4,010	4,600	
	0350	Tee, straight, 1-1/2″ pipe size	Q-1	10	1.600		252	59		311	365	
	0370	2″ pipe size		9	1.778		147	65.50		212.50	260	
	0380	2-1/2″ pipe size		8	2		213	73.50		286.50	345	
	0390	3″ pipe size		7	2.286		150	84		234	291	
	0400	4″ pipe size		5	3.200		228	118		346	425	
	0410	5″ pipe size		4	4		610	147		757	890	
	0420	6″ pipe size	Q-2	6	4		330	153		483	595	
	0430	8″ pipe size		5	4.800		565	183		748	895	
	0440	10″ pipe size		4	6		1,525	229		1,754	2,025	
	0450	12″ pipe size		3	8		2,375	305		2,680	3,075	
	0459	Tee, reducing on outlet										
	0460	2-1/2″ by 2″ pipe size	Q-1	8	2	Ea.	300	73.50		373.50	440	
	0461	3″ by 2-1/2″ pipe size		7	2.286		350	84		434	505	
	0462	4″ by 3″ pipe size		5	3.200		470	118		588	690	
	0463	5″ by 4″ pipe size		4	4		985	147		1,132	1,300	
	0464	6″ by 4″ pipe size	Q-2	6	4		450	153		603	725	
	0465	8″ by 6″ pipe size		5	4.800		705	183		888	1,050	
	0466	10″ by 8″ pipe size		4	6		1,600	229		1,829	2,125	
	0467	12″ by 10″ pipe size		3	8		2,775	305		3,080	3,525	
	0476	Reducer, concentric										
	0477	3″ by 2-1/2″	Q-1	12	1.333	Ea.	430	49		479	545	
	0478	4″ by 3″		9	1.778		300	65.50		365.50	430	
	0479	5″ by 4″		8	2		465	73.50		538.50	620	
	0480	6″ by 4″	Q-2	10	2.400		405	91.50		496.50	585	
	0481	8″ by 6″		9	2.667		510	102		612	715	
	0482	10″ by 8″		8	3		1,050	114		1,164	1,325	
	0483	12″ by 10″		7	3.429		1,800	131		1,931	2,175	
	0492	Reducer, eccentric										
	0493	4″ by 3″	Q-1	8	2	Ea.	490	73.50		563.50	650	
	0494	5″ by 4″	″	7	2.286		725	84		809	920	
	0495	6″ by 4″	Q-2	9	2.667		460	102		562	665	
	0496	8″ by 6″		8	3		585	114		699	815	
	0497	10″ by 8″		7	3.429		1,500	131		1,631	1,850	
	0498	12″ by 10″		6	4		2,025	153		2,178	2,450	
	0500	For galvanized elbows and tees, add					100%					
	0520	For extra heavy weight elbows and tees, add					140%					
	0620	Gasket and bolt set, 150#, 1/2″ pipe size	1 Plum	20	.400		2.36	16.35		18.71	27	
	0622	3/4″ pipe size		19	.421		2.36	17.20		19.56	28.50	
	0624	1″ pipe size		18	.444		2.42	18.15		20.57	30	
	0626	1-1/4″ pipe size		17	.471		2.54	19.20		21.74	32	
	0628	1-1/2″ pipe size		15	.533		2.71	22		24.71	36	
	0630	2″ pipe size		13	.615		4.33	25		29.33	43	
	0640	2-1/2″ pipe size		12	.667		4.58	27		31.58	46	
	0650	3″ pipe size		11	.727		4.80	29.50		34.30	50	
	0660	3-1/2″ pipe size		9	.889		8.70	36.50		45.20	64	

15107 | Metal Pipe & Fittings		CREW	DAILY OUTPUT	LABOR-HOURS	UNIT	2005 BARE COSTS MAT.	LABOR	EQUIP.	TOTAL	TOTAL INCL O&P
660 0670	4" pipe size	1 Plum	8	1	Ea.	8.70	41		49.70	71
0680	5" pipe size		7	1.143		12.90	46.50		59.40	84
0690	6" pipe size		6	1.333		13.55	54.50		68.05	97
0700	8" pipe size		5	1.600		14.45	65.50		79.95	114
0710	10" pipe size		4.50	1.778		30	72.50		102.50	142
0720	12" pipe size		4.20	1.905		32	78		110	153
0730	14" pipe size		4	2		45.50	81.50		127	173
0740	16" pipe size		3	2.667		59.50	109		168.50	230
0750	18" pipe size		2.70	2.963		83	121		204	274
0760	20" pipe size		2.30	3.478		111	142		253	335
0780	24" pipe size		1.90	4.211		149	172		321	425
0790	26" pipe size		1.60	5		168	204		372	490
0810	30" pipe size		1.40	5.714		208	233		441	580
0830	36" pipe size	↓	1.10	7.273	↓	256	297		553	725
0850	For 300 lb gasket set, add					40%				
2000	Flanged unions, 125 lb., black, 1/2" pipe size	1 Plum	17	.471	Ea.	34	19.20		53.20	66.50
2040	3/4" pipe size		17	.471		45	19.20		64.20	78.50
2050	1" pipe size	↓	16	.500		43.50	20.50		64	78.50
2060	1-1/4" pipe size	Q-1	28	.571		51.50	21		72.50	88.50
2070	1-1/2" pipe size		27	.593		47.50	22		69.50	85.50
2080	2" pipe size		26	.615		56	22.50		78.50	95.50
2090	2-1/2" pipe size		24	.667		76	24.50		100.50	121
2100	3" pipe size		22	.727		86	27		113	135
2110	3-1/2" pipe size		18	.889		146	32.50		178.50	209
2120	4" pipe size		16	1		117	37		154	184
2130	5" pipe size	↓	14	1.143		272	42		314	360
2140	6" pipe size	Q-2	19	1.263		255	48		303	355
2150	8" pipe size	"	16	1.500		570	57		627	710
2200	For galvanized unions, add				↓	150%				
2290	Threaded flange									
2300	Cast iron									
2310	Black, 125 lb., per flange									
2320	1" pipe size	1 Plum	27	.296	Ea.	19.50	12.10		31.60	39.50
2330	1-1/4" pipe size	Q-1	44	.364		21.50	13.35		34.85	44
2340	1-1/2" pipe size		40	.400		23.50	14.70		38.20	47.50
2350	2" pipe size		36	.444		21.50	16.35		37.85	48
2360	2-1/2" pipe size		28	.571		25	21		46	59
2370	3" pipe size		20	.800		32.50	29.50		62	79.50
2380	3-1/2" pipe size		16	1		45.50	37		82.50	106
2390	4" pipe size		12	1.333		44	49		93	122
2400	5" pipe size	↓	10	1.600		62	59		121	157
2410	6" pipe size	Q-2	14	1.714		70	65.50		135.50	176
2420	8" pipe size		12	2		110	76.50		186.50	236
2430	10" pipe size		10	2.400		196	91.50		287.50	355
2440	12" pipe size	↓	8	3	↓	420	114		534	630
2460	For galvanized flanges, add					95%				
2580	Forged steel,									
2590	Black 150 lb., per flange									
2600	1/2" pipe size	1 Plum	30	.267	Ea.	13.35	10.90		24.25	31
2610	3/4" pipe size		28	.286		13.35	11.65		25	32
2620	1" pipe size	↓	27	.296		13.35	12.10		25.45	33
2630	1-1/4" pipe size	Q-1	44	.364		13.35	13.35		26.70	34.50
2640	1-1/2" pipe size		40	.400		13.35	14.70		28.05	36.50
2650	2" pipe size		36	.444		16.25	16.35		32.60	42.50
2660	2-1/2" pipe size		28	.571		20	21		41	53.50
2670	3" pipe size	↓	20	.800	↓	21	29.50		50.50	67

Important: See the Reference Section for critical supporting data - Reference Nos., Crews, & City Cost Indexes

15107 | Metal Pipe & Fittings

			CREW	DAILY OUTPUT	LABOR-HOURS	UNIT	2005 BARE COSTS MAT.	LABOR	EQUIP.	TOTAL	TOTAL INCL O&P	
660	2690	4" pipe size	Q-1	12	1.333	Ea.	24.50	49		73.50	101	660
	2700	5" pipe size	↓	10	1.600		36.50	59		95.50	129	
	2710	6" pipe size	Q-2	14	1.714		41	65.50		106.50	144	
	2720	8" pipe size		12	2		69.50	76.50		146	192	
	2730	10" pipe size	↓	10	2.400	↓	126	91.50		217.50	277	
	2860	Black 300 lb., per flange										
	2870	1/2" pipe size	1 Plum	30	.267	Ea.	20	10.90		30.90	38.50	
	2880	3/4" pipe size		28	.286		20	11.65		31.65	39.50	
	2890	1" pipe size	↓	27	.296		20	12.10		32.10	40	
	2900	1-1/4" pipe size	Q-1	44	.364		20	13.35		33.35	42	
	2910	1-1/2" pipe size		40	.400		20	14.70		34.70	44	
	2920	2" pipe size		36	.444		27.50	16.35		43.85	54.50	
	2930	2-1/2" pipe size		28	.571		26	21		47	60.50	
	2940	3" pipe size		20	.800		29	29.50		58.50	76	
	2960	4" pipe size	↓	12	1.333		47.50	49		96.50	126	
	2970	6" pipe size	Q-2	14	1.714	↓	81	65.50		146.50	188	
	3000	Weld joint, butt, carbon steel, standard weight										
	3040	90° elbow, long radius										
	3050	1/2" pipe size	Q-15	16	1	Ea.	18.90	37	5.05	60.95	82	
	3060	3/4" pipe size		16	1		18.90	37	5.05	60.95	82	
	3070	1" pipe size		16	1		9.60	37	5.05	51.65	71.50	
	3080	1-1/4" pipe size		14	1.143		9.60	42	5.75	57.35	80	
	3090	1-1/2" pipe size		13	1.231		9.60	45.50	6.20	61.30	85.50	
	3100	2" pipe size		10	1.600		10.55	59	8.05	77.60	109	
	3110	2-1/2" pipe size		8	2		12.80	73.50	10.10	96.40	136	
	3120	3" pipe size		7	2.286		15.35	84	11.50	110.85	156	
	3130	4" pipe size		5	3.200		25.50	118	16.15	159.65	223	
	3136	5" pipe size	↓	4	4		54	147	20	221	300	
	3140	6" pipe size	Q-16	5	4.800		59	183	16.15	258.15	360	
	3150	8" pipe size		3.75	6.400		111	244	21.50	376.50	510	
	3160	10" pipe size		3	8		203	305	27	535	715	
	3170	12" pipe size		2.50	9.600		300	365	32.50	697.50	915	
	3180	14" pipe size		2	12		395	460	40.50	895.50	1,175	
	3190	16" pipe size		1.50	16		545	610	54	1,209	1,575	
	3191	18" pipe size		1.25	19.200		720	730	64.50	1,514.50	1,950	
	3192	20" pipe size		1.15	20.870		1,025	795	70	1,890	2,400	
	3194	24" pipe size	↓	1.02	23.529	↓	1,375	895	79	2,349	2,925	
	3200	45° Elbow, long										
	3210	1/2" pipe size	Q-15	16	1	Ea.	26.50	37	5.05	68.55	90	
	3220	3/4" pipe size		16	1		26.50	37	5.05	68.55	90	
	3230	1" pipe size		16	1		9.90	37	5.05	51.95	72	
	3240	1-1/4" pipe size		14	1.143		9.90	42	5.75	57.65	80.50	
	3250	1-1/2" pipe size		13	1.231		9.90	45.50	6.20	61.60	85.50	
	3260	2" pipe size		10	1.600		9.90	59	8.05	76.95	108	
	3270	2-1/2" pipe size		8	2		11.85	73.50	10.10	95.45	135	
	3280	3" pipe size		7	2.286		12.80	84	11.50	108.30	153	
	3290	4" pipe size		5	3.200		22.50	118	16.15	156.65	220	
	3296	5" pipe size	↓	4	4		34.50	147	20	201.50	281	
	3300	6" pipe size	Q-16	5	4.800		47.50	183	16.15	246.65	345	
	3310	8" pipe size		3.75	6.400		79.50	244	21.50	345	475	
	3320	10" pipe size		3	8		144	305	27	476	650	
	3330	12" pipe size		2.50	9.600		203	365	32.50	600.50	810	
	3340	14" pipe size		2	12		273	460	40.50	773.50	1,025	
	3341	16" pipe size		1.50	16		460	610	54	1,124	1,475	
	3342	18" pipe size		1.25	19.200		655	730	64.50	1,449.50	1,900	
	3343	20" pipe size	↓	1.15	20.870	↓	680	795	70	1,545	2,025	

15107 | Metal Pipe & Fittings

			CREW	DAILY OUTPUT	LABOR-HOURS	UNIT	2005 BARE COSTS MAT.	LABOR	EQUIP.	TOTAL	TOTAL INCL O&P
660	3345	24" pipe size	Q-16	1.05	22.857	Ea.	1,025	870	77	1,972	2,500
	3346	26" pipe size		.85	28.235		1,125	1,075	95	2,295	2,975
	3347	30" pipe size		.45	53.333		1,250	2,025	179	3,454	4,625
	3349	36" pipe size		.38	63.158		1,375	2,400	212	3,987	5,350
	3350	Tee, straight									
	3352	For reducing tees and concentrics see starting line 4600									
	3360	1/2" pipe size	Q-15	10	1.600	Ea.	48.50	59	8.05	115.55	151
	3370	3/4" pipe size		10	1.600		48.50	59	8.05	115.55	151
	3380	1" pipe size		10	1.600		26	59	8.05	93.05	126
	3390	1-1/4" pipe size		9	1.778		30	65.50	8.95	104.45	141
	3400	1-1/2" pipe size		8	2		30	73.50	10.10	113.60	155
	3410	2" pipe size		6	2.667		25.50	98	13.45	136.95	190
	3420	2-1/2" pipe size		5	3.200		35	118	16.15	169.15	233
	3430	3" pipe size		4	4		39	147	20	206	286
	3440	4" pipe size		3	5.333		55.50	196	27	278.50	385
	3446	5" pipe size		2.50	6.400		91	235	32.50	358.50	490
	3450	6" pipe size	Q-16	3	8		94.50	305	27	426.50	595
	3460	8" pipe size		2.50	9.600		165	365	32.50	562.50	770
	3470	10" pipe size		2	12		297	460	40.50	797.50	1,050
	3480	12" pipe size		1.60	15		415	570	50.50	1,035.50	1,375
	3481	14" pipe size		1.30	18.462		675	705	62	1,442	1,850
	3482	16" pipe size		1	24		815	915	80.50	1,810.50	2,350
	3483	18" pipe size		.80	30		1,300	1,150	101	2,551	3,250
	3484	20" pipe size		.75	32		2,000	1,225	108	3,333	4,150
	3486	24" pipe size		.70	34.286		2,575	1,300	115	3,990	4,925
	3487	26" pipe size		.55	43.636		2,825	1,675	147	4,647	5,775
	3488	30" pipe size		.30	80		3,125	3,050	269	6,444	8,300
	3490	36" pipe size		.20	120		3,425	4,575	405	8,405	11,100
	3491	Eccentric reducer, 1-1/2" pipe size	Q-15	14	1.143		40.50	42	5.75	88.25	114
	3492	2" pipe size		11	1.455		57.50	53.50	7.35	118.35	152
	3493	2-1/2" pipe size		9	1.778		69	65.50	8.95	143.45	184
	3494	3" pipe size		8	2		78	73.50	10.10	161.60	208
	3495	4" pipe size		6	2.667		84.50	98	13.45	195.95	255
	3496	6" pipe size	Q-16	5	4.800		148	183	16.15	347.15	455
	3497	8" pipe size		4	6		170	229	20	419	555
	3498	10" pipe size		3	8		179	305	27	511	685
	3499	12" pipe size		2.50	9.600		256	365	32.50	653.50	865
	3501	Cap, 1-1/2" pipe size	Q-15	28	.571		11.20	21	2.88	35.08	47
	3502	2" pipe size		22	.727		10.90	27	3.67	41.57	56
	3503	2-1/2" pipe size		18	.889		11.50	32.50	4.48	48.48	66.50
	3504	3" pipe size		16	1		11.50	37	5.05	53.55	73.50
	3505	4" pipe size		12	1.333		16.65	49	6.70	72.35	99
	3506	6" pipe size	Q-16	10	2.400		29	91.50	8.05	128.55	178
	3507	8" pipe size		8	3		43.50	114	10.10	167.60	231
	3508	10" pipe size		6	4		71.50	153	13.45	237.95	320
	3509	12" pipe size		5	4.800		108	183	16.15	307.15	410
	3511	14" pipe size		4	6		144	229	20	393	525
	3512	16" pipe size		4	6		167	229	20	416	550
	3513	18" pipe size		3	8		229	305	27	561	740
	3517	Weld joint, butt, carbon steel, extra strong									
	3519	90° elbow, long									
	3520	1/2" pipe size	Q-15	13	1.231	Ea.	24	45.50	6.20	75.70	101
	3530	3/4" pipe size		12	1.333		24	49	6.70	79.70	107
	3540	1" pipe size		11	1.455		21.50	53.50	7.35	82.35	112
	3550	1-1/4" pipe size		10	1.600		21.50	59	8.05	88.55	121
	3560	1-1/2" pipe size		9	1.778		21.50	65.50	8.95	95.95	132

	15107 | Metal Pipe & Fittings		CREW	DAILY OUTPUT	LABOR-HOURS	UNIT	2005 BARE COSTS MAT.	LABOR	EQUIP.	TOTAL	TOTAL INCL O&P	
660	3570	2" pipe size	Q-15	8	2	Ea.	33	73.50	10.10	116.60	159	660
	3580	2-1/2" pipe size		7	2.286		33	84	11.50	128.50	175	
	3590	3" pipe size		6	2.667		33	98	13.45	144.45	198	
	3600	4" pipe size		4	4		55	147	20	222	305	
	3606	5" pipe size		3.50	4.571		114	168	23	305	405	
	3610	6" pipe size	Q-16	4.50	5.333		128	203	17.90	348.90	465	
	3620	8" pipe size		3.50	6.857		276	261	23	560	725	
	3630	10" pipe size		2.50	9.600		375	365	32.50	772.50	1,000	
	3640	12" pipe size		2.25	10.667		450	405	36	891	1,150	
	3650	45° Elbow, long										
	3660	1/2" pipe size	Q-15	13	1.231	Ea.	16.95	45.50	6.20	68.65	93.50	
	3670	3/4" pipe size		12	1.333		16.95	49	6.70	72.65	99.50	
	3680	1" pipe size		11	1.455		8.95	53.50	7.35	69.80	98.50	
	3690	1-1/4" pipe size		10	1.600		8.95	59	8.05	76	107	
	3700	1-1/2" pipe size		9	1.778		8.95	65.50	8.95	83.40	118	
	3710	2" pipe size		8	2		8.95	73.50	10.10	92.55	132	
	3720	2-1/2" pipe size		7	2.286		13.10	84	11.50	108.60	153	
	3730	3" pipe size		6	2.667		16.30	98	13.45	127.75	180	
	3740	4" pipe size		4	4		27.50	147	20	194.50	274	
	3746	5" pipe size		3.50	4.571		59.50	168	23	250.50	345	
	3750	6" pipe size	Q-16	4.50	5.333		66	203	17.90	286.90	400	
	3760	8" pipe size		3.50	6.857		125	261	23	409	560	
	3770	10" pipe size		2.50	9.600		224	365	32.50	621.50	830	
	3780	12" pipe size		2.25	10.667		315	405	36	756	995	
	3800	Tee, straight										
	3810	1/2" pipe size	Q-15	9	1.778	Ea.	57.50	65.50	8.95	131.95	172	
	3820	3/4" pipe size		8.50	1.882		57.50	69	9.50	136	178	
	3830	1" pipe size		8	2		32	73.50	10.10	115.60	157	
	3840	1-1/4" pipe size		7	2.286		32	84	11.50	127.50	174	
	3850	1-1/2" pipe size		6	2.667		36	98	13.45	147.45	201	
	3860	2" pipe size		5	3.200		32	118	16.15	166.15	230	
	3870	2-1/2" pipe size		4	4		51	147	20	218	300	
	3880	3" pipe size		3.50	4.571		49.50	168	23	240.50	335	
	3890	4" pipe size		2.50	6.400		80	235	32.50	347.50	480	
	3896	5" pipe size		2.25	7.111		163	262	36	461	615	
	3900	6" pipe size	Q-16	2.25	10.667		136	405	36	577	800	
	3910	8" pipe size		2	12		255	460	40.50	755.50	1,025	
	3920	10" pipe size		1.75	13.714		415	525	46	986	1,300	
	3930	12" pipe size		1.50	16		600	610	54	1,264	1,625	
	4000	Eccentric reducer, 1-1/2" pipe size	Q-15	10	1.600		41.50	59	8.05	108.55	143	
	4010	2" pipe size		9	1.778		107	65.50	8.95	181.45	226	
	4020	2-1/2" pipe size		8	2		73.50	73.50	10.10	157.10	203	
	4030	3" pipe size		7	2.286		94	84	11.50	189.50	242	
	4040	4" pipe size		5	3.200		131	118	16.15	265.15	340	
	4046	5" pipe size		4.70	3.404		291	125	17.15	433.15	525	
	4050	6" pipe size	Q-16	4.50	5.333		184	203	17.90	404.90	525	
	4060	8" pipe size		3.50	6.857		214	261	23	498	655	
	4070	10" pipe size		2.50	9.600		246	365	32.50	643.50	855	
	4080	12" pipe size		2.25	10.667		257	405	36	698	935	
	4090	14" pipe size		2.10	11.429		500	435	38.50	973.50	1,250	
	4100	16" pipe size		1.90	12.632		1,050	480	42.50	1,572.50	1,925	
	4151	Cap, 1-1/2" pipe size	Q-15	24	.667		13.45	24.50	3.36	41.31	55.50	
	4152	2" pipe size		18	.889		10.90	32.50	4.48	47.88	66	
	4153	2-1/2" pipe size		16	1		15.05	37	5.05	57.10	77.50	
	4154	3" pipe size		14	1.143		16.65	42	5.75	64.40	87.50	
	4155	4" pipe size		10	1.600		23	59	8.05	90.05	123	

15107 | Metal Pipe & Fittings

			CREW	DAILY OUTPUT	LABOR-HOURS	UNIT	2005 BARE COSTS MAT.	LABOR	EQUIP.	TOTAL	TOTAL INCL O&P
660	4156	6″ pipe size	Q-16	9	2.667	Ea.	46.50	102	8.95	157.45	214
	4157	8″ pipe size		7	3.429		70	131	11.50	212.50	286
	4158	10″ pipe size		5	4.800		97.50	183	16.15	296.65	400
	4159	12″ pipe size		4	6		130	229	20	379	510
	4190	Weld fittings, reducing, standard weight									
	4200	Welding ring w/spacer pins, 2″ pipe size				Ea.	1.95			1.95	2.15
	4210	2-1/2″ pipe size					2.35			2.35	2.59
	4220	3″ pipe size					2.44			2.44	2.68
	4230	4″ pipe size					2.54			2.54	2.79
	4236	5″ pipe size					2.99			2.99	3.29
	4240	6″ pipe size					2.99			2.99	3.29
	4250	8″ pipe size					3.53			3.53	3.88
	4260	10″ pipe size					4.32			4.32	4.75
	4270	12″ pipe size					4.95			4.95	5.45
	4280	14″ pipe size					5.75			5.75	6.35
	4290	16″ pipe size					6.50			6.50	7.15
	4300	18″ pipe size					7.50			7.50	8.25
	4310	20″ pipe size					7.95			7.95	8.75
	4330	24″ pipe size					9.25			9.25	10.20
	4340	26″ pipe size					12.35			12.35	13.60
	4350	30″ pipe size					12.35			12.35	13.60
	4370	36″ pipe size					17.75			17.75	19.55
	4600	Tee, reducing on outlet									
	4601	2-1/2″ x 2″ pipe size	Q-15	5	3.200	Ea.	47	118	16.15	181.15	246
	4602	3″ x 2-1/2″ pipe size		4	4		51.50	147	20	218.50	300
	4603	3-1/2″ x 3″ pipe size		3.50	4.571		64.50	168	23	255.50	350
	4604	4″ x 3″ pipe size		3	5.333		105	196	27	328	440
	4605	5″ x 4″ pipe size		2.50	6.400		125	235	32.50	392.50	530
	4606	6″ x 5″ pipe size	Q-16	3	8		340	305	27	672	865
	4607	8″ x 6″ pipe size		2.50	9.600		215	365	32.50	612.50	825
	4608	10″ x 8″ pipe size		2	12		365	460	40.50	865.50	1,125
	4609	12″ x 10″ pipe size		1.60	15		535	570	50.50	1,155.50	1,500
	4610	16″ x 12″ pipe size		1.50	16		955	610	54	1,619	2,025
	4611	14″ x 12″ pipe size		1.52	15.789		730	600	53	1,383	1,775
	4618	Reducer, concentric									
	4619	2-1/2″ by 2″ pipe size	Q-15	10	1.600	Ea.	17.60	59	8.05	84.65	117
	4620	3″ by 2-1/2″ pipe size		9	1.778		21	65.50	8.95	95.45	131
	4621	3-1/2″ by 3″ pipe size		8	2		37.50	73.50	10.10	121.10	163
	4622	4″ by 2-1/2″ pipe size		7	2.286		22.50	84	11.50	118	164
	4623	5″ by 3″ pipe size		7	2.286		47.50	84	11.50	143	191
	4624	6″ by 4″ pipe size	Q-16	6	4		39.50	153	13.45	205.95	287
	4625	8″ by 6″ pipe size		5	4.800		52	183	16.15	251.15	350
	4626	10″ by 8″ pipe size		4	6		91	229	20	340	465
	4627	12″ by 10″ pipe size		3	8		113	305	27	445	615
	4660	Reducer, eccentric									
	4662	3″ x 2″ pipe size	Q-15	8	2	Ea.	24.50	73.50	10.10	108.10	149
	4664	4″ x 3″ pipe size		6	2.667		30	98	13.45	141.45	194
	4666	4″ x 2″ pipe size		6	2.667		37	98	13.45	148.45	202
	4670	6″ x 4″ pipe size	Q-16	5	4.800		62	183	16.15	261.15	360
	4672	6″ x 3″ pipe size		5	4.800		97.50	183	16.15	296.65	400
	4676	8″ x 6″ pipe size		4	6		79	229	20	328	455
	4678	8″ x 4″ pipe size		4	6		145	229	20	394	525
	4682	10″ x 8″ pipe size		3	8		93.50	305	27	425.50	595
	4684	10″ x 6″ pipe size		3	8		165	305	27	497	670
	4688	12″ x 10″ pipe size		2.50	9.600		147	365	32.50	544.50	745
	4690	12″ x 8″ pipe size		2.50	9.600		230	365	32.50	627.50	840

15107 | Metal Pipe & Fittings		CREW	DAILY OUTPUT	LABOR-HOURS	UNIT	2005 BARE COSTS MAT.	LABOR	EQUIP.	TOTAL	TOTAL INCL O&P
660 4691	14" x 12" pipe size	Q-16	2.20	10.909	Ea.	258	415	36.50	709.50	950
4693	16" x 14" pipe size		1.80	13.333		350	510	45	905	1,200
4694	16" x 12" pipe size		2	12		465	460	40.50	965.50	1,250
4696	18" x 16" pipe size	↓	1.60	15	↓	605	570	50.50	1,225.50	1,575
5000	Weld joint, socket, forged steel, 3000 lb., schedule 40 pipe									
5010	90° elbow, straight									
5020	1/4" pipe size	Q-15	22	.727	Ea.	10	27	3.67	40.67	55
5030	3/8" pipe size		22	.727		11.90	27	3.67	42.57	57
5040	1/2" pipe size		20	.800		8.60	29.50	4.03	42.13	58
5050	3/4" pipe size		20	.800		8.80	29.50	4.03	42.33	58
5060	1" pipe size		20	.800		11.35	29.50	4.03	44.88	61
5070	1-1/4" pipe size		18	.889		22	32.50	4.48	58.98	78
5080	1-1/2" pipe size		16	1		23.50	37	5.05	65.55	87
5090	2" pipe size		12	1.333		38	49	6.70	93.70	122
5100	2-1/2" pipe size		10	1.600		107	59	8.05	174.05	215
5110	3" pipe size		8	2		191	73.50	10.10	274.60	330
5120	4" pipe size	↓	6	2.667	↓	490	98	13.45	601.45	700
5130	45° Elbow, straight									
5134	1/4" pipe size	Q-15	22	.727	Ea.	16.25	27	3.67	46.92	62
5135	3/8" pipe size		22	.727		16.25	27	3.67	46.92	62
5136	1/2" pipe size		20	.800		12.15	29.50	4.03	45.68	62
5137	3/4" pipe size		20	.800		14.15	29.50	4.03	47.68	64
5140	1" pipe size		20	.800		18.50	29.50	4.03	52.03	69
5150	1-1/4" pipe size		18	.889		21.50	32.50	4.48	58.48	78
5160	1-1/2" pipe size		16	1		31	37	5.05	73.05	95
5170	2" pipe size		12	1.333		47	49	6.70	102.70	133
5180	2-1/2" pipe size		10	1.600		122	59	8.05	189.05	231
5190	3" pipe size		8	2		205	73.50	10.10	288.60	345
5200	4" pipe size	↓	6	2.667	↓	405	98	13.45	516.45	605
5250	Tee, straight									
5254	1/4" pipe size	Q-15	15	1.067	Ea.	11.55	39	5.40	55.95	77.50
5255	3/8" pipe size		15	1.067		14.15	39	5.40	58.55	80.50
5256	1/2" pipe size		13	1.231		11.05	45.50	6.20	62.75	87
5257	3/4" pipe size		13	1.231		12.95	45.50	6.20	64.65	89
5260	1" pipe size		13	1.231		17.85	45.50	6.20	69.55	94.50
5270	1-1/4" pipe size		12	1.333		27	49	6.70	82.70	110
5280	1-1/2" pipe size		11	1.455		36	53.50	7.35	96.85	128
5290	2" pipe size		8	2		54	73.50	10.10	137.60	182
5300	2-1/2" pipe size		6	2.667		150	98	13.45	261.45	325
5310	3" pipe size		5	3.200		360	118	16.15	494.15	590
5320	4" pipe size	↓	4	4	↓	575	147	20	742	880
5350	For reducing sizes, add					60%				
5450	Couplings									
5451	1/4" pipe size	Q-15	23	.696	Ea.	8.20	25.50	3.51	37.21	51.50
5452	3/8" pipe size		23	.696		7.55	25.50	3.51	36.56	50.50
5453	1/2" pipe size		21	.762		5.20	28	3.84	37.04	52
5454	3/4" pipe size		21	.762		5.55	28	3.84	37.39	52.50
5460	1" pipe size		20	.800		6.20	29.50	4.03	39.73	55
5470	1-1/4" pipe size		20	.800		11.65	29.50	4.03	45.18	61
5480	1-1/2" pipe size		18	.889		12.90	32.50	4.48	49.88	68
5490	2" pipe size		14	1.143		20	42	5.75	67.75	91.50
5500	2-1/2" pipe size		12	1.333		48	49	6.70	103.70	133
5510	3" pipe size		9	1.778		99.50	65.50	8.95	173.95	217
5520	4" pipe size		7	2.286		150	84	11.50	245.50	305
5570	Union, 1/4" pipe size		21	.762		16.90	28	3.84	48.74	65
5571	3/8" pipe size	↓	21	.762	↓	16.90	28	3.84	48.74	65

15107 | Metal Pipe & Fittings

			CREW	DAILY OUTPUT	LABOR-HOURS	UNIT	2005 BARE COSTS MAT.	LABOR	EQUIP.	TOTAL	TOTAL INCL O&P	
660	5572	1/2" pipe size	Q-15	19	.842	Ea.	16.90	31	4.24	52.14	70	660
	5573	3/4" pipe size		19	.842		19.15	31	4.24	54.39	72	
	5574	1" pipe size		19	.842		24.50	31	4.24	59.74	78	
	5575	1-1/4" pipe size		17	.941		52	34.50	4.74	91.24	115	
	5576	1-1/2" pipe size		15	1.067		55.50	39	5.40	99.90	126	
	5577	2" pipe size		11	1.455		76.50	53.50	7.35	137.35	173	
	5600	Reducer, 1/4" pipe size		23	.696		21	25.50	3.51	50.01	66	
	5601	3/8" pipe size		23	.696		21	25.50	3.51	50.01	66	
	5602	1/2" pipe size		21	.762		14.40	28	3.84	46.24	62	
	5603	3/4" pipe size		21	.762		16.15	28	3.84	47.99	64	
	5604	1" pipe size		21	.762		17.65	28	3.84	49.49	65.50	
	5605	1-1/4" pipe size		19	.842		32	31	4.24	67.24	86	
	5607	1-1/2" pipe size		17	.941		34.50	34.50	4.74	73.74	95	
	5608	2" pipe size		13	1.231		38	45.50	6.20	89.70	116	
	5612	Cap, 1/4" pipe size		46	.348		8.35	12.80	1.75	22.90	30.50	
	5613	3/8" pipe size		46	.348		8.35	12.80	1.75	22.90	30.50	
	5614	1/2" pipe size		42	.381		5.70	14	1.92	21.62	29.50	
	5615	3/4" pipe size		42	.381		6.60	14	1.92	22.52	30.50	
	5616	1" pipe size		42	.381		10.15	14	1.92	26.07	34.50	
	5617	1-1/4" pipe size		38	.421		11.85	15.50	2.12	29.47	39	
	5618	1-1/2" pipe size		34	.471		16.20	17.30	2.37	35.87	46.50	
	5619	2" pipe size		26	.615		22	22.50	3.10	47.60	61.50	
	5630	T-O-L, 1/4" pipe size, nozzle		23	.696		4.20	25.50	3.51	33.21	47	
	5631	3/8" pipe size, nozzle		23	.696		4.20	25.50	3.51	33.21	47	
	5632	1/2" pipe size, nozzle		22	.727		4.20	27	3.67	34.87	48.50	
	5633	3/4" pipe size, nozzle		21	.762		4.93	28	3.84	36.77	51.50	
	5634	1" pipe size, nozzle		20	.800		5.65	29.50	4.03	39.18	54.50	
	5635	1-1/4" pipe size, nozzle		18	.889		8.20	32.50	4.48	45.18	63	
	5636	1-1/2" pipe size, nozzle		16	1		8.75	37	5.05	50.80	70.50	
	5637	2" pipe size, nozzle		12	1.333		9.90	49	6.70	65.60	92	
	5638	2-1/2" pipe size, nozzle		10	1.600		32	59	8.05	99.05	132	
	5639	4" pipe size, nozzle		6	2.667		72	98	13.45	183.45	241	
	5640	W-O-L, 1/4" pipe size, nozzle		23	.696		8.30	25.50	3.51	37.31	51.50	
	5641	3/8" pipe size, nozzle		23	.696		8.30	25.50	3.51	37.31	51.50	
	5642	1/2" pipe size, nozzle		22	.727		8.30	27	3.67	38.97	53	
	5643	3/4" pipe size, nozzle		21	.762		8.35	28	3.84	40.19	55.50	
	5644	1" pipe size, nozzle		20	.800		8.90	29.50	4.03	42.43	58	
	5645	1-1/4" pipe size, nozzle		18	.889		10.95	32.50	4.48	47.93	66	
	5646	1-1/2" pipe size, nozzle		16	1		10.95	37	5.05	53	73	
	5647	2" pipe size, nozzle		12	1.333		11.10	49	6.70	66.80	93	
	5648	2-1/2" pipe size, nozzle		10	1.600		25	59	8.05	92.05	125	
	5649	3" pipe size, nozzle		8	2		29	73.50	10.10	112.60	154	
	5650	4" pipe size, nozzle		6	2.667		36.50	98	13.45	147.95	202	
	5651	5" pipe size, nozzle		5	3.200		85.50	118	16.15	219.65	289	
	5652	6" pipe size, nozzle		4	4		103	147	20	270	355	
	5653	8" pipe size, nozzle		3	5.333		197	196	27	420	540	
	5654	10" pipe size, nozzle		2.60	6.154		350	226	31	607	760	
	5655	12" pipe size, nozzle		2.20	7.273		640	267	36.50	943.50	1,150	
	5656	14" pipe size, nozzle		2	8		1,175	294	40.50	1,509.50	1,775	
	5674	S-O-L, 1/4" pipe size, outlet		23	.696		4.20	25.50	3.51	33.21	47	
	5675	3/8" pipe size, outlet		23	.696		4.20	25.50	3.51	33.21	47	
	5676	1/2" pipe size, outlet		22	.727		4.20	27	3.67	34.87	48.50	
	5677	3/4" pipe size, outlet		21	.762		4.93	28	3.84	36.77	51.50	
	5678	1" pipe size, outlet		20	.800		5.65	29.50	4.03	39.18	54.50	
	5679	1-1/4" pipe size, outlet		18	.889		8.40	32.50	4.48	45.38	63	
	5680	1-1/2" pipe size, outlet	↓	16	1	↓	8.75	37	5.05	50.80	70.50	

Important: See the Reference Section for critical supporting data - Reference Nos., Crews, & City Cost Indexes

15107 | Metal Pipe & Fittings

		Description	Crew	Daily Output	Labor-Hours	Unit	2005 Bare Costs Mat.	Labor	Equip.	Total	Total Incl O&P	
660	5681	2" pipe size, outlet	Q-15	12	1.333	Ea.	9.90	49	6.70	65.60	92	660
	6000	Weld-on flange, forged steel										
	6020	Slip-on, 150 lb. flange (welded front and back)										
	6050	1/2" pipe size	Q-15	18	.889	Ea.	10.75	32.50	4.48	47.73	66	
	6060	3/4" pipe size		18	.889		10.75	32.50	4.48	47.73	66	
	6070	1" pipe size		17	.941		10.75	34.50	4.74	49.99	69	
	6080	1-1/4" pipe size		16	1		10.75	37	5.05	52.80	73	
	6090	1-1/2" pipe size		15	1.067		10.75	39	5.40	55.15	77	
	6100	2" pipe size		12	1.333		10	49	6.70	65.70	92	
	6110	2-1/2" pipe size		10	1.600		15.60	59	8.05	82.65	115	
	6120	3" pipe size		9	1.778		19.15	65.50	8.95	93.60	129	
	6130	3-1/2" pipe size		7	2.286		25.50	84	11.50	121	167	
	6140	4" pipe size		6	2.667		24	98	13.45	135.45	188	
	6150	5" pipe size		5	3.200		28	118	16.15	162.15	226	
	6160	6" pipe size	Q-16	6	4		34	153	13.45	200.45	281	
	6170	8" pipe size		5	4.800		58	183	16.15	257.15	355	
	6180	10" pipe size		4	6		85.50	229	20	334.50	460	
	6190	12" pipe size		3	8		133	305	27	465	635	
	6191	14" pipe size		2.50	9.600		180	365	32.50	577.50	785	
	6192	16" pipe size		1.80	13.333		271	510	45	826	1,125	
	6200	300 lb. flange										
	6210	1/2" pipe size	Q-15	17	.941	Ea.	14.35	34.50	4.74	53.59	73	
	6220	3/4" pipe size		17	.941		14.35	34.50	4.74	53.59	73	
	6230	1" pipe size		16	1		14.35	37	5.05	56.40	77	
	6240	1-1/4" pipe size		13	1.231		14.35	45.50	6.20	66.05	90.50	
	6250	1-1/2" pipe size		12	1.333		14.35	49	6.70	70.05	96.50	
	6260	2" pipe size		11	1.455		15.45	53.50	7.35	76.30	106	
	6270	2-1/2" pipe size		9	1.778		22	65.50	8.95	96.45	132	
	6280	3" pipe size		7	2.286		30	84	11.50	125.50	172	
	6290	4" pipe size		6	2.667		43.50	98	13.45	154.95	210	
	6300	5" pipe size		4	4		56	147	20	223	305	
	6310	6" pipe size	Q-16	5	4.800		72	183	16.15	271.15	370	
	6320	8" pipe size		4	6		117	229	20	366	495	
	6330	10" pipe size		3.40	7.059		165	269	23.50	457.50	615	
	6340	12" pipe size		2.80	8.571		204	325	29	558	745	
	6400	Welding neck, 150 lb. flange										
	6410	1/2" pipe size	Q-15	40	.400	Ea.	15.40	14.70	2.02	32.12	41	
	6420	3/4" pipe size		36	.444		15.40	16.35	2.24	33.99	44	
	6430	1" pipe size		32	.500		15.40	18.40	2.52	36.32	47	
	6440	1-1/4" pipe size		29	.552		15.40	20.50	2.78	38.68	50.50	
	6450	1-1/2" pipe size		26	.615		15.40	22.50	3.10	41	54.50	
	6460	2" pipe size		20	.800		17.40	29.50	4.03	50.93	67.50	
	6470	2-1/2" pipe size		16	1		19.30	37	5.05	61.35	82.50	
	6480	3" pipe size		14	1.143		23	42	5.75	70.75	95	
	6500	4" pipe size		10	1.600		35	59	8.05	102.05	136	
	6510	5" pipe size		8	2		41	73.50	10.10	124.60	167	
	6520	6" pipe size	Q-16	10	2.400		50	91.50	8.05	149.55	202	
	6530	8" pipe size		7	3.429		82.50	131	11.50	225	300	
	6540	10" pipe size		6	4		97	153	13.45	263.45	350	
	6550	12" pipe size		5	4.800		175	183	16.15	374.15	485	
	6551	14" pipe size		4.50	5.333		244	203	17.90	464.90	595	
	6552	16" pipe size		3	8		289	305	27	621	810	
	6553	18" pipe size		2.50	9.600		445	365	32.50	842.50	1,075	
	6554	20" pipe size		2.30	10.435		535	400	35	970	1,225	
	6556	24" pipe size		2	12		725	460	40.50	1,225.50	1,525	
	6557	26" pipe size		1.70	14.118		830	540	47.50	1,417.50	1,775	

15107 | Metal Pipe & Fittings

	15107 Metal Pipe & Fittings	CREW	DAILY OUTPUT	LABOR-HOURS	UNIT	2005 BARE COSTS MAT.	LABOR	EQUIP.	TOTAL	TOTAL INCL O&P
660 6558	30" pipe size	Q-16	.90	26.667	Ea.	950	1,025	89.50	2,064.50	2,675
6559	36" pipe size		.75	32		1,100	1,225	108	2,433	3,150
6560	300 lb. flange									
6570	1/2" pipe size	Q-15	36	.444	Ea.	19	16.35	2.24	37.59	48
6580	3/4" pipe size		34	.471		19	17.30	2.37	38.67	49.50
6590	1" pipe size		30	.533		19	19.60	2.69	41.29	53.50
6600	1-1/4" pipe size		28	.571		19	21	2.88	42.88	55.50
6610	1-1/2" pipe size		24	.667		19	24.50	3.36	46.86	61.50
6620	2" pipe size		18	.889		20.50	32.50	4.48	57.48	76.50
6630	2-1/2" pipe size		14	1.143		27.50	42	5.75	75.25	99.50
6640	3" pipe size		12	1.333		35	49	6.70	90.70	119
6650	4" pipe size		8	2		53.50	73.50	10.10	137.10	181
6660	5" pipe size		7	2.286		65	84	11.50	160.50	210
6670	6" pipe size	Q-16	9	2.667		97.50	102	8.95	208.45	270
6680	8" pipe size		6	4		142	153	13.45	308.45	400
6690	10" pipe size		5	4.800		214	183	16.15	413.15	530
6700	12" pipe size		4	6		298	229	20	547	690
6710	14" pipe size		3.50	6.857		410	261	23	694	870
6720	16" pipe size		2	12		760	460	40.50	1,260.50	1,575
7740	Plain ends for plain end pipe, mechanically coupled									
7750	Cplg & labor required at joints not included add 1 per									
7760	joint for installed price, see line 9180									
7770	Malleable iron, painted, unless noted otherwise									
7800	90° Elbow 1"				Ea.	16.75			16.75	18.45
7810	1-1/2"					32			32	35.50
7820	2"					32			32	35.50
7830	2-1/2"					32			32	35.50
7840	3"					38.50			38.50	42.50
7860	4"					56			56	61.50
7870	5" welded steel					101			101	112
7880	6"					80.50			80.50	88.50
7890	8" welded steel					170			170	187
7900	10" welded steel					214			214	235
7910	12" welded steel					272			272	300
7970	45° Elbow 1"					16.75			16.75	18.45
7980	1-1/2"					32			32	35.50
7990	2"					32			32	35.50
8000	2-1/2"					32			32	35.50
8010	3"					38.50			38.50	42.50
8030	4"					56			56	61.50
8040	5" welded steel					101			101	112
8050	6"					80.50			80.50	88.50
8060	8"					93.50			93.50	103
8070	10" welded steel					131			131	144
8080	12" welded steel					169			169	186
8140	Tee, straight 1"					21			21	23
8150	1-1/2"					43			43	47.50
8160	2"					43			43	47.50
8170	2-1/2"					43			43	47.50
8180	3"					62			62	68.50
8200	4"					86.50			86.50	95
8210	5" welded steel					99			99	109
8220	6"					122			122	134
8230	8" welded steel					169			169	186
8240	10" welded steel					258			258	284
8250	12" welded steel					350			350	385

15107 | Metal Pipe & Fittings

			CREW	DAILY OUTPUT	LABOR-HOURS	UNIT	2005 BARE COSTS MAT.	LABOR	EQUIP.	TOTAL	TOTAL INCL O&P	
660	8340	Segmentally welded steel, painted										660
	8390	Wye 2"				Ea.	66			66	72.50	
	8400	2-1/2"					66			66	72.50	
	8410	3"					74			74	81.50	
	8430	4"					110			110	121	
	8440	5"					134			134	147	
	8450	6"					184			184	203	
	8460	8"					239			239	263	
	8470	10"					350			350	385	
	8480	12"					545			545	595	
	8540	Wye, lateral 2"					70.50			70.50	77.50	
	8550	2-1/2"					70.50			70.50	77.50	
	8560	3"					79			79	86.50	
	8580	4"					119			119	131	
	8590	5"					155			155	170	
	8600	6"					198			198	218	
	8610	8"					288			288	315	
	8620	10"					405			405	450	
	8630	12"					550			550	605	
	8690	Cross, 2"					75			75	82.50	
	8700	2-1/2"					75			75	82.50	
	8710	3"					89			89	98	
	8730	4"					122			122	134	
	8740	5"					174			174	191	
	8750	6"					229			229	252	
	8760	8"					294			294	325	
	8770	10"					430			430	470	
	8780	12"					620			620	680	
	8800	Tees, reducing 2" x 1"					44.50			44.50	49	
	8810	2" x 1-1/2"					44.50			44.50	49	
	8820	3" x 1"					44.50			44.50	49	
	8830	3" x 1-1/2"					44.50			44.50	49	
	8840	3" x 2"					44.50			44.50	49	
	8850	4" x 1"					65			65	71.50	
	8860	4" x 1-1/2"					65			65	71.50	
	8870	4" x 2"					65			65	71.50	
	8880	4" x 2-1/2"					66.50			66.50	73.50	
	8890	4" x 3"					66.50			66.50	73.50	
	8900	6" x 2"					108			108	119	
	8910	6" x 3"					114			114	125	
	8920	6" x 4"					114			114	125	
	8930	8" x 2"					139			139	152	
	8940	8" x 3"					148			148	163	
	8950	8" x 4"					148			148	163	
	8960	8" x 5"					152			152	167	
	8970	8" x 6"					152			152	167	
	8980	10" x 4"					218			218	240	
	8990	10" x 6"					231			231	254	
	9000	10" x 8"					231			231	254	
	9010	12" x 6"					335			335	370	
	9020	12" x 8"					335			335	370	
	9030	12" x 10"					350			350	385	
	9080	Adapter nipples 3" long										
	9090	1"				Ea.	5.85			5.85	6.40	
	9100	1-1/2"					5.85			5.85	6.40	
	9110	2"					5.85			5.85	6.40	

15107 | Metal Pipe & Fittings

			CREW	DAILY OUTPUT	LABOR-HOURS	UNIT	MAT.	LABOR	EQUIP.	TOTAL	TOTAL INCL O&P	
660	9120	2-1/2"				Ea.	6.80			6.80	7.50	660
	9130	3"					8.45			8.45	9.30	
	9140	4"					14			14	15.40	
	9150	6"					38			38	41.50	
	9180	Coupling, mechanical, plain end pipe to plain end pipe or fitting										
	9190	1"	Q-1	29	.552	Ea.	21.50	20.50		42	54	
	9200	1-1/2"		28	.571		21.50	21		42.50	55	
	9210	2"		27	.593		21.50	22		43.50	56.50	
	9220	2-1/2"		26	.615		21.50	22.50		44	57.50	
	9230	3"		25	.640		32	23.50		55.50	70.50	
	9240	3-1/2"		24	.667		37	24.50		61.50	77.50	
	9250	4"		22	.727		37	27		64	80.50	
	9260	5"	Q-2	28	.857		52.50	32.50		85	107	
	9270	6"		24	1		64	38		102	128	
	9280	8"		19	1.263		114	48		162	198	
	9290	10"		16	1.500		148	57		205	249	
	9300	12"		12	2		186	76.50		262.50	320	
	9310	Outlets for precut holes through pipe wall										
	9331	Strapless type, with gasket										
	9332	4" to 8" pipe x 1/2"	1 Plum	13	.615	Ea.	21	25		46	61	
	9333	4" to 8" pipe x 3/4"		13	.615		22.50	25		47.50	63	
	9334	10" pipe and larger x 1/2"		11	.727		21	29.50		50.50	67.50	
	9335	10" pipe and larger x 3/4"		11	.727		22.50	29.50		52	69.50	
	9341	Thermometer wells with gasket										
	9342	4" to 8" pipe, 6" stem	1 Plum	14	.571	Ea.	32	23.50		55.50	70.50	
	9343	8" pipe and larger, 6" stem	"	13	.615	"	32	25		57	73.50	
	9400	Mechanical joint ends for plain end pipe										
	9410	Malleable iron, black										
	9420	90° Elbows, 1-1/4"	Q-1	29	.552	Ea.	9.90	20.50		30.40	41.50	
	9430	1-1/2"		27	.593		11.30	22		33.30	45.50	
	9440	2"		24	.667		12.40	24.50		36.90	50.50	
	9490	Tee, reducing outlet										
	9510	1-1/4" x 1/2"	Q-1	18	.889	Ea.	5.90	32.50		38.40	55.50	
	9520	1-1/4" x 3/4"		18	.889		5.90	32.50		38.40	55.50	
	9530	1-1/4" x 1"		18	.889		5.90	32.50		38.40	55.50	
	9540	1-1/2" x 1/2"		17	.941		6.25	34.50		40.75	59	
	9550	1-1/2" x 3/4"		17	.941		6.25	34.50		40.75	59	
	9560	1-1/2" x 1"		17	.941		6.25	34.50		40.75	59	
	9570	2" x 1/2"		15	1.067		7.65	39		46.65	67.50	
	9580	2" x 3/4"		15	1.067		7.65	39		46.65	67.50	
	9590	2" x 1"		15	1.067		7.65	39		46.65	67.50	
	9640	Tee, reducing run and outlet										
	9660	1-1/4" x 1" x 1/2"	Q-1	18	.889	Ea.	6.30	32.50		38.80	56	
	9670	1-1/4" x 1" x 3/4"		18	.889		6.30	32.50		38.80	56	
	9680	1-1/4" x 1" x 1"		18	.889		6.30	32.50		38.80	56	
	9690	1-1/2" x 1-1/4" x 1/2"		17	.941		6.70	34.50		41.20	59.50	
	9700	1-1/2" x 1-1/4" x 3/4"		17	.941		6.70	34.50		41.20	59.50	
	9710	1-1/2" x 1-1/4" x 1"		17	.941		6.70	34.50		41.20	59.50	
	9720	2" x 1-1/2" x 1/2"		15	1.067		8.10	39		47.10	68	
	9730	2" x 1-1/2" x 3/4"		15	1.067		8.10	39		47.10	68	
	9740	2" x 1-1/2" x 1"		15	1.067		8.10	39		47.10	68	
	9790	Tee, outlet										
	9810	3" x 1-1/4"	Q-1	29	.552	Ea.	14.25	20.50		34.75	46	
	9820	3" x 1-1/2"		28	.571		14.25	21		35.25	47	
	9830	3" x 2"		26	.615		15.90	22.50		38.40	51.50	
	9840	4" x 1-1/4"		28	.571		17.70	21		38.70	51	

(Column group: 2005 BARE COSTS = MAT., LABOR, EQUIP., TOTAL)

	15107 | Metal Pipe & Fittings		CREW	DAILY OUTPUT	LABOR-HOURS	UNIT	2005 BARE COSTS MAT.	LABOR	EQUIP.	TOTAL	TOTAL INCL O&P	
660	9850	4″ x 1-1/2″	Q-1	26	.615	Ea.	18.60	22.50		41.10	54.50	660
	9860	4″ x 2″	↓	24	.667	↓	18.60	24.50		43.10	57.50	
	9940	For galvanized fittings for plain end pipe, add					20%					
690	0010	**PIPE, GROOVED-JOINT STEEL FITTINGS & VALVES**										690
	0012	Fittings are ductile iron. Steel fittings noted.										
	0020	Pipe includes coupling & clevis type hanger 10′ O.C.										
	0500	Schedule 10, black										
	0550	2″ diameter	1 Plum	43	.186	L.F.	4.10	7.60		11.70	15.90	
	0560	2-1/2″ diameter	Q-1	61	.262		5.10	9.65		14.75	20	
	0570	3″ diameter		55	.291		5.90	10.70		16.60	22.50	
	0580	3-1/2″ diameter		53	.302		7.30	11.10		18.40	25	
	0590	4″ diameter		49	.327		7.05	12		19.05	26	
	0600	5″ diameter	↓	40	.400		13.45	14.70		28.15	37	
	0610	6″ diameter	Q-2	46	.522		13.75	19.90		33.65	45	
	0620	8″ diameter	″	41	.585	↓	27	22.50		49.50	63	
	0700	To delete couplings & hangers, subtract										
	0710	2″ diam. to 5″ diam.					25%	20%				
	0720	6″ diam. to 8″ diam.					27%	15%				
	1000	Schedule 40, black										
	1040	3/4″ diameter	1 Plum	71	.113	L.F.	1.98	4.60		6.58	9.10	
	1050	1″ diameter		63	.127		1.85	5.20		7.05	9.85	
	1060	1-1/4″ diameter		58	.138		2.37	5.65		8.02	11.05	
	1070	1-1/2″ diameter		51	.157		2.21	6.40		8.61	12.10	
	1080	2″ diameter	↓	40	.200		3.30	8.15		11.45	15.95	
	1090	2-1/2″ diameter	Q-1	57	.281		4.80	10.30		15.10	21	
	1100	3″ diameter		50	.320		6	11.75		17.75	24.50	
	1110	4″ diameter		45	.356		10.95	13.10		24.05	31.50	
	1120	5″ diameter	↓	37	.432		19.95	15.90		35.85	46	
	1130	6″ diameter	Q-2	42	.571		25.50	22		47.50	61	
	1140	8″ diameter		37	.649		38	24.50		62.50	79	
	1150	10″ diameter		31	.774		58.50	29.50		88	109	
	1160	12″ diameter		27	.889		79.50	34		113.50	139	
	1170	14″ diameter		20	1.200		91	46		137	169	
	1180	16″ diameter		17	1.412		99.50	54		153.50	191	
	1190	18″ diameter		14	1.714		160	65.50		225.50	275	
	1200	20″ diameter		12	2		190	76.50		266.50	325	
	1210	24″ diameter	↓	10	2.400	↓	224	91.50		315.50	385	
	1740	To delete coupling & hanger, subtract										
	1750	3/4″ diam. to 2″ diam.					65%	27%				
	1760	2-1/2″ diam. to 5″ diam.					41%	18%				
	1770	6″ diam. to 12″ diam.					31%	13%				
	1780	14″ diam. to 24″ diam.					35%	10%				
	1800	Galvanized										
	1840	3/4″ diameter	1 Plum	71	.113	L.F.	2.19	4.60		6.79	9.30	
	1850	1″ diameter		63	.127		2.13	5.20		7.33	10.15	
	1860	1-1/4″ diameter		58	.138		2.76	5.65		8.41	11.50	
	1870	1-1/2″ diameter		51	.157		3.18	6.40		9.58	13.15	
	1880	2″ diameter	↓	40	.200		3.91	8.15		12.06	16.60	
	1890	2-1/2″ diameter	Q-1	57	.281		5.75	10.30		16.05	22	
	1900	3″ diameter		50	.320		7.25	11.75		19	25.50	
	1910	4″ diameter		45	.356		10	13.10		23.10	30.50	
	1920	5″ diameter	↓	37	.432		26	15.90		41.90	52.50	
	1930	6″ diameter	Q-2	42	.571		33.50	22		55.50	70	
	1940	8″ diameter		37	.649		49.50	24.50		74	91.50	
	1950	10″ diameter	↓	31	.774	↓	75	29.50		104.50	127	

15107 | Metal Pipe & Fittings

			CREW	DAILY OUTPUT	LABOR-HOURS	UNIT	2005 BARE COSTS MAT.	LABOR	EQUIP.	TOTAL	TOTAL INCL O&P	
690	1960	12" diameter	Q-2	27	.889	L.F.	103	34		137	164	690
	2540	To delete coupling & hanger, subtract										
	2550	3/4" diam. to 2" diam.					36%	27%				
	2560	2-1/2" diam. to 5" diam.					19%	18%				
	2570	6" diam. to 12" diam.					14%	13%				
	2600	Schedule 80, black										
	2610	3/4" diameter	1 Plum	65	.123	L.F.	3.99	5.05		9.04	11.95	
	2650	1" diameter		61	.131		3.95	5.35		9.30	12.40	
	2660	1-1/4" diameter		55	.145		5.30	5.95		11.25	14.80	
	2670	1-1/2" diameter		49	.163		6.45	6.65		13.10	17.05	
	2680	2" diameter	↓	38	.211		7.45	8.60		16.05	21	
	2690	2-1/2" diameter	Q-1	54	.296		9.50	10.90		20.40	27	
	2700	3" diameter		48	.333		12.35	12.25		24.60	32	
	2710	4" diameter		44	.364		24	13.35		37.35	46	
	2720	5" diameter	↓	35	.457		27	16.80		43.80	55	
	2730	6" diameter	Q-2	40	.600		44.50	23		67.50	83.50	
	2740	8" diameter		35	.686		86.50	26		112.50	135	
	2750	10" diameter		29	.828		113	31.50		144.50	172	
	2760	12" diameter	↓	24	1	↓	151	38		189	224	
	3240	To delete coupling & hanger, subtract										
	3250	3/4" diam. to 2" diam.					30%	25%				
	3260	2-1/2" diam. to 5" diam.					14%	17%				
	3270	6" diam. to 12" diam.					12%	12%				
	3300	Galvanized										
	3310	3/4" diameter	1 Plum	65	.123	L.F.	5.20	5.05		10.25	13.25	
	3350	1" diameter		61	.131		6.95	5.35		12.30	15.70	
	3360	1-1/4" diameter		55	.145		8.25	5.95		14.20	18.05	
	3370	1-1/2" diameter		46	.174		12.30	7.10		19.40	24.50	
	3380	2" diameter	↓	38	.211		13.15	8.60		21.75	27.50	
	3390	2-1/2" diameter	Q-1	54	.296		14.95	10.90		25.85	33	
	3400	3" diameter		48	.333		18.95	12.25		31.20	39.50	
	3410	4" diameter		44	.364		29.50	13.35		42.85	52.50	
	3420	5" diameter	↓	35	.457		36.50	16.80		53.30	65.50	
	3430	6" diameter	Q-2	40	.600		56	23		79	96	
	3440	8" diameter		35	.686		93.50	26		119.50	143	
	3450	10" diameter		29	.828		110	31.50		141.50	169	
	3460	12" diameter	↓	24	1	↓	147	38		185	220	
	3920	To delete coupling & hanger, subtract										
	3930	3/4" diam. to 2" diam.					30%	25%				
	3940	2-1/2" diam. to 5" diam.					15%	17%				
	3950	6" diam. to 12" diam.					11%	12%				
	3990	Fittings: cplg. & labor required at joints not incl. in fitting										
	3994	price. Add 1 per joint for installed price.										
	4000	Elbow, 90° or 45°, painted										
	4030	3/4" diameter	1 Plum	50	.160	Ea.	19.10	6.55		25.65	31	
	4040	1" diameter		50	.160		10.15	6.55		16.70	21	
	4050	1-1/4" diameter		40	.200		10.15	8.15		18.30	23.50	
	4060	1-1/2" diameter		33	.242		10.15	9.90		20.05	26	
	4070	2" diameter	↓	25	.320		10.15	13.05		23.20	31	
	4080	2-1/2" diameter	Q-1	40	.400		10.15	14.70		24.85	33	
	4090	3" diameter		33	.485		18	17.85		35.85	47	
	4100	4" diameter		25	.640		19.60	23.50		43.10	57	
	4110	5" diameter	↓	20	.800		47	29.50		76.50	96	
	4120	6" diameter	Q-2	25	.960		55.50	36.50		92	116	
	4130	8" diameter		21	1.143		116	43.50		159.50	194	
	4140	10" diameter	↓	18	1.333	↓	211	51		262	310	

15107 | Metal Pipe & Fittings

			CREW	DAILY OUTPUT	LABOR-HOURS	UNIT	2005 BARE COSTS MAT.	LABOR	EQUIP.	TOTAL	TOTAL INCL O&P	
690	4150	12" diameter	Q-2	15	1.600	Ea.	340	61		401	460	690
	4170	14" diameter		12	2		605	76.50		681.50	780	
	4180	16" diameter		11	2.182		785	83		868	990	
	4190	18" diameter	Q-3	15	2.133		1,000	83		1,083	1,225	
	4200	20" diameter		13	2.462		1,325	96		1,421	1,600	
	4210	24" diameter		11	2.909		1,900	113		2,013	2,275	
	4250	For galvanized elbows, add					26%					
	4690	Tee, painted										
	4700	3/4" diameter	1 Plum	38	.211	Ea.	20.50	8.60		29.10	35.50	
	4740	1" diameter		33	.242		15.60	9.90		25.50	32	
	4750	1-1/4" diameter		27	.296		15.60	12.10		27.70	35.50	
	4760	1-1/2" diameter		22	.364		15.60	14.85		30.45	39.50	
	4770	2" diameter		17	.471		15.60	19.20		34.80	46	
	4780	2-1/2" diameter	Q-1	27	.593		15.60	22		37.60	50	
	4790	3" diameter		22	.727		22	27		49	64	
	4800	4" diameter		17	.941		33	34.50		67.50	88.50	
	4810	5" diameter		13	1.231		76	45.50		121.50	152	
	4820	6" diameter	Q-2	17	1.412		90	54		144	180	
	4830	8" diameter		14	1.714		198	65.50		263.50	315	
	4840	10" diameter		12	2		410	76.50		486.50	570	
	4850	12" diameter		10	2.400		575	91.50		666.50	770	
	4851	14" diameter		9	2.667		610	102		712	825	
	4852	16" diameter		8	3		685	114		799	925	
	4853	18" diameter	Q-3	11	2.909		860	113		973	1,125	
	4854	20" diameter		10	3.200		1,225	125		1,350	1,525	
	4855	24" diameter		8	4		1,875	156		2,031	2,300	
	4900	For galvanized tees, add					24%					
	4906	Couplings, rigid style, painted										
	4908	1" diameter	1 Plum	100	.080	Ea.	7.45	3.27		10.72	13.10	
	4909	1-1/4" diameter		100	.080		7.45	3.27		10.72	13.10	
	4910	1-1/2" diameter		67	.119		7.45	4.88		12.33	15.55	
	4912	2" diameter		50	.160		7.65	6.55		14.20	18.25	
	4914	2-1/2" diameter	Q-1	80	.200		8.80	7.35		16.15	21	
	4916	3" diameter		67	.239		10.30	8.80		19.10	24.50	
	4918	4" diameter		50	.320		14.60	11.75		26.35	34	
	4920	5" diameter		40	.400		18.90	14.70		33.60	43	
	4922	6" diameter	Q-2	50	.480		25.50	18.30		43.80	55.50	
	4924	8" diameter		42	.571		40	22		62	76.50	
	4926	10" diameter		35	.686		71	26		97	118	
	4928	12" diameter		32	.750		79.50	28.50		108	131	
	4930	14" diameter		24	1		104	38		142	172	
	4931	16" diameter		20	1.200		136	46		182	218	
	4932	18" diameter		18	1.333		157	51		208	250	
	4933	20" diameter		16	1.500		179	57		236	283	
	4934	24" diameter	Q-9	13	1.231		275	44.50		319.50	370	
	4940	Flexible, standard, painted										
	4950	3/4" diameter	1 Plum	100	.080	Ea.	5.40	3.27		8.67	10.80	
	4960	1" diameter		100	.080		5.40	3.27		8.67	10.80	
	4970	1-1/4" diameter		80	.100		7.15	4.09		11.24	14	
	4980	1-1/2" diameter		67	.119		7.80	4.88		12.68	15.95	
	4990	2" diameter		50	.160		8.25	6.55		14.80	18.90	
	5000	2-1/2" diameter	Q-1	80	.200		9.85	7.35		17.20	22	
	5010	3" diameter		67	.239		10.90	8.80		19.70	25	
	5020	3-1/2" diameter		57	.281		15.85	10.30		26.15	33	
	5030	4" diameter		50	.320		12.95	11.75		24.70	32	
	5040	5" diameter		40	.400		24.50	14.70		39.20	48.50	

15107 | Metal Pipe & Fittings

			CREW	DAILY OUTPUT	LABOR-HOURS	UNIT	2005 BARE COSTS MAT.	LABOR	EQUIP.	TOTAL	TOTAL INCL O&P
690	5050	6″ diameter	Q-2	50	.480	Ea.	28.50	18.30		46.80	59
	5070	8″ diameter		42	.571		46.50	22		68.50	84.50
	5090	10″ diameter		35	.686		77.50	26		103.50	125
	5110	12″ diameter		32	.750		88.50	28.50		117	140
	5120	14″ diameter		24	1		107	38		145	176
	5130	16″ diameter		20	1.200		141	46		187	224
	5140	18″ diameter		18	1.333		165	51		216	258
	5150	20″ diameter		16	1.500		259	57		316	370
	5160	24″ diameter		13	1.846		276	70.50		346.50	410
	5176	Lightweight style, painted									
	5178	1-1/2″ diameter	1 Plum	67	.119	Ea.	6.80	4.88		11.68	14.85
	5180	2″ diameter	″	50	.160		6.95	6.55		13.50	17.45
	5182	2-1/2″ diameter	Q-1	80	.200		8.10	7.35		15.45	19.95
	5184	3″ diameter		67	.239		9.35	8.80		18.15	23.50
	5186	3-1/2″ diameter		57	.281		13.30	10.30		23.60	30
	5188	4″ diameter		50	.320		13.30	11.75		25.05	32.50
	5190	5″ diameter		40	.400		19.10	14.70		33.80	43
	5192	6″ diameter	Q-2	50	.480		23	18.30		41.30	53
	5194	8″ diameter		42	.571		36	22		58	73
	5196	10″ diameter		35	.686		97	26		123	147
	5198	12″ diameter		32	.750		108	28.50		136.50	162
	5200	For galvanized couplings, add					33%				
	5220	Tee, reducing, painted									
	5225	2″ x 1-1/2″ diameter	Q-1	38	.421	Ea.	33.50	15.50		49	60.50
	5226	2-1/2″ x 2″ diameter		28	.571		33.50	21		54.50	68.50
	5227	3″ x 2-1/2″ diameter		23	.696		29.50	25.50		55	71
	5228	4″ x 3″ diameter		18	.889		40	32.50		72.50	93
	5229	5″ x 4″ diameter		15	1.067		86	39		125	154
	5230	6″ x 4″ diameter	Q-2	18	1.333		95	51		146	181
	5231	8″ x 6″ diameter		15	1.600		198	61		259	310
	5232	10″ x 8″ diameter		13	1.846		260	70.50		330.50	390
	5233	12″ x 10″ diameter		11	2.182		395	83		478	560
	5234	14″ x 12″ diameter		10	2.400		445	91.50		536.50	625
	5235	16″ x 12″ diameter		9	2.667		550	102		652	760
	5236	18″ x 12″ diameter	Q-3	12	2.667		660	104		764	880
	5237	18″ x 16″ diameter		11	2.909		830	113		943	1,075
	5238	20″ x 16″ diameter		10	3.200		1,075	125		1,200	1,375
	5239	24″ x 20″ diameter		9	3.556		1,725	138		1,863	2,075
	5240	Reducer, concentric, painted									
	5241	2-1/2″ x 2″ diameter	Q-1	43	.372	Ea.	11.80	13.70		25.50	33.50
	5242	3″ x 2-1/2″ diameter		35	.457		14.25	16.80		31.05	41
	5243	4″ x 3″ diameter		29	.552		17.20	20.50		37.70	49.50
	5244	5″ x 4″ diameter		22	.727		24	27		51	66.50
	5245	6″ x 4″ diameter	Q-2	26	.923		28	35		63	83.50
	5246	8″ x 6″ diameter		23	1.043		72.50	40		112.50	140
	5247	10″ x 8″ diameter		20	1.200		148	46		194	232
	5248	12″ x 10″ diameter		16	1.500		266	57		323	380
	5255	Eccentric, painted									
	5256	2-1/2″ x 2″ diameter	Q-1	42	.381	Ea.	25	14		39	48.50
	5257	3″ x 2-1/2″ diameter		34	.471		28.50	17.30		45.80	57.50
	5258	4″ x 3″ diameter		28	.571		35	21		56	70
	5259	5″ x 4″ diameter		21	.762		47.50	28		75.50	94.50
	5260	6″ x 4″ diameter	Q-2	25	.960		55.50	36.50		92	116
	5261	8″ x 6″ diameter		22	1.091		112	41.50		153.50	186
	5262	10″ x 8″ diameter		19	1.263		310	48		358	415
	5263	12″ x 10″ diameter		15	1.600		425	61		486	555

15107 | Metal Pipe & Fittings

		Description	Crew	Daily Output	Labor-Hours	Unit	2005 Bare Costs Mat.	Labor	Equip.	Total	Total Incl O&P	
690	5270	Coupling, reducing, painted										690
	5272	2" x 1-1/2" diameter	1 Plum	52	.154	Ea.	11.75	6.30		18.05	22.50	
	5274	2-1/2" x 2" diameter	Q-1	82	.195		15.35	7.20		22.55	27.50	
	5276	3" x 2" diameter		69	.232		17.60	8.55		26.15	32	
	5278	4" x 2" diameter		52	.308		28	11.30		39.30	48	
	5280	5" x 4" diameter		42	.381		31.50	14		45.50	56	
	5282	6" x 4" diameter	Q-2	52	.462		48	17.60		65.60	79	
	5284	8" x 6" diameter	"	44	.545		72	21		93	111	
	5290	Outlet coupling, painted										
	5294	1-1/2" x 1" pipe size	1 Plum	65	.123	Ea.	15	5.05		20.05	24	
	5296	2" x 1" pipe size	"	48	.167		15.40	6.80		22.20	27	
	5298	2-1/2" x 1" pipe size	Q-1	78	.205		24	7.55		31.55	38	
	5300	2-1/2" x 1" pipe size	1 Plum	70	.114		27	4.67		31.67	37	
	5302	3" x 1" pipe size	Q-1	65	.246		23	9.05		32.05	39	
	5304	4" x 3/4" pipe size		48	.333		34	12.25		46.25	56	
	5306	4" x 1-1/2" pipe size		46	.348		48.50	12.80		61.30	72.50	
	5308	6" x 1-1/2" pipe size	Q-2	44	.545		68.50	21		89.50	107	
	5750	Flange, w/groove gasket, black steel										
	5752	See 15107-660-0620 for gasket & bolt set										
	5760	ANSI class 125 and 150, painted										
	5780	2" pipe size	1 Plum	23	.348	Ea.	34	14.20		48.20	59	
	5790	2-1/2" pipe size	Q-1	37	.432		42.50	15.90		58.40	70.50	
	5800	3" pipe size		31	.516		45.50	19		64.50	79	
	5820	4" pipe size		23	.696		61.50	25.50		87	106	
	5830	5" pipe size		19	.842		71	31		102	125	
	5840	6" pipe size	Q-2	23	1.043		77.50	40		117.50	146	
	5850	8" pipe size		17	1.412		87.50	54		141.50	178	
	5860	10" pipe size		14	1.714		139	65.50		204.50	252	
	5870	12" pipe size		12	2		182	76.50		258.50	315	
	5880	14" pipe size		10	2.400		415	91.50		506.50	600	
	5890	16" pipe size		9	2.667		485	102		587	690	
	5900	18" pipe size		6	4		595	153		748	885	
	5910	20" pipe size		5	4.800		720	183		903	1,075	
	5920	24" pipe size		4.50	5.333		920	203		1,123	1,300	
	5940	ANSI class 350, painted										
	5946	2" pipe size	1 Plum	23	.348	Ea.	42.50	14.20		56.70	68.50	
	5948	2-1/2" pipe size	Q-1	37	.432		49.50	15.90		65.40	78.50	
	5950	3" pipe size		31	.516		67.50	19		86.50	103	
	5952	4" pipe size		23	.696		90.50	25.50		116	138	
	5954	5" pipe size		19	.842		103	31		134	160	
	5956	6" pipe size	Q-2	23	1.043		119	40		159	191	
	5958	8" pipe size		17	1.412		137	54		191	232	
	5960	10" pipe size		14	1.714		219	65.50		284.50	340	
	5962	12" pipe size	1 Plum	12	.667		233	27		260	298	
	6100	Coupling, for PVC plastic pipe										
	6110	2" diameter	1 Plum	50	.160	Ea.	9.20	6.55		15.75	19.90	
	6112	2-1/2" diameter	Q-1	80	.200		12.85	7.35		20.20	25	
	6114	3" diameter		67	.239		15.60	8.80		24.40	30.50	
	6116	4" diameter		50	.320		20.50	11.75		32.25	40	
	6118	6" diameter	Q-2	50	.480		33.50	18.30		51.80	64.50	
	6120	8" diameter		42	.571		55.50	22		77.50	94	
	6122	10" diameter		35	.686		91	26		117	140	
	6124	12" diameter		32	.750		115	28.50		143.50	170	
	7400	Suction diffuser										
	7402	Grooved end inlet x flanged outlet										
	7410	3" x 3"	Q-1	50	.320	Ea.	355	11.75		366.75	410	

15107 | Metal Pipe & Fittings

	15107 Metal Pipe & Fittings	CREW	DAILY OUTPUT	LABOR-HOURS	UNIT	2005 BARE COSTS MAT.	LABOR	EQUIP.	TOTAL	TOTAL INCL O&P	
690	7412 4" x 4"	Q-1	38	.421	Ea.	480	15.50		495.50	555	690
	7414 5" x 5"		30	.533		560	19.60		579.60	650	
	7416 6" x 6"	Q-2	38	.632		710	24		734	815	
	7418 8" x 8"		27	.889		1,325	34		1,359	1,500	
	7420 10" x 10"		20	1.200		1,800	46		1,846	2,050	
	7422 12" x 12"		16	1.500		2,950	57		3,007	3,325	
	7424 14" x 14"		15	1.600		3,725	61		3,786	4,200	
	7426 16" x 14"		14	1.714		3,825	65.50		3,890.50	4,325	
	7500 Strainer, tee type, painted										
	7506 2" pipe size	1 Plum	38	.211	Ea.	221	8.60		229.60	256	
	7508 2-1/2" pipe size	Q-1	62	.258		232	9.50		241.50	269	
	7510 3" pipe size		50	.320		260	11.75		271.75	305	
	7512 4" pipe size		38	.421		295	15.50		310.50	350	
	7514 5" pipe size		30	.533		425	19.60		444.60	500	
	7516 6" pipe size	Q-2	38	.632		465	24		489	545	
	7518 8" pipe size		27	.889		710	34		744	830	
	7520 10" pipe size		20	1.200		1,050	46		1,096	1,225	
	7522 12" pipe size		16	1.500		1,350	57		1,407	1,550	
	7524 14" pipe size		15	1.600		4,775	61		4,836	5,375	
	7526 16" pipe size		14	1.714		5,950	65.50		6,015.50	6,650	
	7570 Expansion joint, max. 3" travel										
	7572 2" diameter	1 Plum	38	.211	Ea.	181	8.60		189.60	212	
	7574 3" diameter	Q-1	50	.320		213	11.75		224.75	252	
	7576 4" diameter	"	38	.421		266	15.50		281.50	315	
	7578 6" diameter	Q-2	38	.632		465	24		489	545	
	7800 Ball valve w/handle, carbon steel trim										
	7810 1-1/2" pipe size	1 Plum	50	.160	Ea.	63	6.55		69.55	79	
	7812 2" pipe size	"	38	.211		69.50	8.60		78.10	89.50	
	7814 2-1/2" pipe size	Q-1	62	.258		152	9.50		161.50	181	
	7816 3" pipe size		50	.320		246	11.75		257.75	289	
	7818 4" pipe size		38	.421		380	15.50		395.50	445	
	7820 6" pipe size	Q-2	30	.800		1,175	30.50		1,205.50	1,350	
	7830 With gear operator										
	7834 2-1/2" pipe size	Q-1	62	.258	Ea.	310	9.50		319.50	355	
	7836 3" pipe size		50	.320		435	11.75		446.75	500	
	7838 4" pipe size		38	.421		555	15.50		570.50	640	
	7840 6" pipe size	Q-2	30	.800		1,325	30.50		1,355.50	1,500	
	7870 Check valve										
	7874 2-1/2" pipe size	Q-1	62	.258	Ea.	109	9.50		118.50	134	
	7876 3" pipe size		50	.320		128	11.75		139.75	159	
	7878 4" pipe size		38	.421		136	15.50		151.50	173	
	7880 5" pipe size		30	.533		226	19.60		245.60	279	
	7882 6" pipe size	Q-2	38	.632		268	24		292	330	
	7884 8" pipe size		27	.889		365	34		399	455	
	7886 10" pipe size		20	1.200		1,050	46		1,096	1,250	
	7888 12" pipe size		16	1.500		1,250	57		1,307	1,450	
	7900 Plug valve, balancing, w/lever operator										
	7906 3" pipe size	Q-1	50	.320	Ea.	245	11.75		256.75	288	
	7908 4" pipe size	"	38	.421		300	15.50		315.50	355	
	7909 6" pipe size	Q-2	30	.800		465	30.50		495.50	555	
	7916 With gear operator										
	7920 3" pipe size	Q-1	50	.320	Ea.	445	11.75		456.75	510	
	7922 4" pipe size	"	38	.421		460	15.50		475.50	530	
	7924 6" pipe size	Q-2	38	.632		620	24		644	715	
	7926 8" pipe size		27	.889		840	34		874	975	
	7928 10" pipe size		20	1.200		1,300	46		1,346	1,500	

	15107 | Metal Pipe & Fittings		CREW	DAILY OUTPUT	LABOR-HOURS	UNIT	2005 BARE COSTS MAT.	LABOR	EQUIP.	TOTAL	TOTAL INCL O&P	
690	7930	12″ pipe size	Q-2	16	1.500	Ea.	1,975	57		2,032	2,250	690
	8000	Butterfly valve, 2 position handle, with standard trim										
	8010	1-1/2″ pipe size	1 Plum	50	.160	Ea.	96	6.55		102.55	116	
	8020	2″ pipe size	″	38	.211		96	8.60		104.60	119	
	8030	3″ pipe size	Q-1	50	.320		138	11.75		149.75	170	
	8050	4″ pipe size	″	38	.421		171	15.50		186.50	212	
	8070	6″ pipe size	Q-2	38	.632		305	24		329	370	
	8080	8″ pipe size		27	.889		435	34		469	525	
	8090	10″ pipe size		20	1.200		625	46		671	760	
	8200	With stainless steel trim										
	8240	1-1/2″ pipe size	1 Plum	50	.160	Ea.	122	6.55		128.55	144	
	8250	2″ pipe size	″	38	.211		122	8.60		130.60	147	
	8270	3″ pipe size	Q-1	50	.320		164	11.75		175.75	198	
	8280	4″ pipe size	″	38	.421		178	15.50		193.50	220	
	8300	6″ pipe size	Q-2	38	.632		330	24		354	400	
	8310	8″ pipe size		27	.889		505	34		539	610	
	8320	10″ pipe size		20	1.200		725	46		771	870	
	8322	12″ pipe size		16	1.500		1,150	57		1,207	1,325	
	8324	14″ pipe size		15	1.600		1,500	61		1,561	1,750	
	8326	16″ pipe size		14	1.714		2,050	65.50		2,115.50	2,375	
	8328	18″ pipe size	Q-3	12	2.667		2,525	104		2,629	2,950	
	8330	20″ pipe size		11	2.909		3,425	113		3,538	3,950	
	8332	24″ pipe size		10	3.200		4,375	125		4,500	4,975	
	8336	Note: sizes 8″ up w/manual gear operator										
	9000	Cut one groove, labor										
	9010	3/4″ pipe size	Q-1	152	.105	Ea.		3.87		3.87	5.80	
	9020	1″ pipe size		140	.114			4.20		4.20	6.30	
	9030	1-1/4″ pipe size		124	.129			4.75		4.75	7.15	
	9040	1-1/2″ pipe size		114	.140			5.15		5.15	7.75	
	9050	2″ pipe size		104	.154			5.65		5.65	8.50	
	9060	2-1/2″ pipe size		96	.167			6.15		6.15	9.20	
	9070	3″ pipe size		88	.182			6.70		6.70	10.05	
	9080	3-1/2″ pipe size		83	.193			7.10		7.10	10.65	
	9090	4″ pipe size		78	.205			7.55		7.55	11.35	
	9100	5″ pipe size		72	.222			8.15		8.15	12.30	
	9110	6″ pipe size		70	.229			8.40		8.40	12.65	
	9120	8″ pipe size		54	.296			10.90		10.90	16.40	
	9130	10″ pipe size		38	.421			15.50		15.50	23.50	
	9140	12″ pipe size		30	.533			19.60		19.60	29.50	
	9150	14″ pipe size		20	.800			29.50		29.50	44	
	9160	16″ pipe size		19	.842			31		31	46.50	
	9170	18″ pipe size		18	.889			32.50		32.50	49	
	9180	20″ pipe size		17	.941			34.50		34.50	52	
	9190	24″ pipe size		15	1.067			39		39	59	
	9210	Roll one groove										
	9220	3/4″ pipe size	Q-1	266	.060	Ea.		2.21		2.21	3.33	
	9230	1″ pipe size		228	.070			2.58		2.58	3.88	
	9240	1-1/4″ pipe size		200	.080			2.94		2.94	4.42	
	9250	1-1/2″ pipe size		178	.090			3.31		3.31	4.97	
	9260	2″ pipe size		116	.138			5.05		5.05	7.60	
	9270	2-1/2″ pipe size		110	.145			5.35		5.35	8.05	
	9280	3″ pipe size		100	.160			5.90		5.90	8.85	
	9290	3-1/2″ pipe size		94	.170			6.25		6.25	9.40	
	9300	4″ pipe size		86	.186			6.85		6.85	10.30	
	9310	5″ pipe size		84	.190			7		7	10.55	
	9320	6″ pipe size		80	.200			7.35		7.35	11.05	

15107 | Metal Pipe & Fittings

		15107 Metal Pipe & Fittings	CREW	DAILY OUTPUT	LABOR-HOURS	UNIT	2005 BARE COSTS MAT.	LABOR	EQUIP.	TOTAL	TOTAL INCL O&P	
690	9330	8″ pipe size	Q-1	66	.242	Ea.		8.90		8.90	13.40	690
	9340	10″ pipe size		58	.276			10.15		10.15	15.25	
	9350	12″ pipe size		46	.348			12.80		12.80	19.25	
	9360	14″ pipe size		30	.533			19.60		19.60	29.50	
	9370	16″ pipe size		28	.571			21		21	31.50	
	9380	18″ pipe size		27	.593			22		22	33	
	9390	20″ pipe size		25	.640			23.50		23.50	35.50	
	9400	24″ pipe size		23	.696			25.50		25.50	38.50	
920	0010	**PIPE, STAINLESS STEEL**										920
	0020	Welded, with clevis type hangers 10′ O.C.										
	0500	Schedule 5, type 304										
	0540	1/2″ diameter	Q-15	128	.125	L.F.	4.89	4.60	.63	10.12	12.95	
	0550	3/4″ diameter		116	.138		5.60	5.05	.70	11.35	14.50	
	0560	1″ diameter		103	.155		6.90	5.70	.78	13.38	17.05	
	0570	1-1/4″ diameter		93	.172		7.95	6.35	.87	15.17	19.20	
	0580	1-1/2″ diameter		85	.188		9.20	6.90	.95	17.05	21.50	
	0590	2″ diameter		69	.232		11.10	8.55	1.17	20.82	26.50	
	0600	2-1/2″ diameter		53	.302		15.60	11.10	1.52	28.22	35.50	
	0610	3″ diameter		48	.333		18.85	12.25	1.68	32.78	41.50	
	0620	4″ diameter		44	.364		24	13.35	1.83	39.18	48.50	
	0630	5″ diameter		36	.444		49	16.35	2.24	67.59	81	
	0640	6″ diameter	Q-16	42	.571		45	22	1.92	68.92	84.50	
	0650	8″ diameter		34	.706		69.50	27	2.37	98.87	120	
	0660	10″ diameter		26	.923		98.50	35	3.10	136.60	164	
	0670	12″ diameter		21	1.143		130	43.50	3.84	177.34	213	
	0700	To delete hangers, subtract										
	0710	1/2″ diam. to 1-1/2″ diam.					8%	19%				
	0720	2″ diam. to 5″ diam.					4%	9%				
	0730	6″ diam. to 12″ diam.					3%	4%				
	0750	For small quantities, add				L.F.	10%					
	1250	Schedule 5, type 316										
	1290	1/2″ diameter	Q-15	128	.125	L.F.	6.25	4.60	.63	11.48	14.50	
	1300	3/4″ diameter		116	.138		7.30	5.05	.70	13.05	16.40	
	1310	1″ diameter		103	.155		8.35	5.70	.78	14.83	18.65	
	1320	1-1/4″ diameter		93	.172		10.10	6.35	.87	17.32	21.50	
	1330	1-1/2″ diameter		85	.188		11.50	6.90	.95	19.35	24	
	1340	2″ diameter		69	.232		14.05	8.55	1.17	23.77	29.50	
	1350	2-1/2″ diameter		53	.302		20.50	11.10	1.52	33.12	41	
	1360	3″ diameter		48	.333		26	12.25	1.68	39.93	49	
	1370	4″ diameter		44	.364		32	13.35	1.83	47.18	57	
	1380	5″ diameter		36	.444		61	16.35	2.24	79.59	94.50	
	1390	6″ diameter	Q-16	42	.571		59.50	22	1.92	83.42	101	
	1400	8″ diameter		34	.706		90	27	2.37	119.37	142	
	1410	10″ diameter		26	.923		131	35	3.10	169.10	200	
	1420	12″ diameter		21	1.143		168	43.50	3.84	215.34	255	
	1490	For small quantities, add					10%					
	1940	To delete hanger, subtract										
	1950	1/2″ diam. to 1-1/2″ diam.					5%	19%				
	1960	2″ diam. to 5″ diam.					3%	9%				
	1970	6″ diam. to 12″ diam.					2%	4%				
	2000	Schedule 10, type 304										
	2040	1/4″ diameter	Q-15	131	.122	L.F.	3.78	4.49	.62	8.89	11.60	
	2050	3/8″ diameter		128	.125		4.01	4.60	.63	9.24	12	
	2060	1/2″ diameter		125	.128		5	4.71	.65	10.36	13.30	
	2070	3/4″ diameter		113	.142		5.85	5.20	.71	11.76	15.10	
	2080	1″ diameter		100	.160		8.60	5.90	.81	15.31	19.20	

15107 | Metal Pipe & Fittings

15107	Metal Pipe & Fittings	CREW	DAILY OUTPUT	LABOR-HOURS	UNIT	2005 BARE COSTS MAT.	LABOR	EQUIP.	TOTAL	TOTAL INCL O&P
920 2090	1-1/4" diameter	Q-15	91	.176	L.F.	10.45	6.45	.89	17.79	22
2100	1-1/2" diameter		83	.193		11.50	7.10	.97	19.57	24.50
2110	2" diameter		67	.239		14.35	8.80	1.20	24.35	30.50
2120	2-1/2" diameter		51	.314		17.85	11.55	1.58	30.98	38.50
2130	3" diameter		46	.348		22	12.80	1.75	36.55	45
2140	4" diameter		42	.381		29	14	1.92	44.92	54.50
2150	5" diameter		35	.457		39	16.80	2.30	58.10	70.50
2160	6" diameter	Q-16	40	.600		45	23	2.02	70.02	86
2170	8" diameter		33	.727		69.50	27.50	2.44	99.44	121
2180	10" diameter		25	.960		98.50	36.50	3.23	138.23	167
2190	12" diameter		21	1.143		123	43.50	3.84	170.34	206
2250	For small quantities, add					10%				
2650	To delete hanger, subtract									
2660	1/4" diam. to 3/4" diam.					9%	22%			
2670	1" diam. to 2" diam.					4%	15%			
2680	2-1/2" diam. to 5" diam.					3%	8%			
2690	6" diam. to 12" diam.					3%	4%			
2750	Schedule 10, type 316									
2790	1/4" diameter	Q-15	131	.122	L.F.	4.30	4.49	.62	9.41	12.15
2800	3/8" diameter		128	.125		4.83	4.60	.63	10.06	12.90
2810	1/2" diameter		125	.128		6.25	4.71	.65	11.61	14.65
2820	3/4" diameter		113	.142		7.35	5.20	.71	13.26	16.70
2830	1" diameter		100	.160		9.75	5.90	.81	16.46	20.50
2840	1-1/4" diameter		91	.176		13.25	6.45	.89	20.59	25
2850	1-1/2" diameter		83	.193		14.85	7.10	.97	22.92	28
2860	2" diameter		67	.239		18.40	8.80	1.20	28.40	34.50
2870	2-1/2" diameter		51	.314		24	11.55	1.58	37.13	45.50
2880	3" diameter		46	.348		30.50	12.80	1.75	45.05	54.50
2890	4" diameter		42	.381		37	14	1.92	52.92	63.50
2900	5" diameter		35	.457		55	16.80	2.30	74.10	88.50
2910	6" diameter	Q-16	40	.600		59.50	23	2.02	84.52	102
2920	8" diameter		33	.727		98	27.50	2.44	127.94	152
2930	10" diameter		25	.960		131	36.50	3.23	170.73	203
2940	12" diameter		21	1.143		164	43.50	3.84	211.34	250
2990	For small quantities, add					10%				
3430	To delete hanger, subtract									
3440	1/4" diam. to 3/4" diam.					6%	22%			
3450	1" diam. to 2" diam.					3%	15%			
3460	2-1/2" diam. to 5" diam.					2%	8%			
3470	6" diam. to 12" diam.					2%	4%			
3500	Threaded, couplings and hangers 10' O.C.									
3520	Schedule 40, type 304									
3540	1/4" diameter	1 Plum	54	.148	L.F.	5.25	6.05		11.30	14.90
3550	3/8" diameter		53	.151		6.10	6.15		12.25	15.95
3560	1/2" diameter		52	.154		7.20	6.30		13.50	17.35
3570	3/4" diameter		51	.157		9.60	6.40		16	20
3580	1" diameter		45	.178		12.05	7.25		19.30	24
3590	1-1/4" diameter	Q-1	76	.211		15.60	7.75		23.35	29
3600	1-1/2" diameter		69	.232		18.25	8.55		26.80	33
3610	2" diameter		57	.281		24.50	10.30		34.80	42.50
3620	2-1/2" diameter		44	.364		42	13.35		55.35	66
3630	3" diameter		38	.421		55	15.50		70.50	84
3640	4" diameter	Q-2	51	.471		76.50	17.95		94.45	111
3740	For small quantities, add					10%				
4200	To delete couplings & hangers, subtract									
4210	1/4" diam. to 3/4" diam.					15%	56%			

15107 | Metal Pipe & Fittings

			Crew	Daily Output	Labor-Hours	Unit	2005 Bare Costs Mat.	Labor	Equip.	Total	Total Incl O&P	
920	4220	1" diam. to 2" diam.					18%	49%				920
	4230	2-1/2" diam. to 4" diam.					34%	40%				
	4250	Schedule 40, type 316										
	4290	1/4" diameter	1 Plum	54	.148	L.F.	6.15	6.05		12.20	15.85	
	4300	3/8" diameter		53	.151		7.35	6.15		13.50	17.35	
	4310	1/2" diameter		52	.154		9.50	6.30		15.80	19.95	
	4320	3/4" diameter		51	.157		11.80	6.40		18.20	22.50	
	4330	1" diameter		45	.178		15.45	7.25		22.70	28	
	4340	1-1/4" diameter	Q-1	76	.211		19.90	7.75		27.65	33.50	
	4350	1-1/2" diameter		69	.232		23.50	8.55		32.05	38.50	
	4360	2" diameter		57	.281		31.50	10.30		41.80	50	
	4370	2-1/2" diameter		44	.364		54	13.35		67.35	79.50	
	4380	3" diameter		38	.421		73.50	15.50		89	104	
	4390	4" diameter	Q-2	51	.471		100	17.95		117.95	137	
	4490	For small quantities, add					10%					
	4900	To delete couplings & hangers, subtract										
	4910	1/4" diam. to 3/4" diam.					12%	56%				
	4920	1" diam. to 2" diam.					14%	49%				
	4930	2-1/2" diam. to 4" diam.					27%	40%				
	5000	Schedule 80, type 304										
	5040	1/4" diameter	1 Plum	53	.151	L.F.	7.05	6.15		13.20	17.05	
	5050	3/8" diameter		52	.154		8.10	6.30		14.40	18.35	
	5060	1/2" diameter		51	.157		10.15	6.40		16.55	21	
	5070	3/4" diameter		48	.167		13.20	6.80		20	25	
	5080	1" diameter		43	.186		18.30	7.60		25.90	31.50	
	5090	1-1/4" diameter	Q-1	73	.219		27	8.05		35.05	41.50	
	5100	1-1/2" diameter		67	.239		31	8.80		39.80	47	
	5110	2" diameter		54	.296		39.50	10.90		50.40	60	
	5190	For small quantities, add					10%					
	5700	To delete couplings & hangers, subtract										
	5710	1/4" diam. to 3/4" diam.					10%	53%				
	5720	1" diam. to 2" diam.					14%	47%				
	5750	Schedule 80, type 316										
	5790	1/4" diameter	1 Plum	53	.151	L.F.	8.75	6.15		14.90	18.90	
	5800	3/8" diameter		52	.154		10	6.30		16.30	20.50	
	5810	1/2" diameter		51	.157		14.45	6.40		20.85	25.50	
	5820	3/4" diameter		48	.167		16.55	6.80		23.35	28.50	
	5830	1" diameter		43	.186		23	7.60		30.60	37	
	5840	1-1/4" diameter	Q-1	73	.219		34.50	8.05		42.55	50	
	5850	1-1/2" diameter		67	.239		39.50	8.80		48.30	56.50	
	5860	2" diameter		54	.296		50.50	10.90		61.40	72.50	
	5950	For small quantities, add					10%					
	7000	To delete couplings & hangers, subtract										
	7010	1/4" diam. to 3/4" diam.					9%	53%				
	7020	1" diam. to 2" diam.					14%	47%				
	8000	Weld joints with clevis type hangers 10' O.C.										
	8010	Schedule 40, type 304										
	8050	1/8" pipe size	Q-15	126	.127	L.F.	4.35	4.67	.64	9.66	12.50	
	8060	1/4" pipe size		125	.128		4.46	4.71	.65	9.82	12.70	
	8070	3/8" pipe size		122	.131		5.10	4.82	.66	10.58	13.60	
	8080	1/2" pipe size		118	.136		6.25	4.99	.68	11.92	15.10	
	8090	3/4" pipe size		109	.147		7.90	5.40	.74	14.04	17.60	
	8100	1" pipe size		95	.168		10	6.20	.85	17.05	21	
	8110	1-1/4" pipe size		86	.186		12.40	6.85	.94	20.19	25	
	8120	1-1/2" pipe size		78	.205		14.50	7.55	1.03	23.08	28.50	
	8130	2" pipe size		62	.258		18.75	9.50	1.30	29.55	36	

15107 | Metal Pipe & Fittings

			CREW	DAILY OUTPUT	LABOR-HOURS	UNIT	2005 BARE COSTS MAT.	LABOR	EQUIP.	TOTAL	TOTAL INCL O&P	
920	8140	2-1/2" pipe size	Q-15	49	.327	L.F.	28.50	12	1.65	42.15	51	920
	8150	3" pipe size		44	.364		36.50	13.35	1.83	51.68	62	
	8160	3-1/2" pipe size		44	.364		43	13.35	1.83	58.18	69	
	8170	4" pipe size		39	.410		49	15.10	2.07	66.17	79	
	8180	5" pipe size	↓	32	.500		71.50	18.40	2.52	92.42	109	
	8190	6" pipe size	Q-16	37	.649		85.50	24.50	2.18	112.18	133	
	8200	8" pipe size		29	.828		138	31.50	2.78	172.28	203	
	8210	10" pipe size		24	1		236	38	3.36	277.36	320	
	8220	12" pipe size	↓	20	1.200	↓	267	46	4.03	317.03	365	
	8300	Schedule 40, type 316										
	8310	1/8" pipe size	Q-15	126	.127	L.F.	4.31	4.67	.64	9.62	12.45	
	8320	1/4" pipe size		125	.128		5.15	4.71	.65	10.51	13.45	
	8330	3/8" pipe size		122	.131		6.20	4.82	.66	11.68	14.85	
	8340	1/2" pipe size		118	.136		8.35	4.99	.68	14.02	17.45	
	8350	3/4" pipe size		109	.147		9.75	5.40	.74	15.89	19.60	
	8360	1" pipe size		95	.168		12.95	6.20	.85	20	24.50	
	8370	1-1/4" pipe size		86	.186		16.05	6.85	.94	23.84	29	
	8380	1-1/2" pipe size		78	.205		18.80	7.55	1.03	27.38	33	
	8390	2" pipe size		62	.258		24.50	9.50	1.30	35.30	42.50	
	8400	2-1/2" pipe size		49	.327		37.50	12	1.65	51.15	61.50	
	8410	3" pipe size		44	.364		51	13.35	1.83	66.18	78	
	8420	3-1/2" pipe size		44	.364		59	13.35	1.83	74.18	87	
	8430	4" pipe size		39	.410		67	15.10	2.07	84.17	99	
	8440	5" pipe size	↓	32	.500		102	18.40	2.52	122.92	142	
	8450	6" pipe size	Q-16	37	.649		114	24.50	2.18	140.68	164	
	8460	8" pipe size		29	.828		213	31.50	2.78	247.28	286	
	8470	10" pipe size		24	1		287	38	3.36	328.36	375	
	8480	12" pipe size	↓	20	1.200	↓	390	46	4.03	440.03	505	
	8500	Schedule 80, type 304										
	8510	1/4" pipe size	Q-15	110	.145	L.F.	6.15	5.35	.73	12.23	15.60	
	8520	3/8" pipe size		109	.147		7	5.40	.74	13.14	16.60	
	8530	1/2" pipe size		106	.151		8.90	5.55	.76	15.21	18.95	
	8540	3/4" pipe size		96	.167		11.50	6.15	.84	18.49	23	
	8550	1" pipe size		87	.184		15.45	6.75	.93	23.13	28	
	8560	1-1/4" pipe size		81	.198		20	7.25	1	28.25	34	
	8570	1-1/2" pipe size		74	.216		23	7.95	1.09	32.04	38	
	8580	2" pipe size		58	.276		28.50	10.15	1.39	40.04	48.50	
	8590	2-1/2" pipe size		46	.348		59	12.80	1.75	73.55	86	
	8600	3" pipe size		41	.390		79.50	14.35	1.97	95.82	111	
	8610	4" pipe size	↓	33	.485		124	17.85	2.44	144.29	167	
	8630	6" pipe size	Q-16	30	.800	↓	274	30.50	2.69	307.19	350	
	8640	Schedule 80, type 316										
	8650	1/4" pipe size	Q-15	110	.145	L.F.	7.80	5.35	.73	13.88	17.40	
	8660	3/8" pipe size		109	.147		8.95	5.40	.74	15.09	18.70	
	8670	1/2" pipe size		106	.151		12.95	5.55	.76	19.26	23.50	
	8680	3/4" pipe size		96	.167		14.35	6.15	.84	21.34	26	
	8690	1" pipe size		87	.184		19.30	6.75	.93	26.98	32	
	8700	1-1/4" pipe size		81	.198		26.50	7.25	1	34.75	41	
	8710	1-1/2" pipe size		74	.216		30	7.95	1.09	39.04	46	
	8720	2" pipe size		58	.276		37.50	10.15	1.39	49.04	58	
	8730	2-1/2" pipe size		46	.348		97.50	12.80	1.75	112.05	128	
	8740	3" pipe size		41	.390		114	14.35	1.97	130.32	149	
	8760	4" pipe size	↓	33	.485		158	17.85	2.44	178.29	204	
	8770	6" pipe size	Q-16	30	.800	↓	425	30.50	2.69	458.19	520	
	8790	Schedule 160, type 304										
	8800	1/2" pipe size	Q-15	96	.167	L.F.	12.70	6.15	.84	19.69	24	

15107 | Metal Pipe & Fittings

	Line	Description	CREW	DAILY OUTPUT	LABOR-HOURS	UNIT	2005 BARE COSTS MAT.	LABOR	EQUIP.	TOTAL	TOTAL INCL O&P	
920	8810	3/4" pipe size	Q-15	88	.182	L.F.	14.95	6.70	.92	22.57	27.50	920
	8820	1" pipe size		80	.200		25.50	7.35	1.01	33.86	40	
	8840	1-1/2" pipe size		67	.239		35	8.80	1.20	45	53.50	
	8850	2" pipe size		53	.302		51	11.10	1.52	63.62	74.50	
	8870	3" pipe size		37	.432		83.50	15.90	2.18	101.58	118	
	8900	Schedule 160, type 316										
	8910	1/2" pipe size	Q-15	96	.167	L.F.	19	6.15	.84	25.99	31	
	8920	3/4" pipe size		88	.182		20	6.70	.92	27.62	33	
	8930	1" pipe size		80	.200		28.50	7.35	1.01	36.86	43.50	
	8940	1-1/4" pipe size		73	.219		41	8.05	1.10	50.15	58.50	
	8950	1-1/2" pipe size		67	.239		47	8.80	1.20	57	66	
	8960	2" pipe size		53	.302		65	11.10	1.52	77.62	90	
	8990	4" pipe size		30	.533		164	19.60	2.69	186.29	213	
	9100	Threading pipe labor, sst, one end, schedules 40 & 80										
	9110	1/4" through 3/4" pipe size	1 Plum	61.50	.130	Ea.		5.30		5.30	8	
	9120	1" through 2" pipe size		55.90	.143			5.85		5.85	8.80	
	9130	2-1/2" pipe size		41.50	.193			7.85		7.85	11.85	
	9140	3" pipe size		38.50	.208			8.50		8.50	12.75	
	9150	3-1/2" pipe size	Q-1	68.40	.234			8.60		8.60	12.95	
	9160	4" pipe size		73	.219			8.05		8.05	12.10	
	9170	5" pipe size		40.70	.393			14.45		14.45	21.50	
	9180	6" pipe size		35.40	.452			16.60		16.60	25	
	9190	8" pipe size		22.30	.717			26.50		26.50	39.50	
	9200	10" pipe size		16.10	.994			36.50		36.50	55	
	9210	12" pipe size		12.30	1.301			48		48	72	
	9250	Welding labor per joint for stainless steel										
	9260	Schedule 5 and 10										
	9270	1/4" pipe size	Q-15	36	.444	Ea.		16.35	2.24	18.59	27	
	9280	3/8" pipe size		35	.457			16.80	2.30	19.10	28	
	9290	1/2" pipe size		35	.457			16.80	2.30	19.10	28	
	9300	3/4" pipe size		28	.571			21	2.88	23.88	34.50	
	9310	1" pipe size		25	.640			23.50	3.23	26.73	39	
	9320	1-1/4" pipe size		22	.727			27	3.67	30.67	44	
	9330	1-1/2" pipe size		21	.762			28	3.84	31.84	46	
	9340	2" pipe size		18	.889			32.50	4.48	36.98	54	
	9350	2-1/2" pipe size		12	1.333			49	6.70	55.70	81	
	9360	3" pipe size		9.73	1.644			60.50	8.30	68.80	100	
	9370	4" pipe size		7.37	2.171			80	10.95	90.95	132	
	9380	5" pipe size		6.15	2.602			95.50	13.10	108.60	158	
	9390	6" pipe size		5.71	2.802			103	14.10	117.10	171	
	9400	8" pipe size		3.69	4.336			159	22	181	264	
	9410	10" pipe size		2.91	5.498			202	27.50	229.50	335	
	9420	12" pipe size		2.31	6.926			255	35	290	425	
	9500	Schedule 40										
	9510	1/4" pipe size	Q-15	28	.571	Ea.		21	2.88	23.88	34.50	
	9520	3/8" pipe size		27	.593			22	2.99	24.99	36.50	
	9530	1/2" pipe size		25.40	.630			23	3.17	26.17	38.50	
	9540	3/4" pipe size		22.22	.720			26.50	3.63	30.13	44	
	9550	1" pipe size		20.25	.790			29	3.98	32.98	48	
	9560	1-1/4" pipe size		18.82	.850			31.50	4.28	35.78	51.50	
	9570	1-1/2" pipe size		17.78	.900			33	4.54	37.54	55	
	9580	2" pipe size		15.09	1.060			39	5.35	44.35	64.50	
	9590	2-1/2" pipe size		7.96	2.010			74	10.15	84.15	122	
	9600	3" pipe size		6.43	2.488			91.50	12.55	104.05	152	
	9610	4" pipe size		4.88	3.279			121	16.50	137.50	199	
	9620	5" pipe size		4.26	3.756			138	18.95	156.95	229	

	15107	Metal Pipe & Fittings	CREW	DAILY OUTPUT	LABOR-HOURS	UNIT	2005 BARE COSTS MAT.	LABOR	EQUIP.	TOTAL	TOTAL INCL O&P
920	9630	6″ pipe size	Q-15	3.77	4.244	Ea.		156	21.50	177.50	259
	9640	8″ pipe size		2.44	6.557			241	33	274	395
	9650	10″ pipe size		1.92	8.333			305	42	347	505
	9660	12″ pipe size		1.52	10.526			385	53	438	640
	9750	Schedule 80									
	9760	1/4″ pipe size	Q-15	21.55	.742	Ea.		27.50	3.74	31.24	45
	9770	3/8″ pipe size		20.75	.771			28.50	3.89	32.39	47
	9780	1/2″ pipe size		19.54	.819			30	4.13	34.13	50
	9790	3/4″ pipe size		17.09	.936			34.50	4.72	39.22	57
	9800	1″ pipe size		15.58	1.027			38	5.20	43.20	62.50
	9810	1-1/4″ pipe size		14.48	1.105			40.50	5.55	46.05	67
	9820	1-1/2″ pipe size		13.68	1.170			43	5.90	48.90	71
	9830	2″ pipe size		11.61	1.378			50.50	6.95	57.45	83.50
	9840	2-1/2″ pipe size		6.12	2.614			96	13.20	109.20	160
	9850	3″ pipe size		4.94	3.239			119	16.30	135.30	197
	9860	4″ pipe size		3.75	4.267			157	21.50	178.50	260
	9870	5″ pipe size		3.27	4.893			180	24.50	204.50	297
	9880	6″ pipe size		2.90	5.517			203	28	231	335
	9890	8″ pipe size		1.87	8.556			315	43	358	525
	9900	10″ pipe size		1.48	10.811			400	54.50	454.50	660
	9910	12″ pipe size		1.17	13.675			505	69	574	830
	9920	Schedule 160, 1/2″ pipe size		17	.941			34.50	4.74	39.24	57
	9930	3/4″ pipe size		14.81	1.080			39.50	5.45	44.95	65.50
	9940	1″ pipe size		13.50	1.185			43.50	5.95	49.45	72
	9950	1-1/4″ pipe size		12.55	1.275			47	6.45	53.45	77.50
	9960	1-1/2″ pipe size		11.85	1.350			49.50	6.80	56.30	82
	9970	2″ pipe size		10	1.600			59	8.05	67.05	97.50
	9980	3″ pipe size		4.28	3.738			138	18.85	156.85	228
	9990	4″ pipe size		3.25	4.923			181	25	206	300
960	0010	**PIPE, STAINLESS STEEL, FITTINGS**									
	0100	Butt weld joint, schedule 5, type 304									
	0120	90° Elbow, long									
	0140	1/2″	Q-15	17.50	.914	Ea.	11.05	33.50	4.61	49.16	67.50
	0150	3/4″		14	1.143		17.35	42	5.75	65.10	88.50
	0160	1″		12.50	1.280		12.70	47	6.45	66.15	92
	0170	1-1/4″		11	1.455		20.50	53.50	7.35	81.35	111
	0180	1-1/2″		10.50	1.524		15.90	56	7.70	79.60	110
	0190	2″		9	1.778		21	65.50	8.95	95.45	131
	0200	2-1/2″		6	2.667		30.50	98	13.45	141.95	195
	0210	3″		4.86	3.292		38.50	121	16.60	176.10	242
	0220	3-1/2″		4.27	3.747		203	138	18.90	359.90	450
	0230	4″		3.69	4.336		71	159	22	252	340
	0240	5″		3.08	5.195		505	191	26	722	875
	0250	6″	Q-16	4.29	5.594		203	213	18.80	434.80	565
	0260	8″		2.76	8.696		345	330	29	704	910
	0270	10″		2.18	11.009		710	420	37	1,167	1,450
	0280	12″		1.73	13.873		1,125	530	46.50	1,701.50	2,075
	0320	For schedule 5, type 316, add					30%				
	0600	45° Elbow, long									
	0620	1/2″	Q-15	17.50	.914	Ea.	15.90	33.50	4.61	54.01	73
	0630	3/4″		14	1.143		15.90	42	5.75	63.65	87
	0640	1″		12.50	1.280		17.95	47	6.45	71.40	98
	0650	1-1/4″		11	1.455		30.50	53.50	7.35	91.35	122
	0660	1-1/2″		10.50	1.524		16.35	56	7.70	80.05	110
	0670	2″		9	1.778		18.75	65.50	8.95	93.20	129

15107 | Metal Pipe & Fittings		CREW	DAILY OUTPUT	LABOR-HOURS	UNIT	2005 BARE COSTS MAT.	LABOR	EQUIP.	TOTAL	TOTAL INCL O&P
960 0680	2-1/2"	Q-15	6	2.667	Ea.	31	98	13.45	142.45	196 960
0690	3"		4.86	3.292		32	121	16.60	169.60	235
0700	3-1/2"		4.27	3.747		203	138	18.90	359.90	450
0710	4"		3.69	4.336		56.50	159	22	237.50	325
0720	5"		3.08	5.195		355	191	26	572	705
0730	6"	Q-16	4.29	5.594		122	213	18.80	353.80	475
0740	8"		2.76	8.696		212	330	29	571	765
0750	10"		2.18	11.009		475	420	37	932	1,200
0760	12"		1.73	13.873		760	530	46.50	1,336.50	1,675
0800	For schedule 5, type 316, add					25%				
1100	Tee, straight									
1130	1/2"	Q-15	11.66	1.372	Ea.	44.50	50.50	6.90	101.90	133
1140	3/4"		9.33	1.715		44.50	63	8.65	116.15	154
1150	1"		8.33	1.921		42.50	70.50	9.70	122.70	164
1160	1-1/4"		7.33	2.183		73	80.50	11	164.50	213
1170	1-1/2"		7	2.286		42.50	84	11.50	138	186
1180	2"		6	2.667		44.50	98	13.45	155.95	211
1190	2-1/2"		4	4		127	147	20	294	385
1200	3"		3.24	4.938		83	182	25	290	390
1210	3-1/2"		2.85	5.614		222	206	28.50	456.50	585
1220	4"		2.46	6.504		118	239	33	390	525
1230	5"		2	8		475	294	40.50	809.50	1,000
1240	6"	Q-16	2.85	8.421		355	320	28.50	703.50	905
1250	8"		1.84	13.043		810	495	44	1,349	1,700
1260	10"		1.45	16.552		1,275	630	55.50	1,960.50	2,400
1270	12"		1.15	20.870		1,600	795	70	2,465	3,050
1320	For schedule 5, type 316, add					25%				
2000	Butt weld joint, schedule 10, type 304									
2020	90° elbow, long									
2040	1/2"	Q-15	17	.941	Ea.	10.40	34.50	4.74	49.64	68.50
2050	3/4"		14	1.143		10.40	42	5.75	58.15	81
2060	1"		12.50	1.280		10.40	47	6.45	63.85	89.50
2070	1-1/4"		11	1.455		19.45	53.50	7.35	80.30	110
2080	1-1/2"		10.50	1.524		15.15	56	7.70	78.85	109
2090	2"		9	1.778		20.50	65.50	8.95	94.95	131
2100	2-1/2"		6	2.667		45	98	13.45	156.45	211
2110	3"		4.86	3.292		41.50	121	16.60	179.10	246
2120	3-1/2"		4.27	3.747		176	138	18.90	332.90	420
2130	4"		3.69	4.336		52.50	159	22	233.50	320
2140	5"		3.08	5.195		128	191	26	345	455
2150	6"	Q-16	4.29	5.594		110	213	18.80	341.80	460
2160	8"		2.76	8.696		137	330	29	496	685
2170	10"		2.18	11.009		395	420	37	852	1,100
2180	12"		1.73	13.873		495	530	46.50	1,071.50	1,400
2500	45° elbow, long									
2520	1/2"	Q-15	17.50	.914	Ea.	5.25	33.50	4.61	43.36	61.50
2530	3/4"		14	1.143		5.25	42	5.75	53	75
2540	1"		12.50	1.280		6.75	47	6.45	60.20	85.50
2550	1-1/4"		11	1.455		10.50	53.50	7.35	71.35	100
2560	1-1/2"		10.50	1.524		15.15	56	7.70	78.85	109
2570	2"		9	1.778		17.40	65.50	8.95	91.85	128
2580	2-1/2"		6	2.667		12	98	13.45	123.45	175
2590	3"		4.86	3.292		13.50	121	16.60	151.10	215
2600	3-1/2"		4.27	3.747		67	138	18.90	223.90	300
2610	4"		3.69	4.336		21	159	22	202	287
2620	5"		3.08	5.195		104	191	26	321	430

Important: See the Reference Section for critical supporting data - Reference Nos., Crews, & City Cost Indexes

	15107 | Metal Pipe & Fittings	CREW	DAILY OUTPUT	LABOR-HOURS	UNIT	2005 BARE COSTS MAT.	LABOR	EQUIP.	TOTAL	TOTAL INCL O&P	
960	2630	6"	Q-16	4.29	5.594	Ea.	46.50	213	18.80	278.30	390	960
	2640	8"		2.76	8.696		76.50	330	29	435.50	615	
	2650	10"		2.18	11.009		179	420	37	636	870	
	2660	12"		1.73	13.873		273	530	46.50	849.50	1,150	
	3000	Tee, straight										
	3030	1/2"	Q-15	11.66	1.372	Ea.	13.50	50.50	6.90	70.90	98.50	
	3040	3/4"		9.33	1.715		13.50	63	8.65	85.15	119	
	3050	1"		8.33	1.921		36.50	70.50	9.70	116.70	157	
	3060	1-1/4"		7.33	2.183		21	80.50	11	112.50	156	
	3070	1-1/2"		7	2.286		40	84	11.50	135.50	183	
	3080	2"		6	2.667		40.50	98	13.45	151.95	206	
	3090	2-1/2"		4	4		38.50	147	20	205.50	285	
	3100	3"		3.24	4.938		27	182	25	234	330	
	3110	3-1/2"		2.85	5.614		40	206	28.50	274.50	385	
	3120	4"		2.46	6.504		111	239	33	383	520	
	3130	5"		2	8		128	294	40.50	462.50	625	
	3140	6"	Q-16	2.85	8.421		107	320	28.50	455.50	635	
	3150	8"		1.84	13.043		233	495	44	772	1,050	
	3151	10"		1.45	16.552		385	630	55.50	1,070.50	1,425	
	3152	12"		1.15	20.870		535	795	70	1,400	1,850	
	3154	For schedule 10, type 316, add					25%					
	3281	Butt weld joint, schedule 40, type 304										
	3284	90° Elbow, long, 1/2"	Q-15	12.70	1.260	Ea.	5.25	46.50	6.35	58.10	82.50	
	3288	3/4"		11.10	1.441		5.25	53	7.25	65.50	93.50	
	3289	1"		10.13	1.579		6	58	7.95	71.95	103	
	3290	1-1/4"		9.40	1.702		8.25	62.50	8.60	79.35	113	
	3300	1-1/2"		8.89	1.800		6.75	66	9.05	81.80	117	
	3310	2"		7.55	2.119		9.75	78	10.70	98.45	140	
	3320	2-1/2"		3.98	4.020		19.50	148	20.50	188	266	
	3330	3"		3.21	4.984		25	183	25	233	330	
	3340	3-1/2"		2.83	5.654		104	208	28.50	340.50	460	
	3350	4"		2.44	6.557		46.50	241	33	320.50	450	
	3360	5"		2.13	7.512		164	276	38	478	635	
	3370	6"	Q-16	2.83	8.481		145	325	28.50	498.50	675	
	3380	8"		1.83	13.115		244	500	44	788	1,075	
	3390	10"		1.44	16.667		510	635	56	1,201	1,575	
	3400	12"		1.14	21.053		720	805	70.50	1,595.50	2,075	
	3410	For schedule 40, type 316, add					25%					
	3460	45° Elbow, long, 1/2"	Q-15	12.70	1.260	Ea.	6.75	46.50	6.35	59.60	84	
	3470	3/4"		11.10	1.441		6.75	53	7.25	67	95	
	3480	1"		10.13	1.579		6.75	58	7.95	72.70	104	
	3490	1-1/4"		9.40	1.702		10.50	62.50	8.60	81.60	115	
	3500	1-1/2"		8.89	1.800		8.25	66	9.05	83.30	119	
	3510	2"		7.55	2.119		15.75	78	10.70	104.45	146	
	3520	2-1/2"		3.98	4.020		19.50	148	20.50	188	266	
	3530	3"		3.21	4.984		18	183	25	226	325	
	3540	3-1/2"		2.83	5.654		106	208	28.50	342.50	465	
	3550	4"		2.44	6.557		33	241	33	307	435	
	3560	5"		2.13	7.512		114	276	38	428	580	
	3570	6"	Q-16	2.83	8.481		85.50	325	28.50	439	610	
	3580	8"		1.83	13.115		162	500	44	706	975	
	3590	10"		1.44	16.667		305	635	56	996	1,350	
	3600	12"		1.14	21.053		415	805	70.50	1,290.50	1,725	
	3610	For schedule 40, type 316, add					25%					
	3660	Tee, straight 1/2"	Q-15	8.46	1.891	Ea.	14.25	69.50	9.55	93.30	131	
	3670	3/4"		7.40	2.162		14.25	79.50	10.90	104.65	148	

15107 | Metal Pipe & Fittings

			CREW	DAILY OUTPUT	LABOR-HOURS	UNIT	2005 BARE COSTS MAT.	LABOR	EQUIP.	TOTAL	TOTAL INCL O&P	
960	3680	1"	Q-15	6.74	2.374	Ea.	14.25	87.50	11.95	113.70	160	960
	3690	1-1/4"		6.27	2.552		24	94	12.85	130.85	182	
	3700	1-1/2"		5.92	2.703		14.25	99.50	13.60	127.35	180	
	3710	2"		5.03	3.181		15.75	117	16.05	148.80	211	
	3720	2-1/2"		2.65	6.038		40	222	30.50	292.50	410	
	3730	3"		2.14	7.477		32.50	275	37.50	345	490	
	3740	3-1/2"		1.88	8.511		25.50	315	43	383.50	545	
	3750	4"		1.62	9.877		25.50	365	50	440.50	630	
	3760	5"	↓	1.42	11.268		49	415	57	521	740	
	3770	6"	Q-16	1.88	12.766		170	485	43	698	965	
	3780	8"		1.22	19.672		340	750	66	1,156	1,575	
	3790	10"		.96	25		715	955	84	1,754	2,300	
	3800	12"	↓	.76	31.579	↓	2,525	1,200	106	3,831	4,700	
	3810	For schedule 40, type 316, add					25%					
	3820	Tee, reducing on outlet, 3/4" x 1/2"	Q-15	7.73	2.070	Ea.	55.50	76	10.45	141.95	187	
	3822	1" x 1/2"		7.24	2.210		77	81.50	11.15	169.65	219	
	3824	1" x 3/4"		6.96	2.299		64.50	84.50	11.60	160.60	210	
	3826	1-1/4" x 1"		6.43	2.488		203	91.50	12.55	307.05	375	
	3828	1-1/2" x 1/2"		6.58	2.432		107	89.50	12.25	208.75	265	
	3830	1-1/2" x 3/4"		6.35	2.520		107	92.50	12.70	212.20	271	
	3832	1-1/2" x 1"		6.18	2.589		77	95	13.05	185.05	242	
	3834	2" x 1"		5.50	2.909		75	107	14.65	196.65	260	
	3836	2" x 1-1/2"		5.30	3.019		70.50	111	15.20	196.70	262	
	3838	2-1/2" x 2"		3.15	5.079		182	187	25.50	394.50	510	
	3840	3" x 1-1/2"		2.72	5.882		155	216	29.50	400.50	530	
	3842	3" x 2"		2.65	6.038		131	222	30.50	383.50	515	
	3844	4" x 2"		2.10	7.619		236	280	38.50	554.50	720	
	3846	4" x 3"		1.77	9.040		193	330	45.50	568.50	760	
	3848	5" x 4"	↓	1.48	10.811		610	400	54.50	1,064.50	1,325	
	3850	6" x 3"	Q-16	2.19	10.959		485	420	37	942	1,200	
	3852	6" x 4"		2.04	11.765		440	450	39.50	929.50	1,200	
	3854	8" x 4"		1.46	16.438		810	625	55	1,490	1,900	
	3856	10" x 8"		.69	34.783		2,250	1,325	117	3,692	4,600	
	3858	12" x 10"	↓	.55	43.636		2,775	1,675	147	4,597	5,700	
	3950	Reducer, concentric, 3/4" x 1/2"	Q-15	11.85	1.350		25.50	49.50	6.80	81.80	111	
	3952	1" x 3/4"		10.60	1.509		30	55.50	7.60	93.10	125	
	3954	1-1/4" x 3/4"		10.19	1.570		85.50	58	7.90	151.40	190	
	3956	1-1/4" x 1"		9.76	1.639		46	60.50	8.25	114.75	150	
	3958	1-1/2" x 3/4"		9.88	1.619		53.50	59.50	8.15	121.15	157	
	3960	1-1/2" x 1"		9.47	1.690		17.90	62	8.50	88.40	123	
	3962	2" x 1"		8.65	1.850		21.50	68	9.30	98.80	136	
	3964	2" x 1-1/2"		8.16	1.961		17.90	72	9.90	99.80	139	
	3966	2-1/2" x 1"		5.71	2.802		78.50	103	14.10	195.60	257	
	3968	2-1/2" x 2"		5.21	3.071		64.50	113	15.50	193	258	
	3970	3" x 1"		4.88	3.279		85.50	121	16.50	223	293	
	3972	3" x 1-1/2"		4.72	3.390		39.50	125	17.10	181.60	249	
	3974	3" x 2"		4.51	3.548		32	130	17.90	179.90	251	
	3976	4" x 2"		3.69	4.336		44	159	22	225	315	
	3978	4" x 3"		2.77	5.776		38.50	212	29	279.50	395	
	3980	5" x 3"		2.56	6.250		320	230	31.50	581.50	735	
	3982	5" x 4"	↓	2.27	7.048		257	259	35.50	551.50	710	
	3984	6" x 3"	Q-16	3.57	6.723		122	256	22.50	400.50	545	
	3986	6" x 4"		3.19	7.524		102	287	25.50	414.50	570	
	3988	8" x 4"		2.44	9.836		385	375	33	793	1,025	
	3990	8" x 6"		2.22	10.811		247	410	36.50	693.50	930	
	3992	10" x 6"	↓	1.91	12.565	↓	515	480	42	1,037	1,325	

15107 | Metal Pipe & Fittings			CREW	DAILY OUTPUT	LABOR-HOURS	UNIT	2005 BARE COSTS MAT.	LABOR	EQUIP.	TOTAL	TOTAL INCL O&P	
960	3994	10" x 8"	Q-16	1.61	14.907	Ea.	305	570	50	925	1,250	960
	3995	12" x 6"		1.63	14.724		770	560	49.50	1,379.50	1,750	
	3996	12" x 8"		1.41	17.021		705	650	57	1,412	1,825	
	3997	12" x 10"	↓	1.27	18.898	↓	425	720	63.50	1,208.50	1,600	
	4000	Socket weld joint, 3000 lb., type 304										
	4100	90° Elbow										
	4140	1/4"	Q-15	13.47	1.188	Ea.	29	43.50	6	78.50	104	
	4150	3/8"		12.97	1.234		37.50	45.50	6.20	89.20	116	
	4160	1/2"		12.21	1.310		39	48	6.60	93.60	123	
	4170	3/4"		10.68	1.498		42	55	7.55	104.55	138	
	4180	1"		9.74	1.643		71	60.50	8.30	139.80	178	
	4190	1-1/4"		9.05	1.768		124	65	8.90	197.90	243	
	4200	1-1/2"		8.55	1.871		150	69	9.45	228.45	278	
	4210	2"	↓	7.26	2.204	↓	243	81	11.10	335.10	400	
	4300	45° Elbow										
	4340	1/4"	Q-15	13.47	1.188	Ea.	54	43.50	6	103.50	132	
	4350	3/8"		12.97	1.234		54	45.50	6.20	105.70	134	
	4360	1/2"		12.21	1.310		54	48	6.60	108.60	139	
	4370	3/4"		10.68	1.498		61.50	55	7.55	124.05	159	
	4380	1"		9.74	1.643		89	60.50	8.30	157.80	198	
	4390	1-1/4"		9.05	1.768		147	65	8.90	220.90	269	
	4400	1-1/2"		8.55	1.871		148	69	9.45	226.45	275	
	4410	2"	↓	7.26	2.204	↓	268	81	11.10	360.10	430	
	4500	Tee										
	4540	1/4"	Q-15	8.97	1.784	Ea.	38	65.50	9	112.50	150	
	4550	3/8"		8.64	1.852		45.50	68	9.35	122.85	162	
	4560	1/2"		8.13	1.968		55.50	72.50	9.90	137.90	181	
	4570	3/4"		7.12	2.247		64.50	82.50	11.35	158.35	207	
	4580	1"		6.48	2.469		86.50	91	12.45	189.95	245	
	4590	1-1/4"		6.03	2.653		153	97.50	13.35	263.85	330	
	4600	1-1/2"		5.69	2.812		220	103	14.15	337.15	415	
	4610	2"	↓	4.83	3.313	↓	335	122	16.70	473.70	565	
	5000	Socket weld joint, 3000 lb., type 316										
	5100	90° Elbow										
	5140	1/4"	Q-15	13.47	1.188	Ea.	34	43.50	6	83.50	109	
	5150	3/8"		12.97	1.234		39.50	45.50	6.20	91.20	118	
	5160	1/2"		12.21	1.310		47.50	48	6.60	102.10	132	
	5170	3/4"		10.68	1.498		62.50	55	7.55	125.05	160	
	5180	1"		9.74	1.643		89	60.50	8.30	157.80	198	
	5190	1-1/4"		9.05	1.768		158	65	8.90	231.90	281	
	5200	1-1/2"		8.55	1.871		179	69	9.45	257.45	310	
	5210	2"	↓	7.26	2.204	↓	305	81	11.10	397.10	470	
	5300	45° Elbow										
	5340	1/4"	Q-15	13.47	1.188	Ea.	68	43.50	6	117.50	147	
	5350	3/8"		12.97	1.234		68	45.50	6.20	119.70	150	
	5360	1/2"		12.21	1.310		68	48	6.60	122.60	155	
	5370	3/4"		10.68	1.498		75.50	55	7.55	138.05	174	
	5380	1"		9.74	1.643		114	60.50	8.30	182.80	225	
	5390	1-1/4"		9.05	1.768		163	65	8.90	236.90	286	
	5400	1-1/2"		8.55	1.871		183	69	9.45	261.45	315	
	5410	2"	↓	7.26	2.204	↓	273	81	11.10	365.10	435	
	5500	Tee										
	5540	1/4"	Q-15	8.97	1.784	Ea.	46	65.50	9	120.50	159	
	5550	3/8"		8.64	1.852		56	68	9.35	133.35	174	
	5560	1/2"		8.13	1.968		62.50	72.50	9.90	144.90	189	
	5570	3/4"	↓	7.12	2.247	↓	77	82.50	11.35	170.85	221	

15107 | Metal Pipe & Fittings

			CREW	DAILY OUTPUT	LABOR-HOURS	UNIT	2005 BARE COSTS MAT.	LABOR	EQUIP.	TOTAL	TOTAL INCL O&P	
960	5580	1"	Q-15	6.48	2.469	Ea.	118	91	12.45	221.45	279	960
	5590	1-1/4"		6.03	2.653		187	97.50	13.35	297.85	370	
	5600	1-1/2"		5.69	2.812		264	103	14.15	381.15	460	
	5610	2"		4.83	3.313		420	122	16.70	558.70	665	
	5700	For socket weld joint, 6000 lb., type 304 and 316, add					100%					
	6000	Threaded companion flange										
	6010	Stainless steel, 150 lb., type 304										
	6020	1/2" diam.	1 Plum	30	.267	Ea.	58	10.90		68.90	80	
	6030	3/4" diam.		28	.286		64.50	11.65		76.15	88.50	
	6040	1" diam.		27	.296		72	12.10		84.10	97	
	6050	1-1/4" diam.	Q-1	44	.364		98	13.35		111.35	128	
	6060	1-1/2" diam.		40	.400		100	14.70		114.70	132	
	6070	2" diam.		36	.444		112	16.35		128.35	148	
	6080	2-1/2" diam.		28	.571		171	21		192	220	
	6090	3" diam.		20	.800		186	29.50		215.50	249	
	6110	4" diam.		12	1.333		293	49		342	400	
	6130	6" diam.	Q-2	14	1.714		560	65.50		625.50	715	
	6140	8" diam.	"	12	2		1,050	76.50		1,126.50	1,275	
	6150	For type 316 add					40%					
	6260	Weld flanges, stainless steel, type 304										
	6270	Slip on, 150 lb. (welded, front and back)										
	6280	1/2" diam.	Q-15	12.70	1.260	Ea.	58	46.50	6.35	110.85	140	
	6290	3/4" diam.		11.11	1.440		64.50	53	7.25	124.75	159	
	6300	1" diam.		10.13	1.579		40	58	7.95	105.95	140	
	6310	1-1/4" diam.		9.41	1.700		98	62.50	8.55	169.05	211	
	6320	1-1/2" diam.		8.89	1.800		59	66	9.05	134.05	174	
	6330	2" diam.		7.55	2.119		77	78	10.70	165.70	213	
	6340	2-1/2" diam.		3.98	4.020		171	148	20.50	339.50	435	
	6350	3" diam.		3.21	4.984		186	183	25	394	510	
	6370	4" diam.		2.44	6.557		162	241	33	436	575	
	6390	6" diam.	Q-16	1.89	12.698		255	485	42.50	782.50	1,050	
	6400	8" diam.	"	1.22	19.672		1,050	750	66	1,866	2,350	
	6410	For type 316, add					40%					
	6530	Weld neck 150 lb.										
	6540	1/2" diam.	Q-15	25.40	.630	Ea.	37.50	23	3.17	63.67	80	
	6550	3/4" diam.		22.22	.720		42.50	26.50	3.63	72.63	91	
	6560	1" diam.		20.25	.790		50.50	29	3.98	83.48	103	
	6570	1-1/4" diam.		18.82	.850		61	31.50	4.28	96.78	119	
	6580	1-1/2" diam.		17.78	.900		71	33	4.54	108.54	133	
	6590	2" diam.		15.09	1.060		91	39	5.35	135.35	164	
	6600	2-1/2" diam.		7.96	2.010		152	74	10.15	236.15	289	
	6610	3" diam.		6.43	2.488		172	91.50	12.55	276.05	340	
	6630	4" diam.		4.88	3.279		223	121	16.50	360.50	445	
	6640	5" diam.		4.26	3.756		296	138	18.95	452.95	555	
	6650	6" diam.	Q-16	5.66	4.240		345	162	14.25	521.25	640	
	6652	8" diam.		3.65	6.575		485	251	22	758	930	
	6654	10" diam.		2.88	8.333		665	320	28	1,013	1,250	
	6656	12" diam.		2.28	10.526		1,050	400	35.50	1,485.50	1,800	
	6670	For type 316 add					23%					
	7000	Threaded joint, 150 lb., type 304										
	7030	90° elbow										
	7040	1/8"	1 Plum	13	.615	Ea.	16.60	25		41.60	56.50	
	7050	1/4"		13	.615		16.60	25		41.60	56.50	
	7070	3/8"		13	.615		18.95	25		43.95	59	
	7080	1/2"		12	.667		12.70	27		39.70	55	
	7090	3/4"		11	.727		14.40	29.50		43.90	60.50	

15 MECHANICAL

	15107 | Metal Pipe & Fittings		CREW	DAILY OUTPUT	LABOR-HOURS	UNIT	2005 BARE COSTS MAT.	LABOR	EQUIP.	TOTAL	TOTAL INCL O&P	
960	7100	1"	1 Plum	10	.800	Ea.	19.15	32.50		51.65	70	960
	7110	1-1/4"	Q-1	17	.941		39.50	34.50		74	95.50	
	7120	1-1/2"		16	1		32.50	37		69.50	91.50	
	7130	2"		14	1.143		47	42		89	115	
	7140	2-1/2"		11	1.455		174	53.50		227.50	272	
	7150	3"		8	2		300	73.50		373.50	440	
	7160	4"	Q-2	11	2.182		320	83		403	480	
	7180	45° elbow										
	7190	1/8"	1 Plum	13	.615	Ea.	24.50	25		49.50	64.50	
	7200	1/4"		13	.615		24.50	25		49.50	64.50	
	7210	3/8"		13	.615		24	25		49	64.50	
	7220	1/2"		12	.667		18.20	27		45.20	61	
	7230	3/4"		11	.727		19.70	29.50		49.20	66	
	7240	1"		10	.800		21.50	32.50		54	73	
	7250	1-1/4"	Q-1	17	.941		37.50	34.50		72	93	
	7260	1-1/2"		16	1		36	37		73	95.50	
	7270	2"		14	1.143		51.50	42		93.50	120	
	7280	2-1/2"		11	1.455		245	53.50		298.50	350	
	7290	3"		8	2		360	73.50		433.50	505	
	7300	4"	Q-2	11	2.182		485	83		568	660	
	7320	Tee, straight										
	7330	1/8"	1 Plum	9	.889	Ea.	25.50	36.50		62	82.50	
	7340	1/4"		9	.889		25.50	36.50		62	82.50	
	7350	3/8"		9	.889		27	36.50		63.50	84	
	7360	1/2"		8	1		19.05	41		60.05	82.50	
	7370	3/4"		7	1.143		20.50	46.50		67	92.50	
	7380	1"		6.50	1.231		24	50.50		74.50	102	
	7390	1-1/4"	Q-1	11	1.455		54	53.50		107.50	140	
	7400	1-1/2"		10	1.600		49.50	59		108.50	143	
	7410	2"		9	1.778		61.50	65.50		127	166	
	7420	2-1/2"		7	2.286		250	84		334	400	
	7430	3"		5	3.200		380	118		498	595	
	7440	4"	Q-2	7	3.429		705	131		836	970	
	7460	Coupling, straight										
	7470	1/8"	1 Plum	19	.421	Ea.	6.85	17.20		24.05	33.50	
	7480	1/4"		19	.421		8.10	17.20		25.30	35	
	7490	3/8"		19	.421		9.70	17.20		26.90	36.50	
	7500	1/2"		19	.421		9.65	17.20		26.85	36.50	
	7510	3/4"		18	.444		17.15	18.15		35.30	46.50	
	7520	1"		15	.533		21	22		43	56	
	7530	1-1/4"	Q-1	26	.615		32	22.50		54.50	69	
	7540	1-1/2"		24	.667		37.50	24.50		62	78	
	7550	2"		21	.762		59	28		87	107	
	7560	2-1/2"		18	.889		137	32.50		169.50	200	
	7570	3"		14	1.143		186	42		228	268	
	7580	4"	Q-2	16	1.500		275	57		332	385	
	7600	Reducer, concentric, 1/2"	1 Plum	12	.667		10.35	27		37.35	52.50	
	7610	3/4"		11	.727		13.35	29.50		42.85	59	
	7612	1"		10	.800		22	32.50		54.50	73.50	
	7614	1-1/4"	Q-1	17	.941		47	34.50		81.50	104	
	7616	1-1/2"		16	1		51.50	37		88.50	113	
	7618	2"		14	1.143		80.50	42		122.50	152	
	7620	2-1/2"		11	1.455		200	53.50		253.50	300	
	7622	3"		8	2		222	73.50		295.50	355	
	7624	4"	Q-2	11	2.182		370	83		453	530	
	7710	Union										

15107 | Metal Pipe & Fittings

	15107 Metal Pipe & Fittings	CREW	DAILY OUTPUT	LABOR-HOURS	UNIT	2005 BARE COSTS MAT.	LABOR	EQUIP.	TOTAL	TOTAL INCL O&P
960 7720	1/8"	1 Plum	12	.667	Ea.	32	27		59	76
7730	1/4"		12	.667		32	27		59	76
7740	3/8"		12	.667		37	27		64	82
7750	1/2"		11	.727		34.50	29.50		64	82.50
7760	3/4"		10	.800		47.50	32.50		80	102
7770	1"		9	.889		69	36.50		105.50	131
7780	1-1/4"	Q-1	16	1		188	37		225	263
7790	1-1/2"		15	1.067		175	39		214	251
7800	2"		13	1.231		221	45.50		266.50	310
7810	2-1/2"		10	1.600		555	59		614	700
7820	3"		7	2.286		740	84		824	935
7830	4"	Q-2	10	2.400		1,025	91.50		1,116.50	1,275
7838	Caps									
7840	1/2"	1 Plum	24	.333	Ea.	6.45	13.60		20.05	27.50
7841	3/4"		22	.364		9.40	14.85		24.25	33
7842	1"		20	.400		15	16.35		31.35	41
7843	1-1/2"	Q-1	32	.500		36	18.40		54.40	67
7844	2"	"	28	.571		45.50	21		66.50	81.50
7845	4"	Q-2	22	1.091		178	41.50		219.50	259
7850	For 150 lb., type 316, add					25%				
8750	Threaded joint, 2000 lb., type 304									
8770	90° Elbow									
8780	1/8"	1 Plum	13	.615	Ea.	21	25		46	61.50
8790	1/4"		13	.615		21	25		46	61.50
8800	3/8"		13	.615		26.50	25		51.50	67
8810	1/2"		12	.667		35.50	27		62.50	80
8820	3/4"		11	.727		41.50	29.50		71	90
8830	1"		10	.800		54.50	32.50		87	109
8840	1-1/4"	Q-1	17	.941		89	34.50		123.50	150
8850	1-1/2"		16	1		143	37		180	213
8860	2"		14	1.143		201	42		243	284
8880	45° Elbow									
8890	1/8"	1 Plum	13	.615	Ea.	44	25		69	86.50
8900	1/4"		13	.615		44	25		69	86.50
8910	3/8"		13	.615		55.50	25		80.50	99
8920	1/2"		12	.667		55.50	27		82.50	102
8930	3/4"		11	.727		61	29.50		90.50	112
8940	1"		10	.800		73	32.50		105.50	129
8950	1-1/4"	Q-1	17	.941		127	34.50		161.50	192
8960	1-1/2"		16	1		170	37		207	243
8970	2"		14	1.143		196	42		238	279
8990	Tee, straight									
9000	1/8"	1 Plum	9	.889	Ea.	27.50	36.50		64	84.50
9010	1/4"		9	.889		27.50	36.50		64	84.50
9020	3/8"		9	.889		35	36.50		71.50	93.50
9030	1/2"		8	1		45.50	41		86.50	112
9040	3/4"		7	1.143		51.50	46.50		98	127
9050	1"		6.50	1.231		71	50.50		121.50	154
9060	1-1/4"	Q-1	11	1.455		119	53.50		172.50	212
9070	1-1/2"		10	1.600		199	59		258	310
9080	2"		9	1.778		276	65.50		341.50	405
9100	For couplings and unions use 3000 lb., type 304									
9120	2000 lb., type 316									
9130	90° Elbow									
9140	1/8"	1 Plum	13	.615	Ea.	25.50	25		50.50	66
9150	1/4"		13	.615		25.50	25		50.50	66

15107 | Metal Pipe & Fittings

		CREW	DAILY OUTPUT	LABOR-HOURS	UNIT	2005 BARE COSTS MAT.	LABOR	EQUIP.	TOTAL	TOTAL INCL O&P
960 9160	3/8"	1 Plum	13	.615	Ea.	29.50	25		54.50	70.50 960
9170	1/2"		12	.667		38.50	27		65.50	83
9180	3/4"		11	.727		47	29.50		76.50	96
9190	1"		10	.800		70	32.50		102.50	126
9200	1-1/4"	Q-1	17	.941		127	34.50		161.50	191
9210	1-1/2"		16	1		146	37		183	216
9220	2"		14	1.143		248	42		290	335
9240	45° Elbow									
9250	1/8"	1 Plum	13	.615	Ea.	53.50	25		78.50	97
9260	1/4"		13	.615		53.50	25		78.50	97
9270	3/8"		13	.615		53.50	25		78.50	97
9280	1/2"		12	.667		53.50	27		80.50	100
9300	3/4"		11	.727		60.50	29.50		90	111
9310	1"		10	.800		91	32.50		123.50	149
9320	1-1/4"	Q-1	17	.941		130	34.50		164.50	195
9330	1-1/2"		16	1		178	37		215	252
9340	2"		14	1.143		218	42		260	305
9360	Tee, straight									
9370	1/8"	1 Plum	9	.889	Ea.	32	36.50		68.50	90
9380	1/4"		9	.889		32	36.50		68.50	90
9390	3/8"		9	.889		39	36.50		75.50	97.50
9400	1/2"		8	1		49	41		90	116
9410	3/4"		7	1.143		60.50	46.50		107	137
9420	1"		6.50	1.231		92.50	50.50		143	178
9430	1-1/4"	Q-1	11	1.455		150	53.50		203.50	246
9440	1-1/2"		10	1.600		215	59		274	325
9450	2"		9	1.778		345	65.50		410.50	475
9470	For couplings and unions use 3000 lb., type 316									
9490	3000 lb., type 304									
9510	Coupling									
9520	1/8"	1 Plum	19	.421	Ea.	8.30	17.20		25.50	35
9530	1/4"		19	.421		9.10	17.20		26.30	36
9540	3/8"		19	.421		10.60	17.20		27.80	37.50
9550	1/2"		19	.421		12.80	17.20		30	40
9560	3/4"		18	.444		17.05	18.15		35.20	46.50
9570	1"		15	.533		29	22		51	65
9580	1-1/4"	Q-1	26	.615		70	22.50		92.50	111
9590	1-1/2"		24	.667		82	24.50		106.50	127
9600	2"		21	.762		108	28		136	161
9620	Union									
9630	1/8"	1 Plum	12	.667	Ea.	64	27		91	112
9640	1/4"		12	.667		64	27		91	112
9650	3/8"		12	.667		68.50	27		95.50	117
9660	1/2"		11	.727		68.50	29.50		98	120
9670	3/4"		10	.800		84	32.50		116.50	142
9680	1"		9	.889		129	36.50		165.50	197
9690	1-1/4"	Q-1	16	1		239	37		276	320
9700	1-1/2"		15	1.067		270	39		309	355
9710	2"		13	1.231		360	45.50		405.50	465
9730	3000 lb., type 316									
9750	Coupling									
9770	1/8"	1 Plum	19	.421	Ea.	8.80	17.20		26	35.50
9780	1/4"		19	.421		9.90	17.20		27.10	37
9790	3/8"		19	.421		10.60	17.20		27.80	37.50
9800	1/2"		19	.421		15.05	17.20		32.25	42.50
9810	3/4"		18	.444		22	18.15		40.15	52

15100 | Building Services Piping

	15107 | Metal Pipe & Fittings		CREW	DAILY OUTPUT	LABOR-HOURS	UNIT	2005 BARE COSTS MAT.	LABOR	EQUIP.	TOTAL	TOTAL INCL O&P	
960	9820	1″	1 Plum	15	.533	Ea.	37	22		59	74	960
	9830	1-1/4″	Q-1	26	.615		83	22.50		105.50	126	
	9840	1-1/2″		24	.667		97	24.50		121.50	143	
	9850	2″		21	.762		134	28		162	189	
	9870	Union										
	9880	1/8″	1 Plum	12	.667	Ea.	65.50	27		92.50	113	
	9890	1/4″		12	.667		65.50	27		92.50	113	
	9900	3/8″		12	.667		76	27		103	125	
	9910	1/2″		11	.727		76	29.50		105.50	128	
	9920	3/4″		10	.800		99	32.50		131.50	158	
	9930	1″		9	.889		156	36.50		192.50	227	
	9940	1-1/4″	Q-1	16	1		273	37		310	355	
	9950	1-1/2″		15	1.067		340	39		379	430	
	9960	2″		13	1.231		430	45.50		475.50	545	

	15108 | Plastic Pipe & Fittings		CREW	DAILY OUTPUT	LABOR-HOURS	UNIT	MAT.	LABOR	EQUIP.	TOTAL	TOTAL INCL O&P	
520	0010	**PIPE, PLASTIC**										520
	0020	Fiberglass reinforced, couplings 10′ O.C., hangers 3 per 10′										
	0080	General service										
	0120	2″ diameter	Q-1	59	.271	L.F.	9.60	9.95		19.55	25.50	
	0140	3″ diameter		52	.308		12.10	11.30		23.40	30.50	
	0150	4″ diameter		48	.333		13.95	12.25		26.20	34	
	0160	6″ diameter		39	.410		21.50	15.10		36.60	46	
	0170	8″ diameter	Q-2	49	.490		33.50	18.70		52.20	65	
	0180	10″ diameter		41	.585		55.50	22.50		78	94.50	
	0190	12″ diameter		36	.667		71	25.50		96.50	117	
	0200	High strength										
	0240	2″ diameter	Q-1	58	.276	L.F.	9.20	10.15		19.35	25.50	
	0260	3″ diameter		51	.314		13.10	11.55		24.65	32	
	0280	4″ diameter		47	.340		15.85	12.50		28.35	36.50	
	0300	6″ diameter		38	.421		24.50	15.50		40	50.50	
	0320	8″ diameter	Q-2	48	.500		39	19.05		58.05	71	
	0340	10″ diameter		40	.600		61.50	23		84.50	103	
	0360	12″ diameter		36	.667		76	25.50		101.50	122	
	0550	To delete coupling & hangers, subtract										
	0560	2″ diam. to 6″ diam.					33%	56%				
	0570	8″ diam. to 12″ diam.					31%	52%				
	0600	PVC, high impact/pressure, cplgs. 10′ O.C., hangers 3 per 10′										
	1020	Schedule 80										
	1040	1/4″ diameter	1 Plum	58	.138	L.F.	1.40	5.65		7.05	10	
	1060	3/8″ diameter		55	.145		1.59	5.95		7.54	10.70	
	1070	1/2″ diameter		50	.160		1.28	6.55		7.83	11.20	
	1080	3/4″ diameter		47	.170		1.46	6.95		8.41	12.05	
	1090	1″ diameter		43	.186		1.76	7.60		9.36	13.35	
	1100	1-1/4″ diameter		39	.205		2.17	8.40		10.57	15	
	1110	1-1/2″ diameter		34	.235		2.43	9.60		12.03	17.15	
	1120	2″ diameter	Q-1	55	.291		3.01	10.70		13.71	19.40	
	1130	2-1/2″ diameter		52	.308		4.21	11.30		15.51	21.50	
	1140	3″ diameter		50	.320		5.70	11.75		17.45	24	
	1150	4″ diameter		46	.348		7.85	12.80		20.65	28	
	1160	5″ diameter		42	.381		10.95	14		24.95	33	
	1170	6″ diameter		38	.421		14.70	15.50		30.20	39.50	
	1180	8″ diameter	Q-2	47	.511		23	19.45		42.45	54.50	
	1190	10″ diameter		42	.571		39	22		61	76	
	1200	12″ diameter		38	.632		51	24		75	92	
	1730	To delete coupling & hangers, subtract										

15100 | Building Services Piping

15108 | Plastic Pipe & Fittings

				DAILY	LABOR-		2005 BARE COSTS				TOTAL	
			CREW	OUTPUT	HOURS	UNIT	MAT.	LABOR	EQUIP.	TOTAL	INCL O&P	
520	1740	1/4" diam. to 1/2" diam.					62%	80%				520
	1750	3/4" diam. to 1-1/4" diam.					58%	73%				
	1760	1-1/2" diam. to 6" diam.					40%	57%				
	1770	8" diam. to 12" diam.					34%	50%				
	1800	PVC, couplings 10' O.C., hangers 3 per 10'										
	1820	Schedule 40										
	1860	1/2" diameter	1 Plum	54	.148	L.F.	.98	6.05		7.03	10.20	
	1870	3/4" diameter		51	.157		1.06	6.40		7.46	10.80	
	1880	1" diameter		46	.174		1.17	7.10		8.27	12	
	1890	1-1/4" diameter		42	.190		1.34	7.80		9.14	13.20	
	1900	1-1/2" diameter		36	.222		1.44	9.10		10.54	15.25	
	1910	2" diameter	Q-1	59	.271		1.63	9.95		11.58	16.80	
	1920	2-1/2" diameter		56	.286		2.20	10.50		12.70	18.20	
	1930	3" diameter		53	.302		2.87	11.10		13.97	19.85	
	1940	4" diameter		48	.333		3.67	12.25		15.92	22.50	
	1950	5" diameter		43	.372		4.86	13.70		18.56	26	
	1960	6" diameter		39	.410		6.50	15.10		21.60	29.50	
	1970	8" diameter	Q-2	48	.500		10.45	19.05		29.50	40	
	1980	10" diameter		43	.558		26	21.50		47.50	61	
	1990	12" diameter		42	.571		34.50	22		56.50	71	
	2000	14" diameter		31	.774		55	29.50		84.50	105	
	2010	16" diameter		23	1.043		69	40		109	136	
	2340	To delete coupling & hangers, subtract										
	2360	1/2" diam. to 1-1/4" diam.					65%	74%				
	2370	1-1/2" diam. to 6" diam.					44%	57%				
	2380	8" diam. to 12" diam.					41%	53%				
	2390	14" diam. to 16" diam.					48%	45%				
	2420	Schedule 80										
	2440	1/4" diameter	1 Plum	58	.138	L.F.	.85	5.65		6.50	9.40	
	2450	3/8" diameter		55	.145		.85	5.95		6.80	9.90	
	2460	1/2" diameter		50	.160		1.02	6.55		7.57	10.90	
	2470	3/4" diameter		47	.170		1.12	6.95		8.07	11.70	
	2480	1" diameter		43	.186		1.26	7.60		8.86	12.80	
	2490	1-1/4" diameter		39	.205		1.47	8.40		9.87	14.20	
	2500	1-1/2" diameter		34	.235		1.60	9.60		11.20	16.20	
	2510	2" diameter	Q-1	55	.291		1.87	10.70		12.57	18.15	
	2520	2-1/2" diameter		52	.308		2.37	11.30		13.67	19.60	
	2530	3" diameter		50	.320		3.35	11.75		15.10	21.50	
	2540	4" diameter		46	.348		4.42	12.80		17.22	24	
	2550	5" diameter		42	.381		7.15	14		21.15	29	
	2560	6" diameter		38	.421		8.15	15.50		23.65	32.50	
	2570	8" diameter	Q-2	47	.511		13	19.45		32.45	44	
	2580	10" diameter		42	.571		31.50	22		53.50	68	
	2590	12" diameter		38	.632		41	24		65	81.50	
	2830	To delete coupling & hangers, subtract										
	2840	1/4" diam. to 1/2" diam.					66%	80%				
	2850	3/4" diam. to 1-1/4" diam.					61%	73%				
	2860	1-1/2" diam. to 6" diam.					41%	57%				
	2870	8" diam. to 12" diam.					31%	50%				
	2900	Schedule 120										
	2910	1/2" diameter	1 Plum	50	.160	L.F.	1.41	6.55		7.96	11.35	
	2950	3/4" diameter		47	.170		1.62	6.95		8.57	12.25	
	2960	1" diameter		43	.186		2	7.60		9.60	13.60	
	2970	1-1/4" diameter		39	.205		2.51	8.40		10.91	15.35	
	2980	1-1/2" diameter		33	.242		2.85	9.90		12.75	18.05	
	2990	2" diameter	Q-1	54	.296		3.64	10.90		14.54	20.50	

15108 | Plastic Pipe & Fittings

			CREW	DAILY OUTPUT	LABOR-HOURS	UNIT	2005 BARE COSTS MAT.	LABOR	EQUIP.	TOTAL	TOTAL INCL O&P	
520	3000	2-1/2" diameter	Q-1	52	.308	L.F.	5.35	11.30		16.65	23	520
	3010	3" diameter		49	.327		6.95	12		18.95	25.50	
	3020	4" diameter		45	.356		10.45	13.10		23.55	31	
	3030	6" diameter		37	.432		19.60	15.90		35.50	45.50	
	3240	To delete coupling & hangers, subtract										
	3250	1/2" diam. to 1-1/4" diam.					52%	74%				
	3260	1-1/2" diam. to 4" diam.					30%	57%				
	3270	6" diam.					17%	50%				
	3300	PVC, pressure, couplings 10' O.C., hangers 3 per 10'										
	3310	SDR 26, 160 psi										
	3350	1-1/4" diameter	1 Plum	42	.190	L.F.	1.17	7.80		8.97	13	
	3360	1-1/2" diameter	"	36	.222		1.25	9.10		10.35	15.05	
	3370	2" diameter	Q-1	59	.271		1.40	9.95		11.35	16.55	
	3380	2-1/2" diameter		56	.286		1.83	10.50		12.33	17.80	
	3390	3" diameter		53	.302		2.47	11.10		13.57	19.40	
	3400	4" diameter		48	.333		3.27	12.25		15.52	22	
	3420	6" diameter		39	.410		6.25	15.10		21.35	29.50	
	3430	8" diameter	Q-2	48	.500		10.45	19.05		29.50	40	
	3660	To delete coupling & hangers, subtract										
	3670	1-1/4" diam.					63%	68%				
	3680	1-1/2" diam. to 4" diam.					48%	57%				
	3690	6" diam. to 8" diam.					60%	54%				
	3720	SDR 21, 200 psi, 1/2" diameter	1 Plum	54	.148	L.F.	.94	6.05		6.99	10.15	
	3740	3/4" diameter		51	.157		.98	6.40		7.38	10.75	
	3750	1" diameter		46	.174		1.05	7.10		8.15	11.85	
	3760	1-1/4" diameter		42	.190		1.20	7.80		9	13	
	3770	1-1/2" diameter		36	.222		1.29	9.10		10.39	15.05	
	3780	2" diameter	Q-1	59	.271		1.48	9.95		11.43	16.60	
	3790	2-1/2" diameter		56	.286		2.29	10.50		12.79	18.30	
	3800	3" diameter		53	.302		2.63	11.10		13.73	19.60	
	3810	4" diameter		48	.333		3.53	12.25		15.78	22.50	
	3830	6" diameter		39	.410		6.80	15.10		21.90	30	
	3840	8" diameter	Q-2	48	.500		11.60	19.05		30.65	41.50	
	4000	To delete coupling & hangers, subtract										
	4010	1/2" diam. to 3/4" diam.					71%	77%				
	4020	1" diam. to 1-1/4" diam.					63%	70%				
	4030	1-1/2" diam. to 6" diam.					44%	57%				
	4040	8" diam.					46%	54%				
	4100	DWV type, schedule 40, couplings 10' O.C., hangers 3 per 10'										
	4120	ABS										
	4140	1-1/4" diameter	1 Plum	42	.190	L.F.	1.08	7.80		8.88	12.90	
	4150	1-1/2" diameter	"	36	.222		1.09	9.10		10.19	14.85	
	4160	2" diameter	Q-1	59	.271		1.19	9.95		11.14	16.30	
	4170	3" diameter		53	.302		2.08	11.10		13.18	19	
	4180	4" diameter		48	.333		2.78	12.25		15.03	21.50	
	4190	6" diameter		39	.410		6.80	15.10		21.90	30	
	4360	To delete coupling & hangers, subtract										
	4370	1-1/4" diam.					64%	68%				
	4380	1-1/2" diam. to 6" diam.					54%	57%				
	4400	PVC										
	4410	1-1/4" diameter	1 Plum	42	.190	L.F.	1.19	7.80		8.99	13	
	4420	1-1/2" diameter	"	36	.222		1.20	9.10		10.30	14.95	
	4460	2" diameter	Q-1	59	.271		1.35	9.95		11.30	16.50	
	4470	3" diameter		53	.302		2.40	11.10		13.50	19.35	
	4480	4" diameter		48	.333		3.13	12.25		15.38	22	
	4490	6" diameter		39	.410		5.95	15.10		21.05	29	

15108 | Plastic Pipe & Fittings		CREW	DAILY OUTPUT	LABOR-HOURS	UNIT	2005 BARE COSTS MAT.	LABOR	EQUIP.	TOTAL	TOTAL INCL O&P	
520	4500	8″ diameter	Q-2	48	.500	L.F.	15.10	19.05		34.15	45	520
	4750	To delete coupling & hangers, subtract										
	4760	1-1/4″ diam. to 1-1/2″ diam.					71%	64%				
	4770	2″ diam. to 8″ diam.					60%	57%				
	4800	PVC, clear pipe, cplgs. 10′ O.C., hangers 3 per 10′, Sched. 40										
	4840	1/4″ diameter	1 Plum	59	.136	L.F.	1.34	5.55		6.89	9.85	
	4850	3/8″ diameter		56	.143		1.54	5.85		7.39	10.45	
	4860	1/2″ diameter		54	.148		1.92	6.05		7.97	11.20	
	4870	3/4″ diameter		51	.157		2.33	6.40		8.73	12.20	
	4880	1″ diameter		46	.174		3.12	7.10		10.22	14.15	
	4890	1-1/4″ diameter		42	.190		3.97	7.80		11.77	16.05	
	4900	1-1/2″ diameter	↓	36	.222		4.59	9.10		13.69	18.70	
	4910	2″ diameter	Q-1	59	.271		5.95	9.95		15.90	21.50	
	4920	2-1/2″ diameter		56	.286		9.10	10.50		19.60	26	
	4930	3″ diameter		53	.302		11.75	11.10		22.85	29.50	
	4940	3-1/2″ diameter		50	.320		15.80	11.75		27.55	35	
	4950	4″ diameter	↓	48	.333	↓	17.30	12.25		29.55	37.50	
	5250	To delete coupling & hangers, subtract										
	5260	1/4″ diam. to 3/8″ diam.					60%	81%				
	5270	1/2″ diam. to 3/4″ diam.					41%	77%				
	5280	1″ diam. to 1-1/2″ diam.					26%	67%				
	5290	2″ diam. to 4″ diam.					16%	58%				
	5360	CPVC, couplings 10′ O.C., hangers 3 per 10′										
	5380	Schedule 40										
	5460	1/2″ diameter	1 Plum	54	.148	L.F.	2.31	6.05		8.36	11.65	
	5470	3/4″ diameter		51	.157		3.05	6.40		9.45	13	
	5480	1″ diameter		46	.174		3.74	7.10		10.84	14.80	
	5490	1-1/4″ diameter		42	.190		4.35	7.80		12.15	16.50	
	5500	1-1/2″ diameter	↓	36	.222		4.86	9.10		13.96	19	
	5510	2″ diameter	Q-1	59	.271		6	9.95		15.95	21.50	
	5520	2-1/2″ diameter		56	.286		9.55	10.50		20.05	26.50	
	5530	3″ diameter		53	.302		11.75	11.10		22.85	29.50	
	5540	4″ diameter		48	.333		19.05	12.25		31.30	39.50	
	5550	6″ diameter	↓	43	.372	↓	31	13.70		44.70	55	
	5730	To delete coupling & hangers, subtract										
	5740	1/2″ diam. to 3/4″ diam.					37%	77%				
	5750	1″ diam. to 1-1/4″ diam.					27%	70%				
	5760	1-1/2″ diam. to 3″ diam.					21%	57%				
	5770	4″ diam. to 6″ diam.					16%	57%				
	5800	Schedule 80										
	5860	1/2″ diameter	1 Plum	50	.160	L.F.	2.73	6.55		9.28	12.80	
	5870	3/4″ diameter		47	.170		3.26	6.95		10.21	14.05	
	5880	1″ diameter		43	.186		4.02	7.60		11.62	15.80	
	5890	1-1/4″ diameter		39	.205		4.79	8.40		13.19	17.85	
	5900	1-1/2″ diameter	↓	34	.235		5.45	9.60		15.05	20.50	
	5910	2″ diameter	Q-1	55	.291		6.90	10.70		17.60	23.50	
	5920	2-1/2″ diameter		52	.308		10.75	11.30		22.05	29	
	5930	3″ diameter		50	.320		13.50	11.75		25.25	32.50	
	5940	4″ diameter		46	.348		22	12.80		34.80	43.50	
	5950	6″ diameter	↓	38	.421		37.50	15.50		53	65	
	5960	8″ diameter	Q-2	47	.511	↓	68	19.45		87.45	105	
	6060	To delete couplings & hangers, subtract										
	6070	1/2″ diam. to 3/4″ diam.					44%	77%				
	6080	1″ diam. to 1-1/4″ diam.					32%	71%				
	6090	1-1/2″ diam. to 4″ diam.					25%	58%				
	6100	6″ diam. to 8″ diam.					20%	53%				

	15108 | Plastic Pipe & Fittings		CREW	DAILY OUTPUT	LABOR-HOURS	UNIT	2005 BARE COSTS MAT.	LABOR	EQUIP.	TOTAL	TOTAL INCL O&P	
520	6240	CTS, 1/2" diameter	1 Plum	54	.148	L.F.	1.25	6.05		7.30	10.45	520
	6250	3/4" diameter		51	.157		1.70	6.40		8.10	11.50	
	6260	1" diameter		46	.174		2.74	7.10		9.84	13.70	
	6270	1 1/4"		42	.190		3.71	7.80		11.51	15.80	
	6280	1 1/2" diameter		36	.222		4.70	9.10		13.80	18.80	
	6290	2" diameter	Q-1	59	.271		7.45	9.95		17.40	23	
	6370	To delete coupling & hangers, subtract										
	6380	1/2" diam.					51%	79%				
	6390	3/4" diam.					40%	76%				
	6392	1" thru 2" diam.					72%	68%				
	7280	Polyethylene, flexible, no couplings or hangers										
	7282	Note: For labor costs add 25% to the couplings and fittings labor total.										
	7300	SDR 15, 100 psi										
	7310	3/4" diameter				L.F.	.24			.24	.26	
	7350	1" diameter					.32			.32	.35	
	7360	1-1/4" diameter					.57			.57	.63	
	7370	1-1/2" diameter					.69			.69	.76	
	7380	2" diameter					1.13			1.13	1.24	
	7700	SIDR 9, 160 psi										
	7710	1/2" diameter				L.F.	.30			.30	.33	
	7750	3/4" diameter					.31			.31	.34	
	7760	1" diameter					.48			.48	.53	
	7770	1-1/4" diameter					.84			.84	.92	
	7780	1-1/2" diameter					1.17			1.17	1.29	
	7790	2" diameter					1.94			1.94	2.13	
	8120	SDR 9, 200 psi										
	8150	3/4" diameter				L.F.	.28			.28	.31	
	8160	1" diameter					.46			.46	.51	
	8170	1-1/4" diameter					.69			.69	.76	
	8420	SDR 7, 250 psi										
	8440	3/4" diameter				L.F.	.40			.40	.44	
	8450	1" diameter					.64			.64	.70	
	8460	1-1/4" diameter					1.13			1.13	1.24	
	8800	PVC, type PSP, drain & sewer, belled end gasket jnt., no hngr.										
	8840	3" diameter				L.F.	1.25			1.25	1.38	
	8850	4" diameter					1.55			1.55	1.71	
	8860	6" diameter					2.99			2.99	3.29	
	9000	Perforated										
	9040	4" diameter				L.F.	1.55			1.55	1.71	
560	0010	**PIPE, PLASTIC, FITTINGS**										560
	0030	Epoxy resin, fiberglass reinforced, general service										
	0090	Elbow, 90°, 2"	Q-1	33.10	.483	Ea.	45.50	17.80		63.30	76.50	
	0100	3"		20.80	.769		64.50	28.50		93	113	
	0110	4"		16.50	.970		88	35.50		123.50	151	
	0120	6"		10.10	1.584		134	58.50		192.50	236	
	0130	8"	Q-2	9.30	2.581		248	98.50		346.50	420	
	0140	10"		8.50	2.824		310	108		418	500	
	0150	12"		7.60	3.158		445	120		565	670	
	0160	45° Elbow, same as 90°										
	0170	Elbow, 90°, flanged										
	0172	2"	Q-1	23	.696	Ea.	100	25.50		125.50	149	
	0173	3"		16	1		116	37		153	183	
	0174	4"		13	1.231		151	45.50		196.50	235	
	0176	6"		8	2		274	73.50		347.50	410	
	0177	8"	Q-2	9	2.667		490	102		592	695	

	15108 | Plastic Pipe & Fittings		CREW	DAILY OUTPUT	LABOR-HOURS	UNIT	2005 BARE COSTS MAT.	LABOR	EQUIP.	TOTAL	TOTAL INCL O&P	
560	0178	10"	Q-2	7	3.429	Ea.	675	131		806	940	560
	0179	12"		5	4.800		915	183		1,098	1,275	
	0186	Elbow, 45°, flanged										
	0188	2"	Q-1	23	.696	Ea.	101	25.50		126.50	150	
	0189	3"		16	1		116	37		153	183	
	0190	4"		13	1.231		151	45.50		196.50	234	
	0192	6"		8	2		276	73.50		349.50	415	
	0193	8"	Q-2	9	2.667		495	102		597	695	
	0194	10"		7	3.429		675	131		806	940	
	0195	12"		5	4.800		915	183		1,098	1,275	
	0290	Tee, 2"	Q-1	20	.800		55	29.50		84.50	105	
	0300	3"		13.90	1.151		64.50	42.50		107	134	
	0310	4"		11	1.455		76.50	53.50		130	165	
	0320	6"		6.70	2.388		205	88		293	360	
	0330	8"	Q-2	6.20	3.871		237	148		385	480	
	0340	10"		5.70	4.211		380	161		541	655	
	0350	12"		5.10	4.706		460	179		639	775	
	0352	Tee, flanged										
	0354	2"	Q-1	17	.941	Ea.	136	34.50		170.50	202	
	0355	3"		10	1.600		183	59		242	290	
	0356	4"		8	2		203	73.50		276.50	335	
	0358	6"		5	3.200		400	118		518	620	
	0359	8"	Q-2	6	4		685	153		838	985	
	0360	10"		5	4.800		995	183		1,178	1,375	
	0361	12"		4	6		1,275	229		1,504	1,750	
	0365	Wye, flanged										
	0367	2"	Q-1	17	.941	Ea.	350	34.50		384.50	435	
	0368	3"		10	1.600		390	59		449	520	
	0369	4"		8	2		485	73.50		558.50	645	
	0371	6"		5	3.200		580	118		698	815	
	0372	8"	Q-2	6	4		755	153		908	1,050	
	0373	10"		5	4.800		1,275	183		1,458	1,675	
	0374	12"		4	6		1,550	229		1,779	2,050	
	0380	Couplings										
	0410	2"	Q-1	33.10	.483	Ea.	9.85	17.80		27.65	37.50	
	0420	3"		20.80	.769		16.95	28.50		45.45	61	
	0430	4"		16.50	.970		22	35.50		57.50	78	
	0440	6"		10.10	1.584		42	58.50		100.50	134	
	0450	8"	Q-2	9.30	2.581		64	98.50		162.50	219	
	0460	10"		8.50	2.824		96	108		204	268	
	0470	12"		7.60	3.158		129	120		249	325	
	0473	High corrosion resistant couplings, add					30%					
	0474	Reducer, concentric, flanged										
	0475	2" x 1-1/2"	Q-1	30	.533	Ea.	145	19.60		164.60	190	
	0476	3" x 2"		24	.667		165	24.50		189.50	219	
	0477	4" x 3"		19	.842		174	31		205	238	
	0479	6" x 4"		15	1.067		227	39		266	310	
	0480	8" x 6"	Q-2	16	1.500		320	57		377	435	
	0481	10" x 8"		13	1.846		425	70.50		495.50	575	
	0482	12" x 10"		11	2.182		610	83		693	795	
	0486	Adapter, bell x male or female										
	0488	2"	Q-1	28	.571	Ea.	17.70	21		38.70	51	
	0489	3"		20	.800		22.50	29.50		52	69	
	0491	4"		17	.941		31	34.50		65.50	86	
	0492	6"		12	1.333		61.50	49		110.50	141	
	0493	8"	Q-2	15	1.600		73	61		134	172	

15108 | Plastic Pipe & Fittings		CREW	DAILY OUTPUT	LABOR-HOURS	UNIT	2005 BARE COSTS MAT.	LABOR	EQUIP.	TOTAL	TOTAL INCL O&P
560 0494	10"	Q-2	11	2.182	Ea.	105	83		188	241
0528	Flange									
0532	2"	Q-1	46	.348	Ea.	19.15	12.80		31.95	40.50
0533	3"		32	.500		23.50	18.40		41.90	53
0534	4"		26	.615		32.50	22.50		55	69.50
0536	6"		16	1		55	37		92	116
0537	8"	Q-2	18	1.333		90	51		141	176
0538	10"		14	1.714		126	65.50		191.50	237
0539	12"		10	2.400		177	91.50		268.50	330
2100	PVC schedule 80, socket joint									
2110	90° elbow, 1/2"	1 Plum	30.30	.264	Ea.	.59	10.80		11.39	16.85
2130	3/4"		26	.308		.76	12.55		13.31	19.75
2140	1"		22.70	.352		1.72	14.40		16.12	23.50
2150	1-1/4"		20.20	.396		1.63	16.20		17.83	26.50
2160	1-1/2"		18.20	.440		1.74	17.95		19.69	29
2170	2"	Q-1	33.10	.483		2.11	17.80		19.91	29
2180	3"		20.80	.769		5.55	28.50		34.05	48.50
2190	4"		16.50	.970		6.30	35.50		41.80	60.50
2200	6"		10.10	1.584		63.50	58.50		122	158
2210	8"	Q-2	9.30	2.581		175	98.50		273.50	340
2250	45° elbow, 1/2"	1 Plum	30.30	.264		1.12	10.80		11.92	17.45
2270	3/4"		26	.308		1.70	12.55		14.25	21
2280	1"		22.70	.352		2.55	14.40		16.95	24.50
2290	1-1/4"		20.20	.396		3.25	16.20		19.45	28
2300	1-1/2"		18.20	.440		3.84	17.95		21.79	31
2310	2"	Q-1	33.10	.483		4.97	17.80		22.77	32
2320	3"		20.80	.769		12.70	28.50		41.20	56.50
2330	4"		16.50	.970		23	35.50		58.50	78.50
2340	6"		10.10	1.584		29	58.50		87.50	119
2350	8"	Q-2	9.30	2.581		166	98.50		264.50	330
2400	Tee, 1/2"	1 Plum	20.20	.396		1.67	16.20		17.87	26.50
2420	3/4"		17.30	.462		1.75	18.90		20.65	30.50
2430	1"		15.20	.526		2.18	21.50		23.68	35
2440	1-1/4"		13.50	.593		6	24		30	43
2450	1-1/2"		12.10	.661		6	27		33	47
2460	2"	Q-1	20	.800		7.50	29.50		37	52.50
2470	3"		13.90	1.151		8.15	42.50		50.65	72.50
2480	4"		11	1.455		11.80	53.50		65.30	93.50
2490	6"		6.70	2.388		40.50	88		128.50	177
2500	8"	Q-2	6.20	3.871		237	148		385	480
2510	Flange, socket, 150 lb., 1/2"	1 Plum	55.60	.144		3.23	5.90		9.13	12.40
2514	3/4"		47.60	.168		3.46	6.85		10.31	14.10
2518	1"		41.70	.192		3.85	7.85		11.70	16.05
2522	1-1/2"		33.30	.240		4.05	9.80		13.85	19.20
2526	2"	Q-1	60.60	.264		5.40	9.70		15.10	20.50
2530	4"		30.30	.528		11.65	19.40		31.05	42
2534	6"		18.50	.865		18.30	32		50.30	68
2538	8"	Q-2	17.10	1.404		87	53.50		140.50	176
2550	Coupling, 1/2"	1 Plum	30.30	.264		1.07	10.80		11.87	17.40
2570	3/4"		26	.308		1.45	12.55		14	20.50
2580	1"		22.70	.352		1.49	14.40		15.89	23
2590	1-1/4"		20.20	.396		2.27	16.20		18.47	27
2600	1-1/2"		18.20	.440		2.44	17.95		20.39	29.50
2610	2"	Q-1	33.10	.483		2.62	17.80		20.42	29.50
2620	3"		20.80	.769		7.40	28.50		35.90	50.50
2630	4"		16.50	.970		9.25	35.50		44.75	63.50

15108 | Plastic Pipe & Fittings

560		CREW	DAILY OUTPUT	LABOR-HOURS	UNIT	2005 BARE COSTS MAT.	LABOR	EQUIP.	TOTAL	TOTAL INCL O&P
2640	6"	Q-1	10.10	1.584	Ea.	19.95	58.50		78.45	110
2650	8"	Q-2	9.30	2.581		37.50	98.50		136	189
2660	10"		8.50	2.824		40.50	108		148.50	207
2670	12"		7.60	3.158		47	120		167	233
2700	PVC (white), schedule 40, socket joints									
2760	90° elbow, 1/2"	1 Plum	33.30	.240	Ea.	.29	9.80		10.09	15.05
2770	3/4"		28.60	.280		.32	11.45		11.77	17.50
2780	1"		25	.320		.57	13.05		13.62	20.50
2790	1-1/4"		22.20	.360		1	14.70		15.70	23
2800	1-1/2"		20	.400		1.07	16.35		17.42	25.50
2810	2"	Q-1	36.40	.440		1.68	16.15		17.83	26.50
2820	2-1/2"		26.70	.599		5.10	22		27.10	38.50
2830	3"		22.90	.699		6.10	25.50		31.60	45
2840	4"		18.20	.879		10.95	32.50		43.45	60.50
2850	5"		12.10	1.322		28.50	48.50		77	104
2860	6"		11.10	1.441		35	53		88	118
2870	8"	Q-2	10.30	2.330		89	89		178	232
2980	45° elbow, 1/2"	1 Plum	33.30	.240		.46	9.80		10.26	15.25
2990	3/4"		28.60	.280		.71	11.45		12.16	17.95
3000	1"		25	.320		.86	13.05		13.91	20.50
3010	1-1/4"		22.20	.360		1.20	14.70		15.90	23.50
3020	1-1/2"		20	.400		1.51	16.35		17.86	26
3030	2"	Q-1	36.40	.440		1.97	16.15		18.12	26.50
3040	2-1/2"		26.70	.599		5.10	22		27.10	38.50
3050	3"		22.90	.699		8.25	25.50		33.75	47.50
3060	4"		18.20	.879		14.25	32.50		46.75	64
3070	5"		12.10	1.322		28.50	48.50		77	104
3080	6"		11.10	1.441		35	53		88	118
3090	8"	Q-2	10.30	2.330		100	89		189	244
3180	Tee, 1/2"	1 Plum	22.20	.360		.35	14.70		15.05	22.50
3190	3/4"		19	.421		.40	17.20		17.60	26.50
3200	1"		16.70	.479		.75	19.55		20.30	30.50
3210	1-1/4"		14.80	.541		1.18	22		23.18	34.50
3220	1-1/2"		13.30	.601		1.43	24.50		25.93	38.50
3230	2"	Q-1	24.20	.661		2.07	24.50		26.57	39
3240	2-1/2"		17.80	.899		6.85	33		39.85	57
3250	3"		15.20	1.053		13.55	38.50		52.05	73
3260	4"		12.10	1.322		16.25	48.50		64.75	91
3270	5"		8.10	1.975		39	72.50		111.50	152
3280	6"		7.40	2.162		54.50	79.50		134	180
3290	8"	Q-2	6.80	3.529		65	135		200	274
3380	Coupling, 1/2"	1 Plum	33.30	.240		.19	9.80		9.99	14.95
3390	3/4"		28.60	.280		.26	11.45		11.71	17.45
3400	1"		25	.320		.44	13.05		13.49	20
3410	1-1/4"		22.20	.360		.61	14.70		15.31	22.50
3420	1-1/2"		20	.400		.65	16.35		17	25
3430	2"	Q-1	36.40	.440		1.01	16.15		17.16	25.50
3440	2-1/2"		26.70	.599		2.22	22		24.22	35.50
3450	3"		22.90	.699		3.49	25.50		28.99	42.50
3460	4"		18.20	.879		5	32.50		37.50	54
3470	5"		12.10	1.322		12.25	48.50		60.75	86.50
3480	6"		11.10	1.441		15.90	53		68.90	97
3490	8"	Q-2	10.30	2.330		39.50	89		128.50	178
3710	Reducing insert, schedule 40, socket weld									
3712	3/4"	1 Plum	31.50	.254	Ea.	.29	10.35		10.64	15.90
3713	1"		27.50	.291		.53	11.90		12.43	18.45

15108 | Plastic Pipe & Fittings

	15108 Plastic Pipe & Fittings	CREW	DAILY OUTPUT	LABOR-HOURS	UNIT	2005 BARE COSTS MAT.	LABOR	EQUIP.	TOTAL	TOTAL INCL O&P
560 3715	1-1/2"	1 Plum	22	.364	Ea.	.71	14.85		15.56	23.50 560
3716	2"	Q-1	40	.400		1.24	14.70		15.94	23.50
3717	4"		20	.800		6.60	29.50		36.10	51.50
3718	6"		12.20	1.311		16.30	48		64.30	90.50
3719	8"	Q-2	11.30	2.124		57	81		138	185
3730	Reducing insert, socket weld x female/male thread									
3732	1/2"	1 Plum	38.30	.209	Ea.	1.32	8.55		9.87	14.30
3733	3/4"		32.90	.243		.82	9.95		10.77	15.85
3734	1"		28.80	.278		1.14	11.35		12.49	18.30
3736	1-1/2"		23	.348		2.06	14.20		16.26	24
3737	2"	Q-1	41.90	.382		2.20	14.05		16.25	23.50
3738	4"	"	20.90	.766		6.15	28		34.15	49.50
3742	Male adapter, socket weld x male thread									
3744	1/2"	1 Plum	38.30	.209	Ea.	.26	8.55		8.81	13.15
3745	3/4"		32.90	.243		.29	9.95		10.24	15.25
3746	1"		28.80	.278		.50	11.35		11.85	17.60
3748	1-1/2"		23	.348		.82	14.20		15.02	22.50
3749	2"	Q-1	41.90	.382		1.07	14.05		15.12	22
3750	4"	"	20.90	.766		5.90	28		33.90	49
3754	Female adapter, socket weld x female thread									
3756	1/2"	1 Plum	38.30	.209	Ea.	.31	8.55		8.86	13.20
3757	3/4"		32.90	.243		.40	9.95		10.35	15.40
3758	1"		28.80	.278		.46	11.35		11.81	17.55
3760	1-1/2"		23	.348		.82	14.20		15.02	22.50
3761	2"	Q-1	41.90	.382		1.10	14.05		15.15	22
3762	4"	"	20.90	.766		6.15	28		34.15	49.50
3800	PVC, schedule 80, socket joints									
3810	Reducing insert									
3812	3/4"	1 Plum	28.60	.280	Ea.	.62	11.45		12.07	17.85
3813	1"		25	.320		2.40	13.05		15.45	22.50
3815	1-1/2"		20	.400		5.15	16.35		21.50	30
3816	2"	Q-1	36.40	.440		7.35	16.15		23.50	32.50
3817	4"		18.20	.879		28	32.50		60.50	79.50
3818	6"		11.10	1.441		39	53		92	123
3819	8"	Q-2	10.20	2.353		83.50	89.50		173	227
3830	Reducing insert, socket weld x female/male thread									
3832	1/2"	1 Plum	34.80	.230	Ea.	5.15	9.40		14.55	19.75
3833	3/4"		29.90	.268		3.92	10.95		14.87	21
3834	1"		26.10	.307		6.30	12.50		18.80	26
3836	1-1/2"		20.90	.383		6.35	15.65		22	30.50
3837	2"	Q-1	38	.421		12.45	15.50		27.95	37
3838	4"	"	19	.842		44	31		75	95
3844	Adapter, male socket x male thread									
3846	1/2"	1 Plum	34.80	.230	Ea.	2.68	9.40		12.08	17.05
3847	3/4"		29.90	.268		2.96	10.95		13.91	19.70
3848	1"		26.10	.307		5.15	12.50		17.65	24.50
3850	1-1/2"		20.90	.383		8.60	15.65		24.25	33
3851	2"	Q-1	38	.421		12.45	15.50		27.95	37
3852	4"	"	19	.842		28	31		59	77.50
3860	Adapter, female socket x female thread									
3862	1/2"	1 Plum	34.80	.230	Ea.	2.19	9.40		11.59	16.50
3863	3/4"		29.90	.268		3.62	10.95		14.57	20.50
3864	1"		26.10	.307		5.35	12.50		17.85	24.50
3866	1-1/2"		20.90	.383		10.60	15.65		26.25	35
3867	2"	Q-1	38	.421		18.55	15.50		34.05	44
3868	4"	"	19	.842		56.50	31		87.50	109

	15108 | Plastic Pipe & Fittings	CREW	DAILY OUTPUT	LABOR-HOURS	UNIT	2005 BARE COSTS MAT.	LABOR	EQUIP.	TOTAL	TOTAL INCL O&P	
560	3872 Union, socket joints										560
	3874 1/2"	1 Plum	25.80	.310	Ea.	5.35	12.65		18	25	
	3875 3/4"		22.10	.362		6.80	14.80		21.60	29.50	
	3876 1"		19.30	.415		7.75	16.95		24.70	34	
	3878 1-1/2"	↓	15.50	.516		17.45	21		38.45	50.50	
	3879 2"	Q-1	28.10	.569	↓	23.50	21		44.50	57.50	
	3888 Cap										
	3890 1/2"	1 Plum	54.50	.147	Ea.	2.56	6		8.56	11.80	
	3891 3/4"		46.70	.171		2.69	7		9.69	13.45	
	3892 1"		41	.195		4.78	7.95		12.73	17.25	
	3894 1-1/2"	↓	32.80	.244		5.75	9.95		15.70	21.50	
	3895 2"	Q-1	59.50	.269		15.15	9.90		25.05	31.50	
	3896 4"		30	.533		46	19.60		65.60	80	
	3897 6"	↓	18.20	.879		112	32.50		144.50	172	
	3898 8"	Q-2	16.70	1.437	↓	143	55		198	241	
	4500 DWV, ABS, non pressure, socket joints										
	4540 1/4 Bend, 1-1/4"	1 Plum	20.20	.396	Ea.	2.57	16.20		18.77	27.50	
	4560 1-1/2"	"	18.20	.440		1.97	17.95		19.92	29	
	4570 2"	Q-1	33.10	.483		3.04	17.80		20.84	30	
	4580 3"		20.80	.769		7.30	28.50		35.80	50.50	
	4590 4"		16.50	.970		10.95	35.50		46.45	65.50	
	4600 6"	↓	10.10	1.584	↓	58.50	58.50		117	152	
	4650 1/8 Bend, same as 1/4 Bend										
	4800 Tee, sanitary										
	4820 1-1/4"	1 Plum	13.50	.593	Ea.	2.91	24		26.91	39.50	
	4830 1-1/2"	"	12.10	.661		2.68	27		29.68	43.50	
	4840 2"	Q-1	20	.800		3.92	29.50		33.42	48.50	
	4850 3"		13.90	1.151		8.80	42.50		51.30	73	
	4860 4"		11	1.455		16.10	53.50		69.60	98	
	4862 Tee, sanitary, reducing, 2" x 1-1/2"		22	.727		3.78	27		30.78	44	
	4864 3" x 2"		15.30	1.046		6.80	38.50		45.30	65.50	
	4868 4" x 3"	↓	12.10	1.322	↓	18	48.50		66.50	93	
	4870 Combination Y and 1/8 bend										
	4872 1-1/2"	1 Plum	12.10	.661	Ea.	5.45	27		32.45	46.50	
	4874 2"	Q-1	20	.800		6.60	29.50		36.10	51.50	
	4876 3"		13.90	1.151		14.65	42.50		57.15	79.50	
	4878 4"		11	1.455		27.50	53.50		81	111	
	4880 3" x 1-1/2"		15.50	1.032		13.60	38		51.60	72	
	4882 4" x 3"	↓	12.10	1.322		20.50	48.50		69	95.50	
	4900 Wye, 1-1/4"	1 Plum	13.50	.593		3.36	24		27.36	40	
	4902 1-1/2"	"	12.10	.661		3.43	27		30.43	44.50	
	4904 2"	Q-1	20	.800		4.82	29.50		34.32	49.50	
	4906 3"		13.90	1.151		11.95	42.50		54.45	76.50	
	4908 4"		11	1.455		21	53.50		74.50	104	
	4910 6"		6.70	2.388		72.50	88		160.50	212	
	4918 3" x 1-1/2"		15.50	1.032		8	38		46	66	
	4920 4" x 3"		12.10	1.322		15.30	48.50		63.80	90	
	4922 6" x 4"	↓	6.90	2.319		60	85.50		145.50	194	
	4930 Double Wye, 1-1/2"	1 Plum	9.10	.879		7.95	36		43.95	63	
	4932 2"	Q-1	16.60	.964		9.35	35.50		44.85	64	
	4934 3"		10.40	1.538		24	56.50		80.50	112	
	4936 4"		8.25	1.939		49	71.50		120.50	161	
	4940 2" x 1-1/2"		16.80	.952		9.35	35		44.35	63	
	4942 3" x 2"		10.60	1.509		17.95	55.50		73.45	103	
	4944 4" x 3"		8.45	1.893		38.50	69.50		108	148	
	4946 6" x 4"	↓	7.25	2.207	↓	81	81		162	211	

560	15108 Plastic Pipe & Fittings	CREW	DAILY OUTPUT	LABOR-HOURS	UNIT	2005 BARE COSTS MAT.	LABOR	EQUIP.	TOTAL	TOTAL INCL O&P	560
	4950 Reducer bushing, 2" x 1-1/2"	Q-1	36.40	.440	Ea.	1.29	16.15		17.44	26	
	4952 3" x 1-1/2"		27.30	.586		4.33	21.50		25.83	37.50	
	4954 4" x 2"		18.20	.879		9.95	32.50		42.45	59.50	
	4956 6" x 4"		11.10	1.441		26.50	53		79.50	109	
	4960 Couplings, 1-1/2"	1 Plum	18.20	.440		.82	17.95		18.77	28	
	4962 2"	Q-1	33.10	.483		1.22	17.80		19.02	28	
	4963 3"		20.80	.769		3.13	28.50		31.63	46	
	4964 4"		16.50	.970		4.90	35.50		40.40	59	
	4966 6"		10.10	1.584		22	58.50		80.50	112	
	4970 2" x 1-1/2"		33.30	.480		2.12	17.65		19.77	29	
	4972 3" x 1-1/2"		21	.762		5.95	28		33.95	48.50	
	4974 4" x 3"		16.70	.958		10.55	35		45.55	64.50	
	4978 Closet flange, 4"	1 Plum	32	.250		5.20	10.20		15.40	21	
	4980 4" x 3"	"	34	.235		5.20	9.60		14.80	20	
	5000 DWV, PVC, schedule 40, socket joints										
	5040 1/4 bend, 1-1/4"	1 Plum	20.20	.396	Ea.	3.25	16.20		19.45	28	
	5060 1-1/2"	"	18.20	.440		1.24	17.95		19.19	28.50	
	5070 2"	Q-1	33.10	.483		1.93	17.80		19.73	28.50	
	5080 3"		20.80	.769		5.55	28.50		34.05	48.50	
	5090 4"		16.50	.970		9.05	35.50		44.55	63.50	
	5100 6"		10.10	1.584		40.50	58.50		99	133	
	5105 8"	Q-2	9.30	2.581		82.50	98.50		181	239	
	5106 10"	"	8.50	2.824		75	108		183	245	
	5110 1/4 bend, long sweep, 1-1/2"	1 Plum	18.20	.440		3.05	17.95		21	30.50	
	5112 2"	Q-1	33.10	.483		2.93	17.80		20.73	29.50	
	5114 3"		20.80	.769		6.85	28.50		35.35	50	
	5116 4"		16.50	.970		12.80	35.50		48.30	67.50	
	5150 1/8 bend, 1-1/4"	1 Plum	20.20	.396		1.95	16.20		18.15	26.50	
	5170 1-1/2"	"	18.20	.440		1.21	17.95		19.16	28.50	
	5180 2"	Q-1	33.10	.483		1.76	17.80		19.56	28.50	
	5190 3"		20.80	.769		4.81	28.50		33.31	48	
	5200 4"		16.50	.970		7.85	35.50		43.35	62	
	5210 6"		10.10	1.584		37.50	58.50		96	129	
	5215 8"	Q-2	9.30	2.581		66.50	98.50		165	221	
	5216 10"		8.50	2.824		109	108		217	282	
	5217 12"		7.60	3.158		150	120		270	345	
	5250 Tee, sanitary 1-1/4"	1 Plum	13.50	.593		3.36	24		27.36	40	
	5254 1-1/2"	"	12.10	.661		2.13	27		29.13	43	
	5255 2"	Q-1	20	.800		3.18	29.50		32.68	47.50	
	5256 3"		13.90	1.151		7.05	42.50		49.55	71.50	
	5257 4"		11	1.455		12.30	53.50		65.80	94	
	5259 6"		6.70	2.388		60.50	88		148.50	199	
	5261 8"	Q-2	6.20	3.871		181	148		329	420	
	5264 2" x 1-1/2"	Q-1	22	.727		2.94	27		29.94	43	
	5266 3" x 1-1/2"		15.50	1.032		4.47	38		42.47	62	
	5268 4" x 3"		12.10	1.322		19.20	48.50		67.70	94	
	5271 6" x 4"		6.90	2.319		65	85.50		150.50	200	
	5276 Tee, sanitary, reducing										
	5281 2" x 1-1/2" x 1-1/2"	Q-1	23	.696	Ea.	2.59	25.50		28.09	41.50	
	5282 2" x 1-1/2" x 2"		22	.727		3.56	27		30.56	44	
	5283 2" x 2" x 1-1/2"		22	.727		2.65	27		29.65	43	
	5284 3" x 3" x 1-1/2"		15.50	1.032		4.47	38		42.47	62	
	5285 3" x 3" x 2"		15.30	1.046		5.30	38.50		43.80	64	
	5286 4" x 4" x 1-1/2"		12.30	1.301		14.30	48		62.30	88	
	5287 4" x 4" x 2"		12.20	1.311		14.05	48		62.05	88	
	5288 4" x 4" x 3"		12.10	1.322		19.15	48.50		67.65	94	

15108 | Plastic Pipe & Fittings

			CREW	DAILY OUTPUT	LABOR-HOURS	UNIT	2005 BARE COSTS MAT.	LABOR	EQUIP.	TOTAL	TOTAL INCL O&P	
560	5291	6″ x 6″ x 4″	Q-1	6.90	2.319	Ea.	58.50	85.50		144	192	560
	5294	Tee, double sanitary										
	5295	1-1/2″	1 Plum	9.10	.879	Ea.	4.32	36		40.32	59	
	5296	2″	Q-1	16.60	.964		6.30	35.50		41.80	60.50	
	5297	3″		10.40	1.538		17.70	56.50		74.20	104	
	5298	4″		8.25	1.939		28.50	71.50		100	139	
	5303	Wye, reducing										
	5304	2″ x 1-1/2″ x 1-1/2″	Q-1	23	.696	Ea.	5.05	25.50		30.55	44	
	5305	2″ x 2″ x 1-1/2″		22	.727		5.35	27		32.35	46	
	5306	3″ x 3″ x 2″		15.30	1.046		14.20	38.50		52.70	73.50	
	5307	4″ x 4″ x 2″		12.20	1.311		9.75	48		57.75	83	
	5309	4″ x 4″ x 3″		12.10	1.322		12.75	48.50		61.25	87	
	5314	Combination Y & 1/8 bend, 1-1/2″	1 Plum	12.10	.661		4.59	27		31.59	45.50	
	5315	2″	Q-1	20	.800		5.95	29.50		35.45	50.50	
	5317	3″		13.90	1.151		11.70	42.50		54.20	76.50	
	5318	4″		11	1.455		21	53.50		74.50	104	
	5319	6″		6.70	2.388		76.50	88		164.50	217	
	5320	8″	Q-2	6.20	3.871		122	148		270	355	
	5324	Combination Y & 1/8 bend, reducing										
	5325	2″ x 2″ x 1-1/2″	Q-1	22	.727	Ea.	6.55	27		33.55	47.50	
	5327	3″ x 3″ x 1-1/2″		15.50	1.032		11.40	38		49.40	69.50	
	5328	3″ x 3″ x 2″		15.30	1.046		8.45	38.50		46.95	67.50	
	5329	4″ x 4″ x 2″		12.20	1.311		15.85	48		63.85	90	
	5331	Wye, 1-1/4″	1 Plum	13.50	.593		4.29	24		28.29	41	
	5332	1-1/2″	″	12.10	.661		2.86	27		29.86	43.50	
	5333	2″	Q-1	20	.800		3.65	29.50		33.15	48	
	5334	3″		13.90	1.151		9.90	42.50		52.40	74.50	
	5335	4″		11	1.455		16	53.50		69.50	98	
	5336	6″		6.70	2.388		62	88		150	200	
	5337	8″	Q-2	6.20	3.871		73.50	148		221.50	305	
	5338	10″		5.70	4.211		143	161		304	400	
	5339	12″		5.10	4.706		232	179		411	525	
	5341	2″ x 1-1/2″	Q-1	22	.727		5.05	27		32.05	45.50	
	5342	3″ x 1-1/2″		15.50	1.032		6.70	38		44.70	64.50	
	5343	4″ x 3″		12.10	1.322		12.75	48.50		61.25	87	
	5344	6″ x 4″		6.90	2.319		44.50	85.50		130	177	
	5345	8″ x 6″	Q-2	6.40	3.750		124	143		267	350	
	5347	Double wye, 1-1/2″	1 Plum	9.10	.879		6.10	36		42.10	60.50	
	5348	2″	Q-1	16.60	.964		7.85	35.50		43.35	62	
	5349	3″		10.40	1.538		20	56.50		76.50	107	
	5350	4″		8.25	1.939		41	71.50		112.50	152	
	5354	2″ x 1-1/2″		16.80	.952		7.15	35		42.15	60.50	
	5355	3″ x 2″		10.60	1.509		15.05	55.50		70.55	100	
	5356	4″ x 3″		8.45	1.893		32.50	69.50		102	141	
	5357	6″ x 4″		7.25	2.207		67.50	81		148.50	197	
	5374	Coupling, 1-1/4″	1 Plum	20.20	.396		1.80	16.20		18	26.50	
	5376	1-1/2″	″	18.20	.440		.63	17.95		18.58	27.50	
	5378	2″	Q-1	33.10	.483		.78	17.80		18.58	27.50	
	5380	3″		20.80	.769		2.71	28.50		31.21	45.50	
	5390	4″		16.50	.970		4.45	35.50		39.95	58.50	
	5400	6″		10.10	1.584		16.30	58.50		74.80	105	
	5402	8″	Q-2	9.30	2.581		36.50	98.50		135	189	
	5404	2″ x 1-1/2″	Q-1	33.30	.480		1.60	17.65		19.25	28.50	
	5406	3″ x 1-1/2″		21	.762		5.10	28		33.10	47.50	
	5408	4″ x 3″		16.70	.958		9	35		44	63	
	5410	Reducer bushing, 2″ x 1-1/4″		36.50	.438		.97	16.10		17.07	25	

15108 | Plastic Pipe & Fittings

		Description	CREW	DAILY OUTPUT	LABOR-HOURS	UNIT	2005 BARE COSTS MAT.	LABOR	EQUIP.	TOTAL	TOTAL INCL O&P	
560	5411	2″ x 1-1/2″	Q-1	36.40	.440	Ea.	1.08	16.15		17.23	25.50	560
	5412	3″ x 1-1/2″		27.30	.586		4.75	21.50		26.25	38	
	5413	3″ x 2″		27.10	.590		2.34	21.50		23.84	35	
	5414	4″ x 2″		18.20	.879		9.25	32.50		41.75	58.50	
	5415	4″ x 3″		16.70	.958		5.10	35		40.10	58.50	
	5416	6″ x 4″		11.10	1.441		26.50	53		79.50	109	
	5418	8″ x 6″	Q-2	10.20	2.353		53	89.50		142.50	194	
	5425	Closet flange 4″	Q-1	32	.500		5.40	18.40		23.80	33.50	
	5426	4″ x 3″	″	34	.471		8.25	17.30		25.55	35	
	5450	Solvent cement for PVC, industrial grade, per quart				Qt.	11.65			11.65	12.80	
	5500	CPVC, Schedule 80, threaded joints										
	5540	90° Elbow, 1/4″	1 Plum	32	.250	Ea.	7.85	10.20		18.05	24	
	5560	1/2″		30.30	.264		4.55	10.80		15.35	21	
	5570	3/4″		26	.308		6.80	12.55		19.35	26.50	
	5580	1″		22.70	.352		9.55	14.40		23.95	32	
	5590	1-1/4″		20.20	.396		18.40	16.20		34.60	45	
	5600	1-1/2″		18.20	.440		19.80	17.95		37.75	49	
	5610	2″	Q-1	33.10	.483		26.50	17.80		44.30	55.50	
	5620	2-1/2″		24.20	.661		78.50	24.50		103	123	
	5630	3″		20.80	.769		88.50	28.50		117	140	
	5640	4″		16.50	.970		138	35.50		173.50	206	
	5650	6″		10.10	1.584		163	58.50		221.50	267	
	5700	45° Elbow same as 90° Elbow										
	5850	Tee, 1/4″	1 Plum	22	.364	Ea.	15.25	14.85		30.10	39.50	
	5870	1/2″		20.20	.396		15.25	16.20		31.45	41.50	
	5880	3/4″		17.30	.462		22	18.90		40.90	52.50	
	5890	1″		15.20	.526		23.50	21.50		45	58.50	
	5900	1-1/4″		13.50	.593		24	24		48	62.50	
	5910	1-1/2″		12.10	.661		25	27		52	68	
	5920	2″	Q-1	20	.800		27.50	29.50		57	74.50	
	5930	2-1/2″		16.20	.988		136	36.50		172.50	205	
	5940	3″		13.90	1.151		158	42.50		200.50	238	
	5950	4″		11	1.455		375	53.50		428.50	490	
	5960	6″		6.70	2.388		203	88		291	355	
	6000	Coupling, 1/4″	1 Plum	32	.250		10	10.20		20.20	26.50	
	6020	1/2″		30.30	.264		8.20	10.80		19	25.50	
	6030	3/4″		26	.308		13.30	12.55		25.85	33.50	
	6040	1″		22.70	.352		15.10	14.40		29.50	38	
	6050	1-1/4″		20.20	.396		16	16.20		32.20	42	
	6060	1-1/2″		18.20	.440		17.20	17.95		35.15	46	
	6070	2″	Q-1	33.10	.483		20.50	17.80		38.30	49	
	6080	2-1/2″		24.20	.661		36.50	24.50		61	76.50	
	6090	3″		20.80	.769		42.50	28.50		71	89	
	6100	4″		16.50	.970		86	35.50		121.50	149	
	6110	6″		10.10	1.584		124	58.50		182.50	224	
	6120	8″	Q-2	9.30	2.581		252	98.50		350.50	425	
	6200	CTS, 100 psi at 180°F, hot and cold water										
	6230	90° Elbow, 1/2″	1 Plum	20	.400	Ea.	.10	16.35		16.45	24.50	
	6250	3/4″		19	.421		.17	17.20		17.37	26	
	6251	1″		16	.500		.58	20.50		21.08	31	
	6252	1-1/4″		15	.533		.99	22		22.99	34	
	6253	1-1/2″		14	.571		1.59	23.50		25.09	37	
	6254	2″	Q-1	23	.696		3.43	25.50		28.93	42.50	
	6260	45° Elbow, 1/2″	1 Plum	20	.400		.19	16.35		16.54	24.50	
	6280	3/4″		19	.421		.23	17.20		17.43	26.50	
	6281	1″		16	.500		.53	20.50		21.03	31	

Important: See the Reference Section for critical supporting data - Reference Nos., Crews, & City Cost Indexes

15108 | Plastic Pipe & Fittings

		Description	Crew	Daily Output	Labor-Hours	Unit	2005 Bare Costs Mat.	Labor	Equip.	Total	Total Incl O&P	
560	6282	1-1/4"	1 Plum	15	.533	Ea.	1.04	22		23.04	34	560
	6283	1-1/2"		14	.571		1.65	23.50		25.15	37	
	6284	2"	Q-1	23	.696		3.48	25.50		28.98	42.50	
	6290	Tee, 1/2"	1 Plum	13	.615		.14	25		25.14	38	
	6310	3/4"		12	.667		.26	27		27.26	41.50	
	6311	1"		11	.727		1.20	29.50		30.70	46	
	6312	1-1/4"		10	.800		1.60	32.50		34.10	51	
	6313	1-1/2"		10	.800		2.60	32.50		35.10	52	
	6314	2"	Q-1	17	.941		4.22	34.50		38.72	56.50	
	6320	Coupling, 1/2"	1 Plum	22	.364		.10	14.85		14.95	22.50	
	6340	3/4"		21	.381		.14	15.55		15.69	23.50	
	6341	1"		18	.444		.50	18.15		18.65	28	
	6342	1-1/4"		17	.471		.62	19.20		19.82	29.50	
	6343	1-1/2"		16	.500		.88	20.50		21.38	31.50	
	6344	2"	Q-1	28	.571		1.75	21		22.75	33.50	
	6360	Solvent cement for CPVC, commercial grade, per quart				Qt.	17.85			17.85	19.65	
	7990	Polybutyl/polyethyl pipe, for copper fittings see 15107-460-7000										
	8000	Compression type, PVC, 160 psi cold water										
	8010	Coupling, 3/4" CTS	1 Plum	21	.381	Ea.	2.36	15.55		17.91	26	
	8020	1" CTS		18	.444		2.92	18.15		21.07	30.50	
	8030	1-1/4" CTS		17	.471		4.08	19.20		23.28	33.50	
	8040	1-1/2" CTS		16	.500		5.60	20.50		26.10	36.50	
	8050	2" CTS		15	.533		7.85	22		29.85	41.50	
	8060	Female adapter, 3/4" FPT x 3/4" CTS		23	.348		3.76	14.20		17.96	25.50	
	8070	3/4" FPT x 1" CTS		21	.381		5.60	15.55		21.15	29.50	
	8080	1" FPT x 1" CTS		20	.400		5.65	16.35		22	30.50	
	8090	1-1/4" FPT x 1-1/4" CTS		18	.444		7.40	18.15		25.55	35.50	
	8100	1-1/2" FPT x 1-1/2" CTS		16	.500		8.65	20.50		29.15	40	
	8110	2" FPT x 2" CTS		13	.615		12.80	25		37.80	52	
	8130	Male adapter, 3/4" MPT x 3/4" CTS		23	.348		3.39	14.20		17.59	25	
	8140	3/4" MPT x 1" CTS		21	.381		4.17	15.55		19.72	28	
	8150	1" MPT x 1" CTS		20	.400		4.26	16.35		20.61	29	
	8160	1-1/4" MPT x 1-1/4" CTS		18	.444		6.55	18.15		24.70	34.50	
	8170	1-1/2" MPT x 1-1/2" CTS		16	.500		7.55	20.50		28.05	39	
	8180	2" MPT x 2" CTS		13	.615		11.75	25		36.75	51	
	8200	Spigot adapter, 3/4" IPS x 3/4" CTS		23	.348		3.39	14.20		17.59	25	
	8210	3/4" IPS x 1" CTS		21	.381		3.39	15.55		18.94	27	
	8220	1" IPS x 1" CTS		20	.400		3.66	16.35		20.01	28.50	
	8230	1-1/4" IPS x 1-1/4" CTS		18	.444		6.55	18.15		24.70	34.50	
	8240	1-1/2" IPS x 1-1/2" CTS		16	.500		7.55	20.50		28.05	39	
	8250	2" IPS x 2" CTS		13	.615		12.10	25		37.10	51.50	
	8270	Price includes insert stiffeners										
	8280	250 psi is same price as 160 psi										
	8300	Insert type, nylon, 160 & 250 psi, cold water										
	8310	Clamp ring stainless steel, 3/4" IPS	1 Plum	115	.070	Ea.	1.49	2.84		4.33	5.90	
	8320	1" IPS		107	.075		1.52	3.05		4.57	6.25	
	8330	1-1/4" IPS		101	.079		1.54	3.24		4.78	6.55	
	8340	1-1/2" IPS		95	.084		2.06	3.44		5.50	7.40	
	8350	2" IPS		85	.094		2.37	3.84		6.21	8.40	
	8370	Coupling, 3/4" IPS		22	.364		.71	14.85		15.56	23.50	
	8390	1-1/4" IPS		18	.444		1.68	18.15		19.83	29.50	
	8400	1-1/2" IPS		17	.471		2.19	19.20		21.39	31.50	
	8410	2" IPS		16	.500		2.63	20.50		23.13	33.50	
	8430	Elbow, 90°, 3/4" IPS		22	.364		1.03	14.85		15.88	23.50	
	8440	1" IPS		19	.421		1.13	17.20		18.33	27	
	8450	1-1/4" IPS		18	.444		1.27	18.15		19.42	29	

		15108 Plastic Pipe & Fittings	Crew	Daily Output	Labor-Hours	Unit	2005 Bare Costs Mat.	Labor	Equip.	Total	Total Incl O&P	
560	8460	1-1/2" IPS	1 Plum	17	.471	Ea.	1.50	19.20		20.70	30.50	560
	8470	2" IPS		16	.500		2.10	20.50		22.60	33	
	8490	Male adapter, 3/4" IPS x 3/4" MPT		25	.320		.71	13.05		13.76	20.50	
	8500	1" IPS x 1" MPT		21	.381		.94	15.55		16.49	24.50	
	8510	1-1/4" IPS x 1-1/4" MPT		20	.400		1.55	16.35		17.90	26	
	8520	1-1/2" IPS x 1-1/2" MPT		18	.444		2.05	18.15		20.20	30	
	8530	2" IPS x 2" MPT		15	.533		2.63	22		24.63	36	
	8550	Tee, 3/4" IPS		14	.571		1.34	23.50		24.84	36.50	
	8560	1" IPS		13	.615		2.21	25		27.21	40.50	
	8570	1-1/4" IPS		12	.667		3.96	27		30.96	45.50	
	8580	1-1/2" IPS		11	.727		4.97	29.50		34.47	50	
	8590	2" IPS		10	.800		6.70	32.50		39.20	56.50	
	8610	Insert type, PVC, 100 psi @ 180°F, hot & cold water										
	8620	Coupler, male, 3/8" CTS x 3/8" MPT	1 Plum	29	.276	Ea.	.40	11.25		11.65	17.40	
	8630	3/8" CTS x 1/2" MPT		28	.286		.47	11.65		12.12	18.05	
	8640	1/2" CTS x 1/2" MPT		27	.296		.43	12.10		12.53	18.65	
	8650	1/2" CTS x 3/4" MPT		26	.308		1.20	12.55		13.75	20	
	8660	3/4" CTS x 1/2" MPT		25	.320		1.20	13.05		14.25	21	
	8670	3/4" CTS x 3/4" MPT		25	.320		1.26	13.05		14.31	21	
	8700	Coupling, 3/8" CTS x 1/2" CTS		25	.320		2.15	13.05		15.20	22	
	8710	1/2" CTS		23	.348		2.50	14.20		16.70	24.50	
	8720	1/2" CTS x stub		23	.348		2.52	14.20		16.72	24.50	
	8730	3/4" CTS		22	.364		3.96	14.85		18.81	27	
	8750	Elbow 90°, 3/8" CTS		25	.320		2.55	13.05		15.60	22.50	
	8760	1/2" CTS		23	.348		3.01	14.20		17.21	25	
	8770	3/4" CTS		22	.364		4.61	14.85		19.46	27.50	
	8800	Rings, crimp, copper, 3/8" CTS		120	.067		.11	2.72		2.83	4.21	
	8810	1/2" CTS		117	.068		.12	2.79		2.91	4.33	
	8820	3/4" CTS		115	.070		.15	2.84		2.99	4.44	
	8850	Reducer tee, 3/8" x 3/8" x 1/2" CTS		17	.471		1.46	19.20		20.66	30.50	
	8860	1/2" x 3/8" x 1/2" CTS		15	.533		1.25	22		23.25	34.50	
	8870	3/4" x 1/2" x 1/2" CTS		14	.571		1.31	23.50		24.81	36.50	
	8890	3/4" x 3/4" x 1/2" CTS		14	.571		1.24	23.50		24.74	36.50	
	8900	3/4" x 1/2" x 3/8" CTS		14	.571		1.09	23.50		24.59	36	
	8930	Tee, 3/8" CTS		17	.471		1.05	19.20		20.25	30	
	8940	1/2" CTS		15	.533		1.19	22		23.19	34.50	
	8950	3/4" CTS		14	.571		1.29	23.50		24.79	36.50	
	8960	Copper rings included in fitting price										
	9000	Flare type, assembled, acetal, hot & cold water										
	9010	Coupling, 1/4" & 3/8" CTS	1 Plum	24	.333	Ea.	2.14	13.60		15.74	23	
	9020	1/2" CTS		22	.364		2.45	14.85		17.30	25	
	9030	3/4" CTS		21	.381		3.96	15.55		19.51	28	
	9040	1" CTS		18	.444		3.96	18.15		22.11	32	
	9050	Elbow 90°, 1/4" CTS		26	.308		2.38	12.55		14.93	21.50	
	9060	3/8" CTS		24	.333		2.55	13.60		16.15	23.50	
	9070	1/2" CTS		22	.364		3.01	14.85		17.86	26	
	9080	3/4" CTS		21	.381		4.61	15.55		20.16	28.50	
	9090	1" CTS		18	.444		5.80	18.15		23.95	34	
	9110	Tee ,1/4" & 3/8" CTS		15	.533		2.62	22		24.62	36	
	9120	1/2" CTS		14	.571		3.44	23.50		26.94	39	
	9130	3/4" CTS		13	.615		5.25	25		30.25	44	
	9140	1" CTS		12	.667		7	27		34	48.50	
	9550	For plastic hangers see 15060-300-8000										
	9560	For copper/brass fittings see 15107-460-7000										
590	0010	**PIPE, HIGH DENSITY POLYETHYLENE PLASTIC (HDPE)**										590
	0020	Not incl. hangers, trenching, backfill, hoisting or digging equipment.										

15108 | Plastic Pipe & Fittings

		15108 Plastic Pipe & Fittings	CREW	DAILY OUTPUT	LABOR-HOURS	UNIT	2005 BARE COSTS MAT.	LABOR	EQUIP.	TOTAL	TOTAL INCL O&P	
590	0030	Standard length is 40′, add a weld for each joint.										590
	0040	Single wall										
	0050	Straight										
	0054	1″ diameter DR 11				L.F.	.37			.37	.41	
	0058	1-1/2″ diameter DR 11					.81			.81	.89	
	0062	2″ diameter DR 11					1.05			1.05	1.16	
	0066	3″ diameter DR 11					2.27			2.27	2.50	
	0070	3″ diameter DR 17					1.52			1.52	1.67	
	0074	4″ diameter DR 11					3.75			3.75	4.13	
	0078	4″ diameter DR 17					2.52			2.52	2.77	
	0082	6″ diameter DR 11					8.10			8.10	8.95	
	0086	6″ diameter DR 17					5.45			5.45	6	
	0090	8″ diameter DR 11					13.75			13.75	15.15	
	0094	8″ diameter DR 26					6.20			6.20	6.80	
	0098	10″ diameter DR 11					21.50			21.50	23.50	
	0102	10″ diameter DR 26					9.60			9.60	10.55	
	0106	12″ diameter DR 11					30			30	33	
	0110	12″ diameter DR 26					13.50			13.50	14.85	
	0114	16″ diameter DR 11					44.50			44.50	49	
	0118	16″ diameter DR 26					20			20	22	
	0122	18″ diameter DR 11					56.50			56.50	62	
	0126	18″ diameter DR 26					25.50			25.50	28	
	0130	20″ diameter DR 11					70			70	77	
	0134	20″ diameter DR 26					31.50			31.50	34.50	
	0138	22″ diameter DR 11					84.50			84.50	93	
	0142	22″ diameter DR 26					38			38	41.50	
	0146	24″ diameter DR 11					101			101	111	
	0150	24″ diameter DR 26					45			45	49.50	
	0154	28″ diameter DR 17					92			92	101	
	0158	28″ diameter DR 26					61.50			61.50	67.50	
	0162	30″ diameter DR 21					86.50			86.50	95	
	0166	30″ diameter DR 26					57			57	62.50	
	0170	36″ diameter DR 26					57			57	62.50	
	0174	42″ diameter DR 26					138			138	152	
	0178	48″ diameter DR 26					175			175	192	
	0182	54″ diameter DR 26					228			228	251	
	0300	90° Elbow										
	0304	1″ diameter DR 11				Ea.	8.50			8.50	9.35	
	0308	1-1/2″ diameter DR 11					14.20			14.20	15.60	
	0312	2″ diameter DR 11					15.85			15.85	17.40	
	0316	3″ diameter DR 11					30.50			30.50	33.50	
	0320	3″ diameter DR 17					30.50			30.50	33.50	
	0324	4″ diameter DR 11					35.50			35.50	39	
	0328	4″ diameter DR 17					35.50			35.50	39	
	0332	6″ diameter DR 11					104			104	115	
	0336	6″ diameter DR 17					104			104	115	
	0340	8″ diameter DR 11					345			345	380	
	0344	8″ diameter DR 26					234			234	257	
	0348	10″ diameter DR 11					850			850	935	
	0352	10″ diameter DR 26					430			430	470	
	0356	12″ diameter DR 11					850			850	935	
	0360	12″ diameter DR 26					470			470	515	
	0364	16″ diameter DR 11					950			950	1,050	
	0368	16″ diameter DR 26					765			765	845	
	0372	18″ diameter DR 11					995			995	1,100	
	0376	18″ diameter DR 26					995			995	1,100	

	15108 | Plastic Pipe & Fittings	CREW	DAILY OUTPUT	LABOR-HOURS	UNIT	2005 BARE COSTS MAT.	LABOR	EQUIP.	TOTAL	TOTAL INCL O&P	
590	0380 20″ diameter DR 11				Ea.	1,475			1,475	1,625	590
	0384 20″ diameter DR 26					1,475			1,475	1,625	
	0388 22″ diameter DR 11					1,850			1,850	2,025	
	0392 22″ diameter DR 26					1,850			1,850	2,025	
	0396 24″ diameter DR 11					2,125			2,125	2,350	
	0400 24″ diameter DR 26					2,125			2,125	2,350	
	0404 28″ diameter DR 17					3,225			3,225	3,550	
	0408 28″ diameter DR 26					3,225			3,225	3,550	
	0412 30″ diameter DR 17					2,975			2,975	3,275	
	0416 30″ diameter DR 26					2,975			2,975	3,275	
	0420 36″ diameter DR 26					4,550			4,550	5,000	
	0424 42″ diameter DR 26					5,100			5,100	5,625	
	0428 48″ diameter DR 26					5,400			5,400	5,925	
	0432 54″ diameter DR 26				↓	12,500			12,500	13,800	
	0500 45° Elbow										
	0512 2″ diameter DR 11				Ea.	31.50			31.50	35	
	0516 3″ diameter DR 11					27.50			27.50	30.50	
	0520 3″ diameter DR 17					45			45	49.50	
	0524 4″ diameter DR 11					46			46	51	
	0528 4″ diameter DR 17					55.50			55.50	61	
	0532 6″ diameter DR 11					97.50			97.50	107	
	0536 6″ diameter DR 17					92.50			92.50	102	
	0540 8″ diameter DR 11					231			231	254	
	0544 8″ diameter DR 26					139			139	152	
	0548 10″ diameter DR 11					645			645	710	
	0552 10″ diameter DR 26					187			187	206	
	0556 12″ diameter DR 11					780			780	860	
	0560 12″ diameter DR 26					296			296	325	
	0564 16″ diameter DR 11					370			370	410	
	0568 16″ diameter DR 26					370			370	410	
	0572 18″ diameter DR 11					545			545	600	
	0576 18″ diameter DR 26					545			545	600	
	0580 20″ diameter DR 11					850			850	935	
	0584 20″ diameter DR 26					850			850	935	
	0588 22″ diameter DR 11					995			995	1,100	
	0592 22″ diameter DR 26					995			995	1,100	
	0596 24″ diameter DR 11					1,225			1,225	1,350	
	0600 24″ diameter DR 26					1,225			1,225	1,350	
	0604 28″ diameter DR 17					1,800			1,800	1,975	
	0608 28″ diameter DR 26					1,800			1,800	1,975	
	0612 30″ diameter DR 17					1,725			1,725	1,875	
	0616 30″ diameter DR 26					1,725			1,725	1,875	
	0620 36″ diameter DR 26					2,800			2,800	3,075	
	0624 42″ diameter DR 26					3,425			3,425	3,775	
	0628 48″ diameter DR 26					3,525			3,525	3,875	
	0632 54″ diameter DR 26				↓	5,750			5,750	6,325	
	0700 Tee										
	0704 1″ diameter DR 11				Ea.	8.50			8.50	9.35	
	0708 1-1/2″ diameter DR 11					14.20			14.20	15.60	
	0712 2″ diameter DR 11					19.80			19.80	22	
	0716 3″ diameter DR 11					31.50			31.50	35	
	0720 3″ diameter DR 17					31.50			31.50	35	
	0724 4″ diameter DR 11					50			50	55	
	0728 4″ diameter DR 17					50			50	55	
	0732 6″ diameter DR 11					124			124	136	
	0736 6″ diameter DR 17				↓	124			124	136	

15 MECHANICAL

15108 | Plastic Pipe & Fittings

			CREW	DAILY OUTPUT	LABOR-HOURS	UNIT	2005 BARE COSTS MAT.	LABOR	EQUIP.	TOTAL	TOTAL INCL O&P	
590	0740	8″ diameter DR 11				Ea.	286			286	315	590
	0744	8″ diameter DR 17					260			260	286	
	0748	10″ diameter DR 11					895			895	985	
	0752	10″ diameter DR 17					320			320	350	
	0756	12″ diameter DR 11					1,275			1,275	1,425	
	0760	12″ diameter DR 17					365			365	400	
	0764	16″ diameter DR 11					855			855	940	
	0768	16″ diameter DR 17					855			855	940	
	0772	18″ diameter DR 11					905			905	995	
	0776	18″ diameter DR 17					905			905	995	
	0780	20″ diameter DR 11					1,125			1,125	1,225	
	0784	20″ diameter DR 17					1,125			1,125	1,225	
	0788	22″ diameter DR 11					1,550			1,550	1,725	
	0792	22″ diameter DR 17					1,550			1,550	1,725	
	0796	24″ diameter DR 11					1,550			1,550	1,725	
	0800	24″ diameter DR 17					1,550			1,550	1,725	
	0804	28″ diameter DR 17					2,775			2,775	3,050	
	0812	30″ diameter DR 17					3,075			3,075	3,400	
	0820	36″ diameter DR 17					6,150			6,150	6,775	
	0824	42″ diameter DR 26					6,775			6,775	7,450	
	0828	48″ diameter DR 26				▼	7,925			7,925	8,725	
	1000	Flange adptr, w/back-up ring and 1/2 cost of plated bolt set										
	1004	1″ diameter DR 11				Ea.	78			78	85.50	
	1008	1-1/2″ diameter DR 11					79.50			79.50	87.50	
	1012	2″ diameter DR 11					44			44	48.50	
	1016	3″ diameter DR 11					54.50			54.50	59.50	
	1020	3″ diameter DR 17					54.50			54.50	59.50	
	1024	4″ diameter DR 11					71			71	78	
	1028	4″ diameter DR 17					71			71	78	
	1032	6″ diameter DR 11					103			103	114	
	1036	6″ diameter DR 17					103			103	114	
	1040	8″ diameter DR 11					148			148	162	
	1044	8″ diameter DR 26					148			148	162	
	1048	10″ diameter DR 11					222			222	244	
	1052	10″ diameter DR 26					222			222	244	
	1056	12″ diameter DR 11					320			320	350	
	1060	12″ diameter DR 26					320			320	350	
	1064	16″ diameter DR 11					685			685	750	
	1068	16″ diameter DR 26					685			685	750	
	1072	18″ diameter DR 11					880			880	965	
	1076	18″ diameter DR 26					880			880	965	
	1080	20″ diameter DR 11					1,250			1,250	1,375	
	1084	20″ diameter DR 26					1,250			1,250	1,375	
	1088	22″ diameter DR 11					1,475			1,475	1,625	
	1092	22″ diameter DR 26					1,475			1,475	1,625	
	1096	24″ diameter DR 17					1,575			1,575	1,725	
	1100	24″ diameter DR 32.5					1,575			1,575	1,725	
	1104	28″ diameter DR 15.5					2,675			2,675	2,925	
	1108	28″ diameter DR 32.5					2,600			2,600	2,875	
	1112	30″ diameter DR 11					2,875			2,875	3,150	
	1116	30″ diameter DR 21					2,500			2,500	2,750	
	1120	36″ diameter DR 26					3,700			3,700	4,050	
	1124	42″ diameter DR 26					4,175			4,175	4,575	
	1128	48″ diameter DR 26					5,875			5,875	6,475	
	1132	54″ diameter DR 26				▼	6,575			6,575	7,225	
	1200	Reducer										

15108 | Plastic Pipe & Fittings

590

	15108 Plastic Pipe & Fittings	CREW	DAILY OUTPUT	LABOR-HOURS	UNIT	2005 BARE COSTS MAT.	LABOR	EQUIP.	TOTAL	TOTAL INCL O&P
1204	1-1/2" x 1" diameter DR 11				Ea.	9.10			9.10	10
1208	2" x 1-1/2" diameter DR 11					14.20			14.20	15.60
1212	3" x 2" diameter DR 11					18.50			18.50	20.50
1216	4" x 2" diameter DR 11					27.50			27.50	30.50
1220	4" x 3" diameter DR 11					27.50			27.50	30.50
1224	6" x 4" diameter DR 11					64.50			64.50	71
1228	8" x 6" diameter DR 11					114			114	125
1232	10" x 8" diameter DR 11					152			152	167
1236	12" x 8" diameter DR 11					284			284	310
1240	12" x 10" diameter DR 11					203			203	224
1244	14" x 12" diameter DR 11					228			228	251
1248	16" x 14" diameter DR 11					275			275	300
1252	18" x 16" diameter DR 11					550			550	605
1256	20" x 18" diameter DR 11					745			745	820
1260	22" x 20" diameter DR 11					1,525			1,525	1,675
1264	24" x 22" diameter DR 11					975			975	1,075
1268	26" x 24" diameter DR 11					1,150			1,150	1,250
1272	28" x 24" diameter DR 11					1,300			1,300	1,450
1276	32" x 28" diameter DR 17					1,575			1,575	1,725
1280	36" x 32" diameter DR 17					1,875			1,875	2,050
4000	Welding labor per joint, not including welding machine									
4010	Pipe joint size (cost based on thickest wall for each dia.)									
4030	1" pipe size	4 Skwk	273	.117	Ea.		4.09		4.09	6.35
4040	1-1/2" pipe size		175	.183			6.35		6.35	9.90
4050	2" pipe size		128	.250			8.70		8.70	13.55
4060	3" pipe size		100	.320			11.15		11.15	17.35
4070	4" pipe size		77	.416			14.50		14.50	22.50
4080	6" pipe size	5 Skwk	63	.635			22		22	34.50
4090	8" pipe size		48	.833			29		29	45
4100	10" pipe size		40	1			35		35	54
4110	12" pipe size	6 Skwk	41	1.171			41		41	63.50
4120	16" pipe size		34	1.412			49		49	76.50
4130	18" pipe size		32	1.500			52.50		52.50	81.50
4140	20" pipe size	8 Skwk	37	1.730			60.50		60.50	94
4150	22" pipe size		35	1.829			63.50		63.50	99
4160	24" pipe size		34	1.882			65.50		65.50	102
4170	28" pipe size		33	1.939			67.50		67.50	105
4180	30" pipe size		32	2			69.50		69.50	108
4190	36" pipe size		31	2.065			72		72	112
4200	42" pipe size		30	2.133			74.50		74.50	116
4210	48" pipe size	9 Skwk	33	2.182			76		76	118
4220	54" pipe size	"	31	2.323			81		81	126
4300	Note: Cost for set up each time welder is moved.									
4301	Add 50% of a weld cost									
4310	Welder usually remains stationary with pipe moved through it.									
4340	Weld machine, rental per day based on dia. capacity									
4350	1" thru 2" diameter				Ea.			35	35	38.50
4360	3" thru 4" diameter							40	40	44
4370	6" thru 8" diameter							90	90	99
4380	10" thru 12" diameter							155	155	170.50
4390	16" thru 18" diameter							225	225	247.50
4400	20" thru 24" diameter							435	435	478.50
4410	28" thru 32" diameter							475	475	525
4420	36" diameter							495	495	545
4430	42" thru 54" diameter							775	775	855
5000	Dual wall contained pipe									

15108 | Plastic Pipe & Fittings		CREW	DAILY OUTPUT	LABOR-HOURS	UNIT	2005 BARE COSTS MAT.	LABOR	EQUIP.	TOTAL	TOTAL INCL O&P	
590	5040 Straight										590
	5054 1" DR 11 x 3" DR 11				L.F.	7.60			7.60	8.35	
	5058 1" DR 11 x 4" DR 11					8.30			8.30	9.15	
	5062 1-1/2" DR 11 x 4" DR 17					8.80			8.80	9.65	
	5066 2" DR 11 x 4" DR 17					8.90			8.90	9.75	
	5070 2" DR 11 x 6" DR 17					13.85			13.85	15.25	
	5074 3" DR 11 x 6" DR 17					15.25			15.25	16.80	
	5078 3" DR 11 x 6" DR 26					12.25			12.25	13.45	
	5082 3" DR 17 x 8" DR 11					25			25	27.50	
	5086 3" DR 17 x 8" DR 17					19.90			19.90	22	
	5090 4" DR 11 x 8" DR 17					22.50			22.50	25	
	5094 4" DR 17 x 8" DR 26					17.80			17.80	19.55	
	5098 6" DR 11 x 10" DR 17					36			36	39.50	
	5102 6" DR 17 x 10" DR 26					27.50			27.50	30	
	5106 6" DR 26 x 10" DR 26					25.50			25.50	28	
	5110 8" DR 17 x 12" DR 26					39.50			39.50	43.50	
	5114 8" DR 26 x 12" DR 32.5					33			33	36	
	5118 10" DR 17 x 14" DR 26					52			52	57	
	5122 10" DR 17 x 16" DR 26					58.50			58.50	64.50	
	5126 10" DR 26 x 16" DR 26					53			53	58.50	
	5130 12" DR 26 x 16" DR 26					60.50			60.50	66.50	
	5134 12" DR 17 x 18" DR 26					77			77	84.50	
	5138 12" DR 26 x 18" DR 26					69			69	76	
	5142 14" DR 26 x 20" DR 32.5					75.50			75.50	83	
	5146 16" DR 26 x 22" DR 32.5					114			114	126	
	5150 18" DR 26 x 24" DR 32.5					105			105	115	
	5154 20" DR 32.5 x 28" DR 32.5					136			136	150	
	5158 22" DR 32.5 x 30" DR 32.5					136			136	150	
	5162 24" DR 32.5 x 32" DR 32.5					173			173	191	
	5166 36" DR 32.5 x 42" DR 32.5				↓	293			293	320	
	5300 Force transfer coupling										
	5354 1" DR 11 x 3" DR 11				Ea.	273			273	300	
	5358 1" DR 11 x 4" DR 17					287			287	315	
	5362 1-1/2" DR 11 x 4" DR 17					300			300	330	
	5366 2" DR 11 x 4" DR 17					315			315	345	
	5370 2" DR 11 x 6" DR 17					455			455	500	
	5374 3" DR 11 x 6" DR 17					455			455	500	
	5378 3" DR 11 x 6" DR 26					455			455	500	
	5382 3" DR 11 x 8" DR 11					480			480	530	
	5386 3" DR 11 x 8" DR 17					480			480	530	
	5390 4" DR 11 x 8" DR 17					480			480	530	
	5394 4" DR 17 x 8" DR 26					425			425	465	
	5398 6" DR 11 x 10" DR 17					610			610	675	
	5402 6" DR 17 x 10" DR 26					505			505	555	
	5406 6" DR 26 x 10" DR 26					505			505	555	
	5410 8" DR 17 x 12" DR 26					670			670	735	
	5414 8" DR 26 x 12" DR 32.5					540			540	595	
	5418 10" DR 17 x 14" DR 26					875			875	965	
	5422 10" DR 17 x 16" DR 26					875			875	965	
	5426 10" DR 26 x 16" DR 26					875			875	965	
	5430 12" DR 26 x 16" DR 26					1,100			1,100	1,200	
	5434 12" DR 17 x 18" DR 26					1,125			1,125	1,250	
	5438 12" DR 26 x 18" DR 26					1,125			1,125	1,250	
	5442 14" DR 26 x 20" DR 32.5					1,200			1,200	1,300	
	5446 16" DR 26 x 22" DR 32.5					1,225			1,225	1,350	
	5450 18" DR 26 x 24" DR 32.5				↓	1,325			1,325	1,450	

	15108 | Plastic Pipe & Fittings		CREW	DAILY OUTPUT	LABOR-HOURS	UNIT	2005 BARE COSTS MAT.	LABOR	EQUIP.	TOTAL	TOTAL INCL O&P	
590	5454	20″ DR 32.5 x 28″ DR 32.5				Ea.	1,625			1,625	1,775	590
	5458	22″ DR 32.5 x 30″ DR 32.5					1,800			1,800	1,975	
	5462	24″ DR 32.5 x 32″ DR 32.5					1,925			1,925	2,125	
	5466	36″ DR 32.5 x 42″ DR 32.5					3,200			3,200	3,525	
	5600	90° Elbow										
	5654	1″ DR 11 x 3″ DR 11				Ea.	226			226	248	
	5658	1″ DR 11 x 4″ DR 17					216			216	238	
	5662	1-1/2″ DR 11 x 4″ DR 17					236			236	260	
	5666	2″ DR 11 x 4″ DR 17					250			250	275	
	5670	2″ DR 11 x 6″ DR 17					320			320	350	
	5674	3″ DR 11 x 6″ DR 17					365			365	400	
	5678	3″ DR 17 x 6″ DR 26					291			291	320	
	5682	3″ DR 17 x 8″ DR 11					590			590	645	
	5686	3″ DR 17 x 8″ DR 17					455			455	505	
	5690	4″ DR 11 x 8″ DR 17					540			540	590	
	5694	4″ DR 17 x 8″ DR 26					425			425	470	
	5698	6″ DR 11 x 10″ DR 17					715			715	790	
	5702	6″ DR 17 x 10″ DR 26					550			550	605	
	5706	6″ DR 26 x 10″ DR 26					515			515	565	
	5710	8″ DR 17 x 12″ DR 26					815			815	895	
	5714	8″ DR 26 x 12″ DR 32.5					755			755	830	
	5718	10″ DR 17 x 14″ DR 26					1,275			1,275	1,400	
	5722	10″ DR 17 x 16″ DR 26					1,225			1,225	1,350	
	5726	10″ DR 26 x 16″ DR 26					1,125			1,125	1,250	
	5730	12″ DR 26 x 16″ DR 26					1,250			1,250	1,400	
	5734	12″ DR 17 x 18″ DR 26					1,500			1,500	1,650	
	5738	12″ DR 26 x 18″ DR 26					1,375			1,375	1,525	
	5742	14″ DR 26 x 20″ DR 32.5					1,825			1,825	2,000	
	5746	16″ DR 26 x 22″ DR 32.5					1,825			1,825	2,025	
	5750	18″ DR 26 x 24″ DR 32.5					2,250			2,250	2,475	
	5754	20″ DR 32.5 x 28″ DR 32.5					2,600			2,600	2,850	
	5758	22″ DR 32.5 x 30″ DR 32.5					3,525			3,525	3,900	
	5762	24″ DR 32.5 x 32″ DR 32.5					3,425			3,425	3,775	
	5766	36″ DR 32.5 x 42″ DR 32.5					5,575			5,575	6,150	
	5800	45° Elbow										
	5804	1″ DR 11 x 3″ DR 11				Ea.	135			135	149	
	5808	1″ DR 11 x 4″ DR 17					134			134	148	
	5812	1-1/2″ DR 11 x 4″ DR 17					144			144	158	
	5816	2″ DR 11 x 4″ DR 17					157			157	173	
	5820	2″ DR 11 x 6″ DR 17					197			197	217	
	5824	3″ DR 11 x 6″ DR 17					219			219	241	
	5828	3″ DR 17 x 6″ DR 26					181			181	200	
	5832	3″ DR 17 x 8″ DR 11					335			335	370	
	5836	3″ DR 17 x 8″ DR 17					268			268	295	
	5840	4″ DR 11 x 8″ DR 17					315			315	345	
	5844	4″ DR 17 x 8″ DR 26					256			256	281	
	5848	6″ DR 11 x 10″ DR 17					415			415	455	
	5852	6″ DR 17 x 10″ DR 26					325			325	355	
	5856	6″ DR 26 x 10″ DR 26					305			305	335	
	5860	8″ DR 17 x 12″ DR 26					495			495	540	
	5864	8″ DR 26 x 12″ DR 32.5					460			460	505	
	5868	10″ DR 17 x 14″ DR 26					770			770	850	
	5872	10″ DR 17 x 16″ DR 26					755			755	830	
	5876	10″ DR 26 x 16″ DR 26					710			710	780	
	5880	12″ DR 26 x 16″ DR 26					800			800	880	
	5884	12″ DR 17 x 18″ DR 26					940			940	1,025	

15100 | Building Services Piping

15108 | Plastic Pipe & Fittings

			CREW	DAILY OUTPUT	LABOR-HOURS	UNIT	2005 BARE COSTS MAT.	LABOR	EQUIP.	TOTAL	TOTAL INCL O&P	
590	5888	12" DR 26 x 18" DR 26				Ea.	875			875	965	590
	5892	14" DR 26 x 20" DR 32.5					1,125			1,125	1,250	
	5896	16" DR 26 x 22" DR 32.5					1,175			1,175	1,300	
	5900	18" DR 26 x 24" DR 32.5					1,400			1,400	1,550	
	5904	20" DR 32.5 x 28" DR 32.5					1,725			1,725	1,875	
	5908	22" DR 32.5 x 30" DR 32.5					2,175			2,175	2,400	
	5912	24" DR 32.5 x 32" DR 32.5					2,175			2,175	2,375	
	5916	36" DR 32.5 x 42" DR 32.5					3,375			3,375	3,700	
	6000	Access port with 4" riser										
	6050	1" DR 11 x 4" DR 17				Ea.	295			295	325	
	6054	1-1/2" DR 11 x 4" DR 17					300			300	330	
	6058	2" DR 11 x 6" DR 17					365			365	405	
	6062	3" DR 11 x 6" DR 17					370			370	405	
	6066	3" DR 17 x 6" DR 26					360			360	400	
	6070	3" DR 17 x 8" DR 11					415			415	455	
	6074	3" DR 17 x 8" DR 17					395			395	435	
	6078	4" DR 11 x 8" DR 17					410			410	450	
	6082	4" DR 17 x 8" DR 26					390			390	430	
	6086	6" DR 11 x 10" DR 17					510			510	565	
	6090	6" DR 17 x 10" DR 26					470			470	520	
	6094	6" DR 26 x 10" DR 26					460			460	510	
	6098	8" DR 17 x 12" DR 26					495			495	545	
	6102	8" DR 26 x 12" DR 32.5					470			470	520	
	6200	End termination with vent plug										
	6204	1" DR 11 x 3" DR 11				Ea.	244			244	269	
	6208	1" DR 11 x 4" DR 17					261			261	287	
	6212	1-1/2" DR 11 x 4" DR 17					261			261	287	
	6216	2" DR 11 x 4" DR 17					284			284	310	
	6220	2" DR 11 x 6" DR 17					325			325	355	
	6224	3" DR 11 x 6" DR 17					325			325	355	
	6228	3" DR 17 x 6" DR 26					325			325	355	
	6232	3" DR 17 x 8" DR 11					450			450	490	
	6236	3" DR 17 x 8" DR 17					450			450	490	
	6240	4" DR 11 x 8" DR 17					450			450	490	
	6244	4" DR 17 x 8" DR 26					390			390	430	
	6248	6" DR 11 x 10" DR 17					530			530	585	
	6252	6" DR 17 x 10" DR 26					435			435	480	
	6256	6" DR 26 x 10" DR 26					435			435	480	
	6260	8" DR 17 x 12" DR 26					645			645	710	
	6264	8" DR 26 x 12" DR 32.5					515			515	570	
	6268	10" DR 17 x 14" DR 26					670			670	740	
	6272	10" DR 17 x 16" DR 26					670			670	740	
	6276	10" DR 26 x 16" DR 26					670			670	740	
	6280	12" DR 26 x 16" DR 26					670			670	740	
	6284	12" DR 17 x 18" DR 26					885			885	975	
	6288	12" DR 26 x 18" DR 26					885			885	975	
	6292	14" DR 26 x 20" DR 32.5					900			900	990	
	6296	16" DR 26 x 22" DR 32.5					1,050			1,050	1,150	
	6300	18" DR 26 x 24" DR 32.5					1,075			1,075	1,175	
	6304	20" DR 32.5 x 28" DR 32.5					1,375			1,375	1,525	
	6308	22" DR 32.5 x 30" DR 32.5					1,525			1,525	1,675	
	6312	24" DR 32.5 x 32" DR 32.5					1,675			1,675	1,850	
	6316	36" DR 32.5 x 42" DR 32.5					2,450			2,450	2,700	
	6600	Tee										
	6604	1" DR 11 x 3" DR 11				Ea.	249			249	274	
	6608	1" DR 11 x 4" DR 17					310			310	340	

15108 | Plastic Pipe & Fittings		CREW	DAILY OUTPUT	LABOR-HOURS	UNIT	2005 BARE COSTS MAT.	LABOR	EQUIP.	TOTAL	TOTAL INCL O&P
590 6612	1-1/2" DR 11 x 4" DR 17				Ea.	345			345	380 590
6616	2" DR 11 x 4" DR 17					375			375	415
6620	2" DR 11 x 6" DR 17					455			455	500
6624	3" DR 11 x 6" DR 17					525			525	575
6628	3" DR 17 x 6" DR 26					510			510	560
6632	3" DR 17 x 8" DR 11					620			620	680
6636	3" DR 17 x 8" DR 17					590			590	645
6640	4" DR 11 x 8" DR 17					640			640	705
6644	4" DR 17 x 8" DR 26					615			615	675
6648	6" DR 11 x 10" DR 17					800			800	880
6652	6" DR 17 x 10" DR 26					760			760	835
6656	6" DR 26 x 10" DR 26					750			750	825
6660	8" DR 17 x 12" DR 26					750			750	825
6664	8" DR 26 x 12" DR 32.5					980			980	1,075
6668	10" DR 17 x 14" DR 26					1,125			1,125	1,250
6672	10" DR 17 x 16" DR 26					1,200			1,200	1,300
6676	10" DR 26 x 16" DR 26					1,200			1,200	1,300
6680	12" DR 26 x 16" DR 26					1,275			1,275	1,400
6684	12" DR 17 x 18" DR 26					1,275			1,275	1,400
6688	12" DR 26 x 18" DR 26					1,425			1,425	1,575
6692	14" DR 26 x 20" DR 32.5					1,875			1,875	2,075
6696	16" DR 26 x 22" DR 32.5					2,300			2,300	2,525
6700	18" DR 26 x 24" DR 32.5					2,500			2,500	2,750
6704	20" DR 32.5 x 28" DR 32.5					3,150			3,150	3,475
6708	22" DR 32.5 x 30" DR 32.5					3,750			3,750	4,125
6712	24" DR 32.5 x 32" DR 32.5					4,500			4,500	4,950
6716	36" DR 32.5 x 42" DR 32.5					9,150			9,150	10,100
6800	Wye									
6816	2" DR 11 x 4" DR 17				Ea.	475			475	520
6820	2" DR 11 x 6" DR 17					535			535	590
6824	3" DR 11 x 6" DR 17					545			545	595
6828	3" DR 17 x 6" DR 26					525			525	575
6832	3" DR 17 x 8" DR 11					660			660	725
6836	3" DR 17 x 8" DR 17					615			615	680
6840	4" DR 11 x 8" DR 17					670			670	740
6844	4" DR 17 x 8" DR 26					640			640	700
6848	6" DR 11 x 10" DR 17					830			830	915
6852	6" DR 17 x 10" DR 26					775			775	850
6856	6" DR 26 x 10" DR 26					770			770	845
6860	8" DR 17 x 12" DR 26					1,100			1,100	1,200
6864	8" DR 26 x 12" DR 32.5					1,075			1,075	1,175
6868	10" DR 17 x 14" DR 26					1,325			1,325	1,450
6872	10" DR 17 x 16" DR 26					1,400			1,400	1,550
6876	10" DR 26 x 16" DR 26					1,400			1,400	1,525
6880	12" DR 26 x 16" DR 26					1,550			1,550	1,725
6884	12" DR 17 x 18" DR 26					1,750			1,750	1,925
6888	12" DR 26 x 18" DR 26					1,775			1,775	1,950
6892	14" DR 26 x 20" DR 32.5					2,175			2,175	2,375
6896	16" DR 26 x 22" DR 32.5					3,125			3,125	3,425
6900	18" DR 26 x 24" DR 32.5					3,675			3,675	4,025
6904	20" DR 32.5 x 28" DR 32.5					4,600			4,600	5,050
6908	22" DR 32.5 x 30" DR 32.5					5,225			5,225	5,750
6912	24" DR 32.5 x 32" DR 32.5					5,950			5,950	6,550
9000	Welding labor per joint, not including welding machine									
9010	Pipe joint size, outer pipe (cost based on the thickest walls)									
9020	Straight pipe									

15108 | Plastic Pipe & Fittings

	Line	Description	Crew	Daily Output	Labor-Hours	Unit	2005 Bare Costs Mat.	Labor	Equip.	Total	Total Incl O&P	
590	9050	3″ pipe size	4 Skwk	96	.333	Ea.		11.60		11.60	18.05	590
	9060	4″ pipe size	″	77	.416			14.50		14.50	22.50	
	9070	6″ pipe size	5 Skwk	60	.667			23		23	36	
	9080	8″ pipe size	″	40	1			35		35	54	
	9090	10″ pipe size	6 Skwk	41	1.171			41		41	63.50	
	9100	12″ pipe size		39	1.231			43		43	66.50	
	9110	14″ pipe size		38	1.263			44		44	68.50	
	9120	16″ pipe size		35	1.371			48		48	74.50	
	9130	18″ pipe size	8 Skwk	45	1.422			49.50		49.50	77	
	9140	20″ pipe size		42	1.524			53		53	82.50	
	9150	22″ pipe size		40	1.600			56		56	86.50	
	9160	24″ pipe size		38	1.684			58.50		58.50	91.50	
	9170	28″ pipe size		37	1.730			60.50		60.50	94	
	9180	30″ pipe size		36	1.778			62		62	96.50	
	9190	32″ pipe size		35	1.829			63.50		63.50	99	
	9200	42″ pipe size		32	2			69.50		69.50	108	
	9300	Note: Cost for set up each time welder is moved.										
	9301	Add 100% of weld labor cost										
	9310	For handling between fitting welds add 50% of weld labor cost										
	9320	Welder usually remains stationary with pipe moved through it										
	9360	Weld machine, rental per day based on dia. capacity										
	9380	3″ thru 4″ diameter				Ea.			55	55	60.50	
	9390	6″ thru 8″ diameter							180	180	198	
	9400	10″ thru 12″ diameter							270	270	297	
	9410	14″ thru 18″ diameter							365	365	402	
	9420	20″ thru 24″ diameter							650	650	715	
	9430	28″ thru 32″ diameter							765	765	840	
	9440	42″ thru 58″ diameter							790	790	870	

15110 | Valves

	Line	Description	Crew	Daily Output	Labor-Hours	Unit	Mat.	Labor	Equip.	Total	Total Incl O&P	
100	0010	**VALVES, BRASS**										100
	0030	For motorized valves, see Division 13838-200										
	0500	Gas cocks, threaded										
	0510	1/4″	1 Plum	26	.308	Ea.	5.95	12.55		18.50	25.50	
	0520	3/8″		24	.333		5.95	13.60		19.55	27	
	0530	1/2″		24	.333		5.75	13.60		19.35	27	
	0540	3/4″		22	.364		10.80	14.85		25.65	34.50	
	0550	1″		19	.421		11.35	17.20		28.55	38.50	
	0560	1-1/4″		15	.533		28	22		50	64	
	0570	1-1/2″		13	.615		29.50	25		54.50	70.50	
	0580	2″		11	.727		55.50	29.50		85	106	
	0670	For larger sizes use lubricated plug valve, Section 15110-600										
160	0010	**VALVES, BRONZE** R15100-090										160
	1020	Angle, 150 lb., rising stem, threaded										
	1030	1/8″	1 Plum	24	.333	Ea.	49.50	13.60		63.10	75	
	1040	1/4″		24	.333		49.50	13.60		63.10	75	
	1050	3/8″		24	.333		49.50	13.60		63.10	75	
	1060	1/2″		22	.364		49.50	14.85		64.35	77	
	1070	3/4″		20	.400		67	16.35		83.35	98	
	1080	1″		19	.421		95.50	17.20		112.70	131	
	1090	1-1/4″		15	.533		124	22		146	169	
	1100	1-1/2″		13	.615		161	25		186	215	
	1110	2″		11	.727		260	29.50		289.50	330	
	1380	Ball, 150 psi, threaded										
	1400	1/4″	1 Plum	24	.333	Ea.	7	13.60		20.60	28	
	1430	3/8″		24	.333		7	13.60		20.60	28	

15110 | Valves

				DAILY	LABOR-		2005 BARE COSTS				TOTAL
			CREW	OUTPUT	HOURS	UNIT	MAT.	LABOR	EQUIP.	TOTAL	INCL O&P
160	1450	1/2" R15100 -090	1 Plum	22	.364	Ea.	7	14.85		21.85	30
	1460	3/4"		20	.400		11.55	16.35		27.90	37
	1470	1"		19	.421		14.55	17.20		31.75	42
	1480	1-1/4"		15	.533		17.15	22		39.15	52
	1490	1-1/2"		13	.615		26.50	25		51.50	67
	1500	2"		11	.727		33	29.50		62.50	81
	1510	2-1/2"		9	.889		156	36.50		192.50	227
	1520	3"		8	1		255	41		296	345
	1600	Butterfly, 175 psi, full port, solder or threaded ends									
	1610	Stainless steel disc and stem									
	1620	1/4"	1 Plum	24	.333	Ea.	6.25	13.60		19.85	27.50
	1630	3/8"		24	.333		6.25	13.60		19.85	27.50
	1640	1/2"		22	.364		6.65	14.85		21.50	30
	1650	3/4"		20	.400		10.75	16.35		27.10	36.50
	1660	1"		19	.421		13.15	17.20		30.35	40.50
	1670	1-1/4"		15	.533		21	22		43	56.50
	1680	1-1/2"		13	.615		27.50	25		52.50	68
	1690	2"		11	.727		34	29.50		63.50	82
	1750	Check, swing, class 150, regrinding disc, threaded									
	1800	1/8"	1 Plum	24	.333	Ea.	25	13.60		38.60	48
	1830	1/4"		24	.333		25	13.60		38.60	48
	1840	3/8"		24	.333		26	13.60		39.60	49
	1850	1/2"		24	.333		26	13.60		39.60	49
	1860	3/4"		20	.400		35	16.35		51.35	63
	1870	1"		19	.421		52	17.20		69.20	83
	1880	1-1/4"		15	.533		73	22		95	113
	1890	1-1/2"		13	.615		85.50	25		110.50	132
	1900	2"		11	.727		125	29.50		154.50	183
	1910	2-1/2"	Q-1	15	1.067		265	39		304	350
	1920	3"	"	13	1.231		395	45.50		440.50	505
	2000	For 200 lb, add					5%	10%			
	2040	For 300 lb, add					15%	15%			
	2060	Check swing, 300#, sweat, 3/8" size	1 Plum	24	.333	Ea.	16	13.60		29.60	38
	2070	1/2"		24	.333		16	13.60		29.60	38
	2080	3/4"		20	.400		20.50	16.35		36.85	47
	2090	1"		19	.421		28.50	17.20		45.70	57.50
	2100	1-1/4"		15	.533		37	22		59	73.50
	2110	1-1/2 "		13	.615		49	25		74	91.50
	2120	2"		11	.727		73.50	29.50		103	126
	2130	2-1/2"	Q-1	15	1.067		175	39		214	251
	2140	3"	"	13	1.231		233	45.50		278.50	325
	2350	Check, lift, class 150, horizontal composition disc, threaded									
	2430	1/4"	1 Plum	24	.333	Ea.	65.50	13.60		79.10	92.50
	2440	3/8"		24	.333		65.50	13.60		79.10	92.50
	2450	1/2"		24	.333		72.50	13.60		86.10	100
	2460	3/4"		20	.400		88.50	16.35		104.85	122
	2470	1"		19	.421		125	17.20		142.20	164
	2480	1-1/4"		15	.533		169	22		191	219
	2490	1-1/2"		13	.615		201	25		226	260
	2500	2"		11	.727		335	29.50		364.50	415
	2850	Gate, N.R.S., soldered, 125 psi									
	2900	3/8"	1 Plum	24	.333	Ea.	18.30	13.60		31.90	40.50
	2920	1/2"		24	.333		16	13.60		29.60	38
	2940	3/4"		20	.400		25	16.35		41.35	52
	2950	1"		19	.421		34.50	17.20		51.70	64
	2960	1-1/4"		15	.533		46.50	22		68.50	84

15 MECHANICAL

15110 | Valves

			CREW	DAILY OUTPUT	LABOR-HOURS	UNIT	2005 BARE COSTS MAT.	LABOR	EQUIP.	TOTAL	TOTAL INCL O&P	
160	2970	1-1/2" R15100-090	1 Plum	13	.615	Ea.	52.50	25		77.50	95.50	160
	2980	2"		11	.727		83.50	29.50		113	137	
	2990	2-1/2"	Q-1	15	1.067		195	39		234	274	
	3000	3"	"	13	1.231		256	45.50		301.50	350	
	3350	Threaded, class 150										
	3410	1/4"	1 Plum	24	.333	Ea.	31	13.60		44.60	55	
	3420	3/8"		24	.333		31	13.60		44.60	55	
	3430	1/2"		24	.333		28.50	13.60		42.10	52	
	3440	3/4"		20	.400		32	16.35		48.35	59.50	
	3450	1"		19	.421		42	17.20		59.20	72	
	3460	1-1/4" size		15	.533		55.50	22		77.50	94	
	3470	1-1/2"		13	.615		71.50	25		96.50	117	
	3480	2"		11	.727		95	29.50		124.50	150	
	3490	2-1/2"	Q-1	15	1.067		222	39		261	305	
	3500	3" size	"	13	1.231		315	45.50		360.50	415	
	3600	Gate, flanged, 150 lb.										
	3610	1"	1 Plum	7	1.143	Ea.	540	46.50		586.50	660	
	3620	1-1/2"		6	1.333		670	54.50		724.50	815	
	3630	2"		5	1.600		970	65.50		1,035.50	1,175	
	3634	2-1/2"	Q-1	5	3.200		1,525	118		1,643	1,850	
	3640	3"	"	4.50	3.556		1,775	131		1,906	2,150	
	3850	Rising stem, soldered, 300 psi										
	3900	3/8"	1 Plum	24	.333	Ea.	43	13.60		56.60	67.50	
	3920	1/2"		24	.333		38.50	13.60		52.10	62.50	
	3940	3/4"		20	.400		49	16.35		65.35	78	
	3950	1"		19	.421		59.50	17.20		76.70	91	
	3960	1-1/4"		15	.533		88	22		110	130	
	3970	1-1/2"		13	.615		107	25		132	156	
	3980	2"		11	.727		161	29.50		190.50	222	
	3990	2-1/2"	Q-1	15	1.067		390	39		429	485	
	4000	3"	"	13	1.231		455	45.50		500.50	570	
	4250	Threaded, class 150										
	4310	1/4"	1 Plum	24	.333	Ea.	27	13.60		40.60	50.50	
	4320	3/8"		24	.333		27	13.60		40.60	50.50	
	4330	1/2"		24	.333		25.50	13.60		39.10	48.50	
	4340	3/4"		20	.400		30	16.35		46.35	57.50	
	4350	1"		19	.421		40.50	17.20		57.70	70.50	
	4360	1-1/4"		15	.533		53.50	22		75.50	91.50	
	4370	1-1/2"		13	.615		67.50	25		92.50	113	
	4380	2"		11	.727		91.50	29.50		121	146	
	4390	2-1/2"	Q-1	15	1.067		214	39		253	294	
	4400	3"	"	13	1.231		299	45.50		344.50	400	
	4500	For 300 psi, threaded, add					100%	15%				
	4540	For chain operated type, add					15%					
	4850	Globe, class 150, rising stem, threaded										
	4920	1/4"	1 Plum	24	.333	Ea.	35.50	13.60		49.10	60	
	4940	3/8" size		24	.333		39	13.60		52.60	63.50	
	4950	1/2"		24	.333		39	13.60		52.60	63.50	
	4960	3/4"		20	.400		53.50	16.35		69.85	83	
	4970	1"		19	.421		83.50	17.20		100.70	118	
	4980	1-1/4"		15	.533		131	22		153	177	
	4990	1-1/2"		13	.615		160	25		185	214	
	5000	2"		11	.727		239	29.50		268.50	310	
	5010	2-1/2"	Q-1	15	1.067		480	39		519	590	
	5020	3"	"	13	1.231		605	45.50		650.50	735	
	5120	For 300 lb threaded, add					50%	15%				

15110 | Valves

			CREW	DAILY OUTPUT	LABOR-HOURS	UNIT	2005 BARE COSTS MAT.	LABOR	EQUIP.	TOTAL	TOTAL INCL O&P
160	5130	Globe, 300 lb., sweat, 3/8" size R15100-090	1 Plum	24	.333	Ea.	36.50	13.60		50.10	61
	5140	1/2"		24	.333		36	13.60		49.60	60
	5150	3/4"		20	.400		52	16.35		68.35	81.50
	5160	1"		19	.421		78.50	17.20		95.70	113
	5170	1-1/4"		15	.533		122	22		144	167
	5180	1-1/2"		13	.615		145	25		170	197
	5190	2"		11	.727		217	29.50		246.50	283
	5200	2-1/2"	Q-1	15	1.067		465	39		504	570
	5210	3"	"	13	1.231		675	45.50		720.50	815
	5600	Relief, pressure & temperature, self-closing, ASME, threaded									
	5640	3/4"	1 Plum	28	.286	Ea.	81.50	11.65		93.15	107
	5650	1"		24	.333		118	13.60		131.60	151
	5660	1-1/4"		20	.400		237	16.35		253.35	286
	5670	1-1/2"		18	.444		455	18.15		473.15	530
	5680	2"		16	.500		495	20.50		515.50	575
	5950	Pressure, poppet type, threaded									
	6000	1/2"	1 Plum	30	.267	Ea.	13.55	10.90		24.45	31.50
	6040	3/4"	"	28	.286	"	45	11.65		56.65	67
	6400	Pressure, water, ASME, threaded									
	6440	3/4"	1 Plum	28	.286	Ea.	47	11.65		58.65	69.50
	6450	1"		24	.333		93.50	13.60		107.10	124
	6460	1-1/4"		20	.400		153	16.35		169.35	193
	6470	1-1/2"		18	.444		211	18.15		229.15	260
	6480	2"		16	.500		305	20.50		325.50	365
	6490	2-1/2"		15	.533		1,550	22		1,572	1,725
	6900	Reducing, water pressure									
	6920	300 psi to 25-75 psi, threaded or sweat									
	6940	1/2"	1 Plum	24	.333	Ea.	147	13.60		160.60	182
	6950	3/4"		20	.400		147	16.35		163.35	186
	6960	1"		19	.421		227	17.20		244.20	276
	6970	1-1/4"		15	.533		405	22		427	480
	6980	1-1/2"		13	.615		615	25		640	715
	6990	2"		11	.727		885	29.50		914.50	1,025
	7100	For built-in by-pass or 10-35 psi, add					7.30			7.30	8
	7700	High capacity, 250 psi to 25-75 psi, threaded									
	7740	1/2"	1 Plum	24	.333	Ea.	147	13.60		160.60	182
	7780	3/4"		20	.400		147	16.35		163.35	186
	7790	1"		19	.421		227	17.20		244.20	276
	7800	1-1/4"		15	.533		405	22		427	480
	7810	1-1/2"		13	.615		615	25		640	715
	7820	2"		11	.727		890	29.50		919.50	1,025
	7830	2-1/2"		9	.889		1,075	36.50		1,111.50	1,225
	7840	3"		8	1		1,175	41		1,216	1,350
	7850	3" flanged (iron body)	Q-1	10	1.600		1,275	59		1,334	1,500
	7860	4" flanged (iron body)	"	8	2		2,125	73.50		2,198.50	2,450
	7920	For higher pressure, add					25%				
	8000	Silent check, bronze trim									
	8010	Compact wafer type, for 125 or 150 lb. flanges									
	8020	1-1/2"	1 Plum	11	.727	Ea.	198	29.50		227.50	263
	8021	2"	"	9	.889		212	36.50		248.50	289
	8022	2-1/2"	Q-1	9	1.778		224	65.50		289.50	345
	8023	3"		8	2		250	73.50		323.50	385
	8024	4"		5	3.200		425	118		543	645
	8025	5"	Q-2	6	4		590	153		743	875
	8026	6"		5	4.800		705	183		888	1,050
	8027	8"		4.50	5.333		1,425	203		1,628	1,875

15110 | Valves

	Line	Description	Crew	Daily Output	Labor-Hours	Unit	2005 Bare Costs Mat.	Labor	Equip.	Total	Total Incl O&P	
160	8028	10″ (R15100-090)	Q-2	4	6	Ea.	2,525	229		2,754	3,125	160
	8029	12″		3	8		5,900	305		6,205	6,925	
	8050	For 250 or 300 lb flanges, thru 6″ no change										
	8051	For 8″ and 10″, add				Ea.	40%	10%				
	8060	Full flange wafer type, 150 lb.										
	8061	1″	1 Plum	14	.571	Ea.	211	23.50		234.50	267	
	8062	1-1/4″		12	.667		221	27		248	284	
	8063	1-1/2″		11	.727		266	29.50		295.50	340	
	8064	2″		9	.889		355	36.50		391.50	450	
	8065	2-1/2″	Q-1	9	1.778		400	65.50		465.50	545	
	8066	3″		8	2		445	73.50		518.50	600	
	8067	4″		5	3.200		660	118		778	900	
	8068	5″	Q-2	6	4		850	153		1,003	1,175	
	8069	6″	″	5	4.800		1,125	183		1,308	1,500	
	8080	For 300 lb., add					40%	10%				
	8100	Globe type, 150 lb.										
	8110	2″	1 Plum	9	.889	Ea.	415	36.50		451.50	515	
	8111	2-1/2″	Q-1	9	1.778		515	65.50		580.50	665	
	8112	3″		8	2		615	73.50		688.50	785	
	8113	4″		5	3.200		850	118		968	1,100	
	8114	5″	Q-2	6	4		1,050	153		1,203	1,375	
	8115	6″	″	5	4.800		1,450	183		1,633	1,875	
	8130	For 300 lb., add					20%	10%				
	8140	Screwed end type, 250 lb.										
	8141	1/2″	1 Plum	24	.333	Ea.	37	13.60		50.60	61.50	
	8142	3/4″		20	.400		37	16.35		53.35	65.50	
	8143	1″		19	.421		43	17.20		60.20	73.50	
	8144	1-1/4″		15	.533		59	22		81	98	
	8145	1-1/2″		13	.615		66	25		91	111	
	8146	2″		11	.727		89	29.50		118.50	143	
	8350	Tempering, water, sweat connections										
	8400	1/2″	1 Plum	24	.333	Ea.	52.50	13.60		66.10	78	
	8440	3/4″	″	20	.400	″	64	16.35		80.35	95	
	8650	Threaded connections										
	8700	1/2″	1 Plum	24	.333	Ea.	79	13.60		92.60	107	
	8740	3/4″		20	.400		244	16.35		260.35	294	
	8750	1″		19	.421		269	17.20		286.20	320	
	8760	1-1/4″		15	.533		430	22		452	505	
	8770	1-1/2″		13	.615		465	25		490	555	
	8780	2″		11	.727		700	29.50		729.50	815	
200	0010	**VALVES, IRON BODY** (R15100-090)										200
	0020	For grooved joint, see Division 15107-690										
	0100	Angle, 125 lb.										
	0110	Flanged										
	0116	2″	1 Plum	5	1.600	Ea.	660	65.50		725.50	830	
	0118	4″	Q-1	3	5.333		1,100	196		1,296	1,500	
	0120	6″	Q-2	3	8		2,150	305		2,455	2,800	
	0122	8″	″	2.50	9.600		3,825	365		4,190	4,775	
	0560	Butterfly, lug type, pneumatic operator, 2″ size	1 Stpi	14	.571		230	23.50		253.50	288	
	0570	3″	Q-1	8	2		238	73.50		311.50	375	
	0580	4″	″	5	3.200		260	118		378	465	
	0590	6″	Q-2	5	4.800		350	183		533	660	
	0600	8″		4.50	5.333		420	203		623	770	
	0610	10″		4	6		530	229		759	930	
	0620	12″		3	8		715	305		1,020	1,250	
	0630	14″		2.30	10.435		1,025	400		1,425	1,725	

15110 | Valves

		Description	Crew	Daily Output	Labor-Hours	Unit	2005 Bare Costs Mat.	2005 Bare Costs Labor	2005 Bare Costs Equip.	2005 Bare Costs Total	Total Incl O&P	
200	0640	18" R15100-090	Q-2	1.50	16	Ea.	1,775	610		2,385	2,875	200
	0650	20"		1	24		2,125	915		3,040	3,725	
	0790	Butterfly, lug type, gear operated, 2" size, 200 lb. except noted	1 Plum	14	.571		181	23.50		204.50	234	
	0800	2-1/2"	Q-1	9	1.778		182	65.50		247.50	299	
	0810	3"		8	2		191	73.50		264.50	320	
	0820	4"		5	3.200		216	118		334	415	
	0830	5"	Q-2	5	4.800		269	183		452	570	
	0840	6"		5	4.800		298	183		481	605	
	0850	8"		4.50	5.333		385	203		588	730	
	0860	10"		4	6		510	229		739	905	
	0870	12"		3	8		740	305		1,045	1,275	
	0880	14", 150 lb.		2.30	10.435		845	400		1,245	1,525	
	0890	16", 150 lb.		1.75	13.714		1,200	525		1,725	2,100	
	0900	18", 150 lb.		1.50	16		1,550	610		2,160	2,625	
	0910	20", 150 lb.		1	24		1,850	915		2,765	3,400	
	0930	24", 150 lb.		.75	32		2,575	1,225		3,800	4,650	
	1020	Butterfly, wafer type, gear actuator, 200 lb.										
	1030	2"	1 Plum	14	.571	Ea.	113	23.50		136.50	159	
	1040	2-1/2"	Q-1	9	1.778		116	65.50		181.50	227	
	1050	3"		8	2		121	73.50		194.50	244	
	1060	4"		5	3.200		151	118		269	345	
	1070	5"	Q-2	5	4.800		182	183		365	475	
	1080	6"		5	4.800		205	183		388	500	
	1090	8"		4.50	5.333		274	203		477	605	
	1100	10"		4	6		345	229		574	725	
	1110	12"		3	8		480	305		785	985	
	1200	Wafer type, lever actuator, 200 lb.										
	1220	2"	1 Plum	14	.571	Ea.	79.50	23.50		103	123	
	1230	2-1/2"	Q-1	9	1.778		82.50	65.50		148	190	
	1240	3"		8	2		87	73.50		160.50	207	
	1250	4"		5	3.200		107	118		225	294	
	1260	5"	Q-2	5	4.800		148	183		331	440	
	1270	6"		5	4.800		178	183		361	470	
	1280	8"		4.50	5.333		263	203		466	595	
	1290	10"		4	6		475	229		704	865	
	1300	12"		3	8		670	305		975	1,200	
	1600	Gate, threaded, 125 lb.										
	1603	1-1/2"	1 Plum	13	.615	Ea.	236	25		261	298	
	1604	2"		11	.727		625	29.50		654.50	730	
	1605	2-1/2"		10	.800		675	32.50		707.50	790	
	1606	3"		8	1		815	41		856	955	
	1607	4"		5	1.600		1,125	65.50		1,190.50	1,350	
	1650	Gate, 125 lb., N.R.S.										
	2150	Flanged										
	2200	2"	1 Plum	5	1.600	Ea.	320	65.50		385.50	450	
	2240	2-1/2"	Q-1	5	3.200		330	118		448	535	
	2260	3"		4.50	3.556		365	131		496	600	
	2280	4"		3	5.333		525	196		721	875	
	2290	5"	Q-2	3.40	7.059		895	269		1,164	1,400	
	2300	6"		3	8		895	305		1,200	1,450	
	2320	8"		2.50	9.600		1,525	365		1,890	2,250	
	2340	10"		2.20	10.909		2,700	415		3,115	3,600	
	2360	12"		1.70	14.118		3,725	540		4,265	4,900	
	2370	14"		1.30	18.462		5,950	705		6,655	7,600	
	2380	16"		1	24		8,300	915		9,215	10,500	
	2420	For 250 lb flanged, add					200%	10%				

	15110 | Valves	CREW	DAILY OUTPUT	LABOR-HOURS	UNIT	2005 BARE COSTS MAT.	LABOR	EQUIP.	TOTAL	TOTAL INCL O&P	
200	3500 OS&Y, 125 lb., (225# noted), threaded R15100-090										200
	3504 3/4" (225 lb.)	1 Plum	18	.444	Ea.	385	18.15		403.15	455	
	3505 1" (225 lb.)		16	.500		440	20.50		460.50	510	
	3506 1-1/2" (225 lb.)		13	.615		640	25		665	740	
	3507 2"		11	.727		560	29.50		589.50	660	
	3508 2-1/2"		9	.889		575	36.50		611.50	685	
	3510 4"		7	1.143		925	46.50		971.50	1,100	
	3550 OS&Y, 125 lb., flanged										
	3600 2"	1 Plum	5	1.600	Ea.	226	65.50		291.50	345	
	3640 2-1/2"	Q-1	5	3.200		234	118		352	435	
	3660 3"		4.50	3.556		263	131		394	485	
	3670 3-1/2"		3	5.333		410	196		606	745	
	3680 4"		3	5.333		375	196		571	705	
	3690 5"	Q-2	3.40	7.059		615	269		884	1,075	
	3700 6"		3	8		615	305		920	1,150	
	3720 8"		2.50	9.600		1,100	365		1,465	1,750	
	3740 10"		2.20	10.909		2,000	415		2,415	2,825	
	3760 12"		1.70	14.118		2,675	540		3,215	3,750	
	3770 14"		1.30	18.462		5,100	705		5,805	6,675	
	3780 16"		1	24		7,650	915		8,565	9,800	
	3790 18"		.80	30		10,300	1,150		11,450	13,000	
	3800 20"		.60	40		14,300	1,525		15,825	18,100	
	3830 24"		.50	48		21,300	1,825		23,125	26,200	
	3900 For 175 lb, flanged, add					200%	10%				
	4350 Globe, OS&Y										
	4540 Class 125, flanged										
	4550 2"	1 Plum	5	1.600	Ea.	465	65.50		530.50	610	
	4560 2-1/2"	Q-1	5	3.200		490	118		608	715	
	4570 3"		4.50	3.556		570	131		701	820	
	4580 4"		3	5.333		810	196		1,006	1,200	
	4590 5"	Q-2	3.40	7.059		1,475	269		1,744	2,025	
	4600 6"		3	8		1,475	305		1,780	2,075	
	4610 8"		2.50	9.600		2,900	365		3,265	3,725	
	4612 10"		2.20	10.909		4,525	415		4,940	5,600	
	4614 12"		1.70	14.118		4,850	540		5,390	6,125	
	5040 Class 250, flanged										
	5050 2"	1 Plum	4.50	1.778	Ea.	630	72.50		702.50	805	
	5060 2-1/2"	Q-1	4.50	3.556		820	131		951	1,100	
	5070 3"		4	4		850	147		997	1,150	
	5080 4"		2.70	5.926		1,250	218		1,468	1,700	
	5090 5"	Q-2	3	8		2,225	305		2,530	2,900	
	5100 6"		2.70	8.889		2,225	340		2,565	2,950	
	5110 8"		2.20	10.909		3,775	415		4,190	4,775	
	5120 10"		2	12		6,175	460		6,635	7,500	
	5130 12"		1.60	15		9,275	570		9,845	11,100	
	5240 Valve sprocket rim w/chain, for 2" valve	1 Stpi	30	.267		74.50	10.90		85.40	98.50	
	5250 2-1/2" valve		27	.296		74.50	12.15		86.65	100	
	5260 3-1/2" valve		25	.320		74.50	13.10		87.60	102	
	5270 6" valve		20	.400		74.50	16.40		90.90	107	
	5280 8" valve		18	.444		112	18.20		130.20	152	
	5290 12" valve		16	.500		112	20.50		132.50	155	
	5300 16" valve		12	.667		152	27.50		179.50	209	
	5310 20" valve		10	.800		152	33		185	217	
	5320 36" valve		8	1		283	41		324	370	
	5450 Swing check, 125 lb., threaded										
	5470 1"	1 Plum	13	.615	Ea.	310	25		335	380	

15110 | Valves

		Description	CREW	DAILY OUTPUT	LABOR-HOURS	UNIT	2005 BARE COSTS MAT.	LABOR	EQUIP.	TOTAL	TOTAL INCL O&P	
200	5500	2″ (R15100-090)	1 Plum	11	.727	Ea.	400	29.50		429.50	485	200
	5540	2-1/2″	Q-1	15	1.067		455	39		494	560	
	5550	3″		13	1.231		510	45.50		555.50	630	
	5560	4″		10	1.600		815	59		874	985	
	5950	Flanged										
	5994	1″	1 Plum	7	1.143	Ea.	87	46.50		133.50	166	
	5998	1-1/2″		6	1.333		132	54.50		186.50	227	
	6000	2″		5	1.600		156	65.50		221.50	269	
	6040	2-1/2″	Q-1	5	3.200		189	118		307	385	
	6050	3″		4.50	3.556		202	131		333	420	
	6060	4″		3	5.333		320	196		516	645	
	6070	6″	Q-2	3	8		545	305		850	1,050	
	6080	8″		2.50	9.600		1,025	365		1,390	1,675	
	6090	10″		2.20	10.909		1,750	415		2,165	2,550	
	6100	12″		1.70	14.118		2,725	540		3,265	3,800	
	6110	18″		1.30	18.462		18,500	705		19,205	21,500	
	6114	24″		.75	32		34,400	1,225		35,625	39,600	
	6160	For 250 lb flanged, add					200%	20%				
	6600	Silent check, bronze trim										
	6610	Compact wafer type, for 125 or 150 lb. flanges										
	6630	1-1/2″	1 Plum	11	.727	Ea.	91.50	29.50		121	146	
	6640	2″	″	9	.889		109	36.50		145.50	174	
	6650	2-1/2″	Q-1	9	1.778		119	65.50		184.50	230	
	6660	3″		8	2		128	73.50		201.50	252	
	6670	4″		5	3.200		163	118		281	355	
	6680	5″	Q-2	6	4		214	153		367	465	
	6690	6″		6	4		286	153		439	545	
	6700	8″		4.50	5.333		510	203		713	865	
	6710	10″		4	6		885	229		1,114	1,325	
	6720	12″		3	8		1,675	305		1,980	2,300	
	6740	For 250 or 300 lb. flanges, thru 6″ no change										
	6741	For 8″ and 10″, add				Ea.	11%	10%				
	6750	Twin disc										
	6752	2″	1 Plum	9	.889	Ea.	152	36.50		188.50	222	
	6754	4″	Q-1	5	3.200		248	118		366	450	
	6756	6″	Q-2	5	4.800		365	183		548	680	
	6758	8″		4.50	5.333		565	203		768	925	
	6760	10″		4	6		900	229		1,129	1,325	
	6762	12″		3	8		1,175	305		1,480	1,725	
	6764	18″		1.50	16		5,050	610		5,660	6,475	
	6766	24″		.75	32		7,100	1,225		8,325	9,650	
	6800	Full flange type, 150 lb.										
	6810	1″	1 Plum	14	.571	Ea.	107	23.50		130.50	152	
	6811	1-1/4″		12	.667		113	27		140	165	
	6812	1-1/2″		11	.727		130	29.50		159.50	188	
	6813	2″		9	.889		150	36.50		186.50	220	
	6814	2-1/2″	Q-1	9	1.778		181	65.50		246.50	298	
	6815	3″		8	2		188	73.50		261.50	320	
	6816	4″		5	3.200		232	118		350	435	
	6817	5″	Q-2	6	4		365	153		518	635	
	6818	6″		5	4.800		480	183		663	805	
	6819	8″		4.50	5.333		815	203		1,018	1,200	
	6820	10″		4	6		1,025	229		1,254	1,475	
	6840	For 250 lb., add					30%	10%				
	6900	Globe type, 125 lb.										
	6910	2″	1 Plum	9	.889	Ea.	208	36.50		244.50	283	

Important: See the Reference Section for critical supporting data - Reference Nos., Crews, & City Cost Indexes

15110 | Valves

		15110 Valves	CREW	DAILY OUTPUT	LABOR-HOURS	UNIT	MAT.	LABOR	EQUIP.	TOTAL	TOTAL INCL O&P	
							2005 BARE COSTS					
200	6911	2-1/2" (R15100-090)	Q-1	9	1.778	Ea.	228	65.50		293.50	350	200
	6912	3"		8	2		246	73.50		319.50	380	
	6913	4"		5	3.200		330	118		448	540	
	6914	5"	Q-2	6	4		425	153		578	700	
	6915	6"		5	4.800		535	183		718	860	
	6916	8"		4.50	5.333		975	203		1,178	1,375	
	6917	10"		4	6		1,200	229		1,429	1,675	
	6918	12"		3	8		2,050	305		2,355	2,700	
	6919	14"		2.30	10.435		2,775	400		3,175	3,675	
	6920	16"		1.75	13.714		4,025	525		4,550	5,200	
	6921	18"		1.50	16		5,800	610		6,410	7,300	
	6922	20"		1	24		7,050	915		7,965	9,125	
	6923	24"		.75	32		10,300	1,225		11,525	13,100	
	6940	For 250 lb., add					40%	10%				
	6980	Screwed end type, 125 lb.										
	6981	1"	1 Plum	19	.421	Ea.	39	17.20		56.20	69	
	6982	1-1/4"		15	.533		50.50	22		72.50	89	
	6983	1-1/2"		13	.615		62.50	25		87.50	107	
	6984	2"		11	.727		85	29.50		114.50	138	
400	0010	**MULTIPURPOSE VALVES**										400
	0100	Functions as a shut off, balancing, check & metering valve										
	1000	Cast iron body										
	1010	Threaded										
	1020	1-1/2" size	1 Stpi	11	.727	Ea.	105	30		135	160	
	1030	2" size	"	8	1		355	41		396	450	
	1040	2-1/2" size	Q-5	5	3.200		385	118		503	600	
	1050	3" size	"	4.50	3.556		410	131		541	645	
	1200	Flanged										
	1210	3" size	Q-5	4.20	3.810	Ea.	445	140		585	700	
	1220	4" size	"	3	5.333		895	197		1,092	1,275	
	1230	5" size	Q-6	3.80	6.316		1,050	241		1,291	1,550	
	1240	6" size		3	8		1,325	305		1,630	1,900	
	1250	8" size		2.50	9.600		1,950	365		2,315	2,700	
	1260	10" size		2.20	10.909		2,775	415		3,190	3,700	
	1270	12" size		2.10	11.429		7,600	435		8,035	9,000	
	1280	14" size		2	12		10,500	460		10,960	12,300	
500	0010	**VALVES, PLASTIC** (R15100-090)										500
	1100	Angle, PVC, threaded										
	1110	1/4"	1 Plum	26	.308	Ea.	49.50	12.55		62.05	73.50	
	1120	1/2"		26	.308		49.50	12.55		62.05	73.50	
	1130	3/4"		25	.320		58.50	13.05		71.55	84	
	1140	1"		23	.348		70	14.20		84.20	98.50	
	1150	Ball, PVC, socket or threaded, single union										
	1200	1/4"	1 Plum	26	.308	Ea.	24	12.55		36.55	45.50	
	1220	3/8"		26	.308		24	12.55		36.55	45.50	
	1230	1/2"		26	.308		24	12.55		36.55	45.50	
	1240	3/4"		25	.320		28.50	13.05		41.55	51	
	1250	1"		23	.348		34.50	14.20		48.70	59.50	
	1260	1-1/4"		21	.381		57	15.55		72.55	86	
	1270	1-1/2"		20	.400		57	16.35		73.35	87	
	1280	2"		17	.471		82	19.20		101.20	119	
	1290	2-1/2"	Q-1	26	.615		204	22.50		226.50	258	
	1300	3"		24	.667		204	24.50		228.50	261	
	1310	4"		20	.800		278	29.50		307.50	350	

15110 | Valves

			Crew	Daily Output	Labor-Hours	Unit	2005 Bare Costs Mat.	Labor	Equip.	Total	Total Incl O&P
500	1360	For PVC, flanged, add (R15100-090)				Ea.	100%	15%			
	1450	Double union 1/2"	1 Plum	26	.308		13.80	12.55		26.35	34
	1460	3/4"		25	.320		16.20	13.05		29.25	37.50
	1470	1"		23	.348		17.75	14.20		31.95	41
	1480	1-1/4"		21	.381		27	15.55		42.55	53
	1490	1-1/2"		20	.400		30	16.35		46.35	57.50
	1500	2"		17	.471		42.50	19.20		61.70	75.50
	1650	CPVC, socket or threaded, single union									
	1700	1/2"	1 Plum	26	.308	Ea.	35	12.55		47.55	57.50
	1720	3/4"		25	.320		44	13.05		57.05	68
	1730	1"		23	.348		52.50	14.20		66.70	79.50
	1750	1-1/4"		21	.381		88	15.55		103.55	121
	1760	1-1/2"		20	.400		88	16.35		104.35	121
	1770	2"		17	.471		122	19.20		141.20	163
	1780	3"	Q-1	24	.667		365	24.50		389.50	435
	1840	For CPVC, flanged, add					65%	15%			
	1880	For true union, socket or threaded, add					50%	5%			
	2050	Polypropylene, threaded									
	2100	1/4"	1 Plum	26	.308	Ea.	36	12.55		48.55	58.50
	2120	3/8"		26	.308		36	12.55		48.55	58.50
	2130	1/2"		26	.308		36	12.55		48.55	58.50
	2140	3/4"		25	.320		45.50	13.05		58.55	69.50
	2150	1"		23	.348		53.50	14.20		67.70	80.50
	2160	1-1/4"		21	.381		77.50	15.55		93.05	109
	2170	1-1/2"		20	.400		89.50	16.35		105.85	123
	2180	2"		17	.471		122	19.20		141.20	164
	2190	3"	Q-1	24	.667		325	24.50		349.50	390
	2200	4"	"	20	.800		540	29.50		569.50	635
	2550	PVC, three way, socket or threaded									
	2600	1/2"	1 Plum	26	.308	Ea.	44.50	12.55		57.05	68
	2640	3/4"		25	.320		50.50	13.05		63.55	75
	2650	1"		23	.348		54.50	14.20		68.70	81.50
	2660	1-1/2"		20	.400		111	16.35		127.35	147
	2670	2"		17	.471		148	19.20		167.20	192
	2680	3"	Q-1	24	.667		320	24.50		344.50	390
	2740	For flanged, add					60%	15%			
	3150	Ball check, PVC, socket or threaded									
	3200	1/4"	1 Plum	26	.308	Ea.	30.50	12.55		43.05	52.50
	3220	3/8"		26	.308		30.50	12.55		43.05	52.50
	3240	1/2"		26	.308		30.50	12.55		43.05	52.50
	3250	3/4"		25	.320		34	13.05		47.05	57
	3260	1"		23	.348		43	14.20		57.20	68.50
	3270	1-1/4"		21	.381		71.50	15.55		87.05	103
	3280	1-1/2"		20	.400		71.50	16.35		87.85	104
	3290	2"		17	.471		170	19.20		189.20	216
	3310	3"	Q-1	24	.667		271	24.50		295.50	335
	3320	4"	"	20	.800		385	29.50		414.50	465
	3360	For PVC, flanged, add					50%	15%			
	3750	CPVC, socket or threaded									
	3800	1/2"	1 Plum	26	.308	Ea.	44.50	12.55		57.05	68
	3840	3/4"		25	.320		53.50	13.05		66.55	78
	3850	1"		23	.348		63	14.20		77.20	90.50
	3860	1-1/2"		20	.400		108	16.35		124.35	144
	3870	2"		17	.471		146	19.20		165.20	189
	3880	3"	Q-1	24	.667		395	24.50		419.50	470
	3920	4"	"	20	.800		530	29.50		559.50	630

Important: See the Reference Section for critical supporting data - Reference Nos., Crews, & City Cost Indexes

15110 | Valves

	Line	Description	Crew	Daily Output	Labor-Hours	Unit	2005 Bare Costs Mat.	Labor	Equip.	Total	Total Incl O&P	
500	3930	For CPVC, flanged, add R15100-090				Ea.	40%	15%				500
	4340	Polypropylene, threaded										
	4360	1/2"	1 Plum	26	.308	Ea.	29.50	12.55		42.05	51.50	
	4400	3/4"		25	.320		34.50	13.05		47.55	57.50	
	4440	1"		23	.348		44	14.20		58.20	70	
	4450	1-1/2"		20	.400		85	16.35		101.35	118	
	4460	2"	↓	17	.471		106	19.20		125.20	146	
	4500	For polypropylene flanged, add				↓	200%	15%				
	4850	Foot valve, PVC, socket or threaded										
	4900	1/2"	1 Plum	34	.235	Ea.	49.50	9.60		59.10	69	
	4930	3/4"		32	.250		56	10.20		66.20	77	
	4940	1"		28	.286		73	11.65		84.65	97.50	
	4950	1-1/4"		27	.296		140	12.10		152.10	172	
	4960	1-1/2"		26	.308		140	12.55		152.55	173	
	4970	2"		24	.333		162	13.60		175.60	199	
	4980	3"		20	.400		385	16.35		401.35	450	
	4990	4"	↓	18	.444		680	18.15		698.15	775	
	5000	For flanged, add				↓	25%	10%				
	5050	CPVC, socket or threaded										
	5060	1/2"	1 Plum	34	.235	Ea.	48.50	9.60		58.10	67.50	
	5070	3/4"		32	.250		55.50	10.20		65.70	77	
	5080	1"		28	.286		69	11.65		80.65	93	
	5090	1-1/4"		27	.296		110	12.10		122.10	139	
	5100	1-1/2"		26	.308		110	12.55		122.55	140	
	5110	2"		24	.333		141	13.60		154.60	176	
	5120	3"		20	.400		288	16.35		304.35	340	
	5130	4"	↓	18	.444		525	18.15		543.15	605	
	5140	For flanged, add				↓	25%	10%				
	5280	Needle valve, PVC, threaded										
	5300	1/4"	1 Plum	26	.308	Ea.	35	12.55		47.55	57.50	
	5340	3/8"		26	.308		41	12.55		53.55	64	
	5360	1/2"	↓	26	.308		41	12.55		53.55	64	
	5380	For polypropylene, add				↓	10%					
	5800	Y check, PVC, socket or threaded										
	5820	1/2"	1 Plum	26	.308	Ea.	55	12.55		67.55	79.50	
	5840	3/4"		25	.320		59.50	13.05		72.55	85	
	5850	1"		23	.348		65	14.20		79.20	93	
	5860	1-1/4"		21	.381		102	15.55		117.55	136	
	5870	1-1/2"		20	.400		111	16.35		127.35	147	
	5880	2"		17	.471		138	19.20		157.20	180	
	5890	2-1/2"	↓	15	.533		201	22		223	254	
	5900	3"	Q-1	24	.667		273	24.50		297.50	335	
	5910	4"	"	20	.800		475	29.50		504.50	570	
	5960	For PVC flanged, add				↓	45%	15%				
	6350	Y sediment strainer, PVC, socket or threaded										
	6400	1/2"	1 Plum	26	.308	Ea.	36	12.55		48.55	58.50	
	6440	3/4"		24	.333		39	13.60		52.60	63.50	
	6450	1"		23	.348		47	14.20		61.20	73.50	
	6460	1-1/4"		21	.381		77.50	15.55		93.05	109	
	6470	1-1/2"		20	.400		77.50	16.35		93.85	110	
	6480	2"		17	.471		95	19.20		114.20	134	
	6490	2-1/2"	↓	15	.533		233	22		255	289	
	6500	3"	Q-1	24	.667		233	24.50		257.50	293	
	6510	4"	"	20	.800		390	29.50		419.50	470	
	6560	For PVC, flanged, add				↓	55%	15%				

15110 | Valves

			CREW	DAILY OUTPUT	LABOR-HOURS	UNIT	2005 BARE COSTS MAT.	LABOR	EQUIP.	TOTAL	TOTAL INCL O&P	
600	0010	**VALVES, SEMI-STEEL** R15100-090										600
	1020	Lubricated plug valve, threaded, 200 psi										
	1030	1/2"	1 Plum	18	.444	Ea.	61.50	18.15		79.65	95	
	1040	3/4"		16	.500		61.50	20.50		82	98	
	1050	1"		14	.571		79	23.50		102.50	122	
	1060	1-1/4"		12	.667		94.50	27		121.50	145	
	1070	1-1/2"		11	.727		102	29.50		131.50	157	
	1080	2"		8	1		120	41		161	194	
	1090	2-1/2"	Q-1	5	3.200		186	118		304	380	
	1100	3"	"	4.50	3.556		228	131		359	450	
	6990	Flanged, 200 psi										
	7000	2"	1 Plum	8	1	Ea.	145	41		186	221	
	7010	2-1/2"	Q-1	5	3.200		218	118		336	415	
	7020	3"		4.50	3.556		263	131		394	485	
	7030	4"		3	5.333		335	196		531	660	
	7036	5"		2.50	6.400		500	235		735	905	
	7040	6"	Q-2	3	8		650	305		955	1,175	
	7050	8"		2.50	9.600		1,175	365		1,540	1,850	
	7060	10"		2.20	10.909		1,725	415		2,140	2,525	
	7070	12"		1.70	14.118		2,975	540		3,515	4,050	
700	0010	**VALVES, STEEL** R15100-090										700
	0800	Cast										
	1350	Check valve, swing type, 150 lb., flanged										
	1370	1"	1 Plum	10	.800	Ea.	365	32.50		397.50	455	
	1400	2"	"	8	1		515	41		556	625	
	1440	2-1/2"	Q-1	5	3.200		505	118		623	730	
	1450	3"		4.50	3.556		595	131		726	850	
	1460	4"		3	5.333		875	196		1,071	1,250	
	1470	6"	Q-2	3	8		1,350	305		1,655	1,950	
	1480	8"		2.50	9.600		2,225	365		2,590	3,000	
	1490	10"		2.20	10.909		3,300	415		3,715	4,250	
	1500	12"		1.70	14.118		4,725	540		5,265	6,000	
	1510	14"		1.30	18.462		7,400	705		8,105	9,175	
	1520	16"		1	24		9,100	915		10,015	11,400	
	1548	For 600 lb., flanged, add					110%	20%				
	1571	300 lb., 2"	1 Plum	7.40	1.081		545	44		589	665	
	1572	2-1/2"	Q-1	4.20	3.810		825	140		965	1,125	
	1573	3"		4	4		825	147		972	1,125	
	1574	4"		2.80	5.714		900	210		1,110	1,300	
	1575	6"	Q-2	2.90	8.276		1,450	315		1,765	2,075	
	1576	8"		2.40	10		2,375	380		2,755	3,175	
	1577	10"		2.10	11.429		3,500	435		3,935	4,500	
	1578	12"		1.60	15		5,000	570		5,570	6,350	
	1579	14"		1.20	20		7,600	765		8,365	9,500	
	1581	16"		.90	26.667		8,650	1,025		9,675	11,100	
	1950	Gate valve, 150 lb., flanged										
	2000	2"	1 Plum	8	1	Ea.	540	41		581	655	
	2040	2-1/2"	Q-1	5	3.200		765	118		883	1,025	
	2050	3"		4.50	3.556		765	131		896	1,025	
	2060	4"		3	5.333		930	196		1,126	1,325	
	2070	6"	Q-2	3	8		1,475	305		1,780	2,075	
	2080	8"		2.50	9.600		2,350	365		2,715	3,150	
	2090	10"		2.20	10.909		3,700	415		4,115	4,700	
	2100	12"		1.70	14.118		5,050	540		5,590	6,350	
	2110	14"		1.30	18.462		7,400	705		8,105	9,200	
	2120	16"		1	24		10,200	915		11,115	12,600	

Important: See the Reference Section for critical supporting data - Reference Nos., Crews, & City Cost Indexes

15110 | Valves

			CREW	DAILY OUTPUT	LABOR-HOURS	UNIT	2005 BARE COSTS MAT.	LABOR	EQUIP.	TOTAL	TOTAL INCL O&P	
700	2130	18" R15100-090	Q-2	.80	30	Ea.	12,000	1,150		13,150	14,900	700
	2140	20"	↓	.60	40	↓	14,300	1,525		15,825	18,000	
	2650	300 lb., flanged										
	2700	2"	1 Plum	7.40	1.081	Ea.	740	44		784	875	
	2740	2-1/2"	Q-1	4.20	3.810		995	140		1,135	1,300	
	2750	3"		4	4		995	147		1,142	1,325	
	2760	4"	↓	2.80	5.714		1,375	210		1,585	1,850	
	2770	6"	Q-2	2.90	8.276		2,350	315		2,665	3,050	
	2780	8"		2.40	10		3,675	380		4,055	4,600	
	2790	10"		2.10	11.429		4,925	435		5,360	6,075	
	2800	12"		1.60	15		6,500	570		7,070	8,000	
	2810	14"		1.20	20		12,100	765		12,865	14,500	
	2820	16"		.90	26.667		15,500	1,025		16,525	18,600	
	2830	18"		.70	34.286		19,700	1,300		21,000	23,600	
	2840	20"	↓	.50	48	↓	23,800	1,825		25,625	29,000	
	3650	Globe valve, 150 lb., flanged										
	3700	2"	1 Plum	8	1	Ea.	710	41		751	840	
	3740	2-1/2"	Q-1	5	3.200		910	118		1,028	1,175	
	3750	3"		4.50	3.556		910	131		1,041	1,200	
	3760	4"	↓	3	5.333		1,325	196		1,521	1,750	
	3770	6"	Q-2	3	8		2,100	305		2,405	2,750	
	3780	8"		2.50	9.600		3,850	365		4,215	4,775	
	3790	10"		2.20	10.909		9,375	415		9,790	10,900	
	3800	12"	↓	1.70	14.118	↓	12,000	540		12,540	14,000	
	4080	300 lb., flanged										
	4100	2"	1 Plum	7.40	1.081	Ea.	990	44		1,034	1,175	
	4140	2-1/2"	Q-1	4.20	3.810		1,350	140		1,490	1,675	
	4150	3"		4	4		1,350	147		1,497	1,700	
	4160	4"	↓	2.80	5.714		1,850	210		2,060	2,375	
	4170	6"	Q-2	2.90	8.276		3,350	315		3,665	4,175	
	4180	8"		2.40	10		5,500	380		5,880	6,625	
	4190	10"		2.10	11.429		12,600	435		13,035	14,600	
	4200	12"	↓	1.60	15	↓	14,900	570		15,470	17,300	
	4680	600 lb., flanged										
	4700	2"	1 Plum	7	1.143	Ea.	1,400	46.50		1,446.50	1,625	
	4740	2-1/2"	Q-1	4	4		2,100	147		2,247	2,550	
	4750	3"		3.60	4.444		2,100	163		2,263	2,575	
	4760	4"	↓	2.50	6.400		3,400	235		3,635	4,075	
	4770	6"	Q-2	2.60	9.231		7,050	350		7,400	8,300	
	4780	8"	"	2.10	11.429	↓	11,500	435		11,935	13,300	
	4800	Silent check, 316 S.S. trim										
	4810	Full flange type, 150 lb.										
	4811	1"	1 Plum	14	.571	Ea.	320	23.50		343.50	390	
	4812	1-1/4"		12	.667		320	27		347	395	
	4813	1-1/2"		11	.727		350	29.50		379.50	430	
	4814	2"	↓	9	.889		425	36.50		461.50	520	
	4815	2-1/2"	Q-1	9	1.778		510	65.50		575.50	660	
	4816	3"		8	2		595	73.50		668.50	765	
	4817	4"	↓	5	3.200		810	118		928	1,075	
	4818	5"	Q-2	6	4		1,025	153		1,178	1,350	
	4819	6"		5	4.800		1,300	183		1,483	1,725	
	4820	8"		4.50	5.333		1,900	203		2,103	2,400	
	4821	10"	↓	4	6		2,675	229		2,904	3,300	
	4840	For 300 lb., add					20%	10%				
	4860	For 600 lb., add				↓	50%	15%				
	4900	Globe type, 150 lb.										

15110 | Valves

	Line	Description	Crew	Daily Output	Labor-Hours	Unit	2005 Bare Costs Mat.	2005 Bare Costs Labor	2005 Bare Costs Equip.	2005 Bare Costs Total	Total Incl O&P	
700	4910	2″ (R15100-090)	1 Plum	9	.889	Ea.	490	36.50		526.50	595	700
	4911	2-1/2″	Q-1	9	1.778		605	65.50		670.50	765	
	4912	3″		8	2		675	73.50		748.50	850	
	4913	4″		5	3.200		895	118		1,013	1,150	
	4914	5″	Q-2	6	4		1,125	153		1,278	1,450	
	4915	6″		5	4.800		1,375	183		1,558	1,800	
	4916	8″		4.50	5.333		1,950	203		2,153	2,425	
	4917	10″		4	6		3,350	229		3,579	4,050	
	4918	12″		3	8		4,125	305		4,430	5,000	
	4919	14″		2.30	10.435		6,250	400		6,650	7,475	
	4920	16″		1.75	13.714		8,725	525		9,250	10,400	
	4921	18″		1.50	16		9,175	610		9,785	11,000	
	4940	For 300 lb., add					30%	10%				
	4960	For 600 lb., add					60%	15%				
	5150	Forged										
	5340	Ball valve, 800 psi, threaded, 1/4″ size	1 Plum	24	.333	Ea.	40.50	13.60		54.10	65	
	5350	3/8″		24	.333		40.50	13.60		54.10	65	
	5360	1/2″		24	.333		50	13.60		63.60	75.50	
	5370	3/4″		20	.400		66	16.35		82.35	97.50	
	5380	1″		19	.421		82.50	17.20		99.70	117	
	5390	1-1/4″		15	.533		110	22		132	154	
	5400	1-1/2″		13	.615		144	25		169	196	
	5410	2″		11	.727		190	29.50		219.50	254	
	5460	Ball valve, 800 lb., socket weld, 1/4″ size	Q-15	19	.842		44	31	4.24	79.24	99.50	
	5470	3/8″		19	.842		44	31	4.24	79.24	99.50	
	5480	1/2″		19	.842		52.50	31	4.24	87.74	109	
	5490	3/4″		19	.842		57.50	31	4.24	92.74	114	
	5500	1″		15	1.067		67.50	39	5.40	111.90	139	
	5510	1-1/4″		13	1.231		90	45.50	6.20	141.70	174	
	5520	1-1/2″		11	1.455		118	53.50	7.35	178.85	219	
	5530	2″		8.50	1.882		164	69	9.50	242.50	295	
	5550	Ball valve, 150 lb., flanged										
	5560	4″	Q-1	3	5.333	Ea.	435	196		631	770	
	5570	6″	Q-2	3	8		1,050	305		1,355	1,600	
	5580	8″	″	2.50	9.600		1,925	365		2,290	2,675	
	5650	Check valve, class 800, horizontal, socket										
	5652	1/4″	Q-15	19	.842	Ea.	61	31	4.24	96.24	119	
	5654	3/8″		19	.842		61	31	4.24	96.24	119	
	5656	1/2″		19	.842		61	31	4.24	96.24	119	
	5658	3/4″		19	.842		67	31	4.24	102.24	125	
	5660	1″		15	1.067		78	39	5.40	122.40	150	
	5662	1-1/4″		13	1.231		153	45.50	6.20	204.70	244	
	5664	1-1/2″		11	1.455		153	53.50	7.35	213.85	258	
	5666	2″		8.50	1.882		214	69	9.50	292.50	350	
	5698	Threaded										
	5700	1/4″	1 Plum	24	.333	Ea.	61	13.60		74.60	88	
	5720	3/8″		24	.333		61	13.60		74.60	88	
	5730	1/2″		24	.333		61	13.60		74.60	88	
	5740	3/4″		20	.400		67	16.35		83.35	98	
	5750	1″		19	.421		78	17.20		95.20	112	
	5760	1-1/4″		15	.533		153	22		175	202	
	5770	1-1/2″		13	.615		153	25		178	207	
	5780	2″		11	.727		214	29.50		243.50	280	
	5840	For class 150, flanged, add					100%	15%				
	5860	For class 300, flanged, add					120%	20%				
	6100	Gate, class 800, OS&Y, socket										

15100 | Building Services Piping

15110 | Valves

			CREW	DAILY OUTPUT	LABOR-HOURS	UNIT	2005 BARE COSTS MAT.	LABOR	EQUIP.	TOTAL	TOTAL INCL O&P	
700	6102	3/8" R15100-090	Q-15	19	.842	Ea.	44.50	31	4.24	79.74	100	700
	6103	1/2"		19	.842		44.50	31	4.24	79.74	100	
	6104	3/4"		19	.842		49.50	31	4.24	84.74	106	
	6105	1"		15	1.067		60	39	5.40	104.40	130	
	6106	1-1/4"		13	1.231		114	45.50	6.20	165.70	200	
	6107	1-1/2"		11	1.455		114	53.50	7.35	174.85	214	
	6108	2"		8.50	1.882		148	69	9.50	226.50	276	
	6118	Threaded										
	6120	3/8"	1 Plum	24	.333	Ea.	44.50	13.60		58.10	69.50	
	6130	1/2"		24	.333		44.50	13.60		58.10	69.50	
	6140	3/4"		20	.400		49.50	16.35		65.85	79	
	6150	1"		19	.421		60	17.20		77.20	91.50	
	6160	1-1/4"		15	.533		114	22		136	158	
	6170	1-1/2"		13	.615		114	25		139	163	
	6180	2"		11	.727		148	29.50		177.50	207	
	6260	For OS&Y, flanged, add					100%	20%				
	6700	Globe, OS&Y, class 800, socket										
	6710	1/4"	Q-15	19	.842	Ea.	67	31	4.24	102.24	125	
	6720	3/8"		19	.842		67	31	4.24	102.24	125	
	6730	1/2"		19	.842		67	31	4.24	102.24	125	
	6740	3/4"		19	.842		75.50	31	4.24	110.74	134	
	6750	1"		15	1.067		98.50	39	5.40	142.90	174	
	6760	1-1/4"		13	1.231		194	45.50	6.20	245.70	288	
	6770	1-1/2"		11	1.455		194	53.50	7.35	254.85	300	
	6780	2"		8.50	1.882		248	69	9.50	326.50	385	
	6860	For OS&Y, flanged, add					300%	20%				
	6880	Threaded										
	6882	1/4"	1 Plum	24	.333	Ea.	67	13.60		80.60	94	
	6884	3/8"		24	.333		67	13.60		80.60	94	
	6886	1/2"		24	.333		67	13.60		80.60	94	
	6888	3/4"		20	.400		75.50	16.35		91.85	108	
	6890	1"		19	.421		98.50	17.20		115.70	135	
	6892	1-1/4"		15	.533		194	22		216	246	
	6894	1-1/2"		13	.615		194	25		219	251	
	6896	2"		11	.727		248	29.50		277.50	320	
800	0010	**VALVES, STAINLESS STEEL** R15100-090										800
	1610	Ball, threaded 1/4"	1 Stpi	24	.333	Ea.	30.50	13.65		44.15	54	
	1620	3/8"		24	.333		30.50	13.65		44.15	54	
	1630	1/2"		22	.364		30.50	14.90		45.40	56	
	1640	3/4"		20	.400		50.50	16.40		66.90	80	
	1650	1"		19	.421		61.50	17.25		78.75	93.50	
	1660	1-1/4		15	.533		117	22		139	162	
	1670	1-1/2"		13	.615		122	25		147	173	
	1680	2"		11	.727		159	30		189	220	
	1700	Check, 200 lb., threaded										
	1710	1/4"	1 Plum	24	.333	Ea.	88	13.60		101.60	117	
	1720	1/2"		22	.364		88	14.85		102.85	119	
	1730	3/4"		20	.400		96	16.35		112.35	130	
	1750	1"		19	.421		122	17.20		139.20	161	
	1760	1-1/2"		13	.615		235	25		260	296	
	1770	2"		11	.727		395	29.50		424.50	480	
	1800	150 lb., flanged										
	1810	2-1/2"	Q-1	5	3.200	Ea.	945	118		1,063	1,225	
	1820	3"		4.50	3.556		945	131		1,076	1,250	
	1830	4"		3	5.333		1,400	196		1,596	1,850	

15110 | Valves		CREW	DAILY OUTPUT	LABOR-HOURS	UNIT	2005 BARE COSTS MAT.	LABOR	EQUIP.	TOTAL	TOTAL INCL O&P
800 1840	6″ (R15100-090)	Q-2	3	8	Ea.	2,525	305		2,830	3,225 800
1850	8″	″	2.50	9.600		4,200	365		4,565	5,175
2100	Gate, OS&Y, 150 lb., flanged									
2120	1/2″	1 Plum	18	.444	Ea.	277	18.15		295.15	335
2140	3/4″		16	.500		296	20.50		316.50	355
2150	1″		14	.571		360	23.50		383.50	435
2155	1-1/4″		12	.667		505	27		532	600
2160	1-1/2″		11	.727		505	29.50		534.50	605
2170	2″		8	1		570	41		611	685
2180	2-1/2″	Q-1	5	3.200		940	118		1,058	1,200
2190	3″		4.50	3.556		940	131		1,071	1,225
2200	4″		3	5.333		1,375	196		1,571	1,825
2205	5″		2.80	5.714		2,400	210		2,610	2,975
2210	6″	Q-2	3	8		2,400	305		2,705	3,100
2220	8″		2.50	9.600		3,975	365		4,340	4,925
2230	10″		2.30	10.435		6,325	400		6,725	7,550
2240	12″		1.90	12.632		8,075	480		8,555	9,625
2260	For 300 lb., flanged, add					120%	15%			
2600	600 lb., flanged									
2620	1/2″	1 Plum	16	.500	Ea.	850	20.50		870.50	965
2640	3/4″		14	.571		925	23.50		948.50	1,050
2650	1″		12	.667		1,175	27		1,202	1,325
2660	1-1/2″		10	.800		1,650	32.50		1,682.50	1,850
2670	2″		7	1.143		2,350	46.50		2,396.50	2,675
2680	2-1/2″	Q-1	4	4		3,650	147		3,797	4,250
2690	3″	″	3.60	4.444		3,900	163		4,063	4,525
3100	Globe, OS&Y, 150 lb., flanged									
3120	1/2″	1 Plum	18	.444	Ea.	370	18.15		388.15	435
3140	3/4″		16	.500		410	20.50		430.50	480
3150	1″		14	.571		505	23.50		528.50	590
3160	1-1/2″		11	.727		750	29.50		779.50	870
3170	2″		8	1		910	41		951	1,050
3180	2-1/2″	Q-1	5	3.200		1,325	118		1,443	1,625
3190	3″		4.50	3.556		1,625	131		1,756	1,975
3200	4″		3	5.333		2,475	196		2,671	3,025
3210	6″	Q-2	3	8		3,100	305		3,405	3,850
5000	Silent check, 316 S.S. body and trim									
5010	Compact wafer type, 300 lb.									
5020	1″	1 Plum	14	.571	Ea.	585	23.50		608.50	675
5021	1-1/4″		12	.667		575	27		602	670
5022	1-1/2″		11	.727		595	29.50		624.50	695
5023	2″		9	.889		680	36.50		716.50	800
5024	2-1/2″	Q-1	9	1.778		930	65.50		995.50	1,125
5025	3″	″	8	2		1,075	73.50		1,148.50	1,275
5100	Full flange wafer type, 150 lb.									
5110	1″	1 Plum	14	.571	Ea.	375	23.50		398.50	445
5111	1-1/4″		12	.667		400	27		427	480
5112	1-1/2″		11	.727		460	29.50		489.50	550
5113	2″		9	.889		600	36.50		636.50	715
5114	2-1/2″	Q-1	9	1.778		705	65.50		770.50	875
5115	3″		8	2		805	73.50		878.50	995
5116	4″		5	3.200		1,200	118		1,318	1,500
5117	5″	Q-2	6	4		1,400	153		1,553	1,750
5118	6″		5	4.800		1,750	183		1,933	2,200
5119	8″		4.50	5.333		2,900	203		3,103	3,500
5120	10″		4	6		4,025	229		4,254	4,775

15100 | Building Services Piping

	15110 | Valves		CREW	DAILY OUTPUT	LABOR-HOURS	UNIT	2005 BARE COSTS MAT.	LABOR	EQUIP.	TOTAL	TOTAL INCL O&P	
800	5200	Globe type, 300 lb. R15100-090										800
	5210	2"	1 Plum	9	.889	Ea.	685	36.50		721.50	805	
	5211	2-1/2"	Q-1	9	1.778		840	65.50		905.50	1,025	
	5212	3"		8	2		1,025	73.50		1,098.50	1,250	
	5213	4"		5	3.200		1,450	118		1,568	1,750	
	5214	5"	Q-2	6	4		2,025	153		2,178	2,450	
	5215	6"		5	4.800		2,275	183		2,458	2,775	
	5216	8"		4.50	5.333		3,225	203		3,428	3,850	
	5217	10"		4	6		4,575	229		4,804	5,375	
	5300	Screwed end type, 300 lb.										
	5310	1/2"	1 Plum	24	.333	Ea.	58	13.60		71.60	84	
	5311	3/4"		20	.400		69	16.35		85.35	101	
	5312	1"		19	.421		80	17.20		97.20	115	
	5313	1-1/4"		15	.533		106	22		128	150	
	5314	1-1/2"		13	.615		114	25		139	164	
	5315	2"		11	.727		142	29.50		171.50	201	
	15120 | Piping Specialties											
120	0010	**AIR CONTROL**										120
	0030	Air separator, with strainer										
	0040	2" diameter	Q-5	6	2.667	Ea.	555	98.50		653.50	760	
	0080	2-1/2" diameter		5	3.200		630	118		748	870	
	0100	3" diameter		4	4		960	147		1,107	1,275	
	0120	4" diameter		3	5.333		1,400	197		1,597	1,825	
	0130	5" diameter	Q-6	3.60	6.667		1,775	255		2,030	2,325	
	0140	6" diameter		3.40	7.059		2,125	270		2,395	2,725	
	0160	8" diameter		3	8		3,175	305		3,480	3,925	
	0180	10" diameter		2.20	10.909		4,925	415		5,340	6,025	
	0200	12" diameter		1.70	14.118		8,250	540		8,790	9,875	
	0210	14" diameter		1.30	18.462		9,825	705		10,530	11,900	
	0220	16" diameter		1	24		14,800	915		15,715	17,600	
	0230	18" diameter		.80	30		19,400	1,150		20,550	23,100	
	0240	20" diameter		.60	40		21,700	1,525		23,225	26,200	
	0300	Without strainer										
	0310	2" diameter	Q-5	6	2.667	Ea.	430	98.50		528.50	625	
	0320	2-1/2" diameter		5	3.200		515	118		633	740	
	0330	3" diameter		4	4		725	147		872	1,025	
	0340	4" diameter		3	5.333		1,125	197		1,322	1,550	
	0350	5" diameter		2.40	6.667		1,525	246		1,771	2,075	
	0360	6" diameter	Q-6	3.40	7.059		1,775	270		2,045	2,350	
	0370	8" diameter		3	8		2,425	305		2,730	3,125	
	0380	10" diameter		2.20	10.909		3,575	415		3,990	4,550	
	0390	12" diameter		1.70	14.118		5,500	540		6,040	6,850	
	0400	14" diameter		1.30	18.462		8,225	705		8,930	10,100	
	0410	16" diameter		1	24		11,500	915		12,415	14,100	
	0420	18" diameter		.80	30		15,000	1,150		16,150	18,200	
	0430	20" diameter		.60	40		18,000	1,525		19,525	22,100	
	1000	Micro-bubble separator for total air removal, closed loop system										
	1010	Requires bladder type tank in system.										
	1020	Water (hot or chilled) or glycol system										
	1030	Threaded										
	1040	3/4" diameter	1 Stpi	20	.400	Ea.	69	16.40		85.40	100	
	1050	1" diameter		19	.421		77	17.25		94.25	111	
	1060	1-1/4" diameter		16	.500		107	20.50		127.50	148	
	1070	1-1/2" diameter		13	.615		139	25		164	191	
	1080	2" diameter		11	.727		675	30		705	785	

		15120 | Piping Specialties	CREW	DAILY OUTPUT	LABOR-HOURS	UNIT	2005 BARE COSTS MAT.	LABOR	EQUIP.	TOTAL	TOTAL INCL O&P	
120	1090	2-1/2" diameter	Q-5	15	1.067	Ea.	750	39.50		789.50	885	120
	1100	3" diameter		13	1.231		1,075	45.50		1,120.50	1,250	
	1110	4" diameter		10	1.600		1,150	59		1,209	1,375	
	1230	Flanged										
	1250	2" diameter	1 Stpi	8	1	Ea.	865	41		906	1,000	
	1260	2-1/2" diameter	Q-5	5	3.200		930	118		1,048	1,200	
	1270	3" diameter		4.50	3.556		1,250	131		1,381	1,575	
	1280	4" diameter		3	5.333		1,400	197		1,597	1,825	
	1290	5" diameter	Q-6	3.40	7.059		2,225	270		2,495	2,850	
	1300	6" diameter	"	3	8		2,700	305		3,005	3,425	
	1400	Larger sizes available										
	1590	With extended tank dirt catcher										
	1600	Flanged										
	1620	2" diameter	1 Stpi	7.40	1.081	Ea.	1,125	44.50		1,169.50	1,325	
	1630	2-1/2" diameter	Q-5	4.20	3.810		1,200	140		1,340	1,500	
	1640	3" diameter		4	4		1,550	147		1,697	1,925	
	1650	4" diameter		2.80	5.714		1,700	211		1,911	2,175	
	1660	5" diameter	Q-6	3.10	7.742		2,475	296		2,771	3,175	
	1670	6" diameter	"	2.90	8.276		3,050	315		3,365	3,825	
	1800	With drain/dismantle/cleaning access flange										
	1810	Flanged										
	1820	2" diameter	1 Stpi	7	1.143	Ea.	1,775	47		1,822	2,025	
	1830	2-1/2" diameter	Q-5	4	4		1,900	147		2,047	2,325	
	1840	3" diameter		3.60	4.444		2,575	164		2,739	3,075	
	1850	4" diameter		2.50	6.400		2,850	236		3,086	3,475	
	1860	5" diameter	Q-6	2.80	8.571		4,250	330		4,580	5,175	
	1870	6" diameter	"	2.60	9.231		5,525	355		5,880	6,625	
	2000	Boiler air fitting, separator										
	2010	1-1/4"	Q-5	22	.727	Ea.	126	27		153	180	
	2020	1-1/2"		20	.800		245	29.50		274.50	315	
	2021	2"		18	.889		425	33		458	520	
	2400	Compression tank air fitting										
	2410	For tanks 9" to 24" diameter	1 Stpi	15	.533	Ea.	37.50	22		59.50	74	
	2420	For tanks 100 gallon or larger	"	12	.667	"	151	27.50		178.50	207	
140	0010	**AIR PURGING SCOOP** with tappings										140
	0020	for air vent and expansion tank connection										
	0100	1" pipe size, threaded	1 Stpi	19	.421	Ea.	13.90	17.25		31.15	41.50	
	0110	1-1/4" pipe size, threaded		15	.533		13.70	22		35.70	48	
	0120	1-1/2" pipe size, threaded		13	.615		30.50	25		55.50	71.50	
	0130	2" pipe size, threaded		11	.727		35	30		65	83.50	
	0140	2-1/2" pipe size, threaded	Q-5	15	1.067		74	39.50		113.50	141	
	0150	3" pipe size, threaded		13	1.231		97	45.50		142.50	175	
	0160	4" 150 lb. flanges		3	5.333		218	197		415	535	
160	0010	**AUTOMATIC AIR VENT**										160
	0020	Cast iron body, stainless steel internals, float type										
	0060	1/2" NPT inlet, 300 psi	1 Stpi	12	.667	Ea.	73	27.50		100.50	121	
	0140	3/4" NPT inlet, 300 psi		12	.667		73	27.50		100.50	121	
	0180	1/2" NPT inlet, 250 psi		10	.800		234	33		267	305	
	0220	3/4" NPT inlet, 250 psi		10	.800		234	33		267	305	
	0260	1" NPT inlet, 250 psi		10	.800		345	33		378	430	
	0340	1-1/2" NPT inlet, 250 psi	Q-5	12	1.333		725	49		774	875	
	0380	2" NPT inlet, 250 psi	"	12	1.333		725	49		774	875	
	0600	Forged steel body, stainless steel internals, float type										
	0640	1/2" NPT inlet, 750 psi	1 Stpi	12	.667	Ea.	725	27.50		752.50	835	
	0680	3/4" NPT inlet, 750 psi		12	.667		725	27.50		752.50	835	

15120 | Piping Specialties

				DAILY	LABOR-		2005 BARE COSTS				TOTAL	
			CREW	OUTPUT	HOURS	UNIT	MAT.	LABOR	EQUIP.	TOTAL	INCL O&P	
160	0760	3/4" NPT inlet, 1000 psi	1 Stpi	10	.800	Ea.	1,075	33		1,108	1,250	160
	0800	1" NPT inlet, 1000 psi	Q-5	12	1.333		1,050	49		1,099	1,225	
	0880	1-1/2" NPT inlet, 1000 psi		10	1.600		3,050	59		3,109	3,450	
	0920	2" NPT inlet, 1000 psi		10	1.600		3,050	59		3,109	3,450	
	1100	Formed steel body, non corrosive										
	1110	1/8" NPT inlet 150 psi	1 Stpi	32	.250	Ea.	7.05	10.25		17.30	23	
	1120	1/4" NPT inlet 150 psi		32	.250		23.50	10.25		33.75	41.50	
	1130	3/4" NPT inlet 150 psi		32	.250		23.50	10.25		33.75	41.50	
	1300	Chrome plated brass, automatic/manual, for radiators										
	1310	1/8" NPT inlet, nickel plated brass	1 Stpi	32	.250	Ea.	4.10	10.25		14.35	19.90	
180	0010	**CIRCUIT SENSOR** Flow meter										180
	0020	Metering stations										
	0040	Wafer orifice insert type										
	0060	2-1/2" pipe size	Q-5	12	1.333	Ea.	130	49		179	216	
	0100	3" pipe size		11	1.455		144	53.50		197.50	239	
	0140	4" pipe size		8	2		161	73.50		234.50	288	
	0180	5" pipe size		7.30	2.192		207	81		288	350	
	0220	6" pipe size		6.40	2.500		245	92		337	410	
	0260	8" pipe size	Q-6	5.30	4.528		345	173		518	635	
	0280	10" pipe size		4.60	5.217		400	199		599	740	
	0360	12" pipe size		4.20	5.714		660	218		878	1,050	
	2000	In-line probe type										
	2200	Copper, soldered										
	2210	1/2" size	1 Stpi	24	.333	Ea.	109	13.65		122.65	141	
	2220	3/4" size		20	.400		113	16.40		129.40	150	
	2230	1" size		19	.421		128	17.25		145.25	167	
	2240	1-1/4" size		15	.533		135	22		157	182	
	2250	1-1/2" size		13	.615		158	25		183	212	
	2260	2" size		11	.727		179	30		209	241	
	2270	2-1/2" size	Q-5	15	1.067		188	39.50		227.50	266	
	2280	3" size	″	13	1.231		204	45.50		249.50	292	
	2500	Brass, threaded										
	2510	1" size	1 Stpi	19	.421	Ea.	135	17.25		152.25	175	
	2520	1-1/4" size		15	.533		149	22		171	197	
	2530	1-1/2" size		13	.615		179	25		204	234	
	2540	2" size		11	.727		188	30		218	252	
	2550	2-1/2" size	Q-5	15	1.067		204	39.50		243.50	284	
	2560	3" size	″	13	1.231		209	45.50		254.50	298	
	2700	Steel, threaded										
	2710	1" size	1 Stpi	19	.421	Ea.	129	17.25		146.25	168	
	2720	1-1/4" size		15	.533		141	22		163	188	
	2730	1-1/2" size		13	.615		159	25		184	213	
	2740	2" size		11	.727		177	30		207	240	
	2750	2-1/2" size	Q-5	15	1.067		188	39.50		227.50	266	
	2760	3" size	″	13	1.231		209	45.50		254.50	297	
	3000	Pitot tube probe type										
	3100	Weld-on mounting with probe										
	3110	2" size	Q-17	18	.889	Ea.	146	33	4.48	183.48	215	
	3120	2-1/2" size		18	.889		149	33	4.48	186.48	218	
	3130	3" size		18	.889		158	33	4.48	195.48	228	
	3140	4" size		18	.889		209	33	4.48	246.48	284	
	3150	5" size		17	.941		234	34.50	4.74	273.24	315	
	3160	6" size		17	.941		305	34.50	4.74	344.24	390	
	3170	8" size		16	1		430	37	5.05	472.05	535	
	3180	10" size		15	1.067		465	39.50	5.40	509.90	580	
	3190	12" size		15	1.067		500	39.50	5.40	544.90	615	

15120 | Piping Specialties

			CREW	DAILY OUTPUT	LABOR-HOURS	UNIT	2005 BARE COSTS MAT.	LABOR	EQUIP.	TOTAL	TOTAL INCL O&P	
180	3200	14" size	Q-17	14	1.143	Ea.	575	42	5.75	622.75	700	180
	3210	16" size		14	1.143		675	42	5.75	722.75	815	
	3220	18" size		13	1.231		720	45.50	6.20	771.70	870	
	3230	20" size		13	1.231		780	45.50	6.20	831.70	935	
	3240	24" size	↓	12	1.333	↓	875	49	6.70	930.70	1,050	
	3400	Weld-on wet tap with probe										
	3410	2" size, 5/16" probe dia.	Q-17	14	1.143	Ea.	355	42	5.75	402.75	460	
	3420	2-1/2" size, 5/16" probe dia.		14	1.143		360	42	5.75	407.75	470	
	3430	3" size, 5/16" probe dia.		14	1.143		380	42	5.75	427.75	490	
	3440	4" size, 5/16" probe dia.		14	1.143		390	42	5.75	437.75	500	
	3450	5" size, 3/8" probe dia.		13	1.231		410	45.50	6.20	461.70	525	
	3460	6" size, 3/8" probe dia.		13	1.231		435	45.50	6.20	486.70	555	
	3470	8" size, 1/2" probe dia.		12	1.333		450	49	6.70	505.70	575	
	3480	10" size, 3/4" probe dia.		11	1.455		480	53.50	7.35	540.85	620	
	3490	12" size, 3/4" probe dia.		11	1.455		525	53.50	7.35	585.85	670	
	3500	14" size, 1" probe dia.		10	1.600		595	59	8.05	662.05	750	
	3510	16" size, 1" probe dia.		10	1.600		710	59	8.05	777.05	875	
	3520	18" size, 1" probe dia.		9	1.778		760	65.50	8.95	834.45	945	
	3530	20" size, 1" probe dia.		8	2		815	73.50	10.10	898.60	1,025	
	3540	24" size, 1" probe dia.	↓	8	2	↓	910	73.50	10.10	993.60	1,125	
	3700	Clamp-on wet tap with probe										
	3710	2" size, 5/16" probe dia.	Q-5	16	1	Ea.	415	37		452	510	
	3720	2-1/2" size, 5/16" probe dia.		16	1		430	37		467	530	
	3730	3" size, 5/16" probe dia.		15	1.067		440	39.50		479.50	545	
	3740	4" size, 5/16" probe dia.		15	1.067		510	39.50		549.50	620	
	3750	5" size, 3/8" probe dia.		14	1.143		540	42		582	660	
	3760	6" size, 3/8" probe dia.		14	1.143		570	42		612	695	
	3770	8" size, 1/2" probe dia.		13	1.231		630	45.50		675.50	765	
	3780	10" size, 3/4" probe dia.		12	1.333		660	49		709	800	
	3790	12" size, 3/4" probe dia.		12	1.333		735	49		784	885	
	3800	14" size, 1" probe dia.		11	1.455		995	53.50		1,048.50	1,175	
	3810	16" size, 1" probe dia.		11	1.455		1,150	53.50		1,203.50	1,350	
	3820	18" size, 1" probe dia.		10	1.600		1,225	59		1,284	1,450	
	3830	20" size, 1" probe dia		9	1.778		1,400	65.50		1,465.50	1,650	
	3840	24" size, 1" probe dia.	↓	9	1.778	↓	1,450	65.50		1,515.50	1,700	
	4000	Wet tap drills										
	4010	3/8" dia. for 5/16" probe				Ea.	85			85	93.50	
	4020	1/2" dia. for 3/8" probe					88			88	97	
	4030	5/8" dia for 1/2" probe					105			105	116	
	4040	7/8" dia for 3/4" probe					137			137	150	
	4050	1-3/16" dia for 1" probe				↓	146			146	160	
	4100	Wet tap punch										
	4110	5/16" dia for 5/16" probe				Ea.	88			88	97	
	4120	3/8" dia for 3/8" probe					95			95	104	
	4130	1/2" dia for 1/2" probe				↓	113			113	125	
	4150	Note: labor for wet tap drill or punch										
	4160	is included with the mounting and probe assembly										
	9000	Readout instruments										
	9200	Gauges										
	9220	GPM gauge				Ea.	1,425			1,425	1,575	
	9230	Dual gauge kit				"	1,475			1,475	1,625	
200	0010	**CIRCUIT SETTER** Balance valve										200
	0012	Bronze body, soldered										
	0013	1/2" size	1 Stpi	24	.333	Ea.	43.50	13.65		57.15	68.50	
	0014	3/4" size	"	20	.400	"	46	16.40		62.40	75	

		15120 | Piping Specialties	CREW	DAILY OUTPUT	LABOR-HOURS	UNIT	2005 BARE COSTS MAT.	LABOR	EQUIP.	TOTAL	TOTAL INCL O&P	
200	0018	Threaded										200
	0019	1/2" pipe size	1 Stpi	22	.364	Ea.	45	14.90		59.90	72	
	0020	3/4" pipe size		20	.400		48.50	16.40		64.90	77.50	
	0040	1" pipe size		18	.444		62.50	18.20		80.70	96	
	0050	1-1/4" pipe size		15	.533		92	22		114	134	
	0060	1-1/2" pipe size		12	.667		109	27.50		136.50	160	
	0080	2" pipe size	↓	10	.800		155	33		188	220	
	0100	2-1/2" pipe size	Q-5	15	1.067		350	39.50		389.50	445	
	0120	3" pipe size	"	10	1.600	↓	495	59		554	635	
	0130	Cast iron body, flanged										
	0136	3" pipe size, flanged	Q-5	4	4	Ea.	535	147		682	805	
	0140	4" pipe size	"	3	5.333		740	197		937	1,100	
	0200	For differential meter, accurate to 1%, add					600			600	660	
	0300	For differential meter 400 psi/200° F continuous, add				↓	1,475			1,475	1,625	
220	0010	**COCKS, DRAINS & SPECIALTIES**										220
	1000	Boiler drain										
	1010	Pipe thread to hose										
	1020	Bronze										
	1030	1/2" size	1 Stpi	36	.222	Ea.	4.10	9.10		13.20	18.20	
	1040	3/4" size	"	34	.235	"	4.44	9.65		14.09	19.40	
	1100	Solder to hose										
	1110	Bronze										
	1120	1/2" size	1 Stpi	46	.174	Ea.	3.87	7.10		10.97	14.95	
	1130	3/4" size	"	44	.182	"	3.68	7.45		11.13	15.25	
	1600	With built-in vacuum breaker										
	1610	1/2" IP or solder	1 Stpi	36	.222	Ea.	16.45	9.10		25.55	32	
	1630	With tamper proof vacuum breaker										
	1640	1/2" IP or solder	1 Stpi	36	.222	Ea.	19.50	9.10		28.60	35	
	1650	3/4" IP or solder	"	34	.235	"	19.80	9.65		29.45	36.50	
	3000	Cocks										
	3010	Air, lever or tee handle										
	3020	Bronze, single thread										
	3030	1/8" size	1 Stpi	52	.154	Ea.	5.25	6.30		11.55	15.25	
	3040	1/4" size		46	.174		5.60	7.10		12.70	16.85	
	3050	3/8" size		40	.200		5.65	8.20		13.85	18.50	
	3060	1/2" size	↓	36	.222	↓	6.65	9.10		15.75	21	
	3100	Bronze, double thread										
	3110	1/8" size	1 Stpi	26	.308	Ea.	6.60	12.60		19.20	26.50	
	3120	1/4" size		22	.364		6.95	14.90		21.85	30	
	3130	3/8" size		18	.444		7.20	18.20		25.40	35.50	
	3140	1/2" size	↓	15	.533	↓	8.85	22		30.85	42.50	
	4000	Steam										
	4010	Bronze										
	4020	1/8" size	1 Stpi	26	.308	Ea.	16.30	12.60		28.90	37	
	4030	1/4" size		22	.364		14.50	14.90		29.40	38.50	
	4040	3/8" size		18	.444		13.40	18.20		31.60	42.50	
	4050	1/2" size		15	.533		12.60	22		34.60	47	
	4060	3/4" size		14	.571		14.60	23.50		38.10	51	
	4070	1" size		13	.615		27	25		52	67.50	
	4080	1-1/4" size		11	.727		48.50	30		78.50	98.50	
	4090	1-1/2" size		10	.800		63	33		96	119	
	4100	2" size	↓	9	.889	↓	96.50	36.50		133	161	
	4300	Bronze, 3 way										
	4320	1/4" size	1 Stpi	15	.533	Ea.	16.85	22		38.85	51.50	
	4330	3/8" size		12	.667		16.85	27.50		44.35	59.50	
	4340	1/2" size	↓	10	.800	↓	16.85	33		49.85	67.50	

15120 | Piping Specialties

			CREW	DAILY OUTPUT	LABOR-HOURS	UNIT	MAT.	LABOR	EQUIP.	TOTAL	TOTAL INCL O&P	
220	4350	3/4″ size	1 Stpi	9.50	.842	Ea.	29	34.50		63.50	84	220
	4360	1″ size		8.50	.941		35	38.50		73.50	96.50	
	4370	1-1/4″ size		7.50	1.067		52.50	43.50		96	123	
	4380	1-1/2″ size		6.50	1.231		61	50.50		111.50	143	
	4390	2″ size	↓	6	1.333	↓	90	54.50		144.50	181	
	4500	Gauge cock, brass										
	4510	1/4″ FPT	1 Stpi	24	.333	Ea.	5.80	13.65		19.45	27	
	4512	1/4″ MPT	″	24	.333	″	7.30	13.65		20.95	28.50	
	4600	Pigtail, steam syphon										
	4604	1/4″	1 Stpi	24	.333	Ea.	11.20	13.65		24.85	33	
	4650	Snubber valve										
	4654	1/4″	1 Stpi	22	.364	Ea.	9.75	14.90		24.65	33.50	
	4660	Nipple, black steel										
	4664	1/4″ x 3″	1 Stpi	37	.216	Ea.	.58	8.85		9.43	13.95	
250	0010	**DIELECTRIC UNIONS** Standard gaskets for water and air										250
	0020	250 psi maximum pressure										
	0280	Female IPT to sweat, straight										
	0300	1/2″ pipe size	1 Plum	24	.333	Ea.	3.15	13.60		16.75	24	
	0340	3/4″ pipe size		20	.400		3.15	16.35		19.50	28	
	0360	1″ pipe size		19	.421		6.10	17.20		23.30	32.50	
	0380	1-1/4″ pipe size		15	.533		10	22		32	44	
	0400	1-1/2″ pipe size		13	.615		15	25		40	54.50	
	0420	2″ pipe size	↓	11	.727	↓	21	29.50		50.50	67.50	
	0580	Female IPT to brass pipe thread, straight										
	0600	1/2″ pipe size	1 Plum	24	.333	Ea.	5.95	13.60		19.55	27	
	0640	3/4″ pipe size		20	.400		7.60	16.35		23.95	33	
	0660	1″ pipe size		19	.421		11.40	17.20		28.60	38.50	
	0680	1-1/4″ pipe size		15	.533		17.05	22		39.05	52	
	0700	1-1/2″ pipe size		13	.615		23.50	25		48.50	64	
	0720	2″ pipe size	↓	11	.727	↓	30.50	29.50		60	78.50	
	0780	Female IPT to female IPT, straight										
	0800	1/2″ pipe size	1 Plum	24	.333	Ea.	4.75	13.60		18.35	26	
	0840	3/4″ pipe size		20	.400		5.30	16.35		21.65	30.50	
	0860	1″ pipe size		19	.421		7	17.20		24.20	33.50	
	0880	1-1/4″ pipe size		15	.533		9.75	22		31.75	44	
	0900	1-1/2″ pipe size		13	.615		14.95	25		39.95	54.50	
	0920	2″ pipe size	↓	11	.727	↓	21.50	29.50		51	68	
	2000	175 psi maximum pressure										
	2180	Female IPT to sweat										
	2200	1-1/2″ pipe size	1 Plum	11	.727	Ea.	38.50	29.50		68	86.50	
	2240	2″ pipe size	″	9	.889		44	36.50		80.50	103	
	2260	2-1/2″ pipe size	Q-1	15	1.067		48	39		87	112	
	2280	3″ pipe size		14	1.143		65.50	42		107.50	135	
	2300	4″ pipe size	↓	11	1.455		174	53.50		227.50	273	
	2320	5″ pipe size	Q-2	14	1.714		284	65.50		349.50	410	
	2340	6″ pipe size		12	2		330	76.50		406.50	475	
	2360	8″ pipe size	↓	10	2.400	↓	910	91.50		1,001.50	1,150	
	2480	Female IPT to brass pipe										
	2500	1-1/2″ pipe size	1 Plum	11	.727	Ea.	58	29.50		87.50	108	
	2540	2″ pipe size	″	9	.889		69	36.50		105.50	131	
	2560	2-1/2″ pipe size	Q-1	15	1.067		100	39		139	169	
	2580	3″ pipe size		14	1.143		116	42		158	191	
	2600	4″ pipe size	↓	11	1.455		199	53.50		252.50	300	
	2620	5″ pipe size	Q-2	14	1.714		224	65.50		289.50	345	
	2640	6″ pipe size		12	2		267	76.50		343.50	410	
	2680	8″ pipe size	↓	10	2.400	↓	435	91.50		526.50	620	

(Column group MAT., LABOR, EQUIP., TOTAL: 2005 BARE COSTS)

	15120 | Piping Specialties		CREW	DAILY OUTPUT	LABOR-HOURS	UNIT	2005 BARE COSTS MAT.	LABOR	EQUIP.	TOTAL	TOTAL INCL O&P	
250	2880	Female IPT to female IPT										250
	2900	1-1/2″ pipe size	1 Plum	11	.727	Ea.	36	29.50		65.50	84	
	2940	2″ pipe size	″	9	.889		38.50	36.50		75	97	
	2960	2-1/2″ pipe size	Q-1	15	1.067		48	39		87	112	
	2980	3″ pipe size		14	1.143		60	42		102	130	
	3000	4″ pipe size		11	1.455		91	53.50		144.50	181	
	3020	5″ pipe size	Q-2	14	1.714		105	65.50		170.50	215	
	3040	6″ pipe size		12	2		141	76.50		217.50	270	
	3060	8″ pipe size		10	2.400		192	91.50		283.50	350	
	3380	Copper to copper										
	3400	1-1/2″ pipe size	1 Plum	11	.727	Ea.	47.50	29.50		77	97	
	3440	2″ pipe size	″	9	.889		54	36.50		90.50	114	
	3460	2-1/2″ pipe size	Q-1	15	1.067		58	39		97	123	
	3480	3″ pipe size		14	1.143		72.50	42		114.50	143	
	3500	4″ pipe size		11	1.455		230	53.50		283.50	335	
	3520	5″ pipe size	Q-2	14	1.714		370	65.50		435.50	505	
	3540	6″ pipe size		12	2		450	76.50		526.50	610	
	3560	8″ pipe size		10	2.400		1,200	91.50		1,291.50	1,475	
280	0010	**EXPANSION COUPLINGS** Hydronic										280
	0100	Copper to copper, sweat										
	0120	3/4″ diameter	1 Stpi	20	.400	Ea.	23	16.40		39.40	49.50	
	0140	1″ diameter		19	.421		28.50	17.25		45.75	57.50	
	0160	1-1/4″ diameter		15	.533		39	22		61	76	
	1000	Baseboard riser fitting, 5″ stub by coupling 12″ long										
	1020	1/2″ diameter	1 Stpi	24	.333	Ea.	10.90	13.65		24.55	32.50	
	1040	3/4″ diameter		20	.400		15.50	16.40		31.90	41.50	
	1060	1″ diameter		19	.421		21.50	17.25		38.75	49.50	
	1080	1-1/4″ diameter		15	.533		28	22		50	64	
	1180	9″ Stub by tubing 8″ long										
	1200	1/2″ diameter	1 Stpi	24	.333	Ea.	9.80	13.65		23.45	31.50	
	1220	3/4″ diameter		20	.400		12.95	16.40		29.35	39	
	1240	1″ diameter		19	.421		18	17.25		35.25	46	
	1260	1-1/4″ diameter		15	.533		25	22		47	60.50	
300	0010	**EXPANSION JOINTS**										300
	0100	Bellows type, neoprene cover, flanged spool										
	0140	6″ face to face, 1-1/4″ diameter	1 Stpi	11	.727	Ea.	210	30		240	276	
	0160	1-1/2″ diameter	″	10.60	.755		210	31		241	278	
	0180	2″ diameter	Q-5	13.30	1.203		214	44.50		258.50	300	
	0190	2-1/2″ diameter		12.40	1.290		222	47.50		269.50	315	
	0200	3″ diameter		11.40	1.404		252	51.50		303.50	355	
	0210	4″ diameter		8.40	1.905		269	70		339	400	
	0220	5″ diameter		7.60	2.105		330	77.50		407.50	480	
	0230	6″ diameter		6.80	2.353		340	86.50		426.50	505	
	0240	8″ diameter		5.40	2.963		390	109		499	595	
	0250	10″ diameter		5	3.200		520	118		638	745	
	0260	12″ diameter		4.60	3.478		595	128		723	850	
	0480	10″ face to face, 2″ diameter		13	1.231		315	45.50		360.50	415	
	0500	2-1/2″ diameter		12	1.333		330	49		379	440	
	0520	3″ diameter		11	1.455		335	53.50		388.50	450	
	0540	4″ diameter		8	2		375	73.50		448.50	525	
	0560	5″ diameter		7	2.286		455	84		539	625	
	0580	6″ diameter		6	2.667		465	98.50		563.50	665	
	0600	8″ diameter		5	3.200		530	118		648	755	
	0620	10″ diameter		4.60	3.478		590	128		718	845	
	0640	12″ diameter		4	4		675	147		822	965	

	15120 | Piping Specialties		CREW	DAILY OUTPUT	LABOR-HOURS	UNIT	2005 BARE COSTS MAT.	LABOR	EQUIP.	TOTAL	TOTAL INCL O&P	
300	0660	14″ diameter	Q-5	3.80	4.211	Ea.	850	155		1,005	1,175	300
	0680	16″ diameter		2.90	5.517		975	203		1,178	1,375	
	0700	18″ diameter		2.50	6.400		1,100	236		1,336	1,550	
	0720	20″ diameter		2.10	7.619		1,150	281		1,431	1,700	
	0740	24″ diameter		1.80	8.889		1,325	330		1,655	1,950	
	0760	26″ diameter		1.40	11.429		1,475	420		1,895	2,225	
	0780	30″ diameter		1.20	13.333		1,625	490		2,115	2,525	
	0800	36″ diameter		1	16		2,200	590		2,790	3,275	
	1000	Bellows with internal sleeves and external covers										
	1010	Stainless steel, 150 lb.										
	1020	With male threads										
	1030	1/2″ diameter	1 Stpi	22	.364	Ea.	105	14.90		119.90	138	
	1040	3/4″ diameter		20	.400		105	16.40		121.40	140	
	1050	1″ diameter		19	.421		110	17.25		127.25	148	
	1060	1-1/4″ diameter		16	.500		115	20.50		135.50	158	
	1070	1-1/2″ diameter		13	.615		141	25		166	193	
	1080	2″ diameter		11	.727		180	30		210	243	
	1110	With flanged ends										
	1120	1-1/4″ diameter	1 Stpi	12	.667	Ea.	365	27.50		392.50	445	
	1130	1-1/2″ diameter		11	.727		365	30		395	450	
	1140	2″ diameter		9	.889		585	36.50		621.50	700	
	1150	3″ diameter	Q-5	8	2		590	73.50		663.50	760	
	1160	4″ diameter	″	5	3.200		630	118		748	870	
	1170	5″ diameter	Q-6	6	4		675	153		828	975	
	1180	6″ diameter		5	4.800		740	183		923	1,100	
	1190	8″ diameter		4.50	5.333		885	204		1,089	1,275	
	1200	10″ diameter		4	6		1,075	229		1,304	1,550	
	1210	12″ diameter		3.60	6.667		1,375	255		1,630	1,900	
320	0010	**EXPANSION TANKS**										320
	1400	Plastic, corrosion resistant, see Plumbing Cost Data										
	1410	Div. 15220										
	1505	Fiberglass and steel single / double wall storage, see Div 13201										
	1510	Tank leak detection systems, see Div 13851-350										
	2000	Steel, liquid expansion, ASME, painted, 15 gallon capacity	Q-5	17	.941	Ea.	365	34.50		399.50	450	
	2020	24 gallon capacity		14	1.143		370	42		412	470	
	2040	30 gallon capacity		12	1.333		405	49		454	520	
	2060	40 gallon capacity		10	1.600		460	59		519	595	
	2080	60 gallon capacity		8	2		555	73.50		628.50	720	
	2100	80 gallon capacity		7	2.286		570	84		654	750	
	2120	100 gallon capacity		6	2.667		735	98.50		833.50	960	
	2130	120 gallon capacity		5	3.200		815	118		933	1,075	
	2140	135 gallon capacity		4.50	3.556		850	131		981	1,125	
	2150	175 gallon capacity		4	4		1,175	147		1,322	1,525	
	2160	220 gallon capacity		3.60	4.444		1,350	164		1,514	1,725	
	2170	240 gallon capacity		3.30	4.848		1,500	179		1,679	1,925	
	2180	305 gallon capacity		3	5.333		2,025	197		2,222	2,525	
	2190	400 gallon capacity		2.80	5.714		2,275	211		2,486	2,825	
	2360	Galvanized										
	2370	15 gallon capacity	Q-5	17	.941	Ea.	625	34.50		659.50	740	
	2380	24 gallon capacity		14	1.143		670	42		712	805	
	2390	30 gallon capacity		12	1.333		785	49		834	935	
	2400	40 gallon capacity		10	1.600		895	59		954	1,075	
	2410	60 gallon capacity		8	2		1,025	73.50		1,098.50	1,225	
	2420	80 gallon capacity		7	2.286		1,175	84		1,259	1,425	
	2430	100 gallon capacity		6	2.667		1,425	98.50		1,523.50	1,725	
	2440	120 gallon capacity		5	3.200		1,675	118		1,793	2,000	

Important: See the Reference Section for critical supporting data - Reference Nos., Crews, & City Cost Indexes

		15120 | Piping Specialties	CREW	DAILY OUTPUT	LABOR-HOURS	UNIT	2005 BARE COSTS MAT.	LABOR	EQUIP.	TOTAL	TOTAL INCL O&P	
320	2450	135 gallon capacity	Q-5	4.50	3.556	Ea.	1,775	131		1,906	2,150	320
	2460	175 gallon capacity		4	4		2,175	147		2,322	2,625	
	2470	220 gallon capacity		3.60	4.444		2,525	164		2,689	3,025	
	2480	240 gallon capacity		3.30	4.848		2,625	179		2,804	3,150	
	2490	305 gallon capacity		3	5.333		3,550	197		3,747	4,200	
	2500	400 gallon capacity		2.80	5.714		4,325	211		4,536	5,075	
	3000	Steel ASME expansion, rubber diaphragm, 19 gal. cap. accept.		12	1.333		1,425	49		1,474	1,650	
	3020	31 gallon capacity		8	2		1,600	73.50		1,673.50	1,850	
	3040	61 gallon capacity		6	2.667		2,250	98.50		2,348.50	2,625	
	3060	79 gallon capacity		5	3.200		2,275	118		2,393	2,675	
	3080	119 gallon capacity		4	4		2,425	147		2,572	2,875	
	3100	158 gallon capacity		3.80	4.211		3,350	155		3,505	3,925	
	3120	211 gallon capacity		3.30	4.848		3,875	179		4,054	4,525	
	3140	317 gallon capacity		2.80	5.714		5,075	211		5,286	5,900	
	3160	422 gallon capacity		2.60	6.154		7,500	227		7,727	8,600	
	3180	528 gallon capacity	↓	2.40	6.667	↓	8,200	246		8,446	9,400	
350	0010	**FLEXIBLE CONNECTORS**, Corrugated, 7/8" O.D., 1/2" I.D.										350
	0050	Gas, seamless brass, steel fittings										
	0200	12" long	1 Plum	36	.222	Ea.	11.95	9.10		21.05	27	
	0220	18" long		36	.222		14.85	9.10		23.95	30	
	0240	24" long		34	.235		17.55	9.60		27.15	34	
	0260	30" long		34	.235		18.95	9.60		28.55	35.50	
	0280	36" long		32	.250		21	10.20		31.20	38.50	
	0320	48" long		30	.267		26.50	10.90		37.40	46	
	0340	60" long		30	.267		31.50	10.90		42.40	51.50	
	0360	72" long	↓	30	.267	↓	36.50	10.90		47.40	56.50	
	2000	Water, copper tubing, dielectric separators										
	2100	12" long	1 Plum	36	.222	Ea.	9.40	9.10		18.50	24	
	2220	15" long		36	.222		10.45	9.10		19.55	25	
	2240	18" long		36	.222		11.40	9.10		20.50	26	
	2260	24" long	↓	34	.235	↓	13.95	9.60		23.55	30	
370	0010	**FLEXIBLE METAL HOSE** Connectors, standard lengths										370
	0100	Bronze braided, bronze ends										
	0120	3/8" diameter x 12"	1 Stpi	26	.308	Ea.	15.40	12.60		28	36	
	0140	1/2" diameter x 12"		24	.333		14.65	13.65		28.30	36.50	
	0160	3/4" diameter x 12"		20	.400		21.50	16.40		37.90	48	
	0180	1" diameter x 18"		19	.421		27.50	17.25		44.75	56.50	
	0200	1-1/2" diameter x 18"		13	.615		44	25		69	86.50	
	0220	2" diameter x 18"	↓	11	.727	↓	53	30		83	103	
	1000	Carbon steel ends										
	1020	1/4" diameter x 12"	1 Stpi	28	.286	Ea.	12.65	11.70		24.35	31.50	
	1040	3/8" diameter x 12"		26	.308		13	12.60		25.60	33.50	
	1060	1/2" diameter x 12"		24	.333		12.65	13.65		26.30	34.50	
	1080	1/2" diameter x 24"		24	.333		16.20	13.65		29.85	38.50	
	1100	1/2" diameter x 36"		24	.333		19.80	13.65		33.45	42.50	
	1120	3/4" diameter x 12"		20	.400		16.80	16.40		33.20	43	
	1140	3/4" diameter x 24"		20	.400		20.50	16.40		36.90	47	
	1160	3/4" diameter x 36"		20	.400		24.50	16.40		40.90	51.50	
	1180	1" diameter x 18"		19	.421		22	17.25		39.25	50.50	
	1200	1" diameter x 30"		19	.421		26	17.25		43.25	54.50	
	1220	1" diameter x 36"		19	.421		28	17.25		45.25	57	
	1240	1-1/4" diameter x 18"		15	.533		22.50	22		44.50	58	
	1260	1-1/4" diameter x 36"		15	.533		30	22		52	66	
	1280	1-1/2" diameter x 18"		13	.615		24.50	25		49.50	64.50	
	1300	1-1/2" diameter x 36"	↓	13	.615	↓	32.50	25		57.50	73.50	

15120 | Piping Specialties

	Line	Description	CREW	DAILY OUTPUT	LABOR-HOURS	UNIT	2005 BARE COSTS MAT.	LABOR	EQUIP.	TOTAL	TOTAL INCL O&P	
370	1320	2″ diameter x 24″	1 Stpi	11	.727	Ea.	32	30		62	80	370
	1340	2″ diameter x 36″		11	.727		39.50	30		69.50	88.50	
	1360	2-1/2″ diameter x 24″		9	.889		45	36.50		81.50	104	
	1380	2-1/2″ diameter x 36″		9	.889		53	36.50		89.50	113	
	1400	3″ diameter x 24″		7	1.143		53.50	47		100.50	130	
	1420	3″ diameter x 36″		7	1.143		62	47		109	139	
	2000	Carbon steel braid, carbon steel solid ends										
	2100	1/2″ diameter x 12″	1 Stpi	24	.333	Ea.	12.65	13.65		26.30	34.50	
	2120	3/4″ diameter x 12″		20	.400		16.80	16.40		33.20	43	
	2140	1″ diameter x 12″		19	.421		20	17.25		37.25	48	
	2160	1-1/4″ diameter x 12″		15	.533		20	22		42	55	
	2180	1-1/2″ diameter x 12″		13	.615		21.50	25		46.50	62	
	3000	Stainless steel braid, welded on carbon steel ends										
	3100	1/2″ diameter x 12″	1 Stpi	24	.333	Ea.	18.50	13.65		32.15	41	
	3120	3/4″ diameter x 12″		20	.400		26.50	16.40		42.90	53.50	
	3140	3/4″ diameter x 24″		20	.400		32	16.40		48.40	59.50	
	3160	3/4″ diameter x 36″		20	.400		37.50	16.40		53.90	65.50	
	3180	1″ diameter x 12″		19	.421		30	17.25		47.25	59	
	3200	1″ diameter x 24″		19	.421		36	17.25		53.25	66	
	3220	1″ diameter x 36″		19	.421		42.50	17.25		59.75	73	
	3240	1-1/4″ diameter x 12″		15	.533		32.50	22		54.50	69	
	3260	1-1/4″ diameter x 24″		15	.533		41.50	22		63.50	78.50	
	3280	1-1/4″ diameter x 36″		15	.533		50	22		72	88	
	3300	1-1/2″ diameter x 12″		13	.615		36	25		61	77.50	
	3320	1-1/2″ diameter x 24″		13	.615		46	25		71	88.50	
	3340	1-1/2″ diameter x 36″		13	.615		56	25		81	99.50	
	3400	Metal sst braid, over corrugated stainless steel, flanged ends										
	3410	150 PSI										
	3420	1/2″ diameter x 12″	1 Stpi	24	.333	Ea.	30	13.65		43.65	53.50	
	3430	1″ diameter x 12″		20	.400		40	16.40		56.40	68.50	
	3440	1-1/2″ diameter x 12″		15	.533		50	22		72	88	
	3450	2-1/2″ diameter x 9″		12	.667		54	27.50		81.50	101	
	3460	3″ diameter x 9″		9	.889		65	36.50		101.50	126	
	3470	4″ diameter x 9″		7	1.143		80	47		127	159	
	3480	4″ diameter x 30″		5	1.600		200	65.50		265.50	320	
	3490	4″ diameter x 36″		4.80	1.667		220	68.50		288.50	345	
	3500	6″ diameter x 11″		5	1.600		132	65.50		197.50	244	
	3510	6″ diameter x 36″		3.80	2.105		296	86		382	455	
	3520	8″ diameter x 12″		4	2		281	82		363	435	
	3530	10″ diameter x 13″		3	2.667		380	109		489	585	
	3540	12″ diameter x 14″	Q-5	4	4		570	147		717	850	
	6000	Molded rubber with helical wire reinforcement										
	6010	150 PSI										
	6020	1-1/2″ diameter x 12″	1 Stpi	15	.533	Ea.	65	22		87	105	
	6030	2″ diameter x 12″		12	.667		162	27.50		189.50	219	
	6040	3″ diameter x 12″		8	1		176	41		217	256	
	6050	4″ diameter x 12″		6	1.333		225	54.50		279.50	330	
	6060	6″ diameter x 18″		4	2		330	82		412	490	
	6070	8″ diameter x 24″		3	2.667		455	109		564	665	
	6080	10″ diameter x 24″		2	4		545	164		709	845	
	6090	12″ diameter x 24″	Q-5	3	5.333		625	197		822	985	
	7000	Molded teflon with stainless steel flanges										
	7010	150 PSI										
	7020	2-1/2″ diameter x 3-3/16″	Q-1	7.80	2.051	Ea.	980	75.50		1,055.50	1,200	
	7030	3″ diameter x 3-5/8″		6.50	2.462		1,150	90.50		1,240.50	1,400	
	7040	4″ diameter x 3-5/8″		5	3.200		1,500	118		1,618	1,825	

15120 | Piping Specialties

		15120 Piping Specialties	CREW	DAILY OUTPUT	LABOR-HOURS	UNIT	2005 BARE COSTS MAT.	LABOR	EQUIP.	TOTAL	TOTAL INCL O&P	
370	7050	6″ diameter x 4″	Q-1	4.30	3.721	Ea.	2,100	137		2,237	2,500	370
	7060	8″ diameter x 6″	↓	3.80	4.211	↓	3,325	155		3,480	3,900	
400	0010	**FLOAT VALVES**										400
	0020	With ball and bracket										
	0030	Single seat, threaded										
	0040	Brass body										
	0050	1/2″	1 Stpi	11	.727	Ea.	36.50	30		66.50	85.50	
	0060	3/4″		9	.889		41.50	36.50		78	101	
	0070	1″		7	1.143		61	47		108	138	
	0080	1-1/2″		4.50	1.778		101	73		174	220	
	0090	2″	↓	3.60	2.222	↓	102	91		193	249	
	0300	For condensate receivers, CI, in-line mount										
	0320	1″ inlet	1 Stpi	7	1.143	Ea.	104	47		151	185	
	0360	For condensate receiver, CI, external float, flanged tank mount										
	0370	3/4″ inlet	1 Stpi	5	1.600	Ea.	87	65.50		152.50	195	
420	0010	**FLOW CHECK CONTROL**										420
	0100	Bronze body, soldered										
	0110	3/4″ size	1 Stpi	20	.400	Ea.	36.50	16.40		52.90	64.50	
	0120	1″ size	″	19	.421	″	43.50	17.25		60.75	74	
	0200	Cast iron body, threaded										
	0210	3/4″ size	1 Stpi	20	.400	Ea.	29	16.40		45.40	56.50	
	0220	1″ size		19	.421		32.50	17.25		49.75	62	
	0230	1-1/4″ size		15	.533		39.50	22		61.50	76.50	
	0240	1-1/2″ size		13	.615		60.50	25		85.50	105	
	0250	2″ size	↓	11	.727	↓	87	30		117	141	
	0300	Flanged inlet, threaded outlet										
	0310	2-1/2″ size	1 Stpi	8	1	Ea.	193	41		234	274	
	0320	3″ size	Q-5	9	1.778		229	65.50		294.50	350	
	0330	4″ size	″	7	2.286	↓	445	84		529	615	
520	0010	**HYDRONIC HEATING CONTROL VALVES**										520
	0050	Hot water, nonelectric, thermostatic										
	0100	Radiator supply, 1/2″ diameter	1 Stpi	24	.333	Ea.	40	13.65		53.65	64.50	
	0120	3/4″ diameter		20	.400		40.50	16.40		56.90	69.50	
	0140	1″ diameter		19	.421		53	17.25		70.25	84	
	0160	1-1/4″ diameter	↓	15	.533	↓	74	22		96	115	
	0500	For low pressure steam, add					25%					
	1000	Manual, radiator supply										
	1010	1/2″ pipe size, angle union	1 Stpi	24	.333	Ea.	30	13.65		43.65	53.50	
	1020	3/4″ pipe size, angle union		20	.400		28	16.40		44.40	55.50	
	1030	1″ pipe size, angle union	↓	19	.421	↓	36.50	17.25		53.75	66	
	1100	Radiator, balancing, straight, sweat connections										
	1110	1/2″ pipe size	1 Stpi	24	.333	Ea.	8.50	13.65		22.15	30	
	1120	3/4″ pipe size		20	.400		11.85	16.40		28.25	37.50	
	1130	1″ pipe size	↓	19	.421	↓	57	17.25		74.25	88.50	
	1200	Steam, radiator, supply										
	1210	1/2″ pipe size, angle union	1 Stpi	24	.333	Ea.	22	13.65		35.65	44.50	
	1220	3/4″ pipe size, angle union		20	.400		23.50	16.40		39.90	50.50	
	1230	1″ pipe size, angle union		19	.421		26	17.25		43.25	54.50	
	1240	1-1/4″ pipe size, angle union	↓	15	.533	↓	35	22		57	71.50	
	8000	System balancing and shut-off										
	8020	Butterfly, quarter turn, calibrated, threaded or solder										
	8040	Bronze, -30° F to +350° F, pressure to 175 psi										
	8060	1/2″ size	1 Stpi	22	.364	Ea.	6	14.90		20.90	29	
	8070	3/4″ size		20	.400		9.70	16.40		26.10	35	
	8080	1″ size	↓	19	.421	↓	11.90	17.25		29.15	39	

15120 | Piping Specialties

			CREW	DAILY OUTPUT	LABOR-HOURS	UNIT	2005 BARE COSTS MAT.	LABOR	EQUIP.	TOTAL	TOTAL INCL O&P	
520	8090	1-1/4″ size	1 Stpi	15	.533	Ea.	19.15	22		41.15	54	520
	8100	1-1/2″ size		13	.615		24.50	25		49.50	65	
	8110	2″ size		11	.727		31	30		61	79	
550	0010	**LIQUID DRAINERS**										550
	0100	Guided lever type										
	0110	Cast iron body, threaded										
	0120	1/2″ pipe size	1 Stpi	22	.364	Ea.	73	14.90		87.90	103	
	0130	3/4″ pipe size		20	.400		195	16.40		211.40	240	
	0140	1″ pipe size		19	.421		310	17.25		327.25	365	
	0150	1-1/2″ pipe size		11	.727		680	30		710	790	
	0160	2″ pipe size		8	1		680	41		721	805	
	0200	Forged steel body, threaded										
	0210	1/2″ pipe size	1 Stpi	18	.444	Ea.	685	18.20		703.20	780	
	0220	3/4″ pipe size		16	.500		685	20.50		705.50	780	
	0230	1″ pipe size		14	.571		1,025	23.50		1,048.50	1,150	
	0240	1-1/2″ pipe size	Q-5	10	1.600		2,975	59		3,034	3,375	
	0250	2″ pipe size	″	8	2		2,975	73.50		3,048.50	3,375	
	0300	Socket weld										
	0310	1/2″ pipe size	Q-17	16	1	Ea.	785	37	5.05	827.05	925	
	0320	3/4″ pipe size		14	1.143		785	42	5.75	832.75	935	
	0330	1″ pipe size		12	1.333		1,150	49	6.70	1,205.70	1,325	
	0340	1-1/2″ pipe size		9	1.778		3,100	65.50	8.95	3,174.45	3,500	
	0350	2″ pipe size		7	2.286		3,100	84	11.50	3,195.50	3,550	
	0400	Flanged										
	0410	1/2″ pipe size	1 Stpi	12	.667	Ea.	1,375	27.50		1,402.50	1,550	
	0420	3/4″ pipe size		10	.800		1,375	33		1,408	1,550	
	0430	1″ pipe size		7	1.143		1,775	47		1,822	2,025	
	0440	1-1/2″ pipe size	Q-5	9	1.778		3,900	65.50		3,965.50	4,400	
	0450	2″ pipe size	″	7	2.286		3,900	84		3,984	4,425	
	1000	Fixed lever and snap action type										
	1100	Cast iron body, threaded										
	1110	1/2″ pipe size	1 Stpi	22	.364	Ea.	114	14.90		128.90	149	
	1120	3/4″ pipe size		20	.400		305	16.40		321.40	360	
	1130	1″ pipe size		19	.421		305	17.25		322.25	360	
	1200	Forged steel body, threaded										
	1210	1/2″ pipe size	1 Stpi	18	.444	Ea.	740	18.20		758.20	845	
	1220	3/4″ pipe size		16	.500		740	20.50		760.50	845	
	1230	1″ pipe size		14	.571		1,825	23.50		1,848.50	2,025	
	1240	1-1/4″ pipe size		12	.667		1,825	27.50		1,852.50	2,050	
	2000	High pressure high leverage type										
	2100	Forged steel body, threaded										
	2110	1/2″ pipe size	1 Stpi	18	.444	Ea.	1,450	18.20		1,468.20	1,625	
	2120	3/4″ pipe size		16	.500		1,450	20.50		1,470.50	1,625	
	2130	1″ pipe size		14	.571		2,425	23.50		2,448.50	2,700	
	2140	1-1/4″ pipe size		12	.667		2,425	27.50		2,452.50	2,725	
	2150	1-1/2″ pipe size	Q-5	10	1.600		3,675	59		3,734	4,125	
	2160	2″ pipe size	″	8	2		3,675	73.50		3,748.50	4,125	
	2200	Socket weld										
	2210	1/2″ pipe size	Q-17	16	1	Ea.	1,575	37	5.05	1,617.05	1,775	
	2220	3/4″ pipe size		14	1.143		1,575	42	5.75	1,622.75	1,800	
	2230	1″ pipe size		12	1.333		2,550	49	6.70	2,605.70	2,900	
	2240	1-1/4″ pipe size		11	1.455		2,550	53.50	7.35	2,610.85	2,925	
	2250	1-1/2″ pipe size		9	1.778		3,775	65.50	8.95	3,849.45	4,250	
	2260	2″ pipe size		7	2.286		3,775	84	11.50	3,870.50	4,300	
	2300	Flanged										

15120 | Piping Specialties

			CREW	DAILY OUTPUT	LABOR-HOURS	UNIT	2005 BARE COSTS MAT.	LABOR	EQUIP.	TOTAL	TOTAL INCL O&P	
550	2310	1/2" pipe size	1 Stpi	12	.667	Ea.	2,175	27.50		2,202.50	2,450	550
	2320	3/4" pipe size		10	.800		2,175	33		2,208	2,450	
	2330	1" pipe size		7	1.143		3,375	47		3,422	3,775	
	2340	1-1/4" pipe size	↓	6	1.333		3,375	54.50		3,429.50	3,775	
	2350	1-1/2" pipe size	Q-5	9	1.778		4,550	65.50		4,615.50	5,125	
	2360	2" pipe size	"	7	2.286	↓	4,550	84		4,634	5,150	
	3000	Guided lever dual gravity type										
	3100	Cast iron body, threaded										
	3110	1/2" pipe size	1 Stpi	22	.364	Ea.	281	14.90		295.90	335	
	3120	3/4" pipe size		20	.400		281	16.40		297.40	335	
	3130	1" pipe size		19	.421		410	17.25		427.25	475	
	3140	1-1/2" pipe size		11	.727		790	30		820	915	
	3150	2" pipe size	↓	8	1	↓	790	41		831	930	
	3200	Forged steel body, threaded										
	3210	1/2" pipe size	1 Stpi	18	.444	Ea.	795	18.20		813.20	900	
	3220	3/4" pipe size		16	.500		795	20.50		815.50	900	
	3230	1" pipe size	↓	14	.571		1,175	23.50		1,198.50	1,325	
	3240	1-1/2" pipe size	Q-5	10	1.600		3,125	59		3,184	3,550	
	3250	2" pipe size	"	8	2	↓	3,125	73.50		3,198.50	3,550	
	3300	Socket weld										
	3310	1/2" pipe size	Q-17	16	1	Ea.	895	37	5.05	937.05	1,050	
	3320	3/4" pipe size		14	1.143		895	42	5.75	942.75	1,050	
	3330	1" pipe size		12	1.333		1,300	49	6.70	1,355.70	1,500	
	3340	1-1/2" pipe size		9	1.778		3,250	65.50	8.95	3,324.45	3,675	
	3350	2" pipe size	↓	7	2.286	↓	3,250	84	11.50	3,345.50	3,725	
	3400	Flanged										
	3410	1/2" pipe size	1 Stpi	12	.667	Ea.	1,475	27.50		1,502.50	1,675	
	3420	3/4" pipe size		10	.800		1,475	33		1,508	1,675	
	3430	1" pipe size	↓	7	1.143		1,925	47		1,972	2,200	
	3440	1-1/2" pipe size	Q-5	9	1.778		4,050	65.50		4,115.50	4,550	
	3450	2" pipe size	"	7	2.286	↓	4,050	84		4,134	4,575	
580	0010	**MIXING VALVE** Automatic water tempering										580
	0040	1/2" size	1 Stpi	19	.421	Ea.	395	17.25		412.25	460	
	0050	3/4" size		18	.444		395	18.20		413.20	465	
	0100	1" size		16	.500		630	20.50		650.50	725	
	0120	1-1/4" size		13	.615		820	25		845	940	
	0140	1-1/2" size		10	.800		980	33		1,013	1,125	
	0160	2" size		8	1		1,225	41		1,266	1,400	
	0170	2-1/2" size		6	1.333		1,225	54.50		1,279.50	1,425	
	0180	3" size		4	2		2,850	82		2,932	3,250	
	0190	4" size	↓	3	2.667	↓	2,850	109		2,959	3,300	
610	0010	**MONOFLOW TEE FITTING**										610
	1100	For one pipe hydronic, supply and return										
	1110	Copper, soldered										
	1120	3/4" x 1/2" size	1 Stpi	13	.615	Ea.	13.50	25		38.50	53	
	1130	1" x 1/2" size		12	.667		13.50	27.50		41	56	
	1140	1" x 3/4" size		11	.727		13.50	30		43.50	60	
	1150	1-1/4" x 1/2" size		11	.727		14.25	30		44.25	60.50	
	1160	1-1/4" x 3/4" size		10	.800		14.25	33		47.25	64.50	
	1170	1-1/2" x 3/4" size		10	.800		19.90	33		52.90	71	
	1180	1-1/2" x 1" size		9	.889		19.90	36.50		56.40	76.50	
	1190	2" x 3/4" size		9	.889		30.50	36.50		67	88	
	1200	2" x 1" size	↓	8	1	↓	30.50	41		71.50	95	
640	0010	**PRESSURE REDUCING VALVE** Steam, pilot operated										640
	0100	Threaded, iron body										

		15120 | Piping Specialties	CREW	DAILY OUTPUT	LABOR-HOURS	UNIT	2005 BARE COSTS MAT.	LABOR	EQUIP.	TOTAL	TOTAL INCL O&P	
640	0200	1-1/2″ size	1 Stpi	8	1	Ea.	650	41		691	775	640
	0220	2″ size	″	5	1.600	″	835	65.50		900.50	1,025	
	1000	Flanged, iron body, 125 lb. flanges										
	1020	2″ size	1 Stpi	8	1	Ea.	855	41		896	1,000	
	1040	2-1/2″ size	″	4	2		1,125	82		1,207	1,375	
	1060	3″ size	Q-5	4.50	3.556		1,150	131		1,281	1,475	
	1080	4″ size	″	3	5.333		1,450	197		1,647	1,900	
	1500	For 250 lb. flanges, add					5%					
670	0010	PRESSURE REGULATOR										670
	0100	Gas appliance regulators										
	0106	Main burner and pilot applications										
	0108	Rubber seat poppet type										
	0109	1/8″ pipe size	1 Stpi	24	.333	Ea.	10.85	13.65		24.50	32.50	
	0110	1/4″ pipe size		24	.333		12.60	13.65		26.25	34.50	
	0112	3/8″ pipe size		24	.333		13.40	13.65		27.05	35.50	
	0113	1/2″ pipe size		24	.333		14.80	13.65		28.45	37	
	0114	3/4″ pipe size		20	.400		17.50	16.40		33.90	44	
	0122	Lever action type										
	0123	3/8″ pipe size	1 Stpi	24	.333	Ea.	20.50	13.65		34.15	43	
	0124	1/2″ pipe size		24	.333		20.50	13.65		34.15	43	
	0125	3/4″ pipe size		20	.400		41.50	16.40		57.90	70.50	
	0126	1″ pipe size		19	.421		41.50	17.25		58.75	72	
	0132	Double diaphragm type										
	0133	3/8″ pipe size	1 Stpi	24	.333	Ea.	26	13.65		39.65	49	
	0134	1/2″ pipe size		24	.333		38	13.65		51.65	62.50	
	0135	3/4″ pipe size		20	.400		65	16.40		81.40	96	
	0136	1″ pipe size		19	.421		79	17.25		96.25	113	
	0137	1-1/4″ pipe size		15	.533		256	22		278	315	
	0138	1-1/2″ pipe size		13	.615		475	25		500	565	
	0139	2″ pipe size		11	.727		475	30		505	570	
	0140	2-1/2″ pipe size	Q-5	15	1.067		930	39.50		969.50	1,075	
	0141	3″ pipe size		13	1.231		930	45.50		975.50	1,100	
	0142	4″ pipe size (flanged)		8	2		1,725	73.50		1,798.50	2,000	
	0160	Main burner only										
	0162	Straight-thru-flow design										
	0163	1/2″ pipe size	1 Stpi	24	.333	Ea.	28.50	13.65		42.15	51.50	
	0164	3/4″ pipe size		20	.400		39	16.40		55.40	67	
	0165	1″ pipe size		19	.421		59	17.25		76.25	90.50	
	0166	1-1/4″ pipe size		15	.533		59	22		81	97.50	
	0200	Oil, light, hot water, ordinary steam, threaded										
	0220	Bronze body, 1/4″ size	1 Stpi	24	.333	Ea.	102	13.65		115.65	133	
	0230	3/8″ size		24	.333		110	13.65		123.65	142	
	0240	1/2″ size		24	.333		136	13.65		149.65	171	
	0250	3/4″ size		20	.400		163	16.40		179.40	204	
	0260	1″ size		19	.421		247	17.25		264.25	298	
	0270	1-1/4″ size		15	.533		335	22		357	400	
	0280	1-1/2″ size		13	.615		335	25		360	410	
	0290	2″ size		11	.727		655	30		685	765	
	0320	Iron body, 1/4″ size		24	.333		79.50	13.65		93.15	108	
	0330	3/8″ size		24	.333		93.50	13.65		107.15	124	
	0340	1/2″ size		24	.333		99.50	13.65		113.15	130	
	0350	3/4″ size		20	.400		121	16.40		137.40	158	
	0360	1″ size		19	.421		157	17.25		174.25	199	
	0370	1-1/4″ size		15	.533		218	22		240	273	
	0380	1-1/2″ size		13	.615		243	25		268	305	
	0390	2″ size		11	.727		340	30		370	420	

15120 | Piping Specialties

		15120 Piping Specialties	CREW	DAILY OUTPUT	LABOR-HOURS	UNIT	2005 BARE COSTS MAT.	LABOR	EQUIP.	TOTAL	TOTAL INCL O&P	
670	0500	Oil, heavy, viscous fluids, threaded										670
	0520	Bronze body, 3/8″ size	1 Stpi	24	.333	Ea.	178	13.65		191.65	217	
	0530	1/2″ size		24	.333		225	13.65		238.65	269	
	0540	3/4″ size		20	.400		250	16.40		266.40	300	
	0550	1″ size		19	.421		310	17.25		327.25	365	
	0560	1-1/4″ size		15	.533		440	22		462	520	
	0570	1-1/2″ size		13	.615		505	25		530	595	
	0600	Iron body, 3/8″ size		24	.333		139	13.65		152.65	174	
	0620	1/2″ size		24	.333		169	13.65		182.65	207	
	0630	3/4″ size		20	.400		191	16.40		207.40	236	
	0640	1″ size		19	.421		226	17.25		243.25	275	
	0650	1-1/4″ size		15	.533		325	22		347	395	
	0660	1-1/2″ size		13	.615		350	25		375	425	
	0800	Process steam, wet or super heated, monel trim, threaded										
	0820	Bronze body, 1/4″ size	1 Stpi	24	.333	Ea.	270	13.65		283.65	320	
	0830	3/8″ size		24	.333		270	13.65		283.65	320	
	0840	1/2″ size		24	.333		297	13.65		310.65	345	
	0850	3/4″ size		20	.400		325	16.40		341.40	385	
	0860	1″ size		19	.421		415	17.25		432.25	485	
	0870	1-1/4″ size		15	.533		505	22		527	590	
	0880	1-1/2″ size		13	.615		605	25		630	705	
	0890	2″ size		11	.727		880	30		910	1,025	
	0920	Iron body, max 125 PSIG press out, 1/4″ size		24	.333		79.50	13.65		93.15	108	
	0930	3/8″ size		24	.333		93.50	13.65		107.15	124	
	0940	1/2″ size		24	.333		99.50	13.65		113.15	130	
	0950	3/4″ size		20	.400		121	16.40		137.40	158	
	0960	1″ size		19	.421		157	17.25		174.25	199	
	0970	1-1/4″ size		15	.533		218	22		240	273	
	0980	1-1/2″ size		13	.615		243	25		268	305	
	0990	2″ size		11	.727		350	30		380	430	
	1000	Flanged, Class 125										
	1006	2-1/2″ size	Q-5	10	1.600	Ea.	1,125	59		1,184	1,350	
	1008	3″ size		9	1.778		1,150	65.50		1,215.50	1,375	
	1010	4″ size		8	2		1,450	73.50		1,523.50	1,700	
	1020	6″ size		7	2.286		3,700	84		3,784	4,200	
	3000	Steam, high capacity, bronze body, stainless steel trim										
	3020	Threaded, 1/2″ diameter	1 Stpi	24	.333	Ea.	940	13.65		953.65	1,050	
	3030	3/4″ diameter		24	.333		975	13.65		988.65	1,100	
	3040	1″ diameter		19	.421		1,100	17.25		1,117.25	1,225	
	3060	1-1/4″ diameter		15	.533		1,150	22		1,172	1,300	
	3080	1-1/2″ diameter		13	.615		1,325	25		1,350	1,500	
	3100	2″ diameter		11	.727		1,625	30		1,655	1,825	
	3120	2-1/2″ diameter	Q-5	12	1.333		2,025	49		2,074	2,300	
	3140	3″ diameter	″	11	1.455		2,325	53.50		2,378.50	2,625	
	3500	Flanged connection, iron body, 125 lb. W.S.P.										
	3520	3″ diameter	Q-5	11	1.455	Ea.	2,550	53.50		2,603.50	2,875	
	3540	4″ diameter	″	5	3.200	″	3,200	118		3,318	3,700	
	9000	For water pressure regulators, see Div. 15110-160										
700	0010	**PRESSURE & TEMPERATURE SAFETY PLUG**										700
	1000	3/4″ external thread, 3/8″ diam. element										
	1020	Carbon steel										
	1050	7-1/2″ insertion	1 Stpi	32	.250	Ea.	59.50	10.25		69.75	80.50	
	1120	304 stainless steel										
	1150	7-1/2″ insertion	1 Stpi	32	.250	Ea.	54	10.25		64.25	75	
	1220	316 Stainless steel										
	1250	7-1/2″ insertion	1 Stpi	32	.250	Ea.	57	10.25		67.25	78.50	

15120 | Piping Specialties

		15120 Piping Specialties	Crew	Daily Output	Labor-Hours	Unit	2005 Bare Costs Mat.	Labor	Equip.	Total	Total Incl O&P	
730	0010	SLEEVES AND ESCUTCHEONS										730
	0100	Pipe sleeve										
	0110	Steel, w/water stop, 12" long, with link seal										
	0120	2" diam. for 1/2" carrier pipe	1 Plum	8.40	.952	Ea.	26.50	39		65.50	87.50	
	0130	2-1/2" diam. for 3/4" carrier pipe		8	1		31.50	41		72.50	96	
	0140	2-1/2" diam. for 1" carrier pipe		8	1		29	41		70	93.50	
	0150	3" diam. for 1-1/4" carrier pipe		7.20	1.111		41	45.50		86.50	114	
	0160	3-1/2" diam. for 1-1/2" carrier pipe		6.80	1.176		42.50	48		90.50	119	
	0170	4" diam. for 2" carrier pipe		6	1.333		47	54.50		101.50	134	
	0180	4" diam. for 2-1/2" carrier pipe		6	1.333		46.50	54.50		101	133	
	0190	5" diam. for 3" carrier pipe		5.40	1.481		57	60.50		117.50	154	
	0200	6" diam. for 4" carrier pipe		4.80	1.667		68	68		136	177	
	0210	10" diam. for 6" carrier pipe	Q-1	8	2		124	73.50		197.50	247	
	0220	12" diam. for 8" carrier pipe		7.20	2.222		182	81.50		263.50	325	
	0230	14" diam. for 10" carrier pipe		6.40	2.500		192	92		284	350	
	0240	16" diam. for 12" carrier pipe		5.80	2.759		224	101		325	400	
	0250	18" diam. for 14" carrier pipe		5.20	3.077		246	113		359	440	
	0260	24" diam. for 18" carrier pipe		4	4		410	147		557	670	
	0270	24" diam. for 20" carrier pipe		4	4		330	147		477	585	
	0280	30" diam. for 24" carrier pipe		3.20	5		735	184		919	1,075	
	0500	Wall sleeve										
	0510	Ductile iron with rubber gasket seal										
	0520	3"	1 Plum	8.40	.952	Ea.	155	39		194	230	
	0530	4"		7.20	1.111		210	45.50		255.50	299	
	0540	6"		6	1.333		320	54.50		374.50	430	
	0550	8"		4	2		380	81.50		461.50	545	
	0560	10"		3	2.667		455	109		564	665	
	0570	12"		2.40	3.333		274	136		410	505	
	5000	Escutcheon										
	5100	Split ring, pipe										
	5110	Chrome plated										
	5120	1/2"	1 Plum	160	.050	Ea.	1.01	2.04		3.05	4.18	
	5130	3/4"		160	.050		1.05	2.04		3.09	4.23	
	5140	1"		135	.059		1.12	2.42		3.54	4.87	
	5150	1-1/2"		115	.070		1.69	2.84		4.53	6.15	
	5160	2"		100	.080		1.77	3.27		5.04	6.85	
	5170	4"		80	.100		3.93	4.09		8.02	10.45	
	5180	6"		68	.118		6.55	4.81		11.36	14.40	
760	0010	STEAM TRAP										760
	0030	Cast iron body, threaded										
	0040	Inverted bucket										
	0050	1/2" pipe size	1 Stpi	12	.667	Ea.	107	27.50		134.50	158	
	0070	3/4" pipe size		10	.800		185	33		218	253	
	0100	1" pipe size		9	.889		286	36.50		322.50	370	
	0120	1-1/4" pipe size		8	1		430	41		471	535	
	0130	1-1/2" pipe size		7	1.143		510	47		557	635	
	0140	2" pipe size		6	1.333		715	54.50		769.50	870	
	0200	With thermic vent & check valve										
	0202	1/2" pipe size	1 Stpi	12	.667	Ea.	124	27.50		151.50	178	
	0204	3/4" pipe size		10	.800		211	33		244	281	
	0206	1" pipe size		9	.889		310	36.50		346.50	395	
	0208	1-1/4" pipe size		8	1		465	41		506	570	
	0210	1-1/2" pipe size		7	1.143		515	47		562	640	
	0212	2" pipe size		6	1.333		790	54.50		844.50	945	
	1000	Float & thermostatic, 15 psi										
	1010	3/4" pipe size	1 Stpi	16	.500	Ea.	87	20.50		107.50	127	

Important: See the Reference Section for critical supporting data - Reference Nos., Crews, & City Cost Indexes

15120 | Piping Specialties

			CREW	DAILY OUTPUT	LABOR-HOURS	UNIT	2005 BARE COSTS MAT.	LABOR	EQUIP.	TOTAL	TOTAL INCL O&P	
760	1020	1" pipe size	1 Stpi	15	.533	Ea.	104	22		126	147	760
	1030	1-1/4" pipe size		13	.615		127	25		152	178	
	1040	1-1/2" pipe size		9	.889		185	36.50		221.50	259	
	1060	2" pipe size		6	1.333		340	54.50		394.50	455	
	1100	For vacuum breaker, add					47			47	52	
	1110	Differential condensate controller										
	1120	With union and manual metering valve										
	1130	1/2" pipe size	1 Stpi	24	.333	Ea.	385	13.65		398.65	445	
	1140	3/4" pipe size		20	.400		450	16.40		466.40	520	
	1150	1" pipe size		19	.421		605	17.25		622.25	690	
	1160	1-1/4" pipe size		15	.533		815	22		837	930	
	1170	1-1/2" pipe size		13	.615		985	25		1,010	1,125	
	1180	2" pipe size		11	.727		1,275	30		1,305	1,450	
	1290	Brass body, threaded										
	1300	Thermostatic, angle union, 25 psi										
	1310	1/2" pipe size	1 Stpi	24	.333	Ea.	42	13.65		55.65	66.50	
	1320	3/4" pipe size		20	.400		74.50	16.40		90.90	106	
	1330	1" pipe size		19	.421		109	17.25		126.25	146	
	1990	Carbon steel body, threaded										
	2000	Controlled disc, integral strainer & blow down										
	2010	Threaded										
	2020	3/8" pipe size	1 Stpi	24	.333	Ea.	170	13.65		183.65	208	
	2030	1/2" pipe size		24	.333		170	13.65		183.65	208	
	2040	3/4" pipe size		20	.400		228	16.40		244.40	276	
	2050	1" pipe size		19	.421		293	17.25		310.25	345	
	2100	Socket weld										
	2110	3/8" pipe size	Q-17	17	.941	Ea.	203	34.50	4.74	242.24	280	
	2120	1/2" pipe size		16	1		203	37	5.05	245.05	284	
	2130	3/4" pipe size		14	1.143		264	42	5.75	311.75	360	
	2140	1" pipe size		12	1.333		330	49	6.70	385.70	445	
	2170	Flanged, 600 lb. ASA										
	2180	3/8" pipe size	1 Stpi	12	.667	Ea.	470	27.50		497.50	555	
	2190	1/2" pipe size		12	.667		470	27.50		497.50	555	
	2200	3/4" pipe size		10	.800		560	33		593	665	
	2210	1" pipe size		7	1.143		660	47		707	795	
	5000	Forged steel body										
	5010	Inverted bucket										
	5020	Threaded										
	5030	1/2" pipe size	1 Stpi	12	.667	Ea.	305	27.50		332.50	375	
	5040	3/4" pipe size		10	.800		525	33		558	625	
	5050	1" pipe size		9	.889		785	36.50		821.50	920	
	5060	1-1/4" pipe size		8	1		1,150	41		1,191	1,300	
	5070	1-1/2" pipe size		7	1.143		1,625	47		1,672	1,850	
	5080	2" pipe size		6	1.333		2,850	54.50		2,904.50	3,200	
	5100	Socket weld										
	5110	1/2" pipe size	Q-17	16	1	Ea.	365	37	5.05	407.05	460	
	5120	3/4" pipe size		14	1.143		585	42	5.75	632.75	715	
	5130	1" pipe size		12	1.333		850	49	6.70	905.70	1,025	
	5140	1-1/4" pipe size		10	1.600		1,200	59	8.05	1,267.05	1,425	
	5150	1-1/2" pipe size		9	1.778		1,675	65.50	8.95	1,749.45	1,950	
	5160	2" pipe size		7	2.286		2,925	84	11.50	3,020.50	3,350	
	5200	Flanged										
	5210	1/2" pipe size	1 Stpi	12	.667	Ea.	575	27.50		602.50	670	
	5220	3/4" pipe size		10	.800		925	33		958	1,075	
	5230	1" pipe size		7	1.143		1,300	47		1,347	1,525	
	5240	1-1/4" pipe size		6	1.333		1,825	54.50		1,879.50	2,075	

	15120 | Piping Specialties		CREW	DAILY OUTPUT	LABOR-HOURS	UNIT	2005 BARE COSTS MAT.	LABOR	EQUIP.	TOTAL	TOTAL INCL O&P	
760	5250	1-1/2" pipe size	Q-5	9	1.778	Ea.	2,275	65.50		2,340.50	2,600	760
	5260	2" pipe size	"	7	2.286		4,775	84		4,859	5,375	
	6000	Stainless steel body										
	6010	Inverted bucket										
	6020	Threaded										
	6030	1/2" pipe size	1 Stpi	12	.667	Ea.	99.50	27.50		127	150	
	6040	3/4" pipe size		10	.800		131	33		164	193	
	6050	1" pipe size		9	.889		465	36.50		501.50	565	
	6100	Threaded with thermic vent										
	6110	1/2" pipe size	1 Stpi	12	.667	Ea.	113	27.50		140.50	165	
	6120	3/4" pipe size		10	.800		146	33		179	210	
	6130	1" pipe size		9	.889		485	36.50		521.50	585	
	6200	Threaded with check valve										
	6210	1/2" pipe size	1 Stpi	12	.667	Ea.	142	27.50		169.50	198	
	6220	3/4" pipe size	"	10	.800	"	195	33		228	264	
	6300	Socket weld										
	6310	1/2" pipe size	Q-17	16	1	Ea.	115	37	5.05	157.05	188	
	6320	3/4" pipe size	"	14	1.143	"	149	42	5.75	196.75	234	
790	0010	**STRAINERS, BASKET TYPE** Perforated stainless steel basket										790
	0100	Brass or monel available										
	2000	Simplex style										
	2300	Bronze body										
	2320	Screwed, 3/8" pipe size	1 Stpi	22	.364	Ea.	103	14.90		117.90	136	
	2340	1/2" pipe size		20	.400		105	16.40		121.40	141	
	2360	3/4" pipe size		17	.471		108	19.25		127.25	148	
	2380	1" pipe size		15	.533		141	22		163	188	
	2400	1-1/4" pipe size		13	.615		212	25		237	271	
	2420	1-1/2" pipe size		12	.667		214	27.50		241.50	277	
	2440	2" pipe size		10	.800		325	33		358	405	
	2460	2-1/2" pipe size	Q-5	15	1.067		465	39.50		504.50	570	
	2480	3" pipe size	"	14	1.143		670	42		712	805	
	2600	Flanged, 2" pipe size	1 Stpi	6	1.333		490	54.50		544.50	620	
	2620	2-1/2" pipe size	Q-5	4.50	3.556		770	131		901	1,050	
	2640	3" pipe size		3.50	4.571		865	168		1,033	1,200	
	2660	4" pipe size		3	5.333		1,475	197		1,672	1,925	
	2680	5" pipe size	Q-6	3.40	7.059		2,200	270		2,470	2,825	
	2700	6" pipe size		3	8		2,725	305		3,030	3,425	
	2710	8" pipe size		2.50	9.600		4,650	365		5,015	5,675	
	3600	Iron body										
	3700	Screwed, 3/8" pipe size	1 Stpi	22	.364	Ea.	117	14.90		131.90	152	
	3720	1/2" pipe size		20	.400		117	16.40		133.40	154	
	3740	3/4" pipe size		17	.471		155	19.25		174.25	199	
	3760	1" pipe size		15	.533		155	22		177	203	
	3780	1-1/4" pipe size		13	.615		207	25		232	265	
	3800	1-1/2" pipe size		12	.667		227	27.50		254.50	290	
	3820	2" pipe size		10	.800		271	33		304	345	
	3840	2-1/2" pipe size	Q-5	15	1.067		390	39.50		429.50	485	
	3860	3" pipe size	"	14	1.143		440	42		482	550	
	4000	Flanged, 2" pipe size	1 Stpi	6	1.333		370	54.50		424.50	485	
	4020	2-1/2" pipe size	Q-5	4.50	3.556		500	131		631	745	
	4040	3" pipe size		3.50	4.571		530	168		698	840	
	4060	4" pipe size		3	5.333		800	197		997	1,175	
	4080	5" pipe size	Q-6	3.40	7.059		1,250	270		1,520	1,775	
	4100	6" pipe size		3	8		1,550	305		1,855	2,150	
	4120	8" pipe size		2.50	9.600		2,875	365		3,240	3,700	
	4140	10" pipe size		2.20	10.909		4,300	415		4,715	5,350	

Important: See the Reference Section for critical supporting data - Reference Nos., Crews, & City Cost Indexes

15120 | Piping Specialties

		15120 Piping Specialties	CREW	DAILY OUTPUT	LABOR-HOURS	UNIT	2005 BARE COSTS MAT.	LABOR	EQUIP.	TOTAL	TOTAL INCL O&P	
790	4160	12" pipe size	Q-6	1.70	14.118	Ea.	5,125	540		5,665	6,450	790
	4180	14" pipe size		1.40	17.143		6,450	655		7,105	8,075	
	4200	16" pipe size		1	24		6,750	915		7,665	8,800	
	6000	Cast steel body										
	6400	Screwed, 1" pipe size	1 Stpi	15	.533	Ea.	253	22		275	310	
	6410	1-1/4" pipe size		13	.615		380	25		405	460	
	6420	1-1/2" pipe size		12	.667		380	27.50		407.50	460	
	6440	2" pipe size		10	.800		540	33		573	645	
	6460	2-1/2" pipe size	Q-5	15	1.067		760	39.50		799.50	895	
	6480	3" pipe size	"	14	1.143		995	42		1,037	1,175	
	6560	Flanged, 2" pipe size	1 Stpi	6	1.333		785	54.50		839.50	945	
	6580	2-1/2" pipe size	Q-5	4.50	3.556		1,150	131		1,281	1,475	
	6600	3" pipe size		3.50	4.571		1,250	168		1,418	1,625	
	6620	4" pipe size		3	5.333		1,750	197		1,947	2,225	
	6640	6" pipe size	Q-6	3	8		3,200	305		3,505	3,975	
	6660	8" pipe size	"	2.50	9.600		5,200	365		5,565	6,275	
	7000	Stainless steel body										
	7200	Screwed, 1" pipe size	1 Stpi	15	.533	Ea.	415	22		437	490	
	7210	1-1/4" pipe size		13	.615		605	25		630	705	
	7220	1-1/2" pipe size		12	.667		605	27.50		632.50	705	
	7240	2" pipe size		10	.800		885	33		918	1,025	
	7260	2-1/2" pipe size	Q-5	15	1.067		1,475	39.50		1,514.50	1,675	
	7280	3" pipe size	"	14	1.143		1,950	42		1,992	2,225	
	7400	Flanged, 2" pipe size	1 Stpi	6	1.333		1,225	54.50		1,279.50	1,400	
	7420	2-1/2" pipe size	Q-5	4.50	3.556		2,200	131		2,331	2,625	
	7440	3" pipe size		3.50	4.571		2,275	168		2,443	2,750	
	7460	4" pipe size		3	5.333		3,550	197		3,747	4,200	
	7480	6" pipe size	Q-6	3	8		6,150	305		6,455	7,225	
	7500	8" pipe size	"	2.50	9.600		10,400	365		10,765	12,100	
	8100	Duplex style										
	8200	Bronze body										
	8240	Screwed, 3/4" pipe size	1 Stpi	16	.500	Ea.	1,100	20.50		1,120.50	1,225	
	8260	1" pipe size		14	.571		1,100	23.50		1,123.50	1,225	
	8280	1-1/4" pipe size		12	.667		2,150	27.50		2,177.50	2,425	
	8300	1-1/2" pipe size		11	.727		2,150	30		2,180	2,425	
	8320	2" pipe size		9	.889		3,400	36.50		3,436.50	3,800	
	8340	2-1/2" pipe size	Q-5	14	1.143		3,400	42		3,442	3,825	
	8420	Flanged, 2" pipe size	1 Stpi	6	1.333		3,850	54.50		3,904.50	4,325	
	8440	2-1/2" pipe size	Q-5	4.50	3.556		3,950	131		4,081	4,550	
	8460	3" pipe size		3.50	4.571		5,450	168		5,618	6,250	
	8480	4" pipe size		3	5.333		8,625	197		8,822	9,775	
	8500	5" pipe size	Q-6	3.40	7.059		15,000	270		15,270	16,900	
	8520	6" pipe size	"	3	8		18,900	305		19,205	21,300	
	8700	Iron body										
	8740	Screwed, 3/4" pipe size	1 Stpi	16	.500	Ea.	575	20.50		595.50	660	
	8760	1" pipe size		14	.571		575	23.50		598.50	665	
	8780	1-1/4" pipe size		12	.667		995	27.50		1,022.50	1,150	
	8800	1-1/2" pipe size		11	.727		995	30		1,025	1,150	
	8820	2" pipe size		9	.889		1,700	36.50		1,736.50	1,925	
	8840	2-1/2" pipe size	Q-5	14	1.143		1,700	42		1,742	1,950	
	9000	Flanged, 2" pipe size	1 Stpi	6	1.333		1,775	54.50		1,829.50	2,050	
	9020	2-1/2" pipe size	Q-5	4.50	3.556		1,850	131		1,981	2,250	
	9040	3" pipe size		3.50	4.571		2,375	168		2,543	2,850	
	9060	4" pipe size		3	5.333		3,750	197		3,947	4,400	
	9080	5" pipe size	Q-6	3.40	7.059		7,500	270		7,770	8,650	
	9100	6" pipe size		3	8		7,500	305		7,805	8,700	

15120 | Piping Specialties

		15120 Piping Specialties	CREW	DAILY OUTPUT	LABOR-HOURS	UNIT	2005 BARE COSTS MAT.	LABOR	EQUIP.	TOTAL	TOTAL INCL O&P	
790	9120	8″ pipe size	Q-6	2.50	9.600	Ea.	14,400	365		14,765	16,500	790
	9140	10″ pipe size		2.20	10.909		20,100	415		20,515	22,700	
	9160	12″ pipe size		1.70	14.118		22,300	540		22,840	25,300	
	9170	14″ pipe size		1.40	17.143		27,000	655		27,655	30,700	
	9180	16″ pipe size		1	24		33,900	915		34,815	38,700	
	9300	Cast steel body										
	9340	Screwed, 1″ pipe size	1 Stpi	14	.571	Ea.	1,600	23.50		1,623.50	1,800	
	9360	1-1/2″ pipe size		11	.727		2,350	30		2,380	2,650	
	9380	2″ pipe size		9	.889		4,000	36.50		4,036.50	4,450	
	9460	Flanged, 2″ pipe size		6	1.333		4,025	54.50		4,079.50	4,500	
	9480	2-1/2″ pipe size	Q-5	4.50	3.556		4,450	131		4,581	5,100	
	9500	3″ pipe size		3.50	4.571		5,875	168		6,043	6,700	
	9520	4″ pipe size		3	5.333		8,600	197		8,797	9,750	
	9540	6″ pipe size	Q-6	3	8		14,700	305		15,005	16,700	
	9560	8″ pipe size	″	2.50	9.600		29,100	365		29,465	32,600	
	9700	Stainless steel body										
	9740	Screwed, 1″ pipe size	1 Stpi	14	.571	Ea.	2,900	23.50		2,923.50	3,225	
	9760	1-1/2″ pipe size		11	.727		4,400	30		4,430	4,900	
	9780	2″ pipe size		9	.889		7,100	36.50		7,136.50	7,875	
	9860	Flanged, 2″ pipe size		6	1.333		7,400	54.50		7,454.50	8,225	
	9880	2-1/2″ pipe size	Q-5	4.50	3.556		7,775	131		7,906	8,750	
	9900	3″ pipe size		3.50	4.571		12,200	168		12,368	13,700	
	9920	4″ pipe size		3	5.333		14,900	197		15,097	16,700	
	9940	6″ pipe size	Q-6	3	8		26,300	305		26,605	29,400	
	9960	8″ pipe size	″	2.50	9.600		48,700	365		49,065	54,000	
820	0010	**STRAINERS, Y TYPE** Bronze body										820
	0050	Screwed, 150 lb., 1/4″ pipe size	1 Stpi	24	.333	Ea.	12.05	13.65		25.70	34	
	0070	3/8″ pipe size		24	.333		15.90	13.65		29.55	38	
	0100	1/2″ pipe size		20	.400		15.90	16.40		32.30	42	
	0120	3/4″ pipe size		19	.421		17.40	17.25		34.65	45	
	0140	1″ pipe size		17	.471		18.70	19.25		37.95	49.50	
	0150	1-1/4″ pipe size		15	.533		42	22		64	79	
	0160	1-1/2″ pipe size		14	.571		40.50	23.50		64	79.50	
	0180	2″ pipe size		13	.615		53.50	25		78.50	97	
	0182	3″ pipe size		12	.667		340	27.50		367.50	415	
	0200	300 lb., 2-1/2″ pipe size	Q-5	17	.941		286	34.50		320.50	365	
	0220	3″ pipe size		16	1		565	37		602	680	
	0240	4″ pipe size		15	1.067		1,300	39.50		1,339.50	1,475	
	0500	For 300 lb rating 1/4″ thru 2″, add					15%					
	1000	Flanged, 150 lb., 1-1/2″ pipe size	1 Stpi	11	.727	Ea.	296	30		326	370	
	1020	2″ pipe size	″	8	1		345	41		386	440	
	1030	2-1/2″ pipe size	Q-5	5	3.200		530	118		648	760	
	1040	3″ pipe size		4.50	3.556		655	131		786	915	
	1060	4″ pipe size		3	5.333		990	197		1,187	1,400	
	1080	5″ pipe size	Q-6	3.40	7.059		2,025	270		2,295	2,625	
	1100	6″ pipe size		3	8		2,650	305		2,955	3,350	
	1106	8″ pipe size		2.60	9.231		2,750	355		3,105	3,550	
	1500	For 300 lb rating, add					40%					
840	0010	**STRAINERS, Y TYPE** Iron body										840
	0050	Screwed, 250 lb., 1/4″ pipe size	1 Stpi	20	.400	Ea.	6.65	16.40		23.05	32	
	0070	3/8″ pipe size		20	.400		6.65	16.40		23.05	32	
	0100	1/2″ pipe size		20	.400		6.65	16.40		23.05	32	
	0120	3/4″ pipe size		18	.444		7.95	18.20		26.15	36.50	
	0140	1″ pipe size		16	.500		10.75	20.50		31.25	43	

15120 | Piping Specialties

		15120 Piping Specialties	CREW	DAILY OUTPUT	LABOR-HOURS	UNIT	2005 BARE COSTS MAT.	LABOR	EQUIP.	TOTAL	TOTAL INCL O&P	
840	0150	1-1/4″ pipe size	1 Stpi	15	.533	Ea.	14.75	22		36.75	49	840
	0160	1-1/2″ pipe size		12	.667		17.75	27.50		45.25	60.50	
	0180	2″ pipe size		8	1		27.50	41		68.50	91.50	
	0200	2-1/2″ pipe size	Q-5	12	1.333		140	49		189	228	
	0220	3″ pipe size		11	1.455		151	53.50		204.50	247	
	0240	4″ pipe size		5	3.200		255	118		373	455	
	0500	For galvanized body, add					50%					
	1000	Flanged, 125 lb., 1-1/2″ pipe size	1 Stpi	11	.727	Ea.	86	30		116	140	
	1020	2″ pipe size	″	8	1		64	41		105	132	
	1030	2-1/2″ pipe size	Q-5	5	3.200		72.50	118		190.50	257	
	1040	3″ pipe size		4.50	3.556		84.50	131		215.50	290	
	1060	4″ pipe size		3	5.333		155	197		352	465	
	1080	5″ pipe size	Q-6	3.40	7.059		242	270		512	670	
	1100	6″ pipe size		3	8		295	305		600	785	
	1120	8″ pipe size		2.50	9.600		495	365		860	1,100	
	1140	10″ pipe size		2	12		950	460		1,410	1,750	
	1160	12″ pipe size		1.70	14.118		1,425	540		1,965	2,375	
	1170	14″ pipe size		1.30	18.462		2,625	705		3,330	3,950	
	1180	16″ pipe size		1	24		3,725	915		4,640	5,475	
	1500	For 250 lb rating, add					20%					
	2000	For galvanized body, add					50%					
	2500	For steel body, add					40%					
870	0010	**SUCTION DIFFUSERS**										870
	0100	Cast iron body with integral straightening vanes, strainer										
	1000	Flanged										
	1010	2″ inlet, 1-1/2″ pump side	1 Stpi	6	1.333	Ea.	224	54.50		278.50	330	
	1020	2″ pump side	″	5	1.600		232	65.50		297.50	355	
	1030	3″ inlet, 2″ pump side	Q-5	6.50	2.462		284	90.50		374.50	445	
	1040	2-1/2″ pump side		5.30	3.019		415	111		526	620	
	1050	3″ pump side		4.50	3.556		420	131		551	655	
	1060	4″ inlet, 3″ pump side		3.50	4.571		490	168		658	795	
	1070	4″ pump side		3	5.333		570	197		767	920	
	1080	5″ inlet, 4″ pump side	Q-6	3.80	6.316		650	241		891	1,075	
	1090	5″ pump side		3.40	7.059		775	270		1,045	1,250	
	1100	6″ inlet, 4″ pump side		3.30	7.273		675	278		953	1,150	
	1110	5″ pump side		3.10	7.742		810	296		1,106	1,325	
	1120	6″ pump side		3	8		855	305		1,160	1,400	
	1130	8″ inlet, 6″ pump side		2.70	8.889		920	340		1,260	1,500	
	1140	8″ pump side		2.50	9.600		1,600	365		1,965	2,300	
	1150	10″ inlet, 6″ pump side		2.40	10		2,200	380		2,580	3,000	
	1160	8″ pump side		2.30	10.435		1,850	400		2,250	2,625	
	1170	10″ pump side		2.20	10.909		2,650	415		3,065	3,550	
	1180	12″ inlet, 8″ pump side		2	12		2,825	460		3,285	3,800	
	1190	10″ pump side		1.80	13.333		3,125	510		3,635	4,200	
	1200	12″ pump side		1.70	14.118		3,500	540		4,040	4,650	
	1210	14″ inlet, 12″ pump side		1.50	16		4,550	610		5,160	5,925	
	1220	14″ pump side		1.30	18.462		4,725	705		5,430	6,250	
900	0010	**THERMOFLO INDICATOR** For balancing										900
	1000	Sweat connections, 1-1/4″ pipe size	1 Stpi	12	.667	Ea.	375	27.50		402.50	450	
	1020	1-1/2″ pipe size		10	.800		380	33		413	470	
	1040	2″ pipe size		8	1		405	41		446	505	
	1060	2-1/2″ pipe size		7	1.143		610	47		657	740	
	2000	Flange connections, 3″ pipe size	Q-5	5	3.200		730	118		848	975	
	2020	4″ pipe size		4	4		875	147		1,022	1,175	
	2030	5″ pipe size		3.50	4.571		1,100	168		1,268	1,450	

	15120 | Piping Specialties		CREW	DAILY OUTPUT	LABOR-HOURS	UNIT	2005 BARE COSTS MAT.	LABOR	EQUIP.	TOTAL	TOTAL INCL O&P	
900	2040	6" pipe size	Q-5	3	5.333	Ea.	1,175	197		1,372	1,600	900
	2060	8" pipe size	↓	2	8	↓	1,425	295		1,720	2,025	
920	0010	**VENTURI FLOW** Measuring device										920
	0050	1/2" diameter	1 Stpi	24	.333	Ea.	175	13.65		188.65	214	
	0100	3/4" diameter		20	.400		163	16.40		179.40	204	
	0120	1" diameter		19	.421		169	17.25		186.25	212	
	0140	1-1/4" diameter		15	.533		203	22		225	256	
	0160	1-1/2" diameter		13	.615		206	25		231	265	
	0180	2" diameter	↓	11	.727		222	30		252	289	
	0200	2-1/2" diameter	Q-5	16	1		330	37		367	420	
	0220	3" diameter		14	1.143		380	42		422	485	
	0240	4" diameter	↓	11	1.455		595	53.50		648.50	735	
	0260	5" diameter	Q-6	4	6		720	229		949	1,125	
	0280	6" diameter		3.50	6.857		780	262		1,042	1,250	
	0300	8" diameter		3	8		1,075	305		1,380	1,625	
	0320	10" diameter		2	12		2,050	460		2,510	2,975	
	0330	12" diameter		1.80	13.333		3,400	510		3,910	4,500	
	0340	14" diameter		1.60	15		3,650	575		4,225	4,875	
	0350	16" diameter	↓	1.40	17.143		3,950	655		4,605	5,325	
	0500	For meter, add				↓	1,175			1,175	1,300	
940	0010	**WATER SUPPLY METERS**										940
	1000	Detector, serves dual systems such as fire and domestic or										
	1020	process water, wide range cap., UL and FM approved										
	1100	3" mainline x 2" by-pass, 400 GPM	Q-1	3.60	4.444	Ea.	5,150	163		5,313	5,900	
	1140	4" mainline x 2" by-pass, 700 GPM	"	2.50	6.400		5,150	235		5,385	6,000	
	1180	6" mainline x 3" by-pass, 1600 GPM	Q-2	2.60	9.231		7,775	350		8,125	9,075	
	1220	8" mainline x 4" by-pass, 2800 GPM		2.10	11.429		11,500	435		11,935	13,300	
	1260	10" mainline x 6" by-pass, 4400 GPM		2	12		15,700	460		16,160	18,000	
	1300	10"x12" mainlines x 6" by-pass,5400 GPM	↓	1.70	14.118	↓	20,700	540		21,240	23,600	
	2000	Domestic/commercial, bronze										
	2020	Threaded										
	2060	5/8" diameter, to 20 GPM	1 Plum	16	.500	Ea.	40	20.50		60.50	74.50	
	2080	3/4" diameter, to 30 GPM		14	.571		67.50	23.50		91	110	
	2100	1" diameter, to 50 GPM	↓	12	.667	↓	94	27		121	144	
	2300	Threaded/flanged										
	2340	1-1/2" diameter, to 100 GPM	1 Plum	8	1	Ea.	330	41		371	425	
	2360	2" diameter, to 160 GPM	"	6	1.333	"	415	54.50		469.50	535	
	2600	Flanged, compound										
	2640	3" diameter, 320 GPM	Q-1	3	5.333	Ea.	1,950	196		2,146	2,425	
	2660	4" diameter, to 500 GPM		1.50	10.667		3,025	390		3,415	3,925	
	2680	6" diameter, to 1,000 GPM		1	16		4,350	590		4,940	5,650	
	2700	8" diameter, to 1,800 GPM	↓	.80	20	↓	8,600	735		9,335	10,600	
	7000	Turbine										
	7260	Flanged										
	7300	2" diameter, to 160 GPM	1 Plum	7	1.143	Ea.	455	46.50		501.50	570	
	7320	3" diameter, to 450 GPM	Q-1	3.60	4.444		745	163		908	1,050	
	7340	4" diameter, to 650 GPM	"	2.50	6.400		1,350	235		1,585	1,825	
	7360	6" diameter, to 1800 GPM	Q-2	2.60	9.231		2,550	350		2,900	3,325	
	7380	8" diameter, to 2500 GPM		2.10	11.429		4,050	435		4,485	5,100	
	7400	10" diameter, to 5500 GPM	↓	1.70	14.118	↓	5,400	540		5,940	6,750	
960	0010	**WELD END BALL JOINTS** Steel										960
	0050	2-1/2" diameter	Q-17	13	1.231	Ea.	515	45.50	6.20	566.70	640	
	0100	3" diameter		12	1.333		595	49	6.70	650.70	730	
	0120	4" diameter	↓	11	1.455		915	53.50	7.35	975.85	1,100	
	0140	5" diameter	Q-18	14	1.714		1,200	65.50	5.75	1,271.25	1,425	
	0160	6" diameter	↓	12	2	↓	1,600	76.50	6.70	1,683.20	1,900	

Important: See the Reference Section for critical supporting data - Reference Nos., Crews, & City Cost Indexes

15120 | Piping Specialties

	Line	Description	Crew	Daily Output	Labor-Hours	Unit	2005 Bare Costs Mat.	Labor	Equip.	Total	Total Incl O&P	
960	0180	8″ diameter	Q-18	9	2.667	Ea.	2,175	102	8.95	2,285.95	2,550	960
	0200	10″ diameter		8	3		3,075	115	10.10	3,200.10	3,575	
	0220	12″ diameter		6	4		4,050	153	13.45	4,216.45	4,700	
	0240	14″ diameter		4	6		4,825	229	20	5,074	5,700	
980	0010	**ZONE VALVES**										980
	1000	Bronze body										
	1010	2 way, 125 psi										
	1020	Standard, 65′ pump head										
	1030	1/2″ soldered	1 Stpi	24	.333	Ea.	73	13.65		86.65	101	
	1040	3/4″ soldered		20	.400		73	16.40		89.40	105	
	1050	1″ soldered		19	.421		80.50	17.25		97.75	115	
	1060	1-1/4″ soldered		15	.533		91	22		113	133	
	1200	High head, 125′ pump head										
	1210	1/2″ soldered	1 Stpi	24	.333	Ea.	98.50	13.65		112.15	130	
	1220	3/4″ soldered		20	.400		102	16.40		118.40	137	
	1230	1″ soldered		19	.421		112	17.25		129.25	149	
	1300	Geothermal, 150′ pump head										
	1310	3/4″ soldered	1 Stpi	20	.400	Ea.	118	16.40		134.40	154	
	1320	1″ soldered		19	.421		125	17.25		142.25	163	
	1330	3/4″ threaded		20	.400		75.50	16.40		91.90	108	
	3000	3 way, 125 psi										
	3200	By-pass, 65′ pump head										
	3210	1/2″ soldered	1 Stpi	24	.333	Ea.	103	13.65		116.65	135	
	3220	3/4″ soldered		20	.400		107	16.40		123.40	142	
	3230	1″ soldered		19	.421		121	17.25		138.25	160	
	9000	Transformer, for up to 3 zone valves		20	.400		15	16.40		31.40	41	

15140 | Domestic Water Piping

	Line	Description	Crew	Daily Output	Labor-Hours	Unit	2005 Bare Costs Mat.	Labor	Equip.	Total	Total Incl O&P	
100	0010	**BACKFLOW PREVENTER** Includes valves										100
	0020	and four test cocks, corrosion resistant, automatic operation										
	1000	Double check principle										
	1010	Threaded, with ball valves										
	1020	3/4″ pipe size	1 Plum	16	.500	Ea.	127	20.50		147.50	170	
	1030	1″ pipe size		14	.571		140	23.50		163.50	189	
	1040	1-1/2″ pipe size		10	.800		271	32.50		303.50	350	
	1050	2″ pipe size		7	1.143		315	46.50		361.50	415	
	1080	Threaded, with gate valves										
	1100	3/4″ pipe size	1 Plum	16	.500	Ea.	500	20.50		520.50	575	
	1120	1″ pipe size		14	.571		500	23.50		523.50	590	
	1140	1-1/2″ pipe size		10	.800		645	32.50		677.50	760	
	1160	2″ pipe size		7	1.143		790	46.50		836.50	940	
	1200	Flanged, valves are gate										
	1210	3″ pipe size	Q-1	4.50	3.556	Ea.	1,375	131		1,506	1,700	
	1220	4″ pipe size	″	3	5.333		1,675	196		1,871	2,150	
	1230	6″ pipe size	Q-2	3	8		2,475	305		2,780	3,175	
	1240	8″ pipe size		2	12		4,500	460		4,960	5,650	
	1250	10″ pipe size		1	24		6,675	915		7,590	8,725	
	1300	Flanged, valves are OS&Y										
	1370	1″ pipe size	1 Plum	5	1.600	Ea.	545	65.50		610.50	700	
	1374	1-1/2″ pipe size		5	1.600		695	65.50		760.50	865	
	1378	2″ pipe size		4.80	1.667		840	68		908	1,025	
	1380	3″ pipe size	Q-1	4.50	3.556		1,550	131		1,681	1,925	
	1400	4″ pipe size	″	3	5.333		2,175	196		2,371	2,700	
	1420	6″ pipe size	Q-2	3	8		3,425	305		3,730	4,225	
	1430	8″ pipe size	″	2	12		6,350	460		6,810	7,700	
	4000	Reduced pressure principle										

	15140 | Domestic Water Piping		CREW	DAILY OUTPUT	LABOR-HOURS	UNIT	2005 BARE COSTS MAT.	LABOR	EQUIP.	TOTAL	TOTAL INCL O&P	
100	4100	Threaded, bronze, valves are ball										100
	4120	3/4" pipe size	1 Plum	16	.500	Ea.	188	20.50		208.50	238	
	4140	1" pipe size		14	.571		203	23.50		226.50	258	
	4150	1-1/4" pipe size		12	.667		345	27		372	420	
	4160	1-1/2" pipe size		10	.800		380	32.50		412.50	470	
	4180	2" pipe size		7	1.143		425	46.50		471.50	540	
	5000	Flanged, valves are OS&Y										
	5060	2-1/2" pipe size	Q-1	5	3.200	Ea.	2,175	118		2,293	2,575	
	5080	3" pipe size		4.50	3.556		2,300	131		2,431	2,725	
	5100	4" pipe size		3	5.333		2,900	196		3,096	3,475	
	5120	6" pipe size	Q-2	3	8		4,175	305		4,480	5,050	
	5200	Flanged, iron, valves are gate										
	5210	2-1/2" pipe size	Q-1	5	3.200	Ea.	1,550	118		1,668	1,875	
	5220	3" pipe size		4.50	3.556		1,600	131		1,731	1,975	
	5230	4" pipe size		3	5.333		2,175	196		2,371	2,700	
	5240	6" pipe size	Q-2	3	8		3,075	305		3,380	3,825	
	5250	8" pipe size		2	12		5,500	460		5,960	6,750	
	5260	10" pipe size		1	24		7,750	915		8,665	9,900	
	5600	Flanged, iron, valves are OS&Y										
	5660	2-1/2" pipe size	Q-1	5	3.200	Ea.	1,750	118		1,868	2,100	
	5680	3" pipe size		4.50	3.556		1,850	131		1,981	2,225	
	5700	4" pipe size		3	5.333		2,375	196		2,571	2,925	
	5720	6" pipe size	Q-2	3	8		3,350	305		3,655	4,125	
	5740	8" pipe size		2	12		5,875	460		6,335	7,175	
	5760	10" pipe size		1	24		7,875	915		8,790	10,000	
600	0010	**VACUUM BREAKERS** Hot or cold water										600
	1030	Anti-siphon, brass										
	1040	1/4" size	1 Plum	24	.333	Ea.	21	13.60		34.60	43.50	
	1050	3/8" size		24	.333		21	13.60		34.60	43.50	
	1060	1/2" size		24	.333		16.35	13.60		29.95	38.50	
	1080	3/4" size		20	.400		28	16.35		44.35	55	
	1100	1" size		19	.421		43.50	17.20		60.70	74	
	1120	1-1/4" size		15	.533		76.50	22		98.50	117	
	1140	1-1/2" size		13	.615		89.50	25		114.50	137	
	1160	2" size		11	.727		140	29.50		169.50	199	
	1180	2-1/2" size		9	.889		400	36.50		436.50	495	
	1200	3" size		7	1.143		535	46.50		581.50	655	
	1300	For polished chrome, (1/4" thru 1"), add					50%					
	1900	Vacuum relief, water service, bronze										
	2000	1/2" size	1 Plum	30	.267	Ea.	20.50	10.90		31.40	39	
	2040	3/4" size	"	28	.286	"	25.50	11.65		37.15	46	
700	0010	**VACUUM BREAKERS**										700
	0011	See also backflow preventers 15140-100										
	1000	Anti-siphon continuous pressure type										
	1010	Max. 150 PSI - 210°F										
	1020	Bronze body										
	1030	1/2" size	1 Stpi	24	.333	Ea.	92.50	13.65		106.15	123	
	1040	3/4" size		20	.400		92.50	16.40		108.90	127	
	1050	1" size		19	.421		96	17.25		113.25	131	
	1060	1-1/4" size		15	.533		189	22		211	241	
	1070	1-1/2" size		13	.615		233	25		258	294	
	1080	2" size		11	.727		240	30		270	310	
	1200	Max. 125 PSI with atmospheric vent										
	1210	Brass, in-line construction										
	1220	1/4" size	1 Stpi	24	.333	Ea.	38	13.65		51.65	62.50	

		15140 Domestic Water Piping	CREW	DAILY OUTPUT	LABOR-HOURS	UNIT	2005 BARE COSTS MAT.	LABOR	EQUIP.	TOTAL	TOTAL INCL O&P	
700	1230	3/8" size	1 Stpi	24	.333	Ea.	38	13.65		51.65	62.50	700
	1260	For polished chrome finish, add				↓	13%					
	2000	Anti-siphon, non-continuous pressure type										
	2010	Hot or cold water 125 PSI - 210° F										
	2020	Bronze body										
	2030	1/4" size	1 Stpi	24	.333	Ea.	23	13.65		36.65	46	
	2040	3/8" size		24	.333		23	13.65		36.65	46	
	2050	1/2" size		24	.333		26	13.65		39.65	49	
	2060	3/4" size		20	.400		31	16.40		47.40	58.50	
	2070	1" size		19	.421		48	17.25		65.25	79	
	2080	1-1/4" size		15	.533		84.50	22		106.50	126	
	2090	1-1/2" size		13	.615		99	25		124	147	
	2100	2" size		11	.727		154	30		184	215	
	2110	2-1/2" size		8	1		445	41		486	550	
	2120	3" size	↓	6	1.333	↓	590	54.50		644.50	730	
	2150	For polished chrome finish, add					50%					
800	0010	**WATER HAMMER ARRESTORS / SHOCK ABSORBERS**										800
	0490	Copper										
	0500	3/4" male I.P.S. For 1 to 11 fixtures	1 Plum	12	.667	Ea.	14.50	27		41.50	57	
	0600	1" male I.P.S., For 12 to 32 fixtures		8	1		37	41		78	103	
	0700	1-1/4" male I.P.S. For 33 to 60 fixtures		8	1		43	41		84	109	
	0800	1-1/2" male I.P.S. For 61 to 113 fixtures		8	1		58.50	41		99.50	126	
	0900	2" male I.P.S.For 114 to 154 fixtures		8	1		94	41		135	165	
	1000	2-1/2" male I.P.S. For 155 to 330 fixtures	↓	4	2	↓	266	81.50		347.50	415	
	4000	Bellows type										
	4010	3/4" FNPT, to 11 fixture units	1 Plum	10	.800	Ea.	90.50	32.50		123	149	
	4020	1" FNPT, to 32 fixture units		8.80	.909		182	37		219	256	
	4030	1" FNPT, to 60 fixture units		8.80	.909		273	37		310	355	
	4040	1" FNPT, to 113 fixture units		8.80	.909		685	37		722	805	
	4050	1" FNPT, to 154 fixture units		6.60	1.212		770	49.50		819.50	925	
	4060	1-1/2" FNPT, to 300 fixture units	↓	5	1.600	↓	950	65.50		1,015.50	1,150	

		15150 Sanitary Waste and Vent Piping	CREW	DAILY OUTPUT	LABOR-HOURS	UNIT	MAT.	LABOR	EQUIP.	TOTAL	TOTAL INCL O&P	
500	0010	**SEPARATORS** Entrainment eliminator, steel body, 150 PSIG										500
	0100	1/4" size	1 Stpi	24	.333	Ea.	169	13.65		182.65	207	
	0120	1/2" size		24	.333		177	13.65		190.65	216	
	0140	3/4" size		20	.400		189	16.40		205.40	233	
	0160	1" size		19	.421		257	17.25		274.25	310	
	0180	1-1/4" size		15	.533		291	22		313	355	
	0200	1-1/2" size		13	.615		320	25		345	395	
	0220	2" size	↓	11	.727		355	30		385	440	
	0240	2-1/2" size	Q-5	15	1.067		1,450	39.50		1,489.50	1,650	
	0260	3" size		13	1.231		1,725	45.50		1,770.50	1,975	
	0280	4" size		10	1.600		1,950	59		2,009	2,225	
	0300	5" size		6	2.667		2,275	98.50		2,373.50	2,650	
	0320	6" size	↓	3	5.333		2,500	197		2,697	3,050	
	0340	8" size	Q-6	4.40	5.455		3,125	208		3,333	3,750	
	0360	10" size	"	4	6		4,525	229		4,754	5,325	
	1000	For 300 PSIG, add				↓	15%					
800	0010	**TRAPS**										800
	4700	Copper, drainage, drum trap										
	4800	3" x 5" solid, 1-1/2" pipe size	1 Plum	16	.500	Ea.	38	20.50		58.50	72.50	
	4840	3" x 6" swivel, 1-1/2" pipe size	"	16	.500	"	60.50	20.50		81	97	
	5100	P trap, standard pattern										
	5200	1-1/4" pipe size	1 Plum	18	.444	Ea.	28	18.15		46.15	58.50	

	15150 | Sanitary Waste and Vent Piping		CREW	DAILY OUTPUT	LABOR-HOURS	UNIT	2005 BARE COSTS MAT.	LABOR	EQUIP.	TOTAL	TOTAL INCL O&P	
800	5240	1-1/2″ pipe size	1 Plum	17	.471	Ea.	27	19.20		46.20	59	800
	5260	2″ pipe size		15	.533		42	22		64	79	
	5280	3″ pipe size		11	.727		101	29.50		130.50	156	
	5340	With cleanout and slip joint										
	5360	1-1/4″ pipe size	1 Plum	18	.444	Ea.	27	18.15		45.15	57	
	5400	1-1/2″ pipe size		17	.471		41.50	19.20		60.70	74.50	
	5420	2″ pipe size		15	.533		67	22		89	107	
	5460	For swivel, add					4.13			4.13	4.54	
	6710	ABS DWV P trap, solvent weld joint										
	6720	1-1/2″ pipe size	1 Plum	18	.444	Ea.	4.31	18.15		22.46	32	
	6722	2″ pipe size		17	.471		6.15	19.20		25.35	36	
	6724	3″ pipe size		15	.533		22.50	22		44.50	57.50	
	6726	4″ pipe size		14	.571		20	23.50		43.50	57	
	6732	PVC DWV P trap, solvent weld joint										
	6733	1-1/2″ pipe size	1 Plum	18	.444	Ea.	4.34	18.15		22.49	32.50	
	6734	2″ pipe size		17	.471		6.45	19.20		25.65	36	
	6735	3″ pipe size		15	.533		22.50	22		44.50	57.50	
	6736	4″ pipe size		14	.571		51	23.50		74.50	91	
	6760	PP DWV, dilution trap, 1-1/2″ pipe size		16	.500		190	20.50		210.50	240	
	6770	P trap, 1-1/2″ pipe size		17	.471		51	19.20		70.20	85	
	6780	2″ pipe size		16	.500		69.50	20.50		90	107	
	6790	3″ pipe size		14	.571		161	23.50		184.50	212	
	6800	4″ pipe size		13	.615		205	25		230	263	
	6810	Running trap, 1-1/2″ pipe size		16	.500		28.50	20.50		49	62	
	6820	2″ pipe size		15	.533		40	22		62	77	
	6830	S trap, 1-1/2″ pipe size		16	.500		29	20.50		49.50	62.50	
	6840	2″ pipe size		15	.533		36.50	22		58.50	73	
	6850	Universal trap, 1-1/2″ pipe size		14	.571		82.50	23.50		106	126	
	6860	PVC DWV hub x hub, basin trap, 1-1/4″ pipe size		18	.444		6.10	18.15		24.25	34	
	6870	Sink P trap, 1-1/2″ pipe size		18	.444		6.10	18.15		24.25	34	
	6880	Tubular S trap, 1-1/2″ pipe size		17	.471		11.25	19.20		30.45	41.50	
	6890	PVC sch. 40 DWV, drum trap										
	6900	1-1/2″ pipe size	1 Plum	16	.500	Ea.	16.15	20.50		36.65	48.50	
	6910	P trap, 1-1/2″ pipe size		18	.444		4.34	18.15		22.49	32.50	
	6920	2″ pipe size		17	.471		6.45	19.20		25.65	36	
	6930	3″ pipe size		15	.533		22.50	22		44.50	57.50	
	6940	4″ pipe size		14	.571		51	23.50		74.50	91	
	6950	P trap w/clean out, 1-1/2″ pipe size		18	.444		7.40	18.15		25.55	35.50	
	6960	2″ pipe size		17	.471		12.55	19.20		31.75	43	
	6970	P trap adjustable, 1-1/2″ pipe size		17	.471		6.25	19.20		25.45	36	
	6980	P trap adj. w/union & cleanout, 1-1/2″ pipe size		16	.500		12.50	20.50		33	44.50	
900	0010	**VENT FLASHING, CAPS**										900
	0120	Vent caps										
	0140	Cast iron										
	0180	2-1/2″ - 3-5/8″ pipe	1 Plum	21	.381	Ea.	35	15.55		50.55	61.50	
	0190	4″ - 4-1/8″ pipe	″	19	.421	″	41.50	17.20		58.70	72	
	0900	Vent flashing										
	1000	Aluminum with lead ring										
	1020	1-1/4″ pipe	1 Plum	20	.400	Ea.	6.65	16.35		23	32	
	1030	1-1/2″ pipe		20	.400		7.15	16.35		23.50	32.50	
	1040	2″ pipe		18	.444		7.15	18.15		25.30	35.50	
	1050	3″ pipe		17	.471		7.90	19.20		27.10	37.50	
	1060	4″ pipe		16	.500		9.55	20.50		30.05	41	
	1350	Copper with neoprene ring										
	1400	1-1/4″ pipe	1 Plum	20	.400	Ea.	14.05	16.35		30.40	40	

15150 | Sanitary Waste and Vent Piping

		15150 Sanitary Waste and Vent Piping	CREW	DAILY OUTPUT	LABOR-HOURS	UNIT	2005 BARE COSTS MAT.	LABOR	EQUIP.	TOTAL	TOTAL INCL O&P	
900	1430	1-1/2" pipe	1 Plum	20	.400	Ea.	14.05	16.35		30.40	40	900
	1440	2" pipe		18	.444		14.85	18.15		33	44	
	1450	3" pipe		17	.471		17.40	19.20		36.60	48	
	1460	4" pipe		16	.500		19.20	20.50		39.70	51.50	
	2000	Galvanized with neoprene ring										
	2020	1-1/4" pipe	1 Plum	20	.400	Ea.	5.30	16.35		21.65	30.50	
	2030	1-1/2" pipe		20	.400		5.30	16.35		21.65	30.50	
	2040	2" pipe		18	.444		5.60	18.15		23.75	33.50	
	2050	3" pipe		17	.471		6.10	19.20		25.30	35.50	
	2060	4" pipe		16	.500		7.90	20.50		28.40	39	
	2980	Neoprene, one piece										
	3000	1-1/4" pipe	1 Plum	24	.333	Ea.	5.15	13.60		18.75	26	
	3030	1-1/2" pipe		24	.333		5.40	13.60		19	26.50	
	3040	2" pipe		23	.348		5.40	14.20		19.60	27.50	
	3050	3" pipe		21	.381		6.55	15.55		22.10	31	
	3060	4" pipe		20	.400		8.90	16.35		25.25	34.50	

15180 | Heating and Cooling Piping

		15180 Heating and Cooling Piping	CREW	DAILY OUTPUT	LABOR-HOURS	UNIT	MAT.	LABOR	EQUIP.	TOTAL	TOTAL INCL O&P	
100	0010	**ANTI-FREEZE** Inhibited										100
	0900	Ethylene glycol concentrated										
	1000	55 gallon drums, small quantities				Gal.	7.70			7.70	8.50	
	1200	Large quantities					7.25			7.25	8	
	2000	Propylene glycol, for solar heat, small quantities					8.75			8.75	9.65	
	2100	Large quantities					8.65			8.65	9.50	
200	0010	**PUMPS, CIRCULATING** Heated or chilled water application										200
	0600	Bronze, sweat connections, 1/40 HP, in line										
	0640	3/4" size	Q-1	16	1	Ea.	120	37		157	188	
	1000	Flange connection, 3/4" to 1-1/2" size										
	1040	1/12 HP	Q-1	6	2.667	Ea.	325	98		423	505	
	1060	1/8 HP		6	2.667		560	98		658	760	
	1100	1/3 HP		6	2.667		630	98		728	835	
	1140	2" size, 1/6 HP		5	3.200		805	118		923	1,050	
	1180	2-1/2" size, 1/4 HP		5	3.200		1,050	118		1,168	1,325	
	1220	3" size, 1/4 HP		4	4		1,100	147		1,247	1,425	
	1260	1/3 HP		4	4		1,325	147		1,472	1,700	
	1300	1/2 HP		4	4		1,375	147		1,522	1,725	
	1340	3/4 HP		4	4		1,500	147		1,647	1,875	
	1380	1 HP		4	4		2,400	147		2,547	2,875	
	2000	Cast iron, flange connection										
	2040	3/4" to 1-1/2" size, in line, 1/12 HP	Q-1	6	2.667	Ea.	213	98		311	380	
	2060	1/8 HP		6	2.667		355	98		453	535	
	2100	1/3 HP		6	2.667		395	98		493	580	
	2140	2" size, 1/6 HP		5	3.200		435	118		553	655	
	2180	2-1/2" size, 1/4 HP		5	3.200		570	118		688	800	
	2220	3" size, 1/4 HP		4	4		580	147		727	855	
	2260	1/3 HP		4	4		780	147		927	1,075	
	2300	1/2 HP		4	4		810	147		957	1,100	
	2340	3/4 HP		4	4		935	147		1,082	1,250	
	2380	1 HP		4	4		1,350	147		1,497	1,700	
	2600	For non-ferrous impeller, add					3%					
	3000	High head, bronze impeller										
	3030	1-1/2" size 1/2 HP	Q-1	5	3.200	Ea.	680	118		798	920	
	3040	1-1/2" size 3/4 HP		5	3.200		725	118		843	975	
	3050	2" size 1 HP		4	4		870	147		1,017	1,175	
	3090	2" size 1-1/2 HP		4	4		1,000	147		1,147	1,325	
	4000	Close coupled, end suction, bronze impeller										

15100 | Building Services Piping

15180 | Heating and Cooling Piping

				DAILY	LABOR-		2005 BARE COSTS				TOTAL	
			CREW	OUTPUT	HOURS	UNIT	MAT.	LABOR	EQUIP.	TOTAL	INCL O&P	
200	4040	1-1/2" size, 1-1/2 HP, to 40 GPM	Q-1	3	5.333	Ea.	1,125	196		1,321	1,550	200
	4090	2" size, 2 HP, to 50 GPM		3	5.333		1,150	196		1,346	1,550	
	4100	2" size, 3 HP, to 90 GPM		2.30	6.957		1,250	256		1,506	1,750	
	4190	2-1/2" size, 3 HP, to 150 GPM		2	8		1,350	294		1,644	1,925	
	4300	3" size, 5 HP, to 225 GPM		1.80	8.889		1,475	325		1,800	2,100	
	4410	3" size, 10 HP, to 350 GPM		1.60	10		2,125	370		2,495	2,875	
	4420	4" size, 7-1/2 HP, to 350 GPM		1.60	10		2,050	370		2,420	2,800	
	4520	4" size, 10 HP, to 600 GPM	Q-2	1.70	14.118		2,525	540		3,065	3,575	
	4530	5" size, 15 HP, to 1000 GPM		1.70	14.118		3,025	540		3,565	4,125	
	4610	5" size, 20 HP, to 1350 GPM		1.50	16		3,350	610		3,960	4,600	
	4620	5" size, 25 HP, to 1550 GPM		1.50	16		3,975	610		4,585	5,300	
	5000	Base mounted, bronze impeller, coupling guard										
	5040	1-1/2" size, 1-1/2 HP, to 40 GPM	Q-1	2.30	6.957	Ea.	1,750	256		2,006	2,300	
	5090	2" size, 2 HP, to 50 GPM		2.30	6.957		1,825	256		2,081	2,375	
	5100	2" size, 3 HP, to 90 GPM		2	8		1,825	294		2,119	2,475	
	5190	2-1/2" size, 3 HP, to 150 GPM		1.80	8.889		1,925	325		2,250	2,625	
	5300	3" size, 5 HP, to 225 GPM		1.60	10		2,000	370		2,370	2,750	
	5410	4" size, 5 HP, to 350 GPM		1.50	10.667		2,175	390		2,565	3,000	
	5420	4" size, 7-1/2 HP, to 350 GPM		1.50	10.667		2,500	390		2,890	3,350	
	5520	5" size, 10 HP, to 600 GPM	Q-2	1.60	15		3,175	570		3,745	4,350	
	5530	5" size, 15 HP, to 1000 GPM		1.60	15		3,425	570		3,995	4,625	
	5610	6" size, 20 HP, to 1350 GPM		1.40	17.143		3,800	655		4,455	5,150	
	5620	6" size, 25 HP, to 1550 GPM		1.40	17.143		4,275	655		4,930	5,700	
	5800	The above pump capacities are based on 1800 RPM,										
	5810	at a 60 foot head. Increasing the RPM										
	5820	or decreasing the head will increase the GPM.										
300	0010	**PUMPS, CONDENSATE RETURN SYSTEM**										300
	2000	Simplex										
	2010	With pump, motor, CI receiver, float switch										
	2020	3/4 HP, 15 GPM	Q-1	1.80	8.889	Ea.	3,725	325		4,050	4,600	
	2100	Duplex										
	2110	With 2 pumps and motors, CI receiver, float switch, alternator										
	2120	3/4 HP, 15 GPM, 15 Gal CI rcvr	Q-1	1.40	11.429	Ea.	4,075	420		4,495	5,125	
	2130	1 HP, 25 GPM		1.20	13.333		4,450	490		4,940	5,625	
	2140	1-1/2 HP, 45 GPM		1	16		5,575	590		6,165	7,000	
	2150	1-1/2 HP, 60 GPM		1	16		6,275	590		6,865	7,775	
700	0010	**REFRIGERATION SPECIALTIES**										700
	0600	Accumulator										
	0610	3/4"	1 Stpi	8.80	.909	Ea.	84	37		121	149	
	0614	7/8"		6.40	1.250		84	51		135	170	
	0618	1-1/8"		4.80	1.667		114	68.50		182.50	228	
	0622	1-3/8"		4	2		159	82		241	298	
	0626	1-5/8"		3.20	2.500		187	102		289	360	
	0630	2-1/8"		2.40	3.333		470	137		607	725	
	1000	Filter dryer										
	1010	Replaceable core type, solder										
	1020	1/2"	1 Stpi	20	.400	Ea.	102	16.40		118.40	138	
	1030	5/8"		19	.421		115	17.25		132.25	152	
	1040	7/8"		18	.444		115	18.20		133.20	154	
	1050	1-1/8"		15	.533		120	22		142	165	
	1060	1-3/8"		14	.571		144	23.50		167.50	193	
	1070	1-5/8"		12	.667		144	27.50		171.50	199	
	1080	2-1/8"		10	.800		236	33		269	310	
	1090	2-5/8"		9	.889		273	36.50		309.50	355	
	1100	3-1/8"		8	1		283	41		324	370	
	1200	Sealed in-line, solder										

	15180 | Heating and Cooling Piping		CREW	DAILY OUTPUT	LABOR-HOURS	UNIT	2005 BARE COSTS MAT.	LABOR	EQUIP.	TOTAL	TOTAL INCL O&P	
700	1210	1/4", 3 cubic inches	1 Stpi	22	.364	Ea.	8.10	14.90		23	31.50	700
	1220	3/8", 5 cubic inches		21	.381		7.40	15.60		23	31.50	
	1230	1/2", 9 cubic inches		20	.400		12	16.40		28.40	37.50	
	1240	1/2", 16 cubic inches		20	.400		15.60	16.40		32	41.50	
	1250	5/8", 16 cubic inches		19	.421		16.35	17.25		33.60	44	
	1260	5/8", 30 cubic inches		19	.421		26.50	17.25		43.75	55	
	1270	7/8", 30 cubic inches		18	.444		28	18.20		46.20	58.50	
	1280	7/8", 41 cubic inches		17	.471		37.50	19.25		56.75	70.50	
	1290	1-1/8", 60 cubic inches		15	.533		52	22		74	90.50	
	4000	P-Trap, suction line, solder										
	4010	5/8"	1 Stpi	19	.421	Ea.	22.50	17.25		39.75	50.50	
	4020	3/4"		19	.421		16.70	17.25		33.95	44.50	
	4030	7/8"		18	.444		19.05	18.20		37.25	48.50	
	4040	1-1/8"		15	.533		46	22		68	84	
	4050	1-3/8"		14	.571		70	23.50		93.50	112	
	4060	1-5/8"		12	.667		70	27.50		97.50	118	
	4070	2-1/8"		10	.800		144	33		177	207	
	4420	Refrigerant, R-22, 50 lb. disposable cylinder				Lb.	1.31			1.31	1.44	
	4440	Refrigerant, R-507, 25 lb. disposable cylinder				"	10			10	11	
	5000	Sightglass										
	5010	Moisture and liquid indicator, solder										
	5020	1/4"	1 Stpi	22	.364	Ea.	11.70	14.90		26.60	35.50	
	5030	3/8"		21	.381		11.90	15.60		27.50	36.50	
	5040	1/2"		20	.400		14.95	16.40		31.35	41	
	5050	5/8"		19	.421		15.25	17.25		32.50	43	
	5060	7/8"		18	.444		20.50	18.20		38.70	50	
	5070	1-1/8"		15	.533		21.50	22		43.50	57	
	5080	1-3/8"		14	.571		26	23.50		49.50	63.50	
	5090	1-5/8"		12	.667		30	27.50		57.50	74	
	5100	2-1/8"		10	.800		37.50	33		70.50	90	
	7400	Vacuum pump set										
	7410	Two stage, high vacuum continuous duty	1 Stpi	8	1	Ea.	2,300	41		2,341	2,600	
	8000	Valves										
	8100	Check valve, soldered										
	8110	5/8"	1 Stpi	36	.222	Ea.	22	9.10		31.10	38	
	8114	7/8"		26	.308		57.50	12.60		70.10	82	
	8118	1-1/8"		18	.444		159	18.20		177.20	203	
	8122	1-3/8"		14	.571		214	23.50		237.50	271	
	8126	1-5/8"		13	.615		291	25		316	360	
	8130	2-1/8"		12	.667		440	27.50		467.50	525	
	8134	2-5/8"	Q-5	22	.727		730	27		757	845	
	8138	3-1/8"	"	20	.800		900	29.50		929.50	1,025	
	8500	Refrigeration valve, packless, soldered										
	8510	1/2"	1 Stpi	38	.211	Ea.	25.50	8.60		34.10	41	
	8514	5/8"		36	.222		30.50	9.10		39.60	47.50	
	8518	7/8"		26	.308		76.50	12.60		89.10	103	
	8520	Packed, soldered										
	8522	1-1/8"	1 Stpi	18	.444	Ea.	108	18.20		126.20	147	
	8526	1-3/8"		14	.571		259	23.50		282.50	320	
	8530	1-5/8"		13	.615		375	25		400	455	
	8534	2-1/8"		12	.667		475	27.50		502.50	565	
	8538	2-5/8"	Q-5	22	.727		830	27		857	955	
	8542	3-1/8"		20	.800		935	29.50		964.50	1,075	
	8546	4-1/8"		18	.889		1,325	33		1,358	1,500	
	8600	Solenoid valve, flange/solder										
	8610	1/2"	1 Stpi	38	.211	Ea.	130	8.60		138.60	156	

15180 | Heating and Cooling Piping

			CREW	DAILY OUTPUT	LABOR-HOURS	UNIT	2005 BARE COSTS MAT.	LABOR	EQUIP.	TOTAL	TOTAL INCL O&P	
700	8614	5/8"	1 Stpi	36	.222	Ea.	161	9.10		170.10	191	700
	8618	3/4"		30	.267		202	10.90		212.90	238	
	8622	7/8"		26	.308		255	12.60		267.60	300	
	8626	1-1/8"		18	.444		335	18.20		353.20	395	
	8630	1-3/8"		14	.571		405	23.50		428.50	480	
	8634	1-5/8"		13	.615		570	25		595	670	
	8638	2-1/8"	↓	12	.667	↓	625	27.50		652.50	730	
	8800	Thermostatic valve, flange/solder										
	8810	1/2 - 3 ton, 3/8" x 5/8"	1 Stpi	9	.889	Ea.	116	36.50		152.50	183	
	8814	4 - 5 ton, 1/2" x 7/8"		7	1.143		116	47		163	199	
	8818	6 - 8 ton, 5/8" x 7/8"		5	1.600		116	65.50		181.50	227	
	8822	7 - 12 ton, 7/8" x 1-1/8"		4	2		136	82		218	272	
	8826	15 - 20 ton, 7/8" x 1-3/8"	↓	3.20	2.500	↓	136	102		238	305	
800	0010	STEAM CONDENSATE METER										800
	0100	500 lb. per hour	1 Stpi	14	.571	Ea.	2,300	23.50		2,323.50	2,550	
	0140	1500 lb. per hour		7	1.143		2,475	47		2,522	2,800	
	0160	3000 lb. per hour	↓	5	1.600		3,325	65.50		3,390.50	3,750	
	0200	12,000 lb. per hour	Q-5	3.50	4.571	↓	4,450	168		4,618	5,125	

15190 | Fuel Piping

			CREW	DAILY OUTPUT	LABOR-HOURS	UNIT	MAT.	LABOR	EQUIP.	TOTAL	TOTAL INCL O&P	
490	0010	FUEL OIL SPECIALTIES										490
	0020	Foot valve, single poppet, metal to metal construction										
	0040	Bevel seat, 1/2" diameter	1 Stpi	20	.400	Ea.	36	16.40		52.40	64	
	0060	3/4" diameter		18	.444		37.50	18.20		55.70	69	
	0080	1" diameter		16	.500		28	20.50		48.50	61.50	
	0100	1-1/4" diameter		15	.533		45	22		67	82.50	
	0120	1-1/2" diameter		13	.615		48.50	25		73.50	91.50	
	0140	2" diameter	↓	11	.727	↓	54	30		84	105	
	0160	Foot valve, double poppet, metal to metal construction										
	0164	1" diameter	1 Stpi	15	.533	Ea.	80	22		102	122	
	0166	1-1/2" diameter	"	12	.667	"	111	27.50		138.50	163	
	0400	Fuel fill box, flush type										
	0408	Nonlocking, watertight										
	0410	1-1/2" diameter	1 Stpi	12	.667	Ea.	13.60	27.50		41.10	56	
	0440	2" diameter		10	.800		14.65	33		47.65	65	
	0450	2-1/2" diameter		9	.889		49	36.50		85.50	109	
	0460	3" diameter		7	1.143		57.50	47		104.50	134	
	0470	4" diameter	↓	5	1.600	↓	65	65.50		130.50	170	
	0500	Locking inner cover										
	0510	2" diameter	1 Stpi	8	1	Ea.	71	41		112	140	
	0520	2-1/2" diameter		7	1.143		95.50	47		142.50	176	
	0530	3" diameter		5	1.600		102	65.50		167.50	211	
	0540	4" diameter	↓	4	2	↓	131	82		213	267	
	0600	Fuel system components										
	0620	Spill container	1 Stpi	4	2	Ea.	485	82		567	660	
	0640	Fill adapter, 4", straight drop		8	1		27	41		68	91	
	0680	Fill cap, 4"		30	.267		21.50	10.90		32.40	40	
	0700	Extractor fitting, 4" x 1-1/2"		8	1		340	41		381	435	
	0740	Vapor hose adapter, 4"	↓	8	1		57.50	41		98.50	125	
	0760	Wood gage stick, 10'					9.60			9.60	10.55	
	1000	Oil filters, 3/8" IPT., 20 gal. per hour	1 Stpi	20	.400		23	16.40		39.40	49.50	
	1020	32 gal. per hour		18	.444		34.50	18.20		52.70	65.50	
	1040	40 gal. per hour		16	.500		39.50	20.50		60	74.50	
	1060	50 gal. per hour	↓	14	.571	↓	41	23.50		64.50	80	

15190 Fuel Piping			CREW	DAILY OUTPUT	LABOR-HOURS	UNIT	2005 BARE COSTS MAT.	LABOR	EQUIP.	TOTAL	TOTAL INCL O&P	
490	1800	Pump and motor sets										490
	1810	Light fuel and diesel oils, 100 PSI										
	1820	25 GPH 1/4 HP	Q-5	6	2.667	Ea.	795	98.50		893.50	1,025	
	1830	45 GPH, 1/4 HP		6	2.667		800	98.50		898.50	1,025	
	1840	90 GPH, 1/4 HP		5	3.200		860	118		978	1,125	
	1850	160 GPH, 1/3 HP		4	4		885	147		1,032	1,200	
	1860	325 GPH, 3/4 HP		4	4		970	147		1,117	1,300	
	1870	700 GPH, 1-1/2 HP		3	5.333		1,750	197		1,947	2,225	
	1880	1000 GPH, 2 HP		3	5.333		1,800	197		1,997	2,275	
	1890	1800 GPH, 5 HP	↓	1.80	8.889	↓	2,750	330		3,080	3,550	
	2000	Remote tank gauging system, self contained										
	2100	Gage and one probe	1 Stpi	2.50	3.200	Ea.	2,600	131		2,731	3,050	
	2120	Gage with two probes		2	4		3,500	164		3,664	4,100	
	3000	Valve, ball check, globe type, 3/8" diameter		24	.333		9.05	13.65		22.70	30.50	
	3500	Fusible, 3/8" diameter		24	.333		7.20	13.65		20.85	28.50	
	3600	1/2" diameter		24	.333		18.65	13.65		32.30	41	
	3610	3/4" diameter		20	.400		40	16.40		56.40	68	
	3620	1" diameter		19	.421		118	17.25		135.25	155	
	4000	Nonfusible, 3/8" diameter	↓	24	.333	↓	14.50	13.65		28.15	36.50	
	4500	Shutoff, gate type, lever handle, spring-fusible kit										
	4520	1/4" diameter	1 Stpi	14	.571	Ea.	20	23.50		43.50	57	
	4540	3/8" diameter		12	.667		21.50	27.50		49	64.50	
	4560	1/2" diameter		10	.800		27.50	33		60.50	79.50	
	4570	3/4" diameter	↓	8	1	↓	48	41		89	115	
	4580	Lever handle, requires weight and fusible kit										
	4600	1" diameter	1 Stpi	9	.889	Ea.	284	36.50		320.50	365	
	4620	1-1/4" diameter		8	1		294	41		335	385	
	4640	1-1/2" diameter		7	1.143		400	47		447	510	
	4660	2" diameter	↓	6	1.333		420	54.50		474.50	540	
	4680	For fusible link, weight and braided wire, add					5%					
	5000	Vent alarm, whistling signal					26			26	28.50	
	5500	Vent protector/breather, 1-1/4" diameter	1 Stpi	32	.250		7.20	10.25		17.45	23.50	
	5520	1-1/2" diameter		32	.250		8	10.25		18.25	24	
	5540	2" diameter		32	.250		16	10.25		26.25	33	
	5560	3" diameter		28	.286		40	11.70		51.70	61.50	
	5580	4" diameter		24	.333		47	13.65		60.65	72.50	
	5600	Dust cap, breather, 2"	↓	40	.200	↓	10.20	8.20		18.40	23.50	
	8000	Fuel oil and tank heaters										
	8020	Electric, capacity rated at 230 volts										
	8040	Immersion element in steel manifold										
	8060	96 GPH at 50°F rise	Q-5	6.40	2.500	Ea.	630	92		722	830	
	8070	128 GPH at 50°F rise		6.20	2.581		665	95		760	875	
	8080	160 GPH at 50°F rise		5.90	2.712		695	100		795	915	
	8090	192 GPH at 50°F rise		5.50	2.909		740	107		847	975	
	8100	240 GPH at 50°F rise		5.10	3.137		890	116		1,006	1,150	
	8110	288 GPH at 50°F rise		4.60	3.478		1,050	128		1,178	1,375	
	8120	384 GPH at 50°F rise		3.10	5.161		1,200	190		1,390	1,600	
	8130	480 GPH at 50°F rise		2.30	6.957		1,500	256		1,756	2,025	
	8140	576 GPH at 50°F rise	↓	2.10	7.619	↓	1,625	281		1,906	2,225	
	8300	Suction stub, immersion type										
	8320	75" long, 750 watts	1 Stpi	14	.571	Ea.	530	23.50		553.50	620	
	8330	99" long, 2000 watts		12	.667		745	27.50		772.50	860	
	8340	123" long, 3000 watts	↓	10	.800	↓	805	33		838	935	
	8660	Steam, cross flow, rated at 5 PSIG										
	8680	42 GPH	Q-5	7	2.286	Ea.	1,225	84		1,309	1,475	
	8690	73 GPH	↓	6.70	2.388	↓	1,425	88		1,513	1,700	

15100 | Building Services Piping

15190 | Fuel Piping

		15190 Fuel Piping	CREW	DAILY OUTPUT	LABOR-HOURS	UNIT	2005 BARE COSTS MAT.	LABOR	EQUIP.	TOTAL	TOTAL INCL O&P	
490	8700	112 GPH	Q-5	6.20	2.581	Ea.	1,475	95		1,570	1,775	490
	8710	158 GPH		5.80	2.759		1,625	102		1,727	1,925	
	8720	187 GPH		4	4		2,225	147		2,372	2,675	
	8730	270 GPH		3.60	4.444		2,400	164		2,564	2,875	
	8740	365 GPH	Q-6	4.90	4.898		2,800	187		2,987	3,375	
	8750	635 GPH		3.70	6.486		3,550	248		3,798	4,275	
	8760	845 GPH		2.50	9.600		5,200	365		5,565	6,275	
	8770	1420 GPH		1.60	15		8,475	575		9,050	10,200	
	8780	2100 GPH		1.10	21.818		11,600	835		12,435	14,100	
605	0010	METERS										605
	4000	Residential										
	4010	Gas meter, residential, 3/4" pipe size	1 Plum	14	.571	Ea.	117	23.50		140.50	163	
	4020	Gas meter, residential, 1" pipe size		12	.667		166	27		193	224	
	4030	Gas meter, residential, 1-1/4" pipe size		10	.800		176	32.50		208.50	243	

15200 | Process Piping

15210 | Process Air/Gas Piping

		15210 Process Air/Gas Piping	CREW	DAILY OUTPUT	LABOR-HOURS	UNIT	2005 BARE COSTS MAT.	LABOR	EQUIP.	TOTAL	TOTAL INCL O&P	
100	0010	AIR COMPRESSORS										100
	5250	Air, reciprocating air cooled, splash lubricated, tank mounted										
	5300	Single stage, 1 phase, 140 psi										
	5303	1/2 HP, 30 gal tank	1 Stpi	3	2.667	Ea.	925	109		1,034	1,200	
	5305	3/4 HP, 30 gal tank		2.60	3.077		955	126		1,081	1,250	
	5307	1 HP, 30 gal tank		2.20	3.636		1,175	149		1,324	1,525	
	5309	2 HP, 30 gal tank	Q-5	4	4		1,550	147		1,697	1,950	
	5310	3 HP, 30 gal tank		3.60	4.444		2,125	164		2,289	2,600	
	5314	3 HP, 60 gal tank		3.50	4.571		2,450	168		2,618	2,925	
	5320	5 HP, 60 gal tank		3.20	5		2,825	184		3,009	3,375	
	5330	5 HP, 80 gal tank		3	5.333		2,975	197		3,172	3,575	
	5340	7.5 HP, 80 gal tank		2.60	6.154		4,550	227		4,777	5,375	
	5600	2 stage pkg., 3 phase										
	5650	6 CFM at 125 psi 1-1/2 HP, 60 gal tank	Q-5	3	5.333	Ea.	2,050	197		2,247	2,575	
	5670	10.9 CFM at 125 psi, 3 HP, 80 gal tank		1.50	10.667		2,375	395		2,770	3,200	
	5680	38.7 CFM at 125 psi, 10 HP, 120 gal tank		.60	26.667		4,775	985		5,760	6,725	
	5690	105 CFM at 125 psi, 25 HP, 250 gal tank	Q-6	.60	40		9,200	1,525		10,725	12,400	
	5800	With single stage pump										
	5850	8.3 CFM at 125 psi, 2 HP, 80 gal tank	Q-6	3.50	6.857	Ea.	2,575	262		2,837	3,250	
	5860	38.7 CFM at 125 psi, 10 HP, 120 gal tank	"	.90	26.667	"	5,450	1,025		6,475	7,525	
	6000	Reciprocating, 2 stage, tank mtd, 3 Ph., Cap. rated @175 PSIG										
	6050	Pressure lubcatd, hvy duty, 9.7 CFM, 3 HP, 120 Gal. tank	Q-5	1.30	12.308	Ea.	4,050	455		4,505	5,125	
	6054	5 CFM, 1-1/2 HP, 80 gal tank		2.80	5.714		2,375	211		2,586	2,950	
	6056	6.4 CFM, 2 HP, 80 gal tank		2	8		2,500	295		2,795	3,200	
	6058	8.1 CFM, 3 HP, 80 gal tank		1.70	9.412		2,625	345		2,970	3,400	
	6059	14.8 CFM, 5 HP, 80 gal tank		1	16		2,700	590		3,290	3,850	
	6060	16.5 CFM, 5 HP, 120 gal. tank		1	16		4,100	590		4,690	5,375	
	6063	13 CFM, 6 HP, 80 gal tank		.90	17.778		4,325	655		4,980	5,725	
	6066	19.8 CFM, 7.5 HP, 80 gal tank		.80	20		4,550	735		5,285	6,100	
	6070	25.8 CFM, 7-1/2 HP, 120 gal. tank		.80	20		6,100	735		6,835	7,825	

15210 | Process Air/Gas Piping

			CREW	DAILY OUTPUT	LABOR-HOURS	UNIT	2005 BARE COSTS MAT.	LABOR	EQUIP.	TOTAL	TOTAL INCL O&P	
100	6078	34.8 CFM, 10 HP, 80 gal. tank	Q-5	.70	22.857	Ea.	5,775	840		6,615	7,625	100
	6080	34.8 CFM, 10 HP, 120 gal. tank	↓	.60	26.667		6,075	985		7,060	8,150	
	6090	53.7 CFM, 15 HP, 120 gal. tank	Q-6	.80	30		6,300	1,150		7,450	8,650	
	6100	76.7 CFM, 20 HP, 120 gal. tank		.70	34.286		7,600	1,300		8,900	10,400	
	6104	76.7 CFM, 20 HP, 240 gal tank		.68	35.294		9,325	1,350		10,675	12,300	
	6110	90.1 CFM, 25 HP, 120 gal. tank		.63	38.095		9,400	1,450		10,850	12,500	
	6120	101 CFM, 30 HP, 120 gal. tank		.57	42.105		9,425	1,600		11,025	12,800	
	6130	101 CFM, 30 HP, 250 gal. tank	↓	.52	46.154		11,500	1,775		13,275	15,300	
	6200	Oil-less, 13.6 CFM, 5 HP, 120 gal. tank	Q-5	.88	18.182		12,600	670		13,270	14,900	
	6210	13.6 CFM, 5 HP, 250 gal. tank		.80	20		13,500	735		14,235	16,000	
	6220	18.2 CFM, 7.5 HP, 120 gal. tank		.73	21.918		12,800	810		13,610	15,300	
	6230	18.2 CFM, 7.5 HP, 250 gal. tank		.67	23.881		13,700	880		14,580	16,400	
	6250	30.5 CFM, 10 HP, 120 gal. tank		.57	28.070		15,000	1,025		16,025	18,000	
	6260	30.5 CFM, 10 HP, 250 gal. tank	↓	.53	30.189		15,800	1,100		16,900	19,100	
	6270	41.3 CFM, 15 HP, 120 gal. tank	Q-6	.70	34.286		16,200	1,300		17,500	19,800	
	6280	41.3 CFM, 15 HP, 250 gal. tank	"	.67	35.821	↓	17,100	1,375		18,475	20,900	
200	0010	**COMPRESSOR ACCESSORIES**										200
	1700	Refrigerated air dryers with ambient air filters										
	1710	10 CFM	Q-5	8	2	Ea.	800	73.50		873.50	985	
	1720	25 CFM		6.60	2.424		1,125	89.50		1,214.50	1,350	
	1730	50 CFM		6.20	2.581		1,775	95		1,870	2,100	
	1740	75 CFM		5.80	2.759		2,800	102		2,902	3,225	
	1750	100 CFM	↓	5.60	2.857	↓	3,125	105		3,230	3,575	
	4000	Couplers, air line, sleeve type										
	4010	Female, connection size NPT										
	4020	1/4"	1 Stpi	38	.211	Ea.	4.16	8.60		12.76	17.55	
	4030	3/8"		36	.222		5.20	9.10		14.30	19.40	
	4040	1/2"		35	.229		10.90	9.35		20.25	26	
	4050	3/4"	↓	34	.235	↓	12.40	9.65		22.05	28	
	4100	Male										
	4110	1/4"	1 Stpi	38	.211	Ea.	4.11	8.60		12.71	17.45	
	4120	3/8"		36	.222		5.20	9.10		14.30	19.40	
	4130	1/2"		35	.229		10.25	9.35		19.60	25.50	
	4140	3/4"	↓	34	.235	↓	12.10	9.65		21.75	28	
	4150	Coupler, combined male and female halves										
	4160	1/2"	1 Stpi	17	.471	Ea.	21	19.25		40.25	52.50	
	4170	3/4"	"	15	.533	"	24.50	22		46.50	60	
500	0010	**MEDICAL GAS SYSTEM SPECIALTIES**										500
	0100	Vacuum system										
	0110	Vacuum outlet alarm panel	1 Plum	3.20	2.500	Ea.	810	102		912	1,050	
	1000	Nitrogen or oxygen system										
	1010	Cylinder manifold										
	1020	5 cylinder	1 Plum	.80	10	Ea.	4,350	410		4,760	5,400	
	1026	10 cylinder	"	.40	20	"	5,275	815		6,090	7,025	
	3000	Outlets and valves										
	3010	Recessed, wall mounted										
	3012	Single outlet	1 Plum	3.20	2.500	Ea.	51	102		153	210	
	3100	Ceiling outlet										
	3200	Zone valve with box										
	3210	2"	1 Plum	3.20	2.500	Ea.	360	102		462	555	
	4000	Alarm panel, medical gases and vacuum										
	4010	Alarm panel	1 Plum	3.20	2.500	Ea.	810	102		912	1,050	
900	0010	**VACUUM PUMPS**										900
	0100	Medical, with receiver										

	Line	15210 | Process Air/Gas Piping	Crew	Daily Output	Labor-Hours	Unit	2005 Bare Costs Mat.	Labor	Equip.	Total	Total Incl O&P	
900	0110	Duplex										900
	0120	20 SCFM	Q-1	1.14	14.035	Ea.	14,200	515		14,715	16,400	
	0130	60 SCFM, 10 HP	Q-2	1.20	20	″	25,500	765		26,265	29,300	
	0200	Triplex										
	0220	180 SCFM	Q-2	.86	27.907	Ea.	46,400	1,075		47,475	52,500	
	0300	Dental oral										
	0310	Duplex										
	0330	165 SCFM with 77 Gal separator	Q-2	1.30	18.462	Ea.	42,000	705		42,705	47,300	
		15230 | Industrial Process Piping										
500	0010	**PUMPS, GENERAL UTILITY** With motor										500
	0200	Multi-stage, horizontal split, for boiler feed applications										
	0300	Two stage, 3″ discharge x 4″ suction, 75 HP	Q-7	.30	106	Ea.	15,700	4,175		19,875	23,500	
	0340	Four stage, 3″ discharge x 4″ suction, 150 HP	″	.18	177	″	25,800	6,950		32,750	38,700	
	2000	Single stage										
	2060	End suction, 1″D. x 2″S., 3 HP	Q-1	.50	32	Ea.	3,825	1,175		5,000	5,975	
	2100	1-1/2″D. x 3″S., 10 HP	″	.40	40		4,125	1,475		5,600	6,725	
	2140	2″D. x 3″S., 15 HP	Q-2	.60	40		4,475	1,525		6,000	7,225	
	2180	3″D. x 4″S., 20 HP		.50	48		4,775	1,825		6,600	8,000	
	2220	4″D. x 6″S., 30 HP		.40	60		5,175	2,300		7,475	9,150	
	3000	Double suction, 2″D. x 2-1/2″S., 10 HP	Q-1	.30	53.333		5,300	1,950		7,250	8,775	
	3060	3″D. x 4″S., 15 HP	Q-2	.46	52.174		5,475	2,000		7,475	9,000	
	3100	4″D. x 5″S., 30 HP		.40	60		7,425	2,300		9,725	11,600	
	3140	5″D. x 6″S., 50 HP		.33	72.727		7,950	2,775		10,725	12,900	
	3180	6″D. x 8″S., 60 HP	Q-3	.30	106		8,900	4,150		13,050	16,100	
	3190	75 HP, to 2500 GPM		.28	114		11,100	4,450		15,550	18,900	
	3220	100 HP, to 3000 GPM		.26	123		14,100	4,800		18,900	22,700	
	3240	150 HP, to 4000 GPM		.24	133		18,900	5,200		24,100	28,600	
	4000	Centrifugal, end suction, mounted on base										
	4010	Horizontal mounted, with drip proof motor, rated @ 100′ head										
	4020	Vertical split case, single stage										
	4040	100 GPM, 5 HP, 1-1/2″ discharge	Q-1	1.70	9.412	Ea.	2,025	345		2,370	2,750	
	4050	200 GPM, 10 HP, 2″ discharge		1.30	12.308		2,450	455		2,905	3,375	
	4060	250 GPM, 10 HP, 3″ discharge		1.28	12.500		2,500	460		2,960	3,450	
	4070	300 GPM, 15 HP, 2″ discharge	Q-2	1.56	15.385		2,800	585		3,385	3,950	
	4080	500 GPM, 20 HP, 4″ discharge		1.44	16.667		3,275	635		3,910	4,550	
	4090	750 GPM, 30 HP, 4″ discharge		1.20	20		4,375	765		5,140	5,975	
	4100	1050 GPM, 40 HP, 5″ discharge		1	24		4,725	915		5,640	6,575	
	4110	1500 GPM, 60 HP, 6″ discharge		.60	40		6,875	1,525		8,400	9,850	
	4120	2000 GPM, 75 HP, 6″ discharge		.50	48		7,900	1,825		9,725	11,400	
	4130	3000 GPM, 100 HP, 8″ discharge		.40	60		11,900	2,300		14,200	16,500	
	4200	Horizontal split case, single stage										
	4210	100 GPM, 7.5 HP, 1-1/2″ discharge	Q-1	1.70	9.412	Ea.	3,350	345		3,695	4,225	
	4220	250 GPM, 15 HP, 2-1/2″ discharge	″	1.30	12.308		4,450	455		4,905	5,550	
	4230	500 GPM, 20 HP, 4″ discharge	Q-2	1.60	15		5,375	570		5,945	6,750	
	4240	750 GPM, 25 HP, 5″ discharge		1.54	15.584		5,550	595		6,145	7,000	
	4250	1000 GPM, 40 HP, 5″ discharge		1.20	20		6,600	765		7,365	8,400	
	4260	1500 GPM, 50 HP, 6″ discharge	Q-3	1.42	22.535		8,675	880		9,555	10,900	
	4270	2000 GPM, 75 HP, 8″ discharge		1.14	28.070		12,400	1,100		13,500	15,300	
	4280	3000 GPM, 100 HP, 10″ discharge		.96	33.333		13,700	1,300		15,000	17,100	
	4290	3500 GPM, 150 HP, 10″ discharge		.86	37.209		17,100	1,450		18,550	21,000	
	4300	4000 GPM, 200 HP, 10″ discharge		.66	48.485		18,900	1,900		20,800	23,600	
	4330	Horizontal split case, two stage, 500′ head										
	4340	100 GPM, 40 HP, 1-1/2″ discharge	Q-2	1.70	14.118	Ea.	4,400	540		4,940	5,625	
	4350	200 GPM, 50 HP, 1-1/2″ discharge	″	1.44	16.667		4,900	635		5,535	6,325	
	4360	300 GPM, 75 HP, 2″ discharge	Q-3	1.57	20.382		5,200	795		5,995	6,925	

Important: See the Reference Section for critical supporting data - Reference Nos., Crews, & City Cost Indexes

15230 | Industrial Process Piping

		15230 Industrial Process Piping	CREW	DAILY OUTPUT	LABOR-HOURS	UNIT	2005 BARE COSTS MAT.	LABOR	EQUIP.	TOTAL	TOTAL INCL O&P
500	4370	400 GPM, 100 HP, 3" discharge	Q-3	1.14	28.070	Ea.	6,550	1,100		7,650	8,850
	4380	800 GPM, 200 HP, 4" discharge		.86	37.209		15,300	1,450		16,750	19,100
	5000	Centrifugal, in-line									
	5006	Vertical mount, iron body, 125 lb. flgd, 3550 RPM TEFC mtr									
	5010	Single stage									
	5020	50 GPM, 3 HP, 1-1/2" discharge	Q-1	2.30	6.957	Ea.	590	256		846	1,025
	5030	75 GPM, 5 HP, 1-1/2" discharge		1.60	10		995	370		1,365	1,650
	5040	100 GPM, 7.5 HP, 1-1/2" discharge		1.30	12.308		1,075	455		1,530	1,850
	5050	125 GPM, 10 HP, 1-1/2" discharge	Q-2	1.70	14.118		1,325	540		1,865	2,275
	5060	150 GPM, 15 HP, 1-1/2" discharge		1.60	15		1,725	570		2,295	2,725
	5070	200 GPM, 30 HP, 1-1/2" discharge		1.50	16		2,800	610		3,410	4,000
	5080	250 GPM, 40 HP, 2" discharge		1.40	17.143		3,700	655		4,355	5,050
	5090	300 GPM, 50 HP, 3" discharge		1.30	18.462		4,475	705		5,180	5,975
	5100	400 GPM, 75 HP, 3" discharge	Q-3	.60	53.333		7,300	2,075		9,375	11,200
	5110	600 GPM, 100 HP, 3" discharge	"	.50	64		9,800	2,500		12,300	14,600
	8000	Vertical submerged, with non-submerged motor									
	8060	1"D., 3 HP	Q-1	.50	32	Ea.	5,275	1,175		6,450	7,575
	8100	1-1/2"D., 10 HP	"	.30	53.333		5,525	1,950		7,475	9,025
	8140	2"D., 15 HP	Q-2	.40	60		5,675	2,300		7,975	9,700
	8180	3"D., 25 HP		.30	80		5,800	3,050		8,850	11,000
	8220	4"D., 30 HP		.20	120		6,500	4,575		11,075	14,000
600	0010	**PUMPS MISCELLANEOUS**									
	0020	Water pump, portable, gasoline powered									
	0100	6000 GPH, 2" discharge	Q-1	11	1.455	Ea.	1,000	53.50		1,053.50	1,175
	0110	8000 GPH, 2" discharge		10.50	1.524		1,175	56		1,231	1,375
	0120	10,000 GPH, 2" discharge		10	1.600		1,800	59		1,859	2,075
	0500	Pump, propylene body, housing and impeller									
	0510	22 GPM, 1/3 HP, 40' HD	1 Plum	5	1.600	Ea.	470	65.50		535.50	615
	0520	33 GPM, 1/2 HP, 40' HD	Q-1	5	3.200		480	118		598	700
	0530	53 GPM, 3/4 HP, 40' HD	"	4	4		700	147		847	990
	0600	Rotary pump, CI									
	0610	10 GPM, 1/2 HP, 1" discharge	Q-1	5	3.200	Ea.	965	118		1,083	1,225
	0614	10 GPM, 3/4 HP, 1" discharge		4.50	3.556		965	131		1,096	1,250
	0618	10 GPM 1 HP, 1" discharge		4	4		1,000	147		1,147	1,325
	0620	25 GPM, 1 HP, 1-1/4" discharge		3.80	4.211		1,000	155		1,155	1,325
	0624	25 GPM, 1.5 HP, 1-1/4" discharge		3.60	4.444		1,100	163		1,263	1,450
	0628	25 GPM, 2 HP, 1-1/4" discharge		3.20	5		1,100	184		1,284	1,475
	0700	Fuel oil pump, 2 stage, 3450 RPM									
	0710	1/4 HP, 100 PSI	Q-1	8	2	Ea.	1,625	73.50		1,698.50	1,900
	0800	Hydraulic oil pump									
	0810	75 HP, 3 PH, 480 V	Q-3	1.20	26.667	Ea.	5,275	1,050		6,325	7,350
	1000	Turbine pump, CI									
	1010	50 GPM, 2 HP, 3" discharge	Q-1	.80	20	Ea.	1,100	735		1,835	2,325
	1020	100 GPM, 3 HP, 4" discharge	Q-2	.96	25		1,750	955		2,705	3,350
	1030	250 GPM, 15 HP, 6" discharge	"	.94	25.532		3,500	975		4,475	5,325
	1040	500 GPM, 25 HP, 6" discharge	Q-3	1.22	26.230		5,850	1,025		6,875	7,950
	1050	1000 GPM, 50 HP, 8" discharge		1.14	28.070		9,350	1,100		10,450	12,000
	1060	2000 GPM, 100 HP, 10" discharge		1	32		10,500	1,250		11,750	13,500
	1070	3000 GPM, 150 HP, 10" discharge		.80	40		11,000	1,550		12,550	14,500
	1080	4000 GPM, 200 HP, 12" discharge		.70	45.714		17,500	1,775		19,275	22,000
	1090	6000 GPM, 300 HP, 14" discharge		.60	53.333		17,900	2,075		19,975	22,800
	1100	10,000 GPM, 300 HP, 18" discharge		.58	55.172		19,500	2,150		21,650	24,600

	15410 | Plumbing Fixtures		CREW	DAILY OUTPUT	LABOR-HOURS	UNIT	MAT.	LABOR	EQUIP.	TOTAL	TOTAL INCL O&P	
300	0010	FAUCETS/FITTINGS										300
	5000	Sillcock, compact, brass, IPS or copper to hose	1 Plum	24	.333	Ea.	4.74	13.60		18.34	25.50	
	6000	Stop and waste valves, bronze										
	6100	Angle, solder end 1/2"	1 Plum	24	.333	Ea.	4.18	13.60		17.78	25	
	6110	3/4"		20	.400		4.59	16.35		20.94	29.50	
	6300	Straightway, solder end 3/8"		24	.333		3.59	13.60		17.19	24.50	
	6310	1/2"		24	.333		2.32	13.60		15.92	23	
	6320	3/4"		20	.400		2.68	16.35		19.03	27.50	
	6330	1"		19	.421		4.04	17.20		21.24	30.50	
	6400	Straightway, threaded 3/8"		24	.333		3.30	13.60		16.90	24	
	6410	1/2"		24	.333		3.13	13.60		16.73	24	
	6420	3/4"		20	.400		3.73	16.35		20.08	28.50	
	6430	1"		19	.421		7.55	17.20		24.75	34.50	

	15411 | Commercial/Indust Fixtures		CREW	DAILY OUTPUT	LABOR-HOURS	UNIT	MAT.	LABOR	EQUIP.	TOTAL	TOTAL INCL O&P	
500	0010	HYDRANTS										500
	0050	Wall type, moderate climate, bronze, encased										
	0200	3/4" IPS connection	1 Plum	16	.500	Ea.	300	20.50		320.50	360	
	0300	1" IPS connection	"	14	.571		299	23.50		322.50	365	
	0500	Anti-siphon type					259			259	285	
	1000	Non-freeze, bronze, exposed										
	1100	3/4" IPS connection, 4" to 9" thick wall	1 Plum	14	.571	Ea.	203	23.50		226.50	258	
	1120	10" to 14" thick wall		12	.667		220	27		247	283	
	1140	15" to 19" thick wall		12	.667		245	27		272	310	
	1160	20" to 24" thick wall		10	.800		265	32.50		297.50	340	
	1200	For 1" IPS connection, add					15%	10%				
	1240	For 3/4" adapter type vacuum breaker, add				Ea.	25			25	27.50	
	1280	For anti-siphon type, add				"	54			54	59.50	
	2000	Non-freeze bronze, encased, anti-siphon type										
	2100	3/4" IPS connection, 5" to 9" thick wall	1 Plum	14	.571	Ea.	271	23.50		294.50	335	
	2120	10" to 14" thick wall		12	.667		365	27		392	445	
	2140	15" to 19" thick wall		12	.667		395	27		422	470	
	2160	20" to 24" thick wall		10	.800		410	32.50		442.50	500	
	2200	For 1" IPS connection, add					10%	10%				
	3000	Ground box type, bronze frame, 3/4" IPS connection										
	3080	Non-freeze, all bronze, polished face, set flush										
	3100	2 feet depth of bury	1 Plum	8	1	Ea.	375	41		416	470	
	3120	3 feet depth of bury		8	1		400	41		441	500	
	3140	4 feet depth of bury		8	1		430	41		471	530	
	3160	5 feet depth of bury		7	1.143		460	46.50		506.50	575	
	3180	6 feet depth of bury		7	1.143		490	46.50		536.50	605	
	3200	7 feet depth of bury		6	1.333		515	54.50		569.50	645	
	3220	8 feet depth of bury		5	1.600		545	65.50		610.50	700	
	3240	9 feet depth of bury		4	2		575	81.50		656.50	755	
	3260	10 feet depth of bury		4	2		600	81.50		681.50	785	
	3400	For 1" IPS connection, add					15%	10%				
	3450	For 1-1/4" IPS connection, add					325%	14%				
	3500	For 1-1/2" connection, add					370%	18%				
	3550	For 2" connection, add					445%	24%				
	3600	For tapped drain port in box, add					34.50			34.50	37.50	
	4000	Non-freeze, CI body, bronze frame & scoriated cover										
	4010	with hose storage										
	4100	2 feet depth of bury	1 Plum	7	1.143	Ea.	705	46.50		751.50	845	
	4120	3 feet depth of bury		7	1.143		735	46.50		781.50	880	
	4140	4 feet depth of bury		7	1.143		760	46.50		806.50	905	

Important: See the Reference Section for critical supporting data - Reference Nos., Crews, & City Cost Indexes

	15411 | Commercial/Indust Fixtures		CREW	DAILY OUTPUT	LABOR-HOURS	UNIT	2005 BARE COSTS MAT.	LABOR	EQUIP.	TOTAL	TOTAL INCL O&P	
500	4160	5 feet depth of bury	1 Plum	6.50	1.231	Ea.	775	50.50		825.50	925	500
	4180	6 feet depth of bury		6	1.333		790	54.50		844.50	945	
	4200	7 feet depth of bury		5.50	1.455		815	59.50		874.50	985	
	4220	8 feet depth of bury		5	1.600		845	65.50		910.50	1,025	
	4240	9 feet depth of bury		4.50	1.778		870	72.50		942.50	1,075	
	4260	10 feet depth of bury		4	2		900	81.50		981.50	1,125	
	4280	For 1" IPS connection, add					155			155	171	
	4300	For tapped drain port in box, add					34.50			34.50	37.50	
	5000	Moderate climate, all bronze, polished face										
	5020	and scoriated cover, set flush										
	5100	3/4" IPS connection	1 Plum	16	.500	Ea.	265	20.50		285.50	325	
	5120	1" IPS connection	"	14	.571		265	23.50		288.50	325	
	5200	For tapped drain port in box, add					34.50			34.50	37.50	
	6000	Ground post type, all non-freeze, all bronze, aluminum casing										
	6010	guard, exposed head, 3/4" IPS connection										
	6100	2 feet depth of bury	1 Plum	8	1	Ea.	375	41		416	470	
	6120	3 feet depth of bury		8	1		405	41		446	505	
	6140	4 feet depth of bury		8	1		435	41		476	540	
	6160	5 feet depth of bury		7	1.143		465	46.50		511.50	585	
	6180	6 feet depth of bury		7	1.143		500	46.50		546.50	620	
	6200	7 feet depth of bury		6	1.333		535	54.50		589.50	665	
	6220	8 feet depth of bury		5	1.600		565	65.50		630.50	725	
	6240	9 feet depth of bury		4	2		600	81.50		681.50	785	
	6260	10 feet depth of bury		4	2		635	81.50		716.50	825	
	6300	For 1" IPS connection, add					40%	10%				
	6350	For 1-1/4" IPS connection, add					140%	14%				
	6400	For 1-1/2" IPS connection, add					225%	18%				
	6450	For 2" IPS connection, add					315%	24%				

	15440 | Plumbing Pumps		CREW	DAILY OUTPUT	LABOR-HOURS	UNIT	MAT.	LABOR	EQUIP.	TOTAL	TOTAL INCL O&P	
240	0010	**PUMPS, PRESSURE BOOSTER SYSTEM**										240
	0200	Pump system, with diaphragm tank, control, press. switch										
	0300	1 HP pump	Q-1	1.30	12.308	Ea.	4,000	455		4,455	5,075	
	0400	1-1/2 HP pump		1.25	12.800		4,050	470		4,520	5,150	
	0420	2 HP pump		1.20	13.333		4,150	490		4,640	5,300	
	0440	3 HP pump		1.10	14.545		4,200	535		4,735	5,425	
	0460	5 HP pump	Q-2	1.50	16		4,650	610		5,260	6,050	
	0480	7-1/2 HP pump		1.42	16.901		5,175	645		5,820	6,675	
	0500	10 HP pump		1.34	17.910		5,425	685		6,110	7,000	
	1000	Pump/ energy storage system, diaphragm tank, 3 HP pump										
	1100	motor, PRV, switch, gauge, control center, flow switch										
	1200	125 lb. working pressure	Q-2	.70	34.286	Ea.	11,700	1,300		13,000	14,800	
	1300	250 lb. working pressure	"	.64	37.500	"	12,800	1,425		14,225	16,200	
400	0010	**PUMPS, GRINDER SYSTEM** Complete, incl. check valve, tank, std.										400
	0020	controls incl. alarm/disconnect panel w/wire. Excavation not included										
	0260	Simplex, 9 GPM at 60 PSIG, 70 gal. tank				Ea.	2,425			2,425	2,675	
	0300	For manway, 26" I.D., 18" high, add					310			310	340	
	0340	26" I.D., 36" high, add					425			425	470	
	0380	43" I.D., 4' high, add					475			475	525	
	0600	Simplex, 9 GPM at 60 PSIG, 150 gal. tank					2,625			2,625	2,875	
	0660	For manway, add										
	0700	26" I.D., 36" high, add				Ea.	565			565	625	
	0740	26" I.D., 4' high, add					645			645	710	
	2000	Duplex, 18 GPM at 60 PSIG, 150 gal. tank					1,725			1,725	1,900	
	2060	For manway 43" I.D., 4' high, add					1,425			1,425	1,550	

	15440 | Plumbing Pumps	CREW	DAILY OUTPUT	LABOR-HOURS	UNIT	2005 BARE COSTS MAT.	LABOR	EQUIP.	TOTAL	TOTAL INCL O&P	
400	2400 For core only				Ea.	1,300			1,300	1,425	400
800	0010 **PUMPS, SEWAGE EJECTOR** With operating and level controls										800
	0100 Simplex system incl. tank, cover, pump 15′ head										
	0500 37 gal PE tank, 12 GPM, 1/2 HP, 2″ discharge	Q-1	3.20	5	Ea.	375	184		559	690	
	0510 3″ discharge		3.10	5.161		405	190		595	730	
	0530 87 GPM, .7 HP, 2″ discharge		3.20	5		570	184		754	905	
	0540 3″ discharge		3.10	5.161		620	190		810	965	
	0600 45 gal. coated stl tank, 12 GPM, 1/2 HP, 2″ discharge		3	5.333		670	196		866	1,025	
	0610 3″ discharge		2.90	5.517		695	203		898	1,075	
	0630 87 GPM, .7 HP, 2″ discharge		3	5.333		855	196		1,051	1,225	
	0640 3″ discharge		2.90	5.517		905	203		1,108	1,300	
	0660 134 GPM, 1 HP, 2″ discharge		2.80	5.714		925	210		1,135	1,350	
	0680 3″ discharge		2.70	5.926		975	218		1,193	1,400	
	0700 70 gal. PE tank, 12 GPM, 1/2 HP, 2″ discharge		2.60	6.154		735	226		961	1,150	
	0710 3″ discharge		2.40	6.667		785	245		1,030	1,225	
	0730 87 GPM, 0.7 HP, 2″ discharge		2.50	6.400		940	235		1,175	1,375	
	0740 3″ discharge		2.30	6.957		1,000	256		1,256	1,475	
	0760 134 GPM, 1 HP, 2″ discharge		2.20	7.273		1,025	267		1,292	1,525	
	0770 3″ discharge		2	8		1,100	294		1,394	1,650	
	0800 75 gal. coated stl. tank, 12 GPM, 1/2 HP, 2 ″ discharge		2.40	6.667		760	245		1,005	1,200	
	0810 3″ discharge		2.20	7.273		790	267		1,057	1,275	
	0830 87 GPM, .7 HP, 2″ discharge		2.30	6.957		955	256		1,211	1,425	
	0840 3″ discharge		2.10	7.619		1,000	280		1,280	1,525	
	0860 134 GPM, 1 HP, 2″ discharge		2	8		1,025	294		1,319	1,575	
	0880 3″ discharge	↓	1.80	8.889	↓	1,075	325		1,400	1,675	
	1040 Duplex system incl. tank, covers, pumps										
	1060 110 gal. fiberglass tank, 24 GPM, 1/2 HP, 2″ discharge	Q-1	1.60	10	Ea.	1,425	370		1,795	2,125	
	1080 3″ discharge		1.40	11.429		1,500	420		1,920	2,275	
	1100 174 GPM, .7 HP, 2″ discharge		1.50	10.667		1,875	390		2,265	2,650	
	1120 3″ discharge		1.30	12.308		1,950	455		2,405	2,825	
	1140 268 GPM, 1 HP, 2″ discharge		1.20	13.333		2,050	490		2,540	2,975	
	1160 3″ discharge	↓	1	16		2,100	590		2,690	3,200	
	1260 135 gal. coated stl. tank, 24 GPM, 1/2 HP, 2″ discharge	Q-2	1.70	14.118		1,500	540		2,040	2,450	
	2000 3″ discharge		1.60	15		1,575	570		2,145	2,575	
	2640 174 GPM, .7 HP, 2″ discharge		1.60	15		1,950	570		2,520	3,000	
	2660 3″ discharge		1.50	16		2,050	610		2,660	3,175	
	2700 268 GPM, 1 HP, 2″ discharge		1.30	18.462		2,125	705		2,830	3,375	
	3040 3″ discharge		1.10	21.818		2,250	830		3,080	3,725	
	3060 275 gal. coated stl. tank, 24 GPM, 1/2 HP, 2″ discharge		1.50	16		1,850	610		2,460	2,975	
	3080 3″ discharge		1.40	17.143		1,875	655		2,530	3,050	
	3100 174 GPM, .7 HP, 2″ discharge		1.40	17.143		2,400	655		3,055	3,625	
	3120 3″ discharge		1.30	18.462		2,425	705		3,130	3,700	
	3140 268 GPM, 1 HP, 2″ discharge		1.10	21.818		2,625	830		3,455	4,150	
	3160 3″ discharge	↓	.90	26.667	↓	2,725	1,025		3,750	4,525	
	3260 Pump system accessories, add										
	3300 Alarm horn and lights, 115V mercury switch	Q-1	8	2	Ea.	82.50	73.50		156	202	
	3340 Switch, mag. contactor, alarm bell, light, 3 level control		5	3.200		590	118		708	825	
	3380 Alternator, mercury switch activated	↓	4	4	↓	750	147		897	1,050	
900	0010 **PUMPS, PEDESTAL SUMP** With float control										900
	0400 Molded PVC base, 21 GPM at 15′ head, 1/3 HP	1 Plum	5	1.600	Ea.	87	65.50		152.50	194	
	0800 Iron base, 21 GPM at 15′ head, 1/3 HP		5	1.600		110	65.50		175.50	219	
	1200 Solid brass, 21 GPM at 15′ head, 1/3 HP	↓	5	1.600	↓	179	65.50		244.50	295	
940	0010 **PUMPS, SUBMERSIBLE** Sump										940
	7000 Sump pump, automatic										

15400 | Plumbing Fixtures & Equipment

15440 | Plumbing Pumps

			CREW	DAILY OUTPUT	LABOR-HOURS	UNIT	2005 BARE COSTS MAT.	LABOR	EQUIP.	TOTAL	TOTAL INCL O&P	
940	7100	Plastic, 1-1/4" discharge, 1/4 HP	1 Plum	6	1.333	Ea.	105	54.50		159.50	198	940
	7140	1/3 HP		5	1.600		127	65.50		192.50	238	
	7160	1/2 HP		5	1.600		153	65.50		218.50	266	
	7180	1-1/2" discharge, 1/2 HP		4	2		171	81.50		252.50	310	
	7500	Cast iron, 1-1/4" discharge, 1/4 HP		6	1.333		122	54.50		176.50	216	
	7540	1/3 HP		6	1.333		144	54.50		198.50	240	
	7560	1/2 HP		5	1.600		170	65.50		235.50	285	

15450 | Potable Water Storage Tanks

			CREW	DAILY OUTPUT	LABOR-HOURS	UNIT	MAT.	LABOR	EQUIP.	TOTAL	TOTAL INCL O&P	
900	0010	**WATER HEATER STORAGE TANKS** 125 psi ASME										900
	2000	Galvanized steel, 15 gal., 14" diam., 26" LOA	1 Plum	12	.667	Ea.	535	27		562	630	
	2060	30 gal., 14" diam. x 49" LOA		11	.727		640	29.50		669.50	745	
	2080	80 gal., 20" diam. x 64" LOA		9	.889		990	36.50		1,026.50	1,150	
	2100	135 gal., 24" diam. x 75" LOA		6	1.333		2,150	54.50		2,204.50	2,450	
	2120	240 gal., 30" diam. x 86" LOA		4	2		2,525	81.50		2,606.50	2,900	
	2140	300 gal., 36" diam. x 76" LOA		3	2.667		4,100	109		4,209	4,700	
	2160	400 gal., 36" diam. x 100" LOA	Q-1	4	4		5,050	147		5,197	5,775	
	2180	500 gal., 36" diam., x 126" LOA	"	3	5.333		6,250	196		6,446	7,175	
	3000	Glass lined, P.E., 80 gal., 20" diam. x 60" LOA	1 Plum	9	.889		1,725	36.50		1,761.50	1,925	
	3060	140 gal., 24" diam. x 80" LOA		6	1.333		2,450	54.50		2,504.50	2,775	
	3080	225 gal., 30" diam. x 78" LOA		4	2		2,800	81.50		2,881.50	3,200	
	3100	325 gal., 36" diam. x 81" LOA		3	2.667		3,575	109		3,684	4,100	
	3120	460 gal., 42" diam. x 84" LOA	Q-1	4	4		3,750	147		3,897	4,350	
	3140	605 gal., 48" diam. x 87" LOA		3	5.333		7,175	196		7,371	8,200	
	3160	740 gal., 54" diam. x 91" LOA		3	5.333		8,075	196		8,271	9,200	
	3180	940 gal., 60" diam. x 93" LOA		2.50	6.400		9,325	235		9,560	10,700	
	3200	1330 gal., 66" diam. x 107" LOA		2	8		12,800	294		13,094	14,500	
	3220	1615 gal., 72" diam. x 110" LOA		1.50	10.667		14,500	390		14,890	16,600	
	3240	2285 gal., 84" diam. x 128" LOA		1	16		18,000	590		18,590	20,700	
	3260	3440 gal., 96" diam. x 157" LOA	Q-2	1.50	16		26,600	610		27,210	30,200	

15480 | Domestic Water Heaters

			CREW	DAILY OUTPUT	LABOR-HOURS	UNIT	MAT.	LABOR	EQUIP.	TOTAL	TOTAL INCL O&P	
900	0010	**HEAT TRANSFER PACKAGES** Complete, controls,										900
	0020	expansion tank, converter, air separator										
	1000	Hot water, 180°F enter, 200°F leaving, 15# steam										
	1010	One pump system, 28 GPM	Q-6	.75	32	Ea.	11,600	1,225		12,825	14,700	
	1020	35 GPM		.70	34.286		12,800	1,300		14,100	16,100	
	1040	55 GPM		.65	36.923		14,600	1,400		16,000	18,200	
	1060	130 GPM		.55	43.636		18,100	1,675		19,775	22,400	
	1080	255 GPM		.40	60		24,900	2,300		27,200	30,900	
	1100	550 GPM		.30	80		31,700	3,050		34,750	39,400	
	1120	800 GPM		.25	96		41,200	3,675		44,875	51,000	
	1220	Two pump system, 28 GPM		.70	34.286		15,400	1,300		16,700	19,000	
	1240	35 GPM		.65	36.923		18,500	1,400		19,900	22,400	
	1260	55 GPM		.60	40		19,000	1,525		20,525	23,200	
	1280	130 GPM		.50	48		24,300	1,825		26,125	29,500	
	1300	255 GPM		.35	68.571		33,300	2,625		35,925	40,600	
	1320	550 GPM		.25	96		42,100	3,675		45,775	52,000	
	1340	800 GPM		.20	120		56,000	4,575		60,575	68,500	

15490 | Pool and Fountain Equipment

			CREW	DAILY OUTPUT	LABOR-HOURS	UNIT	MAT.	LABOR	EQUIP.	TOTAL	TOTAL INCL O&P	
100	0010	**FOUNTAINS/AERATORS**										100
	0100	Pump w/controls										

15400 | Plumbing Fixtures & Equipment

15490 | Pool and Fountain Equipment

			CREW	DAILY OUTPUT	LABOR-HOURS	UNIT	2005 BARE COSTS MAT.	LABOR	EQUIP.	TOTAL	TOTAL INCL O&P	
100	0200	Single phase, 100′ cord, 1/2 H.P. pump	2 Skwk	4.40	3.636	Ea.	2,675	127		2,802	3,150	100
	0300	3/4 H.P. pump		4.30	3.721		3,075	130		3,205	3,575	
	0400	1 H.P. pump		4.20	3.810		3,100	133		3,233	3,600	
	0500	1-1/2 H.P. pump		4.10	3.902		3,200	136		3,336	3,725	
	0600	2 H.P. pump		4	4		3,250	139		3,389	3,800	
	0700	Three phase, 200′ cord, 5 H.P. pump		3.90	4.103		7,850	143		7,993	8,875	
	0800	7-1/2 H.P. pump		3.80	4.211		8,875	147		9,022	9,975	
	0900	10 H.P. pump		3.70	4.324		9,900	151		10,051	11,100	
	1000	15 H.P. pump		3.60	4.444		11,700	155		11,855	13,100	
	1100	Nozzles, minimum		8	2		197	69.50		266.50	325	
	1200	Maximum		8	2		259	69.50		328.50	395	
	1300	Lights w/mounting kits, 200 watt		18	.889		325	31		356	410	
	1400	300 watt		18	.889		360	31		391	445	
	1500	500 watt		18	.889		395	31		426	485	
	1600	Color blender	↓	12	1.333	↓	305	46.50		351.50	410	

15500 | Heat Generation Equipment

15510 | Heating Boilers and Accessories

			CREW	DAILY OUTPUT	LABOR-HOURS	UNIT	2005 BARE COSTS MAT.	LABOR	EQUIP.	TOTAL	TOTAL INCL O&P	
110	0010	**BOILER BLOWDOWN SYSTEMS**										110
	1010	Boiler blowdown, auto/manual to 2000 MBH	Q-5	3.75	4.267	Ea.	3,025	157		3,182	3,550	
	1020	7300 MBH	″	3	5.333	″	3,475	197		3,672	4,125	
120	0010	**BURNERS**										120
	0990	Residential, conversion, gas fired, LP or natural										
	1000	Gun type, atmospheric input 72 to 200 MBH	Q-1	2.50	6.400	Ea.	655	235		890	1,075	
	1020	120 to 360 MBH		2	8		725	294		1,019	1,250	
	1040	280 to 800 MBH	↓	1.70	9.412	↓	1,400	345		1,745	2,075	
	2000	Commercial and industrial, gas/oil, input										
	2050	400 MBH	Q-1	1.50	10.667	Ea.	5,625	390		6,015	6,800	
	2080	For mounting plate, add		35	.457		154	16.80		170.80	195	
	2090	670 MBH		1.40	11.429		8,025	420		8,445	9,450	
	2110	For mounting plate, add		35	.457		154	16.80		170.80	195	
	2140	1155 MBH		1.30	12.308		10,000	455		10,455	11,800	
	2170	For mounting plate, add		35	.457		154	16.80		170.80	195	
	2200	1800 MBH		1.20	13.333		14,300	490		14,790	16,500	
	2230	For mounting plate, add		35	.457		154	16.80		170.80	195	
	2260	3000 MBH		1.10	14.545		19,600	535		20,135	22,400	
	2290	For mounting plate, add		35	.457		154	16.80		170.80	195	
	2320	4100 MBH		1	16		20,300	590		20,890	23,200	
	2350	For mounting plate, add	↓	35	.457	↓	154	16.80		170.80	195	
	2500	Impinged jet, rectangular, burner only, 3-1/2″ WC,										
	2520	Gas fired, input, 420 to 640 MBH	Q-1	2	8	Ea.	660	294		954	1,175	
	2540	560 to 860 MBH		2	8		920	294		1,214	1,475	
	2560	630 to 950 MBH		1.90	8.421		985	310		1,295	1,550	
	2580	950 to 1200 MBH		1.80	8.889		1,375	325		1,700	2,025	
	2600	1120 to 1700 MBH		1.80	8.889		1,850	325		2,175	2,525	
	2620	1700 to 1900 MBH		1.70	9.412		2,075	345		2,420	2,800	
	2640	1400 to 2100 MBH	↓	1.60	10	↓	2,300	370		2,670	3,075	

15510 | Heating Boilers and Accessories

		15510 Heating Boilers and Accessories	Crew	Daily Output	Labor-Hours	Unit	2005 Bare Costs Mat.	Labor	Equip.	Total	Total Incl O&P	
120	2660	1600 to 2400 MBH	Q-1	1.60	10	Ea.	2,475	370		2,845	3,275	120
	2680	1700 to 2500 MBH		1.50	10.667		2,775	390		3,165	3,650	
	2700	1880 to 2880 MBH		1.40	11.429		2,950	420		3,370	3,875	
	2720	2260 to 3680 MBH		1.40	11.429		3,450	420		3,870	4,425	
	2740	2590 to 4320 MBH		1.30	12.308		4,150	455		4,605	5,225	
	2760	3020 to 5040 MBH		1.20	13.333		4,475	490		4,965	5,650	
	2780	3450 to 5760 MBH		1.20	13.333		5,525	490		6,015	6,800	
	2850	Burner pilot		50	.320		15.90	11.75		27.65	35	
	2860	Thermocouple		50	.320		16.40	11.75		28.15	35.50	
	2870	Thermocouple & pilot bracket		45	.356		43.50	13.10		56.60	67.50	
	2900	For 7" W.C. pressure, increase MBH 40%										
	3000	Flame retention oil fired assembly, input										
	3020	.50 to 2.25 GPH	Q-1	2.40	6.667	Ea.	291	245		536	690	
	3040	2.0 to 5.0 GPH		2	8		525	294		819	1,025	
	3060	3.0 to 7.0 GPH		1.80	8.889		535	325		860	1,075	
	3080	6.0 to 12.0 GPH		1.60	10		835	370		1,205	1,475	
	4600	Gas safety, shut off valve, 3/4" threaded	1 Stpi	20	.400		165	16.40		181.40	207	
	4610	1" threaded		19	.421		166	17.25		183.25	209	
	4620	1-1/4" threaded		15	.533		181	22		203	232	
	4630	1-1/2" threaded		13	.615		184	25		209	240	
	4640	2" threaded		11	.727		219	30		249	286	
	4650	2-1/2" threaded	Q-1	15	1.067		251	39		290	335	
	4660	3" threaded		13	1.231		345	45.50		390.50	450	
	4670	4" flanged		3	5.333		2,375	196		2,571	2,900	
	4680	6" flanged	Q-2	3	8		5,275	305		5,580	6,250	
300	0010	**BOILERS, ELECTRIC, ASME** Standard controls and trim R15500-030										300
	1000	Steam, 6 KW, 20.5 MBH	Q-19	1.20	20	Ea.	2,725	765		3,490	4,150	
	1040	9 KW, 30.7 MBH D3020 102		1.20	20		2,775	765		3,540	4,200	
	1060	18 KW, 61.4 MBH		1.20	20		2,850	765		3,615	4,300	
	1080	24 KW, 81.8 MBH D3020 104		1.10	21.818		3,325	830		4,155	4,900	
	1120	36 KW, 123 MBH		1.10	21.818		3,650	830		4,480	5,275	
	1160	60 KW, 205 MBH		1	24		4,600	915		5,515	6,425	
	1220	112 KW, 382 MBH		.75	32		6,300	1,225		7,525	8,750	
	1240	148 KW, 505 MBH		.65	36.923		6,625	1,400		8,025	9,400	
	1260	168 KW, 573 MBH		.60	40		9,100	1,525		10,625	12,300	
	1280	222 KW, 758 MBH		.55	43.636		10,400	1,675		12,075	14,000	
	1300	296 KW, 1010 MBH		.45	53.333		10,500	2,025		12,525	14,600	
	1320	300 KW, 1023 MBH		.40	60		13,500	2,300		15,800	18,300	
	1340	370 KW, 1263 MBH		.35	68.571		14,500	2,625		17,125	19,900	
	1360	444 KW, 1515 MBH		.30	80		15,500	3,050		18,550	21,700	
	1380	518 KW, 1768 MBH	Q-21	.36	88.889		17,600	3,450		21,050	24,500	
	1400	592 KW, 2020 MBH		.34	94.118		19,900	3,650		23,550	27,400	
	1420	666 KW, 2273 MBH		.32	100		20,900	3,875		24,775	28,800	
	1460	740 KW, 2526 MBH		.28	114		22,000	4,450		26,450	30,900	
	1480	814 KW, 2778 MBH		.25	128		24,600	4,975		29,575	34,600	
	1500	962 KW, 3283 MBH		.22	145		25,200	5,650		30,850	36,300	
	1520	1036 KW, 3536 MBH		.20	160		26,500	6,225		32,725	38,500	
	1540	1110 KW, 3788 MBH		.19	168		27,800	6,550		34,350	40,400	
	1560	2070 KW, 7063 MBH		.18	177		39,000	6,900		45,900	53,500	
	1580	2250 KW, 7677 MBH		.17	188		45,400	7,325		52,725	61,000	
	1600	2,340 KW, 7984 MBH		.16	200		50,500	7,775		58,275	67,000	
	2000	Hot water, 7.5 KW, 25.6 MBH	Q-19	1.30	18.462		2,900	705		3,605	4,225	
	2020	15 KW, 51.2 MBH		1.30	18.462		2,925	705		3,630	4,275	
	2040	30 KW, 102 MBH		1.20	20		3,050	765		3,815	4,525	
	2060	45 KW, 164 MBH		1.20	20		3,450	765		4,215	4,925	

15510 | Heating Boilers and Accessories

			CREW	DAILY OUTPUT	LABOR-HOURS	UNIT	2005 BARE COSTS MAT.	LABOR	EQUIP.	TOTAL	TOTAL INCL O&P	
300	2070	60 KW, 205 MBH R15500-030	Q-19	1.20	20	Ea.	3,650	765		4,415	5,150	300
	2080	75 KW, 256 MBH		1.10	21.818		3,900	830		4,730	5,550	
	2100	90 KW, 307 MBH		1.10	21.818		4,225	830		5,055	5,900	
	2120	105 KW, 358 MBH		1	24		4,500	915		5,415	6,325	
	2140	120 KW, 410 MBH		.90	26.667		4,675	1,025		5,700	6,675	
	2160	135 KW, 461 MBH		.75	32		5,075	1,225		6,300	7,425	
	2180	150 KW, 512 MBH		.65	36.923		5,275	1,400		6,675	7,900	
	2200	165 KW, 563 MBH		.60	40		5,450	1,525		6,975	8,250	
	2220	296 KW, 1010 MBH		.55	43.636		9,200	1,675		10,875	12,600	
	2280	370 KW, 1263 MBH		.40	60		10,200	2,300		12,500	14,600	
	2300	444 KW, 1515 MBH		.35	68.571		12,400	2,625		15,025	17,500	
	2340	518 KW, 1768 MBH	Q-21	.44	72.727		13,600	2,825		16,425	19,200	
	2360	592 KW, 2020 MBH		.43	74.419		15,100	2,900		18,000	20,900	
	2400	666 KW, 2273 MBH		.40	80		16,100	3,100		19,200	22,400	
	2420	740 KW, 2526 MBH		.39	82.051		16,500	3,200		19,700	23,000	
	2440	814 KW, 2778 MBH		.38	84.211		17,000	3,275		20,275	23,600	
	2460	888 KW, 3031 MBH		.37	86.486		17,900	3,350		21,250	24,700	
	2480	962 KW, 3283 MBH		.36	88.889		20,100	3,450		23,550	27,300	
	2500	1,036 KW, 3536 MBH		.34	94.118		21,400	3,650		25,050	29,000	
	2520	1110 KW, 3788 MBH		.33	96.970		23,600	3,775		27,375	31,700	
	2540	1440 KW, 4915 MBH		.32	100		27,100	3,875		30,975	35,700	
	2560	1560 KW, 5323 MBH		.31	103		30,500	4,000		34,500	39,500	
	2580	1680 KW, 5733 MBH		.30	106		32,600	4,150		36,750	42,000	
	2600	1800 KW, 6143 MBH		.29	110		34,300	4,275		38,575	44,100	
	2620	1980 KW, 6757 MBH		.28	114		36,600	4,450		41,050	47,000	
	2640	2100 KW, 7167 MBH		.27	118		40,800	4,600		45,400	52,000	
	2660	2220 KW, 7576 MBH		.26	123		42,200	4,775		46,975	53,500	
	2680	2,400 KW, 8191 MBH		.25	128		44,500	4,975		49,475	56,500	
	2700	2610 KW, 8905 MBH		.24	133		47,900	5,175		53,075	60,500	
	2720	2790 KW, 9519 MBH		.23	139		50,000	5,400		55,400	63,000	
	2740	2970 KW, 10133 MBH		.21	152		51,000	5,925		56,925	65,500	
	2760	3150 KW, 10748 MBH		.19	168		56,000	6,550		62,550	71,500	
	2780	3240 KW, 11055 MBH		.18	177		57,000	6,900		63,900	73,500	
	2800	3420 KW, 11669 MBH		.17	188		60,500	7,325		67,825	77,500	
	2820	3,600 KW, 12,283 MBH		.16	200		63,000	7,775		70,775	80,500	
400	0010	**BOILERS, GAS FIRED** Natural or propane, standard controls										400
	1000	Cast iron, with insulated jacket										
	2000	Steam, gross output, 81 MBH	Q-7	1.40	22.857	Ea.	1,450	890		2,340	2,950	
	2020	102 MBH		1.30	24.615		1,650	960		2,610	3,250	
	2040	122 MBH		1	32		1,775	1,250		3,025	3,825	
	2060	163 MBH		.90	35.556		2,050	1,400		3,450	4,325	
	2080	203 MBH		.90	35.556		2,375	1,400		3,775	4,675	
	2100	240 MBH		.85	37.647		2,525	1,475		4,000	4,975	
	2120	280 MBH		.80	40		2,750	1,550		4,300	5,375	
	2140	320 MBH		.70	45.714		3,000	1,775		4,775	5,975	
	2160	360 MBH		.63	51.200		3,375	2,000		5,375	6,725	
	2180	400 MBH		.56	56.838		3,600	2,225		5,825	7,275	
	2200	440 MBH		.51	62.500		3,850	2,450		6,300	7,900	
	2220	544 MBH		.45	71.588		6,450	2,800		9,250	11,300	
	2240	765 MBH		.43	74.419		7,800	2,900		10,700	13,000	
	2260	892 MBH		.38	84.211		9,050	3,275		12,325	14,900	
	2280	1275 MBH		.34	94.118		10,300	3,675		13,975	16,900	
	2300	1530 MBH		.32	100		10,900	3,900		14,800	17,900	
	2320	1,875 MBH		.30	106		11,800	4,175		15,975	19,300	
	2340	2170 MBH		.26	122		13,500	4,775		18,275	22,000	

15510 | Heating Boilers and Accessories

		15510 Heating Boilers and Accessories	CREW	DAILY OUTPUT	LABOR-HOURS	UNIT	2005 BARE COSTS MAT.	LABOR	EQUIP.	TOTAL	TOTAL INCL O&P	
400	2360	2675 MBH	Q-7	.20	163	Ea.	15,400	6,375		21,775	26,600	400
	2380	3060 MBH		.19	172		16,700	6,725		23,425	28,500	
	2400	3570 MBH		.18	181		18,700	7,100		25,800	31,200	
	2420	4207 MBH		.16	205		20,600	8,000		28,600	34,700	
	2440	4,720 MBH		.15	207		22,700	8,100		30,800	37,100	
	2460	5660 MBH		.15	220		48,800	8,625		57,425	66,500	
	2480	6,100 MBH		.13	246		56,000	9,600		65,600	76,000	
	2500	6390 MBH		.12	266		59,000	10,400		69,400	80,500	
	2520	6680 MBH		.11	290		60,000	11,400		71,400	83,000	
	2540	6,970 MBH		.10	320		62,500	12,500		75,000	88,000	
	3000	Hot water, gross output, 80 MBH		1.46	21.918		1,550	855		2,405	2,975	
	3020	100 MBH		1.35	23.704		1,750	925		2,675	3,325	
	3040	122 MBH		1.10	29.091		1,875	1,125		3,000	3,750	
	3060	163 MBH		1	32		2,150	1,250		3,400	4,250	
	3080	203 MBH		1	32		2,200	1,250		3,450	4,300	
	3100	240 MBH		.95	33.684		2,375	1,325		3,700	4,600	
	3120	280 MBH		.90	35.556		2,625	1,400		4,025	4,975	
	3140	320 MBH		.80	40		2,800	1,550		4,350	5,425	
	3160	360 MBH		.71	45.070		3,175	1,750		4,925	6,150	
	3180	400 MBH		.64	50		3,400	1,950		5,350	6,650	
	3200	440 MBH		.58	54.983		3,650	2,150		5,800	7,225	
	3220	544 MBH		.51	62.992		6,250	2,450		8,700	10,600	
	3240	765 MBH		.46	70.022		6,525	2,725		9,250	11,300	
	3260	1,088 MBH		.40	80		6,825	3,125		9,950	12,200	
	3280	1,275 MBH		.36	89.888		9,300	3,500		12,800	15,500	
	3300	1,530 MBH		.31	104		9,950	4,100		14,050	17,200	
	3320	2,000 MBH		.26	125		12,300	4,875		17,175	20,800	
	3340	2,312 MBH		.22	148		14,200	5,775		19,975	24,400	
	3360	2,856 MBH		.20	160		17,500	6,250		23,750	28,600	
	3380	3,264 MBH		.18	179		17,700	7,025		24,725	30,000	
	3400	3,808 MBH		.16	195		19,600	7,625		27,225	33,000	
	3420	4,488 MBH		.15	210		21,800	8,225		30,025	36,200	
	3440	4,720 MBH		.15	220		49,400	8,625		58,025	67,500	
	3460	5,520 MBH		.14	228		55,500	8,925		64,425	75,000	
	3480	6,100 MBH		.13	250		56,000	9,750		65,750	76,500	
	3500	6,390 MBH		.11	285		57,500	11,200		68,700	80,500	
	3520	6,680 MBH		.10	310		61,000	12,100		73,100	85,000	
	3540	6,970 MBH		.09	359		63,000	14,000		77,000	90,500	
	4000	Steel, insulating jacket										
	4500	Steam, not including burner, gross output										
	5500	1,440 MBH	Q-6	.35	68.571	Ea.	19,100	2,625		21,725	25,000	
	5520	1630 MBH		.30	80		19,800	3,050		22,850	26,400	
	5540	1800 MBH		.26	92.308		20,500	3,525		24,025	27,900	
	5560	2520 MBH		.20	120		25,400	4,575		29,975	34,900	
	5580	3,065 MBH		.18	133		27,200	5,100		32,300	37,600	
	5600	3600 MBH		.16	150		28,900	5,725		34,625	40,300	
	5620	5400 MBH		.14	171		40,200	6,550		46,750	54,000	
	5640	7,200 MBH	Q-7	.18	177		49,900	6,950		56,850	65,500	
	5660	10,800 MBH		.17	188		62,000	7,350		69,350	79,500	
	5680	12,600 MBH		.15	213		69,000	8,325		77,325	88,500	
	5700	16,190 MBH		.12	266		87,500	10,400		97,900	111,500	
	5720	17,990 MBH		.10	320		93,000	12,500		105,500	121,500	
	6000	Hot water, including burner & one zone valve, gross output										
	6010	51.2 MBH	Q-6	2	12	Ea.	1,650	460		2,110	2,525	
	6020	72 MBH		2	12		2,025	460		2,485	2,925	
	6040	89 MBH		1.90	12.632		2,075	485		2,560	3,000	

	15510 | Heating Boilers and Accessories		CREW	DAILY OUTPUT	LABOR-HOURS	UNIT	2005 BARE COSTS MAT.	LABOR	EQUIP.	TOTAL	TOTAL INCL O&P	
400	6060	105 MBH	Q-6	1.80	13.333	Ea.	2,325	510		2,835	3,325	400
	6080	132 MBH		1.70	14.118		2,650	540		3,190	3,725	
	6100	155 MBH		1.50	16		3,075	610		3,685	4,300	
	6110	186 MBH		1.40	17.143		3,700	655		4,355	5,050	
	6120	227 MBH		1.30	18.462		4,725	705		5,430	6,225	
	6140	292 MBH		1.20	20		5,300	765		6,065	7,000	
	6160	400 MBH		.80	30		6,725	1,150		7,875	9,100	
	6180	480 MBH		.70	34.286		7,675	1,300		8,975	10,400	
	6200	640 MBH		.60	40		9,125	1,525		10,650	12,300	
	6220	800 MBH		.50	48		10,600	1,825		12,425	14,500	
	6240	960 MBH		.45	53.333		13,200	2,050		15,250	17,600	
	6260	1200 MBH		.40	60		16,200	2,300		18,500	21,300	
	6280	1440 MBH		.35	68.571		18,400	2,625		21,025	24,200	
	6300	1960 MBH		.30	80		24,300	3,050		27,350	31,400	
	6320	2400 MBH		.20	120		29,700	4,575		34,275	39,500	
	6340	3,000 MBH		.15	160		36,700	6,125		42,825	49,600	
	7000	For tankless water heater on smaller gas units, add					10%					
	7050	For additional zone valves up to 312 MBH add				Ea.	117			117	129	
	7990	Special feature gas fired boilers										
	8000	Pulse combustion, standard controls / trim										
	8050	88,000 BTU	Q-5	1.40	11.429	Ea.	3,600	420		4,020	4,575	
	8080	134,000 BTU	"	1.20	13.333	"	4,300	490		4,790	5,475	
460	0010	**BOILERS, GAS/OIL** Combination with burners and controls										460
	1000	Cast iron with insulated jacket										
	2000	Steam, gross output, 720 MBH	Q-7	.43	74.074	Ea.	6,950	2,900		9,850	12,000	
	2020	810 MBH		.38	83.989		7,425	3,275		10,700	13,100	
	2040	1,084 MBH		.34	93.023		8,550	3,625		12,175	14,900	
	2060	1,360 MBH		.33	98.160		10,100	3,825		13,925	16,900	
	2080	1,600 MBH		.30	107		11,400	4,175		15,575	18,900	
	2100	2,040 MBH		.25	130		14,000	5,075		19,075	23,000	
	2120	2,450 MBH		.20	156		15,400	6,100		21,500	26,100	
	2140	2,700 MBH		.19	165		16,300	6,475		22,775	27,600	
	2160	3,000 MBH		.18	175		17,600	6,850		24,450	29,600	
	2180	3,270 MBH		.17	183		18,900	7,175		26,075	31,600	
	2200	3,770 MBH		.17	191		46,100	7,475		53,575	61,500	
	2220	4,070 MBH		.16	200		48,500	7,800		56,300	65,000	
	2240	4,650 MBH		.15	210		51,000	8,225		59,225	68,500	
	2260	5,230 MBH		.14	223		56,500	8,725		65,225	75,500	
	2280	5,520 MBH		.14	235		61,500	9,175		70,675	82,000	
	2300	5,810 MBH		.13	248		62,500	9,675		72,175	83,000	
	2320	6,100 MBH		.12	260		62,500	10,200		72,700	84,500	
	2340	6,390 MBH		.11	296		66,000	11,600		77,600	90,000	
	2360	6,680 MBH		.10	320		69,500	12,500		82,000	95,500	
	2380	6,970 MBH		.09	372		70,000	14,500		84,500	99,000	
	2900	Hot water, gross output										
	2910	200 MBH	Q-6	.61	39.024	Ea.	5,700	1,500		7,200	8,525	
	2920	300 MBH		.49	49.080		5,700	1,875		7,575	9,100	
	2930	400 MBH		.41	57.971		6,675	2,225		8,900	10,700	
	2940	500 MBH		.36	67.039		7,200	2,550		9,750	11,800	
	3000	584 MBH	Q-7	.44	72.072		8,800	2,825		11,625	13,900	
	3020	876 MBH		.40	79.012		10,500	3,075		13,575	16,100	
	3040	1,168 MBH		.31	103		11,200	4,050		15,250	18,400	
	3060	1,460 MBH		.28	113		14,300	4,425		18,725	22,300	
	3080	2,044 MBH		.26	122		17,900	4,775		22,675	26,900	
	3100	2,628 MBH		.21	150		21,300	5,875		27,175	32,200	
	3120	3,210 MBH		.18	174		26,200	6,825		33,025	39,100	

15510 | Heating Boilers and Accessories

		15510 Heating Boilers and Accessories	CREW	DAILY OUTPUT	LABOR-HOURS	UNIT	2005 BARE COSTS MAT.	LABOR	EQUIP.	TOTAL	TOTAL INCL O&P	
460	3140	3,796 MBH	Q-7	.17	186	Ea.	29,300	7,250		36,550	43,200	460
	3160	4,088 MBH		.16	195		33,400	7,625		41,025	48,100	
	3180	4,672 MBH		.16	203		36,400	7,950		44,350	52,000	
	3200	5,256 MBH		.15	217		40,800	8,500		49,300	57,500	
	3220	6,000 MBH, 179 BHP		.13	256		58,500	10,000		68,500	79,000	
	3240	7,130 MBH, 213 BHP		.08	385		61,000	15,000		76,000	90,000	
	3260	9,800 MBH, 286 BHP		.06	533		70,500	20,800		91,300	109,000	
	3280	10,900 MBH, 325.6 BHP		.05	592		85,500	23,100		108,600	129,000	
	3290	12,200 MBH, 364.5 BHP		.05	666		93,000	26,000		119,000	141,500	
	3300	13,500 MBH, 403.3 BHP	↓	.04	727	↓	103,000	28,400		131,400	156,000	
	4000	Steel, insulated jacket, skid base, tubeless										
	4100	Steam, 15 psi, gross output 517 MBH, 15 BHP	Q-6	.41	57.971	Ea.	14,800	2,225		17,025	19,600	
	4110	690 MBH, 20 BHP		.38	62.992		15,200	2,400		17,600	20,300	
	4120	862 MBH, 25 BHP		.36	67.039		16,200	2,550		18,750	21,700	
	4130	1,004 MBH, 30 BHP		.32	74.074		19,100	2,825		21,925	25,300	
	4140	1,340 MBH, 40 BHP		.31	78.431		20,900	3,000		23,900	27,500	
	4150	1,675 MBH, 50 BHP		.28	85.714		21,700	3,275		24,975	28,800	
	4160	2,175 MBH, 65 BHP		.23	104		22,400	4,000		26,400	30,600	
	4170	2,510 MBH, 75 BHP		.19	125		24,000	4,775		28,775	33,600	
	4180	3,015 MBH, 90 BHP	↓	.17	140		27,300	5,375		32,675	38,100	
	4190	3,350 MBH, 100 BHP	Q-7	.22	146		31,100	5,725		36,825	42,900	
	4200	3,880 MBH, 116 BHP		.21	153		33,600	6,000		39,600	46,000	
	4210	4,685 MBH, 140 BHP		.19	168		35,800	6,575		42,375	49,300	
	4220	5,025 MBH, 150 BHP		.18	177		37,000	6,950		43,950	51,000	
	4230	5,550 MBH, 166 BHP		.17	189		48,000	7,400		55,400	64,000	
	4240	6,700 MBH, 200 BHP		.13	256		52,500	10,000		62,500	72,500	
	4250	7,760 MBH, 232 BHP		.11	296		58,500	11,600		70,100	82,000	
	4260	8,870 MBH, 265 BHP		.09	336		63,500	13,100		76,600	90,000	
	4270	11,100 MBH, 322 BHP		.07	426		68,000	16,700		84,700	100,000	
	4280	13,400 MBH, 400 BHP		.06	516		78,000	20,100		98,100	116,000	
	4290	17,750 MBH, 530 BHP		.05	680		97,500	26,600		124,100	147,500	
	4300	20,750 MBH, 620 BHP	↓	.04	800		109,000	31,200		140,200	167,000	
	4500	Steam, 150 psi gross output, 335 MBH, 10 BHP	Q-6	.54	44.037		12,000	1,675		13,675	15,700	
	4520	502 MBH, 15 BHP		.40	60		13,600	2,300		15,900	18,500	
	4560	670 MBH, 20 BHP		.36	65.934		14,800	2,525		17,325	20,100	
	4600	1,005 MBH, 30 BHP		.31	77.922		19,400	2,975		22,375	25,800	
	4640	1,339 MBH, 40 BHP		.29	81.911		23,500	3,125		26,625	30,500	
	4660	1,674 MBH, 50 BHP		.24	100		26,200	3,825		30,025	34,600	
	4680	2,009 MBH, 60 BHP		.21	114		31,100	4,375		35,475	40,800	
	4720	2,511 MBH, 75 BHP		.17	141		32,000	5,400		37,400	43,300	
	5000	Hot water, gross output, 525 MBH		.45	52.980		11,400	2,025		13,425	15,700	
	5020	630 MBH		.43	55.944		11,600	2,150		13,750	15,900	
	5040	735 MBH		.41	58.968		14,300	2,250		16,550	19,100	
	5060	840 MBH		.38	64		15,800	2,450		18,250	21,100	
	5080	1,050 MBH		.35	67.989		18,400	2,600		21,000	24,200	
	5100	1,365 MBH		.33	72.727		23,000	2,775		25,775	29,500	
	5120	1,680 MBH		.29	82.759		26,400	3,175		29,575	33,800	
	5140	2,310 MBH		.20	120		34,300	4,575		38,875	44,600	
	5160	2,835 MBH		.18	129		41,400	4,950		46,350	53,000	
	5180	3,150 MBH	↓	.16	146	↓	45,700	5,600		51,300	58,500	
500	0010	**BOILERS, OIL FIRED** Standard controls, flame retention burner										500
	1000	Cast iron, with insulated flush jacket										
	2000	Steam, gross output, 109 MBH	Q-7	1.20	26.667	Ea.	1,925	1,050		2,975	3,700	
	2020	144 MBH		1.10	29.091		1,975	1,125		3,100	3,875	
	2040	173 MBH		1	32		2,650	1,250		3,900	4,800	
	2060	207 MBH	↓	.90	35.556	↓	2,775	1,400		4,175	5,125	

MECHANICAL 15

15510 | Heating Boilers and Accessories

			CREW	DAILY OUTPUT	LABOR-HOURS	UNIT	2005 BARE COSTS MAT.	LABOR	EQUIP.	TOTAL	TOTAL INCL O&P	
500	2080	236 MBH	Q-7	.85	37.647	Ea.	2,900	1,475		4,375	5,400	500
	2100	300 MBH		.70	45.714		2,975	1,775		4,750	5,950	
	2120	480 MBH		.50	64		3,775	2,500		6,275	7,900	
	2140	665 MBH		.45	71.111		5,150	2,775		7,925	9,825	
	2160	794 MBH		.41	78.049		6,100	3,050		9,150	11,300	
	2180	1,084 MBH		.38	85.106		7,025	3,325		10,350	12,700	
	2200	1,360 MBH		.33	98.160		8,225	3,825		12,050	14,800	
	2220	1,600 MBH		.26	122		9,300	4,775		14,075	17,400	
	2240	2,175 MBH		.24	133		11,800	5,225		17,025	20,900	
	2260	2,480 MBH		.20	156		13,700	6,100		19,800	24,300	
	2280	3,000 MBH		.19	170		15,800	6,650		22,450	27,400	
	2300	3,550 MBH		.17	187		17,700	7,300		25,000	30,400	
	2320	3,820 MBH		.16	200		28,600	7,800		36,400	43,200	
	2340	4,360 MBH		.15	214		38,800	8,375		47,175	55,500	
	2360	4,940 MBH		.14	225		44,000	8,800		52,800	61,500	
	2380	5,520 MBH		.14	235		52,000	9,175		61,175	71,000	
	2400	6,100 MBH		.13	256		59,500	10,000		69,500	80,500	
	2420	6,390 MBH		.11	290		62,500	11,400		73,900	86,000	
	2440	6,680 MBH		.10	313		64,500	12,200		76,700	89,000	
	2460	6,970 MBH		.09	363		66,500	14,200		80,700	94,500	
	3000	Hot water, same price as steam										
	4000	For tankless coil in smaller sizes, add					15%					
	5000	Steel, insulated jacket, burner										
	6000	Steam, full water leg construction, gross output										
	6020	144 MBH	Q-6	1.33	18.045	Ea.	4,050	690		4,740	5,475	
	6040	198 MBH		1.20	20		4,400	765		5,165	6,000	
	6060	252 MBH		1.13	21.295		5,475	815		6,290	7,250	
	6080	324 MBH		1.04	23.011		6,250	880		7,130	8,200	
	6100	396 MBH		.78	30.691		6,800	1,175		7,975	9,250	
	6120	468 MBH		.70	34.483		7,450	1,325		8,775	10,200	
	6140	648 MBH		.52	45.977		7,925	1,750		9,675	11,400	
	6160	792 MBH		.44	55.046		9,325	2,100		11,425	13,500	
	6180	1,008 MBH		.39	61.381		11,600	2,350		13,950	16,200	
	6200	1,260 MBH		.35	68.966		12,800	2,625		15,425	18,100	
	6220	1,512 MBH		.31	78.689		15,300	3,000		18,300	21,300	
	6240	1,800 MBH		.29	83.624		16,800	3,200		20,000	23,300	
	6260	2,100 MBH		.23	106		19,800	4,050		23,850	27,900	
	6280	2,400 MBH		.19	125		22,100	4,775		26,875	31,500	
	6400	Larger sizes are same as steel, gas fired										
	7000	Hot water, gross output, 103 MBH	Q-6	1.60	15	Ea.	1,250	575		1,825	2,250	
	7020	122 MBH		1.45	16.506		2,475	630		3,105	3,675	
	7040	137 MBH		1.36	17.595		2,725	670		3,395	4,000	
	7060	168 MBH		1.30	18.405		3,125	705		3,830	4,475	
	7080	225 MBH		1.22	19.704		3,225	755		3,980	4,650	
	7100	315 MBH		.96	25.105		3,725	960		4,685	5,550	
	7120	420 MBH		.70	34.483		4,950	1,325		6,275	7,400	
	7140	525 MBH		.57	42.403		5,375	1,625		7,000	8,350	
	7160	630 MBH		.52	45.977		6,450	1,750		8,200	9,750	
	7180	735 MBH		.48	50.104		6,450	1,925		8,375	9,975	
	7200	840 MBH		.43	55.172		10,000	2,100		12,100	14,200	
	7220	1,050 MBH		.36	65.753		10,200	2,525		12,725	15,100	
	7240	1,365 MBH		.32	74.534		11,900	2,850		14,750	17,400	
	7260	1,680 MBH		.29	83.624		13,800	3,200		17,000	20,000	
	7280	2,310 MBH		.21	114		15,200	4,400		19,600	23,300	
	7320	3,150 MBH		.13	184		19,900	7,050		26,950	32,500	
	7340	For tankless coil in steam or hot water, add					7%					

15510 | Heating Boilers and Accessories

			CREW	DAILY OUTPUT	LABOR-HOURS	UNIT	2005 BARE COSTS				TOTAL
							MAT.	LABOR	EQUIP.	TOTAL	INCL O&P
700	0010	**BOILERS, PACKAGED SCOTCH MARINE** Steam or hot water									
	1000	Packaged fire tube, #2 oil, gross output									
	1006	15 PSI steam									
	1010	1005 MBH, 30 HP	Q-6	.33	72.072	Ea.	26,100	2,750		28,850	32,900
	1014	1675 MBH, 50 HP	"	.26	93.023		32,700	3,550		36,250	41,400
	1020	3348 MBH, 100 HP	Q-7	.21	152		38,600	5,950		44,550	51,500
	1030	5025 MBH, 150 HP		.17	192		48,800	7,525		56,325	65,000
	1040	6696 MBH, 200 HP		.14	223		55,500	8,725		64,225	74,000
	1050	8375 MBH, 250 HP		.14	231		62,500	9,050		71,550	82,000
	1060	10,044 MBH, 300 HP		.13	251		80,000	9,825		89,825	103,000
	1080	16,740 MBH, 500 HP		.08	380		114,500	14,900		129,400	148,000
	1100	23,435 MBH, 700 HP		.07	484		140,000	18,900		158,900	182,500
	1140	To fire #6, and gas, add		.42	76.190		12,500	2,975		15,475	18,200
	1180	For duplex package feed system									
	1200	To 3348 MBH boiler, add	Q-7	.54	59.259	Ea.	6,700	2,325		9,025	10,900
	1220	To 6696 MBH boiler, add		.41	78.049		7,950	3,050		11,000	13,300
	1240	To 10,044 MBH boiler, add		.38	84.211		9,125	3,275		12,400	15,000
	1260	To 16,740 MBH boiler, add		.28	114		10,400	4,450		14,850	18,100
	1280	To 23,435 MBH boiler, add		.25	128		12,300	5,000		17,300	21,000
	1300	150 PSI steam									
	1310	1005 MBH, 30 HP	Q-6	.33	72.072	Ea.	27,300	2,750		30,050	34,200
	1320	1675 MBH, 50 HP	"	.26	93.023		34,100	3,550		37,650	42,900
	1330	3350 MBH, 100 HP	Q-7	.21	152		46,500	5,950		52,450	60,000
	1340	5025 MBH, 150 HP		.17	192		59,500	7,525		67,025	77,000
	1350	6700 MBH, 200 HP		.14	223		66,000	8,725		74,725	85,500
	1360	10,050 MBH, 300 HP		.13	251		97,500	9,825		107,325	122,000
	1400	Packaged scotch marine, #6 oil									
	1410	15 PSI steam									
	1420	1675 MBH, 50 HP	Q-6	.25	97.166	Ea.	40,200	3,725		43,925	49,800
	1430	3350 MBH, 100 HP	Q-7	.20	159		47,700	6,225		53,925	62,000
	1440	5025 MBH, 150 HP		.16	202		60,000	7,900		67,900	78,000
	1450	6700 MBH, 200 HP		.14	233		65,500	9,125		74,625	85,500
	1460	8375 MBH, 250 HP		.13	244		83,500	9,525		93,025	106,500
	1470	10,050 MBH, 300 HP		.12	262		98,500	10,200		108,700	123,500
	1480	13,400 MBH, 400 HP		.09	351		112,500	13,700		126,200	144,500
	1500	150 PSI steam									
	1510	1675 MBH, 50 HP	Q-6	.25	97.166	Ea.	47,700	3,725		51,425	58,000
	1520	3350 MBH, 100 HP	Q-7	.20	159		54,000	6,225		60,225	69,000
	1530	5025 MBH, 150 HP		.16	202		67,500	7,900		75,400	86,000
	1540	6700 MBH, 200 HP		.14	233		72,000	9,125		81,125	92,500
	1550	8375 MBH, 250 HP		.13	244		91,500	9,525		101,025	115,500
	1560	10,050 MBH, 300 HP		.12	262		104,000	10,200		114,200	129,500
	1570	13,400 MBH, 400 HP		.09	351		120,500	13,700		134,200	153,000
	2000	Packaged water tube, #2 oil, steam or hot water, gross output									
	2010	200 MBH	Q-6	1.50	16	Ea.	8,375	610		8,985	10,100
	2014	275 MBH		1.40	17.143		9,100	655		9,755	11,000
	2018	360 MBH		1.10	21.818		10,200	835		11,035	12,500
	2022	520 MBH		.65	36.923		10,400	1,400		11,800	13,600
	2026	600 MBH		.60	40		10,600	1,525		12,125	14,000
	2030	720 MBH		.55	43.636		11,100	1,675		12,775	14,700
	2034	960 MBH		.48	50		12,200	1,900		14,100	16,300
	2040	1200 MBH	Q-7	.50	64		13,900	2,500		16,400	19,000
	2044	1440 MBH		.45	71.111		18,200	2,775		20,975	24,200
	2060	1600 MBH		.40	80		18,600	3,125		21,725	25,200
	2068	1920 MBH		.35	91.429		20,200	3,575		23,775	27,600
	2072	2160 MBH		.33	96.970		22,400	3,775		26,175	30,300

		15510 | Heating Boilers and Accessories	CREW	DAILY OUTPUT	LABOR-HOURS	UNIT	2005 BARE COSTS MAT.	LABOR	EQUIP.	TOTAL	TOTAL INCL O&P	
700	2080	2400 MBH	Q-7	.30	106	Ea.	25,000	4,175		29,175	33,800	700
	2100	3200 MBH		.25	128		28,600	5,000		33,600	39,000	
	2120	4800 MBH		.20	160		37,400	6,250		43,650	50,500	
	2200	Gas fired										
	2204	200 MBH	Q-6	1.50	16	Ea.	6,350	610		6,960	7,925	
	2208	275 MBH		1.40	17.143		6,600	655		7,255	8,250	
	2212	360 MBH		1.10	21.818		6,600	835		7,435	8,525	
	2216	520 MBH		.65	36.923		7,800	1,400		9,200	10,700	
	2220	600 MBH		.60	40		7,800	1,525		9,325	10,900	
	2224	720 MBH		.55	43.636		8,600	1,675		10,275	12,000	
	2228	960 MBH		.48	50		10,100	1,900		12,000	14,000	
	2232	1220 MBH	Q-7	.50	64		10,900	2,500		13,400	15,800	
	2236	1440 MBH		.45	71.111		11,600	2,775		14,375	17,000	
	2240	1680 MBH		.40	80		13,000	3,125		16,125	19,000	
	2244	1920 MBH		.35	91.429		13,700	3,575		17,275	20,500	
	2248	2160 MBH		.33	96.970		14,800	3,775		18,575	22,000	
	2252	2400 MBH		.30	106		16,600	4,175		20,775	24,600	
760	0010	**BOILERS, SOLID FUEL**										760
	0020	Wood, coal, wood/paper waste, peat. Steel, 3" diam. horiz.										
	0040	fire tubes, low pressure steam or hot water, natural or										
	0060	mechanical draft. Not including dry base.										
	1000	Natural draft, gross output, 1445 MBH	Q-6	.45	53.333	Ea.	14,600	2,050		16,650	19,200	
	1050	1635 MBH		.43	55.814		15,200	2,125		17,325	20,000	
	1100	1805 MBH		.41	58.537		15,900	2,225		18,125	20,900	
	1150	2160 MBH		.38	63.158		17,100	2,425		19,525	22,400	
	1200	2520 MBH		.32	75		20,600	2,875		23,475	27,000	
	1250	3065 MBH	Q-7	.40	80		22,200	3,125		25,325	29,100	
	1300	3600 MBH		.38	84.211		23,700	3,275		26,975	31,100	
	1350	4500 MBH		.29	110		30,700	4,300		35,000	40,300	
	1400	5400 MBH		.26	123		33,100	4,800		37,900	43,600	
	1450	6295 MBH		.23	139		38,400	5,425		43,825	50,500	
	1500	7200 MBH		.21	152		43,400	5,950		49,350	56,500	
	1550	9000 MBH		.19	168		47,200	6,575		53,775	62,000	
	1600	10,800 MBH		.17	188		54,000	7,350		61,350	70,500	
	1650	12,600 MBH		.15	213		59,000	8,325		67,325	77,500	
	1700	14,390 MBH		.13	246		71,000	9,600		80,600	93,000	
	1750	16,190 MBH		.12	266		77,500	10,400		87,900	100,500	
	1800	17,990 MBH		.10	320		86,000	12,500		98,500	113,500	
	3000	Stoker fired (coal) cast iron with flush jacket and										
	3080	insulation, steam or water, gross output, 148 MBH	Q-6	1.50	16	Ea.	4,125	610		4,735	5,450	
	3100	200 MBH		1.30	18.462		4,725	705		5,430	6,225	
	3120	247 MBH		1.20	20		5,275	765		6,040	6,950	
	3140	300 MBH		1	24		5,900	915		6,815	7,850	
	3160	410 MBH		.96	25		6,750	955		7,705	8,850	
	3180	495 MBH		.88	27.273		7,550	1,050		8,600	9,875	
	3200	580 MBH		.80	30		8,350	1,150		9,500	10,900	
	3400	1280 MBH		.36	66.667		20,600	2,550		23,150	26,500	
	3420	1460 MBH		.30	80		22,600	3,050		25,650	29,500	
	3440	1640 MBH		.28	85.714		24,100	3,275		27,375	31,400	
	3460	1820 MBH		.26	92.308		25,900	3,525		29,425	33,700	
	3480	2000 MBH		.25	96		27,500	3,675		31,175	35,800	
	3500	2360 MBH		.23	104		31,100	4,000		35,100	40,200	
	3540	2725 MBH		.20	120		34,600	4,575		39,175	44,900	
	3800	2950 MBH	Q-7	.16	200		58,000	7,800		65,800	75,000	
	3820	3210 MBH		.15	213		61,000	8,325		69,325	79,500	

Important: See the Reference Section for critical supporting data - Reference Nos., Crews, & City Cost Indexes

15510 | Heating Boilers and Accessories

			CREW	DAILY OUTPUT	LABOR-HOURS	UNIT	2005 BARE COSTS MAT.	LABOR	EQUIP.	TOTAL	TOTAL INCL O&P	
760	3840	3480 MBH	Q-7	.14	228	Ea.	63,500	8,925		72,425	83,500	760
	3860	3745 MBH		.14	228		66,500	8,925		75,425	86,500	
	3880	4000 MBH		.13	246		69,000	9,600		78,600	90,500	
	3900	4200 MBH		.13	246		72,500	9,600		82,100	94,000	
	3920	4400 MBH		.12	266		75,000	10,400		85,400	98,000	
	3940	4600 MBH	↓	.12	266	↓	78,000	10,400		88,400	101,500	
880	0010	**SWIMMING POOL HEATERS** Not including wiring, external										880
	0020	piping, base or pad,										
	0060	Gas fired, input, 115 MBH	Q-6	3	8	Ea.	1,850	305		2,155	2,500	
	0100	135 MBH		2	12		2,100	460		2,560	3,000	
	0160	155 MBH		1.50	16		2,200	610		2,810	3,350	
	0180	175 MBH		1.30	18.462		2,450	705		3,155	3,750	
	0200	190 MBH		1	24		2,875	915		3,790	4,525	
	0220	235 MBH		.70	34.286		3,500	1,300		4,800	5,825	
	0240	285 MBH		.60	40		4,500	1,525		6,025	7,250	
	0260	365 MBH		.50	48		5,075	1,825		6,900	8,325	
	0280	500 MBH		.40	60		6,950	2,300		9,250	11,100	
	0300	600 MBH		.35	68.571		8,000	2,625		10,625	12,800	
	0320	800 MBH		.33	72.727		9,525	2,775		12,300	14,700	
	0360	1000 MBH		.22	109		11,100	4,175		15,275	18,600	
	0370	1,200 MBH		.21	114		13,900	4,375		18,275	21,800	
	0380	1,500 MBH		.19	126		17,100	4,825		21,925	26,100	
	0400	1,800 MBH		.14	171		19,400	6,550		25,950	31,200	
	0410	2,450 MBH		.13	184		25,800	7,050		32,850	39,000	
	0420	3,000 MBH		.11	218		31,500	8,350		39,850	47,100	
	0440	3,750 MBH		.09	266		39,100	10,200		49,300	58,500	
	1000	Oil fired, gross output, 97 MBH		1.70	14.118		2,125	540		2,665	3,125	
	1020	118 MBH		1.70	14.118		2,175	540		2,715	3,200	
	1040	134 MBH		1.60	15		2,300	575		2,875	3,400	
	1060	161 MBH		1.30	18.462		2,800	705		3,505	4,150	
	1080	187 MBH		1.20	20		2,975	765		3,740	4,400	
	1100	214 MBH		1.10	21.818		3,375	835		4,210	4,975	
	1120	262 MBH		.89	26.966		3,825	1,025		4,850	5,750	
	1140	340 MBH		.68	35.294		5,050	1,350		6,400	7,575	
	1160	420 MBH		.58	41.379		5,800	1,575		7,375	8,750	
	1180	525 MBH		.46	52.174		7,425	2,000		9,425	11,200	
	1200	630 MBH		.41	58.537		8,275	2,225		10,500	12,500	
	1220	735 MBH		.35	68.571		10,200	2,625		12,825	15,200	
	1240	840 MBH		.33	72.727		11,600	2,775		14,375	17,000	
	1260	1,050 MBH		.27	88.889		14,200	3,400		17,600	20,700	
	1280	1,365 MBH		.21	114		18,200	4,375		22,575	26,700	
	1300	1,680 MBH		.18	133		21,400	5,100		26,500	31,300	
	1320	2,310 MBH		.13	184		27,600	7,050		34,650	41,000	
	1340	2,835 MBH		.11	218		34,400	8,350		42,750	50,500	
	1360	3,150 MBH	↓	.09	266		38,600	10,200		48,800	58,000	
	2000	Electric, 12 KW, 4,800 gallon pool	Q-19	3	8		1,725	305		2,030	2,350	
	2020	15 KW, 7,200 gallon pool		2.80	8.571		1,725	325		2,050	2,400	
	2040	24 KW, 9,600 gallon pool		2.40	10		2,325	380		2,705	3,150	
	2060	30 KW, 12,000 gallon pool		2	12		2,425	460		2,885	3,325	
	2080	35 KW, 14,400 gallon pool		1.60	15		2,425	570		2,995	3,525	
	2100	55 KW, 24,000 gallon pool	↓	1.20	20	↓	3,325	765		4,090	4,800	
	9000	To select pool heater: 12 BTUH x S.F. pool area										
	9010	X temperature differential =required output										
	9050	For electric, KW = gallons x 2.5 divided by 1000										
	9100	For family home type pool, double the										
	9110	Rated gallon capacity = 1/2°F rise per hour										

	15520 | Feedwater Equipment		CREW	DAILY OUTPUT	LABOR-HOURS	UNIT	2005 BARE COSTS MAT.	LABOR	EQUIP.	TOTAL	TOTAL INCL O&P	
050	0010	**SHOT CHEMICAL FEEDER**										050
	0200	Shot chem feeder, by pass, in line mount, 125 PSIG, 1.7 gallon	2 Stpi	1.80	8.889	Ea.	320	365		685	900	
	0220	Floor mount, 175 PSIG, 5 gallon		1.24	12.903		405	530		935	1,250	
	0230	12 gallon		.93	17.204		645	705		1,350	1,750	
	0300	150 lb., ASME, 5 gallon		1.24	12.903		1,225	530		1,755	2,150	
	0320	10 gallon		.93	17.204		1,325	705		2,030	2,500	
	0400	300 lb., ASME, 5 gallon		1.24	12.903		3,275	530		3,805	4,400	
	0420	10 gallon		.93	17.204		4,100	705		4,805	5,550	
	15530 | Furnaces											
200	0010	**FURNACE COMPONENTS AND COMBINATIONS**										200
	0080	Coils, A/C evaporator, for gas or oil furnaces										
	0090	Add-on, with holding charge										
	0100	Upflow										
	0120	1-1/2 ton cooling	Q-5	4	4	Ea.	127	147		274	360	
	0130	2 ton cooling		3.70	4.324		153	159		312	410	
	0140	3 ton cooling		3.30	4.848		191	179		370	480	
	0150	4 ton cooling		3	5.333		270	197		467	590	
	0160	5 ton cooling		2.70	5.926		345	218		563	710	
	0300	Downflow										
	0330	2-1/2 ton cooling	Q-5	3	5.333	Ea.	178	197		375	490	
	0340	3-1/2 ton cooling		2.60	6.154		239	227		466	605	
	0350	5 ton cooling		2.20	7.273		345	268		613	785	
	0600	Horizontal										
	0630	2 ton cooling	Q-5	3.90	4.103	Ea.	180	151		331	425	
	0640	3 ton cooling		3.50	4.571		206	168		374	480	
	0650	4 ton cooling		3.20	5		259	184		443	560	
	0660	5 ton cooling		2.90	5.517		345	203		548	685	
	2000	Cased evaporator coils for air handlers										
	2100	1-1/2 ton cooling	Q-5	4.40	3.636	Ea.	191	134		325	410	
	2110	2 ton cooling		4.10	3.902		194	144		338	430	
	2120	2-1/2 ton cooling		3.90	4.103		218	151		369	465	
	2130	3 ton cooling		3.70	4.324		256	159		415	520	
	2140	3-1/2 ton cooling		3.50	4.571		240	168		408	515	
	2150	4 ton cooling		3.20	5		288	184		472	590	
	2160	5 ton cooling		2.90	5.517		330	203		533	670	
	3010	Air handler, modular										
	3100	With cased evaporator cooling coil										
	3120	1-1/2 ton cooling	Q-5	3.80	4.211	Ea.	495	155		650	780	
	3130	2 ton cooling		3.50	4.571		520	168		688	830	
	3140	2-1/2 ton cooling		3.30	4.848		570	179		749	895	
	3150	3 ton cooling		3.10	5.161		620	190		810	965	
	3160	3-1/2 ton cooling		2.90	5.517		765	203		968	1,150	
	3170	4 ton cooling		2.50	6.400		940	236		1,176	1,375	
	3180	5 ton cooling		2.10	7.619		1,025	281		1,306	1,550	
	3500	With no cooling coil										
	3520	1-1/2 ton coil size	Q-5	12	1.333	Ea.	350	49		399	455	
	3530	2 ton coil size		10	1.600		375	59		434	500	
	3540	2-1/2 ton coil size		10	1.600		415	59		474	545	
	3554	3 ton coil size		9	1.778		460	65.50		525.50	605	
	3560	3-1/2 ton coil size		9	1.778		515	65.50		580.50	665	
	3570	4 ton coil size		8.50	1.882		660	69.50		729.50	830	
	3580	5 ton coil size		8	2		710	73.50		783.50	890	
	4000	With heater										
	4120	5 kW, 17.1 MBH	Q-5	16	1	Ea.	315	37		352	400	
	4130	7.5 kW, 25.6 MBH		15.60	1.026		335	38		373	425	

Important: See the Reference Section for critical supporting data - Reference Nos., Crews, & City Cost Indexes

	15530 | Furnaces		CREW	DAILY OUTPUT	LABOR-HOURS	UNIT	2005 BARE COSTS MAT.	LABOR	EQUIP.	TOTAL	TOTAL INCL O&P	
200	4140	10 kW, 34.2 MBH	Q-5	15.20	1.053	Ea.	390	39		429	490	200
	4150	12.5 KW, 42.7 MBH		14.80	1.081		440	40		480	540	
	4160	15 KW, 51.2 MBH		14.40	1.111		525	41		566	640	
	4170	25 KW, 85.4 MBH		14	1.143		595	42		637	715	
	4180	30 KW, 102 MBH	↓	13	1.231	↓	695	45.50		740.50	835	
400	0010	**FURNACES** Hot air heating, blowers, standard controls										400
	0020	not including gas, oil or flue piping										
	1000	Electric, UL listed										
	1020	10.2 MBH	Q-20	5	4	Ea.	315	148		463	570	
	1040	17.1 MBH		4.80	4.167		330	155		485	595	
	1060	27.3 MBH		4.60	4.348		395	161		556	680	
	1100	34.1 MBH		4.40	4.545		405	169		574	705	
	1120	51.6 MBH		4.20	4.762		505	177		682	825	
	1140	68.3 MBH		4	5		610	185		795	955	
	1160	85.3 MBH	↓	3.80	5.263	↓	670	195		865	1,025	
	3000	Gas, AGA certified, upflow, direct drive models										
	3020	45 MBH input	Q-9	4	4	Ea.	435	145		580	705	
	3040	60 MBH input		3.80	4.211		580	152		732	875	
	3060	75 MBH input		3.60	4.444		615	161		776	925	
	3100	100 MBH input		3.20	5		655	181		836	1,000	
	3120	125 MBH input		3	5.333		755	193		948	1,125	
	3130	150 MBH input		2.80	5.714		875	207		1,082	1,275	
	3140	200 MBH input		2.60	6.154		2,000	223		2,223	2,550	
	3160	300 MBH input		2.30	6.957		2,075	252		2,327	2,675	
	3180	400 MBH input	↓	2	8	↓	2,700	289		2,989	3,425	
	3200	Gas fired wall furnace										
	3204	Horizontal flow										
	3208	7.7 MBH	Q-9	7	2.286	Ea.	400	82.50		482.50	565	
	3212	14 MBH		6.50	2.462		370	89		459	540	
	3216	24 MBH		5	3.200		390	116		506	605	
	3220	49 MBH		4	4		530	145		675	810	
	3224	65 MBH	↓	3.60	4.444	↓	695	161		856	1,000	
	3240	Up-flow										
	3244	7.7 MBH	Q-9	7	2.286	Ea.	370	82.50		452.50	530	
	3248	14 MBH		6.50	2.462		370	89		459	540	
	3252	24 MBH		5	3.200		370	116		486	590	
	3256	49 MBH		4	4		715	145		860	1,000	
	3260	65 MBH	↓	3.60	4.444	↓	790	161		951	1,100	
	3500	Gas furnace										
	3510	Up-flow										
	3514	1200 CFM, 45 MBH	Q-9	4	4	Ea.	685	145		830	980	
	3518	75 MBH		3.60	4.444		755	161		916	1,075	
	3522	100 MBH		3.20	5		865	181		1,046	1,225	
	3526	125 MBH		3	5.333		1,000	193		1,193	1,400	
	3530	150 MBH		2.80	5.714		1,225	207		1,432	1,675	
	3540	2000 CFM, 75 MBH		3.60	4.444		830	161		991	1,150	
	3544	100 MBH		3.20	5		940	181		1,121	1,300	
	3548	120 MBH		3	5.333		1,075	193		1,268	1,500	
	3552	150 MBH	↓	2.80	5.714	↓	1,300	207		1,507	1,750	
	3560	Horizontal flow										
	3564	1200 CFM, 45 MBH	Q-9	4	4	Ea.	690	145		835	980	
	3568	60 MBH		3.80	4.211		790	152		942	1,100	
	3572	70 MBH		3.60	4.444		800	161		961	1,125	
	3576	2000 CFM, 90 MBH		3.30	4.848		865	175		1,040	1,225	
	3580	110 MBH	↓	3.10	5.161	↓	905	187		1,092	1,275	

		15530 | Furnaces	CREW	DAILY OUTPUT	LABOR-HOURS	UNIT	2005 BARE COSTS MAT.	LABOR	EQUIP.	TOTAL	TOTAL INCL O&P	
400	6000	Oil, UL listed, atomizing gun type burner										400
	6020	56 MBH output	Q-9	3.60	4.444	Ea.	745	161		906	1,075	
	6030	84 MBH output		3.50	4.571		775	165		940	1,100	
	6040	95 MBH output		3.40	4.706		790	170		960	1,125	
	6060	134 MBH output		3.20	5		1,100	181		1,281	1,475	
	6080	151 MBH output		3	5.333		1,225	193		1,418	1,625	
	6100	200 MBH input		2.60	6.154		2,125	223		2,348	2,675	
	6120	300 MBH input		2.30	6.957		2,325	252		2,577	2,950	
	6140	400 MBH input		2	8		2,700	289		2,989	3,425	
	6200	Up flow										
	6210	105 MBH	Q-9	3.40	4.706	Ea.	1,275	170		1,445	1,650	
	6220	140 MBH		3.30	4.848		1,525	175		1,700	1,950	
	6230	168 MBH		2.90	5.517		2,000	200		2,200	2,500	
	6260	Lo-boy style										
	6270	105 MBH	Q-9	3.40	4.706	Ea.	1,400	170		1,570	1,800	
	6280	140 MBH		3.30	4.848		1,500	175		1,675	1,925	
	6290	189 MBH		2.80	5.714		1,675	207		1,882	2,175	
	6320	Down flow										
	6330	105 MBH	Q-9	3.40	4.706	Ea.	1,275	170		1,445	1,650	
	6340	140 MBH		3.30	4.848		1,525	175		1,700	1,950	
	6350	168 MBH		2.90	5.517		1,550	200		1,750	2,025	
	6380	Horizontal										
	6390	105 MBH	Q-9	3.40	4.706	Ea.	1,250	170		1,420	1,625	
	6400	140 MBH		3.30	4.848		1,450	175		1,625	1,875	
	6410	189 MBH		2.80	5.714		1,575	207		1,782	2,050	

		15540 | Fuel-Fired Heaters	CREW	DAILY OUTPUT	LABOR-HOURS	UNIT	MAT.	LABOR	EQUIP.	TOTAL	TOTAL INCL O&P	
300	0010	**DUCT FURNACES** Includes burner, controls, stainless steel										300
	0020	heat exchanger. Gas fired, electric ignition										
	0030	Indoor installation										
	0080	100 MBH output	Q-5	5	3.200	Ea.	1,200	118		1,318	1,475	
	0100	120 MBH output		4	4		1,425	147		1,572	1,800	
	0130	200 MBH output		2.70	5.926		1,825	218		2,043	2,350	
	0140	240 MBH output		2.30	6.957		1,975	256		2,231	2,550	
	0160	280 MBH output		2	8		2,200	295		2,495	2,875	
	0180	320 MBH output		1.60	10		2,375	370		2,745	3,175	
	0300	For powered venter and adapter, add					325			325	360	
	0500	For required flue pipe, see 15550										
	1000	Outdoor installation, with vent cap										
	1020	75 MBH output	Q-5	4	4	Ea.	1,850	147		1,997	2,250	
	1040	94 MBH output		4	4		2,075	147		2,222	2,500	
	1060	120 MBH output		4	4		2,375	147		2,522	2,825	
	1080	157 MBH output		3.50	4.571		2,750	168		2,918	3,275	
	1100	187 MBH output		3	5.333		3,375	197		3,572	4,025	
	1120	225 MBH output		2.50	6.400		3,500	236		3,736	4,200	
	1140	300 MBH output		1.80	8.889		4,400	330		4,730	5,350	
	1160	375 MBH output		1.60	10		4,675	370		5,045	5,675	
	1180	450 MBH output		1.40	11.429		4,900	420		5,320	6,000	
	1200	600 MBH output		1	16		5,950	590		6,540	7,425	
	1300	Aluminized exchanger, subtract					15%					
	1400	For powered venter, add					10%					
	1500	For two stage gas valve, add					180			180	198	
900	0010	**SPACE HEATERS** Cabinet, grilles, fan, controls, burner,										900
	0020	thermostat, no piping. For flue see 15550										

15540 | Fuel-Fired Heaters

			CREW	DAILY OUTPUT	LABOR-HOURS	UNIT	2005 BARE COSTS MAT.	LABOR	EQUIP.	TOTAL	TOTAL INCL O&P	
900	1000	Gas fired, floor mounted										900
	1100	60 MBH output	Q-5	10	1.600	Ea.	545	59		604	690	
	1120	80 MBH output		9	1.778		555	65.50		620.50	710	
	1140	100 MBH output		8	2		595	73.50		668.50	765	
	1160	120 MBH output		7	2.286		670	84		754	865	
	1180	180 MBH output		6	2.667		815	98.50		913.50	1,050	
	1500	Rooftop mounted, gravity vent, stainless steel exchanger										
	1520	75 MBH output	Q-6	4	6	Ea.	3,725	229		3,954	4,450	
	1540	95 MBH output		3.60	6.667		3,775	255		4,030	4,525	
	1560	120 MBH output		3.30	7.273		4,050	278		4,328	4,875	
	1580	159 MBH output		3	8		4,175	305		4,480	5,050	
	1600	190 MBH output		2.60	9.231		4,700	355		5,055	5,700	
	1620	225 MBH output		2.30	10.435		4,775	400		5,175	5,850	
	1640	300 MBH output		1.90	12.632		5,700	485		6,185	7,000	
	1660	375 MBH output		1.40	17.143		6,475	655		7,130	8,100	
	1680	450 MBH output		1.20	20		6,600	765		7,365	8,400	
	1700	600 MBH output		1	24		7,950	915		8,865	10,100	
	1720	750 MBH output		.80	30		13,600	1,150		14,750	16,600	
	1740	900 MBH output		.60	40		15,200	1,525		16,725	19,100	
	1760	1200 MBH output		.30	80		18,500	3,050		21,550	25,000	
	1800	For power vent, add					10%					
	1900	For aluminized steel exchanger, subtract					10%					
	2000	Suspension mounted, propeller fan, 20 MBH output	Q-5	8.50	1.882		450	69.50		519.50	600	
	2020	40 MBH output		7.50	2.133		495	78.50		573.50	665	
	2040	60 MBH output		7	2.286		555	84		639	735	
	2060	80 MBH output		6	2.667		635	98.50		733.50	850	
	2080	100 MBH output		5.50	2.909		735	107		842	970	
	2100	130 MBH output		5	3.200		825	118		943	1,075	
	2120	140 MBH output		4.50	3.556		910	131		1,041	1,200	
	2140	160 MBH output		4	4		915	147		1,062	1,225	
	2160	180 MBH output		3.50	4.571		1,025	168		1,193	1,400	
	2180	200 MBH output		3	5.333		1,075	197		1,272	1,475	
	2200	240 MBH output		2.70	5.926		1,275	218		1,493	1,725	
	2220	280 MBH output		2.30	6.957		1,550	256		1,806	2,075	
	2240	320 MBH output		2	8		1,700	295		1,995	2,325	
	2500	For powered venter and adapter, add					282			282	310	
	3000	Suspension mounted, blower type, 40 MBH output	Q-5	6.80	2.353		730	86.50		816.50	935	
	3020	60 MBH output		6.60	2.424		810	89.50		899.50	1,025	
	3040	84 MBH output		5.80	2.759		865	102		967	1,100	
	3060	104 MBH output		5.20	3.077		1,000	113		1,113	1,275	
	3080	140 MBH output		4.30	3.721		1,200	137		1,337	1,525	
	3100	180 MBH output		3.30	4.848		1,250	179		1,429	1,650	
	3120	240 MBH output		2.50	6.400		1,825	236		2,061	2,375	
	3140	280 MBH output		2	8		2,050	295		2,345	2,700	
	4000	Suspension mounted, sealed combustion system,										
	4020	Aluminized steel exchanger, powered vent										
	4040	100 MBH output	Q-5	5	3.200	Ea.	1,450	118		1,568	1,775	
	4060	120 MBH output		4.70	3.404		1,625	125		1,750	1,975	
	4080	160 MBH output		3.70	4.324		1,775	159		1,934	2,200	
	4100	200 MBH output		2.90	5.517		2,125	203		2,328	2,650	
	4120	240 MBH output		2.50	6.400		2,275	236		2,511	2,850	
	4140	320 MBH output		1.70	9.412		2,725	345		3,070	3,525	
	5000	Wall furnace, 17.5 MBH output		6	2.667		570	98.50		668.50	775	
	5020	24 MBH output		5	3.200		595	118		713	830	
	5040	35 MBH output		4	4		805	147		952	1,100	
	6000	Oil fired, suspension mounted, 94 MBH output		4	4		2,175	147		2,322	2,625	

	15540 | Fuel-Fired Heaters		CREW	DAILY OUTPUT	LABOR-HOURS	UNIT	2005 BARE COSTS MAT.	LABOR	EQUIP.	TOTAL	TOTAL INCL O&P	
900	6040	140 MBH output	Q-5	3	5.333	Ea.	2,325	197		2,522	2,850	900
	6060	184 MBH output		3	5.333		2,500	197		2,697	3,050	
	15550 | Breechings, Chimneys & Stacks											
200	0010	DRAFT CONTROL DEVICES										200
	1000	Barometric,gas fired system only, 6″ size for 5″ and 6″ pipes	1 Shee	20	.400	Ea.	41	16.10		57.10	69.50	
	1020	7″ size, for 6″ and 7″ pipes		19	.421		52	16.95		68.95	83.50	
	1040	8″ size, for 7″ and 8″ pipes		18	.444		60.50	17.85		78.35	94	
	1060	9″ size, for 8″ and 9″ pipes		16	.500		70.50	20		90.50	109	
	2000	All fuel, oil, oil/gas, coal										
	2020	10″ for 9″ and 10″ pipes	1 Shee	15	.533	Ea.	96	21.50		117.50	139	
	2040	12″ for 11″ and 12″ pipes		15	.533		126	21.50		147.50	171	
	2060	14″ for 13″ and 14″ pipes		14	.571		160	23		183	212	
	2080	16″ for 15″ and 16″ pipes		13	.615		224	24.50		248.50	284	
	2100	18″ for 17″ and 18″ pipes		12	.667		292	27		319	360	
	2120	20″ for 19″ and 21″ pipes		10	.800		350	32		382	435	
	2140	24″ for 22″ and 25″ pipes	Q-9	12	1.333		435	48		483	555	
	2160	28″ for 26″ and 30″ pipes		10	1.600		520	58		578	665	
	2180	32″ for 31″ and 34″ pipes		8	2		680	72.50		752.50	855	
	3260	For thermal switch for above, add	1 Shee	24	.333		39.50	13.40		52.90	63.50	
	5000	Vent damper, bi-metal, gas, 3″ diameter	Q-9	24	.667		23.50	24		47.50	62.50	
	5010	4″ diameter		24	.667		23.50	24		47.50	62.50	
	5020	5″ diameter		23	.696		23.50	25		48.50	64	
	5030	6″ diameter		22	.727		23.50	26.50		50	66	
	5040	7″ diameter		21	.762		25	27.50		52.50	70	
	5050	8″ diameter		20	.800		27	29		56	74.50	
	5101	Electric, automatic, gas, 4″ diameter		24	.667		145	24		169	196	
	5110	5″ diameter		23	.696		148	25		173	202	
	5121	6″ diameter		22	.727		150	26.50		176.50	206	
	5130	7″ diameter		21	.762		153	27.50		180.50	211	
	5140	8″ diameter		20	.800		156	29		185	216	
	5150	9″ diameter		20	.800		159	29		188	220	
	5160	10″ diameter		19	.842		164	30.50		194.50	227	
	5170	12″ diameter		19	.842		167	30.50		197.50	230	
	5250	Automatic, oil, 4″ diameter		24	.667		162	24		186	215	
	5260	5″ diameter		23	.696		165	25		190	220	
	5270	6″ diameter		22	.727		169	26.50		195.50	227	
	5280	7″ diameter		21	.762		172	27.50		199.50	232	
	5290	8″ diameter		20	.800		179	29		208	242	
	5300	9″ diameter		20	.800		182	29		211	245	
	5310	10″ diameter		19	.842		182	30.50		212.50	248	
	5320	12″ diameter		19	.842		185	30.50		215.50	251	
	5330	14″ diameter		18	.889		248	32		280	320	
	5340	16″ diameter		17	.941		288	34		322	370	
	5350	18″ diameter		16	1		320	36		356	410	
440	0010	**VENT CHIMNEY** Prefab metal, U.L. listed										440
	0020	Gas, double wall, galvanized steel										
	0080	3″ diameter	Q-9	72	.222	V.L.F.	3.77	8.05		11.82	16.50	
	0100	4″ diameter		68	.235		4.71	8.50		13.21	18.30	
	0120	5″ diameter		64	.250		5.50	9.05		14.55	19.95	
	0140	6″ diameter		60	.267		6.40	9.65		16.05	22	
	0160	7″ diameter		56	.286		9.40	10.35		19.75	26.50	
	0180	8″ diameter		52	.308		10.45	11.15		21.60	28.50	
	0200	10″ diameter		48	.333		22	12.05		34.05	43	
	0220	12″ diameter		44	.364		29.50	13.15		42.65	52.50	

	15550 | Breechings, Chimneys & Stacks	CREW	DAILY OUTPUT	LABOR-HOURS	UNIT	2005 BARE COSTS MAT.	LABOR	EQUIP.	TOTAL	TOTAL INCL O&P	
440	0240 14" diameter	Q-9	42	.381	V.L.F.	49.50	13.80		63.30	75.50	440
	0260 16" diameter		40	.400		67	14.45		81.45	96.50	
	0280 18" diameter	↓	38	.421		86.50	15.25		101.75	119	
	0300 20" diameter	Q-10	36	.667		102	25		127	151	
	0320 22" diameter		34	.706		129	26.50		155.50	183	
	0340 24" diameter		32	.750		159	28		187	219	
	0360 26" diameter		31	.774		191	29		220	256	
	0380 28" diameter		30	.800		202	30		232	268	
	0400 30" diameter		28	.857		213	32		245	285	
	0420 32" diameter		27	.889		248	33.50		281.50	325	
	0440 34" diameter		26	.923		282	34.50		316.50	365	
	0460 36" diameter		25	.960		298	36		334	385	
	0480 38" diameter		24	1		325	37.50		362.50	420	
	0500 40" diameter		23	1.043		365	39		404	460	
	0520 42" diameter		22	1.091		380	41		421	485	
	0540 44" diameter		21	1.143		420	43		463	530	
	0560 46" diameter		20	1.200		465	45		510	580	
	0580 48" diameter	↓	19	1.263	↓	515	47.50		562.50	640	
	0600 For 4", 5" and 6" oval, add					50%					
	0650 Gas, double wall, galvanized steel, fittings										
	0660 Elbow 45°, 3" diameter	Q-9	36	.444	Ea.	9.35	16.10		25.45	35	
	0670 4" diameter		34	.471		11.05	17.05		28.10	38	
	0680 5" diameter		32	.500		12.95	18.10		31.05	42.50	
	0690 6" diameter		30	.533		15.55	19.30		34.85	46.50	
	0700 7" diameter		28	.571		22.50	20.50		43	56.50	
	0710 8" diameter		26	.615		26	22.50		48.50	63	
	0720 10" diameter		24	.667		64	24		88	108	
	0730 12" diameter		22	.727		81	26.50		107.50	130	
	0740 14" diameter		21	.762		113	27.50		140.50	167	
	0750 16" diameter		20	.800		142	29		171	201	
	0760 18" diameter	↓	19	.842		191	30.50		221.50	257	
	0770 20" diameter	Q-10	18	1.333		216	50		266	315	
	0780 22" diameter		17	1.412		325	53		378	435	
	0790 24" diameter		16	1.500		415	56.50		471.50	540	
	0800 26" diameter		16	1.500		815	56.50		871.50	980	
	0810 28" diameter		15	1.600		835	60		895	1,025	
	0820 30" diameter		14	1.714		880	64.50		944.50	1,075	
	0830 32" diameter		14	1.714		990	64.50		1,054.50	1,200	
	0840 34" diameter		13	1.846		1,100	69.50		1,169.50	1,300	
	0850 36" diameter		12	2		1,125	75		1,200	1,350	
	0860 38" diameter		12	2		1,200	75		1,275	1,450	
	0870 40" diameter		11	2.182		1,275	82		1,357	1,525	
	0880 42" diameter		11	2.182		1,350	82		1,432	1,625	
	0890 44" diameter		10	2.400		1,425	90		1,515	1,725	
	0900 46" diameter		10	2.400		1,525	90		1,615	1,825	
	0910 48" diameter	↓	10	2.400	↓	1,575	90		1,665	1,900	
	0916 Adjustable length										
	0918 3" diameter, to 12"	Q-9	36	.444	Ea.	7.65	16.10		23.75	33	
	0920 4" diameter, to 12"		34	.471		9.05	17.05		26.10	36	
	0924 6" diameter, to 12"		30	.533		13.90	19.30		33.20	45	
	0928 8" diameter, to 12"		26	.615		17.15	22.50		39.65	53.50	
	0930 10" diameter, to 18"		24	.667		64.50	24		88.50	108	
	0932 12" diameter, to 18"		22	.727		78.50	26.50		105	127	
	0936 16" diameter, to 18"		20	.800		151	29		180	211	
	0938 18" diameter, to 18"	↓	19	.842		180	30.50		210.50	246	
	0944 24" diameter, to 18"	Q-10	16	1.500	↓	325	56.50		381.50	440	

15550 | Breechings, Chimneys & Stacks

			Description	CREW	DAILY OUTPUT	LABOR-HOURS	UNIT	2005 BARE COSTS MAT.	LABOR	EQUIP.	TOTAL	TOTAL INCL O&P	
440	0950		Elbow 90°, adjustable, 3″ diameter	Q-9	36	.444	Ea.	15.55	16.10		31.65	41.50	440
	0960		4″ diameter		34	.471		18.95	17.05		36	47	
	0970		5″ diameter		32	.500		22	18.10		40.10	52.50	
	0980		6″ diameter		30	.533		26.50	19.30		45.80	58.50	
	0990		7″ diameter		28	.571		42	20.50		62.50	78	
	1010		8″ diameter		26	.615		44.50	22.50		67	83.50	
	1020		Wall thimble, 4 to 7″ adjustable, 3″ diameter		36	.444		9.85	16.10		25.95	35.50	
	1022		4″ diameter		34	.471		11	17.05		28.05	38	
	1024		5″ diameter		32	.500		12.35	18.10		30.45	41.50	
	1026		6″ diameter		30	.533		15.10	19.30		34.40	46	
	1028		7″ diameter		28	.571		17.60	20.50		38.10	51.50	
	1030		8″ diameter		26	.615		18.20	22.50		40.70	54.50	
	1040		Roof flashing, 3″ diameter		36	.444		5.60	16.10		21.70	30.50	
	1050		4″ diameter		34	.471		6.35	17.05		23.40	33	
	1060		5″ diameter		32	.500		9.85	18.10		27.95	39	
	1070		6″ diameter		30	.533		13.80	19.30		33.10	44.50	
	1080		7″ diameter		28	.571		17.40	20.50		37.90	51	
	1090		8″ diameter		26	.615		18.90	22.50		41.40	55.50	
	1100		10″ diameter		24	.667		51.50	24		75.50	94	
	1110		12″ diameter		22	.727		66	26.50		92.50	113	
	1120		14″ diameter		20	.800		81	29		110	134	
	1130		16″ diameter		18	.889		100	32		132	160	
	1140		18″ diameter		16	1		136	36		172	205	
	1150		20″ diameter	Q-10	18	1.333		176	50		226	270	
	1160		22″ diameter		14	1.714		223	64.50		287.50	345	
	1170		24″ diameter		12	2		268	75		343	410	
	1200		Tee, 3″ diameter	Q-9	27	.593		23	21.50		44.50	58.50	
	1210		4″ diameter		26	.615		24	22.50		46.50	61	
	1220		5″ diameter		25	.640		26	23		49	64	
	1230		6″ diameter		24	.667		29	24		53	68.50	
	1240		7″ diameter		23	.696		31	25		56	72.50	
	1250		8″ diameter		22	.727		42.50	26.50		69	87	
	1260		10″ diameter		21	.762		115	27.50		142.50	169	
	1270		12″ diameter		20	.800		138	29		167	197	
	1280		14″ diameter		18	.889		225	32		257	298	
	1290		16″ diameter		16	1		320	36		356	405	
	1300		18″ diameter		14	1.143		400	41.50		441.50	505	
	1310		20″ diameter	Q-10	17	1.412		540	53		593	675	
	1320		22″ diameter		13	1.846		670	69.50		739.50	845	
	1330		24″ diameter		12	2		790	75		865	980	
	1340		26″ diameter		12	2		1,125	75		1,200	1,350	
	1350		28″ diameter		11	2.182		1,250	82		1,332	1,500	
	1360		30″ diameter		10	2.400		1,375	90		1,465	1,650	
	1370		32″ diameter		10	2.400		1,575	90		1,665	1,900	
	1380		34″ diameter		9	2.667		1,600	100		1,700	1,900	
	1390		36″ diameter		9	2.667		1,900	100		2,000	2,250	
	1400		38″ diameter		8	3		1,900	113		2,013	2,275	
	1410		40″ diameter		7	3.429		2,150	129		2,279	2,575	
	1420		42″ diameter		6	4		2,275	150		2,425	2,725	
	1430		44″ diameter		5	4.800		2,475	180		2,655	3,000	
	1440		46″ diameter		5	4.800		2,725	180		2,905	3,275	
	1450		48″ diameter		5	4.800		2,900	180		3,080	3,475	
	1460		Tee cap, 3″ diameter	Q-9	45	.356		1.63	12.85		14.48	21.50	
	1470		4″ diameter		42	.381		1.85	13.80		15.65	23	
	1480		5″ diameter		40	.400		2.47	14.45		16.92	25	
	1490		6″ diameter		37	.432		2.86	15.65		18.51	27	

Important: See the Reference Section for critical supporting data - Reference Nos., Crews, & City Cost Indexes

15550 | Breechings, Chimneys & Stacks

			CREW	DAILY OUTPUT	LABOR-HOURS	UNIT	2005 BARE COSTS MAT.	LABOR	EQUIP.	TOTAL	TOTAL INCL O&P	
440	1500	7″ diameter	Q-9	35	.457	Ea.	4.65	16.55		21.20	30.50	440
	1510	8″ diameter		34	.471		5.45	17.05		22.50	32	
	1520	10″ diameter		32	.500		10.35	18.10		28.45	39.50	
	1530	12″ diameter		30	.533		13.85	19.30		33.15	45	
	1540	14″ diameter		28	.571		28.50	20.50		49	63.50	
	1550	16″ diameter		25	.640		33	23		56	72	
	1560	18″ diameter		24	.667		37.50	24		61.50	78	
	1570	20″ diameter	Q-10	27	.889		40	33.50		73.50	95.50	
	1580	22″ diameter		22	1.091		46	41		87	114	
	1590	24″ diameter		21	1.143		51.50	43		94.50	123	
	1600	26″ diameter		20	1.200		89.50	45		134.50	168	
	1610	28″ diameter		19	1.263		106	47.50		153.50	189	
	1620	30″ diameter		18	1.333		122	50		172	211	
	1630	32″ diameter		17	1.412		142	53		195	238	
	1640	34″ diameter		16	1.500		158	56.50		214.50	261	
	1650	36″ diameter		15	1.600		176	60		236	286	
	1660	38″ diameter		14	1.714		191	64.50		255.50	310	
	1670	40″ diameter		14	1.714		209	64.50		273.50	330	
	1680	42″ diameter		13	1.846		222	69.50		291.50	350	
	1690	44″ diameter		12	2		240	75		315	380	
	1700	46″ diameter		12	2		247	75		322	385	
	1710	48″ diameter		11	2.182		286	82		368	440	
	1750	Top, 3″ diameter	Q-9	46	.348		8.65	12.60		21.25	29	
	1760	4″ diameter		44	.364		9.05	13.15		22.20	30	
	1770	5″ diameter		42	.381		12.35	13.80		26.15	34.50	
	1780	6″ diameter		40	.400		15.85	14.45		30.30	40	
	1790	7″ diameter		38	.421		23.50	15.25		38.75	49.50	
	1800	8″ diameter		36	.444		30.50	16.10		46.60	58	
	1810	10″ diameter		34	.471		71	17.05		88.05	105	
	1820	12″ diameter		32	.500		107	18.10		125.10	146	
	1830	14″ diameter		30	.533		123	19.30		142.30	165	
	1840	16″ diameter		28	.571		221	20.50		241.50	275	
	1850	18″ diameter		26	.615		390	22.50		412.50	465	
	1860	20″ diameter	Q-10	28	.857		435	32		467	525	
	1870	22″ diameter		22	1.091		590	41		631	715	
	1880	24″ diameter		20	1.200		735	45		780	875	
	1882	26″ diameter		19	1.263		1,100	47.50		1,147.50	1,300	
	1884	28″ diameter		18	1.333		1,275	50		1,325	1,475	
	1886	30″ diameter		17	1.412		1,425	53		1,478	1,625	
	1888	32″ diameter		16	1.500		1,500	56.50		1,556.50	1,725	
	1890	34″ diameter		15	1.600		1,600	60		1,660	1,850	
	1892	36″ diameter		14	1.714		1,700	64.50		1,764.50	1,975	
	1900	Gas, double wall, galvanized steel, oval										
	1904	4″ x 1′	Q-9	68	.235	V.L.F.	13.50	8.50		22	28	
	1906	5″		64	.250		31.50	9.05		40.55	48.50	
	1908	5″/6″ x 1′		60	.267		29	9.65		38.65	47	
	1910	Oval fittings										
	1912	Adjustable length										
	1914	4″ diameter to 12″ lg	Q-9	34	.471	Ea.	12.80	17.05		29.85	40	
	1916	5″ diameter to 12″ lg		32	.500		35.50	18.10		53.60	67	
	1918	5″/6″ diameter to 12″		30	.533		32.50	19.30		51.80	65	
	1920	Elbow 45°										
	1922	4″	Q-9	34	.471	Ea.	23.50	17.05		40.55	51.50	
	1924	5″		32	.500		44.50	18.10		62.60	76.50	
	1926	5″/6″		30	.533		42	19.30		61.30	75.50	
	1930	Elbow 45°, flat										

440

	15550 | Breechings, Chimneys & Stacks	CREW	DAILY OUTPUT	LABOR-HOURS	UNIT	2005 BARE COSTS MAT.	LABOR	EQUIP.	TOTAL	TOTAL INCL O&P
1932	4″	Q-9	34	.471	Ea.	23.50	17.05		40.55	51.50
1934	5″		32	.500		44.50	18.10		62.60	76.50
1936	5″/6″		30	.533		42	19.30		61.30	75.50
1940	Top									
1942	4″	Q-9	44	.364	Ea.	17.40	13.15		30.55	39
1944	5″		42	.381		42.50	13.80		56.30	67.50
1946	5″/6″		40	.400		26.50	14.45		40.95	51.50
1950	Adjustable flashing									
1952	4″	Q-9	34	.471	Ea.	10.75	17.05		27.80	38
1954	5″		32	.500		31	18.10		49.10	62
1956	5″/6″		30	.533		21	19.30		40.30	52.50
1960	Tee									
1962	4″	Q-9	26	.615	Ea.	34.50	22.50		57	72.50
1964	5″		25	.640		73	23		96	116
1966	5″/6″		24	.667		67.50	24		91.50	111
1970	Tee with short snout									
1972	4″	Q-9	26	.615	Ea.	34.50	22.50		57	72.50
3200	All fuel, pressure tight, double wall, U.L. listed, 1400°F.									
3210	304 stainless steel liner, aluminized steel outer jacket									
3220	6″ diameter	Q-9	60	.267	L.F.	36.50	9.65		46.15	55
3221	8″ diameter		52	.308		45	11.15		56.15	66.50
3222	10″ diameter		48	.333		50	12.05		62.05	73.50
3223	12″ diameter		44	.364		57.50	13.15		70.65	83
3224	14″ diameter		42	.381		65.50	13.80		79.30	93
3225	16″ diameter		40	.400		69.50	14.45		83.95	99
3226	18″ diameter		38	.421		78.50	15.25		93.75	110
3227	20″ diameter	Q-10	36	.667		89.50	25		114.50	137
3228	24″ diameter		32	.750		115	28		143	170
3229	28″ diameter		30	.800		132	30		162	191
3230	32″ diameter		27	.889		148	33.50		181.50	215
3231	36″ diameter		25	.960		169	36		205	242
3232	42″ diameter		22	1.091		194	41		235	276
3233	48″ diameter		19	1.263		227	47.50		274.50	325
3260	For 316 stainless steel liner add					6%				
3280	All fuel, pressure tight, double wall fittings									
3284	304 stainless steel inner, aluminized steel jacket									
3288	Adjustable 18″/30″ section									
3291	5″ diameter	Q-9	32	.500	Ea.	94	18.10		112.10	132
3292	6″ diameter		30	.533		119	19.30		138.30	161
3293	8″ diameter		26	.615		137	22.50		159.50	185
3294	10″ diameter		24	.667		154	24		178	206
3295	12″ diameter		22	.727		175	26.50		201.50	233
3296	14″ diameter		21	.762		202	27.50		229.50	265
3297	16″ diameter		20	.800		210	29		239	276
3298	18″ diameter		19	.842		259	30.50		289.50	330
3299	20″ diameter	Q-10	18	1.333		298	50		348	405
3300	24″ diameter		16	1.500		385	56.50		441.50	510
3301	28″ diameter		15	1.600		430	60		490	565
3302	32″ diameter		14	1.714		505	64.50		569.50	655
3303	36″ diameter		12	2		505	75		580	670
3304	42″ diameter		11	2.182		580	82		662	760
3305	48″ diameter		10	2.400		680	90		770	885
3326	For 316 stainless steel liner, add					15%				
3350	Elbow 90° fixed									
3353	5″ diameter	Q-9	32	.500	Ea.	176	18.10		194.10	221
3354	6″ diameter		30	.533		206	19.30		225.30	257

15550 | Breechings, Chimneys & Stacks

	Line	Description	Crew	Daily Output	Labor-Hours	Unit	2005 Bare Costs Mat.	2005 Bare Costs Labor	2005 Bare Costs Equip.	2005 Bare Costs Total	Total Incl O&P	
440	3355	8" diameter	Q-9	26	.615	Ea.	228	22.50		250.50	286	440
	3356	10" diameter		24	.667		262	24		286	325	
	3357	12" diameter		22	.727		296	26.50		322.50	365	
	3358	14" diameter		21	.762		345	27.50		372.50	425	
	3359	16" diameter		20	.800		385	29		414	470	
	3360	18" diameter		19	.842		450	30.50		480.50	540	
	3361	20" diameter	Q-10	18	1.333		505	50		555	630	
	3362	24" diameter		16	1.500		625	56.50		681.50	770	
	3363	28" diameter		15	1.600		780	60		840	950	
	3364	32" diameter		14	1.714		1,025	64.50		1,089.50	1,250	
	3365	36" diameter		12	2		1,225	75		1,300	1,475	
	3366	42" diameter		11	2.182		1,850	82		1,932	2,175	
	3367	48" diameter		10	2.400		2,350	90		2,440	2,725	
	3380	For 316 stainless steel liner, add					20%					
	3400	Elbow 45°										
	3404	6" diameter	Q-9	30	.533	Ea.	130	19.30		149.30	173	
	3405	8" diameter		26	.615		148	22.50		170.50	197	
	3406	10" diameter		24	.667		166	24		190	219	
	3407	12" diameter		22	.727		189	26.50		215.50	249	
	3408	14" diameter		21	.762		217	27.50		244.50	282	
	3409	16" diameter		20	.800		245	29		274	315	
	3410	18" diameter		19	.842		277	30.50		307.50	350	
	3411	20" diameter	Q-10	18	1.333		315	50		365	425	
	3412	24" diameter		16	1.500		405	56.50		461.50	530	
	3413	28" diameter		15	1.600		460	60		520	600	
	3414	32" diameter		14	1.714		615	64.50		679.50	780	
	3415	36" diameter		12	2		870	75		945	1,075	
	3416	42" diameter		11	2.182		1,175	82		1,257	1,425	
	3417	48" diameter		10	2.400		1,600	90		1,690	1,900	
	3430	For 316 stainless steel liner, add					20%					
	3450	Tee 90°										
	3454	6" diameter	Q-9	24	.667	Ea.	136	24		160	187	
	3455	8" diameter		22	.727		157	26.50		183.50	213	
	3456	10" diameter		21	.762		212	27.50		239.50	277	
	3457	12" diameter		20	.800		214	29		243	280	
	3458	14" diameter		18	.889		253	32		285	330	
	3459	16" diameter		16	1		284	36		320	365	
	3460	18" diameter		14	1.143		335	41.50		376.50	435	
	3461	20" diameter	Q-10	17	1.412		375	53		428	495	
	3462	24" diameter		12	2		465	75		540	625	
	3463	28" diameter		11	2.182		580	82		662	765	
	3464	32" diameter		10	2.400		795	90		885	1,025	
	3465	36" diameter		9	2.667		925	100		1,025	1,175	
	3466	42" diameter		6	4		1,475	150		1,625	1,850	
	3467	48" diameter		5	4.800		1,875	180		2,055	2,350	
	3480	For Tee Cap, add					35%	20%				
	3500	For 316 stainless steel liner, add					22%					
	3520	Plate support, galvanized										
	3524	6" diameter	Q-9	26	.615	Ea.	74.50	22.50		97	117	
	3525	8" diameter		22	.727		84	26.50		110.50	133	
	3526	10" diameter		20	.800		92	29		121	146	
	3527	12" diameter		18	.889		95.50	32		127.50	155	
	3528	14" diameter		17	.941		116	34		150	180	
	3529	16" diameter		16	1		122	36		158	190	
	3530	18" diameter		15	1.067		128	38.50		166.50	201	
	3531	20" diameter	Q-10	16	1.500		136	56.50		192.50	237	

15550 | Breechings, Chimneys & Stacks		CREW	DAILY OUTPUT	LABOR-HOURS	UNIT	2005 BARE COSTS MAT.	LABOR	EQUIP.	TOTAL	TOTAL INCL O&P
440	3532 24" diameter	Q-10	14	1.714	Ea.	143	64.50		207.50	257
	3533 28" diameter		13	1.846		176	69.50		245.50	300
	3534 32" diameter		12	2		229	75		304	365
	3535 36" diameter		10	2.400		285	90		375	455
	3536 42" diameter		9	2.667		505	100		605	710
	3537 48" diameter		8	3		610	113		723	845
	3570 Bellows, lined, 316 stainless steel only									
	3574 6" diameter	Q-9	30	.533	Ea.	315	19.30		334.30	375
	3575 8" diameter		26	.615		340	22.50		362.50	410
	3576 10" diameter		24	.667		375	24		399	445
	3577 12" diameter		22	.727		415	26.50		441.50	495
	3578 14" diameter		21	.762		465	27.50		492.50	560
	3579 16" diameter		20	.800		515	29		544	615
	3580 18" diameter		19	.842		575	30.50		605.50	680
	3581 20" diameter	Q-10	18	1.333		650	50		700	790
	3582 24" diameter		16	1.500		815	56.50		871.50	980
	3583 28" diameter		15	1.600		915	60		975	1,100
	3584 32" diameter		14	1.714		1,050	64.50		1,114.50	1,250
	3585 36" diameter		12	2		1,200	75		1,275	1,450
	3586 42" diameter		11	2.182		1,975	82		2,057	2,300
	3587 48" diameter		10	2.400		2,475	90		2,565	2,875
	3600 Ventilated roof thimble, 304 stainless steel									
	3620 6" diameter	Q-9	26	.615	Ea.	183	22.50		205.50	236
	3624 8" diameter		22	.727		196	26.50		222.50	257
	3625 10" diameter		20	.800		212	29		241	279
	3626 12" diameter		18	.889		229	32		261	300
	3627 14" diameter		17	.941		247	34		281	325
	3628 16" diameter		16	1		262	36		298	345
	3629 18" diameter		15	1.067		282	38.50		320.50	370
	3630 20" diameter	Q-10	16	1.500		300	56.50		356.50	420
	3631 24" diameter		14	1.714		340	64.50		404.50	475
	3632 28" diameter		13	1.846		390	69.50		459.50	535
	3633 32" diameter		12	2		440	75		515	600
	3634 36" diameter		10	2.400		505	90		595	695
	3635 42" diameter		9	2.667		670	100		770	890
	3636 48" diameter		8	3		845	113		958	1,100
	3650 For 316 stainless steel, add					10%				
	3670 Exit cone, 316 stainless steel only									
	3674 6" diameter	Q-9	46	.348	Ea.	129	12.60		141.60	161
	3675 8" diameter		42	.381		136	13.80		149.80	170
	3676 10" diameter		40	.400		146	14.45		160.45	184
	3677 12" diameter		38	.421		154	15.25		169.25	193
	3678 14" diameter		37	.432		170	15.65		185.65	211
	3679 16" diameter		36	.444		188	16.10		204.10	232
	3680 18" diameter		35	.457		231	16.55		247.55	280
	3681 20" diameter	Q-10	28	.857		275	32		307	350
	3682 24" diameter		26	.923		355	34.50		389.50	445
	3683 28" diameter		25	.960		430	36		466	530
	3684 32" diameter		24	1		535	37.50		572.50	645
	3685 36" diameter		22	1.091		640	41		681	770
	3686 42" diameter		21	1.143		820	43		863	965
	3687 48" diameter		20	1.200		1,050	45		1,095	1,250
	3720 Roof support assy., incl. 30" pipe sect, 304 st. st.									
	3724 6" diameter	Q-9	25	.640	Ea.	211	23		234	268
	3725 8" diameter		21	.762		230	27.50		257.50	296
	3726 10" diameter		19	.842		252	30.50		282.50	325

15550 | Breechings, Chimneys & Stacks

				DAILY	LABOR-		2005 BARE COSTS				TOTAL	
			CREW	OUTPUT	HOURS	UNIT	MAT.	LABOR	EQUIP.	TOTAL	INCL O&P	
440	3727	12″ diameter	Q-9	17	.941	Ea.	268	34		302	345	440
	3728	14″ diameter		16	1		300	36		336	385	
	3729	16″ diameter		15	1.067		320	38.50		358.50	410	
	3730	18″ diameter		14	1.143		345	41.50		386.50	445	
	3731	20″ diameter	Q-10	15	1.600		365	60		425	495	
	3732	24″ diameter		13	1.846		405	69.50		474.50	555	
	3733	28″ diameter		12	2		480	75		555	640	
	3734	32″ diameter		11	2.182		575	82		657	760	
	3735	36″ diameter		9	2.667		695	100		795	920	
	3736	42″ diameter		8	3		1,100	113		1,213	1,375	
	3737	48″ diameter		7	3.429		1,350	129		1,479	1,675	
	3750	For 316 stainless steel, add					20%					
	3770	Stack cap, 316 stainless steel only										
	3773	5″ diameter	Q-9	48	.333	Ea.	123	12.05		135.05	154	
	3774	6″ diameter		46	.348		144	12.60		156.60	177	
	3775	8″ diameter		42	.381		167	13.80		180.80	204	
	3776	10″ diameter		40	.400		196	14.45		210.45	239	
	3777	12″ diameter		38	.421		227	15.25		242.25	274	
	3778	14″ diameter		37	.432		268	15.65		283.65	320	
	3779	16″ diameter		36	.444		305	16.10		321.10	360	
	3780	18″ diameter		35	.457		345	16.55		361.55	405	
	3781	20″ diameter	Q-10	28	.857		395	32		427	485	
	3782	24″ diameter		26	.923		475	34.50		509.50	580	
	3783	28″ diameter		25	.960		520	36		556	625	
	3784	32″ diameter		24	1		595	37.50		632.50	715	
	3785	36″ diameter		22	1.091		730	41		771	865	
	3786	42″ diameter		21	1.143		900	43		943	1,050	
	3787	48″ diameter		20	1.200		1,050	45		1,095	1,225	
	3800	Round storm collar										
	3806	5″ diameter	Q-9	32	.500	Ea.	21.50	18.10		39.60	52	
	7800	All fuel, double wall, stainless steel, 6″ diameter		60	.267	V.L.F.	32.50	9.65		42.15	50.50	
	7802	7″ diameter		56	.286		42	10.35		52.35	62	
	7804	8″ diameter		52	.308		49	11.15		60.15	71	
	7806	10″ diameter		48	.333		71.50	12.05		83.55	97	
	7808	12″ diameter		44	.364		96	13.15		109.15	125	
	7810	14″ diameter		42	.381		126	13.80		139.80	159	
	8000	All fuel, double wall, stainless steel fittings										
	8010	Roof support 6″ diameter	Q-9	30	.533	Ea.	80.50	19.30		99.80	118	
	8020	7″ diameter		28	.571		91	20.50		111.50	132	
	8030	8″ diameter		26	.615		102	22.50		124.50	147	
	8040	10″ diameter		24	.667		117	24		141	166	
	8050	12″ diameter		22	.727		141	26.50		167.50	196	
	8060	14″ diameter		21	.762		179	27.50		206.50	239	
	8100	Elbow 15°, 6″ diameter		30	.533		129	19.30		148.30	172	
	8120	7″ diameter		28	.571		145	20.50		165.50	192	
	8140	8″ diameter		26	.615		165	22.50		187.50	217	
	8160	10″ diameter		24	.667		216	24		240	275	
	8180	12″ diameter		22	.727		262	26.50		288.50	330	
	8200	14″ diameter		21	.762		315	27.50		342.50	390	
	8300	Insulated tee with insulated tee cap, 6″ diameter		30	.533		122	19.30		141.30	164	
	8340	7″ diameter		28	.571		160	20.50		180.50	208	
	8360	8″ diameter		26	.615		180	22.50		202.50	233	
	8380	10″ diameter		24	.667		253	24		277	315	
	8400	12″ diameter		22	.727		355	26.50		381.50	430	
	8420	14″ diameter		21	.762		465	27.50		492.50	560	
	8500	Joist shield, 6″ diameter		30	.533		37	19.30		56.30	70.50	

15550 | Breechings, Chimneys & Stacks		CREW	DAILY OUTPUT	LABOR-HOURS	UNIT	2005 BARE COSTS MAT.	LABOR	EQUIP.	TOTAL	TOTAL INCL O&P
440 8510	7″ diameter	Q-9	28	.571	Ea.	40.50	20.50		61	76.50 440
8520	8″ diameter		26	.615		49.50	22.50		72	89
8530	10″ diameter		24	.667		67	24		91	111
8540	12″ diameter		22	.727		84	26.50		110.50	133
8550	14″ diameter		21	.762		104	27.50		131.50	157
8600	Round top, 6″ diameter		30	.533		41	19.30		60.30	75
8620	7″ diameter		28	.571		55	20.50		75.50	92.50
8640	8″ diameter		26	.615		115	22.50		137.50	161
8660	10″ diameter		24	.667		136	24		160	186
8680	12″ diameter		22	.727		194	26.50		220.50	254
8700	14″ diameter		21	.762		259	27.50		286.50	330
8800	Adjustable roof flashing, 6″ diameter		30	.533		49	19.30		68.30	83.50
8820	7″ diameter		28	.571		56	20.50		76.50	93.50
8840	8″ diameter		26	.615		61	22.50		83.50	102
8860	10″ diameter		24	.667		78.50	24		102.50	123
8880	12″ diameter		22	.727		101	26.50		127.50	152
8900	14″ diameter		21	.762		126	27.50		153.50	182
9000	High temp. (2000° F), steel jacket, acid resist. refractory lining									
9010	11 ga. galvanized jacket, U.L. listed									
9020	Straight section, 48″ long, 10″ diameter	Q-10	13.30	1.805	Ea.	355	67.50		422.50	495
9030	12″ diameter		11.20	2.143		385	80.50		465.50	550
9040	18″ diameter		7.40	3.243		545	122		667	785
9050	24″ diameter		4.60	5.217		745	196		941	1,125
9060	30″ diameter		3.70	6.486		950	243		1,193	1,425
9070	36″ diameter		2.70	8.889		1,150	335		1,485	1,800
9080	42″ diameter	Q-11	3.10	10.323		1,450	395		1,845	2,175
9090	48″ diameter		2.70	11.852		1,700	455		2,155	2,575
9100	54″ diameter		2.20	14.545		2,050	555		2,605	3,125
9110	60″ diameter		2	16		2,700	615		3,315	3,925
9120	Tee section, 10″ diameter	Q-10	4.40	5.455		755	205		960	1,150
9130	12″ diameter		3.70	6.486		795	243		1,038	1,250
9140	18″ diameter		2.40	10		1,075	375		1,450	1,750
9150	24″ diameter		1.50	16		1,475	600		2,075	2,550
9160	30″ diameter		1.20	20		2,100	750		2,850	3,450
9170	36″ diameter		.80	30		3,525	1,125		4,650	5,600
9180	42″ diameter	Q-11	1	32		4,625	1,225		5,850	6,975
9190	48″ diameter		.90	35.556		6,000	1,350		7,350	8,700
9200	54″ diameter		.70	45.714		8,750	1,750		10,500	12,300
9210	60″ diameter		.60	53.333		10,600	2,050		12,650	14,800
9220	Cleanout pier section, 10″ diameter	Q-10	3.50	6.857		580	257		837	1,025
9230	12″ diameter		2.50	9.600		610	360		970	1,225
9240	18″ diameter		1.90	12.632		780	475		1,255	1,600
9250	24″ diameter		1.30	18.462		990	695		1,685	2,175
9260	30″ diameter		1	24		1,200	900		2,100	2,700
9270	36″ diameter		.75	32		1,425	1,200		2,625	3,425
9280	42″ diameter	Q-11	.90	35.556		1,825	1,350		3,175	4,100
9290	48″ diameter		.75	42.667		2,175	1,625		3,800	4,925
9300	54″ diameter		.60	53.333		2,625	2,050		4,675	6,025
9310	60″ diameter		.50	64		3,325	2,450		5,775	7,450
9320	For drain, add					53%				
9330	Elbow, 30° and 45°, 10″ diameter	Q-10	6.60	3.636		495	136		631	750
9340	12″ diameter		5.60	4.286		530	161		691	825
9350	18″ diameter		3.70	6.486		795	243		1,038	1,250
9360	24″ diameter		2.30	10.435		1,200	390		1,590	1,900
9370	30″ diameter		1.85	12.973		1,625	485		2,110	2,550
9380	36″ diameter		1.30	18.462		1,975	695		2,670	3,250

15550 Breechings, Chimneys & Stacks		CREW	DAILY OUTPUT	LABOR-HOURS	UNIT	2005 BARE COSTS MAT.	LABOR	EQUIP.	TOTAL	TOTAL INCL O&P
440	9390 42" diameter	Q-11	1.60	20	Ea.	2,700	765		3,465	4,150
	9400 48" diameter		1.35	23.704		3,450	910		4,360	5,200
	9410 54" diameter		1.10	29.091		4,475	1,125		5,600	6,625
	9420 60" diameter	↓	1	32		5,450	1,225		6,675	7,850
	9430 For 60° and 90° elbow, add					112%				
	9440 End cap, 10" diameter	Q-10	26	.923		845	34.50		879.50	985
	9450 12" diameter		22	1.091		965	41		1,006	1,125
	9460 18" diameter		15	1.600		1,200	60		1,260	1,425
	9470 24" diameter		9	2.667		1,475	100		1,575	1,750
	9480 30" diameter		7.30	3.288		1,800	123		1,923	2,200
	9490 36" diameter	↓	5	4.800		2,100	180		2,280	2,575
	9500 42" diameter	Q-11	6.30	5.079		2,600	195		2,795	3,150
	9510 48" diameter		5.30	6.038		3,075	231		3,306	3,725
	9520 54" diameter		4.40	7.273		3,500	279		3,779	4,275
	9530 60" diameter	↓	3.90	8.205		3,925	315		4,240	4,775
	9540 Increaser (1 diameter), 10" diameter	Q-10	6.60	3.636		420	136		556	670
	9550 12" diameter		5.60	4.286		450	161		611	740
	9560 18" diameter		3.70	6.486		655	243		898	1,100
	9570 24" diameter		2.30	10.435		940	390		1,330	1,625
	9580 30" diameter		1.85	12.973		1,150	485		1,635	2,025
	9590 36" diameter	↓	1.30	18.462		1,275	695		1,970	2,475
	9592 42" diameter	Q-11	1.60	20		1,400	765		2,165	2,725
	9594 48" diameter		1.35	23.704		1,700	910		2,610	3,275
	9596 54" diameter	↓	1.10	29.091	↓	2,050	1,125		3,175	3,975
	9600 For expansion joints, add to straight section					7%				
	9610 For 1/4" hot rolled steel jacket, add					157%				
	9620 For 2950°F very high temperature, add					89%				
	9630 26 ga. aluminized jacket, straight section, 48" long									
	9640 10" diameter	Q-10	15.30	1.569	V.L.F.	210	59		269	320
	9650 12" diameter		12.90	1.860		223	70		293	350
	9660 18" diameter		8.50	2.824		325	106		431	520
	9670 24" diameter		5.30	4.528		460	170		630	765
	9680 30" diameter		4.30	5.581		590	209		799	970
	9690 36" diameter	↓	3.10	7.742	↓	745	290		1,035	1,275
	9700 Accessories (all models)									
	9710 Guy band, 10" diameter	Q-10	32	.750	Ea.	65	28		93	115
	9720 12" diameter		30	.800		69	30		99	122
	9730 18" diameter		26	.923		89.50	34.50		124	152
	9740 24" diameter		24	1		132	37.50		169.50	203
	9750 30" diameter		20	1.200		152	45		197	238
	9760 36" diameter	↓	18	1.333		169	50		219	263
	9770 42" diameter	Q-11	22	1.455		305	55.50		360.50	425
	9780 48" diameter		20	1.600		335	61.50		396.50	465
	9790 54" diameter		16	2		370	76.50		446.50	530
	9800 60" diameter	↓	12	2.667		400	102		502	600
	9810 Draw band, galv. stl., 11 gauge, 10" diameter	Q-10	32	.750		57	28		85	106
	9820 12" diameter		30	.800		61	30		91	113
	9830 18" diameter		26	.923		76	34.50		110.50	138
	9840 24" diameter		24	1		96.50	37.50		134	164
	9850 30" diameter		20	1.200		113	45		158	194
	9860 36" diameter	↓	18	1.333		129	50		179	219
	9870 42" diameter	Q-11	22	1.455		173	55.50		228.50	276
	9880 48" diameter		20	1.600		222	61.50		283.50	340
	9890 54" diameter		16	2		258	76.50		334.50	400
	9900 60" diameter	↓	12	2.667		325	102		427	515
	9910 Draw band, aluminized stl., 26 gauge, 10" diameter	Q-10	32	.750	↓	16.25	28		44.25	61.50

15550 | Breechings, Chimneys & Stacks

			CREW	DAILY OUTPUT	LABOR-HOURS	UNIT	2005 BARE COSTS MAT.	LABOR	EQUIP.	TOTAL	TOTAL INCL O&P	
440	9920	12" diameter	Q-10	30	.800	Ea.	16.25	30		46.25	64	440
	9930	18" diameter		26	.923		20.50	34.50		55	76	
	9940	24" diameter		24	1		24.50	37.50		62	84.50	
	9950	30" diameter		20	1.200		28.50	45		73.50	101	
	9960	36" diameter		18	1.333		32.50	50		82.50	113	
	9962	42" diameter	Q-11	22	1.455		36.50	55.50		92	126	
	9964	48" diameter		20	1.600		41.50	61.50		103	141	
	9966	54" diameter		16	2		44.50	76.50		121	167	
	9968	60" diameter		12	2.667		49	102		151	211	
600	0010	**INDUCED DRAFT FANS**										600
	1000	Breeching installation										
	1800	Hot gas, 600°F, variable pitch pulley and motor										
	1840	6" diam. inlet, 1/4 H.P., 1 phase, 400 CFM	Q-9	6	2.667	Ea.	1,025	96.50		1,121.50	1,275	
	1850	7" diam. inlet, 1/4 H.P., 1 phase, 800 CFM		5	3.200		1,100	116		1,216	1,375	
	1860	8" diam. inlet, 1/4 H.P., 1phase, 1120 CFM		4	4		1,600	145		1,745	1,975	
	1870	9" diam. inlet, 3/4 H.P., 1 phase, 1440 CFM		3.60	4.444		1,875	161		2,036	2,325	
	1880	10" diam. inlet, 3/4 H.P., 1 phase, 2000 CFM		3.30	4.848		2,125	175		2,300	2,600	
	1900	12" diam. inlet, 3/4 H.P., 3 phase, 2960 CFM		3	5.333		2,175	193		2,368	2,700	
	1910	14" diam. inlet, 1 H.P., 3 phase, 4160 CFM		2.60	6.154		2,575	223		2,798	3,175	
	1920	16" diam. inlet, 2 H.P., 3 phase, 6720 CFM		2.30	6.957		2,800	252		3,052	3,450	
	1940	18" diam. inlet, 3 H.P., 3 phase, 9120 CFM		2	8		3,375	289		3,664	4,175	
	1950	20" diam. inlet, 3 H.P., 3 phase, 9760 CFM		1.50	10.667		3,675	385		4,060	4,650	
	1960	22" diam. inlet, 5 H.P., 3 phase, 13,360 CFM		1	16		4,325	580		4,905	5,650	
	1980	24" diam. inlet, 7-1/2 H.P., 3 phase, 17,760 CFM		.80	20		5,300	725		6,025	6,975	
	2300	For multi-blade damper at fan inlet, add					20%					
	3600	Chimneytop installation										
	3700	6" size	1 Shee	8	1	Ea.	495	40		535	605	
	3740	8" size		7	1.143		500	46		546	620	
	3780	13" size		6	1.333		680	53.50		733.50	835	
	3880	For speed control switch, add					41			41	45	
	3920	For thermal fan control, add					41			41	45	
	5500	Flue shutter damper for draft control,										
	5510	parallel blades										
	5550	8" size	Q-9	8	2	Ea.	360	72.50		432.50	505	
	5560	9" size		7.50	2.133		390	77		467	545	
	5570	10" size		7	2.286		400	82.50		482.50	565	
	5580	12" size		6.50	2.462		440	89		529	615	
	5590	14" size		6	2.667		500	96.50		596.50	700	
	5600	16" size		5.50	2.909		540	105		645	755	
	5610	18" size		5	3.200		595	116		711	835	
	5620	20" size		4.50	3.556		655	129		784	925	
	5630	22" size		4	4		725	145		870	1,025	
	5640	24" size		3.50	4.571		790	165		955	1,125	
	5650	27" size		3	5.333		810	193		1,003	1,175	
	5660	30" size		2.50	6.400		840	232		1,072	1,275	
	5670	32" size		2	8		880	289		1,169	1,425	
	5680	36" size		1.50	10.667		950	385		1,335	1,650	

Important: See the Reference Section for critical supporting data - Reference Nos., Crews, & City Cost Indexes

15600 | Refrigeration Equipment

15610 | Refrigeration Compressors

		15610 Refrigeration Compressors	CREW	DAILY OUTPUT	LABOR-HOURS	UNIT	2005 BARE COSTS MAT.	LABOR	EQUIP.	TOTAL	TOTAL INCL O&P	
400	0010	ROTARY/RECIPROCATING REFRIGERANT COMPRESSOR										400
	0100	Refrigeration, hermetic, switches and protective devices										
	0200	High back pressure types for R-22										
	0210	1.25 ton	1 Stpi	3	2.667	Ea.	340	109		449	540	
	0220	1.42 ton		3	2.667		350	109		459	550	
	0230	1.68 ton		2.80	2.857		395	117		512	610	
	0240	2.00 ton		2.60	3.077		485	126		611	725	
	0250	2.37 ton		2.50	3.200		515	131		646	760	
	0260	2.67 ton		2.40	3.333		565	137		702	825	
	0270	3.53 ton		2.30	3.478		590	142		732	865	
	0280	4.43 ton		2.20	3.636		625	149		774	915	
	0290	5.08 ton		2.10	3.810		775	156		931	1,075	
	0300	5.22 ton		2	4		775	164		939	1,100	
	0310	7.31 ton	Q-5	3	5.333		1,525	197		1,722	1,975	
	0320	9.95 ton		2.60	6.154		1,675	227		1,902	2,175	
	0330	11.6 ton		2.50	6.400		1,725	236		1,961	2,250	
	0340	15.25 ton		2.30	6.957		2,225	256		2,481	2,800	
	0350	17.7 ton		2.10	7.619		2,475	281		2,756	3,150	
	0990	Refrigeration, recip. hermetic, switches & protective devices										
	1000	(Ratings are ARI standard 515 group IV using R-22) 10 ton	Q-5	1	16	Ea.	6,350	590		6,940	7,875	
	1100	20 ton	Q-6	.72	33.333		9,525	1,275		10,800	12,400	
	1200	30 ton		.64	37.500		11,500	1,425		12,925	14,900	
	1300	40 ton		.44	54.545		12,400	2,075		14,475	16,700	
	1400	50 ton		.20	120		13,100	4,575		17,675	21,300	
	1500	75 ton	Q-7	.27	118		14,500	4,625		19,125	22,900	
	1600	130 ton	"	.21	152		16,500	5,950		22,450	27,100	
800	0010	SCROLL REFRIGERANT COMPRESSORS										800
	1800	Refrigeration, scroll type										
	1810	1.9 ton	1 Stpi	2.70	2.963	Ea.	425	121		546	650	
	1820	2.35 ton		2.50	3.200		490	131		621	730	
	1830	2.82 ton		2.35	3.404		525	139		664	785	
	1840	3.3 ton		2.30	3.478		550	142		692	820	
	1850	3.83 ton		2.25	3.556		630	146		776	915	
	1860	4.1 ton		2.20	3.636		730	149		879	1,025	
	1870	4.8 ton		2.10	3.810		750	156		906	1,050	
	1880	5 ton		2.07	3.865		760	158		918	1,075	

15620 | Packaged Water Chillers

		15620 Packaged Water Chillers	CREW	DAILY OUTPUT	LABOR-HOURS	UNIT	MAT.	LABOR	EQUIP.	TOTAL	TOTAL INCL O&P	
100	0010	ABSORPTION WATER CHILLERS										100
	0020	Steam or hot water, water cooled										
	0050	100 ton	Q-7	.13	240	Ea.	119,000	9,375		128,375	145,000	
	0100	148 ton		.12	258		122,000	10,100		132,100	149,000	
	0200	200 ton		.12	275		135,500	10,800		146,300	165,500	
	0240	250 ton		.11	290		149,500	11,400		160,900	181,500	
	0300	354 ton		.10	304		186,500	11,900		198,400	223,000	
	0400	420 ton		.10	323		204,000	12,600		216,600	243,500	
	0500	665 ton		.09	340		286,500	13,300		299,800	335,000	
	0600	750 ton		.09	363		334,500	14,200		348,700	389,500	
	0700	850 ton		.08	385		357,000	15,000		372,000	415,000	
	0800	955 ton		.08	410		381,500	16,000		397,500	444,000	
	0900	1125 ton		.08	421		440,500	16,400		456,900	509,500	
	1000	1250 ton		.07	444		474,000	17,300		491,300	547,500	
	1100	1465 ton		.07	463		562,000	18,100		580,100	645,000	
	1200	1660 ton		.07	477		661,000	18,600		679,600	755,000	
	2000	For two stage unit, add					80%	25%				
	3000	Gas fired, air cooled										

	15620	Packaged Water Chillers	CREW	DAILY OUTPUT	LABOR-HOURS	UNIT	2005 BARE COSTS MAT.	LABOR	EQUIP.	TOTAL	TOTAL INCL O&P	
100	3180	3 ton	Q-5	1.30	12.308	Ea.	6,625	455		7,080	7,975	100
	3220	5 ton		.60	26.667		7,900	985		8,885	10,200	
	3270	10 ton		.40	40		22,300	1,475		23,775	26,800	
	4000	Water cooled, duplex										
	4130	100 ton	Q-7	.13	246	Ea.	129,000	9,600		138,600	156,500	
	4140	200 ton		.11	283		179,000	11,100		190,100	213,000	
	4150	300 ton		.11	299		207,000	11,700		218,700	245,000	
	4160	400 ton		.10	316		269,000	12,400		281,400	314,500	
	4170	500 ton		.10	329		338,500	12,900		351,400	392,000	
	4180	600 ton		.09	340		400,000	13,300		413,300	460,000	
	4190	700 ton		.09	359		443,500	14,000		457,500	509,000	
	4200	800 ton		.08	380		535,500	14,900		550,400	611,500	
	4210	900 ton		.08	400		602,500	15,600		618,100	686,000	
	4220	1000 ton		.08	421		669,500	16,400		685,900	760,500	
600	0010	**CENTRIFUGAL/SCREW/RECIP. WATER CHILLERS,** W/ standard controls										600
	0020	Centrifugal liquid chiller, water cooled										
	0030	not including water tower										
	0100	Open drive, 2000 ton	Q-7	.07	477	Ea.	537,000	18,600		555,600	618,500	
	0110	Screw, liquid chiller, air cooled, insulated evaporator										
	0120	130 ton	Q-7	.14	228	Ea.	61,000	8,925		69,925	80,500	
	0124	160 ton		.13	246		75,000	9,600		84,600	97,000	
	0128	180 ton		.13	250		84,500	9,750		94,250	107,000	
	0132	210 ton		.12	258		93,000	10,100		103,100	117,000	
	0136	270 ton		.12	266		106,000	10,400		116,400	132,000	
	0140	320 ton		.12	275		133,000	10,800		143,800	162,500	
	0200	Packaged unit, water cooled, not incl. tower										
	0210	80 ton	Q-7	.14	223	Ea.	34,000	8,725		42,725	50,500	
	0220	100 ton		.14	230		39,600	8,975		48,575	57,000	
	0230	150 ton		.13	240		56,500	9,400		65,900	76,500	
	0240	200 ton		.13	251		65,500	9,825		75,325	87,000	
	0250	250 ton		.12	260		69,500	10,200		79,700	92,000	
	0260	300 ton		.12	266		79,500	10,400		89,900	102,500	
	0270	350 ton		.12	275		111,000	10,800		121,800	138,000	
	0274	Centrifugal, packaged unit, water cooled, not incl. tower										
	0280	400 ton	Q-7	.11	283	Ea.	122,000	11,100		133,100	150,500	
	0282	450 ton		.11	290		136,000	11,400		147,400	167,000	
	0290	500 ton		.11	296		158,000	11,600		169,600	191,500	
	0292	550 ton		.10	304		174,000	11,900		185,900	209,500	
	0300	600 ton		.10	310		187,000	12,100		199,100	223,500	
	0302	650 ton		.10	320		202,500	12,500		215,000	242,000	
	0304	700 ton		.10	326		218,000	12,700		230,700	259,000	
	0306	750 ton		.10	333		233,000	13,000		246,000	275,500	
	0310	800 ton		.09	340		243,500	13,300		256,800	288,000	
	0312	850 ton		.09	351		263,000	13,700		276,700	310,000	
	0316	900 ton		.09	359		289,500	14,000		303,500	339,500	
	0318	950 ton		.09	363		305,500	14,200		319,700	358,000	
	0320	1000 ton		.09	372		319,500	14,500		334,000	373,500	
	0324	1100 ton		.08	385		354,000	15,000		369,000	412,000	
	0330	1200 ton		.08	395		386,000	15,400		401,400	448,000	
	0340	1300 ton		.08	410		410,500	16,000		426,500	476,000	
	0350	1400 ton		.08	421		450,500	16,400		466,900	520,000	
	0360	1500 ton		.07	426		471,500	16,700		488,200	543,500	
	0370	1600 ton		.07	438		512,000	17,100		529,100	588,500	
	0380	1700 ton		.07	450		544,000	17,600		561,600	624,500	
	0390	1800 ton		.07	457		576,000	17,800		593,800	660,500	
	0400	1900 ton		.07	470		608,000	18,400		626,400	696,000	

Important: See the Reference Section for critical supporting data - Reference Nos., Crews, & City Cost Indexes

	15620 | Packaged Water Chillers	CREW	DAILY OUTPUT	LABOR-HOURS	UNIT	2005 BARE COSTS MAT.	LABOR	EQUIP.	TOTAL	TOTAL INCL O&P	
600	0490 Reciprocating, packaged w/integral air cooled condenser, 15 ton cool	Q-7	.37	86.486	Ea.	13,300	3,375		16,675	19,700	600
	0500 20 ton cooling		.34	94.955		17,600	3,700		21,300	25,000	
	0510 25 ton cooling		.33	97.859		18,400	3,825		22,225	26,100	
	0515 30 ton cooling		.31	101		19,800	3,975		23,775	27,800	
	0517 35 ton cooling		.31	104		22,100	4,100		26,200	30,500	
	0520 40 ton cooling		.30	108		23,900	4,225		28,125	32,700	
	0528 45 ton cooling		.29	109		25,400	4,300		29,700	34,400	
	0536 50 ton cooling		.28	113		27,800	4,425		32,225	37,200	
	0538 60 ton cooling		.28	115		34,600	4,500		39,100	44,800	
	0540 65 ton cooling		.27	118		34,800	4,600		39,400	45,200	
	0546 70 ton cooling		.27	119		39,600	4,675		44,275	50,500	
	0552 80 ton cooling		.26	123		42,400	4,800		47,200	54,000	
	0554 90 ton cooling		.25	125		46,000	4,925		50,925	58,000	
	0600 100 ton cooling		.25	129		51,500	5,025		56,525	64,000	
	0620 110 ton cooling		.24	132		56,000	5,150		61,150	69,500	
	0630 130 ton cooling		.24	135		62,000	5,275		67,275	76,000	
	0640 150 ton cooling		.23	137		71,500	5,375		76,875	87,000	
	0650 175 ton cooling		.23	140		76,500	5,500		82,000	92,500	
	0654 190 ton cooling		.22	144		86,500	5,625		92,125	103,500	
	0660 210 ton cooling		.22	148		94,500	5,775		100,275	112,500	
	0662 240 ton cooling		.21	151		113,500	5,925		119,425	134,000	
	0664 270 ton cooling		.20	156		127,500	6,100		133,600	149,500	
	0666 300 ton cooling		.20	160		138,000	6,250		144,250	161,000	
	0668 330 ton cooling		.19	164		153,000	6,400		159,400	177,500	
	0670 360 ton cooling		.19	168		173,000	6,575		179,575	200,000	
	0672 390 ton cooling		.18	172		181,500	6,750		188,250	209,500	
	0674 420 ton cooling		.18	177		189,500	6,950		196,450	219,000	
	0680 Water cooled, single compressor, semi-hermetic, tower not incl.										
	0700 2 to 5 ton cooling	Q-5	.57	28.021	Ea.	6,375	1,025		7,400	8,575	
	0720 6 ton cooling		.42	38.005		9,400	1,400		10,800	12,500	
	0740 8 ton cooling		.31	52.117		10,100	1,925		12,025	14,000	
	0760 10 ton cooling	Q-6	.36	67.039		10,300	2,550		12,850	15,200	
	0780 15 ton cooling	″	.33	72.727		10,700	2,775		13,475	16,000	
	0800 20 ton cooling	Q-7	.38	83.989		11,300	3,275		14,575	17,400	
	0820 30 ton cooling		.33	96.096		12,900	3,750		16,650	19,800	
	0840 35 ton cooling		.33	97.859		14,000	3,825		17,825	21,200	
	0980 Water cooled, multiple compress., semi-hermetic, tower not incl.										
	1000 15 ton cooling	Q-6	.36	65.934	Ea.	11,000	2,525		13,525	15,900	
	1020 20 ton cooling	Q-7	.41	78.049		12,000	3,050		15,050	17,800	
	1040 25 ton cooling		.36	89.888		13,500	3,500		17,000	20,100	
	1060 30 ton cooling		.31	101		14,200	3,975		18,175	21,600	
	1080 40 ton cooling		.30	108		16,500	4,225		20,725	24,500	
	1100 50 ton cooling		.28	113		25,400	4,450		29,850	34,700	
	1120 60 ton cooling		.25	125		28,000	4,925		32,925	38,200	
	1130 70 ton cooling		.23	139		37,900	5,425		43,325	49,800	
	1140 80 ton cooling		.21	151		39,500	5,925		45,425	52,500	
	1150 90 ton cooling		.19	164		42,400	6,450		48,850	56,500	
	1160 100 ton cooling		.18	179		45,300	7,025		52,325	60,500	
	1170 110 ton cooling		.17	190		47,800	7,425		55,225	63,500	
	1180 120 ton cooling		.16	196		51,500	7,650		59,150	68,000	
	1200 140 ton cooling		.16	202		60,500	7,900		68,400	78,500	
	1210 160 ton cooling		.15	210		63,500	8,225		71,725	82,500	
	1300 Water cooled, single compressor, direct drive, tower not incl.										
	1320 40 ton cooling,	Q-7	.30	108	Ea.	23,700	4,225		27,925	32,500	
	1340 50 ton cooling		.28	113		28,800	4,450		33,250	38,400	
	1360 60 ton cooling		.25	125		31,400	4,925		36,325	41,900	

15620 | Packaged Water Chillers

			CREW	DAILY OUTPUT	LABOR-HOURS	UNIT	2005 BARE COSTS MAT.	LABOR	EQUIP.	TOTAL	TOTAL INCL O&P	
600	1380	80 ton cooling	Q-7	.21	151	Ea.	38,600	5,925		44,525	51,500	600
	1450	Water cooled, dual compressors, direct drive, tower not incl.										
	1500	80 ton cooling	Q-7	.14	222	Ea.	47,800	8,675		56,475	65,500	
	1520	100 ton cooling		.14	228		55,500	8,925		64,425	74,500	
	1540	120 ton cooling		.14	231		61,500	9,050		70,550	81,000	
	1560	135 ton cooling		.14	235		70,000	9,175		79,175	91,000	
	1580	150 ton cooling		.13	240		74,500	9,400		83,900	96,000	
	1600	175 ton cooling		.13	244		85,500	9,525		95,025	108,500	
	1620	200 ton cooling		.13	250		91,000	9,750		100,750	114,500	
	1640	225 ton cooling		.13	256		99,000	10,000		109,000	124,000	
	1660	250 ton cooling		.12	260		105,000	10,200		115,200	131,000	
	4000	Packaged chiller, remote air cooled condensers not incl.										
	4020	15 ton cooling	Q-7	.30	108	Ea.	11,000	4,225		15,225	18,500	
	4030	20 ton cooling		.28	115		12,000	4,525		16,525	20,000	
	4040	25 ton cooling		.25	125		13,500	4,925		18,425	22,200	
	4050	30 ton cooling		.24	133		14,200	5,225		19,425	23,500	
	4060	40 ton cooling		.22	144		16,500	5,625		22,125	26,600	
	4070	50 ton cooling		.21	153		20,200	6,000		26,200	31,200	
	4080	60 ton cooling		.19	164		23,500	6,400		29,900	35,500	
	4090	70 ton cooling		.18	173		37,900	6,800		44,700	52,000	
	4100	80 ton cooling		.17	183		39,500	7,175		46,675	54,500	
	4110	90 ton cooling		.16	193		42,400	7,575		49,975	58,000	
	4120	100 ton cooling		.16	203		45,300	7,950		53,250	62,000	
	4130	110 ton cooling		.15	213		47,800	8,325		56,125	65,000	
	4140	120 ton cooling		.14	223		51,500	8,725		60,225	69,500	
	4150	140 ton cooling		.14	233		60,500	9,125		69,625	80,000	
	8000	Direct expansion, shell and tube type, for built up systems										
	8020	1 ton	Q-5	2	8	Ea.	4,875	295		5,170	5,800	
	8030	5 ton		1.90	8.421		8,075	310		8,385	9,350	
	8040	10 ton		1.70	9.412		10,000	345		10,345	11,500	
	8050	20 ton		1.50	10.667		11,400	395		11,795	13,200	
	8060	30 ton		1	16		14,600	590		15,190	16,900	
	8070	50 ton		.90	17.778		23,400	655		24,055	26,700	
	8080	100 ton	Q-6	.90	26.667		41,600	1,025		42,625	47,300	

15640 | Packaged Cooling Towers

			CREW	DAILY OUTPUT	LABOR-HOURS	UNIT	MAT.	LABOR	EQUIP.	TOTAL	TOTAL INCL O&P	
400	0010	**COOLING TOWERS** Packaged units D3030 115										400
	0070	Galvanized steel										
	0080	Induced draft, crossflow D3030 210										
	0100	Vertical, belt drive, 61 tons	Q-6	90	.267	TonAC	82.50	10.20		92.70	106	
	0150	100 ton		100	.240		65.50	9.15		74.65	86	
	0200	115 ton		109	.220		64	8.40		72.40	83	
	0250	131 ton		120	.200		56.50	7.65		64.15	73.50	
	0260	162 ton		132	.182		47.50	6.95		54.45	63	
	1000	For higher capacities, use multiples										
	1500	Induced air, double flow										
	1900	Vertical, gear drive, 167 ton	Q-6	126	.190	TonAC	88	7.30		95.30	108	
	2000	297 ton		129	.186		63	7.10		70.10	80	
	2100	582 ton		132	.182		52	6.95		58.95	68	
	2150	849 ton		142	.169		51.50	6.45		57.95	66	
	2200	1016 ton		150	.160		49	6.10		55.10	63	
	2500	Blow through, centrifugal type										
	2510	50 ton	Q-6	2.28	10.526	Ea.	7,300	400		7,700	8,625	
	2520	75 ton		1.52	15.789		9,425	605		10,030	11,300	
	2524	100 ton		1.22	19.672		11,100	750		11,850	13,400	
	2528	125 ton	Q-7	1.31	24.427		13,000	955		13,955	15,700	

15 MECHANICAL

15640 | Packaged Cooling Towers

			CREW	DAILY OUTPUT	LABOR-HOURS	UNIT	2005 BARE COSTS MAT.	LABOR	EQUIP.	TOTAL	TOTAL INCL O&P
400	2532	200 ton (D3030 115)	Q-7	.82	39.216	Ea.	17,000	1,525		18,525	21,000
	2536	250 ton		.65	49.231		20,300	1,925		22,225	25,300
	2540	300 ton (D3030 210)		.54	59.259		23,400	2,325		25,725	29,200
	2544	350 ton		.47	68.085		26,400	2,650		29,050	33,100
	2548	400 ton		.41	78.049		27,500	3,050		30,550	34,900
	2552	450 ton		.36	88.154		33,100	3,450		36,550	41,600
	2556	500 ton		.32	100		35,300	3,900		39,200	44,700
	2560	550 ton		.30	107		38,000	4,200		42,200	48,100
	2564	600 ton		.27	117		39,800	4,600		44,400	50,500
	2568	650 ton		.25	127		44,300	4,975		49,275	56,000
	2572	700 ton		.23	137		52,000	5,350		57,350	65,000
	2576	750 ton		.22	146		56,000	5,725		61,725	70,000
	2580	800 ton		.21	152		60,500	5,950		66,450	75,500
	2584	850 ton		.20	156		64,000	6,125		70,125	79,500
	2588	900 ton		.20	161		68,000	6,300		74,300	84,000
	2592	950 ton		.19	170		71,500	6,650		78,150	89,000
	2596	1000 ton	↓	.18	180	↓	77,000	7,050		84,050	95,500
	2700	Axial fan, induced draft									
	2710	50 ton	Q-6	2.28	10.526	Ea.	6,175	400		6,575	7,400
	2720	75 ton		1.52	15.789		7,275	605		7,880	8,900
	2724	100 ton	↓	1.22	19.672		7,700	750		8,450	9,600
	2728	125 ton	Q-7	1.31	24.427		9,475	955		10,430	11,800
	2732	200 ton		.82	39.216		16,100	1,525		17,625	20,000
	2736	250 ton		.65	49.231		18,100	1,925		20,025	22,800
	2740	300 ton		.54	59.259		21,200	2,325		23,525	26,800
	2744	350 ton		.47	68.085		22,700	2,650		25,350	28,900
	2748	400 ton		.41	78.049		26,000	3,050		29,050	33,200
	2752	450 ton		.36	88.154		28,800	3,450		32,250	36,900
	2756	500 ton		.32	100		30,600	3,900		34,500	39,600
	2760	550 ton		.30	107		34,200	4,200		38,400	44,000
	2764	600 ton		.27	117		36,100	4,600		40,700	46,600
	2768	650 ton		.25	127		39,400	4,975		44,375	51,000
	2772	700 ton		.23	137		44,800	5,350		50,150	57,500
	2776	750 ton		.22	146		46,600	5,725		52,325	60,000
	2780	800 ton		.21	152		48,600	5,950		54,550	62,500
	2784	850 ton		.20	156		49,000	6,125		55,125	63,000
	2788	900 ton		.20	161		50,500	6,300		56,800	65,000
	2792	950 ton		.19	170		53,500	6,650		60,150	69,000
	2796	1000 ton	↓	.18	180	↓	56,500	7,050		63,550	72,500
	3000	For higher capacities, use multiples									
	3500	For pumps and piping, add	Q-6	38	.632	TonAC	41.50	24		65.50	82
	4000	For absorption systems, add				"	75%	75%			
	4100	Cooling water chemical feeder	Q-5	3	5.333	Ea.	286	197		483	610
	4500	For rigging, see division 01590-600									
	5000	Fiberglass									
	5010	Draw thru									
	5100	60 ton	Q-6	1.50	16	Ea.	3,175	610		3,785	4,425
	5120	125 ton		.99	24.242		6,500	925		7,425	8,550
	5140	300 ton		.43	55.814		15,200	2,125		17,325	20,000
	5160	600 ton		.22	109		27,400	4,175		31,575	36,400
	5180	1000 ton	↓	.15	160	↓	47,000	6,125		53,125	60,500
	6000	Stainless steel									
	6010	Induced draft, crossflow, horizontal, belt drive									
	6100	57 ton	Q-6	1.50	16	Ea.	8,350	610		8,960	10,100
	6120	91 ton		.99	24.242		12,000	925		12,925	14,600
	6140	111 ton	↓	.43	55.814	↓	14,200	2,125		16,325	18,800

15640 | Packaged Cooling Towers

		15640 Packaged Cooling Towers	CREW	DAILY OUTPUT	LABOR-HOURS	UNIT	2005 BARE COSTS MAT.	LABOR	EQUIP.	TOTAL	TOTAL INCL O&P	
400	6160	126 ton (D3030 115)	Q-6	.22	109	Ea.	15,300	4,175		19,475	23,200	400
	6170	Induced draft, crossflow, vertical, gear drive										
	6172	167 ton (D3030 210)	Q-6	.75	32	Ea.	24,700	1,225		25,925	29,000	
	6174	297 ton		.43	55.814		31,700	2,125		33,825	38,000	
	6176	582 ton		.23	104		49,900	4,000		53,900	61,000	
	6178	849 ton		.17	141		80,000	5,400		85,400	96,000	
	6180	1016 ton		.15	160		87,500	6,125		93,625	105,000	

15660 | Liquid Coolers/Evap Condensers

		15660 Liquid Coolers/Evap Condensers	CREW	DAILY OUTPUT	LABOR-HOURS	UNIT	2005 BARE COSTS MAT.	LABOR	EQUIP.	TOTAL	TOTAL INCL O&P	
100	0010	**CONDENSERS** Ratings are for 30°F TD, R-22 (D3030 110)										100
	0080	Air cooled, belt drive, propeller fan										
	0220	45 ton	Q-6	.70	34.286	Ea.	10,200	1,300		11,500	13,200	
	0240	50 ton		.69	34.985		10,500	1,325		11,825	13,600	
	0260	54 ton		.63	37.795		10,900	1,450		12,350	14,200	
	0280	59 ton		.58	41.308		11,700	1,575		13,275	15,300	
	0300	65 ton		.53	45.541		12,200	1,750		13,950	16,000	
	0320	73 ton		.47	51.173		13,800	1,950		15,750	18,200	
	0340	81 ton		.42	56.738		15,700	2,175		17,875	20,500	
	0360	86 ton		.40	60.302		16,400	2,300		18,700	21,500	
	0380	88 ton		.39	61.697		17,500	2,350		19,850	22,900	
	0400	101 ton	Q-7	.45	70.640		20,600	2,750		23,350	26,900	
	0500	159 ton		.31	102		30,800	4,025		34,825	39,800	
	0600	228 ton		.22	148		44,500	5,775		50,275	57,500	
	0700	314 ton		.16	203		66,500	7,950		74,450	85,500	
	0800	471 ton		.11	283		76,000	11,100		87,100	100,500	
	1500	May be specified single or multi-circuit										
	1550	Air cooled, direct drive, propeller fan										
	1590	1 ton	Q-5	3.80	4.211	Ea.	540	155		695	830	
	1600	1-1/2 ton		3.60	4.444		640	164		804	950	
	1620	2 ton		3.20	5		710	184		894	1,050	
	1630	3 ton		2.40	6.667		790	246		1,036	1,250	
	1640	5 ton		2	8		1,450	295		1,745	2,025	
	1650	8 ton		1.80	8.889		1,825	330		2,155	2,500	
	1660	10 ton		1.40	11.429		2,225	420		2,645	3,075	
	1670	12 ton		1.30	12.308		2,500	455		2,955	3,425	
	1680	14 ton		1.20	13.333		2,850	490		3,340	3,900	
	1690	16 ton		1.10	14.545		3,125	535		3,660	4,250	
	1700	21 ton		1	16		3,600	590		4,190	4,850	
	1720	26 ton		.84	19.002		4,050	700		4,750	5,500	
	1740	30 ton		.70	22.792		6,750	840		7,590	8,675	
	1760	41 ton	Q-6	.77	31.008		7,425	1,175		8,600	9,950	
	1780	52 ton		.66	36.419		10,600	1,400		12,000	13,800	
	1800	63 ton		.54	44.037		11,700	1,675		13,375	15,400	
	1820	76 ton		.45	52.980		13,300	2,025		15,325	17,700	
	1840	86 ton		.40	60		16,400	2,300		18,700	21,600	
	1860	97 ton		.35	67.989		18,700	2,600		21,300	24,500	
	1880	105 ton	Q-7	.43	73.563		21,600	2,875		24,475	28,000	
	1890	118 ton		.39	82.687		23,200	3,225		26,425	30,500	
	1900	126 ton		.36	88.154		27,000	3,450		30,450	34,900	
	1910	136 ton		.34	95.238		30,200	3,725		33,925	38,800	
	1920	142 ton		.32	99.379		31,000	3,875		34,875	39,900	
	3400	Evaporative, copper coil, pump, fan motor										
	3440	10 ton	Q-5	.54	29.630	Ea.	4,300	1,100		5,400	6,375	
	3460	15 ton		.50	32		4,475	1,175		5,650	6,700	
	3480	20 ton		.47	34.043		4,650	1,250		5,900	7,000	

Important: See the Reference Section for critical supporting data - Reference Nos., Crews, & City Cost Indexes

15660 | Liquid Coolers/Evap Condensers

			CREW	DAILY OUTPUT	LABOR-HOURS	UNIT	MAT.	LABOR	EQUIP.	TOTAL	TOTAL INCL O&P	
							2005 BARE COSTS					
100	3500	25 ton D3030 110	Q-5	.45	35.556	Ea.	4,975	1,300		6,275	7,450	100
	3520	30 ton	↓	.42	38.095		5,300	1,400		6,700	7,925	
	3540	40 ton	Q-6	.49	48.980		7,025	1,875		8,900	10,600	
	3560	50 ton		.39	61.538		8,800	2,350		11,150	13,200	
	3580	65 ton		.35	68.571		9,750	2,625		12,375	14,700	
	3600	80 ton		.33	72.727		10,600	2,775		13,375	15,900	
	3620	90 ton	↓	.29	82.759		11,600	3,175		14,775	17,600	
	3640	100 ton	Q-7	.36	88.889		12,700	3,475		16,175	19,200	
	3660	110 ton		.33	96.970		14,000	3,775		17,775	21,100	
	3680	125 ton		.30	106		15,200	4,175		19,375	23,100	
	3700	135 ton		.28	114		16,300	4,450		20,750	24,700	
	3720	150 ton		.25	128		18,000	5,000		23,000	27,300	
	3740	165 ton		.23	139		19,800	5,425		25,225	30,000	
	3760	185 ton	↓	.22	145		20,900	5,675		26,575	31,500	
	3860	For fan damper control, add	Q-5	2	8	↓	450	295		745	940	
	8000	Oils										
	8100	Lubricating										
	8120	Oil, lubricating				Oz.	.17			.17	.19	
	8500	Refrigeration										
	8520	Oil, refrigeration				Gal.	13.20			13.20	14.50	

15670 | Refrigerant Condensing Units

			CREW	DAILY OUTPUT	LABOR-HOURS	UNIT	MAT.	LABOR	EQUIP.	TOTAL	TOTAL INCL O&P	
300	0010	**CONDENSING UNITS**										300
	0030	Air cooled, compressor, standard controls										
	0050	1.5 ton	Q-5	2.50	6.400	Ea.	555	236		791	965	
	0100	2 ton		2.10	7.619		585	281		866	1,050	
	0200	2.5 ton		1.70	9.412		705	345		1,050	1,300	
	0300	3 ton		1.30	12.308		830	455		1,285	1,600	
	0350	3.5 ton		1.10	14.545		935	535		1,470	1,825	
	0400	4 ton		.90	17.778		1,075	655		1,730	2,150	
	0500	5 ton		.60	26.667		1,300	985		2,285	2,925	
	0550	7.5 ton		.55	29.091		2,525	1,075		3,600	4,375	
	0560	8.5 ton		.53	30.189		3,075	1,100		4,175	5,075	
	0600	10 ton		.50	32		3,350	1,175		4,525	5,475	
	0620	11 ton		.48	33.333		5,175	1,225		6,400	7,525	
	0650	15 ton	↓	.40	40		5,600	1,475		7,075	8,400	
	0700	20 ton	Q-6	.40	60		7,975	2,300		10,275	12,200	
	0720	25 ton		.35	68.571		11,600	2,625		14,225	16,700	
	0750	30 ton		.30	80		13,200	3,050		16,250	19,100	
	0800	40 ton		.20	120		16,000	4,575		20,575	24,500	
	0840	50 ton		.18	133		19,000	5,100		24,100	28,600	
	0860	60 ton		.16	150		21,900	5,725		27,625	32,700	
	0900	75 ton		.14	171		25,000	6,550		31,550	37,400	
	1000	80 ton		.12	200		27,800	7,650		35,450	42,100	
	1100	100 ton	↓	.09	266	↓	39,100	10,200		49,300	58,500	
	2000	Water cooled, compressor, heat exchanger, controls										
	2100	5 ton	Q-5	.70	22.857	Ea.	8,500	840		9,340	10,600	
	2110	10 ton		.60	26.667		9,825	985		10,810	12,300	
	2200	15 ton	↓	.50	32		13,400	1,175		14,575	16,600	
	2300	20 ton	Q-6	.40	60		15,000	2,300		17,300	20,000	
	2310	30 ton		.30	80		17,800	3,050		20,850	24,200	
	2400	40 ton		.20	120		20,300	4,575		24,875	29,200	
	2410	60 ton		.18	133		30,800	5,100		35,900	41,600	
	2420	80 ton		.14	171		35,600	6,550		42,150	49,100	
	2500	100 ton		.11	218		43,600	8,350		51,950	60,500	
	2510	120 ton	↓	.10	240	↓	70,000	9,175		79,175	91,500	

15680 | Evaporators, DX Coils

			CREW	DAILY OUTPUT	LABOR-HOURS	UNIT	2005 BARE COSTS MAT.	LABOR	EQUIP.	TOTAL	TOTAL INCL O&P	
300	0010	**EVAPORATORS** DX coils, remote compressors not included										300
	1000	Coolers, reach-in type, above freezing temperatures										
	1300	Shallow depth, wall mount, 7 fins per inch, air defrost										
	1310	600 BTUH, 8" fan, rust-proof core	Q-5	3.80	4.211	Ea.	276	155		431	540	
	1320	900 BTUH, 8" fan, rust-proof core		3.30	4.848		340	179		519	645	
	1330	1200 BTUH, 8" fan, rust-proof core		3	5.333		400	197		597	740	
	1340	1800 BTUH, 8" fan, rust-proof core		2.40	6.667		470	246		716	885	
	1350	2500 BTUH, 10" fan, rust-proof core		2.20	7.273		505	268		773	960	
	1360	3500 BTUH, 10" fan		2	8		595	295		890	1,100	
	1370	4500 BTUH, 10" fan		1.90	8.421		625	310		935	1,150	
	1390	For pan drain, add					38			38	41.50	
	1600	Undercounter refrigerators, ceiling or wall mount,										
	1610	8 fins per inch, air defrost, rust-proof core										
	1630	800 BTUH, one 6" fan	Q-5	5	3.200	Ea.	185	118		303	380	
	1640	1300 BTUH, two 6" fans		4	4		225	147		372	470	
	1650	1700 BTUH, two 6" fans		3.60	4.444		252	164		416	525	
	2000	Coolers, reach-in and walk-in types, above freezing temperatures										
	2600	Two-way discharge, ceiling mount, 150-4100 CFM,										
	2610	above 34°F applications, air defrost										
	2630	900 BTUH, 7 fins per inch, 8" fan	Q-5	4	4	Ea.	197	147		344	440	
	2660	2500 BTUH, 7 fins per inch, 10" fan		2.60	6.154		370	227		597	745	
	2690	5500 BTUH, 7 fins per inch, 12" fan		1.70	9.412		735	345		1,080	1,325	
	2720	8500 BTUH, 8 fins per inch, 16" fan		1.20	13.333		955	490		1,445	1,800	
	2750	15,000 BTUH, 7 fins per inch, 18" fan		1.10	14.545		1,325	535		1,860	2,250	
	2770	24,000 BTUH, 7 fins per inch, two 16" fans		1	16		2,000	590		2,590	3,075	
	2790	30,000 BTUH, 7 fins per inch, two 18" fans		.90	17.778		2,475	655		3,130	3,700	
	2850	Two-way discharge, low profile, ceiling mount,										
	2860	8 fins per inch, 200-570 CFM, air defrost										
	2880	800 BTUH, one 6" fan	Q-5	4	4	Ea.	224	147		371	470	
	2890	1300 BTUH, two 6" fans		3.50	4.571		266	168		434	545	
	2900	1800 BTUH, two 6" fans		3.30	4.848		281	179		460	580	
	2910	2700 BTUH, three 6" fans		2.70	5.926		330	218		548	695	
	3000	Coolers, walk-in type, above freezing temperatures										
	3300	General use, ceiling mount, 108-2080 CFM										
	3320	600 BTUH, 7 fins per inch, 6" fan	Q-5	5.20	3.077	Ea.	176	113		289	365	
	3340	1200 BTUH, 7 fins per inch, 8" fan		3.70	4.324		243	159		402	505	
	3360	1800 BTUH, 7 fins per inch, 10" fan		2.70	5.926		340	218		558	705	
	3380	3500 BTUH, 7 fins per inch, 12" fan		2.40	6.667		500	246		746	920	
	3400	5500 BTUH, 8 fins per inch, 12" fan		2	8		560	295		855	1,050	
	3430	8500 BTUH, 8 fins per inch, 16" fan		1.40	11.429		830	420		1,250	1,550	
	3460	15,000 BTUH, 7 fins per inch, two 16" fans		1.10	14.545		1,325	535		1,860	2,275	
	3640	Low velocity, high latent load, ceiling mount,										
	3650	1050-2420 CFM, air defrost, 6 fins per inch										
	3670	6700 BTUH, two 10" fans	Q-5	1.30	12.308	Ea.	1,100	455		1,555	1,875	
	3680	10,000 BTUH, three 10" fans		1	16		1,275	590		1,865	2,275	
	3690	13,500 BTUH, three 10" fans		1	16		1,525	590		2,115	2,575	
	3700	18,000 BTUH, four 10" fans		1	16		2,025	590		2,615	3,125	
	3710	26,500 BTUH, four 10" fans		.80	20		2,525	735		3,260	3,875	
	3730	For electric defrost, add					17%					
	5000	Freezers and coolers, reach-in type, above 34°F										
	5030	to sub-freezing temperature range, low latent load,										
	5050	air defrost, 7 fins per inch										
	5070	1200 BTUH, 8" fan, rustproof core	Q-5	4.40	3.636	Ea.	229	134		363	455	
	5080	1500 BTUH, 8" fan, rustproof core		4.20	3.810		236	140		376	470	
	5090	1800 BTUH, 8" fan, rustproof core		3.70	4.324		247	159		406	510	
	5100	2500 BTUH, 10" fan, rustproof core		2.90	5.517		282	203		485	615	

Important: See the Reference Section for critical supporting data - Reference Nos., Crews, & City Cost Indexes

15680 | Evaporators, DX Coils

			CREW	DAILY OUTPUT	LABOR-HOURS	UNIT	2005 BARE COSTS MAT.	LABOR	EQUIP.	TOTAL	TOTAL INCL O&P	
300	5110	3500 BTUH, 10" fan, rustproof core	Q-5	2.50	6.400	Ea.	390	236		626	780	300
	5120	4500 BTUH, 12" fan, rustproof core	↓	2.10	7.619	↓	515	281		796	985	
	6000	Freezers and coolers, walk-in type										
	6050	1960-18,000 CFM, medium profile										
	6060	Standard motor, 6 fins per inch, to -30°F, all aluminum										
	6080	10,500 BTUH, air defrost, one 18" fan	Q-5	1.40	11.429	Ea.	985	420		1,405	1,700	
	6090	12,500 BTUH, air defrost, one 18" fan		1.40	11.429		1,075	420		1,495	1,800	
	6100	16,400 BTUH, air defrost, one 18" fan		1.30	12.308		1,100	455		1,555	1,875	
	6110	20,900 BTUH, air defrost, one 18" fans		1	16		1,275	590		1,865	2,275	
	6120	27,000 BTUH, air defrost, two 18" fans	↓	1	16		1,900	590		2,490	2,950	
	6130	32,900 BTUH, air defrost, two 20" fans	Q-6	1.30	18.462		2,125	705		2,830	3,375	
	6140	39,000 BTUH, air defrost, two 20" fans		1.25	19.200		2,450	735		3,185	3,800	
	6150	44,100 BTUH, air defrost, three 20" fans		1	24		2,675	915		3,590	4,325	
	6155	53,000 BTUH, air defrost, three 20" fans		1	24		2,800	915		3,715	4,450	
	6160	66,200 BTUH, air defrost, four 20" fans		1	24		3,175	915		4,090	4,875	
	6170	78,000 BTUH, air defrost, four 20" fans		.80	30		3,950	1,150		5,100	6,075	
	6180	88,200 BTUH, air defrost, four 24" fans		.60	40		4,200	1,525		5,725	6,925	
	6190	110,000 BTUH, air defrost, four 24" fans	↓	.50	48		4,775	1,825		6,600	8,000	
	6330	Hot gas defrost, standard motor units, add					35%					
	6370	Electric defrost, 230V, standard motor, add					16%					
	6410	460V, standard or high capacity motor, add				↓	25%					
	6800	Eight fins per inch, increases BTUH 75%										
	6810	Standard capacity, add				Ea.	5%					
	6830	Hot gas & electric defrost not recommended										
	7000	800-4150 CFM, 12" fans										
	7010	Four fins per inch										
	7030	3400 BTUH, 1 fan	Q-5	2.80	5.714	Ea.	345	211		556	695	
	7040	4200 BTUH, 1 fan		2.40	6.667		350	246		596	760	
	7050	5300 BTUH, 1 fan		1.90	8.421		410	310		720	915	
	7060	6800 BTUH, 1 fan		1.70	9.412		460	345		805	1,025	
	7070	8400 BTUH, 2 fans		1.60	10		610	370		980	1,225	
	7080	10,500 BTUH, 2 fans		1.40	11.429		665	420		1,085	1,375	
	7090	13,000 BTUH, 3 fans		1.30	12.308		860	455		1,315	1,625	
	7100	17,000 BTUH, 4 fans		1.20	13.333		1,075	490		1,565	1,950	
	7110	21,500 BTUH, 5 fans	↓	1	16		1,350	590		1,940	2,350	
	7160	Six fins per inch increases BTUH 35%, add				↓	5%					
	7180	Eight fins per inch increases BTUH 53%,										
	7190	not recommended for below 35°F, add				Ea.	8%					
	7210	For 12°F temperature differential, add					20%					
	7230	For adjustable thermostat control, add				↓	73			73	80.50	
	7240	For hot gas defrost, except 8 fin models										
	7250	on applications below 35°F, add				Ea.	58%					
	7260	For electric defrost, except 8 fin models										
	7270	on applications below 35°F, add				Ea.	36%					
	8000	Freezers, pass-thru door uprights, walk-in storage										
	8300	Low temperature, thin profile, electric defrost										
	8310	138-1800 CFM, 5 fins per inch										
	8320	900 BTUH, one 6" fan	Q-5	3.30	4.848	Ea.	260	179		439	555	
	8340	1500 BTUH, two 6" fans		2.20	7.273		390	268		658	835	
	8360	2600 BTUH, three 6" fans		1.90	8.421		560	310		870	1,075	
	8370	3300 BTUH, four 6" fans		1.70	9.412		625	345		970	1,200	
	8380	4400 BTUH, five 6" fans		1.40	11.429		880	420		1,300	1,600	
	8400	7000 BTUH, three 10" fans		1.30	12.308		1,050	455		1,505	1,850	
	8410	8700 BTUH, four 10" fans	↓	1.10	14.545		1,200	535		1,735	2,125	
	8450	For air defrost, deduct				↓	130			130	143	

15705 | Curbs/Pads/Stands Prefab

			CREW	DAILY OUTPUT	LABOR-HOURS	UNIT	2005 BARE COSTS MAT.	LABOR	EQUIP.	TOTAL	TOTAL INCL O&P	
600	0010	**CURBS/PADS PREFABRICATED**										600
	6000	Pad, fiberglass reinforced concrete with polystyrene foam core										
	6050	Condenser, 2″ thick, 20″ x 38″	1 Shee	8	1	Ea.	8.85	40		48.85	71.50	
	6070	24″ x 24″		16	.500		7.85	20		27.85	39.50	
	6090	24″ x 36″		12	.667		12.15	27		39.15	54.50	
	6110	24″ x 42″		8	1		14.10	40		54.10	77.50	
	6150	26″ x 36″	Q-9	8	2		13.30	72.50		85.80	126	
	6170	28″ x 38″	1 Shee	8	1		15.05	40		55.05	78.50	
	6190	30″ x 30″		12	.667		13.10	27		40.10	55.50	
	6220	30″ x 36″		8	1		15.05	40		55.05	78.50	
	6240	30″ x 40″	Q-9	8	2		16.40	72.50		88.90	129	
	6260	32″ x 32″		7	2.286		14.65	82.50		97.15	143	
	6280	36″ x 36″		8	2		17.50	72.50		90	130	
	6300	36″ x 40″		7	2.286		19.90	82.50		102.40	149	
	6320	36″ x 48″		7	2.286		23.50	82.50		106	153	
	6340	36″ x 54″		6	2.667		28.50	96.50		125	179	
	6360	36″ x 60″ x 3″		7	2.286		33.50	82.50		116	164	
	6400	41″ x 47″ x 3″		6	2.667		30.50	96.50		127	182	
	6490	26″ round		10	1.600		8.15	58		66.15	98	
	6550	30″ round		9	1.778		11.35	64.50		75.85	111	
	6600	36″ round		8	2		23.50	72.50		96	137	
	6700	For 3″ thick pad, add					20%	5%				
	8800	Stand, corrosion-resistant plastic										
	8850	Heat pump, 6″ high	1 Shee	32	.250	Ea.	7	10.05		17.05	23	
	8870	12″ high	″	32	.250	″	9.50	10.05		19.55	26	

15710 | Heat Exchangers

			CREW	DAILY OUTPUT	LABOR-HOURS	UNIT	MAT.	LABOR	EQUIP.	TOTAL	TOTAL INCL O&P	
900	0010	**HEAT EXCHANGERS**										900
	0016	Shell & tube type, 2 or 4 pass, 3/4″ O.D. copper tubes,										
	0020	C.I. heads, C.I. tube sheet, steel shell										
	0100	Hot water 40°F to 180°F, by steam at 10 PSI										
	0120	8 GPM	Q-5	6	2.667	Ea.	960	98.50		1,058.50	1,200	
	0140	10 GPM		5	3.200		1,450	118		1,568	1,775	
	0160	40 GPM		4	4		2,250	147		2,397	2,700	
	0180	64 GPM		2	8		3,425	295		3,720	4,225	
	0200	96 GPM		1	16		4,600	590		5,190	5,925	
	0220	120 GPM	Q-6	1.50	16		6,050	610		6,660	7,575	
	0240	168 GPM		1	24		7,400	915		8,315	9,525	
	0260	240 GPM		.80	30		11,600	1,150		12,750	14,400	
	0300	600 GPM		.70	34.286		24,900	1,300		26,200	29,300	
	0500	For bronze head and tube sheet, add					50%					
	1000	Hot water 40°F to 140°F, by water at 200°F										
	1020	7 GPM	Q-5	6	2.667	Ea.	1,175	98.50		1,273.50	1,450	
	1040	16 GPM		5	3.200		1,675	118		1,793	2,025	
	1060	34 GPM		4	4		2,525	147		2,672	3,000	
	1080	55 GPM		3	5.333		3,675	197		3,872	4,350	
	1100	74 GPM		1.50	10.667		4,600	395		4,995	5,650	
	1120	86 GPM		1.40	11.429		6,150	420		6,570	7,375	
	1140	112 GPM	Q-6	2	12		7,650	460		8,110	9,100	
	1160	126 GPM		1.80	13.333		9,550	510		10,060	11,300	
	1180	152 GPM		1	24		12,200	915		13,115	14,800	
	3000	Plate type,										
	3100	400 GPM	Q-6	.80	30	Ea.	18,100	1,150		19,250	21,600	
	3120	800 GPM	″	.50	48		31,300	1,825		33,125	37,300	
	3140	1200 GPM	Q-7	.34	94.118		38,000	3,675		41,675	47,300	

15710 | Heat Exchangers

			CREW	DAILY OUTPUT	LABOR-HOURS	UNIT	2005 BARE COSTS MAT.	LABOR	EQUIP.	TOTAL	TOTAL INCL O&P	
900	3160	1800 GPM	Q-7	.24	133	Ea.	56,500	5,200		61,700	70,500	900
	8000	Heat pipe type, glycol, 50% efficient										
	8010	100 MBH, 1700 CFM	1 Stpi	.80	10	Ea.	2,425	410		2,835	3,300	
	8020	160 MBH, 2700 CFM		.60	13.333		3,275	545		3,820	4,425	
	8030	620 MBH, 4000 CFM		.40	20		4,925	820		5,745	6,650	

15720 | Air Handling Units

			CREW	DAILY OUTPUT	LABOR-HOURS	UNIT	MAT.	LABOR	EQUIP.	TOTAL	TOTAL INCL O&P	
100	0010	**AIR HANDLING UNIT** Built-up										100
	0100	With cooling/heating coil section, filters, mixing box										
	0880	Single zone, horizontal / vertical										
	0890	Constant volume										
	0900	1600 CFM	Q-5	1.20	13.333	Ea.	2,500	490		2,990	3,500	
	0906	2000 CFM		1.10	14.545		3,125	535		3,660	4,225	
	0910	3000 CFM		1	16		3,900	590		4,490	5,175	
	0916	4000 CFM		.95	16.842		5,200	620		5,820	6,650	
	0920	5000 CFM	Q-6	1.40	17.143		6,500	655		7,155	8,125	
	0926	6500 CFM		1.30	18.462		8,450	705		9,155	10,400	
	0930	7500 CFM		1.20	20		9,750	765		10,515	11,900	
	0936	9200 CFM		1.10	21.818		12,000	835		12,835	14,500	
	0940	11,500 CFM		1	24		13,200	915		14,115	15,900	
	0946	13,200 CFM		.90	26.667		15,100	1,025		16,125	18,100	
	0950	16,500 CFM		.80	30		18,900	1,150		20,050	22,500	
	0960	19,500 CFM		.70	34.286		22,300	1,300		23,600	26,500	
	0970	22,000 CFM		.60	40		25,200	1,525		26,725	30,000	
	0980	27,000 CFM		.50	48		28,100	1,825		29,925	33,700	
	0990	34,000 CFM		.40	60		35,400	2,300		37,700	42,400	
	1000	40,000 CFM		.30	80		38,500	3,050		41,550	47,000	
	1010	47,000 CFM		.20	120		48,900	4,575		53,475	61,000	
	1140	60,000 CFM	Q-7	.18	177		53,500	6,950		60,450	69,500	
	1160	75,000 CFM	"	.16	200		67,000	7,800		74,800	85,000	
	2000	Variable air volume										
	2100	75,000 CFM	Q-7	.12	266	Ea.	68,500	10,400		78,900	91,000	
	2140	100,000 CFM		.11	290		97,000	11,400		108,400	124,000	
	2160	150,000 CFM		.10	320		146,500	12,500		159,000	180,000	
	2300	Multi-zone, horizontal or vertical, blow-thru fan										
	2310	3000 CFM	Q-5	1	16	Ea.	1,700	590		2,290	2,725	
	2314	4000 CFM	"	.95	16.842		2,250	620		2,870	3,400	
	2318	5000 CFM	Q-6	1.40	17.143		2,825	655		3,480	4,075	
	2322	6500 CFM		1.30	18.462		3,650	705		4,355	5,075	
	2326	7500 CFM		1.20	20		4,225	765		4,990	5,800	
	2330	9200 CFM		1.10	21.818		5,175	835		6,010	6,950	
	2334	11,500 CFM		1	24		6,475	915		7,390	8,500	
	2336	13,200 CFM		.90	26.667		7,425	1,025		8,450	9,700	
	2340	16,500 CFM		.80	30		9,300	1,150		10,450	11,900	
	2344	19,500 CFM		.70	34.286		11,000	1,300		12,300	14,100	
	2348	22,000 CFM		.60	40		12,400	1,525		13,925	15,900	
	2352	27,000 CFM		.50	48		15,200	1,825		17,025	19,500	
	2356	34,000 CFM		.40	60		19,100	2,300		21,400	24,600	
	2360	40,000 CFM		.30	80		22,500	3,050		25,550	29,400	
	2364	47,000 CFM		.20	120		26,500	4,575		31,075	36,000	
	3000	Rooftop										
	3010	Constant volume										
	3100	2000 CFM	Q-5	1	16	Ea.	4,025	590		4,615	5,300	
	3140	5000 CFM	Q-6	.79	30.380		7,200	1,150		8,350	9,675	
	3150	10,000 CFM		.52	46.154		9,125	1,775		10,900	12,700	

	15720 Air Handling Units		CREW	DAILY OUTPUT	LABOR-HOURS	UNIT	MAT.	LABOR	EQUIP.	TOTAL	TOTAL INCL O&P	
100	3160	15,000 CFM	Q-6	.39	61.538	Ea.	13,800	2,350		16,150	18,700	100
	3170	20,000 CFM	↓	.30	80	↓	17,200	3,050		20,250	23,500	
	3200	Variable air volume										
	3240	5000 CFM	Q-6	.70	34.286	Ea.	8,825	1,300		10,125	11,700	
	3250	10,000 CFM		.50	48		11,100	1,825		12,925	15,000	
	3260	15,000 CFM		.40	60		15,700	2,300		18,000	20,800	
	3270	20,000 CFM		.29	82.759		20,500	3,175		23,675	27,300	
	3280	30,000 CFM	↓	.20	120	↓	34,500	4,575		39,075	44,900	
200	0010	**CENTRAL STATION AIR-HANDLING UNIT** Chilled water										200
	2300	Variable air volume, also incl. heating coil										
	2340	5000 CFM	Q-6	.70	34.286	Ea.	8,275	1,300		9,575	11,100	
	2350	10,000 CFM		.50	48		10,400	1,825		12,225	14,200	
	2360	15,000 CFM		.40	60		14,700	2,300		17,000	19,600	
	2370	20,000 CFM		.29	82.759		19,200	3,175		22,375	25,900	
	2380	30,000 CFM		.20	120		32,200	4,575		36,775	42,400	
	3000	Packaged, 3000 CFM, 7.5 ton		1.70	14.118		1,900	540		2,440	2,900	
	3050	3200 CFM, 8 ton		1.60	15		2,125	575		2,700	3,200	
	3100	4000 CFM, 10 ton		1.40	17.143		2,250	655		2,905	3,450	
	3150	4400 CFM, 11 ton		1.35	17.778		2,675	680		3,355	3,950	
	3200	6000 CFM, 15 ton		1	24		2,900	915		3,815	4,550	
	3250	7000 CFM, 17.5 ton		.90	26.667		3,500	1,025		4,525	5,375	
	3300	10,000 CFM, 25 ton		.65	36.923		5,325	1,400		6,725	8,000	
	3400	12,000 CFM, 30 ton	↓	.50	48		5,725	1,825		7,550	9,050	
	3500	For hot water heat coil, add				↓	30%	5%				
300	0010	**EVAPORATIVE COOLERS** Ducted, not incl. duct.										300
	0100	Side discharge style, capacities at .25" S.P.										
	0120	1785 CFM, 1/3 HP, 115 V	Q-9	5	3.200	Ea.	365	116		481	580	
	0140	2740 CFM, 1/3 HP, 115 V		4.50	3.556		445	129		574	690	
	0160	3235 CFM, 1/2 HP, 115 V		4	4		465	145		610	735	
	0180	3615 CFM, 1/2 HP, 230 V		3.60	4.444		465	161		626	760	
	0200	4215 CFM, 3/4 HP, 230 V		3.20	5		595	181		776	935	
	0220	5255 CFM. 1 HP, 115/230 V		3	5.333		975	193		1,168	1,375	
	0240	6090 CFM, 1 HP, 230/460 V		2.80	5.714		1,375	207		1,582	1,825	
	0260	8300 CFM, 1-1/2 HP, 230/460 V		2.60	6.154		1,375	223		1,598	1,850	
	0280	8360 CFM, 1-1/2 HP, 230/460 V		2.20	7.273		1,800	263		2,063	2,375	
	0300	9725 CFM, 2 HP, 230/460 V		1.80	8.889		1,850	320		2,170	2,525	
	0320	11,715 CFM, 3 HP, 230/460 V		1.40	11.429		1,875	415		2,290	2,700	
	0340	14,410 CFM, 5 HP, 230/460 V	↓	1	16	↓	2,025	580		2,605	3,125	
	0400	For two-speed motor, add					5%					
	0500	For down discharge style, add					10%					
	1000	High capacities at .4" S.P.										
	1040	15,000 CFM, 5 HP, 230/460 V	Q-9	.54	29.630	Ea.	5,975	1,075		7,050	8,200	
	1080	20,000 CFM, 10HP, 230/460 V	"	.48	33.333		7,650	1,200		8,850	10,300	
	1100	25,000 CFM, 20 HP, 230/460 V	Q-10	.54	44.444		8,425	1,675		10,100	11,900	
	1140	30,000 CFM, 20 HP, 230/460 V		.46	52.174		9,725	1,950		11,675	13,700	
	1160	35,000 CFM, 20 HP, 230/460 V		.44	54.545		10,600	2,050		12,650	14,900	
	1180	40,000 CFM, 30 HP, 230/460 V	↓	.42	57.143	↓	11,100	2,150		13,250	15,600	
500	0010	**MAKE-UP AIR UNIT**										500
	0020	Indoor suspension, natural/LP gas, direct fired,										
	0030	standard control. For flue see division 15550-440										
	0040	70°F temperature rise, MBH is input										
	0100	2000 CFM, 168 MBH	Q-6	3	8	Ea.	6,425	305		6,730	7,525	
	0120	3000 CFM, 252 MBH		2	12		6,600	460		7,060	7,975	
	0140	4000 CFM, 336 MBH		1.80	13.333		6,725	510		7,235	8,150	
	0160	6000 CFM, 502 MBH	↓	1.50	16	↓	8,100	610		8,710	9,825	

Column headings: DAILY OUTPUT, LABOR-HOURS; 2005 BARE COSTS (MAT., LABOR, EQUIP., TOTAL).

15700 | Heating/Ventilating/Air Conditioning Equipment

	15720 Air Handling Units		CREW	DAILY OUTPUT	LABOR-HOURS	UNIT	2005 BARE COSTS MAT.	LABOR	EQUIP.	TOTAL	TOTAL INCL O&P	
500	0180	8000 CFM, 670 MBH	Q-6	1.40	17.143	Ea.	8,875	655		9,530	10,700	500
	0200	10,000 CFM, 838 MBH		1.20	20		9,975	765		10,740	12,200	
	0220	12,000 CFM, 1005 MBH		1	24		10,100	915		11,015	12,500	
	0240	14,000 CFM, 1180 MBH		.94	25.532		10,200	975		11,175	12,700	
	0260	18,000 CFM, 1340 MBH		.86	27.907		10,300	1,075		11,375	12,900	
	0280	20,000 CFM, 1675 MBH		.76	31.579		12,700	1,200		13,900	15,800	
	0300	24,000 CFM, 2007 MBH	Q-7	1	32		13,100	1,250		14,350	16,300	
	0320	30,000 CFM, 2510 MBH		.96	33.333		13,500	1,300		14,800	16,800	
	0340	35,000 CFM, 2930 MBH		.92	34.783		17,000	1,350		18,350	20,800	
	0360	40,000 CFM, 3350 MBH		.88	36.364		17,400	1,425		18,825	21,200	
	0380	45,000 CFM, 3770 MBH		.84	38.095		17,800	1,475		19,275	21,800	
	0400	50,000 CFM, 4180 MBH		.80	40		18,000	1,550		19,550	22,200	
	0420	55,000 CFM, 4600 MBH		.75	42.667		23,000	1,675		24,675	27,800	
	0440	60,000 CFM, 5020 MBH		.70	45.714		23,900	1,775		25,675	29,000	
	0460	65,000 CFM, 5435 MBH		.60	53.333		24,100	2,075		26,175	29,600	
	0480	75,000 CFM, 6275 MBH		.50	64		25,300	2,500		27,800	31,700	
	0600	For discharge louver assembly, add					10%					
	0700	For filters, add					20%					
	0800	For air shut-off damper section, add					10%					
	0900	For vertical unit, add					10%					
	1000	Rooftop unit, natural gas, gravity vent, S.S. exchanger										
	1010	70°F temperature rise, MBH is input										
	1020	995 CFM, 75 MBH	Q-6	4	6	Ea.	4,325	229		4,554	5,100	
	1040	1245 CFM, 95 MBH		3.60	6.667		4,375	255		4,630	5,175	
	1060	1590 CFM, 120 MBH		3.30	7.273		4,700	278		4,978	5,600	
	1080	2090 CFM, 159 MBH		3	8		4,850	305		5,155	5,775	
	1100	2490 CFM, 190 MBH		2.60	9.231		5,550	355		5,905	6,625	
	1120	2985 CFM, 225 MBH		2.30	10.435		5,625	400		6,025	6,800	
	1140	3975 CFM, 300 MBH		1.90	12.632		6,675	485		7,160	8,050	
	1160	4970 CFM, 375 MBH		1.40	17.143		7,725	655		8,380	9,475	
	1180	5950 CFM, 450 MBH		1.20	20		8,150	765		8,915	10,100	
	1200	7940 CFM, 600 MBH		1	24		9,450	915		10,365	11,800	
	1220	9930 CFM, 750 MBH		.80	30		16,100	1,150		17,250	19,500	
	1240	11,900 CFM, 900 MBH		.60	40		18,000	1,525		19,525	22,100	
	1260	15,880 CFM, 1200 MBH		.30	80		21,700	3,050		24,750	28,500	
	1500	For power vent, add					8%					
	1600	For cleanable filters, add				Ea.	125			125	138	
	1700	For electric modulating gas control, add				"	1,350			1,350	1,475	
	1800	For aluminized steel exchanger, deduct					10%					
	8000	Rooftop unit, electric, with curb, controls										
	8010	33 kW	Q-5	4	4	Ea.	3,400	147		3,547	3,975	
	8020	100 kW		3.60	4.444		5,650	164		5,814	6,475	
	8030	150 kW		3.20	5		6,550	184		6,734	7,475	
	8040	250 kW		2.80	5.714		8,525	211		8,736	9,700	
	8050	400 kW		2.20	7.273		12,400	268		12,668	14,100	

	15730 Unitary Air Conditioning Equip		CREW	DAILY OUTPUT	LABOR-HOURS	UNIT	MAT.	LABOR	EQUIP.	TOTAL	TOTAL INCL O&P	
200	0010	COMPUTER ROOM UNITS (D3030 222)										200
	1000	Air cooled, includes remote condenser but not										
	1020	interconnecting tubing or refrigerant										
	1080	3 ton	Q-5	.50	32	Ea.	8,725	1,175		9,900	11,400	
	1120	5 ton		.45	35.556		11,700	1,300		13,000	14,900	
	1160	6 ton		.30	53.333		21,500	1,975		23,475	26,700	
	1200	8 ton		.27	59.259		24,400	2,175		26,575	30,100	
	1240	10 ton		.25	64		25,400	2,350		27,750	31,500	

	15730 | Unitary Air Conditioning Equip		CREW	DAILY OUTPUT	LABOR-HOURS	UNIT	2005 BARE COSTS MAT.	LABOR	EQUIP.	TOTAL	TOTAL INCL O&P	
200	1280	15 ton	Q-5	.22	72.727	Ea.	28,300	2,675		30,975	35,100	200
	1290	18 ton		.20	80		32,300	2,950		35,250	39,900	
	1320	20 ton	Q-6	.29	82.759		33,800	3,175		36,975	42,000	
	1360	23 ton	"	.28	85.714		35,500	3,275		38,775	43,900	
	2200	Chilled water, for connection to										
	2220	existing chiller system of adequate capacity										
	2260	5 ton	Q-5	.74	21.622	Ea.	9,850	795		10,645	12,000	
	2280	6 ton		.52	30.769		13,100	1,125		14,225	16,100	
	2300	8 ton		.50	32		14,100	1,175		15,275	17,300	
	2320	10 ton		.49	32.653		15,600	1,200		16,800	19,000	
	2330	12 ton		.48	32.990		16,200	1,225		17,425	19,600	
	2360	15 ton		.48	33.333		16,800	1,225		18,025	20,300	
	2400	20 ton		.46	34.783		17,200	1,275		18,475	20,800	
	2440	23 ton		.42	38.095		18,000	1,400		19,400	21,900	
	4000	Glycol system, complete except for interconnecting tubing										
	4060	3 ton	Q-5	.40	40	Ea.	10,700	1,475		12,175	14,000	
	4100	5 ton		.38	42.105		13,900	1,550		15,450	17,600	
	4120	6 ton		.25	64		27,900	2,350		30,250	34,300	
	4140	8 ton		.23	69.565		30,100	2,575		32,675	37,100	
	4160	10 ton		.21	76.190		32,700	2,800		35,500	40,100	
	4200	15 ton	Q-6	.26	92.308		39,600	3,525		43,125	48,800	
	4240	20 ton		.24	100		42,400	3,825		46,225	52,500	
	4280	23 ton		.22	109		46,200	4,175		50,375	57,500	
	8000	Water cooled system, not including condenser,										
	8020	water supply or cooling tower										
	8060	3 ton	Q-5	.62	25.806	Ea.	7,875	950		8,825	10,100	
	8100	5 ton		.54	29.630		11,100	1,100		12,200	13,900	
	8120	6 ton		.35	45.714		20,800	1,675		22,475	25,400	
	8140	8 ton		.33	48.485		22,900	1,775		24,675	27,900	
	8160	10 ton		.31	51.613		24,400	1,900		26,300	29,700	
	8200	15 ton		.27	59.259		28,000	2,175		30,175	34,100	
	8240	20 ton	Q-6	.38	63.158		33,100	2,425		35,525	40,000	
	8280	23 ton	"	.34	70.588		35,100	2,700		37,800	42,700	
500	0010	**PACKAGED TERMINAL AIR CONDITIONER** Cabinet, wall sleeve,										500
	0100	louver, electric heat, thermostat, manual changeover, 208 V										
	0200	6,000 BTUH cooling, 8800 BTU heat	Q-5	6	2.667	Ea.	955	98.50		1,053.50	1,200	
	0220	9,000 BTUH cooling, 13,900 BTU heat		5	3.200		1,000	118		1,118	1,275	
	0240	12,000 BTUH cooling, 13,900 BTU heat		4	4		1,100	147		1,247	1,450	
	0260	15,000 BTUH cooling, 13,900 BTU heat		3	5.333		1,300	197		1,497	1,750	
	0280	18,000 BTUH cooling, 10 KW heat		2.40	6.667		1,350	246		1,596	1,875	
	0300	24,000 BTUH cooling, 10 KW heat		1.90	8.421		1,375	310		1,685	1,975	
	0320	30,000 BTUH cooling, 10 KW heat		1.40	11.429		1,575	420		1,995	2,350	
	0340	36,000 BTUH cooling, 10 KW heat		1.25	12.800		1,625	470		2,095	2,500	
	0360	42,000 BTUH cooling, 10 KW heat		1	16		2,050	590		2,640	3,150	
	0380	48,000 BTUH cooling, 10 KW heat		.90	17.778		2,225	655		2,880	3,425	
	0500	For hot water coil, increase heat by 10%, add					5%	10%				
	1000	For steam, increase heat output by 30%, add					8%	10%				
600	0010	**ROOF TOP AIR CONDITIONERS** Standard controls, curb, economizer										600
	1000	Single zone, electric cool, gas heat										
	1090	2 ton cooling, 55 MBH heating (D3030 202)	Q-5	.93	17.204	Ea.	2,550	635		3,185	3,750	
	1100	3 ton cooling, 60 MBH heating		.70	22.857		2,550	840		3,390	4,075	
	1120	4 ton cooling, 95 MBH heating (D3030 206)		.61	26.403		3,700	975		4,675	5,525	
	1140	5 ton cooling, 112 MBH heating		.56	28.520		4,050	1,050		5,100	6,025	
	1145	6 ton cooling, 140 MBH heating (R15700 -010)		.52	30.769		4,750	1,125		5,875	6,925	
	1150	7.5 ton cooling, 170 MBH heating		.50	32.258		5,900	1,200		7,100	8,275	

15730 | Unitary Air Conditioning Equip

			CREW	DAILY OUTPUT	LABOR-HOURS	UNIT	2005 BARE COSTS MAT.	LABOR	EQUIP.	TOTAL	TOTAL INCL O&P	
600	1160	10 ton cooling, 200 MBH heating (R15700-010)	Q-6	.67	35.982	Ea.	7,600	1,375		8,975	10,400	600
	1170	12.5 ton cooling, 230 MBH heating		.63	37.975		8,650	1,450		10,100	11,700	
	1180	15 ton cooling, 270 MBH heating (R15700-020)		.57	42.032		11,900	1,600		13,500	15,500	
	1190	18 ton cooling, 330 MBH heating		.52	45.889		14,100	1,750		15,850	18,200	
	1200	20 ton cooling, 360 MBH heating	Q-7	.67	47.976		15,800	1,875		17,675	20,200	
	1210	25 ton cooling, 450 MBH heating		.56	57.554		19,500	2,250		21,750	24,800	
	1220	30 ton cooling, 540 MBH heating		.47	68.376		24,600	2,675		27,275	31,000	
	1240	40 ton cooling, 675 MBH heating		.35	91.168		32,100	3,550		35,650	40,700	
	1260	50 ton cooling, 810 MBH heating		.28	113		35,500	4,450		39,950	45,800	
	1265	60 ton cooling, 900 MBH heating		.23	136		44,200	5,325		49,525	56,500	
	1270	80 ton cooling, 1000 MBH heating		.20	160		56,000	6,250		62,250	71,500	
	1275	90 ton cooling, 1200 MBH heating		.17	188		59,000	7,350		66,350	76,000	
	1280	100 ton cooling, 1350 MBH heating		.14	228		67,500	8,925		76,425	87,500	
	1300	Electric cool, electric heat										
	1310	5 ton	Q-5	.63	25.600	Ea.	2,850	945		3,795	4,575	
	1314	10 ton	Q-6	.74	32.432		5,200	1,250		6,450	7,600	
	1318	15 ton	"	.63	37.795		7,800	1,450		9,250	10,800	
	1322	20 ton	Q-7	.74	43.185		10,400	1,675		12,075	13,900	
	1326	25 ton		.62	51.696		13,000	2,025		15,025	17,300	
	1330	30 ton		.52	61.657		15,600	2,400		18,000	20,800	
	1334	40 ton		.39	82.051		20,800	3,200		24,000	27,700	
	1338	50 ton		.31	102		24,700	4,000		28,700	33,200	
	1342	60 ton		.26	123		29,700	4,800		34,500	39,800	
	1346	75 ton		.21	153		37,100	6,000		43,100	49,800	
	2000	Multizone, electric cool, gas heat, economizer										
	2100	15 ton cooling, 360 MBH heating	Q-7	.61	52.545	Ea.	73,500	2,050		75,550	84,000	
	2120	20 ton cooling, 360 MBH heating		.53	60.038		84,500	2,350		86,850	96,500	
	2140	25 ton cooling, 450 MBH heating		.44	71.910		96,000	2,800		98,800	110,000	
	2160	28 ton cooling, 450 MBH heating		.40	79.012		100,500	3,075		103,575	115,000	
	2180	30 ton cooling, 540 MBH heating		.37	85.562		109,000	3,350		112,350	125,000	
	2200	40 ton cooling, 540 MBH heating		.28	113		114,000	4,450		118,450	132,000	
	2210	50 ton cooling, 540 MBH heating		.22	142		121,500	5,550		127,050	142,000	
	2220	70 ton cooling, 1500 MBH heating		.16	198		131,000	7,750		138,750	156,000	
	2240	80 ton cooling, 1500 MBH heating		.14	228		150,000	8,925		158,925	178,500	
	2260	90 ton cooling, 1500 MBH heating		.13	256		159,000	10,000		169,000	189,500	
	2280	105 ton cooling, 1500 MBH heating		.11	290		173,500	11,400		184,900	208,000	
	2400	For hot water heat coil, deduct					5%					
	2500	For steam heat coil, deduct					2%					
	2600	For electric heat, deduct					3%	5%				
	4000	Single zone, electric cool, gas heat, variable air volume										
	4100	12.5 ton cooling, 230 MBH heating	Q-6	.55	43.716	Ea.	10,600	1,675		12,275	14,100	
	4120	18 ton cooling, 330 MBH heating	"	.45	52.747		16,400	2,025		18,425	21,000	
	4140	25 ton cooling, 450 MBH heating	Q-7	.48	66.116		22,800	2,575		25,375	29,000	
	4160	40 ton cooling, 675 MBH heating		.31	104		38,900	4,100		43,000	49,000	
	4180	60 ton cooling, 900 MBH heating		.20	157		55,500	6,150		61,650	70,500	
	4200	80 ton cooling, 1000 MBH heating		.15	209		67,500	8,175		75,675	87,000	
	5000	Single zone, electric cool only										
	5050	3 ton cooling	Q-5	.88	18.203	Ea.	2,525	670		3,195	3,775	
	5060	4 ton cooling		.76	21.108		2,825	780		3,605	4,275	
	5070	5 ton cooling		.70	22.792		3,225	840		4,065	4,775	
	5080	6 ton cooling		.65	24.502		3,675	905		4,580	5,375	
	5090	7.5 ton cooling		.62	25.806		4,775	950		5,725	6,675	
	5100	8.5 ton cooling		.59	26.981		5,275	995		6,270	7,300	
	5110	10 ton cooling	Q-6	.83	28.812		5,950	1,100		7,050	8,200	
	5120	12 ton cooling		.80	30		7,425	1,150		8,575	9,875	
	5130	15 ton cooling		.71	33.613		10,200	1,275		11,475	13,100	

MECHANICAL 15

	15730 | Unitary Air Conditioning Equip		CREW	DAILY OUTPUT	LABOR-HOURS	UNIT	2005 BARE COSTS MAT.	LABOR	EQUIP.	TOTAL	TOTAL INCL O&P	
600	5140	18 ton cooling R15700-010	Q-6	.65	36.697	Ea.	12,200	1,400		13,600	15,500	600
	5150	20 ton cooling	Q-7	.83	38.415		14,100	1,500		15,600	17,800	
	5160	25 ton cooling R15700-020		.70	45.977		17,300	1,800		19,100	21,800	
	5170	30 ton cooling		.58	54.701		21,700	2,125		23,825	27,000	
	5180	40 ton cooling		.44	73.059		28,400	2,850		31,250	35,500	
	5400	For low heat, add					7%					
	5410	For high heat, add					10%					
	6000	Single zone electric cooling with variable volume distribution										
	6020	20 ton cooling	Q-7	.72	44.199	Ea.	17,500	1,725		19,225	21,800	
	6030	25 ton cooling		.60	52.893		20,700	2,075		22,775	25,900	
	6040	30 ton cooling		.51	62.868		25,000	2,450		27,450	31,200	
	6050	40 ton cooling		.38	83.989		31,600	3,275		34,875	39,700	
	6060	50 ton cooling		.31	104		35,100	4,100		39,200	44,800	
	6070	60 ton cooling		.25	125		39,300	4,925		44,225	50,500	
	6200	For low heat, add					7%					
	6210	For high heat, add					10%					
	7000	Multizone, cool/heat, variable volume distribution										
	7100	50 ton cooling	Q-7	.19	164	Ea.	77,000	6,400		83,400	94,500	
	7110	70 ton cooling		.14	228		108,000	8,925		116,925	132,500	
	7120	90 ton cooling		.11	296		124,000	11,600		135,600	154,000	
	7130	105 ton cooling		.10	333		139,000	13,000		152,000	172,500	
	7140	120 ton cooling		.08	380		159,000	14,900		173,900	197,000	
	7150	140 ton cooling		.07	444		170,000	17,300		187,300	213,000	
	7300	Penthouse unit, cool/heat, variable volume distribution										
	7400	150 ton cooling	Q-7	.10	310	Ea.	248,000	12,100		260,100	291,000	
	7410	170 ton cooling		.10	320		270,000	12,500		282,500	316,000	
	7420	200 ton cooling		.10	326		317,500	12,700		330,200	368,500	
	7430	225 ton cooling		.09	340		347,500	13,300		360,800	402,000	
	7440	250 ton cooling		.09	359		358,500	14,000		372,500	415,000	
	7450	270 ton cooling		.09	363		387,000	14,200		401,200	447,000	
	7460	300 ton cooling		.09	372		430,000	14,500		444,500	495,000	
	9400	Sub assemblies for assembly systems										
	9410	Duct work per ton roof top 1 zone units	Q-9	.96	16.667	Ton	283	605		888	1,250	
	9420	Ductwork per ton rooftop multizone units		.50	32		405	1,150		1,555	2,225	
	9430	Ductwork, VAV cooling/ton rooftop multizone, 1 zone/ton		.32	50		595	1,800		2,395	3,425	
	9440	Duct work per ton packaged water & air cooled units		1.26	12.698		90.50	460		550.50	805	
	9450	Duct work per ton split system remote condensing units		1.32	12.121		85.50	440		525.50	770	
800	0010	**WINDOW UNIT AIR CONDITIONERS**										800
	4000	Portable/window, 15 amp 125V grounded receptacle required										
	4060	5000 BTUH	1 Carp	8	1	Ea.	194	34.50		228.50	267	
	4340	6000 BTUH		8	1		204	34.50		238.50	278	
	4480	8000 BTUH		6	1.333		290	45.50		335.50	390	
	4500	10,000 BTUH		6	1.333		390	45.50		435.50	495	
	4520	12,000 BTUH	L-2	8	2		475	60		535	615	
	4600	Window/thru-the-wall, 15 amp 230V grounded receptacle required										
	4780	17,000 BTUH	L-2	6	2.667	Ea.	615	79.50		694.50	800	
	4940	25,000 BTUH		4	4		780	120		900	1,050	
	4960	29,000 BTUH		4	4		840	120		960	1,100	
840	0010	**SELF-CONTAINED SINGLE PACKAGE** D3030 210										840
	0100	Air cooled, for free blow or duct, not incl. remote condenser										
	0110	Constant volume D3030 214										
	0200	3 ton cooling	Q-5	1	16	Ea.	2,575	590		3,165	3,725	
	0210	4 ton cooling	"	.80	20		2,800	735		3,535	4,200	
	0220	5 ton cooling	Q-6	1.20	20		3,050	765		3,815	4,525	

15730 | Unitary Air Conditioning Equip			DAILY OUTPUT	LABOR-HOURS	UNIT	2005 BARE COSTS MAT.	LABOR	EQUIP.	TOTAL	TOTAL INCL O&P
		CREW								
840 0230	7.5 ton cooling (D3030 210)	Q-6	.90	26.667	Ea.	3,950	1,025		4,975	5,875
0240	10 ton cooling	Q-7	1	32		5,250	1,250		6,500	7,650
0250	15 ton cooling (D3030 214)		.95	33.684		7,300	1,325		8,625	10,000
0260	20 ton cooling		.90	35.556		12,300	1,400		13,700	15,600
0270	25 ton cooling		.85	37.647		12,400	1,475		13,875	15,900
0280	30 ton cooling		.80	40		12,800	1,550		14,350	16,500
0300	40 ton cooling		.60	53.333		17,800	2,075		19,875	22,600
0320	50 ton cooling		.50	64		21,800	2,500		24,300	27,800
0340	60 ton cooling	Q-8	.40	80		25,500	3,125	202	28,827	32,900
0380	With hot water heat, variable air volume									
0390	10 ton cooling	Q-7	.96	33.333	Ea.	7,700	1,300		9,000	10,400
0400	20 ton cooling		.86	37.209		16,800	1,450		18,250	20,700
0410	30 ton cooling		.76	42.105		18,700	1,650		20,350	23,100
0420	40 ton cooling		.58	55.172		26,400	2,150		28,550	32,200
0430	50 ton cooling		.48	66.667		31,500	2,600		34,100	38,500
0440	60 ton cooling	Q-8	.38	84.211		39,300	3,275	212	42,787	48,400
0490	For duct mounting no price change									
0500	For steam heating coils, add				Ea.	10%	10%			
0550	Packaged, with electric heat									
0552	Constant volume									
0560	5 ton cooling	Q-6	1.20	20	Ea.	5,525	765		6,290	7,225
0570	10 ton cooling	Q-7	1	32		9,025	1,250		10,275	11,800
0580	20 ton cooling		.90	35.556		17,700	1,400		19,100	21,500
0590	30 ton cooling		.80	40		21,900	1,550		23,450	26,500
0600	40 ton cooling		.60	53.333		27,500	2,075		29,575	33,400
0610	50 ton cooling		.50	64		35,400	2,500		37,900	42,800
0700	Variable air volume									
0710	10 ton cooling	Q-7	1	32	Ea.	11,000	1,250		12,250	14,000
0720	20 ton cooling		.90	35.556		21,000	1,400		22,400	25,200
0730	30 ton cooling		.80	40		26,500	1,550		28,050	31,600
0740	40 ton cooling		.60	53.333		34,400	2,075		36,475	40,900
0750	50 ton cooling		.50	64		42,900	2,500		45,400	51,000
0760	60 ton cooling	Q-8	.40	80		49,000	3,125	202	52,327	59,000
1000	Water cooled for free blow or duct, not including tower									
1010	Constant volume									
1100	3 ton cooling	Q-6	1	24	Ea.	2,450	915		3,365	4,075
1120	5 ton cooling		1	24		3,175	915		4,090	4,875
1130	7.5 ton cooling		.80	30		4,475	1,150		5,625	6,625
1140	10 ton cooling	Q-7	.90	35.556		6,275	1,400		7,675	8,975
1150	15 ton cooling		.85	37.647		9,075	1,475		10,550	12,200
1160	20 ton cooling		.80	40		21,100	1,550		22,650	25,600
1170	25 ton cooling		.75	42.667		24,900	1,675		26,575	29,900
1180	30 ton cooling		.70	45.714		28,000	1,775		29,775	33,400
1200	40 ton cooling		.40	80		35,400	3,125		38,525	43,700
1220	50 ton cooling		.30	106		44,100	4,175		48,275	55,000
1240	60 ton cooling	Q-8	.30	106		51,500	4,175	269	55,944	63,000
1300	For hot water or steam heat coils, add					12%	10%			
2000	Water cooled with electric heat, not including tower									
2010	Constant volume									
2100	5 ton cooling	Q-6	1.20	20	Ea.	5,650	765		6,415	7,375
2120	10 ton cooling	Q-7	1	32		10,100	1,250		11,350	13,000
2130	15 ton cooling	″	.90	35.556		14,500	1,400		15,900	18,000
2400	Variable air volume									
2440	10 ton cooling	Q-7	1	32	Ea.	12,000	1,250		13,250	15,100
2450	15 ton cooling	″	.90	35.556	″	17,800	1,400		19,200	21,700
2600	Water cooled, hot water coils, not including tower									

	15730 | Unitary Air Conditioning Equip	CREW	DAILY OUTPUT	LABOR-HOURS	UNIT	2005 BARE COSTS MAT.	LABOR	EQUIP.	TOTAL	TOTAL INCL O&P	
840	2610 Variable air volume (D3030 210)										840
	2640 10 ton cooling	Q-7	1	32	Ea.	8,825	1,250		10,075	11,600	
	2650 15 ton cooling (D3030 214)	"	.80	40	"	13,300	1,550		14,850	17,000	
900	0010 **SPLIT DUCTLESS SYSTEM**										900
	0100 Cooling only, single zone										
	0110 Wall mount										
	0120 3/4 ton cooling	Q-5	2	8	Ea.	1,125	295		1,420	1,675	
	0130 1 ton cooling		1.80	8.889		1,225	330		1,555	1,850	
	0140 1-1/2 ton cooling		1.60	10		1,575	370		1,945	2,300	
	0150 2 ton cooling	↓	1.40	11.429	↓	2,275	420		2,695	3,125	
	1000 Ceiling mount										
	1020 2 ton cooling	Q-5	1.40	11.429	Ea.	1,100	420		1,520	1,850	
	1030 3 ton cooling	"	1.20	13.333	"	3,300	490		3,790	4,375	
	2000 T-Bar mount										
	2010 2 ton cooling	Q-5	1.40	11.429	Ea.	2,450	420		2,870	3,325	
	2020 3 ton cooling		1.20	13.333		2,950	490		3,440	4,000	
	2030 3-1/2 ton cooling	↓	1.10	14.545	↓	3,550	535		4,085	4,700	
	3000 Multizone										
	3010 Wall mount										
	3020 2 @ 3/4 ton cooling	Q-5	1.80	8.889	Ea.	1,075	330		1,405	1,700	
	5000 Cooling / Heating										
	5010 Wall mount										
	5110 1 ton cooling	Q-5	1.70	9.412	Ea.	795	345		1,140	1,400	
	5120 1-1/2 ton cooling	"	1.50	10.667	"	1,275	395		1,670	2,000	
	5300 Ceiling mount										
	5310 3 ton cooling	Q-5	1	16	Ea.	3,825	590		4,415	5,075	
	7000 Accessories for all split ductless systems										
	7010 Add for ambient frost control	Q-5	8	2	Ea.	120	73.50		193.50	243	
	7020 Add for tube / wiring kit										
	7030 15′ kit	Q-5	32	.500	Ea.	28.50	18.45		46.95	58.50	
	7040 35′ kit	"	24	.667	"	91.50	24.50		116	137	

	15740 | Heat Pumps	CREW	DAILY OUTPUT	LABOR-HOURS	UNIT	MAT.	LABOR	EQUIP.	TOTAL	TOTAL INCL O&P	
100	0010 **AIR-SOURCE HEAT PUMPS** (Not including interconnecting tubing)										100
	1000 Air to air, split system, not including curbs, pads, or ductwork										
	1010 For curbs/pads see division 15705										
	1015 1.5 ton cooling, 7 MBH heat @ 0°F	Q-5	1.22	13.115	Ea.	1,375	485		1,860	2,225	
	1020 2 ton cooling, 8.5 MBH heat @ 0°F		1.20	13.333		1,425	490		1,915	2,300	
	1030 2.5 ton cooling, 10 MBH heat @ 0°F		1	16		1,575	590		2,165	2,625	
	1040 3 ton cooling, 13 MBH heat @ 0°F		.80	20		1,725	735		2,460	3,000	
	1050 3.5 ton cooling, 18 MBH heat @ 0°F		.75	21.333		1,975	785		2,760	3,350	
	1054 4 ton cooling, 24 MBH heat @ 0°F		.60	26.667		2,150	985		3,135	3,850	
	1060 5 ton cooling, 27 MBH heat @ 0°F		.50	32		2,450	1,175		3,625	4,475	
	1080 7.5 ton cooling, 33 MBH heat @ 0°F	↓	.30	53.333		5,450	1,975		7,425	8,925	
	1100 10 ton cooling, 50 MBH heat @ 0°F	Q-6	.38	63.158		7,300	2,425		9,725	11,700	
	1120 15 ton cooling, 64 MBH heat @ 0°F		.26	92.308		10,400	3,525		13,925	16,700	
	1130 20 ton cooling, 85 MBH heat @ 0°F		.20	120		13,600	4,575		18,175	21,900	
	1140 25 ton cooling, 119 MBH heat @ 0°F	↓	.20	120	↓	16,400	4,575		20,975	24,900	
	1300 Supplementary electric heat coil, included										
	1500 Single package, not including curbs, pads, or plenums										
	1502 1/2 ton cooling, supplementary heat not incl.	Q-5	8	2	Ea.	1,025	73.50		1,098.50	1,225	
	1504 3/4 ton cooling, supplementary heat not incl.		6	2.667		1,100	98.50		1,198.50	1,350	
	1506 1 ton cooling, supplementary heat not incl.		4	4		1,250	147		1,397	1,600	
	1510 1.5 ton cooling, 5 MBH heat @ 0°F		1.55	10.323		1,925	380		2,305	2,700	
	1520 2 ton cooling, 6.5 MBH heat @ 0°F	↓	1.50	10.667	↓	2,200	395		2,595	3,000	

	15740 | Heat Pumps		CREW	DAILY OUTPUT	LABOR-HOURS	UNIT	2005 BARE COSTS MAT.	LABOR	EQUIP.	TOTAL	TOTAL INCL O&P	
100	1540	2.5 ton cooling, 8 MBH heat @ 0°F	Q-5	1.40	11.429	Ea.	2,375	420		2,795	3,250	100
	1560	3 ton cooling, 10 MBH heat @ 0°F		1.20	13.333		2,625	490		3,115	3,625	
	1570	3.5 ton cooling, 11 MBH heat @ 0°F		1	16		2,850	590		3,440	4,025	
	1580	4 ton cooling, 13 MBH heat @ 0°F		.96	16.667		3,075	615		3,690	4,325	
	1620	5 ton cooling, 27 MBH heat @ 0°F		.65	24.615		3,325	905		4,230	5,025	
	1640	7.5 ton cooling, 35 MBH heat @ 0°F		.40	40		5,075	1,475		6,550	7,825	
	1648	10 ton cooling, 45 MBH heat @ 0°F	Q-6	.40	60		6,725	2,300		9,025	10,900	
	1652	12 ton cooling, 50 MBH heat @ 0°F	"	.36	66.667		8,725	2,550		11,275	13,400	
	1696	Supplementary electric heat coil incl., except as noted										
800	0010	**WATER-SOURCE HEAT PUMPS** (Not including interconnecting tubing)										800
	2000	Water source to air, single package										
	2100	1 ton cooling, 13 MBH heat @ 75°F	Q-5	2	8	Ea.	990	295		1,285	1,550	
	2120	1.5 ton cooling, 17 MBH heat @ 75°F		1.80	8.889		1,075	330		1,405	1,700	
	2140	2 ton cooling, 19 MBH heat @ 75°F		1.70	9.412		1,125	345		1,470	1,750	
	2160	2.5 ton cooling, 25 MBH heat @ 75°F		1.60	10		1,200	370		1,570	1,875	
	2180	3 ton cooling, 27 MBH heat @ 75°F		1.40	11.429		1,275	420		1,695	2,050	
	2190	3.5 ton cooling, 29 MBH heat @ 75°F		1.30	12.308		1,325	455		1,780	2,150	
	2200	4 ton cooling, 31 MBH heat @ 75°F		1.20	13.333		1,475	490		1,965	2,375	
	2220	5 ton cooling, 29 MBH heat @ 75°F		.90	17.778		1,725	655		2,380	2,875	
	2240	7.5 ton cooling, 35 MBH heat @ 75°F		.60	26.667		5,000	985		5,985	6,975	
	2250	8.5 ton cooling, 40 MBH heat @ 75°F		.58	27.586		5,400	1,025		6,425	7,450	
	2260	10 ton cooling, 50 MBH heat @ 75°F		.53	30.189		5,450	1,100		6,550	7,675	
	2280	15 ton cooling, 64 MBH heat @ 75°F	Q-6	.47	51.064		10,200	1,950		12,150	14,200	
	2300	20 ton cooling, 100 MBH heat @ 75°F		.41	58.537		11,200	2,225		13,425	15,700	
	2310	25 ton cooling, 100 MBH heat @ 75°F		.32	75		15,200	2,875		18,075	21,000	
	2320	30 ton cooling		.23	102		16,700	3,900		20,600	24,300	
	2340	40 ton cooling		.20	117		23,600	4,475		28,075	32,600	
	2360	50 ton cooling		.15	160		26,600	6,125		32,725	38,500	
	3960	For supplementary heat coil, add					10%					
	4000	For increase in capacity thru use										
	4020	of solar collector, size boiler at 60%										
	15750 | Humidity Control Equipment											
300	0010	**DEHUMIDIFIERS**										300
	6000	Self contained with filters and standard controls										
	6040	1.5 lb/hr, 50 cfm	1 Plum	8	1	Ea.	3,250	41		3,291	3,625	
	6060	3 lb/hr, 150 cfm	Q-1	12	1.333		3,700	49		3,749	4,150	
	6065	6 lb/hr, 150 cfm		9	1.778		6,850	65.50		6,915.50	7,625	
	6070	16 to 20 lb/hr, 600 cfm		5	3.200		13,600	118		13,718	15,200	
	6080	30 to 40 lb/hr, 1125 cfm		4	4		22,500	147		22,647	25,000	
	6090	60 to 75 lb/hr, 2250 cfm		3	5.333		29,600	196		29,796	32,800	
	6100	120 to 155 lb/hr, 4500 cfm		2	8		54,000	294		54,294	60,000	
	6110	240 to 310 lb/hr, 9000 cfm		1.50	10.667		76,000	390		76,390	84,000	
	6120	400 to 515 lb/hr, 15,000 cfm	Q-2	1.60	15		103,500	570		104,070	114,500	
	6130	530 to 690 lb/hr, 20,000 cfm		1.40	17.143		112,000	655		112,655	124,500	
	6140	800 to 1030 lb/hr, 30,000 cfm		1.20	20		127,500	765		128,265	141,500	
	6150	1060 to 1375 lb/hr, 40,000 cfm		1	24		178,000	915		178,915	197,500	
500	0010	**HUMIDIFIERS**										500
	0520	Steam, room or duct, filter, regulators, auto. controls, 220 V										
	0540	11 lb. per hour	Q-5	6	2.667	Ea.	2,175	98.50		2,273.50	2,550	
	0560	22 lb. per hour		5	3.200		2,400	118		2,518	2,800	
	0580	33 lb. per hour		4	4		2,450	147		2,597	2,925	
	0600	50 lb. per hour		4	4		3,025	147		3,172	3,550	
	0620	100 lb. per hour		3	5.333		3,600	197		3,797	4,275	
	0640	150 lb. per hour		2.50	6.400		4,775	236		5,011	5,600	

	15750 | Humidity Control Equipment		CREW	DAILY OUTPUT	LABOR-HOURS	UNIT	2005 BARE COSTS MAT.	LABOR	EQUIP.	TOTAL	TOTAL INCL O&P	
500	0660	200 lb. per hour	Q-5	2	8	Ea.	5,925	295		6,220	6,950	500
	0700	With blower										
	0720	11 lb. per hour	Q-5	5.50	2.909	Ea.	3,025	107		3,132	3,475	
	0740	22 lb. per hour		4.75	3.368		3,250	124		3,374	3,750	
	0760	33 lb. per hour		3.75	4.267		3,325	157		3,482	3,875	
	0780	50 lb. per hour		3.50	4.571		4,000	168		4,168	4,650	
	0800	100 lb. per hour		2.75	5.818		4,500	214		4,714	5,275	
	0820	150 lb. per hour		2	8		6,675	295		6,970	7,800	
	0840	200 lb. per hour		1.50	10.667		7,825	395		8,220	9,200	
	1000	Steam for duct installation, manifold with pneum. controls										
	1010	3 - 191 lb/hr, 24" x 24"	Q-5	5	3.200	Ea.	1,125	118		1,243	1,400	
	1020	3 - 191 lb/hr, 48" x 48"		4.60	3.478		2,175	128		2,303	2,575	
	1030	65 - 334 lb/hr, 36" x 36"		4.80	3.333		2,225	123		2,348	2,625	
	1040	65 - 334 lb/hr, 72" x 72"		4	4		2,400	147		2,547	2,850	
	1050	100 - 690 lb/hr, 48" x 48"		4	4		3,125	147		3,272	3,675	
	1060	100 - 690 lb/hr, 84" x 84"		3.60	4.444		3,325	164		3,489	3,925	
	1070	220 - 2000 lb/hr, 72" x 72"		3.80	4.211		3,375	155		3,530	3,950	
	1080	220 - 2000 lb/hr, 144" x 144"		3.40	4.706		3,975	173		4,148	4,625	
	1090	For electrically controlled unit, add					700			700	770	
	1092	For additional manifold, add					25%	15%				
	1200	Electric steam, self generating, with manifold										
	1210	1 - 96 lb/hr, 12" x 72"	Q-5	3.60	4.444	Ea.	5,300	164		5,464	6,075	
	1220	20 - 168 lb/hr, 12" x 72"		3.60	4.444		8,550	164		8,714	9,650	
	1290	For room mounting, add		16	1		153	37		190	224	
	5000	Furnace type, wheel bypass										
	5020	10 GPD	1 Stpi	4	2	Ea.	129	82		211	265	
	5040	14 GPD		3.80	2.105		175	86		261	325	
	5060	19 GPD		3.60	2.222		204	91		295	360	

	15760 | Terminal Heating & Cooling Units		CREW	DAILY OUTPUT	LABOR-HOURS	UNIT	MAT.	LABOR	EQUIP.	TOTAL	TOTAL INCL O&P	
100	0010	**COILS, FLANGED**										100
	0100	Basic water, DX, or condenser coils										
	0110	Copper tubes, alum. fins, galv. end sheets										
	0112	H is finned height, L is finned length										
	0120	3/8" x .016 tube, .0065 aluminum fins										
	0130	2 row, 8 fins per inch										
	0140	4" H x 12" L	Q-5	48	.333	Ea.	310	12.30		322.30	365	
	0150	4" H x 24" L		38.38	.417		320	15.35		335.35	380	
	0160	4" H x 48" L		19.25	.831		340	30.50		370.50	420	
	0170	4" H x 72" L		12.80	1.250		480	46		526	600	
	0180	6" H x 12" L		48	.333		325	12.30		337.30	375	
	0190	6" H x 24" L		25.60	.625		335	23		358	405	
	0200	6" H x 48" L		12.80	1.250		360	46		406	465	
	0210	6" H x 72" L		8.53	1.876		505	69		574	665	
	0220	10" H x 12" L		30.73	.521		345	19.20		364.20	410	
	0230	10" H x 24" L		15.33	1.044		360	38.50		398.50	455	
	0240	10" H x 48" L		7.69	2.081		395	76.50		471.50	550	
	0250	10" H x 72" L		5.12	3.125		560	115		675	790	
	0260	12" H x 12" L		25.60	.625		355	23		378	425	
	0270	12" H x 24" L		12.80	1.250		375	46		421	480	
	0280	12" H x 48" L		6.40	2.500		410	92		502	595	
	0290	12" H x 72" L		4.27	3.747		585	138		723	855	
	0300	16" H x 16" L		14.38	1.113		385	41		426	485	
	0310	16" H x 24" L		9.59	1.668		400	61.50		461.50	535	
	0320	16" H x 48" L		4.80	3.333		510	123		633	750	
	0330	16" H x 72" L		3.20	5		640	184		824	975	

	15760 | Terminal Heating & Cooling Units		CREW	DAILY OUTPUT	LABOR-HOURS	UNIT	2005 BARE COSTS MAT.	LABOR	EQUIP.	TOTAL	TOTAL INCL O&P	
100	0340	20" H x 24" L	Q-5	7.69	2.081	Ea.	425	76.50		501.50	585	100
	0350	20" H x 48" L		3.84	4.167		550	154		704	835	
	0360	20" H x 72" L		2.56	6.250		690	230		920	1,100	
	0370	24" H x 24" L		6.40	2.500		450	92		542	640	
	0380	24" H x 48" L		3.20	5		590	184		774	925	
	0390	24" H x 72" L		2.13	7.512		740	277		1,017	1,225	
	0400	30" H x 28" L		4.39	3.645		560	134		694	820	
	0410	30" H x 48" L		2.56	6.250		655	230		885	1,075	
	0420	30" H x 72" L		1.71	9.357		820	345		1,165	1,425	
	0430	30" H x 84" L		1.46	10.959		910	405		1,315	1,600	
	0440	34" H x 32" L		3.39	4.720		615	174		789	935	
	0450	34" H x 48" L		2.26	7.080		695	261		956	1,150	
	0460	34" H x 72" L		1.51	10.596		875	390		1,265	1,550	
	0470	34" H x 84" L	↓	1.29	12.403	↓	970	455		1,425	1,750	
	0600	For 10 fins per inch, add					1%					
	0610	For 12 fins per inch, add					3%					
	0620	For 4 row, add				Ea.	30%	50%				
	0630	For 6 row, add					60%	100%				
	0640	For 8 row, add				↓	80%	150%				
	1100	1/2" x .017 tube, .0065 aluminum fins										
	1110	2 row, 8 fins per inch										
	1120	5" H x 5" L	Q-5	40	.400	Ea.	310	14.75		324.75	360	
	1130	5" H x 20" L		37.10	.431		325	15.90		340.90	385	
	1140	5" H x 45" L		16.41	.975		350	36		386	440	
	1150	5" H x 90" L		8.21	1.949		570	72		642	735	
	1160	7.5" H x 5" L		38.50	.416		320	15.30		335.30	380	
	1170	7.5" H x 20" L		24.62	.650		340	24		364	410	
	1180	7.5" H x 45" L		10.94	1.463		370	54		424	490	
	1190	7.5" H x 90" L		5.46	2.930		610	108		718	830	
	1200	12.5" H x 10" L		29.49	.543		355	20		375	420	
	1210	12.5" H x 20" L		14.75	1.085		370	40		410	470	
	1220	12.5" H x 45" L		6.55	2.443		415	90		505	590	
	1230	12.5" H x 90" L		3.28	4.878		685	180		865	1,025	
	1240	15" H x 15" L		16.41	.975		375	36		411	470	
	1250	15" H x 30" L		8.21	1.949		405	72		477	560	
	1260	15" H x 45" L		5.46	2.930		435	108		543	640	
	1270	15" H x 90" L		2.73	5.861		725	216		941	1,125	
	1280	20" H x 20" L		9.22	1.735		420	64		484	555	
	1290	20" H x 45" L		4.10	3.902		550	144		694	820	
	1300	20" H x 90" L		2.05	7.805		805	288		1,093	1,325	
	1310	25" H x 25" L		5.90	2.712		465	100		565	660	
	1320	25" H x 45" L		3.28	4.878		590	180		770	920	
	1330	25" H x 90" L		1.64	9.756		885	360		1,245	1,500	
	1340	30" H x 30" L		4.10	3.902		575	144		719	850	
	1350	30" H x 45" L		2.73	5.861		650	216		866	1,050	
	1360	30" H x 90" L		1.37	11.679		965	430		1,395	1,700	
	1370	35" H x 35" L		3.01	5.316		645	196		841	1,000	
	1380	35" H x 45" L		2.34	6.838		700	252		952	1,150	
	1390	35" H x 90" L		1.17	13.675		1,050	505		1,555	1,900	
	1400	42.5" H x 45" L		1.93	8.290		775	305		1,080	1,300	
	1410	42.5" H x 75" L		1.16	13.793		1,025	510		1,535	1,900	
	1420	42.5" H x 105" L	↓	.83	19.277		1,250	710		1,960	2,450	
	1600	For 10 fins per inch, add					2%					
	1610	For 12 fins per inch, add					5%					
	1620	For 4 row, add					35%	50%				
	1630	For 6 row, add				↓	65%	100%				

15760	Terminal Heating & Cooling Units	CREW	DAILY OUTPUT	LABOR-HOURS	UNIT	2005 BARE COSTS MAT.	LABOR	EQUIP.	TOTAL	TOTAL INCL O&P
100 1640	For 8 row, add				Ea.	90%	150%			
2010	5/8" x .020" tube, .0065 aluminum fins									
2020	2 row, 8 fins per inch									
2030	6" H x 6" L	Q-5	38	.421	Ea.	315	15.50		330.50	375
2040	6" H x 24" L		24.96	.641		340	23.50		363.50	405
2050	6" H x 48" L		12.48	1.282		365	47		412	475
2060	6" H x 96" L		6.24	2.564		610	94.50		704.50	810
2070	12" H x 12" L		24.96	.641		320	23.50		343.50	385
2080	12" H x 24" L		12.48	1.282		345	47		392	450
2090	12" H x 48" L		6.24	2.564		425	94.50		519.50	610
2100	12" H x 96" L		3.12	5.128		720	189		909	1,075
2110	18" H x 24" L		8.32	1.923		420	71		491	570
2120	18" H x 48" L		4.16	3.846		555	142		697	825
2130	18" H x 96" L		2.08	7.692		825	283		1,108	1,325
2140	24" H x 24" L		6.24	2.564		465	94.50		559.50	650
2150	24" H x 48" L		3.12	5.128		550	189		739	890
2160	24" H x 96" L		1.56	10.256		935	380		1,315	1,600
2170	30" H x 30" L		3.99	4.010		590	148		738	870
2180	30" H x 48" L		2.50	6.400		615	236		851	1,025
2190	30" H x 96" L		1.25	12.800		1,050	470		1,520	1,850
2200	30" H x 120" L		1	16		1,225	590		1,815	2,225
2210	36" H x 36" L		2.77	5.776		680	213		893	1,075
2220	36" H x 48" L		2.08	7.692		755	283		1,038	1,250
2230	36" H x 90" L		1.11	14.414		1,125	530		1,655	2,025
2240	36" H x 120" L		.83	19.277		1,350	710		2,060	2,575
2250	42" H x 48" L		1.78	8.989		820	330		1,150	1,400
2260	42" H x 90" L		.95	16.842		1,225	620		1,845	2,275
2270	42" H x 120" L		.71	22.535		1,475	830		2,305	2,875
2280	51" H x 48" L		1.47	10.884		940	400		1,340	1,625
2290	51" H x 96" L		.73	21.918		1,450	810		2,260	2,825
2300	51" H x 120" L		.59	27.119		1,700	1,000		2,700	3,375
2500	For 10 fins per inch, add					3%				
2510	For 12 fins per inch, add					6%				
2520	For 4 row, add				Ea.	40%	50%			
2530	For 6 row, add					80%	100%			
2540	For 8 row, add					110%	150%			
3000	Hot water booster coils									
3010	Copper tubes, alum. fins, galv. end sheets									
3012	H is finned height, L is finned length									
3020	1/2" x .017" tube, .0065 Al fins									
3030	1 row, 10 fins per inch									
3040	5"H x 10"L	Q-5	40	.400	Ea.	221	14.75		235.75	265
3050	5"H x 25"L		33.18	.482		230	17.75		247.75	280
3060	10"H x 10"L		39	.410		230	15.10		245.10	276
3070	10"H x 20"L		20.73	.772		238	28.50		266.50	305
3080	10"H x 30"L		13.83	1.157		246	42.50		288.50	335
3090	15"H x 15"L		18.44	.868		244	32		276	315
3110	15"H x 20"L		13.83	1.157		250	42.50		292.50	340
3120	15"H x 30"L		9.22	1.735		261	64		325	385
3130	20"H x 20"L		10.37	1.543		261	57		318	375
3140	20"H x 30"L		6.91	2.315		275	85.50		360.50	430
3150	20"H x 40"L		5.18	3.089		289	114		403	490
3160	25"H x 25"L		6.64	2.410		281	89		370	445
3170	25"H x 35"L		4.74	3.376		355	124		479	575
3180	25"H x 45"L		3.69	4.336		385	160		545	660
3190	30"H x 30"L		4.61	3.471		360	128		488	585

Important: See the Reference Section for critical supporting data - Reference Nos., Crews, & City Cost Indexes

15760 | Terminal Heating & Cooling Units

		15760 Terminal Heating & Cooling Units	CREW	DAILY OUTPUT	LABOR-HOURS	UNIT	2005 BARE COSTS MAT.	LABOR	EQUIP.	TOTAL	TOTAL INCL O&P	
100	3200	30"H x 40"L	Q-5	3.46	4.624	Ea.	395	170		565	690	100
	3210	30"H x 50"L		2.76	5.797		425	214		639	785	
	3300	For 2 row, add					20%	50%				
	3400	1/2" x .020" tube, .008 Al fins										
	3410	2 row, 10 fins per inch										
	3420	20" H x 27" L	Q-5	7.27	2.201	Ea.	320	81		401	470	
	3430	20" H x 36" L		5.44	2.941		340	108		448	540	
	3440	20" H x 45" L		4.35	3.678		425	136		561	675	
	3450	25" H x 45" L		3.49	4.585		470	169		639	770	
	3460	25" H x 54" L		2.90	5.517		560	203		763	920	
	3470	30" H x 54" L		2.42	6.612		605	244		849	1,025	
	3480	35" H x 54" L		2.07	7.729		655	285		940	1,150	
	3490	35" H x 63" L		1.78	8.989		705	330		1,035	1,275	
	3500	40" H x 72" L		1.36	11.765		750	435		1,185	1,475	
	3510	55" H x 63" L		1.13	14.159		930	520		1,450	1,800	
	3520	55" H x 72" L		.99	16.162		1,000	595		1,595	2,000	
	3530	55" H x 81" L		.88	18.182		1,100	670		1,770	2,225	
	3540	75" H x 81" L		.64	25		1,450	920		2,370	2,975	
	3550	75" H x 90" L		.58	27.586		1,475	1,025		2,500	3,150	
	3560	75" H x 108" L		.48	33.333		1,700	1,225		2,925	3,725	
	3800	5/8" x .020 tubing, .0065 aluminum fins										
	3810	1 row, 10 fins per inch										
	3820	6" H x 12" L	Q-5	38	.421	Ea.	225	15.50		240.50	272	
	3830	6" H x 24" L		28.80	.556		234	20.50		254.50	288	
	3840	6" H x 30" L		23.04	.694		238	25.50		263.50	300	
	3850	12" H x 12" L		28.80	.556		237	20.50		257.50	291	
	3860	12" H x 24" L		14.40	1.111		250	41		291	335	
	3870	12" H x 30" L		11.52	1.389		256	51		307	360	
	3880	18" H x 18" L		12.80	1.250		257	46		303	350	
	3890	18" H x 24" L		9.60	1.667		266	61.50		327.50	385	
	3900	18" H x 36" L		6.40	2.500		284	92		376	450	
	3910	24" H x 24" L		7.20	2.222		283	82		365	435	
	3920	24" H x 30" L		4.80	3.333		320	123		443	535	
	3930	24" H x 36" L		5.05	3.168		390	117		507	600	
	3940	30" H x 30" L		4.61	3.471		390	128		518	620	
	3950	30" H x 36" L		3.85	4.159		415	153		568	685	
	3960	30" H x 42" L		3.29	4.863		445	179		624	760	
	3970	36" H x 36" L		3.20	5		450	184		634	770	
	3980	36" H x 42" L		2.74	5.839		480	215		695	855	
	4100	For 2 row, add					30%	50%				
	4400	Steam and hot water heating coils										
	4410	Copper tubes, alum. fins. galv. end sheets										
	4412	H is finned height, L is finned length										
	4420	5/8" tubing, to 175 psig working pressure										
	4430	1 row, 8 fins per inch										
	4440	6"H x 6"L	Q-5	38	.421	Ea.	345	15.50		360.50	405	
	4450	6"H x 12"L		37	.432		350	15.95		365.95	410	
	4460	6"H x 24"L		24.48	.654		365	24		389	435	
	4470	6"H x 36"L		16.32	.980		380	36		416	475	
	4480	6"H x 48"L		12.24	1.307		395	48		443	510	
	4490	6"H x 84"L		6.99	2.289		740	84.50		824.50	940	
	4500	6" H x 108" L		5.44	2.941		850	108		958	1,100	
	4510	12" H x 12" L		24.48	.654		395	24		419	470	
	4520	12" H x 24" L		12.24	1.307		415	48		463	535	
	4530	12" H x 36" L		8.16	1.961		440	72.50		512.50	595	
	4540	12" H x 48" L		6.12	2.614		525	96.50		621.50	725	

	15760 | Terminal Heating & Cooling Units	CREW	DAILY OUTPUT	LABOR-HOURS	UNIT	2005 BARE COSTS MAT.	LABOR	EQUIP.	TOTAL	TOTAL INCL O&P	
100	4550	12″ H x 84″ L	Q-5	3.50	4.571	Ea.	885	168		1,053	1,225	100
	4560	12″ H x 108″ L		2.72	5.882		1,050	217		1,267	1,475	
	4570	18″ H x18″ L		10.88	1.471		450	54		504	580	
	4580	18″ H x 24″ L		8.16	1.961		470	72.50		542.50	625	
	4590	18″ H x 48″ L		4.08	3.922		605	145		750	880	
	4600	18″ H x 84″ L		2.33	6.867		1,025	253		1,278	1,500	
	4610	18″ H x 108″ L		1.81	8.840		1,225	325		1,550	1,850	
	4620	24″ H x 24″ L		6.12	2.614		575	96.50		671.50	775	
	4630	24″ H x 48″ L		3.06	5.229		685	193		878	1,050	
	4640	24″ H x 84″ L		1.75	9.143		1,175	335		1,510	1,800	
	4650	24″ H x 108″ L		1.36	11.765		1,425	435		1,860	2,200	
	4660	36″ H x 36″ L		2.72	5.882		770	217		987	1,175	
	4670	36″ H x 48″ L		2.04	7.843		845	289		1,134	1,375	
	4680	36″ H x 84″ L		1.17	13.675		1,450	505		1,955	2,350	
	4690	36″ H x 108″ L		.91	17.582		1,800	650		2,450	2,950	
	4700	48″H x 48″L		1.53	10.458		1,000	385		1,385	1,675	
	4710	48″H x 84″L		.87	18.391		1,775	680		2,455	2,975	
	4720	48″H x 108″L		.68	23.529		2,200	865		3,065	3,700	
	4730	48″H x 126″L	↓	.58	27.586	↓	2,350	1,025		3,375	4,100	
	4910	For 10 fins per inch, add					2%					
	4920	For 12 fins per inch, add					5%					
	4930	For 2 row, add					30%	50%				
	4940	For 3 row, add					50%	100%				
	4950	For 4 row, add					80%	123%				
	4960	For 230 PSIG heavy duty, add					20%	30%				
	6000	Steam, heavy duty, 200 PSIG										
	6010	Copper tubes, alum. fins, galv. end sheets										
	6012	H is finned height, L is finned length										
	6020	1″ x .035″ tubing, .010 aluminum fins										
	6030	1 row, 8 fins per inch										
	6040	12″H x 12″L	Q-5	24.48	.654	Ea.	560	24		584	650	
	6050	12″H x 24″L		12.24	1.307		605	48		653	740	
	6060	12″H x 36″L		8.16	1.961		655	72.50		727.50	830	
	6070	12″H x 48″L		6.12	2.614		765	96.50		861.50	985	
	6080	12″H x 84″L		3.50	4.571		1,025	168		1,193	1,375	
	6090	12″H x 108″L		2.72	5.882		1,200	217		1,417	1,650	
	6100	18″H x 24″L		8.16	1.961		715	72.50		787.50	900	
	6110	18″H x 36″L		5.44	2.941		845	108		953	1,100	
	6120	18″H x 48″L		4.08	3.922		920	145		1,065	1,250	
	6130	18″H x 84″L		2.33	6.867		1,250	253		1,503	1,750	
	6140	18″H x 108″L		1.81	8.840		1,500	325		1,825	2,150	
	6150	24″H x 24″L		6.12	2.614		880	96.50		976.50	1,100	
	6160	24″H x 36″L		4.08	3.922		980	145		1,125	1,300	
	6170	24″H x 48″L		3.06	5.229		1,075	193		1,268	1,500	
	6180	24″H x 84″L		1.75	9.143		1,500	335		1,835	2,125	
	6190	24″H x 108″L		1.36	11.765		1,775	435		2,210	2,600	
	6200	30″H x 36″L		3.26	4.908		1,125	181		1,306	1,500	
	6210	30″H x 48″L		2.45	6.531		1,250	241		1,491	1,725	
	6220	30″H x 84″L		1.40	11.429		1,725	420		2,145	2,500	
	6230	30″H x 108″L		1.09	14.679		2,075	540		2,615	3,100	
	6240	36″H x 36″L		2.72	5.882		1,250	217		1,467	1,700	
	6250	36″H x 48″L		2.04	7.843		1,400	289		1,689	1,975	
	6260	36″H x 84″L		1.17	13.675		1,950	505		2,455	2,875	
	6270	36″H x 108″L		.91	17.582		2,350	650		3,000	3,550	
	6280	36″H x 120″L		.82	19.512		2,500	720		3,220	3,825	
	6290	48″H x 48″L	↓	1.53	10.458	↓	1,725	385		2,110	2,475	

15 MECHANICAL

		15760 | Terminal Heating & Cooling Units	CREW	DAILY OUTPUT	LABOR-HOURS	UNIT	2005 BARE COSTS MAT.	LABOR	EQUIP.	TOTAL	TOTAL INCL O&P	
100	6300	48"H x 84"L	Q-5	.87	18.391	Ea.	2,400	680		3,080	3,675	100
	6310	48"H x 108"L		.68	23.529		2,925	865		3,790	4,525	
	6320	48"H x 120"L	↓	.61	26.230		3,125	965		4,090	4,875	
	6400	For 10 fins per inch, add					4%					
	6410	For 12 fins per inch, add				↓	10%					
200	0010	**DUCT HEATERS** Electric, 480 V, 3 Ph										200
	0020	Finned tubular insert, 500°F										
	0100	8" wide x 6" high, 4.0 kW	Q-20	16	1.250	Ea.	540	46.50		586.50	665	
	0120	12" high, 8.0 kW		15	1.333		895	49.50		944.50	1,050	
	0140	18" high, 12.0 kW		14	1.429		1,250	53		1,303	1,450	
	0160	24" high, 16.0 kW		13	1.538		1,625	57		1,682	1,850	
	0180	30" high, 20.0 kW		12	1.667		1,975	62		2,037	2,275	
	0300	12" wide x 6" high, 6.7 kW		15	1.333		575	49.50		624.50	705	
	0320	12" high, 13.3 kW		14	1.429		925	53		978	1,100	
	0340	18" high, 20.0 kW		13	1.538		1,300	57		1,357	1,500	
	0360	24" high, 26.7 kW		12	1.667		1,675	62		1,737	1,925	
	0380	30" high, 33.3 kW		11	1.818		2,050	67.50		2,117.50	2,350	
	0500	18" wide x 6" high, 13.3 kW		14	1.429		605	53		658	750	
	0520	12" high, 26.7 kW		13	1.538		1,050	57		1,107	1,250	
	0540	18" high, 40.0 kW		12	1.667		1,400	62		1,462	1,650	
	0560	24" high, 53.3 kW		11	1.818		1,875	67.50		1,942.50	2,175	
	0580	30" high, 66.7 kW		10	2		2,350	74		2,424	2,700	
	0700	24" wide x 6" high, 17.8 kW		13	1.538		670	57		727	820	
	0720	12" high, 35.6 kW		12	1.667		1,150	62		1,212	1,375	
	0740	18" high, 53.3 kW		11	1.818		1,600	67.50		1,667.50	1,850	
	0760	24" high, 71.1 kW		10	2		2,075	74		2,149	2,425	
	0780	30" high, 88.9 kW		9	2.222		2,550	82.50		2,632.50	2,925	
	0900	30" wide x 6" high, 22.2 kW		12	1.667		710	62		772	875	
	0920	12" high, 44.4 kW		11	1.818		1,225	67.50		1,292.50	1,425	
	0940	18" high, 66.7 kW		10	2		1,675	74		1,749	1,975	
	0960	24" high, 88.9 kW		9	2.222		2,175	82.50		2,257.50	2,525	
	0980	30" high, 111.0 kW	↓	8	2.500	↓	2,700	92.50		2,792.50	3,125	
	1400	Note decreased kW available for										
	1410	each duct size at same cost										
	1420	See line 5000 for modifications and accessories										
	2000	Finned tubular flange with insulated										
	2020	terminal box, 500°F										
	2100	12" wide x 36" high, 54 kW	Q-20	10	2	Ea.	2,650	74		2,724	3,050	
	2120	40" high, 60 kW		9	2.222		3,050	82.50		3,132.50	3,500	
	2200	24" wide x 36" high, 118.8 kW		9	2.222		2,925	82.50		3,007.50	3,350	
	2220	40" high, 132 kW		8	2.500		3,350	92.50		3,442.50	3,825	
	2400	36" wide x 8" high, 40 kW		11	1.818		1,350	67.50		1,417.50	1,575	
	2420	16" high, 80 kW		10	2		1,800	74		1,874	2,125	
	2440	24" high, 120 kW		9	2.222		2,375	82.50		2,457.50	2,725	
	2460	32" high, 160 kW		8	2.500		3,025	92.50		3,117.50	3,475	
	2480	36" high, 180 kW		7	2.857		3,725	106		3,831	4,250	
	2500	40" high, 200 kW		6	3.333		4,125	124		4,249	4,750	
	2600	40" wide x 8" high, 45 kW		11	1.818		1,400	67.50		1,467.50	1,650	
	2620	16" high, 90 kW		10	2		1,950	74		2,024	2,275	
	2640	24" high, 135 kW		9	2.222		2,475	82.50		2,557.50	2,850	
	2660	32" high, 180 kW		8	2.500		3,300	92.50		3,392.50	3,775	
	2680	36" high, 202.5 kW		7	2.857		3,800	106		3,906	4,325	
	2700	40" high, 225 kW		6	3.333		4,325	124		4,449	4,950	
	2800	48" wide x 8" high, 54.8 kW		10	2		1,475	74		1,549	1,750	
	2820	16" high, 109.8 kW	↓	9	2.222	↓	2,050	82.50		2,132.50	2,375	

	15760 | Terminal Heating & Cooling Units		CREW	DAILY OUTPUT	LABOR-HOURS	UNIT	2005 BARE COSTS MAT.	LABOR	EQUIP.	TOTAL	TOTAL INCL O&P	
200	2840	24" high, 164.4 kW	Q-20	8	2.500	Ea.	2,625	92.50		2,717.50	3,025	200
	2860	32" high, 219.2 kW		7	2.857		3,500	106		3,606	4,000	
	2880	36" high, 246.6 kW		6	3.333		4,025	124		4,149	4,625	
	2900	40" high, 274 kW		5	4		4,550	148		4,698	5,225	
	3000	56" wide x 8" high, 64 kW		9	2.222		1,700	82.50		1,782.50	1,975	
	3020	16" high, 128 kW		8	2.500		2,350	92.50		2,442.50	2,725	
	3040	24" high, 192 kW		7	2.857		2,825	106		2,931	3,275	
	3060	32" high, 256 kW		6	3.333		3,975	124		4,099	4,550	
	3080	36" high, 288kW		5	4		4,550	148		4,698	5,225	
	3100	40" high, 320kW		4	5		5,025	185		5,210	5,800	
	3200	64" wide x 8" high, 74kW		8	2.500		1,725	92.50		1,817.50	2,050	
	3220	16" high, 148kW		7	2.857		2,400	106		2,506	2,800	
	3240	24" high, 222kW		6	3.333		3,075	124		3,199	3,575	
	3260	32" high, 296kW		5	4		4,100	148		4,248	4,750	
	3280	36" high, 333kW		4	5		4,900	185		5,085	5,650	
	3300	40" high, 370kW	↓	3	6.667	↓	5,400	247		5,647	6,325	
	3800	Note decreased kW available for										
	3820	each duct size at same cost										
	5000	Duct heater modifications and accessories										
	5120	T.C.O. limit auto or manual reset	Q-20	42	.476	Ea.	88.50	17.65		106.15	125	
	5140	Thermostat		28	.714		380	26.50		406.50	455	
	5160	Overheat thermocouple (removable)		7	2.857		535	106		641	750	
	5180	Fan interlock relay		18	1.111		129	41		170	205	
	5200	Air flow switch		20	1		112	37		149	180	
	5220	Split terminal box cover	↓	100	.200	↓	37	7.40		44.40	52	
	8000	To obtain BTU multiply kW by 3413										
250	0010	**ELECTRIC HEATING**, not incl. conduit or feed wiring										250
	1100	Rule of thumb: Baseboard units, including control	1 Elec	4.40	1.818	kW	77	74		151	195	
	1300	Baseboard heaters, 2' long, 375 watt		8	1	Ea.	29	41		70	92.50	
	1400	3' long, 500 watt		8	1		34.50	41		75.50	98.50	
	1600	4' long, 750 watt		6.70	1.194		41	48.50		89.50	118	
	1800	5' long, 935 watt		5.70	1.404		48.50	57		105.50	139	
	2000	6' long, 1125 watt		5	1.600		54	65		119	157	
	2200	7' long, 1310 watt		4.40	1.818		59	74		133	175	
	2400	8' long, 1500 watt		4	2		68	81.50		149.50	196	
	2600	9' long, 1680 watt		3.60	2.222		76	90.50		166.50	219	
	2800	10' long, 1875 watt	↓	3.30	2.424	↓	84.50	99		183.50	240	
	2950	Wall heaters with fan, 120 to 277 volt										
	3160	Recessed, residential, 750 watt	1 Elec	6	1.333	Ea.	89.50	54.50		144	180	
	3170	1000 watt		6	1.333		92	54.50		146.50	182	
	3180	1250 watt		5	1.600		101	65		166	208	
	3190	1500 watt		4	2		101	81.50		182.50	232	
	3210	2000 watt		4	2		106	81.50		187.50	238	
	3230	2500 watt		3.50	2.286		229	93		322	390	
	3240	3000 watt		3	2.667		345	109		454	540	
	3250	4000 watt		2.70	2.963		350	121		471	565	
	3260	Commercial, 750 watt		6	1.333		146	54.50		200.50	241	
	3270	1000 watt		6	1.333		146	54.50		200.50	241	
	3280	1250 watt		5	1.600		146	65		211	257	
	3290	1500 watt		4	2		146	81.50		227.50	281	
	3300	2000 watt		4	2		146	81.50		227.50	281	
	3310	2500 watt		3.50	2.286		146	93		239	299	
	3320	3000 watt		3	2.667		257	109		366	445	
	3330	4000 watt		2.70	2.963		257	121		378	465	
	3600	Thermostats, integral		16	.500		18	20.50		38.50	50.50	
	3800	Line voltage, 1 pole	↓	8	1	↓	22.50	41		63.50	85	

	15760 | Terminal Heating & Cooling Units		CREW	DAILY OUTPUT	LABOR-HOURS	UNIT	2005 BARE COSTS MAT.	LABOR	EQUIP.	TOTAL	TOTAL INCL O&P	
250	3810	2 pole	1 Elec	8	1	Ea.	29	41		70	92.50	250
	3820	Low voltage, 1 pole		8	1		25	41		66	88	
	4000	Heat trace system, 400 degree										
	4020	115V, 2.5 watts per L.F.	1 Elec	530	.015	L.F.	6.70	.61		7.31	8.25	
	4030	5 watts per L.F.		530	.015		6.70	.61		7.31	8.25	
	4050	10 watts per L.F.		530	.015		6.70	.61		7.31	8.25	
	4060	208V, 5 watts per L.F.		530	.015		6.70	.61		7.31	8.25	
	4080	480V, 8 watts per L.F.		530	.015		6.70	.61		7.31	8.25	
	4200	Heater raceway										
	4260	Heat transfer cement										
	4280	1 gallon				Ea.	53.50			53.50	58.50	
	4300	5 gallon				"	218			218	240	
	4320	Snap band, clamp										
	4340	3/4" pipe size	1 Elec	470	.017	Ea.		.69		.69	1.03	
	4360	1" pipe size		444	.018			.73		.73	1.09	
	4380	1-1/4" pipe size		400	.020			.82		.82	1.21	
	4400	1-1/2" pipe size		355	.023			.92		.92	1.37	
	4420	2" pipe size		320	.025			1.02		1.02	1.52	
	4440	3" pipe size		160	.050			2.04		2.04	3.03	
	4460	4" pipe size		100	.080			3.26		3.26	4.85	
	4480	Thermostat NEMA 3R, 22 amp, 0-150 Deg, 10′ cap.		8	1		182	41		223	261	
	4500	Thermostat NEMA 4X, 25 amp, 40 Deg, 5-1/2′ cap.		7	1.143		182	46.50		228.50	270	
	4520	Thermostat NEMA 4X, 22 amp, 25-325 Deg, 10′ cap.		7	1.143		490	46.50		536.50	610	
	4540	Thermostat NEMA 4X, 22 amp, 15-140 Deg,		6	1.333		410	54.50		464.50	535	
	4580	Thermostat NEMA 4,7,9, 22 amp, 25-325 Deg, 10′ cap.		3.60	2.222		605	90.50		695.50	800	
	4600	Thermostat NEMA 4,7,9, 22 amp, 15-140 Deg,		3	2.667		600	109		709	820	
	4720	Fiberglass application tape, 36 yard roll		11	.727		55	29.50		84.50	105	
	5000	Radiant heating ceiling panels, 2′ x 4′, 500 watt		16	.500		203	20.50		223.50	254	
	5050	750 watt		16	.500		223	20.50		243.50	277	
	5200	For recessed plaster frame, add		32	.250		42	10.20		52.20	61.50	
	5300	Infrared quartz heaters, 120 volts, 1000 watts		6.70	1.194		125	48.50		173.50	211	
	5350	1500 watt		5	1.600		125	65		190	235	
	5400	240 volts, 1500 watt		5	1.600		125	65		190	235	
	5450	2000 watt		4	2		125	81.50		206.50	259	
	5500	3000 watt		3	2.667		145	109		254	320	
	5550	4000 watt		2.60	3.077		145	125		270	345	
	5570	Modulating control		.80	10		72.50	410		482.50	685	
	5600	Unit heaters, heavy duty, with fan & mounting bracket										
	5650	Single phase, 208-240-277 volt, 3 kW	1 Elec	3.20	2.500	Ea.	335	102		437	515	
	5750	5 kW		2.40	3.333		345	136		481	580	
	5800	7 kW		1.90	4.211		530	172		702	840	
	5850	10 kW		1.30	6.154		605	251		856	1,050	
	5950	15 kW		.90	8.889		975	360		1,335	1,625	
	6000	480 volt, 3 kW		3.30	2.424		345	99		444	525	
	6020	4 kW		3	2.667		375	109		484	570	
	6040	5 kW		2.60	3.077		385	125		510	605	
	6060	7 kW		2	4		570	163		733	870	
	6080	10 kW		1.40	5.714		625	233		858	1,025	
	6100	13 kW		1.10	7.273		975	296		1,271	1,525	
	6120	15 kW		1	8		975	325		1,300	1,550	
	6140	20 kW		.90	8.889		1,275	360		1,635	1,950	
	6300	3 phase, 208-240 volt, 5 kW		2.40	3.333		325	136		461	555	
	6320	7 kW		1.90	4.211		500	172		672	805	
	6340	10 kW		1.30	6.154		540	251		791	970	
	6360	15 kW		.90	8.889		905	360		1,265	1,550	
	6380	20 kW		.70	11.429		1,300	465		1,765	2,125	

	15760 | Terminal Heating & Cooling Units		CREW	DAILY OUTPUT	LABOR-HOURS	UNIT	2005 BARE COSTS MAT.	LABOR	EQUIP.	TOTAL	TOTAL INCL O&P	
250	6400	25 kW	1 Elec	.50	16	Ea.	1,525	650		2,175	2,650	250
	6500	480 volt, 5 kW		2.60	3.077		455	125		580	685	
	6520	7 kW		2	4		570	163		733	875	
	6540	10 kW		1.40	5.714		605	233		838	1,000	
	6560	13 kW		1.10	7.273		975	296		1,271	1,525	
	6580	15 kW		1	8		975	325		1,300	1,550	
	6600	20 kW		.90	8.889		1,075	360		1,435	1,750	
	6620	25 kW		.60	13.333		1,250	545		1,795	2,175	
	6630	30 kW		.70	11.429		1,350	465		1,815	2,175	
	6640	40 kW		.60	13.333		2,250	545		2,795	3,275	
	6650	50 kW		.50	16		2,700	650		3,350	3,950	
	6800	Vertical discharge heaters, with fan										
	6820	Single phase, 208-240-277 volt, 10 kW	1 Elec	1.30	6.154	Ea.	560	251		811	990	
	6840	15 kW		.90	8.889		940	360		1,300	1,575	
	6900	3 phase, 208-240 volt, 10 kW		1.30	6.154		540	251		791	970	
	6920	15 kW		.90	8.889		905	360		1,265	1,525	
	6940	20 kW		.70	11.429		1,300	465		1,765	2,125	
	6960	25 kW		.50	16		1,525	650		2,175	2,650	
	6980	30 kW		.40	20		1,750	815		2,565	3,150	
	7000	40 kW		.36	22.222		2,250	905		3,155	3,825	
	7020	50 kW		.32	25		2,700	1,025		3,725	4,500	
	7100	480 volt, 10 kW		1.40	5.714		605	233		838	1,000	
	7120	15 kW		1	8		975	325		1,300	1,550	
	7140	20 kW		.90	8.889		1,275	360		1,635	1,950	
	7160	25 kW		.60	13.333		1,525	545		2,070	2,475	
	7180	30 kW		.50	16		1,750	650		2,400	2,900	
	7200	40 kW		.40	20		2,250	815		3,065	3,700	
	7220	50 kW		.35	22.857		2,700	930		3,630	4,350	
	7900	Cabinet convector heaters, 240 volt										
	7920	3′ long, 2000 watt	1 Elec	5.30	1.509	Ea.	1,500	61.50		1,561.50	1,750	
	7940	3000 watt		5.30	1.509		1,575	61.50		1,636.50	1,825	
	7960	4000 watt		5.30	1.509		1,650	61.50		1,711.50	1,925	
	7980	6000 watt		4.60	1.739		1,725	71		1,796	2,000	
	8000	8000 watt		4.60	1.739		1,725	71		1,796	1,975	
	8020	4′ long, 4000 watt		4.60	1.739		1,650	71		1,721	1,900	
	8040	6000 watt		4	2		1,725	81.50		1,806.50	2,025	
	8060	8000 watt		4	2		1,825	81.50		1,906.50	2,125	
	8080	10,000 watt		4	2		1,900	81.50		1,981.50	2,225	
	8100	Available also in 208 or 277 volt										
	8200	Cabinet unit heaters, 120 to 277 volt, 1 pole,										
	8220	wall mounted, 2 kW	1 Elec	4.60	1.739	Ea.	1,500	71		1,571	1,750	
	8230	3 kW		4.60	1.739		1,550	71		1,621	1,825	
	8240	4 kW		4.40	1.818		1,600	74		1,674	1,875	
	8250	5 kW		4.40	1.818		1,675	74		1,749	1,925	
	8260	6 kW		4.20	1.905		1,675	77.50		1,752.50	1,975	
	8270	8 kW		4	2		1,750	81.50		1,831.50	2,050	
	8280	10 kW		3.80	2.105		1,775	86		1,861	2,100	
	8290	12 kW		3.50	2.286		1,825	93		1,918	2,150	
	8300	13.5 kW		2.90	2.759		1,850	112		1,962	2,225	
	8310	16 kW		2.70	2.963		1,900	121		2,021	2,275	
	8320	20 kW		2.30	3.478		2,850	142		2,992	3,350	
	8330	24 kW		1.90	4.211		2,900	172		3,072	3,450	
	8350	Recessed, 2 kW		4.40	1.818		1,500	74		1,574	1,725	
	8370	3 kW		4.40	1.818		1,550	74		1,624	1,825	
	8380	4 kW		4.20	1.905		1,600	77.50		1,677.50	1,900	
	8390	5 kW		4.20	1.905		1,675	77.50		1,752.50	1,950	

15760 | Terminal Heating & Cooling Units

			CREW	DAILY OUTPUT	LABOR-HOURS	UNIT	2005 BARE COSTS MAT.	LABOR	EQUIP.	TOTAL	TOTAL INCL O&P	
250	8400	6 kW	1 Elec	4	2	Ea.	1,700	81.50		1,781.50	2,000	250
	8410	8 kW		3.80	2.105		1,775	86		1,861	2,075	
	8420	10 kW		3.50	2.286		2,200	93		2,293	2,575	
	8430	12 kW		2.90	2.759		2,200	112		2,312	2,600	
	8440	13.5 kW		2.70	2.963		2,225	121		2,346	2,625	
	8450	16 kW		2.30	3.478		2,250	142		2,392	2,675	
	8460	20 kW		1.90	4.211		2,700	172		2,872	3,225	
	8470	24 kW		1.60	5		2,775	204		2,979	3,350	
	8490	Ceiling mounted, 2 kW		3.20	2.500		1,500	102		1,602	1,775	
	8510	3 kW		3.20	2.500		1,550	102		1,652	1,875	
	8520	4 kW		3	2.667		1,625	109		1,734	1,925	
	8530	5 kW		3	2.667		1,675	109		1,784	1,975	
	8540	6 kW		2.80	2.857		1,675	116		1,791	2,025	
	8550	8 kW		2.40	3.333		1,750	136		1,886	2,125	
	8560	10 kW		2.20	3.636		1,775	148		1,923	2,200	
	8570	12 kW		2	4		1,775	163		1,938	2,225	
	8580	13.5 kW		1.50	5.333		1,850	217		2,067	2,375	
	8590	16 kW		1.30	6.154		1,900	251		2,151	2,475	
	8600	20 kW		.90	8.889		2,700	360		3,060	3,525	
	8610	24 kW	↓	.60	13.333	↓	2,750	545		3,295	3,825	
	8630	208 to 480 V, 3 pole										
	8650	Wall mounted, 2 kW	1 Elec	4.60	1.739	Ea.	1,500	71		1,571	1,725	
	8670	3 kW		4.60	1.739		1,550	71		1,621	1,825	
	8680	4 kW		4.40	1.818		1,600	74		1,674	1,875	
	8690	5 kW		4.40	1.818		1,675	74		1,749	1,925	
	8700	6 kW		4.20	1.905		1,675	77.50		1,752.50	1,975	
	8710	8 kW		4	2		1,750	81.50		1,831.50	2,050	
	8720	10 kW		3.80	2.105		1,775	86		1,861	2,100	
	8730	12 kW		3.50	2.286		1,775	93		1,868	2,125	
	8740	13.5 kW		2.90	2.759		1,850	112		1,962	2,225	
	8750	16 kW		2.70	2.963		2,050	121		2,171	2,425	
	8760	20 kW		2.30	3.478		2,700	142		2,842	3,175	
	8770	24 kW		1.90	4.211		2,750	172		2,922	3,275	
	8790	Recessed, 2 kW		4.40	1.818		1,500	74		1,574	1,725	
	8810	3 kW		4.40	1.818		1,550	74		1,624	1,825	
	8820	4 kW		4.20	1.905		1,600	77.50		1,677.50	1,900	
	8830	5 kW		4.20	1.905		1,675	77.50		1,752.50	1,950	
	8840	6 kW		4	2		1,675	81.50		1,756.50	1,975	
	8850	8 kW		3.80	2.105		1,750	86		1,836	2,050	
	8860	10 kW		3.50	2.286		1,775	93		1,868	2,100	
	8870	12 kW		2.90	2.759		1,775	112		1,887	2,125	
	8880	13.5 kW		2.70	2.963		1,900	121		2,021	2,275	
	8890	16 kW		2.30	3.478		1,900	142		2,042	2,300	
	8900	20 kW		1.90	4.211		2,700	172		2,872	3,225	
	8920	24 kW		1.60	5		2,750	204		2,954	3,325	
	8940	Ceiling mount, 2 kW		3.20	2.500		1,500	102		1,602	1,775	
	8950	3 kW		3.20	2.500		1,550	102		1,652	1,875	
	8960	4 kW		3	2.667		1,600	109		1,709	1,925	
	8970	5 kW		3	2.667		1,675	109		1,784	1,975	
	8980	6 kW		2.80	2.857		1,675	116		1,791	2,025	
	8990	8 kW		2.40	3.333		1,750	136		1,886	2,125	
	9000	10 kW		2.20	3.636		1,775	148		1,923	2,200	
	9020	13.5 kW		1.50	5.333		1,850	217		2,067	2,375	
	9030	16 kW		1.30	6.154		1,900	251		2,151	2,475	
	9040	20 kW		.90	8.889		2,700	360		3,060	3,525	
	9060	24 kW	↓	.60	13.333	↓	2,750	545		3,295	3,825	

15760 | Terminal Heating & Cooling Units

		15760 Terminal Heating & Cooling Units	CREW	DAILY OUTPUT	LABOR-HOURS	UNIT	2005 BARE COSTS MAT.	LABOR	EQUIP.	TOTAL	TOTAL INCL O&P	
300	0010	**FAN COIL AIR CONDITIONING** Cabinet mounted, filters, controls										300
	0100	Chilled water, 1/2 ton cooling	Q-5	8	2	Ea.	520	73.50		593.50	685	
	0110	3/4 ton cooling		7	2.286		550	84		634	730	
	0120	1 ton cooling		6	2.667		625	98.50		723.50	840	
	0140	1.5 ton cooling		5.50	2.909		715	107		822	950	
	0150	2 ton cooling		5.25	3.048		915	112		1,027	1,175	
	0160	2.5 ton cooling		5	3.200		1,350	118		1,468	1,650	
	0180	3 ton cooling		4	4		1,500	147		1,647	1,850	
	0190	7.5 ton cooling		2.70	5.926		1,575	218		1,793	2,075	
	0262	For hot water coil, add					40%	10%				
	0300	Console, 2 pipe with electric heat										
	0310	1/2 ton cooling	Q-5	8	2	Ea.	960	73.50		1,033.50	1,150	
	0315	3/4 ton cooling		7	2.286		1,050	84		1,134	1,275	
	0320	1 ton cooling		6	2.667		1,125	98.50		1,223.50	1,400	
	0330	1.5 ton cooling		5.50	2.909		1,400	107		1,507	1,675	
	0340	2 ton cooling		4.70	3.404		1,850	125		1,975	2,225	
	0345	2-1/2 ton cooling		4.70	3.404		2,675	125		2,800	3,150	
	0350	3 ton cooling		4	4		3,525	147		3,672	4,100	
	0360	4 ton cooling		4	4		3,675	147		3,822	4,275	
	0940	Direct expansion, for use w/air cooled condensing, 1.5 ton cooling		5	3.200		410	118		528	625	
	0950	2 ton cooling		4.80	3.333		435	123		558	665	
	0960	2.5 ton cooling		4.40	3.636		500	134		634	750	
	0970	3 ton cooling		3.80	4.211		525	155		680	810	
	0980	3.5 ton cooling		3.60	4.444		635	164		799	945	
	0990	4 ton cooling		3.40	4.706		675	173		848	1,000	
	1000	5 ton cooling		3	5.333		760	197		957	1,125	
	1020	7.5 ton cooling		3	5.333		1,775	197		1,972	2,250	
	1040	10 ton cooling	Q-6	2.60	9.231		2,125	355		2,480	2,875	
	1042	11 ton cooling		2.30	10.435		2,750	400		3,150	3,625	
	1050	15 ton cooling		1.60	15		2,825	575		3,400	3,950	
	1060	20 ton cooling		.70	34.286		3,750	1,300		5,050	6,100	
	1070	25 ton cooling		.65	36.923		4,700	1,400		6,100	7,300	
	1080	30 ton cooling		.60	40		5,175	1,525		6,700	8,000	
	1500	For hot water coil, add					40%	10%				
	1510	For condensing unit add see division 15670.										
	3000	Chilled water, horizontal unit, housing, 2 pipe, controls										
	3100	1/2 ton cooling	Q-5	8	2	Ea.	895	73.50		968.50	1,100	
	3110	1 ton cooling		6	2.667		1,075	98.50		1,173.50	1,350	
	3120	1.5 ton cooling		5.50	2.909		1,300	107		1,407	1,575	
	3130	2 ton cooling		5.25	3.048		1,550	112		1,662	1,875	
	3140	3 ton cooling		4	4		1,800	147		1,947	2,200	
	3150	3.5 ton cooling		3.80	4.211		1,975	155		2,130	2,400	
	3160	4 ton cooling		3.80	4.211		1,975	155		2,130	2,400	
	3170	5 ton cooling		3.40	4.706		2,300	173		2,473	2,775	
	3180	6 ton cooling		3.40	4.706		2,300	173		2,473	2,775	
	3190	7 ton cooling		2.90	5.517		2,500	203		2,703	3,050	
	3200	8 ton cooling		2.90	5.517		2,500	203		2,703	3,050	
	3210	10 ton cooling	Q-6	2.80	8.571		2,775	330		3,105	3,550	
	3240	For hot water coil, add					40%	10%				
	4000	With electric heat, 2 pipe										
	4100	1/2 ton cooling	Q-5	8	2	Ea.	1,100	73.50		1,173.50	1,300	
	4105	3/4 ton cooling		7	2.286		1,200	84		1,284	1,450	
	4110	1 ton cooling		6	2.667		1,325	98.50		1,423.50	1,600	
	4120	1.5 ton cooling		5.50	2.909		1,475	107		1,582	1,775	
	4130	2 ton cooling		5.25	3.048		1,850	112		1,962	2,225	
	4135	2.5 ton cooling		4.80	3.333		2,550	123		2,673	3,000	

	15760 | Terminal Heating & Cooling Units		CREW	DAILY OUTPUT	LABOR-HOURS	UNIT	2005 BARE COSTS MAT.	LABOR	EQUIP.	TOTAL	TOTAL INCL O&P
300	4140	3 ton cooling	Q-5	4	4	Ea.	3,400	147		3,547	3,950
	4150	3.5 ton cooling		3.80	4.211		4,075	155		4,230	4,700
	4160	4 ton cooling		3.60	4.444		4,675	164		4,839	5,400
	4170	5 ton cooling		3.40	4.706		5,225	173		5,398	6,000
	4180	6 ton cooling		3.20	5		5,225	184		5,409	6,025
	4190	7 ton cooling		2.90	5.517		5,525	203		5,728	6,375
	4200	8 ton cooling		2.20	7.273		6,225	268		6,493	7,250
	4210	10 ton cooling	Q-6	2.80	8.571		6,600	330		6,930	7,775
500	0010	**HEATING & VENTILATING UNITS** Classroom									
	0020	Includes filter, heating/cooling coils, standard controls									
	0080	750 CFM, 2 tons cooling	Q-6	2	12	Ea.	3,150	460		3,610	4,150
	0100	1000 CFM, 2-1/2 tons cooling		1.60	15		3,700	575		4,275	4,925
	0120	1250 CFM, 3 tons cooling		1.40	17.143		3,825	655		4,480	5,200
	0140	1500 CFM, 4 tons cooling		.80	30		4,100	1,150		5,250	6,225
	0160	2000 CFM 5 tons cooling		.50	48		4,825	1,825		6,650	8,050
	0500	For electric heat, add					35%				
	1000	For no cooling, deduct					25%	10%			
600	0010	**HYDRONIC HEATING** Terminal units, not incl. main supply pipe									
	1000	Radiation									
	1100	Panel, baseboard, C.I., including supports, no covers	Q-5	46	.348	L.F.	24.50	12.80		37.30	46.50
	1150	Fin tube, wall hung, 14" slope top cover, with damper									
	1200	1-1/4" copper tube, 4-1/4" alum. fin	Q-5	38	.421	L.F.	28	15.50		43.50	54
	1250	1-1/4" steel tube, 4-1/4" steel fin		36	.444		25.50	16.40		41.90	52.50
	1255	2" steel tube, 4-1/4" steel fin		32	.500		27.50	18.45		45.95	58
	1260	21", two tier, slope top, w/damper									
	1262	1-1/4" copper tube, 4-1/4" alum. fin	Q-5	30	.533	L.F.	33	19.65		52.65	65.50
	1264	1-1/4" steel tube, 4-1/4" steel fin		28	.571		31	21		52	65.50
	1266	2" steel tube, 4-1/4" steel fin		25	.640		33.50	23.50		57	72.50
	1270	10", flat top, wtih damper									
	1272	1-1/4" copper tube, 4-1/4" alum. fin	Q-5	38	.421	L.F.	28	15.50		43.50	54
	1274	1-1/4" steel tube, 4-1/4" steel fin		36	.444		28	16.40		44.40	55
	1276	2" steel tube, 4-1/4" steel fin		32	.500		27.50	18.45		45.95	58
	1280	17", two tier, flat top, w/damper									
	1282	1-1/4" copper tube, 4-1/4" alum. fin	Q-5	30	.533	L.F.	32.50	19.65		52.15	65.50
	1284	1-1/4" steel tube, 4-1/4" steel fin		28	.571		31	21		52	65.50
	1286	2" steel tube, 4-1/4" steel fin		25	.640		32.50	23.50		56	71
	1301	Rough in wall hung, steel fin w/supply & balance valves	1 Stpi	.91	8.791	Ea.	144	360		504	700
	1310	Baseboard, pkgd, 1/2" copper tube, alum. fin, 7" high	Q-5	60	.267	L.F.	5.35	9.85		15.20	20.50
	1320	3/4" copper tube, alum. fin, 7" high		58	.276		5.65	10.15		15.80	21.50
	1340	1" copper tube, alum. fin, 8-7/8" high		56	.286		11.70	10.55		22.25	28.50
	1360	1-1/4" copper tube, alum. fin, 8-7/8" high		54	.296		17.30	10.90		28.20	35.50
	1380	1-1/4" IPS steel tube with steel fins		52	.308		17.25	11.35		28.60	36
	1381	Rough in baseboard panel & fin tube, supply & balance valves	1 Stpi	1.06	7.547	Ea.	107	310		417	585
	1500	Note: fin tube may also require corners, caps, etc.									
	1990	Convector unit, floor recessed, flush, with trim									
	2000	for under large glass wall areas, no damper	Q-5	20	.800	L.F.	25.50	29.50		55	72.50
	2100	For unit with damper				"	36			36	40
	3000	Radiators, cast iron									
	3100	Free standing or wall hung, 6 tube, 25" high	Q-5	96	.167	Section	24	6.15		30.15	36
	3150	4 tube 25" high		96	.167		17.90	6.15		24.05	29
	3200	4 tube, 19" high		96	.167		16.65	6.15		22.80	27.50
	3250	Adj. brackets, 2 per wall radiator up to 30 sections	1 Stpi	32	.250	Ea.	17.10	10.25		27.35	34
	3500	Recessed, 20" high x 5" deep, without grille	Q-5	60	.267	Section	19.20	9.85		29.05	36
	3600	For inlet grille, add				"	2.06			2.06	2.27
	9500	To convert SFR to BTU rating: Hot water, 150 x SFR									

	15760 | Terminal Heating & Cooling Units		Crew	Daily Output	Labor-Hours	Unit	2005 Bare Costs Mat.	Labor	Equip.	Total	Total Incl O&P	
600	9510	Forced hot water, 180 x SFR; steam, 240 x SFR										600
700	0010	**INFRARED UNIT**										700
	0020	Gas fired, unvented, electric ignition, 100% shutoff.										
	0030	Piping and wiring not included										
	0060	Input, 15 MBH	Q-5	7	2.286	Ea.	360	84		444	525	
	0100	30 MBH		6	2.667		375	98.50		473.50	560	
	0120	45 MBH		5	3.200		430	118		548	650	
	0140	50 MBH		4.50	3.556		450	131		581	690	
	0160	60 MBH		4	4		465	147		612	730	
	0180	75 MBH		3	5.333		540	197		737	890	
	0200	90 MBH		2.50	6.400		560	236		796	970	
	0220	105 MBH		2	8		685	295		980	1,200	
	0240	120 MBH		2	8		895	295		1,190	1,425	
	2000	Electric, single or three phase										
	2050	6 kW, 20,478 BTU	1 Elec	2.30	3.478	Ea.	370	142		512	620	
	2100	13.5 KW, 40,956 BTU		2.20	3.636		585	148		733	865	
	2150	24 KW, 81,912 BTU		2	4		1,250	163		1,413	1,625	
	3000	Oil fired, pump, controls, fusible valve, oil supply tank										
	3050	91,000 BTU	Q-5	2.50	6.400	Ea.	5,700	236		5,936	6,625	
	3080	105.000 BTU		2.25	7.111		5,975	262		6,237	6,950	
	3110	119,000 BTU		2	8		6,550	295		6,845	7,650	
800	0010	**UNIT HEATERS**										800
	3950	Unit heaters, propeller, 115 V 2 psi steam, 60°F entering air										
	4000	Horizontal, 12 MBH	Q-5	12	1.333	Ea.	257	49		306	355	
	4020	28.2 MBH		10	1.600		355	59		414	480	
	4040	36.5 MBH		8	2		415	73.50		488.50	565	
	4060	43.9 MBH		8	2		415	73.50		488.50	565	
	4080	56.5 MBH		7.50	2.133		460	78.50		538.50	625	
	4100	65.6 MBH		7	2.286		520	84		604	695	
	4120	87.6 MBH		6.50	2.462		575	90.50		665.50	770	
	4140	96.8 MBH		6	2.667		600	98.50		698.50	810	
	4160	133.3 MBH		5	3.200		615	118		733	850	
	4180	157.6 MBH		4	4		770	147		917	1,075	
	4200	197.7 MBH		3	5.333		910	197		1,107	1,300	
	4220	257.2 MBH		2.50	6.400		1,250	236		1,486	1,725	
	4240	286.9 MBH		2	8		1,300	295		1,595	1,875	
	4250	326.0 MBH		1.90	8.421		1,500	310		1,810	2,125	
	4260	364 MBH		1.80	8.889		1,850	330		2,180	2,550	
	4270	404 MBH		1.60	10		2,025	370		2,395	2,800	
	4280	Rough in unit heater, 72.7 MBH, piping, supports & controls	1 Stpi	.47	17.021	Set	350	695		1,045	1,425	
	4300	For vertical diffuser, add				Ea.	143			143	157	
	4310	Vertical flow, 40 MBH	Q-5	11	1.455		410	53.50		463.50	530	
	4314	58.5 MBH		8	2		465	73.50		538.50	620	
	4318	92 MBH		7	2.286		560	84		644	740	
	4322	109.7 MBH		6	2.667		650	98.50		748.50	865	
	4326	131.0 MBH		4	4		650	147		797	935	
	4330	160.0 MBH		3	5.333		770	197		967	1,150	
	4334	194.0 MBH		2.20	7.273		825	268		1,093	1,300	
	4338	212.0 MBH		2.10	7.619		950	281		1,231	1,475	
	4342	247.0 MBH		1.96	8.163		1,000	300		1,300	1,550	
	4346	297.0 MBH		1.80	8.889		1,025	330		1,355	1,625	
	4350	333.0 MBH	Q-6	1.90	12.632		1,525	485		2,010	2,400	
	4354	420 MBH,(460 V)		1.80	13.333		1,550	510		2,060	2,475	
	4358	500 MBH, (460 V)		1.71	14.035		2,200	535		2,735	3,225	
	4362	570 MBH, (460 V)		1.40	17.143		3,025	655		3,680	4,300	

15760 | Terminal Heating & Cooling Units

		15760 Terminal Heating & Cooling Units	CREW	DAILY OUTPUT	LABOR-HOURS	UNIT	2005 BARE COSTS MAT.	LABOR	EQUIP.	TOTAL	TOTAL INCL O&P	
800	4366	620 MBH, (460 V)	Q-6	1.30	18.462	Ea.	3,475	705		4,180	4,875	800
	4370	960 MBH, (460 V)		1.10	21.818		6,100	835		6,935	7,950	
	4400	Unit heaters, propeller, hot water, 115 V										
	4404	Horizontal										
	4408	6 MBH	Q-5	11	1.455	Ea.	257	53.50		310.50	365	
	4412	12 MBH		11	1.455		264	53.50		317.50	370	
	4416	16 MBH		10	1.600		278	59		337	395	
	4420	24 MBH		9	1.778		355	65.50		420.50	490	
	4424	29 MBH		8	2		415	73.50		488.50	565	
	4428	47 MBH		7	2.286		465	84		549	635	
	4432	63 MBH		6	2.667		490	98.50		588.50	690	
	4436	81 MBH		5.50	2.909		595	107		702	815	
	4440	90 MBH		5	3.200		615	118		733	850	
	4460	133 MBH		4	4		770	147		917	1,075	
	4464	139 MBH		3.50	4.571		910	168		1,078	1,250	
	4468	198 MBH		2.50	6.400		1,250	236		1,486	1,725	
	4472	224 MBH		2	8		1,300	295		1,595	1,875	
	4476	273 MBH		1.50	10.667		1,375	395		1,770	2,125	
	4500	Vertical										
	4504	8 MBH	Q-5	13	1.231	Ea.	410	45.50		455.50	520	
	4508	12 MBH		13	1.231		410	45.50		455.50	520	
	4512	16 MBH		13	1.231		410	45.50		455.50	520	
	4516	30 MBH		12	1.333		410	49		459	525	
	4520	43 MBH		11	1.455		465	53.50		518.50	590	
	4524	57 MBH		8	2		520	73.50		593.50	685	
	4528	68 MBH		7	2.286		560	84		644	740	
	4540	105 MBH		6	2.667		650	98.50		748.50	865	
	4544	123 MBH		5	3.200		770	118		888	1,025	
	4550	140 MBH		5	3.200		840	118		958	1,100	
	4554	156 MBH		4	4		950	147		1,097	1,275	
	4560	210 MBH		4	4		1,000	147		1,147	1,325	
	4564	223 MBH		3	5.333		1,525	197		1,722	1,975	
	4568	257 MBH		2	8		1,525	295		1,820	2,125	
	5400	Cabinet, horizontal, hot water, blower type										
	5420	20 MBH	Q-5	10	1.600	Ea.	695	59		754	855	
	5430	60 MBH		7	2.286		1,200	84		1,284	1,450	
	5440	100 MBH		5	3.200		1,275	118		1,393	1,575	
	5450	120 MBH		4	4		1,325	147		1,472	1,675	
	6000	Valance units, complete with 1/2″ cooling coil, enclosure										
	6020	2 tube	Q-5	18	.889	L.F.	31	33		64	83	
	6040	3 tube		16	1		37	37		74	96	
	6060	4 tube		16	1		42	37		79	102	
	6080	5 tube		15	1.067		48	39.50		87.50	112	
	6100	6 tube		15	1.067		53	39.50		92.50	118	
	6120	8 tube		14	1.143		77	42		119	148	
	6200	For 3/4″ cooling coil, add					10%					

15770 | Floor-Heating & Snow-Melting Eq.

		15770 Floor-Heating & Snow-Melting Eq.	CREW	DAILY OUTPUT	LABOR-HOURS	UNIT	MAT.	LABOR	EQUIP.	TOTAL	TOTAL INCL O&P	
500	0010	**RADIANT FLOOR HEATING**										500
	0100	Tubing, PEX (cross-linked polyethylene)										
	0110	Oxygen barrier type for systems with ferrous materials										
	0120	1/2″	Q-5	800	.020	L.F.	.43	.74		1.17	1.58	
	0130	3/4″		535	.030		.55	1.10		1.65	2.27	
	0140	1″		400	.040		1.25	1.47		2.72	3.60	
	0200	Non barrier type for ferrous free systems										
	0210	1/2″	Q-5	800	.020	L.F.	.23	.74		.97	1.36	

	15770 | Floor-Heating & Snow-Melting Eq.	CREW	DAILY OUTPUT	LABOR-HOURS	UNIT	2005 BARE COSTS MAT.	LABOR	EQUIP.	TOTAL	TOTAL INCL O&P	
500	0220 3/4"	Q-5	535	.030	L.F.	.38	1.10		1.48	2.08	500
	0230 1"	↓	400	.040	↓	.66	1.47		2.13	2.95	
	1000 Manifolds										
	1110 Brass										
	1120 Valved										
	1130 1", 2 circuit	Q-5	13	1.231	Ea.	38.50	45.50		84	110	
	1140 1", 3 circuit		11	1.455		52	53.50		105.50	138	
	1150 1", 4 circuit		9	1.778		70.50	65.50		136	176	
	1160 1-1/4", 3 circuit		10	1.600		54	59		113	148	
	1170 1-1/4", 4 circuit	↓	8	2	↓	72	73.50		145.50	190	
	1300 Valveless										
	1310 1", 2 circuit	Q-5	14	1.143	Ea.	19.50	42		61.50	85	
	1320 1", 3 circuit		12	1.333		26.50	49		75.50	103	
	1330 1", 4 circuit		10	1.600		34	59		93	126	
	1340 1-1/4", 3 circuit		11	1.455		32.50	53.50		86	116	
	1350 1-1/4", 4 circuit	↓	9	1.778	↓	37	65.50		102.50	139	
	1400 Manifold connector kit										
	1410 Mounting brackets, couplings, end cap, air vent, valve, plug										
	1420 1"	Q-5	8	2	Ea.	66	73.50		139.50	184	
	1430 1-1/4"	"	7	2.286	"	79	84		163	214	
	1610 Copper, (cut to size)										
	1620 1" x 1/2" x 72" Lg, 2" sweat drops, 24 circuit	Q-5	3.33	4.805	Ea.	69	177		246	340	
	1630 1-1/4" x 1/2" x 72" Lg, 2" sweat drops, 24 circuit		3.20	5		81	184		265	365	
	1640 1-1/2" x 1/2" x 72" Lg, 2" sweat drops, 24 circuit		3	5.333		99	197		296	405	
	1650 1-1/4" x 3/4" x 72" Lg, 2" sweat drops, 24 circuit		3.10	5.161		86.50	190		276.50	380	
	1660 1-1/2" x 3/4" x 72" Lg, 2" sweat drops, 24 circuit	↓	2.90	5.517	↓	104	203		307	420	
	1710 Copper, (modular)										
	1720 1" x 1/2", 2" sweat drops, 2 circuit	Q-5	17.50	.914	Ea.	6.75	33.50		40.25	58	
	1730 1" x 1/2", 2" sweat drops, 3 circuit		14.50	1.103		9.75	40.50		50.25	72	
	1740 1" x 1/2", 2" sweat drops, 4 circuit		12.30	1.301		11.25	48		59.25	84.50	
	1750 1-1/4" x 1/2", 2" sweat drops, 2 circuit		17	.941		7.50	34.50		42	60.50	
	1760 1-1/4" x 1/2", 2" sweat drops, 3 circuit		14	1.143		11.25	42		53.25	76	
	1770 1-1/4" x 1/2", 2" sweat drops, 4 circuit	↓	11.80	1.356	↓	15	50		65	91.50	
	3000 Valves										
	3110 Motorized zone valve operator, 24V	Q-5	30	.533	Ea.	40.50	19.65		60.15	74	
	3120 Motorized zone valve with operator complete, 24V										
	3130 3/4"	Q-5	35	.457	Ea.	70	16.85		86.85	102	
	3140 1"		32	.500		79.50	18.45		97.95	115	
	3150 1-1/4"		29.60	.541		97.50	19.90		117.40	137	
	3200 Flow control/isolation valve, 1/2"	↓	32	.500	↓	21	18.45		39.45	50.50	
	3300 Thermostatic mixing valves										
	3310 1/2"	Q-5	26.60	.601	Ea.	73	22		95	114	
	3320 3/4"		25	.640		88.50	23.50		112	133	
	3330 1"	↓	21.40	.748	↓	94.50	27.50		122	146	
	3400 3 Way mixing / diverting valve, manual, brass										
	3410 1/2"	Q-5	21.30	.751	Ea.	96	27.50		123.50	148	
	3420 3/4"		20	.800		99	29.50		128.50	154	
	3430 1"	↓	17.80	.899	↓	115	33		148	176	
	3500 4 Way mixing / diverting valve, manual, brass										
	3510 1/2"	Q-5	16	1	Ea.	101	37		138	167	
	3520 3/4"		15	1.067		112	39.50		151.50	182	
	3530 1"		13.30	1.203		125	44.50		169.50	205	
	3540 1-1/4"		11.40	1.404		137	51.50		188.50	229	
	3550 1-1/2"		10.60	1.509		185	55.50		240.50	287	
	3560 2"		10	1.600		199	59		258	310	
	3570 2-1/2"	↓	8	2	↓	550	73.50		623.50	710	

 Important: See the Reference Section for critical supporting data - Reference Nos., Crews, & City Cost Indexes

15700 | Heating/Ventilating/Air Conditioning Equipment

	15770 Floor-Heating & Snow-Melting Eq.		CREW	DAILY OUTPUT	LABOR-HOURS	UNIT	2005 BARE COSTS MAT.	LABOR	EQUIP.	TOTAL	TOTAL INCL O&P	
500	3800	Motor control, 3 or 4 way, for valves up to 1-1/2"	Q-5	30	.533	Ea.	138	19.65		157.65	182	500
	5000	Radiant floor heating, zone control panel										
	5110	6 Zone control										
	5120	For zone valves	Q-5	16	1	Ea.	131	37		168	200	
	5130	For circulators	"	16	1	"	185	37		222	260	
	7000	PEX tubing fittings, compression type										
	7110	PEX tubing to brass valve conn., 1/2"	Q-5	54	.296	Ea.	2.25	10.90		13.15	18.90	
	7200	PEX x male NPT										
	7210	1/2" x 1/2"	Q-5	54	.296	Ea.	4.37	10.90		15.27	21	
	7220	3/4" x 3/4"		46	.348		7.30	12.80		20.10	27.50	
	7230	1" x 1"	↓	40	.400	↓	17.65	14.75		32.40	41.50	
	7300	PEX coupling										
	7310	1/2" x 1/2"	Q-5	50	.320	Ea.	5.25	11.80		17.05	23.50	
	7320	3/4" x 3/4"		42	.381		15.20	14.05		29.25	37.50	
	7330	1" x 1"	↓	36	.444	↓	20.50	16.40		36.90	47	
	7400	PEX x copper sweat										
	7410	1/2" x 1/2"	Q-5	38	.421	Ea.	3.88	15.50		19.38	28	
	7420	3/4" x 3/4"		34	.471		8.05	17.35		25.40	35	
	7430	1" x 1"	↓	29	.552	↓	17.65	20.50		38.15	50	

	15780 Energy Recovery Equipment		CREW	DAILY OUTPUT	LABOR-HOURS	UNIT	MAT.	LABOR	EQUIP.	TOTAL	TOTAL INCL O&P	
100	0010	HEAT RECOVERY PACKAGES										100
	0100	Air to air										
	4000	Enthalpy recovery wheel										
	4010	1000 max CFM	Q-9	1.20	13.333	Ea.	4,425	480		4,905	5,600	
	4020	2000 max CFM		1	16		5,175	580		5,755	6,575	
	4030	4000 max CFM		.80	20		5,975	725		6,700	7,700	
	4040	6000 max CFM	↓	.70	22.857		7,000	825		7,825	8,950	
	4050	8000 max CFM	Q-10	1	24		7,725	900		8,625	9,875	
	4060	10,000 max CFM		.90	26.667		9,250	1,000		10,250	11,800	
	4070	20,000 max CFM		.80	30		16,700	1,125		17,825	20,100	
	4080	25,000 max CFM		.70	34.286		20,500	1,275		21,775	24,500	
	4090	30,000 max CFM		.50	48		22,700	1,800		24,500	27,800	
	4100	40,000 max CFM		.45	53.333		31,300	2,000		33,300	37,600	
	4110	50,000 max CFM	↓	.40	60	↓	36,400	2,250		38,650	43,500	

MECHANICAL 15

15800 | Air Distribution

	15810 Ducts		CREW	DAILY OUTPUT	LABOR-HOURS	UNIT	2005 BARE COSTS MAT.	LABOR	EQUIP.	TOTAL	TOTAL INCL O&P	
100	0010	METAL DUCTWORK										100
	0020	Fabricated rectangular, includes fittings, joints, supports, (R15810-050)										
	0030	allowance for flexible connections, no insulation										
	0031	NOTE: Fabrication and installation are combined (R15810-070)										
	0040	as LABOR cost. Approx. 25% fittings assumed.										
	0042	Fabrication/Inst. is to commercial quality standards (R15810-100)										
	0043	(SMACNA or equiv.) for structure, sealing, leak testing, etc.										
	0050	Add to labor for elevated installation										
	0051	of fabricated ductwork										
	0052	10' to 15' high						6%				

	15810 Ducts	CREW	DAILY OUTPUT	LABOR-HOURS	UNIT	2005 BARE COSTS MAT.	LABOR	EQUIP.	TOTAL	TOTAL INCL O&P	
100	0053 15′ to 20′ high R15810-050						12%				100
	0054 20′ to 25′ high						15%				
	0055 25′ to 30′ high R15810-070						21%				
	0056 30′ to 35′ high						24%				
	0057 35′ to 40′ high R15810-100						30%				
	0058 Over 40′ high						33%				
	0070 For duct insulation and lining see 15080-200-3000										
	0100 Aluminum, alloy 3003-H14, under 100 lb.	Q-10	75	.320	Lb.	2.50	12		14.50	21	
	0110 100 to 500 lb.		80	.300		2	11.25		13.25	19.50	
	0120 500 to 1,000 lb.		95	.253		1.70	9.50		11.20	16.45	
	0140 1,000 to 2,000 lb.		120	.200		1.60	7.50		9.10	13.30	
	0150 2,000 to 5,000 lb.		130	.185		1.53	6.95		8.48	12.35	
	0160 Over 5,000 lb.		145	.166		1.50	6.20		7.70	11.20	
	0500 Galvanized steel, under 200 lb.		235	.102		1.04	3.83		4.87	7.05	
	0520 200 to 500 lb.		245	.098		.84	3.68		4.52	6.55	
	0540 500 to 1,000 lb.		255	.094		.74	3.53		4.27	6.25	
	0560 1,000 to 2,000 lb.		265	.091		.59	3.40		3.99	5.90	
	0570 2,000 to 5,000 lb.		275	.087		.56	3.27		3.83	5.65	
	0580 Over 5,000 lb.	↓	285	.084	↓	.54	3.16		3.70	5.45	
	0600 For large quantities special prices available from supplier										
	1000 Stainless steel, type 304, under 100 lb.	Q-10	165	.145	Lb.	2.47	5.45		7.92	11.10	
	1020 100 to 500 lb.		175	.137		1.97	5.15		7.12	10.05	
	1030 500 to 1,000 lb.		190	.126		1.67	4.74		6.41	9.15	
	1040 1,000 to 2,000 lb.		200	.120		1.57	4.50		6.07	8.70	
	1050 2,000 to 5,000 lb.		225	.107		1.50	4		5.50	7.80	
	1060 Over 5,000 lb.	↓	235	.102	↓	1.47	3.83		5.30	7.50	
	1080 Note: Minimum order cost exceeds per lb. cost for min. wt.										
	1100 For medium pressure ductwork, add						15%				
	1200 For high pressure ductwork, add						40%				
	1276 Duct tape, 2″ x 60 yd roll				Ea.	5.20			5.20	5.70	
	1280 Add to labor for elevated installation										
	1282 of prefabricated (purchased) ductwork										
	1283 10′ to 15′ high						10%				
	1284 15′ to 20′ high						20%				
	1285 20′ to 25′ high						25%				
	1286 25′ to 30′ high						35%				
	1287 30′ to 35′ high						40%				
	1288 35′ to 40′ high						50%				
	1289 Over 40′ high						55%				
	5400 Spiral preformed, steel, galv., straight lengths, Max 10″ spwg.										
	5410 4″ diameter, 26 ga.	Q-9	360	.044	L.F.	.85	1.61		2.46	3.41	
	5416 5″ diameter, 26 ga.		320	.050		1.08	1.81		2.89	3.97	
	5420 6″ diameter, 26 ga.		280	.057		1.23	2.07		3.30	4.53	
	5425 7″ diameter, 26 ga.		240	.067		1.46	2.41		3.87	5.30	
	5430 8″ diameter, 26 ga.		200	.080		1.62	2.89		4.51	6.25	
	5440 10″ diameter, 26 ga.		160	.100		2	3.62		5.62	7.75	
	5450 12″ diameter, 26 ga.		120	.133		2.46	4.82		7.28	10.10	
	5460 14″ diameter, 26 ga.		80	.200		2.85	7.25		10.10	14.30	
	5480 16″ diameter, 24 ga.		60	.267		4.24	9.65		13.89	19.50	
	5490 18″ diameter, 24 ga.	↓	50	.320		4.70	11.60		16.30	23	
	5500 20″ diameter, 24 ga.	Q-10	65	.369		5.25	13.85		19.10	27.50	
	5510 22″ diameter, 24 ga.		60	.400		5.80	15		20.80	29.50	
	5520 24″ diameter, 24 ga.		55	.436		6.30	16.35		22.65	32	
	5540 30″ diameter, 22 ga.		45	.533		9.55	20		29.55	41.50	
	5600 36″ diameter, 22 ga.	↓	40	.600	↓	11.45	22.50		33.95	47	
	5800 Connector, 4″ diameter	Q-9	100	.160	Ea.	1.58	5.80		7.38	10.65	

15 MECHANICAL

Important: See the Reference Section for critical supporting data - Reference Nos., Crews, & City Cost Indexes

	15810 | Ducts		Crew	Daily Output	Labor-Hours	Unit	2005 Bare Costs Mat.	Labor	Equip.	Total	Total Incl O&P	
100	5810	5″ diameter R15810-050	Q-9	94	.170	Ea.	1.69	6.15		7.84	11.30	100
	5820	6″ diameter		88	.182		1.82	6.60		8.42	12.10	
	5840	8″ diameter R15810-070		78	.205		2.35	7.40		9.75	14	
	5860	10″ diameter		70	.229		3.71	8.25		11.96	16.80	
	5880	12″ diameter R15810-100		50	.320		4.47	11.60		16.07	22.50	
	5900	14″ diameter		44	.364		5.20	13.15		18.35	25.50	
	5920	16″ diameter		40	.400		5.80	14.45		20.25	29	
	5930	18″ diameter		37	.432		6.50	15.65		22.15	31	
	5940	20″ diameter		34	.471		7.05	17.05		24.10	34	
	5950	22″ diameter		31	.516		7.70	18.65		26.35	37	
	5960	24″ diameter		28	.571		8.25	20.50		28.75	41	
	5980	30″ diameter		22	.727		10.75	26.50		37.25	52.50	
	6000	36″ diameter		18	.889		16.10	32		48.10	67.50	
	6300	Elbow, 45°, 4″ diameter		60	.267		1.79	9.65		11.44	16.80	
	6310	5″ diameter		52	.308		2.30	11.15		13.45	19.65	
	6320	6″ diameter		44	.364		2.51	13.15		15.66	23	
	6340	8″ diameter		28	.571		3.31	20.50		23.81	35.50	
	6360	10″ diameter		18	.889		5.05	32		37.05	55	
	6380	12″ diameter		13	1.231		6.50	44.50		51	75.50	
	6400	14″ diameter		11	1.455		12.20	52.50		64.70	94.50	
	6420	16″ diameter		10	1.600		20	58		78	111	
	6430	18″ diameter		9.60	1.667		24.50	60.50		85	120	
	6440	20″ diameter	Q-10	14	1.714		31	64.50		95.50	133	
	6450	22″ diameter		13	1.846		34.50	69.50		104	145	
	6460	24″ diameter		12	2		41.50	75		116.50	161	
	6480	30″ diameter		9	2.667		64	100		164	225	
	6500	36″ diameter		7	3.429		78	129		207	284	
	6600	Elbow, 90°, 4″ diameter	Q-9	60	.267		3.05	9.65		12.70	18.20	
	6610	5″ diameter		52	.308		3.51	11.15		14.66	21	
	6620	6″ diameter		44	.364		4.02	13.15		17.17	24.50	
	6625	7″ diameter		36	.444		4.25	16.10		20.35	29	
	6630	8″ diameter		28	.571		4.93	20.50		25.43	37.50	
	6640	10″ diameter		18	.889		7.05	32		39.05	57.50	
	6650	12″ diameter		13	1.231		9.30	44.50		53.80	79	
	6660	14″ diameter		11	1.455		17.45	52.50		69.95	100	
	6670	16″ diameter		10	1.600		27.50	58		85.50	120	
	6676	18″ diameter		9.60	1.667		36	60.50		96.50	133	
	6680	20″ diameter	Q-10	14	1.714		48	64.50		112.50	152	
	6684	22″ diameter		13	1.846		74.50	69.50		144	189	
	6690	24″ diameter		12	2		83.50	75		158.50	207	
	6700	30″ diameter		9	2.667		128	100		228	295	
	6710	36″ diameter		7	3.429		156	129		285	370	
	6720	End cap										
	6722	4″	Q-9	110	.145	Ea.	1.58	5.25		6.83	9.85	
	6724	6″		94	.170		2.10	6.15		8.25	11.75	
	6725	8″		88	.182		3.01	6.60		9.61	13.40	
	6726	10″		78	.205		4.37	7.40		11.77	16.20	
	6727	12″		70	.229		5.80	8.25		14.05	19.10	
	6728	14″		50	.320		8.05	11.60		19.65	26.50	
	6729	16″		44	.364		9.65	13.15		22.80	30.50	
	6730	18″		40	.400		11.30	14.45		25.75	35	
	6731	20″		37	.432		12.30	15.65		27.95	37.50	
	6732	22″		34	.471		13.70	17.05		30.75	41	
	6733	24″		30	.533		16.10	19.30		35.40	47	
	6800	Reducing coupling, 6″ x 4″		46	.348		4.55	12.60		17.15	24.50	
	6820	8″ x 6″		40	.400		5.60	14.45		20.05	28.50	

15810 | Ducts		CREW	DAILY OUTPUT	LABOR-HOURS	UNIT	2005 BARE COSTS MAT.	LABOR	EQUIP.	TOTAL	TOTAL INCL O&P
100 6840	10″ x 8″ R15810-050	Q-9	32	.500	Ea.	7.05	18.10		25.15	36
6860	12″ x 10″		24	.667		10.10	24		34.10	48
6880	14″ x 12″ R15810-070		20	.800		16.65	29		45.65	63
6900	16″ x 14″		18	.889		24.50	32		56.50	76
6920	18″ x 16″ R15810-100	↓	16	1		30	36		66	88.50
6940	20″ x 18″	Q-10	24	1		36.50	37.50		74	98
6950	22″ x 20″		23	1.043		43	39		82	107
6960	24″ x 22″		22	1.091		49	41		90	117
6980	30″ x 28″		18	1.333		73	50		123	158
7000	36″ x 34″	↓	16	1.500		98	56.50		154.50	195
7100	Tee, 90°, 4″ diameter	Q-9	40	.400		3.99	14.45		18.44	27
7110	5″ diameter		34.60	.462		5.30	16.75		22.05	31.50
7120	6″ diameter		30	.533		5.75	19.30		25.05	36
7130	8″ diameter		19	.842		7.50	30.50		38	55.50
7140	10″ diameter		12	1.333		10.70	48		58.70	86
7150	12″ diameter		9	1.778		13.65	64.50		78.15	114
7160	14″ diameter		7	2.286		26	82.50		108.50	156
7170	16″ diameter		6.70	2.388		36.50	86.50		123	173
7176	18″ diameter	↓	6	2.667		50.50	96.50		147	204
7180	20″ diameter	Q-10	8.70	2.759		56	104		160	221
7190	24″ diameter		7	3.429		66.50	129		195.50	271
7200	30″ diameter		6	4		82	150		232	320
7210	36″ diameter	↓	5	4.800	↓	104	180		284	390
7350	Lateral 45°, reducing									
7354	6″ x 6″ x 4″	Q-9	30	.533	Ea.	7.35	19.30		26.65	37.50
7358	8″ x 8″ x 4″		19	.842		9.05	30.50		39.55	57
7362	10″ x 10″ x 4″		12	1.333		12.70	48		60.70	88
7366	12″ x 12″ x 4″		9	1.778		14.35	64.50		78.85	115
7370	14″ x 14″ x 4″		7	2.286		20.50	82.50		103	150
7374	16″ x 16″ x 4″		6.70	2.388		32.50	86.50		119	169
7378	18″ x 18″ x 4″	↓	6	2.667		41	96.50		137.50	193
7382	20″ x 20″ x 4″	Q-10	8.70	2.759		44.50	104		148.50	208
7386	22″ x 22″ x 4″		8	3		50	113		163	228
7390	24″ x 24″ x 4″	↓	7	3.429	↓	53.50	129		182.50	257
7400	Steel, PVC coated both sides, straight lengths									
7410	4″ diameter, 26 ga.	Q-9	360	.044	L.F.	1.19	1.61		2.80	3.78
7416	5″ diameter, 26 ga.		240	.067		1.48	2.41		3.89	5.35
7420	6″ diameter, 26 ga.		280	.057		1.65	2.07		3.72	5
7440	8″ diameter, 26 ga.		200	.080		2.28	2.89		5.17	6.95
7460	10″ diameter, 26 ga.		160	.100		3.07	3.62		6.69	8.95
7480	12″ diameter, 26 ga.		120	.133		3.80	4.82		8.62	11.60
7500	14″ diameter, 26 ga.		80	.200		5.20	7.25		12.45	16.85
7520	16″ diameter, 24 ga.		60	.267		7.35	9.65		17	23
7540	18″ diameter, 24 ga.	↓	45	.356		7.95	12.85		20.80	28.50
7560	20″ diameter, 24 ga.	Q-10	65	.369		9.35	13.85		23.20	32
7580	24″ diameter, 24 ga.		55	.436		11.90	16.35		28.25	38
7600	30″ diameter, 22 ga.		45	.533		17.25	20		37.25	50
7620	36″ diameter, 22 ga.	↓	40	.600	↓	21	22.50		43.50	57.50
7890	Connector, 4″ diameter	Q-9	100	.160	Ea.	2.49	5.80		8.29	11.65
7896	5″ diameter		83	.193		2.66	6.95		9.61	13.70
7900	6″ diameter		88	.182		2.80	6.60		9.40	13.20
7920	8″ diameter		78	.205		3.54	7.40		10.94	15.30
7940	10″ diameter		70	.229		4.87	8.25		13.12	18.05
7960	12″ diameter		50	.320		6.25	11.60		17.85	24.50
7980	14″ diameter		44	.364		8	13.15		21.15	29
8000	16″ diameter	↓	40	.400	↓	8.90	14.45		23.35	32.50

	15810 | Ducts		CREW	DAILY OUTPUT	LABOR-HOURS	UNIT	2005 BARE COSTS MAT.	LABOR	EQUIP.	TOTAL	TOTAL INCL O&P	
100	8020	18″ diameter R15810 -050	Q-9	37	.432	Ea.	10	15.65		25.65	35	100
	8040	20″ diameter		34	.471		10.75	17.05		27.80	38	
	8060	24″ diameter R15810 -070		28	.571		12.05	20.50		32.55	45.50	
	8080	30″ diameter		22	.727		15.75	26.50		42.25	58	
	8100	36″ diameter R15810 -100		18	.889		21.50	32		53.50	73	
	8390	Elbow, 45°, 4″ diameter		60	.267		3.14	9.65		12.79	18.30	
	8396	5″ diameter		36	.444		3.79	16.10		19.89	28.50	
	8400	6″ diameter		44	.364		4.12	13.15		17.27	24.50	
	8420	8″ diameter		28	.571		5.45	20.50		25.95	38	
	8440	10″ diameter		18	.889		7.90	32		39.90	58	
	8460	12″ diameter		13	1.231		11.65	44.50		56.15	81.50	
	8480	14″ diameter		11	1.455		21	52.50		73.50	104	
	8500	16″ diameter		10	1.600		30	58		88	122	
	8520	18″ diameter		9	1.778		34.50	64.50		99	137	
	8540	20″ diameter	Q-10	13	1.846		46.50	69.50		116	158	
	8560	24″ diameter		11	2.182		59	82		141	191	
	8580	30″ diameter		9	2.667		90.50	100		190.50	254	
	8600	36″ diameter		7	3.429		110	129		239	320	
	8800	Elbow, 90°, 4″ diameter	Q-9	60	.267		5.20	9.65		14.85	20.50	
	8804	5″ diameter		52	.308		5.95	11.15		17.10	23.50	
	8806	6″ diameter		44	.364		6.50	13.15		19.65	27	
	8808	8″ diameter		28	.571		8.10	20.50		28.60	41	
	8810	10″ diameter		18	.889		10.95	32		42.95	61.50	
	8812	12″ diameter		13	1.231		15.10	44.50		59.60	85	
	8814	14″ diameter		11	1.455		27	52.50		79.50	111	
	8816	16″ diameter		10	1.600		42	58		100	135	
	8818	18″ diameter		9.60	1.667		51.50	60.50		112	150	
	8820	20″ diameter	Q-10	14	1.714		65.50	64.50		130	171	
	8822	24″ diameter		12	2		118	75		193	244	
	8824	30″ diameter		9	2.667		181	100		281	355	
	8826	36″ diameter		7	3.429		220	129		349	440	
	8850	Tee, 90°, 4″ diameter	Q-9	40	.400		5.85	14.45		20.30	29	
	8854	5″ diameter		34.60	.462		7.70	16.75		24.45	34	
	8856	6″ diameter		30	.533		8.55	19.30		27.85	39	
	8858	8″ diameter		19	.842		10.95	30.50		41.45	59	
	8860	10″ diameter		12	1.333		14.30	48		62.30	90	
	8862	12″ diameter		9	1.778		20.50	64.50		85	122	
	8864	14″ diameter		7	2.286		35.50	82.50		118	166	
	8866	16″ diameter		6.70	2.388		51.50	86.50		138	190	
	8868	18″ diameter		6	2.667		74.50	96.50		171	230	
	8870	20″ diameter	Q-10	8.70	2.759		80.50	104		184.50	248	
	8872	24″ diameter		7	3.429		111	129		240	320	
	8874	30″ diameter		6	4		138	150		288	385	
	8876	36″ diameter		5	4.800		175	180		355	470	
	9000	Reducing coupling, tapered offset										
	9020	Diameter x any smaller diameter										
	9040	6″ x	Q-9	46	.348	Ea.	6.50	12.60		19.10	26.50	
	9060	8″ x		40	.400		8	14.45		22.45	31.50	
	9080	10″ x		32	.500		9.65	18.10		27.75	38.50	
	9100	12″ x		24	.667		13.45	24		37.45	52	
	9120	14″ x		20	.800		21.50	29		50.50	68.50	
	9140	16″ x		18	.889		33	32		65	85.50	
	9160	18″ x		16	1		40	36		76	99.50	
	9180	20″ x	Q-10	24	1		49	37.50		86.50	112	
	9200	24″ x		22	1.091		66.50	41		107.50	136	
	9220	30″ x		18	1.333		94	50		144	180	

	15810 | Ducts		CREW	DAILY OUTPUT	LABOR-HOURS	UNIT	2005 BARE COSTS MAT.	LABOR	EQUIP.	TOTAL	TOTAL INCL O&P
100	9240	36″ x R15810-050	Q-10	16	1.500	Ea.	122	56.50		178.50	221
	9400	Lateral 45°, straight barrel									
	9410	4″ size R15810-070	Q-9	40	.400	Ea.	6.75	14.45		21.20	30
	9416	5″ size		34.60	.462		8.85	16.75		25.60	35.50
	9420	6″ size R15810-100		30	.533		9.80	19.30		29.10	40.50
	9440	8″ size		19	.842		12.60	30.50		43.10	61
	9460	10″ size		12	1.333		16.45	48		64.45	92
	9480	12″ size		9	1.778		20.50	64.50		85	122
	9500	14″ size		7	2.286		27	82.50		109.50	157
	9520	16″ size		6.70	2.388		51	86.50		137.50	189
	9540	18″ size		6	2.667		72	96.50		168.50	227
	9560	20″ size	Q-10	8.70	2.759		85.50	104		189.50	253
	9580	24″ size		7	3.429		119	129		248	330
	9600	30″ size		6	4		139	150		289	385
	9620	36″ size		5	4.800		181	180		361	475
	9640	For reducing lateral, add					40%				
	9650	Lateral 45°, reducing									
	9654	6″ x 6″ x 4″	Q-9	30	.533	Ea.	14.55	19.30		33.85	45.50
	9658	8″ x 8″ x 4″		19	.842		18.40	30.50		48.90	67.50
	9662	10″ x 10″ x 4″		12	1.333		24	48		72	101
	9666	12″ x 12″ x 4″		9	1.778		28	64.50		92.50	130
	9670	14″ x 14″ x 4″		7	2.286		46.50	82.50		129	178
	9674	16″ x 16″ x 4″		6.70	2.388		58	86.50		144.50	197
	9678	18″ x 18″ x 4″		6	2.667		91	96.50		187.50	248
	9682	20″ x 20″ x 4″	Q-10	8.70	2.759		116	104		220	287
	9686	22″ x 22″ x 4″		8	3		132	113		245	320
	9690	24″ x 24″ x 4″		7	3.429		142	129		271	355
	9700	Register boot, elbow or perimeter									
	9710	6″ duct diameter	Q-9	60	.267	Ea.	24	9.65		33.65	41.50
	9720	8″ duct diameter		56	.286		29	10.35		39.35	48
	9730	10″ duct diameter		45	.356		40	12.85		52.85	63.50
	9740	12″ duct diameter		36	.444		40	16.10		56.10	68
	9800	Steel, stainless, straight lengths									
	9810	3″ diameter, 24 Ga.	Q-9	400	.040	L.F.	4.28	1.45		5.73	6.95
	9820	4″ diameter, 24 Ga.		360	.044		5.40	1.61		7.01	8.40
	9830	5″ diameter, 24 Ga.		320	.050		6.65	1.81		8.46	10.10
	9840	6″ diameter, 24 Ga.		280	.057		7.80	2.07		9.87	11.80
	9850	7″ diameter, 24 Ga.		240	.067		8.95	2.41		11.36	13.55
	9860	8″ diameter, 24 Ga.		200	.080		10.15	2.89		13.04	15.60
	9870	9″ diameter, 24 Ga.		180	.089		11.30	3.22		14.52	17.40
	9880	10″ diameter, 24 Ga.		160	.100		12.45	3.62		16.07	19.25
	9890	12″ diameter, 24 Ga.		120	.133		14.85	4.82		19.67	23.50
	9900	14″ diameter, 24 Ga.		80	.200		17.15	7.25		24.40	30
	9910	16″ diameter, 24 Ga.		60	.267		19.50	9.65		29.15	36.50
	9920	18″ diameter, 24 Ga.		45	.356		22	12.85		34.85	44
	9930	20″ diameter, 24 Ga.	Q-10	65	.369		24	13.85		37.85	48
	9940	24″ diameter, 24 Ga.		55	.436		29	16.35		45.35	57
	9950	30″ diameter, 22 Ga.		45	.533		44	20		64	79.50
	9960	36″ diameter, 22 Ga.		40	.600		53.50	22.50		76	93.50
	9970	40″ diameter, 20 Ga.		35	.686		71	25.50		96.50	118
300	0010	**FIBEROUS GLASS DUCTWORK** R15810-050									
	1280	Add to labor for elevated installation									
	1282	of prefabricated (purchased) ductwork									
	1283	10′ to 15′ high						10%			
	1284	15′ to 20′ high						20%			
	1285	20′ to 25′ high						25%			

15 MECHANICAL

	15810 | Ducts		Crew	Daily Output	Labor-Hours	Unit	2005 Bare Costs Mat.	Labor	Equip.	Total	Total Incl O&P	
300	1286	25′ to 30′ high (R15810-050)						35%				300
	1287	30′ to 35′ high						40%				
	1288	35′ to 40′ high						50%				
	1289	Over 40′ high						55%				
	3490	Rigid fiberglass duct board, foil reinf. kraft facing										
	3500	Rectangular, 1″ thick, alum. faced, (FRK), std. weight	Q-10	350	.069	SF Surf	.66	2.57		3.23	4.69	
500	0010	FLEXIBLE DUCTS (R15810-050)										500
	1280	Add to labor for elevated installation										
	1282	of prefabricated (purchased) ductwork										
	1283	10′ to 15′ high						10%				
	1284	15′ to 20′ high						20%				
	1285	20′ to 25′ high						25%				
	1286	25′ to 30′ high						35%				
	1287	30′ to 35′ high						40%				
	1288	35′ to 40′ high						50%				
	1289	Over 40′ high						55%				
	1300	Flexible, coated fiberglass fabric on corr. resist. metal helix										
	1400	pressure to 12″ (WG) UL-181										
	1500	Non-insulated, 3″ diameter	Q-9	400	.040	L.F.	1.03	1.45		2.48	3.36	
	1520	4″ diameter		360	.044		1.03	1.61		2.64	3.60	
	1540	5″ diameter		320	.050		1.29	1.81		3.10	4.20	
	1560	6″ diameter		280	.057		1.52	2.07		3.59	4.85	
	1580	7″ diameter		240	.067		1.81	2.41		4.22	5.70	
	1600	8″ diameter		200	.080		2.09	2.89		4.98	6.75	
	1620	9″ diameter		180	.089		2.47	3.22		5.69	7.65	
	1640	10″ diameter		160	.100		2.63	3.62		6.25	8.45	
	1660	12″ diameter		120	.133		3.18	4.82		8	10.90	
	1680	14″ diameter		80	.200		3.86	7.25		11.11	15.40	
	1700	16″ diameter		60	.267		5.70	9.65		15.35	21	
	1800	For adjustable clamps, add				Ea.	.74			.74	.81	
	1900	Insulated, 1″ thick, PE jacket, 3″ diameter	Q-9	380	.042	L.F.	1.14	1.52		2.66	3.59	
	1910	4″ diameter		340	.047		1.23	1.70		2.93	3.97	
	1920	5″ diameter		300	.053		1.35	1.93		3.28	4.46	
	1940	6″ diameter		260	.062		1.56	2.23		3.79	5.15	
	1960	7″ diameter		220	.073		1.76	2.63		4.39	6	
	1980	8″ diameter		180	.089		1.96	3.22		5.18	7.10	
	2000	9″ diameter		160	.100		2.15	3.62		5.77	7.90	
	2020	10″ diameter		140	.114		2.40	4.14		6.54	9	
	2040	12″ diameter		100	.160		2.95	5.80		8.75	12.15	
	2060	14″ diameter		80	.200		3.35	7.25		10.60	14.85	
	2080	16″ diameter		60	.267		3.70	9.65		13.35	18.90	
	2100	18″ diameter		45	.356		4.20	12.85		17.05	24.50	
	2120	20″ diameter	Q-10	65	.369		8.65	13.85		22.50	31	
	2500	Insulated, heavy duty, coated fiberglass fabric										
	2520	4″ diameter	Q-9	340	.047	L.F.	3.18	1.70		4.88	6.10	
	2540	5″ diameter		300	.053		3.41	1.93		5.34	6.70	
	2560	6″ diameter		260	.062		3.61	2.23		5.84	7.40	
	2580	7″ diameter		220	.073		4.36	2.63		6.99	8.85	
	2600	8″ diameter		180	.089		4.49	3.22		7.71	9.90	
	2620	9″ diameter		160	.100		5.10	3.62		8.72	11.15	
	2640	10″ diameter		140	.114		5.25	4.14		9.39	12.15	
	2660	12″ diameter		100	.160		6.60	5.80		12.40	16.15	
	2680	14″ diameter		80	.200		8	7.25		15.25	19.95	
	2700	16″ diameter		60	.267		9.95	9.65		19.60	26	
	2720	18″ diameter		45	.356		12.60	12.85		25.45	33.50	
	2800	Flexible, alum., pressure to 12″ (WG) UL-181										

	15810 | Ducts		CREW	DAILY OUTPUT	LABOR-HOURS	UNIT	2005 BARE COSTS MAT.	LABOR	EQUIP.	TOTAL	TOTAL INCL O&P	
500	2820	Non-insulated R15810-050										500
	2830	3" diameter	Q-9	400	.040	L.F.	1.07	1.45		2.52	3.41	
	2831	4" diameter		360	.044		1.24	1.61		2.85	3.83	
	2832	5" diameter		320	.050		1.59	1.81		3.40	4.53	
	2833	6" diameter		280	.057		2.02	2.07		4.09	5.40	
	2834	7" diameter		240	.067		2.46	2.41		4.87	6.40	
	2835	8" diameter		200	.080		2.80	2.89		5.69	7.55	
	2836	9" diameter		180	.089		3.32	3.22		6.54	8.60	
	2837	10" diameter		160	.100		3.90	3.62		7.52	9.85	
	2838	12" diameter		120	.133		4.80	4.82		9.62	12.70	
	2839	14" diameter		80	.200		5.90	7.25		13.15	17.65	
	2840	15" diameter		70	.229		6.90	8.25		15.15	20.50	
	2841	16" diameter		60	.267		9.20	9.65		18.85	25	
	2842	18" diameter	↓	45	.356		10.25	12.85		23.10	31	
	2843	20" diameter	Q-10	65	.369	↓	11.35	13.85		25.20	34	
	2880	Insulated, 1" thick with 3/4 lb., PE jacket										
	2890	3" diameter	Q-9	380	.042	L.F.	1.79	1.52		3.31	4.31	
	2891	4" diameter		340	.047		2.05	1.70		3.75	4.88	
	2892	5" diameter		300	.053		2.83	1.93		4.76	6.10	
	2893	6" diameter		260	.062		3.13	2.23		5.36	6.85	
	2894	7" diameter		220	.073		3.85	2.63		6.48	8.30	
	2895	8" diameter		180	.089		4.13	3.22		7.35	9.50	
	2896	9" diameter		160	.100		4.67	3.62		8.29	10.70	
	2897	10" diameter		140	.114		5.75	4.14		9.89	12.65	
	2898	12" diameter		100	.160		6.30	5.80		12.10	15.80	
	2899	14" diameter		80	.200		7.70	7.25		14.95	19.60	
	2900	15" diameter		70	.229		10.05	8.25		18.30	24	
	2901	16" diameter		60	.267		11.95	9.65		21.60	28	
	2902	18" diameter	↓	45	.356	↓	13.35	12.85		26.20	34.50	
	3000	Transitions, 3" deep, square to round										
	3010	6" x 6" to round	Q-9	44	.364	Ea.	7.80	13.15		20.95	28.50	
	3014	6" x 12" to round		42	.381		13.65	13.80		27.45	36	
	3018	8" x 8" to round		40	.400		8.45	14.45		22.90	32	
	3022	9" x 9" to round		38	.421		9.75	15.25		25	34.50	
	3026	9" x 12" to round		37	.432		17.55	15.65		33.20	43.50	
	3030	9" x 15" to round		36	.444		19.45	16.10		35.55	46	
	3034	10" x 10" to round		34	.471		11.05	17.05		28.10	38	
	3038	10" x 22" to round		32	.500		17.55	18.10		35.65	47.50	
	3042	12" x 12" to round		30	.533		11.70	19.30		31	42.50	
	3046	12" x 18" to round		28	.571		24.50	20.50		45	59	
	3050	12" x 24" to round		26	.615		24.50	22.50		47	61.50	
	3054	15" x 15" to round		25	.640		14.95	23		37.95	52	
	3058	15" x 18" to round		24	.667		23.50	24		47.50	62.50	
	3062	18" x 18" to round		22	.727		17.55	26.50		44.05	60	
	3066	21" x 21" to round		21	.762		31	27.50		58.50	77	
	3070	22" x 22" to round		20	.800		31	29		60	79	
	3074	24" x 24" to round ↓	↓	18	.889	↓	31	32		63	84	
700	0010	**GLASS FIBER-REINFORCED PLASTIC DUCTS** R15810-050										700
	1280	Add to labor for elevated installation										
	1282	of prefabricated (purchased) ductwork										
	1283	10' to 15' high						10%				
	1284	15' to 20' high						20%				
	1285	20' to 25' high						25%				
	1286	25' to 30' high						35%				
	1287	30' to 35' high ↓						40%				

15 MECHANICAL

	15810 | Ducts		CREW	DAILY OUTPUT	LABOR-HOURS	UNIT	2005 BARE COSTS MAT.	LABOR	EQUIP.	TOTAL	TOTAL INCL O&P
700	1288	35′ to 40′ high R15810-050						50%			
	1289	Over 40′ high						55%			
	3550	Rigid fiberglass reinforced plastic, FM approved									
	3552	for acid fume and smoke exhaust system, nonflammable									
	3554	Straight, 4″ diameter	Q-9	200	.080	L.F.	9.20	2.89		12.09	14.55
	3555	6″ diameter		146	.110		12.05	3.97		16.02	19.35
	3556	8″ diameter		106	.151		14.60	5.45		20.05	24.50
	3557	10″ diameter		85	.188		17.55	6.80		24.35	30
	3558	12″ diameter		68	.235		20.50	8.50		29	35.50
	3559	14″ diameter		50	.320		30.50	11.60		42.10	51.50
	3560	16″ diameter		40	.400		34	14.45		48.45	59.50
	3561	18″ diameter	↓	31	.516		37.50	18.65		56.15	70
	3562	20″ diameter	Q-10	44	.545		41.50	20.50		62	77
	3563	22″ diameter		40	.600		45	22.50		67.50	84
	3564	24″ diameter		37	.649		48.50	24.50		73	90.50
	3565	26″ diameter		34	.706		52	26.50		78.50	98
	3566	28″ diameter		28	.857		56	32		88	111
	3567	30″ diameter		24	1		59.50	37.50		97	123
	3568	32″ diameter		22.60	1.062		63.50	40		103.50	131
	3569	34″ diameter		21.60	1.111		66.50	41.50		108	138
	3570	36″ diameter		20.60	1.165		70.50	43.50		114	145
	3571	38″ diameter		19.80	1.212		109	45.50		154.50	189
	3572	42″ diameter		18.30	1.311		121	49		170	210
	3573	46″ diameter		17	1.412		134	53		187	230
	3574	50″ diameter		15.90	1.509		146	56.50		202.50	248
	3575	54″ diameter		14.90	1.611		161	60.50		221.50	270
	3576	60″ diameter		13.60	1.765		184	66		250	305
	3577	64″ diameter		12.80	1.875		269	70.50		339.50	405
	3578	68″ diameter		12.20	1.967		286	74		360	430
	3579	72″ diameter	↓	11.90	2.017	↓	310	75.50		385.50	455
	3584	Note: joints are cemented with									
	3586	fiberglass resin, included in material cost.									
	3590	Elbow, 90°, 4″ diameter	Q-9	20.30	.788	Ea.	44.50	28.50		73	93
	3591	6″ diameter		13.80	1.159		62	42		104	133
	3592	8″ diameter		10	1.600		68.50	58		126.50	164
	3593	10″ diameter		7.80	2.051		90	74		164	213
	3594	12″ diameter		6.50	2.462		114	89		203	262
	3595	14″ diameter		5.50	2.909		139	105		244	315
	3596	16″ diameter		4.80	3.333		166	121		287	370
	3597	18″ diameter	↓	4.40	3.636		238	132		370	465
	3598	20″ diameter	Q-10	5.90	4.068		251	153		404	510
	3599	22″ diameter		5.40	4.444		310	167		477	595
	3600	24″ diameter		4.90	4.898		355	184		539	675
	3601	26″ diameter		4.60	5.217		360	196		556	695
	3602	28″ diameter		4.20	5.714		450	214		664	825
	3603	30″ diameter		3.90	6.154		500	231		731	905
	3604	32″ diameter		3.70	6.486		555	243		798	985
	3605	34″ diameter		3.45	6.957		655	261		916	1,125
	3606	36″ diameter		3.20	7.500		750	281		1,031	1,275
	3607	38″ diameter		3.07	7.818		1,075	293		1,368	1,625
	3608	42″ diameter		2.70	8.889		1,275	335		1,610	1,925
	3609	46″ diameter		2.54	9.449		1,500	355		1,855	2,200
	3610	50″ diameter		2.33	10.300		1,725	385		2,110	2,500
	3611	54″ diameter		2.14	11.215		2,000	420		2,420	2,825
	3612	60″ diameter		1.94	12.371		2,650	465		3,115	3,650
	3613	64″ diameter	↓	1.82	13.187	↓	3,850	495		4,345	4,975

15810 | Ducts

			CREW	DAILY OUTPUT	LABOR-HOURS	UNIT	2005 BARE COSTS MAT.	LABOR	EQUIP.	TOTAL	TOTAL INCL O&P	
700	3614	68" diameter (R15810-050)	Q-10	1.71	14.035	Ea.	4,275	525		4,800	5,500	700
	3615	72" diameter		1.62	14.815		4,725	555		5,280	6,050	
	3626	Elbow, 45°, 4" diameter	Q-9	22.30	.717		43	26		69	87	
	3627	6" diameter		15.20	1.053		59	38		97	124	
	3628	8" diameter		11	1.455		63	52.50		115.50	150	
	3629	10" diameter		8.60	1.860		66.50	67.50		134	177	
	3630	12" diameter		7.20	2.222		72.50	80.50		153	204	
	3631	14" diameter		6.10	2.623		89	95		184	244	
	3632	16" diameter		5.28	3.030		106	110		216	286	
	3633	18" diameter		4.84	3.306		124	120		244	320	
	3634	20" diameter	Q-10	6.50	3.692		144	139		283	370	
	3635	22" diameter		5.94	4.040		196	152		348	450	
	3636	24" diameter		5.40	4.444		223	167		390	500	
	3637	26" diameter		5.06	4.743		253	178		431	550	
	3638	28" diameter		4.62	5.195		274	195		469	600	
	3639	30" diameter		4.30	5.581		315	209		524	670	
	3640	32" diameter		4.07	5.897		350	221		571	725	
	3641	34" diameter		3.80	6.316		415	237		652	820	
	3642	36" diameter		3.52	6.818		480	256		736	920	
	3643	38" diameter		3.38	7.101		670	266		936	1,150	
	3644	42" diameter		2.97	8.081		805	305		1,110	1,350	
	3645	46" diameter		2.80	8.571		940	320		1,260	1,525	
	3646	50" diameter		2.56	9.375		1,100	350		1,450	1,750	
	3647	54" diameter		2.35	10.213		1,250	385		1,635	1,975	
	3648	60" diameter		2.13	11.268		1,675	425		2,100	2,475	
	3649	64" diameter		2	12		2,425	450		2,875	3,350	
	3650	68" diameter		1.88	12.766		2,675	480		3,155	3,675	
	3651	72" diameter		1.78	13.483		2,950	505		3,455	4,025	
	3660	Tee, 90°, 4" diameter	Q-9	14.06	1.138		24.50	41		65.50	90.50	
	3661	6" diameter		9.52	1.681		37	61		98	135	
	3662	8" diameter		6.98	2.292		51	83		134	184	
	3663	10" diameter		5.45	2.936		66.50	106		172.50	236	
	3664	12" diameter		4.60	3.478		83	126		209	285	
	3665	14" diameter		3.86	4.145		121	150		271	365	
	3666	16" diameter		3.40	4.706		144	170		314	420	
	3667	18" diameter		3.05	5.246		170	190		360	480	
	3668	20" diameter	Q-10	4.12	5.825		199	219		418	555	
	3669	22" diameter		3.78	6.349		227	238		465	615	
	3670	24" diameter		3.46	6.936		257	260		517	680	
	3671	26" diameter		3.20	7.500		291	281		572	755	
	3672	28" diameter		2.96	8.108		325	305		630	825	
	3673	30" diameter		2.75	8.727		365	325		690	905	
	3674	32" diameter		2.54	9.449		400	355		755	985	
	3675	34" diameter		2.40	10		440	375		815	1,050	
	3676	36" diameter		2.27	10.573		485	395		880	1,150	
	3677	38" diameter		2.15	11.163		725	420		1,145	1,450	
	3678	42" diameter		1.91	12.565		875	470		1,345	1,700	
	3679	46" diameter		1.78	13.483		1,025	505		1,530	1,900	
	3680	50" diameter		1.63	14.724		1,200	550		1,750	2,175	
	3681	54" diameter		1.50	16		1,400	600		2,000	2,450	
	3682	60" diameter		1.35	17.778		1,725	665		2,390	2,925	
	3683	64" diameter		1.27	18.898		2,600	710		3,310	3,975	
	3684	68" diameter		1.20	20		2,900	750		3,650	4,350	
	3685	72" diameter		1.13	21.239		3,250	795		4,045	4,800	
	3690	For Y @ 45°, add					22%					
	3700	Blast gate, 4" diameter	Q-9	22.30	.717		118	26		144	170	

Important: See the Reference Section for critical supporting data - Reference Nos., Crews, & City Cost Indexes

	15810 | Ducts		CREW	DAILY OUTPUT	LABOR-HOURS	UNIT	2005 BARE COSTS MAT.	LABOR	EQUIP.	TOTAL	TOTAL INCL O&P	
700	3701	6″ diameter R15810-050	Q-9	15.20	1.053	Ea.	133	38		171	206	700
	3702	8″ diameter		11	1.455		180	52.50		232.50	279	
	3703	10″ diameter		8.60	1.860		205	67.50		272.50	330	
	3704	12″ diameter		7.20	2.222		251	80.50		331.50	400	
	3705	14″ diameter		6.10	2.623		278	95		373	450	
	3706	16″ diameter		5.28	3.030		310	110		420	510	
	3707	18″ diameter		4.84	3.306		340	120		460	555	
	3708	20″ diameter	Q-10	6.50	3.692		365	139		504	615	
	3709	22″ diameter		5.94	4.040		395	152		547	670	
	3710	24″ diameter		5.40	4.444		425	167		592	720	
	3711	26″ diameter		5.06	4.743		455	178		633	775	
	3712	28″ diameter		4.62	5.195		485	195		680	835	
	3713	30″ diameter		4.30	5.581		515	209		724	890	
	3714	32″ diameter		4.07	5.897		565	221		786	960	
	3715	34″ diameter		3.80	6.316		615	237		852	1,050	
	3716	36″ diameter		3.52	6.818		665	256		921	1,125	
	3717	38″ diameter		3.38	7.101		900	266		1,166	1,400	
	3718	42″ diameter		2.97	8.081		1,025	305		1,330	1,600	
	3719	46″ diameter		2.80	8.571		1,175	320		1,495	1,800	
	3720	50″ diameter		2.56	9.375		1,350	350		1,700	2,025	
	3721	54″ diameter		2.35	10.213		1,525	385		1,910	2,275	
	3722	60″ diameter		2.13	11.268		1,775	425		2,200	2,600	
	3723	64″ diameter		2	12		1,975	450		2,425	2,875	
	3724	68″ diameter		1.88	12.766		2,175	480		2,655	3,125	
	3725	72″ diameter		1.78	13.483		2,400	505		2,905	3,400	
	3740	Reducers, 4″ diameter	Q-9	24.50	.653		29.50	23.50		53	69	
	3741	6″ diameter		16.70	.958		44	34.50		78.50	102	
	3742	8″ diameter		12	1.333		59.50	48		107.50	140	
	3743	10″ diameter		9.50	1.684		73.50	61		134.50	175	
	3744	12″ diameter		7.90	2.025		88	73.50		161.50	210	
	3745	14″ diameter		6.70	2.388		103	86.50		189.50	246	
	3746	16″ diameter		5.80	2.759		117	100		217	283	
	3747	18″ diameter		5.30	3.019		133	109		242	315	
	3748	20″ diameter	Q-10	7.20	3.333		147	125		272	355	
	3749	22″ diameter		6.50	3.692		237	139		376	475	
	3750	24″ diameter		5.90	4.068		257	153		410	515	
	3751	26″ diameter		5.60	4.286		278	161		439	550	
	3752	28″ diameter		5.10	4.706		305	177		482	605	
	3753	30″ diameter		4.70	5.106		325	192		517	650	
	3754	32″ diameter		4.50	5.333		345	200		545	690	
	3755	34″ diameter		4.20	5.714		370	214		584	735	
	3756	36″ diameter		3.90	6.154		390	231		621	785	
	3757	38″ diameter		3.70	6.486		590	243		833	1,025	
	3758	42″ diameter		3.30	7.273		650	273		923	1,125	
	3759	46″ diameter		3.10	7.742		715	290		1,005	1,225	
	3760	50″ diameter		2.80	8.571		775	320		1,095	1,350	
	3761	54″ diameter		2.60	9.231		840	345		1,185	1,450	
	3762	60″ diameter		2.34	10.256		940	385		1,325	1,625	
	3763	64″ diameter		2.20	10.909		1,450	410		1,860	2,225	
	3764	68″ diameter		2.06	11.650		1,550	435		1,985	2,375	
	3765	72″ diameter		1.96	12.245		1,650	460		2,110	2,500	
	3780	Flange, per each, 4″ diameter	Q-9	40.60	.394		71.50	14.25		85.75	101	
	3781	6″ diameter		27.70	.578		83.50	21		104.50	124	
	3782	8″ diameter		20.50	.780		104	28		132	159	
	3783	10″ diameter		16.30	.982		125	35.50		160.50	192	
	3784	12″ diameter		13.90	1.151		144	41.50		185.50	222	

15810 | Ducts

	Line	Description	Crew	Daily Output	Labor-Hours	Unit	2005 Bare Costs Mat.	2005 Bare Costs Labor	2005 Bare Costs Equip.	2005 Bare Costs Total	Total Incl O&P	
700	3785	14″ diameter (R15810-050)	Q-9	11.30	1.416	Ea.	166	51		217	261	700
	3786	16″ diameter		9.70	1.649		186	59.50		245.50	297	
	3787	18″ diameter		8.30	1.928		207	69.50		276.50	335	
	3788	20″ diameter	Q-10	11.30	2.124		227	79.50		306.50	375	
	3789	22″ diameter		10.40	2.308		248	86.50		334.50	405	
	3790	24″ diameter		9.50	2.526		269	95		364	440	
	3791	26″ diameter		8.80	2.727		291	102		393	475	
	3792	28″ diameter		8.20	2.927		310	110		420	515	
	3793	30″ diameter		7.60	3.158		330	118		448	540	
	3794	32″ diameter		7.20	3.333		350	125		475	575	
	3795	34″ diameter		6.80	3.529		370	132		502	615	
	3796	36″ diameter		6.40	3.750		395	141		536	650	
	3797	38″ diameter		6.10	3.934		415	148		563	680	
	3798	42″ diameter		5.50	4.364		455	164		619	750	
	3799	46″ diameter		5	4.800		495	180		675	820	
	3800	50″ diameter		4.60	5.217		540	196		736	895	
	3801	54″ diameter		4.20	5.714		575	214		789	965	
	3802	60″ diameter		3.80	6.316		640	237		877	1,075	
	3803	64″ diameter		3.60	6.667		680	250		930	1,125	
	3804	68″ diameter		3.40	7.059		720	265		985	1,200	
	3805	72″ diameter		3.20	7.500		765	281		1,046	1,275	
	4000	Rigid plastic, corrosive fume resistant PVC										
	4020	Straight, 6″ diameter	Q-9	220	.073	L.F.	6.65	2.63		9.28	11.40	
	4030	7″ diameter		160	.100		9.90	3.62		13.52	16.45	
	4040	8″ diameter		160	.100		9.90	3.62		13.52	16.45	
	4050	9″ diameter		140	.114		11.20	4.14		15.34	18.65	
	4060	10″ diameter		120	.133		12.50	4.82		17.32	21	
	4070	12″ diameter		100	.160		14.55	5.80		20.35	25	
	4080	14″ diameter		70	.229		18.90	8.25		27.15	33.50	
	4090	16″ diameter		62	.258		22	9.35		31.35	38.50	
	4100	18″ diameter		58	.276		34	10		44	53	
	4110	20″ diameter	Q-10	75	.320		44	12		56	66.50	
	4120	22″ diameter		60	.400		50	15		65	78	
	4130	24″ diameter		55	.436		56	16.35		72.35	87	
	4140	26″ diameter		42	.571		63.50	21.50		85	103	
	4150	28″ diameter		35	.686		68.50	25.50		94	115	
	4160	30″ diameter		28	.857		74	32		106	131	
	4170	36″ diameter		23	1.043		88.50	39		127.50	157	
	4180	42″ diameter		20	1.200		103	45		148	183	
	4200	48″ diameter		18	1.333		118	50		168	207	
	4250	Coupling, 6″ diameter	Q-9	88	.182	Ea.	6.95	6.60		13.55	17.75	
	4260	7″ diameter		82	.195		7.45	7.05		14.50	19.05	
	4270	8″ diameter		78	.205		7.80	7.40		15.20	19.95	
	4280	9″ diameter		74	.216		8.50	7.80		16.30	21.50	
	4290	10″ diameter		70	.229		8.50	8.25		16.75	22	
	4300	12″ diameter		55	.291		9.10	10.55		19.65	26	
	4310	14″ diameter		44	.364		11.05	13.15		24.20	32	
	4320	16″ diameter		40	.400		13.10	14.45		27.55	37	
	4330	18″ diameter		38	.421		15.20	15.25		30.45	40	
	4340	20″ diameter	Q-10	55	.436		26.50	16.35		42.85	54	
	4350	22″ diameter		50	.480		30.50	18		48.50	61	
	4360	24″ diameter		45	.533		34	20		54	68.50	
	4370	26″ diameter		39	.615		36	23		59	75	
	4380	28″ diameter		33	.727		38	27.50		65.50	83.50	
	4390	30″ diameter		27	.889		41.50	33.50		75	97	
	4400	36″ diameter		25	.960		65.50	36		101.50	128	

15810 | Ducts

		15810 Ducts	CREW	DAILY OUTPUT	LABOR-HOURS	UNIT	2005 BARE COSTS MAT.	LABOR	EQUIP.	TOTAL	TOTAL INCL O&P	
700	4410	42″ diameter R15810-050	Q-10	23	1.043	Ea.	73	39		112	141	700
	4420	48″ diameter		21	1.143		85.50	43		128.50	160	
	4470	Elbow, 90°, 6″ diameter	Q-9	44	.364		65.50	13.15		78.65	92	
	4480	7″ diameter		36	.444		70	16.10		86.10	102	
	4490	8″ diameter		28	.571		70	20.50		90.50	109	
	4500	9″ diameter		22	.727		96	26.50		122.50	147	
	4510	10″ diameter		18	.889		113	32		145	175	
	4520	12″ diameter		15	1.067		154	38.50		192.50	230	
	4530	14″ diameter		11	1.455		211	52.50		263.50	315	
	4540	16″ diameter		10	1.600		255	58		313	370	
	4550	18″ diameter	Q-10	15	1.600		276	60		336	400	
	4560	20″ diameter		14	1.714		293	64.50		357.50	420	
	4570	22″ diameter		13	1.846		375	69.50		444.50	515	
	4580	24″ diameter		12	2		420	75		495	575	
	4590	26″ diameter		11	2.182		480	82		562	655	
	4600	28″ diameter		10	2.400		570	90		660	765	
	4610	30″ diameter		9	2.667		670	100		770	895	
	4620	36″ diameter		7	3.429		920	129		1,049	1,225	
	4630	42″ diameter		5	4.800		1,275	180		1,455	1,675	
	4640	48″ diameter		4	6		1,675	225		1,900	2,175	
	4750	Elbow 45°, use 90° and deduct					45%					

15820 | Duct Accessories

		15820 Duct Accessories	CREW	DAILY OUTPUT	LABOR-HOURS	UNIT	MAT.	LABOR	EQUIP.	TOTAL	TOTAL INCL O&P	
300	0010	**DUCT ACCESSORIES**										300
	0050	Air extractors, 12″ x 4″	1 Shee	24	.333	Ea.	15.55	13.40		28.95	37.50	
	0100	8″ x 6″		22	.364		15.55	14.60		30.15	39.50	
	0120	12″ x 6″		21	.381		21.50	15.30		36.80	47	
	0140	16″ x 6″		20	.400		27	16.10		43.10	54.50	
	0160	24″ x 6″		18	.444		45	17.85		62.85	77	
	0180	12″ x 8″		20	.400		24	16.10		40.10	51	
	0200	20″ x 8″		16	.500		35	20		55	70	
	0240	18″ x 10″		14	.571		34.50	23		57.50	73	
	0260	24″ x 10″		12	.667		44.50	27		71.50	89.50	
	0280	24″ x 12″		10	.800		48	32		80	102	
	0300	30″ x 12″		8	1		59.50	40		99.50	128	
	0350	Duct collar, spin-in type										
	0352	Without damper										
	0354	Round										
	0356	4″	1 Shee	32	.250	Ea.	2.02	10.05		12.07	17.65	
	0357	5″		30	.267		2.02	10.70		12.72	18.70	
	0358	6″		28	.286		2.21	11.50		13.71	20	
	0359	7″		26	.308		2.41	12.35		14.76	21.50	
	0360	8″		24	.333		2.54	13.40		15.94	23.50	
	0361	9″		22	.364		2.86	14.60		17.46	25.50	
	0362	10″		20	.400		2.99	16.10		19.09	28	
	0363	12″		17	.471		3.45	18.90		22.35	33	
	1000	Duct access door, insulated, 6″ x 6″		14	.571		11.65	23		34.65	48.50	
	1020	10″ x 10″		11	.727		16.90	29		45.90	63.50	
	1040	12″ x 12″		10	.800		18.20	32		50.20	69.50	
	1050	12″ x 18″		9	.889		22.50	35.50		58	79.50	
	1060	16″ x 12″		9	.889		26	35.50		61.50	83.50	
	1070	18″ x 18″		8	1		32.50	40		72.50	98	
	1074	24″ x 18″		8	1		32	40		72	97	
	1080	24″ x 24″		8	1		34	40		74	99	
	1100	Exhaust vent damper, auto., OB with elect. actuator										

	15820 | Duct Accessories	CREW	DAILY OUTPUT	LABOR-HOURS	UNIT	2005 BARE COSTS MAT.	LABOR	EQUIP.	TOTAL	TOTAL INCL O&P	
300	1110 6" x 6"	1 Shee	24	.333	Ea.	73	13.40		86.40	101	300
	1114 8" x 6"		23	.348		75	14		89	104	
	1116 8" x 8"		22	.364		79	14.60		93.60	110	
	1118 10" x 6"		23	.348		80	14		94	110	
	1120 10" x 10"		22	.364		83	14.60		97.60	114	
	1122 12" x 6"		22	.364		80	14.60		94.60	111	
	1124 12" x 10"		21	.381		87.50	15.30		102.80	120	
	1126 12" x 12"		20	.400		91.50	16.10		107.60	126	
	1128 14" x 6"		21	.381		81	15.30		96.30	113	
	1130 14" x 10"		19	.421		91.50	16.95		108.45	127	
	1132 16" x 6"		21	.381		81	15.30		96.30	113	
	1134 16" x 8"		20	.400		90.50	16.10		106.60	124	
	1136 18" x 6"		20	.400		86.50	16.10		102.60	120	
	1138 18" x 8"		19	.421	↓	91.50	16.95		108.45	127	
	2000 Fabrics for flexible connections, with metal edge		100	.080	L.F.	1.62	3.22		4.84	6.75	
	2100 Without metal edge		160	.050	"	1.07	2.01		3.08	4.27	
	3000 Fire damper, curtain type, 1-1/2 hr rated, vertical, 6" x 6"		24	.333	Ea.	13.40	13.40		26.80	35	
	3020 8" x 6"		22	.364		13.40	14.60		28	37.50	
	3021 8" x 8"		22	.364		13.55	14.60		28.15	37.50	
	3030 10" x 10"		22	.364		14.10	14.60		28.70	38	
	3040 12" x 6"		22	.364		21	14.60		35.60	45.50	
	3044 12" x 12"		20	.400		14.45	16.10		30.55	40.50	
	3060 20" x 6"		18	.444		24.50	17.85		42.35	54.50	
	3080 12" x 8"		22	.364		21	14.60		35.60	45.50	
	3100 24" x 8"		16	.500		30	20		50	64	
	3120 12" x 10"		21	.381		21	15.30		36.30	46.50	
	3140 24" x 10"		15	.533		32.50	21.50		54	69	
	3160 36" x 10"		12	.667		43.50	27		70.50	89	
	3180 16" x 12"		20	.400		26.50	16.10		42.60	54	
	3200 24" x 12"		13	.615		22.50	24.50		47	63	
	3220 48" x 12"		10	.800		49.50	32		81.50	104	
	3238 14" x 14"		19	.421		21.50	16.95		38.45	49.50	
	3240 16" x 14"		18	.444		32.50	17.85		50.35	63.50	
	3260 24" x 14"		12	.667		43.50	27		70.50	89	
	3280 30" x 14"		11	.727		33.50	29		62.50	82	
	3298 16" x 16"		18	.444		21.50	17.85		39.35	51	
	3300 18" x 16"		17	.471		24	18.90		42.90	55.50	
	3320 24" x 16"		11	.727		38.50	29		67.50	87	
	3340 36" x 16"		10	.800		34	32		66	87	
	3356 18" x 18"		16	.500		26	20		46	59.50	
	3360 24" x 18"		10	.800		41.50	32		73.50	95.50	
	3380 48" x 18"		8	1		66.50	40		106.50	135	
	3398 20" x 20"		10	.800		26	32		58	78	
	3400 24" x 20"		8	1		41.50	40		81.50	108	
	3420 36" x 20"		7	1.143		55.50	46		101.50	132	
	3440 24" x 22"		7	1.143		43.50	46		89.50	119	
	3460 30" x 22"		6	1.333		49.50	53.50		103	137	
	3478 24" x 24"		8	1		30	40		70	95	
	3480 26" x 24"	↓	7	1.143		49	46		95	124	
	3484 32" x 24"	Q-9	13	1.231		37.50	44.50		82	110	
	3500 48" x 24"		12	1.333		43.50	48		91.50	122	
	3520 28" x 26"		13	1.231		55.50	44.50		100	130	
	3540 30" x 28"		12	1.333		58.50	48		106.50	139	
	3560 48" x 30"		11	1.455		84.50	52.50		137	174	
	3580 48" x 36"		10	1.600		92.50	58		150.50	191	
	3590 32" x 40"	↓	10	1.600	↓	51	58		109	145	

	15820 | Duct Accessories	CREW	DAILY OUTPUT	LABOR-HOURS	UNIT	2005 BARE COSTS MAT.	LABOR	EQUIP.	TOTAL	TOTAL INCL O&P	
300	3600	44" x 44"	Q-9	10	1.600	Ea.	96	58		154	195	300
	3620	48" x 48"	↓	8	2	↓	88	72.50		160.50	208	
	3700	U.L. label included in above										
	3800	For horizontal operation, add				Ea.	20%					
	3900	For cap for blades out of air stream, add					20%					
	4000	For 10" 22 ga., U.L. approved sleeve, add					35%					
	4100	For 10" 22 ga., U.L. sleeve, 100% free area, add				↓	65%					
	4200	For oversize openings group dampers										
	4210	Fire damper, round, Type A										
	4214	4" diam.	1 Shee	20	.400	Ea.	40.50	16.10		56.60	69	
	4216	5" diam.		16	.500		37.50	20		57.50	72.50	
	4218	6" diam.		14.30	.559		38.50	22.50		61	76.50	
	4220	7" diam.		13.30	.601		38.50	24		62.50	79	
	4222	8" diam.		11.40	.702		38.50	28		66.50	85.50	
	4224	9" diam.		10.80	.741		43.50	30		73.50	94	
	4226	10" diam.		10	.800		44.50	32		76.50	98.50	
	4228	12" diam.		9.50	.842		47.50	34		81.50	104	
	4230	14" diam.		8.20	.976		73	39		112	141	
	4232	16" diam.		7.60	1.053		78.50	42.50		121	152	
	4234	18" diam.		7.20	1.111		93.50	44.50		138	172	
	4236	20" diam.		6.40	1.250		93.50	50.50		144	181	
	4238	22" diam.		6.10	1.311		107	52.50		159.50	198	
	4240	24" diam.		5.70	1.404		111	56.50		167.50	209	
	4242	26" diam.		5.50	1.455		125	58.50		183.50	227	
	4244	28" diam.		5.40	1.481		131	59.50		190.50	236	
	4246	30" diam.	↓	5.30	1.509	↓	137	60.50		197.50	245	
	4500	Fire/smoke combination damper, louver type, UL										
	4506	6" x 6"	1 Shee	24	.333	Ea.	55	13.40		68.40	81	
	4510	8" x 8"		22	.364		126	14.60		140.60	162	
	4520	16" x 8"		20	.400		138	16.10		154.10	177	
	4540	18" x 8"		18	.444		138	17.85		155.85	180	
	4560	20" x 8"		16	.500		148	20		168	193	
	4580	10" x 10"		21	.381		143	15.30		158.30	181	
	4600	24" x 10"		15	.533		168	21.50		189.50	217	
	4620	30" x 10"		12	.667		173	27		200	232	
	4640	12" x 12"		20	.400		149	16.10		165.10	189	
	4660	18" x 12"		18	.444		152	17.85		169.85	195	
	4680	24" x 12"		13	.615		160	24.50		184.50	214	
	4700	30" x 12"		11	.727		166	29		195	228	
	4720	14" x 14"		17	.471		160	18.90		178.90	205	
	4740	16" x 14"		18	.444		165	17.85		182.85	209	
	4760	20" x 14"		14	.571		168	23		191	220	
	4780	24" x 14"		12	.667		172	27		199	230	
	4800	30" x 14"		10	.800		182	32		214	250	
	4820	16" x 16"		16	.500		177	20		197	226	
	4840	20" x 16"		14	.571		182	23		205	236	
	4860	24" x 16"		11	.727		188	29		217	252	
	4880	30" x 16"		8	1		196	40		236	278	
	4900	18" x 18"		15	.533		176	21.50		197.50	226	
	5000	24" x 18"		10	.800		188	32		220	256	
	5020	36" x 18"		7	1.143		205	46		251	297	
	5040	20" x 20"		13	.615		196	24.50		220.50	254	
	5060	24" x 20"		8	1		201	40		241	283	
	5080	30" x 20"		8	1		211	40		251	294	
	5100	36" x 20"		7	1.143		222	46		268	315	
	5120	24" x 24"	↓	8	1	↓	217	40		257	300	

	15820 | Duct Accessories	CREW	DAILY OUTPUT	LABOR-HOURS	UNIT	2005 BARE COSTS MAT.	LABOR	EQUIP.	TOTAL	TOTAL INCL O&P	
300	5130 30″ x 24″	1 Shee	7.60	1.053	Ea.	177	42.50		219.50	260	300
	5140 36″ x 24″	Q-9	12	1.333		237	48		285	335	
	5141 36″ x 30″		10	1.600		209	58		267	320	
	5142 40″ x 36″		8	2		445	72.50		517.50	600	
	5143 48″ x 48″	▼	4	4		870	145		1,015	1,175	
	5150 Damper operator motor, 24 or 120 volt	1 Shee	16	.500		263	20		283	320	
	5990 Multi-blade dampers, opposed blade, 8″ x 6″		24	.333		19.95	13.40		33.35	42.50	
	5994 8″ x 8″		22	.364		21	14.60		35.60	45.50	
	5996 10″ x 10″		21	.381		23.50	15.30		38.80	49	
	6000 12″ x 12″		21	.381		27.50	15.30		42.80	53.50	
	6020 12″ x 18″		18	.444		36.50	17.85		54.35	68	
	6030 14″ x 10″		20	.400		26.50	16.10		42.60	54	
	6031 14″ x 14″		17	.471		32.50	18.90		51.40	64.50	
	6032 16″ x 10″		18	.444		29	17.85		46.85	59.50	
	6033 16″ x 12″		17	.471		32.50	18.90		51.40	64.50	
	6035 16″ x 16″		16	.500		40	20		60	75	
	6036 18″ x 14″		16	.500		40	20		60	75	
	6037 18″ x 16″		15	.533		45	21.50		66.50	82.50	
	6038 18″ x 18″		15	.533		48.50	21.50		70	86	
	6040 18″ x 24″		12	.667		62.50	27		89.50	110	
	6060 18″ x 28″		10	.800		71.50	32		103.50	128	
	6068 20″ x 6″		18	.444		25	17.85		42.85	55	
	6069 20″ x 8″		16	.500		30	20		50	64	
	6070 20″ x 16″		14	.571		48.50	23		71.50	88.50	
	6071 20″ x 18″		13	.615		52.50	24.50		77	95.50	
	6072 20″ x 20″		13	.615		57.50	24.50		82	101	
	6073 22″ x 6″		20	.400		26.50	16.10		42.60	54	
	6074 22″ x 18″		14	.571		57.50	23		80.50	98.50	
	6075 22″ x 20″		12	.667		63	27		90	111	
	6076 24″ x 16″		11	.727		56.50	29		85.50	107	
	6077 22″ x 22″		10	.800		67.50	32		99.50	124	
	6078 24″ x 20″		8	1		67.50	40		107.50	136	
	6080 24″ x 24″		8	1		79	40		119	149	
	6100 24″ x 28″		6	1.333		90.50	53.50		144	183	
	6110 26″ x 26″	▼	6	1.333		91.50	53.50		145	184	
	6120 28″ x 28″	Q-9	11	1.455		107	52.50		159.50	198	
	6130 30″ x 18″		10	1.600		75.50	58		133.50	173	
	6132 30″ x 24″		7	2.286		95	82.50		177.50	231	
	6133 30″ x 30″		6.60	2.424		122	87.50		209.50	270	
	6135 32″ x 32″		6.40	2.500		140	90.50		230.50	293	
	6150 34″ x 34″		6	2.667		157	96.50		253.50	320	
	6151 36″ x 12″		10	1.600		65	58		123	161	
	6152 36″ x 16″		8	2		81.50	72.50		154	201	
	6156 36″ x 32″		6.30	2.540		157	92		249	315	
	6157 36″ x 34″		6.20	2.581		166	93.50		259.50	325	
	6158 36″ x 36″		6	2.667		176	96.50		272.50	340	
	6160 44″ x 28″		5.80	2.759		168	100		268	340	
	6180 48″ x 36″		5.60	2.857		234	103		337	415	
	6200 56″ x 36″		5.40	2.963		273	107		380	465	
	6220 60″ x 36″		5.20	3.077		293	111		404	490	
	6240 60″ x 44″	▼	5	3.200		360	116		476	575	
	7000 Splitter damper assembly, self-locking, 1′ rod	1 Shee	24	.333		17.40	13.40		30.80	39.50	
	7020 3′ rod		22	.364		22.50	14.60		37.10	47.50	
	7040 4′ rod		20	.400		25	16.10		41.10	52	
	7060 6′ rod	▼	18	.444	▼	30.50	17.85		48.35	61	
	7500 Variable volume modulating motorized damper, incl. elect. mtr.										

15820 | Duct Accessories

				DAILY	LABOR-		2005 BARE COSTS				TOTAL
			CREW	OUTPUT	HOURS	UNIT	MAT.	LABOR	EQUIP.	TOTAL	INCL O&P
300	7504	8″ x 6″	1 Shee	15	.533	Ea.	110	21.50		131.50	154
	7506	10″ x 6″		14	.571		110	23		133	157
	7510	10″ x 10″		13	.615		114	24.50		138.50	163
	7520	12″ x 12″		12	.667		116	27		143	169
	7522	12″ x 16″		11	.727		120	29		149	177
	7524	16″ x 10″		12	.667		117	27		144	170
	7526	16″ x 14″		10	.800		122	32		154	185
	7528	16″ x 18″		9	.889		130	35.50		165.50	198
	7540	18″ x 12″		10	.800		124	32		156	186
	7542	18″ x 18″		8	1		135	40		175	210
	7544	20″ x 14″		8	1		137	40		177	213
	7546	20″ x 18″		7	1.143		146	46		192	231
	7560	24″ x 12″		8	1		139	40		179	215
	7562	24″ x 18″		7	1.143		155	46		201	241
	7568	28″ x 10″		7	1.143		231	46		277	325
	7580	28″ x 16″		6	1.333		245	53.50		298.50	350
	7590	30″ x 14″		5	1.600		245	64.50		309.50	370
	7600	30″ x 18″		4	2		260	80.50		340.50	410
	7610	30″ x 24″		3.80	2.105		298	84.50		382.50	460
	7690	48″ x 48″	Q-9	6	2.667		665	96.50		761.50	885
	7694	6′ x 14′ w/2 motors	″	3	5.333		2,600	193		2,793	3,150
	7700	For thermostat, add	1 Shee	8	1		28	40		68	93
	7800	For transformer 40 VA capacity, add	″	16	.500		22.50	20		42.50	55.50
	8000	Multi-blade dampers, parallel blade									
	8100	8″ x 8″	1 Shee	24	.333	Ea.	53.50	13.40		66.90	79.50
	8120	12″ x 8″		22	.364		53.50	14.60		68.10	81.50
	8140	16″ x 10″		20	.400		71	16.10		87.10	103
	8160	18″ x 12″		18	.444		77	17.85		94.85	112
	8180	22″ x 12″		15	.533		82.50	21.50		104	124
	8200	24″ x 16″		11	.727		92	29		121	146
	8220	28″ x 16″		10	.800		105	32		137	165
	8240	30″ x 16″		8	1		105	40		145	177
	8260	30″ x 18″		7	1.143		127	46		173	211
	8300	Relief damper, electronic bypass with tight seal									
	8310	8″ x 6″	1 Shee	22	.364	Ea.	110	14.60		124.60	144
	8314	10″ x 6″		22	.364		124	14.60		138.60	159
	8318	10″ x 10″		21	.381		161	15.30		176.30	201
	8322	12″ x 12″		20	.400		172	16.10		188.10	214
	8326	12″ x 16″		19	.421		179	16.95		195.95	223
	8330	16″ x 10″		20	.400		171	16.10		187.10	214
	8334	16″ x 14″		18	.444		180	17.85		197.85	226
	8338	16″ x 18″		17	.471		198	18.90		216.90	247
	8342	18″ x 12″		18	.444		179	17.85		196.85	225
	8346	18″ x 18″		15	.533		186	21.50		207.50	237
	8350	20″ x 14″		14	.571		186	23		209	240
	8354	20″ x 18″		13	.615		190	24.50		214.50	247
	8358	24″ x 12″		13	.615		199	24.50		223.50	257
	8362	24″ x 18″		10	.800		213	32		245	284
	8363	24″ x 24″		10	.800		222	32		254	294
	8364	24″ x 36″		9	.889		258	35.50		293.50	340
	8365	24″ x 48″		6	1.333		289	53.50		342.50	405
	8366	28″ x 10″		13	.615		186	24.50		210.50	242
	8370	28″ x 16″		10	.800		220	32		252	292
	8374	30″ x 14″		11	.727		215	29		244	281
	8378	30″ x 18″		8	1		216	40		256	300
	8382	30″ x 24″		6	1.333		206	53.50		259.50	310

15820 | Duct Accessories

			CREW	DAILY OUTPUT	LABOR-HOURS	UNIT	2005 BARE COSTS MAT.	LABOR	EQUIP.	TOTAL	TOTAL INCL O&P	
300	8390	46″ x 36″	1 Shee	4	2	Ea.	300	80.50		380.50	455	300
	8394	48″ x 48″		3	2.667		350	107		457	550	
	8396	54″ x 36″		2	4		370	161		531	655	
	8400	Round damper, butterfly, vol. control w/lever lock reg.										
	8410	6″ diam.	1 Shee	22	.364	Ea.	23.50	14.60		38.10	48	
	8412	7″ diam.		21	.381		24	15.30		39.30	50	
	8414	8″ diam.		20	.400		25	16.10		41.10	52	
	8416	9″ diam.		19	.421		26	16.95		42.95	54.50	
	8418	10″ diam.		18	.444		27.50	17.85		45.35	57.50	
	8420	12″ diam.		16	.500		28.50	20		48.50	62	
	8422	14″ diam.		14	.571		36.50	23		59.50	76	
	8424	16″ diam.		13	.615		39	24.50		63.50	81	
	8426	18″ diam.		12	.667		44	27		71	89.50	
	8428	20″ diam.		11	.727		51.50	29		80.50	102	
	8430	24″ diam.		10	.800		59	32		91	115	
	8432	30″ diam.		9	.889		82.50	35.50		118	146	
	8434	36″ diam.		8	1		111	40		151	184	
	8500	Round motor operated damper										
	8510	6″ dia.	1 Shee	22	.364	Ea.	55.50	14.60		70.10	84	
	8512	7″ dia.		21	.381		57.50	15.30		72.80	86.50	
	8514	8″ dia.		20	.400		61.50	16.10		77.60	92	
	8516	10″ dia.		18	.444		67.50	17.85		85.35	102	
	8518	12″ dia.		16	.500		72.50	20		92.50	111	
	8520	14″ dia.		14	.571		82.50	23		105.50	126	
	9000	Silencers, noise control for air flow, duct				MCFM	48			48	53	
	9004	Duct sound trap, packaged										
	9010	12″ x 18″ x 36″, 2000 CFM	1 Shee	12	.667	Ea.	288	27		315	355	
	9011	24″ x 18″ x 36″, 9000 CFM	Q-9	80	.200		325	7.25		332.25	370	
	9012	24″ x 24″ x 36″, 9000 CFM		60	.267		490	9.65		499.65	550	
	9013	24″ x 30″ x 36″, 9000 CFM		50	.320		520	11.60		531.60	595	
	9014	24″ x 36″ x 36″, 9000 CFM		40	.400		650	14.45		664.45	740	
	9015	24″ x 48″ x 36″, 9000 CFM		28	.571		845	20.50		865.50	960	
	9018	36″ x 18″ x 36″, 6300 CFM		42	.381		650	13.80		663.80	735	
	9019	36″ x 36″ x 36″, 6300 CFM		21	.762		1,300	27.50		1,327.50	1,475	
	9020	36″ x 48″ x 36″, 6300 CFM		14	1.143		1,700	41.50		1,741.50	1,925	
	9021	36″ x 60″ x 36″, 6300 CFM		13	1.231		2,150	44.50		2,194.50	2,450	
	9022	48″ x 48″ x 36″, 6300 CFM		12	1.333		2,275	48		2,323	2,575	
	9023	24″ x 24″ x 60″, 6000 CFM		3.60	4.444		750	161		911	1,075	
	9026	24″ x 30″ x 60″, 7000 CFM		3.50	4.571		840	165		1,005	1,175	
	9030	24″ x 36″ x 60″, 8000 CFM		3.40	4.706		930	170		1,100	1,275	
	9034	24″ x 48″ x 60″, 11,000 CFM		3.30	4.848		815	175		990	1,175	
	9038	36″ x 18″ x 60″, 6300 CFM		3.50	4.571		625	165		790	945	
	9042	36″ x 36″ x 60″, 12,000 CFM		3.30	4.848		1,000	175		1,175	1,375	
	9046	36″ x 48″ x 60″, 17,000 CFM		3.10	5.161		1,325	187		1,512	1,725	
	9050	36″ x 60″ x 60″, 20,000 CFM		2.80	5.714		1,600	207		1,807	2,100	
	9054	48″ x 48″ x 60″, 23,000 CFM		2.60	6.154		1,850	223		2,073	2,400	
	9056	48″ x 60″ x 60″, 23,000 CFM		2.10	7.619		2,325	276		2,601	2,975	
	9400	Turning vane components										
	9410	Turning vane rail	1 Shee	160	.050	L.F.	1.24	2.01		3.25	4.45	
	9420	Double thick, factory fab. vane		300	.027		1.51	1.07		2.58	3.31	
	9428	12″ high set		170	.047		8.50	1.89		10.39	12.25	
	9432	14″ high set		160	.050		9.50	2.01		11.51	13.55	
	9434	16″ high set		150	.053		10.50	2.14		12.64	14.85	
	9436	18″ high set		144	.056		11.50	2.23		13.73	16.10	
	9438	20″ high set		138	.058		12.50	2.33		14.83	17.35	
	9440	22″ high set		130	.062		13.50	2.47		15.97	18.65	

	15820 | Duct Accessories	CREW	DAILY OUTPUT	LABOR-HOURS	UNIT	2005 BARE COSTS MAT.	LABOR	EQUIP.	TOTAL	TOTAL INCL O&P	
300	9442 24" high set	1 Shee	124	.065	L.F.	14.50	2.59		17.09	19.95	300
	9444 26" high set		116	.069		15.50	2.77		18.27	21.50	
	9446 30" high set		112	.071		17.50	2.87		20.37	23.50	

	15830 | Fans	CREW	DAILY OUTPUT	LABOR-HOURS	UNIT	MAT.	LABOR	EQUIP.	TOTAL	TOTAL INCL O&P	
100	0010 FANS R15700-040										100
	0020 Air conditioning and process air handling										
	0030 Axial flow, compact, low sound, 2.5" S.P. R15800-080										
	0050 3,800 CFM, 5 HP	Q-20	3.40	5.882	Ea.	3,775	218		3,993	4,475	
	0080 6,400 CFM, 5 HP		2.80	7.143		4,200	265		4,465	5,025	
	0100 10,500 CFM, 7-1/2 HP		2.40	8.333		5,250	310		5,560	6,250	
	0120 15,600 CFM, 10 HP		1.60	12.500		6,600	465		7,065	7,975	
	0140 23,000 CFM, 15 HP		.70	28.571		10,100	1,050		11,150	12,700	
	0160 28,000 CFM, 20 HP		.40	50		11,300	1,850		13,150	15,200	
	0200 In-line centrifugal, supply/exhaust booster										
	0220 aluminum wheel/hub, disconnect switch, 1/4" S.P.										
	0240 500 CFM, 10" diameter connection	Q-20	3	6.667	Ea.	900	247		1,147	1,375	
	0260 1,380 CFM, 12" diameter connection		2	10		955	370		1,325	1,625	
	0280 1,520 CFM, 16" diameter connection		2	10		1,050	370		1,420	1,725	
	0300 2,560 CFM, 18" diameter connection		1	20		1,125	740		1,865	2,375	
	0320 3,480 CFM, 20" diameter connection		.80	25		1,350	925		2,275	2,900	
	0326 5,080 CFM, 20" diameter connection		.75	26.667		1,475	990		2,465	3,100	
	0340 7,500 CFM, 22" diameter connection		.70	28.571		1,750	1,050		2,800	3,525	
	0350 10,000 CFM, 27" diameter connection		.65	30.769		2,100	1,150		3,250	4,075	
	0500 Axial flow, constant speed										
	0505 Direct drive, 1/8" S.P.										
	0510 12", 1060 CFM, 1/6 HP	Q-20	3	6.667	Ea.	440	247		687	865	
	0514 12", 2095 CFM, 1/2 HP		3	6.667		515	247		762	950	
	0518 16", 2490 CFM, 1/3 HP		2.80	7.143		515	265		780	975	
	0522 20", 4130 CFM, 3/4 HP		2.60	7.692		615	285		900	1,125	
	0526 22", 4700 CFM, 3/4 HP		2.60	7.692		915	285		1,200	1,425	
	0530 24", 5850 CFM, 1 HP		2.50	8		945	297		1,242	1,500	
	0534 24", 7925 CFM, 1-1/2 HP		2.40	8.333		870	310		1,180	1,425	
	0538 30", 10640 CFM, 2 HP		2.20	9.091		1,500	335		1,835	2,175	
	0542 30", 14765 CFM, 2-1/2 HP		2.10	9.524		1,750	355		2,105	2,450	
	0546 36", 16780 CFM, 2 HP		2.10	9.524		1,800	355		2,155	2,525	
	0550 36", 22920 CFM, 5 HP		1.80	11.111		1,975	410		2,385	2,800	
	0560 Belt drive, 1/8" S.P.										
	0562 15", 2800 CFM, 1/3 HP	Q-20	3.20	6.250	Ea.	530	232		762	935	
	0564 15", 3400 CFM, 1/2 HP		3	6.667		540	247		787	975	
	0568 18", 3280 CFM, 1/3 HP		2.80	7.143		575	265		840	1,025	
	0572 18", 3900 CFM, 1/2 HP		2.80	7.143		580	265		845	1,050	
	0576 18", 5250 CFM, 1 HP		2.70	7.407		665	275		940	1,150	
	0584 24", 6430 CFM, 1 HP		2.50	8		830	297		1,127	1,375	
	0588 24", 8860 CFM, 2 HP		2.40	8.333		835	310		1,145	1,400	
	0592 30", 9250 CFM, 1 HP		2.20	9.091		1,025	335		1,360	1,650	
	0596 30", 16900 CFM, 5 HP		2	10		1,125	370		1,495	1,825	
	0604 36", 14475 CFM, 2 HP		2.30	8.696		1,175	325		1,500	1,800	
	0608 36", 20080 CFM, 5 HP		1.80	11.111		1,275	410		1,685	2,025	
	0612 36", 14475 CFM, 7-1/2 HP		1.60	12.500		1,425	465		1,890	2,275	
	0616 42", 29000 CFM, 7-1/2 HP		1.40	14.286		1,700	530		2,230	2,675	
	1000 Return air fan										
	1010 9200 CFM	Q-20	1.40	14.286	Ea.	3,075	530		3,605	4,175	
	1020 13,200 CFM		1.30	15.385		4,425	570		4,995	5,725	
	1030 16,500 CFM		1.20	16.667		5,525	620		6,145	7,025	

15830 | Fans

			CREW	DAILY OUTPUT	LABOR-HOURS	UNIT	2005 BARE COSTS MAT.	LABOR	EQUIP.	TOTAL	TOTAL INCL O&P	
100	1040	19,500 CFM R15700-040	Q-20	1	20	Ea.	6,525	740		7,265	8,300	100
	1500	Vaneaxial, low pressure, 2000 CFM, 1/2 HP		3.60	5.556		1,625	206		1,831	2,100	
	1520	4,000 CFM, 1 HP R15800-080		3.20	6.250		1,650	232		1,882	2,175	
	1540	8,000 CFM, 2 HP		2.80	7.143		1,950	265		2,215	2,550	
	1560	16,000 CFM, 5 HP		2.40	8.333		2,575	310		2,885	3,300	
	2000	Blowers, direct drive with motor, complete										
	2020	1045 CFM @ .5" S.P., 1/5 HP	Q-20	18	1.111	Ea.	145	41		186	223	
	2040	1385 CFM @ .5" S.P., 1/4 HP		18	1.111		147	41		188	225	
	2060	1640 CFM @ .5" S.P., 1/3 HP		18	1.111		175	41		216	256	
	2080	1760 CFM @ .5" S.P., 1/2 HP		18	1.111		190	41		231	272	
	2090	4 speed										
	2100	1164 to 1739 CFM @ .5" S.P., 1/3 HP	Q-20	16	1.250	Ea.	251	46.50		297.50	345	
	2120	1467 to 2218 CFM @ 1.0" S.P., 3/4 HP	"	14	1.429	"	325	53		378	440	
	2500	Ceiling fan, right angle, extra quiet, 0.10" S.P.										
	2520	95 CFM	Q-20	20	1	Ea.	143	37		180	214	
	2540	210 CFM		19	1.053		169	39		208	246	
	2560	385 CFM		18	1.111		214	41		255	299	
	2580	885 CFM		16	1.250		420	46.50		466.50	535	
	2600	1,650 CFM		13	1.538		585	57		642	725	
	2620	2,960 CFM		11	1.818		780	67.50		847.50	960	
	2640	For wall or roof cap, add	1 Shee	16	.500		143	20		163	188	
	2660	For straight thru fan, add					10%					
	2680	For speed control switch, add	1 Elec	16	.500		78	20.50		98.50	116	
	3000	Paddle blade air circulator, 3 speed switch										
	3020	42", 5,000 CFM high, 3000 CFM low	1 Elec	2.40	3.333	Ea.	73.50	136		209.50	283	
	3040	52", 6,500 CFM high, 4000 CFM low	"	2.20	3.636	"	74.50	148		222.50	305	
	3100	For antique white motor, same cost										
	3200	For brass plated motor, same cost										
	3300	For light adaptor kit, add				Ea.	27.50			27.50	30	
	3310	Industrial grade, reversable, 4 blade										
	3312	5500 CFM	Q-20	5	4	Ea.	110	148		258	350	
	3314	7600 CFM		5	4		120	148		268	360	
	3316	21,015 CFM		4	5		155	185		340	455	
	3500	Centrifugal, airfoil, motor and drive, complete										
	3520	1000 CFM, 1/2 HP	Q-20	2.50	8	Ea.	905	297		1,202	1,450	
	3540	2,000 CFM, 1 HP		2	10		970	370		1,340	1,650	
	3560	4,000 CFM, 3 HP		1.80	11.111		1,250	410		1,660	2,000	
	3580	8,000 CFM, 7-1/2 HP		1.40	14.286		1,825	530		2,355	2,825	
	3600	12,000 CFM, 10 HP		1	20		2,700	740		3,440	4,100	
	4000	Single width, belt drive, not incl. motor, capacities										
	4020	at 2000 FPM, 2.5" S.P. for indicated motor										
	4040	6900 CFM, 5 HP	Q-9	2.40	6.667	Ea.	2,750	241		2,991	3,400	
	4060	10,340 CFM, 7-1/2" HP		2.20	7.273		3,825	263		4,088	4,625	
	4080	15,320 CFM, 10 HP		2	8		4,525	289		4,814	5,425	
	4100	22,780 CFM, 15 HP		1.80	8.889		6,725	320		7,045	7,900	
	4120	33,840 CFM, 20 HP		1.60	10		8,925	360		9,285	10,400	
	4140	41,400 CFM, 25 HP		1.40	11.429		11,800	415		12,215	13,600	
	4160	50,100 CFM, 30 HP		.80	20		15,000	725		15,725	17,600	
	4200	Double width wheel, 12,420 CFM, 7.5 HP		2.20	7.273		3,925	263		4,188	4,725	
	4220	18,620 CFM, 15 HP		2	8		5,625	289		5,914	6,650	
	4240	27,580 CFM, 20 HP		1.80	8.889		6,875	320		7,195	8,075	
	4260	40,980 CFM, 25 HP		1.50	10.667		10,800	385		11,185	12,500	
	4280	60,920 CFM, 40 HP		1	16		14,500	580		15,080	16,900	
	4300	74,520 CFM, 50 HP		.80	20		17,200	725		17,925	20,000	
	4320	90,160 CFM, 50 HP		.70	22.857		23,600	825		24,425	27,200	
	4340	110,300 CFM, 60 HP		.50	32		31,900	1,150		33,050	36,900	

		15830 | Fans		CREW	DAILY OUTPUT	LABOR-HOURS	UNIT	2005 BARE COSTS MAT.	LABOR	EQUIP.	TOTAL	TOTAL INCL O&P	
100	4360	134,960 CFM, 75 HP	R15700 -040	Q-9	.40	40	Ea.	42,400	1,450		43,850	48,800	100
	4500	Corrosive fume resistant, plastic											
	4600	roof ventilators, centrifugal, V belt drive, motor	R15800 -080										
	4620	1/4" S.P., 250 CFM, 1/4 HP		Q-20	6	3.333	Ea.	2,475	124		2,599	2,925	
	4640	895 CFM, 1/3 HP			5	4		2,700	148		2,848	3,200	
	4660	1630 CFM, 1/2 HP			4	5		3,200	185		3,385	3,800	
	4680	2240 CFM, 1 HP			3	6.667		3,325	247		3,572	4,025	
	4700	3810 CFM, 2 HP			2	10		3,700	370		4,070	4,650	
	4710	5000 CFM, 2 HP			1.80	11.111		7,775	410		8,185	9,175	
	4715	8000 CFM, 5 HP			1.40	14.286		8,650	530		9,180	10,300	
	4720	11760 CFM, 5 HP			1	20		8,800	740		9,540	10,800	
	4740	18810 CFM, 10 HP		↓	.70	28.571	↓	9,100	1,050		10,150	11,600	
	4800	For intermediate capacity, motors may be varied											
	4810	For explosion proof motor, add					Ea.	15%					
	5000	Utility set, centrifugal, V belt drive, motor											
	5020	1/4" S.P., 1200 CFM, 1/4 HP		Q-20	6	3.333	Ea.	2,550	124		2,674	3,000	
	5040	1520 CFM, 1/3 HP			5	4		2,550	148		2,698	3,025	
	5060	1850 CFM, 1/2 HP			4	5		2,550	185		2,735	3,100	
	5080	2180 CFM, 3/4 HP			3	6.667		2,600	247		2,847	3,225	
	5100	1/2" S.P., 3600 CFM, 1 HP			2	10		3,750	370		4,120	4,700	
	5120	4250 CFM, 1-1/2 HP			1.60	12.500		3,800	465		4,265	4,875	
	5140	4800 CFM, 2 HP			1.40	14.286		3,850	530		4,380	5,025	
	5160	6920 CFM, 5 HP			1.30	15.385		3,975	570		4,545	5,250	
	5180	7700 CFM, 7-1/2 HP		↓	1.20	16.667	↓	4,100	620		4,720	5,450	
	5200	For explosion proof motor, add						15%					
	5300	Fume exhauster without hose and intake nozzle											
	5310	630 CFM, 1-1/2 HP		Q-20	2	10	Ea.	705	370		1,075	1,350	
	5320	Hose extension kit 5'		"	20	1	"	49	37		86	111	
	5500	Fans, industrial exher, for air which may contain granular matl.											
	5520	1000 CFM, 1-1/2 HP		Q-20	2.50	8	Ea.	2,100	297		2,397	2,750	
	5540	2000 CFM, 3 HP			2	10		2,525	370		2,895	3,350	
	5560	4000 CFM, 7-1/2 HP			1.80	11.111		4,075	410		4,485	5,100	
	5580	8000 CFM, 15 HP			1.40	14.286		6,375	530		6,905	7,800	
	5600	12,000 CFM, 30 HP		↓	1	20	↓	8,850	740		9,590	10,900	
	6000	Propeller exhaust, wall shutter, 1/4" S.P.											
	6020	Direct drive, two speed											
	6100	375 CFM, 1/10 HP		Q-20	10	2	Ea.	256	74		330	395	
	6120	730 CFM, 1/7 HP			9	2.222		285	82.50		367.50	440	
	6140	1000 CFM, 1/8 HP			8	2.500		395	92.50		487.50	575	
	6160	1890 CFM, 1/4 HP			7	2.857		405	106		511	605	
	6180	3275 CFM, 1/2 HP			6	3.333		410	124		534	645	
	6200	4720 CFM, 1 HP		↓	5	4	↓	635	148		783	925	
	6300	V-belt drive, 3 phase											
	6320	6175 CFM, 3/4 HP		Q-20	5	4	Ea.	535	148		683	810	
	6340	7500 CFM, 3/4 HP			5	4		560	148		708	845	
	6360	10,100 CFM, 1 HP			4.50	4.444		685	165		850	1,000	
	6380	14,300 CFM, 1-1/2 HP			4	5		805	185		990	1,175	
	6400	19,800 CFM, 2 HP			3	6.667		925	247		1,172	1,400	
	6420	26,250 CFM, 3 HP			2.60	7.692		1,100	285		1,385	1,650	
	6440	38,500 CFM, 5 HP			2.20	9.091		1,300	335		1,635	1,975	
	6460	46,000 CFM, 7-1/2 HP			2	10		1,400	370		1,770	2,125	
	6480	51,500 CFM, 10 HP		↓	1.80	11.111	↓	1,450	410		1,860	2,225	
	6490	V-belt drive, 115V., residential, whole house											
	6500	Ceiling-wall, 5200 CFM, 1/4 HP, 30" x 30"		1 Shee	6	1.333	Ea.	279	53.50		332.50	390	
	6510	7500 CFM, 1/3 HP, 36"x36"			5	1.600		292	64.50		356.50	420	
	6520	10,500 CFM, 1/3 HP, 42"x42"		↓	5	1.600	↓	305	64.50		369.50	435	

15830 | Fans			CREW	DAILY OUTPUT	LABOR-HOURS	UNIT	MAT.	LABOR	EQUIP.	TOTAL	TOTAL INCL O&P	
100	6530	13,200 CFM, 1/3 HP, 48" x 48" R15700-040	1 Shee	4	2	Ea.	335	80.50		415.50	495	100
	6540	15,445 CFM, 1/2 HP, 48" x 48"		4	2		350	80.50		430.50	510	
	6550	17,025 CFM, 1/2 HP, 54" x 54" R15800-080		4	2		675	80.50		755.50	870	
	6560	For two speed motor, add					20%					
	6570	Shutter, automatic, ceiling/wall										
	6580	30" x 30"	1 Shee	8	1	Ea.	91	40		131	162	
	6590	36" x 36"		8	1		104	40		144	176	
	6600	42" x 42"		8	1		130	40		170	205	
	6610	48" x 48"		7	1.143		141	46		187	226	
	6620	54" x 54"		6	1.333		186	53.50		239.50	287	
	6630	Timer, shut off, to 12 Hr.		20	.400		30	16.10		46.10	57.50	
	6650	Residential, bath exhaust, grille, back draft damper										
	6660	50 CFM	Q-20	24	.833	Ea.	30	31		61	80	
	6670	110 CFM		22	.909		50	33.50		83.50	107	
	6672	180 CFM		22	.909		86	33.50		119.50	146	
	6673	210 CFM		22	.909		115	33.50		148.50	179	
	6674	260 CFM		22	.909		129	33.50		162.50	194	
	6675	300 CFM		22	.909		176	33.50		209.50	246	
	6680	Light combination, squirrel cage, 100 watt, 70 CFM		24	.833		61.50	31		92.50	115	
	6700	Light/heater combination, ceiling mounted										
	6710	70 CFM, 1450 watt	Q-20	24	.833	Ea.	74.50	31		105.50	129	
	6800	Heater combination, recessed, 70 CFM		24	.833		35.50	31		66.50	86	
	6820	With 2 infrared bulbs		23	.870		53.50	32.50		86	109	
	6840	Wall mount, 170 CFM		22	.909		115	33.50		148.50	179	
	6846	Ceiling mount, 180 CFM		22	.909		117	33.50		150.50	180	
	6900	Kitchen exhaust, grille, complete, 160 CFM		22	.909		62.50	33.50		96	120	
	6910	180 CFM		20	1		73	37		110	137	
	6920	270 CFM		18	1.111		95.50	41		136.50	168	
	6930	350 CFM		16	1.250		105	46.50		151.50	186	
	6940	Residential roof jacks and wall caps										
	6944	Wall cap with back draft damper										
	6946	3" & 4" dia. round duct	1 Shee	11	.727	Ea.	12.80	29		41.80	59	
	6948	6" dia. round duct	"	11	.727	"	31	29		60	79	
	6958	Roof jack with bird screen and back draft damper										
	6960	3" & 4" dia. round duct	1 Shee	11	.727	Ea.	12.30	29		41.30	58.50	
	6962	3-1/4" x 10" rectangular duct	"	10	.800	"	22.50	32		54.50	74.50	
	6980	Transition										
	6982	3-1/4" x 10" to 6" dia. round	1 Shee	20	.400	Ea.	13.85	16.10		29.95	40	
	7000	Roof exhauster, centrifugal, aluminum housing, 12" galvanized										
	7020	curb, bird screen, back draft damper, 1/4" S.P.										
	7100	Direct drive, 320 CFM, 11" sq. damper	Q-20	7	2.857	Ea.	320	106		426	510	
	7120	600 CFM, 11" sq. damper		6	3.333		325	124		449	545	
	7140	815 CFM, 13" sq. damper		5	4		325	148		473	580	
	7160	1450 CFM, 13" sq. damper		4.20	4.762		415	177		592	725	
	7180	2050 CFM, 16" sq. damper		4	5		415	185		600	740	
	7200	V-belt drive, 1650 CFM, 12" sq. damper		6	3.333		725	124		849	985	
	7220	2750 CFM, 21" sq. damper		5	4		840	148		988	1,150	
	7230	3500 CFM, 21" sq. damper		4.50	4.444		925	165		1,090	1,275	
	7240	4910 CFM, 23" sq. damper		4	5		1,150	185		1,335	1,550	
	7260	8525 CFM, 28" sq. damper		3	6.667		1,425	247		1,672	1,950	
	7280	13,760 CFM, 35" sq. damper		2	10		1,975	370		2,345	2,750	
	7300	20,558 CFM, 43" sq. damper		1	20		4,200	740		4,940	5,750	
	7320	For 2 speed winding, add					15%					
	7340	For explosionproof motor, add				Ea.	330			330	360	
	7360	For belt driven, top discharge, add					15%					
	7400	Roof mounted kitchen exhaust, aluminum, centrifugal										

Column group "2005 BARE COSTS" spans MAT., LABOR, EQUIP., TOTAL.

15 MECHANICAL

15830 | Fans

100		CREW	DAILY OUTPUT	LABOR-HOURS	UNIT	2005 BARE COSTS MAT.	LABOR	EQUIP.	TOTAL	TOTAL INCL O&P
7410	Direct drive, 2 speed, temp to 200°F (R15700-040)									
7412	1/12 HP, 9-3/4"	Q-20	8	2.500	Ea.	320	92.50		412.50	495
7414	1/3 HP, 12-5/8" (R15800-080)		7	2.857		395	106		501	595
7416	1/2 HP, 13-1/2"		6	3.333		440	124		564	670
7418	3/4 HP, 15"		5	4		625	148		773	915
7424	Belt drive, temp to 250°F									
7426	3/4 HP, 20"	Q-20	5	4	Ea.	1,375	148		1,523	1,750
7428	1-1/2 HP, 24-1/2"		4	5		1,525	185		1,710	1,950
7430	1-1/2 HP, 30"		4	5		2,000	185		2,185	2,475
7450	Upblast, propeller, w/BDD, BS, + C									
7454	30,300 CFM @ 3/8" S.P., 5 HP	Q-20	1.60	12.500	Ea.	2,325	465		2,790	3,250
7458	36,000 CFM @ 1/2" S.P., 15 HP	″	1.60	12.500	″	3,300	465		3,765	4,350
7500	Utility set, steel construction, pedestal, 1/4" S.P.									
7520	Direct drive, 150 CFM, 1/8 HP	Q-20	6.40	3.125	Ea.	635	116		751	870
7540	485 CFM, 1/6 HP		5.80	3.448		795	128		923	1,075
7560	1950 CFM, 1/2 HP		4.80	4.167		935	155		1,090	1,250
7580	2410 CFM, 3/4 HP		4.40	4.545		1,725	169		1,894	2,150
7600	3328 CFM, 1-1/2 HP		3	6.667		1,925	247		2,172	2,475
7680	V-belt drive, drive cover, 3 phase									
7700	800 CFM, 1/4 HP	Q-20	6	3.333	Ea.	460	124		584	700
7720	1,300 CFM, 1/3 HP		5	4		485	148		633	760
7740	2,000 CFM, 1 HP		4.60	4.348		575	161		736	875
7760	2,900 CFM, 3/4 HP		4.20	4.762		775	177		952	1,125
7780	3,600 CFM, 3/4 HP		4	5		950	185		1,135	1,325
7800	4,800 CFM, 1 HP		3.50	5.714		1,125	212		1,337	1,550
7820	6,700 CFM, 1-1/2 HP		3	6.667		1,375	247		1,622	1,900
7830	7,500 CFM, 2 HP		2.50	8		1,875	297		2,172	2,525
7840	11,000 CFM, 3 HP		2	10		2,525	370		2,895	3,350
7860	13,000 CFM, 3 HP		1.60	12.500		2,550	465		3,015	3,500
7880	15,000 CFM, 5 HP		1	20		2,650	740		3,390	4,025
7900	17,000 CFM, 7-1/2 HP		.80	25		2,825	925		3,750	4,525
7920	20,000 CFM, 7-1/2 HP		.80	25		3,375	925		4,300	5,125
8000	Ventilation, residential									
8020	Attic, roof type									
8030	Aluminum dome, damper & curb									
8040	6" diameter, 300 CFM	1 Elec	16	.500	Ea.	238	20.50		258.50	293
8050	7" diameter, 450 CFM		15	.533		260	21.50		281.50	320
8060	9" diameter, 900 CFM		14	.571		415	23.50		438.50	495
8080	12" diameter, 1000 CFM (gravity)		10	.800		295	32.50		327.50	375
8090	16" diameter, 1500 CFM (gravity)		9	.889		355	36		391	445
8100	20" diameter, 2500 CFM (gravity)		8	1		435	41		476	540
8110	26" diameter, 4000 CFM (gravity)		7	1.143		530	46.50		576.50	650
8120	32" diameter, 6500 CFM (gravity)		6	1.333		725	54.50		779.50	875
8130	38" diameter, 8000 CFM (gravity)		5	1.600		1,075	65		1,140	1,275
8140	50" diameter, 13,000 CFM (gravity)		4	2		1,550	81.50		1,631.50	1,850
8160	Plastic, ABS dome									
8180	1050 CFM	1 Elec	14	.571	Ea.	86.50	23.50		110	130
8200	1600 CFM	″	12	.667	″	130	27		157	184
8240	Attic, wall type, with shutter, one speed									
8250	12" diameter, 1000 CFM	1 Elec	14	.571	Ea.	187	23.50		210.50	241
8260	14" diameter, 1500 CFM		12	.667		203	27		230	264
8270	16" diameter, 2000 CFM		9	.889		230	36		266	305
8290	Whole house, wall type, with shutter, one speed									
8300	30" diameter, 4800 CFM	1 Elec	7	1.143	Ea.	490	46.50		536.50	610
8310	36" diameter, 7000 CFM		6	1.333		535	54.50		589.50	670
8320	42" diameter, 10,000 CFM		5	1.600		600	65		665	755

15830 | Fans

			CREW	DAILY OUTPUT	LABOR-HOURS	UNIT	2005 BARE COSTS MAT.	LABOR	EQUIP.	TOTAL	TOTAL INCL O&P	
100	8330	48" diameter, 16,000 CFM (R15700-040)	1 Elec	4	2	Ea.	745	81.50		826.50	940	100
	8340	For two speed, add					45			45	49.50	
	8350	Whole house, lay-down type, with shutter, one speed (R15800-080)										
	8360	30" diameter, 4500 CFM	1 Elec	8	1	Ea.	525	41		566	635	
	8370	36" diameter, 6500 CFM		7	1.143		560	46.50		606.50	690	
	8380	42" diameter, 9000 CFM		6	1.333		620	54.50		674.50	760	
	8390	48" diameter, 12,000 CFM		5	1.600		700	65		765	865	
	8440	For two speed, add					33.50			33.50	37	
	8450	For 12 hour timer switch, add	1 Elec	32	.250		33.50	10.20		43.70	52	
	8500	Wall exhausters, centrifugal, auto damper, 1/8" S.P.										
	8520	Direct drive, 610 CFM, 1/20 HP	Q-20	14	1.429	Ea.	207	53		260	310	
	8540	796 CFM, 1/12 HP		13	1.538		214	57		271	325	
	8560	822 CFM, 1/6 HP		12	1.667		340	62		402	465	
	8580	1,320 CFM, 1/4 HP		12	1.667		340	62		402	470	
	8600	1756 CFM, 1/4 HP		11	1.818		400	67.50		467.50	545	
	8620	1983 CFM, 1/4 HP		10	2		475	74		549	640	
	8640	2900 CFM, 1/2 HP		9	2.222		575	82.50		657.50	760	
	8660	3307 CFM, 3/4 HP		8	2.500		635	92.50		727.50	840	
	8670	5940 CFM, 1 HP		7	2.857		900	106		1,006	1,150	
	9500	V-belt drive, 3 phase										
	9520	2,800 CFM, 1/4 HP	Q-20	9	2.222	Ea.	940	82.50		1,022.50	1,150	
	9540	3,740 CFM, 1/2 HP		8	2.500		975	92.50		1,067.50	1,225	
	9560	4400 CFM, 3/4 HP		7	2.857		990	106		1,096	1,225	
	9580	5700 CFM, 1-1/2 HP		6	3.333		1,025	124		1,149	1,325	

15840 | Air Terminal Units

			CREW	DAILY OUTPUT	LABOR-HOURS	UNIT	MAT.	LABOR	EQUIP.	TOTAL	TOTAL INCL O&P	
200	0010	**AIR CURTAINS** Incl. motor starters, transformers,										200
	0050	door switches & temperature controls										
	0100	Shipping and receiving doors, unheated, minimal wind stoppage										
	0150	8' high, multiples of 3' wide	2 Shee	6	2.667	L.F.	320	107		427	515	
	0160	5' wide		10	1.600		310	64.50		374.50	445	
	0210	10' high, multiples of 4' wide		8	2		305	80.50		385.50	460	
	0250	12' high, 3'-6" wide		7	2.286		335	92		427	505	
	0260	12' wide		6	2.667		305	107		412	500	
	0350	16' high, 3'-6" wide		7	2.286		320	92		412	490	
	0360	12' wide		6	2.667		355	107		462	555	
	1500	Customer entrance doors, unheated, minimal wind stoppage										
	1550	10' high, multiples of 3' wide	2 Shee	6	2.667	L.F.	282	107		389	475	
	1560	5' wide		10	1.600		305	64.50		369.50	435	
	1650	Maximum wind stoppage, 12' high, multiples of 4' wide		8	2		227	80.50		307.50	375	
	1700	Heated, minimal wind stoppage, electric heat										
	1750	8' high, multiples of 3' wide	2 Shee	6	2.667	L.F.	560	107		667	785	
	1850	10' high, multiples of 3' wide		6	2.667		480	107		587	695	
	1860	Multiples of 5' wide		10	1.600		525	64.50		589.50	675	
	1950	Maximum wind stoppage, steam heat										
	1960	12' high, multiples of 4' wide	2 Shee	8	2	L.F.	510	80.50		590.50	685	
	2000	Walk-in coolers and freezers, ambient air, minimal wind stoppage										
	2050	8' high, multiples of 3' wide	2 Shee	6	2.667	L.F.	214	107		321	400	
	2060	Multiples of 5' wide		10	1.600		231	64.50		295.50	355	
	2250	Maximum wind stoppage, 12' high, multiples of 3' wide		6	2.667		246	107		353	435	
	2450	Conveyor openings or service windows, unheated, 5' high		5	3.200		268	129		397	490	
	2460	Heated, electric, 5' high, 2'-6" wide		5	3.200		194	129		323	410	
500	0010	**CONSTANT VOLUME MIXING BOXES**										500
	5180	Mixing box, includes electric or pneumatic motor										
	5190	Recommend use with attenuator, see 15820-300-9000										
	5200	Constant volume, 150 to 270 CFM	Q-9	12	1.333	Ea.	525	48		573	650	

15840 | Air Terminal Units

		15840 Air Terminal Units	CREW	DAILY OUTPUT	LABOR-HOURS	UNIT	2005 BARE COSTS MAT.	LABOR	EQUIP.	TOTAL	TOTAL INCL O&P	
500	5210	270 to 600 CFM	Q-9	11	1.455	Ea.	540	52.50		592.50	670	500
	5230	550 to 1000 CFM		9	1.778		540	64.50		604.50	690	
	5240	1000 to 1600 CFM		8	2		550	72.50		622.50	715	
	5250	1300 to 1900 CFM		6	2.667		565	96.50		661.50	770	
	5260	550 to 2640 CFM		5.60	2.857		620	103		723	840	
	5270	650 to 3120 CFM		5.20	3.077		660	111		771	895	
700	0010	VARIABLE VOLUME MIXING BOXES										700
	5500	VAV Cool only, pneum. press indep. 300 to 600 CFM	Q-9	11	1.455	Ea.	281	52.50		333.50	390	
	5510	500 to 1000 CFM		9	1.778		293	64.50		357.50	425	
	5520	800 to 1600 CFM		9	1.778		305	64.50		369.50	435	
	5530	1100 to 2000 CFM		8	2		320	72.50		392.50	460	
	5540	1500 to 3000 CFM		7	2.286		335	82.50		417.50	495	
	5550	2000 to 4000 CFM		6	2.667		355	96.50		451.50	540	
	5560	For electric, w/thermostat, press. dependent, add					20			20	22	
	5600	VAV Cool & HW coils, damper, actuator and t'stat										
	5610	200 CFM	Q-9	11	1.455	Ea.	590	52.50		642.50	730	
	5620	400 CFM		10	1.600		595	58		653	745	
	5630	600 CFM		10	1.600		595	58		653	745	
	5640	800 CFM		8	2		610	72.50		682.50	785	
	5650	1000 CFM		8	2		610	72.50		682.50	785	
	5660	1250 CFM		6	2.667		670	96.50		766.50	885	
	5670	1500 CFM		6	2.667		670	96.50		766.50	885	
	5680	2000 CFM		4	4		725	145		870	1,025	
	5684	3000 CFM		3.80	4.211		790	152		942	1,100	
	5700	VAV Cool only, fan powr'd, damper, actuator, t'stat										
	5710	200 CFM	Q-9	10	1.600	Ea.	805	58		863	975	
	5720	400 CFM		9	1.778		835	64.50		899.50	1,025	
	5730	600 CFM		9	1.778		835	64.50		899.50	1,025	
	5740	800 CFM		7	2.286		895	82.50		977.50	1,100	
	5750	1000 CFM		7	2.286		895	82.50		977.50	1,100	
	5760	1250 CFM		5	3.200		965	116		1,081	1,225	
	5770	1500 CFM		5	3.200		965	116		1,081	1,225	
	5780	2000 CFM		4	4		1,075	145		1,220	1,400	
	5800	VAV Fan powr'd, cooling, w/HW coil, damper, actuator, t'stat										
	5810	200 CFM	Q-9	10	1.600	Ea.	970	58		1,028	1,175	
	5820	400 CFM		9	1.778		1,000	64.50		1,064.50	1,200	
	5830	600 CFM		9	1.778		1,000	64.50		1,064.50	1,200	
	5840	800 CFM		7	2.286		1,050	82.50		1,132.50	1,300	
	5850	1000 CFM		7	2.286		1,050	82.50		1,132.50	1,300	
	5860	1250 CFM		5	3.200		1,200	116		1,316	1,500	
	5870	1500 CFM		5	3.200		1,200	116		1,316	1,500	
	5880	2000 CFM		3.50	4.571		1,300	165		1,465	1,675	
800	0010	INDUCTION AIR, TERMINAL MIXING BOXES										800
	5154	Induction air terminal unit, cooling and heating										
	5156	Vertical, cooling w/ electric heating coils										
	5158	60 CFM	Q-20	7.60	2.632	Ea.	930	97.50		1,027.50	1,175	
	5160	90 CFM		7.20	2.778		1,150	103		1,253	1,400	
	5162	120 CFM		6.80	2.941		1,150	109		1,259	1,425	
	5164	150 CFM		6.40	3.125		1,300	116		1,416	1,625	
	5166	170 CFM		6.20	3.226		1,300	120		1,420	1,625	
	5170	Horizontal, cooling w/ electric heating coils										
	5172	60 CFM	Q-20	7.60	2.632	Ea.	755	97.50		852.50	985	
	5174	90 CFM		7.20	2.778		930	103		1,033	1,175	
	5176	120 CFM		6.80	2.941		930	109		1,039	1,200	

15850 | Air Outlets & Inlets

300		CREW	DAILY OUTPUT	LABOR-HOURS	UNIT	2005 BARE COSTS MAT.	LABOR	EQUIP.	TOTAL	TOTAL INCL O&P
0010	**DIFFUSERS** Aluminum, opposed blade damper unless noted R15850-060									
0100	Ceiling, linear, also for sidewall									
0120	2" wide R15810-070	1 Shee	32	.250	L.F.	29	10.05		39.05	47
0140	3" wide		30	.267		33.50	10.70		44.20	53.50
0160	4" wide R15800-080		26	.308		37.50	12.35		49.85	60.50
0180	6" wide		24	.333		45	13.40		58.40	70
0200	8" wide		22	.364		51.50	14.60		66.10	79
0220	10" wide		20	.400		58	16.10		74.10	88
0240	12" wide		18	.444		71.50	17.85		89.35	106
0260	For floor or sill application, add					15%				
0500	Perforated, 24" x 24" lay-in panel size, 6" x 6"	1 Shee	16	.500	Ea.	75.50	20		95.50	114
0520	8" x 8"		15	.533		77.50	21.50		99	118
0530	9" x 9"		14	.571		79	23		102	123
0540	10" x 10"		14	.571		79.50	23		102.50	123
0560	12" x 12"		12	.667		81.50	27		108.50	131
0580	15" x 15"		11	.727		111	29		140	167
0590	16" x 16"		11	.727		96	29		125	150
0600	18" x 18"		10	.800		101	32		133	161
0610	20" x 20"		10	.800		116	32		148	177
0620	24" x 24"		9	.889		133	35.50		168.50	202
1000	Rectangular, 1 to 4 way blow, 6" x 6"		16	.500		40.50	20		60.50	76
1010	8" x 8"		15	.533		48.50	21.50		70	86
1014	9" x 9"		15	.533		52	21.50		73.50	90
1016	10" x 10"		15	.533		61	21.50		82.50	100
1020	12" x 6"		15	.533		67	21.50		88.50	107
1040	12" x 9"		14	.571		73	23		96	116
1060	12" x 12"		12	.667		71.50	27		98.50	120
1070	14" x 6"		13	.615		72	24.50		96.50	118
1074	14" x 14"		12	.667		93.50	27		120.50	144
1080	18" x 12"		11	.727		128	29		157	186
1120	15" x 15"		10	.800		125	32		157	187
1140	18" x 15"		9	.889		151	35.50		186.50	221
1150	18" x 18"		9	.889		137	35.50		172.50	206
1160	21" x 21"		8	1		166	40		206	244
1170	24" x 12"		10	.800		124	32		156	186
1180	24" x 24"		7	1.143		281	46		327	380
1500	Round, butterfly damper, 6" diameter		18	.444		16.85	17.85		34.70	46
1520	8" diameter		16	.500		18.20	20		38.20	51
1540	10" diameter		14	.571		22	23		45	60
1560	12" diameter		12	.667		29.50	27		56.50	73.50
1580	14" diameter		10	.800		38	32		70	91.50
1600	18" diameter		9	.889		90.50	35.50		126	155
1610	20" diameter		9	.889		108	35.50		143.50	174
1620	22" diameter		8	1		122	40		162	196
1630	24" diameter		8	1		134	40		174	210
1640	28" diameter		7	1.143		149	46		195	235
1650	32" diameter		6	1.333		164	53.50		217.50	263
1700	Round, stl, adjustable core, annular segmented damper									
1704	6" diameter	1 Shee	18	.444	Ea.	67	17.85		84.85	101
1708	8" diameter		16	.500		72	20		92	110
1712	10" diameter		14	.571		81	23		104	125
1716	12" diameter		12	.667		90	27		117	140
1720	14" diameter		10	.800		100	32		132	160
1724	18" diameter		9	.889		153	35.50		188.50	223
1728	20" diameter		9	.889		177	35.50		212.50	250
1732	24" diameter		8	1		276	40		316	365

300

15 MECHANICAL

	15850 | Air Outlets & Inlets		CREW	DAILY OUTPUT	LABOR-HOURS	UNIT	2005 BARE COSTS MAT.	LABOR	EQUIP.	TOTAL	TOTAL INCL O&P	
300	1736	30" diameter R15850-060	1 Shee	7	1.143	Ea.	480	46		526	600	300
	1740	36" diameter		6	1.333		585	53.50		638.50	730	
	2000	T bar mounting, 24" x 24" lay-in frame, 6" x 6" R15810-070		16	.500		82	20		102	121	
	2020	9" x 9"		14	.571		90.50	23		113.50	135	
	2040	12" x 12" R15800-080		12	.667		117	27		144	170	
	2060	15" x 15"		11	.727		150	29		179	210	
	2080	18" x 18"		10	.800		157	32		189	222	
	2500	Combination supply and return										
	2520	21" x 21" supply, 15" x 15" return	Q-9	10	1.600	Ea.	198	58		256	305	
	2540	24" x 24" supply, 18" x 18" return		9.50	1.684		250	61		311	370	
	2560	27" x 27" supply, 18" x 18" return		9	1.778		305	64.50		369.50	435	
	2580	30" x 30" supply, 21" x 21" return		8.50	1.882		385	68		453	530	
	2600	33" x 33" supply, 24" x 24" return		8	2		450	72.50		522.50	605	
	2620	36" x 36" supply, 24" x 24" return		7.50	2.133		530	77		607	705	
	3000	Baseboard, white enameled steel										
	3100	18" long	1 Shee	20	.400	Ea.	8.70	16.10		24.80	34	
	3120	24" long		18	.444		17	17.85		34.85	46	
	3140	48" long		16	.500		31	20		51	65.50	
	3400	For matching return, deduct					10%					
	4000	Floor, steel, adjustable pattern										
	4100	2" x 10"	1 Shee	34	.235	Ea.	11.75	9.45		21.20	27.50	
	4120	2" x 12"		32	.250		12.90	10.05		22.95	29.50	
	4140	2" x 14"		30	.267		14.65	10.70		25.35	32.50	
	4200	4" x 10"		28	.286		13.60	11.50		25.10	32.50	
	4220	4" x 12"		26	.308		15.05	12.35		27.40	35.50	
	4240	4" x 14"		25	.320		17	12.85		29.85	38.50	
	4260	6" x 10"		26	.308		20.50	12.35		32.85	41.50	
	4280	6" x 12"		24	.333		21.50	13.40		34.90	44	
	4300	6" x 14"		22	.364		22.50	14.60		37.10	47	
	4500	Linear bar diffuser in wall, sill or ceiling										
	4510	2" wide	1 Shee	32	.250	L.F.	29	10.05		39.05	47	
	4512	3" wide		30	.267		34	10.70		44.70	54	
	4514	4" wide		26	.308		36.50	12.35		48.85	59	
	4516	6" wide		24	.333		42.50	13.40		55.90	67.50	
	4520	10" wide		20	.400		52	16.10		68.10	81.50	
	4522	12" wide		18	.444		68	17.85		85.85	103	
	5000	Sidewall, aluminum, 3 way dispersion										
	5100	8" x 4"	1 Shee	24	.333	Ea.	6.45	13.40		19.85	27.50	
	5110	8" x 6"		24	.333		6.90	13.40		20.30	28	
	5120	10" x 4"		22	.364		6.75	14.60		21.35	30	
	5130	10" x 6"		22	.364		7.25	14.60		21.85	30.50	
	5140	10" x 8"		20	.400		9.90	16.10		26	35.50	
	5160	12" x 4"		18	.444		7.30	17.85		25.15	35.50	
	5170	12" x 6"		17	.471		8.05	18.90		26.95	38	
	5180	12" x 8"		16	.500		10.45	20		30.45	42.50	
	5200	14" x 4"		15	.533		8.15	21.50		29.65	42	
	5220	14" x 6"		14	.571		8.95	23		31.95	45.50	
	5240	14" x 8"		13	.615		11.65	24.50		36.15	51	
	5260	16" x 6"		12.50	.640		11.85	25.50		37.35	52.50	
	6000	For steel diffusers instead of aluminum, deduct					10%					
500	0010	**GRILLES**										500
	0020	Aluminum										
	0100	Air supply, single deflection, adjustable										
	0120	8" x 4"	1 Shee	30	.267	Ea.	12	10.70		22.70	29.50	
	0140	8" x 8"		28	.286		14.55	11.50		26.05	33.50	
	0160	10" x 4"		24	.333		13.25	13.40		26.65	35	

	15850 | Air Outlets & Inlets	CREW	DAILY OUTPUT	LABOR-HOURS	UNIT	2005 BARE COSTS MAT.	LABOR	EQUIP.	TOTAL	TOTAL INCL O&P	
500	0180 10″ x 10″	1 Shee	23	.348	Ea.	16.80	14		30.80	40	500
	0200 12″ x 6″		23	.348		15.65	14		29.65	38.50	
	0220 12″ x 12″		22	.364		20.50	14.60		35.10	45	
	0230 14″ x 6″		23	.348		17.45	14		31.45	40.50	
	0240 14″ x 8″		23	.348		18.65	14		32.65	42	
	0260 14″ x 14″		22	.364		25	14.60		39.60	50	
	0270 18″ x 8″		23	.348		21.50	14		35.50	45	
	0280 18″ x 10″		23	.348		24.50	14		38.50	48.50	
	0300 18″ x 18″		21	.381		38.50	15.30		53.80	66	
	0320 20″ x 12″		22	.364		28	14.60		42.60	53.50	
	0340 20″ x 20″		21	.381		46.50	15.30		61.80	74.50	
	0350 24″ x 8″		20	.400		27	16.10		43.10	54	
	0360 24″ x 14″		18	.444		38	17.85		55.85	69	
	0380 24″ x 24″		15	.533		66	21.50		87.50	106	
	0400 30″ x 8″		20	.400		34.50	16.10		50.60	62.50	
	0420 30″ x 10″		19	.421		38.50	16.95		55.45	68	
	0440 30″ x 12″		18	.444		42	17.85		59.85	73.50	
	0460 30″ x 16″		17	.471		54.50	18.90		73.40	89	
	0480 30″ x 18″		17	.471		61	18.90		79.90	96	
	0500 30″ x 30″		14	.571		109	23		132	156	
	0520 36″ x 12″		17	.471		52	18.90		70.90	86.50	
	0540 36″ x 16″		16	.500		66.50	20		86.50	104	
	0560 36″ x 18″		15	.533		73.50	21.50		95	114	
	0570 36″ x 20″		15	.533		80.50	21.50		102	122	
	0580 36″ x 24″		14	.571		105	23		128	151	
	0600 36″ x 28″		13	.615		128	24.50		152.50	179	
	0620 36″ x 30″		12	.667		141	27		168	196	
	0640 36″ x 32″		12	.667		155	27		182	211	
	0660 36″ x 34″		11	.727		169	29		198	230	
	0680 36″ x 36″	↓	11	.727		183	29		212	246	
	0700 For double deflecting, add					70%					
	1000 Air return, 6″ x 6″	1 Shee	26	.308		13	12.35		25.35	33.50	
	1020 10″ x 6″		24	.333		15.60	13.40		29	37.50	
	1040 14″ x 6″		23	.348		18.20	14		32.20	41.50	
	1060 10″ x 8″		23	.348		15.60	14		29.60	38.50	
	1080 16″ x 8″		22	.364		23.50	14.60		38.10	48	
	1100 12″ x 12″		22	.364		23.50	14.60		38.10	48	
	1120 24″ x 12″		18	.444		41.50	17.85		59.35	73.50	
	1140 30″ x 12″		16	.500		49.50	20		69.50	85.50	
	1160 14″ x 14″		22	.364		28.50	14.60		43.10	54	
	1180 16″ x 16″		22	.364		34	14.60		48.60	59.50	
	1200 18″ x 18″		21	.381		39	15.30		54.30	66.50	
	1220 24″ x 18″		16	.500		49.50	20		69.50	85.50	
	1240 36″ x 18″		15	.533		83	21.50		104.50	125	
	1260 24″ x 24″		15	.533		65	21.50		86.50	105	
	1280 36″ x 24″		14	.571		99	23		122	145	
	1300 48″ x 24″		12	.667		161	27		188	218	
	1320 48″ x 30″		11	.727		208	29		237	274	
	1340 36″ x 36″		13	.615		166	24.50		190.50	221	
	1360 48″ x 36″		11	.727		234	29		263	300	
	1380 48″ x 48″		8	1		310	40		350	405	
	2000 Door grilles, 12″ x 12″		22	.364		34	14.60		48.60	59.50	
	2020 18″ x 12″		22	.364		41.50	14.60		56.10	68.50	
	2040 24″ x 12″		18	.444		49.50	17.85		67.35	82	
	2060 18″ x 18″		18	.444		62.50	17.85		80.35	96	
	2080 24″ x 18″	↓	16	.500	↓	78	20		98	117	

	15850 | Air Outlets & Inlets		CREW	DAILY OUTPUT	LABOR-HOURS	UNIT	2005 BARE COSTS MAT.	LABOR	EQUIP.	TOTAL	TOTAL INCL O&P	
500	2100	24" x 24"	1 Shee	15	.533	Ea.	104	21.50		125.50	147	500
	3000	Filter grille with filter, 12" x 12"		24	.333		39	13.40		52.40	63.50	
	3020	18" x 12"		20	.400		57	16.10		73.10	87.50	
	3040	24" x 18"		18	.444		83	17.85		100.85	119	
	3060	24" x 24"		16	.500		109	20		129	151	
	3080	30" x 24"		14	.571		140	23		163	190	
	3100	30" x 30"		13	.615		177	24.50		201.50	232	
	3950	Eggcrate, framed, 6" x 6" opening		26	.308		14.60	12.35		26.95	35	
	3954	8" x 8" opening		24	.333		15.10	13.40		28.50	37	
	3960	10" x 10" opening		23	.348		23.50	14		37.50	47	
	3970	12" x 12" opening		22	.364		20.50	14.60		35.10	45	
	3980	14" x 14" opening		22	.364		25.50	14.60		40.10	50.50	
	3984	16" x 16" opening		21	.381		32	15.30		47.30	58.50	
	3990	18" x 18" opening		21	.381		37.50	15.30		52.80	65	
	4020	22" x 22" opening		17	.471		60	18.90		78.90	95	
	4040	24" x 24" opening		15	.533		63	21.50		84.50	103	
	4044	28" x 28" opening		14	.571		90	23		113	135	
	4048	36" x 36" opening		12	.667		107	27		134	159	
	4050	48" x 24" opening		12	.667		187	27		214	247	
	4060	Eggcrate, lay-in, T-bar system										
	4070	48" x 24" sheet	1 Shee	40	.200	Ea.	76.50	8.05		84.55	97	
	5000	Transfer grille, vision proof, 8" x 4"		30	.267		22	10.70		32.70	41	
	5020	8" x 6"		28	.286		22	11.50		33.50	42	
	5040	8" x 8"		26	.308		24.50	12.35		36.85	46	
	5060	10" x 6"		24	.333		23	13.40		36.40	46	
	5080	10" x 10"		23	.348		29	14		43	53.50	
	5090	12" x 6"		23	.348		31	14		45	55.50	
	5100	12" x 10"		22	.364		32	14.60		46.60	57.50	
	5110	12" x 12"		22	.364		34.50	14.60		49.10	60.50	
	5120	14" x 10"		22	.364		34.50	14.60		49.10	60.50	
	5140	16" x 8"		21	.381		34.50	15.30		49.80	61.50	
	5160	18" x 12"		20	.400		45	16.10		61.10	74	
	5170	18" x 18"		21	.381		63.50	15.30		78.80	93.50	
	5180	20" x 12"		20	.400		48.50	16.10		64.60	78	
	5200	20" x 20"		19	.421		77	16.95		93.95	111	
	5220	24" x 12"		17	.471		54.50	18.90		73.40	89	
	5240	24" x 24"		16	.500		105	20		125	147	
	5260	30" x 6"		15	.533		48.50	21.50		70	86.50	
	5280	30" x 8"		15	.533		56.50	21.50		78	95	
	5300	30" x 12"		14	.571		70.50	23		93.50	113	
	5320	30" x 16"		13	.615		84	24.50		108.50	130	
	5340	30" x 20"		12	.667		106	27		133	157	
	5360	30" x 24"		11	.727		131	29		160	189	
	5380	30" x 30"		10	.800		172	32		204	239	
	6000	For steel grilles instead of aluminum in above, deduct					10%					
	6200	Plastic, eggcrate, lay-in, T-bar system										
	6210	48" x 24" sheet	1 Shee	50	.160	Ea.	19.50	6.45		25.95	31.50	
	6250	Steel door louver										
	6270	With fire link, steel only										
	6350	12" x 18"	1 Shee	22	.364	Ea.	135	14.60		149.60	172	
	6410	12" x 24"		21	.381		150	15.30		165.30	189	
	6470	18" x 18"		20	.400		150	16.10		166.10	190	
	6530	18" x 24"		19	.421		172	16.95		188.95	215	
	6600	18" x 36"		17	.471		195	18.90		213.90	244	
	6660	24" x 24"		15	.533		181	21.50		202.50	232	
	6720	24" x 30"		15	.533		195	21.50		216.50	248	

		15850 | Air Outlets & Inlets	CREW	DAILY OUTPUT	LABOR-HOURS	UNIT	2005 BARE COSTS MAT.	LABOR	EQUIP.	TOTAL	TOTAL INCL O&P	
500	6780	30″ x 30″	1 Shee	13	.615	Ea.	229	24.50		253.50	290	500
	6840	30″ x 36″		12	.667		265	27		292	335	
	6900	36″ x 36″	↓	10	.800	↓	305	32		337	385	
600	0010	**LOUVERS**										600
	0100	Aluminum, extruded, with screen, mill finish										
	1000	Brick vent, (see also division 04090-860)										
	1100	Standard, 4″ deep, 8″ wide, 5″ high	1 Shee	24	.333	Ea.	24	13.40		37.40	47	
	1200	Modular, 4″ deep, 7-3/4″ wide, 5″ high		24	.333		25	13.40		38.40	48	
	1300	Speed brick, 4″ deep, 11-5/8″ wide, 3-7/8″ high		24	.333		25	13.40		38.40	48	
	1400	Fuel oil brick, 4″ deep, 8″ wide, 5″ high		24	.333	↓	44	13.40		57.40	69	
	2000	Cooling tower and mechanical equip., screens, light weight		40	.200	S.F.	10.45	8.05		18.50	24	
	2020	Standard weight		35	.229		29	9.20		38.20	46	
	2500	Dual combination, automatic, intake or exhaust		20	.400		40	16.10		56.10	68.50	
	2520	Manual operation		20	.400		29.50	16.10		45.60	57	
	2540	Electric or pneumatic operation		20	.400	↓	29.50	16.10		45.60	57	
	2560	Motor, for electric or pneumatic	↓	14	.571	Ea.	350	23		373	420	
	3000	Fixed blade, continuous line										
	3100	Mullion type, stormproof	1 Shee	28	.286	S.F.	29.50	11.50		41	50	
	3200	Stormproof		28	.286		29.50	11.50		41	50	
	3300	Vertical line	↓	28	.286	↓	39	11.50		50.50	60.50	
	3500	For damper to use with above, add					50%	30%				
	3520	Motor, for damper, electric or pneumatic	1 Shee	14	.571	Ea.	350	23		373	420	
	4000	Operating, 45°, manual, electric or pneumatic		24	.333	S.F.	29.50	13.40		42.90	53	
	4100	Motor, for electric or pneumatic		14	.571	Ea.	350	23		373	420	
	4200	Penthouse, roof		56	.143	S.F.	17.15	5.75		22.90	27.50	
	4300	Walls		40	.200		41	8.05		49.05	58	
	5000	Thinline, under 4″ thick, fixed blade	↓	40	.200	↓	16.60	8.05		24.65	30.50	
	5010	Finishes, applied by mfr. at additional cost, available in colors										
	5020	Prime coat only, add				S.F.	2.42			2.42	2.66	
	5040	Baked enamel finish coating, add					4.44			4.44	4.88	
	5060	Anodized finish, add					4.82			4.82	5.30	
	5080	Duranodic finish, add					8.75			8.75	9.60	
	5100	Fluoropolymer finish coating, add					13.65			13.65	15.05	
	9980	For small orders (under 10 pieces), add				↓	25%					
700	0010	**REGISTERS**										700
	0980	Air supply										
	1000	Ceiling/wall, O.B. damper, anodized aluminum										
	1010	One or two way deflection, adj. curved face bars										
	1014	6″ x 6″	1 Shee	24	.333	Ea.	20.50	13.40		33.90	43.50	
	1020	8″ x 4″		26	.308		20.50	12.35		32.85	42	
	1040	8″ x 8″		24	.333		26	13.40		39.40	49	
	1060	10″ x 6″		20	.400		23.50	16.10		39.60	50.50	
	1080	10″ x 10″		19	.421		32	16.95		48.95	61	
	1100	12″ x 6″		19	.421		23.50	16.95		40.45	52	
	1120	12″ x 12″		18	.444		34	17.85		51.85	64.50	
	1140	14″ x 8″		17	.471		33	18.90		51.90	65	
	1160	14″ x 14″		18	.444		39.50	17.85		57.35	71	
	1170	16″ x 16″		17	.471		43.50	18.90		62.40	77	
	1180	18″ x 8″		18	.444		53	17.85		70.85	86	
	1200	18″ x 18″		17	.471		67	18.90		85.90	103	
	1220	20″ x 4″		19	.421		34	16.95		50.95	63	
	1240	20″ x 6″		18	.444		34	17.85		51.85	64.50	
	1260	20″ x 8″		18	.444		37	17.85		54.85	68.50	
	1280	20″ x 20″	↓	17	.471	↓	81.50	18.90		100.40	119	

15850 | Air Outlets & Inlets

			Crew	Daily Output	Labor-Hours	Unit	2005 Bare Costs Mat.	Labor	Equip.	Total	Total Incl O&P	
700	1290	22" x 22"	1 Shee	15	.533	Ea.	94	21.50		115.50	136	700
	1300	24" x 4"		17	.471		41.50	18.90		60.40	75	
	1320	24" x 6"		16	.500		41.50	20		61.50	77	
	1340	24" x 8"		13	.615		45.50	24.50		70	88	
	1350	24" x 18"		12	.667		85	27		112	135	
	1360	24" x 24"		11	.727		115	29		144	172	
	1380	30" x 4"		16	.500		52	20		72	88.50	
	1400	30" x 6"		15	.533		52	21.50		73.50	90.50	
	1420	30" x 8"		14	.571		60	23		83	102	
	1440	30" x 24"		12	.667		149	27		176	205	
	1460	30" x 30"		10	.800		191	32		223	260	
	1504	4 way deflection, adjustable curved face bars										
	1510	6" x 6"	1 Shee	26	.308	Ea.	25	12.35		37.35	46.50	
	1514	8" x 8"		24	.333		31	13.40		44.40	54.50	
	1518	10" x 10"		19	.421		38.50	16.95		55.45	68.50	
	1522	12" x 6"		19	.421		28	16.95		44.95	57	
	1526	12" x 12"		18	.444		40.50	17.85		58.35	72	
	1530	14" x 14"		18	.444		47	17.85		64.85	79.50	
	1534	16" x 16"		17	.471		52.50	18.90		71.40	86.50	
	1538	18" x 18"		17	.471		80.50	18.90		99.40	118	
	1542	22" x 22"		16	.500		113	20		133	155	
	1980	One way deflection, adj. vert. or horiz. face bars										
	1990	6" x 6"	1 Shee	26	.308	Ea.	29	12.35		41.35	51	
	2000	8" x 4"		26	.308		30.50	12.35		42.85	53	
	2020	8" x 8"		24	.333		33.50	13.40		46.90	57.50	
	2040	10" x 6"		20	.400		33.50	16.10		49.60	61.50	
	2060	10" x 10"		19	.421		39	16.95		55.95	68.50	
	2080	12" x 6"		19	.421		36	16.95		52.95	65.50	
	2100	12" x 12"		18	.444		47	17.85		64.85	79.50	
	2120	14" x 8"		17	.471		46	18.90		64.90	79.50	
	2140	14" x 12"		18	.444		51.50	17.85		69.35	84	
	2160	14" x 14"		18	.444		58.50	17.85		76.35	92	
	2180	16" x 6"		18	.444		43	17.85		60.85	75	
	2200	16" x 12"		18	.444		57	17.85		74.85	90	
	2220	16" x 16"		17	.471		72.50	18.90		91.40	109	
	2240	18" x 8"		18	.444		50.50	17.85		68.35	83.50	
	2260	18" x 12"		17	.471		61	18.90		79.90	96	
	2280	18" x 18"		16	.500		89	20		109	129	
	2300	20" x 10"		18	.444		60	17.85		77.85	93.50	
	2320	20" x 16"		16	.500		86.50	20		106.50	126	
	2340	20" x 20"		16	.500		108	20		128	150	
	2360	24" x 12"		15	.533		74.50	21.50		96	115	
	2380	24" x 16"		14	.571		99.50	23		122.50	146	
	2400	24" x 20"		12	.667		124	27		151	178	
	2420	24" x 24"		10	.800		152	32		184	217	
	2440	30" x 12"		13	.615		97	24.50		121.50	144	
	2460	30" x 16"		13	.615		126	24.50		150.50	176	
	2480	30" x 24"		11	.727		196	29		225	261	
	2500	30" x 30"		9	.889		253	35.50		288.50	335	
	2520	36" x 12"		11	.727		118	29		147	175	
	2540	36" x 24"		10	.800		243	32		275	315	
	2560	36" x 36"		8	1		425	40		465	530	
	2600	For 2 way deflect., adj. vert. or horiz. face bars, add					40%					
	2700	Above registers in steel instead of aluminum, deduct					10%					
	3000	Baseboard, hand adj. damper, enameled steel										
	3012	8" x 6"	1 Shee	26	.308	Ea.	12.45	12.35		24.80	32.50	

700	15850 | Air Outlets & Inlets	CREW	DAILY OUTPUT	LABOR-HOURS	UNIT	2005 BARE COSTS MAT.	LABOR	EQUIP.	TOTAL	TOTAL INCL O&P	
700	3020 10" x 6"	1 Shee	24	.333	Ea.	14	13.40		27.40	36	700
	3040 12" x 5"		23	.348		16	14		30	39	
	3060 12" x 6"		23	.348		14.75	14		28.75	38	
	3080 12" x 8"		22	.364		21.50	14.60		36.10	46	
	3100 14" x 6"	↓	20	.400	↓	16	16.10		32.10	42	
	4000 Floor, toe operated damper, enameled steel										
	4020 4" x 8"	1 Shee	32	.250	Ea.	17.75	10.05		27.80	35	
	4040 4" x 12"		26	.308		21	12.35		33.35	42	
	4060 6" x 8"		28	.286		18.10	11.50		29.60	37.50	
	4080 6" x 14"		22	.364		24	14.60		38.60	49	
	4100 8" x 10"		22	.364		21.50	14.60		36.10	46.50	
	4120 8" x 16"		20	.400		36.50	16.10		52.60	64.50	
	4140 10" x 10"		20	.400		25.50	16.10		41.60	52.50	
	4160 10" x 16"		18	.444		39.50	17.85		57.35	71	
	4180 12" x 12"		18	.444		32	17.85		49.85	62.50	
	4200 12" x 24"		16	.500		110	20		130	152	
	4220 14" x 14"		16	.500		90.50	20		110.50	131	
	4240 14" x 20"	↓	15	.533	↓	121	21.50		142.50	166	
	4300 Spiral pipe supply register										
	4310 Aluminum, double deflection, w/damper extractor										
	4320 4" x 12", for 6" thru 10" diameter duct	1 Shee	25	.320	Ea.	62	12.85		74.85	88	
	4330 4" x 18", for 6" thru 10" diameter duct		18	.444		76.50	17.85		94.35	112	
	4340 6" x 12", for 8" thru 12" diameter duct		19	.421		67.50	16.95		84.45	101	
	4350 6" x 18", for 8" thru 12" diameter duct		18	.444		88	17.85		105.85	124	
	4360 6" x 24", for 8" thru 12" diameter duct		16	.500		109	20		129	150	
	4370 6" x 30", for 8" thru 12" diameter duct		15	.533		138	21.50		159.50	185	
	4380 8" x 18", for 10" thru 14" diameter duct		18	.444		93.50	17.85		111.35	131	
	4390 8" x 24", for 10" thru 14" diameter duct		15	.533		117	21.50		138.50	162	
	4400 8" x 30", for 10" thru 14" diameter duct		14	.571		155	23		178	207	
	4410 10" x 24", for 12" thru 18" diameter duct		13	.615		129	24.50		153.50	180	
	4420 10" x 30", for 12" thru 18" diameter duct		12	.667		170	27		197	228	
	4430 10" x 36", for 12" thru 18" diameter duct	↓	11	.727	↓	211	29		240	277	
	4980 Air return										
	5000 Ceiling or wall, fixed 45° face blades										
	5010 Adjustable O.B. damper, anodized aluminum										
	5020 4" x 8"	1 Shee	26	.308	Ea.	23.50	12.35		35.85	44.50	
	5040 6" x 8"		24	.333		23.50	13.40		36.90	46	
	5060 6" x 10"		19	.421		26	16.95		42.95	54.50	
	5070 6" x 12"		19	.421		28.50	16.95		45.45	57.50	
	5080 6" x 16"		18	.444		34	17.85		51.85	64.50	
	5100 8" x 10"		19	.421		28.50	16.95		45.45	57.50	
	5120 8" x 12"		16	.500		31	20		51	65.50	
	5140 10" x 10"		18	.444		31	17.85		48.85	62	
	5160 10" x 16"		17	.471		41.50	18.90		60.40	75	
	5170 12" x 12"		18	.444		39	17.85		56.85	70.50	
	5180 12" x 18"		18	.444		54.50	17.85		72.35	87.50	
	5184 12" x 24"		14	.571		65	23		88	107	
	5200 12" x 30"		12	.667		80.50	27		107.50	130	
	5210 12" x 36"		10	.800		101	32		133	162	
	5220 16" x 16"		17	.471		57	18.90		75.90	92	
	5240 18" x 18"		16	.500		67.50	20		87.50	106	
	5250 18" x 24"		13	.615		86	24.50		110.50	133	
	5254 18" x 30"		11	.727		107	29		136	162	
	5260 18" x 36"		10	.800		135	32		167	199	
	5270 20" x 20"		10	.800		80.50	32		112.50	138	
	5274 20" x 24"	↓	12	.667	↓	94.50	27		121.50	145	

	15850 | Air Outlets & Inlets		CREW	DAILY OUTPUT	LABOR-HOURS	UNIT	2005 BARE COSTS MAT.	LABOR	EQUIP.	TOTAL	TOTAL INCL O&P	
700	5280	24" x 24"	1 Shee	11	.727	Ea.	112	29		141	168	700
	5281	24" x 24"		11	.727		112	29		141	168	
	5290	24" x 30"		10	.800		143	32		175	207	
	5300	24" x 36"		8	1		174	40		214	254	
	5320	24" x 48"		6	1.333		255	53.50		308.50	365	
	5340	30" x 30"		6.50	1.231		180	49.50		229.50	274	
	5344	30" x 120"	Q-9	6	2.667		970	96.50		1,066.50	1,225	
	5360	36" x 36"	1 Shee	6	1.333		286	53.50		339.50	400	
	5370	40" x 32"	"	6	1.333		269	53.50		322.50	380	
	5400	Ceiling or wall, removable-reversible core, single deflection										
	5402	Adjustable O.B. damper, radiused frame with border, aluminum										
	5410	8" x 4"	1 Shee	26	.308	Ea.	32	12.35		44.35	54	
	5414	8" x 6"		24	.333		36	13.40		49.40	60	
	5418	10" x 6"		19	.421		39	16.95		55.95	69	
	5422	10" x 10"		18	.444		49	17.85		66.85	81.50	
	5426	12" x 6"		19	.421		42	16.95		58.95	72	
	5430	12" x 12"		18	.444		63	17.85		80.85	97	
	5434	16" x 16"		16	.500		94	20		114	134	
	5438	18" x 18"		17	.471		109	18.90		127.90	149	
	5442	20" x 20"		16	.500		134	20		154	178	
	5446	24" x 12"		15	.533		92	21.50		113.50	134	
	5450	24" x 18"		13	.615		135	24.50		159.50	187	
	5454	24" x 24"		11	.727		175	29		204	238	
	5458	30" x 12"		13	.615		110	24.50		134.50	159	
	5462	30" x 18"		12.50	.640		158	25.50		183.50	214	
	5466	30" x 24"		11	.727		210	29		239	276	
	5470	30" x 30"		9	.889		262	35.50		297.50	345	
	5474	36" x 12"		11	.727		133	29		162	191	
	5478	36" x 18"		10.50	.762		187	30.50		217.50	253	
	5482	36" x 24"		10	.800		259	32		291	335	
	5486	36" x 30"		9	.889		340	35.50		375.50	425	
	5490	36" x 36"		8	1		425	40		465	525	
	6000	For steel construction instead of aluminum, deduct					10%					
800	0010	**VENTILATORS** Base & damper										800
	1280	Rotary ventilators, wind driven, galvanized										
	1300	4" neck diameter	Q-9	20	.800	Ea.	52	29		81	102	
	1320	5" neck diameter		18	.889		58	32		90	114	
	1340	6" neck diameter		16	1		58	36		94	120	
	1360	8" neck diameter		14	1.143		60.50	41.50		102	130	
	1380	10" neck diameter		12	1.333		77.50	48		125.50	160	
	1400	12" neck diameter		10	1.600		84.50	58		142.50	182	
	1420	14" neck diameter		10	1.600		117	58		175	218	
	1440	16" neck diameter		9	1.778		141	64.50		205.50	255	
	1460	18" neck diameter, 1700 CFM		9	1.778		164	64.50		228.50	280	
	1480	20" neck diameter, 2100 CFM		8	2		182	72.50		254.50	310	
	1500	24" neck diameter, 3,100 CFM		8	2		245	72.50		317.50	380	
	1520	30" neck diameter, 4500 CFM		7	2.286		510	82.50		592.50	685	
	1540	36" neck diameter, 5,500 CFM		6	2.667		720	96.50		816.50	940	
	1600	For aluminum, add					300%					
	1620	For stainless steel, add					600%					
	1630	For copper, add					600%					
	2000	Stationary, gravity, syphon, galvanized										
	2100	3" neck diameter, 40 CFM	Q-9	24	.667	Ea.	47	24		71	88.50	
	2120	4" neck diameter, 50 CFM		20	.800		51.50	29		80.50	102	
	2140	5" neck diameter, 58 CFM		18	.889		52	32		84	107	

	15850 | Air Outlets & Inlets		CREW	DAILY OUTPUT	LABOR-HOURS	UNIT	2005 BARE COSTS MAT.	LABOR	EQUIP.	TOTAL	TOTAL INCL O&P	
800	2160	6" neck diameter, 66 CFM	Q-9	16	1	Ea.	45	36		81	105	800
	2180	7" neck diameter, 86 CFM		15	1.067		56.50	38.50		95	122	
	2200	8" neck diameter, 110 CFM		14	1.143		57	41.50		98.50	126	
	2220	10" neck diameter, 140 CFM		12	1.333		73.50	48		121.50	155	
	2240	12" neck diameter, 160 CFM		10	1.600		94.50	58		152.50	193	
	2260	14" neck diameter, 250 CFM		10	1.600		119	58		177	219	
	2280	16" neck diameter, 380 CFM		9	1.778		158	64.50		222.50	273	
	2300	18" neck diameter, 500 CFM		9	1.778		211	64.50		275.50	330	
	2320	20" neck diameter, 625 CFM		8	2		223	72.50		295.50	355	
	2340	24" neck diameter, 900 CFM		8	2		257	72.50		329.50	395	
	2360	30" neck diameter, 1375 CFM		7	2.286		585	82.50		667.50	770	
	2380	36" neck diameter, 2,000 CFM		6	2.667		805	96.50		901.50	1,025	
	2400	42" neck diameter, 3000 CFM		4	4		1,125	145		1,270	1,475	
	2410	48" neck diameter, 4000 CFM	↓	3	5.333		1,525	193		1,718	1,975	
	2500	For aluminum, add					300%					
	2520	For stainless steel, add					600%					
	2530	For copper, add					600%					
	3000	Rotating chimney cap, galvanized, 4" neck diameter	Q-9	20	.800		40.50	29		69.50	89	
	3020	5" neck diameter		18	.889		41.50	32		73.50	95	
	3040	6" neck diameter		16	1		42	36		78	102	
	3060	7" neck diameter		15	1.067		45.50	38.50		84	110	
	3080	8" neck diameter		14	1.143		47	41.50		88.50	116	
	3100	10" neck diameter		12	1.333		61	48		109	141	
	3600	Stationary chimney rain cap, galvanized, 3" neck diameter		24	.667		35.50	24		59.50	76	
	3620	4" neck diameter		20	.800		36	29		65	84	
	3640	6" neck diameter		16	1		37.50	36		73.50	97	
	3680	8" neck diameter		14	1.143		40.50	41.50		82	108	
	3700	10" neck diameter		12	1.333		51.50	48		99.50	131	
	3720	12" neck diameter		10	1.600		85.50	58		143.50	183	
	3740	14" neck diameter		10	1.600		91.50	58		149.50	190	
	3760	16" neck diameter		9	1.778		112	64.50		176.50	222	
	3770	18" neck diameter		9	1.778		162	64.50		226.50	277	
	3780	20" neck diameter		8	2		216	72.50		288.50	350	
	3790	24" neck diameter		8	2		252	72.50		324.50	390	
	4200	Stationary mushroom, aluminum, 16" orifice diameter		10	1.600		320	58		378	440	
	4220	26" orifice diameter		6.15	2.602		470	94		564	665	
	4230	30" orifice diameter		5.71	2.802		690	101		791	915	
	4240	38" orifice diameter		5	3.200		990	116		1,106	1,250	
	4250	42" orifice diameter		4.70	3.404		1,300	123		1,423	1,650	
	4260	50" orifice diameter	↓	4.44	3.604	↓	1,550	130		1,680	1,925	
	5000	Relief vent										
	5500	Rectangular, aluminum, galvanized curb										
	5510	intake/exhaust, 0.05" SP										
	5600	500 CFM, 12" x 16"	Q-9	8	2	Ea.	390	72.50		462.50	540	
	5620	750 CFM, 12" x 20"		7.20	2.222		430	80.50		510.50	595	
	5640	1000 CFM, 12" x 24"		6.60	2.424		465	87.50		552.50	645	
	5660	1500 CFM, 12" x 36"		5.80	2.759		560	100		660	770	
	5680	3000 CFM, 20" x 42"		4	4		805	145		950	1,100	
	5700	6000 CFM, 20" x 84"		2.60	6.154		1,225	223		1,448	1,700	
	5720	8000 CFM, 24" x 96"		2.30	6.957		1,400	252		1,652	1,925	
	5740	10,000 CFM, 48" x 60"		1.80	8.889		1,725	320		2,045	2,400	
	5760	12,500 CFM, 48" x 72"		1.60	10		2,025	360		2,385	2,775	
	5780	15,000 CFM, 48" x 96"		1.30	12.308		2,325	445		2,770	3,250	
	5800	20,000 CFM, 48" x 120"		1.20	13.333		2,700	480		3,180	3,725	
	5820	25,000 CFM, 60" x 120"		.90	17.778		3,575	645		4,220	4,925	
	5840	30,000 CFM, 72" x 120"	↓	.70	22.857	↓	4,700	825		5,525	6,425	

15800 | Air Distribution

15850 | Air Outlets & Inlets

			CREW	DAILY OUTPUT	LABOR-HOURS	UNIT	2005 BARE COSTS MAT.	LABOR	EQUIP.	TOTAL	TOTAL INCL O&P	
800	5860	40,000 CFM, 96″ x 120″	Q-9	.60	26.667	Ea.	6,025	965		6,990	8,100	800
	5880	50,000 CFM, 96″ x 144″		.50	32		6,400	1,150		7,550	8,800	
	7000	Note: sizes based on exhaust. Intake, with 0.125″ SP										
	7100	loss, approximately twice listed capacity.										
	7500	For power open, spring return damper										
	7600	motor for each S.F. of throat area	Q-9	100	.160	S.F.	47.50	5.80		53.30	61.50	

15860 | Air Cleaning Devices

			CREW	DAILY OUTPUT	LABOR-HOURS	UNIT	MAT.	LABOR	EQUIP.	TOTAL	TOTAL INCL O&P	
100	0010	AIR FILTERS										100
	0050	Activated charcoal type, full flow				MCFM	600			600	660	
	0060	Activated charcoal type, full flow, impregnated media 12″ deep					175			175	193	
	0070	Activated charcoal type, HEPA filter & frame for field erection					175			175	193	
	0080	Activated charcoal type, HEPA filter-diffuser, ceiling install.					250			250	275	
	0500	Chemical media filtration type										
	1100	Industrial air fume & odor scrubber unit w/pump & motor										
	1110	corrosion resistant PVC construction										
	1120	Single pack filter, horizontal type										
	1130	500 CFM	Q-9	14	1.143	Ea.	4,025	41.50		4,066.50	4,500	
	1140	1000 CFM		11	1.455		4,725	52.50		4,777.50	5,275	
	1150	2000 CFM		8	2		5,725	72.50		5,797.50	6,400	
	1160	3000 CFM		7	2.286		6,650	82.50		6,732.50	7,425	
	1170	5000 CFM		5	3.200		8,650	116		8,766	9,700	
	1180	8000 CFM		4	4		12,100	145		12,245	13,500	
	1190	12,000 CFM		3	5.333		16,100	193		16,293	18,000	
	1200	16,000 CFM		2.50	6.400		19,800	232		20,032	22,200	
	1210	20,000 CFM		2	8		24,800	289		25,089	27,700	
	1220	26,000 CFM		1.50	10.667		28,600	385		28,985	32,100	
	1230	30,000 CFM	Q-10	2	12		32,500	450		32,950	36,500	
	1240	40,000 CFM	″	1.50	16		42,600	600		43,200	47,700	
	1250	50,000 CFM	Q-11	2	16		52,000	615		52,615	58,000	
	1260	55,000 CFM		1.80	17.778		56,500	680		57,180	63,000	
	1270	60,000 CFM		1.50	21.333		61,000	815		61,815	68,500	
	1300	Double pack filter, horizontal type										
	1310	500 CFM	Q-9	10	1.600	Ea.	5,250	58		5,308	5,875	
	1320	1000 CFM		8	2		6,300	72.50		6,372.50	7,025	
	1330	2000 CFM		6	2.667		7,575	96.50		7,671.50	8,475	
	1340	3000 CFM		5	3.200		9,050	116		9,166	10,200	
	1350	5000 CFM		4	4		11,500	145		11,645	12,900	
	1360	8000 CFM		3	5.333		15,200	193		15,393	17,000	
	1370	12,000 CFM		2.50	6.400		20,700	232		20,932	23,100	
	1380	16,000 CFM		2	8		26,900	289		27,189	30,000	
	1390	20,000 CFM		1.50	10.667		32,700	385		33,085	36,500	
	1400	26,000 CFM		1	16		37,800	580		38,380	42,500	
	1410	30,000 CFM	Q-10	1.50	16		44,400	600		45,000	49,700	
	1420	40,000 CFM	″	1.30	18.462		57,000	695		57,695	63,500	
	1430	50,000 CFM	Q-11	1.50	21.333		71,000	815		71,815	79,500	
	1440	55,000 CFM		1.30	24.615		77,500	945		78,445	87,000	
	1450	60,000 CFM		1	32		84,000	1,225		85,225	94,500	
	1500	Single pack filter, vertical type										
	1510	500 CFM	Q-9	24	.667	Ea.	3,850	24		3,874	4,250	
	1520	1000 CFM		18	.889		4,550	32		4,582	5,050	
	1530	2000 CFM		12	1.333		5,450	48		5,498	6,075	
	1540	3000 CFM		9	1.778		6,375	64.50		6,439.50	7,125	
	1550	5000 CFM		6	2.667		8,500	96.50		8,596.50	9,500	
	1560	8000 CFM		4	4		12,100	145		12,245	13,500	
	1570	12,000 CFM		3	5.333		16,500	193		16,693	18,400	

	15860 | Air Cleaning Devices		CREW	DAILY OUTPUT	LABOR-HOURS	UNIT	2005 BARE COSTS MAT.	LABOR	EQUIP.	TOTAL	TOTAL INCL O&P	
100	1580	16,000 CFM	Q-9	2	8	Ea.	19,300	289		19,589	21,600	100
	1590	20,000 CFM		1.80	8.889		24,300	320		24,620	27,200	
	1600	24,000 CFM		1.60	10		27,100	360		27,460	30,500	
	1650	Double pack filter, vertical type										
	1660	500 CFM	Q-9	22	.727	Ea.	5,000	26.50		5,026.50	5,550	
	1670	1000 CFM		16	1		5,950	36		5,986	6,600	
	1680	2000 CFM		10	1.600		7,150	58		7,208	7,975	
	1690	3000 CFM		7	2.286		8,425	82.50		8,507.50	9,375	
	1700	5000 CFM		5	3.200		11,000	116		11,116	12,300	
	1710	8000 CFM		3	5.333		15,100	193		15,293	16,900	
	1720	12,000 CFM		2.50	6.400		20,400	232		20,632	22,800	
	1730	16,000 CFM		2	8		25,600	289		25,889	28,600	
	1740	20,000 CFM		1.50	10.667		32,000	385		32,385	35,800	
	1750	24,000 CFM		1	16		35,700	580		36,280	40,100	
	1800	Inlet or outlet transition, horizontal										
	1810	Single pack to 12,000 CFM, add					4%					
	1820	Single pack to 30,000 CFM, add					6%					
	1830	Single pack to 60,000 CFM, add					8%					
	1840	Double pack to 12,000 CFM, add					3%					
	1850	Double pack to 30,000 CFM, add					5%					
	1860	Double pack to 60,000 CFM, add					6%					
	1870	Inlet or outlet transition, vertical										
	1880	Single pack to 5000 CFM, add					2%					
	1890	Single pack to 24,000 CFM, add					3%					
	1900	Double pack to 24,000 CFM, add					2%					
	2000	Electronic air cleaner, duct mounted										
	2150	400 - 1000 CFM	1 Shee	2.30	3.478	Ea.	695	140		835	980	
	2200	1000 - 1400 CFM		2.20	3.636		725	146		871	1,025	
	2250	1400 - 2000 CFM		2.10	3.810		800	153		953	1,125	
	2260	2000 - 2500 CFM		2	4		825	161		986	1,150	
	2950	Mechanical media filtration units										
	3000	High efficiency type, with frame, non-supported				MCFM	45			45	49.50	
	3100	Supported type					55			55	60.50	
	4000	Medium efficiency, extended surface					5			5	5.50	
	4500	Permanent washable					20			20	22	
	5000	Renewable disposable roll					120			120	132	
	5500	Throwaway glass or paper media type				Ea.	4.60			4.60	5.05	
	5800	Filter, bag type										
	5810	90-95% efficiency										
	5820	24″ x 12″ x 29″, .75-1.25 MCFM	1 Shee	1.90	4.211	Ea.	34	169		203	298	
	5830	24″ x 24″ x 29″, 1.5-2.5 MCFM	″	1.90	4.211	″	71.50	169		240.50	340	
	5850	80-85% efficiency										
	5860	24″ x 12″ x 29″, .75-1.25 MCFM	1 Shee	1.90	4.211	Ea.	31	169		200	294	
	5870	24″ x 24″ x 29″, 1.5-2.5 MCFM	″	1.90	4.211	″	54.50	169		223.50	320	
	6000	HEPA filter complete w/particle board,										
	6010	kraft paper frame, separator material										
	6020	95% DOP efficiency										
	6030	12″ x 12″ x 6″, 150 CFM	1 Shee	3.70	2.162	Ea.	31.50	87		118.50	169	
	6034	24″ x 12″ x 6″, 375 CFM		1.90	4.211		33	169		202	297	
	6038	24″ x 18″ x 6″, 450 CFM		1.90	4.211		42.50	169		211.50	305	
	6042	24″ x 24″ x 6″, 700 CFM		1.90	4.211		50	169		219	315	
	6046	12″ x 12″ x 12″, 250 CFM		3.70	2.162		42.50	87		129.50	181	
	6050	24″ x 12″ x 12″, 500 CFM		1.90	4.211		50	169		219	315	
	6054	24″ x 18″ x 12″, 875 CFM		1.90	4.211		67.50	169		236.50	335	
	6058	24″ x 24″ x 12″, 1000 CFM		1.90	4.211		70	169		239	335	
	6100	99% DOP efficiency										

15860 | Air Cleaning Devices

			CREW	DAILY OUTPUT	LABOR-HOURS	UNIT	2005 BARE COSTS MAT.	LABOR	EQUIP.	TOTAL	TOTAL INCL O&P	
100	6110	12" x 12" x 6", 150 CFM	1 Shee	3.70	2.162	Ea.	53	87		140	192	100
	6114	24" x 12" x 6", 325 CFM		1.90	4.211		58	169		227	325	
	6118	24" x 18" x 6", 550 CFM		1.90	4.211		81.50	169		250.50	350	
	6122	24" x 24" x 6", 775 CFM		1.90	4.211		84	169		253	355	
	6124	24" x 48" x 6", 775 CFM		1.30	6.154		168	247		415	565	
	6126	12" x 12" x 12", 250 CFM		3.60	2.222		60	89.50		149.50	203	
	6130	24" x 12" x 12", 500 CFM		1.90	4.211		72.50	169		241.50	340	
	6134	24" x 18" x 12", 775 CFM		1.90	4.211		116	169		285	390	
	6138	24" x 24" x 12", 1100 CFM		1.90	4.211		120	169		289	390	
	6500	HEPA filter housing, 14 ga. galv. sheet metal										
	6510	12" x 12" x 6"	1 Shee	2.50	3.200	Ea.	450	129		579	695	
	6514	24" x 12" x 6"		2	4		500	161		661	795	
	6518	12" x 12" x 12"		2.40	3.333		450	134		584	700	
	6522	14" x 12" x 12"		2.30	3.478		500	140		640	765	
	6526	24" x 18" x 6"		1.90	4.211		640	169		809	965	
	6530	24" x 24" x 6"		1.80	4.444		585	179		764	915	
	6534	24" x 48" x 6"		1.70	4.706		780	189		969	1,150	
	6538	24" x 72" x 6"		1.60	5		1,000	201		1,201	1,400	
	6542	24" x 18" x 12"		1.80	4.444		640	179		819	980	
	6546	24" x 24" x 12"		1.70	4.706		585	189		774	930	
	6550	24" x 48" x 12"		1.60	5		780	201		981	1,175	
	6554	24" x 72" x 12"		1.50	5.333		1,000	214		1,214	1,425	
	6558	48" x 48" x 6"	Q-9	2.80	5.714		1,125	207		1,332	1,550	
	6562	48" x 72" x 6"		2.60	6.154		1,400	223		1,623	1,900	
	6566	48" x 96" x 6"		2.40	6.667		1,575	241		1,816	2,100	
	6570	48" x 48" x 12"		2.70	5.926		1,125	214		1,339	1,550	
	6574	48" x 72" x 12"		2.50	6.400		1,400	232		1,632	1,900	
	6578	48" x 96" x 12"		2.30	6.957		1,575	252		1,827	2,100	
	6582	114" x 72" x 12"		2	8		2,400	289		2,689	3,100	
500	0010	**EXHAUST SYSTEMS** D3090 320										500
	0500	Engine exhaust, garage, in-floor system										
	0510	Single tube outlet assemblies										
	0520	For transite pipe ducting, self-storing tube										
	0530	3" tubing adapter plate	1 Shee	16	.500	Ea.	162	20		182	209	
	0540	4" tubing adapter plate		16	.500		162	20		182	209	
	0550	5" tubing adapter plate		16	.500		162	20		182	209	
	0600	For vitrified tile ducting										
	0610	3" tubing adapter plate, self-storing tube	1 Shee	16	.500	Ea.	162	20		182	209	
	0620	4" tubing adapter plate, self-storing tube		16	.500		162	20		182	209	
	0660	5" tubing adapter plate, self-storing tube		16	.500		162	20		182	209	
	0800	Two tube outlet assemblies										
	0810	For transite pipe ducting, self-storing tube										
	0820	3" tubing, dual exhaust adapter plate	1 Shee	16	.500	Ea.	162	20		182	209	
	0850	For vitrified tile ducting										
	0860	3" tubing, dual exhaust, self-storing tube	1 Shee	16	.500	Ea.	162	20		182	209	
	0870	3" tubing, double outlet, non-storing tubes	"	16	.500	"	162	20		182	209	
	0900	Accessories for metal tubing, (overhead systems also)										
	0910	Adapters, for metal tubing end										
	0920	3" tail pipe type				Ea.	36			36	39.50	
	0930	4" tail pipe type					36			36	39.50	
	0940	5" tail pipe type					36			36	39.50	
	0990	5" diesel stack type					252			252	277	
	1000	6" diesel stack type					252			252	277	
	1100	Bullnose (guide) required for in-floor assemblies										
	1110	3" tubing size				Ea.	18			18	19.80	

15860	Air Cleaning Devices	CREW	DAILY OUTPUT	LABOR-HOURS	UNIT	2005 BARE COSTS MAT.	LABOR	EQUIP.	TOTAL	TOTAL INCL O&P
500 1120	4″ tubing size				Ea.	18			18	19.80
1130	5″ tubing size				↓	18			18	19.80
1150	Plain rings, for tubing end									
1160	3″ tubing size				Ea.	8.55			8.55	9.40
1170	4″ tubing size				″	10.80			10.80	11.90
1200	Tubing, galvanized, flexible, (for overhead systems also)									
1210	3″ ID				L.F.	4.82			4.82	5.30
1220	4″ ID					5.80			5.80	6.40
1230	5″ ID					6.75			6.75	7.45
1240	6″ ID				↓	7.75			7.75	8.50
1250	Stainless steel, flexible, (for overhead sys, also)									
1260	3″ ID				L.F.	10			10	11
1270	4″ ID					13			13	14.30
1280	5″ ID					15.70			15.70	17.30
1290	6″ ID				↓	18.70			18.70	20.50
1500	Engine exhaust, garage, ovrhd. compon., for neoprene tubing									
1510	Alternate metal tubing & accessories see above									
1550	Adapters, for neoprene tubing end									
1560	3″ tail pipe, adjustable, neoprene				Ea.	31.50			31.50	34.50
1570	3″ tail pipe, heavy wall neoprene					40.50			40.50	44.50
1580	4″ tail pipe, heavy wall neoprene					43			43	47.50
1590	5″ tail pipe, heavy wall neoprene				↓	58.50			58.50	64.50
1650	Connectors, tubing									
1660	3″ interior, aluminum				Ea.	8.55			8.55	9.40
1670	4″ interior, aluminum					10.80			10.80	11.90
1710	5″ interior, neoprene					27			27	29.50
1750	3″ spiralock, neoprene					8.55			8.55	9.40
1760	4″ spiralock, neoprene					10.80			10.80	11.90
1780	Y for 3″ ID tubing, neoprene, dual exhaust					22.50			22.50	25
1790	Y for 4″ ID tubing, aluminum, dual exhaust				↓	40.50			40.50	44.50
1850	Elbows, aluminum, splice into tubing for strap									
1860	3″ neoprene tubing size				Ea.	27			27	29.50
1870	4″ neoprene tubing size					31.50			31.50	34.50
1900	Flange assemblies, connect tubing to overhead duct				↓	29.50			29.50	32.50
2000	Hardware and accessories									
2020	Cable, galvanized, 1/8″ diameter				L.F.	.32			.32	.35
2040	Cleat, tie down cable or rope				Ea.	3.60			3.60	3.96
2060	Pulley					4.95			4.95	5.45
2080	Pulley hook, universal				↓	3.60			3.60	3.96
2100	Rope, nylon, 1/4″ diameter				L.F.	.27			.27	.30
2120	Winch, 1″ diameter				Ea.	76.50			76.50	84
2150	Lifting strap, mounts on neoprene									
2160	3″ tubing size				Ea.	18			18	19.80
2170	4″ tubing size					18			18	19.80
2180	5″ tubing size					18			18	19.80
2190	6″ tubing size				↓	18			18	19.80
2200	Tubing, neoprene, 11′ lengths									
2210	3″ ID				L.F.	5.40			5.40	5.95
2220	4″ ID					6.75			6.75	7.45
2230	5″ ID				↓	12.15			12.15	13.35
2500	Engine exhaust, thru-door outlet									
2510	3″ tube size	1 Carp	16	.500	Ea.	31.50	17.15		48.65	61
2530	4″ tube size	″	16	.500	″	31.50	17.15		48.65	61
3000	Tubing, exhaust, flex hose, with									
3010	coupler, damper and tail pipe adapter									
3020	Neoprene									

15 MECHANICAL

15800 | Air Distribution

15860 | Air Cleaning Devices

		15860 Air Cleaning Devices	CREW	DAILY OUTPUT	LABOR-HOURS	UNIT	2005 BARE COSTS MAT.	LABOR	EQUIP.	TOTAL	TOTAL INCL O&P	
500	3040	3″ x 20′	1 Shee	6	1.333	Ea.	171	53.50		224.50	272	500
	3050	4″ x 15′		5.40	1.481		171	59.50		230.50	281	
	3060	4″ x 20′		5	1.600		205	64.50		269.50	325	
	3070	5″ x 15′		4.40	1.818		289	73		362	430	
	3100	Galvanized										
	3110	3″ x 20′	1 Shee	6	1.333	Ea.	155	53.50		208.50	254	
	3120	4″ x 17′		5.60	1.429		162	57.50		219.50	267	
	3130	4″ x 20′		5	1.600		179	64.50		243.50	296	
	3140	5″ x 17′		4.60	1.739		199	70		269	325	
	7500	Welding fume elimination accessories for garage exhaust systems										
	7600	Cut off (blast gate)										
	7610	3″ tubing size, 3″ x 6″ opening	1 Shee	24	.333	Ea.	14.40	13.40		27.80	36.50	
	7620	4″ tubing size, 4″ x 8″ opening		24	.333		16.20	13.40		29.60	38.50	
	7630	5″ tubing size, 5″ x 10″ opening		24	.333		21.50	13.40		34.90	44.50	
	7640	6″ tubing size		24	.333		25	13.40		38.40	48	
	7650	8″ tubing size		24	.333		36	13.40		49.40	60	
	7700	Hoods, magnetic, with handle & screen										
	7710	3″ tubing size, 3″ x 6″ opening	1 Shee	24	.333	Ea.	67.50	13.40		80.90	95	
	7720	4″ tubing size, 4″ x 8″ opening		24	.333		67.50	13.40		80.90	95	
	7730	5″ tubing size, 5″ x 10″ opening		24	.333		67.50	13.40		80.90	95	
	8000	Blower, for tailpipe exhaust system										
	8010	Direct drive										
	8012	495 CFM, 1/3 HP	Q-9	5	3.200	Ea.	1,200	116		1,316	1,500	
	8014	1445 CFM, 3/4 HP		4	4		1,400	145		1,545	1,775	
	8016	1840 CFM, 1-1/2 HP		3	5.333		1,900	193		2,093	2,400	
	8030	Beltdrive										
	8032	1400 CFM, 1 HP	Q-9	4	4	Ea.	1,550	145		1,695	1,925	
	8034	2023 CFM, 1-1/2 HP		3	5.333		2,150	193		2,343	2,675	
	8036	2750 CFM, 2 HP		2	8		2,600	289		2,889	3,300	
	8038	4400 CFM, 3 HP		1.80	8.889		3,950	320		4,270	4,850	
	8040	7060 CFM, 5 HP		1.40	11.429		7,000	415		7,415	8,325	

MECHANICAL 15

15900 | HVAC Instrumentation & Controls

15905 | HVAC Instrumentation

		15905 HVAC Instrumentation	CREW	DAILY OUTPUT	LABOR-HOURS	UNIT	2005 BARE COSTS MAT.	LABOR	EQUIP.	TOTAL	TOTAL INCL O&P	
960	0010	WATER LEVEL CONTROLS										960
	1000	Electric water feeder	1 Stpi	12	.667	Ea.	208	27.50		235.50	270	
	2000	Feeder cut-off combination										
	2100	Steam system up to 5000 sq. ft.	1 Stpi	12	.667	Ea.	405	27.50		432.50	485	
	2200	Steam system above 5000 sq. ft.		12	.667		495	27.50		522.50	585	
	2300	Steam and hot water, high pressure		10	.800		580	33		613	685	
	3000	Low water cut-off for hot water boiler, 50 psi maximum										
	3100	1″ top & bottom equalizing pipes, manual reset	1 Stpi	14	.571	Ea.	224	23.50		247.50	282	
	3200	1″ top & bottom equalizing pipes		14	.571		224	23.50		247.50	282	
	3300	2-1/2″ side connection for nipple-to-boiler		14	.571		202	23.50		225.50	257	
	4000	Low water cut-off for low pressure steam with quick hook-up ftgs.										
	4100	For installation in gauge glass tappings	1 Stpi	16	.500	Ea.	161	20.50		181.50	208	
	4200	Built-in type, 2-1/2″ tap - 3-1/8″ insertion		16	.500		136	20.50		156.50	180	
	4300	Built-in type, 2-1/2″ tap - 1-3/4″ insertion		16	.500		144	20.50		164.50	189	

15900 | HVAC Instrumentation & Controls

	15905 | HVAC Instrumentation		CREW	DAILY OUTPUT	LABOR-HOURS	UNIT	2005 BARE COSTS MAT.	LABOR	EQUIP.	TOTAL	TOTAL INCL O&P	
960	4400	Side connection to 2-1/2" tapping	1 Stpi	16	.500	Ea.	154	20.50		174.50	200	960
	5000	Pump control, low water cut-off and alarm switch	↓	14	.571	↓	425	23.50		448.50	500	
	9000	Water gauges, complete										
	9010	Rough brass, wheel type										
	9020	125 PSI at 350° F										
	9030	3/8" pipe size	1 Stpi	11	.727	Ea.	36	30		66	84.50	
	9040	1/2" pipe size	"	10	.800	"	39	33		72	92	
	9060	200 PSI at 400° F										
	9070	3/8" pipe size	1 Stpi	11	.727	Ea.	44.50	30		74.50	94	
	9080	1/2" pipe size		10	.800		44.50	33		77.50	98	
	9090	3/4" pipe size	↓	9	.889	↓	57	36.50		93.50	117	
	9130	Rough brass, chain lever type										
	9140	250 PSI at 400° F										
	9200	Polished brass, wheel type										
	9210	200 PSI at 400° F										
	9220	3/8" pipe size	1 Stpi	11	.727	Ea.	85.50	30		115.50	139	
	9230	1/2" pipe size		10	.800		87	33		120	145	
	9240	3/4" pipe size	↓	9	.889	↓	92	36.50		128.50	156	
	9260	Polished brass, chain lever type										
	9270	250 PSI at 400° F										
	9280	1/2" pipe size	1 Stpi	10	.800	Ea.	123	33		156	185	
	9290	3/4" pipe size	"	9	.889	"	134	36.50		170.50	202	
	9400	Bronze, high pressure, ASME										
	9410	1/2" pipe size	1 Stpi	10	.800	Ea.	155	33		188	220	
	9420	3/4" pipe size	"	9	.889	"	162	36.50		198.50	233	
	9460	Chain lever type										
	9470	1/2" pipe size	1 Stpi	10	.800	Ea.	200	33		233	269	
	9480	3/4" pipe size	"	9	.889	"	211	36.50		247.50	287	
	9500	316 stainless steel, high pressure, ASME										
	9510	500 PSI at 450° F										
	9520	1/2" pipe size	1 Stpi	10	.800	Ea.	505	33		538	605	
	9530	3/4" pipe size	"	9	.889	"	490	36.50		526.50	595	

15 MECHANICAL

15950 | Testing/Adjusting/Balancing

	15955 | HVAC Test/Adjust/Balance		CREW	DAILY OUTPUT	LABOR-HOURS	UNIT	2005 BARE COSTS MAT.	LABOR	EQUIP.	TOTAL	TOTAL INCL O&P	
100	0010	**BALANCING, AIR** (Subcontractor's quote incl. material & labor) R15050-710										100
	0900	Heating and ventilating equipment										
	1000	Centrifugal fans, utility sets				Ea.					296.88	
	1100	Heating and ventilating unit									445.32	
	1200	In-line fan									445.32	
	1300	Propeller and wall fan									84.12	
	1400	Roof exhaust fan									197.92	
	2000	Air conditioning equipment, central station									643.24	
	2100	Built-up low pressure unit									593.76	
	2200	Built-up high pressure unit									692.72	
	2300	Built-up high pressure dual duct									1,088.56	
	2400	Built-up variable volume									1,286.48	
	2500	Multi-zone A.C. and heating unit									445.32	
	2600	For each zone over one, add				↓					98.96	

	15955 | HVAC Test/Adjust/Balance	CREW	DAILY OUTPUT	LABOR-HOURS	UNIT	2005 BARE COSTS MAT.	LABOR	EQUIP.	TOTAL	TOTAL INCL O&P	
100	2700 Package A.C. unit R15050-710				Ea.					247.40	100
	2800 Rooftop heating and cooling unit									346.36	
	3000 Supply, return, exhaust, registers & diffusers, avg. height ceiling									59.38	
	3100 High ceiling									89.06	
	3200 Floor height									49.48	
	3300 Off mixing box									39.58	
	3500 Induction unit									64.32	
	3600 Lab fume hood									296.88	
	3700 Linear supply									148.44	
	3800 Linear supply high									173.18	
	4000 Linear return									49.48	
	4100 Light troffers									59.38	
	4200 Moduline - master									59.38	
	4300 Moduline - slaves									29.69	
	4400 Regenerators									395.84	
	4500 Taps into ceiling plenums									74.22	
	4600 Variable volume boxes				↓					59.38	
700	0010 **PIPING, TESTING**										700
	0100 Nondestructive testing										
	0110 Nondestructive hydraulic pressure test, isolate & 1 hr. hold										
	0120 1" - 4" pipe										
	0140 0 - 250 L.F.	1 Stpi	1.33	6.015	Ea.		246		246	370	
	0160 250 - 500 L.F.	"	.80	10			410		410	615	
	0180 500 - 1000 L.F.	Q-5	1.14	14.035			515		515	775	
	0200 1000 - 2000 L.F.	"	.80	20	↓		735		735	1,100	
	0300 6" - 10" pipe										
	0320 0 - 250 L.F.	Q-5	1	16	Ea.		590		590	885	
	0340 250 - 500 L.F.		.73	21.918			810		810	1,225	
	0360 500 - 1000 L.F.		.53	30.189			1,100		1,100	1,675	
	0380 1000 - 2000 L.F.	↓	.38	42.105	↓		1,550		1,550	2,325	
	1000 Pneumatic pressure test, includes soaping joints										
	1120 1" - 4" pipe										
	1140 0 - 250 L.F.	Q-5	2.67	5.993	Ea.	5.10	221		226.10	335	
	1160 250 - 500 L.F.		1.33	12.030		10.25	445		455.25	675	
	1180 500 - 1000 L.F.		.80	20		15.35	735		750.35	1,125	
	1200 1000 - 2000 L.F.	↓	.50	32	↓	20.50	1,175		1,195.50	1,800	
	1300 6" - 10" pipe										
	1320 0 - 250 L.F.	Q-5	1.33	12.030	Ea.	5.10	445		450.10	670	
	1340 250 - 500 L.F.		.67	23.881		10.25	880		890.25	1,325	
	1360 500 - 1000 L.F.		.40	40		20.50	1,475		1,495.50	2,250	
	1380 1000 - 2000 L.F.	↓	.25	64	↓	25.50	2,350		2,375.50	3,575	
	2000 X-Ray of welds										
	2110 2" diam.	1 Stpi	8	1	Ea.	8.20	41		49.20	70.50	
	2120 3" diam.		8	1		8.20	41		49.20	70.50	
	2130 4" diam.		8	1		12.30	41		53.30	75	
	2140 6" diam.		8	1		12.30	41		53.30	75	
	2150 8" diam.		6.60	1.212		12.30	49.50		61.80	88	
	2160 10" diam.	↓	6	1.333	↓	16.40	54.50		70.90	100	
	3000 Liquid penetration of welds										
	3110 2" diam.	1 Stpi	14	.571	Ea.	2.30	23.50		25.80	37.50	
	3120 3" diam.		13.60	.588		2.30	24		26.30	38.50	
	3130 4" diam.		13.40	.597		2.30	24.50		26.80	39.50	
	3140 6" diam.		13.20	.606		2.30	25		27.30	40	
	3150 8" diam.		13	.615		3.45	25		28.45	42	
	3160 10" diam.	↓	12.80	.625	↓	3.45	25.50		28.95	42.50	

15955 | HVAC Test/Adjust/Balance			CREW	DAILY OUTPUT	LABOR-HOURS	UNIT	2005 BARE COSTS MAT.	LABOR	EQUIP.	TOTAL	TOTAL INCL O&P	
900	0010	**BALANCING, WATER** (Subcontractor's quote incl. material & labor)										900
	0050	Air cooled condenser				Ea.					172.34	
	0100	Cabinet unit heater R15050-710									59.09	
	0200	Chiller									418.54	
	0300	Convector									49.24	
	0400	Converter									246.20	
	0500	Cooling tower									320.06	
	0600	Fan coil unit, unit ventilator									88.63	
	0700	Fin tube and radiant panels									98.48	
	0800	Main and duct re-heat coils									91.09	
	0810	Heat exchanger									91.09	
	0900	Main balancing cocks									73.86	
	1000	Pumps									216.66	
	1100	Unit heater									68.94	

For information about Means Estimating Seminars, see yellow pages 12 and 13 in back of book

Division 16 Electrical

Estimating Tips

16060 Grounding & Bonding

- When taking off grounding system, identify separately the type and size of wire and list each unique type of ground connection.

16100 Wiring Methods

- Conduit should be taken off in three main categories: power distribution, branch power, and branch lighting, so the estimator can concentrate on systems and components, therefore making it easier to ensure all items have been accounted for.
- For cost modifications for elevated conduit installation, add the percentages to labor according to the height of installation and only the quantities exceeding the different height levels, not to the total conduit quantities.
- Remember that aluminum wiring of equal ampacity is larger in diameter than copper and may require larger conduit.
- If more than three wires at a time are being pulled, deduct percentages from the labor hours of that grouping of wires.
- The estimator should take the weights of materials into consideration when completing a takeoff. Topics to consider include: How will the materials be supported? What methods of support are available? How high will the support structure have to reach? Will the final support structure be able to withstand the total burden? Is the support material included or separate from the fixture, equipment and material specified?

16200 Electrical Power

- Do not overlook the costs for equipment used in the installation. If scaffolding or highlifts are available in the field, contractors may use them in lieu of the proposed ladders and rolling staging.

16400 Low-Voltage Distribution

- Supports and concrete pads may be shown on drawings for the larger equipment, or the support system may be just a piece of plywood for the back of a panelboard. In either case, it must be included in the costs.

16500 Lighting

- Fixtures should be taken off room by room, using the fixture schedule, specifications, and the ceiling plan. For large concentrations of lighting fixtures in the same area deduct the percentages from labor hours.

16700 Communications
16800 Sound & Video

- When estimating material costs for special systems, it is always prudent to obtain manufacturers' quotations for equipment prices and special installation requirements which will affect the total costs.

Reference Numbers

Reference numbers are shown in bold squares at the beginning of some major classifications. These numbers refer to related items in the Reference Section. The reference information may be an estimating procedure, an alternate pricing method or technical information.

Note: Not all subdivisions listed here necessarily appear in this publication.

Note: **i2 Trade Service,** *in part, has been used as a reference source for some of the material prices used in Division 16.*

16132 | Conduit & Tubing				CREW	DAILY OUTPUT	LABOR-HOURS	UNIT	2005 BARE COSTS MAT.	LABOR	EQUIP.	TOTAL	TOTAL INCL O&P	
260	0010	CUTTING AND DRILLING											260
	0100	Hole drilling to 10′ high, concrete wall											
	0110	8″ thick, 1/2″ pipe size		R-31	12	.667	Ea.	3.26	27	4.19	34.45	48.50	
	0120	3/4″ pipe size			12	.667		3.26	27	4.19	34.45	48.50	
	0130	1″ pipe size			9.50	.842		6.70	34.50	5.30	46.50	64	
	0140	1-1/4″ pipe size			9.50	.842		6.70	34.50	5.30	46.50	64	
	0150	1-1/2″ pipe size			9.50	.842		6.70	34.50	5.30	46.50	64	
	0160	2″ pipe size			4.40	1.818		7.25	74	11.40	92.65	131	
	0170	2-1/2″ pipe size			4.40	1.818		7.25	74	11.40	92.65	131	
	0180	3″ pipe size			4.40	1.818		7.25	74	11.40	92.65	131	
	0190	3-1/2″ pipe size			3.30	2.424		7.75	99	15.20	121.95	172	
	0200	4″ pipe size			3.30	2.424		7.75	99	15.20	121.95	172	
	0500	12″ thick, 1/2″ pipe size			9.40	.851		4.98	34.50	5.35	44.83	63	
	0520	3/4″ pipe size			9.40	.851		4.98	34.50	5.35	44.83	63	
	0540	1″ pipe size			7.30	1.096		9.45	44.50	6.90	60.85	84.50	
	0560	1-1/4″ pipe size			7.30	1.096		9.45	44.50	6.90	60.85	84.50	
	0570	1-1/2″ pipe size			7.30	1.096		9.45	44.50	6.90	60.85	84.50	
	0580	2″ pipe size			3.60	2.222		11.10	90.50	13.95	115.55	163	
	0590	2-1/2″ pipe size			3.60	2.222		11.10	90.50	13.95	115.55	163	
	0600	3″ pipe size			3.60	2.222		11.10	90.50	13.95	115.55	163	
	0610	3-1/2″ pipe size			2.80	2.857		12.60	116	17.95	146.55	207	
	0630	4″ pipe size			2.50	3.200		12.60	130	20	162.60	230	
	0650	16″ thick, 1/2″ pipe size			7.60	1.053		6.70	43	6.60	56.30	78.50	
	0670	3/4″ pipe size			7	1.143		6.70	46.50	7.20	60.40	85	
	0690	1″ pipe size			6	1.333		12.25	54.50	8.35	75.10	104	
	0710	1-1/4″ pipe size			5.50	1.455		12.25	59.50	9.15	80.90	112	
	0730	1-1/2″ pipe size			5.50	1.455		12.25	59.50	9.15	80.90	112	
	0750	2″ pipe size			3	2.667		14.95	109	16.75	140.70	197	
	0770	2-1/2″ pipe size			2.70	2.963		14.95	121	18.60	154.55	217	
	0790	3″ pipe size			2.50	3.200		14.95	130	20	164.95	232	
	0810	3-1/2″ pipe size			2.30	3.478		17.45	142	22	181.45	254	
	0830	4″ pipe size			2	4		17.45	163	25	205.45	290	
	0850	20″ thick, 1/2″ pipe size			6.40	1.250		8.40	51	7.85	67.25	94	
	0870	3/4″ pipe size			6	1.333		8.40	54.50	8.35	71.25	99.50	
	0890	1″ pipe size			5	1.600		15	65	10.05	90.05	125	
	0910	1-1/4″ pipe size			4.80	1.667		15	68	10.45	93.45	129	
	0930	1-1/2″ pipe size			4.60	1.739		15	71	10.90	96.90	134	
	0950	2″ pipe size			2.70	2.963		18.75	121	18.60	158.35	221	
	0970	2-1/2″ pipe size			2.40	3.333		18.75	136	21	175.75	246	
	0990	3″ pipe size			2.20	3.636		18.75	148	23	189.75	267	
	1010	3-1/2″ pipe size			2	4		22.50	163	25	210.50	295	
	1030	4″ pipe size			1.70	4.706		22.50	192	29.50	244	340	
	1050	24″ thick, 1/2″ pipe size			5.50	1.455		10.15	59.50	9.15	78.80	109	
	1070	3/4″ pipe size			5.10	1.569		10.15	64	9.85	84	117	
	1090	1″ pipe size			4.30	1.860		17.75	76	11.70	105.45	145	
	1110	1-1/4″ pipe size			4	2		17.75	81.50	12.55	111.80	154	
	1130	1-1/2″ pipe size			4	2		17.75	81.50	12.55	111.80	154	
	1150	2″ pipe size			2.40	3.333		22.50	136	21	179.50	250	
	1170	2-1/2″ pipe size			2.20	3.636		22.50	148	23	193.50	271	
	1190	3″ pipe size			2	4		22.50	163	25	210.50	296	
	1210	3-1/2″ pipe size			1.80	4.444		27	181	28	236	330	
	1230	4″ pipe size			1.50	5.333		27	217	33.50	277.50	390	
	1500	Brick wall, 8″ thick, 1/2″ pipe size			18	.444		3.26	18.10	2.79	24.15	33.50	
	1520	3/4″ pipe size			18	.444		3.26	18.10	2.79	24.15	33.50	
	1540	1″ pipe size			13.30	.601		6.70	24.50	3.78	34.98	48	
	1560	1-1/4″ pipe size		↓	13.30	.601	↓	6.70	24.50	3.78	34.98	48	

16 ELECTRICAL

16132 | Conduit & Tubing		CREW	DAILY OUTPUT	LABOR-HOURS	UNIT	2005 BARE COSTS MAT.	LABOR	EQUIP.	TOTAL	TOTAL INCL O&P
260 1580	1-1/2" pipe size	R-31	13.30	.601	Ea.	6.70	24.50	3.78	34.98	48
1600	2" pipe size		5.70	1.404		7.25	57	8.80	73.05	103
1620	2-1/2" pipe size		5.70	1.404		7.25	57	8.80	73.05	103
1640	3" pipe size		5.70	1.404		7.25	57	8.80	73.05	103
1660	3-1/2" pipe size		4.40	1.818		7.75	74	11.40	93.15	131
1680	4" pipe size		4	2		7.75	81.50	12.55	101.80	143
1700	12" thick, 1/2" pipe size		14.50	.552		4.98	22.50	3.46	30.94	43
1720	3/4" pipe size		14.50	.552		4.98	22.50	3.46	30.94	43
1740	1" pipe size		11	.727		9.45	29.50	4.57	43.52	59.50
1760	1-1/4" pipe size		11	.727		9.45	29.50	4.57	43.52	59.50
1780	1-1/2" pipe size		11	.727		9.45	29.50	4.57	43.52	59.50
1800	2" pipe size		5	1.600		11.10	65	10.05	86.15	120
1820	2-1/2" pipe size		5	1.600		11.10	65	10.05	86.15	120
1840	3" pipe size		5	1.600		11.10	65	10.05	86.15	120
1860	3-1/2" pipe size		3.80	2.105		12.60	86	13.20	111.80	156
1880	4" pipe size		3.30	2.424		12.60	99	15.20	126.80	178
1900	16" thick, 1/2" pipe size		12.30	.650		6.70	26.50	4.08	37.28	51.50
1920	3/4" pipe size		12.30	.650		6.70	26.50	4.08	37.28	51.50
1940	1" pipe size		9.30	.860		12.25	35	5.40	52.65	71.50
1960	1-1/4" pipe size		9.30	.860		12.25	35	5.40	52.65	71.50
1980	1-1/2" pipe size		9.30	.860		12.25	35	5.40	52.65	71.50
2000	2" pipe size		4.40	1.818		14.95	74	11.40	100.35	139
2010	2-1/2" pipe size		4.40	1.818		14.95	74	11.40	100.35	139
2030	3" pipe size		4.40	1.818		14.95	74	11.40	100.35	139
2050	3-1/2" pipe size		3.30	2.424		17.45	99	15.20	131.65	183
2070	4" pipe size		3	2.667		17.45	109	16.75	143.20	200
2090	20" thick, 1/2" pipe size		10.70	.748		8.40	30.50	4.70	43.60	60
2110	3/4" pipe size		10.70	.748		8.40	30.50	4.70	43.60	60
2130	1" pipe size		8	1		15	41	6.30	62.30	84
2150	1-1/4" pipe size		8	1		15	41	6.30	62.30	84
2170	1-1/2" pipe size		8	1		15	41	6.30	62.30	84
2190	2" pipe size		4	2		18.75	81.50	12.55	112.80	155
2210	2-1/2" pipe size		4	2		18.75	81.50	12.55	112.80	155
2230	3" pipe size		4	2		18.75	81.50	12.55	112.80	155
2250	3-1/2" pipe size		3	2.667		22.50	109	16.75	148.25	205
2270	4" pipe size		2.70	2.963		22.50	121	18.60	162.10	225
2290	24" thick, 1/2" pipe size		9.40	.851		10.15	34.50	5.35	50	68.50
2310	3/4" pipe size		9.40	.851		10.15	34.50	5.35	50	68.50
2330	1" pipe size		7.10	1.127		17.75	46	7.10	70.85	96
2350	1-1/4" pipe size		7.10	1.127		17.75	46	7.10	70.85	96
2370	1-1/2" pipe size		7.10	1.127		17.75	46	7.10	70.85	96
2390	2" pipe size		3.60	2.222		22.50	90.50	13.95	126.95	175
2410	2-1/2" pipe size		3.60	2.222		22.50	90.50	13.95	126.95	175
2430	3" pipe size		3.60	2.222		22.50	90.50	13.95	126.95	175
2450	3-1/2" pipe size		2.80	2.857		27	116	17.95	160.95	223
2470	4" pipe size	↓	2.50	3.200	↓	27	130	20	177	246
3000	Knockouts to 8' high, metal boxes & enclosures									
3020	With hole saw, 1/2" pipe size	1 Elec	53	.151	Ea.		6.15		6.15	9.15
3040	3/4" pipe size		47	.170			6.95		6.95	10.30
3050	1" pipe size		40	.200			8.15		8.15	12.15
3060	1-1/4" pipe size		36	.222			9.05		9.05	13.50
3070	1-1/2" pipe size		32	.250			10.20		10.20	15.15
3080	2" pipe size		27	.296			12.05		12.05	17.95
3090	2-1/2" pipe size		20	.400			16.30		16.30	24.50
4010	3" pipe size		16	.500			20.50		20.50	30.50
4030	3-1/2" pipe size	↓	13	.615	↓		25		25	37.50

16132 | Conduit & Tubing

			CREW	DAILY OUTPUT	LABOR-HOURS	UNIT	2005 BARE COSTS MAT.	LABOR	EQUIP.	TOTAL	TOTAL INCL O&P	
260	4050	4" pipe size	1 Elec	11	.727	Ea.		29.50		29.50	44	260
	4070	With hand punch set, 1/2" pipe size		40	.200			8.15		8.15	12.15	
	4090	3/4" pipe size		32	.250			10.20		10.20	15.15	
	4110	1" pipe size		30	.267			10.85		10.85	16.15	
	4130	1-1/4" pipe size		28	.286			11.65		11.65	17.35	
	4150	1-1/2" pipe size		26	.308			12.55		12.55	18.65	
	4170	2" pipe size		20	.400			16.30		16.30	24.50	
	4190	2-1/2" pipe size		17	.471			19.20		19.20	28.50	
	4200	3" pipe size		15	.533			21.50		21.50	32.50	
	4220	3-1/2" pipe size		12	.667			27		27	40.50	
	4240	4" pipe size		10	.800			32.50		32.50	48.50	
	4260	With hydraulic punch, 1/2" pipe size		44	.182			7.40		7.40	11.05	
	4280	3/4" pipe size		38	.211			8.60		8.60	12.75	
	4300	1" pipe size		38	.211			8.60		8.60	12.75	
	4320	1-1/4" pipe size		38	.211			8.60		8.60	12.75	
	4340	1-1/2" pipe size		38	.211			8.60		8.60	12.75	
	4360	2" pipe size		32	.250			10.20		10.20	15.15	
	4380	2-1/2" pipe size		27	.296			12.05		12.05	17.95	
	4400	3" pipe size		23	.348			14.15		14.15	21	
	4420	3-1/2" pipe size		20	.400			16.30		16.30	24.50	
	4440	4" pipe size	↓	18	.444	↓		18.10		18.10	27	

16139 | Residential Wiring

			CREW	DAILY OUTPUT	LABOR-HOURS	UNIT	MAT.	LABOR	EQUIP.	TOTAL	TOTAL INCL O&P	
700	0010	RESIDENTIAL WIRING										700
	0020	20' avg. runs and #14/2 wiring incl. unless otherwise noted										
	1000	Service & panel, includes 24' SE-AL cable, service eye, meter,										
	1010	Socket, panel board, main bkr., ground rod, 15 or 20 amp										
	1020	1-pole circuit breakers, and misc. hardware										
	1100	100 amp, with 10 branch breakers	1 Elec	1.19	6.723	Ea.	445	274		719	895	
	1110	With PVC conduit and wire		.92	8.696		485	355		840	1,050	
	1120	With RGS conduit and wire		.73	10.959		675	445		1,120	1,400	
	1150	150 amp, with 14 branch breakers		1.03	7.767		690	315		1,005	1,225	
	1170	With PVC conduit and wire		.82	9.756		775	400		1,175	1,450	
	1180	With RGS conduit and wire	↓	.67	11.940		1,125	485		1,610	1,975	
	1200	200 amp, with 18 branch breakers	2 Elec	1.80	8.889		910	360		1,270	1,550	
	1220	With PVC conduit and wire		1.46	10.959		990	445		1,435	1,775	
	1230	With RGS conduit and wire	↓	1.24	12.903		1,450	525		1,975	2,375	
	1800	Lightning surge suppressor for above services, add	1 Elec	32	.250	↓	41.50	10.20		51.70	60.50	
	2000	Switch devices										
	2100	Single pole, 15 amp, Ivory, with a 1-gang box, cover plate,										
	2110	Type NM (Romex) cable	1 Elec	17.10	.468	Ea.	8.10	19.05		27.15	37.50	
	2120	Type MC (BX) cable		14.30	.559		20.50	23		43.50	56.50	
	2130	EMT & wire		5.71	1.401		29	57		86	117	
	2150	3-way, #14/3, type NM cable		14.55	.550		12	22.50		34.50	46.50	
	2170	Type MC cable		12.31	.650		29	26.50		55.50	71.50	
	2180	EMT & wire		5	1.600		31	65		96	132	
	2200	4-way, #14/3, type NM cable		14.55	.550		25.50	22.50		48	61.50	
	2220	Type MC cable		12.31	.650		42.50	26.50		69	86.50	
	2230	EMT & wire		5	1.600		45	65		110	146	
	2250	S.P., 20 amp, #12/2, type NM cable		13.33	.600		14.05	24.50		38.55	52	
	2270	Type MC cable		11.43	.700		25	28.50		53.50	70	
	2280	EMT & wire		4.85	1.649		35	67		102	139	
	2290	S.P. rotary dimmer, 600W, no wiring		17	.471		16.20	19.20		35.40	46.50	
	2300	S.P. rotary dimmer, 600W, type NM cable		14.55	.550		19.65	22.50		42.15	55	
	2320	Type MC cable	↓	12.31	.650	↓	32	26.50		58.50	75	

16139 | Residential Wiring		CREW	DAILY OUTPUT	LABOR-HOURS	UNIT	2005 BARE COSTS MAT.	LABOR	EQUIP.	TOTAL	TOTAL INCL O&P
700 2330	EMT & wire	1 Elec	5	1.600	Ea.	41	65		106	143 700
2350	3-way rotary dimmer, type NM cable		13.33	.600		17.35	24.50		41.85	55.50
2370	Type MC cable		11.43	.700		30	28.50		58.50	75.50
2380	EMT & wire		4.85	1.649		39	67		106	143
2400	Interval timer wall switch, 20 amp, 1-30 min., #12/2									
2410	Type NM cable	1 Elec	14.55	.550	Ea.	31	22.50		53.50	67.50
2420	Type MC cable		12.31	.650		39	26.50		65.50	82
2430	EMT & wire		5	1.600		52	65		117	154
2500	Decorator style									
2510	S.P., 15 amp, type NM cable	1 Elec	17.10	.468	Ea.	11.65	19.05		30.70	41.50
2520	Type MC cable		14.30	.559		24	23		47	60.50
2530	EMT & wire		5.71	1.401		32.50	57		89.50	121
2550	3-way, #14/3, type NM cable		14.55	.550		15.55	22.50		38.05	50.50
2570	Type MC cable		12.31	.650		32.50	26.50		59	75.50
2580	EMT & wire		5	1.600		34.50	65		99.50	135
2600	4-way, #14/3, type NM cable		14.55	.550		29	22.50		51.50	65.50
2620	Type MC cable		12.31	.650		46	26.50		72.50	90.50
2630	EMT & wire		5	1.600		48.50	65		113.50	150
2650	S.P., 20 amp, #12/2, type NM cable		13.33	.600		17.60	24.50		42.10	56
2670	Type MC cable		11.43	.700		29	28.50		57.50	74
2680	EMT & wire		4.85	1.649		38.50	67		105.50	143
2700	S.P., slide dimmer, type NM cable		17.10	.468		27	19.05		46.05	58
2720	Type MC cable		14.30	.559		39.50	23		62.50	77.50
2730	EMT & wire		5.71	1.401		48.50	57		105.50	138
2750	S.P., touch dimmer, type NM cable		17.10	.468		23	19.05		42.05	54
2770	Type MC cable		14.30	.559		35.50	23		58.50	73
2780	EMT & wire		5.71	1.401		44.50	57		101.50	134
2800	3-way touch dimmer, type NM cable		13.33	.600		41	24.50		65.50	81.50
2820	Type MC cable		11.43	.700		53.50	28.50		82	102
2830	EMT & wire		4.85	1.649		62.50	67		129.50	169
3000	Combination devices									
3100	S.P. switch/15 amp recpt., Ivory, 1-gang box, plate									
3110	Type NM cable	1 Elec	11.43	.700	Ea.	16.35	28.50		44.85	60.50
3120	Type MC cable		10	.800		29	32.50		61.50	80.50
3130	EMT & wire		4.40	1.818		38	74		112	152
3150	S.P. switch/pilot light, type NM cable		11.43	.700		17	28.50		45.50	61
3170	Type MC cable		10	.800		29.50	32.50		62	81
3180	EMT & wire		4.43	1.806		38.50	73.50		112	153
3190	2-S.P. switches, 2-#14/2, no wiring		14	.571		6.75	23.50		30.25	42
3200	2-S.P. switches, 2-#14/2, type NM cables		10	.800		19.05	32.50		51.55	69.50
3220	Type MC cable		8.89	.900		40	36.50		76.50	98.50
3230	EMT & wire		4.10	1.951		39.50	79.50		119	162
3250	3-way switch/15 amp recpt., #14/3, type NM cable		10	.800		23.50	32.50		56	74
3270	Type MC cable		8.89	.900		40.50	36.50		77	99
3280	EMT & wire		4.10	1.951		42.50	79.50		122	165
3300	2-3 way switches, 2-#14/3, type NM cables		8.89	.900		31.50	36.50		68	89
3320	Type MC cable		8	1		61	41		102	128
3330	EMT & wire		4	2		47	81.50		128.50	173
3350	S.P. switch/20 amp recpt., #12/2, type NM cable		10	.800		27.50	32.50		60	78.50
3370	Type MC cable		8.89	.900		35.50	36.50		72	93.50
3380	EMT & wire		4.10	1.951		48.50	79.50		128	172
3400	Decorator style									
3410	S.P. switch/15 amp recpt., type NM cable	1 Elec	11.43	.700	Ea.	19.90	28.50		48.40	64.50
3420	Type MC cable		10	.800		32.50	32.50		65	84
3430	EMT & wire		4.40	1.818		41.50	74		115.50	156
3450	S.P. switch/pilot light, type NM cable		11.43	.700		20.50	28.50		49	65

16139 | Residential Wiring

			CREW	DAILY OUTPUT	LABOR-HOURS	UNIT	2005 BARE COSTS MAT.	LABOR	EQUIP.	TOTAL	TOTAL INCL O&P	
700	3470	Type MC cable	1 Elec	10	.800	Ea.	33	32.50		65.50	85	700
	3480	EMT & wire		4.40	1.818		42	74		116	157	
	3500	2-S.P. switches, 2-#14/2, type NM cables		10	.800		22.50	32.50		55	73.50	
	3520	Type MC cable		8.89	.900		43.50	36.50		80	103	
	3530	EMT & wire		4.10	1.951		43	79.50		122.50	166	
	3550	3-way/15 amp recpt., #14/3, type NM cable		10	.800		27	32.50		59.50	78	
	3570	Type MC cable		8.89	.900		44	36.50		80.50	103	
	3580	EMT & wire		4.10	1.951		46	79.50		125.50	169	
	3650	2-3 way switches, 2-#14/3, type NM cables		8.89	.900		35	36.50		71.50	93	
	3670	Type MC cable		8	1		64.50	41		105.50	132	
	3680	EMT & wire		4	2		50.50	81.50		132	177	
	3700	S.P. switch/20 amp recpt., #12/2, type NM cable		10	.800		31	32.50		63.50	82.50	
	3720	Type MC cable		8.89	.900		39	36.50		75.50	97	
	3730	EMT & wire		4.10	1.951		52	79.50		131.50	175	
	4000	Receptacle devices										
	4010	Duplex outlet, 15 amp recpt., Ivory, 1-gang box, plate										
	4015	Type NM cable	1 Elec	14.55	.550	Ea.	6.45	22.50		28.95	40.50	
	4020	Type MC cable		12.31	.650		19	26.50		45.50	60.50	
	4030	EMT & wire		5.33	1.501		27	61		88	121	
	4050	With #12/2, type NM cable		12.31	.650		8.05	26.50		34.55	48.50	
	4070	Type MC cable		10.67	.750		19.20	30.50		49.70	66.50	
	4080	EMT & wire		4.71	1.699		29	69		98	135	
	4100	20 amp recpt., #12/2, type NM cable		12.31	.650		15.45	26.50		41.95	56.50	
	4120	Type MC cable		10.67	.750		26.50	30.50		57	75	
	4130	EMT & wire		4.71	1.699		36.50	69		105.50	143	
	4140	For GFI see line 4300 below										
	4150	Decorator style, 15 amp recpt., type NM cable	1 Elec	14.55	.550	Ea.	10	22.50		32.50	44.50	
	4170	Type MC cable		12.31	.650		22.50	26.50		49	64.50	
	4180	EMT & wire		5.33	1.501		31	61		92	125	
	4200	With #12/2, type NM cable		12.31	.650		11.60	26.50		38.10	52.50	
	4220	Type MC cable		10.67	.750		23	30.50		53.50	70.50	
	4230	EMT & wire		4.71	1.699		32.50	69		101.50	139	
	4250	20 amp recpt. #12/2, type NM cable		12.31	.650		19	26.50		45.50	60.50	
	4270	Type MC cable		10.67	.750		30	30.50		60.50	78.50	
	4280	EMT & wire		4.71	1.699		40	69		109	147	
	4300	GFI, 15 amp recpt., type NM cable		12.31	.650		35	26.50		61.50	78	
	4320	Type MC cable		10.67	.750		47.50	30.50		78	98	
	4330	EMT & wire		4.71	1.699		56	69		125	165	
	4350	GFI with #12/2, type NM cable		10.67	.750		36.50	30.50		67	85.50	
	4370	Type MC cable		9.20	.870		47.50	35.50		83	105	
	4380	EMT & wire		4.21	1.900		57.50	77.50		135	179	
	4400	20 amp recpt., #12/2 type NM cable		10.67	.750		38	30.50		68.50	87.50	
	4420	Type MC cable		9.20	.870		49.50	35.50		85	107	
	4430	EMT & wire		4.21	1.900		59.50	77.50		137	180	
	4500	Weather-proof cover for above receptacles, add		32	.250		4.40	10.20		14.60	20	
	4550	Air conditioner outlet, 20 amp-240 volt recpt.										
	4560	30′ of #12/2, 2 pole circuit breaker										
	4570	Type NM cable	1 Elec	10	.800	Ea.	44.50	32.50		77	97	
	4580	Type MC cable		9	.889		59	36		95	119	
	4590	EMT & wire		4	2		64.50	81.50		146	192	
	4600	Decorator style, type NM cable		10	.800		48	32.50		80.50	102	
	4620	Type MC cable		9	.889		63	36		99	123	
	4630	EMT & wire		4	2		68.50	81.50		150	196	
	4650	Dryer outlet, 30 amp-240 volt recpt., 20′ of #10/3										
	4660	2 pole circuit breaker										
	4670	Type NM cable	1 Elec	6.41	1.248	Ea.	54	51		105	135	

Important: See the Reference Section for critical supporting data - Reference Nos., Crews, & City Cost Indexes

700	16139 | Residential Wiring	CREW	DAILY OUTPUT	LABOR-HOURS	UNIT	2005 BARE COSTS MAT.	LABOR	EQUIP.	TOTAL	TOTAL INCL O&P	700
4680	Type MC cable	1 Elec	5.71	1.401	Ea.	65.50	57		122.50	157	
4690	EMT & wire	↓	3.48	2.299	↓	69	93.50		162.50	215	
4700	Range outlet, 50 amp-240 volt recpt., 30′ of #8/3										
4710	Type NM cable	1 Elec	4.21	1.900	Ea.	80.50	77.50		158	204	
4720	Type MC cable		4	2		128	81.50		209.50	262	
4730	EMT & wire		2.96	2.703		86	110		196	259	
4750	Central vacuum outlet, Type NM cable		6.40	1.250		48	51		99	129	
4770	Type MC cable		5.71	1.401		70	57		127	162	
4780	EMT & wire	↓	3.48	2.299	↓	75	93.50		168.50	222	
4800	30 amp-110 volt locking recpt., #10/2 circ. bkr.										
4810	Type NM cable	1 Elec	6.20	1.290	Ea.	58.50	52.50		111	143	
4830	EMT & wire	"	3.20	2.500	"	82.50	102		184.50	243	
4900	Low voltage outlets										
4910	Telephone recpt., 20′ of 4/C phone wire	1 Elec	26	.308	Ea.	7.40	12.55		19.95	27	
4920	TV recpt., 20′ of RG59U coax wire, F type connector	"	16	.500	"	12.65	20.50		33.15	44.50	
4950	Door bell chime, transformer, 2 buttons, 60′ of bellwire										
4970	Economy model	1 Elec	11.50	.696	Ea.	54.50	28.50		83	102	
4980	Custom model		11.50	.696		85.50	28.50		114	136	
4990	Luxury model, 3 buttons	↓	9.50	.842	↓	231	34.50		265.50	305	
6000	Lighting outlets										
6050	Wire only (for fixture), type NM cable	1 Elec	32	.250	Ea.	4.42	10.20		14.62	20	
6070	Type MC cable		24	.333		14.10	13.60		27.70	35.50	
6080	EMT & wire		10	.800		21	32.50		53.50	72	
6100	Box (4″), and wire (for fixture), type NM cable		25	.320		9.40	13.05		22.45	30	
6120	Type MC cable		20	.400		19.10	16.30		35.40	45.50	
6130	EMT & wire	↓	11	.727	↓	26	29.50		55.50	73	
6200	Fixtures (use with lines 6050 or 6100 above)										
6210	Canopy style, economy grade	1 Elec	40	.200	Ea.	26	8.15		34.15	40.50	
6220	Custom grade		40	.200		48	8.15		56.15	65	
6250	Dining room chandelier, economy grade		19	.421		78	17.15		95.15	112	
6260	Custom grade		19	.421		230	17.15		247.15	279	
6270	Luxury grade		15	.533		510	21.50		531.50	595	
6310	Kitchen fixture (fluorescent), economy grade		30	.267		52	10.85		62.85	73	
6320	Custom grade		25	.320		165	13.05		178.05	201	
6350	Outdoor, wall mounted, economy grade		30	.267		27.50	10.85		38.35	46.50	
6360	Custom grade		30	.267		104	10.85		114.85	130	
6370	Luxury grade		25	.320		235	13.05		248.05	278	
6410	Outdoor PAR floodlights, 1 lamp, 150 watt		20	.400		21.50	16.30		37.80	48	
6420	2 lamp, 150 watt each		20	.400		36	16.30		52.30	64	
6430	For infrared security sensor, add		32	.250		90	10.20		100.20	114	
6450	Outdoor, quartz-halogen, 300 watt flood		20	.400		39	16.30		55.30	67.50	
6600	Recessed downlight, round, pre-wired, 50 or 75 watt trim		30	.267		36	10.85		46.85	55.50	
6610	With shower light trim		30	.267		45	10.85		55.85	65.50	
6620	With wall washer trim		28	.286		54	11.65		65.65	77	
6630	With eye-ball trim	↓	28	.286		54	11.65		65.65	77	
6640	For direct contact with insulation, add					1.65			1.65	1.82	
6700	Porcelain lamp holder	1 Elec	40	.200		3.60	8.15		11.75	16.10	
6710	With pull switch		40	.200		3.86	8.15		12.01	16.40	
6750	Fluorescent strip, 1-20 watt tube, wrap around diffuser, 24″		24	.333		53	13.60		66.60	78.50	
6770	2-40 watt tubes, 48″		20	.400		79	16.30		95.30	112	
6780	With residential ballast		20	.400		89.50	16.30		105.80	123	
6800	Bathroom heat lamp, 1-250 watt		28	.286		35	11.65		46.65	56	
6810	2-250 watt lamps	↓	28	.286	↓	56	11.65		67.65	79	
6820	For timer switch, see line 2400										
6900	Outdoor post lamp, incl. post, fixture, 35′ of #14/2										
6910	Type NMC cable	1 Elec	3.50	2.286	Ea.	184	93		277	340	

16139 | Residential Wiring		CREW	DAILY OUTPUT	LABOR-HOURS	UNIT	2005 BARE COSTS MAT.	LABOR	EQUIP.	TOTAL	TOTAL INCL O&P	
700	6920 Photo-eye, add	1 Elec	27	.296	Ea.	30	12.05		42.05	51	700
	6950 Clock dial time switch, 24 hr., w/enclosure, type NM cable	↓	11.43	.700	↓	53.50	28.50		82	102	
	6970 Type MC cable		11	.727		66	29.50		95.50	117	
	6980 EMT & wire	↓	4.85	1.649	↓	74.50	67		141.50	182	
	7000 Alarm systems										
	7050 Smoke detectors, box, #14/3, type NM cable	1 Elec	14.55	.550	Ea.	29.50	22.50		52	66	
	7070 Type MC cable		12.31	.650		43.50	26.50		70	87.50	
	7080 EMT & wire	↓	5	1.600		45.50	65		110.50	147	
	7090 For relay output to security system, add				↓	12.10			12.10	13.30	
	8000 Residential equipment										
	8050 Disposal hook-up, incl. switch, outlet box, 3′ of flex										
	8060 20 amp-1 pole circ. bkr., and 25′ of #12/2										
	8070 Type NM cable	1 Elec	10	.800	Ea.	22.50	32.50		55	73	
	8080 Type MC cable		8	1		35.50	41		76.50	99.50	
	8090 EMT & wire	↓	5	1.600	↓	46.50	65		111.50	149	
	8100 Trash compactor or dishwasher hook-up, incl. outlet box,										
	8110 3′ of flex, 15 amp-1 pole circ. bkr., and 25′ of #14/2										
	8120 Type NM cable	1 Elec	10	.800	Ea.	15.90	32.50		48.40	66	
	8130 Type MC cable		8	1		31	41		72	94.50	
	8140 EMT & wire	↓	5	1.600	↓	40.50	65		105.50	142	
	8150 Hot water sink dispensor hook-up, use line 8100										
	8200 Vent/exhaust fan hook-up, type NM cable	1 Elec	32	.250	Ea.	4.42	10.20		14.62	20	
	8220 Type MC cable		24	.333		14.10	13.60		27.70	35.50	
	8230 EMT & wire	↓	10	.800	↓	21	32.50		53.50	72	
	8250 Bathroom vent fan, 50 CFM (use with above hook-up)										
	8260 Economy model	1 Elec	15	.533	Ea.	22	21.50		43.50	56.50	
	8270 Low noise model		15	.533		31	21.50		52.50	66.50	
	8280 Custom model	↓	12	.667	↓	114	27		141	166	
	8300 Bathroom or kitchen vent fan, 110 CFM										
	8310 Economy model	1 Elec	15	.533	Ea.	58	21.50		79.50	96.50	
	8320 Low noise model	"	15	.533	"	77	21.50		98.50	117	
	8350 Paddle fan, variable speed (w/o lights)										
	8360 Economy model (AC motor)	1 Elec	10	.800	Ea.	103	32.50		135.50	162	
	8370 Custom model (AC motor)		10	.800		178	32.50		210.50	245	
	8380 Luxury model (DC motor)		8	1		350	41		391	445	
	8390 Remote speed switch for above, add	↓	12	.667	↓	25	27		52	68	
	8500 Whole house exhaust fan, ceiling mount, 36″, variable speed										
	8510 Remote switch, incl. shutters, 20 amp-1 pole circ. bkr.										
	8520 30′ of #12/2, type NM cable	1 Elec	4	2	Ea.	615	81.50		696.50	795	
	8530 Type MC cable		3.50	2.286		630	93		723	830	
	8540 EMT & wire	↓	3	2.667	↓	645	109		754	865	
	8600 Whirlpool tub hook-up, incl. timer switch, outlet box										
	8610 3′ of flex, 20 amp-1 pole GFI circ. bkr.										
	8620 30′ of #12/2, type NM cable	1 Elec	5	1.600	Ea.	85	65		150	191	
	8630 Type MC cable		4.20	1.905		93.50	77.50		171	219	
	8640 EMT & wire	↓	3.40	2.353	↓	103	96		199	257	
	8650 Hot water heater hook-up, incl. 1-2 pole circ. bkr., box;										
	8660 3′ of flex, 20′ of #10/2, type NM cable	1 Elec	5	1.600	Ea.	25	65		90	125	
	8670 Type MC cable		4.20	1.905		40.50	77.50		118	161	
	8680 EMT & wire	↓	3.40	2.353	↓	40.50	96		136.50	188	
	9000 Heating/air conditioning										
	9050 Furnace/boiler hook-up, incl. firestat, local on-off switch										
	9060 Emergency switch, and 40′ of type NM cable	1 Elec	4	2	Ea.	45	81.50		126.50	171	
	9070 Type MC cable		3.50	2.286		66	93		159	212	
	9080 EMT & wire	↓	1.50	5.333	↓	82	217		299	415	
	9100 Air conditioner hook-up, incl. local 60 amp disc. switch										

	16139 | Residential Wiring		CREW	DAILY OUTPUT	LABOR-HOURS	UNIT	2005 BARE COSTS MAT.	LABOR	EQUIP.	TOTAL	TOTAL INCL O&P	
700	9110	3′ sealtite, 40 amp, 2 pole circuit breaker										700
	9130	40′ of #8/2, type NM cable	1 Elec	3.50	2.286	Ea.	158	93		251	315	
	9140	Type MC cable		3	2.667		233	109		342	420	
	9150	EMT & wire	↓	1.30	6.154	↓	183	251		434	575	
	9200	Heat pump hook-up, 1-40 & 1-100 amp 2 pole circ. bkr.										
	9210	Local disconnect switch, 3′ sealtite										
	9220	40′ of #8/2 & 30′ of #3/2										
	9230	Type NM cable	1 Elec	1.30	6.154	Ea.	385	251		636	795	
	9240	Type MC cable		1.08	7.407		560	300		860	1,075	
	9250	EMT & wire	↓	.94	8.511	↓	425	345		770	985	
	9500	Thermostat hook-up, using low voltage wire										
	9520	Heating only	1 Elec	24	.333	Ea.	5.25	13.60		18.85	26	
	9530	Heating/cooling	"	20	.400	"	6.35	16.30		22.65	31.50	

	16150 | Wiring Connections		CREW	DAILY OUTPUT	LABOR-HOURS	UNIT	MAT.	LABOR	EQUIP.	TOTAL	TOTAL INCL O&P	
275	0010	**MOTOR CONNECTIONS**										275
	0020	Flexible conduit and fittings, 115 volt, 1 phase, up to 1 HP motor	1 Elec	8	1	Ea.	4.96	41		45.96	66	
	0050	2 HP motor		6.50	1.231		5.05	50		55.05	80	
	0100	3 HP motor		5.50	1.455		8.40	59.50		67.90	97.50	
	0120	230 volt, 10 HP motor, 3 phase		4.20	1.905		9.05	77.50		86.55	126	
	0150	15 HP motor		3.30	2.424		17.35	99		116.35	166	
	0200	25 HP motor		2.70	2.963		19.50	121		140.50	202	
	0400	50 HP motor		2.20	3.636		54.50	148		202.50	281	
	0600	100 HP motor		1.50	5.333		139	217		356	480	
	1500	460 volt, 5 HP motor, 3 phase		8	1		5.50	41		46.50	66.50	
	1520	10 HP motor		8	1		5.50	41		46.50	66.50	
	1530	25 HP motor		6	1.333		10.35	54.50		64.85	92.50	
	1540	30 HP motor		6	1.333		10.35	54.50		64.85	92.50	
	1550	40 HP motor		5	1.600		16.30	65		81.30	115	
	1560	50 HP motor		5	1.600		18.20	65		83.20	117	
	1570	60 HP motor		3.80	2.105		24.50	86		110.50	155	
	1580	75 HP motor		3.50	2.286		32.50	93		125.50	175	
	1590	100 HP motor		2.50	3.200		46.50	130		176.50	245	
	1600	125 HP motor		2	4		60.50	163		223.50	310	
	1610	150 HP motor		1.80	4.444		63	181		244	340	
	1620	200 HP motor		1.50	5.333		109	217		326	445	
	2005	460 Volt, 5 HP motor, 3 Phase, w/sealtite		8	1		11.05	41		52.05	72.50	
	2010	10 HP motor		8	1		11.05	41		52.05	72.50	
	2015	25 HP motor		6	1.333		19.65	54.50		74.15	103	
	2020	30 HP motor		6	1.333		19.65	54.50		74.15	103	
	2025	40 HP motor		5	1.600		37	65		102	138	
	2030	50 HP motor		5	1.600		35	65		100	136	
	2035	60 HP motor		3.80	2.105		53	86		139	187	
	2040	75 HP motor		3.50	2.286		56.50	93		149.50	202	
	2045	100 HP motor		2.50	3.200		75	130		205	277	
	2055	150 HP motor		1.80	4.444		79.50	181		260.50	360	
	2060	200 HP motor	↓	1.50	5.333	↓	520	217		737	900	

16200 | Electrical Power

16220 | Motors & Generators

			CREW	DAILY OUTPUT	LABOR-HOURS	UNIT	2005 BARE COSTS MAT.	LABOR	EQUIP.	TOTAL	TOTAL INCL O&P	
900	0010	**VARIABLE FREQUENCY DRIVES/ADJUSTABLE FREQUENCY DRIVES**										900
	0100	Enclosed (NEMA 1), 460 volt, for 3 HP motor size	1 Elec	.80	10	Ea.	1,500	410		1,910	2,250	
	0110	5 HP motor size		.80	10		1,625	410		2,035	2,400	
	0120	7.5 HP motor size		.67	11.940		1,925	485		2,410	2,850	
	0130	10 HP motor size		.67	11.940		1,925	485		2,410	2,850	
	0140	15 HP motor size	2 Elec	.89	17.978		2,225	735		2,960	3,550	
	0150	20 HP motor size		.89	17.978		3,300	735		4,035	4,725	
	0160	25 HP motor size		.67	23.881		3,825	975		4,800	5,675	
	0170	30 HP motor size		.67	23.881		4,675	975		5,650	6,600	
	0180	40 HP motor size		.67	23.881		6,875	975		7,850	9,000	
	0190	50 HP motor size		.53	30.189		7,475	1,225		8,700	10,100	
	0200	60 HP motor size	R-3	.56	35.714		8,175	1,425	283	9,883	11,400	
	0210	75 HP motor size		.56	35.714		11,200	1,425	283	12,908	14,700	
	0220	100 HP motor size		.50	40		11,400	1,600	315	13,315	15,300	
	0230	125 HP motor size		.50	40		12,700	1,600	315	14,615	16,600	
	0240	150 HP motor size		.50	40		16,200	1,600	315	18,115	20,500	
	0250	200 HP motor size		.42	47.619		20,000	1,900	380	22,280	25,300	
	1100	Custom-engineered, 460 volt, for 3 HP motor size	1 Elec	.56	14.286		2,475	580		3,055	3,600	
	1110	5 HP motor size		.56	14.286		2,600	580		3,180	3,725	
	1120	7.5 HP motor size		.47	17.021		3,125	695		3,820	4,475	
	1130	10 HP motor size		.47	17.021		3,125	695		3,820	4,475	
	1140	15 HP motor size	2 Elec	.62	25.806		3,425	1,050		4,475	5,350	
	1150	20 HP motor size		.62	25.806		4,675	1,050		5,725	6,725	
	1160	25 HP motor size		.47	34.043		5,075	1,375		6,450	7,650	
	1170	30 HP motor size		.47	34.043		6,925	1,375		8,300	9,675	
	1180	40 HP motor size		.47	34.043		7,725	1,375		9,100	10,600	
	1190	50 HP motor size		.37	43.243		7,725	1,750		9,475	11,100	
	1200	60 HP motor size	R-3	.39	51.282		11,500	2,050	405	13,955	16,200	
	1210	75 HP motor size		.39	51.282		13,300	2,050	405	15,755	18,200	
	1220	100 HP motor size		.35	57.143		13,900	2,275	455	16,630	19,200	
	1230	125 HP motor size		.35	57.143		14,700	2,275	455	17,430	20,000	
	1240	150 HP motor size		.35	57.143		16,400	2,275	455	19,130	22,000	
	1250	200 HP motor size		.29	68.966		22,700	2,750	545	25,995	29,700	
	2000	For complex & special design systems to meet specific										
	2010	requirements, obtain quote from vendor.										

16400 | Low-Voltage Distribution

16420 | Enclosed Controllers

			CREW	DAILY OUTPUT	LABOR-HOURS	UNIT	2005 BARE COSTS MAT.	LABOR	EQUIP.	TOTAL	TOTAL INCL O&P	
220	0010	**CONTROL STATIONS**										220
	0050	NEMA 1, heavy duty, stop/start	1 Elec	8	1	Ea.	121	41		162	195	
	0100	Stop/start, pilot light		6.20	1.290		165	52.50		217.50	261	
	0200	Hand/off/automatic		6.20	1.290		90	52.50		142.50	177	
	0400	Stop/start/reverse		5.30	1.509		164	61.50		225.50	272	
	0500	NEMA 7, heavy duty, stop/start		6	1.333		199	54.50		253.50	300	
	0600	Stop/start, pilot light		4	2		244	81.50		325.50	390	

START

STOP

	16440 | Swbds, Panels & Control Centers	CREW	DAILY OUTPUT	LABOR-HOURS	UNIT	2005 BARE COSTS MAT.	LABOR	EQUIP.	TOTAL	TOTAL INCL O&P	
660	0010 MOTOR STARTERS & CONTROLS										660
	0050 Magnetic, FVNR, with enclosure and heaters, 480 volt										
	0080 2 HP, size 00	1 Elec	3.50	2.286	Ea.	173	93		266	330	
	0100 5 HP, size 0		2.30	3.478		209	142		351	440	
	0200 10 HP, size 1		1.60	5		235	204		439	565	
	0300 25 HP, size 2	2 Elec	2.20	7.273		440	296		736	925	
	0400 50 HP, size 3		1.80	8.889		720	360		1,080	1,325	
	0500 100 HP, size 4		1.20	13.333		1,600	545		2,145	2,550	
	0600 200 HP, size 5		.90	17.778		3,725	725		4,450	5,175	
	0610 400 HP, size 6		.80	20		10,900	815		11,715	13,100	
	0620 NEMA 7, 5 HP, size 0	1 Elec	1.60	5		895	204		1,099	1,300	
	0630 10 HP, size 1	"	1.10	7.273		930	296		1,226	1,475	
	0640 25 HP, size 2	2 Elec	1.80	8.889		1,500	360		1,860	2,200	
	0650 50 HP, size 3		1.20	13.333		2,225	545		2,770	3,250	
	0660 100 HP, size 4		.90	17.778		3,650	725		4,375	5,075	
	0670 200 HP, size 5		.50	32		8,500	1,300		9,800	11,300	
	0700 Combination, with motor circuit protectors, 5 HP, size 0	1 Elec	1.80	4.444		680	181		861	1,025	
	0800 10 HP, size 1	"	1.30	6.154		705	251		956	1,150	
	0900 25 HP, size 2	2 Elec	2	8		990	325		1,315	1,550	
	1000 50 HP, size 3		1.32	12.121		1,425	495		1,920	2,300	
	1200 100 HP, size 4		.80	20		3,100	815		3,915	4,650	
	1220 NEMA 7, 5 HP, size 0	1 Elec	1.30	6.154		1,575	251		1,826	2,125	
	1230 10 HP, size 1	"	1	8		1,625	325		1,950	2,250	
	1240 25 HP, size 2	2 Elec	1.32	12.121		2,150	495		2,645	3,100	
	1250 50 HP, size 3		.80	20		3,525	815		4,340	5,125	
	1260 100 HP, size 4		.60	26.667		5,500	1,075		6,575	7,675	
	1270 200 HP, size 5		.40	40		11,900	1,625		13,525	15,500	
	1400 Combination, with fused switch, 5 HP, size 0	1 Elec	1.80	4.444		520	181		701	840	
	1600 10 HP, size 1	"	1.30	6.154		555	251		806	985	
	1800 25 HP, size 2	2 Elec	2	8		900	325		1,225	1,475	
	2000 50 HP, size 3		1.32	12.121		1,525	495		2,020	2,400	
	2200 100 HP, size 4		.80	20		2,650	815		3,465	4,150	
	3500 Magnetic FVNR with NEMA 12, enclosure & heaters, 480 volt										
	3600 5 HP, size 0	1 Elec	2.20	3.636	Ea.	197	148		345	440	
	3700 10 HP, size 1	"	1.50	5.333		297	217		514	650	
	3800 25 HP, size 2	2 Elec	2	8		555	325		880	1,100	
	3900 50 HP, size 3		1.60	10		855	410		1,265	1,550	
	4000 100 HP, size 4		1	16		2,050	650		2,700	3,225	
	4100 200 HP, size 5		.80	20		4,900	815		5,715	6,625	
	4200 Combination, with motor circuit protectors, 5 HP, size 0	1 Elec	1.70	4.706		655	192		847	1,000	
	4300 10 HP, size 1	"	1.20	6.667		680	272		952	1,150	
	4400 25 HP, size 2	2 Elec	1.80	8.889		1,025	360		1,385	1,675	
	4500 50 HP, size 3		1.20	13.333		1,650	545		2,195	2,625	
	4600 100 HP, size 4		.74	21.622		3,725	880		4,605	5,400	
	4700 Combination, with fused switch, 5 HP, size 0	1 Elec	1.70	4.706		630	192		822	980	
	4800 10 HP, size 1	"	1.20	6.667		660	272		932	1,125	
	4900 25 HP, size 2	2 Elec	1.80	8.889		1,000	360		1,360	1,650	
	5000 50 HP, size 3		1.20	13.333		1,600	545		2,145	2,575	
	5100 100 HP, size 4		.74	21.622		3,250	880		4,130	4,875	
	5200 Factory installed controls, adders to size 0 thru 5										
	5300 Start-stop push button	1 Elec	32	.250	Ea.	41.50	10.20		51.70	60.50	
	5400 Hand-off-auto-selector switch		32	.250		41.50	10.20		51.70	60.50	
	5500 Pilot light		32	.250		77.50	10.20		87.70	100	
	5600 Start-stop-pilot		32	.250		119	10.20		129.20	146	
	5700 Auxiliary contact, NO or NC		32	.250		57	10.20		67.20	77.50	
	5800 NO-NC		32	.250		114	10.20		124.20	140	

ELECTRICAL 16

Division Notes

		CREW	DAILY OUTPUT	LABOR-HOURS	UNIT	2005 BARE COSTS MAT.	LABOR	EQUIP.	TOTAL	TOTAL INCL O&P

Division 17
Square Foot & Cubic Foot Costs

Estimating Tips

- The cost figures in Division 17 were derived from approximately 11,200 projects contained in the Means database of completed construction projects, and include the contractor's overhead and profit, but do not generally include architectural fees or land costs. The figures have been adjusted to January of the current year. New projects are added to our files each year, and outdated projects are discarded. For this reason, certain costs may not show a uniform annual progression. In no case are all subdivisions of a project listed.
- These projects were located throughout the U.S. and reflect a tremendous variation in square foot (S.F.) and cubic foot (C.F.) costs. This is due to differences, not only in labor and material costs, but also in individual owners' requirements. For instance, a bank in a large city would have different features than one in a rural area. This is true of all the different types of buildings analyzed. Therefore, caution should be exercised when using Division 17 costs. For example, for court houses, costs in the database are local court house costs and will not apply to the larger, more elaborate federal court houses. As a general rule, the projects in the 1/4 column do not include any site work or equipment, while the projects in the 3/4 column may include both equipment and site work. The median figures do not generally include site work.
- None of the figures "go with" any others. All individual cost items were computed and tabulated separately. Thus the sum of the median figures for Plumbing, HVAC and Electrical will not normally total up to the total Mechanical and Electrical costs arrived at by separate analysis and tabulation of the projects.
- Each building was analyzed as to total and component costs and percentages. The figures were arranged in ascending order with the results tabulated as shown. The 1/4 column shows that 25% of the projects had lower costs, 75% higher. The 3/4 column shows that 75% of the projects had lower costs, 25% had higher. The median column shows that 50% of the projects had lower costs, 50% had higher.
- There are two times when square foot costs are useful. The first is in the conceptual stage when no details are available. Then square foot costs make a useful starting point. The second is after the bids are in and the costs can be worked back into their appropriate units for information purposes. As soon as details become available in the project design, the square foot approach should be discontinued and the project priced as to its particular components. When more precision is required or for estimating the replacement cost of specific buildings, the current edition of *Means Square Foot Costs* should be used.
- In using the figures in Division 17, it is recommended that the median column be used for preliminary figures if no additional information is available. The median figures, when multiplied by the total city construction cost index figures (see City Cost Indexes) and then multiplied by the project size modifier in Reference Number R17100-100, should present a fairly accurate base figure, which would then have to be adjusted in view of the estimator's experience, local economic conditions, code requirements and the owner's particular requirements. There is no need to factor the percentage figures, as these should remain constant from city to city. All tabulations mentioning air conditioning had at least partial air conditioning.
- The editors of this book would greatly appreciate receiving cost figures on one or more of your recent projects which would then be included in the averages for next year. All cost figures received will be kept confidential except that they will be averaged with other similar projects to arrive at S.F. and C.F. cost figures for next year's book. See the last page of the book for details and the discount available for submitting one or more of your projects.

	17100 | S.F. & C.F. Costs			UNIT	UNIT COSTS 1/4	UNIT COSTS MEDIAN	UNIT COSTS 3/4	% OF TOTAL 1/4	% OF TOTAL MEDIAN	% OF TOTAL 3/4	
340	2900	Electrical		S.F.	5.15	8.15	12.25	8.30%	10.55%	14.20%	340
	3100	Total: Mechanical & Electrical			12.10	19.40	29.50	22%	28.50%	34.50%	
360	0010	**FIRE STATIONS**	R17100-100	S.F.	78	107	144				360
	0020	Total project costs		C.F.	4.74	6.50	8.65				
	2720	Plumbing		S.F.	5.15	7.55	11.15	5.85%	7.35%	9.50%	
	2770	Heating, ventilating, air conditioning			4.42	7.05	11	4.86%	7.25%	9.25%	
	2900	Electrical			5.75	9.70	13.05	6.80%	8.75%	10.65%	
	3100	Total: Mechanical & Electrical			28	33.50	39	18.75%	23%	27%	
370	0010	**FRATERNITY HOUSES** and Sorority Houses	R17100-100	S.F.	81	104	143				370
	0020	Total project costs		C.F.	7.75	8.40	10.80				
	2720	Plumbing		S.F.	6.15	7	12.85	6.80%	8%	10.85%	
	2900	Electrical			5.35	11.55	14.15	6.60%	9.90%	10.65%	
	3100	Total: Mechanical & Electrical			14.60	20.50	25	14.60%	15.10%	15.90%	
380	0010	**FUNERAL HOMES**	R17100-100	S.F.	85.50	117	212				380
	0020	Total project costs		C.F.	8.75	9.70	18.65				
	2900	Electrical		S.F.	3.78	6.95	8.15	3.58%	4.44%	5.95%	
	3100	Total: Mechanical & Electrical		"	13.50	20	26.50	12.90%	17.80%	18.80%	
390	0010	**GARAGES, COMMERCIAL** (Service)	R17100-100	S.F.	48	76	103				390
	0020	Total project costs		C.F.	3.26	4.71	6.85				
	2720	Plumbing		S.F.	3.36	5.15	9.45	5.45%	7.85%	10.65%	
	2730	Heating & ventilating			4.41	6	8.30	5.25%	6.85%	8.20%	
	2900	Electrical			4.67	7.15	10.20	7.15%	9.25%	10.85%	
	3100	Total: Mechanical & Electrical			10.45	19.40	29	13.60%	17.40%	27%	
400	0010	**GARAGES, MUNICIPAL** (Repair)	R17100-100	S.F.	71	95	134				400
	0020	Total project costs		C.F.	4.44	5.65	9.20				
	2720	Plumbing		S.F.	3.19	6.15	11.55	3.59%	6.70%	7.95%	
	2730	Heating & ventilating			5.45	7.90	15.25	6.15%	7.45%	13.50%	
	2900	Electrical			5.25	8.25	11.90	6.65%	8.15%	11.15%	
	3100	Total: Mechanical & Electrical			16.90	29.50	49	21.50%	25.50%	28.50%	
410	0010	**GARAGES, PARKING**	R17100-100	S.F.	27.50	40	69.50				410
	0020	Total project costs		C.F.	2.60	3.53	5.15				
	2720	Plumbing		S.F.	.78	1.22	1.88	1.72%	2.70%	3.85%	
	2900	Electrical			1.52	1.84	2.87	4.33%	5.20%	6.25%	
	3100	Total: Mechanical & Electrical			2.42	4.33	5.70	6.90%	8.80%	11.05%	
	9000	Per car, total cost		Car	11,700	14,700	18,700				
430	0010	**GYMNASIUMS**	R17100-100	S.F.	77	102	125				430
	0020	Total project costs		C.F.	3.84	5.20	6.40				
	2720	Plumbing		S.F.	4.87	6	7.45	4.95%	6.75%	7.75%	
	2770	Heating, ventilating, air conditioning		"	5.25	8	16.05	5.80%	9.80%	11.10%	
	2900	Electrical		S.F.	5.30	7.80	9.90	6.60%	8.50%	10.35%	
	3100	Total: Mechanical & Electrical			21	29.50	33.50	19.60%	26%	29.50%	
460	0010	**HOSPITALS**	R17100-100		152	181	275				460
	0020	Total project costs		C.F.	11.20	13.95	19.95				
	2720	Plumbing		S.F.	12.75	17.80	23	7.60%	9.10%	10.85%	
	2770	Heating, ventilating, air conditioning			18.70	24	32.50	7.80%	12.95%	16.65%	
	2900	Electrical			16.15	21	32.50	9.85%	11.75%	14%	
	3100	Total: Mechanical & Electrical			45.50	60	98.50	26.50%	33%	36.50%	
	9000	Per bed or person, total cost		Bed	76,500	171,500	223,500				
480	0010	**HOUSING** For the Elderly	R17100-100	S.F.	72.50	92	113				480
	0020	Total project costs		C.F.	5.20	7.20	9.20				
	2720	Plumbing		S.F.	5.40	6.90	9	8.15%	9.55%	10.50%	
	2730	Heating, ventilating, air conditioning			2.77	3.92	5.85	3.30%	5.60%	7.25%	
	2900	Electrical			5.40	7.35	9.45	7.30%	8.50%	10.25%	
	3100	Total: Mechanical & Electrical			18.65	22	29.50	18.10%	22.50%	29%	
	9000	Per rental unit, total cost		Unit	67,500	79,000	88,000				
	9500	Total: Mechanical & Electrical		"	14,900	17,300	20,200				

See R17100-100 and City Cost Indexes in the Reference Section

17100 | S.F., C.F. and % of Total Costs

	17100 | S.F. & C.F. Costs			UNIT COSTS			% OF TOTAL			
			UNIT	1/4	MEDIAN	3/4	1/4	MEDIAN	3/4	
170	2900	Electrical	S.F.	5.55	8.65	10.45	6.50%	9.50%	10.55%	170
	3100	Total: Mechanical & Electrical	↓	15.55	27	29.50	21%	23%	37%	
180	0010	**CLUBS, Y.M.C.A.** R17100-100	S.F.	88	119	145				180
	0020	Total project costs	C.F.	4.05	6.80	10.10				
	2720	Plumbing	S.F.	5.55	11.05	12.40	5.65%	7.60%	10.85%	
	2900	Electrical		6.65	9.10	12.95	6.25%	8.65%	10.20%	
	3100	Total: Mechanical & Electrical	↓	19.70	25	33.50	18.40%	22.50%	29.50%	
190	0010	**COLLEGES** Classrooms & Administration R17100-100	S.F.	99	131	175				190
	0020	Total project costs	C.F.	7.15	10.10	15.90				
	2720	Plumbing	S.F.	4.81	9.65	17.30	5.10%	6.60%	8.95%	
	2900	Electrical		8.05	12.40	14.90	7.70%	9.85%	12%	
	3100	Total: Mechanical & Electrical	↓	19.80	35.50	49	24%	28%	31.50%	
210	0010	**COLLEGES** Science, Engineering, Laboratories R17100-100	S.F.	165	193	239				210
	0020	Total project costs	C.F.	9.45	13.80	15.70				
	2900	Electrical	S.F.	13.60	19.30	29.50	7.10%	9.40%	12.10%	
	3100	Total: Mechanical & Electrical	"	50.50	60	93	28.50%	31.50%	41%	
230	0010	**COLLEGES** Student Unions R17100-100	S.F.	105	148	173				230
	0020	Total project costs	C.F.	5.90	7.70	9.95				
	3100	Total: Mechanical & Electrical	S.F.	27.50	42.50	50.50	23.50%	26%	29%	
250	0010	**COMMUNITY CENTERS** R17100-100	"	82.50	107	144				250
	0020	Total project costs	C.F.	5.70	8.10	10.50				
	2720	Plumbing	S.F.	4.22	7.20	10.50	4.94%	7%	9.10%	
	2770	Heating, ventilating, air conditioning		6.95	10	13.75	6.95%	10.65%	13.05%	
	2900	Electrical		6.95	9.40	14.30	7.35%	9.10%	10.85%	
	3100	Total: Mechanical & Electrical	↓	26	31	44.50	22.50%	26.50%	32.50%	
280	0010	**COURT HOUSES** R17100-100	S.F.	124	144	166				280
	0020	Total project costs	C.F.	9.60	11.50	14.50				
	2720	Plumbing	S.F.	5.95	8.35	12	5.95%	7.45%	8.20%	
	2900	Electrical		11.70	14	17.50	8.55%	9.95%	11.50%	
	3100	Total: Mechanical & Electrical	↓	30	34	46.50	22.50%	29.50%	30.50%	
300	0010	**DEPARTMENT STORES** R17100-100	S.F.	46.50	62.50	79				300
	0020	Total project costs	C.F.	2.50	3.37	4.74				
	2720	Plumbing	S.F.	1.51	1.83	2.77	1.82%	4.21%	5.90%	
	2770	Heating, ventilating, air conditioning		4.22	6.50	9.80	8.20%	9.10%	14.80%	
	2900	Electrical		5.35	7.30	8.60	9.05%	12.15%	14.95%	
	3100	Total: Mechanical & Electrical	↓	9.35	12	21	13.20%	21.50%	50%	
310	0010	**DORMITORIES** Low Rise (1 to 3 story) R17100-100	S.F.	76	111	135				310
	0020	Total project costs	C.F.	5.25	8.05	12.05				
	2720	Plumbing	S.F.	5.30	7.10	8.95	8.05%	9%	9.65%	
	2770	Heating, ventilating, air conditioning		5.60	6.70	8.95	4.61%	8.05%	10%	
	2900	Electrical		5.65	8.55	10.75	6.55%	8.90%	9.55%	
	3100	Total: Mechanical & Electrical	↓	29.50	31	32.50	22%	26%	29%	
	9000	Per bed, total cost	Bed	37,300	41,400	89,000				
320	0010	**DORMITORIES** Mid Rise (4 to 8 story) R17100-100	S.F.	108	141	171				320
	0020	Total project costs	C.F.	11.90	13.10	15.65				
	2900	Electrical	S.F.	8	12.55	14.20	8.20%	9.45%	10.35%	
	3100	Total: Mechanical & Electrical	"	24	32	42	19.50%	30.50%	34.50%	
	9000	Per bed, total cost	Bed	15,400	35,100	72,500				
340	0010	**FACTORIES** R17100-100	S.F.	41	61	94				340
	0020	Total project costs	C.F.	2.62	3.91	6.50				
	2720	Plumbing	S.F.	2.29	4.10	6.80	3.73%	6.05%	8.10%	
	2770	Heating, ventilating, air conditioning	↓	4.29	6.15	8.30	5.25%	8.45%	11.35%	

	17100 | S.F. & C.F. Costs		UNIT	UNIT COSTS 1/4	UNIT COSTS MEDIAN	UNIT COSTS 3/4	% OF TOTAL 1/4	% OF TOTAL MEDIAN	% OF TOTAL 3/4	
340	2900	Electrical	S.F.	5.15	8.15	12.25	8.30%	10.55%	14.20%	340
	3100	Total: Mechanical & Electrical		12.10	19.40	29.50	22%	28.50%	34.50%	
360	0010	**FIRE STATIONS** R17100-100	S.F.	78	107	144				360
	0020	Total project costs	C.F.	4.74	6.50	8.65				
	2720	Plumbing	S.F.	5.15	7.55	11.15	5.85%	7.35%	9.50%	
	2770	Heating, ventilating, air conditioning		4.42	7.05	11	4.86%	7.25%	9.25%	
	2900	Electrical		5.75	9.70	13.05	6.80%	8.75%	10.65%	
	3100	Total: Mechanical & Electrical		28	33.50	39	18.75%	23%	27%	
370	0010	**FRATERNITY HOUSES** and Sorority Houses R17100-100	S.F.	81	104	143				370
	0020	Total project costs	C.F.	7.75	8.40	10.80				
	2720	Plumbing	S.F.	6.15	7	12.85	6.80%	8%	10.85%	
	2900	Electrical		5.35	11.55	14.15	6.60%	9.90%	10.65%	
	3100	Total: Mechanical & Electrical		14.60	20.50	25		15.10%	15.90%	
380	0010	**FUNERAL HOMES** R17100-100	S.F.	85.50	117	212				380
	0020	Total project costs	C.F.	8.75	9.70	18.65				
	2900	Electrical	S.F.	3.78	6.95	8.15	3.58%	4.44%	5.95%	
	3100	Total: Mechanical & Electrical	″	13.50	20	26.50	12.90%	12.90%	12.90%	
390	0010	**GARAGES, COMMERCIAL** (Service) R17100-100	S.F.	48	76	103				390
	0020	Total project costs	C.F.	3.26	4.71	6.85				
	2720	Plumbing	S.F.	3.36	5.15	9.45	5.45%	7.85%	10.65%	
	2730	Heating & ventilating		4.41	6	8.30	5.25%	6.85%	8.20%	
	2900	Electrical		4.67	7.15	10.20	7.15%	9.25%	10.85%	
	3100	Total: Mechanical & Electrical		10.45	19.40	29	13.60%	17.40%	27%	
400	0010	**GARAGES, MUNICIPAL** (Repair) R17100-100	S.F.	71	95	134				400
	0020	Total project costs	C.F.	4.44	5.65	9.20				
	2720	Plumbing	S.F.	3.19	6.15	11.55	3.59%	6.70%	7.95%	
	2730	Heating & ventilating		5.45	7.90	15.25	6.15%	7.45%	13.50%	
	2900	Electrical		5.25	8.25	11.90	6.65%	8.15%	11.15%	
	3100	Total: Mechanical & Electrical		16.90	29.50	49	21.50%	25.50%	28.50%	
410	0010	**GARAGES, PARKING** R17100-100	S.F.	27.50	40	69.50				410
	0020	Total project costs	C.F.	2.60	3.53	5.15				
	2720	Plumbing	S.F.	.78	1.22	1.88	1.72%	2.70%	3.85%	
	2900	Electrical		1.52	1.84	2.87	4.33%	5.20%	6.25%	
	3100	Total: Mechanical & Electrical		2.42	4.33	5.70	6.90%	8.80%	11.05%	
	9000	Per car, total cost	Car	11,700	14,700	18,700				
430	0010	**GYMNASIUMS** R17100-100	S.F.	77	102	125				430
	0020	Total project costs	C.F.	3.84	5.20	6.40				
	2720	Plumbing	S.F.	4.87	6	7.45	4.95%	6.75%	7.75%	
	2770	Heating, ventilating, air conditioning	″	5.25	8	16.05	5.80%	9.80%	11.10%	
	2900	Electrical	S.F.	5.30	7.80	9.90	6.60%	8.50%	10.35%	
	3100	Total: Mechanical & Electrical		21	29.50	33.50	19.60%	26%	29.50%	
460	0010	**HOSPITALS** R17100-100		152	181	275				460
	0020	Total project costs	C.F.	11.20	13.95	19.95				
	2720	Plumbing	S.F.	12.75	17.80	23	7.60%	9.10%	10.85%	
	2770	Heating, ventilating, air conditioning		18.70	24	32.50	7.80%	12.95%	16.65%	
	2900	Electrical		16.15	21	32.50	9.85%	11.75%	14%	
	3100	Total: Mechanical & Electrical		45.50	60	98.50	26.50%	33%	36.50%	
	9000	Per bed or person, total cost	Bed	76,500	171,500	223,500				
480	0010	**HOUSING** For the Elderly R17100-100	S.F.	72.50	92	113				480
	0020	Total project costs	C.F.	5.20	7.20	9.20				
	2720	Plumbing	S.F.	5.40	6.90	9	8.15%	9.55%	10.50%	
	2730	Heating, ventilating, air conditioning		2.77	3.92	5.85	3.30%	5.60%	7.25%	
	2900	Electrical		5.40	7.35	9.45	7.30%	8.50%	10.25%	
	3100	Total: Mechanical & Electrical		18.65	22	29.50	18.10%	22.50%	29%	
	9000	Per rental unit, total cost	Unit	67,500	79,000	88,000				
	9500	Total: Mechanical & Electrical	″	14,900	17,300	20,200				

See R17100-100 and City Cost Indexes in the Reference Section

		17100 | S.F. & C.F. Costs	Unit	Unit Costs 1/4	Unit Costs Median	Unit Costs 3/4	% of Total 1/4	% of Total Median	% of Total 3/4	
500	0010	HOUSING Public (Low Rise) R17100-100	S.F.	61	85	111				500
	0020	Total project costs	C.F.	5.45	6.80	8.45				
	2720	Plumbing	S.F.	4.41	5.85	7.35	7.15%	9.05%	11.60%	
	2730	Heating, ventilating, air conditioning		2.22	4.30	4.71	4.26%	6.05%	6.45%	
	2900	Electrical		3.70	5.50	7.65	5.10%	6.55%	8.25%	
	3100	Total: Mechanical & Electrical	↓	17.55	22.50	25.50	14.50%	17.55%	26.50%	
	9000	Per apartment, total cost	Apt.	67,000	76,500	96,000				
	9500	Total: Mechanical & Electrical	"	14,300	17,600	19,600				
510	0010	ICE SKATING RINKS R17100-100	S.F.	54.50	122	134				510
	0020	Total project costs	C.F.	3.84	3.93	4.53				
	2720	Plumbing	S.F.	1.95	3.66	3.74	3.12%	3.23%	5.65%	
	2900	Electrical		5.60	8.60	9.10	6.30%	10.15%	15.05%	
	3100	Total: Mechanical & Electrical	↓	9.35	13.20	16.50	18.95%	18.95%	18.95%	
520	0010	JAILS R17100-100	S.F.	159	205	265				520
	0020	Total project costs	C.F.	15	20	24.50				
	2720	Plumbing	S.F.	16.20	20.50	27	7%	8.90%	13.35%	
	2770	Heating, ventilating, air conditioning		14.35	19.15	37	7.50%	9.45%	17.75%	
	2900	Electrical		17.50	22	27.50	8.20%	11.45%	14.70%	
	3100	Total: Mechanical & Electrical	↓	43.50	79	94	27.50%	30%	34%	
530	0010	LIBRARIES R17100-100	S.F.	99	126	165				530
	0020	Total project costs	C.F.	6.90	8.35	11				
	2720	Plumbing	S.F.	3.71	5.40	7.30	3.43%	4.60%	5.70%	
	2770	Heating, ventilating, air conditioning		8.20	13.90	18.10	7.80%	10.95%	12.80%	
	2900	Electrical		10.10	13	16.65	8.35%	10.40%	11.95%	
	3100	Total: Mechanical & Electrical	↓	30.50	38	47.50	19.65%	23%	26.50%	
540	0010	LIVING, ASSISTED R17100-100	S.F.	94	111	129				540
	0020	Total project costs	C.F.	7.85	9.15	10.40				
	2720	Plumbing	S.F.	7.80	10.45	10.80	6.05%	8.15%	10.60%	
	2770	Heating, ventilating, air conditioning		9.25	9.65	10.60	7.95%	9.35%	9.70%	
	2900	Electrical		9.15	10.25	11.65	9%	10.30%	10.70%	
	3100	Total: Mechanical & Electrical	↓	26.50	30	34.50	26%	29%	31.50%	
550	0010	MEDICAL CLINICS R17100-100	S.F.	93	117	149				550
	0020	Total project costs	C.F.	6.95	9.40	12.10				
	2720	Plumbing	S.F.	6.30	8.85	12	6.15%	8.40%	10.10%	
	2770	Heating, ventilating, air conditioning		7.50	9.85	14.45	6.65%	8.85%	11.35%	
	2900	Electrical		8.15	11.50	15.05	8.10%	10%	12.20%	
	3100	Total: Mechanical & Electrical	↓	26	36	48.50	22.50%	27.50%	33.50%	
570	0010	MEDICAL OFFICES R17100-100	S.F.	89	110	135				570
	0020	Total project costs	C.F.	6.65	9	12.15				
	2720	Plumbing	S.F.	4.92	7.60	10.20	5.60%	6.80%	8.50%	
	2770	Heating, ventilating, air conditioning		5.95	8.75	11.35	6.15%	8.05%	9.70%	
	2900	Electrical		7.10	10.35	14.50	7.65%	9.80%	11.70%	
	3100	Total: Mechanical & Electrical	↓	19.20	27.50	41	19.35%	23.50%	30.50%	
590	0010	MOTELS R17100-100	S.F.	56	82.50	107				590
	0020	Total project costs	C.F.	5	6.70	10.95				
	2720	Plumbing	S.F.	5.70	7.25	8.65	9.45%	10.60%	12.55%	
	2770	Heating, ventilating, air conditioning		3.47	5.15	9.30	5.60%	5.60%	10%	
	2900	Electrical		5.35	6.80	8.65	7.45%	9.20%	10.80%	
	3100	Total: Mechanical & Electrical	↓	18.05	22.50	38.50	18.50%	21%	25.50%	
	9000	Per rental unit, total cost	Unit	28,600	54,500	58,500				
	9500	Total: Mechanical & Electrical	"	5,575	8,425	9,800				
600	0010	NURSING HOMES R17100-100	S.F.	89	114	139				600
	0020	Total project costs	C.F.	6.90	8.65	11.80				
	2720	Plumbing	S.F.	7.85	11.45	13.80	8.75%	10.10%	12.70%	
	2770	Heating, ventilating, air conditioning		7.95	12.05	16	9.70%	11.45%	11.80%	
	2900	Electrical		8.75	11.10	14.85	9.50%	10.60%	12.50%	
	3100	Total: Mechanical & Electrical	↓	21	29	49	26%	29.50%	30.50%	

For expanded coverage of these items see *Means Square Foot Costs 2005*

17100 | S.F. & C.F. Costs

				UNIT COSTS			% OF TOTAL			
			UNIT	1/4	MEDIAN	3/4	1/4	MEDIAN	3/4	
600	9000	Per bed or person, total cost	Bed	39,600	50,000	63,500				600
610	0010	**OFFICES** Low Rise (1 to 4 story) R17100-100	S.F.	73.50	95	124				610
	0020	Total project costs	C.F.	5.30	7.30	9.60				
	2720	Plumbing	S.F.	2.65	4.10	5.95	3.66%	4.50%	6.10%	
	2770	Heating, ventilating, air conditioning		5.85	8.30	11.95	7.20%	10.40%	11.70%	
	2900	Electrical		6.05	8.60	12.15	7.45%	9.65%	11.40%	
	3100	Total: Mechanical & Electrical	↓	15.80	22	33	18.25%	22.50%	27%	
620	0010	**OFFICES** Mid Rise (5 to 10 story) R17100-100	S.F.	78.50	95	129				620
	0020	Total project costs	C.F.	5.55	7.10	10.05				
	2720	Plumbing	S.F.	2.37	3.83	5.30	2.83%	3.74%	4.50%	
	2770	Heating, ventilating, air conditioning		5.95	8.55	13.60	7.65%	9.40%	11%	
	3100	Total: Mechanical & Electrical	↓	14.90	18.80	37.50	19.15%	21%	26%	
630	0010	**OFFICES** High Rise (11 to 20 story) R17100-100	S.F.	96.50	122	150				630
	0020	Total project costs	C.F.	6.75	8.45	12.10				
	2900	Electrical	S.F.	5.85	7.15	10.60	5.80%	7.85%	10.50%	
	3100	Total: Mechanical & Electrical	"	18.85	25.50	42.50	16.90%	23.50%	34%	
640	0010	**POLICE STATIONS** R17100-100	S.F.	116	152	192				640
	0020	Total project costs	C.F.	9.20	11.30	15.45				
	2720	Plumbing	S.F.	6.70	12.90	16.05	5.65%	6.90%	10.75%	
	2770	Heating, ventilating, air conditioning		10.10	13.45	18.25	5.85%	10.55%	11.70%	
	2900	Electrical		12.60	18.90	24	9.80%	11.85%	14.80%	
	3100	Total: Mechanical & Electrical	↓	41.50	50	67.50	28.50%	32%	32.50%	
650	0010	**POST OFFICES** R17100-100	S.F.	87.50	113	140				650
	0020	Total project costs	C.F.	5.50	6.95	8.40				
	2720	Plumbing	S.F.	4.11	5.10	6.45	4.24%	5.30%	5.60%	
	2770	Heating, ventilating, air conditioning		6.45	7.95	8.85	6.65%	7.15%	9.35%	
	2900	Electrical		7.55	10.60	12.60	7.25%	9%	11%	
	3100	Total: Mechanical & Electrical	↓	22	28.50	32.50	16.25%	18.80%	22%	
660	0010	**POWER PLANTS** R17100-100	S.F.	635	835	1,550				660
	0020	Total project costs	C.F.	17.45	38	81.50				
	2900	Electrical	S.F.	45	95	142	9.30%	12.75%	21.50%	
	8100	Total: Mechanical & Electrical	"	112	365	815	32.50%	32.50%	52.50%	
670	0010	**RELIGIOUS EDUCATION** R17100-100	S.F.	73.50	94	118				670
	0020	Total project costs	C.F.	4.09	5.85	7.55				
	2720	Plumbing	S.F.	3.08	4.36	6.15	4.40%	5.30%	7.10%	
	2770	Heating, ventilating, air conditioning		7.80	8.80	12.45	10.05%	11.45%	12.35%	
	2900	Electrical		5.85	7.70	10.80	7.60%	9.10%	10.35%	
	3100	Total: Mechanical & Electrical	↓	20	31	33.50	22%	23%	26%	
690	0010	**RESEARCH** Laboratories and facilities R17100-100	S.F.	110	158	230				690
	0020	Total project costs	C.F.	8.60	16.60	19.35				
	2720	Plumbing	S.F.	11.05	14.15	22.50	6.15%	8.30%	10.80%	
	2770	Heating, ventilating, air conditioning		9.90	33.50	39.50	7.25%	16.50%	17.50%	
	2900	Electrical		12.70	21.50	36	9.45%	11.15%	15.40%	
	3100	Total: Mechanical & Electrical	↓	36	71	106	29.50%	37%	45.50%	
700	0010	**RESTAURANTS** R17100-100	S.F.	106	137	179				700
	0020	Total project costs	C.F.	9	11.80	15.45				
	2720	Plumbing	S.F.	8.55	10.25	14.60	6.10%	8.15%	9%	
	2770	Heating, ventilating, air conditioning		11.15	14.85	19.35	9.20%	12%	12.40%	
	2900	Electrical		11.30	13.95	18.10	8.35%	10.55%	11.55%	
	3100	Total: Mechanical & Electrical	↓	35	37	47.50	19.25%	24%	29.50%	
	9000	Per seat unit, total cost	Seat	3,925	5,225	6,175				
	9500	Total: Mechanical & Electrical	"	980	1,300	1,550				
720	0010	**RETAIL STORES** R17100-100	S.F.	50	67	88.50				720
	0020	Total project costs	C.F.	3.38	4.82	6.70				
	2720	Plumbing	S.F.	1.81	3.02	5.15	3.26%	4.60%	6.80%	
	2770	Heating, ventilating, air conditioning	↓	3.91	5.35	8.05	6.75%	8.75%	10.15%	

See R17100-100 and City Cost Indexes in the Reference Section

17100 | S.F., C.F. and % of Total Costs

17100 | S.F. & C.F. Costs

				UNIT COSTS			% OF TOTAL			
			UNIT	1/4	MEDIAN	3/4	1/4	MEDIAN	3/4	
720	2900	Electrical	S.F.	4.48	6.25	8.90	7.30%	9.90%	11.65%	720
	3100	Total: Mechanical & Electrical	↓	11.95	15.35	20.50	17.10%	21.50%	23.50%	
740	0010	**SCHOOLS** Elementary R17100 -100	S.F.	80.50	99.50	120				740
	0020	Total project costs	C.F.	5.35	6.80	8.80				
	2720	Plumbing	S.F.	4.67	6.60	8.85	5.70%	7.15%	9.35%	
	2730	Heating, ventilating, air conditioning		7	11.20	15.60	8.15%	10.80%	15.20%	
	2900	Electrical		7.55	9.90	12.50	8.45%	10.05%	11.80%	
	3100	Total: Mechanical & Electrical	↓	27	33	39.50	24%	27.50%	30.50%	
	9000	Per pupil, total cost	Ea.	7,850	14,000	41,400				
	9500	Total: Mechanical & Electrical	″	2,650	3,350	12,000				
760	0010	**SCHOOLS** Junior High & Middle R17100 -100	S.F.	81.50	102	120				760
	0020	Total project costs	C.F.	5.35	6.90	7.75				
	2720	Plumbing	S.F.	5.35	6.05	7.80	5.50%	6.90%	8.15%	
	2770	Heating, ventilating, air conditioning		6.15	11.80	16.50	8.75%	12.75%	17.45%	
	2900	Electrical		7.75	9.85	12.20	7.90%	9.30%	10.60%	
	3100	Total: Mechanical & Electrical	↓	23.50	33	38.50	22.50%	25.50%	29.50%	
	9000	Per pupil, total cost	Ea.	10,700	14,400	18,800				
780	0010	**SCHOOLS** Senior High R17100 -100	S.F.	85.50	108	136				780
	0020	Total project costs	C.F.	5.55	7.75	12.85				
	2720	Plumbing	S.F.	4.47	5.95	13.45	5.70%	7%	8.35%	
	2770	Heating, ventilating, air conditioning		10	11.45	22	8.95%	11.60%	15%	
	2900	Electrical		8.55	11.15	17.90	8.35%	10.10%	11.95%	
	3100	Total: Mechanical & Electrical	↓	30	34	57	23%	26.50%	28.50%	
	9000	Per pupil, total cost	Ea.	8,225	14,000	21,000				
800	0010	**SCHOOLS** Vocational R17100 -100	S.F.	70.50	100	125				800
	0020	Total project costs	C.F.	4.39	6.30	8.70				
	2720	Plumbing	S.F.	4.60	6.90	10.35	5.40%	6.90%	8.55%	
	2770	Heating, ventilating, air conditioning		6.30	11.75	19.70	8.60%	11.90%	14.65%	
	2900	Electrical		7.20	10.05	14.20	8.45%	10.95%	13.20%	
	3100	Total: Mechanical & Electrical	↓	25.50	28.50	48.50	23.50%	29.50%	31%	
	9000	Per pupil, total cost	Ea.	9,850	26,300	39,300				
830	0010	**SPORTS ARENAS** R17100 -100	S.F.	60	82.50	127				830
	0020	Total project costs	C.F.	3.35	6	7.75				
	2720	Plumbing	S.F.	3.58	5.45	11.50	4.35%	6.35%	9.40%	
	2770	Heating, ventilating, air conditioning		7.70	9.15	12.65	8.80%	10.20%	13.55%	
	2900	Electrical		6.50	8.75	11.30	8.60%	9.90%	12.25%	
	3100	Total: Mechanical & Electrical	↓	16	28.50	37	21.50%	25%	27.50%	
850	0010	**SUPERMARKETS** R17100 -100	S.F.	56	66	80				850
	0020	Total project costs	C.F.	3.18	3.84	5.80				
	2720	Plumbing	S.F.	3.19	4.10	4.67	5.40%	6%	7.45%	
	2770	Heating, ventilating, air conditioning		4.69	6.25	7.60	8.60%	8.65%	9.60%	
	2900	Electrical		7.15	8.20	9.70	10.40%	12.45%	13.60%	
	3100	Total: Mechanical & Electrical	↓	18.35	19.95	28	20.50%	26.50%	31%	
860	0010	**SWIMMING POOLS** R17100 -100	S.F.	92.50	155	305				860
	0020	Total project costs	C.F.	7.40	9.25	10.05				
	2720	Plumbing	S.F.	8.55	9.75	13.60	4.80%	9.70%	20.50%	
	2900	Electrical		6.95	11.25	16.40	6.50%	7.25%	7.60%	
	3100	Total: Mechanical & Electrical	↓	16.95	43	58.50	11.15%	14.10%	23.50%	
870	0010	**TELEPHONE EXCHANGES** R17100 -100	S.F.	123	180	228				870
	0020	Total project costs	C.F.	7.65	12.25	16.85				
	2720	Plumbing	S.F.	5.20	8.20	11.70	4.52%	5.80%	6.90%	
	2770	Heating, ventilating, air conditioning		12	24	30	11.80%	16.05%	18.40%	
	2900	Electrical		12.50	19.80	35	10.90%	14%	17.85%	
	3100	Total: Mechanical & Electrical	↓	37	70	99	29.50%	33.50%	44.50%	

		17100 | S.F. & C.F. Costs		UNIT	UNIT COSTS 1/4	UNIT COSTS MEDIAN	UNIT COSTS 3/4	% OF TOTAL 1/4	% OF TOTAL MEDIAN	% OF TOTAL 3/4	
910	0010	**THEATERS**	R17100-100	S.F.	77	99	146				910
	0020	Total project costs		C.F.	3.56	5.25	7.75				
	2720	Plumbing		S.F.	2.57	2.79	11.40	2.92%	4.70%	6.80%	
	2770	Heating, ventilating, air conditioning			7.50	9.10	11.25	8%	12.25%	13.40%	
	2900	Electrical			6.75	9.10	18.55	8.05%	9.95%	12.25%	
	3100	Total: Mechanical & Electrical		↓	17.35	26	53.50	23%	26.50%	27.50%	
940	0010	**TOWN HALLS** City Halls & Municipal Buildings	R17100-100	S.F.	90.50	116	153				940
	0020	Total project costs		C.F.	7.85	9.70	13.25				
	2720	Plumbing		S.F.	3.60	6.75	12.40	4.31%	5.95%	7.95%	
	2770	Heating, ventilating, air conditioning			6.50	12.90	18.85	7.05%	9.05%	13.45%	
	2900	Electrical			8.20	11.95	15.95	8.05%	9.50%	12.05%	
	3100	Total: Mechanical & Electrical		↓	27.50	32.50	42.50	22%	26.50%	31%	
970	0010	**WAREHOUSES** And Storage Buildings	R17100-100	S.F.	34	49.50	68.50				970
	0020	Total project costs		C.F.	1.85	2.63	4.36				
	2720	Plumbing		S.F.	1.07	1.92	3.67	2.90%	4.80%	6.55%	
	2730	Heating, ventilating, air conditioning			1.29	3.45	4.63	2.41%	5%	8.90%	
	2900	Electrical			2.03	2.69	5.95	5.15%	7.20%	10.10%	
	3100	Total: Mechanical & Electrical		↓	5.30	8.15	17.80	12.75%	18.90%	26%	
990	0010	**WAREHOUSE & OFFICES** Combination	R17100-100	S.F.	39.50	52.50	73				990
	0020	Total project costs		C.F.	2.02	2.94	4.34				
	2720	Plumbing		S.F.	1.53	2.71	4.07	3.74%	4.76%	6.30%	
	2770	Heating, ventilating, air conditioning			2.41	3.77	5.30	5%	5.65%	10.05%	
	2900	Electrical		S.F.	2.65	3.93	6.20	5.85%	8%	10%	
	3100	Total: Mechanical & Electrical		↓	7.30	10.45	17.25	14.40%	19.95%	24.50%	

See R17100-100 and City Cost Indexes in the Reference Section

Assemblies Section

Table of Contents

How to Use the Assemblies Cost Tables

The following is a detailed explanation of a sample Assemblies Cost Table. Most Assembly Tables are separated into three parts: 1) an illustration of the system to be estimated; 2) the components and related costs of a typical system; and 3) the costs for similar systems with dimensional and/or size variations. For costs of the components that comprise these systems or "assemblies" refer to the Unit Price Section. Next to each bold number below is the item being described with the appropriate component of the sample entry following in parenthesis. In most cases, if the work is to be subcontracted, the general contractor will need to add an additional markup (RSMeans suggests using 10%) to the "Total" figures.

System/Line Numbers (D3010 510 1760)

Each Assemblies Cost Line has been assigned a unique identification number based on the UniFormat classification sytem.

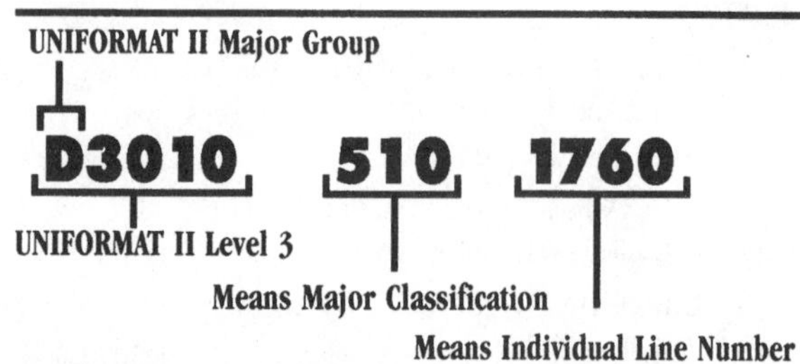

D30 HVAC

D3010 Energy Supply

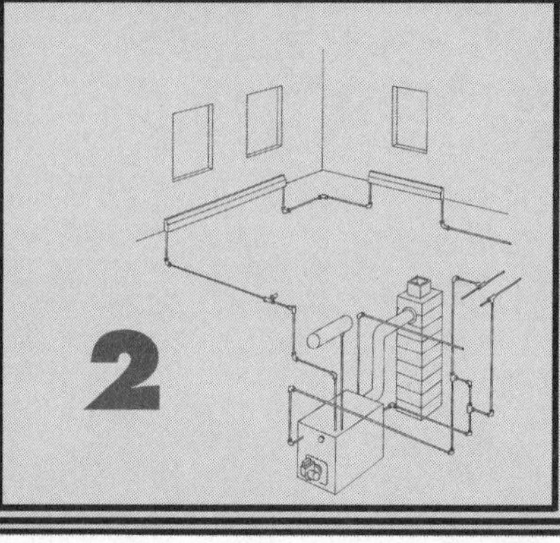

Basis for Heat Loss Estimate, Apartment Type Structures:

1. Masonry walls and flat roof are insulated. U factor is assumed at .08.
2. Window glass area taken as BOCA minimum, 1/10th of floor area. Double insulating glass with 1/4" air space, U = .65.
3. Infiltration = 0.3 C.F. per hour per S.F. of net wall.
4. Concrete floor loss is 2 BTUH per S.F.
5. Temperature difference taken as 70° F.
6. Ventilating or makeup air has not been included and must be added if desired. Air shafts are not used.

System Components	QUANTITY	UNIT	COST EACH MAT.	COST EACH INST.	COST EACH TOTAL
SYSTEM D3010 510 1760 HEATING SYSTEM, FIN TUBE RADIATION, FORCED HOT WATER 1,000 S.F. AREA, 10,000 C.F. VOLUME					
Boiler, oil fired, CI, burner, ctrls/insul/breech/pipe/ftng/valves, 109 MBH	1.000	Ea.	3,718.75	2,756.25	6,475
Circulating pump, CI flange connection, 1/12 HP	1.000	Ea.	234	147	381
Expansion tank, painted steel, ASME 18 Gal capacity	1.000	Ea.	405	63.50	468.50
Storage tank, steel, above ground, 275 Gal capacity w/supports	1.000	Ea.	[illegible]	[illegible]	[illegible]
Copper tubing type L, solder joint, hanger 10' OC, 3/4" diam	100.000	L.F.	[illegible]	[illegible]	[illegible]
Radiation, 3/4" copper tube w/alum fin baseboard pkg, 7" high	30.000	L.F.	[illegible]7.50	[illegible]	[illegible]
Pipe covering, calcium silicate w/cover, 1" wall, 3/4" diam	100.000	L.F.	240	496	736
TOTAL			5,367.25	4,743.75	10,111
COST PER S.F.			5.37	4.74	10.11

D3010 510	Apartment Building Heating - Fin Tube Radiation		COST PER S.F. MAT.	COST PER S.F. INST.	COST PER S.F. TOTAL
1740	Heating systems, fin tube radiation, forced hot water				
1760	1,000 S.F. area, 10,000 C.F. volume		5.38	4.75	10.13
1800	10,000 S.F. area, 100,000 C.F. volume	R15500 -010	1.70	2.86	4.56
1840	20,000 S.F. area, 200,000 C.F. volume		1.72	3.22	4.94
1880	30,000 S.F. area, 300,000 C.F. volume	R15500 -020	1.61	3.11	4.72
1890					

2 *Illustration*

At the top of most assembly pages is an illustration, a brief description, and the design criteria used to develop the cost.

3 *System Components*

The components of a typical system are listed separately to show what has been included in the development of the total system price. The table below contains prices for other similar systems with dimensional and/or size variations.

4 *Quantity*

This is the number of line item units required for one system unit. For example, we assume that it will take 30 linear feet of radiation for the 1000 S.F. area shown.

5 *Unit of Measure for Each Item*

The abbreviated designation indicates the unit of measure, as defined by industry standards, upon which the price of the component is based. For example, baseboard radiation is priced by the linear foot. For a complete listing of abbreviations, see the Reference Section.

6 *Unit of Measure for Each System (Each)*

Costs shown in the three right hand columns have been adjusted by the component quantity and unit of measure for the entire system. In this example, "Cost Each" is the unit of measure for this system or "assembly."

7 *Materials (5,367.25)*

This column contains the Materials Cost of each component. These cost figures are bare costs plus 10% for profit.

8 *Installation (4,743.75)*

Installation includes labor and equipment plus the installing contractor's overhead and profit. Equipment costs are the bare rental costs plus 10% for profit. The labor overhead and profit is defined on the inside back cover of this book.

9 *Total (10,111.00)*

The figure in this column is the sum of the material and installation costs.

Material Cost	+	Installation Cost	=	Total
$5,367.25	+	$4,743.75	=	$10,111.00

A SUBSTRUCTURE

A2020 Basement Walls

A2020 220	Subdrainage Piping	COST PER L.F.		
		MAT.	INST.	TOTAL
2000	Piping, excavation & backfill excluded, PVC, perforated			
2110	3" diameter	2.11	3	5.11
2130	4" diameter	2.11	3	5.11
2140	5" diameter	3.78	3.21	6.99
2150	6" diameter	3.78	3.21	6.99
3000	Metal alum. or steel, perforated asphalt coated			
3150	6" diameter	3.04	6.15	9.19
3160	8" diameter	4.33	6.30	10.63
3170	10" diameter	5.40	6.45	11.85
3180	12" diameter	6.05	8.15	14.20
3220	18" diameter	9.10	11.30	20.40
4000	Porous wall concrete			
4130	4" diameter	2.15	3.35	5.50
4150	6" diameter	2.79	3.57	6.36
4160	8" diameter	3.44	4.88	8.32
4180	12" diameter	7.30	5.30	12.60
4200	15" diameter	8.35	6.60	14.95
4220	18" diameter	11.05	9.15	20.20
5000	Vitrified clay C-211, perforated			
5130	4" diameter	2.20	5.80	8
5150	6" diameter	3.64	7.35	10.99
5160	8" diameter	5.20	8	13.20
5180	12" diameter	10.90	8.45	19.35

For information about Means Estimating Seminars, see yellow pages 12 and 13 in back of book

D SERVICES

D2020 Domestic Water Distribution

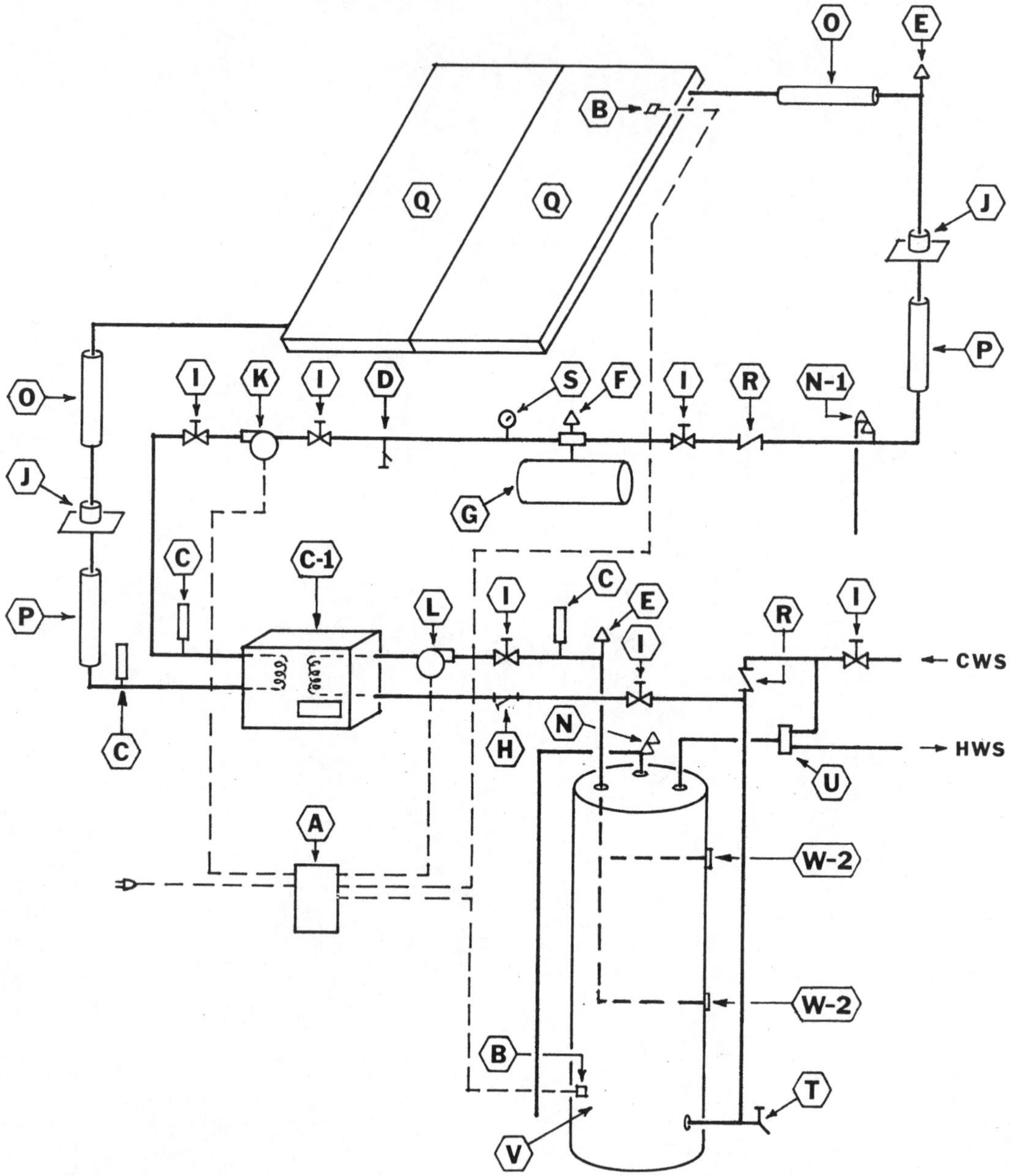

In this closed-loop indirect collection system, fluid with a low freezing temperature, such as propylene glycol, transports heat from the collectors to water storage. The transfer fluid is contained in a closed-loop consisting of collectors, supply and return piping, and a remote heat exchanger. The heat exchanger transfers heat energy from the fluid in the collector loop to potable water circulated in a storage loop. A typical two-or-three panel system contains 5 to 6 gallons of heat transfer fluid.

When the collectors become approximately 20° F warmer than the storage temperature, a controller activates the circulator on the collector and storage loops. The circulators will move the fluid and potable water through the heat exchanger until heat collection no longer occurs. At that point, the system shuts down. Since the heat transfer medium is a fluid with a very low freezing temperature, there is no need for it to be drained from the system between periods of collection.

D2020 Domestic Water Distribution

System Components		QUANTITY	UNIT	COST EACH MAT.	INST.	TOTAL
	SYSTEM D2020 265 2760					
	SOLAR, CLOSED LOOP, ADD-ON HOT WATER SYS., EXTERNAL HEAT EXCHANGER					
	3/4" TUBING, TWO 3'X7' BLACK CHROME COLLECTORS					
A,B,G,L,K,M	Heat exchanger fluid-fluid pkg incl 2 circulators, expansion tank,					
	Check valve, relief valve, controller, hi temp cutoff, & 2 sensors	1.000	Ea.	765	355	1,120
C	Thermometer, 2" dial	3.000	Ea.	66	91.50	157.50
D, T	Fill & drain valve, brass, 3/4" connection	1.000	Ea.	5.20	20.50	25.70
E	Air vent, manual, 1/8" fitting	2.000	Ea.	4.92	30.70	35.62
F	Air purger	1.000	Ea.	47.50	41	88.50
H	Strainer, Y type, bronze body, 3/4" IPS	1.000	Ea.	19.15	26	45.15
I	Valve, gate, bronze, NRS, soldered 3/4" diam	6.000	Ea.	165	147	312
J	Neoprene vent flashing	2.000	Ea.	19.60	49	68.60
N-1, N	Relief valve temp & press, 150 psi 210°F self-closing 3/4" IPS	1.000	Ea.	10	16.35	26.35
O	Pipe covering, urethane, ultraviolet cover, 1" wall 3/4" diam	20.000	L.F.	38.20	91.80	130
P	Pipe covering, fiberglass, all service jacket, 1" wall, 3/4" diam	50.000	L.F.	45.50	183.50	229
Q	Collector panel solar energy blk chrome on copper, 1/8" temp glass 3'x7'	2.000	Ea.	1,200	186	1,386
	Roof clamps for solar energy collector panels	2.000	Set	3.54	25.30	28.84
R	Valve, swing check, bronze, regrinding disc, 3/4" diam	2.000	Ea.	77	49	126
S	Pressure gauge, 60 psi, 2" dial	1.000	Ea.	24.50	15.35	39.85
U	Valve, water tempering, bronze, sweat connections, 3/4" diam	1.000	Ea.	70.50	24.50	95
W-2, V	Tank water storage w/heating element, drain, relief valve, existing	1.000	Ea.			
	Copper tubing type L, solder joint, hanger 10' OC 3/4" diam	20.000	L.F.	48.40	129	177.40
	Copper tubing, type M, solder joint, hanger 10' OC 3/4" diam	70.000	L.F.	133	441	574
	Sensor wire, #22-2 conductor multistranded	.500	C.L.F.	5.68	24.25	29.93
	Solar energy heat transfer fluid, propylene glycol anti-freeze	6.000	Gal.	58.20	105.30	163.50
	Wrought copper fittings & solder, 3/4" diam	76.000	Ea.	88.92	1,976	2,064.92
	TOTAL			2,895.81	4,028.05	6,923.86

D2020 265	Solar, Closed Loop, Add-On Hot Water Systems	COST EACH MAT.	INST.	TOTAL
2550	Solar, closed loop, add-on hot water system, external heat exchanger			
2570	3/8" tubing, 3 ea. 4' x 4'-4" vacuum tube collectors	4,450	3,775	8,225
2580	1/2" tubing, 4 ea 4 x 4'-4" vacuum tube collectors R13600-610	5,350	4,050	9,400
2600	2 ea 3'x7' black chrome collectors	2,675	3,825	6,500
2620	3 ea 3'x7' black chrome collectors	3,275	3,925	7,200
2640	2 ea 3'x7' flat black collectors	2,425	3,850	6,275
2660	3 ea 3'x7' flat black collectors	2,900	3,950	6,850
2700	3/4" tubing, 3 ea 3'x7' black chrome collectors	3,500	4,125	7,625
2720	3 ea 3'x7' flat black absorber plate collectors	3,125	4,150	7,275
2740	2 ea 4'x9' flat black w/plastic glazing collectors	2,775	4,175	6,950
2760	2 ea 3'x7' black chrome collectors	2,900	4,025	6,925
2780	1" tubing,4 ea 2'x9' plastic absorber & glazing collectors	4,175	4,650	8,825
2800	4 ea 3'x7' black chrome absorber collectors	4,400	4,675	9,075
2820	4 ea 3'x7' flat black absorber collectors	3,900	4,700	8,600

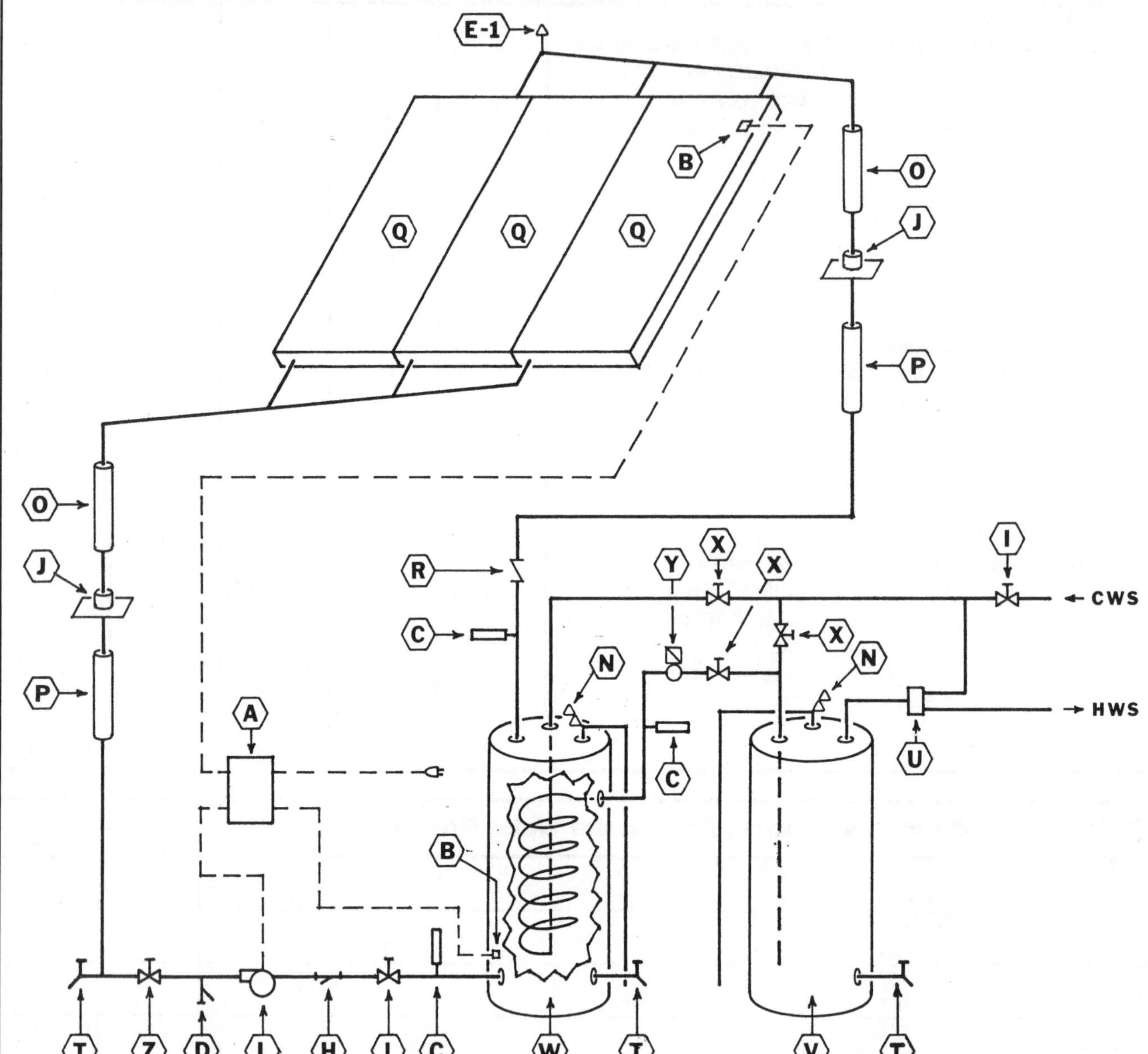

In the drainback indirect-collection system, the heat transfer fluid is distilled water contained in a loop consisting of collectors, supply and return piping, and an unpressurized holding tank. A large heat exchanger containing incoming potable water is immersed in the holding tank. When a controller activates solar collection, the distilled water is pumped through the collectors and heated and pumped back down to the holding tank. When the temperature differential between the water in the collectors and water in storage is such that collection no longer occurs, the pump turns off and gravity causes the distilled water in the collector loop to drain back to the holding tank. All the loop piping is pitched so that the water can drain out of the collectors and piping and not freeze there. As hot water is needed in the home, incoming water first flows through the holding tank with the immersed heat exchanger and is warmed and then flows through a conventional heater for any supplemental heating that is necessary.

D2020 Domestic Water Distribution

System Components		QUANTITY	UNIT	COST EACH MAT.	COST EACH INST.	COST EACH TOTAL
	SYSTEM D2020 270 2760					
	SOLAR, DRAINBACK, ADD ON, HOT WATER, IMMERSED HEAT EXCHANGER					
	3/4" TUBING, THREE EA 3'X7' BLACK CHROME COLLECTOR					
A, B	Differential controller 2 sensors, thermostat, solar energy system	1.000	Ea.	92	41	133
C	Thermometer 2" dial	3.000	Ea.	66	91.50	157.50
D, T	Fill & drain valve, brass, 3/4" connection	1.000	Ea.	5.20	20.50	25.70
E-1	Automatic air vent 1/8" fitting	1.000	Ea.	11.25	15.35	26.60
H	Strainer, Y type, bronze body, 3/4" IPS	1.000	Ea.	19.15	26	45.15
I	Valve, gate, bronze, NRS, soldered 3/4" diam	2.000	Ea.	55	49	104
J	Neoprene vent flashing	2.000	Ea.	19.60	49	68.60
L	Circulator, solar heated liquid, 1/20 HP	1.000	Ea.	170	73.50	243.50
N	Relief valve temp. & press. 150 psi 210°F self-closing 3/4" IPS	1.000	Ea.	10	16.35	26.35
O	Pipe covering, urethane, ultraviolet cover, 1" wall, 3/4" diam	20.000	L.F.	38.20	91.80	130
P	Pipe covering, fiberglass, all service jacket, 1" wall, 3/4" diam	50.000	L.F.	45.50	183.50	229
Q	Collector panel solar energy blk chrome on copper, 1/8" temp glas 3'x7'	3.000	Ea.	1,800	279	2,079
	Roof clamps for solar energy collector panels	3.000	Set	5.31	37.95	43.26
R	Valve, swing check, bronze, regrinding disc, 3/4" diam	1.000	Ea.	38.50	24.50	63
U	Valve, water tempering, bronze sweat connections, 3/4" diam	1.000	Ea.	70.50	24.50	95
V	Tank, water storage w/heating element, drain, relief valve, existing	1.000	Ea.			
W	Tank, water storage immersed heat exchr elec 2"x1/2# insul 120 gal	1.000	Ea.	1,025	350	1,375
X	Valve, globe, bronze, rising stem, 3/4" diam, soldered	3.000	Ea.	175.50	73.50	249
Y	Flow control valve	1.000	Ea.	64.50	22.50	87
Z	Valve, ball, bronze, solder 3/4" diam, solar loop flow control	1.000	Ea.	12.70	24.50	37.20
	Copper tubing, type L, solder joint, hanger 10' OC 3/4" diam	20.000	L.F.	48.40	129	177.40
	Copper tubing, type M, solder joint, hanger 10' OC 3/4" diam	70.000	L.F.	133	441	574
	Sensor wire, #22-2 conductor, multistranded	.500	C.L.F.	5.68	24.25	29.93
	Wrought copper fittings & solder, 3/4" diam	76.000	Ea.	88.92	1,976	2,064.92
	TOTAL			3,999.91	4,064.20	8,064.11

D2020 270	Solar, Drainback, Hot Water Systems	COST EACH MAT.	COST EACH INST.	COST EACH TOTAL
2550	Solar, drainback, hot water, immersed heat exchanger R13600-610			
2560	3/8" tubing, 3 ea. 4' x 4'-4" vacuum tube collectors	4,850	3,675	8,525
2580	1/2" tubing, 4 ea 4' x 4'-4" vacuum tube collectors, 80 gal tank	5,750	3,950	9,700
2600	120 gal tank	5,850	4,000	9,850
2640	2 ea. 3'x7' blk chrome collectors, 80 gal tank	3,075	3,725	6,800
2660	3 ea. 3'x7' blk chrome collectors, 120 gal tank	3,775	3,875	7,650
2700	2 ea. 3'x7' flat blk collectors, 120 gal tank	2,925	3,775	6,700
2720	3 ea. 3'x7' flat blk collectors, 120 gal tank	3,400	3,875	7,275
2760	3/4" tubing, 3 ea 3'x7' black chrome collectors, 120 gal tank	4,000	4,075	8,075
2780	3 ea. 3'x7' flat black absorber collectors, 120 gal tank	3,625	4,075	7,700
2800	2 ea. 4'x9' flat blk w/plastic glazing collectors 120 gal tank	3,275	4,100	7,375
2840	1" tubing, 4 ea. 2'x9' plastic absorber & glazing collectors, 120 gal tank	4,725	4,575	9,300
2860	4 ea. 3'x7' black chrome absorber collectors, 120 gal tank	4,950	4,600	9,550
2880	4 ea. 3'x7' flat black absorber collectors, 120 gal tank	4,450	4,625	9,075

D SERVICES

D2020 Domestic Water Distribution

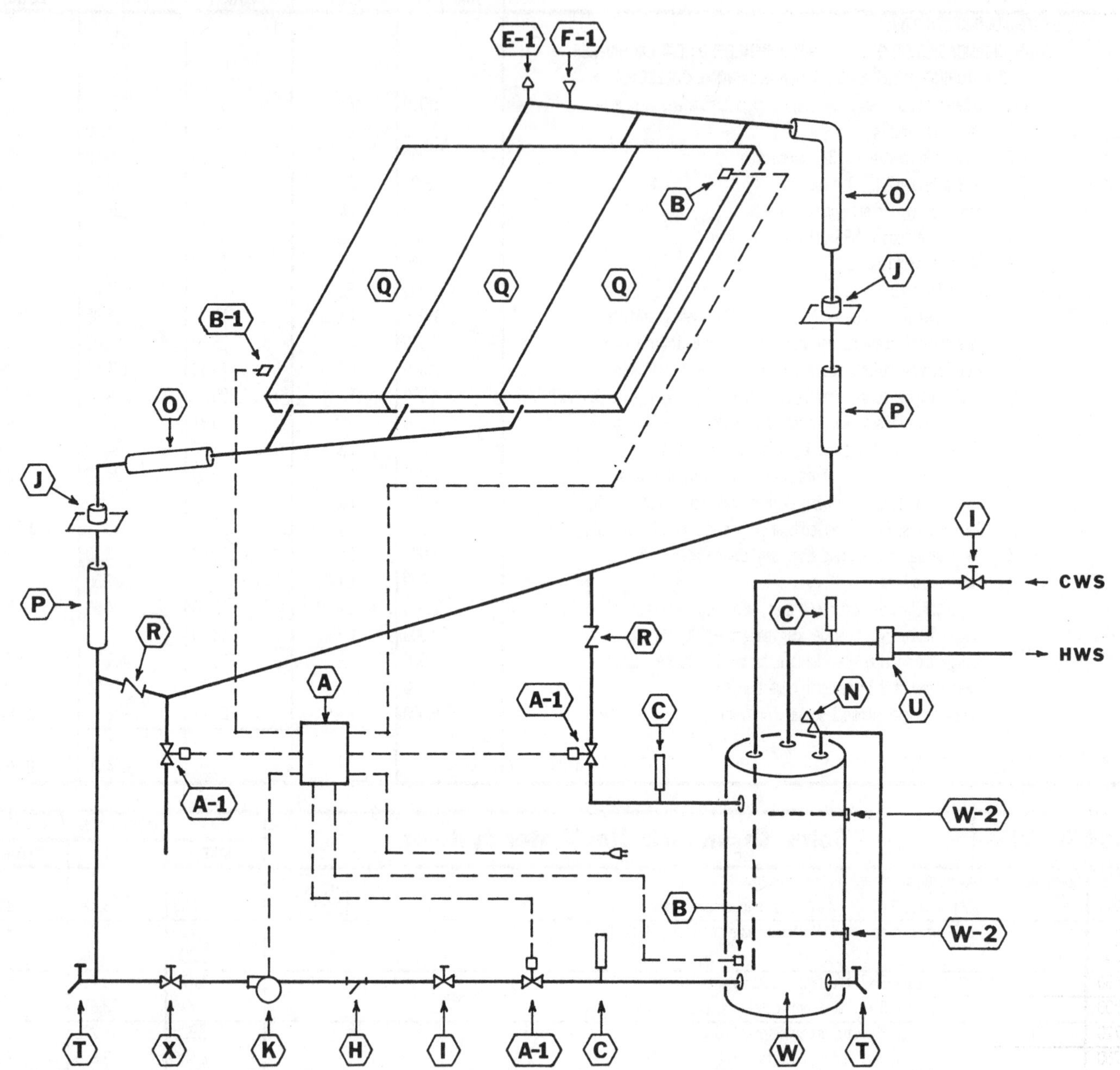

In the draindown direct-collection system, incoming domestic water is heated in the collectors. When the controller activates solar collection, domestic water is first heated as it flows through the collectors and is then pumped to storage. When conditions are no longer suitable for heat collection, the pump shuts off and the water in the loop drains down and out of the system by means of solenoid valves and properly pitched piping.

D2020 Domestic Water Distribution

System Components		QUANTITY	UNIT	COST EACH MAT.	INST.	TOTAL
	SYSTEM D2020 275 2760					
	SOLAR, DRAINDOWN, HOT WATER, DIRECT COLLECTION					
	3/4" TUBING, THREE 3'X7' BLACK CHROME COLLECTORS					
A, B	Differential controller, 2 sensors, thermostat, solar energy system	1.000	Ea.	92	41	133
A-1	Solenoid valve, solar heating loop, brass, 3/4" diam, 24 volts	3.000	Ea.	699	163.50	862.50
B-1	Solar energy sensor, freeze prevention	1.000	Ea.	24	15.35	39.35
C	Thermometer, 2" dial	3.000	Ea.	66	91.50	157.50
E-1	Vacuum relief valve, 3/4" diam	1.000	Ea.	29	15.35	44.35
F-1	Air vent, automatic, 1/8" fitting	1.000	Ea.	11.25	15.35	26.60
H	Strainer, Y type, bronze body, 3/4" IPS	1.000	Ea.	19.15	26	45.15
I	Valve, gate, bronze, NRS, soldered, 3/4" diam	2.000	Ea.	55	49	104
J	Vent flashing neoprene	2.000	Ea.	19.60	49	68.60
K	Circulator, solar heated liquid, 1/25 HP	1.000	Ea.	162	63	225
N	Relief valve temp & press 150 psi 210°F self-closing 3/4" IPS	1.000	Ea.	10	16.35	26.35
O	Pipe covering, urethane, ultraviolet cover, 1" wall, 3/4" diam	20.000	L.F.	38.20	91.80	130
P	Pipe covering, fiberglass, all service jacket, 1" wall, 3/4" diam	50.000	L.F.	45.50	183.50	229
	Roof clamps for solar energy collector panels	3.000	Set	5.31	37.95	43.26
Q	Collector panel solar energy blk chrome on copper, 1/8" temp glass 3'x7'	3.000	Ea.	1,800	279	2,079
R	Valve, swing check, bronze, regrinding disc, 3/4" diam, soldered	2.000	Ea.	77	49	126
T	Drain valve, brass, 3/4" connection	2.000	Ea.	10.40	41	51.40
U	Valve, water tempering, bronze, sweat connections, 3/4" diam	1.000	Ea.	70.50	24.50	95
W-2, W	Tank, water storage elec elem 2"x1/2# insul 120 gal	1.000	Ea.	1,025	350	1,375
X	Valve, globe, bronze, rising stem, 3/4" diam, soldered	1.000	Ea.	58.50	24.50	83
	Copper tubing, type L, solder joints, hangers 10' OC 3/4" diam	20.000	L.F.	48.40	129	177.40
	Copper tubing, type M, solder joints, hangers 10' OC 3/4" diam	70.000	L.F.	133	441	574
	Sensor wire, #22-2 conductor, multistranded	.500	C.L.F.	5.68	24.25	29.93
	Wrought copper fittings & solder, 3/4" diam	76.000	Ea.	88.92	1,976	2,064.92
	TOTAL			4,593.41	4,196.90	8,790.31

D2020 275	Solar, Draindown, Hot Water Systems	COST EACH MAT.	INST.	TOTAL
2550	Solar, draindown, hot water			
2560	3/8" tubing, 3 ea. 4' x 4'-4" vacuum tube collectors, 80 gal tank	5,400	3,975	9,375
2580	1/2" tubing, 4 ea. 4' x 4'-4" vacuum tube collectors, 80 gal tank R13600-610	6,350	4,100	10,450
2600	120 gal tank	6,475	4,125	10,600
2640	2 ea 3'x7' black chrome collectors, 80 gal tank	3,700	3,850	7,550
2660	3 ea 3'x7' black chrome collectors, 120 gal tank	4,400	4,000	8,400
2700	2 ea 3'x7' flat black collectors, 120 gal tank	3,550	3,925	7,475
2720	3 ea 3'x7' flat black collectors, 120 gal tank	4,000	4,000	8,000
2760	3/4" tubing, 3 ea 3'x7' black chrome collectors, 120 gal tank	4,600	4,200	8,800
2780	3 ea 3'x7' flat collectors, 120 gal tank	4,225	4,225	8,450
2800	2 ea. 4'x9' flat black & plastic glazing collectors, 120 gal tank	3,875	4,225	8,100
2840	1" tubing, 4 ea. 2'x9' plastic absorber & glazing collectors, 120 gal tank	5,275	4,700	9,975
2860	4 ea 3'x7' black chrome absorber collectors, 120 gal tank	5,500	4,725	10,225
2880	4 ea 3'x7' flat black absorber collectors, 120 gal tank	5,000	4,750	9,750

D2020 Domestic Water Distribution

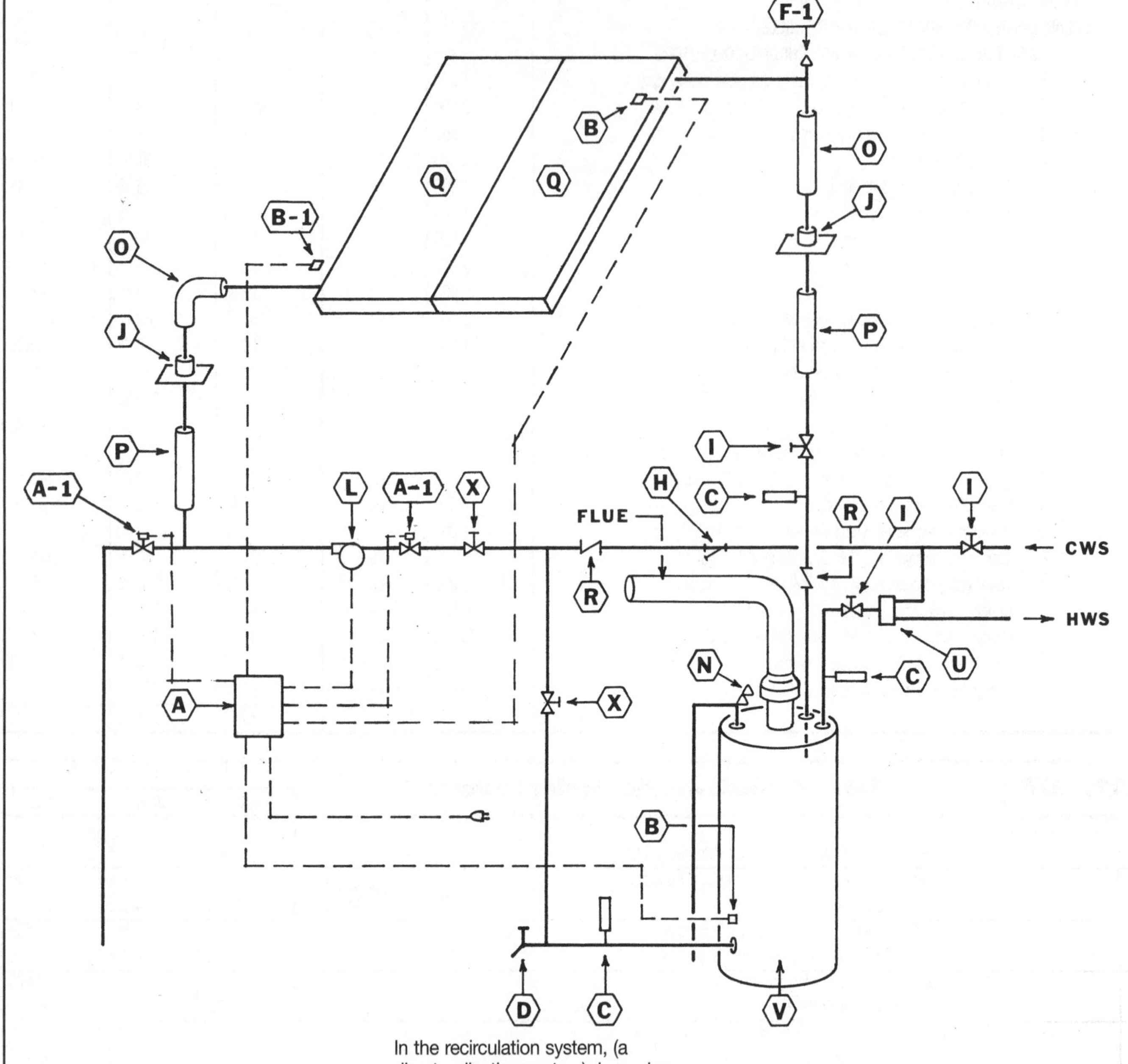

In the recirculation system, (a direct-collection system), incoming domestic water is heated in the collectors. When the controller activates solar collection, domestic water is heated as it flows through the collectors and then it flows back to storage. When conditions are not suitable for heat collection, the pump shuts off and the flow of the water stops. In this type of system, water remains in the collector loop at all times. A “frost sensor” at the collector activates the circulation of warm water from storage through the collectors when protection from freezing is required.

D2020 Domestic Water Distribution

System Components

System Components		QUANTITY	UNIT	COST EACH MAT.	INST.	TOTAL
	SYSTEM D2020 280 2820					
	SOLAR, RECIRCULATION, HOT WATER					
	3/4″ TUBING, TWO 3′X7′ BLACK CHROME COLLECTORS					
A, B	Differential controller 2 sensors, thermostat, for solar energy system	1.000	Ea.	92	41	133
A-1	Solenoid valve, solar heating loop, brass, 3/4″ IPS, 24 volts	2.000	Ea.	466	109	575
B-1	Solar energy sensor freeze prevention	1.000	Ea.	24	15.35	39.35
C	Thermometer, 2″ dial	3.000	Ea.	66	91.50	157.50
D	Drain valve, brass, 3/4″ connection	1.000	Ea.	5.20	20.50	25.70
F-1	Air vent, automatic, 1/8″ fitting	1.000	Ea.	11.25	15.35	26.60
H	Strainer, Y type, bronze body, 3/4″ IPS	1.000	Ea.	19.15	26	45.15
I	Valve, gate, bronze, 125 lb, soldered 3/4″ diam	3.000	Ea.	82.50	73.50	156
J	Vent flashing, neoprene	2.000	Ea.	19.60	49	68.60
L	Circulator, solar heated liquid, 1/20 HP	1.000	Ea.	170	73.50	243.50
N	Relief valve, temp & press 150 psi 210°F self-closing 3/4″ IPS	2.000	Ea.	20	32.70	52.70
O	Pipe covering, urethane, ultraviolet cover, 1″ wall, 3/4″ diam	20.000	L.F.	38.20	91.80	130
P	Pipe covering, fiberglass, all service jacket, 1″ wall 3/4″ diam	50.000	L.F.	45.50	183.50	229
Q	Collector panel solar energy blk chrome on copper, 1/8″ temp glass 3′x7′	2.000	Ea.	1,200	186	1,386
	Roof clamps for solar energy collector panels	2.000	Set	3.54	25.30	28.84
R	Valve, swing check, bronze, 125 lb, regrinding disc, soldered 3/4″ diam	2.000	Ea.	77	49	126
U	Valve, water tempering, bronze, sweat connections, 3/4″ diam	1.000	Ea.	70.50	24.50	95
V	Tank, water storage, w/heating element, drain, relief valve, existing	1.000	Ea.			
X	Valve, globe, bronze, 125 lb, 3/4″ diam	2.000	Ea.	117	49	166
	Copper tubing, type L, solder joints, hangers 10′ OC 3/4″ diam	20.000	L.F.	48.40	129	177.40
	Copper tubing, type M, solder joints, hangers 10′ OC 3/4″ diam	70.000	L.F.	133	441	574
	Wrought copper fittings & solder, 3/4″ diam	76.000	Ea.	88.92	1,976	2,064.92
	Sensor wire, #22-2 conductor, multistranded	.500	C.L.F.	5.68	24.25	29.93
	TOTAL			2,803.44	3,726.75	6,530.19

D2020 280	Solar Recirculation, Domestic Hot Water Systems	COST EACH MAT.	INST.	TOTAL
2550	Solar recirculation, hot water			
2560	3/8″ tubing, 3 ea. 4′ x 4′-4″ vacuum tube collectors	4,350	3,475	7,825
2580	1/2″ tubing, 4 ea. 4′ x 4′-4″ vacuum tube collectors (R13600-610)	5,250	3,775	9,025
2640	2 ea. 3′x7′ black chrome collectors	2,575	3,525	6,100
2660	3 ea. 3′x7′ black chrome collectors	3,200	3,625	6,825
2700	2 ea. 3′x7′ flat black collectors	2,325	3,550	5,875
2720	3 ea. 3′x7′ flat black collectors	2,825	3,650	6,475
2760	3/4″ tubing, 3 ea. 3′x7′ black chrome collectors	3,400	3,825	7,225
2780	3 ea. 3′x7′ flat black absorber plate collectors	3,025	3,850	6,875
2800	2 ea. 4′x9′ flat black w/plastic glazing collectors	2,675	3,875	6,550
2820	2 ea. 3′x7′ black chrome collectors	2,800	3,725	6,525
2840	1″ tubing, 4 ea. 2′x9′ black plastic absorber & glazing collectors	4,125	4,350	8,475
2860	4 ea. 3′x7′ black chrome absorber collectors	4,350	4,375	8,725
2880	4 ea. 3′x7′ flat black absorber collectors	3,850	4,400	8,250

D2020 Domestic Water Distribution

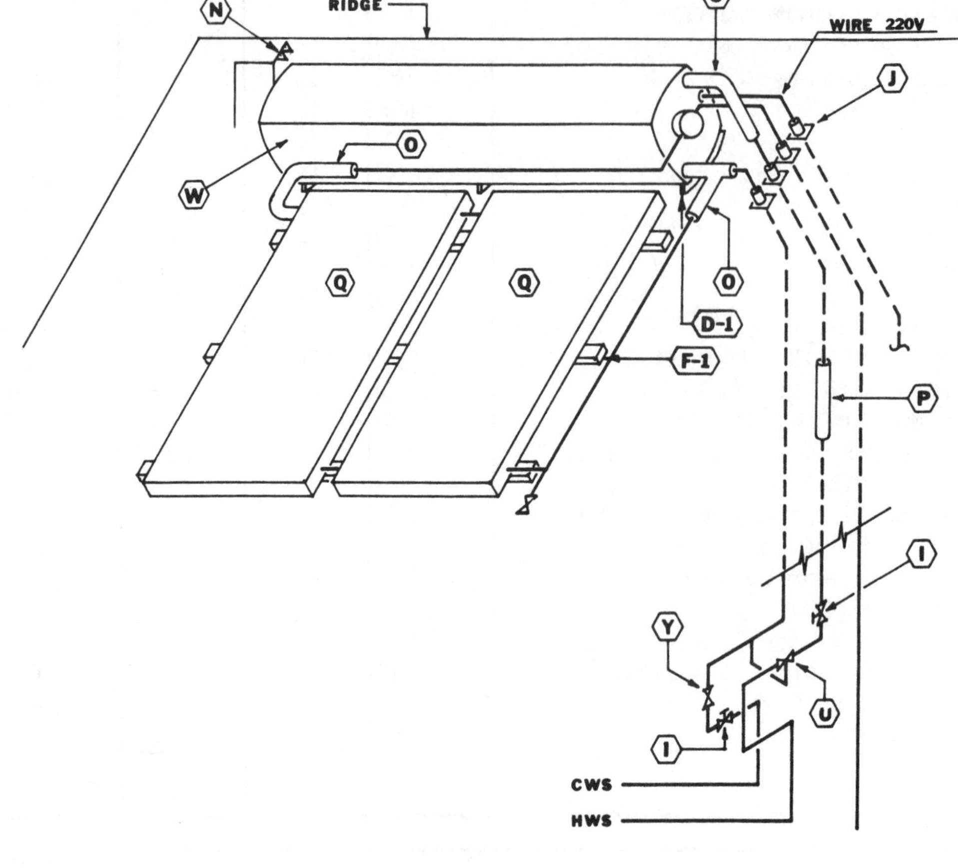

The thermosyphon domestic hot water system, a direct collection system, operates under city water pressure and does not require pumps for system operation. An insulated water storage tank is located above the collectors. As the sun heats the collectors, warm water in them rises by means of natural convection; the colder water in the storage tank flows into the collectors by means of gravity. As long as the sun is shining the water continues to flow through the collectors and to become warmer.

To prevent freezing, the system must be drained or the collectors covered with an insulated lid when the temperature drops below 32°F.

D2020 Domestic Water Distribution

System Components	QUANTITY	UNIT	COST EACH MAT.	COST EACH INST.	COST EACH TOTAL
SYSTEM D2020 285 0960					
SOLAR, THERMOSYPHON, WATER HEATER					
3/4" TUBING, TWO 3'X7' BLACK CHROME COLLECTORS					
D-1 Framing lumber, fir, 2" x 6" x 8', tank cradle	.008	M.B.F.	4.72	5.48	10.20
F-1 Framing lumber, fir, 2" x 4" x 24', sleepers	.016	M.B.F.	8.96	16.40	25.36
I Valve, gate, bronze, 125 lb, soldered 1/2" diam	2.000	Ea.	35.20	41	76.20
J Vent flashing, neoprene	4.000	Ea.	39.20	98	137.20
O Pipe covering, urethane, ultraviolet cover, 1" wall, 1/2" diam	40.000	L.F.	58.40	180	238.40
P Pipe covering fiberglass all service jacket 1" wall 1/2" diam	160.000	L.F.	124.80	561.60	686.40
Q Collector panel solar, blk chrome on copper, 3/16" temp glass 3'-8" x 6'	2.000	Ea.	950	197	1,147
Y Flow control valve, globe, bronze, 125#, soldered, 1/2" diam	1.000	Ea.	43	20.50	63.50
U Valve, water tempering, bronze, sweat connections, 1/2" diam	1.000	Ea.	57.50	20.50	78
W Tank, water storage, solar energy system, 80 Gal, 2" x 1/2 lb insul	1.000	Ea.	920	305	1,225
Copper tubing type L, solder joints, hangers 10' OC 1/2" diam	150.000	L.F.	247.50	907.50	1,155
Copper tubing type M, solder joints, hangers 10' OC 1/2" diam	50.000	L.F.	65	292.50	357.50
Sensor wire, #22-2 gauge multistranded	.500	C.L.F.	5.68	24.25	29.93
Wrought copper fittings & solder, 1/2" diam	75.000	Ea.	39.75	1,837.50	1,877.25
TOTAL			2,599.71	4,507.23	7,106.94

D2020 285	Thermosyphon, Hot Water		MAT.	INST.	TOTAL
0960	Solar, thermosyphon, hot water, two collector system	R13600 -610	2,600	4,500	7,100
0970					

D2020 Domestic Water Distribution

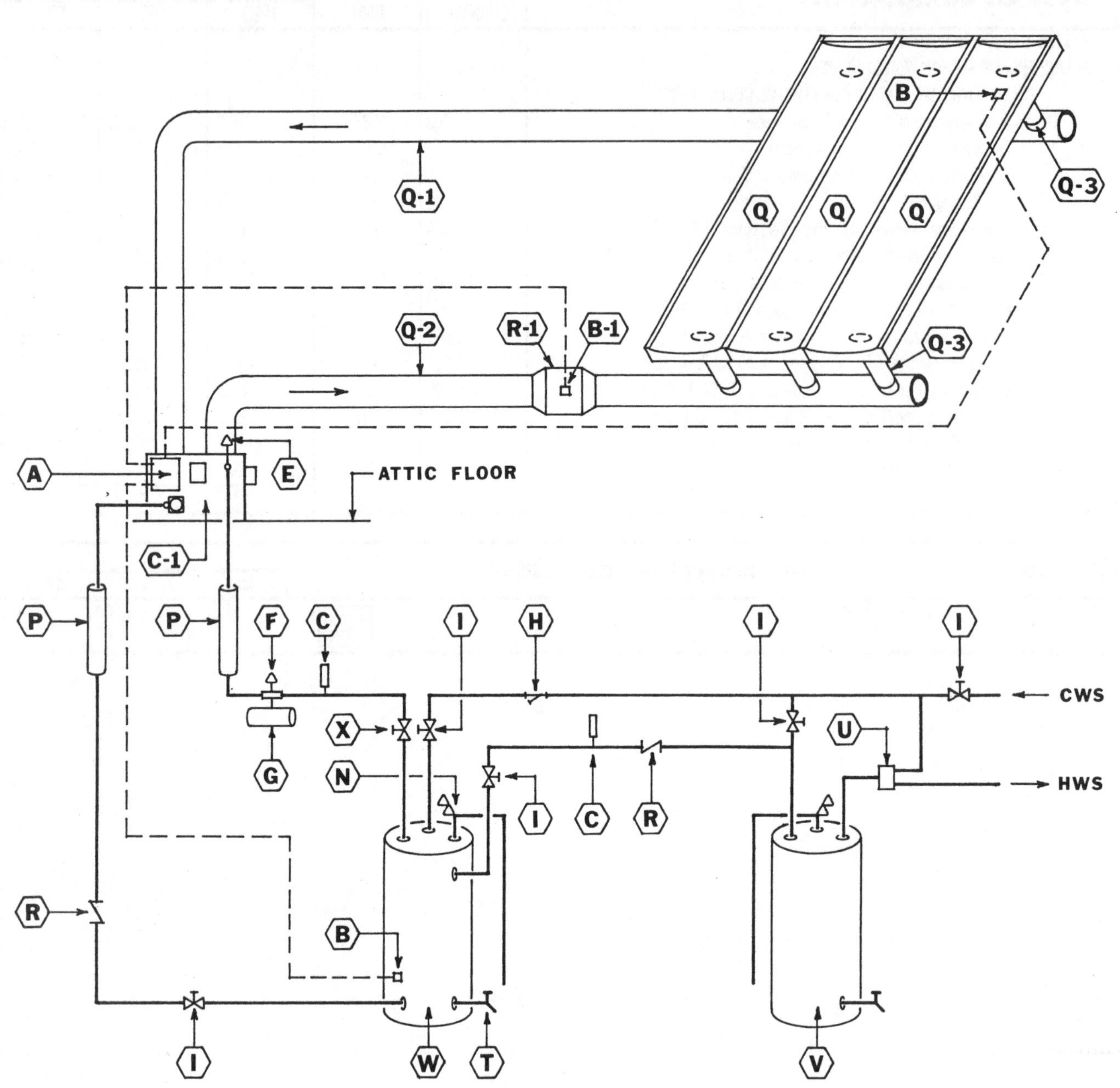

This domestic hot water pre-heat system includes heat exchanger with a circulating pump, blower, air-to-water, coil and controls, mounted in the upper collector manifold. Heat from the hot air coming out of the collectors is transferred through the heat exchanger. For each degree of DHW preheating gained, one degree less heating is needed from the fuel fired water heater. The system is simple, inexpensive to operate and can provide a substantial portion of DHW requirements for modest additional cost.

D2020 Domestic Water Distribution

System Components

System Components		QUANTITY	UNIT	COST EACH MAT.	INST.	TOTAL
	SYSTEM D2020 290 2560					
	SOLAR, HOT WATER, AIR TO WATER HEAT EXCHANGE					
	THREE COLLECTORS, OPTICAL BLACK ON ALUMINUM, 10'X2',80GAL TANK					
A, B	Differential controller, 2 sensors, thermostat, solar energy system	1.000	Ea.	92	41	133
C	Thermometer, 2" dial	2.000	Ea.	44	61	105
C-1	Heat exchanger, air to fluid, up flow 70 MBH	1.000	Ea.	365	253	618
E	Air vent, manual, for solar energy system 1/8" fitting	1.000	Ea.	2.46	15.35	17.81
F	Air purger	1.000	Ea.	47.50	41	88.50
G	Expansion tank	1.000	Ea.	67.50	15.35	82.85
H	Strainer, Y type, bronze body, 1/2" IPS	1.000	Ea.	17.50	24.50	42
I	Valve, gate, bronze, 125 lb, NRS, soldered 1/2" diam	5.000	Ea.	88	102.50	190.50
N	Relief valve, temp & pressure solar 150 psi 210°F self-closing	1.000	Ea.	10	16.35	26.35
P	Pipe covering, fiberglass, all service jacket, 1" wall, 1/2" diam	60.000	L.F.	46.80	210.60	257.40
Q	Collector panel solar energy, air, black on alum plate, flush mount, 10'x2'	30.000	L.F.	1,710	279	1,989
B-1, R-1	Shutter damper wfor solar heater circulator	1.000	Ea.	43	74	117
B-1, R-1	Shutter motor for solar heater circulator	1.000	Ea.	102	55.50	157.50
R	Backflow preventer for solar energy system, 1/2" diam	2.000	Ea.	128	61	189
T	Drain valve, brass, 3/4" connection	1.000	Ea.	5.20	20.50	25.70
U	Valve, water tempering, bronze, sweat connections, 1/2" diam	1.000	Ea.	57.50	20.50	78
W	Tank, water storage, solar energy system, 80 Gal, 2" x 2 lb insul	1.000	Ea.	920	305	1,225
V	Tank, water storage, w/heating element, drain, relief valve, existing	1.000	System			
X	Valve, globe, bronze, 125 lb, rising stem, 1/2" diam	1.000	Ea.	43	20.50	63.50
	Copper tubing type M, solder joints, hangers 10' OC 1/2" diam	50.000	L.F.	65	292.50	357.50
	Copper tubing type L, solder joints, hangers 10' OC 1/2" diam	10.000	L.F.	16.50	60.50	77
	Wrought copper fittings & solder, 1/2" diam	10.000	Ea.	5.30	245	250.30
	Sensor wire, #22-2 conductor multistranded	.500	C.L.F.	5.68	24.25	29.93
Q-1, Q-2	Ductwork, fiberglass, aluminzed jacket, 1-1/2" thick, 8" diam	32.000	S.F.	158.08	158.40	316.48
Q-3	Manifold for flush mount solar energy collector panels, air	6.000	L.F.	177	33.30	210.30
	TOTAL			4,217.02	2,430.60	6,647.62

D2020 290	Solar Hot Water, Air To Water Heat Exchange		COST EACH MAT.	INST.	TOTAL
2550	Solar hot water, air to water heat exchange				
2560	Three collectors, optical black on aluminum, 10' x 2', 80 Gal tank		4,225	2,425	6,650
2580	Four collectors, optical black on aluminum, 10' x 2', 80 Gal tank	R13600 -610	4,900	2,575	7,475
2600	Four collectors, optical black on aluminum, 10' x 2', 120 Gal tank		5,000	2,625	7,625

D2020 Domestic Water Distribution

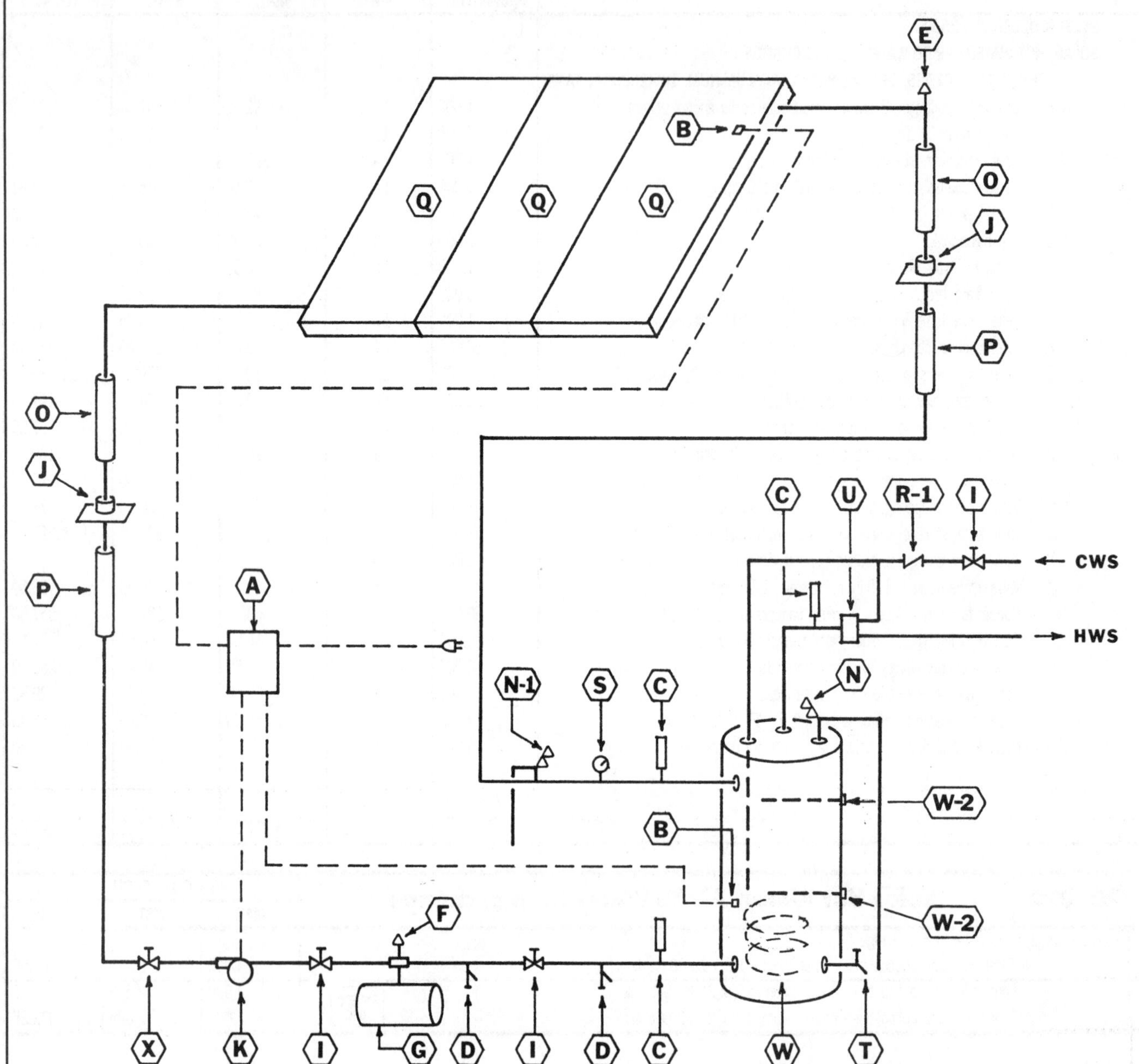

In this closed-loop indirect collection system, fluid with a low freezing temperature, such as propylene glycol, transports heat from the collectors to water storage. The transfer fluid is contained in a closed-loop consisting of collectors, supply and return piping, and a heat exchanger immersed in the storage tank. A typical two-or-three panel system contains 5 to 6 gallons of heat transfer fluid.

When the collectors become approximately 20° F warmer than the storage temperature, a controller activates the circulator. The circulator moves the fluid continuously through the collectors until the temperature difference between the collectors and storage is such that heat collection no longer occurs; at that point, the circulator shuts off. Since the heat transfer fluid has a very low freezing temperature, there is no need for it to be drained from the collectors between periods of collection.

Important: See the Reference Section for critical supporting data - Reference Numbers and City Cost Indexes

D2020 Domestic Water Distribution

System Components	QUANTITY	UNIT	COST EACH MAT.	COST EACH INST.	COST EACH TOTAL
SYSTEM D2020 295 2760					
SOLAR, CLOSED LOOP, HOT WATER SYSTEM, IMMERSED HEAT EXCHANGER					
3/4" TUBING, THREE 3' X 7' BLACK CHROME COLLECTORS					
A, B Differential controller, 2 sensors, thermostat, solar energy system	1.000	Ea.	92	41	133
C Thermometer 2" dial	3.000	Ea.	66	91.50	157.50
D, T Fill & drain valves, brass, 3/4" connection	3.000	Ea.	15.60	61.50	77.10
E Air vent, manual, 1/8" fitting	1.000	Ea.	2.46	15.35	17.81
F Air purger	1.000	Ea.	47.50	41	88.50
G Expansion tank	1.000	Ea.	67.50	15.35	82.85
I Valve, gate, bronze, NRS, soldered 3/4" diam	3.000	Ea.	82.50	73.50	156
J Neoprene vent flashing	2.000	Ea.	19.60	49	68.60
K Circulator, solar heated liquid, 1/25 HP	1.000	Ea.	162	63	225
N-1, N Relief valve, temp & press 150 psi 210°F self-closing 3/4" IPS	2.000	Ea.	20	32.70	52.70
O Pipe covering, urethane ultraviolet cover, 1" wall, 3/4" diam	20.000	L.F.	38.20	91.80	130
P Pipe covering, fiberglass, all service jacket, 1" wall, 3/4" diam	50.000	L.F.	45.50	183.50	229
Roof clamps for solar energy collector panel	3.000	Set	5.31	37.95	43.26
Q Collector panel solar blk chrome on copper, 1/8" temp glass, 3'x7'	3.000	Ea.	1,800	279	2,079
R-1 Valve, swing check, bronze, regrinding disc, 3/4" diam, soldered	1.000	Ea.	38.50	24.50	63
S Pressure gauge, 60 psi, 2-1/2" dial	1.000	Ea.	24.50	15.35	39.85
U Valve, water tempering, bronze, sweat connections, 3/4" diam	1.000	Ea.	70.50	24.50	95
W-2, W Tank, water storage immersed heat exchr elec elem 2"x2# insul 120 Gal	1.000	Ea.	1,025	350	1,375
X Valve, globe, bronze, rising stem, 3/4" diam, soldered	1.000	Ea.	58.50	24.50	83
Copper tubing type L, solder joint, hanger 10' OC 3/4" diam	20.000	L.F.	48.40	129	177.40
Copper tubing, type M, solder joint, hanger 10' OC 3/4" diam	70.000	L.F.	133	441	574
Sensor wire, #22-2 conductor multistranded	.500	C.L.F.	5.68	24.25	29.93
Solar energy heat transfer fluid, propylene glycol, anti-freeze	6.000	Gal.	58.20	105.30	163.50
Wrought copper fittings & solder, 3/4" diam	76.000	Ea.	88.92	1,976	2,064.92
TOTAL			4,015.37	4,190.55	8,205.92

D2020 295	Solar, Closed Loop, Hot Water Systems	COST EACH MAT.	COST EACH INST.	COST EACH TOTAL
2550	Solar, closed loop, hot water system, immersed heat exchanger			
2560	3/8" tubing, 3 ea. 4' x 4'-4" vacuum tube collectors, 80 gal. tank	4,900	3,800	8,700
2580	1/2" tubing, 4 ea. 4' x 4'-4" vacuum tube collectors, 80 gal. tank	5,775	4,075	9,850
2600	120 gal. tank	5,875	4,125	10,000
2640	2 ea. 3'x7' black chrome collectors, 80 gal. tank (R13600-610)	3,150	3,900	7,050
2660	120 gal. tank	3,225	3,900	7,125
2700	2 ea. 3'x7' flat black collectors, 120 gal. tank	2,975	3,900	6,875
2720	3 ea. 3'x7' flat black collectors, 120 gal. tank	3,450	4,025	7,475
2760	3/4" tubing, 3 ea. 3'x7' black chrome collectors, 120 gal. tank	4,025	4,200	8,225
2780	3 ea. 3'x7' flat black collectors, 120 gal. tank	3,650	4,200	7,850
2800	2 ea. 4'x9' flat black w/plastic glazing collectors 120 gal. tank	3,300	4,225	7,525
2840	1" tubing, 4 ea. 2'x9' plastic absorber & glazing collectors 120 gal. tank	4,675	4,675	9,350
2860	4 ea. 3'x7' black chrome collectors, 120 gal. tank	4,875	4,700	9,575
2880	4 ea. 3'x7' flat black absorber collectors, 120 gal. tank	4,375	4,725	9,100

D3010 Energy Supply

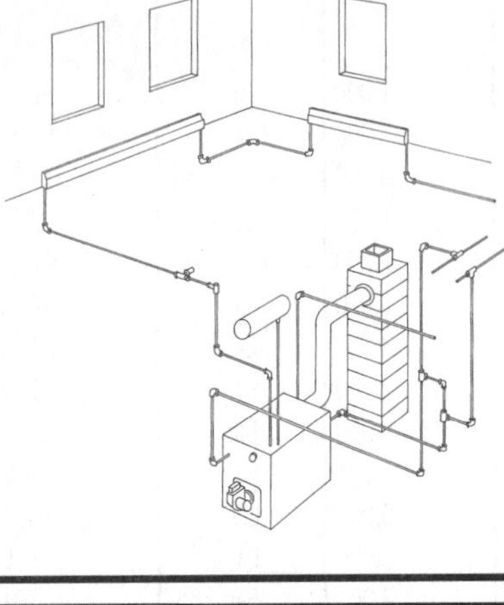

Basis for Heat Loss Estimate, Apartment Type Structures:

1. Masonry walls and flat roof are insulated. U factor is assumed at .08.
2. Window glass area taken as BOCA minimum, 1/10th of floor area. Double insulating glass with 1/4″ air space, U = .65.
3. Infiltration = 0.3 C.F. per hour per S.F. of net wall.
4. Concrete floor loss is 2 BTUH per S.F.
5. Temperature difference taken as 70° F.
6. Ventilating or makeup air has not been included and must be added if desired. Air shafts are not used.

System Components	QUANTITY	UNIT	COST EACH MAT.	COST EACH INST.	COST EACH TOTAL
SYSTEM D3010 510 1760					
HEATING SYSTEM, FIN TUBE RADIATION, FORCED HOT WATER					
1,000 S.F. AREA, 10,000 C.F. VOLUME					
Boiler, oil fired, CI, burner, ctrls/insul/breech/pipe/ftng/valves, 109 MBH	1.000	Ea.	3,718.75	2,756.25	6,475
Circulating pump, CI flange connection, 1/12 HP	1.000	Ea.	234	147	381
Expansion tank, painted steel, ASME 18 Gal capacity	1.000	Ea.	405	63.50	468.50
Storage tank, steel, above ground, 275 Gal capacity w/supports	1.000	Ea.	340	177	517
Copper tubing type L, solder joint, hanger 10′ OC, 3/4″ diam	100.000	L.F.	242	645	887
Radiation, 3/4″ copper tube w/alum fin baseboard pkg, 7″ high	30.000	L.F.	187.50	459	646.50
Pipe covering, calcium silicate w/cover, 1″ wall, 3/4″ diam	100.000	L.F.	240	496	736
TOTAL			5,367.25	4,743.75	10,111
COST PER S.F.			5.37	4.74	10.11

D3010 510	Apartment Building Heating - Fin Tube Radiation		COST PER S.F. MAT.	COST PER S.F. INST.	COST PER S.F. TOTAL
1740	Heating systems, fin tube radiation, forced hot water				
1760	1,000 S.F. area, 10,000 C.F. volume		5.38	4.75	10.13
1800	10,000 S.F. area, 100,000 C.F. volume	R15500 -010	1.70	2.86	4.56
1840	20,000 S.F. area, 200,000 C.F. volume		1.72	3.22	4.94
1880	30,000 S.F. area, 300,000 C.F. volume	R15500 -020	1.61	3.11	4.72
1890					

D3010 Energy Supply

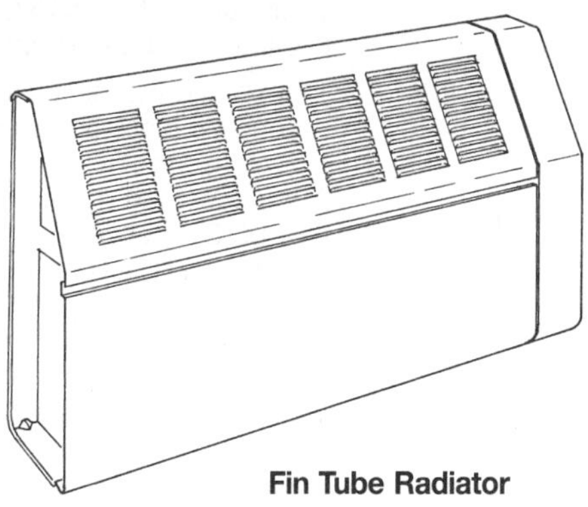

Fin Tube Radiator

Basis for Heat Loss Estimate, Factory or Commercial Type Structures:

1. Walls and flat roof are of lightly insulated concrete block or metal. U factor is assumed at .17.
2. Windows and doors are figured at 1-1/2 S.F. per linear foot of building perimeter. The U factor for flat single glass is 1.13.
3. Infiltration is approximately 5 C.F. per hour per S.F. of wall.
4. Concrete floor loss is 2 BTUH per S.F. of floor.
5. Temperature difference is assumed at 70° F.
6. Ventilation or makeup air has not been included and must be added if desired.

System Components	QUANTITY	UNIT	COST EACH		
			MAT.	INST.	TOTAL
SYSTEM D3010 520 1960					
HEATING SYSTEM, FIN TUBE RADIATION, FORCED HOT WATER					
1,000 S.F. BLDG., ONE FLOOR					
Boiler, oil fired, CI, burner/ctrls/insul/breech/pipe/ftngs/valves, 109 MBH	1.000	Ea.	3,718.75	2,756.25	6,475
Expansion tank, painted steel, ASME 18 Gal capacity	1.000	Ea.	1,575	74	1,649
Storage tank, steel, above ground, 550 Gal capacity w/supports	1.000	Ea.	1,525	330	1,855
Circulating pump, CI flanged, 1/8 HP	1.000	Ea.	390	147	537
Pipe, steel, black, sch. 40, threaded, cplg & hngr 10'OC, 1-1/2" diam.	260.000	L.F.	904.80	2,873	3,777.80
Pipe covering, calcium silicate w/cover , 1" wall, 1-1/2" diam	260.000	L.F.	712.40	1,326	2,038.40
Radiation, steel 1-1/4" tube & 4-1/4" fin w/cover & damper, wall hung	42.000	L.F.	1,176	1,029	2,205
Rough in, steel fin tube radiation w/supply & balance valves	4.000	Set	636	2,160	2,796
TOTAL			10,637.95	10,695.25	21,333.20
COST PER S.F.			10.64	10.70	21.34

D3010 520	Commercial Building Heating - Fin Tube Radiation		COST PER S.F.		
			MAT.	INST.	TOTAL
1940	Heating systems, fin tube radiation, forced hot water				
1960	1,000 S.F. bldg, one floor		10.60	10.70	21.30
2000	10,000 S.F., 100,000 C.F., total two floors	R15500 -010	2.65	4.09	6.74
2040	100,000 S.F., 1,000,000 C.F., total three floors		1.15	1.84	2.99
2080	1,000,000 S.F., 10,000,000 C.F., total five floors	R15500 -020	.56	.98	1.54
2090					

D3010 Energy Supply

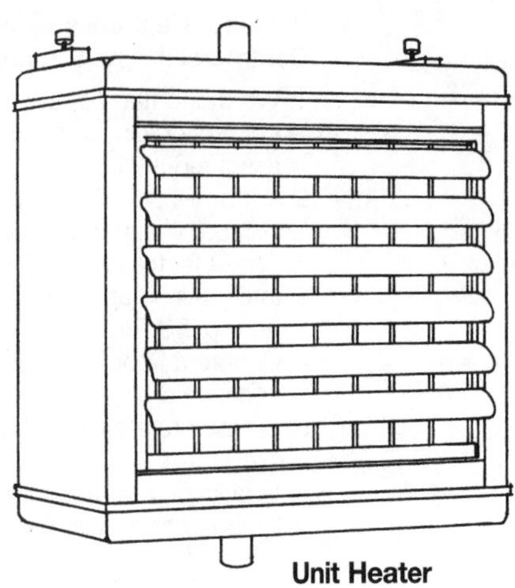
Unit Heater

Basis for Heat Loss Estimate, Factory or Commercial Type Structures:

1. Walls and flat roof are of lightly insulated concrete block or metal. U factor is assumed at .17.
2. Windows and doors are figured at 1-1/2 S.F. per linear foot of building perimeter. The U factor for flat single glass is 1.13.
3. Infiltration is approximately 5 C.F. per hour per S.F. of wall.
4. Concrete floor loss is 2 BTUH per S.F. of floor.
5. Temperature difference is assumed at 70° F.
6. Ventilation or makeup air has not been included and must be added if desired.

System Components	QUANTITY	UNIT	COST EACH		
			MAT.	INST.	TOTAL
SYSTEM D3010 530 1880					
HEATING SYSTEM, TERMINAL UNIT HEATERS, FORCED HOT WATER					
1,000 S.F. BLDG., ONE FLOOR					
Boiler oil fired, CI, burner/ctrls/insul/breech/pipe/ftngs/valves, 109 MBH	1.000	Ea.	3,718.75	2,756.25	6,475
Expansion tank, painted steel, ASME 18 Gal capacity	1.000	Ea.	1,575	74	1,649
Storage tank, steel, above ground, 550 Gal capacity w/supports	1.000	Ea.	1,525	330	1,855
Circulating pump, CI, flanged, 1/8 HP	1.000	Ea.	390	147	537
Pipe, steel, black, schedule 40, threaded, cplg & hngr 10'OC 1-1/2" diam	260.000	L.F.	904.80	2,873	3,777.80
Pipe covering, calcium silicate w/cover, 1" wall, 1-1/2" diam	260.000	L.F.	712.40	1,326	2,038.40
Unit heater, 1 speed propeller, horizontal, 200° EWT, 26.9 MBH	2.000	Ea.	910	222	1,132
Unit heater piping hookup with controls	2.000	Set	770	2,100	2,870
TOTAL			10,505.95	9,828.25	20,334.20
COST PER S.F.			10.51	9.83	20.34

D3010 530	Commercial Bldg. Heating - Terminal Unit Heaters		COST PER S.F.		
			MAT.	INST.	TOTAL
1860	Heating systems, terminal unit heaters, forced hot water				
1880	1,000 S.F. bldg., one floor		10.50	9.83	20.33
1920	10,000 S.F. bldg., 100,000 C.F. total two floors	R15500 -010	2.40	3.39	5.79
1960	100,000 S.F. bldg., 1,000,000 C.F. total three floors		1.19	1.70	2.89
2000	1,000,000 S.F. bldg., 10,000,000 C.F. total five floors	R15500 -020	.76	1.10	1.86
2010					

SERVICES D

Important: See the Reference Section for critical supporting data - Reference Numbers and City Cost Indexes

D3010 Energy Supply

Many styles of active solar energy systems exist. Those shown on the following page represent the majority of systems now being installed in different regions of the country.

The five active domestic hot water (DHW) systems which follow are typically specified as two or three panel systems with additional variations being the type of glazing and size of the storage tanks. Various combinations have been costed for the user's evaluation and comparison. The basic specifications from which the following systems were developed satisfy the construction detail requirements specified in the HUD Intermediate Minimum Property Standards (IMPS). If these standards are not complied with, the renewable energy system's costs could be significantly lower than shown.

To develop the system's specifications and costs it was necessary to make a number of assumptions about the systems. Certain systems are more appropriate to one climatic region than another or the systems may require modifications to be usable in particular locations. Specific instances in which the systems are not appropriate throughout the country as specified, or in which modification will be needed include the following:

- The freeze protection mechanisms provided in the DHW systems vary greatly. In harsh climates, where freeze is a major concern, a closed-loop indirect collection system may be more appropriate than a direct collection system.
- The thermosyphon water heater system described cannot be used when temperatures drop below 32° F.
- In warm climates it may be necessary to modify the systems installed to prevent overheating.

For each renewable resource (solar) system a schematic diagram and descriptive summary of the system is provided along with a list of all the components priced as part of the system. The costs were developed based on these specifications.

Considerations affecting costs which may increase or decrease beyond the estimates presented here include the following:

- Special structural qualities (allowance for earthquake, future expansion, high winds, and unusual spans or shapes);
- Isolated building site or rough terrain that would affect the transportation of personnel, material, or equipment;
- Unusual climatic conditions during the construction process;
- Substitution of other materials or system components for those used in the system specifications.

D3010 Energy Supply

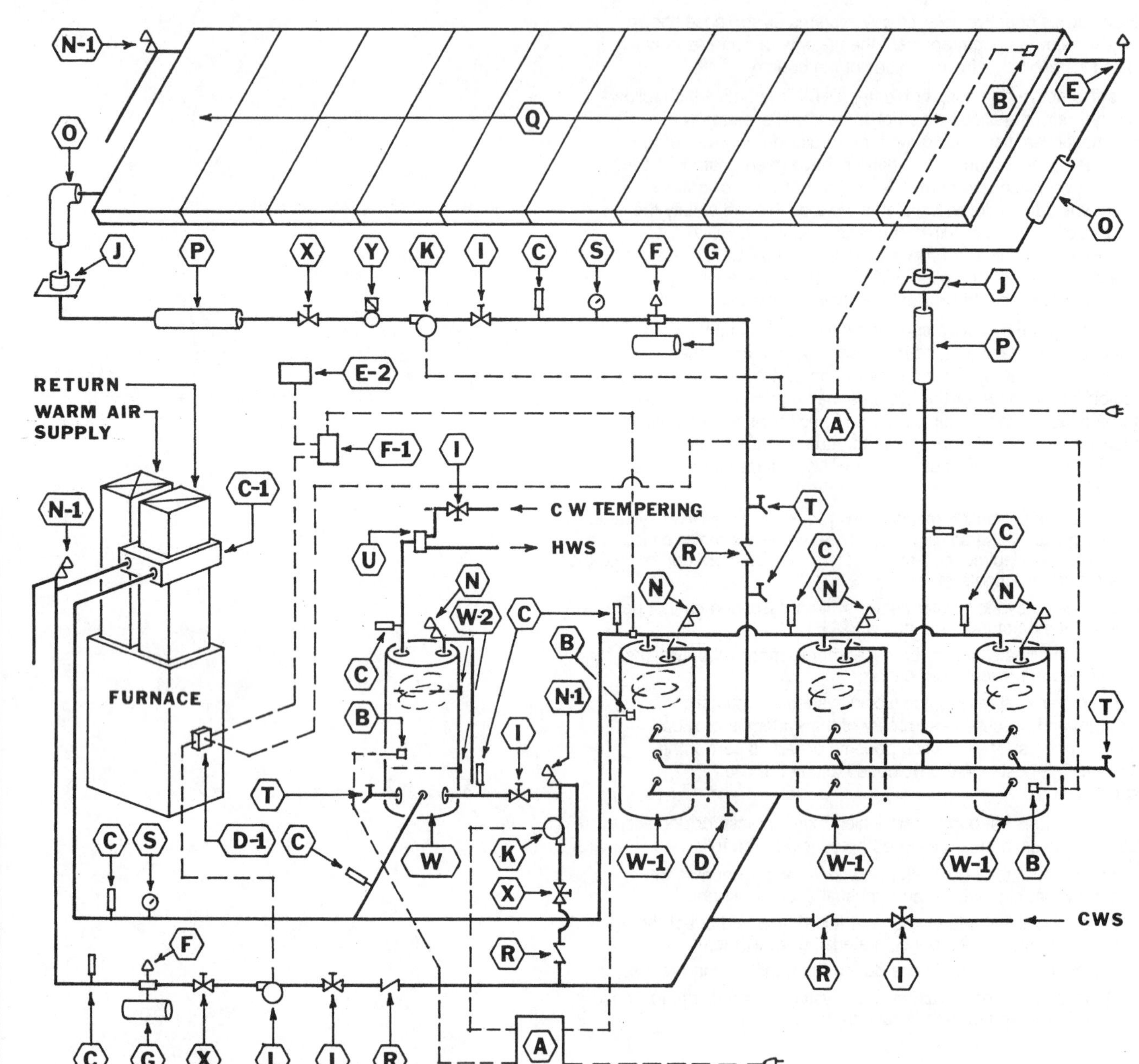

In this closed-loop indirect collection system, fluid with a low freezing temperature, propylene glycol, transports heat from the collectors to water storage. The transfer fluid is contained in a closed loop consisting of collectors, supply and return piping, and a heat exchanger immersed in the storage tank.

When the collectors become approximately 20° F warmer than the storage temperature, the controller activates the circulator. The circulator moves the fluid continuously until the temperature difference between fluid in the collectors and storage is such that the collection will no longer occur and then the circulator turns off. Since the heat transfer fluid has a very low freezing temperature, there is no need for it to be drained from the collectors between periods of collection.

D3010 Energy Supply

System Components	QUANTITY	UNIT	COST EACH MAT.	INST.	TOTAL
SYSTEM D3010 650 2750					
SOLAR, CLOSED LOOP, SPACE/HOT WATER					
1" TUBING, TEN 3'X7' BLK CHROME ON COPPER ABSORBER COLLECTORS					
A, B Differential controller 2 sensors, thermostat, solar energy system	2.000	Ea.	184	82	266
C Thermometer, 2" dial	10.000	Ea.	220	305	525
C-1 Heat exchanger, solar energy system, fluid to air, up flow, 80 MBH	1.000	Ea.	490	295	785
D, T Fill & drain valves, brass, 3/4" connection	5.000	Ea.	26	102.50	128.50
D-1 Fan center	1.000	Ea.	119	108	227
E Air vent, manual, 1/8" fitting	1.000	Ea.	2.46	15.35	17.81
E-2 Thermostat, 2 stage for sensing room temperature	1.000	Ea.	196	62	258
F Air purger	2.000	Ea.	95	82	177
F-1 Controller, liquid temperature, solar energy system	1.000	Ea.	96	98	194
G Expansion tank	2.000	Ea.	135	30.70	165.70
I Valve, gate, bronze, 125 lb, soldered, 1" diam	5.000	Ea.	190	130	320
J Vent flashing, neoprene	2.000	Ea.	19.60	49	68.60
K Circulator, solar heated liquid, 1/25 HP	2.000	Ea.	324	126	450
L Circulator, solar heated liquid, 1/20 HP	1.000	Ea.	170	73.50	243.50
N Relief valve, temp & pressure 150 psi 210°F self-closing	4.000	Ea.	40	65.40	105.40
N-1 Relief valve, pressure poppet, bronze, 30 psi, 3/4" IPS	3.000	Ea.	148.50	52.65	201.15
O Pipe covering, urethane, ultraviolet cover, 1" wall, 1" diam	50.000	L.F.	95.50	229.50	325
P Pipe covering, fiberglass, all service jacket, 1" wall, 1" diam	60.000	L.F.	54.60	220.20	274.80
Q Collector panel solar energy blk chrome on copper 1/8" temp glass 3'x7'	10.000	Ea.	6,000	930	6,930
Roof clamps for solar energy collector panel	10.000	Set	17.70	126.50	144.20
R Valve, swing check, bronze, 125 lb, regrinding disc, 3/4" & 1" diam	4.000	Ea.	228	104	332
S Pressure gage, 0-60 psi, for solar energy system	2.000	Ea.	49	30.70	79.70
U Valve, water tempering, bronze, sweat connections, 3/4" diam	1.000	Ea.	70.50	24.50	95
W-2, W-1, W Tank, water storage immersed heat xchr elec elem 2"x1/2# ins 120 gal	4.000	Ea.	4,100	1,400	5,500
X Valve, globe, bronze, 125 lb, rising stem, 1" diam	3.000	Ea.	274.50	78	352.50
Y Valve, flow control	1.000	Ea.	64.50	22.50	87
Copper tubing type M, solder joint, hanger 10' OC, 1" diam	110.000	L.F.	288.20	770	1,058.20
Copper tubing, type L, solder joint, hanger 10' OC 3/4" diam	20.000	L.F.	48.40	129	177.40
Wrought copper fittings & solder, 3/4" & 1" diam	121.000	Ea.	347.27	3,690.50	4,037.77
Sensor, wire, #22-2 conductor, multistranded	.700	C.L.F.	7.95	33.95	41.90
Ductwork, galvanized steel, for heat exchanger	8.000	Lb.	9.12	47.20	56.32
Solar energy heat transfer fluid propylene glycol, anti-freeze	25.000	Gal.	242.50	438.75	681.25
TOTAL			14,353.30	9,952.40	24,305.70

D3010 650	Solar, Closed Loop, Space/Hot Water Systems	COST EACH MAT.	INST.	TOTAL
2540	Solar, closed loop, space/hot water			
2550	1/2" tubing, 12 ea. 4'x4'4" vacuum tube collectors	19,000	9,300	28,300
2600	3/4" tubing, 12 ea. 4'x4'4" vacuum tube collectors R13600-610	19,600	9,600	29,200
2650	10 ea. 3' x 7' black chrome absorber collectors	14,000	9,325	23,325
2700	10 ea. 3' x 7' flat black absorber collectors	12,700	9,375	22,075
2750	1" tubing, 10 ea. 3' x 7' black chrome absorber collectors	14,400	9,950	24,350
2800	10 ea. 3' x 7' flat black absorber collectors	13,100	10,000	23,100
2850	6 ea. 4' x 9' flat black w/plastic glazing collectors	11,600	9,950	21,550
2900	12 ea. 2' x 9' plastic absorber and glazing collectors	14,900	10,100	25,000

D3010 Energy Supply

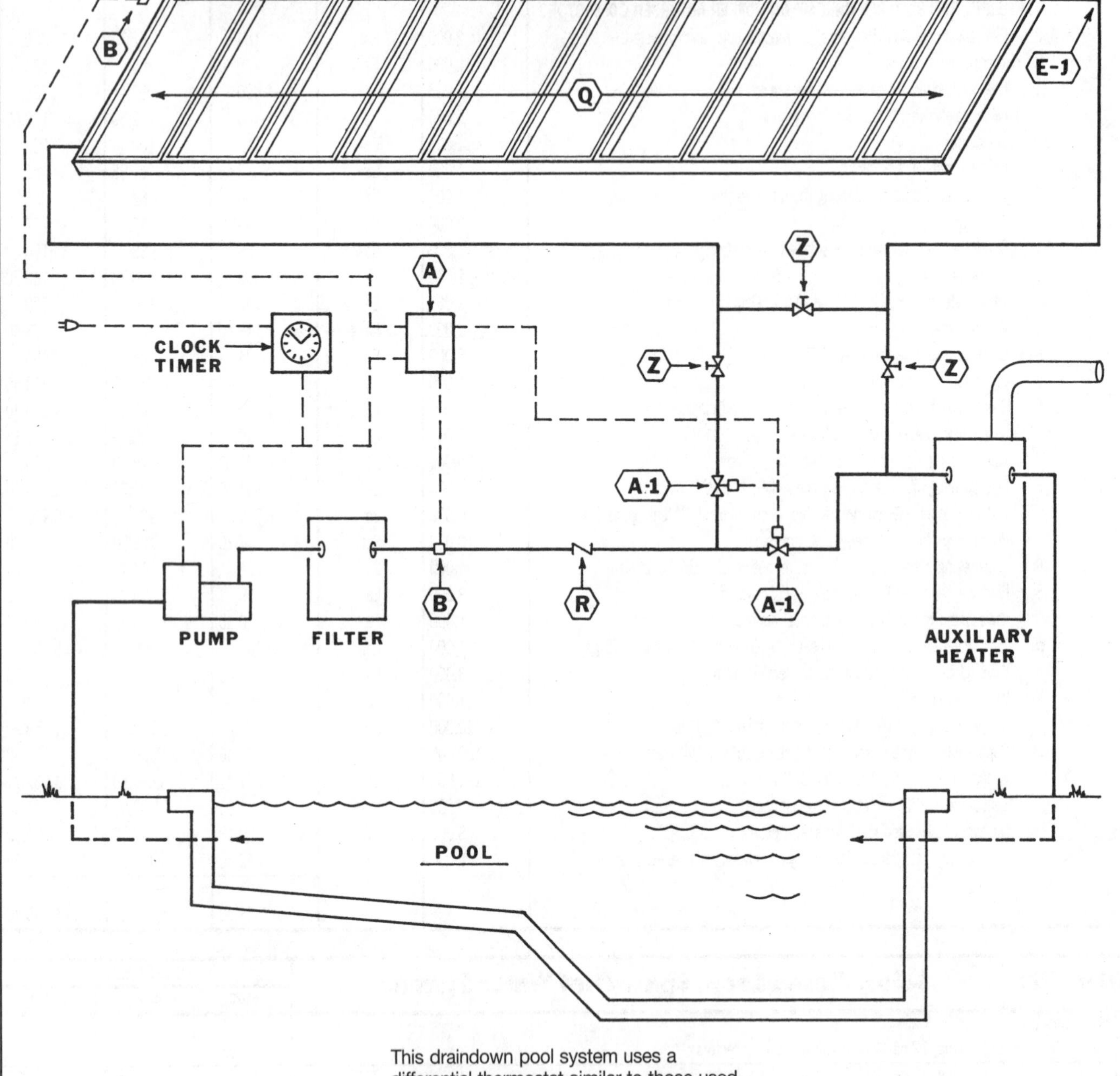

This draindown pool system uses a differential thermostat similar to those used in solar domestic hot water and space heating applications. To heat the pool, the pool water passes through the conventional pump-filter loop and then flows through the collectors. When collection is not possible, or when the pool temperature is reached, all water drains from the solar loop back to the pool through the existing piping. The modes are controlled by solenoid valves or other automatic valves in conjunction with a vacuum breaker relief valve, which facilitates draindown.

D3010 Energy Supply

System Components	QUANTITY	UNIT	COST EACH MAT.	COST EACH INST.	COST EACH TOTAL
SYSTEM D3010 660 2640					
SOLAR SWIMMING POOL HEATER, ROOF MOUNTED COLLECTORS					
TEN 4′ X 10′ FULLY WETTED UNGLAZED PLASTIC ABSORBERS					
A Differential thermostat/controller, 110V, adj pool pump system	1.000	Ea.	305	246	551
A-1 Solenoid valve, PVC, normally 1 open 1 closed (included)	2.000	Ea.			
B Sensor, thermistor type (included)	2.000	Ea.			
E-1 Valve, vacuum relief	1.000	Ea.	29	15.35	44.35
Q Collector panel, solar energy, plastic, liquid full wetted, 4′ x 10′	10.000	Ea.	2,150	1,770	3,920
R Valve, ball check, PVC, socket, 1-1/2″ diam	1.000	Ea.	79	24.50	103.50
Z Valve, ball, PVC, socket, 1-1/2″ diam	3.000	Ea.	187.50	73.50	261
Pipe, PVC, sch 40, 1-1/2″ diam	80.000	L.F.	127.20	1,092	1,219.20
Pipe fittings, PVC sch 40, socket joint, 1-1/2″ diam	10.000	Ea.	11.80	245	256.80
Sensor wire, #22-2 conductor, multistranded	.500	C.L.F.	5.68	24.25	29.93
Roof clamps for solar energy collector panels	10.000	Set	17.70	126.50	144.20
Roof strap, teflon for solar energy collector panels	26.000	L.F.	271.70	62.40	334.10
TOTAL			3,184.58	3,679.50	6,864.08

D3010 660	Solar Swimming Pool Heater Systems	COST EACH MAT.	COST EACH INST.	COST EACH TOTAL
2530	Solar swimming pool heater systems, roof mounted collectors			
2540	10 ea. 3′x7′ black chrome absorber, 1/8″ temp. glass	7,025	2,850	9,875
2560	10 ea. 4′x8′ black chrome absorber, 3/16″ temp. glass (R13600 -610)	8,125	3,375	11,500
2580	10 ea. 3′8″x6′ flat black absorber, 3/16″ temp. glass	5,775	2,900	8,675
2600	10 ea. 4′x9′ flat black absorber, plastic glazing	6,425	3,525	9,950
2620	10 ea. 2′x9′ rubber absorber, plastic glazing	12,000	3,800	15,800
2640	10 ea. 4′x10′ fully wetted unglazed plastic absorber	3,175	3,675	6,850
2660	Ground mounted collectors			
2680	10 ea. 3′x7′ black chrome absorber, 1/8″ temp. glass	7,250	3,275	10,525
2700	10 ea. 4′x8′ black chrome absorber, 3/16″ temp glass	8,350	3,800	12,150
2720	10 ea. 3′8″x6′ flat blk absorber, 3/16″ temp. glass	6,000	3,325	9,325
2740	10 ea. 4′x9′ flat blk absorber, plastic glazing	6,650	3,950	10,600
2760	10 ea. 2′x9′ rubber absorber, plastic glazing	12,100	4,100	16,200
2780	10 ea. 4′x10′ fully wetted unglazed plastic absorber	3,400	4,100	7,500

D3010 Energy Supply

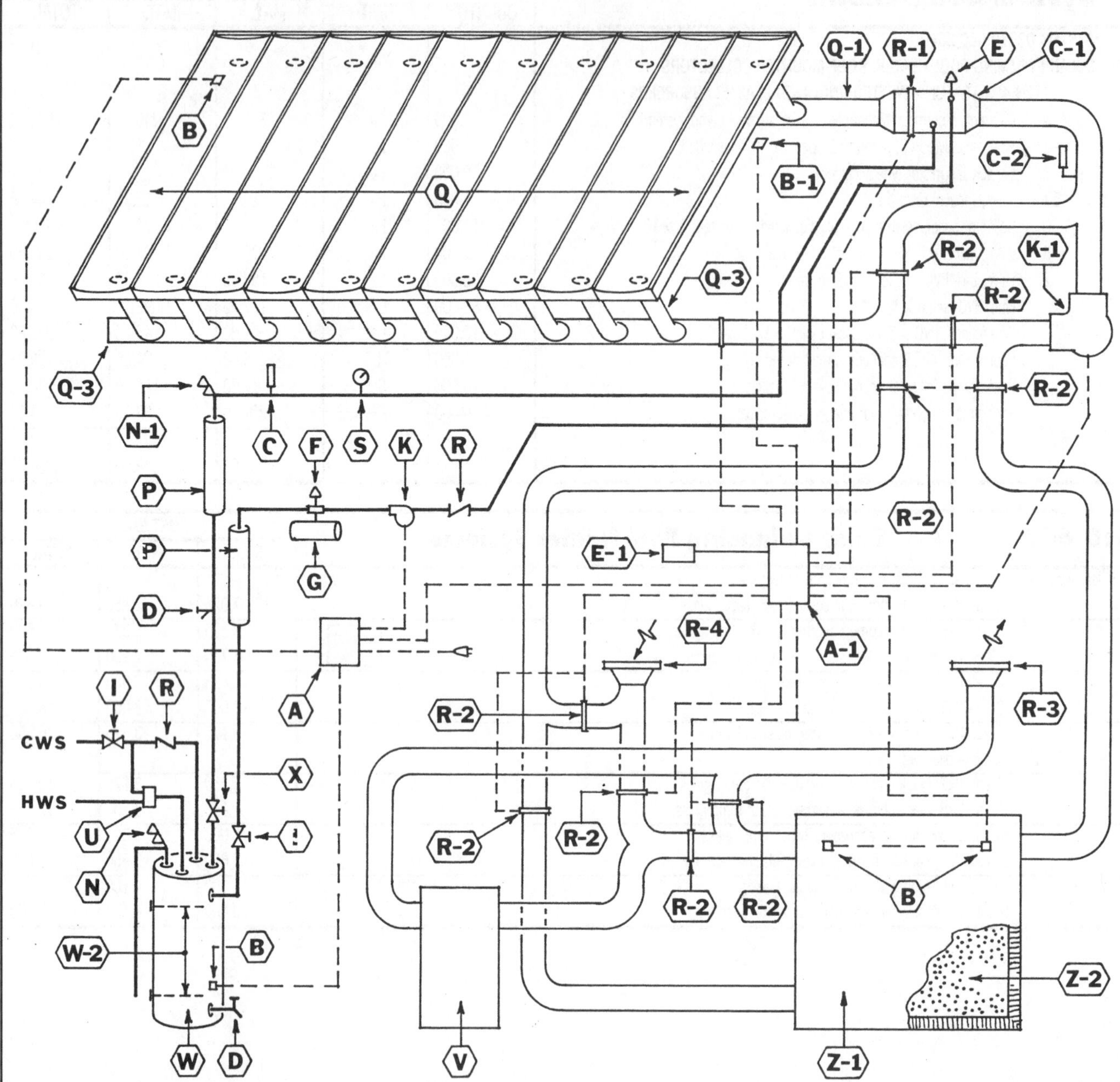

The complete Solar Air Heating System provides maximum savings of conventional fuel with both space heating and year-round domestic hot water heating. It allows for the home air conditioning to operate simultaneously and independently from the solar domestic water heating in summer. The system's modes of operation are:

Mode 1: The building is heated directly from the collectors with air circulated by the Solar Air Mover.

Mode 2: When heat is not needed in the building, the dampers change within the air mover to circulate the air from the collectors to the rock storage bin.

Mode 3: When heat is not available from the collector array and is available in rock storage, the air mover draws heated air from rock storage and directs it into the building. When heat is not available from the collectors or the rock storage bin, the auxiliary heating unit will provide heat for the building. The size of the collector array is typically 25% the size of the main floor area.

D3010 Energy Supply

System Components	QUANTITY	UNIT	COST EACH MAT.	INST.	TOTAL
SYSTEM D3010 675 1210					
SOLAR, SPACE/HOT WATER, AIR TO WATER HEAT EXCHANGE					
A, B Differential controller 2 sensors thermos., solar energy sys liquid loop	1.000	Ea.	92	41	133
A-1, B Differential controller 2 sensors 6 station solar energy sys air loop	1.000	Ea.	239	164	403
B-1 Solar energy sensor, freeze prevention	1.000	Ea.	24	15.35	39.35
C Thermometer for solar energy system, 2″ dial	2.000	Ea.	44	61	105
C-1 Heat exchanger, solar energy system, air to fluid, up flow, 70 MBH	1.000	Ea.	365	253	618
D Drain valve, brass, 3/4″ connection	2.000	Ea.	10.40	41	51.40
E Air vent, manual, for solar energy system 1/8″ fitting	1.000	Ea.	2.46	15.35	17.81
E-1 Thermostat, 2 stage for sensing room temperature	1.000	Ea.	196	62	258
F Air purger	1.000	Ea.	47.50	41	88.50
G Expansion tank, for solar energy system	1.000	Ea.	67.50	15.35	82.85
I Valve, gate, bronze, 125 lb, NRS, soldered 3/4″ diam	2.000	Ea.	55	49	104
K Circulator, solar heated liquid, 1/25 HP	1.000	Ea.	162	63	225
N Relief valve temp & press 150 psi 210°F self-closing, 3/4″ IPS	1.000	Ea.	10	16.35	26.35
N-1 Relief valve, pressure, poppet, bronze, 30 psi, 3/4″ IPS	1.000	Ea.	49.50	17.55	67.05
P Pipe covering, fiberglass, all service jacket, 1″ wall, 3/4″ diam	60.000	L.F.	54.60	220.20	274.80
Q-3 Manifold for flush mount solar energy collector panels	20.000	L.F.	590	111	701
Q Collector panel solar energy, air, black on alum. plate, flush mount 10′x2′	100.000	L.F.	5,700	930	6,630
R Valve, swing check, bronze, 125 lb, regrinding disc, 3/4″ diam	2.000	Ea.	77	49	126
S Pressure gage, 2″ dial, for solar energy system	1.000	Ea.	24.50	15.35	39.85
U Valve, water tempering, bronze, sweat connections, 3/4″ diam	1.000	Ea.	70.50	24.50	95
W-2, W Tank, water storage, solar, elec element 2″x1/2# insul, 80 Gal	1.000	Ea.	920	305	1,225
X Valve, globe, bronze, 125 lb, soldered, 3/4″ diam	1.000	Ea.	58.50	24.50	83
Copper tubing type L, solder joints, hangers 10′ OC 3/4″ diam	10.000	L.F.	24.20	64.50	88.70
Copper tubing type M, solder joints, hangers 10′ OC 3/4″ diam	60.000	L.F.	114	378	492
Wrought copper fittings & solder, 3/4″ diam	26.000	Ea.	30.42	676	706.42
Sensor wire, #22-2 conductor multistranded	1.200	C.L.F.	13.62	58.20	71.82
Q-1 Duct work, rigid fiberglass, rectangular	400.000	S.F.	292	1,584	1,876
Ductwork, transition pieces	8.000	Ea.	118.40	296	414.40
R-2 Shutter/damper for solar heater circulator	9.000	Ea.	387	666	1,053
K-1 Shutter motor for solar heater circulator blower	9.000	Ea.	918	499.50	1,417.50
K-1 Fan, solar energy heated air circulator, space & DHW system	1.000	Ea.	1,625	1,775	3,400
Z-1 Tank, solar energy air storage, 6′-3″H 7′x7′ = 306 CF/2000 Gal	1.000	Ea.	10,400	1,150	11,550
Z-2 Crushed stone 1-1/2″	11.000	C.Y.	330	60.72	390.72
C-2 Themometer, remote probe, 2″ dial	1.000	Ea.	33.50	61.50	95
R-1 Solenoid valve	1.000	Ea.	233	54.50	287.50
V, R-4, R-3 Furnace, supply diffusers, return grilles, existing	1.000	Ea.	10,400	1,150	11,550
TOTAL			33,778.60	11,008.42	44,787.02

D3010 675	Air To Water Heat Exchange		COST EACH MAT.	INST.	TOTAL
1210	Solar, air to water heat exchange, for space / hot water heating	R13600 -610	33,800	11,000	44,800
1220					

D SERVICES

D3020 Heat Generating Systems

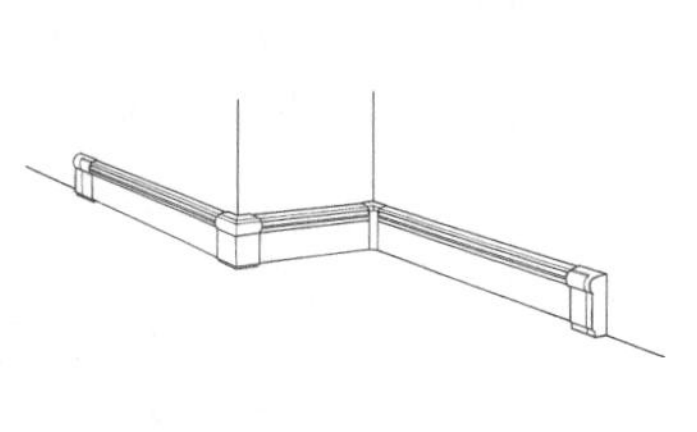

Boiler

Baseboard Radiation

Small Electric Boiler System Considerations:

1. Terminal units are fin tube baseboard radiation rated at 720 BTU/hr with 200° water temperature or 820 BTU/hr steam.
2. Primary use being for residential or smaller supplementary areas, the floor levels are based on 7-1/2′ ceiling heights.
3. All distribution piping is copper for boilers through 205 MBH. All piping for larger systems is steel pipe.

System Components	QUANTITY	UNIT	COST EACH		
			MAT.	INST.	TOTAL
SYSTEM D3020 102 1120					
SMALL HEATING SYSTEM, HYDRONIC, ELECTRIC BOILER					
1,480 S.F., 61 MBH, STEAM, 1 FLOOR					
Boiler, electric steam, std cntrls, trim, ftngs and valves, 18 KW, 61.4 MBH	1.000	Ea.	3,465	1,265	4,730
Copper tubing type L, solder joint, hanger 10′OC, 1-1/4″ diam	160.000	L.F.	748.80	1,352	2,100.80
Radiation, 3/4″ copper tube w/alum fin baseboard pkg 7″ high	60.000	L.F.	375	918	1,293
Rough in baseboard panel or fin tube with valves & traps	10.000	Set	1,180	4,650	5,830
Pipe covering, calcium silicate w/cover 1″ wall 1-1/4″ diam	160.000	L.F.	417.60	816	1,233.60
Low water cut-off, quick hookup, in gage glass tappings	1.000	Ea.	177	31	208
TOTAL			6,363.40	9,032	15,395.40
COST PER S.F.			4.30	6.10	10.40

D3020 102	Small Heating Systems, Hydronic, Electric Boilers		COST PER S.F.		
			MAT.	INST.	TOTAL
1100	Small heating systems, hydronic, electric boilers				
1120	Steam, 1 floor, 1480 S.F., 61 M.B.H.		4.31	6.10	10.41
1160	3,000 S.F., 123 M.B.H.	R15500 -010	3.20	5.35	8.55
1200	5,000 S.F., 205 M.B.H.		2.73	4.92	7.65
1240	2 floors, 12,400 S.F., 512 M.B.H.	R15500 -020	2.17	4.89	7.06
1280	3 floors, 24,800 S.F., 1023 M.B.H.		2.24	4.83	7.07
1320	34,750 S.F., 1,433 M.B.H.	R15500 -030	2.03	4.70	6.73
1360	Hot water, 1 floor, 1,000 S.F., 41 M.B.H.		7.25	3.41	10.66
1400	2,500 S.F., 103 M.B.H.		4.94	6.10	11.04
1440	2 floors, 4,850 S.F., 205 M.B.H.		4.61	7.35	11.96
1480	3 floors, 9,700 S.F., 410 M.B.H.		4.55	7.55	12.10

D3020 Heat Generating Systems

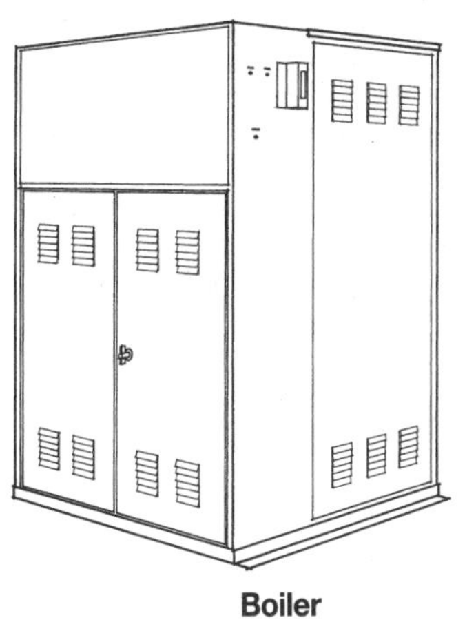

Boiler

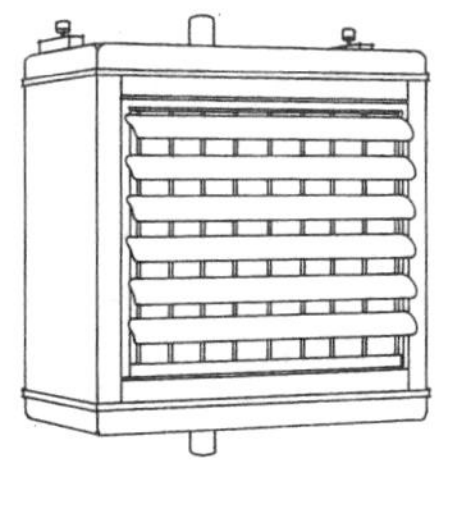

Unit Heater

Large Electric Boiler System Considerations:

1. Terminal units are all unit heaters of the same size. Quantities are varied to accommodate total requirements.
2. All air is circulated through the heaters a minimum of three times per hour.
3. As the capacities are adequate for commercial use, floor levels are based on 10′ ceiling heights.
4. All distribution piping is black steel pipe.

System Components	QUANTITY	UNIT	COST EACH MAT.	COST EACH INST.	COST EACH TOTAL
SYSTEM D3020 104 1240					
LARGE HEATING SYSTEM, HYDRONIC, ELECTRIC BOILER					
9,280 S.F., 135 KW, 461 MBH, 1 FLOOR					
Boiler, electric hot water, std ctrls, trim, ftngs, valves, 135 KW, 461 MBH	1.000	Ea.	8,400	2,737.50	11,137.50
Expansion tank, painted steel, 60 Gal capacity ASME	1.000	Ea.	2,475	148	2,623
Circulating pump, CI, close cpld, 50 GPM, 2 HP, 2″ pipe conn	1.000	Ea.	1,250	295	1,545
Unit heater, 1 speed propeller, horizontal, 200° EWT, 72.7 MBH	7.000	Ea.	4,585	1,127	5,712
Unit heater piping hookup with controls	7.000	Set	2,695	7,350	10,045
Pipe, steel, black, schedule 40, welded, 2-1/2″ diam	380.000	L.F.	2,242	7,862.20	10,104.20
Pipe covering, calcium silicate w/cover, 1″ wall, 2-1/2″ diam	380.000	L.F.	1,124.80	1,995	3,119.80
TOTAL			22,771.80	21,514.70	44,286.50
COST PER S.F.			2.45	2.32	4.77

D3020 104	Large Heating Systems, Hydronic, Electric Boilers		COST PER S.F. MAT.	COST PER S.F. INST.	COST PER S.F. TOTAL
1230	Large heating systems, hydronic, electric boilers				
1240	9,280 S.F., 135 K.W., 461 M.B.H., 1 floor		2.46	2.32	4.78
1280	14,900 S.F., 240 K.W., 820 M.B.H., 2 floors	R15500 -010	3.01	3.80	6.81
1320	18,600 S.F., 296 K.W., 1,010 M.B.H., 3 floors		2.93	4.12	7.05
1360	26,100 S.F., 420 K.W., 1,432 M.B.H., 4 floors	R15500 -020	2.91	4.04	6.95
1400	39,100 S.F., 666 K.W., 2,273 M.B.H., 4 floors		2.52	3.38	5.90
1440	57,700 S.F., 900 K.W., 3,071 M.B.H., 5 floors	R15500 -030	2.37	3.32	5.69
1480	111,700 S.F., 1,800 K.W., 6,148 M.B.H., 6 floors		2.17	2.85	5.02
1520	149,000 S.F., 2,400 K.W., 8,191 M.B.H., 8 floors		2.15	2.84	4.99
1560	223,300 S.F., 3,600 K.W., 12,283 M.B.H., 14 floors		2.26	3.23	5.49

D3020 Heat Generating Systems

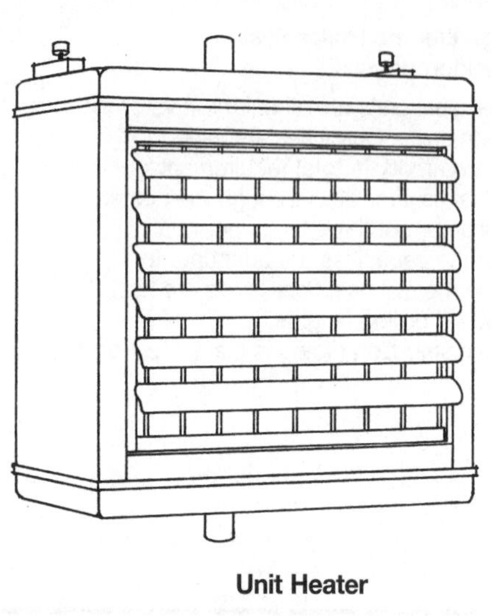

Unit Heater

Fossil Fuel Boiler System Considerations:

1. Terminal units are horizontal unit heaters. Quantities are varied to accommodate total heat loss per building.
2. Unit heater selection was determined by their capacity to circulate the building volume a minimum of three times per hour in addition to the BTU output.
3. Systems shown are forced hot water. Steam boilers cost slightly more than hot water boilers. However, this is compensated for by the smaller size or fewer terminal units required with steam.
4. Floor levels are based on 10′ story heights.
5. MBH requirements are gross boiler output.

System Components	QUANTITY	UNIT	COST EACH MAT.	COST EACH INST.	COST EACH TOTAL
SYSTEM D3020 108 1280					
HEATING SYSTEM, HYDRONIC, FOSSIL FUEL, TERMINAL UNIT HEATERS					
CAST IRON BOILER, GAS, 80 MBH, 1,070 S.F. BUILDING					
Boiler, gas, hot water, CI, burner, controls, insulation, breeching, 80 MBH	1.000	Ea.	1,785	1,338.75	3,123.75
Pipe, steel, black, schedule 40, threaded, cplg & hngr 10′OC, 2″ diam	200.000	L.F.	910	2,760	3,670
Unit heater, 1 speed propeller, horizontal, 200° EWT, 72.7 MBH	2.000	Ea.	1,310	322	1,632
Unit heater piping hookup with controls	2.000	Set	770	2,100	2,870
Expansion tank, painted steel, ASME, 18 Gal capacity	1.000	Ea.	1,575	74	1,649
Circulating pump, CI, flange connection, 1/12 HP	1.000	Ea.	234	147	381
Pipe covering, calcium silicate w/cover, 1″ wall, 2″ diam	200.000	L.F.	556	1,050	1,606
TOTAL			7,140	7,791.75	14,931.75
COST PER S.F.			6.67	7.28	13.95

D3020 108	Heating Systems, Unit Heaters		COST PER S.F. MAT.	COST PER S.F. INST.	COST PER S.F. TOTAL
1260	Heating systems, hydronic, fossil fuel, terminal unit heaters,				
1280	Cast iron boiler, gas, 80 M.B.H., 1,070 S.F. bldg.		6.68	7.29	13.97
1320	163 M.B.H., 2,140 S.F. bldg.	R15500-010	4.40	4.94	9.34
1360	544 M.B.H., 7,250 S.F. bldg.		3.12	3.56	6.68
1400	1,088 M.B.H., 14,500 S.F. bldg.	R15500-020	2.43	3.37	5.80
1440	3,264 M.B.H., 43,500 S.F. bldg.		2.06	2.53	4.59
1480	5,032 M.B.H., 67,100 S.F. bldg.	R15500-030	2.62	2.62	5.24
1520	Oil, 109 M.B.H., 1,420 S.F. bldg.		6.85	6.50	13.35
1560	235 M.B.H., 3,150 S.F. bldg.		4.38	4.70	9.08
1600	940 M.B.H., 12,500 S.F. bldg.		3.14	3.08	6.22
1640	1,600 M.B.H., 21,300 S.F. bldg.		3	2.95	5.95
1680	2,480 M.B.H., 33,100 S.F. bldg.		3.04	2.66	5.70
1720	3,350 M.B.H., 44,500 S.F. bldg.		2.60	2.72	5.32
1800	300 M.B.H., 4,000 S.F. bldg.		4.13	3.32	7.45
1840	2,360 M.B.H., 31,500 S.F. bldg.		2.87	2.72	5.59
1880	Steel boiler, gas, 72 M.B.H., 1,020 S.F. bldg.		6.35	5.40	11.75
1920	240 M.B.H., 3,200 S.F. bldg.		4.30	4.30	8.60
1960	480 M.B.H., 6,400 S.F. bldg.		3.52	3.19	6.71
2000	800 M.B.H., 10,700 S.F. bldg.		3.02	2.84	5.86
2040	1,960 M.B.H., 26,100 S.F. bldg.		2.69	2.57	5.26
2080	3,000 M.B.H., 40,000 S.F. bldg.		2.67	2.61	5.28

D3020 Heat Generating Systems

D3020 108	Heating Systems, Unit Heaters	COST PER S.F.		
		MAT.	INST.	TOTAL
2120	Oil, 97 M.B.H., 1,300 S.F. bldg.	5.80	6.15	11.95
2160	315 M.B.H., 4,550 S.F. bldg.	3.68	3.13	6.81
2200	525 M.B.H., 7,000 S.F. bldg.	4.10	3.18	7.28
2240	1,050 M.B.H., 14,000 S.F. bldg.	3.41	3.21	6.62
2280	2,310 M.B.H. 30,800 S.F. bldg.	3	2.79	5.79
2320	3,150 M.B.H., 42,000 S.F. bldg.	2.71	2.82	5.53

D3020 Heat Generating Systems

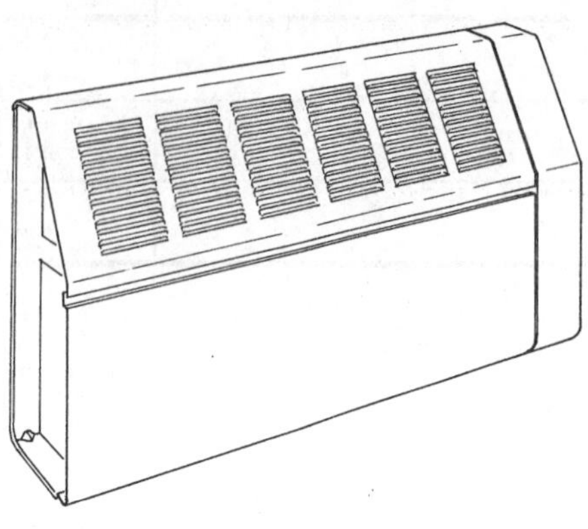

Fin Tube Radiator

Fossil Fuel Boiler System Considerations:

1. Terminal units are commercial steel fin tube radiation. Quantities are varied to accommodate total heat loss per building.
2. Systems shown are forced hot water. Steam boilers cost slightly more than hot water boilers. However, this is compensated for by the smaller size or fewer terminal units required with steam.
3. Floor levels are based on 10′ story heights.
4. MBH requirements are gross boiler output.

System Components	QUANTITY	UNIT	COST EACH		
			MAT.	INST.	TOTAL
SYSTEM D3020 110 3240					
HEATING SYSTEM, HYDRONIC, FOSSIL FUEL, FIN TUBE RADIATION					
CAST IRON BOILER, GAS, 80 MBH, 1,070 S.F. BLDG.					
Boiler, gas, hot water, CI, burner, controls, insulation, breeching, 80 MBH	1.000	Ea.	1,785	1,338.75	3,123.75
Pipe, steel, black, schedule 40, threaded, 2″ diam	200.000	L.F.	910	2,760	3,670
Radiation,steel 1-1/4″ tube & 4-1/4″ fin w/cover & damper, wall hung	80.000	L.F.	2,240	1,960	4,200
Rough-in, wall hung steel fin radiation with supply & balance valves	8.000	Set	1,272	4,320	5,592
Expansion tank, painted steel, ASME, 18 Gal capacity	1.000	Ea.	1,575	74	1,649
Circulating pump, CI flanged, 1/12 HP	1.000	Ea.	234	147	381
Pipe covering, calcium silicate w/cover, 1″ wall, 2″ diam	200.000	L.F.	556	1,050	1,606
TOTAL			8,572	11,649.75	20,221.75
COST PER S.F.			8.01	10.89	18.90

D3020 110	Heating System, Fin Tube Radiation		COST PER S.F.		
			MAT.	INST.	TOTAL
3230	Heating systems, hydronic, fossil fuel, fin tube radiation				
3240	Cast iron boiler, gas, 80 MBH, 1,070 S.F. bldg.		8.01	10.84	18.85
3280	169 M.B.H., 2,140 S.F. bldg.	R15500 -010	4.98	6.90	11.88
3320	544 M.B.H., 7,250 S.F. bldg.		4.11	6	10.11
3360	1,088 M.B.H., 14,500 S.F. bldg.	R15500 -020	3.48	5.90	9.38
3400	3,264 M.B.H., 43,500 S.F. bldg.		3.21	5.15	8.36
3440	5,032 M.B.H., 67,100 S.F. bldg.	R15500 -030	3.79	5.20	8.99
3480	Oil, 109 M.B.H., 1,420 S.F. bldg.		9.30	11.90	21.20
3520	235 M.B.H., 3,150 S.F. bldg.		5.35	7.10	12.45
3560	940 M.B.H., 12,500 S.F. bldg.		4.23	5.60	9.83
3600	1,600 M.B.H., 21,300 S.F. bldg.		4.17	5.55	9.72
3640	2,480 M.B.H., 33,100 S.F. bldg.		4.22	5.30	9.52
3680	3,350 M.B.H., 44,500 S.F. bldg.		3.77	5.30	9.07
3720	Coal, 148 M.B.H., 1,975 S.F. bldg.		6.35	6.60	12.95
3760	300 M.B.H., 4,000 S.F. bldg.		5.05	5.65	10.70
3800	2,360 M.B.H., 31,500 S.F. bldg.		4	5.25	9.25
3840	Steel boiler, gas, 72 M.B.H., 1,020 S.F. bldg.		8.75	10.35	19.10
3880	240 M.B.H., 3,200 S.F. bldg.		5.35	6.80	12.15
3920	480 M.B.H., 6,400 S.F. bldg.		4.54	5.60	10.14
3960	800 M.B.H., 10,700 S.F. bldg.		4.43	5.95	10.38
4000	1,960 M.B.H., 26,100 S.F. bldg.		3.82	5.15	8.97
4040	3,000 M.B.H., 40,000 S.F. bldg.		3.77	5.15	8.92
4080	Oil, 97 M.B.H., 1,300 S.F. bldg.		6.85	9.80	16.65
4120	315 M.B.H., 4,550 S.F. bldg.		4.60	5.35	9.95

Important: See the Reference Section for critical supporting data - Reference Numbers and City Cost Indexes

D3020 Heat Generating Systems

D3020 110	Heating System, Fin Tube Radiation	COST PER S.F.		
		MAT.	INST.	TOTAL
4160	525 M.B.H., 7,000 S.F. bldg.	5.20	5.70	10.90
4200	1,050 M.B.H., 14,000 S.F. bldg.	4.52	5.75	10.27
4240	2,310 M.B.H., 30,800 S.F. bldg.	4.11	5.30	9.41
4280	3,150 M.B.H., 42,000 S.F. bldg.	3.83	5.40	9.23

D3020 Heat Generating Systems

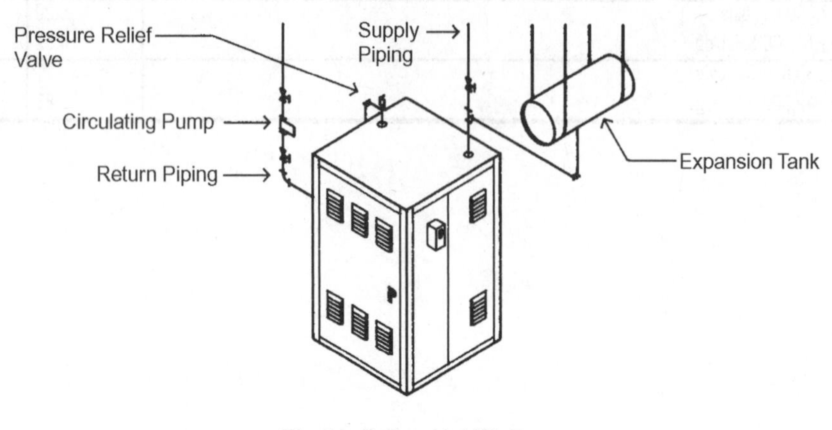

Electric Boiler, Hot Water

System Components	QUANTITY	UNIT	COST EACH MAT.	COST EACH INST.	COST EACH TOTAL
SYSTEM D3020 126 1010					
BOILER, ELECTRIC, HOT WATER, 15 kW, 52 MBH					
Boilers, electric, ASME, hot water, 15 KW, 52 MBH	1.000	Ea.	3,225	1,050	4,275
Pipe, black steel, Sch 40, threaded, W/coupling & hangers, 10′ OC, 3/4″ dia	10.000	L.F.	19.44	92.58	112.02
Pipe, black steel, Sch 40, threaded, W/coupling & hangers, 10′ OC, 1″ dia	20.000	L.F.	51.17	198.88	250.05
Elbow, 90°, black, straight, 3/4″ dia.	6.000	Ea.	11.04	210	221.04
Elbow, 90°, black, straight, 1″ dia.	6.000	Ea.	19.14	228	247.14
Tee, black, straight, 3/4″ dia.	1.000	Ea.	2.92	54.50	57.42
Tee, black, straight, 1″ dia.	2.000	Ea.	9.94	123	132.94
Tee, black, reducing, 1″ dia.	2.000	Ea.	11.40	123	134.40
Union, black with brass seat, 3/4″ dia.	1.000	Ea.	7.70	38	45.70
Union, black with brass seat, 1″ dia.	2.000	Ea.	22	82	104
Pipe nipples, 3/4″ diam.	3.000	Ea.	2.54	12.08	14.62
Pipe nipples, 1″ diam.	3.000	Ea.	3.57	13.88	17.45
Valves, bronze, gate, N.R.S., threaded, class 150, 3/4″ size	2.000	Ea.	70	49	119
Valves, bronze, gate, N.R.S., threaded, class 150, 1″ size	2.000	Ea.	92	52	144
Thermometer, stem type, 9″ case, 8″ stem, 3/4″ NPT	1.000	Ea.	260	35.20	295.20
Tank, steel, liquid expansion, ASME, painted, 15 gallon capacity	1.000	Ea.	400	52	452
Pump, circulating, bronze, flange connection, 3/4″ to 1-1/2″ size, 1/8 HP	1.000	Ea.	615	147	762
Insulation, fiberglass pipe covering, 1″ wall, 1″ IPS	20.000	L.F.	18.20	76.60	94.80
TOTAL			4,841.06	2,637.72	7,478.78

D3020 126	Electric Boiler, Hot Water	COST EACH MAT.	COST EACH INST.	COST EACH TOTAL
1010	Boiler, electric, hot water, 15 kW, 52 MBH	4,850	2,650	7,500
1020	30 kW, 103 MBH	5,475	2,800	8,275
1030	60 kW, 205 MBH	5,550	2,875	8,425
1040	120 kW, 410 MBH	7,325	3,450	10,775
1050	150 kW, 510 MBH	11,100	6,750	17,850
1060	210 kW, 716 MBH	13,200	7,400	20,600
1070	296 kW, 1010 MBH	17,600	7,625	25,225
1080	444 kW, 1515 MBH	21,600	10,300	31,900
1090	666 kW, 2273 MBH	25,900	11,100	37,000
1100	900 kW, 3071 MBH	31,700	12,900	44,600
1110	1320 kW, 4505 MBH	38,000	13,500	51,500
1120	2100 kW, 7167 MBH	58,000	16,400	74,400

Important: See the Reference Section for critical supporting data - Reference Numbers and City Cost Indexes

D3020 Heat Generating Systems

D3020 126	Electric Boiler, Hot Water	COST EACH		
		MAT.	INST.	TOTAL
1130	2610 kW, 8905 MBH	71,500	19,800	91,300
1140	3600 kW, 12,283 MBH	88,000	23,600	111,600

D3020 128	Electric Boiler, Steam	COST EACH		
		MAT.	INST.	TOTAL
1010	Boiler, electric, steam, 18 kW, 61.4 MBH	3,800	2,550	6,350
1020	36 kW, 123 MBH	4,750	2,700	7,450
1030	60 kW, 205 MBH	11,500	5,475	16,975
1040	120 kW, 409 MBH	13,400	6,000	19,400
1050	150 kW, 512 MBH	15,700	9,175	24,875
1060	210 kW, 716 MBH	20,400	9,850	30,250
1070	300 kW, 1023 MBH	23,800	10,800	34,600
1080	510 kW, 1740 MBH	29,500	14,000	43,500
1090	720 kW, 2456 MBH	33,200	14,700	47,900
1100	1080 kW, 3685 MBH	38,800	17,400	56,200
1110	1260 kW, 4300 MBH	40,700	18,500	59,200
1120	1620 kW, 5527 MBH	42,800	21,000	63,800
1130	2070 kW, 7063 MBH	56,500	22,000	78,500
1140	2340 kW, 7984 MBH	73,000	24,900	97,900

D3020 Heat Generating Systems

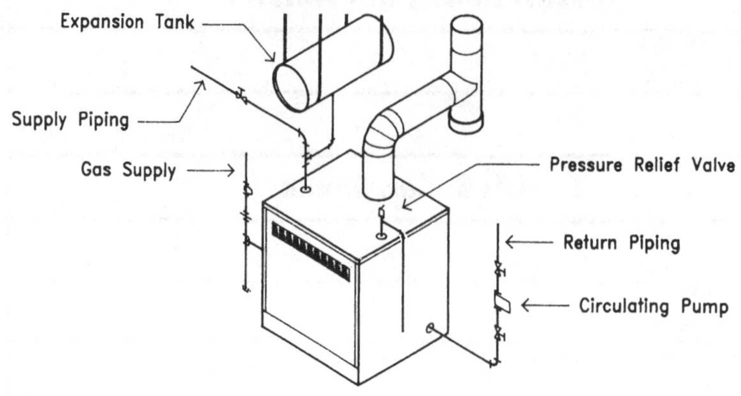

Cast Iron Boiler, Hot Water, Gas Fired

System Components	QUANTITY	UNIT	COST EACH MAT.	COST EACH INST.	COST EACH TOTAL
SYSTEM D3020 130 1010					
BOILER, CAST IRON, GAS, HOT WATER, 100 MBH					
Boilers, gas fired, std controls, CI, insulated, HW, gross output 100 MBH	1.000	Ea.	1,925	1,400	3,325
Pipe, black steel, Sch 40, threaded, W/coupling & hangers, 10′ OC, 3/4″ dia	20.000	L.F.	38.03	181.13	219.16
Pipe, black steel, Sch 40, threaded, W/coupling & hangers, 10′ OC, 1″ dia.	20.000	L.F.	51.17	198.88	250.05
Elbow, 90°, black, straight, 3/4″ dia.	9.000	Ea.	16.56	315	331.56
Elbow, 90°, black, straight, 1″ dia.	6.000	Ea.	19.14	228	247.14
Tee, black, straight, 3/4″ dia.	2.000	Ea.	5.84	109	114.84
Tee, black, straight, 1″ dia.	2.000	Ea.	9.94	123	132.94
Tee, black, reducing, 1″ dia.	2.000	Ea.	11.40	123	134.40
Pipe cap, black, 3/4″ dia.	1.000	Ea.	2.07	15.35	17.42
Union, black with brass seat, 3/4″ dia.	2.000	Ea.	15.40	76	91.40
Union, black with brass seat, 1″ dia.	2.000	Ea.	22	82	104
Pipe nipples, black, 3/4″ dia.	5.000	Ea.	4.23	20.13	24.36
Pipe nipples, black, 1″ dia.	3.000	Ea.	3.57	13.88	17.45
Valves, bronze, gate, N.R.S., threaded, class 150, 3/4″ size	2.000	Ea.	70	49	119
Valves, bronze, gate, N.R.S., threaded, class 150, 1″ size	2.000	Ea.	92	52	144
Gas cock, brass, 3/4″ size	1.000	Ea.	11.90	22.50	34.40
Thermometer, stem type, 9″ case, 8″ stem, 3/4″ NPT	2.000	Ea.	260	35.20	295.20
Tank, steel, liquid expansion, ASME, painted, 15 gallon capacity	1.000	Ea.	400	52	452
Pump, circulating, bronze, flange connection, 3/4″ to 1-1/2″ size, 1/8 HP	1.000	Ea.	615	147	762
Vent chimney, all fuel, pressure tight, double wall, SS, 6″ dia.	20.000	L.F.	800	297	1,097
Vent chimney, elbow, 90° fixed, 6″ dia.	2.000	Ea.	454	59	513
Vent chimney, Tee, 6″ dia.	2.000	Ea.	300	74	374
Vent chimney, ventilated roof thimble, 6″ dia.	1.000	Ea.	201	34.50	235.50
Vent chimney, adjustable roof flashing, 6″ dia.	1.000	Ea.	54	29.50	83.50
Vent chimney, stack cap, 6″ diameter	1.000	Ea.	158	19.35	177.35
Insulation, fiberglass pipe covering, 1″ wall, 1″ IPS	20.000	L.F.	18.20	76.60	94.80
TOTAL			5,558.45	3,833.02	9,391.47

D3020 130	Boiler, Cast Iron, Hot Water, Gas	COST EACH MAT.	COST EACH INST.	COST EACH TOTAL
1010	Boiler, cast iron, gas, hot water, 100 MBH	5,550	3,825	9,375
1020	200 MBH	6,150	4,375	10,525
1030	320 MBH	6,400	4,750	11,150
1040	440 MBH	8,550	6,125	14,675
1050	544 MBH	14,800	9,725	24,525
1060	765 MBH	17,500	10,400	27,900

Important: See the Reference Section for critical supporting data - Reference Numbers and City Cost Indexes

D3020 Heat Generating Systems

D3020 130	Boiler, Cast Iron, Hot Water, Gas	COST EACH		
		MAT.	INST.	TOTAL
1070	1088 MBH	17,900	11,100	29,000
1080	1530 MBH	22,600	14,800	37,400
1090	2312 MBH	27,300	17,300	44,600
1100	2856 MBH	34,400	18,600	53,000
1110	3808 MBH	37,700	21,800	59,500
1120	4720 MBH	71,500	24,400	95,900
1130	6100 MBH	79,500	27,200	106,700
1140	6970 MBH	87,000	33,400	120,400

D3020 134	Boiler, Cast Iron, Steam, Gas	COST EACH		
		MAT.	INST.	TOTAL
1010	Boiler, cast iron, gas, steam, 100 MBH	4,500	3,700	8,200
1020	200 MBH	11,000	6,950	17,950
1030	320 MBH	12,100	7,800	19,900
1040	544 MBH	17,900	12,400	30,300
1050	765 MBH	20,200	12,900	33,100
1060	1275 MBH	23,100	14,100	37,200
1070	2675 MBH	30,600	20,600	51,200
1080	3570 MBH	36,100	23,100	59,200
1090	4720 MBH	40,500	24,600	65,100
1100	6100 MBH	79,000	28,500	107,500
1110	6970 MBH	87,000	33,300	120,300

D3020 136	Boiler, Cast Iron, Hot Water, Gas/Oil	COST EACH		
		MAT.	INST.	TOTAL
1010	Boiler, cast iron, gas & oil, hot water, 584 MBH	17,700	10,800	28,500
1020	876 MBH	22,000	11,500	33,500
1030	1168 MBH	22,800	13,000	35,800
1040	1460 MBH	27,300	15,700	43,000
1050	2044 MBH	31,300	16,300	47,600
1060	2628 MBH	39,500	19,700	59,200
1070	3210 MBH	44,900	21,200	66,100
1080	3796 MBH	48,400	21,800	70,200
1090	4672 MBH	56,000	22,900	78,900
1100	5256 MBH	68,500	27,500	96,000
1110	6000 MBH	81,500	28,000	109,500
1120	9800 MBH	102,500	47,700	150,200
1130	12,200 MBH	127,500	55,500	183,000
1140	13,500 MBH	138,500	59,000	197,500

D3020 138	Boiler, Cast Iron, Steam, Gas/Oil	COST EACH		
		MAT.	INST.	TOTAL
1010	Boiler, cast iron, gas & oil, steam, 810 MBH	19,900	14,000	33,900
1020	1360 MBH	22,900	14,900	37,800
1030	2040 MBH	29,100	19,200	48,300
1040	2700 MBH	31,600	21,300	52,900
1050	3270 MBH	36,600	23,800	60,400
1060	3770 MBH	66,500	24,200	90,700
1070	4650 MBH	74,000	27,400	101,400
1080	5520 MBH	86,000	28,900	114,900
1090	6100 MBH	87,000	30,400	117,400
1100	6970 MBH	95,000	36,900	131,900

D3020 Heat Generating Systems

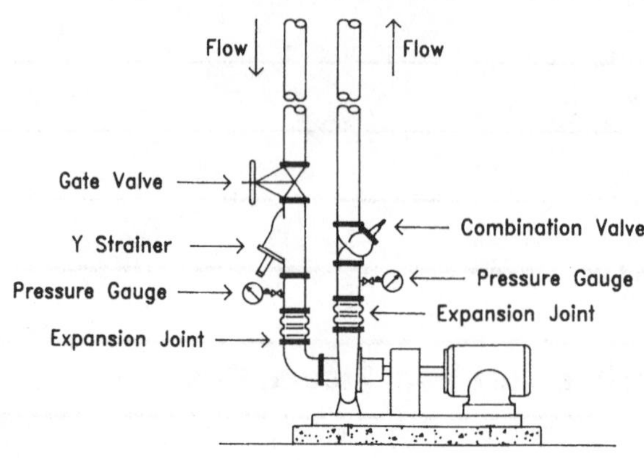

Base Mounted End–Suction Pump

System Components	QUANTITY	UNIT	COST EACH MAT.	COST EACH INST.	COST EACH TOTAL
SYSTEM D3020 330 1010					
PUMP, BASE MTD. WITH MOTOR, END-SUCTION, 2-1/2" SIZE, 3 HP, TO 150 GPM					
Pump, circulating, CI, base mounted, 2-1/2" size, 3 HP, to 150 GPM	1.000	Ea.	2,125	490	2,615
Pipe, black steel, Sch. 40, on yoke & roll hangers, 10′ O.C., 2-1/2" dia	12.000	L.F.	70.80	248.28	319.08
Elbow, 90°, weld joint, steel, 2-1/2" pipe size	1.000	Ea.	14.10	122.10	136.20
Flange, weld neck, 150 LB, 2-1/2" pipe size	8.000	Ea.	172	488.40	660.40
Valve, iron body, gate, 125 lb., N.R.S., flanged, 2-1/2" size	1.000	Ea.	360	177	537
Strainer, Y type, iron body, flanged, 125 lb., 2-1/2" pipe size	1.000	Ea.	80	177	257
Multipurpose valve, CI body, 2" size	1.000	Ea.	390	61.50	451.50
Expansion joint, flanged spool, 6" F to F, 2-1/2" dia	2.000	Ea.	488	143	631
T-O-L, weld joint, socket, 1/4" pipe size, nozzle	2.000	Ea.	9.24	84.70	93.94
T-O-L, weld joint, socket, 1/2" pipe size, nozzle	2.000	Ea.	9.24	88.06	97.30
Control gauges, pressure or vacuum, 3-1/2" diameter dial	2.000	Ea.	44	30.80	74.80
Pressure/temperature relief plug, 316 SS, 3/4" OD, 7-1/2" insertion	2.000	Ea.	126	30.80	156.80
Insulation, fiberglass pipe covering, 1-1/2" wall, 2-1/2" IPS	12.000	L.F.	28.80	56.16	84.96
Pump control system	1.000	Ea.	995	455	1,450
Pump balancing	1.000	Ea.		217	217
TOTAL			4,912.18	2,869.80	7,781.98

D3020 330	Circulating Pump Systems, End Suction	COST EACH MAT.	COST EACH INST.	COST EACH TOTAL
1010	Pump, base mtd with motor, end-suction, 2-1/2" size, 3 HP, to 150 GPM	4,900	2,875	7,775
1020	3" size, 5 HP, to 225 GPM	5,450	3,275	8,725
1030	4" size, 7-1/2 HP, to 350 GPM	6,800	3,950	10,750
1040	5" size, 15 HP, to 1000 GPM	9,675	6,000	15,675
1050	6" size, 25 HP, to 1550 GPM	12,900	7,300	20,200

D3020 340	Circulating Pump Systems, Double Suction	COST EACH MAT.	COST EACH INST.	COST EACH TOTAL
1010	Pump, base mtd w/ motor, double suction, 6" size, 50 HP, to 1200 GPM	16,900	10,500	27,400
1020	8" size, 75 HP, to 2500 GPM	24,000	14,200	38,200
1030	8" size, 100 HP, to 3000 GPM	35,400	15,800	51,200
1040	10" size, 150 HP, to 4000 GPM	48,800	17,900	66,700

Important: See the Reference Section for critical supporting data - Reference Numbers and City Cost Indexes

D3030 Cooling Generating Systems

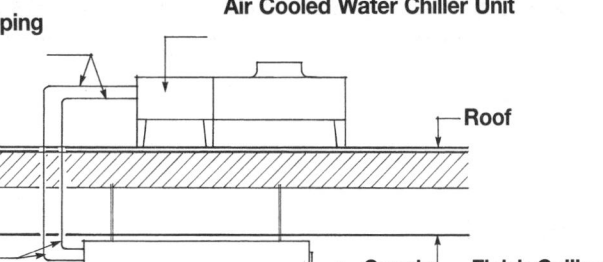

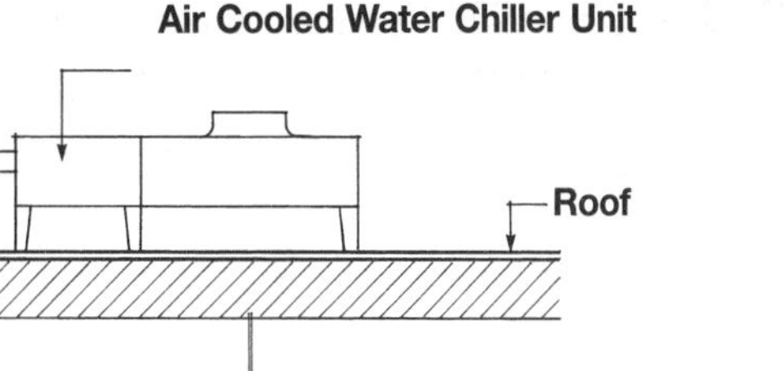

Design Assumptions: The chilled water, air cooled systems priced, utilize reciprocating hermetic compressors and propeller-type condenser fans. Piping with pumps and expansion tanks is included based on a two pipe system. No ducting is included and the fan-coil units are cooling only. Water treatment and balancing are not included. Chilled water piping is insulated. Area distribution is through the use of multiple fan coil units. Fewer but larger fan coil units with duct distribution would be approximately the same S.F. cost.

System Components	QUANTITY	UNIT	COST EACH		
			MAT.	INST.	TOTAL
SYSTEM D3030 110 1200					
PACKAGED CHILLER, AIR COOLED, WITH FAN COIL UNIT					
APARTMENT CORRIDORS, 3,000 S.F., 5.50 TON					
Fan coil air conditioning unit, cabinet mounted & filters chilled water	1.000	Ea.	2,978.63	406.93	3,385.56
Water chiller,air conditioning unit, reciprocating, air cooled,	1.000	Ea.	5,335	1,533.13	6,868.13
Chilled water unit coil connections	1.000	Ea.	610	1,100	1,710
Chilled water distribution piping	440.000	L.F.	4,928	14,960	19,888
TOTAL			13,851.63	18,000.06	31,851.69
COST PER S.F.			4.62	6	10.62
*Cooling requirements would lead to choosing a water cooled unit					

D3030 110	Chilled Water, Air Cooled Condenser Systems	COST PER S.F.		
		MAT.	INST.	TOTAL
1180	Packaged chiller, air cooled, with fan coil unit			
1200	Apartment corridors, 3,000 S.F., 5.50 ton	4.63	6	10.63
1240	6,000 S.F., 11.00 ton	3.80	4.80	8.60
1280	10,000 S.F., 18.33 ton	3.38	3.65	7.03
1320	20,000 S.F., 36.66 ton	2.52	2.64	5.16
1360	40,000 S.F., 73.33 ton	3.41	2.68	6.09
1440	Banks and libraries, 3,000 S.F., 12.50 ton	7.35	7.10	14.45
1480	6,000 S.F., 25.00 ton	6.85	5.75	12.60
1520	10,000 S.F., 41.66 ton	5.10	4.02	9.12
1560	20,000 S.F., 83.33 ton	5.90	3.65	9.55
1600	40,000 S.F., 167 ton*			
1680	Bars and taverns, 3,000 S.F., 33.25 ton	12.70	8.40	21.10
1720	6,000 S.F., 66.50 ton	11.50	6.50	18
1760	10,000 S.F., 110.83 ton	9.80	2.08	11.88
1800	20,000 S.F., 220 ton*			
1840	40,000 S.F., 440 ton*			
1920	Bowling alleys, 3,000 S.F., 17.00 ton	9.40	7.95	17.35
1960	6,000 S.F., 34.00 ton	6.95	5.60	12.55
2000	10,000 S.F., 56.66 ton	6.10	4.05	10.15
2040	20,000 S.F., 113.33 ton	6.70	3.69	10.39
2080	40,000 S.F., 227 ton*			
2160	Department stores, 3,000 S.F., 8.75 ton	6.60	6.70	13.30
2200	6,000 S.F., 17.50 ton	5.15	5.15	10.30
2240	10,000 S.F., 29.17 ton	3.96	3.74	7.70

D3030 Cooling Generating Systems

D3030 110	Chilled Water, Air Cooled Condenser Systems	COST PER S.F.		
		MAT.	INST.	TOTAL
2280	20,000 S.F., 58.33 ton	3.39	2.77	6.16
2320	40,000 S.F., 116.66 ton	4.29	2.76	7.05
2400	Drug stores, 3,000 S.F., 20.00 ton	10.75	8.20	18.95
2440	6,000 S.F., 40.00 ton	8.20	6.10	14.30
2480	10,000 S.F., 66.66 ton	9	5.25	14.25
2520	20,000 S.F., 133.33 ton	7.95	3.89	11.84
2560	40,000 S.F., 267 ton*			
2640	Factories, 2,000 S.F., 10.00 ton	6.30	6.80	13.10
2680	6,000 S.F., 20.00 ton	5.85	5.45	11.30
2720	10,000 S.F., 33.33 ton	4.34	3.83	8.17
2760	20,000 S.F., 66.66 ton	5.10	3.46	8.56
2800	40,000 S.F., 133.33 ton	4.66	2.83	7.49
2880	Food supermarkets, 3,000 S.F., 8.50 ton	6.50	6.65	13.15
2920	6,000 S.F., 17.00 ton	5.05	5.15	10.20
2960	10,000 S.F., 28.33 ton	3.77	3.63	7.40
3000	20,000 S.F., 56.66 ton	3.27	2.70	5.97
3040	40,000 S.F., 113.33 ton	4.20	2.76	6.96
3120	Medical centers, 3,000 S.F., 7.00 ton	5.75	6.50	12.25
3160	6,000 S.F., 14.00 ton	4.41	4.97	9.38
3200	10,000 S.F., 23.33 ton	4.04	3.80	7.84
3240	20,000 S.F., 46.66 ton	3.05	2.79	5.84
3280	40,000 S.F., 93.33 ton	3.81	2.71	6.52
3360	Offices, 3,000 S.F., 9.50 ton	5.80	6.60	12.40
3400	6,000 S.F., 19.00 ton	5.65	5.40	11.05
3440	10,000 S.F., 31.66 ton	4.19	3.79	7.98
3480	20,000 S.F., 63.33 ton	5.10	3.53	8.63
3520	40,000 S.F., 126.66 ton	4.58	2.86	7.44
3600	Restaurants, 3,000 S.F., 15.00 ton	8.35	7.35	15.70
3640	6,000 S.F., 30.00 ton	6.50	5.55	12.05
3680	10,000 S.F., 50.00 ton	5.75	4.11	9.86
3720	20,000 S.F., 100.00 ton	6.75	3.86	10.61
3760	40,000 S.F., 200 ton*			
3840	Schools and colleges, 3,000 S.F., 11.50 ton	6.90	6.95	13.85
3880	6,000 S.F., 23.00 ton	6.45	5.65	12.10
3920	10,000 S.F., 38.33 ton	4.80	3.95	8.75
3960	20,000 S.F., 76.66 ton	5.70	3.69	9.39
4000	40,000 S.F., 153 ton*			

D3030 Cooling Generating Systems

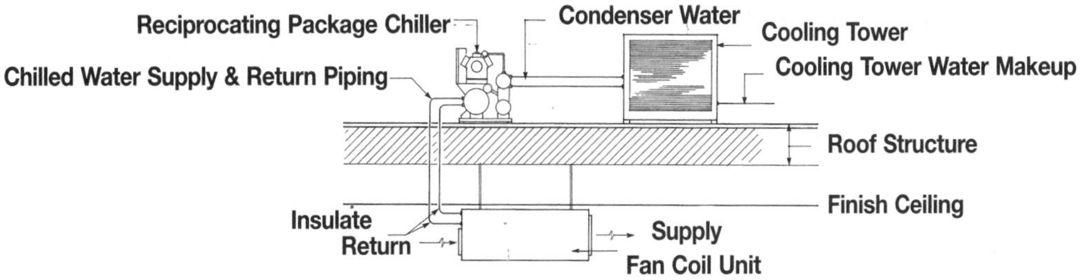

Reciprocating Package Chiller
Chilled Water Supply & Return Piping
Condenser Water
Cooling Tower
Cooling Tower Water Makeup
Roof Structure
Finish Ceiling
Insulate
Return
Supply
Fan Coil Unit

General: Water cooled chillers are available in the same sizes as air cooled units. They are also available in larger capacities.

Design Assumptions: The chilled water systems with water cooled condenser include reciprocating hermetic compressors, water cooling tower, pumps, piping and expansion tanks and are based on a two pipe system. Chilled water piping is insulated. No ducts are included and fan-coil units are cooling only. Area distribution is through use of multiple fan coil units. Fewer but larger fan coil units with duct distribution would be approximately the same S.F. cost. Water treatment and balancing are not included.

System Components	QUANTITY	UNIT	COST EACH MAT.	COST EACH INST.	COST EACH TOTAL
SYSTEM D3030 115 1320					
PACKAGED CHILLER, WATER COOLED, WITH FAN COIL UNIT					
APARTMENT CORRIDORS, 4,000 S.F., 7.33 TON					
Fan coil air conditioner unit, cabinet mounted & filters, chilled water	2.000	Ea.	3,969.88	542.35	4,512.23
Water chiller, reciprocating, water cooled, 1 compressor semihermetic	1.000	Ea.	10,167.60	2,633.50	12,801.10
Cooling tower, draw thru single flow, belt drive	1.000	Ea.	667.03	112.15	779.18
Cooling tower pumps & piping	1.000	System	333.52	267.55	601.07
Chilled water unit coil connections	2.000	Ea.	1,220	2,200	3,420
Chilled water distribution piping	520.000	L.F.	5,824	17,680	23,504
TOTAL			22,182.03	23,435.55	45,617.58
COST PER S.F.			5.55	5.86	11.41
*Cooling requirements would lead to choosing a water cooled unit					

D3030 115	Chilled Water, Cooling Tower Systems	COST PER S.F. MAT.	COST PER S.F. INST.	COST PER S.F. TOTAL
1300	Packaged chiller, water cooled, with fan coil unit			
1320	Apartment corridors, 4,000 S.F., 7.33 ton	5.55	5.85	11.40
1360	6,000 S.F., 11.00 ton R15700-020	4.43	5.05	9.48
1400	10,000 S.F., 18.33 ton	3.17	3.75	6.92
1440	20,000 S.F., 26.66 ton	2.54	2.81	5.35
1480	40,000 S.F., 73.33 ton	3.78	2.88	6.66
1520	60,000 S.F., 110.00 ton	3.74	2.95	6.69
1600	Banks and libraries, 4,000 S.F., 16.66 ton	6.90	6.45	13.35
1640	6,000 S.F., 25.00 ton	5.85	5.70	11.55
1680	10,000 S.F., 41.66 ton	4.83	4.23	9.06
1720	20,000 S.F., 83.33 ton	6.70	4.06	10.76
1760	40,000 S.F., 166.66 ton	6.55	4.98	11.53
1800	60,000 S.F., 250.00 ton	6.50	5.30	11.80
1880	Bars and taverns, 4,000 S.F., 44.33 ton	11.55	8.15	19.70
1920	6,000 S.F., 66.50 ton	14.75	8.45	23.20
1960	10,000 S.F., 110.83 ton	14.10	6.70	20.80
2000	20,000 S.F., 221.66 ton	13.15	7.15	20.30
2040	40,000 S.F., 440 ton*			
2080	60,000 S.F., 660 ton*			
2160	Bowling alleys, 4,000 S.F., 22.66 ton	8.15	7.10	15.25
2200	6,000 S.F., 34.00 ton	7.10	6.25	13.35
2240	10,000 S.F., 56.66 ton	6.80	4.59	11.39
2280	20,000 S.F., 113.33 ton	7.85	4.29	12.14
2320	40,000 S.F., 226.66 ton	7.60	5.15	12.75

D3030 Cooling Generating Systems

D3030 115	Chilled Water, Cooling Tower Systems	COST PER S.F.		
		MAT.	INST.	TOTAL
2360	60,000 S.F., 340 ton			
2440	Department stores, 4,000 S.F., 11.66 ton	6.35	6.35	12.70
2480	6,000 S.F., 17.50 ton	5.05	5.25	10.30
2520	10,000 S.F., 29.17 ton	3.85	3.92	7.77
2560	20,000 S.F., 58.33 ton	3.52	2.92	6.44
2600	40,000 S.F., 116.66 ton	4.65	3.03	7.68
2640	60,000 S.F., 175.00 ton	5.35	4.93	10.28
2720	Drug stores, 4,000 S.F., 26.66 ton	8.75	7.30	16.05
2760	6,000 S.F., 40.00 ton	7.65	6.25	13.90
2800	10,000 S.F., 66.66 ton	9.40	5.70	15.10
2840	20,000 S.F., 133.33 ton	8.90	4.47	13.37
2880	40,000 S.F., 266.67 ton	8.75	5.60	14.35
2920	60,000 S.F., 400 ton*			
3000	Factories, 4,000 S.F., 13.33 ton	5.90	6.15	12.05
3040	6,000 S.F., 20.00 ton	5.15	5.30	10.45
3080	10,000 S.F., 33.33 ton	4.22	4.04	8.26
3120	20,000 S.F., 66.66 ton	5.25	3.64	8.89
3160	40,000 S.F., 133.33 ton	5.10	3.12	8.22
3200	60,000 S.F., 200.00 ton	5.75	5.05	10.80
3280	Food supermarkets, 4,000 S.F., 11.33 ton	6.20	6.30	12.50
3320	6,000 S.F., 17.00 ton	4.54	5.15	9.69
3360	10,000 S.F., 28.33 ton	3.77	3.88	7.65
3400	20,000 S.F., 56.66 ton	3.58	2.92	6.50
3440	40,000 S.F., 113.33 ton	4.60	3.02	7.62
3480	60,000 S.F., 170.00 ton	5.30	4.91	10.21
3560	Medical centers, 4.000 S.F., 9.33 ton	5.30	5.80	11.10
3600	6,000 S.F., 14.00 ton	4.35	5.05	9.40
3640	10,000 S.F., 23.33 ton	3.48	3.77	7.25
3680	20,000 S.F., 46.66 ton	3.15	2.86	6.01
3720	40,000 S.F., 93.33 ton	4.23	2.97	7.20
3760	60,000 S.F., 140.00 ton	4.93	4.88	9.81
3840	Offices, 4,000 S.F., 12.66 ton	5.70	6.10	11.80
3880	6,000 S.F., 19.00 ton	5.10	5.40	10.50
3920	10,000 S.F., 31.66 ton	4.19	4.09	8.28
3960	20,000 S.F., 63.33 ton	5.15	3.66	8.81
4000	40,000 S.F., 126.66 ton	5.55	4.77	10.32
4040	60,000 S.F., 190.00 ton	5.55	5	10.55
4120	Restaurants, 4,000 S.F., 20.00 ton	7.30	6.55	13.85
4160	6,000 S.F., 30.00 ton	6.40	5.85	12.25
4200	10,000 S.F., 50.00 ton	6.25	4.40	10.65
4240	20,000 S.F., 100.00 ton	7.50	4.26	11.76
4280	40,000 S.F., 200.00 ton	6.80	4.84	11.64
4320	60,000 S.F., 300.00 ton	7.25	5.50	12.75
4400	Schools and colleges, 4,000 S.F., 15.33 ton	6.50	6.30	12.80
4440	6,000 S.F., 23.00 ton	5.50	5.60	11.10
4480	10,000 S.F., 38.33 ton	4.53	4.15	8.68
4520	20,000 S.F., 76.66 ton	6.35	4.01	10.36
4560	40,000 S.F., 153.33 ton	6.15	4.88	11.03
4600	60,000 S.F., 230.00 ton	6	5.10	11.10

D3030 Cooling Generating Systems

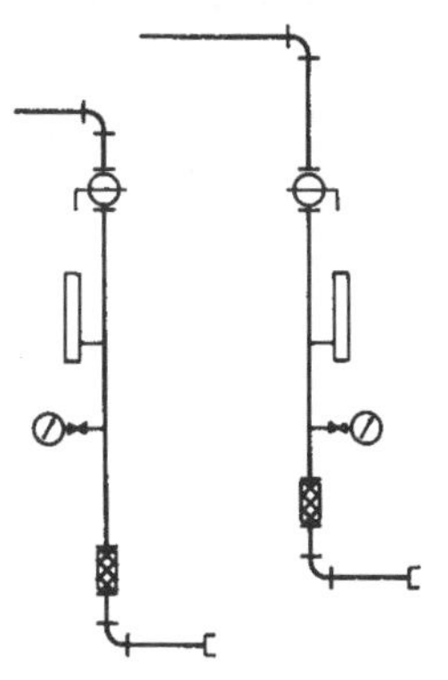
Chilled Water Piping

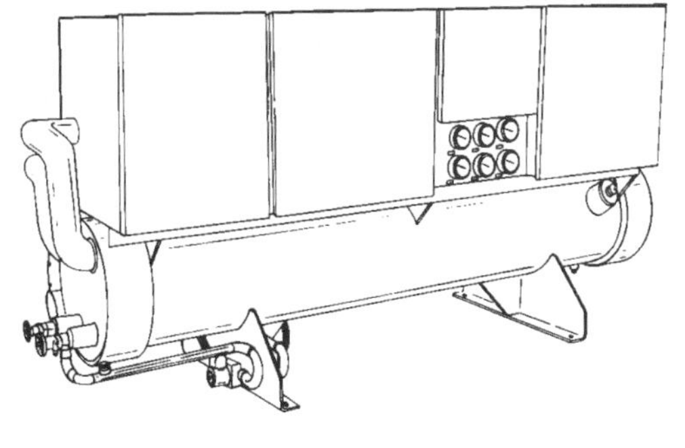
Reciprocating Chiller, Water Cooled

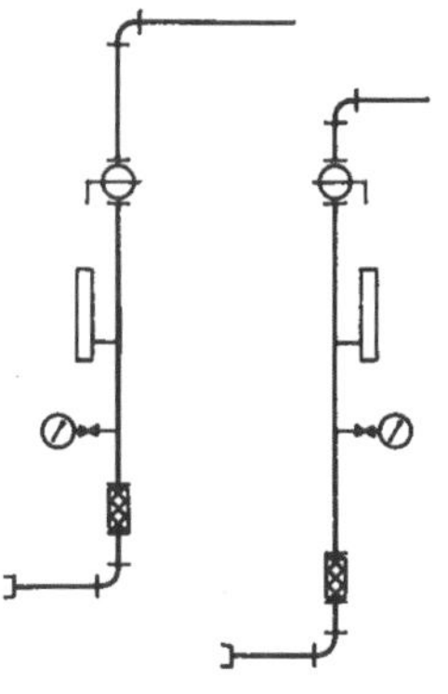
Condenser Water Piping

System Components	QUANTITY	UNIT	COST EACH MAT.	COST EACH INST.	COST EACH TOTAL
SYSTEM D3030 130 1010					
CHILLER, RECIPROCATING, WATER COOLED, STD. CONTROLS, 60 TON					
Water chiller, reciprocating, water cooled, 60 ton cooling	1.000	Ea.	30,800	7,400	38,200
Pipe, black steel, welded, Sch. 40, on yoke & roll hgrs, 10′ O.C., 4″ dia	40.000	L.F.	528	1,056	1,584
Elbow, 90°, weld joint, butt, 4″ pipe size	8.000	Ea.	224	1,558	1,782
Flange, welding neck, 150 lb., 4″ pipe size	16.000	Ea.	616	1,557.60	2,173.60
Valves, iron body, butterfly, lug type, lever actuator, 4″ size	4.000	Ea.	468	708	1,176
Expansion joints, bellows type, flange spool, 6″ F to F, 4″ dia	4.000	Ea.	1,184	420	1,604
T-O-L socket, weld joint, 1/4″ pipe size, nozzle	4.000	Ea.	18.48	169.40	187.88
T-O-L socket, weld joint, 3/4″ pipe size, nozzle	4.000	Ea.	21.60	184.88	206.48
Thermometers, stem type, 9″ case, 8″ stem, 3/4″ NPT	4.000	Ea.	520	70.40	590.40
Gauges, pressure or vacuum, 3-1/2″ diameter dial	4.000	Ea.	88	61.60	149.60
Insulation, fiberglass pipe covering, 1-1/2″ wall, 4″ IPS	20.000	L.F.	56.80	120	176.80
Chiller balancing	1.000	Ea.		420	420
TOTAL			34,524.88	13,725.88	48,250.76

D3030 130	Reciprocating Chiller, Water Cooled	COST EACH MAT.	COST EACH INST.	COST EACH TOTAL
1010	Chiller, reciprocating, water cooled, std. controls, 60 Ton	34,500	13,700	48,200
1020	100 Ton	65,500	20,600	86,100
1030	150 Ton	87,500	22,600	110,100
1040	200 Ton	105,500	23,800	129,300

D3030 135	Reciprocating Chiller, Air Cooled	COST EACH MAT.	COST EACH INST.	COST EACH TOTAL
1010	Chiller, reciprocating, air cooled, std. controls, 20 Ton	20,800	8,075	28,875
1020	30 Ton	23,200	8,475	31,675
1030	40 Ton	27,800	9,050	36,850
1040	60 Ton	39,900	10,200	50,100
1050	70 Ton	45,500	10,500	56,000
1060	80 Ton	48,500	10,700	59,200
1070	90 Ton	52,500	10,800	63,300
1080	100 Ton	58,500	11,000	69,500
1090	110 Ton	64,000	11,800	75,800
1100	130 Ton	70,500	12,000	82,500

D3030 Cooling Generating Systems

D3030 135	Reciprocating Chiller, Air Cooled	COST EACH		
		MAT.	INST.	TOTAL
1110	150 Ton	81,500	12,200	93,700
1120	175 Ton	86,500	12,400	98,900
1130	210 Ton	107,000	13,500	120,500

D3030 Cooling Generating Systems

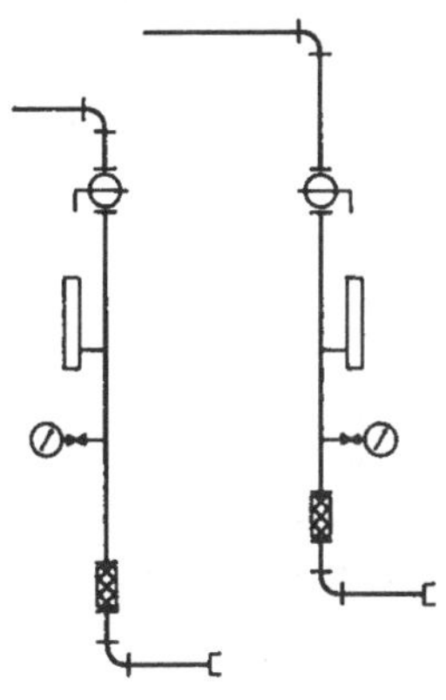
Chilled Water Piping

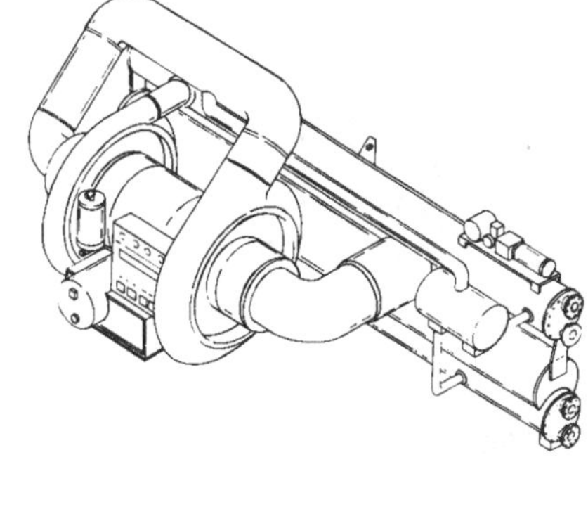
Centrifugal Chiller, Water Cooled, Hermetic

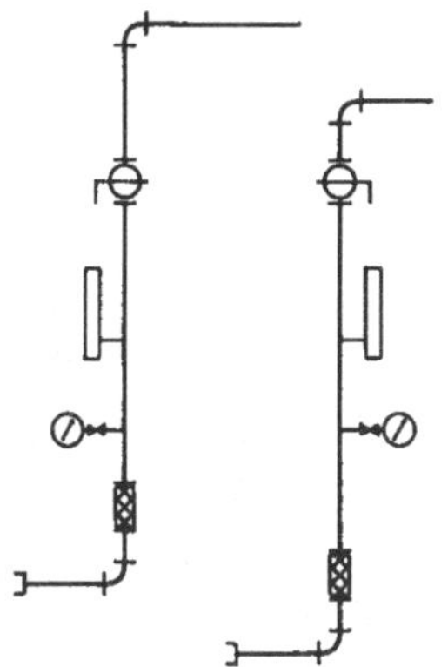
Condenser Water Piping

System Components	QUANTITY	UNIT	COST EACH MAT.	COST EACH INST.	COST EACH TOTAL
SYSTEM D3030 140 1010					
CHILLER, CENTRIF., WATER COOLED, PKGD. HERMETIC, STD. CONTROLS, 200 TON					
Water chiller, screw, water cooled, not incl tower, 200 ton	1.000	Ea.	72,000	14,800	86,800
Pipe, black steel, welded, Sch. 40, on yoke & roll hgrs, 10′ o.c., 6″ dia	40.000	L.F.	1,180	1,618.80	2,798.80
Elbow, 90°, weld joint, 6″ pipe size	8.000	Ea.	520	2,342	2,862
Flange, welding neck, 150 lb., 6″ pipe size	16.000	Ea.	880	2,350.40	3,230.40
Valves, iron body, butterfly, lug type, lever actuator, 6″ size	4.000	Ea.	784	1,100	1,884
Expansion joints, bellows type, flange spool, 6″ F to F, 6″ dia	4.000	Ea.	1,500	520	2,020
T-O-L socket, weld jiont, 1/4″ pipe size, nozzle	4.000	Ea.	18.48	169.40	187.88
T-O-L socket, weld joint, 3/4″ pipe size, nozzle	4.000	Ea.	21.60	184.88	206.48
Thermometers, stem type, 9″ case, 8″ stem, 3/4″ NPT	4.000	Ea.	520	70.40	590.40
Gauges, pressure or vacuum, 3-1/2″ diameter dial	4.000	Ea.	88	61.60	149.60
Insulation, fiberglass pipe covering, 1-1/2″ wall, 6″ IPS	40.000	L.F.	130.40	306	436.40
Chiller balancing	1.000	Ea.		420	420
TOTAL			77,642.48	23,943.48	101,585.96

D3030 140	Centrifugal Chiller, Water Cooled	COST EACH MAT.	COST EACH INST.	COST EACH TOTAL
1010	Chiller, centrif., water cooled, pkgd. hermetic, std controls, 200 Ton	77,500	23,900	101,400
1020	400 Ton	144,000	32,900	176,900
1030	1,000 Ton	371,000	49,600	420,600
1040	1,500 Ton	545,500	61,000	606,500

D3030 Cooling Generating Systems

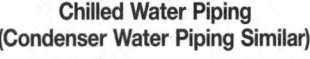

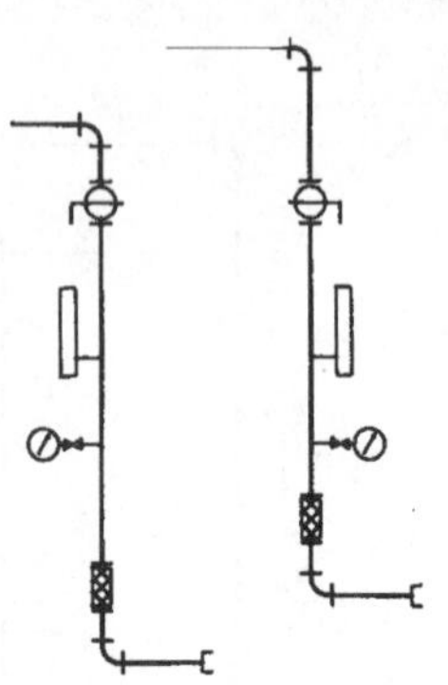
Chilled Water Piping
(Condenser Water Piping Similar)

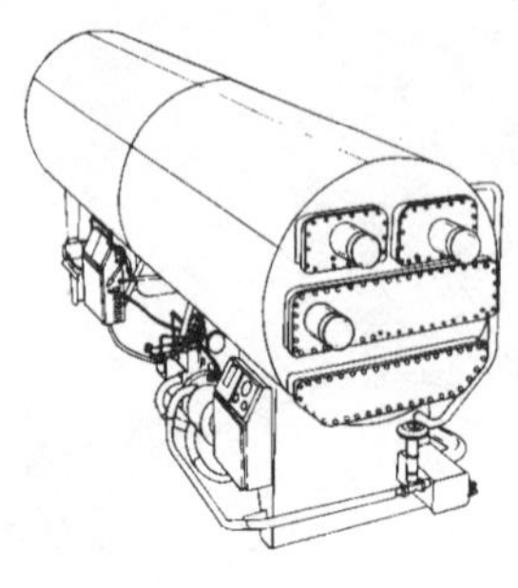
Gas Absorption Chiller, Water Cooled

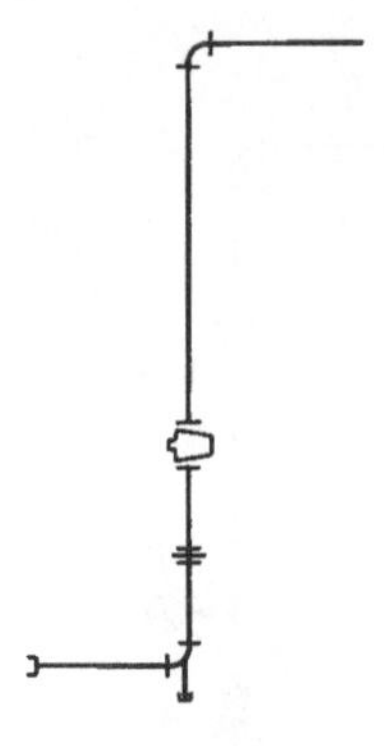
Gas Piping

System Components	QUANTITY	UNIT	COST EACH MAT.	COST EACH INST.	COST EACH TOTAL
SYSTEM D3030 145 1010					
CHILLER, GAS FIRED ABSORPTION, WATER COOLED, STD. CONTROLS, 800 TON					
Absorption water chiller, gas fired, water cooled, 800 ton	1.000	Ea.	589,000	22,300	611,300
Pipe, black steel, weld, Sch. 40, on yoke & roll hangers, 10′ O.C., 10″ dia	20.000	L.F.	1,410	1,224	2,634
Pipe, black steel, welded, Sch. 40, (no hgrs incl), 14″ dia.	44.000	L.F.	1,738	4,285.60	6,023.60
Pipe, blacksteel, welded, Sch. 40, on yoke & roll hangers, 10′ O.C., 4″ dia	20.000	L.F.	264	528	792
Elbow, 90°, weld joint, 10″ pipe size	4.000	Ea.	896	1,958	2,854
Elbow, 90°, weld joint, 14″ pipe size	6.000	Ea.	2,610	4,407	7,017
Elbow, 90°, weld joint, 4″ pipe size	3.000	Ea.	84	584.25	668.25
Flange, welding neck, 150 lb., 10″ pipe size	8.000	Ea.	856	1,950.40	2,806.40
Flange, welding neck, 150 lb., 14″ pipe size	14.000	Ea.	3,752	4,546.50	8,298.50
Flange, welding neck, 150 lb., 4″ pipe size	3.000	Ea.	115.50	292.05	407.55
Valves, iron body, butterfly, lug type, gear operated, 10″ size	2.000	Ea.	1,120	690	1,810
Valves, iron body, butterfly, lug type, gear operated, 14″ size	5.000	Ea.	4,650	3,000	7,650
Valves, semi-steel, lubricated plug valve, flanged, 4″ pipe size	1.000	Ea.	365	295	660
Expansion joints, bellows type, flange spool, 6″ F to F, 10″ dia	2.000	Ea.	1,140	354	1,494
Expansion joints, bellows type, flange spool, 10″ F to F, 14″ dia	2.000	Ea.	1,870	466	2,336
T-O-L socket, weld joint, 1/4″ pipe size, nozzle	4.000	Ea.	18.48	169.40	187.88
T-O-L socket, weld joint, 3/4″ pipe size, nozzle	4.000	Ea.	21.60	184.88	206.48
Thermometers, stem type, 9″ case, 8″ stem, 3/4″ NPT	4.000	Ea.	520	70.40	590.40
Gauges, pressure or vacuum, 3-1/2″ diameter dial	4.000	Ea.	88	61.60	149.60
Insulation, fiberglass pipe covering, 2″ wall, 10″ IPS	20.000	L.F.	148	241	389
Vent chimney, all fuel, pressure tight, double wall, SS, 12″ dia	20.000	L.F.	1,260	400	1,660
Chiller balancing	1.000	Ea.		420	420
TOTAL			611,926.58	48,428.08	660,354.66

D3030 145	Gas Absorption Chiller, Water Cooled	MAT.	INST.	TOTAL
1010	Chiller, gas fired absorption, water cooled, std. ctrls, 800 Ton	612,000	48,500	660,500
1020	1000 Ton	772,000	62,500	834,500

D3030 150	Steam Absorption Chiller, Water Cooled	COST EACH MAT.	COST EACH INST.	COST EACH TOTAL
1010	Chiller, steam absorption, water cooled, std. ctrls, 750 Ton	401,000	57,000	458,000
1020	955 Ton	464,000	68,500	532,500
1030	1465 Ton	669,000	79,000	748,000
1040	1660 Ton	797,500	90,000	887,500

SERVICES D

D3030 Cooling Generating Systems

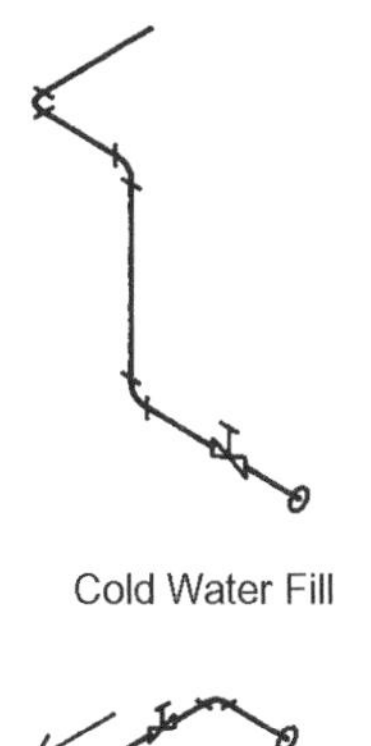

Cold Water Fill

Drain To Waste Piping

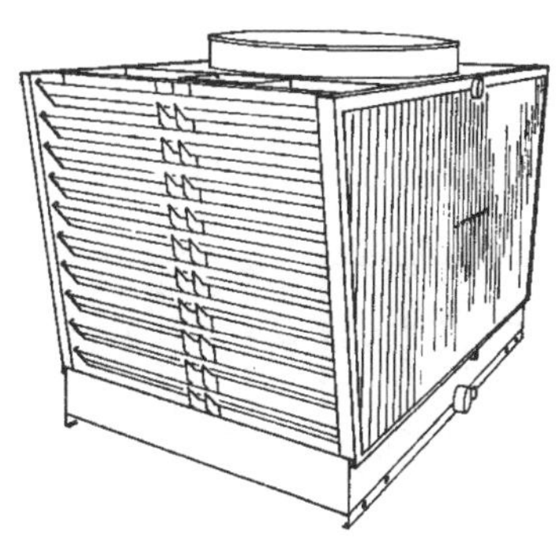

Cooling Tower

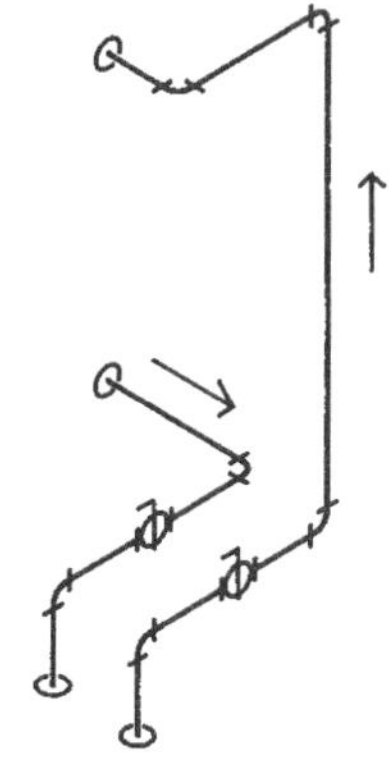

Condenser Water Piping

System Components	QUANTITY	UNIT	COST EACH MAT.	COST EACH INST.	COST EACH TOTAL
SYSTEM D3030 310 1010					
COOLING TOWER, GALVANIZED STEEL, PACKAGED UNIT, DRAW THRU, 60 TON					
Cooling towers, draw thru, single flow, belt drive, 60 tons	1.000	Ea.	5,460	918	6,378
Pipe, black steel, welded, Sch. 40, on yoke & roll hangers, 10′ OC, 4″ dia.	40.000	L.F.	528	1,056	1,584
Tubing, copper, Type L, couplings & hangers 10′ OC, 1-1/2″ dia.	20.000	L.F.	119	189	308
Elbow, 90°, copper, cu x cu, 1-1/2″ dia.	3.000	Ea.	18.30	114	132.30
Elbow, 90°, weld joint, 4″ pipe size	6.000	Ea.	168	1,168.50	1,336.50
Flange, welding neck, 150 lb., 4″ pipe size	6.000	Ea.	231	584.10	815.10
Valves, bronze, gate, N.R.S., soldered, 125 psi, 1-1/2″ size	1.000	Ea.	57.50	38	95.50
Valves, iron body, butterfly, lug type, lever actuator, 4″ size	2.000	Ea.	234	354	588
Electric heat trace system, 400 degree, 115v, 5 watts per L.F.	160.000	L.F.	1,176	147.20	1,323.20
Insulation, fiberglass pipe covering, , 1-1/2″ wall, 4″ IPS	40.000	L.F.	113.60	240	353.60
Insulation, fiberglass pipe covering, 1″ wall, 1-1/2″ IPS	20.000	L.F.	22.40	80.40	102.80
Cooling tower condenser control system	1.000	Ea.	4,300	2,050	6,350
Cooling tower balancing	1.000	Ea.		320	320
TOTAL			12,427.80	7,259.20	19,687

D3030 310	Galvanized Draw Through Cooling Tower	COST EACH MAT.	COST EACH INST.	COST EACH TOTAL
1010	Cooling tower, galvanized steel, packaged unit, draw thru, 60 Ton	12,400	7,250	19,650
1020	110 Ton	15,400	8,525	23,925
1030	300 Ton	30,400	12,800	43,200
1040	600 Ton	49,100	19,100	68,200
1050	1000 Ton	70,500	28,100	98,600

D3030 320	Fiberglass Draw Through Cooling Tower	COST EACH MAT.	COST EACH INST.	COST EACH TOTAL
1010	Cooling tower, fiberglass, packaged unit, draw thru, 60 Ton	10,500	7,250	17,750
1020	125 Ton	14,800	8,550	23,350
1030	300 Ton	26,300	12,700	39,000
1040	600 Ton	44,700	19,100	63,800
1050	1000 Ton	68,000	28,100	96,100

D3030 330	Stainless Steel Draw Through Cooling Tower	COST EACH MAT.	COST EACH INST.	COST EACH TOTAL
1010	Cooling tower, stainless steel, packaged unit, draw thru, 60 Ton	16,200	7,250	23,450
1020	110 Ton	20,900	8,550	29,450
1030	300 Ton	44,300	12,700	57,000
1040	600 Ton	69,500	18,800	88,300
1050	1000 Ton	112,500	28,100	140,600

D3040 Distribution Systems

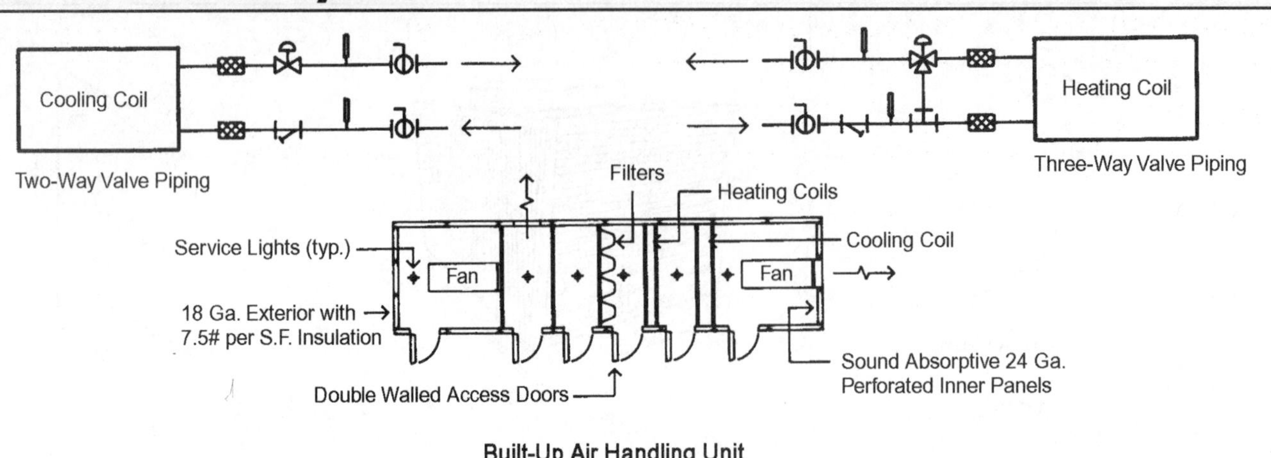

Built-Up Air Handling Unit

System Components	QUANTITY	UNIT	COST EACH MAT.	COST EACH INST.	COST EACH TOTAL
SYSTEM D3040 106 1010					
AHU, FIELD FAB., BLT. UP, COOL/HEAT COILS, FLTRS., CONST. VOL., 40,000 CFM					
AHU, Built-up, cool/heat coils, filters, mix box, constant volume	1.000	Ea.	42,400	4,600	47,000
Pipe, black steel, welded, Sch. 40, on yoke & roll hgrs, 10′ OC, 3″ dia.	26.000	L.F.	184.60	586.56	771.16
Pipe, black steel, welded, Sch. 40, on yoke & roll hgrs, 10′ OC, 4″ dia	20.000	L.F.	264	528	792
Elbow, 90°, weld joint, 3″ pipe size	5.000	Ea.	84.50	693.25	777.75
Elbow, 90°, weld joint, 4″ pipe size	4.000	Ea.	112	779	891
Tee, welded, reducing on outlet, 3″ by 2-1/2″ pipe	3.000	Ea.	169.50	729	898.50
Tee, welded, reducing on outlet, 4″ by 3″ pipe	2.000	Ea.	232	649	881
Reducer, eccentric, weld joint, 3″ pipe size	4.000	Ea.	342	488.40	830.40
Reducer, eccentric, weld joint, 4″ pipe size	4.000	Ea.	372	647	1,019
Flange, weld neck, 150 LB, 2-1/2″ pipe size	3.000	Ea.	64.50	183.15	247.65
Flange, weld neck, 150 lb., 3″ pipe size	14.000	Ea.	357	970.90	1,327.90
Flange, weld neck, 150 lb., 4″ pipe size	10.000	Ea.	385	973.50	1,358.50
Valves, iron body, butterfly, lug type, lever actuator, 3″ size	2.000	Ea.	191	222	413
Valves, iron body, butterfly, lug type, lever actuator, 4″ size	2.000	Ea.	234	354	588
Strainer, Y type, iron body, flanged, 125 lb., 3″ pipe size	1.000	Ea.	93	197	290
Strainer, Y type, iron body, flanged, 125 lb., 4″ pipe size	1.000	Ea.	170	295	465
Valve, electric motor actuated, iron body, 3 way, flanged, 2-1/2″ pipe	1.000	Ea.	885	245	1,130
Valve, electric motor actuated, iron body, two way, flanged, 3″ pipe size	1.000	Ea.	525	245	770
Expansion joints, bellows type, flange spool, 6″ F to F, 3″ dia	2.000	Ea.	554	155	709
Expansion joints, bellows type, flange spool, 6″ F to F, 4″ dia	2.000	Ea.	592	210	802
T-O-L socket, weld joint, 1/4″ pipe size, nozzle	4.000	Ea.	18.48	169.40	187.88
T-O-L socket, weld joint, 1/2″ pipe size, nozzle	12.000	Ea.	55.44	528.36	583.80
Thermometers, stem type, 9″ case, 8″ stem, 3/4″ NPT	4.000	Ea.	520	70.40	590.40
Pressure/temperature relief plug, 316 SS, 3/4″OD, 7-1/2″ insertion	4.000	Ea.	252	61.60	313.60
Insulation, fiberglass pipe covering, 1-1/2″ wall, 3″ IPS	26.000	L.F.	64.48	128.96	193.44
Insulation, fiberglass pipe covering, 1-1/2″ wall, 4″ IPS	20.000	L.F.	56.80	120	176.80
Fan coil control system	1.000	Ea.	970	455	1,425
Re-heat coil balancing	2.000	Ea.		182	182
Air conditioner balancing	1.000	Ea.		595	595
TOTAL			50,148.30	16,061.48	66,209.78

D3040 106	Field Fabricated Air Handling Unit	COST EACH MAT.	COST EACH INST.	COST EACH TOTAL
1010	AHU, Field fab, blt up, cool/heat coils, fltrs, const vol, 40,000 CFM	50,000	16,000	66,000
1020	60,000 CFM	69,500	24,700	94,200
1030	75,000 CFM	84,000	26,000	110,000

SERVICES D

D3040 Distribution Systems

D3040 108	Field Fabricated VAV Air Handling Unit	COST EACH		
		MAT.	INST.	TOTAL
1010	AHU, Field fab, blt up, cool/heat coils, fltrs, VAV, 75,000 CFM	86,000	29,800	115,800
1020	100,000 CFM	135,000	39,500	174,500
1030	150,000 CFM	220,500	63,000	283,500

D3040 Distribution Systems

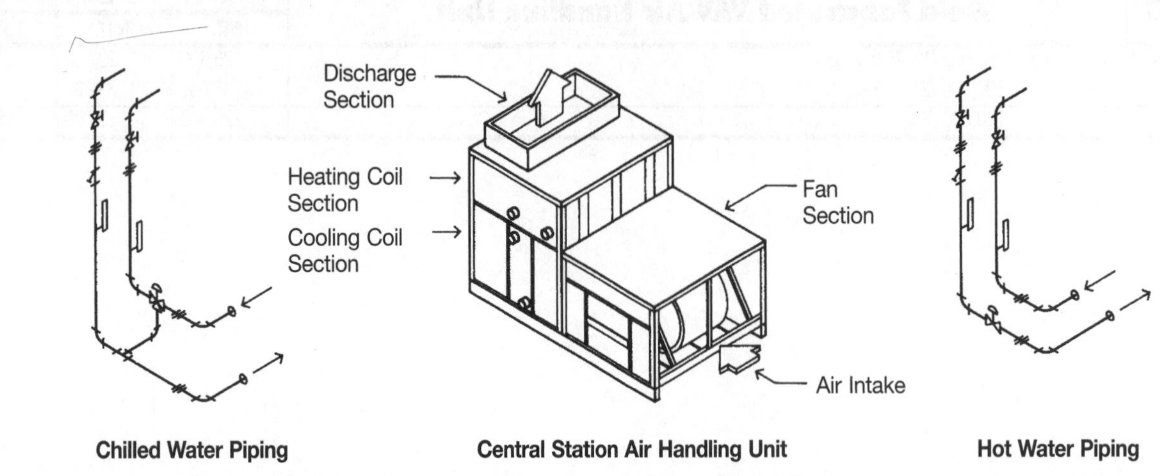

System Components	QUANTITY	UNIT	COST EACH MAT.	COST EACH INST.	COST EACH TOTAL
SYSTEM D3040 110 1010					
AHU, CENTRAL STATION, COOL/HEAT COILS, CONST. VOL., FLTRS., 2,000 CFM					
Central station air handling unit, chilled water, 1900 CFM	1.000	Ea.	3,910	925.75	4,835.75
Pipe, black steel, Sch 40, threaded, W/cplgs & hangers, 10′ OC, 1″ dia.	26.000	L.F.	77.35	300.63	377.98
Pipe, black steel, Sch 40, threaded, W/cplg & hangers, 10′ OC, 1-1/4″ dia.	20.000	L.F.	78.52	258.70	337.22
Elbow, 90°, black, straight, 1″ dia.	4.000	Ea.	12.76	152	164.76
Elbow, 90°, black, straight, 1-1/4″ dia.	4.000	Ea.	21	160	181
Tee, black, reducing, 1″ dia.	8.000	Ea.	45.60	492	537.60
Tee, black, reducing, 1-1/4″ dia.	8.000	Ea.	78.80	504	582.80
Reducer, concentric, black, 1″ dia.	4.000	Ea.	23.20	132	155.20
Reducer, concentric, black, 1-1/4″ dia.	4.000	Ea.	26	136	162
Union, black with brass seat, 1″ dia.	4.000	Ea.	44	164	208
Union, black with brass seat, 1-1/4″ dia.	2.000	Ea.	30.20	84	114.20
Pipe nipple, black, 1″ dia	13.000	Ea.	15.47	60.13	75.60
Pipe nipple, black, 1-1/4″ dia	12.000	Ea.	18.12	59.70	77.82
Valves, bronze, gate, N.R.S., threaded, class 150, 1″ size	2.000	Ea.	92	52	144
Valves, bronze, gate, N.R.S., threaded, class 150, 1-1/4″ size	2.000	Ea.	122	66	188
Strainer, Y type, bronze body, screwed, 150 lb., 1″ pipe size	1.000	Ea.	20.50	29	49.50
Strainer, Y type, bronze body, screwed, 150 lb., 1-1/4″ pipe size	1.000	Ea.	46	33	79
Valve, electric motor actuated, brass, 3 way, screwed, 3/4″ pipe	1.000	Ea.	273	27	300
Valve, electric motor actuated, brass, 2 way, screwed, 1″ pipe size	1.000	Ea.	370	26	396
Thermometers, stem type, 9″ case, 8″ stem, 3/4″ NPT	4.000	Ea.	520	70.40	590.40
Pressure/temperature relief plug, 316 SS, 3/4″OD, 7-1/2″ insertion	4.000	Ea.	252	61.60	313.60
Insulation, fiberglass pipe covering, 1″ wall, 1″ IPS	26.000	L.F.	23.66	99.58	123.24
Insulation, fiberglass pipe covering, 1″ wall, 1-1/4″ IPS	20.000	L.F.	21.20	80.40	101.60
Fan coil control system	1.000	Ea.	970	455	1,425
Re-heat coil balancing	2.000	Ea.		182	182
Multizone A/C balancing	1.000	Ea.		297	297
TOTAL			7,091.38	4,907.89	11,999.27

D3040 110	Central Station Air Handling Unit	COST EACH MAT.	COST EACH INST.	COST EACH TOTAL
1010	AHU, Central station, cool/heat coils, constant vol, fltrs, 2,000 CFM	7,100	4,900	12,000
1020	5,000 CFM	10,900	6,450	17,350
1030	10,000 CFM	14,500	9,175	23,675
1040	15,000 CFM	20,500	12,600	33,100
1050	20,000 CFM	24,600	14,800	39,400

Important: See the Reference Section for critical supporting data - Reference Numbers and City Cost Indexes

D3040 Distribution Systems

D3040 112	Central Station Air Handling Unit, VAV	COST EACH		
		MAT.	INST.	TOTAL
1010	AHU, Central station, cool/heat coils, VAV, fltrs, 5,000 CFM	13,000	6,450	19,450
1020	10,000 CFM	17,000	9,000	26,000
1030	15,000 CFM	23,200	11,900	35,100
1040	20,000 CFM	29,000	14,400	43,400
1050	30,000 CFM	43,400	16,500	59,900

D3040 Distribution Systems

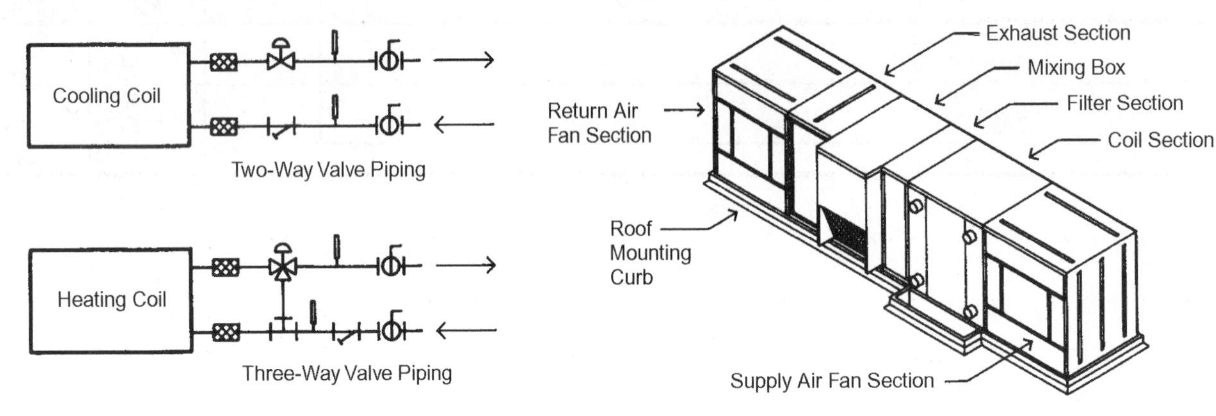

Rooftop Air Handling Unit

System Components	QUANTITY	UNIT	COST EACH MAT.	COST EACH INST.	COST EACH TOTAL
SYSTEM D3040 114 1010					
AHU, ROOFTOP, COOL/HEAT COILS, FLTRS., CONST. VOL., 2,000 CFM					
AHU, Built-up, cool/heat coils, filter, mix box, rooftop, constant volume	1.000	Ea.	4,425	885	5,310
Pipe, black steel, Sch 40, threaded, W/cplgs & hangers, 10′ OC, 1″ dia.	26.000	L.F.	77.35	300.63	377.98
Pipe, black steel, Sch 40, threaded, W/cplg & hangers, 10′ OC, 1-1/4″ dia.	20.000	L.F.	78.52	258.70	337.22
Elbow, 90°, black, straight, 1″ dia.	4.000	Ea.	12.76	152	164.76
Elbow, 90°, black, straight, 1-1/4″ dia.	4.000	Ea.	21	160	181
Tee, black, reducing, 1″ dia.	8.000	Ea.	45.60	492	537.60
Tee, black, reducing, 1-1/4″ dia.	8.000	Ea.	78.80	504	582.80
Reducer, black, concentric, 1″ dia.	4.000	Ea.	23.20	132	155.20
Reducer, black, concentric, 1-1/4″ dia.	4.000	Ea.	26	136	162
Union, black with brass seat, 1″ dia.	4.000	Ea.	44	164	208
Union, black with brass seat, 1-1/4″ dia.	2.000	Ea.	30.20	84	114.20
Pipe nipples, 1″ dia	13.000	Ea.	15.47	60.13	75.60
Pipe nipples, 1-1/4″ dia	12.000	Ea.	18.12	59.70	77.82
Valves, bronze, gate, N.R.S., threaded, class 150, 1″ size	2.000	Ea.	92	52	144
Valves, bronze, gate, N.R.S., threaded, class 150, 1-1/4″ size	2.000	Ea.	122	66	188
Strainer, Y type, bronze body, screwed, 150 lb., 1″ pipe size	1.000	Ea.	20.50	29	49.50
Strainer, Y type, bronze body, screwed, 150 lb., 1-1/4″ pipe size	1.000	Ea.	46	33	79
Valve, electric motor actuated, brass, 3 way, screwed, 3/4″ pipe	1.000	Ea.	273	27	300
Valve, electric motor actuated, brass, 2 way, screwed, 1″ pipe size	1.000	Ea.	370	26	396
Thermometers, stem type, 9″ case, 8″ stem, 3/4″ NPT	4.000	Ea.	520	70.40	590.40
Pressure/temperature relief plug, 316 SS, 3/4″OD, 7-1/2″ insertion	4.000	Ea.	252	61.60	313.60
Insulation, fiberglass pipe covering, 1″ wall, 1″ IPS	26.000	L.F.	23.66	99.58	123.24
Insulation, fiberglass pipe covering, 1″ wall, 1-1/4″ IPS	20.000	L.F.	21.20	80.40	101.60
Insulation pipe covering .016″ aluminum jacket finish, add	45.000	S.F.	33.75	189.90	223.65
Fan coil control system	1.000	Ea.	970	455	1,425
Re-heat coil balancing	2.000	Ea.		182	182
Rooftop unit heat/cool balancing	1.000	Ea.		345	345
TOTAL			7,640.13	5,105.04	12,745.17

D3040 114	Rooftop Air Handling Unit	COST EACH MAT.	COST EACH INST.	COST EACH TOTAL
1010	AHU, Rooftop, cool/heat coils, constant vol., fltrs, 2,000 CFM	7,650	5,100	12,750
1020	5,000 CFM	11,800	6,500	18,300

SERVICES D

D3040 Distribution Systems

D3040 114	Rooftop Air Handling Unit	COST EACH		
		MAT.	INST.	TOTAL
1030	10,000 CFM	15,700	9,150	24,850
1040	15,000 CFM	22,300	12,400	34,700
1050	20,000 CFM	26,900	14,700	41,600

D3040 116	Rooftop Air Handling Unit, VAV	COST EACH		
		MAT.	INST.	TOTAL
1010	AHU, Rooftop, cool/heat coils, VAV, fltrs, 5,000 CFM	13,600	6,725	20,325
1020	10,000 CFM	17,900	9,250	27,150
1030	15,000 CFM	24,400	12,300	36,700
1040	20,000 CFM	30,500	14,800	45,300
1050	30,000 CFM	46,000	17,000	63,000

D3040 Distribution Systems

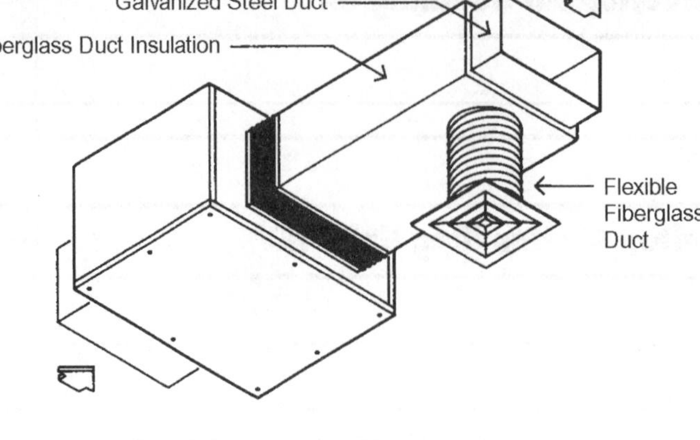

Horizontal Fan Coil Air Conditioning System

System Components	QUANTITY	UNIT	COST EACH MAT.	COST EACH INST.	COST EACH TOTAL
SYSTEM D3040 124 1010					
FAN COIL A/C SYSTEM, HORIZONTAL, W/HOUSING, CONTROLS, 2 PIPE, 1/2 TON					
Fan coil A/C, horizontal housing, filters, chilled water, 1/2 ton cooling	1.000	Ea.	985	111	1,096
Pipe, black steel, Sch 40, threaded, W/cplgs & hangers, 10' OC, 3/4" dia.	20.000	L.F.	36.34	173.08	209.42
Elbow, 90°, black, straight, 3/4" dia.	6.000	Ea.	11.04	210	221.04
Tee, black, straight, 3/4" dia.	2.000	Ea.	5.84	109	114.84
Union, black with brass seat, 3/4" dia.	2.000	Ea.	15.40	76	91.40
Pipe nipples, 3/4" diam	3.000	Ea.	2.54	12.08	14.62
Valves, bronze, gate, N.R.S., threaded, class 150, 3/4" size	2.000	Ea.	70	49	119
Circuit setter, balance valve, bronze body, threaded, 3/4" pipe size	1.000	Ea.	53	24.50	77.50
Insulation, fiberglass pipe covering, 1" wall, 3/4" IPS	20.000	L.F.	18.20	73.40	91.60
Ductwork, 12" x 8" fabricated, galvanized steel, 12 LF	55.000	Lb.	62.70	324.50	387.20
Insulation, ductwork, blanket type, fiberglass, 1" thk, 1-1/2 LB density	40.000	S.F.	22	96.40	118.40
Diffusers, aluminum, OB damper, ceiling, perf, 24"x24" panel size, 6"x6"	2.000	Ea.	166	62	228
Ductwork, flexible, fiberglass fabric, insulated, 1"thk, PE jacket, 6" dia	16.000	L.F.	27.52	54.72	82.24
Round volume control damper 6" dia.	2.000	Ea.	51	45	96
Fan coil unit balancing	1.000	Ea.		64.50	64.50
Re-heat coil balancing	1.000	Ea.		91	91
Diffuser/register, high, balancing	2.000	Ea.		178	178
TOTAL			1,526.58	1,754.18	3,280.76

D3040 118	Fan Coil A/C Unit, Two Pipe	COST EACH MAT.	COST EACH INST.	COST EACH TOTAL
1010	Fan coil A/C system, cabinet mounted, controls, 2 pipe, 1/2 Ton	785	925	1,710
1020	1 Ton	975	1,050	2,025
1030	1-1/2 Ton	1,075	1,050	2,125
1040	2 Ton	1,675	1,275	2,950
1050	3 Ton	2,300	1,325	3,625

D3040 120	Fan Coil A/C Unit, Two Pipe, Electric Heat	COST EACH MAT.	COST EACH INST.	COST EACH TOTAL
1010	Fan coil A/C system, cabinet mntd, elect. ht, controls, 2 pipe, 1/2 Ton	1,250	925	2,175
1020	1 Ton	1,525	1,050	2,575
1030	1-1/2 Ton	1,800	1,050	2,850
1040	2 Ton	2,700	1,300	4,000
1050	3 Ton	4,550	1,325	5,875

D3040 Distribution Systems

D3040 122	Fan Coil A/C Unit, Four Pipe	COST EACH		
		MAT.	INST.	TOTAL
1010	Fan coil A/C system, cabinet mounted, controls, 4 pipe, 1/2 Ton	1,250	1,800	3,050
1020	1 Ton	1,475	1,925	3,400
1030	1-1/2 Ton	1,600	1,925	3,525
1040	2 Ton	2,000	2,000	4,000
1050	3 Ton	2,950	2,150	5,100

D3040 124	Fan Coil A/C, Horizontal, Duct Mount, 2 Pipe	COST EACH		
		MAT.	INST.	TOTAL
1010	Fan coil A/C system, horizontal w/housing, controls, 2 pipe, 1/2 Ton	1,525	1,750	3,275
1020	1 Ton	2,025	2,550	4,575
1030	1-1/2 Ton	2,525	3,325	5,850
1040	2 Ton	3,225	4,000	7,225
1050	3 Ton	3,950	5,300	9,250
1060	3-1/2 Ton	4,150	5,500	9,650
1070	4 Ton	4,300	6,225	10,525
1080	5 Ton	5,075	7,950	13,025
1090	6 Ton	5,175	8,525	13,700
1100	7 Ton	5,200	8,850	14,050
1110	8 Ton	5,200	9,025	14,225
1120	10 Ton	5,675	9,375	15,050

D3040 126	Fan Coil A/C, Horiz., Duct Mount, 2 Pipe, Elec. Ht.	COST EACH		
		MAT.	INST.	TOTAL
1010	Fan coil A/C system, horiz. hsng, elect. ht, ctrls, 2 pipe, 1/2 Ton	1,750	1,750	3,500
1020	1 Ton	2,275	2,550	4,825
1030	1-1/2 Ton	2,725	3,325	6,050
1040	2 Ton	3,575	4,000	7,575
1050	3 Ton	5,700	5,475	11,175
1060	3-1/2 Ton	6,450	5,500	11,950
1070	4 Ton	7,275	6,250	13,525
1080	5 Ton	8,300	7,950	16,250
1090	6 Ton	8,400	8,550	16,950
1100	7 Ton	8,525	8,500	17,025
1110	8 Ton	9,300	8,600	17,900
1120	10 Ton	9,900	8,850	18,750

D3040 128	Fan Coil A/C, Horizontal, Duct Mount, 4 Pipe	COST EACH		
		MAT.	INST.	TOTAL
1010	Fan coil A/C system, horiz. w/cabinet, controls, 4 pipe, 1/2 Ton	2,125	2,625	4,750
1020	1 Ton	2,725	3,525	6,250
1030	1-1/2 Ton	3,300	4,225	7,525
1040	2 Ton	3,850	4,725	8,575
1050	3 Ton	4,750	6,300	11,050
1060	3.5 Ton	5,425	7,725	13,150
1070	4 Ton	5,450	7,800	13,250
1080	5 Ton	6,625	10,800	17,425
1090	6 Ton	7,000	12,700	19,700
1100	7 Ton	7,900	14,300	22,200
1110	8 Ton	8,025	15,100	23,125
1120	10 Ton	9,700	18,500	28,200

D3040 Distribution Systems

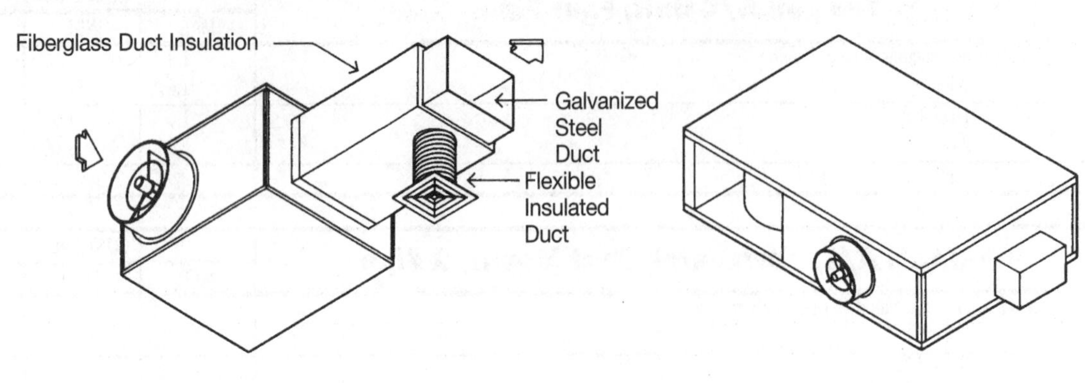

VAV Terminal

Fan Powered VAV Terminal

System Components	QUANTITY	UNIT	COST EACH MAT.	COST EACH INST.	COST EACH TOTAL
SYSTEM D3040 132 1010					
VAV TERMINAL, COOLING ONLY, WITH ACTUATOR / CONTROLS, 200 CFM					
Mixing box variable volume 300 to 600CFM, cool only	1.000	Ea.	341	89.10	430.10
Ductwork, 12" x 8" fabricated, galvanized steel, 12 LF	55.000	Lb.	62.70	324.50	387.20
Insulation, ductwork, blanket type, fiberglass, 1" thk, 1-1/2 LB density	40.000	S.F.	22	96.40	118.40
Diffusers, aluminum, OB damper, ceiling, perf, 24"x24" panel size, 6"x6"	2.000	Ea.	166	62	228
Ductwork, flexible, fiberglass fabric, insulated, 1"thk, PE jacket, 6" dia	16.000	L.F.	27.52	54.72	82.24
Round volume control damper 6" dia.	2.000	Ea.	51	45	96
VAV box balancing	1.000	Ea.		59.50	59.50
Diffuser/register, high, balancing	2.000	Ea.		178	178
TOTAL			670.22	909.22	1,579.44

D3040 132	VAV Terminal, Cooling Only	COST EACH MAT.	COST EACH INST.	COST EACH TOTAL
1010	VAV Terminal, cooling only, with actuator / controls, 200 CFM	670	910	1,580
1020	400 CFM	880	1,600	2,480
1030	600 CFM	1,175	2,375	3,550
1040	800 CFM	1,275	2,900	4,175
1050	1000 CFM	1,475	3,350	4,825
1060	1250 CFM	1,675	4,250	5,925
1070	1500 CFM	1,850	5,025	6,875
1080	2000 CFM	2,400	7,150	9,550

D3040 134	VAV Terminal, Hot Water Reheat	COST EACH MAT.	COST EACH INST.	COST EACH TOTAL
1010	VAV Terminal, cooling, HW reheat, with actuator / controls, 200 CFM	1,550	1,675	3,225
1020	400 CFM	1,750	2,300	4,050
1030	600 CFM	2,025	3,125	5,150
1040	800 CFM	2,150	3,675	5,825
1050	1000 CFM	2,350	4,125	6,475
1060	1250 CFM	2,600	5,075	7,675
1070	1500 CFM	2,875	6,000	8,875
1080	2000 CFM	3,475	8,175	11,650

D3040 Distribution Systems

D3040 136	Fan Powered VAV Terminal, Cooling Only	COST EACH		
		MAT.	INST.	TOTAL
1010	VAV Terminal, cooling, fan powrd, with actuator / controls, 200 CFM	1,225	910	2,135
1020	400 CFM	1,425	1,500	2,925
1030	600 CFM	1,725	2,375	4,100
1040	800 CFM	1,850	2,850	4,700
1050	1000 CFM	2,025	3,275	5,300
1060	1250 CFM	2,350	4,325	6,675
1070	1500 CFM	2,500	5,050	7,550
1080	2000 CFM	3,150	7,350	10,500

D3040 138	Fan Powered VAV Terminal, Hot Water Reheat	COST EACH		
		MAT.	INST.	TOTAL
1010	VAV Terminal, cool, HW reht, fan powrd, with actuator/ctrls, 200 CFM	1,975	1,675	3,650
1020	400 CFM	2,200	2,300	4,500
1030	600 CFM	2,475	3,150	5,625
1040	800 CFM	2,600	3,625	6,225
1050	1000 CFM	2,775	4,075	6,850
1060	1250 CFM	3,200	5,100	8,300
1070	1500 CFM	3,475	6,050	9,525
1080	2000 CFM	4,100	8,400	12,500

D3040 Distribution Systems

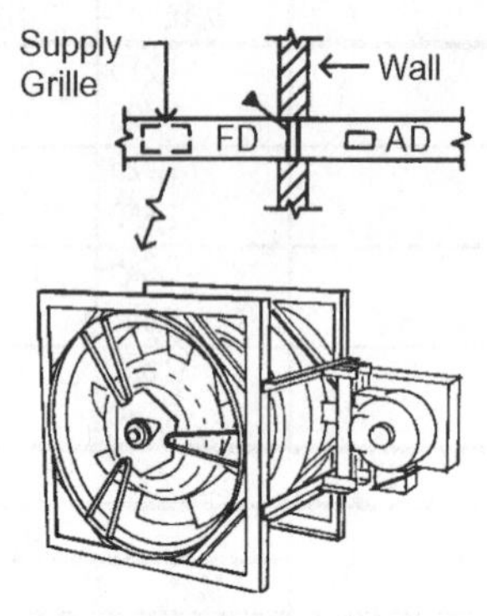

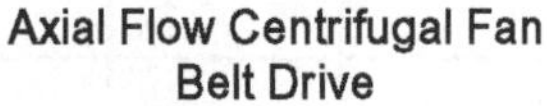
Axial Flow Centrifugal Fan Belt Drive

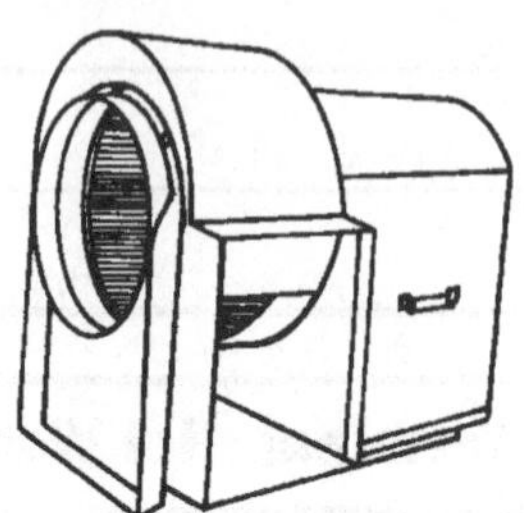
Belt Drive Utility Set

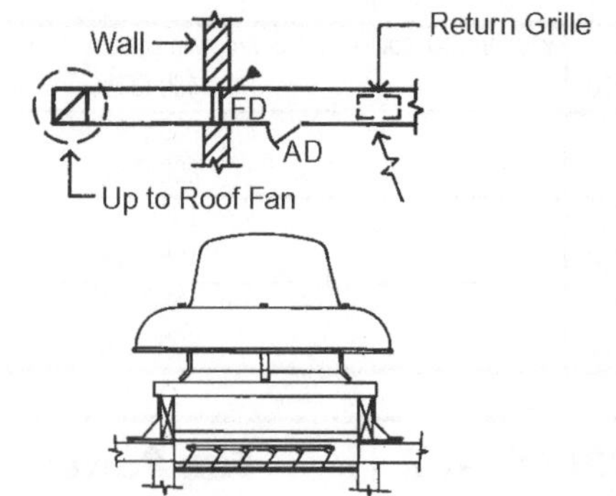

Centrifugal Roof Exhaust Fan Direct Drive

System Components	QUANTITY	UNIT	COST EACH MAT.	COST EACH INST.	COST EACH TOTAL
SYSTEM D3040 220 1010					
FAN SYSTEM, IN-LINE CENTRIFUGAL, 500 CFM					
Fans, in-line centrifugal, supply/exhaust, 500 CFM, 10" dia conn	1.000	Ea.	990	380	1,370
Louver, alum., W/screen, damper, mill fin, fxd blade, cont line, stormproof	4.050	S.F.	131.63	71.48	203.11
Motor for louver damper, electric or pneumatic	1.000	Ea.	385	35.50	420.50
Ductwork, 12" x 8" fabricated, galvanized steel, 60 LF	200.000	Lb.	239.20	1,469	1,708.20
Grilles, aluminum, air supply, single deflection, adjustable, 12" x 6"	1.000	Ea.	51.60	64.50	116.10
Fire damper, curtain type, vertical, 12" x 8"	1.000	Ea.	23	22.50	45.50
Duct access door, insulated, 10" x 10"	1.000	Ea.	18.60	45	63.60
Duct access door, insulated, 16" x 12"	1.000	Ea.	28.50	55	83.50
Pressure controller/switch, air flow controller	1.000	Ea.	126	41	167
Fan control switch and mounting base	1.000	Ea.	73.50	31	104.50
Fan balancing	1.000	Ea.		297	297
Diffuser/register balancing	3.000	Ea.		178.50	178.50
TOTAL			2,067.03	2,690.48	4,757.51

D3040 220	Centrifugal In Line Fan Systems	COST EACH MAT.	COST EACH INST.	COST EACH TOTAL
1010	Fan system, in-line centrifugal, 500 CFM	2,075	2,700	4,775
1020	1300 CFM	2,500	4,675	7,175
1030	1500 CFM	2,825	5,400	8,225
1040	2500 CFM	4,425	16,500	20,925
1050	3500 CFM	5,600	22,200	27,800
1060	5000 CFM	7,275	32,900	40,175
1070	7500 CFM	8,400	33,600	42,000
1080	10,000 CFM	10,200	39,400	49,600

D3040 230	Utility Set Fan Systems	COST EACH MAT.	COST EACH INST.	COST EACH TOTAL
1010	Utility fan set system, belt drive, 2000 CFM	1,875	8,725	10,600
1020	3500 CFM	3,250	15,400	18,650
1030	5000 CFM	4,150	21,400	25,550
1040	7500 CFM	5,575	24,900	30,475
1050	10,000 CFM	7,375	32,500	39,875
1060	15,000 CFM	7,950	33,200	41,150
1070	20,000 CFM	8,800	37,700	46,500

Important: See the Reference Section for critical supporting data - Reference Numbers and City Cost Indexes

D3040 Distribution Systems

D3040 240	Roof Exhaust Fan Systems	COST EACH		
		MAT.	INST.	TOTAL
1010	Roof vent. system, centrifugal, alum., galv curb, BDD, 500 CFM	745	1,775	2,520
1020	800 CFM	1,025	3,400	4,425
1030	1500 CFM	1,325	5,075	6,400
1040	2750 CFM	2,575	11,500	14,075
1050	3500 CFM	3,200	14,700	17,900
1060	5000 CFM	4,650	24,300	28,950
1070	8500 CFM	5,825	31,100	36,925
1080	13,800 CFM	8,325	45,200	53,525

D3040 Controls & Instrumentation

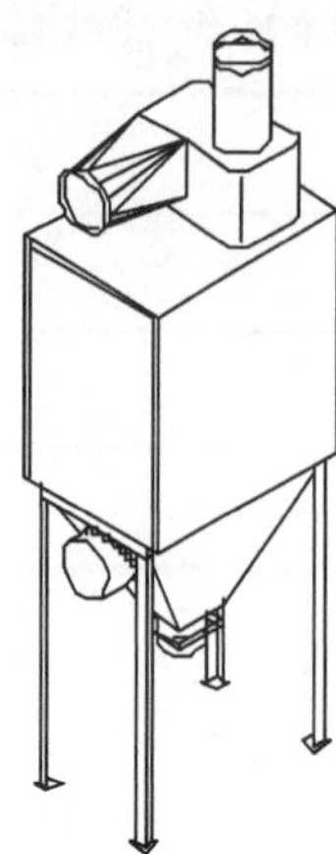

Commercial/Industrial Dust Collection System

System Components	QUANTITY	UNIT	COST EACH MAT.	COST EACH INST.	COST EACH TOTAL
SYSTEM D3040 250 1010					
COMMERCIAL/INDUSTRIAL VACUUM DUST COLLECTION SYSTEM, 500 CFM					
Dust collection central vac unit inc. stand, filter & shaker, 500 CFM 2 HP	1.000	Ea.	3,825	470	4,295
Galvanized tubing, 16 ga., 4″ OD	80.000	L.F.	345.60	197.60	543.20
Galvanized tubing, 90° ell, 4″ dia.	3.000	Ea.	73.50	53.40	126.90
Galvanized tubing, 45° ell, 4″ dia.	4.000	Ea.	86	71.20	157.20
Galvanized tubing, TY slip fit, 4″ dia.	4.000	Ea.	208	118	326
Flexible rubber hose, 4″ dia.	32.000	L.F.	278.40	79.04	357.44
Galvanized air gate valve, 4″ dia.	4.000	Ea.	624	154	778
Galvanized compression coupling, 4″ dia.	36.000	Ea.	738	972	1,710
Pipe hangers, steel, 4″ pipe size	16.000	Ea.	184.80	128.80	313.60
TOTAL			6,363.30	2,244.04	8,607.34

D3040 250	Commercial/Industrial Vacuum Dust Collection	COST EACH MAT.	COST EACH INST.	COST EACH TOTAL
1010	Commercial / industrial, vacuum dust collection system, 500 CFM	6,375	2,250	8,625
1020	1000 CFM	8,450	2,550	11,000
1030	1500 CFM	10,200	2,825	13,025
1040	3000 CFM	20,200	3,950	24,150
1050	5000 CFM	30,000	5,925	35,925

D3040 Controls & Instrumentation

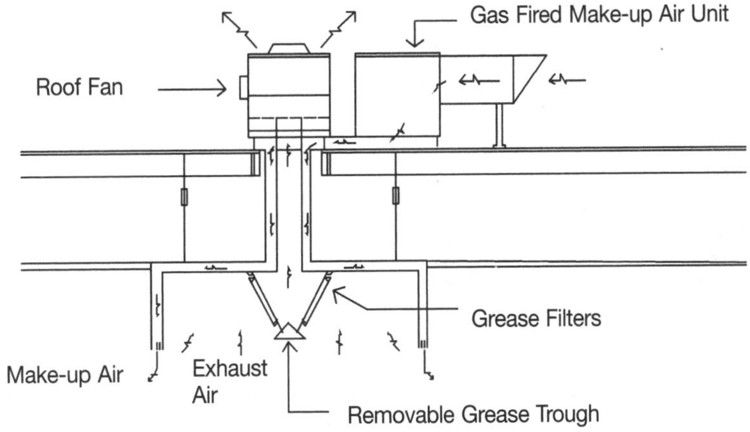

Commercial Kitchen Exhaust/Make-up Air Rooftop System

System Components	QUANTITY	UNIT	COST EACH MAT.	COST EACH INST.	COST EACH TOTAL
SYSTEM D3040 260 1010					
COMMERCIAL KITCHEN EXHAUST/MAKE-UP AIR SYSTEM, ROOFTOP, GAS, 2000 CFM					
Make-up air rooftop unit, gas, 2090 CFM, 159 MBH	1.000	Ea.	5,751	496.80	6,247.80
Pipe, black steel, Sch 40, thrded, nipples, W/cplngs/hngrs, 10′ OC, 1″ dia.	20.000	L.F.	52.36	203.50	255.86
Elbow, 90°, black steel, straight, 3/4″ dia.	2.000	Ea.	3.68	70	73.68
Elbow, 90°, black steel, straight, 1″ dia.	3.000	Ea.	9.57	114	123.57
Tee, black steel, reducing, 1″ dia.	1.000	Ea.	5.70	61.50	67.20
Union, black with brass seat, 1″ dia.	1.000	Ea.	11	41	52
Pipe nipples, black, 3/4″ dia	2.000	Ea.	3.38	16.10	19.48
Pipe cap, malleable iron, black, 1″ dia.	1.000	Ea.	2.51	16.35	18.86
Gas cock, brass, 1″ size	1.000	Ea.	12.50	26	38.50
Ductwork, 20″ x 14″ fabricated, galv. steel, 30 LF	220.000	Lb.	178.20	1,199	1,377.20
Ductwork, 18″ x 10″ fabricated, galv. steel, 12 LF	72.000	Lb.	58.32	392.40	450.72
Ductwork, 30″ x 10″ fabricated, galv. steel, 30 LF	260.000	Lb.	210.60	1,417	1,627.60
Ductwork, 16″ x 10″ fabricated, galv. steel, 12 LF	70.000	Lb.	56.70	381.50	438.20
Kitchen ventilation, island style, water wash, Stainless Steel	5.000	L.F.	9,500	975	10,475
Control system, Make-up Air Unit / Exhauster	1.000	Ea.	2,150	1,725	3,875
Rooftop unit heat/cool balancing	1.000	Ea.		345	345
Roof fan balancing	1.000	Ea.		198	198
TOTAL			18,005.52	7,678.15	25,683.67

D3040 260	Kitchen Exhaust/Make-Up Air	COST EACH MAT.	COST EACH INST.	COST EACH TOTAL
1010	Commercial kitchen exhaust/make-up air system, rooftop, gas, 2000 CFM	18,000	7,675	25,675
1020	3000 CFM	19,300	10,600	29,900
1030	5000 CFM	24,000	12,700	36,700
1040	8000 CFM	38,200	19,000	57,200
1050	12,000 CFM	56,000	23,000	79,000
1060	16,000 CFM	69,500	36,600	106,100

D3040 Distribution Systems

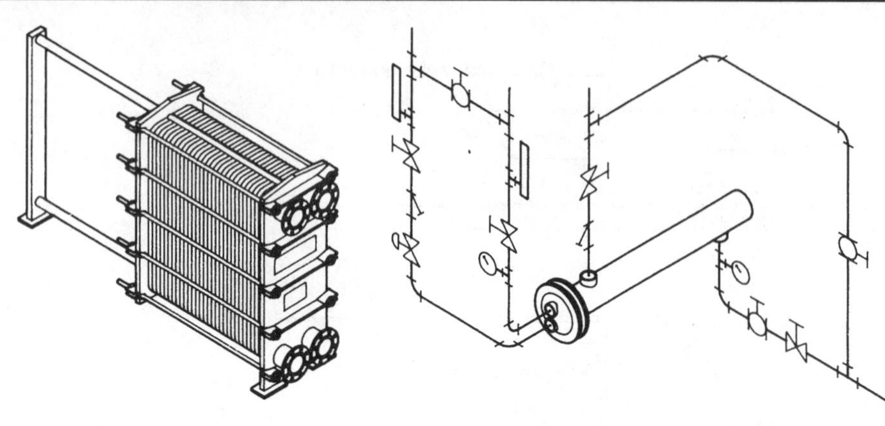

Plate Heat Exchanger

Shell and Tube Heat Exchanger

System Components	QUANTITY	UNIT	COST EACH MAT.	COST EACH INST.	COST EACH TOTAL
SYSTEM D3040 620 1010					
SHELL & TUBE HEAT EXCHANGER, 40 GPM					
Heat exchanger, 4 pass, 3/4" O.D. copper tubes, by steam at 10 psi, 40 GPM	1.000	Ea.	2,475	222	2,697
Pipe, black steel, Sch 40, threaded, W/couplings & hangers, 10' OC, 2" dia.	40.000	L.F.	229.78	696.90	926.68
Elbow, 90°, straight, 2" dia.	16.000	Ea.	190.40	784	974.40
Reducer, black steel, concentric, 2" dia.	2.000	Ea.	26.80	84	110.80
Valves, bronze, gate, rising stem, threaded, class 150, 2" size	6.000	Ea.	606	267	873
Valves, bronze, globe, class 150, rising stem, threaded, 2" size	2.000	Ea.	526	89	615
Strainers, Y type, bronze body, screwed, 150 lb., 2" pipe size	2.000	Ea.	118	76	194
Valve, electric motor actuated, brass, 2 way, screwed, 1-1/2" pipe size	1.000	Ea.	415	38.50	453.50
Union, black with brass seat, 2" dia.	8.000	Ea.	184	416	600
Tee, black, straight, 2" dia.	9.000	Ea.	134.55	724.50	859.05
Tee, black, reducing run and outlet, 2" dia.	4.000	Ea.	108	322	430
Pipe nipple, black, 2" dia.	21.000	Ea.	2.28	6.90	9.18
Thermometers, stem type, 9" case, 8" stem, 3/4" NPT	2.000	Ea.	260	35.20	295.20
Gauges, pressure or vacuum, 3-1/2" diameter dial	2.000	Ea.	44	30.80	74.80
Insulation, fiberglass pipe covering, 1" wall, 2" IPS	40.000	L.F.	49.20	168.80	218
Heat exchanger control system	1.000	Ea.	2,100	1,600	3,700
Coil balancing	1.000	Ea.		91	91
TOTAL			7,469.01	5,652.60	13,121.61

D3040 610	Heat Exchanger, Plate Type	COST EACH MAT.	COST EACH INST.	COST EACH TOTAL
1010	Plate heat exchanger, 400 GPM	29,100	11,700	40,800
1020	800 GPM	46,200	14,700	60,900
1030	1200 GPM	59,500	19,900	79,400
1040	1800 GPM	87,500	24,700	112,200

D3040 620	Heat Exchanger, Shell & Tube	COST EACH MAT.	COST EACH INST.	COST EACH TOTAL
1010	Shell & tube heat exchanger, 40 GPM	7,475	5,650	13,125
1020	96 GPM	12,900	10,300	23,200
1030	240 GPM	23,200	13,900	37,100
1040	600 GPM	43,900	19,400	63,300

Important: See the Reference Section for critical supporting data - Reference Numbers and City Cost Indexes

D3050 Terminal & Package Units

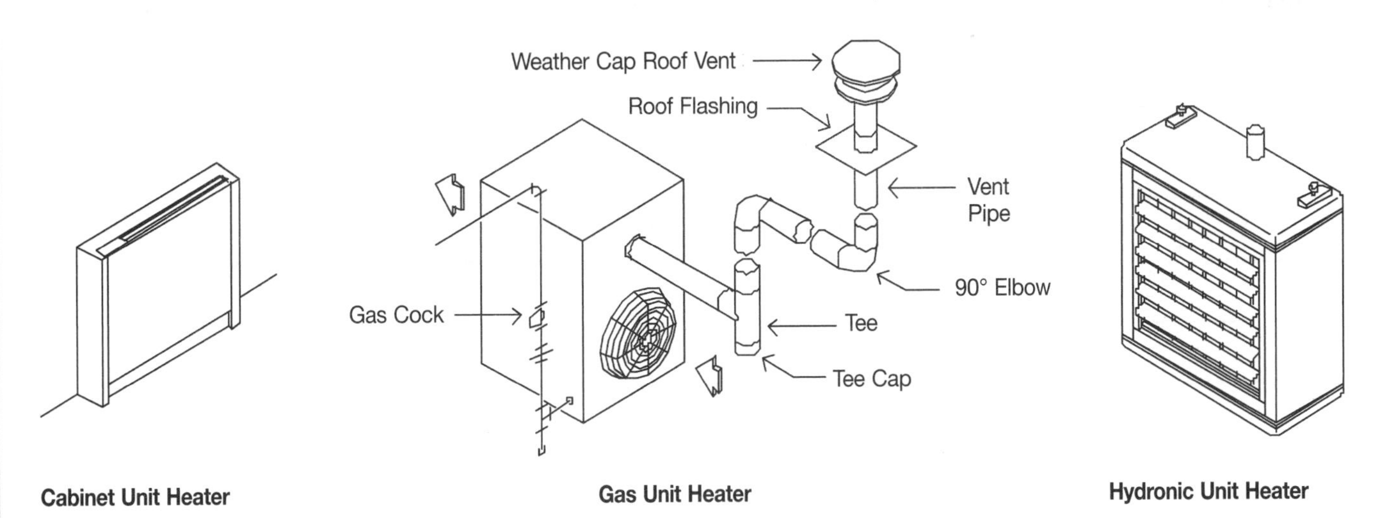

System Components	QUANTITY	UNIT	COST EACH MAT.	COST EACH INST.	COST EACH TOTAL
SYSTEM D3050 120 1010					
SPACE HEATER, SUSPENDED, GAS FIRED, PROPELLER FAN, 20 MBH					
Space heater, propeller fan, 20 MBH output	1.000	Ea.	495	104	599
Pipe, black steel, Sch 40, threaded, W/coupling & hangers, 10′ OC, 3/4″ dia	20.000	L.F.	35.49	169.05	204.54
Elbow, 90°, black steel, straight, 1/2″ dia.	3.000	Ea.	4.56	99	103.56
Elbow, 90°, black steel, straight, 3/4″ dia.	3.000	Ea.	5.52	105	110.52
Tee, black steel, reducing, 3/4″ dia.	1.000	Ea.	4.50	54.50	59
Union, black with brass seat, 3/4″ dia.	1.000	Ea.	7.70	38	45.70
Pipe nipples, black, 1/2″ dia	2.000	Ea.	2.82	15.60	18.42
Pipe nipples, black, 3/4″ dia	2.000	Ea.	1.69	8.05	9.74
Pipe cap, black, 3/4″ dia.	1.000	Ea.	2.07	15.35	17.42
Gas cock, brass, 3/4″ size	1.000	Ea.	11.90	22.50	34.40
Thermostat, 1 set back, electric, timed	1.000	Ea.	89	62	151
Wiring, thermostat hook-up	1.000	Ea.	5.75	20	25.75
Vent chimney, prefab metal, U.L. listed, gas, double wall, galv. st, 4″ dia	12.000	L.F.	62.40	157.20	219.60
Vent chimney, gas, double wall, galv. steel, elbow 90°, 4″ dia.	2.000	Ea.	42	52	94
Vent chimney, gas, double wall, galv. steel, Tee, 4″ dia.	1.000	Ea.	26.50	34.50	61
Vent chimney, gas, double wall, galv. steel, T cap, 4″ dia.	1.000	Ea.	2.04	21	23.04
Vent chimney, gas, double wall, galv. steel, roof flashing, 4″ dia.	1.000	Ea.	6.95	26	32.95
Vent chimney, gas, double wall, galv. steel, top, 4″ dia.	1.000	Ea.	10	20	30
TOTAL			815.89	1,023.75	1,839.64

D3050 120	Unit Heaters, Gas	COST EACH MAT.	COST EACH INST.	COST EACH TOTAL
1010	Space heater, suspended, gas fired, propeller fan, 20 MBH	815	1,025	1,840
1020	60 MBH	960	1,075	2,035
1030	100 MBH	1,200	1,125	2,325
1040	160 MBH	1,400	1,250	2,650
1050	200 MBH	1,725	1,375	3,100
1060	280 MBH	2,250	1,450	3,700
1070	320 MBH	2,975	1,650	4,625

D3050 130	Unit Heaters, Hydronic	COST EACH MAT.	COST EACH INST.	COST EACH TOTAL
1010	Space heater, suspended, horiz. mount, HW, prop. fan, 20 MBH	1,400	1,075	2,475
1020	60 MBH	1,800	1,225	3,025

D3050 Terminal & Package Units

D3050 130	Unit Heaters, Hydronic	COST EACH		
		MAT.	INST.	TOTAL
1030	100 MBH	2,125	1,325	3,450
1040	150 MBH	2,675	1,475	4,150
1050	200 MBH	2,850	1,650	4,500
1060	300 MBH	3,125	1,950	5,075

D3050 140	Cabinet Unit Heaters, Hydronic	COST EACH		
		MAT.	INST.	TOTAL
1010	Unit heater, cabinet type, horizontal blower, hot water, 20 MBH	1,700	1,075	2,775
1020	60 MBH	2,450	1,175	3,625
1030	100 MBH	2,700	1,300	4,000
1040	120 MBH	2,750	1,325	4,075

D3050 Terminal & Package Units

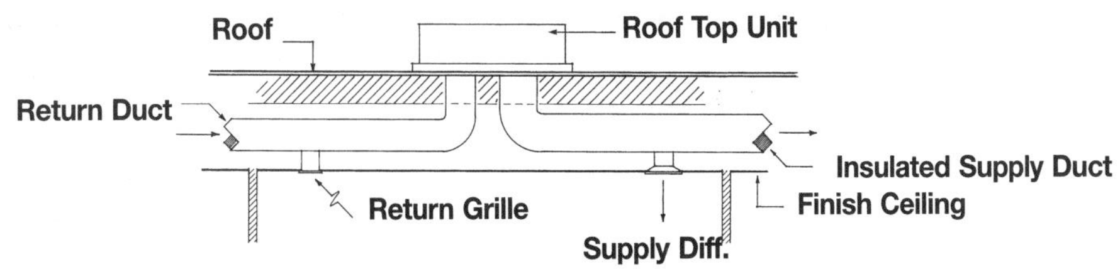

System Description: Rooftop single zone units are electric cooling and gas heat. Duct systems are low velocity, galvanized steel supply and return. Price variations between sizes are due to several factors. Jumps in the cost of the rooftop unit occur when the manufacturer shifts from the largest capacity unit on a small frame to the smallest capacity on the next larger frame, or changes from one compressor to two. As the unit capacity increases for larger areas the duct distribution grows in proportion. For most applications there is a tradeoff point where it is less expensive and more efficient to utilize smaller units with short simple distribution systems. Larger units also require larger initial supply and return ducts which can create a space problem. Supplemental heat may be desired in colder locations. The table below is based on one unit supplying the area listed. The 10,000 S.F. unit for bars and taverns is not listed because a nominal 110 ton unit would be required and this is above the normal single zone rooftop capacity.

System Components	QUANTITY	UNIT	COST EACH MAT.	COST EACH INST.	COST EACH TOTAL
SYSTEM D3050 150 1280					
ROOFTOP, SINGLE ZONE, AIR CONDITIONER					
APARTMENT CORRIDORS, 500 S.F., .92 TON					
Rooftop air-conditioner, 1 zone, electric cool, standard controls, curb	1.000	Ea.	1,288	586.50	1,874.50
Ductwork package for rooftop single zone units	1.000	System	285.20	855.60	1,140.80
TOTAL			1,573.20	1,442.10	3,015.30
COST PER S.F.			3.15	2.88	6.03
*Size would suggest multiple units					

D3050 150	Rooftop Single Zone Unit Systems	COST PER S.F. MAT.	COST PER S.F. INST.	COST PER S.F. TOTAL
1260	Rooftop, single zone, air conditioner			
1280	Apartment corridors, 500 S.F., .92 ton	3.15	2.90	6.05
1320	1,000 S.F., 1.83 ton (R15700-020)	3.13	2.87	6
1360	1500 S.F., 2.75 ton	2.44	2.36	4.80
1400	3,000 S.F., 5.50 ton	2.20	2.29	4.49
1440	5,000 S.F., 9.17 ton	2.10	2.09	4.19
1480	10,000 S.F., 18.33 ton	2.17	1.96	4.13
1560	Banks or libraries, 500 S.F., 2.08 ton	7.10	6.50	13.60
1600	1,000 S.F., 4.17 ton	5.55	5.40	10.95
1640	1,500 S.F., 6.25 ton	5	5.20	10.20
1680	3,000 S.F., 12.50 ton	4.77	4.74	9.51
1720	5,000 S.F., 20.80 ton	4.91	4.46	9.37
1760	10,000 S.F., 41.67 ton	4.96	4.44	9.40
1840	Bars and taverns, 500 S.F. 5.54 ton	11.60	8.65	20.25
1880	1,000 S.F., 11.08 ton	10.95	7.45	18.40
1920	1,500 S.F., 16.62 ton	11.40	6.95	18.35
1960	3,000 S.F., 33.25 ton	11.70	6.65	18.35
2000	5,000 S.F., 55.42 ton	10.40	6.65	17.05
2040	10,000 S.F., 110.83 ton*			
2080	Bowling alleys, 500 S.F., 2.83 ton	7.50	7.30	14.80
2120	1,000 S.F., 5.67 ton	6.80	7.05	13.85
2160	1,500 S.F., 8.50 ton	6.50	6.45	12.95
2200	3,000 S.F., 17.00 ton	6.70	6.20	12.90
2240	5,000 S.F., 28.33 ton	6.85	6.05	12.90
2280	10,000 S.F., 56.67 ton	6.20	6.05	12.25
2360	Department stores, 500 S.F., 1.46 ton	4.99	4.58	9.57

D3050 Terminal & Package Units

D3050 150	Rooftop Single Zone Unit Systems	COST PER S.F. MAT.	COST PER S.F. INST.	COST PER S.F. TOTAL
2400	1,000 S.F., 2.92 ton	3.88	3.78	7.66
2440	1,500 S.F., 4.37 ton	3.50	3.63	7.13
2480	3,000 S.F., 8.75 ton	3.34	3.32	6.66
2520	5,000 S.F., 14.58 ton	3.45	3.18	6.63
2560	10,000 S.F., 29.17 ton	3.52	3.10	6.62
2640	Drug stores, 500 S.F., 3.33 ton	8.85	8.60	17.45
2680	1,000 S.F., 6.67 ton	8	8.30	16.30
2720	1,500 S.F., 10.00 ton	7.65	7.60	15.25
2760	3,000 S.F., 20.00 ton	7.85	7.15	15
2800	5,000 S.F., 33.33 ton	8.05	7.10	15.15
2840	10,000 S.F., 66.67 ton	7.30	7.10	14.40
2920	Factories, 500 S.F., 1.67 ton	5.70	5.25	10.95
2960	1,000 S.F., 3.33 ton	4.42	4.31	8.73
3000	1,500 S.F., 5.00 ton	4	4.15	8.15
3040	3,000 S.F., 10.00 ton	3.81	3.79	7.60
3080	5,000 S.F., 16.67 ton	3.94	3.64	7.58
3120	10,000 S.F., 33.33 ton	4.03	3.54	7.57
3200	Food supermarkets, 500 S.F., 1.42 ton	4.84	4.45	9.29
3240	1,000 S.F., 2.83 ton	3.73	3.65	7.38
3280	1,500 S.F., 4.25 ton	3.40	3.53	6.93
3320	3,000 S.F., 8.50 ton	3.25	3.23	6.48
3360	5,000 S.F., 14.17 ton	3.35	3.10	6.45
3400	10,000 S.F., 28.33 ton	3.43	3.01	6.44
3480	Medical centers, 500 S.F., 1.17 ton	3.99	3.67	7.66
3520	1,000 S.F., 2.33 ton	3.98	3.66	7.64
3560	1,500 S.F., 3.50 ton	3.10	3.02	6.12
3600	3,000 S.F., 7.00 ton	2.80	2.91	5.71
3640	5,000 S.F., 11.67 ton	2.67	2.65	5.32
3680	10,000 S.F., 23.33 ton	2.75	2.50	5.25
3760	Offices, 500 S.F., 1.58 ton	5.40	4.96	10.36
3800	1,000 S.F., 3.17 ton	4.21	4.10	8.31
3840	1,500 S.F., 4.75 ton	3.80	3.95	7.75
3880	3,000 S.F., 9.50 ton	3.62	3.61	7.23
3920	5,000 S.F., 15.83 ton	3.75	3.45	7.20
3960	10,000 S.F., 31.67 ton	3.83	3.37	7.20
4000	Restaurants, 500 S.F., 2.50 ton	8.55	7.85	16.40
4040	1,000 S.F., 5.00 ton	6	6.25	12.25
4080	1,500 S.F., 7.50 ton	5.75	5.70	11.45
4120	3,000 S.F., 15.00 ton	5.90	5.45	11.35
4160	5,000 S.F., 25.00 ton	5.90	5.35	11.25
4200	10,000 S.F., 50.00 ton	5.45	5.30	10.75
4240	Schools and colleges, 500 S.F., 1.92 ton	6.55	6	12.55
4280	1,000 S.F., 3.83 ton	5.10	4.95	10.05
4320	1,500 S.F., 5.75 ton	4.60	4.78	9.38
4360	3,000 S.F., 11.50 ton	4.39	4.37	8.76
4400	5,000 S.F., 19.17 ton	4.52	4.11	8.63
4440	10,000 S.F., 38.33 ton	4.57	4.07	8.64
5000				
5100				
5200				
9000	Components of ductwork packages for above systems, per ton of cooling:			
9010	Ductwork; galvanized steel, 120 pounds			
9020	Insulation; fiberglass 2" thick, FRK faced, 52 SF			
9030	Diffusers; aluminum, 24" x 12", one			
9040	Registers; aluminum, return with dampers, one			

Important: See the Reference Section for critical supporting data - Reference Numbers and City Cost Indexes

D3050 Terminal & Package Units

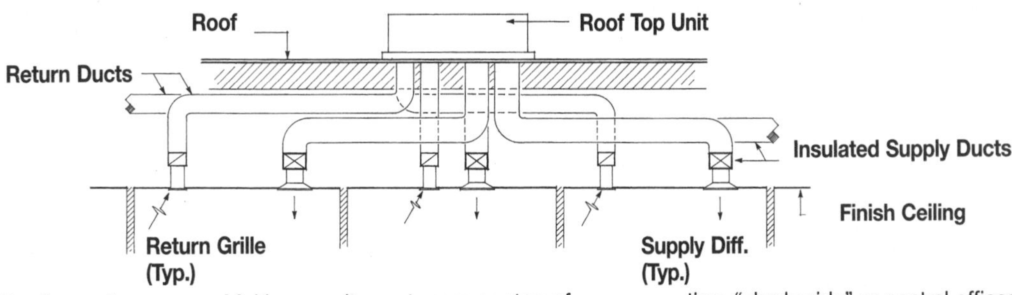

System Description: Rooftop units are multizone with up to 12 zones, and include electric cooling, gas heat, thermostats, filters, supply and return fans complete. Duct systems are low velocity, galvanized steel supply and return with insulated supplies.

Multizone units cost more per ton of cooling than single zone. However, they offer flexibility where load conditions are varied due to heat generating areas or exposure to radiational heating. For example, perimeter offices on the "sunny side" may require cooling at the same time "shady side" or central offices may require heating. It is possible to accomplish similar results using duct heaters in branches of the single zone unit. However, heater location could be a problem and total system operating energy efficiency could be lower.

System Components	QUANTITY	UNIT	COST EACH MAT.	COST EACH INST.	COST EACH TOTAL
SYSTEM D3050 155 1280					
ROOFTOP, MULTIZONE, AIR CONDITIONER					
APARTMENT CORRIDORS, 3,000 S.F., 5.50 TON					
Rooftop multizone unit, standard controls, curb	1.000	Ea.	35,640	1,353	36,993
Ductwork package for rooftop multizone units	1.000	System	2,447.50	9,762.50	12,210
TOTAL			38,087.50	11,115.50	49,203
COST PER S.F.			12.70	3.71	16.41
Note A: Small single zone unit recommended					
Note B: A combination of multizone units recommended					

D3050 155	Rooftop Multizone Unit Systems		COST PER S.F. MAT.	COST PER S.F. INST.	COST PER S.F. TOTAL
1240	Rooftop, multizone, air conditioner				
1260	Apartment corridors, 1,500 S.F., 2.75 ton. See Note A.				
1280	3,000 S.F., 5.50 ton	R15700 -020	12.70	3.70	16.40
1320	10,000 S.F., 18.30 ton		9.35	3.57	12.92
1360	15,000 S.F., 27.50 ton		8.05	3.55	11.60
1400	20,000 S.F., 36.70 ton		8.15	3.57	11.72
1440	25,000 S.F., 45.80 ton		7.05	3.58	10.63
1520	Banks or libraries, 1,500 S.F., 6.25 ton		29	8.45	37.45
1560	3,000 S.F., 12.50 ton		24.50	8.25	32.75
1600	10,000 S.F., 41.67 ton		16	8.15	24.15
1640	15,000 S.F., 62.50 ton		10.45	8.10	18.55
1680	20,000 S.F., 83.33 ton		10.45	8.10	18.55
1720	25,000 S.F., 104.00 ton		9.40	8.05	17.45
1800	Bars and taverns, 1,500 S.F., 16.62 ton		62.50	12.10	74.60
1840	3,000 S.F., 33.24 ton		47	11.70	58.70
1880	10,000 S.F., 110.83 ton		22.50	11.65	34.15
1920	15,000 S.F., 165 ton, See Note B				
1960	20,000 S.F., 220 ton, See Note B				
2000	25,000 S.F., 275 ton, See Note B				
2080	Bowling alleys, 1,500 S.F., 8.50 ton		39	11.45	50.45
2120	3,000 S.F., 17.00 ton		33	11.20	44.20
2160	10,000 S.F., 56.70 ton		22	11.10	33.10
2200	15,000 S.F., 85.00 ton		14.20	11	25.20
2240	20,000 S.F., 113.00 ton		12.80	10.95	23.75
2280	25,000 S.F., 140.00 ton see Note B				
2360	Department stores, 1,500 S.F., 4.37 ton, See Note A.				

D3050 Terminal & Package Units

D3050 155	Rooftop Multizone Unit Systems	COST PER S.F.		
		MAT.	INST.	TOTAL
2400	3,000 S.F., 8.75 ton	20	5.90	25.90
2440	10,000 S.F., 29.17 ton	12.95	5.65	18.60
2480	15,000 S.F., 43.75 ton	11.20	5.70	16.90
2520	20,000 S.F., 58.33 ton	7.30	5.65	12.95
2560	25,000 S.F., 72.92 ton	7.30	5.65	12.95
2640	Drug stores, 1,500 S.F., 10.00 ton	46	13.45	59.45
2680	3,000 S.F., 20.00 ton	34	13	47
2720	10,000 S.F., 66.66 ton	16.75	12.95	29.70
2760	15,000 S.F., 100.00 ton	15.10	12.90	28
2800	20,000 S.F., 135 ton, See Note B			
2840	25,000 S.F., 165 ton, See Note B			
2920	Factories, 1,500 S.F., 5 ton, See Note A			
2960	3,000 S.F., 10.00 ton	23	6.75	29.75
3000	10,000 S.F., 33.33 ton	14.80	6.50	21.30
3040	15,000 S.F., 50.00 ton	12.80	6.50	19.30
3080	20,000 S.F., 66.66 ton	8.35	6.50	14.85
3120	25,000 S.F., 83.33 ton	8.35	6.50	14.85
3200	Food supermarkets, 1,500 S.F., 4.25 ton, See Note A			
3240	3,000 S.F., 8.50 ton	19.60	5.75	25.35
3280	10,000 S.F., 28.33 ton	12.45	5.50	17.95
3320	15,000 S.F., 42.50 ton	10.85	5.55	16.40
3360	20,000 S.F., 56.67 ton	7.10	5.50	12.60
3400	25,000 S.F., 70.83 ton	7.10	5.50	12.60
3480	Medical centers, 1,500 S.F., 3.5 ton, See Note A			
3520	3,000 S.F., 7.00 ton	16.15	4.71	20.86
3560	10,000 S.F., 23.33 ton	10.95	4.53	15.48
3600	15,000 S.F., 35.00 ton	10.35	4.53	14.88
3640	20,000 S.F., 46.66 ton	8.95	4.56	13.51
3680	25,000 S.F., 58.33 ton	5.85	4.53	10.38
3760	Offices, 1,500 S.F., 4.75 ton, See Note A			
3800	3,000 S.F., 9.50 ton	22	6.40	28.40
3840	10,000 S.F., 31.66 ton	14.10	6.15	20.25
3880	15,000 S.F., 47.50 ton	12.15	6.20	18.35
3920	20,000 S.F., 63.33 ton	7.95	6.15	14.10
3960	25,000 S.F., 79.16 ton	7.95	6.15	14.10
4000	Restaurants, 1,500 S.F., 7.50 ton	34.50	10.10	44.60
4040	3,000 S.F., 15.00 ton	29	9.90	38.90
4080	10,000 S.F., 50.00 ton	19.20	9.80	29
4120	15,000 S.F., 75.00 ton	12.55	9.70	22.25
4160	20,000 S.F., 100.00 ton	11.30	9.70	21
4200	25,000 S.F., 125 ton, See Note B			
4240	Schools and colleges, 1,500 S.F., 5.75 ton	26.50	7.75	34.25
4280	3,000 S.F., 11.50 ton	22.50	7.60	30.10
4320	10,000 S.F., 38.33 ton	14.70	7.50	22.20
4360	15,000 S.F., 57.50 ton	9.60	7.45	17.05
4400	20,000 S.F., 76.66 ton	9.60	7.45	17.05
4440	25,000 S.F., 95.83 ton	9.15	7.45	16.60
4450				
9000	Components of ductwork packages for above systems, per ton of cooling:			
9010	Ductwork; galvanized steel, 240 pounds			
9020	Insulation; fiberglass, 2" thick, FRK faced, 104 SF			
9030	Diffusers; aluminum, 24" x 12", two			
9040	Registers; aluminum, return with dampers, one			

Important: See the Reference Section for critical supporting data - Reference Numbers and City Cost Indexes

D3050 Terminal & Package Units

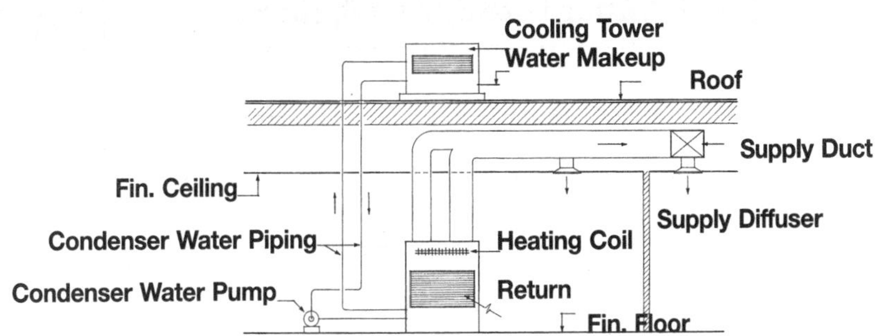

System Description: Self-contained, single package water cooled units include cooling tower, pump, piping allowance. Systems for 1000 S.F. and up include duct and diffusers to provide for even distribution of air. Smaller units distribute air through a supply air plenum, which is integral with the unit.

Returns are not ducted and supplies are not insulated.

Hot water or steam heating coils are included but piping to boiler and the boiler itself is not included.

Where local codes or conditions permit single pass cooling for the smaller units, deduct 10%.

System Components	QUANTITY	UNIT	COST EACH MAT.	COST EACH INST.	COST EACH TOTAL
SYSTEM D3050 160 1300					
SELF-CONTAINED, WATER COOLED UNIT					
APARTMENT CORRIDORS, 500 S.F., .92 TON					
Self contained, water cooled, single package air conditioner unit	1.000	Ea.	907.20	462	1,369.20
Ductwork package for water or air cooled packaged units	1.000	System	92	648.60	740.60
Cooling tower, draw thru single flow, belt drive	1.000	Ea.	83.72	14.08	97.80
Cooling tower pumps & piping	1.000	System	41.86	33.58	75.44
TOTAL			1,124.78	1,158.26	2,283.04
COST PER S.F.			2.25	2.32	4.57

D3050 160	Self-contained, Water Cooled Unit Systems	COST PER S.F. MAT.	COST PER S.F. INST.	COST PER S.F. TOTAL
1280	Self-contained, water cooled unit	2.24	2.32	4.56
1300	Apartment corridors, 500 S.F., .92 ton	2.25	2.30	4.55
1320	1,000 S.F., 1.83 ton (R15700 -020)	2.24	2.31	4.55
1360	3,000 S.F., 5.50 ton	1.84	1.94	3.78
1400	5,000 S.F., 9.17 ton	1.82	1.81	3.63
1440	10,000 S.F., 18.33 ton	2.77	1.63	4.40
1520	Banks or libraries, 500 S.F., 2.08 ton	4.68	2.31	6.99
1560	1,000 S.F., 4.17 ton	4.20	4.41	8.61
1600	3,000 S.F., 12.50 ton	4.15	4.10	8.25
1640	5,000 S.F., 20.80 ton	6.30	3.68	9.98
1680	10,000 S.F., 41.66 ton	5.45	3.69	9.14
1760	Bars and taverns, 500 S.F., 5.54 ton	10	3.90	13.90
1800	1,000 S.F., 11.08 ton	10.40	7	17.40
1840	3,000 S.F., 33.25 ton	14.45	5.55	20
1880	5,000 S.F., 55.42 ton	13.55	5.80	19.35
1920	10,000 S.F., 110.00 ton	13.20	5.75	18.95
2000	Bowling alleys, 500 S.F., 2.83 ton	6.40	3.15	9.55
2040	1,000 S.F., 5.66 ton	5.70	6	11.70
2080	3,000 S.F., 17.00 ton	8.60	5.05	13.65
2120	5,000 S.F., 28.33 ton	7.70	4.85	12.55
2160	10,000 S.F., 56.66 ton	7.20	4.97	12.17
2200	Department stores, 500 S.F., 1.46 ton	3.29	1.62	4.91
2240	1,000 S.F., 2.92 ton	2.94	3.09	6.03
2280	3,000 S.F., 8.75 ton	2.90	2.88	5.78
2320	5,000 S.F., 14.58 ton	4.41	2.59	7
2360	10,000 S.F., 29.17 ton	3.97	2.50	6.47

D3050 Terminal & Package Units

D3050 160	Self-contained, Water Cooled Unit Systems	COST PER S.F. MAT.	INST.	TOTAL
2440	Drug stores, 500 S.F., 3.33 ton	7.50	3.70	11.20
2480	1,000 S.F., 6.66 ton	6.70	7.05	13.75
2520	3,000 S.F., 20.00 ton	10.10	5.90	16
2560	5,000 S.F., 33.33 ton	10.60	5.80	16.40
2600	10,000 S.F., 66.66 ton	8.50	5.85	14.35
2680	Factories, 500 S.F., 1.66 ton	3.74	1.85	5.59
2720	1,000 S.F. 3.37 ton	3.39	3.57	6.96
2760	3,000 S.F., 10.00 ton	3.31	3.28	6.59
2800	5,000 S.F., 16.66 ton	5.05	2.95	8
2840	10,000 S.F., 33.33 ton	4.53	2.85	7.38
2920	Food supermarkets, 500 S.F., 1.42 ton	3.19	1.57	4.76
2960	1,000 S.F., 2.83 ton	3.47	3.57	7.04
3000	3,000 S.F., 8.50 ton	2.85	3	5.85
3040	5,000 S.F., 14.17 ton	2.82	2.79	5.61
3080	10,000 S.F., 28.33 ton	3.86	2.42	6.28
3160	Medical centers, 500 S.F., 1.17 ton	2.63	1.30	3.93
3200	1,000 S.F., 2.33 ton	2.86	2.95	5.81
3240	3,000 S.F., 7.00 ton	2.35	2.49	4.84
3280	5,000 S.F., 11.66 ton	2.32	2.30	4.62
3320	10,000 S.F., 23.33 ton	3.53	2.07	5.60
3400	Offices, 500 S.F., 1.58 ton	3.56	1.77	5.33
3440	1,000 S.F., 3.17 ton	3.88	4	7.88
3480	3,000 S.F., 9.50 ton	3.15	3.12	6.27
3520	5,000 S.F., 15.83 ton	4.79	2.81	7.60
3560	10,000 S.F., 31.67 ton	4.31	2.71	7.02
3640	Restaurants, 500 S.F., 2.50 ton	5.65	2.78	8.43
3680	1,000 S.F., 5.00 ton	5.05	5.30	10.35
3720	3,000 S.F., 15.00 ton	4.99	4.93	9.92
3760	5,000 S.F., 25.00 ton	7.55	4.44	11.99
3800	10,000 S.F., 50.00 ton	4.57	4.08	8.65
3880	Schools and colleges, 500 S.F., 1.92 ton	4.31	2.13	6.44
3920	1,000 S.F., 3.83 ton	3.85	4.06	7.91
3960	3,000 S.F., 11.50 ton	3.81	3.77	7.58
4000	5,000 S.F., 19.17 ton	5.80	3.40	9.20
4040	10,000 S.F., 38.33 ton	5	3.40	8.40
4050				
9000	Components of ductwork packages for above systems, per ton of cooling:			
9010	Ductwork; galvanized steel, 108 pounds			
9020	Diffusers, aluminum, 24″ x 12″, two			

D3050 Terminal & Package Units

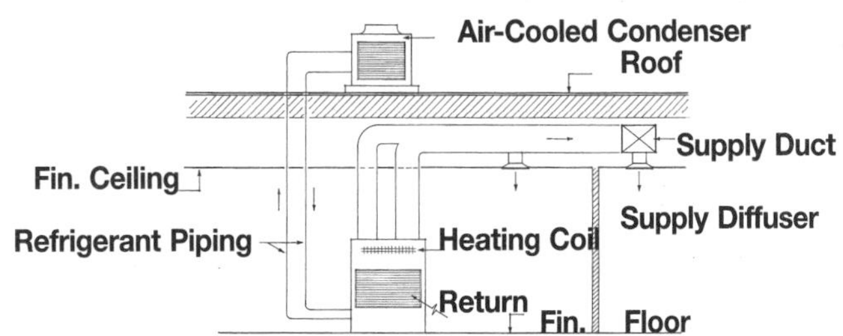

System Description: Self-contained air cooled units with remote air cooled condenser and interconnecting tubing. Systems for 1000 S.F. and up include duct and diffusers. Smaller units distribute air directly.

Returns are not ducted and supplies are not insulated.

Potential savings may be realized by using a single zone rooftop system or through-the-wall unit, especially in the smaller capacities, if the application permits.

Hot water or steam heating coils are included but piping to boiler and the boiler itself is not included.

Condenserless models are available for 15% less where remote refrigerant source is available.

System Components	QUANTITY	UNIT	COST EACH		
			MAT.	INST.	TOTAL
SYSTEM D3050 165 1320					
SELF-CONTAINED, AIR COOLED UNIT					
APARTMENT CORRIDORS, 500 S.F., .92 TON					
Air cooled, package unit	1.000	Ea.	969	300.90	1,269.90
Ductwork package for water or air cooled packaged units	1.000	System	92	648.60	740.60
Refrigerant piping	1.000	System	170.20	409.40	579.60
Air cooled condenser, direct drive, propeller fan	1.000	Ea.	269.70	114.70	384.40
TOTAL			1,500.90	1,473.60	2,974.50
COST PER S.F.			3	2.95	5.95

D3050 165	Self-contained, Air Cooled Unit Systems	COST PER S.F.		
		MAT.	INST.	TOTAL
1300	Self-contained, air cooled unit			
1320	Apartment corridors, 500 S.F., .92 ton	3	2.95	5.95
1360	1,000 S.F., 1.83 ton R15700 -020	2.96	2.92	5.88
1400	3,000 S.F., 5.50 ton	2.46	2.73	5.19
1440	5,000 S.F., 9.17 ton	2.14	2.61	4.75
1480	10,000 S.F., 18.33 ton	2.23	2.40	4.63
1560	Banks or libraries, 500 S.F., 2.08 ton	6.30	3.70	10
1600	1,000 S.F., 4.17 ton	5.55	6.20	11.75
1640	3,000 S.F., 12.50 ton	4.87	5.90	10.77
1680	5,000 S.F., 20.80 ton	5.10	5.45	10.55
1720	10,000 S.F., 41.66 ton	4.25	5.35	9.60
1800	Bars and taverns, 500 S.F., 5.54 ton	12.50	8.10	20.60
1840	1,000 S.F., 11.08 ton	12.35	11.85	24.20
1880	3,000 S.F., 33.25 ton	11.10	10.25	21.35
1920	5,000 S.F., 55.42 ton	10.55	10.30	20.85
1960	10,000 S.F., 110.00 ton	10.70	10.25	20.95
2040	Bowling alleys, 500 S.F., 2.83 ton	8.60	5.05	13.65
2080	1,000 S.F., 5.66 ton	7.60	8.45	16.05
2120	3,000 S.F., 17.00 ton	6.90	7.40	14.30
2160	5,000 S.F., 28.33 ton	5.95	7.25	13.20
2200	10,000 S.F., 56.66 ton	5.70	7.25	12.95
2240	Department stores, 500 S.F., 1.46 ton	4.43	2.61	7.04
2280	1,000 S.F., 2.92 ton	3.92	4.36	8.28
2320	3,000 S.F., 8.75 ton	3.39	4.15	7.54

D3050 Terminal & Package Units

D3050 165	Self-contained, Air Cooled Unit Systems	COST PER S.F. MAT.	INST.	TOTAL
2360	5,000 S.F., 14.58 ton	3.39	4.15	7.54
2400	10,000 S.F., 29.17 ton	3.06	3.73	6.79
2480	Drug stores, 500 S.F., 3.33 ton	10.10	5.95	16.05
2520	1,000 S.F., 6.66 ton	8.95	9.95	18.90
2560	3,000 S.F., 20.00 ton	8.20	8.75	16.95
2600	5,000 S.F., 33.33 ton	7	8.50	15.50
2640	10,000 S.F., 66.66 ton	6.75	8.55	15.30
2720	Factories, 500 S.F., 1.66 ton	5.15	3	8.15
2760	1,000 S.F., 3.33 ton	4.51	4.98	9.49
2800	3,000 S.F., 10.00 ton	3.89	4.73	8.62
2840	5,000 S.F., 16.66 ton	4.04	4.35	8.39
2880	10,000 S.F., 33.33 ton	3.49	4.26	7.75
2960	Food supermarkets, 500 S.F., 1.42 ton	4.30	2.53	6.83
3000	1,000 S.F., 2.83 ton	4.56	4.52	9.08
3040	3,000 S.F., 8.50 ton	3.79	4.23	8.02
3080	5,000 S.F., 14.17 ton	3.29	4.03	7.32
3120	10,000 S.F., 28.33 ton	2.96	3.62	6.58
3200	Medical centers, 500 S.F., 1.17 ton	3.56	2.09	5.65
3240	1,000 S.F., 2.33 ton	3.79	3.73	7.52
3280	3,000 S.F., 7.00 ton	3.13	3.49	6.62
3320	5,000 S.F., 16.66 ton	2.72	3.31	6.03
3360	10,000 S.F., 23.33 ton	2.84	3.05	5.89
3440	Offices, 500 S.F., 1.58 ton	4.82	2.83	7.65
3480	1,000 S.F., 3.16 ton	5.15	5.05	10.20
3520	3,000 S.F., 9.50 ton	3.71	4.50	8.21
3560	5,000 S.F., 15.83 ton	3.89	4.14	8.03
3600	10,000 S.F., 31.66 ton	3.34	4.04	7.38
3680	Restaurants, 500 S.F., 2.50 ton	7.55	4.45	12
3720	1,000 S.F., 5.00 ton	6.70	7.50	14.20
3760	3,000 S.F., 15.00 ton	5.85	7.10	12.95
3800	5,000 S.F., 25.00 ton	5.25	6.40	11.65
3840	10,000 S.F., 50.00 ton	5.20	6.35	11.55
3920	Schools and colleges, 500 S.F., 1.92 ton	5.80	3.42	9.22
3960	1,000 S.F., 3.83 ton	5.15	5.75	10.90
4000	3,000 S.F., 11.50 ton	4.47	5.45	9.92
4040	5,000 S.F., 19.17 ton	4.67	5	9.67
4080	10,000 S.F., 38.33 ton	3.92	4.91	8.83
4090				
9000	Components of ductwork packages for above systems, per ton of cooling:			
9010	Ductwork; galvanized steel, 102 pounds			
9020	Diffusers; aluminum, 24" x 12", two			

SERVICES D

D3050 Terminal & Package Units

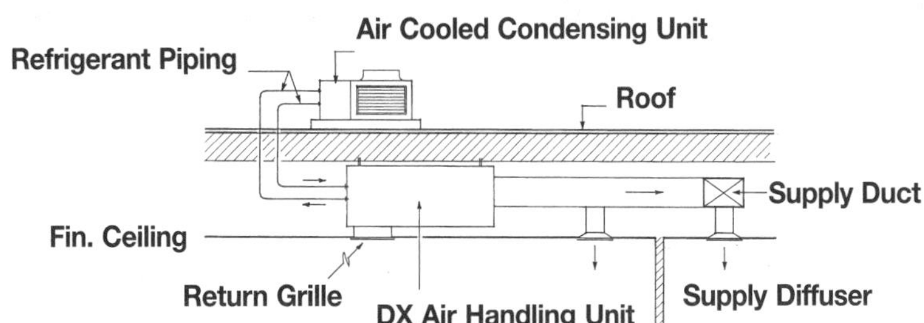

General: Split systems offer several important advantages which should be evaluated when a selection is to be made. They provide a greater degree of flexibility in component selection which permits an accurate match-up of the proper equipment size and type with the particular needs of the building. This allows for maximum use of modern energy saving concepts in heating and cooling. Outdoor installation of the air cooled condensing unit allows space savings in the building and also isolates the equipment operating sounds from building occupants.

Design Assumptions: The systems below are comprised of a direct expansion air handling unit and air cooled condensing unit with interconnecting copper tubing. Ducts and diffusers are also included for distribution of air. Systems are priced for cooling only. Heat can be added as desired either by putting hot water/steam coils into the air unit or into the duct supplying the particular area of need. Gas fired duct furnaces are also available. Refrigerant liquid line is insulated.

System Components	QUANTITY	UNIT	COST EACH		
			MAT.	INST.	TOTAL
SYSTEM D3050 170 1280					
SPLIT SYSTEM, AIR COOLED CONDENSING UNIT					
APARTMENT CORRIDORS, 1,000 S.F., 1.80 TON					
Fan coil AC unit, cabinet mntd & filters direct expansion air cool	1.000	Ea.	307.44	107.97	415.41
Ductwork package, for split system, remote condensing unit	1.000	System	86.01	617.63	703.64
Refrigeration piping	1.000	System	150.98	677.10	828.08
Condensing unit, air cooled, incls compressor & standard controls	1.000	Ea.	744.20	433.10	1,177.30
TOTAL			1,288.63	1,835.80	3,124.43
COST PER S.F.			1.29	1.84	3.13
*Cooling requirements would lead to choosing a water cooled unit					

D3050 170	Split Systems With Air Cooled Condensing Units	COST PER S.F.		
		MAT.	INST.	TOTAL
1260	Split system, air cooled condensing unit			
1280	Apartment corridors, 1,000 S.F., 1.83 ton	1.30	1.83	3.13
1320	2,000 S.F., 3.66 ton (R15700 -020)	1.09	1.86	2.95
1360	5,000 S.F., 9.17 ton	1.43	2.35	3.78
1400	10,000 S.F., 18.33 ton	1.57	2.56	4.13
1440	20,000 S.F., 36.66 ton	1.65	2.59	4.24
1520	Banks and libraries, 1,000 S.F., 4.17 ton	2.46	4.23	6.69
1560	2,000 S.F., 8.33 ton	3.25	5.30	8.55
1600	5,000 S.F., 20.80 ton	3.58	5.80	9.38
1640	10,000 S.F., 41.66 ton	3.75	5.90	9.65
1680	20,000 S.F., 83.32 ton	4.32	6.10	10.42
1760	Bars and taverns, 1,000 S.F., 11.08 ton	7.70	8.35	16.05
1800	2,000 S.F., 22.16 ton	10.15	9.95	20.10
1840	5,000 S.F., 55.42 ton	8.60	9.30	17.90
1880	10,000 S.F., 110.84 ton	10.40	9.70	20.10
1920	20,000 S.F., 220 ton*			
2000	Bowling alleys, 1,000 S.F., 5.66 ton	3.59	7.90	11.49
2040	2,000 S.F., 11.33 ton	4.43	7.25	11.68
2080	5,000 S.F., 28.33 ton	4.88	7.90	12.78
2120	10,000 S.F., 56.66 ton	5.10	8	13.10
2160	20,000 S.F., 113.32 ton	6.45	8.65	15.10
2320	Department stores, 1,000 S.F., 2.92 ton	1.76	2.89	4.65
2360	2,000 S.F., 5.83 ton	1.85	4.08	5.93
2400	5,000 S.F., 14.58 ton	2.28	3.72	6

D3050 Terminal & Package Units

D3050 170	Split Systems With Air Cooled Condensing Units	COST PER S.F.		
		MAT.	INST.	TOTAL
2440	10,000 S.F., 29.17 ton	2.50	4.06	6.56
2480	20,000 S.F., 58.33 ton	2.62	4.12	6.74
2560	Drug stores, 1,000 S.F., 6.66 ton	4.23	9.30	13.53
2600	2,000 S.F., 13.32 ton	5.20	8.50	13.70
2640	5,000 S.F., 33.33 ton	6	9.40	15.40
2680	10,000 S.F., 66.66 ton	6.05	9.85	15.90
2720	20,000 S.F., 133.32 ton*			
2800	Factories, 1,000 S.F., 3.33 ton	2.01	3.30	5.31
2840	2,000 S.F., 6.66 ton	2.11	4.66	6.77
2880	5,000 S.F., 16.66 ton	2.86	4.63	7.49
2920	10,000 S.F., 33.33 ton	3	4.70	7.70
2960	20,000 S.F., 66.66 ton	3.01	4.93	7.94
3040	Food supermarkets, 1,000 S.F., 2.83 ton	1.70	2.82	4.52
3080	2,000 S.F., 5.66 ton	1.80	3.97	5.77
3120	5,000 S.F., 14.66 ton	2.22	3.61	5.83
3160	10,000 S.F., 28.33 ton	2.43	3.94	6.37
3200	20,000 S.F., 56.66 ton	2.55	4	6.55
3280	Medical centers, 1,000 S.F., 2.33 ton	1.41	2.28	3.69
3320	2,000 S.F., 4.66 ton	1.48	3.26	4.74
3360	5,000 S.F., 11.66 ton	1.82	2.96	4.78
3400	10,000 S.F., 23.33 ton	2	3.24	5.24
3440	20,000 S.F., 46.66 ton	2.10	3.28	5.38
3520	Offices, 1,000 S.F., 3.17 ton	1.91	3.15	5.06
3560	2,000 S.F., 6.33 ton	2.01	4.43	6.44
3600	5,000 S.F., 15.83 ton	2.47	4.04	6.51
3640	10,000 S.F., 31.66 ton	2.72	4.41	7.13
3680	20,000 S.F., 63.32 ton	2.86	4.69	7.55
3760	Restaurants, 1,000 S.F., 5.00 ton	3.17	7	10.17
3800	2,000 S.F., 10.00 ton	3.91	6.40	10.31
3840	5,000 S.F., 25.00 ton	4.30	6.95	11.25
3880	10,000 S.F., 50.00 ton	4.50	7.05	11.55
3920	20,000 S.F., 100.00 ton	5.70	7.65	13.35
4000	Schools and colleges, 1,000 S.F., 3.83 ton	2.27	3.88	6.15
4040	2,000 S.F., 7.66 ton	3	4.89	7.89
4080	5,000 S.F., 19.17 ton	3.29	5.35	8.64
4120	10,000 S.F., 38.33 ton	3.45	5.40	8.85
4160	20,000 S.F., 76.66 ton	3.49	5.65	9.14
5000				
5100				
5200				
9000	Components of ductwork packages for above systems, per ton of cooling:			
9010	Ductwork; galvanized steel, 102 pounds			
9020	Diffusers; aluminum, 24″ x 12″, two			

D3050 Terminal & Package Units

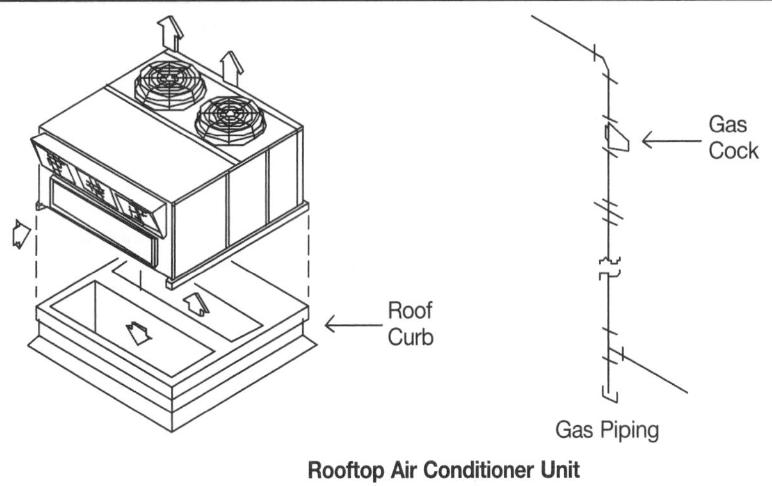

Rooftop Air Conditioner Unit

System Components	QUANTITY	UNIT	COST EACH MAT.	COST EACH INST.	COST EACH TOTAL
SYSTEM D3050 175 1010					
A/C, ROOFTOP, DX COOL, GAS HEAT, CURB, ECONOMIZER, FILTERS., 5 TON					
Roof top A/C, curb, economizer, sgl zone, elec cool, gas ht, 5 ton, 112 MBH	1.000	Ea.	4,450	1,575	6,025
Pipe, black steel, Sch 40, threaded, W/cplgs & hangers, 10′ OC, 1″ dia	20.000	L.F.	49.98	194.25	244.23
Elbow, 90°, black, straight, 3/4″ dia.	3.000	Ea.	5.52	105	110.52
Elbow, 90°, black, straight, 1″ dia.	3.000	Ea.	9.57	114	123.57
Tee, black, reducing, 1″ dia.	1.000	Ea.	5.70	61.50	67.20
Union, black with brass seat, 1″ dia.	1.000	Ea.	11	41	52
Pipe nipple, black, 3/4″ dia	2.000	Ea.	3.38	16.10	19.48
Pipe nipple, black, 1″ dia	2.000	Ea.	2.38	9.25	11.63
Cap, black, 1″ dia.	1.000	Ea.	2.51	16.35	18.86
Gas cock, brass, 1″ size	1.000	Ea.	12.50	26	38.50
Control system, pneumatic, Rooftop A/C unit	1.000	Ea.	2,150	1,725	3,875
Rooftop unit heat/cool balancing	1.000	Ea.		345	345
TOTAL			6,702.54	4,228.45	10,930.99

D3050 175	Rooftop Air Conditioner, Const. Volume	COST EACH MAT.	COST EACH INST.	COST EACH TOTAL
1010	A/C, Rooftop, DX cool, gas heat, curb, economizer, fltrs, 5 Ton	6,700	4,225	10,925
1020	7-1/2 Ton	8,750	4,400	13,150
1030	12-1/2 Ton	11,800	4,825	16,625
1040	18 Ton	17,900	5,275	23,175
1050	25 Ton	23,700	6,025	29,725
1060	40 Ton	37,600	8,050	45,650

D3050 180	Rooftop Air Conditioner, Variable Air Volume	COST EACH MAT.	COST EACH INST.	COST EACH TOTAL
1010	A/C, Rooftop, DX cool, gas heat, curb, ecmizr, fltrs, VAV, 12-1/2 Ton	13,900	5,150	19,050
1020	18 Ton	20,300	5,675	25,975
1030	25 Ton	27,400	6,525	33,925
1040	40 Ton	45,100	8,850	53,950
1050	60 Ton	63,500	12,100	75,600
1060	80 Ton	77,000	15,100	92,100

D3050 Terminal & Package Units

Computer rooms impose special requirements on air conditioning systems. A prime requirement is reliability, due to the potential monetary loss that could be incurred by a system failure. A second basic requirement is the tolerance of control with which temperature and humidity are regulated, and dust eliminated. As the air conditioning system reliability is so vital, the additional cost of reserve capacity and redundant components is often justified.

System Descriptions: Computer areas may be environmentally controlled by one of three methods as follows:

1. **Self-contained Units**
   These are units built to higher standards of performance and reliability. They usuallycontain alarms and controls to indicate component operation failure, filter change, etc. It should be remembered that these units in the room will occupy space that is relatively expensive to build and that all alterations and service of the equipment will also have to be accomplished within the computer area.
2. **Decentralized Air Handling Units** In operation these are similar to the self-contained units except that their cooling capability comes from remotely located refrigeration equipment as refrigerant or chilled water. As no compressors or refrigerating equipment are required in the air units, they are smaller and require less service than self-contained units. An added plus for this type of system occurs if some of the computer components themselves also require chilled water for cooling.
3. **Central System Supply** Cooling is obtained from a central source which, since it is not located within the computer room, may have excess capacity and permit greater flexibility without interfering with the computer components. System performance criteria must still be met.

Note: The costs shown below do not include an allowance for ductwork or piping.

D3050 185	Computer Room Cooling Units	COST EACH		
		MAT.	INST.	TOTAL
0560	Computer room unit, air cooled, includes remote condenser			
0580	3 ton	9,600	1,775	11,375
0600	5 ton	12,900	1,975	14,875
0620	8 ton	26,800	3,275	30,075
0640	10 ton	27,900	3,550	31,450
0660	15 ton	31,100	4,025	35,125
0680	20 ton	37,200	4,750	41,950
0700	23 ton	39,000	4,925	43,925
0800	Chilled water, for connection to existing chiller system			
0820	5 ton	10,800	1,200	12,000
0840	8 ton	15,500	1,775	17,275
0860	10 ton	17,200	1,800	19,000
0880	15 ton	18,400	1,850	20,250
0900	20 ton	18,900	1,925	20,825
0920	23 ton	19,800	2,100	21,900
1000	Glycol system, complete except for interconnecting tubing			
1020	3 ton	11,800	2,225	14,025
1040	5 ton	15,300	2,325	17,625
1060	8 ton	33,200	3,850	37,050
1080	10 ton	35,900	4,225	40,125
1100	15 ton	43,500	5,300	48,800
1120	20 ton	46,600	5,750	52,350
1140	23 ton	51,000	6,275	57,275
1240	Water cooled, not including condenser water supply or cooling tower			
1260	3 ton	8,675	1,425	10,100
1280	5 ton	12,200	1,650	13,850
1300	8 ton	25,200	2,675	27,875
1320	15 ton	30,800	3,275	34,075
1340	20 ton	36,400	3,625	40,025
1360	23 ton	38,600	4,050	42,650

D3050 201	Packaged A/C, Elec. Ht., Const. Volume	COST EACH		
		MAT.	INST.	TOTAL
1010	A/C Packaged, DX, air cooled, electric heat, constant vol., 5 Ton	6,350	3,100	9,450
1020	10 Ton	10,300	4,250	14,550

SERVICES D

D3050 Terminal & Package Units

D3050 201	Packaged A/C, Elec. Ht., Const. Volume	COST EACH		
		MAT.	INST.	TOTAL
1030	20 Ton	20,000	5,600	25,600
1040	30 Ton	24,700	6,200	30,900
1050	40 Ton	30,900	7,125	38,025
1060	50 Ton	39,600	8,000	47,600

D3050 202	Packaged Air Conditioner, Elect. Heat, VAV	COST EACH		
		MAT.	INST.	TOTAL
1010	A/C Packaged, DX, air cooled, electric heat, VAV, 10 Ton	12,800	5,325	18,125
1020	20 Ton	24,200	7,025	31,225
1030	30 Ton	30,300	7,675	37,975
1040	40 Ton	39,000	8,775	47,775
1050	50 Ton	48,500	9,750	58,250
1060	60 Ton	55,500	11,500	67,000

D3050 203	Packaged A/C, Hot Wtr. Heat, Const. Volume	COST EACH		
		MAT.	INST.	TOTAL
1010	A/C Packaged, DX, air cooled, H/W heat, constant vol., 5 Ton	5,200	4,550	9,750
1020	10 Ton	7,750	5,950	13,700
1030	20 Ton	17,000	8,025	25,025
1040	30 Ton	18,700	10,100	28,800
1050	40 Ton	24,900	11,300	36,200
1060	50 Ton	30,400	12,300	42,700

D3050 204	Packaged A/C, Hot Wtr. Heat, VAV	COST EACH		
		MAT.	INST.	TOTAL
1010	A/C Packaged, DX, air cooled, H/W heat, VAV, 10 Ton	10,300	6,900	17,200
1020	20 Ton	21,200	9,075	30,275
1030	30 Ton	24,300	11,400	35,700
1040	40 Ton	33,100	12,300	45,400
1050	50 Ton	39,200	14,200	53,400
1060	60 Ton	48,000	16,000	64,000

D3050 Terminal & Package Units

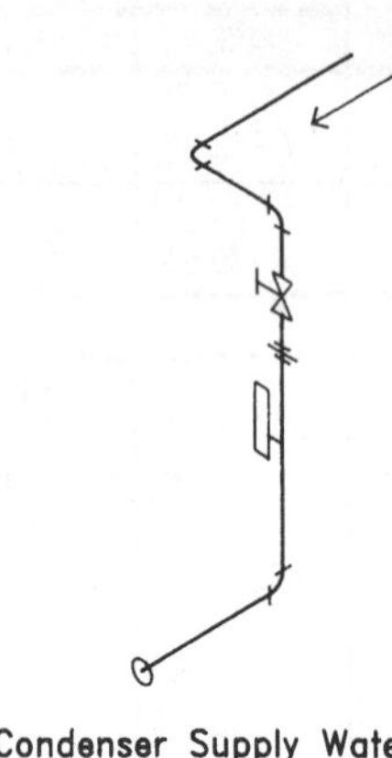

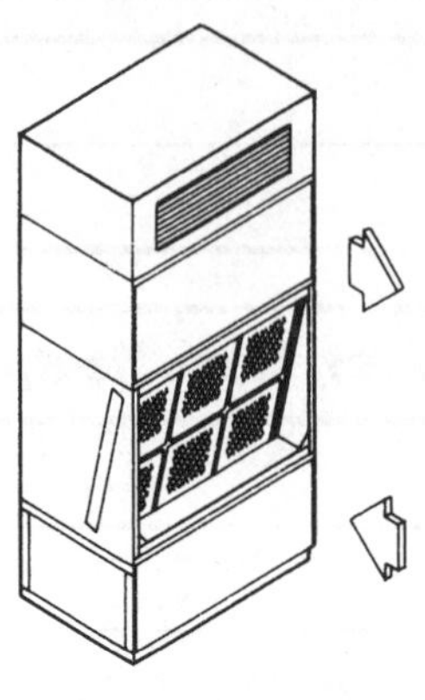

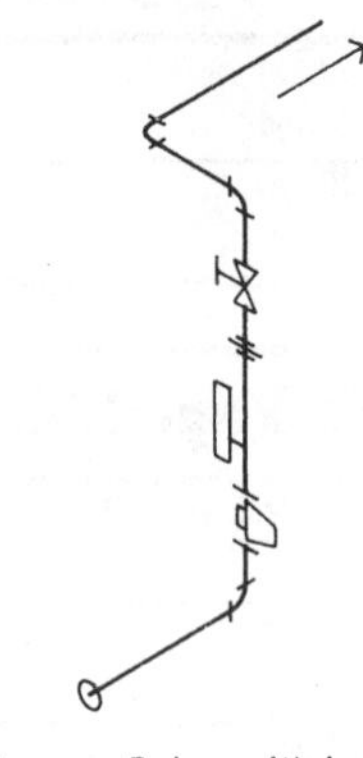

Self–Contained Air Conditioner Unit, Water Cooled

System Components	QUANTITY	UNIT	COST EACH MAT.	COST EACH INST.	COST EACH TOTAL
SYSTEM D3050 210 1010					
A/C, SELF CONTAINED, SINGLE PKG., WATER COOLED, ELECT. HEAT, 5 TON					
Self-contained, water cooled, elect. heat, not inc tower, 5 ton, const vol	1.000	Ea.	6,225	1,150	7,375
Pipe, black steel, Sch 40, threaded, W/cplg & hangers, 10′ OC, 1-1/4″ dia.	20.000	L.F.	64.93	213.93	278.86
Elbow, 90°, black, straight, 1-1/4″ dia.	6.000	Ea.	31.50	240	271.50
Tee, black, straight, 1-1/4″ dia.	2.000	Ea.	16.10	126	142.10
Tee, black, reducing, 1-1/4″ dia.	2.000	Ea.	19.70	126	145.70
Thermometers, stem type, 9″ case, 8″ stem, 3/4″ NPT	2.000	Ea.	260	35.20	295.20
Union, black with brass seat, 1-1/4″ dia.	2.000	Ea.	30.20	84	114.20
Pipe nipple, black, 1-1/4″ dia	3.000	Ea.	4.53	14.93	19.46
Valves, bronze, gate, N.R.S., threaded, class 150, 1-1/4″ size	2.000	Ea.	122	66	188
Circuit setter, bal valve, bronze body, threaded, 1-1/4″ pipe size	1.000	Ea.	101	33	134
Insulation, fiberglass pipe covering, 1″ wall, 1-1/4″ IPS	20.000	L.F.	21.20	80.40	101.60
Control system, pneumatic, A/C Unit with heat	1.000	Ea.	2,150	1,725	3,875
Re-heat coil balancing	1.000	Ea.		91	91
Rooftop unit heat/cool balancing	1.000	Ea.		345	345
TOTAL			9,046.16	4,330.46	13,376.62

D3050 210	AC Unit, Package, Elec. Ht., Water Cooled	COST EACH MAT.	COST EACH INST.	COST EACH TOTAL
1010	A/C, Self contained, single pkg., water cooled, elect. heat, 5 Ton	9,050	4,325	13,375
1020	10 Ton	14,200	5,325	19,525
1030	20 Ton	19,300	6,875	26,175
1040	30 Ton	27,600	7,500	35,100
1050	40 Ton	33,800	8,275	42,075
1060	50 Ton	43,000	9,900	52,900

D3050 215	AC Unit, Package, Elec. Ht., Water Cooled, VAV	COST EACH MAT.	COST EACH INST.	COST EACH TOTAL
1010	A/C, Self contained, single pkg., water cooled, elect. ht, VAV, 10 Ton	16,300	5,325	21,625
1020	20 Ton	23,000	6,875	29,875
1030	30 Ton	32,700	7,500	40,200
1040	40 Ton	41,300	8,275	49,575
1050	50 Ton	51,000	9,900	60,900
1060	60 Ton	58,500	12,200	70,700

D3050 Terminal & Package Units

D3050 220	AC Unit, Package, Hot Water Coil, Water Cooled	COST EACH		
		MAT.	INST.	TOTAL
1010	A/C, Self contn'd, single pkg., water cool, H/W ht, const. vol, 5 Ton	7,675	6,000	13,675
1020	10 Ton	11,400	7,150	18,550
1030	20 Ton	30,500	9,075	39,575
1040	30 Ton	40,000	11,600	51,600
1050	40 Ton	49,400	14,100	63,500
1060	50 Ton	61,000	17,300	78,300

D3050 225	AC Unit, Package, HW Coil, Water Cooled, VAV	COST EACH		
		MAT.	INST.	TOTAL
1010	A/C, Self contn'd, single pkg., water cool, H/W ht, VAV, 10 Ton	13,600	6,750	20,350
1020	20 Ton	19,600	8,850	28,450
1030	30 Ton	26,800	11,100	37,900
1040	40 Ton	35,500	12,100	47,600
1050	50 Ton	42,400	14,300	56,700
1060	60 Ton	51,000	15,500	66,500

D3050 Terminal & Package Units

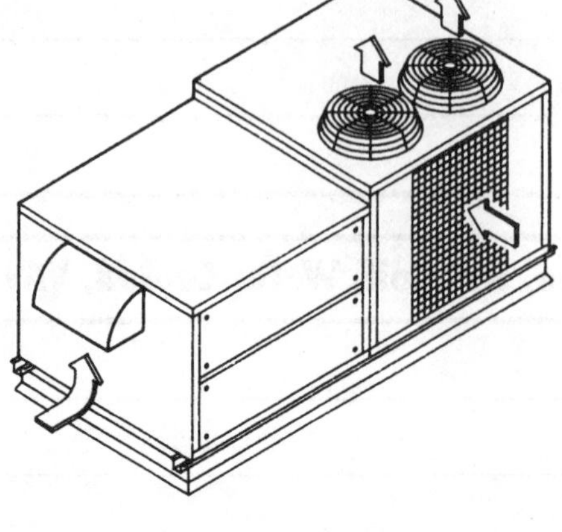

Heat Pump, Air/Air, Rooftop w/ Economizer

System Components	QUANTITY	UNIT	COST EACH MAT.	COST EACH INST.	COST EACH TOTAL
SYSTEM D3050 245 1010					
HEAT PUMP, ROOF TOP, AIR/AIR, CURB., ECONOMIZER., SUP. ELECT. HEAT, 10 TON					
Heat pump, air to air, rooftop, curb, economizer, elect ht, 10 ton cooling	1.000	Ea.	41,700	3,450	45,150
Ductwork, 32" x 12" fabricated, galvanized steel, 16 LF	176.000	Lb.	200.64	1,038.40	1,239.04
Insulation, ductwork, blanket type, fiberglass, 1" thk, 1-1/2 LB density	117.000	S.F.	64.35	281.97	346.32
Rooftop unit heat/cool balancing	1.000	Ea.		345	345
TOTAL			41,964.99	5,115.37	47,080.36

D3050 230	Heat Pump, Water Source, Central Station	COST EACH MAT.	COST EACH INST.	COST EACH TOTAL
1010	Heat pump, central station, water source, constant vol., 5 Ton	2,575	2,450	5,025
1020	10 Ton	6,950	3,400	10,350
1030	20 Ton	13,600	6,425	20,025
1040	30 Ton	19,800	9,300	29,100
1050	40 Ton	27,300	10,100	37,400
1060	50 Ton	31,100	13,700	44,800

D3050 235	Heat Pump, Water Source, Console	COST EACH MAT.	COST EACH INST.	COST EACH TOTAL
1020	Heat pump, console, water source, 1 Ton	1,375	1,350	2,725
1030	1-1/2 Ton	1,475	1,375	2,850
1040	2 Ton	1,900	1,625	3,525
1050	3 Ton	2,100	1,750	3,850
1060	3-1/2 Ton	2,150	1,800	3,950

D3050 240	Heat Pump, Horizontal, Duct Mounted	COST EACH MAT.	COST EACH INST.	COST EACH TOTAL
1020	Heat pump, horizontal, ducted, water source, 1 Ton	1,900	2,900	4,800
1030	1-1/2 Ton	2,225	3,650	5,875
1040	2 Ton	2,650	4,350	7,000
1050	3 Ton	3,300	5,725	9,025
1060	3-1/2 Ton	3,375	5,900	9,275

Important: See the Reference Section for critical supporting data - Reference Numbers and City Cost Indexes

D3050 Terminal & Package Units

D3050 245	Heat Pump, Roof Top, Air/Air	COST EACH		
		MAT.	INST.	TOTAL
1010	Heat pump, roof top, air/air, curb, economizer, sup. elect heat, 10 Ton	42,000	5,125	47,125
1020	20 Ton	60,000	8,700	68,700
1030	30 Ton	81,000	12,100	93,100
1040	40 Ton	90,000	13,900	103,900
1050	50 Ton	105,000	15,700	120,700
1060	60 Ton	111,000	17,800	128,800

D3050 Terminal & Package Units

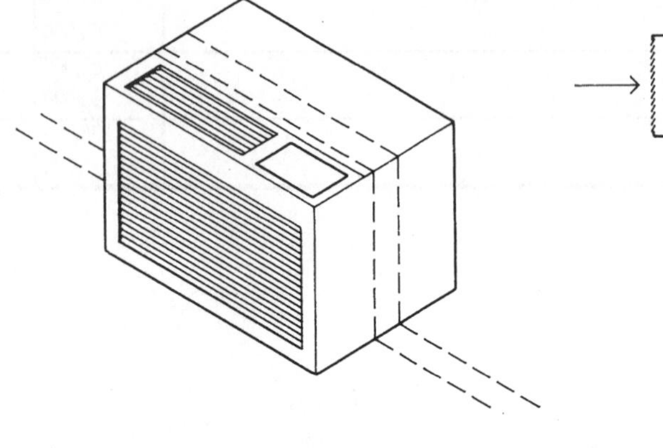

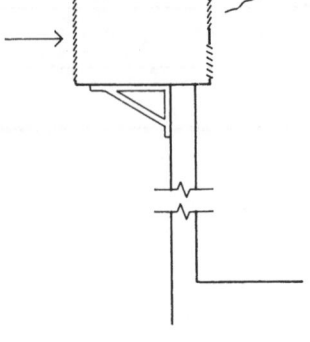

Thru–Wall Heat Pump   Thru–Wall Air Conditioner

System Components	QUANTITY	UNIT	COST EACH MAT.	COST EACH INST.	COST EACH TOTAL
SYSTEM D3050 255 1010 **A/C UNIT, THRU-THE-WALL, SUP. ELECT. HEAT, CABINET, LOUVER, 1/2 TON**					
A/C unit, thru-wall, electric heat., cabinet, louver, 1/2 ton	1.000	Ea.	1,050	148	1,198
TOTAL			1,050	148	1,198

D3050 255	Thru-Wall A/C Unit	COST EACH MAT.	COST EACH INST.	COST EACH TOTAL
1010	A/C Unit, thru-the-wall, sup. elect. heat, cabinet, louver, 1/2 Ton	1,050	148	1,198
1020	3/4 Ton	1,100	177	1,277
1030	1 Ton	1,225	222	1,447
1040	1-1/2 Ton	1,500	370	1,870
1050	2 Ton	1,500	465	1,965

D3050 260	Thru-Wall Heat Pump	COST EACH MAT.	COST EACH INST.	COST EACH TOTAL
1010	Heat pump, thru-the-wall, cabinet, louver, 1/2 Ton	1,125	111	1,236
1020	3/4 Ton	1,200	148	1,348
1030	1 Ton	1,375	222	1,597
1040	Sup. elect. heat, 1-1/2 Ton	2,125	570	2,695
1050	2 Ton	2,400	590	2,990

D3060 Controls & Instrumentation

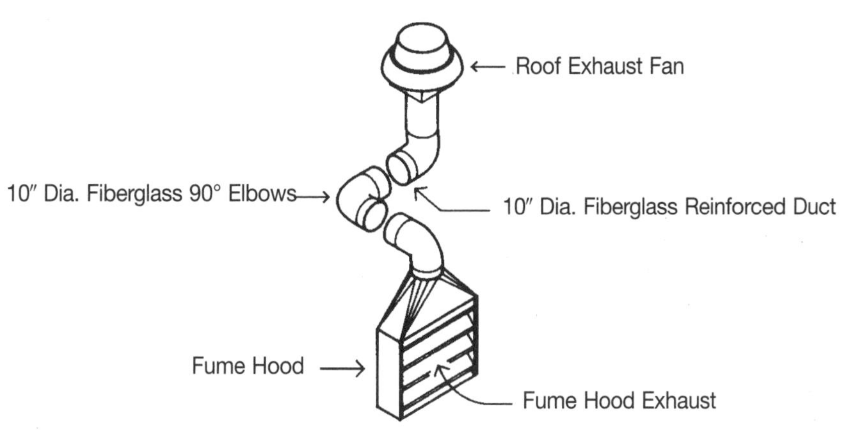

System Components	QUANTITY	UNIT	COST EACH MAT.	COST EACH INST.	COST EACH TOTAL
SYSTEM D3060 510 1010					
FUME HOOD EXHAUST SYSTEM, 3′ LONG, 1000 CFM					
Fume hood	3.000	L.F.	4,050	1,065	5,115
Ductwork, plastic, FM approved for acid fume & smoke, straight, 10″ dia	60.000	L.F.	1,158	630	1,788
Ductwork, plastic, FM approved for acid fume & smoke, elbow 90°, 10″ dia	3.000	Ea.	297	342	639
Fan, corrosive fume resistant plastic, 1630 CFM, 1/2 HP	1.000	Ea.	3,525	283	3,808
Fan control switch & mounting base	1.000	Ea.	73.50	31	104.50
Roof fan balancing	1.000	Ea.		198	198
TOTAL			9,103.50	2,549	11,652.50

D3060 510	Fume Hood Exhaust Systems	COST EACH MAT.	COST EACH INST.	COST EACH TOTAL
1010	Fume hood exhaust system, 3′ long, 1000 CFM	9,100	2,550	11,650
1020	4′ long, 2000 CFM	10,800	3,225	14,025
1030	6′ long, 3500 CFM	15,000	4,825	19,825
1040	6′ long, 5000 CFM	20,600	5,650	26,250
1050	10′ long, 8000 CFM	27,700	7,750	35,450

D3090 Other HVAC Systems/Equip

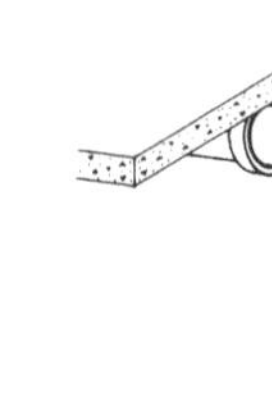

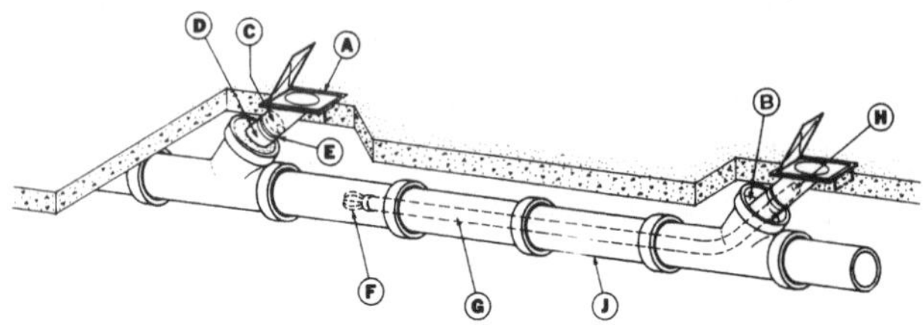

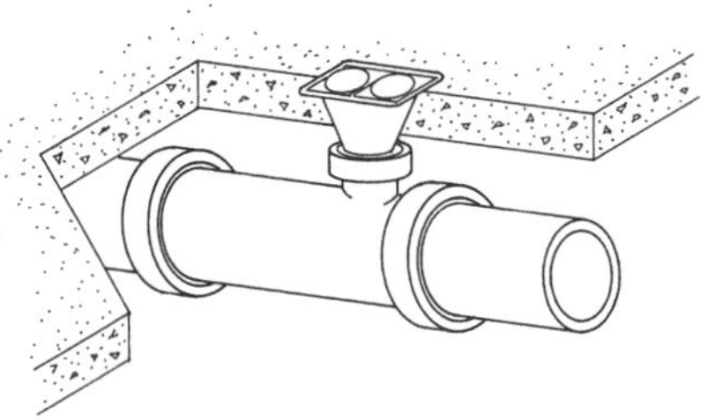

Vitrified Clay Garage Exhaust System

Dual Exhaust System

System Components	QUANTITY	UNIT	COST EACH MAT.	COST EACH INST.	COST EACH TOTAL
SYSTEM D3090 320 1040					
GARAGE, EXHAUST, SINGLE OUTLET, 3″ EXHAUST, CARS & LIGHT TRUCKS					
A Outlet top assy. for engine exhaust system with adapters and ftngs, 3″ diam	1.000	Ea.	178	31	209
F Bullnose (guide) for engine exhaust system, 3″ diam	1.000	Ea.	19.80		19.80
G Galvanized flexible tubing for engine exhaust system, 3″ diam	8.000	L.F.	42.40		42.40
H Adapter for metal tubing end of engine exhaust system, 3″ tail pipe	1.000	Ea.	39.50		39.50
J Pipe, sewer, vitrified clay, premium joint, 8″ diam.	18.000	L.F.	79.74	136.26	216
Excavate utility trench w/chain trencher, 12 H.P., oper. walking	18.000	L.F.		23.22	23.22
Backfill utility trench by hand, incl. compaction, 8″ wide 24″ deep	18.000	L.F.		33.30	33.30
Stand for blower, concrete over polystyrene core, 6″ high	1.000	Ea.	7.70	15.45	23.15
AC&V duct spiral reducer 10″x8″	1.000	Ea.	7.80	28	35.80
AC&V duct spiral reducer 12″x10″	1.000	Ea.	11.10	37	48.10
AC&V duct spiral preformed 45° elbow 8″ diam	1.000	Ea.	3.64	32	35.64
AC&V utility fan, belt drive, 3 phase, 2000 CFM, 1 HP	2.000	Ea.	1,260	492	1,752
Safety switch, heavy duty fused, 240V, 3 pole, 30 amp	1.000	Ea.	138	152	290
TOTAL			1,787.68	980.23	2,767.91

D3090 320	Garage Exhaust Systems	COST PER BAY MAT.	COST PER BAY INST.	COST PER BAY TOTAL
1040	Garage, single exhaust, 3″ outlet, cars & light trucks, one bay	1,800	980	2,780
1060	Additional bays up to seven bays	325	138	463
1500	4″ outlet, trucks, one bay	1,800	980	2,780
1520	Additional bays up to six bays	335	138	473
1600	5″ outlet, diesel trucks, one bay	2,050	980	3,030
1650	Additional single bays up to six	590	160	750
1700	Two adjoining bays	2,050	980	3,030
2000	Dual exhaust, 3″ outlets, pair of adjoining bays	2,100	1,075	3,175
2100	Additional pairs of adjoining bays	605	160	765

G BUILDING SITEWORK

G10 Site Preparation

G1030 Site Earthwork

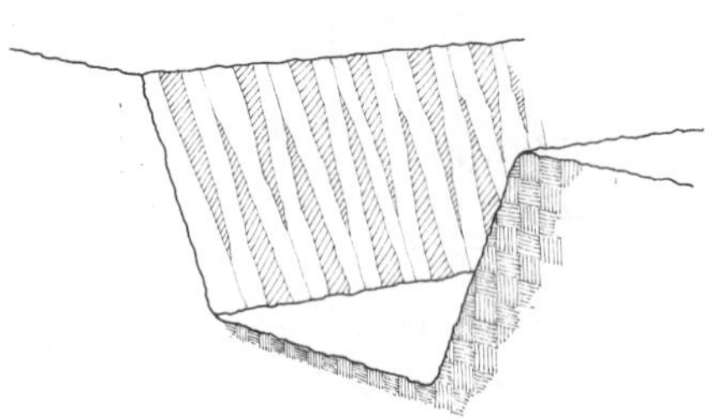

Trenching Systems are shown on a cost per linear foot basis. The systems include: excavation; backfill and removal of spoil; and compaction for various depths and trench bottom widths. The backfill has been reduced to accommodate a pipe of suitable diameter and bedding.

The slope for trench sides varies from 0:1 to 2:1.

The Expanded System Listing shows Trenching Systems that range from 2′ to 12′ in width. Depths range from 2′ to 25′.

System Components	QUANTITY	UNIT	COST PER L.F.		
			EQUIP.	LABOR	TOTAL
SYSTEM G1030 805 1310					
TRENCHING, BACKHOE, 0 TO 1 SLOPE, 2′ WIDE, 2′ DP, 3/8 C.Y. BUCKET					
Excavation, trench, hyd. backhoe, track mtd., 3/8 C.Y. bucket	.174	C.Y.	.27	.87	1.14
Backfill and load spoil, from stockpile	.174	C.Y.	.09	.25	.34
Compaction by rammer tamper, 8″ lifts, 4 passes	.014	C.Y.		.04	.04
Remove excess spoil, 6 C.Y. dump truck, 2 mile roundtrip	.160	C.Y.	.57	.54	1.11
TOTAL			.93	1.70	2.63

G1030 805	Trenching	COST PER L.F.		
		EQUIP.	LABOR	TOTAL
1310	Trenching, backhoe, 0 to 1 slope, 2′ wide, 2′ deep, 3/8 C.Y. bucket	.93	1.70	2.63
1320	3′ deep, 3/8 C.Y. bucket	1.15	2.49	3.64
1330	4′ deep, 3/8 C.Y. bucket	1.36	3.28	4.64
1340	6′ deep, 3/8 C.Y. bucket	1.69	4.18	5.87
1350	8′ deep, 1/2 C.Y. bucket	1.98	5.30	7.28
1360	10′ deep, 1 C.Y. bucket	3.01	6.30	9.31
1400	4′ wide, 2′ deep, 3/8 C.Y. bucket	1.98	3.42	5.40
1410	3′ deep, 3/8 C.Y. bucket	2.80	5.10	7.90
1420	4′ deep, 1/2 C.Y. bucket	2.79	5.10	7.89
1430	6′ deep, 1/2 C.Y. bucket	3.56	7.75	11.31
1440	8′ deep, 1/2 C.Y. bucket	5.60	10	15.60
1450	10′ deep, 1 C.Y. bucket	6.25	12.40	18.65
1460	12′ deep, 1 C.Y. bucket	7.95	15.90	23.85
1470	15′ deep, 1-1/2 C.Y. bucket	6.75	13.70	20.45
1480	18′ deep, 2-1/2 C.Y. bucket	9.30	19.50	28.80
1520	6′ wide, 6′ deep, 5/8 C.Y. bucket	7.25	10.75	18
1530	8′ deep, 3/4 C.Y. bucket	8.85	12.90	21.75
1540	10′ deep, 1 C.Y. bucket	9.50	15.50	25
1550	12′ deep, 1-1/4 C.Y. bucket	8.35	12.30	20.65
1560	16′ deep, 2 C.Y. bucket	10.95	14.05	25
1570	20′ deep, 3-1/2 C.Y. bucket	14.90	25	39.90
1580	24′ deep, 3-1/2 C.Y. bucket	24	41.50	65.50
1640	8′ wide, 12′ deep, 1-1/4 C.Y. bucket	12.40	16.80	29.20
1650	15′ deep, 1-1/2 C.Y. bucket	14.60	20.50	35.10
1660	18′ deep, 2-1/2 C.Y. bucket	17	21	38
1680	24′ deep, 3-1/2 C.Y. bucket	33	56	89
1730	10′ wide, 20′ deep, 3-1/2 C.Y. bucket	27	42	69
1740	24′ deep, 3-1/2 C.Y. bucket	42	70	112
1780	12′ wide, 20′ deep, 3-1/2 C.Y. bucket	32.50	50.50	83
1790	25′ deep, bucket	56	93.50	149.50
1800	1/2 to 1 slope, 2′ wide, 2′ deep, 3/8 C.Y. bucket	1.27	2.18	3.45
1810	3′ deep, 3/8 C.Y. bucket	1.75	3.60	5.35
1820	4′ deep, 3/8 C.Y. bucket	2.22	5.25	7.47
1840	6′ deep, 3/8 C.Y. bucket	3.23	8.10	11.33

G1030 Site Earthwork

G1030 805	Trenching	COST PER L.F.		
		EQUIP.	LABOR	TOTAL
1860	8' deep, 1/2 C.Y. bucket	4.49	12.35	16.84
1880	10' deep, 1 C.Y. bucket	8.35	17	25.35
2300	4' wide, 2' deep, 3/8 C.Y. bucket	2.11	3.63	5.74
2310	3' deep, 3/8 C.Y. bucket	3.48	6	9.48
2320	4' deep, 1/2 C.Y. bucket	3.66	6.25	9.91
2340	6' deep, 1/2 C.Y. bucket	5.10	10.85	15.95
2360	8' deep, 1/2 C.Y. bucket	8.65	14.70	23.35
2380	10' deep, 1 C.Y. bucket	11.35	21.50	32.85
2400	12' deep, 1 C.Y. bucket	11.95	23.50	35.45
2430	15' deep, 1-1/2 C.Y. bucket	15.20	29	44.20
2460	18' deep, 2-1/2 C.Y. bucket	21	35	56
2840	6' wide, 6' deep, 5/8 C.Y. bucket	8.60	11.85	20.45
2860	8' deep, 3/4 C.Y. bucket	12.25	17.75	30
2880	10' deep, 1 C.Y. bucket	14.30	23.50	37.80
2900	12' deep, 1-1/4 C.Y. bucket	13.45	19.55	33
2940	16' deep, 2 C.Y. bucket	21	31	52
2980	20' deep, 3-1/2 C.Y. bucket	32.50	53.50	86
3020	24' deep, 3-1/2 C.Y. bucket	58.50	100	158.50
3100	8' wide, 12' deep, 1-1/4 C.Y. bucket	18.05	24	42.05
3120	15' deep, 1-1/2 C.Y. bucket	23	32	55
3140	18' deep, 2-1/2 C.Y. bucket	29	35	64
3180	24' deep, 3-1/2 C.Y. bucket	67.50	112	179.50
3270	10' wide, 20' deep, 3-1/2 C.Y. bucket	42	67	109
3280	24' deep, 3-1/2 C.Y. bucket	77	126	203
3370	12' wide, 20' deep, 3-1/2 C.Y. bucket	47.50	74	121.50
3380	25' deep, 3-1/2 C.Y. bucket	90	147	237
3500	1 to 1 slope, 2' wide, 2' deep, 3/8 C.Y. bucket	1.48	2.55	4.03
3520	3' deep, 3/8 C.Y. bucket	2.27	4.46	6.73
3540	4' deep, 3/8 C.Y. bucket	2.92	6.70	9.62
3560	6' deep, 3/8 C.Y. bucket	4.40	10.70	15.10
3580	8' deep, 1/2 C.Y. bucket	6.35	16.85	23.20
3600	10' deep, 1 C.Y. bucket	13.90	27	40.90
3800	4' wide, 2' deep, 3/8 C.Y. bucket	2.52	4.35	6.87
3820	3' deep, 3/8 C.Y. bucket	4.44	7.65	12.09
3840	4' deep, 1/2 C.Y. bucket	4.45	7	11.45
3860	6' deep, 1/2 C.Y. bucket	6.35	13	19.35
3880	8' deep, 1/2 C.Y. bucket	11.65	19.40	31.05
3900	10' deep, 1 C.Y. bucket	15.15	28	43.15
3920	12' deep, 1 C.Y. bucket	22	40.50	62.50
3940	15' deep, 1-1/2 C.Y. bucket	21.50	38	59.50
3960	18' deep, 2-1/2 C.Y. bucket	30	45.50	75.50
4030	6' wide, 6' deep, 5/8 C.Y. bucket	10.45	14.40	24.85
4040	8' deep, 3/4 C.Y. bucket	15.30	23	38.30
4050	10' deep, 1 C.Y. bucket	18.45	32	50.45
4060	12' deep, 1-1/4 C.Y. bucket	17.95	29	46.95
4070	16' deep, 2 C.Y. bucket	28.50	41	69.50
4080	20' deep, 3-1/2 C.Y. bucket	46.50	85	131.50
4090	24' deep, 3-1/2 C.Y. bucket	87	165	252
4500	8' wide, 12' deep, 1-1/4 C.Y. bucket	23	32.50	55.50
4550	15' deep, 1-1/2 C.Y. bucket	30.50	46.50	77
4600	18' deep, 2-1/2 C.Y. bucket	39.50	54	93.50
4650	24' deep, 3-1/2 C.Y. bucket	96	176	272
4800	10' wide, 20' deep, 3-1/2 C.Y. bucket	58.50	98.50	157
4850	24' deep, 3-1/2 C.Y. bucket	104	187	291
4950	12' wide, 20' deep, 3-1/2 C.Y. bucket	63.50	105	168.50
4980	25' deep, 3-1/2 C.Y. bucket	126	227	353
5000	1-1/2 to 1 slope, 2' wide, 2' deep, 3/8 C.Y. bucket	1.71	2.95	4.66
5020	3' deep, 3/8 C.Y. bucket	2.77	5.25	8.02

G1030 Site Earthwork

G1030 805	Trenching	COST PER L.F.		
		EQUIP.	LABOR	TOTAL
5040	4′ deep, 3/8 C.Y. bucket	3.58	7.85	11.43
5060	6′ deep, 3/8 C.Y. bucket	5.45	12.50	17.95
5080	8′ deep, 1/2 C.Y. bucket	7.90	19.80	27.70
5100	10′ deep, 1 C.Y. bucket	15.55	28	43.55
5300	4′ wide, 2′ deep, 3/8 C.Y. bucket	2.39	4.10	6.49
5320	3′ deep, 3/8 C.Y. bucket	4.32	7.45	11.77
5340	4′ deep, 1/2 C.Y. bucket	5.30	8.20	13.50
5360	6′ deep, 1/2 C.Y. bucket	7.55	14.75	22.30
5380	8′ deep, 1/2 C.Y. bucket	14	22	36
5400	10′ deep, 1 C.Y. bucket	18.40	31.50	49.90
5420	12′ deep, 1 C.Y. bucket	27	47	74
5450	15′ deep, 1-1/2 C.Y. bucket	26.50	42	68.50
5480	18′ deep, 2-1/2 C.Y. bucket	37.50	49.50	87
5660	6′ wide, 6′ deep, 5/8 C.Y. bucket	12.10	16.10	28.20
5680	8′ deep, 3/4 C.Y. bucket	17.90	25.50	43.40
5700	10′ deep, 1 C.Y. bucket	22	35.50	57.50
5720	12′ deep, 1-1/4 C.Y. bucket	21	30.50	51.50
5760	16′ deep, 2 C.Y. bucket	34.50	43.50	78
5800	20′ deep, 3-1/2 C.Y. bucket	57	96	153
5840	24′ deep, 3-1/2 C.Y. bucket	109	187	296
6020	8′ wide, 12′ deep, 1-1/4 C.Y. bucket	27.50	36	63.50
6050	15′ deep, 1-1/2 C.Y. bucket	36	51	87
6080	18′ deep, 2-1/2 C.Y. bucket	48	58	106
6140	24′ deep, 3-1/2 C.Y. bucket	118	198	316
6300	10′ wide, 20′ deep, 3-1/2 C.Y. bucket	70	109	179
6350	24′ deep, 3-1/2 C.Y. bucket	125	207	332
6450	12′ wide, 20′ deep, 3-1/2 C.Y. bucket	76	116	192
6480	25′ deep, 3-1/2 C.Y. bucket	152	250	402
6600	2 to 1 slope, 2′ wide, 2′ deep, 3/8 C.Y. bucket	2.70	2.68	5.38
6620	3′ deep, 3/8 C.Y. bucket	3.19	5.85	9.04
6640	4′ deep, 3/8 C.Y. bucket	4.10	8.75	12.85
6660	6′ deep, 3/8 C.Y. bucket	5.85	13.25	19.10
6680	8′ deep, 1/2 C.Y. bucket	9	22.50	31.50
6700	10′ deep, 1 C.Y. bucket	17.65	31.50	49.15
6900	4′ wide, 2′ deep, 3/8 C.Y. bucket	2.38	4.08	6.46
6920	3′ deep, 3/8 C.Y. bucket	4.44	7.65	12.09
6940	4′ deep, 1/2 C.Y. bucket	5.95	8.75	14.70
6960	6′ deep, 1/2 C.Y. bucket	8.45	16	24.45
6980	8′ deep, 1/2 C.Y. bucket	15.55	24	39.55
7000	10′ deep, 1 C.Y. bucket	20.50	35	55.50
7020	12′ deep, 1 C.Y. bucket	30.50	52.50	83
7050	15′ deep, 1-1/2 C.Y. bucket	29.50	47	76.50
7080	18′ deep, 2-1/2 C.Y. bucket	42.50	56	98.50
7260	6′ wide, 6′ deep, 5/8 C.Y. bucket	13.35	17.20	30.55
7280	8′ deep, 3/4 C.Y. bucket	19.70	27.50	47.20
7300	10′ deep, 3/4 C.Y. bucket	26	39.50	65.50
7320	12′ deep, 1-1/4 C.Y. bucket	23.50	34.50	58
7360	16′ deep, 2 C.Y. bucket	38.50	48.50	87
7400	20′ deep, 3-1/2 C.Y. bucket	64	107	171
7440	24′ deep, 3-1/2 C.Y. bucket	122	210	332
7620	8′ wide, 12′ deep, 1-1/4 C.Y. bucket	30.50	39	69.50
7650	15′ deep, 1-1/2 C.Y. bucket	40.50	56	96.50
7680	18′ deep, 2-1/2 C.Y. bucket	53	64	117
7740	24′ deep, 3-1/2 C.Y. bucket	132	220	352
7920	10′ wide, 20′ deep, 3-1/2 C.Y. bucket	77	120	197
7940	24′ deep, 3-1/2 C.Y. bucket	139	229	368
8060	12′ wide, 20′ deep, 3-1/2 C.Y. bucket	84	126	210
8080	25′ deep, 3-1/2 C.Y. bucket	168	274	442

G1030 Site Earthwork

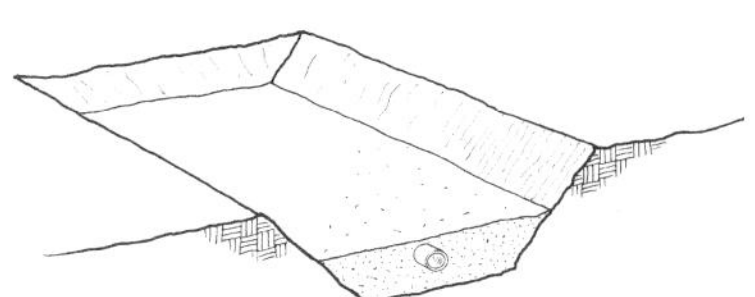

The Pipe Bedding System is shown for various pipe diameters. Compacted bank sand is used for pipe bedding and to fill 12″ over the pipe. No backfill is included. Various side slopes are shown to accommodate different soil conditions. Pipe sizes vary from 6″ to 84″ diameter.

System Components	QUANTITY	UNIT	COST PER L.F. MAT.	COST PER L.F. INST.	COST PER L.F. TOTAL
SYSTEM G1030 815 1440					
PIPE BEDDING, SIDE SLOPE 0 TO 1, 1′ WIDE, PIPE SIZE 6″ DIAMETER					
Borrow, bank sand, 2 mile haul, machine spread	.067	C.Y.	.30	.37	.67
Compaction, vibrating plate	.067	C.Y.		.12	.12
TOTAL			.30	.49	.79

G1030 815	Pipe Bedding	COST PER L.F. MAT.	COST PER L.F. INST.	COST PER L.F. TOTAL
1440	Pipe bedding, side slope 0 to 1, 1′ wide, pipe size 6″ diameter	.30	.49	.79
1460	2′ wide, pipe size 8″ diameter	.67	1.09	1.76
1480	Pipe size 10″ diameter	.69	1.11	1.80
1500	Pipe size 12″ diameter	.71	1.15	1.86
1520	3′ wide, pipe size 14″ diameter	1.17	1.89	3.06
1540	Pipe size 15″ diameter	1.18	1.90	3.08
1560	Pipe size 16″ diameter	1.19	1.93	3.12
1580	Pipe size 18″ diameter	1.22	1.97	3.19
1600	4′ wide, pipe size 20″ diameter	1.76	2.83	4.59
1620	Pipe size 21″ diameter	1.78	2.86	4.64
1640	Pipe size 24″ diameter	1.83	2.95	4.78
1660	Pipe size 30″ diameter	1.87	3.01	4.88
1680	6′ wide, pipe size 32″ diameter	3.26	5.25	8.51
1700	Pipe size 36″ diameter	3.35	5.40	8.75
1720	7′ wide, pipe size 48″ diameter	4.36	7.05	11.41
1740	8′ wide, pipe size 60″ diameter	5.45	8.80	14.25
1760	10′ wide, pipe size 72″ diameter	7.85	12.70	20.55
1780	12′ wide, pipe size 84″ diameter	10.70	17.25	27.95
2140	Side slope 1/2 to 1, 1′ wide, pipe size 6″ diameter	.64	1.02	1.66
2160	2′ wide, pipe size 8″ diameter	1.06	1.71	2.77
2180	Pipe size 10″ diameter	1.15	1.86	3.01
2200	Pipe size 12″ diameter	1.23	1.99	3.22
2220	3′ wide, pipe size 14″ diameter	1.77	2.84	4.61
2240	Pipe size 15″ diameter	1.81	2.92	4.73
2260	Pipe size 16″ diameter	1.87	3.01	4.88
2280	Pipe size 18″ diameter	1.97	3.18	5.15
2300	4′ wide, pipe size 20″ diameter	2.61	4.20	6.81
2320	Pipe size 21″ diameter	2.67	4.31	6.98
2340	Pipe size 24″ diameter	2.86	4.62	7.48
2360	Pipe size 30″ diameter	3.21	5.20	8.41
2380	6′ wide, pipe size 32″ diameter	4.73	7.65	12.38
2400	Pipe size 36″ diameter	5.05	8.15	13.20
2420	7′ wide, pipe size 48″ diameter	6.90	11.15	18.05
2440	8′ wide, pipe size 60″ diameter	9	14.50	23.50
2460	10′ wide, pipe size 72″ diameter	12.60	20.50	33.10
2480	12′ wide, pipe size 84″ diameter	16.75	27	43.75
2620	Side slope 1 to 1, 1′ wide, pipe size 6″ diameter	.98	1.58	2.56
2640	2′ wide,pipe size 8″ diameter	1.47	2.38	3.85

G10 Site Preparation

G1030 Site Earthwork

G1030 815	Pipe Bedding	COST PER L.F.		
		MAT.	INST.	TOTAL
2660	Pipe size 10" diameter	1.61	2.59	4.20
2680	Pipe size 12" diameter	1.77	2.84	4.61
2700	3' wide, pipe size 14" diameter	2.37	3.81	6.18
2720	Pipe size 15" diameter	2.45	3.95	6.40
2740	Pipe size 16" diameter	2.54	4.10	6.64
2760	Pipe size 18" diameter	2.73	4.40	7.13
2780	4' wide, pipe size 20" diameter	3.45	5.55	9
2800	Pipe size 21" diameter	3.55	5.75	9.30
2820	Pipe size 24" diameter	3.89	6.25	10.14
2840	Pipe size 30" diameter	4.54	7.30	11.84
2860	6' wide, pipe size 32" diameter	6.20	9.95	16.15
2880	Pipe size 36" diameter	6.75	10.90	17.65
2900	7' wide, pipe size 48" diameter	9.45	15.25	24.70
2920	8' wide, pipe size 60" diameter	12.55	20.50	33.05
2940	10' wide, pipe size 72" diameter	17.30	28	45.30
2960	12' wide, pipe size 84" diameter	23	37	60
3000	Side slope 1-1/2 to 1, 1' wide, pipe size 6" diameter	1.31	2.11	3.42
3020	2' wide, pipe size 8" diameter	1.86	3.01	4.87
3040	Pipe size 10" diameter	2.06	3.32	5.38
3060	Pipe size 12" diameter	2.29	3.69	5.98
3080	3' wide, pipe size 14" diameter	2.96	4.77	7.73
3100	Pipe size 15" diameter	3.08	4.98	8.06
3120	Pipe size 16" diameter	3.21	5.20	8.41
3140	Pipe size 18" diameter	3.49	5.65	9.14
3160	4' wide, pipe size 20" diameter	4.30	6.95	11.25
3180	Pipe size 21" diameter	4.44	7.15	11.59
3200	Pipe size 24" diameter	4.92	7.90	12.82
3220	Pipe size 30" diameter	5.90	9.50	15.40
3240	6' wide, pipe size 32" diameter	7.65	12.30	19.95
3260	Pipe size 36" diameter	8.45	13.65	22.10
3280	7' wide, pipe size 48" diameter	12	19.35	31.35
3300	8' wide, pipe size 60" diameter	16.10	26	42.10
3320	10' wide, pipe size 72" diameter	22	35.50	57.50
3340	12' wide, pipe size 84" diameter	29	46.50	75.50
3400	Side slope 2 to 1, 1' wide, pipe size 6" diameter	1.68	2.70	4.38
3420	2' wide, pipe size 8" diameter	2.26	3.64	5.90
3440	Pipe size 10" diameter	2.52	4.06	6.58
3460	Pipe size 12" diameter	2.81	4.53	7.34
3480	3' wide, pipe size 14" diameter	3.55	5.75	9.30
3500	Pipe size 15" diameter	3.70	5.95	9.65
3520	Pipe size 16" diameter	3.88	6.25	10.13
3540	Pipe size 18" diameter	4.24	6.85	11.09
3560	4' wide, pipe size 20" diameter	5.15	8.30	13.45
3580	Pipe size 21" diameter	5.35	8.60	13.95
3600	Pipe size 24" diameter	5.95	9.60	15.55
3620	Pipe size 30" diameter	7.25	11.70	18.95
3640	6' wide, pipe size 32" diameter	9.10	14.65	23.75
3660	Pipe size 36" diameter	10.15	16.40	26.55
3680	7' wide, pipe size 48" diameter	14.55	23.50	38.05
3700	8' wide, pipe size 60" diameter	19.65	31.50	51.15
3720	10' wide, pipe size 72" diameter	27	43	70
3740	12' wide, pipe size 84" diameter	35	56.50	91.50

Important: See the Reference Section for critical supporting data - Reference Numbers and City Cost Indexes

G3010 Water Supply

G3010 110	Water Distribution Piping	COST PER L.F.		
		MAT.	INST.	TOTAL
2000	Piping, excav. & backfill excl., ductile iron class 250, mech. joint			
2130	4″ diameter	13.75	12.85	26.60
2150	6″ diameter	15.95	16.05	32
2160	8″ diameter	17.65	19.20	36.85
2170	10″ diameter	24	22.50	46.50
2180	12″ diameter	29.50	24.50	54
2210	16″ diameter	41	35	76
2220	18″ diameter	51.50	37	88.50
3000	Tyton joint			
3130	4″ diameter	8.10	6.40	14.50
3150	6″ diameter	9.20	7.65	16.85
3160	8″ diameter	12.65	12.85	25.50
3170	10″ diameter	20	14.10	34.10
3180	12″ diameter	21	16.05	37.05
3210	16″ diameter	32.50	22.50	55
3220	18″ diameter	36	25.50	61.50
3230	20″ diameter	39.50	29	68.50
3250	24″ diameter	51.50	33.50	85
4000	Copper tubing, type K			
4050	3/4″ diameter	2.97	2.21	5.18
4060	1″ diameter	3.93	2.76	6.69
4080	1-1/2″ diameter	6.45	3.34	9.79
4090	2″ diameter	10	3.85	13.85
4110	3″ diameter	20.50	6.60	27.10
4130	4″ diameter	34	9.30	43.30
4150	6″ diameter	102	14.50	116.50
5000	Polyvinyl chloride class 160, S.D.R. 26			
5130	1-1/2″ diameter	.50	1.50	2
5150	2″ diameter	1.20	1.64	2.84
5160	4″ diameter	4.16	3	7.16
5170	6″ diameter	8.95	3.56	12.51
5180	8″ diameter	15.20	4.32	19.52
6000	Polyethylene 160 psi, S.D.R. 7			
6050	3/4″ diameter	.34	2.14	2.48
6060	1″ diameter	.43	2.32	2.75
6080	1-1/2″ diameter	.78	2.50	3.28
6090	2″ diameter	1.20	3.08	4.28

G3020 Sanitary Sewer

G3020 110	Drainage & Sewage Piping	COST PER L.F.		
		MAT.	INST.	TOTAL
2000	Piping, excavation & backfill excluded, PVC, plain			
2130	4″ diameter	2.11	3	5.11
2150	6″ diameter	3.78	3.21	6.99
2160	8″ diameter	6.40	3.35	9.75
2900	Box culvert, precast, 8′ long			
3000	6′ x 3′	176	23	199
3020	6′ x 7′	267	26	293
3040	8′ x 3′	243	24.50	267.50
3060	8′ x 8′	330	32.50	362.50
3080	10′ x 3′	360	29.50	389.50
3100	10′ x 8′	410	40.50	450.50
3120	12′ x 3′	350	32.50	382.50
3140	12′ x 8′	600	48.50	648.50
4000	Concrete, nonreinforced			
4150	6″ diameter	4.54	8.75	13.29
4160	8″ diameter	5	10.35	15.35
4170	10″ diameter	5.55	10.75	16.30
4180	12″ diameter	6.80	11.65	18.45
4200	15″ diameter	7.95	12.90	20.85
4220	18″ diameter	9.75	16.15	25.90
4250	24″ diameter	14.75	23.50	38.25
4400	Reinforced, no gasket			
4580	12″ diameter	12.35	15.50	27.85
4600	15″ diameter	15.80	15.50	31.30
4620	18″ diameter	16.60	17.60	34.20
4650	24″ diameter	24.50	23.50	48
4670	30″ diameter	33.50	35.50	69
4680	36″ diameter	48	44	92
4690	42″ diameter	65.50	48	113.50
4700	48″ diameter	81	54.50	135.50
4720	60″ diameter	129	72	201
4730	72″ diameter	183	86.50	269.50
4740	84″ diameter	310	109	419
4800	With gasket			
4980	12″ diameter	13.50	9.05	22.55
5000	15″ diameter	16.20	9.50	25.70
5020	18″ diameter	20.50	9.95	30.45
5050	24″ diameter	30.50	11.15	41.65
5070	30″ diameter	40.50	35.50	76
5080	36″ diameter	60.50	44	104.50
5090	42″ diameter	79	43.50	122.50
5100	48″ diameter	99	54.50	153.50
5120	60″ diameter	146	69	215
5130	72″ diameter	256	86.50	342.50
5140	84″ diameter	370	144	514
5700	Corrugated metal, alum. or galv. bit. coated			
5760	8″ diameter	9.85	7	16.85
5770	10″ diameter	11.80	8.90	20.70
5780	12″ diameter	14.15	11.10	25.25
5800	15″ diameter	17.20	11.65	28.85
5820	18″ diameter	22	12.25	34.25
5850	24″ diameter	27	14.55	41.55
5870	30″ diameter	35.50	26	61.50
5880	36″ diameter	52.50	26	78.50
5900	48″ diameter	79.50	32	111.50
5920	60″ diameter	103	46	149
5930	72″ diameter	154	77	231
6000	Plain			

G3020 Sanitary Sewer

G3020 110	Drainage & Sewage Piping	COST PER L.F.		
		MAT.	INST.	TOTAL
6060	8" diameter	7.60	6.50	14.10
6070	10" diameter	8.35	8.30	16.65
6080	12" diameter	9.55	10.60	20.15
6100	15" diameter	12.10	10.60	22.70
6120	18" diameter	16.05	11.30	27.35
6140	24" diameter	23	13.25	36.25
6170	30" diameter	29.50	24.50	54
6180	36" diameter	48.50	24.50	73
6200	48" diameter	65	28.50	93.50
6220	60" diameter	102	44.50	146.50
6230	72" diameter	151	109	260
6300	Steel or alum. oval arch, coated & paved invert			
6400	15" equivalent diameter	29	11.65	40.65
6420	18" equivalent diameter	37.50	15.50	53
6450	24" equivalent diameter	53.50	18.60	72.10
6470	30" equivalent diameter	65.50	23.50	89
6480	36" equivalent diameter	97.50	32	129.50
6490	42" equivalent diameter	117	35	152
6500	48" equivalent diameter	131	42	173
6600	Plain			
6700	15" equivalent diameter	15.75	10.30	26.05
6720	18" equivalent diameter	18.65	13.25	31.90
6750	24" equivalent diameter	30	15.50	45.50
6770	30" equivalent diameter	37.50	29.50	67
6780	36" equivalent diameter	62	29.50	91.50
6790	42" equivalent diameter	73.50	34.50	108
6800	48" equivalent diameter	64.50	42	106.50
8000	Polyvinyl chloride SDR 35			
8130	4" diameter	2.11	3	5.11
8150	6" diameter	3.78	3.21	6.99
8160	8" diameter	6.40	3.35	9.75
8170	10" diameter	9.65	4.59	14.24
8180	12" diameter	10.75	4.74	15.49
8200	15" diameter	16.20	7.95	24.15

G3030 Storm Sewer

Manhole

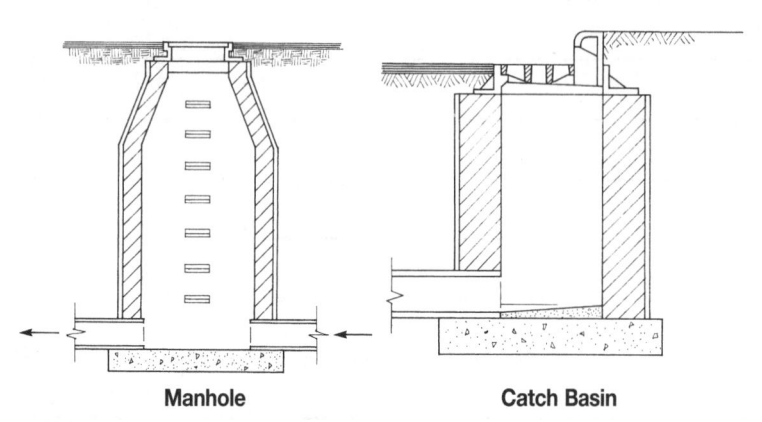

Catch Basin

The Manhole and Catch Basin System includes: excavation with a backhoe; a formed concrete footing; frame and cover; cast iron steps and compacted backfill.

The Expanded System Listing shows manholes that have a 4′, 5′ and 6′ inside diameter riser. Depths range from 4′ to 14′. Construction material shown is either concrete, concrete block, precast concrete, or brick.

System Components	QUANTITY	UNIT	COST PER EACH		
			MAT.	INST.	TOTAL
SYSTEM G3030 210 1920					
MANHOLE/CATCH BASIN, BRICK, 4′ I.D. RISER, 4′ DEEP					
Excavation, hydraulic backhoe, 3/8 C.Y. bucket	14.815	C.Y.		84.74	84.74
Trim sides and bottom of excavation	64.000	S.F.		45.44	45.44
Forms in place, manhole base, 4 uses	20.000	SFCA	11.80	78	89.80
Reinforcing in place footings, #4 to #7	.019	Ton	15.87	18.15	34.02
Concrete, 3000 psi	.925	C.Y.	82.33		82.33
Place and vibrate concrete, footing, direct chute	.925	C.Y.		35.58	35.58
Catch basin or MH, brick, 4′ ID, 4′ deep	1.000	Ea.	350	760	1,110
Catch basin or MH steps; heavy galvanized cast iron	1.000	Ea.	13.45	10.75	24.20
Catch basin or MH frame and cover	1.000	Ea.	212	166.50	378.50
Fill, granular	12.954	C.Y.	112.05		112.05
Backfill, spread with wheeled front end loader	12.954	C.Y.		22.93	22.93
Backfill compaction, 12″ lifts, air tamp	12.954	C.Y.		83.55	83.55
TOTAL			797.50	1,305.64	2,103.14

G3030 210	Manholes & Catch Basins	COST PER EACH		
		MAT.	INST.	TOTAL
1920	Manhole/catch basin, brick, 4′ I.D. riser, 4′ deep	800	1,325	2,125
1940	6′ deep	1,050	1,800	2,850
1960	8′ deep	1,350	2,500	3,850
1980	10′ deep	1,850	3,075	4,925
3000	12′ deep	2,425	3,350	5,775
3020	14′ deep	3,075	4,675	7,750
3200	Block, 4′ I.D. riser, 4′ deep	730	1,050	1,780
3220	6′ deep	940	1,475	2,415
3240	8′ deep	1,175	2,050	3,225
3260	10′ deep	1,375	2,525	3,900
3280	12′ deep	1,675	3,200	4,875
3300	14′ deep	2,025	3,900	5,925
4620	Concrete, cast-in-place, 4′ I.D. riser, 4′ deep	940	1,800	2,740
4640	6′ deep	1,275	2,425	3,700
4660	8′ deep	1,750	3,525	5,275
4680	10′ deep	2,100	4,375	6,475
4700	12′ deep	2,550	5,400	7,950
4720	14′ deep	3,025	6,475	9,500
5820	Concrete, precast, 4′ I.D. riser, 4′ deep	1,200	965	2,165
5840	6′ deep	1,525	1,300	2,825

G3030 Storm Sewer

G3030 210	Manholes & Catch Basins	COST PER EACH		
		MAT.	INST.	TOTAL
5860	8′ deep	1,850	1,825	3,675
5880	10′ deep	2,275	2,250	4,525
5900	12′ deep	2,800	2,750	5,550
5920	14′ deep	3,375	3,500	6,875
6000	5′ I.D. riser, 4′ deep	1,275	1,050	2,325
6020	6′ deep	1,700	1,475	3,175
6040	8′ deep	2,150	1,925	4,075
6060	10′ deep	2,700	2,475	5,175
6080	12′ deep	3,275	3,150	6,425
6100	14′ deep	3,925	3,850	7,775
6200	6′ I.D. riser, 4′ deep	1,875	1,375	3,250
6220	6′ deep	2,400	1,825	4,225
6240	8′ deep	3,000	2,575	5,575
6260	10′ deep	3,750	3,250	7,000
6280	12′ deep	4,575	4,075	8,650
6300	14′ deep	5,425	4,950	10,375

G3060 Fuel Distribution

G3060 110	Gas Service Piping	COST PER L.F.		
		MAT.	INST.	TOTAL
2000	Piping, excavation & backfill excluded, polyethylene			
2070	1-1/4″ diam, SDR 10	.96	2.89	3.85
2090	2″ diam, SDR 11	1.19	3.23	4.42
2110	3″ diam, SDR 11.5	2.48	3.86	6.34
2130	4″ diam,SDR 11	5.70	7.30	13
2150	6″ diam, SDR 21	17.70	7.80	25.50
2160	8″ diam, SDR 21	24	9.45	33.45
3000	Steel, schedule 40, plain end, tarred & wrapped			
3060	1″ diameter	2.90	6.55	9.45
3090	2″ diameter	4.56	7	11.56
3110	3″ diameter	7.55	7.55	15.10
3130	4″ diameter	9.80	11.85	21.65
3140	5″ diameter	14.25	13.70	27.95
3150	6″ diameter	17.40	16.80	34.20
3160	8″ diameter	27.50	21.50	49

Important: See the Reference Section for critical supporting data - Reference Numbers and City Cost Indexes

G3060 Fuel Distribution

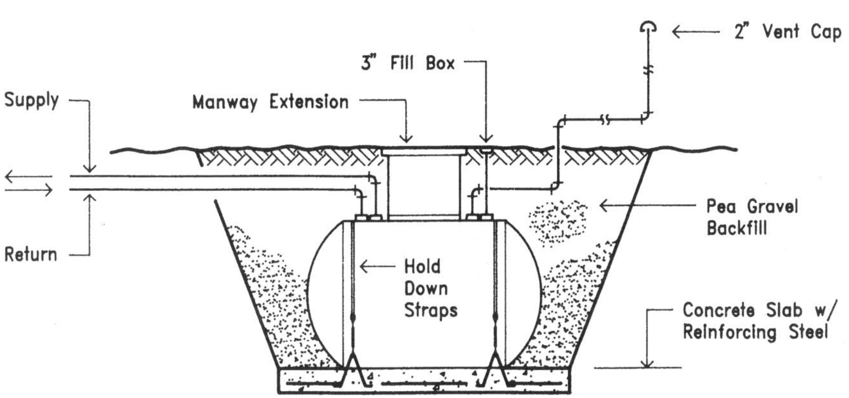

Fiberglass Underground Fuel Storage Tank, Single Wall

System Components	QUANTITY	UNIT	COST EACH MAT.	COST EACH INST.	COST EACH TOTAL
SYSTEM G3060 305 1010					
STORAGE TANK, FUEL, UNDERGROUND, SINGLE WALL FIBERGLASS, 550 GAL.					
Tank, fiberglass, underground, single wall, U.L. listed, 550 gal	1.000	Ea.	1,650	330	1,980
Tank, fiberglass, manways, add	1.000	Ea.	935		935
Tank, for hold-downs 500-4000 gal, add	1.000	Ea.	260	117	377
Foot valve, single poppet, 3/4" dia	1.000	Ea.	41.50	27.50	69
Tubing, copper, Type L, 3/4" dia	120.000	L.F.	290.40	774	1,064.40
Elbow 90°, copper, wrought, cu x cu, 3/4" dia.	8.000	Ea.	9.36	208	217.36
Fuel oil specialties, valve, ball chk, globe type, fusible, 3/4" dia	2.000	Ea.	87	49	136
Elbow 90°, black steel, straight, 2" dia.	3.000	Ea.	35.70	147	182.70
Union, black steel, with brass seat, 2" dia.	1.000	Ea.	23	52	75
Pipe and nipples, steel, Sch. 40, threaded, black, 2" dia	2.000	Ea.	172.90	524.40	697.30
Vent protector / breather, 2" dia.	1.000	Ea.	17.60	15.40	33
Pipe, black steel, welded, Sch. 40, 3" dia.	30.000	L.F.	213	676.80	889.80
Elbow 90°, steel, weld joint, butt, 3" dia.	3.000	Ea.	50.70	415.95	466.65
Flange, weld neck, 150 lb., 3" pipe size	1.000	Ea.	25.50	69.35	94.85
Fuel fill box, locking inner cover, 3" dia.	1.000	Ea.	112	98.50	210.50
Remote tank gauging system, 30', 5" pointer travel	1.000	Ea.	3,850	246	4,096
Tank leak detector system, 8 channel, external monitoring	1.000	Ea.	1,800		1,800
Tank leak detection system, probes, well monitoring, liquid phase detection	2.000	Ea.	1,610		1,610
Excavating, trench, NO sheeting or dewatering, 6'-10' D,3/4 CY hyd backhoe	18.000	C.Y.		102.96	102.96
Concrete hold down pad, 6' x 5' x 8" thick	30.000	C.Y.	70.80	33.30	104.10
Reinforcing in hold down pad, #3 to #7	120.000	Lb.	52.80	57.60	110.40
Stone back-fill, hand spread, pea gravel	9.000	C.Y.	553.50	328.50	882
Corrosion resistance, wrap & coat, small diam pipe, 1" diam, add	120.000	L.F.	145.20		145.20
Corrosion resistance, wrap & coat, small diameter pipe, 2" diam, add	36.000	L.F.	47.88		47.88
Corrosion resistance wrap & coat, 4" diam, add	30.000	L.F.	48.30		48.30
TOTAL			12,102.14	4,273.26	16,375.40

G3060 305	Fiberglass Fuel Tank, Single Wall	COST EACH MAT.	COST EACH INST.	COST EACH TOTAL
1010	Storage tank, fuel, underground, single wall fiberglass, 550 Gal.	12,100	4,275	16,375
1020	2000 Gal.	15,200	5,800	21,000
1030	4000 Gal.	20,400	7,925	28,325
1040	6000 Gal.	21,900	8,850	30,750
1050	8000 Gal.	26,700	10,800	37,500
1060	10,000 Gal.	30,000	11,800	41,800

G30 Site Mechanical Utilities

G3060 Fuel Distribution

G3060 305	Fiberglass Fuel Tank, Single Wall	COST EACH		
		MAT.	INST.	TOTAL
1070	15,000 Gal.	34,800	14,500	49,300
1080	20,000 Gal.	44,200	17,700	61,900
1090	25,000 Gal.	55,000	20,600	75,600
1100	30,000 Gal.	62,500	22,200	84,700
1110	40,000 Gal.	79,000	28,000	107,000
1120	48,000 Gal.	107,000	32,200	139,200

G3060 310	Fiberglass Fuel Tank, Double Wall	COST EACH		
		MAT.	INST.	TOTAL
1010	Storage tank, fuel, underground, double wall fiberglass, 550 Gal.	14,400	4,875	19,275
1020	2500 Gal.	20,800	6,775	27,575
1030	4000 Gal.	26,600	8,400	35,000
1040	6000 Gal.	28,600	9,475	38,075
1050	8000 Gal.	33,900	11,400	45,300
1060	10,000 Gal.	37,900	13,200	51,100
1070	15,000 Gal.	49,300	15,200	64,500
1080	20,000 Gal.	60,000	18,400	78,400
1090	25,000 Gal.	72,000	21,300	93,300
1100	30,000 Gal.	81,500	22,700	104,200

G3060 Fuel Distribution

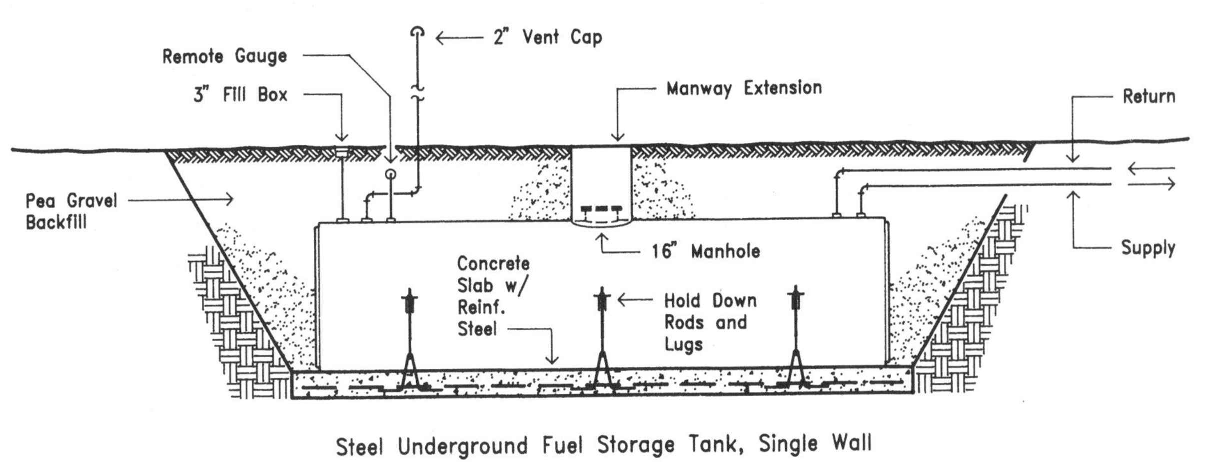

Steel Underground Fuel Storage Tank, Single Wall

System Components	QUANTITY	UNIT	COST EACH MAT.	COST EACH INST.	COST EACH TOTAL
SYSTEM G3060 315 1010					
STORAGE TANK, FUEL, UNDERGROUND, SINGLE WALL STEEL, 550 GAL.					
Tank, steel, underground, sti-P3, set in place, 500 gal, 7 Ga. shell	1.000	Ea.	1,100	330	1,430
Tank, for manways, add	1.000	Ea.	705		705
Tanks, for hold-downs, 500-2000 gal, add	1.000	Ea.	410	117	527
Foot valve, single poppet, 3/4" dia	1.000	Ea.	41.50	27.50	69
Tubing, copper, Type L, 3/4" dia.	120.000	L.F.	290.40	774	1,064.40
Elbow 90°, copper, wrought, cu x cu, 3/4" dia.	8.000	Ea.	9.36	208	217.36
Fuel oil valve, ball chk, globe type, fusible, 3/4" dia.	2.000	Ea.	87	49	136
Elbow 90°, black steel, straight, 2" dia.	3.000	Ea.	35.70	147	182.70
Union, black steel, with brass seat, 3/4" dia.	1.000	Ea.	7.70	38	45.70
Pipe and nipples, steel, Sch. 40, threaded, black, 2" dia	2.000	Ea.	172.90	524.40	697.30
Vent protector / breather, 2" dia.	1.000	Ea.	17.60	15.40	33
Pipe, black steel, welded, Sch. 40, 3" dia.	30.000	L.F.	213	676.80	889.80
Elbow 90°, steel, weld joint, butt, 90° elbow, 3" dia.	3.000	Ea.	50.70	415.95	466.65
Flange, steel, weld neck, 150 lb., 3" pipe size	1.000	Ea.	25.50	69.35	94.85
Fuel fill box, locking inner cover, 3" dia.	1.000	Ea.	112	98.50	210.50
Remote tank gauging system, 30', 5" pointer travel	1.000	Ea.	3,850	246	4,096
Tank leak detector system, 8 channel, external monitoring	1.000	Ea.	1,800		1,800
Tank leak detection system, probes, well monitoring, liquid phase detection	2.000	Ea.	1,610		1,610
Excavation, NO sheeting or dewatering,6'-10' D,3/4 CY hyd backhoe	18.000	C.Y.		102.96	102.96
Concrete hold down pad, 6' x 5' x 8" thick	30.000	C.Y.	70.80	33.30	104.10
Reinforcing in hold down pad, #3 to #7	120.000	Lb.	52.80	57.60	110.40
Stone back-fill, hand spread, pea gravel	9.000	C.Y.	553.50	328.50	882
Corrosion resistance, wrap & coat, small diam pipe, 1" dia., add	120.000	L.F.	145.20		145.20
Corrosion resistance, wrap & coat, small diameter pipe, 2" dia., add	36.000	L.F.	47.88		47.88
Corrosion resistance wrap & coat, 4" dia., add	30.000	L.F.	48.30		48.30
TOTAL			11,456.84	4,259.26	15,716.10

G3060 315	Steel Fuel Tank, Single Wall	COST EACH MAT.	COST EACH INST.	COST EACH TOTAL
1010	Storage tank, fuel, underground, single wall steel, 550 Gal.	11,500	4,250	15,750
1020	2000 Gal.	15,700	5,800	21,500
1030	5000 Gal.	24,300	8,225	32,525
1040	10,000 Gal.	36,800	12,500	49,300
1050	15,000 Gal.	41,500	14,500	56,000
1060	20,000 Gal.	52,000	17,500	69,500

G3060 Fuel Distribution

G3060 315	Steel Fuel Tank, Single Wall	COST EACH		
		MAT.	INST.	TOTAL
1070	25,000 Gal.	60,000	20,400	80,400
1080	30,000 Gal.	70,000	21,900	91,900
1090	40,000 Gal.	86,500	27,600	114,100
1100	50,000 Gal.	105,500	33,400	138,900

G3060 320	Steel Fuel Tank, Double Wall	COST EACH		
		MAT.	INST.	TOTAL
1010	Storage tank, fuel, underground, double wall steel, 500 Gal.	14,700	5,000	19,700
1020	2000 Gal.	19,800	6,525	26,325
1030	4000 Gal.	26,800	8,500	35,300
1040	6000 Gal.	32,400	9,400	41,800
1050	8000 Gal.	38,100	11,600	49,700
1060	10,000 Gal.	42,800	13,400	56,200
1070	15,000 Gal.	50,500	15,500	66,000
1080	20,000 Gal.	61,000	18,700	79,700
1090	25,000 Gal.	81,500	21,400	102,900
1100	30,000 Gal.	93,000	22,800	115,800
1110	40,000 Gal.	115,000	28,600	143,600
1120	50,000 Gal.	143,000	34,300	177,300

G3060 325	Steel Tank, Above Ground, Single Wall	COST EACH		
		MAT.	INST.	TOTAL
1010	Storage tank, fuel, above ground, single wall steel, 550 Gal.	8,375	3,575	11,950
1020	2000 Gal.	11,000	3,650	14,650
1030	5000 Gal.	13,500	5,575	19,075
1040	10,000 Gal.	24,800	5,950	30,750
1050	15,000 Gal.	25,800	6,100	31,900
1060	20,000 Gal.	30,300	6,300	36,600
1070	25,000 Gal.	33,600	6,450	40,050
1080	30,000 Gal.	38,700	6,700	45,400

G3060 330	Steel Tank, Above Ground, Double Wall	COST EACH		
		MAT.	INST.	TOTAL
1010	Storage tank, fuel, above ground, double wall steel, 500 Gal.	11,600	3,625	15,225
1020	2000 Gal.	15,100	3,700	18,800
1030	4000 Gal.	20,200	3,775	23,975
1040	6000 Gal.	23,200	5,775	28,975
1050	8000 Gal.	27,100	5,950	33,050
1060	10,000 Gal.	28,800	6,050	34,850
1070	15,000 Gal.	38,300	6,250	44,550
1080	20,000 Gal.	42,100	6,450	48,550
1090	25,000 Gal.	48,900	6,625	55,525
1100	30,000 Gal.	52,500	6,875	59,375

Reference Section

All the reference information is in one section, making it easy to find what you need to know . . . and easy to use the book on a daily basis. This section is visually identified by a vertical gray bar on the edge of pages.

In the reference number information that follows, you'll see the background that relates to the "reference numbers" that appeared in the Unit Price Section. You'll find reference tables, explanations and estimating information that support how we arrived at the unit price data. Also included are alternate pricing methods, technical data and estimating procedures along with information on design and economy in construction.

Also in this Reference Section, we've included Change Orders, information on pricing changes to contract documents; Crew Listings, a full listing of all the crews, equipment and their costs; Historical Cost Indexes for cost comparisons over time; City Cost Indexes and Location Factors for adjusting costs to the region you are in, and an explanation of all abbreviations used in the book.

Table of Contents

R01100-005 Tips for Accurate Estimating

1. Use pre-printed or columnar forms for orderly sequence of dimensions and locations and for recording telephone quotations.
2. Use only the front side of each paper or form except for certain pre-printed summary forms.
3. Be consistent in listing dimensions: For example, length x width x height. This helps in rechecking to ensure that, the total length of partitions is appropriate for the building area.
4. Use printed (rather than measured) dimensions where given.
5. Add up multiple printed dimensions for a single entry where possible.
6. Measure all other dimensions carefully.
7. Use each set of dimensions to calculate multiple related quantities.
8. Convert foot and inch measurements to decimal feet when listing. Memorize decimal equivalents to .01 parts of a foot (1/8″ equals approximately .01′).
9. Do not "round off" quantities until the final summary.
10. Mark drawings with different colors as items are taken off.
11. Keep similar items together, different items separate.
12. Identify location and drawing numbers to aid in future checking for completeness.
13. Measure or list everything on the drawings or mentioned in the specifications.
14. It may be necessary to list items not called for to make the job complete.
15. Be alert for: Notes on plans such as N.T.S. (not to scale); changes in scale throughout the drawings; reduced size drawings; discrepancies between the specifications and the drawings.
16. Develop a consistent pattern of performing an estimate. For example:
    a. Start the quantity takeoff at the lower floor and move to the next higher floor.
    b. Proceed from the main section of the building to the wings.
    c. Proceed from south to north or vice versa, clockwise or counterclockwise.
    d. Take off floor plan quantities first, elevations next, then detail drawings.
17. List all gross dimensions that can be either used again for different quantities, or used as a rough check of other quantities for verification (exterior perimeter, gross floor area, individual floor areas, etc.).
18. Utilize design symmetry or repetition (repetitive floors, repetitive wings, symmetrical design around a center line, similar room layouts, etc.). Note: Extreme caution is needed here so as not to omit or duplicate an area.
19. Do not convert units until the final total is obtained. For instance, when estimating concrete work, keep all units to the nearest cubic foot, then summarize and convert to cubic yards.
20. When figuring alternatives, it is best to total all items involved in the basic system, then total all items involved in the alternates. Therefore you work with positive numbers in all cases. When adds and deducts are used, it is often confusing whether to add or subtract a portion of an item; especially on a complicated or involved alternate.

R01100-040 Builder's Risk Insurance

Builder's Risk Insurance is insurance on a building during construction. Premiums are paid by the owner or the contractor. Blasting, collapse and underground insurance would raise total insurance costs above those listed. Floater policy for materials delivered to the job runs $.75 to $1.25 per $100 value. Contractor equipment insurance runs $.50 to $1.50 per $100 value. Insurance for miscellaneous tools to $1,500 value runs from $3.00 to $7.50 per $100 value.

Tabulated below are New England Builder's Risk insurance rates in dollars per $100 value for $1,000 deductible. For $25,000 deductible, rates can be reduced 13% to 34%. On contracts over $1,000,000, rates may be lower than those tabulated. Policies are written annually for the total completed value in place. For "all risk" insurance (excluding flood, earthquake and certain other perils) add $.025 to total rates below.

Coverage	Frame Construction (Class 1)		Brick Construction (Class 4)		Fire Resistive (Class 6)	
	Range	Average	Range	Average	Range	Average
Fire Insurance	$.350 to $ .850	$.600	$.158 to $.189	$.174	$.052 to $.080	$.070
Extended Coverage	.115 to .200	.158	.080 to .105	.101	.081 to .105	.100
Vandalism	.012 to .016	.014	.008 to .011	.011	.008 to .011	.010
Total Annual Rate	$.477 to $1.066	$.772	$.246 to $.305	$.286	$.141 to $.196	$.180

R01100-060 Workers' Compensation Insurance Rates by Trade

The table below tabulates the national averages for Workers' Compensation insurance rates by trade and type of building. The average "Insurance Rate" is multiplied by the "% of Building Cost" for each trade. This produces the "Workers' Compensation Cost" by % of total labor cost, to be added for each trade by building type to determine the weighted average Workers' Compensation rate for the building types analyzed.

Trade	Insurance Rate (% Labor Cost)		% of Building Cost			Workers' Compensation		
	Range	Average	Office Bldgs.	Schools & Apts.	Mfg.	Office Bldgs.	Schools & Apts.	Mfg.
Excavation, Grading, etc.	4.2 % to 19.5%	10.4%	4.8%	4.9%	4.5%	.50%	.51%	.47%
Piles & Foundations	7.7 to 76.7	22.5	7.1	5.2	8.7	1.60	1.17	1.96
Concrete	5.2 to 35.8	16.1	5.0	14.8	3.7	.81	2.38	.60
Masonry	5.1 to 28.3	15.1	6.9	7.5	1.9	1.04	1.13	.29
Structural Steel	7.8 to 103.8	38.6	10.7	3.9	17.6	4.13	1.51	6.79
Miscellaneous & Ornamental Metals	4.9 to 24.8	12.6	2.8	4.0	3.6	.35	.50	.45
Carpentry & Millwork	7.0 to 53.2	18.4	3.7	4.0	0.5	.68	.74	.09
Metal or Composition Siding	5.3 to 32.1	17.0	2.3	0.3	4.3	.39	.05	.73
Roofing	7.8 to 78.9	33.1	2.3	2.6	3.1	.76	.86	1.03
Doors & Hardware	4.5 to 24.9	11.3	0.9	1.4	0.4	.10	.16	.05
Sash & Glazing	4.6 to 33.7	13.9	3.5	4.0	1.0	.49	.56	.14
Lath & Plaster	4.3 to 40.2	14.9	3.3	6.9	0.8	.49	1.03	.12
Tile, Marble & Floors	2.4 to 31.8	9.7	2.6	3.0	0.5	.25	.29	.05
Acoustical Ceilings	2.4 to 29.7	10.9	2.4	0.2	0.3	.26	.02	.03
Painting	4.5 to 33.7	13.1	1.5	1.6	1.6	.20	.21	.21
Interior Partitions	7.0 to 53.2	18.4	3.9	4.3	4.4	.72	.79	.81
Miscellaneous Items	2.6 to 137.6	16.9	5.2	3.7	9.7	.88	.63	1.64
Elevators	2.8 to 42.8	7.6	2.1	1.1	2.2	.16	.08	.17
Sprinklers	3.2 to 22.1	8.8	0.5	—	2.0	.04	—	.18
Plumbing	2.7 to 12.4	8.0	4.9	7.2	5.2	.39	.58	.42
Heat., Vent., Air Conditioning	4.0 to 27.2	11.5	13.5	11.0	12.9	1.55	1.27	1.48
Electrical	2.6 to 12.7	6.5	10.1	8.4	11.1	.66	.55	.72
Total	2.4 % to 137.6%	—	100.0%	100.0%	100.0%	16.45%	15.02%	18.43%
		Overall Weighted Average 16.63%						

Workers' Compensation Insurance Rates by States

The table below lists the weighted average Workers' Compensation base rate for each state with a factor comparing this with the national average of 16.2%.

State	Weighted Average	Factor	State	Weighted Average	Factor	State	Weighted Average	Factor
Alabama	25.2%	156	Kentucky	17.4%	107	North Dakota	11.4%	70
Alaska	19.7	122	Louisiana	26.2	162	Ohio	12.8	79
Arizona	7.4	46	Maine	19.3	119	Oklahoma	13.7	85
Arkansas	13.7	85	Maryland	17.7	109	Oregon	12.6	78
California	17.4	107	Massachusetts	12.4	77	Pennsylvania	13.4	83
Colorado	12.3	76	Michigan	17.6	109	Rhode Island	19.6	121
Connecticut	23.1	143	Minnesota	25.1	155	South Carolina	12.9	80
Delaware	14.2	88	Mississippi	16.1	99	South Dakota	14.2	88
District of Columbia	17.3	107	Missouri	17.6	109	Tennessee	17.9	110
Florida	25.1	155	Montana	18.9	117	Texas	13.5	83
Georgia	21.3	131	Nebraska	18.8	116	Utah	11.4	70
Hawaii	15.7	97	Nevada	13.9	86	Vermont	18.6	115
Idaho	10.4	64	New Hampshire	21.3	131	Virginia	10.8	67
Illinois	17.2	106	New Jersey	10.5	65	Washington	10.4	64
Indiana	5.4	33	New Mexico	16.3	101	West Virginia	12.4	77
Iowa	11.8	73	New York	13.4	83	Wisconsin	15.2	94
Kansas	8.5	52	North Carolina	13.2	81	Wyoming	7.9	49
			Weighted Average for U.S. is	15.5% of payroll = 100%				

Rates in the following table are the base or manual costs per $100 of payroll for Workers' Compensation in each state. Rates are usually applied to straight time wages only and not to premium time wages and bonuses.

The weighted average skilled worker rate for 35 trades is 16.2%. For bidding purposes, apply the full value of Workers' Compensation directly to total labor costs, or if labor is 38%, materials 42% and overhead and profit 20% of total cost, carry 38/80 x 16.2% =7.7% of cost (before overhead and profit) into overhead. Rates vary not only from state to state but also with the experience rating of the contractor.

Rates are the most current available at the time of publication.

R01100-060 Workers' Compensation Insurance Rates by Trade and State (cont.)

State	Carpentry — 3 stories or less 5651	Carpentry — interior cab. work 5437	Carpentry — general 5403	Concrete Work — NOC 5213	Concrete Work — flat (flr., sdwk.) 5221	Electrical Wiring — inside 5190	Excavation — earth NOC 6217	Excavation — rock 6217	Glaziers 5462	Insulation Work 5479	Lathing 5443	Masonry 5022	Painting & Decorating 5474	Pile Driving 6003	Plastering 5480	Plumbing 5183	Roofing 5551	Sheet Metal Work (HVAC) 5538	Steel Erection — door & sash 5102	Steel Erection — inter., ornam. 5102	Steel Erection — structure 5040	Steel Erection — NOC 5057	Tile Work — (interior ceramic) 5348	Waterproofing 9014	Wrecking 5701
AL	26.93	15.28	34.89	12.39	9.90	8.95	13.69	13.69	33.66	17.12	12.78	27.57	33.72	34.63	40.23	11.91	60.85	27.15	23.11	23.11	51.16	39.09	13.16	6.65	51.16
AK	17.02	14.74	13.84	12.07	11.18	12.70	19.54	19.54	19.86	25.41	9.78	24.13	16.28	59.80	19.42	11.79	38.40	12.13	12.24	12.24	42.33	18.60	8.48	10.79	42.33
AZ	7.71	5.37	12.63	6.33	3.44	4.18	4.83	4.83	7.59	9.87	4.70	6.40	5.23	9.47	15.76	3.77	11.96	4.71	7.92	7.92	16.58	7.75	2.37	2.55	50.31
AR	13.04	11.01	16.99	13.77	6.64	6.11	9.42	9.42	15.49	33.95	8.65	11.34	10.50	12.82	12.93	5.69	21.95	10.28	7.70	7.70	37.63	28.19	7.10	3.75	37.63
CA	28.28	9.07	28.28	13.18	13.18	9.93	7.88	7.88	16.37	23.34	10.90	14.55	19.45	21.18	18.12	11.59	39.79	15.33	13.55	13.55	23.93	21.91	7.74	19.45	21.91
CO	17.36	9.17	12.24	14.36	7.69	5.32	10.44	10.44	9.28	13.68	5.57	13.41	10.77	17.70	10.81	8.51	24.47	12.11	7.39	7.39	31.30	15.53	7.99	5.78	31.30
CT	20.96	18.42	31.08	29.77	13.28	8.63	11.18	11.18	15.10	24.22	17.24	25.81	19.56	27.54	27.61	10.37	51.40	14.91	18.42	18.42	70.81	35.84	12.65	6.69	50.86
DE	15.39	15.39	12.92	13.03	7.13	6.79	10.03	10.03	11.96	12.92	13.83	12.33	16.85	21.38	13.83	7.94	27.84	10.69	13.02	13.02	29.92	13.02	10.51	12.33	29.92
DC	13.05	10.16	15.52	17.41	20.87	6.52	10.32	10.32	27.59	11.07	8.44	22.34	10.52	18.70	13.70	12.43	21.23	9.29	17.97	17.97	54.79	20.38	31.80	4.05	54.79
FL	30.55	21.73	31.10	30.26	15.05	10.74	13.79	13.79	23.24	22.05	13.08	23.21	21.38	58.93	36.28	10.71	46.17	18.10	14.78	14.78	58.02	39.86	11.26	9.25	58.02
GA	32.05	17.69	24.16	15.16	10.44	8.97	16.86	16.86	15.69	22.02	17.16	20.30	17.13	31.16	19.94	10.77	46.74	17.52	13.97	13.97	44.46	48.01	9.64	8.55	44.46
HI	17.38	11.62	28.20	13.74	12.27	7.10	7.73	7.73	20.55	21.19	10.94	17.87	11.33	20.15	15.61	6.06	33.24	8.02	11.11	11.11	31.71	21.70	9.78	11.54	31.71
ID	11.48	6.50	13.42	8.40	5.87	5.01	6.24	6.24	9.48	7.60	5.27	9.01	7.40	13.14	9.74	4.67	29.72	9.37	9.80	9.80	30.78	13.76	6.47	4.09	30.78
IL	16.38	13.48	17.63	29.67	11.05	8.51	10.49	10.49	15.50	11.58	9.74	16.53	11.34	28.71	11.92	9.79	28.32	14.36	14.87	14.87	52.98	26.69	14.11	4.36	52.98
IN	5.31	4.50	7.02	5.18	3.17	2.64	4.17	4.17	4.59	4.15	2.43	5.07	4.48	7.71	4.30	2.70	10.93	4.03	4.94	4.94	16.12	9.00	3.08	2.55	16.12
IA	11.67	5.80	10.47	12.55	6.82	4.18	5.98	5.98	13.66	9.22	5.45	8.79	7.73	10.70	9.26	5.92	18.71	7.17	9.92	9.92	51.91	36.28	5.29	4.20	30.67
KS	10.58	8.42	10.47	7.12	6.36	3.72	4.73	4.73	6.49	7.80	4.37	7.45	6.06	9.36	8.73	5.40	20.70	6.76	7.83	7.83	23.76	12.28	4.75	3.48	23.76
KY	15.79	12.91	19.21	20.50	6.16	7.64	17.08	17.08	21.66	24.14	11.20	10.81	11.40	18.49	14.90	6.85	48.37	13.64	11.61	11.61	40.21	30.64	12.56	4.72	40.21
LA	23.14	24.93	53.17	26.43	15.61	9.88	17.49	17.49	20.14	21.43	24.51	28.33	29.58	31.25	22.37	8.64	77.12	21.20	18.98	18.98	51.76	24.21	13.77	13.78	66.41
ME	13.43	10.42	43.59	24.29	11.12	5.03	12.19	12.19	13.35	15.39	14.03	17.26	16.07	31.44	17.95	7.32	32.07	8.82	15.08	15.08	37.84	61.81	11.66	6.06	37.84
MD	13.71	5.95	10.55	18.82	8.70	5.15	12.91	12.91	25.52	20.07	11.37	13.31	9.37	26.22	14.21	8.58	48.72	12.74	13.66	13.66	56.25	37.81	8.03	7.14	26.80
MA	9.93	6.07	16.09	17.95	8.10	3.69	6.49	6.49	7.55	13.48	5.76	13.04	7.54	14.31	5.34	4.55	38.25	7.47	12.35	12.35	35.13	27.39	9.24	3.32	28.82
MI	21.68	13.59	19.80	22.12	9.35	5.71	11.69	11.69	13.87	12.38	13.59	19.33	14.93	39.06	15.59	8.24	41.33	10.41	11.36	11.36	39.06	29.15	11.31	6.40	39.06
MN	21.84	21.96	43.39	16.70	14.64	6.94	16.68	16.68	17.42	11.95	21.02	20.28	17.69	26.83	21.02	10.70	78.89	12.72	14.64	14.64	103.82	38.11	14.59	6.96	132.92
MS	16.92	15.26	19.71	11.11	8.98	6.80	12.22	12.22	12.02	14.11	8.02	13.18	12.86	19.38	22.03	8.15	37.48	21.88	11.67	11.67	33.42	32.27	9.89	6.93	33.42
MO	21.84	12.24	16.70	15.66	11.37	7.09	10.05	10.05	11.75	18.49	12.47	16.14	13.64	21.93	16.25	9.37	33.06	14.20	12.51	12.51	59.20	40.85	9.59	7.02	59.20
MT	22.03	10.79	17.11	12.89	9.98	5.54	17.20	17.20	11.51	16.19	19.02	14.58	12.03	76.71	13.82	8.93	57.20	9.44	9.73	9.73	41.23	17.93	6.54	5.33	41.23
NE	23.15	10.97	20.95	28.60	9.65	8.32	12.20	12.20	15.50	24.47	11.30	18.50	12.32	23.35	14.75	11.17	36.72	13.05	13.15	13.15	52.27	38.35	9.97	6.15	50.40
NV	19.78	7.76	13.18	10.57	11.57	6.18	11.21	11.21	12.58	13.22	7.97	11.38	10.62	14.55	12.30	8.93	21.55	18.17	12.87	12.87	31.33	33.90	6.64	6.15	44.29
NH	21.98	14.43	24.90	35.84	11.10	6.02	17.16	17.16	14.70	33.70	8.60	24.98	15.72	17.80	18.65	10.22	60.89	14.43	14.01	14.01	54.76	25.50	17.84	6.36	54.76
NJ	11.76	8.69	11.76	9.82	7.77	4.22	7.74	7.74	6.91	11.95	10.39	12.07	9.66	13.10	10.39	6.14	29.59	6.69	9.94	9.94	19.67	10.61	4.42	4.65	23.42
NM	28.63	8.03	18.67	17.07	9.38	6.85	9.33	9.33	14.38	12.83	7.20	13.69	12.57	18.59	11.22	8.64	32.00	11.70	24.78	24.78	43.20	25.65	7.98	6.81	43.20
NY	17.66	6.98	15.91	17.16	13.57	6.32	10.02	10.02	12.02	9.09	5.11	19.20	13.75	16.81	8.43	8.53	33.44	17.47	9.63	9.63	18.80	19.14	9.43	6.09	11.29
NC	16.34	11.24	15.16	11.77	7.00	8.63	8.65	8.65	10.30	11.88	8.28	10.32	9.75	13.68	15.22	7.78	26.83	11.26	7.71	7.71	52.96	19.19	5.79	4.27	52.96
ND	9.60	9.60	9.60	5.66	5.66	4.00	5.39	5.39	9.60	9.60	9.04	8.10	6.27	20.12	9.04	5.35	20.74	5.35	20.12	20.12	20.12	20.12	9.60	20.24	14.48
OH	6.32	11.48	9.82	11.66	10.37	5.54	8.42	8.42	9.22	14.76	29.68	12.20	15.20	25.34	4.41	7.14	22.43	8.85	10.53	10.53	25.27	18.05	9.73	5.02	25.27
OK	13.44	9.11	12.70	12.65	6.63	5.82	10.25	10.25	16.10	19.50	8.46	10.98	9.09	20.29	12.21	7.24	22.32	9.49	13.78	13.78	39.78	24.01	6.95	5.88	39.78
OR	19.10	9.61	15.51	13.55	8.43	4.65	9.70	9.70	15.02	8.08	7.48	15.36	12.79	15.81	10.75	5.89	23.26	10.49	9.53	9.53	32.00	13.92	11.23	4.97	32.00
PA	13.85	13.85	12.80	14.90	9.67	6.74	8.44	8.44	9.60	12.80	11.67	12.18	14.33	17.72	11.67	7.51	27.83	8.19	15.17	15.17	26.43	15.17	8.22	12.18	26.43
RI	19.53	11.65	18.07	18.23	16.24	4.43	10.38	10.38	12.85	22.78	11.97	25.11	24.13	37.66	17.25	8.52	33.92	10.29	14.07	14.07	59.49	37.50	14.36	7.70	78.79
SC	19.10	13.03	19.24	12.88	6.19	7.02	8.32	8.32	12.22	10.69	7.27	9.55	11.19	16.64	18.89	6.88	28.98	11.42	9.49	9.49	21.80	23.66	6.23	4.65	21.80
SD	17.83	7.27	14.55	18.00	6.34	5.44	13.24	13.24	10.58	13.24	8.39	8.01	12.63	24.87	11.78	8.53	21.26	9.64	11.53	11.53	53.25	20.15	6.68	4.20	53.25
TN	31.79	14.64	24.26	19.40	9.95	8.72	14.06	14.06	12.09	14.84	13.05	17.02	14.59	17.43	17.47	11.33	38.38	14.51	13.38	13.38	45.26	25.19	9.73	6.57	45.26
TX	16.89	11.33	13.25	12.10	8.85	7.36	10.49	10.49	11.61	15.68	9.53	13.68	9.89	13.95	20.99	8.06	22.80	14.58	11.23	11.23	33.39	14.78	7.29	8.09	17.86
UT	9.45	6.23	10.52	7.68	7.99	7.05	6.38	6.38	9.55	10.95	12.32	13.75	15.29	17.24	10.60	6.41	26.27	6.30	8.99	8.99	23.51	23.51	6.06	5.83	26.53
VT	19.90	11.34	19.46	34.13	12.01	6.10	10.61	10.61	21.25	25.54	10.85	19.53	10.41	23.91	18.23	10.10	28.73	12.62	13.94	13.94	47.05	37.23	8.52	9.59	47.05
VA	11.97	7.81	9.72	10.47	5.32	4.43	6.53	6.53	7.60	7.96	13.79	8.17	9.95	13.55	7.16	5.52	21.62	8.30	11.34	11.34	39.01	16.86	6.32	3.09	39.01
WA	9.13	9.13	9.13	8.17	8.17	3.12	8.54	8.54	13.74	8.92	9.13	13.27	9.39	20.04	11.57	5.04	19.54	4.60	12.96	12.96	9.39	9.59	10.76	10.23	9.59
WV	12.30	12.30	12.30	27.37	27.37	5.86	8.90	8.90	6.99	13.15	13.15	11.94	12.35	8.90	12.35	5.48	14.63	6.99	10.89	10.89	10.89	10.89	13.15	13.15	48.62
WI	12.40	10.86	19.86	13.12	11.66	5.32	6.74	6.74	16.53	14.04	10.60	18.99	14.82	17.24	13.95	6.31	43.57	8.01	14.30	14.30	34.84	21.34	14.42	5.80	34.84
WY	7.77	7.77	7.77	7.77	7.77	7.77	7.77	7.77	7.77	7.77	7.77	7.77	7.77	7.77	7.77	7.77	7.77	7.77	7.77	7.77	7.77	7.77	7.77	7.77	7.77
AVG.	16.96	11.33	18.42	16.07	9.94	6.46	10.43	10.43	13.91	15.53	10.87	15.06	13.12	22.45	14.88	7.96	33.14	11.46	12.57	12.57	38.60	24.71	9.66	7.12	39.48

R01100-060 Workers' Compensation (cont.) (Canada in Canadian dollars)

Province		Alberta	British Columbia	Manitoba	Ontario	New Brunswick	Newfndld. & Labrador	Northwest Territories	Nova Scotia	Prince Edward Island	Quebec	Saskat-chewan	Yukon
Carpentry—3 stories or less	Rate	10.80	6.66	4.57	4.83	4.63	10.05	4.25	7.67	6.15	14.69	7.01	4.36
	Code	42143	721028	40102	723	422	4226	4-41	4226	401	80110	B1317	202
Carpentry—interior cab. work	Rate	2.59	5.94	4.57	4.83	4.21	6.35	4.25	5.74	3.41	14.69	3.65	4.36
	Code	42133	721021	40102	723	427	4270	4-41	4274	402	80110	B11-27	202
CARPENTRY—general	Rate	10.80	6.66	4.57	4.83	4.63	6.35	4.25	7.67	6.15	14.69	7.01	4.36
	Code	42143	721028	40102	723	422	4299	4-41	4226	401	80110	B1317	202
CONCRETE WORK—NOC	Rate	7.26	8.48	7.60	16.47	4.63	10.05	4.25	4.83	6.15	15.84	7.01	4.36
	Code	42104	721010	40110	748	422	4224	4-41	4224	401	80100	B13-14	203
CONCRETE WORK—flat (flr. sidewalk)	Rate	7.26	8.48	7.60	16.47	4.63	10.05	4.25	4.83	6.15	15.84	7.01	4.36
	Code	42104	721010	40110	748	422	4224	4-41	4224	401	80100	B13-14	203
ELECTRICAL Wiring—inside	Rate	2.86	2.67	2.52	3.03	2.43	3.19	3.00	2.23	3.41	7.29	3.65	4.36
	Code	42124	721019	40203	704	426	4261	4-46	4261	402	80170	B11-05	206
EXCAVATION—earth NOC	Rate	3.59	4.29	3.84	4.20	3.69	5.55	3.50	4.11	3.58	8.80	4.02	4.36
	Code	40604	721031	40706	711	421	4214	4-43	4214	404	80030	R11-06	207
EXCAVATION—rock	Rate	3.59	4.29	3.84	4.20	3.69	5.55	3.50	4.11	3.58	8.80	4.02	4.36
	Code	40604	721031	40706	711	421	4214	4-43	4214	404	80030	R11-06	207
GLAZIERS	Rate	4.84	3.00	4.57	8.12	6.12	5.97	4.25	7.67	3.41	14.59	7.01	4.36
	Code	42121	715020	40109	751	423	4233	4-41	4233	402	80150	B13-04	212
INSULATION WORK	Rate	3.67	6.94	4.57	8.12	6.12	5.97	4.25	7.67	6.15	14.69	6.20	4.36
	Code	42184	721029	40102	751	423	4234	4-41	4234	401	80110	B12-07	202
LATHING	Rate	8.30	11.20	4.57	4.83	4.21	6.35	4.25	5.74	3.41	14.69	7.01	4.36
	Code	42135	721033	40102	723	427	4279	4-41	4271	402	80110	B13-16	202
MASONRY	Rate	7.26	11.20	4.57	12.21	6.12	5.97	4.25	7.67	6.15	15.84	7.01	4.36
	Code	42102	721037	40102	741	423	4231	4-41	4231	401	80100	B13-18	202
PAINTING & DECORATING	Rate	5.64	6.94	3.45	6.83	4.21	6.35	4.25	5.74	3.41	14.69	6.20	4.36
	Code	42111	721041	40105	719	427	4275	4-41	4275	402	80110	B12-01	202
PILE DRIVING	Rate	7.26	13.07	3.84	5.84	4.63	10.05	3.50	4.83	6.15	8.80	7.01	4.36
	Code	42159	722004	40706	732	422	4221	4-43	4221	401	80030	B13-10	202
PLASTERING	Rate	8.30	11.20	5.13	6.83	4.21	6.35	4.25	5.74	3.41	14.69	6.20	4.36
	Code	42135	721042	40108	719	427	4271	4-41	4271	402	80110	B12-21	202
PLUMBING	Rate	2.86	5.48	2.88	3.83	3.74	3.89	3.00	2.23	3.41	7.75	3.65	4.36
	Code	42122	721043	40204	707	424	4241	4-46	4241	402	80160	B11-01	214
ROOFING	Rate	11.09	10.45	6.52	12.34	8.00	10.05	4.25	9.49	6.15	23.29	7.01	4.36
	Code	42118	721036	40403	728	430	4236	4-41	4236	401	80130	B13-20	202
SHEET METAL WORK (HVAC)	Rate	2.86	5.48	6.52	3.83	3.74	3.89	3.50	3.23	3.41	7.75	3.65	6.60
	Code	42117	721043	40402	707	424	4244	4-46	4244	402	80160	B11-07	208
STEEL ERECTION—door & sash	Rate	3.67	13.07	9.89	16.47	4.63	10.05	4.25	7.67	6.15	29.23	7.01	4.36
	Code	42106	722005	40502	748	422	4227	4-41	4227	401	80080	B13-22	202
STEEL ERECTION—inter., ornam.	Rate	3.67	13.07	9.89	16.47	4.63	10.05	4.25	7.67	6.15	29.23	7.01	4.36
	Code	42106	722005	40502	748	422	4227	4-41	4227	401	80080	B13-22	202
STEEL ERECTION—structure	Rate	3.67	13.07	9.89	16.47	4.63	10.05	4.25	7.67	6.15	29.23	7.01	4.36
	Code	42106	722005	40502	748	422	4227	4-41	4227	401	80080	B13-22	202
STEEL ERECTION—NOC	Rate	3.67	13.07	9.89	16.47	4.63	10.05	4.25	7.67	6.15	29.23	7.01	4.36
	Code	42106	722005	40502	748	422	4227	4-41	4227	401	80080	B13-22	202
TILE WORK—inter. (ceramic)	Rate	4.86	4.70	2.17	6.83	4.21	6.35	4.25	5.74	3.41	14.69	7.01	4.36
	Code	42113	721054	40103	719	427	4276	4-41	4276	402	80110	B13-01	202
WATERPROOFING	Rate	5.64	6.94	4.57	4.83	6.12	6.35	4.25	7.67	3.41	23.29	6.20	4.36
	Code	42139	721016	40102	723	423	4299	4-41	4239	402	80130	B12-17	202
WRECKING	Rate	3.59	6.53	6.46	16.47	3.69	5.55	3.50	4.11	6.15	14.69	7.01	4.36
	Code	40604	721005	40106	748	421	4211	4-43	4211	401	80110	B13-09	202

R01100-070 Contractor's Overhead & Profit

Below are the **average** installing contractor's percentage mark-ups applied to base labor rates to arrive at typical billing rates.

Column A: Labor rates are based on union wages averaged for 30 major U.S. cities. Base rates including fringe benefits are listed hourly and daily. These figures are the sum of the wage rate and employer-paid fringe benefits such as vacation pay, employer-paid health and welfare costs, pension costs, plus appropriate training and industry advancement funds costs.

Column B: Workers' Compensation rates are the national average of state rates established for each trade.

Column C: Column C lists average fixed overhead figures for all trades. Included are Federal and State Unemployment costs set at 6.2%; Social Security Taxes (FICA) set at 7.65%; Builder's Risk Insurance costs set at 0.44%; and Public Liability costs set at 2.02%. All the percentages except those for Social Security Taxes vary from state to state as well as from company to company.

Columns D and E: Percentages in Columns D and E are based on the presumption that the installing contractor has annual billing of $4,000,000 and up. Overhead percentages may increase with smaller annual billing. The overhead percentages for any given contractor may vary greatly and depend on a number of factors, such as the contractor's annual volume, engineering and logistical support costs, and staff requirements. The figures for overhead and profit will also vary depending on the type of job, the job location, and the prevailing economic conditions. All factors should be examined very carefully for each job.

Column F: Column F lists the total of Columns B, C, D, and E.

Column G: Column G is Column A (hourly base labor rate) multiplied by the percentage in Column F (O&P percentage).

Column H: Column H is the total of Column A (hourly base labor rate) plus Column G (Total O&P).

Column I: Column I is Column H multiplied by eight hours.

		A		B	C	D	E	F	G	H	I
		Base Rate Incl. Fringes		Workers' Comp. Ins.	Average Fixed Overhead	Overhead	Profit	Total Overhead & Profit		Rate with O & P	
Abbr.	Trade	Hourly	Daily					%	Amount	Hourly	Daily
Skwk	Skilled Workers Average (35 trades)	$34.85	$278.80	16.2%	16.3%	13.0%	10.0%	55.5%	$19.35	$54.20	$433.60
	Helpers Average (5 trades)	25.55	204.40	18.2		11.0		55.5	$14.20	39.75	318.00
	Foreman Average, Inside ($0.50 over trade)	35.35	282.80	16.2		13.0		55.5	19.60	54.95	439.60
	Foreman Average, Outside ($2.00 over trade)	36.85	294.80	16.2		13.0		55.5	20.45	57.30	458.40
Clab	Common Building Laborers	26.70	213.60	18.4		11.0		55.7	14.85	41.55	332.40
Asbe	Asbestos/Insulation Workers/Pipe Coverers	37.10	296.80	15.5		16.0		57.8	21.45	58.55	468.40
Boil	Boilermakers	42.90	343.20	13.7		16.0		56.0	24.00	66.90	535.20
Bric	Bricklayers	35.25	282.00	15.1		11.0		52.4	18.45	53.70	429.60
Brhe	Bricklayer Helpers	26.90	215.20	15.1		11.0		52.4	14.10	41.00	328.00
Carp	Carpenters	34.25	274.00	18.4		11.0		55.7	19.10	53.35	426.80
Cefi	Cement Finishers	32.85	262.80	9.9		11.0		47.2	15.50	48.35	386.80
Elec	Electricians	40.75	326.00	6.5		16.0		48.8	19.90	60.65	485.20
Elev	Elevator Constructors	45.05	360.40	7.6		16.0		49.9	22.50	67.55	540.40
Eqhv	Equipment Operators, Crane or Shovel	35.90	287.20	10.4		14.0		50.7	18.20	54.10	432.80
Eqmd	Equipment Operators, Medium Equipment	34.65	277.20	10.4		14.0		50.7	17.55	52.20	417.60
Eqlt	Equipment Operators, Light Equipment	33.05	264.40	10.4		14.0		50.7	16.75	49.80	398.40
Eqol	Equipment Operators, Oilers	30.10	240.80	10.4		14.0		50.7	15.25	45.35	362.80
Eqmm	Equipment Operators, Master Mechanics	36.20	289.60	10.4		14.0		50.7	18.35	54.55	436.40
Glaz	Glaziers	33.00	264.00	13.9		11.0		51.2	16.90	49.90	399.20
Lath	Lathers	31.45	251.60	10.9		11.0		48.2	15.15	46.60	372.80
Marb	Marble Setters	33.20	265.60	15.1		11.0		52.4	17.40	50.60	404.80
Mill	Millwrights	35.50	284.00	10.4		11.0		47.7	16.95	52.45	419.60
Mstz	Mosaic and Terrazzo Workers	32.65	261.20	9.7		11.0		47.0	15.35	48.00	384.00
Pord	Painters, Ordinary	30.60	244.80	13.1		11.0		50.4	15.40	46.00	368.00
Psst	Painters, Structural Steel	31.00	248.00	46.8		11.0		84.1	26.05	57.05	456.40
Pape	Paper Hangers	30.30	242.40	13.1		11.0		50.4	15.25	45.55	364.40
Pile	Pile Drivers	33.30	266.40	22.5		16.0		64.8	21.60	54.90	439.20
Plas	Plasterers	31.35	250.80	14.9		11.0		52.2	16.35	47.70	381.60
Plah	Plasterer Helpers	27.10	216.80	14.9		11.0		52.2	14.15	41.25	330.00
Plum	Plumbers	40.85	326.80	8.0		16.0		50.3	20.55	61.40	491.20
Rodm	Rodmen (Reinforcing)	37.95	303.60	24.7		14.0		65.0	24.65	62.60	500.80
Rofc	Roofers, Composition	29.45	235.60	33.1		11.0		70.4	20.75	50.20	401.60
Rots	Roofers, Tile and Slate	29.70	237.60	33.1		11.0		70.4	20.90	50.60	404.80
Rohe	Roofer Helpers (Composition)	21.60	172.80	33.1		11.0		70.4	15.20	36.80	294.40
Shee	Sheet Metal Workers	40.20	321.60	11.5		16.0		53.8	21.65	61.85	494.80
Spri	Sprinkler Installers	40.75	326.00	8.8		16.0		51.1	20.80	61.55	492.40
Stpi	Steamfitters or Pipefitters	40.95	327.60	8.0		16.0		50.3	20.60	61.55	492.40
Ston	Stone Masons	35.30	282.40	15.1		11.0		52.4	18.50	53.80	430.40
Sswk	Structural Steel Workers	38.15	305.20	38.6		14.0		78.9	30.10	68.25	546.00
Tilf	Tile Layers	32.70	261.60	9.7		11.0		47.0	15.35	48.05	384.40
Tilh	Tile Layer Helpers	25.35	202.80	9.7		11.0		47.0	11.90	37.25	298.00
Trlt	Truck Drivers, Light	26.80	214.40	15.2		11.0		52.5	14.05	40.85	326.80
Trhv	Truck Drivers, Heavy	27.55	220.40	15.2		11.0		52.5	14.45	42.00	336.00
Sswl	Welders, Structural Steel	38.15	305.20	38.6		14.0		78.9	30.10	68.25	546.00
Wrck	*Wrecking	26.70	213.60	39.5	▼	11.0	▼	76.8	20.50	47.20	377.60

*Not included in Averages.

R01100-080 Performance Bond

This table shows the cost of a Performance Bond for a construction job scheduled to be completed in 12 months. Add 1% of the premium cost per month for jobs requiring more than 12 months to complete. The rates are "standard" rates offered to contractors that the bonding company considers financially sound and capable of doing the work. Preferred rates are offered by some bonding companies based upon financial strength of the contractor. Actual rates vary from contractor to contractor and from bonding company to bonding company. Contractors should prequalify through a bonding agency before submitting a bid on a contract that requires a bond.

Contract Amount	Building Construction Class B Projects	Highways & Bridges	
		Class A New Construction	Class A-1 Highway Resurfacing
First $ 100,000 bid	$25.00 per M	$15.00 per M	$9.40 per M
Next 400,000 bid	$ 2,500 plus $15.00 per M	$ 1,500 plus $10.00 per M	$ 940 plus $7.20 per M
Next 2,000,000 bid	8,500 plus 10.00 per M	5,500 plus 7.00 per M	3,820 plus 5.00 per M
Next 2,500,000 bid	28,500 plus 7.50 per M	19,500 plus 5.50 per M	15,820 plus 4.50 per M
Next 2,500,000 bid	47,250 plus 7.00 per M	33,250 plus 5.00 per M	28,320 plus 4.50 per M
Over 7,500,000 bid	64,750 plus 6.00 per M	45,750 plus 4.50 per M	39,570 plus 4.00 per M

R01100-090 Sales Tax by State

State sales tax on materials is tabulated below (5 states have no sales tax). Many states allow local jurisdictions, such as a county or city, to levy additional sales tax .

Some projects may be sales tax exempt, particularly those constructed with public funds.

State	Tax (%)	State	Tax (%)	State	Tax (%)	State	Tax (%)
Alabama	4	Illinois	6.25	Montana	0	Rhode Island	7
Alaska	0	Indiana	6	Nebraska	5.5	South Carolina	5
Arizona	5.6	Iowa	5	Nevada	6.5	South Dakota	4
Arkansas	5.125	Kansas	5.3	New Hampshire	0	Tennessee	7
California	7.25	Kentucky	6	New Jersey	6	Texas	6.25
Colorado	2.9	Louisiana	4	New Mexico	5	Utah	4.75
Connecticut	6	Maine	5	New York	4.25	Vermont	6
Delaware	0	Maryland	5	North Carolina	4.5	Virginia	5
District of Columbia	5.75	Massachusetts	5	North Dakota	5	Washington	6.5
Florida	6	Michigan	6	Ohio	6	West Virginia	6
Georgia	4	Minnesota	6.5	Oklahoma	4.5	Wisconsin	5
Hawaii	4	Mississippi	7	Oregon	0	Wyoming	4
Idaho	6	Missouri	4.225	Pennsylvania	6	Average	4.86 %

R01100-090 Sales Tax by Province (Canada)

GST - a value-added tax, which the government imposes on most goods and services provided in or imported into Canada. PST - a retail sales tax, which five of the provinces impose on the price of most goods and some services. QST - a value-added tax, similar to the federal GST, which Quebec imposes. HST - Three provinces have combined their retail sales tax with the federal GST into one harmonized tax.

Province	PST (%)	QST (%)	GST(%)	HST(%)
Alberta	0	0	7	0
British Columbia	7.5	0	7	0
Manitoba	7	0	7	0
New Brunswick	0	0	0	15
Newfoundland	0	0	0	15
Northwest Territories	0	0	7	0
Nova Scotia	0	0	0	15
Ontario	8	0	7	0
Prince Edward Island	10	0	7	0
Quebec	0	7.5	7	0
Saskatchewan	6	0	7	0
Yukon	0	0	7	0

R01100-100 Unemployment Taxes and Social Security Taxes

Mass. State Unemployment tax ranges from 1.12% to 10.96% plus an experience rating assessment the following year, on the first $14,000 of wages. Federal Unemployment tax is 6.2% of the first $7,000 of wages. This is reduced by a credit for payment to the state. The minimum Federal Unemployment tax is 0.8% after all credits.

Combined rates in Mass. thus vary from 1.92% to 11.76% of the first $14,000 of wages. Combined average U.S. rate is about 6.2% of the first $7,000. Contractors with permanent workers will pay less since the average annual wages for skilled workers is $34.85 x 2,000 hours or about $69,700 per year. The average combined rate for U.S. would thus be 6.2% x $7,000 ÷ $69,700 = 0.6% of total wages for permanent employees.

Rates vary not only from state to state but also with the experience rating of the contractor.

Social Security (FICA) for 2005 is estimated at time of publication to be 7.65% of wages up to $87,900.

R01100-110 Overtime

One way to improve the completion date of a project or eliminate negative float from a schedule is to compress activity duration times. This can be achieved by increasing the crew size or working overtime with the proposed crew.

To determine the costs of working overtime to compress activity duration times, consider the following examples. Below is an overtime efficiency and cost chart based on a five, six, or seven day week with an eight through twelve hour day. Payroll percentage increases for time and one half and double time are shown for the various working days.

Days per Week	Hours per Day	Production Efficiency					Payroll Cost Factors	
		1 Week	2 Weeks	3 Weeks	4 Weeks	Average 4 Weeks	@ 1-1/2 Times	@ 2 Times
5	8	100%	100%	100%	100%	100 %	100 %	100 %
	9	100	100	95	90	96.25	105.6	111.1
	10	100	95	90	85	91.25	110.0	120.0
	11	95	90	75	65	81.25	113.6	127.3
	12	90	85	70	60	76.25	116.7	133.3
6	8	100	100	95	90	96.25	108.3	116.7
	9	100	95	90	85	92.50	113.0	125.9
	10	95	90	85	80	87.50	116.7	133.3
	11	95	85	70	65	78.75	119.7	139.4
	12	90	80	65	60	73.75	122.2	144.4
7	8	100	95	85	75	88.75	114.3	128.6
	9	95	90	80	70	83.75	118.3	136.5
	10	90	85	75	65	78.75	121.4	142.9
	11	85	80	65	60	72.50	124.0	148.1
	12	85	75	60	55	68.75	126.2	152.4

R01100-710 Unit Gross Area Requirements

The figures in the table below indicate typical ranges in square feet as a function of the "occupant" unit. This table is best used in the preliminary design stages to help determine the probable size requirement for the total project. See R17100-100 for the typical total size ranges for various types of buildings.

Building Type	Unit	Gross Area in S.F.		
		1/4	Median	3/4
Apartments	Unit	660	860	1,100
Auditorium & Play Theaters	Seat	18	25	38
Bowling Alleys	Lane		940	
Churches & Synagogues	Seat	20	28	39
Dormitories	Bed	200	230	275
Fraternity & Sorority Houses	Bed	220	315	370
Garages, Parking	Car	325	355	385
Hospitals	Bed	685	850	1,075
Hotels	Rental Unit	475	600	710
Housing for the elderly	Unit	515	635	755
Housing, Public	Unit	700	875	1,030
Ice Skating Rinks	Total	27,000	30,000	36,000
Motels	Rental Unit	360	465	620
Nursing Homes	Bed	290	350	450
Restaurants	Seat	23	29	39
Schools, Elementary	Pupil	65	77	90
Junior High & Middle	↓	85	110	129
Senior High	↓	102	130	145
Vocational	↓	110	135	195
Shooting Ranges	Point		450	
Theaters & Movies	Seat		15	

R01100-720 Floor Area Ratios

Table below lists commonly used gross to net area and net to gross area ratios expressed in % for various building types.

Building Type	Gross to Net Ratio	Net to Gross Ratio	Building Type	Gross to Net Ratio	Net to Gross Ratio
Apartment	156	64	School Buildings (campus type)		
Bank	140	72	Administrative	150	67
Church	142	70	Auditorium	142	70
Courthouse	162	61	Biology	161	62
Department Store	123	81	Chemistry	170	59
Garage	118	85	Classroom	152	66
Hospital	183	55	Dining Hall	138	72
Hotel	158	63	Dormitory	154	65
Laboratory	171	58	Engineering	164	61
Library	132	76	Fraternity	160	63
Office	135	75	Gymnasium	142	70
Restaurant	141	70	Science	167	60
Warehouse	108	93	Service	120	83
			Student Union	172	59

The gross area of a building is the total floor area based on outside dimensions.

The net area of a building is the usable floor area for the function intended and excludes such items as stairways, corridors and mechanical rooms. In the case of a commercial building, it might be considered as the "leasable area."

R01100-730 Occupancy Determinations

Description		S.F. Required per Person		
		BOCA	SBC	UBC
Assembly Areas	Fixed Seats	**	6	7
	Movable Seats		15	15
	Concentrated	7		
	Unconcentrated	15		
	Standing Space	3		
Educational	Unclassified			
	Classrooms	20	40	20
	Shop Areas	50	100	50
Institutional	Unclassified		125	
	In-Patient Areas	240		
	Sleeping Areas	120		
Mercantile	Basement	30	30	20
	Ground Floor	30	30	30
	Upper Floors	60	60	50
Office		100	100	100

BOCA=Building Officials & Code Administrators
SBC=Southern Building Code
UBC=Uniform Building Code

** The occupancy load for assembly area with fixed seats shall be determined by the number of fixed seats installed.

R01107-030 Engineering Fees

Typical **Structural Engineering Fees** based on type of construction and total project size. These fees are included in Architectural Fees.

Type of Construction	Total Project Size (in thousands of dollars)			
	$500	$500-$1,000	$1,000-$5,000	Over $5000
Industrial buildings, factories & warehouses	Technical payroll times 2.0 to 2.5	1.60%	1.25%	1.00%
Hotels, apartments, offices, dormitories, hospitals, public buildings, food stores	↓	2.00%	1.70%	1.20%
Museums, banks, churches and cathedrals	↓	2.00%	1.75%	1.25%
Thin shells, prestressed concrete, earthquake resistive	↓	2.00%	1.75%	1.50%
Parking ramps, auditoriums, stadiums, convention halls, hangars & boiler houses	↓	2.50%	2.00%	1.75%
Special buildings, major alterations, underpinning & future expansion	↓	Add to above 0.5%	Add to above 0.5%	Add to above 0.5%

For complex reinforced concrete or unusually complicated structures, add 20% to 50%.

Typical **Mechanical and Electrical Engineering Fees** are based on the size of the subcontract. The fee structure for Mechanical Engineering is shown below; Electrical Engineering fees range from 2 to 5%. These fees are included in Architectural Fees.

Type of Construction	Subcontract Size							
	$25,000	$50,000	$100,000	$225,000	$350,000	$500,000	$750,000	$1,000,000
Simple structures	6.4%	5.7%	4.8%	4.5%	4.4%	4.3%	4.2%	4.1%
Intermediate structures	8.0	7.3	6.5	5.6	5.1	5.0	4.9	4.8
Complex structures	12.0	9.0	9.0	8.0	7.5	7.5	7.0	7.0

For renovations, add 15% to 25% to applicable fee.

R01250-010 Repair and Remodeling

Cost figures are based on new construction utilizing the most cost-effective combination of labor, equipment and material with the work scheduled in proper sequence to allow the various trades to accomplish their work in an efficient manner.

The costs for repair and remodeling work must be modified due to the following factors that may be present in any given repair and remodeling project.

1. Equipment usage curtailment due to the physical limitations of the project, with only hand-operated equipment being used.
2. Increased requirement for shoring and bracing to hold up the building while structural changes are being made and to allow for temporary storage of construction materials on above-grade floors.
3. Material handling becomes more costly due to having to move within the confines of an enclosed building. For multi-story construction, low capacity elevators and stairwells may be the only access to the upper floors.
4. Large amount of cutting and patching and attempting to match the existing construction is required. It is often more economical to remove entire walls rather than create many new door and window openings. This sort of trade-off has to be carefully analyzed.
5. Cost of protection of completed work is increased since the usual sequence of construction usually cannot be accomplished.
6. Economies of scale usually associated with new construction may not be present. If small quantities of components must be custom fabricated due to job requirements, unit costs will naturally increase. Also, if only small work areas are available at a given time, job scheduling between trades becomes difficult and subcontractor quotations may reflect the excessive start-up and shut-down phases of the job.
7. Work may have to be done on other than normal shifts and may have to be done around an existing production facility which has to stay in production during the course of the repair and remodeling.
8. Dust and noise protection of adjoining non-construction areas can involve substantial special protection and alter usual construction methods.
9. Job may be delayed due to unexpected conditions discovered during demolition or removal. These delays ultimately increase construction costs.
10. Piping and ductwork runs may not be as simple as for new construction. Wiring may have to be snaked through walls and floors.
11. Matching "existing construction" may be impossible because materials may no longer be manufactured. Substitutions may be expensive.
12. Weather protection of existing structure requires additional temporary structures to protect building at openings.
13. On small projects, because of local conditions, it may be necessary to pay a tradesman for a minimum of four hours for a task that is completed in one hour.

All of the above areas can contribute to increased costs for a repair and remodeling project. Each of the above factors should be considered in the planning, bidding and construction stage in order to minimize the increased costs associated with repair and remodeling jobs.

R01540-100 Steel Tubular Scaffolding

On new construction, tubular scaffolding is efficient up to 60′ high or five stories. Above this it is usually better to use a hung scaffolding if construction permits. Swing scaffolding operations may interfere with tenants. In this case, the tubular is more practical at all heights.

In repairing or cleaning the front of an existing building the cost of tubular scaffolding per S.F. of building front increases as the height increases above the first tier. The first tier cost is relatively high due to leveling and alignment.

The minimum efficient crew for erection is three workers. For heights over 50′, a crew of four is more efficient. Use two or more on top and two at the bottom for handing up or hoisting. Four workers can erect and dismantle about nine frames per hour up to five stories. From five to eight stories they will average six frames per hour. With 7′ horizontal spacing this will run about 400 S.F. and 265 S.F. of wall surface, respectively. Time for placing planks must be added to the above. On heights above 50′, five planks can be placed per labor-hour.

The table below shows the number of pieces required to erect tubular steel scaffolding for 1000 S.F. of building frontage. This area is made up of a scaffolding system that is 12 frames (11 bays) long by 2 frames high.

For jobs under twenty-five frames, add 50% to rental cost. Rental rates will be lower for jobs over three months duration. Large quantities for long periods can reduce rental rates by 20%.

Description of Component	CSI Line Item	Number of Pieces for 1000 S.F. of Building Front	Unit
5′ Wide Standard Frame, 6′-4″ High	01540-750-2200	24	Ea.
Leveling Jack & Plate	01540-750-2650	24	↓
Cross Brace	01540-750-2500	44	
Side Arm Bracket, 21″	01540-750-2700	12	
Guardrail Post	01540-750-2550	12	
Guardrail, 7′ section	01540-750-2600	22	
Stairway Section	01540-750-2900	2	
Stairway Starter Bar	01540-750-2910	1	
Stairway Inside Handrail	01540-750-2920	2	
Stairway Outside Handrail	01540-750-2930	2	
Walk-Thru Frame Guardrail	01540-750-2940	2	↓

Scaffolding is often used as falsework over 15′ high during construction of cast-in-place concrete beams and slabs. Two foot wide scaffolding is generally used for heavy beam construction. The span between frames depends upon the load to be carried with a maximum span of 5′.

Heavy duty shoring frames with a capacity of 10,000#/leg can be spaced up to 10′ O.C. depending upon form support design and loading.

Scaffolding used as horizontal shoring requires less than half the material required with conventional shoring.

On new construction, erection is done by carpenters.

Rolling towers supporting horizontal shores can reduce labor and speed the job. For maintenance work, catwalks with spans up to 70′ can be supported by the rolling towers.

R01590-100 Contractor Equipment

Rental Rates shown in Division 01590 pertain to late model high quality machines in excellent working condition, rented from equipment dealers. Rental rates from contractors may be substantially lower than the rental rates from equipment dealers depending upon economic conditions; for older, less productive machines, reduce rates by a maximum of 15%. Any overtime must be added to the base rates. For shift work, rates are lower. Usual rule of thumb is 150% of one shift rate for two shifts; 200% for three shifts.

For periods of less than one week, operated equipment is usually more economical to rent than renting bare equipment and hiring an operator.

Costs to move equipment to a job site (mobilization) or from a job site (demobilization) are not included in rental rates in Division 01590, nor in any Equipment costs on any Unit Price line items or crew listings. These costs can be found in section 02305-250. If a piece of equipment is already at a job site, it is not appropriate to utilize mob/demob costs in an estimate again.

Rental rates vary throughout the country with larger cities generally having lower rates. Lease plans for new equipment are available for periods in excess of six months with a percentage of payments applying toward purchase.

Monthly rental rates vary from 2% to 5% of the cost of the equipment depending on the anticipated life of the equipment and its wearing parts. Weekly rates are about 1/3 the monthly rates and daily rental rates about 1/3 the weekly rate.

The hourly operating costs for each piece of equipment include costs to the user such as fuel, oil, lubrication, normal expendables for the equipment, and a percentage of mechanic's wages chargeable to maintenance. The hourly operating costs listed do not include the operator's wages.

The daily cost for equipment used in the standard crews is figured by dividing the weekly rate by five, then adding eight times the hourly operating cost to give the total daily equipment cost, not including the operator. This figure is in the right hand column of Division 01590 under Crew Equipment Cost/Day.

Pile Driving rates shown for pile hammer and extractor do not include leads, crane, boiler or compressor. Vibratory pile driving requires an added field specialist during set-up and pile driving operation for the electric model. The hydraulic model requires a field specialist for set-up only. Up to 125 reuses of sheet piling are possible using vibratory drivers. For normal conditions, crane capacity for hammer type and size are as follows.

Crane Capacity	Hammer Type and Size		
	Air or Steam	Diesel	Vibratory
25 ton	to 8,750 ft.-lb.		70 H.P.
40 ton	15,000 ft.-lb.	to 32,000 ft.-lb.	170 H.P.
60 ton	25,000 ft.-lb.		300 H.P.
100 ton		112,000 ft.-lb.	

Cranes should be specified for the job by size, building and site characteristics, availability, performance characteristics, and duration of time required.

Backhoes & Shovels rent for about the same as equivalent size cranes but maintenance and operating expense is higher. Crane operators rate must be adjusted for high boom heights. Average adjustments: for 150′ boom add 2% per hour; over 185′, add 4% per hour; over 210′, add 6% per hour; over 250′, add 8% per hour and over 295′, add 12% per hour.

Tower Cranes of the climbing or static type have jibs from 50′ to 200′ and capacities at maximum reach range from 4,000 to 14,000 pounds. Lifting capacities increase up to maximum load as the hook radius decreases.

Typical rental rates, based on purchase price are about 2% to 3% per month.

Erection and dismantling runs between 500 and 2000 labor hours. Climbing operation takes 10 labor hours per 20′ climb. Crane dead time is about 5 hours per 40′ climb. If crane is bolted to side of the building add cost of ties and extra mast sections. Climbing cranes have from 80′ to 180′ of mast while static cranes have 80′ to 800′ of mast.

Truck Cranes can be converted to tower cranes by using tower attachments. Mast heights over 400′ have been used. See Division 01590-600 for rental rates of high boom cranes.

A single 100′ high material **Hoist and Tower** can be erected and dismantled in about 400 labor hours; a double 100′ high hoist and tower in about 600 labor hours. Erection times for additional heights are 3 and 4 labor hours per vertical foot respectively up to 150′, and 4 to 5 labor hours per vertical foot over 150′ high. A 40′ high portable Buck hoist takes about 160 labor hours to erect and dismantle. Additional heights take 2 labor hours per vertical foot to 80′ and 3 labor hours per vertical foot for the next 100′. Most material hoists do not meet local code requirements for carrying personnel.

A 150′ high **Personnel Hoist** requires about 500 to 800 labor hours to erect and dismantle. Budget erection time at 5 labor hours per vertical foot for all trades. Local code requirements or labor scarcity requiring overtime can add up to 50% to any of the above erection costs.

Earthmoving Equipment: The selection of earthmoving equipment depends upon the type and quantity of material, moisture content, haul distance, haul road, time available, and equipment available. Short haul cut and fill operations may require dozers only, while another operation may require excavators, a fleet of trucks, and spreading and compaction equipment. Stockpiled material and granular material are easily excavated with front end loaders. Scrapers are most economically used with hauls between 300′ and 1-1/2 miles if adequate haul roads can be maintained. Shovels are often used for blasted rock and any material where a vertical face of 8′ or more can be excavated. Special conditions may dictate the use of draglines, clamshells, or backhoes. Spreading and compaction equipment must be matched to the soil characteristics, the compaction required and the rate the fill is being supplied.

R02115-200 Underground Storage Tank Removal

Underground Storage Tank Removal can be divided into two categories: Non-Leaking and Leaking. Prior to removing an underground storage tank, tests should be made, with the proper authorities present, to determine whether a tank has been leaking or the surrounding soil has been contaminated.

To safely remove Liquid Underground Storage Tanks:

1. Excavate to the top of the tank.
2. Disconnect all piping.
3. Open all tank vents and access ports.
4. Remove all liquids and/or sludge.
5. Purge the tank with an inert gas.
6. Provide access to the inside of the tank and clean out the interior using proper personal protective equipment (PPE).
7. Excavate soil surrounding the tank using proper PPE for on-site personnel.
8. Pull and properly dispose of the tank.
9. Clean up the site of all contaminated material.
10. Install new tanks or close the excavation.

R02220-510 Demolition Defined

Whole Building Demolition (Division 02220-110) - Demolition of the whole building with no concern for any particular building element, component, or material type being demolished. This type of demolition is accomplished with large pieces of construction equipment that break up the structure, load it into trucks and haul it to a disposal site, but disposal or dump fees are not included (Divisions 02220-320 and 02220-330). Demolition of below-grade foundation elements, such as footings, foundation walls, grade beams, slabs on grade, etc. (Division 02220-130), is not included. Certain mechanical equipment containing flammable liquids or ozone-depleting refrigerants, electric lighting elements, communication equipment components, and other building elements may contain hazardous waste, and must be removed, either selectively or carefully, as hazardous waste before the building can be demolished (Division 13281).

Foundation Demolition (Division 02220-130) - Demolition of below-grade foundation footings, foundation walls, grade beams, and slabs on grade. This type of demolition is accomplished by hand or pneumatic hand tools, and does not include saw cutting (Division 02220-360), or handling, loading, hauling, or disposal of the debris (Divisions 02220-320, 02220-330 and 02220-350).

Gutting (Division 02220-340) - Removal of building interior finishes and electrical/mechanical systems down to the load-bearing and sub-floor elements of the rough building frame, with no concern for any particular building element, component, or material type being demolished. This type of demolition is accomplished by hand or pneumatic hand tools, and includes loading into trucks, but not hauling, disposal or dump fees (Divisions 02220-320 and 02220-330); scaffolding (01540-750); or shoring (03150-600). Certain mechanical equipment containing flammable liquids or ozone-depleting refrigerants, electric lighting elements, communication equipment components, and other building elements may contain hazardous waste, and must be removed, either selectively or carefully, as hazardous waste, before the building is gutted (Division 13281).

Selective Demolition - Demolition of a selected building element, component, or finish, with some concern for surrounding or adjacent elements, components, or finishes (see the first Subdivision (s) at the beginning of Divisions 3 through 16). This type of demolition is accomplished by hand or pneumatic hand tools, and does not include handling, loading, storing, hauling, or disposal of the debris (Divisions 02220-320, 02220-330 and 02220-350); scaffolding (Division 01540-750); or shoring (Division 03150-600). "Gutting" methods may be used in order to save time, but damage that was caused to surrounding or adjacent elements, components, or finishes may have to be repaired at a later time.

Careful Removal - Removal of a piece of service equipment, building element or component, or material type, with great concern for both the removed item and surrounding or adjacent elements, components or finishes. The purpose of careful removal may be to protect the removed item for later re-use, preserve a higher salvage value of the removed item, or replace an item while taking care to protect surrounding or adjacent elements, components, connections, or finishes from cosmetic and/or structural damage. An approximation of the time required to perform this type of removal is 1/3 to 1/2 the time it would take to install a new item of like kind (see Reference Numbers R15050-720 and R16055-300). This type of removal is accomplished by hand or pneumatic hand tools, and does not include loading, hauling, or storing the removed item (Divisions 01840-100 and 02220-350); scaffolding (Division 01540-750); shoring (Division 03150-600); or lifting equipment (Division 01590-600).

Cutout Demolition (Division 02220-310) - Demolition of a small quantity of floor, wall, roof, or other assembly, with concern for the appearance and structural integrity of the surrounding materials. This type of demolition is accomplished by hand or pneumatic hand tools, and does not include saw cutting (Division 02220-360); handling, loading, hauling, or disposal of debris (Divisions 02220-320, 02220-330 and 02220-350); scaffolding (Division 01540-750); or shoring (Division 03150-600).

Rubbish Handling (Division 02220-350) - Work activities that involve handling, loading or hauling of debris. Generally, the cost of rubbish handling must be added to the cost of all types of demolition, with the exception of whole building demolition.

Minor Site Demolition (Division 02220-240) - Demolition of site elements outside the footprint of a building. This type of demolition is accomplished by hand or pneumatic hand tools, or with larger pieces of construction equipment, and may include loading a removed item onto a truck (check the Crew for equipment used). It does not include saw cutting (Division 02220-360), hauling or disposal of debris (Divisions 02220-320 and 02220-330), and, sometimes, handling or loading (Division 02220-350).

R02240-900 Wellpoints

A single stage wellpoint system is usually limited to dewatering an average 15′ depth below normal ground water level. Multi-stage systems are employed for greater depth with the pumping equipment installed only at the lowest header level. Ejectors, with unlimited lift capacity, can be economical when two or more stages of wellpoints can be replaced or when horizontal clearance is restricted, such as in deep trenches or tunneling projects, and where low water flows are expected. Wellpoints are usually spaced on 2-1/2′ to 10′ centers along a header pipe. Wellpoint spacing, header size, and pump size are all determined by the expected flow, as dictated by soil conditions.

In almost all soils encountered in wellpoint dewatering, the wellpoints may be jetted into place. Cemented soils and stiff clays may require sand wicks about 12″ in diameter around each wellpoint to increase efficiency and eliminate weeping into the excavation. These sand wicks require 1/2 to 3 C.Y. of washed filter sand and are installed by using a 12″ diameter steel casing and hole puncher jetted into the ground 2′ deeper than the wellpoint. Rock may require predrilled holes.

Labor required for the complete installation and removal of a single stage wellpoint system is in the range of 3/4 to 2 labor-hours per linear foot of header, depending upon jetting conditions, wellpoint spacing, etc.

Continuous pumping is necessary except in some free draining soil where temporary flooding is permissible (as in trenches which are backfilled after each day's work). Good practice requires provision of a stand-by pump during the continuous pumping operation.

Systems for continuous trenching below the water table should be installed three to four times the length of expected daily progress to insure uninterrupted digging, and header pipe size should not be changed during the job.

For pervious free draining soils, deep wells in place of wellpoints may be economical because of lower installation and maintenance costs. Daily production ranges between two to three wells per day, for 25′ to 40′ depths, to one well per day for depths over 50′.

Detailed analysis and estimating for any dewatering problem is available at no cost from wellpoint manufacturers. Major firms will quote "sufficient equipment" quotes or their affiliates offer lump sum proposals to cover complete dewatering responsibility.

Description for 200′ System with 8″ Header		Quantities	1st Month		Thereafter	
			Unit	Total	Unit	Total
Equipment & Material	Wellpoints 25′ long, 2″ diameter @ 5′ O.C.	40 Each	$ 31.00	$ 1,240.00	$ 23.25	$ 930.00
	Header pipe, 8″ diameter	200 L.F.	5.90	1,180.00	2.95	590.00
	Discharge pipe, 8″ diameter	100 L.F.	3.96	396.00	2.57	257.00
	8″ valves	3 Each	151.00	453.00	98.15	294.45
	Combination Jetting & Wellpoint pump (standby)	1 Each	2,450.00	2,450.00	1,715.00	1,715.00
	Wellpoint pump, 8″ diameter	1 Each	2,400.00	2,400.00	1,680.00	1,680.00
	Transportation to and from site	1 Day	697.00	697.00	—	—
	Fuel 30 days x 60 gal./day	1800 Gallons	2.00	3,600.00	2.00	3,600.00
	Lubricants for 30 days x 16 lbs./day	480 Lbs.	1.50	720.00	1.50	720.00
	Sand for points	40 C.Y.	21.35	854.00	—	—
	Equipment & Materials Subtotal			$13,990.00		$ 9,786.45
Labor	Technician to supervise installation	1 Week	$ 36.20	$ 1,448.00	—	—
	Labor for installation and removal of system	300 Labor-hours	26.70	8,010.00	—	—
	4 Operators straight time 40 hrs./wk. for 4.33 wks.	693 Hrs.	33.05	22,903.65	$ 33.05	$22,903.65
	4 Operators overtime 2 hrs./wk. for 4.33 wks.	35 Hrs.	66.10	2,313.50	66.10	2,313.50
	Bare Labor Subtotal			$34,675.15		$25,217.15
	Total Bare Cost			$48,665.15		$35,003.60
	Monthly Bare Cost/L.F. Header			$ 243.33		$ 175.02

R02250-400 Wood Sheet Piling

Wood sheet piling may be used for depths to 20′ where there is no ground water. If moderate ground water is encountered Tongue & Groove sheeting will help to keep it out. When considerable ground water is present, steel sheeting must be used.

For estimating purposes on trench excavation, sizes are as follows:

Depth	Sheeting	Wales	Braces	B.F. per S.F.
To 8′	3 x 12′s	6 x 8′s, 2 line	6 x 8′s, @ 10′	4.0 @ 8′
8′ x 12′	3 x 12′s	10 x 10′s, 2 line	10 x 10′s, @ 9′	5.0 average
12′ to 20′	3 x 12′s	12 x 12′s, 3 line	12 x 12′s, @ 8′	7.0 average

Sheeting to be toed in at least 2′ depending upon soil conditions. A five person crew with an air compressor and sheeting driver can drive and brace 440 SF/day at 8′ deep, 360 SF/day at 12′ deep, and 320 SF/day at 16′ deep. For normal soils, piling can be pulled in 1/3 the time to install. Pulling difficulty increases with the time in the ground. Production can be increased by high pressure jetting. Figures below assume 50% of lumber is salvaged and includes pulling costs. Some jurisdictions require an equipment operator in addition to Crew B-31.

Sheeting Pulled

Daily Cost Crew B-31	L.H./ Day	Hourly Cost	Daily Cost	8′ Depth, 440 S.F./Day To Drive (1 Day)		8′ Depth To Pull (1/3 Day)	16′ Depth, 320 S.F./Day To Drive (1 Day)		16′ Depth To Pull (1/3 Day)
1 Foreman	8	$28.70	$ 229.60		$ 229.60	$ 76.46		$ 229.60	$ 76.46
3 Laborers	24	26.70	640.80		640.80	213.39		640.80	213.39
1 Carpenter	8	34.25	274.00		274.00	91.24		274.00	91.24
1 Air Compressor			112.40		112.40	37.43		112.40	37.43
1 Sheeting Driver			7.20		7.20	2.40		7.20	2.40
2 -50 Ft. Air Hoses, 1-1/2″ Diam.			25.50		25.50	8.49		25.50	8.49
Lumber (50% salvage)				1.76 MBF	616.00		2.24 MBF	784.00	
Bare Total			$1,289.50		$1,905.50	$429.41		$2,073.50	$429.41
Bare Total/S.F.					$ 4.33	$ .98		$ 4.71	$ 1.34
Bare Total (Drive and Pull)/S.F.					$ 5.31			$ 6.05	

Sheeting Left in Place

Daily Cost	8′ Depth 440 S.F./Day		10′ Depth 400 S.F./Day		12′ Depth 360 S.F./Day		16′ Depth 320 S.F./Day		18′ Depth 305 S.F./Day		20′ Depth 280 S.F./Day	
Crew B-31		$1,273.40		$1,273.40		$1,273.40		$1,273.40		$1,273.40		$1,273.40
Lumber	1.76 MBF	1,232.00	1.8 MBF	1,260.00	1.8 MBF	1,260.00	2.24 MBF	1,568.00	2.1 MBF	1,470.00	1.9 MBF	1,330.00
Bare Total in Place		$2,505.40		$2,533.40		$2,533.40		$2,841.40		$2,743.40		$2,603.40
Bare Total/S.F.		$ 5.69		$ 6.33		$ 7.04		$ 8.88		$ 8.99		$ 9.30
Bare Total/M.B.F.		$1,423.52		$1,407.44		$1,407.44		$1,268.48		$1,306.38		$1,370.21

R02250-450 Steel Sheet Piling

Limiting weights are 22 to 38#/S.F. of wall surface with 27#/S.F. average for usual types and sizes. (Weights of piles themselves are from 30.7#/L.F. to 57#/L.F. but they are 15″ to 21″ wide.) Lightweight sections 12″ to 28″ wide from 3 ga. to 12 ga. thick are also available for shallow excavations. Piles may be driven two at a time with an impact or vibratory hammer (use vibratory to pull) hung from a crane without leads. A reasonable estimate of the life of steel sheet piling is 10 uses with up to 125 uses possible if a vibratory hammer is used. Used piling costs from 50% to 80% of new piling depending on location and market conditions. Sheet piling and H piles can be rented for about 30% of the delivered mill price for the first month and 5% per month thereafter. Allow 1 labor-hour per pile for cleaning and trimming after driving. These costs increase with depth and hydrostatic head. Vibratory drivers are faster in wet granular soils and are excellent for pile extraction. Pulling difficulty increases with the time in the ground and may cost more than driving. It is often economical to abandon the sheet piling, especially if it can be used as the outer wall form. Allow about 1/3 additional length or more, for toeing into ground. Add bracing, waler and strut costs. Waler costs can equal the cost per ton of sheeting.

Cost of Sheet Piling & Production Rate by Ton & S.F.

Depth of Excavation	15′ Depth		20′ Depth		25′ Depth	
Description of Pile	22 psf = 90.9 S.F./Ton		27 psf = 74 S.F./Ton		38 psf = 52.6 S.F./Ton	
Type of operation:	Drive &	Drive &	Drive &	Drive &	Drive &	Drive &
Left in Place or Removed	Left	Extract*	Left	Extract*	Left	Extract*
Labor & Equip. to Drive 1 ton	$ 412.06	$742.40	$ 343.97	$680.06	$ 234.44	$424.23
Piling (75% Salvage)	880.00	220.00	880.00	220.00	880.00	220.00
Bare Cost/Ton in Place	$1,292.06	$962.40	$1,223.97	$900.06	$1,114.44	$644.23
Production Rate, Tons/Day	10.81	6.00	12.95	6.55	19.00	10.50
Bare Cost, S.F. in Place (incl. 33% toe in)	$ 14.21	$ 10.59	$ 16.54	$ 12.16	$ 21.19	$ 12.25
Production Rate, S.F./Day (incl. 33% toe in)	983	545	959	485	1000	553

Crew B-40	Bare Costs Hr.	Bare Costs Daily	Inc. Subs O&P Hr.	Inc. Subs O&P Daily
1 Pile Driver Foreman (Out)	$35.30	$ 282.40	$58.15	$ 465.20
4 Pile Drivers	33.30	1,065.60	54.90	1,756.80
2 Equip. Oper. (crane)	35.90	574.40	54.10	865.60
1 Equip. Oper. Oiler	30.10	240.80	45.35	362.80
1 Crane, 40 Ton		888.20		977.00
Vibratory Hammer & Gen.		1,403.00		1,543.30
64 L.H., Daily Totals		$4,454.40		$5,970.70

Bare Installation Cost per Ton

for 15′ Deep Excavation plus 33% toe-in

$4,454 per day/10.81 tons = $412.03 per ton

Bare Installation Cost per S.F.

$$\frac{2000\#}{22\#/S.F.} = 90.9 \text{ S.F.} \times 10.81 \text{ tons} = 983 \text{ S.F. per day}$$

$4,454 / 983 S.F. per day = $4.53 per S.F.

*For driving & extracting, two mobilizations & demobilizations will be necessary

R02510-800 Piping Designations

There are several systems currently in use to describe pipe and fittings. The following paragraphs will help to identify and clarify classifications of piping systems used for water distribution.

Piping may be classified by schedule. Piping schedules include 5S, 10S, 10, 20, 30, Standard, 40, 60, Extra Strong, 80, 100, 120, 140, 160 and Double Extra Strong. These schedules are dependent upon the pipe wall thickness. The wall thickness of a particular schedule may vary with pipe size.

Ductile iron pipe for water distribution is classified by Pressure Classes such as Class 150, 200, 250, 300 and 350. These classes are actually the rated water working pressure of the pipe in pounds per square inch (psi). The pipe in these pressure classes is designed to withstand the rated water working pressure plus a surge allowance of 100 psi.

The American Water Works Association (AWWA) provides standards for various types of **plastic pipe.** C-900 is the specification for polyvinyl chloride (PVC) piping used for water distribution in sizes ranging from 4″ through 12″. C-901 is the specification for polyethylene (PE) pressure pipe, tubing and fittings used for water distribution in sizes ranging from 1/2″ through 3″. C-905 is the specification for PVC piping sizes 14″ and greater.

PVC pressure-rated pipe is identified using the standard dimensional ratio (SDR) method. This method is defined by the American Society for Testing and Materials (ASTM) Standard D 2241. This pipe is available in SDR numbers 64, 41, 32.5, 26, 21, 17, and 13.5. Pipe with an SDR of 64 will have the thinnest wall while pipe with an SDR of 13.5 will have the thickest wall. When the pressure rating (PR) of a pipe is given in psi, it is based on a line supplying water at 73 degrees F.

The National Sanitation Foundation (NSF) seal of approval is applied to products that can be used with potable water. These products have been tested to ANSI/NSF Standard 14.

Valves and strainers are classified by American National Standards Institute (ANSI) Classes. These Classes are 125, 150, 200, 250, 300, 400, 600, 900, 1500 and 2500. Within each class there is an operating pressure range dependent upon temperature. Design parameters should be compared to the appropriate material dependent, pressure-temperature rating chart for accurate valve selection.

Thermal & Moist. Protection | R078 | Fire & Smoke Protection

R07800-030 Firestopping

Firestopping is the sealing of structural, mechanical, electrical and other penetrations through fire-rated assemblies. The basic components of firestop systems are safing insulation and firestop sealant on both sides of wall penetrations and the top side of floor penetrations.

Pipe penetrations are assumed to be through concrete, grout, or joint compound and can be sleeved or unsleeved. Costs for the penetrations and sleeves are not included. An annular space of 1″ is assumed. Escutcheons are not included.

Metallic pipe is assumed to be copper, aluminum, cast iron or similar metallic material. Insulated metallic pipe is assumed to be covered with a thermal insulating jacket of varying thickness and materials.

Non-metallic pipe is assumed to be PVC, CPVC, FR Polypropylene or similar plastic piping material. Intumescent firestop sealant or wrap strips are included. Collars on both sides of wall penetrations and a sheet metal plate on the underside of floor penetrations are included.

Ductwork is assumed to be sheet metal, stainless steel or similar metallic material. Duct penetrations are assumed to be through concrete, grout or joint compound. Costs for penetrations and sleeves are not included. An annular space of 1/2″ is assumed.

Multi-trade openings include costs for sheet metal forms, firestop mortar, wrap strips, collars and sealants as necessary.

Structural penetrations joints are assumed to be 1/2″ or less. CMU walls are assumed to be within 1-1/2″ of metal deck. Drywall walls are assumed to be tight to the underside of metal decking.

Metal panel, glass or curtain wall systems include a spandrel area of 5′ filled with mineral wool foil-faced insulation. Fasteners and stiffeners are included.

R13281-120 Asbestos Removal Process

Asbestos removal is accomplished by a specialty contractor who understands the federal and state regulations regarding the handling and disposal of the material. The process of asbestos removal is divided into many individual steps. An accurate estimate can be calculated only after all the steps have been priced.

The steps are generally as follows:

1. Obtain an asbestos abatement plan from an industrial hygienist.
2. Monitor the air quality in and around the removal area and along the path of travel between the removal area and transport area. This establishes the background contamination.
3. Construct a two part decontamination chamber at entrance to removal area.
4. Install a HEPA filter to create a negative pressure in the removal area.
5. Install wall, floor and ceiling protection as required by the plan, usually 2 layers of fireproof 6 mil polyethylene.
6. Industrial hygienist visually inspects work area to verify compliance with plan.
7. Provide temporary supports for conduit and piping affected by the removal process.
8. Proceed with asbestos removal and bagging process. Monitor air quality as described in Step #2. Discontinue operations when contaminate levels exceed applicable standards.
9. Document the legal disposal of materials in accordance with EPA standards.
10. Thoroughly clean removal area including all ledges, crevices and surfaces.
11. Post abatement inspection by industrial hygienist to verify plan compliance.
12. Provide a certificate from a licensed industrial hygienist attesting that contaminate levels are within acceptable standards before returning area to regular use.

R13600-610 Solar Heating (Space and Hot Water)

Collectors should face as close to due South as possible, however, variations of up to 20 degrees on either side of true South are acceptable. Local climate and collector type may influence the choice between east or west deviations. Obviously they should be located so they are not shaded from the sun's rays. Incline collectors at a slope of latitude minus 5 degrees for domestic hot water and latitude plus 15 degrees for space heating.

Flat plate collectors consist of a number of components as follows: Insulation to reduce heat loss through the bottom and sides of the collector. The enclosure which contains all the components in this assembly is usually weatherproof and prevents dust, wind and water from coming in contact with the absorber plate. The cover plate usually consists of one or more layers of a variety of glass or plastic and reduces the reradiation by creating an air space which traps the heat between the cover and the absorber plates.

The absorber plate must have a good thermal bond with the fluid passages. The absorber plate is usually metallic and treated with a surface coating which improves absorptivity. Black or dark paints or selective coatings are used for this purpose, and the design of this passage and plate combination helps determine a solar system's effectiveness.

Heat transfer fluid passage tubes are attached above and below or integral with an absorber plate for the purpose of transferring thermal energy from the absorber plate to a heat transfer medium. The heat exchanger is a device for transferring thermal energy from one fluid to another.

Piping and storage tanks should be well insulated to minimize heat losses.

Size domestic water heating storage tanks to hold 20 gallons of water per user, minimum, plus 10 gallons per dishwasher or washing machine. For domestic water heating an optimum collector size is approximately 3/4 square foot of area per gallon of water storage. For space heating of residences and small commercial applications the collector is commonly sized between 30% and 50% of the internal floor area. For space heating of large commercial applications, collector areas less than 30% of the internal floor area can still provide significant heat reductions.

A supplementary heat source is recommended for Northern states for December through February.

The solar energy transmission per square foot of collector surface varies greatly with the material used. Initial cost, heat transmittance and useful life are obviously interrelated.

R15050-710 Subcontractors

On the unit cost pages of the R.S. Means Cost Data books, the last column is entitled "Total Incl. O&P". This is normally the cost of the installing contractor. In Division 15, this is the cost of the mechanical contractor. If the particular work being estimated is to be performed by a sub to the mechanical contractor, the mechanical's profit and handling charge (usually 10%) is added to the total of the last column.

R15050-720 Demolition (Selective vs. Removal for Replacement)

Demolition can be divided into two basic categories.

One type of demolition involves the removal of material with no concern for its replacement. The labor-hours to estimate this work are found in Div. 15055 under "Selective Demolition". It is selective in that individual items or all the material installed as a system or trade grouping such as plumbing or heating systems are removed. This may be accomplished by the easiest way possible, such as sawing, torch cutting, or sledge hammer as well as simple unbolting.

The second type of demolition is the removal of some item for repair or replacement. This removal may involve careful draining, opening of unions, disconnecting and tagging of electrical connections, capping of pipes/ducts to prevent entry of debris or leakage of the material contained as well as transport of the item away from its in-place location to a truck/dumpster. An approximation of the time required to accomplish this type of demolition is to use half of the time indicated as necessary to install a new unit. For example; installation of a new pump might be listed as requiring 6 labor-hours so if we had to estimate the removal of the old pump we would allow an additional 3 hours for a total of 9 hours. That is, the complete replacement of a defective pump with a new pump would be estimated to take 9 labor-hours.

R15100-050 Pipe Material Considerations

1. Malleable fittings should be used for gas service.
2. Malleable fittings are used where there are stresses/strains due to expansion and vibration.
3. Cast fittings may be broken as an aid to disassembling of heating lines frozen by long use, temperature and minerals.
4. Cast iron pipe is extensively used for underground and submerged service.
5. Type M (light wall) copper tubing is available in hard temper only and is used for nonpressure and less severe applications than K and L.
6. Type L (medium wall) copper tubing, available hard or soft for interior service.
7. Type K (heavy wall) copper tubing, available in hard or soft temper for use where conditions are severe. For underground and interior service.
8. Hard drawn tubing requires fewer hangers or supports but should not be bent. Silver brazed fittings are recommended, however soft solder is normally used.
9. Type DMV (very light wall) copper tubing designed for drainage, waste and vent plus other non-critical pressure services.

Domestic/Imported Pipe and Fittings Cost

The prices shown in this publication for steel/cast iron pipe and steel, cast iron, malleable iron fittings are based on domestic production sold at the normal trade discounts. The above listed items of foreign manufacture may be available at prices of 1/3 to 1/2 those shown. Some imported items after minor machining or finishing operations are being sold as domestic to further complicate the system.

Caution: Most pipe prices in this book also include a coupling and pipe hangers which for the larger sizes can add significantly to the per foot cost and should be taken into account when comparing "book cost" with quoted supplier's cost.

R15100-070 Piping to 10′ High

When taking off pipe, it is important to identify the different material types and joining procedures, as well as distances between supports and components required for proper support.

During the takeoff, measure through all fittings. Do not subtract the lengths of the fittings, valves, or strainers, etc. This added length plus the final rounding of the totals will compensate for nipples and waste.

When rounding off totals always increase the actual amount to correspond with manufacturer's shipping lengths.

A. Both red brass and yellow brass pipe are normally furnished in 12′ lengths, plain end. Division 15107-220 includes in the linear foot costs two field threads and one coupling per 10′ length. A carbon steel clevis type hanger assembly every 10′ is also prorated into the linear foot costs, including both material and labor.

B. Cast iron soil pipe is furnished in either 5′ or 10′ lengths. For pricing purposes, Division 15107-320 in *Means Plumbing Cost Data* features 10′ lengths with a joint and a carbon steel clevis hanger assembly every 5′ prorated into the per foot costs of both material and labor.

Three methods of joining are considered in Division 15107-320, lead and oakum poured joints, or push-on gasket type joints for the bell and spigot pipe, and a joint clamp for the no-hub soil pipe. The labor and material costs for each of these individual joining procedures are also prorated into the linear costs per foot.

C. Copper tubing in Division 15107-420 covers types K, L, M, and DWV which are furnished in 20′ lengths. Means pricing data is based on a tubing cut each length and a coupling and two soft soldered joints every 10′. A carbon steel, clevis type hanger assembly every 10′ is also prorated into the per foot costs. The prices for refrigeration tubing are for materials only. Labor for full lengths may be based on the type L labor but short cut measures in tight areas can increase the installation labor-hours from 20 to 40%.

D. Corrosion-resistant piping does not lend itself to one particular standard of hanging or support assembly due to its diversity of application and placement. The several varieties of corrosion-resistant piping do not include any material or labor costs for hanger assemblies (See Division 15060-300 for appropriate selection). The estimator should refer to Division 15107-560 in *Means Plumbing Cost Data* for the compatible corrosion-resistant fittings and piping.

E. Glass pipe, as shown in *Means Plumbing Cost Data* Division 15106-120, is furnished in standard lengths either 5′ or 10′ long, beaded on one end. Special orders for diverse lengths beaded on both ends are also available. For pricing purposes, R.S. Means features 10′ lengths with a coupling and a carbon steel band hanger assembly every 10′ prorated into the per foot linear costs.

Glass pipe is also available with conical ends and standard lengths ranging from 6″ through 3′ in 6″ increments, then up to 10′ in 12″ increments. Special lengths can be customized for particular installation requirements.

For pricing purposes, Means has based the labor and material pricing on 10′ lengths. Included in these costs per linear foot are the prorated costs for a flanged assembly every 10′ consisting of two flanges, a gasket, two insertable seals, and the required number of bolts and nuts. A carbon steel band hanger assembly based on 10′ center lines has also been prorated into the costs per foot for labor and materials.

F. Plastic pipe of several compositions and joining methods are considered in Division 15108-520. Fiberglass reinforced pipe (FRP) is priced, based on 10′ lengths (20′ lengths are also available), with coupling and epoxy joints every 10′. FRP is furnished in both "General Service" and "High Strength." A carbon steel clevis hanger assembly, 3 for every 10′, is built into the prorated labor and material costs on a per foot basis.

The PVC and CPVC pipe schedules 40, 80 and 120, plus SDR ratings are all based on 20′ lengths with a coupling installed every 10′, as well as a carbon steel clevis hanger assembly every 3′. The PVC and ABS type DWV piping is based on 10′ lengths with solvent weld couplings every 10′, and with carbon steel clevis hanger assemblies, 3 for every 10′. The rest of the plastic piping in this section is based on flexible 100′ coils and does not include any coupling or supports.

This section ends with PVC drain and sewer piping based on 10′ lengths with bell and spigot ends and 0-ring type, push-on joints. Material costs only are given, but the complete installation costs should be similar to those shown in line number 02620-270.

G. Stainless steel piping, Division 15107-900, includes both weld end and threaded piping, both in the type 304 and 316 specification and in the following schedules, 5, 10, 40, 80, and 160. Although this piping is usually furnished in 20′ lengths, this cost grouping has a joint (either heli-arc butt-welded or threads and coupling) every 10′. A carbon steel clevis type hanger assembly is also included at 10′ intervals and prorated into the linear foot costs.

H. Division 15107-620 covers carbon steel pipe, both black and galvanized. This section encompasses schedules 40 (standard) and 80 (extra heavy).

Several common methods of joining steel pipe — such as thread and coupled, butt welded, and flanged (150 lb. weld neck flanges) are also included.

For estimating purposes, it is assumed that the piping is purchased in 20′ lengths and that a compatible joint is made up every 10′. These joints are prorated into the labor and material costs per linear foot. The following hanger and support assemblies every 10′ are also included: carbon steel clevis for the T & C pipe, and single rod roll type for both the welded and flanged piping. All of these hangers are oversized to accommodate pipe insulation 3/4″ thick through 5″ pipe size and 1-1/2″ thick from 6″ through 12″ pipe size.

I. Division 15107-690 prices grooved joint steel pipe, both black and galvanized, in schedules 10, 40, and 80, furnished in 20′ lengths. This section describes two joining methods: cut groove and roll groove. The schedule 10 piping is roll-grooved, while the heavier schedules are cut-grooved. The labor and material costs are prorated into per linear foot prices, including a coupled joint every 10′, as well as a carbon steel clevis hanger assembly.

Notes:

The pipe hanger assemblies mentioned in the preceding paragraphs include the described hanger; appropriately sized steel, box-type insert and nut; plus a threaded hanger rod. On average, the distance from the pipe center line to the insert face is 2′.

C clamps are used when the pipe is to be supported from steel shapes rather than anchored in the slab. C clamps are slightly less costly than inserts. However, to save time in estimating, it is advisable to use the given line number cost, rather than substituting a C clamp for the insert.

Add to piping labor for elevated installation:

10′ to 15′ high	10%	30′ to 35′ high	40%
15′ to 20′ high	20%	35′ to 40′ high	50%
20′ to 25′ high	25%	Over 40′ high	55%
25′ to 30′ high	35%		

When using the percentage adds for elevated piping installations as shown above, bear in mind that the given heights are for the pipe supports, even though the insert, anchor, or clamp may be several feet higher than the pipe itself.

An allowance has been included in the piping installation time for testing and minor tightening of leaking joints, fittings, stuffing boxes, packing glands, etc. For extraordinary test requirements such as x-rays, prolonged pressure or demonstration tests, a percentage of the piping labor, based on the estimator's experience, must be added to the labor total. A testing service specializing in weld x-rays should be consulted for pricing if it is an estimate requirement. Equipment installation time includes start-up with associated adjustments.

R15100-080 Valve Materials

VALVE MATERIALS

Bronze:
Bronze is one of the oldest materials used to make valves. It is most commonly used in hot and cold water systems and other non-corrosive services. It is often used as a seating surface in larger iron body valves to ensure tight closure.

Carbon Steel:
Carbon steel is a high strength material. Therefore, valves made from this metal are used in higher pressure services, such as steam lines up to 600 psi at 850°F. Many steel valves are available with butt-weld ends for economy and are generally used in high pressure steam service as well as other higher pressure non-corrosive services.

Forged Steel:
Valves from tough carbon steel are used in service up to 2000 psi and temperatures up to 1000°F in Gate, Globe and Check valves.

Iron:
Valves are normally used in medium to large pipe lines to control non-corrosive fluid and gases, where pressures do not exceed 250 psi at 450° or 500 psi cold water, oil or gas.

Stainless Steel:
Developed steel alloys can be used in over 90% corrosive services.

Plastic PVC:
This is used in a great variety of valves generally in high corrosive service with lower temperatures and pressures.

VALVE SERVICE PRESSURES

Pressure ratings on valves provide an indication of the safe operating pressure for a valve at some elevated temperature. This temperature is dependent upon the materials used and the fabrication of the valve. When specific data is not available, a good "rule-of-thumb" to follow is the temperature of saturated steam on the primary rating indicated on the valve body. Example: The valve has the number 150S printed on the side indicating 150 psi and hence, a maximum operating temperature of 367°F (temperature of saturated steam and 150 psi).

DEFINITIONS

1. "WOG" – Water, oil, gas (cold working pressures).
2. "SWP" – Steam working pressure.
3. 100% area (full port) – means the area through the valve is equal to or greater than the area of standard pipe.
4. "Standard Opening" – means that the area through the valve is less than the area of standard pipe and therefore these valves should be used only where restriction of flow is unimportant.
5. "Round Port" – means the valve has a full round opening through the plug and body, of the same size and area as standard pipe.
6. "Rectangular Port" – valves have rectangular shaped ports through the plug body. The area of the port is either equal to 100% of the area of standard pipe, or restricted (standard opening). In either case it is clearly marked.
7. "ANSI" – American National Standards Institute.

R15100-090 Valve Selection Considerations

INTRODUCTION: In any piping application, valve performance is critical. Valves should be selected to give the best performance at the lowest cost.

The following is a list of performance characteristics generally expected of valves.

1. Stopping flow or starting it.
2. Throttling flow (Modulation).
3. Flow direction changing.
4. Checking backflow (Permitting flow in only one direction).
5. Relieving or regulating pressure.

In order to properly select the right valve, some facts must be determined.

A. What liquid or gas will flow through the valve?
B. Does the fluid contain suspended particles?
C. Does the fluid remain in liquid form at all times?
D. Which metals does fluid corrode?
E. What are the pressure and temperature limits? (As temperature and pressure rise, so will the price of the valve.)
F. Is there constant line pressure?
G. Is the valve merely an on-off valve?
H. Will checking of backflow be required?
I. Will the valve operate frequently or infrequently?

Valves are classified by design type into such classifications as Gate, Globe, Angle, Check, Ball, Butterfly and Plug. They are also classified by end connection, stem, pressure restrictions and material such as bronze, cast iron, etc. Each valve has a specific use. A quality valve used correctly will provide a lifetime of trouble- free service, but a high quality valve installed in the wrong service may require frequent attention.

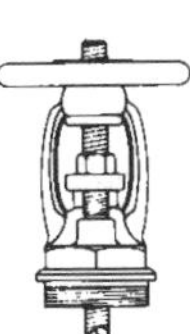

STEM TYPES

(OS & Y)—Rising Stem-Outside Screw and Yoke

Offers a visual indication of whether the valve is open or closed. Recommended where high temperatures, corrosives, and solids in the line might cause damage to inside-valve stem threads. The stem threads are engaged by the yoke bushing so the stem rises through the hand wheel as it is turned.

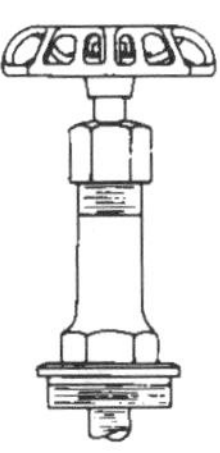

(R.S.)—Rising Stem-Inside Screw

Adequate clearance for operation must be provided because both the hand wheel and the stem rise.
The valve wedge position is indicated by the position of the stem and hand wheel.

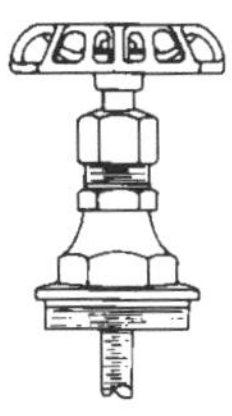

(N.R.S.)—Non-Rising Stem-Inside Screw

A minimum clearance is required for operating this type of valve. Excessive wear or damage to stem threads inside the valve may be caused by heat, corrosion, and solids. Because the hand wheel and stem do not rise, wedge position cannot be visually determined.

VALVE TYPES

Gate Valves

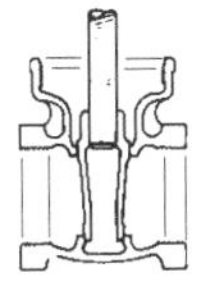

Provide full flow, minute pressure drop, minimum turbulence and minimum fluid trapped in the line.
They are normally used where operation is infrequent.

Globe Valves

Globe valves are designed for throttling and/or frequent operation with positive shut-off. Particular attention must be paid to the several types of seating materials available to avoid unnecessary wear. The seats must be compatible with the fluid in service and may be composition or metal. The configuration of the Globe valve opening causes turbulence which results in increased resistance. Most bronze Globe valves are rising stem-inside screw, but they are also available on O.S. & Y.

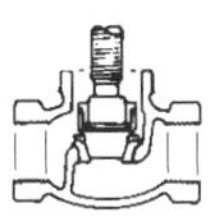

Angle Valves

The fundamental difference between the Angle valve and the Globe valve is the fluid flow through the Angle valve. It makes a 90° turn and offers less resistance to flow than the Globe valve while replacing an elbow. An Angle valve thus reduces the number of joints and installation time.

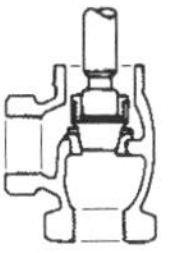

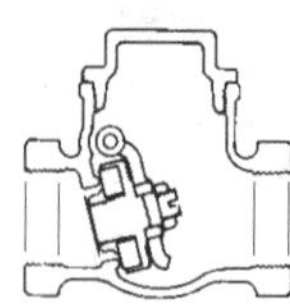

Check Valves

Check valves are designed to prevent backflow by automatically seating when the direction of fluid is reversed.
Swing Check valves are generally installed with Gate-valves, as they provide comparable full flow. Usually recommended for lines where flow velocities are low and should not be used on lines with pulsating flow. Recommended for horizontal installation, or in vertical lines only where flow is upward.

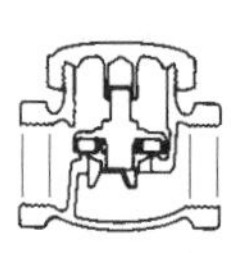
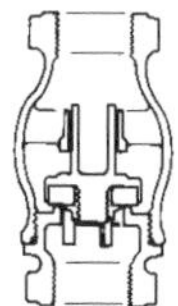

Lift Check Valves

These are commonly used with Globe and Angle valves since they have similar diaphragm seating arrangements and are recommended for preventing backflow of steam, air, gas and water, and on vapor lines with high flow velocities. For horizontal lines, horizontal lift checks should be used and vertical lift checks for vertical lines.

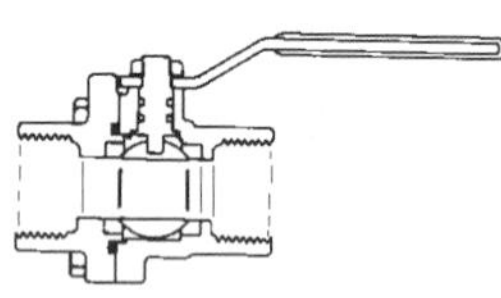

Ball Valves

Ball valves are light and easily installed, yet because of modern elastomeric seats, provide tight closure. Flow is controlled by rotating up to 90° a drilled ball which fits tightly against resilient seals. This ball seats with flow in either direction, and valve handle indicates the degree of opening. Recommended for frequent operation readily adaptable to automation, ideal for installation where space is limited.

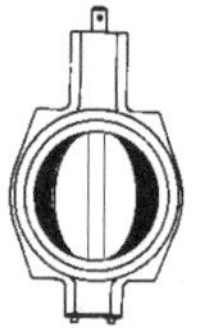

Butterfly Valves

Butterfly valves provide bubble-tight closure with excellent throttling characteristics. They can be used for full-open, closed and for throttling applications.

The Butterfly valve consists of a disc within the valve body which is controlled by a shaft. In its closed position, the valve disc seals against a resilient seat. The disc position throughout the full 90° rotation is visually indicated by the position of the operator.

A Butterfly valve is only a fraction of the weight of a Gate valve and requires no gaskets between flanges in most cases. Recommended for frequent operation and adaptable to automation where space is limited.

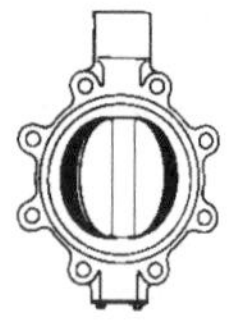

Wafer and Lug type bodies when installed between two pipe flanges, can be easily removed from the line. The pressure of the bolted flanges holds the valve in place.
Locating lugs makes installation easier.

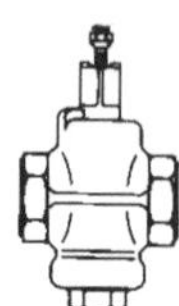

Plug Valves

Lubricated plug valves, because of the wide range of service to which they are adapted, may be classified as all purpose valves. They can be safely used at all pressure and vacuums, and at all temperatures up to the limits of available lubricants. They are the most satisfactory valves for the handling of gritty suspensions and many other destructive, erosive, corrosive and chemical solutions.

R15500-010 Heating Systems

Heating Systems

The basic function of a heating system is to bring an enclosed volume up to a desired temperature and then maintain that temperature within a reasonable range. To accomplish this, the selected system must have sufficient capacity to offset transmission losses resulting from the temperature difference on the interior and exterior of the enclosing walls in addition to losses due to cold air infiltration through cracks, crevices and around doors and windows. The amount of heat to be furnished is dependent upon the building size, construction, temperature difference, air leakage, use, shape, orientation and exposure. Air circulation is also an important consideration. Circulation will prevent stratification which could result in heat losses through uneven temperatures at various levels. For example, the most efficient use of unit heaters can usually be achieved by circulating the space volume through the total number of units once every 20 minutes or 3 times an hour. This general rule must, of course, be adapted for special cases such as large buildings with low ratios of heat transmitting surface to cubical volume. The type of occupancy of a building will have considerable bearing on the number of heat transmitting units and the location selected. It is axiomatic, however, that the basis of any successful heating system is to provide the maximum amount of heat at the points of maximum heat loss such as exposed walls, windows, and doors. Large roof areas, wind direction, and wide doorways create problems of excessive heat loss and require special consideration and treatment.

Heat Transmission

Heat transfer is an important parameter to be considered during selection of the exterior wall style, material and window area. A high rate of transfer will permit greater heat loss during the wintertime with the resultant increase in heating energy costs and a greater rate of heat gain in the summer with proportionally greater cooling cost. Several terms are used to describe various aspects of heat transfer. However, for general estimating purposes this book lists U values for systems of construction materials. U is the "overall heat transfer coefficient." It is defined as the heat flow per hour through one square foot when the temperature difference in the air on either side of the structure wall, roof, ceiling or floor is one degree Fahrenheit. The structural segment may be a single homogeneous material or a composite.

Total heat transfer is found using the following equation:

$Q = AU(T_2 - T_1)$ where

- Q = Heat flow, BTU per hour
- A = Area, square feet
- U = Overall heat transfer coefficient
- $(T_2 - T_1)$ = Difference in temperature of air on each side of the construction component. (Also abbreviated TD)

Note that heat can flow through all surfaces of any building and this flow is in addition to heat gain or loss due to ventilation, infiltration and generation (appliances, machinery, people).

R15500-020 Heating Approximations for Quick Estimating

Oil Piping & Boiler Room Piping	
Small System	20 to 30% of Boiler
Complex System with Pumps, Headers, Etc.	80 to 110% of Boiler
Breeching With Insulation:	
Small	10 to 15% of Boiler
Large	15 to 25% of Boiler
Coils:	15 to 30% of Containing Unit
Balancing (Independent)	1/2% of H.V.A.C. Estimate

Quality/Complexity Adjustment: For all heating installations add these adjustments to the estimate to more closely allow for the equipment and conditions of the particular job under consideration.

Economy installation, add	0 to 5% of System
Good quality, medium complexity, add	5 to 15% of System
Above average quality and complexity, add	15 to 25% of System

R15500-030 The Basics of a Heating System

The function of a heating system is to achieve and maintain a desired temperature in a room or building by replacing the amount of heat being dissipated. There are four kinds of heating systems: hot-water, steam, warm-air and electric resistance. Each has certain essential and similar elements with the exception of electric resistance heating which is covered in Section 16.

The basic elements of a heating system are:

A. A **combustion chamber** in which fuel is burned and heat transferred to a conveying medium.

B. The **"fluid"** used for conveying the heat (water, steam or air).

C. **Conductors** or pipes for transporting the fluid to specific desired locations.

D. A means of disseminating the heat, sometimes called **terminal units.**

A. The **combustion chamber** in a furnace heats air which is then distributed. This is called a warm-air system.

The combustion chamber in a boiler heats water which is either distributed as hot water or steam and this is termed a hydronic system.

The maximum allowable working pressures are limited by ASME "Code for Heating Boilers" to 15 PSI for steam and 160 PSI for hot water heating boilers, with a maximum temperature limitation of 250°F. Hot water boilers are generally rated for a working pressure of 30 PSI. High pressure boilers are governed by the ASME "Code for Power Boilers" which is used almost universally for boilers operating over 15 PSIG. High pressure boilers used for a combination of heating/process loads are usually designed for 150 PSIG.

Boiler ratings are usually indicated as either Gross or Net Output. The Gross Load is equal to the Net Load plus a piping and pickup allowance. When this allowance cannot be determined, divide the gross output rating by 1.25 for a value equal to or greater than the net heat loss requirement of the building.

B. Of the three **fluids** used, steam carries the greatest amount of heat per unit volume. This is due to the fact that it gives up its latent heat of vaporization at a temperature considerably above room temperature. Another advantage is that the pressure to produce a positive circulation is readily available. Piping conducts the steam to terminal units and returns condensate to the boiler.

The **steam system** is well adapted to large buildings because of its positive circulation, its comparatively economical installation and its ability to deliver large quantities of heat. Nearly all large office buildings, stores, hotels, and industrial buildings are so heated, in addition to many residences.

Hot water, when used as the heat carrying fluid, gives up a portion of its sensible heat and then returns to the boiler or heating apparatus for reheating. As the heat conveyed by each pound of water is about one-fiftieth of the heat conveyed by a pound of steam, it is necessary to circulate about fifty times as much water as steam by weight (although only one-thirtieth as much by volume). The hot water system is usually, although not necessarily, designed to operate at temperatures below that of the ordinary steam system and so the amount of heat transfer surface must be correspondingly greater. A temperature of 190°F to 200°F is normally the maximum. Circulation in small buildings may depend on the difference in density between hot water and the cool water returning to the boiler; circulating pumps are normally used to maintain a desired rate of flow. Pumps permit a greater degree of flexibility and better control.

In **warm-air** furnace systems, cool air is taken from one or more points in the building, passed over the combustion chamber and flue gas passages and then distributed through a duct system. A disadvantage of this system is that the ducts take up much more building volume than steam or hot water pipes. Advantages of this system are the relative ease with which humidification can be accomplished by the evaporation of water as the air circulates through the heater, and the lack of need for expensive disseminating units as the warm air simply becomes part of the interior atmosphere of the building.

C. Conductors (pipes and ducts) have been lightly treated in the discussion of conveying fluids. For more detailed information such as sizing and distribution methods, the reader is referred to technical publications such as the American Society of Heating, Refrigerating and Air-Conditioning Engineers "Handbook of Fundamentals."

D. Terminal units come in an almost infinite variety of sizes and styles, but the basic principles of operation are very limited. As previously mentioned, warm-air systems require only a simple register or diffuser to mix heated air with that present in the room. Special application items such as radiant coils and infrared heaters are available to meet particular conditions but are not usually considered for general heating needs. Most heating is accomplished by having air flow over coils or pipes containing the heat transporting medium (steam, hot-water, electricity). These units, while varied, may be separated into two general types, (1) radiator/convectors and (2) unit heaters.

Radiator/convectors may be cast, fin-tube or pipe assemblies. They may be direct, indirect, exposed, concealed or mounted within a cabinet enclosure, upright or baseboard style. These units are often collectively referred to as "radiatiors" or "radiation" although none gives off heat either entirely by radiation or by convection but rather a combination of both. The air flows over the units as a gravity "current." It is necessary to have one or more heat-emitting units in each room. The most efficient placement is low along an outside wall or under a window to counteract the cold coming into the room and achieve an even distribution.

In contrast to radiator/convectors which operate most effectively against the walls of smaller rooms, **unit heaters** utilize a fan to move air over heating coils and are very effective in locations of relatively large volume. Unit heaters, while usually suspended overhead, may be floor mounted. They also may take in fresh outside air for ventilation. The heat distributed by unit heaters may be from a remote source and conveyed by a fluid or it may be from the combustion of fuel in each individual heater. In the latter case the only piping required would be for fuel, however, a vent for the products of combustion would be necessary.

The following list gives may of the advantages of unit heaters for applications other than office or residential:

a. Large capacity so smaller number of units are required,
b. Piping system simplified, **c.** Space saved where they are located overhead out of the way, **d.** Rapid heating directed where needed with effective wide distribution, **e.** Difference between floor and ceiling temperature reduced, **f.** Circulation of air obtained, and ventilation with introduction of fresh air possible, **g.** Heat output flexible and easily controlled.

R15500-050 Factor for Determining Heat Loss for Various Types of Buildings

General: While the most accurate estimates of heating requirements would naturally be based on detailed information about the building being considered, it is possible to arrive at a reasonable approximation using the following procedure:

1. Calculate the cubic volume of the room or building.
2. Select the appropriate factor from Table 1 below. Note that the factors apply only to inside temperatures listed in the first column and to 0°F outside temperature.
3. If the building has bad north and west exposures, multiply the heat loss factor by 1.1.
4. If the outside design temperature is other than 0°F, multiply the factor from Table 1 by the factor from Table 2.
5. Multiply the cubic volume by the factor selected from Table 1. This will give the estimated BTUH heat loss which must be made up to maintain inside temperature.

Table 1 — Building Type	Conditions	Qualifications	Loss Factor*
Factories & Industrial Plants General Office Areas at 70°F	One Story	Skylight in Roof	6.2
		No Skylight in Roof	5.7
	Multiple Story	Two Story	4.6
		Three Story	4.3
		Four Story	4.1
		Five Story	3.9
		Six Story	3.6
	All Walls Exposed	Flat Roof	6.9
		Heated Space Above	5.2
	One Long Warm Common Wall	Flat Roof	6.3
		Heated Space Above	4.7
	Warm Common Walls on Both Long Sides	Flat Roof	5.8
		Heated Space Above	4.1
Warehouses at 60°F	All Walls Exposed	Skylights in Roof	5.5
		No Skylight in Roof	5.1
		Heated Space Above	4.0
	One Long Warm Common Wall	Skylight in Roof	5.0
		No Skylight in Roof	4.9
		Heated Space Above	3.4
	Warm Common Walls on Both Long Sides	Skylight in Roof	4.7
		No Skylight in Roof	4.4
		Heated Space Above	3.0

*Note: This table tends to be conservative particularly for new buildings designed for minimum energy consumption.

Table 2 — Outside Design Temperature Correction Factor (for Degrees Fahrenheit)									
Outside Design Temperature	50	40	30	20	10	0	-10	-20	-30
Correction Factor	.29	.43	.57	.72	.86	1.00	1.14	1.28	1.43

R15500-070 Transmission of Heat

R15500-070 and R15500-080 provide a way to calculate heat transmission of various construction materials from their U values and the TD (Temperature Difference).

1. From the Exterior Enclosure Division or elsewhere, determine U values for the construction desired.
2. Determine the coldest design temperature. The difference between this temperature and the desired interior temperature is the TD (temperature difference).
3. Enter R15500-070 or R15500-080 at correct U Value. Cross horizontally to the intersection with appropriate TD. Read transmission per square foot from bottom of figure.
4. Multiply this value of BTU per hour transmission per square foot of area by the total surface area of that type of construction.

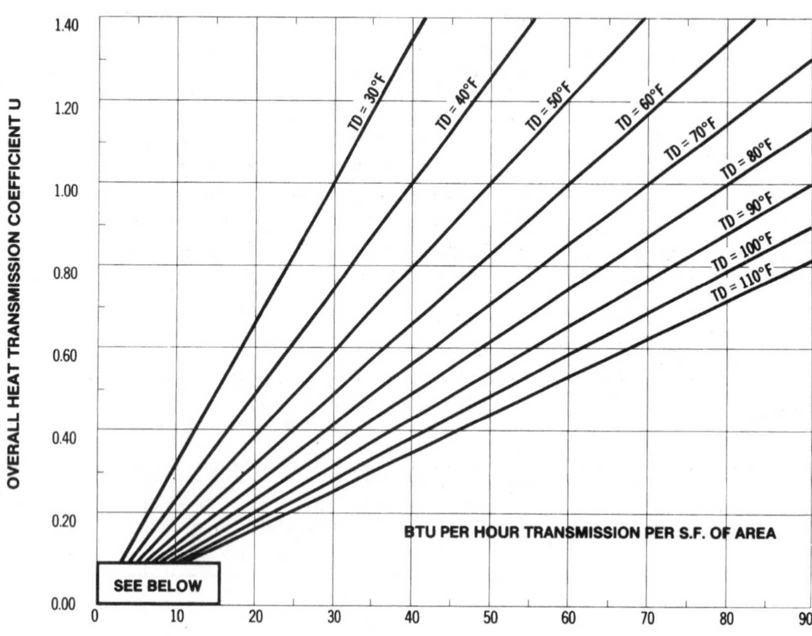

R15500-080 Transmission of Heat (Low Rate)

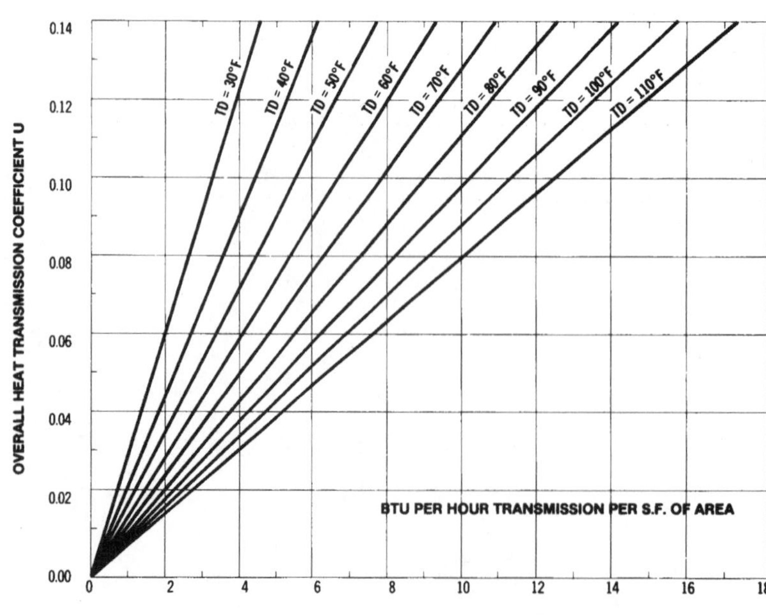

R15700-010 Air Conditioning

General: The purpose of air conditioning is to control the environment of a space so that comfort is provided for the occupants and/or conditions are suitable for the processes or equipment contained therein. The several items which should be evaluated to define system objectives are:

Temperature Control
Humidity Control
Cleanliness
Odor, smoke and fumes
Ventilation

Efforts to control the above parameters must also include consideration of the degree or tolerance of variation, the noise level introduced, the velocity of air motion and the energy requirements to accomplish the desired results.

The variation in **temperature** and **humidity** is a function of the sensor and the controller. The controller reacts to a signal from the sensor and produces the appropriate suitable response in either the terminal unit, the conductor of the transporting medium (air, steam, chilled water, etc.), or the source (boiler, evaporating coils, etc.).

The **noise level** is a by-product of the energy supplied to moving components of the system. Those items which usually contribute the most noise are pumps, blowers, fans, compressors and diffusers. The level of noise can be partially controlled through use of vibration pads, isolators, proper sizing, shields, baffles and sound absorbing liners.

Some **air motion** is necessary to prevent stagnation and stratification. The maximum acceptable velocity varies with the degree of heating or cooling which is taking place. Most people feel air moving past them at velocities in excess of 25 FPM as an annoying draft. However, velocities up to 45 FPM may be acceptable in certain cases. Ventilation, expressed as air changes per hour and percentage of fresh air, is usually an item regulated by local codes.

Selection of the system to be used for a particular application is usually a trade-off. In some cases the building size, style, or room available for mechanical use limits the range of possibilities. Prime factors influencing the decision are first cost and total life (operating, maintenance and replacement costs). The accuracy with which each parameter is determined will be an important measure of the reliability of the decision and subsequent satisfactory operation of the installed system.

Heat delivery may be desired from an air conditioning system. Heating capability usually is added as follows: A gas fired burner or hot water/steam/electric coils may be added to the air handling unit directly and heat all air equally. For limited or localized heat requirements the water/steam/electric coils may be inserted into the duct branch supplying the cold areas. Gas fired duct furnaces are also available.

Note: When water or steam coils are used the cost of the piping and boiler must also be added. For a rough estimate use the cost per square foot of the appropriate sized hydronic system with unit heaters. This will provide a cost for the boiler and piping, and the unit heaters of the system would equate to the approximate cost of the heating coils.

R15700-020 Air Conditioning Requirements

BTU's per hour per S.F. of floor area and S.F. per ton of air conditioning.

Type of Building	BTU per S.F.	S.F. per Ton	Type of Building	BTU per S.F.	S.F. per Ton	Type of Building	BTU per S.F.	S.F. per Ton
Apartments, Individual	26	450	Dormitory, Rooms	40	300	Libraries	50	240
Corridors	22	550	Corridors	30	400	Low Rise Office, Exterior	38	320
Auditoriums & Theaters	40	300/18*	Dress Shops	43	280	Interior	33	360
Banks	50	240	Drug Stores	80	150	Medical Centers	28	425
Barber Shops	48	250	Factories	40	300	Motels	28	425
Bars & Taverns	133	90	High Rise Office—Ext. Rms.	46	263	Office (small suite)	43	280
Beauty Parlors	66	180	Interior Rooms	37	325	Post Office, Individual Office	42	285
Bowling Alleys	68	175	Hospitals, Core	43	280	Central Area	46	260
Churches	36	330/20*	Perimeter	46	260	Residences	20	600
Cocktail Lounges	68	175	Hotel, Guest Rooms	44	275	Restaurants	60	200
Computer Rooms	141	85	Corridors	30	400	Schools & Colleges	46	260
Dental Offices	52	230	Public Spaces	55	220	Shoe Stores	55	220
Dept. Stores, Basement	34	350	Industrial Plants, Offices	38	320	Shop'g. Ctrs., Supermarkets	34	350
Main Floor	40	300	General Offices	34	350	Retail Stores	48	250
Upper Floor	30	400	Plant Areas	40	300	Specialty	60	200

*Persons per ton
12,000 BTU = 1 ton of air conditioning

R15700-030 Psychrometric Table

Dewpoint or Saturation Temperature (F)

Relative humidity (%)	32	35	40	45	50	55	60	65	70	75	80	85	90	95	100
100	32	35	40	45	50	55	60	65	70	75	80	85	90	95	100
90	30	33	37	42	47	52	57	62	67	72	77	82	87	92	97
80	27	30	34	39	44	49	54	58	64	68	73	78	83	88	93
70	24	27	31	36	40	45	50	55	60	64	69	74	79	84	88
60	20	24	28	32	36	41	46	51	55	60	65	69	74	79	83
50	16	20	24	28	33	36	41	46	50	55	60	64	69	73	78
40	12	15	18	23	27	31	35	40	45	49	53	58	62	67	71
30	8	10	14	18	21	25	29	33	37	42	46	50	54	59	62
20	6	7	8	9	13	16	20	24	28	31	35	40	43	48	52
10	4	4	5	5	6	8	9	10	13	17	20	24	27	30	34

Dry bulb temperature (F)

This table shows the relationship between RELATIVE HUMIDITY, DRY BULB TEMPERATURE AND DEWPOINT.

As an example, assume that the thermometer in a room reads 75°F, and we know that the relative humidity is 50%. The chart shows the dewpoint temperature to be 55°. That is, any surface colder than 55°F will "sweat" or collect condensing moisture. This surface could be the outside of an uninsulated chilled water pipe in the summertime, or the inside surface of a wall or deck in the wintertime. After determining the extreme ambient parameters, the table at the left is useful in determining which surfaces need insulation or vapor barrier protection.

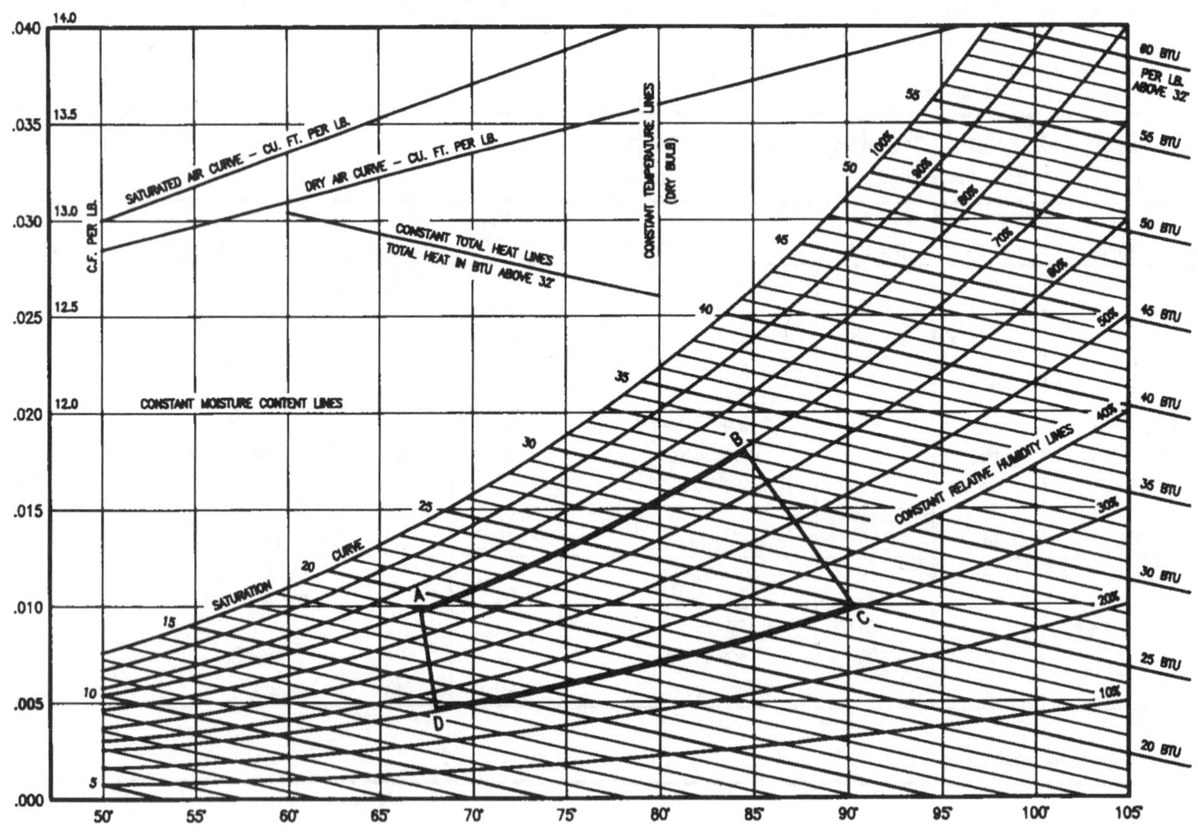

TEMPERATURE DEGREES FAHRENHEIT
TOTAL PRESSURE = 14.696 LB. PER SQ. IN. ABS.

Psychrometric chart showing different variables based on one pound of dry air. Space marked A B C D is temperature-humidity range which is most comfortable for majority of people.

R15700-040 Recommended Ventilation Air Changes

Table below lists range of time in minutes per change for various types of facilities.

Assembly Halls	2-10	Dance Halls	2-10	Laundries	1-3
Auditoriums	2-10	Dining Rooms	3-10	Markets	2-10
Bakeries	2-3	Dry Cleaners	1-5	Offices	2-10
Banks	3-10	Factories	2-5	Pool Rooms	2-5
Bars	2-5	Garages	2-10	Recreation Rooms	2-10
Beauty Parlors	2-5	Generator Rooms	2-5	Sales Rooms	2-10
Boiler Rooms	1-5	Gymnasiums	2-10	Theaters	2-8
Bowling Alleys	2-10	Kitchens-Hospitals	2-5	Toilets	2-5
Churches	5-10	Kitchens-Restaurant	1-3	Transformer Rooms	1-5

CFM air required for changes = Volume of room in cubic feet ÷ Minutes per change.

R15700-090 Quality/Complexity Adjustment for Air Conditioning Systems

Economy installation, add	0 to 5%
Good quality, medium complexity, add	5 to 15%
Above average quality and complexity, add	15 to 25%

Add the above adjustments to the estimate to more closely allow for the equipment and conditions of the particular job under consideration.

R15800-080 Loudness Levels for Moving Air thru Fans, Diffusers, Register, Etc. (Measured in Sones)

Recommended Loudness Levels (Sones)											
Area	Very Quiet	Quiet	Noisy	Area	Very Quiet	Quiet	Noisy	Area	Very Quiet	Quiet	Noisy
Auditoriums				**Hotels**				**Offices**			
Auditorium lobbies	3	4	6	Banquet Rooms	1.5	3	6	Conference rooms	1	1.7	3
Concert and opera halls	0.8	1	1.5	Individual rooms, suites	1	2	4	Drafting	2	4	8
Courtrooms	2	3	4	Kitchens and laundries	4	7	10	General open offices	2	4	8
Lecture halls	1.5	2	3	Lobbies	2	4	8	Halls and corridors	2.5	5	10
Movie theaters	1.5	2	3	**Indoor Sports**				Professional offices	1.5	3	6
Churches and Schools				Bowling alleys	3	4	6	Reception room	1.5	3	6
Kitchens	4	6	8	Gymnasiums	2	4	6	Tabulation & computation	3	6	12
Laboratories	2	4	6	Swimming pools	4	7	10	**Public Buildings**			
Libraries	1.5	2	3	**Manufacturing Areas**				Banks	2	4	6
Recreation halls	2	4	8	Assembly lines	5	12	30	Court houses	2.5	4	6
Sanctuaries	1	1.7	3	Foreman's office	3	5	8	Museums	2	3	4
Schools and classrooms	1.5	2.5	4	Foundries	10	20	40	Planetariums	1.5	2	3
Hospital and Clinics				General storage	5	10	20	Post offices	2.5	4	6
Laboratories	2	4	6	Heavy machinery	10	25	60	Public libraries	1.5	2	4
Lobbies, waiting rooms	2	4	6	Light machinery	5	12	30	Waiting rooms	3	5	8
Halls and corridors	2	4	6	Tool maintenance	4	7	10	**Restaurants**			
Operating rooms	1.5	2.5	4	**Stores**				Cafeterias	3	6	10
Private rooms	1	1.7	3	Department stores	3	6	8	Night clubs	2.5	4	6
Wards	1.5	2.5	4	Supermarkets	4	7	10	Restaurants	2	4	8

R15810-050 Ductwork

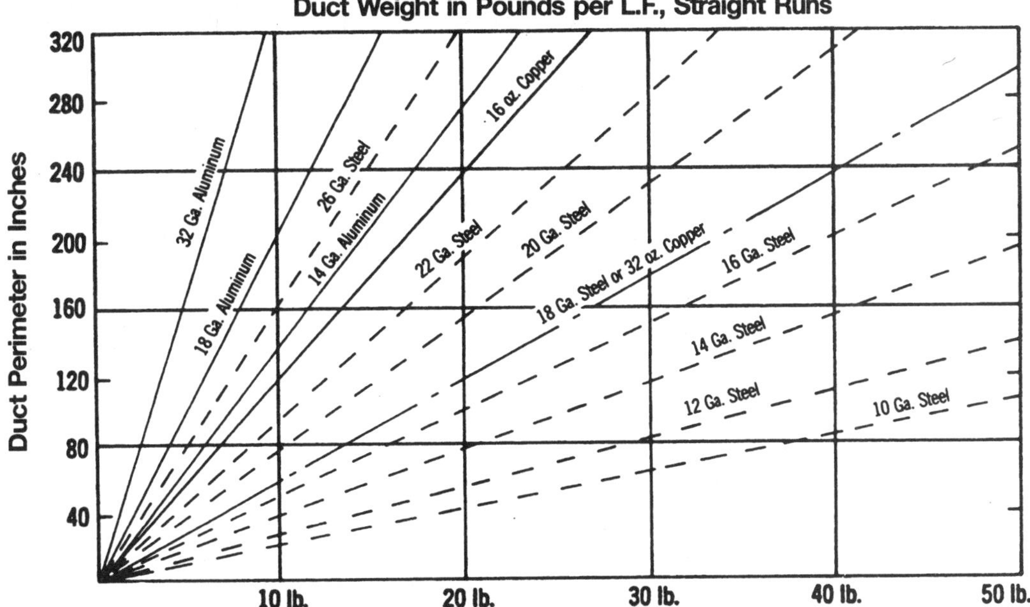

Add to the above for fittings; 90° elbow is 3 L.F.; 45° elbow is 2.5 L.F.; offset is 4 L.F.; transition offset is 6 L.F.; square-to-round transition is 4 L.F.; 90° reducing elbow is 5 L.F. For bracing and waste, add 20% to aluminum and copper, 15% to steel.

R15810-070 Duct Fabrication/Installation

The labor cost for sheet metal duct using lines 15810-100-0070 thru 1060 includes both the cost of fabrication and installation of the duct. The split is approximately 40% for fabrication, 60% for installation. It is for this reason that the percentage add for elevated installation is less than the percentage add for prefabricated duct as found starting on line 15810-100-1283.

Example: assume a piece of duct cost $100 installed (labor only)

Sheet Metal Fabrication = 40% = $40
Installation = 60% = $60

The add for elevated installation is:

Based on total labor (fabrication & installation) = $100 x 6% = $6.00

Based on installation cost only (Material purchased prefabricated) = $60 x 10% = $6.00

The $6.00 markup (10′ to 15′ high) is the same.

R15810-100 Sheet Metal Calculator (Weight in Lb./Ft. of Length)

Gauge	26	24	22	20	18	16	Gauge	26	24	22	20	18	16
Wt.-Lb./S.F.	.906	1.156	1.406	1.656	2.156	2.656	Wt.-Lb./S.F.	.906	1.156	1.406	1.656	2.156	2.656
SMACNA Max. Dimension – Long Side		30″	54″	84″	85″ Up		SMACNA Max. Dimension – Long Side		30″	54″	84″	85″ Up	
Sum-2 sides							Sum-2 Sides						
2	.3	.40	.50	.60	.80	.90	56	9.3	12.0	14.0	16.2	21.3	25.2
3	.5	.65	.80	.90	1.1	1.4	57	9.5	12.3	14.3	16.5	21.7	25.7
4	.7	.85	1.0	1.2	1.5	1.8	58	9.7	12.5	14.5	16.8	22.0	26.1
5	.8	1.1	1.3	1.5	1.9	2.3	59	9.8	12.7	14.8	17.1	22.4	26.6
6	1.0	1.3	1.5	1.7	2.3	2.7	60	10.0	12.9	15.0	17.4	22.8	27.0
7	1.2	1.5	1.8	2.0	2.7	3.2	61	10.2	13.1	15.3	17.7	23.2	27.5
8	1.3	1.7	2.0	2.3	3.0	3.6	62	10.3	13.3	15.5	18.0	23.6	27.9
9	1.5	1.9	2.3	2.6	3.4	4.1	63	10.5	13.5	15.8	18.3	24.0	28.4
10	1.7	2.2	2.5	2.9	3.8	4.5	64	10.7	13.7	16.0	18.6	24.3	28.8
11	1.8	2.4	2.8	3.2	4.2	5.0	65	10.8	13.9	16.3	18.9	24.7	29.3
12	2.0	2.6	3.0	3.5	4.6	5.4	66	11.0	14.1	16.5	19.1	25.1	29.7
13	2.2	2.8	3.3	3.8	4.9	5.9	67	11.2	14.3	16.8	19.4	25.5	30.2
14	2.3	3.0	3.5	4.1	5.3	6.3	68	11.3	14.6	17.0	19.7	25.8	30.6
15	2.5	3.2	3.8	4.4	5.7	6.8	69	11.5	14.8	17.3	20.0	26.2	31.1
16	2.7	3.4	4.0	4.6	6.1	7.2	70	11.7	15.0	17.5	20.3	26.6	31.5
17	2.8	3.7	4.3	4.9	6.5	7.7	71	11.8	15.2	17.8	20.6	27.0	32.0
18	3.0	3.9	4.5	5.2	6.8	8.1	72	12.0	15.4	18.0	20.9	27.4	32.4
19	3.2	4.1	4.8	5.5	7.2	8.6	73	12.2	15.6	18.3	21.2	27.7	32.9
20	3.3	4.3	5.0	5.8	7.6	9.0	74	12.3	15.8	18.5	21.5	28.1	33.3
21	3.5	4.5	5.3	6.1	8.0	9.5	75	12.5	16.1	18.8	21.8	28.5	33.8
22	3.7	4.7	5.5	6.4	8.4	9.9	76	12.7	16.3	19.0	22.0	28.9	34.2
23	3.8	5.0	5.8	6.7	8.7	10.4	77	12.8	16.5	19.3	22.3	29.3	34.7
24	4.0	5.2	6.0	7.0	9.1	10.8	78	13.0	16.7	19.5	22.6	29.6	35.1
25	4.2	5.4	6.3	7.3	9.5	11.3	79	13.2	16.9	19.8	22.9	30.0	35.6
26	4.3	5.6	6.5	7.5	9.9	11.7	80	13.3	17.1	20.0	23.2	30.4	36.0
27	4.5	5.8	6.8	7.8	10.3	12.2	81	13.5	17.3	20.3	23.5	30.8	36.5
28	4.7	6.0	7.0	8.1	10.6	12.6	82	13.7	17.5	20.5	23.8	31.2	36.9
29	4.8	6.2	7.3	8.4	11.0	13.1	83	13.8	17.8	20.8	24.1	31.5	37.4
30	5.0	6.5	7.5	8.7	11.4	13.5	84	14.0	18.0	21.0	24.4	31.9	37.8
31	5.2	6.7	7.8	9.0	11.8	14.0	85	14.2	18.2	21.3	24.7	32.3	38.3
32	5.3	6.9	8.0	9.3	12.2	14.4	86	14.3	18.4	21.5	24.9	32.7	38.7
33	5.5	7.1	8.3	9.6	12.5	14.9	87	14.5	18.6	21.8	25.2	33.1	39.2
34	5.7	7.3	8.5	9.9	12.9	15.3	88	14.7	18.8	22.0	25.5	33.4	39.6
35	5.8	7.5	8.8	10.2	13.3	15.8	89	14.8	19.0	22.3	25.8	33.8	40.1
36	6.0	7.8	9.0	10.4	13.7	16.2	90	15.0	19.3	22.5	26.1	34.2	40.5
37	6.2	8.0	9.3	10.7	14.1	16.7	91	15.2	19.5	22.8	26.4	34.6	41.0
38	6.3	8.2	9.5	11.0	14.4	17.1	92	15.3	19.7	23.0	26.7	35.0	41.4
39	6.5	8.4	9.8	11.3	14.8	17.6	93	15.5	19.9	23.3	27.0	35.3	41.9
40	6.7	8.6	10.0	11.6	15.2	18.0	94	15.7	20.1	23.5	27.3	35.7	42.3
41	6.8	8.8	10.3	11.9	15.6	18.5	95	15.8	20.3	23.8	27.6	36.1	42.8
42	7.0	9.0	10.5	12.2	16.0	18.9	96	16.0	20.5	24.0	27.8	36.5	43.2
43	7.2	9.2	10.8	12.5	16.3	19.4	97	16.2	20.8	24.3	28.1	36.9	43.7
44	7.3	9.5	11.0	12.8	16.7	19.8	98	16.3	21.0	24.5	28.4	37.2	44.1
45	7.5	9.7	11.3	13.1	17.1	20.3	99	16.5	21.2	24.8	28.7	37.6	44.6
46	7.7	9.9	11.5	13.3	17.5	20.7	100	16.7	21.4	25.0	29.0	38.0	45.0
47	7.8	10.1	11.8	13.6	17.9	21.2	101	16.8	21.6	25.3	29.3	38.4	45.5
48	8.0	10.3	12.0	13.9	18.2	21.6	102	17.0	21.8	25.5	29.6	38.8	45.9
49	8.2	10.5	12.3	14.2	18.6	22.1	103	17.2	22.0	25.8	29.9	39.1	46.4
50	8.3	10.7	12.5	14.5	19.0	22.5	104	17.3	22.3	26.0	30.2	39.5	46.8
51	8.5	11.0	12.8	14.8	19.4	23.0	105	17.5	22.5	26.3	30.5	39.9	47.3
52	8.7	11.2	13.0	15.1	19.8	23.4	106	17.7	22.7	26.5	30.7	40.3	47.7
53	8.8	11.4	13.3	15.4	20.1	23.9	107	17.8	22.9	26.8	31.0	40.7	48.2
54	9.0	11.6	13.5	15.7	20.5	24.3	108	18.0	23.1	27.0	31.3	41.0	48.6
55	9.2	11.8	13.8	16.0	20.9	24.8	109	18.2	23.3	27.3	31.6	41.4	49.1
							110	18.3	23.5	27.5	31.9	41.8	49.5

Example: If duct is 34″ x 20″ x 15′ long, 34″ is greater than 30″ maximum, for 24 ga. so must be 22 ga. 34″ + 20″ = 54″ going across from 54″ find 13.5 lb. per foot. 13.5 x 15′ = 202.5 lbs. For S.F. of surface area 202.5 ÷ 1.406 = 144 S.F.

Note: Figures include an allowance for scrap.

R15810-110 Ductwork Packages (per Ton of Cooling)

System	Sheet Metal	Insulation	Diffusers	Return Register
Roof Top Unit Single Zone	120 Lbs.	52 S.F.	1	1
Roof Top Unit Multizone	240 Lbs.	104 S.F.	2	1
Self-contained Air or Water Cooled	108 Lbs.	—	2	—
Split System Air Cooled	102 Lbs.	—	2	—

Systems reflect most common usage.
Refer to system graphics for duct layout.

R15850-060 Diffuser Evaluation

CFM = V × An × K where V = Outlet velocity in feet per minute. An = Neck area in square feet and K = Diffuser delivery factor. An undersized diffuser for a desired CFM will produce a high velocity and noise level. When air moves past people at a velocity in excess of 25 FPM, an annoying draft is felt. An oversized diffuser will result in low velocity with poor mixing. Consideration must be given to avoid vertical stratification or horizontal areas of stagnation.

R17100-100 Square Foot Project Size Modifier

One factor that affects the S.F. cost of a particular building is the size. In general, for buildings built to the same specifications in the same locality, the larger building will have the lower S.F. cost. This is due mainly to the decreasing contribution of the exterior walls plus the economy of scale usually achievable in larger buildings. The Area Conversion Scale shown below will give a factor to convert costs for the typical size building to an adjusted cost for the particular project.

Example: Determine the cost per S.F. for a 100,000 S.F. Mid-rise apartment building.

$$\frac{\text{Proposed building area} = 100{,}000 \text{ S.F.}}{\text{Typical size from below} = 50{,}000 \text{ S.F.}} = 2.00$$

Enter Area Conversion scale at 2.0, intersect curve, read horizontally the appropriate cost multiplier of .94. Size adjusted cost becomes .94 x $85.50 = $80.35 based on national average costs.

Note: For Size Factors less than .50, the Cost Multiplier is 1.1
For Size Factors greater than 3.5, the Cost Multiplier is .90

The Square Foot Base Size lists the median costs, most typical project size in our accumulated data and the range in size of the projects.

The Size Factor for your project is determined by dividing your project area in S.F. by the typical project size for the particular Building Type. With this factor, enter the Area Conversion Scale at the appropriate Size Factor and determine the appropriate cost multiplier for your building size.

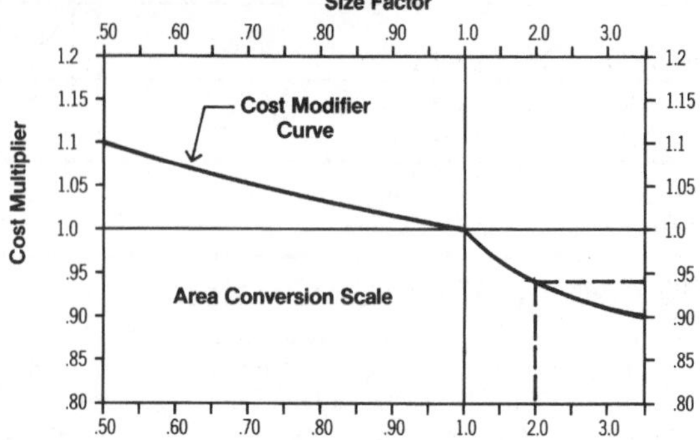

Square Foot Base Size

Building Type	Median Cost per S.F.	Typical Size Gross S.F.	Typical Range Gross S.F.	Building Type	Median Cost per S.F.	Typical Size Gross S.F.	Typical Range Gross S.F.
Apartments, Low Rise	$ 67.50	21,000	9,700 - 37,200	Jails	$205.00	40,000	5,500 - 145,000
Apartments, Mid Rise	85.50	50,000	32,000 - 100,000	Libraries	126.00	12,000	7,000 - 31,000
Apartments, High Rise	97.00	145,000	95,000 - 600,000	Living, Assisted	111.00	32,300	23,500 - 50,300
Auditoriums	113.00	25,000	7,600 - 39,000	Medical Clinics	117.00	7,200	4,200 - 15,700
Auto Sales	84.00	20,000	10,800 - 28,600	Medical Offices	110.00	6,000	4,000 - 15,000
Banks	151.00	4,200	2,500 - 7,500	Motels	82.50	40,000	15,800 - 120,000
Churches	103.00	17,000	2,000 - 42,000	Nursing Homes	114.00	23,000	15,000 - 37,000
Clubs, Country	106.00	6,500	4,500 - 15,000	Offices, Low Rise	95.00	20,000	5,000 - 80,000
Clubs, Social	100.00	10,000	6,000 - 13,500	Offices, Mid Rise	95.00	120,000	20,000 - 300,000
Clubs, YMCA	119.00	28,300	12,800 - 39,400	Offices, High Rise	122.00	260,000	120,000 - 800,000
Colleges (Class)	131.00	50,000	15,000 - 150,000	Police Stations	152.00	10,500	4,000 - 19,000
Colleges (Science Lab)	193.00	45,600	16,600 - 80,000	Post Offices	113.00	12,400	6,800 - 30,000
College (Student Union)	148.00	33,400	16,000 - 85,000	Power Plants	835.00	7,500	1,000 - 20,000
Community Center	107.00	9,400	5,300 - 16,700	Religious Education	94.00	9,000	6,000 - 12,000
Court Houses	144.00	32,400	17,800 - 106,000	Research	158.00	19,000	6,300 - 45,000
Dept. Stores	62.50	90,000	44,000 - 122,000	Restaurants	137.00	4,400	2,800 - 6,000
Dormitories, Low Rise	111.00	25,000	10,000 - 95,000	Retail Stores	67.00	7,200	4,000 - 17,600
Dormitories, Mid Rise	141.00	85,000	20,000 - 200,000	Schools, Elementary	99.50	41,000	24,500 - 55,000
Factories	61.00	26,400	12,900 - 50,000	Schools, Jr. High	102.00	92,000	52,000 - 119,000
Fire Stations	107.00	5,800	4,000 - 8,700	Schools, Sr. High	108.00	101,000	50,500 - 175,000
Fraternity Houses	104.00	12,500	8,200 - 14,800	Schools, Vocational	100.00	37,000	20,500 - 82,000
Funeral Homes	117.00	10,000	4,000 - 20,000	Sports Arenas	82.50	15,000	5,000 - 40,000
Garages, Commercial	76.00	9,300	5,000 - 13,600	Supermarkets	66.00	44,000	12,000 - 60,000
Garages, Municipal	95.00	8,300	4,500 - 12,600	Swimming Pools	155.00	20,000	10,000 - 32,000
Garages, Parking	40.00	163,000	76,400 - 225,300	Telephone Exchange	180.00	4,500	1,200 - 10,600
Gymnasiums	102.00	19,200	11,600 - 41,000	Theaters	99.00	10,500	8,800 - 17,500
Hospitals	181.00	55,000	27,200 - 125,000	Town Halls	116.00	10,800	4,800 - 23,400
House (Elderly)	92.00	37,000	21,000 - 66,000	Warehouses	49.50	25,000	8,000 - 72,000
Housing (Public)	85.00	36,000	14,400 - 74,400	Warehouse & Office	52.50	25,000	8,000 - 72,000
Ice Rinks	122.00	29,000	27,200 - 33,600				

Change Orders

Change Order Considerations

A Change Order is a written document, usually prepared by the design professional, and signed by the owner, the architect/engineer and the contractor. A change order states the agreement of the parties to: an addition, deletion, or revision in the work; an adjustment in the contract sum, if any; or an adjustment in the contract time, if any. Change orders, or "extras" in the construction process occur after execution of the construction contract and impact architects/engineers, contractors and owners.

Change orders that are properly recognized and managed can ensure orderly, professional and profitable progress for all who are involved in the project. There are many causes for change orders and change order requests. In all cases, change orders or change order requests should be addressed promptly and in a precise and prescribed manner. The following paragraphs include information regarding change order pricing and procedures.

The Causes of Change Orders

Reasons for issuing change orders include:

- Unforeseen field conditions that require a change in the work
- Correction of design discrepancies, errors or omissions in the contract documents
- Owner-requested changes, either by design criteria, scope of work, or project objectives
- Completion date changes for reasons unrelated to the construction process
- Changes in building code interpretations, or other public authority requirements that require a change in the work
- Changes in availability of existing or new materials and products

Procedures

Properly written contract documents must include the correct change order procedures for all parties—owners, design professionals and contractors—to follow in order to avoid costly delays and litigation.

Being "in the right" is not always a sufficient or acceptable defense. The contract provisions requiring notification and documentation must be adhered to within a defined or reasonable time frame.

The appropriate method of handling change orders is by a written proposal and acceptance by all parties involved. Prior to starting work on a project, all parties should identify their authorized agents who may sign and accept change orders, as well as any limits placed on their authority.

Time may be a critical factor when the need for a change arises. For such cases, the contractor might be directed to proceed on a "time and materials" basis, rather than wait for all paperwork to be processed—a delay that could impede progress. In this situation, the contractor must still follow the prescribed change order procedures, including but not limited to, notification and documentation.

All forms used for change orders should be dated and signed by the proper authority. Lack of documentation can be very costly, especially if legal judgments are to be made and if certain field personnel are no longer available. For time and material change orders, the contractor should keep accurate daily records of all labor and material allocated to the change. Forms that can be used to document change order work are available in *Means Forms for Building Construction Professionals.*

Owners or awarding authorities who do considerable and continual building construction (such as the federal government) realize the inevitability of change orders for numerous reasons, both predictable and unpredictable. As a result, the federal government, the American Institute of Architects (AIA), the Engineers Joint Contract Documents Committee (EJCDC) and other contractor, legal and technical organizations have developed standards and procedures to be followed by all parties to achieve contract continuance and timely completion, while being financially fair to all concerned.

In addition to the change order standards put forth by industry associations, there are also many books available on the subject.

Pricing Change Orders

When pricing change orders, regardless of their cause, the most significant factor is *when* the change occurs. The need for a change may be perceived in the field or requested by the architect/engineer *before* any of the actual installation has begun, or may evolve or appear *during* construction when the item of work in question is partially installed. In the latter cases, the original sequence of construction is disrupted, along with all contiguous and supporting systems. Change orders cause the greatest impact when they occur *after* the installation has been completed and must be uncovered, or even replaced. Post-completion changes may be caused by necessary design changes, product failure, or changes in the owner's requirements that are not discovered until the building or the systems begin to function.

Specified procedures of notification and record keeping must be adhered to and enforced regardless of the stage of construction: *before, during,* or *after* installation. Some bidding documents anticipate change orders by requiring that unit prices including overhead and profit percentages—for additional as well as deductible changes—be listed. Generally these unit prices do not fully take into account the ripple effect, or impact on other trades, and should be used for general guidance only.

When pricing change orders, it is important to classify the time frame in which the change occurs. There are two basic time frames for change orders: *pre-installation change orders*, which occur before the start of construction, and *post-installation change orders*, which involve reworking after the original installation. Change orders that occur between these stages may be priced according to the extent of work completed using a combination of techniques developed for pricing *pre-* and *post-installation* changes.

The following factors are the basis for a check list to use when preparing a change order estimate.

Factors To Consider When Pricing Change Orders

As an estimator begins to prepare a change order, the following questions should be reviewed to determine their impact on the final price.

General

- Is the change order work *pre-installation* or *post-installation*?

  Change order work costs vary according to how much of the installation has been completed. Once workers have the project scoped in their mind, even though they have not started, it can be difficult to refocus. Consequently they may spend more than the normal amount of time understanding the change. Also, modifications to work in place such as trimming or refitting usually take more time than was initially estimated. The greater the amount of work in place, the more reluctant workers are to change it. Psychologically they may resent the change and as a result the rework takes longer than normal. Post-installation change order estimates must include demolition of existing work as required to accomplish the change. If the work is performed at a later time, additional obstacles such as building finishes may be present which must be protected. Regardless of whether the change occurs pre-installation or post-installation, attempt to isolate the identifiable factors and price them separately. For example, add shipping costs that may be required pre-installation or any demolition required post-installation. Then analyze the potential impact on productivity of psychological and/or learning curve factors and adjust the output rates accordingly. One approach is to break down the typical workday into segments and quantify the impact on each segment. The following chart may be useful as a guide:

Activities (Productivity) Expressed as Percentages of a Workday			
Task	Means Mechanical Cost Data (for New Construction)	Pre-Installation Change Orders	Post-Installation Change Orders
1. Study plans	3%	6%	6%
2. Material procurement	3%	3%	3%
3. Receiving and storing	3%	3%	3%
4. Mobilization	5%	5%	5%
5. Site movement	5%	5%	8%
6. Layout and marking	8%	10%	12%
7. Actual installation	64%	59%	54%
8. Clean-up	3%	3%	3%
9. Breaks—non-productive	6%	6%	6%
Total	100%	100%	100%

Change Order Installation Efficiency

The labor-hours expressed (for new construction) are based on average installation time, using an efficiency level of approximately 60-65%. For change order situations, adjustments to this efficiency level should reflect the daily labor-hour allocation for that particular occurrence.

If any of the specific percentages expressed in the above chart do not apply to a particular project situation, then those percentage points should be reallocated to the appropriate task(s). Example: Using data for new construction, assume there is no new material being utilized. The percentages for Tasks 2 and 3 would therefore be reallocated to other tasks. If the time required for Tasks 2 and 3 can now be applied to installation, we can add the time allocated for *Material Procurement* and *Receiving and Storing* to the *Actual Installation* time for new construction, thereby increasing the Actual Installation percentage.

This chart shows that, due to reduced productivity, labor costs will be higher than those for new construction by 5% to 15% for pre-installation change orders and by 15% to 25% for post-installation change orders. Each job and change order is unique and must be examined individually. Many factors, covered elsewhere in this section, can each have a significant impact on productivity and change order costs. All such factors should be considered in every case.

- Will the change substantially delay the original completion date?

  A significant change in the project may cause the original completion date to be extended. The extended schedule may subject the contractor to new wage rates dictated by relevant labor contracts. Project supervision and other project overhead must also be extended beyond the original completion date. The schedule extension may also put installation into a new weather season. For example, underground piping scheduled for October installation was delayed until January. As a result, frost penetrated the trench area, thereby changing the degree of difficulty of the task. Changes and delays may have a ripple effect throughout the project. This effect must be analyzed and negotiated with the owner.

- What is the net effect of a deduct change order?

  In most cases, change orders resulting in a deduction or credit reflect only bare costs. The contractor may retain the overhead and profit based on the original bid.

Materials

- Will you have to pay more or less for the new material, required by the change order, than you paid for the original purchase?

  The same material prices or discounts will usually apply to materials purchased for change orders as new construction. In some instances, however, the contractor may forfeit the advantages of competitive pricing for change orders. Consider the following example:

  A contractor purchased over $20,000 worth of fan coil units for an installation, and obtained the maximum discount. Some time later it was determined the project required an additional matching unit. The contractor has to purchase this unit from the original supplier to ensure a match. The supplier at this time may not discount the unit because of the small quantity, and the fact that he is no longer in a competitive situation. The impact of quantity on purchase can add between 0% and 25% to material prices and/or subcontractor quotes.

- If materials have been ordered or delivered to the job site, will they be subject to a cancellation charge or restocking fee?

  Check with the supplier to determine if ordered materials are subject to a cancellation charge. Delivered materials not used as result of a change order may be subject to a restocking fee if returned to the supplier. Common restocking charges run between 20% and 40%. Also, delivery charges to return the goods to the supplier must be added.

Labor

- How efficient is the existing crew at the actual installation?

  Is the same crew that performed the initial work going to do the change order? Possibly the change consists of the installation of a unit identical to one already installed; therefore the change should take less time. Be sure to consider this potential productivity increase and modify the productivity rates accordingly.

- If the crew size is increased, what impact will that have on supervision requirements?

  Under most bargaining agreements or management practices, there is a point at which a working foreman is replaced by a nonworking foreman. This replacement increases project overhead by adding a nonproductive worker. If additional workers are added to accelerate the project or to perform changes while maintaining the schedule, be sure to add additional supervision time if warranted. Calculate the hours involved and the additional cost directly if possible.

- What are the other impacts of increased crew size?

  The larger the crew, the greater the potential for productivity to decrease. Some of the factors that cause this productivity loss are: overcrowding (producing restrictive conditions in the working space), and possibly a shortage of any special tools and equipment required. Such factors affect not only the crew working on the elements directly involved in the change order, but other crews whose movement may also be hampered.

  As the crew increases, check its basic composition for changes by the addition or deletion of apprentices or nonworking foreman and quantify the potential effects of equipment shortages or other logistical factors.

- As new crews, unfamiliar with the project, are brought onto the site, how long will it take them to become oriented to the project requirements?

  The orientation time for a new crew to become 100% effective varies with the site and type of project. Orientation is easiest at a new construction site, and most difficult at existing, very restrictive renovation sites. The type of work also affects orientation time. When all elements of the work are exposed, such as concrete or masonry work, orientation is decreased. When the work is concealed or less visible, such as existing electrical systems, orientation takes longer. Usually orientation can be accomplished in one day or less. Costs for added orientation should be itemized and added to the total estimated cost.

- How much actual production can be gained by working overtime?

  Short term overtime can be used effectively to accomplish more work in a day. However, as overtime is scheduled to run beyond several weeks, studies have shown marked decreases in output. The following chart shows the effect of long term overtime on worker efficiency. If the anticipated change requires extended overtime to keep the job on schedule, these factors can be used as a guide to predict the impact on time and cost. Add project overhead, particularly supervision, that may also be incurred.

Days per Week	Hours per Day	Production Efficiency					Payroll Cost Factors	
		1 Week	2 Weeks	3 Weeks	4 Weeks	Average 4 Weeks	@ 1-1/2 Times	@ 2 Times
5	8	100%	100%	100%	100%	100%	100%	100%
	9	100	100	95	90	96.25	105.6	111.1
	10	100	95	90	85	91.25	110.0	120.0
	11	95	90	75	65	81.25	113.6	127.3
	12	90	85	70	60	76.25	116.7	133.3
6	8	100	100	95	90	96.25	108.3	116.7
	9	100	95	90	85	92.50	113.0	125.9
	10	95	90	85	80	87.50	116.7	133.3
	11	95	85	70	65	78.75	119.7	139.4
	12	90	80	65	60	73.75	122.2	144.4
7	8	100	95	85	75	88.75	114.3	128.6
	9	95	90	80	70	83.75	118.3	136.5
	10	90	85	75	65	78.75	121.4	142.9
	11	85	80	65	60	72.50	124.0	148.1
	12	85	75	60	55	68.75	126.2	152.4

Effects of Overtime

Caution: Under many labor agreements, Sundays and holidays are paid at a higher premium than the normal overtime rate.

The use of long-term overtime is counterproductive on almost any construction job; that is, the longer the period of overtime, the lower the actual production rate. Numerous studies have been conducted, and while they have resulted in slightly different numbers, all reach the same conclusion. The figure above tabulates the effects of overtime work on efficiency.

As illustrated, there can be a difference between the *actual* payroll cost per hour and the *effective* cost per hour for overtime work. This is due to the reduced production efficiency with the increase in weekly hours beyond 40. This difference between actual and effective cost results from overtime work over a prolonged period. Short-term overtime work does not result in as great a reduction in efficiency, and in such cases, effective cost may not vary significantly from the actual payroll cost. As the total hours per week are increased on a regular basis, more time is lost because of fatigue, lowered morale, and an increased accident rate.

As an example, assume a project where workers are working 6 days a week, 10 hours per day. From the figure above (based on productivity studies), the average effective productive hours over a four-week period are:

$$0.875 \times 60 = 52.5$$

Depending upon the locale and day of week, overtime hours may be paid at time and a half or double time. For time and a half, the overall (average) *actual* payroll cost (including regular and overtime hours) is determined as follows:

$$\frac{40 \text{ reg. hrs.} + (20 \text{ overtime hrs.} \times 1.5)}{60 \text{ hrs.}} = 1.167$$

Based on 60 hours, the payroll cost per hour will be 116.7% of the normal rate at 40 hours per week. However, because the effective production (efficiency) for 60 hours is reduced to the equivalent of 52.5 hours, the effective cost of overtime is calculated as follows:

For time and a half:

$$\frac{40 \text{ reg. hrs.} + (20 \text{ overtime hrs.} \times 1.5)}{52.5 \text{ hrs.}} = 1.33$$

Installed cost will be 133% of the normal rate (for labor).

Thus, when figuring overtime, the actual cost per unit of work will be higher than the apparent overtime payroll dollar increase, due to the reduced productivity of the longer workweek. These efficiency calculations are true only for those cost factors determined by hours worked. Costs that are applied weekly or monthly, such as equipment rentals, will not be similarly affected.

Equipment

- What equipment is required to complete the change order?

  Change orders may require extending the rental period of equipment already on the job site, or the addition of special equipment brought in to accomplish the change work. In either case, the additional rental charges and operator labor charges must be added.

Summary

The preceding considerations and others you deem appropriate should be analyzed and applied to a change order estimate. The impact of each should be quantified and listed on the estimate to form an audit trail.

Change orders that are properly identified, documented, and managed help to ensure the orderly, professional and profitable progress of the work. They also minimize potential claims or disputes at the end of the project.

Crews

Crew No.	Bare Costs		Incl. Subs O & P		Cost Per Labor-Hour	
Crew A-1	**Hr.**	**Daily**	**Hr.**	**Daily**	**Bare Costs**	**Incl. O&P**
1 Building Laborer	$26.70	$213.60	$41.55	$332.40	$26.70	$41.55
1 Concrete saw, gas manual		49.80		54.80	6.23	6.85
8 L.H., Daily Totals		$263.40		$387.20	$32.93	$48.40
Crew A-1A	**Hr.**	**Daily**	**Hr.**	**Daily**	**Bare Costs**	**Incl. O&P**
1 Skilled Worker	$34.85	$278.80	$54.20	$433.60	$34.85	$54.20
1 Shot Blaster, 20"		338.35		372.20	42.29	46.52
8 L.H., Daily Totals		$617.15		$805.80	$77.14	$100.72
Crew A-1B	**Hr.**	**Daily**	**Hr.**	**Daily**	**Bare Costs**	**Incl. O&P**
1 Building Laborer	$26.70	$213.60	$41.55	$332.40	$26.70	$41.55
1 Concr. saw, gas, self-prop.		114.40		125.85	14.30	15.73
8 L.H., Daily Totals		$328.00		$458.25	$41.00	$57.28
Crew A-1C	**Hr.**	**Daily**	**Hr.**	**Daily**	**Bare Costs**	**Incl. O&P**
1 Building Laborer	$26.70	$213.60	$41.55	$332.40	$26.70	$41.55
1 Brush saw		19.20		21.10	2.40	2.64
8 L.H., Daily Totals		$232.80		$353.50	$29.10	$44.19
Crew A-1D	**Hr.**	**Daily**	**Hr.**	**Daily**	**Bare Costs**	**Incl. O&P**
1 Building Laborer	$26.70	$213.60	$41.55	$332.40	$26.70	$41.55
1 Vibrating plate, gas, 18"		25.20		27.70	3.15	3.47
8 L.H., Daily Totals		$238.80		$360.10	$29.85	$45.02
Crew A-1E	**Hr.**	**Daily**	**Hr.**	**Daily**	**Bare Costs**	**Incl. O&P**
1 Building Laborer	$26.70	$213.60	$41.55	$332.40	$26.70	$41.55
1 Vibrating plate, gas, 21"		36.80		40.50	4.60	5.06
8 L.H., Daily Totals		$250.40		$372.90	$31.30	$46.61
Crew A-1F	**Hr.**	**Daily**	**Hr.**	**Daily**	**Bare Costs**	**Incl. O&P**
1 Building Laborer	$26.70	$213.60	$41.55	$332.40	$26.70	$41.55
1 Rammer/tamper, gas, 8"		34.80		38.30	4.35	4.79
8 L.H., Daily Totals		$248.40		$370.70	$31.05	$46.34
Crew A-1G	**Hr.**	**Daily**	**Hr.**	**Daily**	**Bare Costs**	**Incl. O&P**
1 Building Laborer	$26.70	$213.60	$41.55	$332.40	$26.70	$41.55
1 Rammer/tamper, gas, 8"		34.80		38.30	4.35	4.79
8 L.H., Daily Totals		$248.40		$370.70	$31.05	$46.34
Crew A-1H	**Hr.**	**Daily**	**Hr.**	**Daily**	**Bare Costs**	**Incl. O&P**
1 Building Laborer	$26.70	$213.60	$41.55	$332.40	$26.70	$41.55
1 Pressure washer		55.00		60.50	6.88	7.56
8 L.H., Daily Totals		$268.60		$392.90	$33.58	$49.11
Crew A-1J	**Hr.**	**Daily**	**Hr.**	**Daily**	**Bare Costs**	**Incl. O&P**
1 Building Laborer	$26.70	$213.60	$41.55	$332.40	$26.70	$41.55
1 Rototiller		84.50		92.95	10.56	11.61
8 L.H., Daily Totals		$298.10		$425.35	$37.26	$53.16
Crew A-1K	**Hr.**	**Daily**	**Hr.**	**Daily**	**Bare Costs**	**Incl. O&P**
1 Building Laborer	$26.70	$213.60	$41.55	$332.40	$26.70	$41.55
1 Lawn aerator		23.95		26.35	2.99	3.29
8 L.H., Daily Totals		$237.55		$358.75	$29.69	$44.84
Crew A-1L	**Hr.**	**Daily**	**Hr.**	**Daily**	**Bare Costs**	**Incl. O&P**
1 Building Laborer	$26.70	$213.60	$41.55	$332.40	$26.70	$41.55
1 Power blower/vacuum		23.95		26.35	2.99	3.29
8 L.H., Daily Totals		$237.55		$358.75	$29.69	$44.84

Crew No.	Bare Costs		Incl. Subs O & P		Cost Per Labor-Hour	
Crew A-1M	**Hr.**	**Daily**	**Hr.**	**Daily**	**Bare Costs**	**Incl. O&P**
1 Building Laborer	$26.70	$213.60	$41.55	$332.40	$26.70	$41.55
1 Snow blower		84.50		92.95	10.56	11.61
8 L.H., Daily Totals		$298.10		$425.35	$37.26	$53.16
Crew A-2	**Hr.**	**Daily**	**Hr.**	**Daily**	**Bare Costs**	**Incl. O&P**
2 Laborers	$26.70	$427.20	$41.55	$664.80	$26.73	$41.32
1 Truck Driver (light)	26.80	214.40	40.85	326.80		
1 Light Truck, 1.5 Ton		128.80		141.70	5.37	5.90
24 L.H., Daily Totals		$770.40		$1133.30	$32.10	$47.22
Crew A-2A	**Hr.**	**Daily**	**Hr.**	**Daily**	**Bare Costs**	**Incl. O&P**
2 Laborers	$26.70	$427.20	$41.55	$664.80	$26.73	$41.32
1 Truck Driver (light)	26.80	214.40	40.85	326.80		
1 Light Truck, 1.5 Ton		128.80		141.70		
1 Concrete Saw		114.40		125.85	10.13	11.15
24 L.H., Daily Totals		$884.80		$1259.15	$36.86	$52.47
Crew A-3	**Hr.**	**Daily**	**Hr.**	**Daily**	**Bare Costs**	**Incl. O&P**
1 Truck Driver (heavy)	$27.55	$220.40	$42.00	$336.00	$27.55	$42.00
1 Dump Truck, 12 Ton		326.60		359.25	40.83	44.91
8 L.H., Daily Totals		$547.00		$695.25	$68.38	$86.91
Crew A-3A	**Hr.**	**Daily**	**Hr.**	**Daily**	**Bare Costs**	**Incl. O&P**
1 Truck Driver (light)	$26.80	$214.40	$40.85	$326.80	$26.80	$40.85
1 Pickup Truck (4x4)		85.20		93.70	10.65	11.72
8 L.H., Daily Totals		$299.60		$420.50	$37.45	$52.57
Crew A-3B	**Hr.**	**Daily**	**Hr.**	**Daily**	**Bare Costs**	**Incl. O&P**
1 Equip. Oper. (medium)	$34.65	$277.20	$52.20	$417.60	$31.10	$47.10
1 Truck Driver (heavy)	27.55	220.40	42.00	336.00		
1 Dump Truck, 16 Ton		476.60		524.25		
1 F.E. Loader, 3 C.Y.		305.00		335.50	48.85	53.74
16 L.H., Daily Totals		$1279.20		$1613.35	$79.95	$100.84
Crew A-3C	**Hr.**	**Daily**	**Hr.**	**Daily**	**Bare Costs**	**Incl. O&P**
1 Equip. Oper. (light)	$33.05	$264.40	$49.80	$398.40	$33.05	$49.80
1 Wheeled Skid Steer Loader		197.80		217.60	24.73	27.20
8 L.H., Daily Totals		$462.20		$616.00	$57.78	$77.00
Crew A-3D	**Hr.**	**Daily**	**Hr.**	**Daily**	**Bare Costs**	**Incl. O&P**
1 Truck Driver, Light	$26.80	$214.40	$40.85	$326.80	$26.80	$40.85
1 Pickup Truck (4x4)		85.20		93.70		
1 Flatbed Trailer, 25 Ton		88.00		96.80	21.65	23.82
8 L.H., Daily Totals		$387.60		$517.30	$48.45	$64.67
Crew A-3E	**Hr.**	**Daily**	**Hr.**	**Daily**	**Bare Costs**	**Incl. O&P**
1 Equip. Oper. (crane)	$35.90	$287.20	$54.10	$432.80	$31.73	$48.05
1 Truck Driver (heavy)	27.55	220.40	42.00	336.00		
1 Pickup Truck (4x4)		85.20		93.70	5.33	5.86
16 L.H., Daily Totals		$592.80		$862.50	$37.06	$53.91
Crew A-3F	**Hr.**	**Daily**	**Hr.**	**Daily**	**Bare Costs**	**Incl. O&P**
1 Equip. Oper. (crane)	$35.90	$287.20	$54.10	$432.80	$31.73	$48.05
1 Truck Driver (heavy)	27.55	220.40	42.00	336.00		
1 Pickup Truck (4x4)		85.20		93.70		
1 Tractor, 6x2, 40 Ton Cap.		301.60		331.75		
1 Lowbed Trailer, 75 Ton		170.20		187.20	34.81	38.29
16 L.H., Daily Totals		$1064.60		$1381.45	$66.54	$86.34

Crew No.	Bare Costs		Incl. Subs O & P		Cost Per Labor-Hour	
Crew A-3G	**Hr.**	**Daily**	**Hr.**	**Daily**	**Bare Costs**	**Incl. O&P**
1 Equip. Oper. (crane)	$35.90	$287.20	$54.10	$432.80	$31.73	$48.05
1 Truck Driver (heavy)	27.55	220.40	42.00	336.00		
1 Pickup Truck (4x4)		85.20		93.70		
1 Tractor, 6x4, 45 Ton Cap.		353.60		388.95		
1 Lowbed Trailer, 75 Ton		170.20		187.20	38.06	41.87
16 L.H., Daily Totals		$1116.60		$1438.65	$69.79	$89.92
Crew A-4	**Hr.**	**Daily**	**Hr.**	**Daily**	**Bare Costs**	**Incl. O&P**
2 Carpenters	$34.25	$548.00	$53.35	$853.60	$33.03	$50.90
1 Painter, Ordinary	30.60	244.80	46.00	368.00		
24 L.H., Daily Totals		$792.80		$1221.60	$33.03	$50.90
Crew A-5	**Hr.**	**Daily**	**Hr.**	**Daily**	**Bare Costs**	**Incl. O&P**
2 Laborers	$26.70	$427.20	$41.55	$664.80	$26.71	$41.47
.25 Truck Driver (light)	26.80	53.60	40.85	81.70		
.25 Light Truck, 1.5 Ton		32.20		35.40	1.79	1.97
18 L.H., Daily Totals		$513.00		$781.90	$28.50	$43.44
Crew A-6	**Hr.**	**Daily**	**Hr.**	**Daily**	**Bare Costs**	**Incl. O&P**
1 Instrument Man	$34.85	$278.80	$54.20	$433.60	$33.93	$52.05
1 Rodman/Chainman	33.00	264.00	49.90	399.20		
1 Laser Transit/Level		60.75		66.85	3.79	4.17
16 L.H., Daily Totals		$603.55		$899.65	$37.72	$56.22
Crew A-7	**Hr.**	**Daily**	**Hr.**	**Daily**	**Bare Costs**	**Incl. O&P**
1 Chief Of Party	$42.90	$343.20	$66.90	$535.20	$36.92	$57.00
1 Instrument Man	34.85	278.80	54.20	433.60		
1 Rodman/Chainman	33.00	264.00	49.90	399.20		
1 Laser Transit/Level		60.75		66.85	2.53	2.78
24 L.H., Daily Totals		$946.75		$1434.85	$39.45	$59.78
Crew A-8	**Hr.**	**Daily**	**Hr.**	**Daily**	**Bare Costs**	**Incl. O&P**
1 Chief Of Party	$42.90	$343.20	$66.90	$535.20	$35.94	$55.23
1 Instrument Man	34.85	278.80	54.20	433.60		
2 Rodmen/Chainmen	33.00	528.00	49.90	798.40		
1 Laser Transit/Level		60.75		66.85	1.90	2.09
32 L.H., Daily Totals		$1210.75		$1834.05	$37.84	$57.32
Crew A-9	**Hr.**	**Daily**	**Hr.**	**Daily**	**Bare Costs**	**Incl. O&P**
1 Asbestos Foreman	$37.60	$300.80	$59.35	$474.80	$37.16	$58.65
7 Asbestos Workers	37.10	2077.60	58.55	3278.80		
64 L.H., Daily Totals		$2378.40		$3753.60	$37.16	$58.65
Crew A-10	**Hr.**	**Daily**	**Hr.**	**Daily**	**Bare Costs**	**Incl. O&P**
1 Asbestos Foreman	$37.60	$300.80	$59.35	$474.80	$37.16	$58.65
7 Asbestos Workers	37.10	2077.60	58.55	3278.80		
64 L.H., Daily Totals		$2378.40		$3753.60	$37.16	$58.65
Crew A-10A	**Hr.**	**Daily**	**Hr.**	**Daily**	**Bare Costs**	**Incl. O&P**
1 Asbestos Foreman	$37.60	$300.80	$59.35	$474.80	$37.27	$58.82
2 Asbestos Workers	37.10	593.60	58.55	936.80		
24 L.H., Daily Totals		$894.40		$1411.60	$37.27	$58.82
Crew A-10B	**Hr.**	**Daily**	**Hr.**	**Daily**	**Bare Costs**	**Incl. O&P**
1 Asbestos Foreman	$37.60	$300.80	$59.35	$474.80	$37.23	$58.75
3 Asbestos Workers	37.10	890.40	58.55	1405.20		
32 L.H., Daily Totals		$1191.20		$1880.00	$37.23	$58.75

Crew No.	Bare Costs		Incl. Subs O & P		Cost Per Labor-Hour	
Crew A-10C	**Hr.**	**Daily**	**Hr.**	**Daily**	**Bare Costs**	**Incl. O&P**
3 Asbestos Workers	$37.10	$890.40	$58.55	$1405.20	$37.10	$58.55
1 Flatbed Truck		128.80		141.70	5.37	5.90
24 L.H., Daily Totals		$1019.20		$1546.90	$42.47	$64.45
Crew A-10D	**Hr.**	**Daily**	**Hr.**	**Daily**	**Bare Costs**	**Incl. O&P**
2 Asbestos Workers	$37.10	$593.60	$58.55	$936.80	$35.05	$54.14
1 Equip. Oper. (crane)	35.90	287.20	54.10	432.80		
1 Equip. Oper. Oiler	30.10	240.80	45.35	362.80		
1 Hydraulic Crane, 33 Ton		641.60		705.75	20.05	22.06
32 L.H., Daily Totals		$1763.20		$2438.15	$55.10	$76.20
Crew A-11	**Hr.**	**Daily**	**Hr.**	**Daily**	**Bare Costs**	**Incl. O&P**
1 Asbestos Foreman	$37.60	$300.80	$59.35	$474.80	$37.16	$58.65
7 Asbestos Workers	37.10	2077.60	58.55	3278.80		
2 Chipping Hammers		32.00		35.20	.50	.55
64 L.H., Daily Totals		$2410.40		$3788.80	$37.66	$59.20
Crew A-12	**Hr.**	**Daily**	**Hr.**	**Daily**	**Bare Costs**	**Incl. O&P**
1 Asbestos Foreman	$37.60	$300.80	$59.35	$474.80	$37.16	$58.65
7 Asbestos Workers	37.10	2077.60	58.55	3278.80		
1 Large Prod. Vac. Loader		480.00		528.00	7.50	8.25
64 L.H., Daily Totals		$2858.40		$4281.60	$44.66	$66.90
Crew A-13	**Hr.**	**Daily**	**Hr.**	**Daily**	**Bare Costs**	**Incl. O&P**
1 Equip. Oper. (light)	$33.05	$264.40	$49.80	$398.40	$33.05	$49.80
1 Large Prod. Vac. Loader		480.00		528.00	60.00	66.00
8 L.H., Daily Totals		$744.40		$926.40	$93.05	$115.80
Crew B-1	**Hr.**	**Daily**	**Hr.**	**Daily**	**Bare Costs**	**Incl. O&P**
1 Labor Foreman (outside)	$28.70	$229.60	$44.70	$357.60	$27.37	$42.60
2 Laborers	26.70	427.20	41.55	664.80		
24 L.H., Daily Totals		$656.80		$1022.40	$27.37	$42.60
Crew B-1A	**Hr.**	**Daily**	**Hr.**	**Daily**	**Bare Costs**	**Incl. O&P**
1 Laborer Foreman	$28.70	$229.60	$44.70	$357.60	$27.37	$42.60
2 Laborers	26.70	427.20	41.55	664.80		
2 Cutting Torches		36.00		39.60		
2 Gases		129.60		142.55	6.90	7.59
24 L.H., Daily Totals		$822.40		$1204.55	$34.27	$50.19
Crew B-1B	**Hr.**	**Daily**	**Hr.**	**Daily**	**Bare Costs**	**Incl. O&P**
1 Laborer Foreman	$28.70	$229.60	$44.70	$357.60	$29.50	$45.48
2 Laborers	26.70	427.20	41.55	664.80		
1 Equip. Oper. (crane)	35.90	287.20	54.10	432.80		
2 Cutting Torches		36.00		39.60		
2 Gases		129.60		142.55		
1 Hyd. Crane, 12 Ton		602.60		662.85	24.01	26.41
32 L.H., Daily Totals		$1712.20		$2300.20	$53.51	$71.89
Crew B-2	**Hr.**	**Daily**	**Hr.**	**Daily**	**Bare Costs**	**Incl. O&P**
1 Labor Foreman (outside)	$28.70	$229.60	$44.70	$357.60	$27.10	$42.18
4 Laborers	26.70	854.40	41.55	1329.60		
40 L.H., Daily Totals		$1084.00		$1687.20	$27.10	$42.18

Crews

Crew No.	Bare Costs		Incl. Subs O & P		Cost Per Labor-Hour	
Crew B-3	**Hr.**	**Daily**	**Hr.**	**Daily**	**Bare Costs**	**Incl. O&P**
1 Labor Foreman (outside)	$28.70	$229.60	$44.70	$357.60	$28.64	$44.00
2 Laborers	26.70	427.20	41.55	664.80		
1 Equip. Oper. (med.)	34.65	277.20	52.20	417.60		
2 Truck Drivers (heavy)	27.55	440.80	42.00	672.00		
1 F.E. Loader, T.M., 2.5 C.Y.		788.00		866.80		
2 Dump Trucks, 16 Ton		953.20		1048.50	36.28	39.90
48 L.H., Daily Totals		$3116.00		$4027.30	$64.92	$83.90
Crew B-3A	**Hr.**	**Daily**	**Hr.**	**Daily**	**Bare Costs**	**Incl. O&P**
4 Laborers	$26.70	$854.40	$41.55	$1329.60	$28.29	$43.68
1 Equip. Oper. (med.)	34.65	277.20	52.20	417.60		
1 Hyd. Excavator, 1.5 C.Y.		720.40		792.45	18.01	19.81
40 L.H., Daily Totals		$1852.00		$2539.65	$46.30	$63.49
Crew B-3B	**Hr.**	**Daily**	**Hr.**	**Daily**	**Bare Costs**	**Incl. O&P**
2 Laborers	$26.70	$427.20	$41.55	$664.80	$28.90	$44.33
1 Equip. Oper. (med.)	34.65	277.20	52.20	417.60		
1 Truck Driver (heavy)	27.55	220.40	42.00	336.00		
1 Backhoe Loader, 80 H.P.		256.40		282.05		
1 Dump Truck, 16 Ton		476.60		524.25	22.91	25.20
32 L.H., Daily Totals		$1657.80		$2224.70	$51.81	$69.53
Crew B-3C	**Hr.**	**Daily**	**Hr.**	**Daily**	**Bare Costs**	**Incl. O&P**
3 Laborers	$26.70	$640.80	$41.55	$997.20	$28.69	$44.21
1 Equip. Oper. (med.)	34.65	277.20	52.20	417.60		
1 F.E. Crawler Ldr, 4 C.Y.		1094.00		1203.40	34.19	37.61
32 L.H., Daily Totals		$2012.00		$2618.20	$62.88	$81.82
Crew B-4	**Hr.**	**Daily**	**Hr.**	**Daily**	**Bare Costs**	**Incl. O&P**
1 Labor Foreman (outside)	$28.70	$229.60	$44.70	$357.60	$27.18	$42.15
4 Laborers	26.70	854.40	41.55	1329.60		
1 Truck Driver (heavy)	27.55	220.40	42.00	336.00		
1 Tractor, 4 x 2, 195 H.P.		208.20		229.00		
1 Platform Trailer		119.40		131.35	6.83	7.51
48 L.H., Daily Totals		$1632.00		$2383.55	$34.01	$49.66
Crew B-5	**Hr.**	**Daily**	**Hr.**	**Daily**	**Bare Costs**	**Incl. O&P**
1 Labor Foreman (outside)	$28.70	$229.60	$44.70	$357.60	$29.26	$45.04
4 Laborers	26.70	854.40	41.55	1329.60		
2 Equip. Oper. (med.)	34.65	554.40	52.20	835.20		
1 Air Compr., 250 C.F.M.		112.40		123.65		
2 Air Tools & Accessories		19.60		21.55		
2-50 Ft. Air Hoses, 1.5" Dia.		9.40		10.35		
1 F.E. Loader, T.M., 2.5 C.Y.		788.00		866.80	16.60	18.26
56 L.H., Daily Totals		$2567.80		$3544.75	$45.86	$63.30
Crew B-5A	**Hr.**	**Daily**	**Hr.**	**Daily**	**Bare Costs**	**Incl. O&P**
1 Foreman	$28.70	$229.60	$44.70	$357.60	$28.86	$44.35
6 Laborers	26.70	1281.60	41.55	1994.40		
2 Equip. Oper. (med.)	34.65	554.40	52.20	835.20		
1 Equip. Oper. (light)	33.05	264.40	49.80	398.40		
2 Truck Drivers (heavy)	27.55	440.80	42.00	672.00		
1 Air Compr. 365 C.F.M.		146.80		161.50		
2 Pavement Breakers		19.60		21.55		
8 Air Hoses w/Coup.,1"		29.20		32.10		
2 Dump Trucks, 12 Ton		653.20		718.50	8.84	9.73
96 L.H., Daily Totals		$3619.60		$5191.25	$37.70	$54.08

Crew No.	Bare Costs		Incl. Subs O & P		Cost Per Labor-Hour	
Crew B-5B	**Hr.**	**Daily**	**Hr.**	**Daily**	**Bare Costs**	**Incl. O&P**
1 Powderman	$34.85	$278.80	$54.20	$433.60	$31.13	$47.43
2 Equip. Oper. (med.)	34.65	554.40	52.20	835.20		
3 Truck Drivers (heavy)	27.55	661.20	42.00	1008.00		
1 F.E. Ldr. 2-1/2 CY		305.00		335.50		
3 Dump Trucks, 16 Ton		1429.80		1572.80		
1 Air Compr. 365 C.F.M.		146.80		161.50	39.20	43.12
48 L.H., Daily Totals		$3376.00		$4346.60	$70.33	$90.55
Crew B-5C	**Hr.**	**Daily**	**Hr.**	**Daily**	**Bare Costs**	**Incl. O&P**
3 Laborers	$26.70	$640.80	$41.55	$997.20	$29.48	$45.04
1 Equip. Oper. (medium)	34.65	277.20	52.20	417.60		
2 Truck Drivers (heavy)	27.55	440.80	42.00	672.00		
1 Equip. Oper. (crane)	35.90	287.20	54.10	432.80		
1 Equip. Oper. Oiler	30.10	240.80	45.35	362.80		
2 Dump Trucks, 16 Ton		953.20		1048.50		
1 F.E. Crawler Ldr, 4 C.Y.		1094.00		1203.40		
1 Hyd. Crane, 25 Ton		575.00		632.50	40.97	45.07
64 L.H., Daily Totals		$4509.00		$5766.80	$70.45	$90.11
Crew B-6	**Hr.**	**Daily**	**Hr.**	**Daily**	**Bare Costs**	**Incl. O&P**
2 Laborers	$26.70	$427.20	$41.55	$664.80	$28.82	$44.30
1 Equip. Oper. (light)	33.05	264.40	49.80	398.40		
1 Backhoe Loader, 48 H.P.		215.80		237.40	8.99	9.89
24 L.H., Daily Totals		$907.40		$1300.60	$37.81	$54.19
Crew B-6A	**Hr.**	**Daily**	**Hr.**	**Daily**	**Bare Costs**	**Incl. O&P**
.5 Labor Foreman (outside)	$28.70	$114.80	$44.70	$178.80	$30.28	$46.44
1 Laborer	26.70	213.60	41.55	332.40		
1 Equip. Oper. (med.)	34.65	277.20	52.20	417.60		
1 Vacuum Trk.,5000 Gal.		331.50		364.65	16.58	18.23
20 L.H., Daily Totals		$937.10		$1293.45	$46.86	$64.67
Crew B-6B	**Hr.**	**Daily**	**Hr.**	**Daily**	**Bare Costs**	**Incl. O&P**
2 Labor Foreman (out)	$28.70	$459.20	$44.70	$715.20	$27.37	$42.60
4 Laborers	26.70	854.40	41.55	1329.60		
1 Winch Truck		317.00		348.70		
1 Flatbed Truck		128.80		141.70		
1 Butt Fusion Machine		432.80		476.10	18.30	20.13
48 L.H., Daily Totals		$2192.20		$3011.30	$45.67	$62.73
Crew B-7	**Hr.**	**Daily**	**Hr.**	**Daily**	**Bare Costs**	**Incl. O&P**
1 Labor Foreman (outside)	$28.70	$229.60	$44.70	$357.60	$28.36	$43.85
4 Laborers	26.70	854.40	41.55	1329.60		
1 Equip. Oper. (med.)	34.65	277.20	52.20	417.60		
1 Chipping Machine		164.80		181.30		
1 F.E. Loader, T.M., 2.5 C.Y.		788.00		866.80		
2 Chain Saws, 36"		66.80		73.50	21.24	23.37
48 L.H., Daily Totals		$2380.80		$3226.40	$49.60	$67.22
Crew B-7A	**Hr.**	**Daily**	**Hr.**	**Daily**	**Bare Costs**	**Incl. O&P**
2 Laborers	$26.70	$427.20	$41.55	$664.80	$28.82	$44.30
1 Equip. Oper. (light)	33.05	264.40	49.80	398.40		
1 Rake w/Tractor		191.50		210.65		
2 Chain Saws, 18"		38.40		42.25	9.58	10.54
24 L.H., Daily Totals		$921.50		$1316.10	$38.40	$54.84

Crew No.	Bare Costs		Incl. Subs O & P		Cost Per Labor-Hour	

Crew B-8	Hr.	Daily	Hr.	Daily	Bare Costs	Incl. O&P
1 Labor Foreman (outside)	$28.70	$229.60	$44.70	$357.60	$29.58	$45.19
2 Laborers	26.70	427.20	41.55	664.80		
2 Equip. Oper. (med.)	34.65	554.40	52.20	835.20		
1 Equip. Oper. Oiler	30.10	240.80	45.35	362.80		
2 Truck Drivers (heavy)	27.55	440.80	42.00	672.00		
1 Hyd. Crane, 25 Ton		616.80		678.50		
1 F.E. Loader, T.M., 2.5 C.Y.		788.00		866.80		
2 Dump Trucks, 16 Ton		953.20		1048.50	36.84	40.53
64 L.H., Daily Totals		$4250.80		$5486.20	$66.42	$85.72

Crew B-9	Hr.	Daily	Hr.	Daily	Bare Costs	Incl. O&P
1 Labor Foreman (outside)	$28.70	$229.60	$44.70	$357.60	$27.10	$42.18
4 Laborers	26.70	854.40	41.55	1329.60		
1 Air Compr., 250 C.F.M.		112.40		123.65		
2 Air Tools & Accessories		19.60		21.55		
2-50 Ft. Air Hoses, 1.5" Dia.		9.40		10.35	3.54	3.89
40 L.H., Daily Totals		$1225.40		$1842.75	$30.64	$46.07

Crew B-9A	Hr.	Daily	Hr.	Daily	Bare Costs	Incl. O&P
2 Laborers	$26.70	$427.20	$41.55	$664.80	$26.98	$41.70
1 Truck Driver (heavy)	27.55	220.40	42.00	336.00		
1 Water Tanker		116.80		128.50		
1 Tractor		208.20		229.00		
2-50 Ft. Disch. Hoses		5.00		5.50	13.75	15.13
24 L.H., Daily Totals		$977.60		$1363.80	$40.73	$56.83

Crew B-9B	Hr.	Daily	Hr.	Daily	Bare Costs	Incl. O&P
2 Laborers	$26.70	$427.20	$41.55	$664.80	$26.98	$41.70
1 Truck Driver (heavy)	27.55	220.40	42.00	336.00		
2-50 Ft. Disch. Hoses		5.00		5.50		
1 Water Tanker		116.80		128.50		
1 Tractor		208.20		229.00		
1 Pressure Washer		46.20		50.80	15.68	17.24
24 L.H., Daily Totals		$1023.80		$1414.60	$42.66	$58.94

Crew B-9C	Hr.	Daily	Hr.	Daily	Bare Costs	Incl. O&P
1 Labor Foreman (outside)	$28.70	$229.60	$44.70	$357.60	$27.10	$42.18
4 Laborers	26.70	854.40	41.55	1329.60		
1 Air Compr., 250 C.F.M.		112.40		123.65		
2-50 Ft. Air Hoses, 1.5" Dia.		9.40		10.35		
2 Breaker, Pavement, 60 lb.		19.60		21.55	3.54	3.89
40 L.H., Daily Totals		$1225.40		$1842.75	$30.64	$46.07

Crew B-9D	Hr.	Daily	Hr.	Daily	Bare Costs	Incl. O&P
1 Labor Foreman (Outside)	$28.70	$229.60	$44.70	$357.60	$27.10	$42.18
4 Common Laborers	26.70	854.40	41.55	1329.60		
1 Air Compressor, 250 CFM		112.40		123.65		
2 Air hoses, 1.5" x 50'		9.40		10.35		
2 Air tamper		55.30		60.85	4.43	4.87
40 L.H., Daily Totals		$1261.10		$1882.05	$31.53	$47.05

Crew B-10	Hr.	Daily	Hr.	Daily	Bare Costs	Incl. O&P
1 Equip. Oper. (med.)	$34.65	$277.20	$52.20	$417.60	$32.00	$48.65
.5 Laborer	26.70	106.80	41.55	166.20		
12 L.H., Daily Totals		$384.00		$583.80	$32.00	$48.65

Crew B-10A	Hr.	Daily	Hr.	Daily	Bare Costs	Incl. O&P
1 Equip. Oper. (med.)	$34.65	$277.20	$52.20	$417.60	$32.00	$48.65
.5 Laborer	26.70	106.80	41.55	166.20		
1 Walk behind compactor, 7.5 HP		122.80		135.10	10.23	11.26
12 L.H., Daily Totals		$506.80		$718.90	$42.23	$59.91

Crew B-10B	Hr.	Daily	Hr.	Daily	Bare Costs	Incl. O&P
1 Equip. Oper. (med.)	$34.65	$277.20	$52.20	$417.60	$32.00	$48.65
.5 Laborer	26.70	106.80	41.55	166.20		
1 Dozer, 200 H.P.		919.60		1011.55	76.63	84.30
12 L.H., Daily Totals		$1303.60		$1595.35	$108.63	$132.95

Crew B-10C	Hr.	Daily	Hr.	Daily	Bare Costs	Incl. O&P
1 Equip. Oper. (med.)	$34.65	$277.20	$52.20	$417.60	$32.00	$48.65
.5 Laborer	26.70	106.80	41.55	166.20		
1 Dozer, 200 H.P.		919.60		1011.55		
1 Vibratory Roller, Towed		591.60		650.75	125.93	138.53
12 L.H., Daily Totals		$1895.20		$2246.10	$157.93	$187.18

Crew B-10D	Hr.	Daily	Hr.	Daily	Bare Costs	Incl. O&P
1 Equip. Oper. (med.)	$34.65	$277.20	$52.20	$417.60	$32.00	$48.65
.5 Laborer	26.70	106.80	41.55	166.20		
1 Dozer, 200 H.P.		919.60		1011.55		
1 Sheepsft. Roller, Towed		610.20		671.20	127.48	140.23
12 L.H., Daily Totals		$1913.80		$2266.55	$159.48	$188.88

Crew B-10E	Hr.	Daily	Hr.	Daily	Bare Costs	Incl. O&P
1 Equip. Oper. (med.)	$34.65	$277.20	$52.20	$417.60	$32.00	$48.65
.5 Laborer	26.70	106.80	41.55	166.20		
1 Tandem Roller, 5 Ton		107.40		118.15	8.95	9.85
12 L.H., Daily Totals		$491.40		$701.95	$40.95	$58.50

Crew B-10F	Hr.	Daily	Hr.	Daily	Bare Costs	Incl. O&P
1 Equip. Oper. (med.)	$34.65	$277.20	$52.20	$417.60	$32.00	$48.65
.5 Laborer	26.70	106.80	41.55	166.20		
1 Tandem Roller, 10 Ton		179.20		197.10	14.93	16.43
12 L.H., Daily Totals		$563.20		$780.90	$46.93	$65.08

Crew B-10G	Hr.	Daily	Hr.	Daily	Bare Costs	Incl. O&P
1 Equip. Oper. (med.)	$34.65	$277.20	$52.20	$417.60	$32.00	$48.65
.5 Laborer	26.70	106.80	41.55	166.20		
1 Sheepsft. Roll., 130 H.P.		794.80		874.30	66.23	72.86
12 L.H., Daily Totals		$1178.80		$1458.10	$98.23	$121.51

Crew B-10H	Hr.	Daily	Hr.	Daily	Bare Costs	Incl. O&P
1 Equip. Oper. (med.)	$34.65	$277.20	$52.20	$417.60	$32.00	$48.65
.5 Laborer	26.70	106.80	41.55	166.20		
1 Diaphr. Water Pump, 2"		50.60		55.65		
1-20 Ft. Suction Hose, 2"		2.55		2.80		
2-50 Ft. Disch. Hoses, 2"		3.80		4.20	4.75	5.22
12 L.H., Daily Totals		$440.95		$646.45	$36.75	$53.87

Crew B-10I	Hr.	Daily	Hr.	Daily	Bare Costs	Incl. O&P
1 Equip. Oper. (med.)	$34.65	$277.20	$52.20	$417.60	$32.00	$48.65
.5 Laborer	26.70	106.80	41.55	166.20		
1 Diaphr. Water Pump, 4"		74.40		81.85		
1-20 Ft. Suction Hose, 4"		5.05		5.55		
2-50 Ft. Disch. Hoses, 4"		7.10		7.80	7.21	7.93
12 L.H., Daily Totals		$470.55		$679.00	$39.21	$56.58

Crew B-10J	Hr.	Daily	Hr.	Daily	Bare Costs	Incl. O&P
1 Equip. Oper. (med.)	$34.65	$277.20	$52.20	$417.60	$32.00	$48.65
.5 Laborer	26.70	106.80	41.55	166.20		
1 Centr. Water Pump, 3"		53.60		58.95		
1-20 Ft. Suction Hose, 3"		4.25		4.70		
2-50 Ft. Disch. Hoses, 3"		5.00		5.50	5.24	5.76
12 L.H., Daily Totals		$446.85		$652.95	$37.24	$54.41

Crew No.	Bare Costs		Incl. Subs O & P		Cost Per Labor-Hour	
Crew B-10K	**Hr.**	**Daily**	**Hr.**	**Daily**	**Bare Costs**	**Incl. O&P**
1 Equip. Oper. (med.)	$34.65	$277.20	$52.20	$417.60	$32.00	$48.65
.5 Laborer	26.70	106.80	41.55	166.20		
1 Centr. Water Pump, 6"		217.00		238.70		
1-20 Ft. Suction Hose, 6"		12.80		14.10		
2-50 Ft. Disch. Hoses, 6"		17.80		19.60	20.63	22.70
12 L.H., Daily Totals		$631.60		$856.20	$52.63	$71.35
Crew B-10L	**Hr.**	**Daily**	**Hr.**	**Daily**	**Bare Costs**	**Incl. O&P**
1 Equip. Oper. (med.)	$34.65	$277.20	$52.20	$417.60	$32.00	$48.65
.5 Laborer	26.70	106.80	41.55	166.20		
1 Dozer, 80 H.P.		314.80		346.30	26.23	28.86
12 L.H., Daily Totals		$698.80		$930.10	$58.23	$77.51
Crew B-10M	**Hr.**	**Daily**	**Hr.**	**Daily**	**Bare Costs**	**Incl. O&P**
1 Equip. Oper. (med.)	$34.65	$277.20	$52.20	$417.60	$32.00	$48.65
.5 Laborer	26.70	106.80	41.55	166.20		
1 Dozer, 300 H.P.		1195.00		1314.50	99.58	109.54
12 L.H., Daily Totals		$1579.00		$1898.30	$131.58	$158.19
Crew B-10N	**Hr.**	**Daily**	**Hr.**	**Daily**	**Bare Costs**	**Incl. O&P**
1 Equip. Oper. (med.)	$34.65	$277.20	$52.20	$417.60	$32.00	$48.65
.5 Laborer	26.70	106.80	41.55	166.20		
1 F.E. Loader, T.M., 1.5 C.Y		312.60		343.85	26.05	28.66
12 L.H., Daily Totals		$696.60		$927.65	$58.05	$77.31
Crew B-10O	**Hr.**	**Daily**	**Hr.**	**Daily**	**Bare Costs**	**Incl. O&P**
1 Equip. Oper. (med.)	$34.65	$277.20	$52.20	$417.60	$32.00	$48.65
.5 Laborer	26.70	106.80	41.55	166.20		
1 F.E. Loader, T.M., 2.25 C.Y.		551.80		607.00	45.98	50.58
12 L.H., Daily Totals		$935.80		$1190.80	$77.98	$99.23
Crew B-10P	**Hr.**	**Daily**	**Hr.**	**Daily**	**Bare Costs**	**Incl. O&P**
1 Equip. Oper. (med.)	$34.65	$277.20	$52.20	$417.60	$32.00	$48.65
.5 Laborer	26.70	106.80	41.55	166.20		
1 F.E. Loader, T.M., 2.5 C.Y.		788.00		866.80	65.67	72.23
12 L.H., Daily Totals		$1172.00		$1450.60	$97.67	$120.88
Crew B-10Q	**Hr.**	**Daily**	**Hr.**	**Daily**	**Bare Costs**	**Incl. O&P**
1 Equip. Oper. (med.)	$34.65	$277.20	$52.20	$417.60	$32.00	$48.65
.5 Laborer	26.70	106.80	41.55	166.20		
1 F.E. Loader, T.M., 5 C.Y.		1094.00		1203.40	91.17	100.28
12 L.H., Daily Totals		$1478.00		$1787.20	$123.17	$148.93
Crew B-10R	**Hr.**	**Daily**	**Hr.**	**Daily**	**Bare Costs**	**Incl. O&P**
1 Equip. Oper. (med.)	$34.65	$277.20	$52.20	$417.60	$32.00	$48.65
.5 Laborer	26.70	106.80	41.55	166.20		
1 F.E. Loader, W.M., 1 C.Y.		193.80		213.20	16.15	17.77
12 L.H., Daily Totals		$577.80		$797.00	$48.15	$66.42
Crew B-10S	**Hr.**	**Daily**	**Hr.**	**Daily**	**Bare Costs**	**Incl. O&P**
1 Equip. Oper. (med.)	$34.65	$277.20	$52.20	$417.60	$32.00	$48.65
.5 Laborer	26.70	106.80	41.55	166.20		
1 F.E. Loader, W.M., 1.5 C.Y.		241.00		265.10	20.08	22.09
12 L.H., Daily Totals		$625.00		$848.90	$52.08	$70.74
Crew B-10T	**Hr.**	**Daily**	**Hr.**	**Daily**	**Bare Costs**	**Incl. O&P**
1 Equip. Oper. (med.)	$34.65	$277.20	$52.20	$417.60	$32.00	$48.65
.5 Laborer	26.70	106.80	41.55	166.20		
1 F.E. Loader, W.M.,2.5 C.Y.		305.00		335.50	25.42	27.96
12 L.H., Daily Totals		$689.00		$919.30	$57.42	$76.61

Crew No.	Bare Costs		Incl. Subs O & P		Cost Per Labor-Hour	
Crew B-10U	**Hr.**	**Daily**	**Hr.**	**Daily**	**Bare Costs**	**Incl. O&P**
1 Equip. Oper. (med.)	$34.65	$277.20	$52.20	$417.60	$32.00	$48.65
.5 Laborer	26.70	106.80	41.55	166.20		
1 F.E. Loader, W.M., 5.5 C.Y.		700.40		770.45	58.37	64.20
12 L.H., Daily Totals		$1084.40		$1354.25	$90.37	$112.85
Crew B-10V	**Hr.**	**Daily**	**Hr.**	**Daily**	**Bare Costs**	**Incl. O&P**
1 Equip. Oper. (med.)	$34.65	$277.20	$52.20	$417.60	$32.00	$48.65
.5 Laborer	26.70	106.80	41.55	166.20		
1 Dozer, 700 H.P.		3139.00		3452.90	261.58	287.74
12 L.H., Daily Totals		$3523.00		$4036.70	$293.58	$336.39
Crew B-10W	**Hr.**	**Daily**	**Hr.**	**Daily**	**Bare Costs**	**Incl. O&P**
1 Equip. Oper. (med.)	$34.65	$277.20	$52.20	$417.60	$32.00	$48.65
.5 Laborer	26.70	106.80	41.55	166.20		
1 Dozer, 105 H.P.		453.80		499.20	37.82	41.60
12 L.H., Daily Totals		$837.80		$1083.00	$69.82	$90.25
Crew B-10X	**Hr.**	**Daily**	**Hr.**	**Daily**	**Bare Costs**	**Incl. O&P**
1 Equip. Oper. (med.)	$34.65	$277.20	$52.20	$417.60	$32.00	$48.65
.5 Laborer	26.70	106.80	41.55	166.20		
1 Dozer, 410 H.P.		1524.00		1676.40	127.00	139.70
12 L.H., Daily Totals		$1908.00		$2260.20	$159.00	$188.35
Crew B-10Y	**Hr.**	**Daily**	**Hr.**	**Daily**	**Bare Costs**	**Incl. O&P**
1 Equip. Oper. (med.)	$34.65	$277.20	$52.20	$417.60	$32.00	$48.65
.5 Laborer	26.70	106.80	41.55	166.20		
1 Vibratory Drum Roller		344.80		379.30	28.73	31.61
12 L.H., Daily Totals		$728.80		$963.10	$60.73	$80.26
Crew B-11A	**Hr.**	**Daily**	**Hr.**	**Daily**	**Bare Costs**	**Incl. O&P**
1 Equipment Oper. (med.)	$34.65	$277.20	$52.20	$417.60	$30.68	$46.88
1 Laborer	26.70	213.60	41.55	332.40		
1 Dozer, 200 H.P.		919.60		1011.55	57.48	63.22
16 L.H., Daily Totals		$1410.40		$1761.55	$88.16	$110.10
Crew B-11B	**Hr.**	**Daily**	**Hr.**	**Daily**	**Bare Costs**	**Incl. O&P**
1 Equipment Oper. (light)	$33.05	$264.40	$49.80	$398.40	$29.88	$45.68
1 Laborer	26.70	213.60	41.55	332.40		
1 Air Powered Tamper		27.65		30.40		
1 Air Compr. 365 C.F.M.		146.80		161.50		
2-50 Ft. Air Hoses, 1.5" Dia.		9.40		10.35	11.49	12.64
16 L.H., Daily Totals		$661.85		$933.05	$41.37	$58.32
Crew B-11C	**Hr.**	**Daily**	**Hr.**	**Daily**	**Bare Costs**	**Incl. O&P**
1 Equipment Oper. (med.)	$34.65	$277.20	$52.20	$417.60	$30.68	$46.88
1 Laborer	26.70	213.60	41.55	332.40		
1 Backhoe Loader, 48 H.P.		215.80		237.40	13.49	14.84
16 L.H., Daily Totals		$706.60		$987.40	$44.17	$61.72
Crew B-11J	**Hr.**	**Daily**	**Hr.**	**Daily**	**Bare Costs**	**Incl. O&P**
1 Equipment Oper. (med.)	$34.65	$277.20	$52.20	$417.60	$30.68	$46.88
1 Laborer	26.70	213.60	41.55	332.40		
1 Grader, 30,000 Lbs.		456.60		502.25		
1 Ripper, beam & 1 shank		71.40		78.55	33.00	36.30
16 L.H., Daily Totals		$1018.80		$1330.80	$63.68	$83.18

Crew No.	Bare Costs		Incl. Subs O & P		Cost Per Labor-Hour	
Crew B-11K	**Hr.**	**Daily**	**Hr.**	**Daily**	**Bare Costs**	**Incl. O&P**
1 Equipment Oper. (med.)	$34.65	$277.20	$52.20	$417.60	$30.68	$46.88
1 Laborer	26.70	213.60	41.55	332.40		
1 Trencher, 8' D., 16" W.		1420.00		1562.00	88.75	97.63
16 L.H., Daily Totals		$1910.80		$2312.00	$119.43	$144.51
Crew B-11L	**Hr.**	**Daily**	**Hr.**	**Daily**	**Bare Costs**	**Incl. O&P**
1 Equipment Oper. (med.)	$34.65	$277.20	$52.20	$417.60	$30.68	$46.88
1 Laborer	26.70	213.60	41.55	332.40		
1 Grader, 30,000 Lbs.		456.60		502.25	28.54	31.39
16 L.H., Daily Totals		$947.40		$1252.25	$59.22	$78.27
Crew B-11M	**Hr.**	**Daily**	**Hr.**	**Daily**	**Bare Costs**	**Incl. O&P**
1 Equipment Oper. (med.)	$34.65	$277.20	$52.20	$417.60	$30.68	$46.88
1 Laborer	26.70	213.60	41.55	332.40		
1 Backhoe Loader, 80 H.P.		256.40		282.05	16.03	17.63
16 L.H., Daily Totals		$747.20		$1032.05	$46.71	$64.51
Crew B-11N	**Hr.**	**Daily**	**Hr.**	**Daily**	**Bare Costs**	**Incl. O&P**
1 Labor Foreman	$28.70	$229.60	$44.70	$357.60	$29.26	$44.57
2 Equipment Operators (med.)	34.65	554.40	52.20	835.20		
6 Truck Drivers (hvy.)	27.55	1322.40	42.00	2016.00		
1 F.E. Loader, 5.5 C.Y.		700.40		770.45		
1 Dozer, 400 H.P.		1524.00		1676.40		
6 Off Hwy. Tks. 50 Ton		7086.00		7794.60	129.31	142.24
72 L.H., Daily Totals		$11416.80		$13450.25	$158.57	$186.81
Crew B-11Q	**Hr.**	**Daily**	**Hr.**	**Daily**	**Bare Costs**	**Incl. O&P**
1 Equipment Operator (med.)	$34.65	$277.20	$52.20	$417.60	$32.00	$48.65
.5 Laborer	26.70	106.80	41.55	166.20		
1 Dozer, 140 H.P.		572.00		629.20	47.67	52.43
12 L.H., Daily Totals		$956.00		$1213.00	$79.67	$101.08
Crew B-11R	**Hr.**	**Daily**	**Hr.**	**Daily**	**Bare Costs**	**Incl. O&P**
1 Equipment Operator (med.)	$34.65	$277.20	$52.20	$417.60	$32.00	$48.65
.5 Laborer	26.70	106.80	41.55	166.20		
1 Dozer, 215 H.P.		919.60		1011.55	76.63	84.30
12 L.H., Daily Totals		$1303.60		$1595.35	$108.63	$132.95
Crew B-11S	**Hr.**	**Daily**	**Hr.**	**Daily**	**Bare Costs**	**Incl. O&P**
1 Equipment Operator (med.)	$34.65	$277.20	$52.20	$417.60	$32.00	$48.65
.5 Laborer	26.70	106.80	41.55	166.20		
1 Dozer, 300 H.P.		1195.00		1314.50		
1 Ripper, Beam & 1 Shank		71.40		78.55	105.53	116.09
12 L.H., Daily Totals		$1650.40		$1976.85	$137.53	$164.74
Crew B-11T	**Hr.**	**Daily**	**Hr.**	**Daily**	**Bare Costs**	**Incl. O&P**
1 Equipment Operator (med.)	$34.65	$277.20	$52.20	$417.60	$32.00	$48.65
.5 Laborer	26.70	106.80	41.55	166.20		
1 Dozer, 410 H.P.		1524.00		1676.40		
1 Ripper, Beam & 2 Shanks		80.20		88.20	133.68	147.05
12 L.H., Daily Totals		$1988.20		$2348.40	$165.68	$195.70
Crew B-11U	**Hr.**	**Daily**	**Hr.**	**Daily**	**Bare Costs**	**Incl. O&P**
1 Equipment Operator (med.)	$34.65	$277.20	$52.20	$417.60	$32.00	$48.65
.5 Laborer	26.70	106.80	41.55	166.20		
1 Dozer, 520 H.P.		2035.00		2238.50	169.58	186.54
12 L.H., Daily Totals		$2419.00		$2822.30	$201.58	$235.19

Crew No.	Bare Costs		Incl. Subs O & P		Cost Per Labor-Hour	
Crew B-11V	**Hr.**	**Daily**	**Hr.**	**Daily**	**Bare Costs**	**Incl. O&P**
3 Laborers	$26.70	$640.80	$41.55	$997.20	$26.70	$41.55
1 Walk behind compactor, 7.5 HP		122.80		135.10	5.12	5.63
24 L.H., Daily Totals		$763.60		$1132.30	$31.82	$47.18
Crew B-11W	**Hr.**	**Daily**	**Hr.**	**Daily**	**Bare Costs**	**Incl. O&P**
1 Equipment Operator (med.)	$34.65	$277.20	$52.20	$417.60	$28.07	$42.81
1 Common Laborer	26.70	213.60	41.55	332.40		
10 Truck Drivers, Heavy	27.55	2204.00	42.00	3360.00		
1 Dozer, 200 H.P.		919.60		1011.55		
1 Vib. roller, smth, towed, 23 Ton		591.60		650.75		
10 Dump Truck, 10 Ton		3266.00		3592.60	49.76	54.74
96 L.H., Daily Totals		$7472.00		$9364.90	$77.83	$97.55
Crew B-11Y	**Hr.**	**Daily**	**Hr.**	**Daily**	**Bare Costs**	**Incl. O&P**
1 Labor Foreman (Outside)	$28.70	$229.60	$44.70	$357.60	$29.57	$45.45
5 Common Laborers	26.70	1068.00	41.55	1662.00		
3 Equipment Operator (med.)	34.65	831.60	52.20	1252.80		
1 Dozer, 80 H.P.		314.80		346.30		
2 Walk behind compactor, 7.5 HP		245.60		270.15		
4 Vibratory plate, gas, 21"		147.20		161.90	9.83	10.81
72 L.H., Daily Totals		$2836.80		$4050.75	$39.40	$56.26
Crew B-12A	**Hr.**	**Daily**	**Hr.**	**Daily**	**Bare Costs**	**Incl. O&P**
1 Equip. Oper. (crane)	$35.90	$287.20	$54.10	$432.80	$31.30	$47.83
1 Laborer	26.70	213.60	41.55	332.40		
1 Hyd. Excavator, 1 C.Y.		557.80		613.60	34.86	38.35
16 L.H., Daily Totals		$1058.60		$1378.80	$66.16	$86.18
Crew B-12B	**Hr.**	**Daily**	**Hr.**	**Daily**	**Bare Costs**	**Incl. O&P**
1 Equip. Oper. (crane)	$35.90	$287.20	$54.10	$432.80	$31.30	$47.83
1 Laborer	26.70	213.60	41.55	332.40		
1 Hyd. Excavator, 1.5 C.Y.		720.40		792.45	45.03	49.53
16 L.H., Daily Totals		$1221.20		$1557.65	$76.33	$97.36
Crew B-12C	**Hr.**	**Daily**	**Hr.**	**Daily**	**Bare Costs**	**Incl. O&P**
1 Equip. Oper. (crane)	$35.90	$287.20	$54.10	$432.80	$31.30	$47.83
1 Laborer	26.70	213.60	41.55	332.40		
1 Hyd. Excavator, 2 C.Y.		909.80		1000.80	56.86	62.55
16 L.H., Daily Totals		$1410.60		$1766.00	$88.16	$110.38
Crew B-12D	**Hr.**	**Daily**	**Hr.**	**Daily**	**Bare Costs**	**Incl. O&P**
1 Equip. Oper. (crane)	$35.90	$287.20	$54.10	$432.80	$31.30	$47.83
1 Laborer	26.70	213.60	41.55	332.40		
1 Hyd. Excavator, 3.5 C.Y.		2020.00		2222.00	126.25	138.88
16 L.H., Daily Totals		$2520.80		$2987.20	$157.55	$186.71
Crew B-12E	**Hr.**	**Daily**	**Hr.**	**Daily**	**Bare Costs**	**Incl. O&P**
1 Equip. Oper. (crane)	$35.90	$287.20	$54.10	$432.80	$31.30	$47.83
1 Laborer	26.70	213.60	41.55	332.40		
1 Hyd. Excavator, .5 C.Y.		332.40		365.65	20.78	22.85
16 L.H., Daily Totals		$833.20		$1130.85	$52.08	$70.68
Crew B-12F	**Hr.**	**Daily**	**Hr.**	**Daily**	**Bare Costs**	**Incl. O&P**
1 Equip. Oper. (crane)	$35.90	$287.20	$54.10	$432.80	$31.30	$47.83
1 Laborer	26.70	213.60	41.55	332.40		
1 Hyd. Excavator, .75 C.Y.		474.00		521.40	29.63	32.59
16 L.H., Daily Totals		$974.80		$1286.60	$60.93	$80.42

Crew B-12G	Bare Costs Hr.	Bare Costs Daily	Incl. Subs O & P Hr.	Incl. Subs O & P Daily	Bare Costs	Incl. O&P
1 Equip. Oper. (crane)	$35.90	$287.20	$54.10	$432.80	$31.30	$47.83
1 Laborer	26.70	213.60	41.55	332.40		
1 Power Shovel, .5 C.Y.		508.10		558.90		
1 Clamshell Bucket, .5 C.Y.		33.80		37.20	33.87	37.25
16 L.H., Daily Totals		$1042.70		$1361.30	$65.17	$85.08

Crew B-12H	Hr.	Daily	Hr.	Daily	Bare Costs	Incl. O&P
1 Equip. Oper. (crane)	$35.90	$287.20	$54.10	$432.80	$31.30	$47.83
1 Laborer	26.70	213.60	41.55	332.40		
1 Power Shovel, 1 C.Y.		833.20		916.50		
1 Clamshell Bucket, 1 C.Y.		43.40		47.75	54.79	60.27
16 L.H., Daily Totals		$1377.40		$1729.45	$86.09	$108.10

Crew B-12I	Hr.	Daily	Hr.	Daily	Bare Costs	Incl. O&P
1 Equip. Oper. (crane)	$35.90	$287.20	$54.10	$432.80	$31.30	$47.83
1 Laborer	26.70	213.60	41.55	332.40		
1 Power Shovel, .75 C.Y.		630.45		693.50		
1 Dragline Bucket, .75 C.Y.		18.80		20.70	40.58	44.63
16 L.H., Daily Totals		$1150.05		$1479.40	$71.88	$92.46

Crew B-12J	Hr.	Daily	Hr.	Daily	Bare Costs	Incl. O&P
1 Equip. Oper. (crane)	$35.90	$287.20	$54.10	$432.80	$31.30	$47.83
1 Laborer	26.70	213.60	41.55	332.40		
1 Gradall, 3 Ton, .5 C.Y.		837.80		921.60	52.36	57.60
16 L.H., Daily Totals		$1338.60		$1686.80	$83.66	$105.43

Crew B-12K	Hr.	Daily	Hr.	Daily	Bare Costs	Incl. O&P
1 Equip. Oper. (crane)	$35.90	$287.20	$54.10	$432.80	$31.30	$47.83
1 Laborer	26.70	213.60	41.55	332.40		
1 Gradall, 3 Ton, 1 C.Y.		971.00		1068.10	60.69	66.76
16 L.H., Daily Totals		$1471.80		$1833.30	$91.99	$114.59

Crew B-12L	Hr.	Daily	Hr.	Daily	Bare Costs	Incl. O&P
1 Equip. Oper. (crane)	$35.90	$287.20	$54.10	$432.80	$31.30	$47.83
1 Laborer	26.70	213.60	41.55	332.40		
1 Power Shovel, .5 C.Y.		508.10		558.90		
1 F.E. Attachment, .5 C.Y.		47.40		52.15	34.72	38.19
16 L.H., Daily Totals		$1056.30		$1376.25	$66.02	$86.02

Crew B-12M	Hr.	Daily	Hr.	Daily	Bare Costs	Incl. O&P
1 Equip. Oper. (crane)	$35.90	$287.20	$54.10	$432.80	$31.30	$47.83
1 Laborer	26.70	213.60	41.55	332.40		
1 Power Shovel, .75 C.Y.		630.45		693.50		
1 F.E. Attachment, .75 C.Y.		51.80		57.00	42.64	46.90
16 L.H., Daily Totals		$1183.05		$1515.70	$73.94	$94.73

Crew B-12N	Hr.	Daily	Hr.	Daily	Bare Costs	Incl. O&P
1 Equip. Oper. (crane)	$35.90	$287.20	$54.10	$432.80	$31.30	$47.83
1 Laborer	26.70	213.60	41.55	332.40		
1 Power Shovel, 1 C.Y.		833.20		916.50		
1 F.E. Attachment, 1 C.Y.		58.60		64.45	55.74	61.31
16 L.H., Daily Totals		$1392.60		$1746.15	$87.04	$109.14

Crew B-12O	Hr.	Daily	Hr.	Daily	Bare Costs	Incl. O&P
1 Equip. Oper. (crane)	$35.90	$287.20	$54.10	$432.80	$31.30	$47.83
1 Laborer	26.70	213.60	41.55	332.40		
1 Power Shovel, 1.5 C.Y.		888.20		977.00		
1 F.E. Attachment, 1.5 C.Y.		67.80		74.60	59.75	65.73
16 L.H., Daily Totals		$1456.80		$1816.80	$91.05	$113.56

Crew B-12P	Hr.	Daily	Hr.	Daily	Bare Costs	Incl. O&P
1 Equip. Oper. (crane)	$35.90	$287.20	$54.10	$432.80	$31.30	$47.83
1 Laborer	26.70	213.60	41.55	332.40		
1 Crawler Crane, 40 Ton		888.20		977.00		
1 Dragline Bucket, 1.5 C.Y.		30.60		33.65	57.43	63.17
16 L.H., Daily Totals		$1419.60		$1775.85	$88.73	$111.00

Crew B-12Q	Hr.	Daily	Hr.	Daily	Bare Costs	Incl. O&P
1 Equip. Oper. (crane)	$35.90	$287.20	$54.10	$432.80	$31.30	$47.83
1 Laborer	26.70	213.60	41.55	332.40		
1 Hyd. Excavator, 5/8 C.Y.		432.40		475.65	27.03	29.73
16 L.H., Daily Totals		$933.20		$1240.85	$58.33	$77.56

Crew B-12R	Hr.	Daily	Hr.	Daily	Bare Costs	Incl. O&P
1 Equip. Oper. (crane)	$35.90	$287.20	$54.10	$432.80	$31.30	$47.83
1 Laborer	26.70	213.60	41.55	332.40		
1 Hyd. Excavator, 1.5 C.Y.		720.40		792.45	45.03	49.53
16 L.H., Daily Totals		$1221.20		$1557.65	$76.33	$97.36

Crew B-12S	Hr.	Daily	Hr.	Daily	Bare Costs	Incl. O&P
1 Equip. Oper. (crane)	$35.90	$287.20	$54.10	$432.80	$31.30	$47.83
1 Laborer	26.70	213.60	41.55	332.40		
1 Hyd. Excavator, 2.5 C.Y.		1215.00		1336.50	75.94	83.53
16 L.H., Daily Totals		$1715.80		$2101.70	$107.24	$131.36

Crew B-12T	Hr.	Daily	Hr.	Daily	Bare Costs	Incl. O&P
1 Equip. Oper. (crane)	$35.90	$287.20	$54.10	$432.80	$31.30	$47.83
1 Laborer	26.70	213.60	41.55	332.40		
1 Crawler Crane, 75 Ton		1170.00		1287.00		
1 F.E. Attachment, 3 C.Y.		90.20		99.20	78.76	86.64
16 L.H., Daily Totals		$1761.00		$2151.40	$110.06	$134.47

Crew B-12V	Hr.	Daily	Hr.	Daily	Bare Costs	Incl. O&P
1 Equip. Oper. (crane)	$35.90	$287.20	$54.10	$432.80	$31.30	$47.83
1 Laborer	26.70	213.60	41.55	332.40		
1 Crawler Crane, 75 Ton		1170.00		1287.00		
1 Dragline Bucket, 3 C.Y.		47.40		52.15	76.09	83.70
16 L.H., Daily Totals		$1718.20		$2104.35	$107.39	$131.53

Crew B-13	Hr.	Daily	Hr.	Daily	Bare Costs	Incl. O&P
1 Labor Foreman (outside)	$28.70	$229.60	$44.70	$357.60	$28.79	$44.34
4 Laborers	26.70	854.40	41.55	1329.60		
1 Equip. Oper. (crane)	35.90	287.20	54.10	432.80		
1 Equip. Oper. Oiler	30.10	240.80	45.35	362.80		
1 Hyd. Crane, 25 Ton		616.80		678.50	11.01	12.12
56 L.H., Daily Totals		$2228.80		$3161.30	$39.80	$56.46

Crew B-13A	Hr.	Daily	Hr.	Daily	Bare Costs	Incl. O&P
1 Foreman	$28.70	$229.60	$44.70	$357.60	$29.50	$45.17
2 Laborers	26.70	427.20	41.55	664.80		
2 Equipment Operators	34.65	554.40	52.20	835.20		
2 Truck Drivers (heavy)	27.55	440.80	42.00	672.00		
1 Crane, 75 Ton		1170.00		1287.00		
1 F.E. Lder, 3.75 C.Y.		1094.00		1203.40		
2 Dump Trucks, 12 Ton		653.20		718.50	52.09	57.30
56 L.H., Daily Totals		$4569.20		$5738.50	$81.59	$102.47

Crew No.	Bare Costs		Incl. Subs O & P		Cost Per Labor-Hour	
Crew B-13B	**Hr.**	**Daily**	**Hr.**	**Daily**	**Bare Costs**	**Incl. O&P**
1 Labor Foreman (outside)	$28.70	$229.60	$44.70	$357.60	$28.79	$44.34
4 Laborers	26.70	854.40	41.55	1329.60		
1 Equip. Oper. (crane)	35.90	287.20	54.10	432.80		
1 Equip. Oper. Oiler	30.10	240.80	45.35	362.80		
1 Hyd. Crane, 55 Ton		895.40		984.95	15.99	17.59
56 L.H., Daily Totals		$2507.40		$3467.75	$44.78	$61.93

Crew B-13C	Hr.	Daily	Hr.	Daily	Bare Costs	Incl. O&P
1 Labor Foreman (outside)	$28.70	$229.60	$44.70	$357.60	$28.79	$44.34
4 Laborers	26.70	854.40	41.55	1329.60		
1 Equip. Oper. (crane)	35.90	287.20	54.10	432.80		
1 Equip. Oper. Oiler	30.10	240.80	45.35	362.80		
1 Crawler Crane, 100 Ton		1519.00		1670.90	27.13	29.84
56 L.H., Daily Totals		$3131.00		$4153.70	$55.92	$74.18

Crew B-14	Hr.	Daily	Hr.	Daily	Bare Costs	Incl. O&P
1 Labor Foreman (outside)	$28.70	$229.60	$44.70	$357.60	$28.09	$43.45
4 Laborers	26.70	854.40	41.55	1329.60		
1 Equip. Oper. (light)	33.05	264.40	49.80	398.40		
1 Backhoe Loader, 48 H.P.		215.80		237.40	4.50	4.95
48 L.H., Daily Totals		$1564.20		$2323.00	$32.59	$48.40

Crew B-15	Hr.	Daily	Hr.	Daily	Bare Costs	Incl. O&P
1 Equipment Oper. (med)	$34.65	$277.20	$52.20	$417.60	$29.46	$44.85
.5 Laborer	26.70	106.80	41.55	166.20		
2 Truck Drivers (heavy)	27.55	440.80	42.00	672.00		
2 Dump Trucks, 16 Ton		953.20		1048.50		
1 Dozer, 200 H.P.		919.60		1011.55	66.89	73.57
28 L.H., Daily Totals		$2697.60		$3315.85	$96.35	$118.42

Crew B-16	Hr.	Daily	Hr.	Daily	Bare Costs	Incl. O&P
1 Labor Foreman (outside)	$28.70	$229.60	$44.70	$357.60	$27.41	$42.45
2 Laborers	26.70	427.20	41.55	664.80		
1 Truck Driver (heavy)	27.55	220.40	42.00	336.00		
1 Dump Truck, 16 Ton		476.60		524.25	14.89	16.38
32 L.H., Daily Totals		$1353.80		$1882.65	$42.30	$58.83

Crew B-17	Hr.	Daily	Hr.	Daily	Bare Costs	Incl. O&P
2 Laborers	$26.70	$427.20	$41.55	$664.80	$28.50	$43.73
1 Equip. Oper. (light)	33.05	264.40	49.80	398.40		
1 Truck Driver (heavy)	27.55	220.40	42.00	336.00		
1 Backhoe Loader, 48 H.P.		215.80		237.40		
1 Dump Truck, 12 Ton		326.60		359.25	16.95	18.65
32 L.H., Daily Totals		$1454.40		$1995.85	$45.45	$62.38

Crew B-17A	Hr.	Daily	Hr.	Daily	Bare Costs	Incl. O&P
2 Laborer Foremen	$28.70	$459.20	$44.70	$715.20	$28.93	$45.02
6 Laborers	26.70	1281.60	41.55	1994.40		
1 Skilled Worker Foreman	36.85	294.80	57.30	458.40		
1 Skilled Worker	34.85	278.80	54.20	433.60		
80 L.H., Daily Totals		$2314.40		$3601.60	$28.93	$45.02

Crew B-18	Hr.	Daily	Hr.	Daily	Bare Costs	Incl. O&P
1 Labor Foreman (outside)	$28.70	$229.60	$44.70	$357.60	$27.37	$42.60
2 Laborers	26.70	427.20	41.55	664.80		
1 Vibrating Compactor		36.80		40.50	1.53	1.69
24 L.H., Daily Totals		$693.60		$1062.90	$28.90	$44.29

Crew No.	Bare Costs		Incl. Subs O & P		Cost Per Labor-Hour	
Crew B-19	**Hr.**	**Daily**	**Hr.**	**Daily**	**Bare Costs**	**Incl. O&P**
1 Pile Driver Foreman	$35.30	$282.40	$58.15	$465.20	$33.80	$53.91
4 Pile Drivers	33.30	1065.60	54.90	1756.80		
2 Equip. Oper. (crane)	35.90	574.40	54.10	865.60		
1 Equip. Oper. Oiler	30.10	240.80	45.35	362.80		
1 Crane, 40 Ton & Access.		888.20		977.00		
60 L.F. Pile Leads		75.00		82.50		
1 Hammer, Diesel, 22k Ft-Lb		564.00		620.40	23.91	26.30
64 L.H., Daily Totals		$3690.40		$5130.30	$57.71	$80.21

Crew B-19A	Hr.	Daily	Hr.	Daily	Bare Costs	Incl. O&P
1 Pile Driver Foreman	$35.30	$282.40	$58.15	$465.20	$33.80	$53.91
4 Pile Drivers	33.30	1065.60	54.90	1756.80		
2 Equip. Oper. (crane)	35.90	574.40	54.10	865.60		
1 Equip. Oper. Oiler	30.10	240.80	45.35	362.80		
1 Crawler Crane, 75 Ton		1170.00		1287.00		
60 L.F. Leads, 25K Ft. Lbs.		120.00		132.00		
1 Hammer, Diesel, 41k Ft-Lb		662.60		728.85	30.51	33.56
64 L.H., Daily Totals		$4115.80		$5598.25	$64.31	$87.47

Crew B-20	Hr.	Daily	Hr.	Daily	Bare Costs	Incl. O&P
1 Labor Foreman (out)	$28.70	$229.60	$44.70	$357.60	$30.08	$46.82
1 Skilled Worker	34.85	278.80	54.20	433.60		
1 Laborer	26.70	213.60	41.55	332.40		
24 L.H., Daily Totals		$722.00		$1123.60	$30.08	$46.82

Crew B-20A	Hr.	Daily	Hr.	Daily	Bare Costs	Incl. O&P
1 Labor Foreman	$28.70	$229.60	$44.70	$357.60	$32.24	$49.20
1 Laborer	26.70	213.60	41.55	332.40		
1 Plumber	40.85	326.80	61.40	491.20		
1 Plumber Apprentice	32.70	261.60	49.15	393.20		
32 L.H., Daily Totals		$1031.60		$1574.40	$32.24	$49.20

Crew B-21	Hr.	Daily	Hr.	Daily	Bare Costs	Incl. O&P
1 Labor Foreman (out)	$28.70	$229.60	$44.70	$357.60	$30.91	$47.86
1 Skilled Worker	34.85	278.80	54.20	433.60		
1 Laborer	26.70	213.60	41.55	332.40		
.5 Equip. Oper. (crane)	35.90	143.60	54.10	216.40		
.5 S.P. Crane, 5 Ton		158.50		174.35	5.66	6.23
28 L.H., Daily Totals		$1024.10		$1514.35	$36.57	$54.09

Crew B-21A	Hr.	Daily	Hr.	Daily	Bare Costs	Incl. O&P
1 Labor Foreman	$28.70	$229.60	$44.70	$357.60	$32.97	$50.18
1 Laborer	26.70	213.60	41.55	332.40		
1 Plumber	40.85	326.80	61.40	491.20		
1 Plumber Apprentice	32.70	261.60	49.15	393.20		
1 Equip. Oper. (crane)	35.90	287.20	54.10	432.80		
1 S.P. Crane, 12 Ton		505.80		556.40	12.65	13.91
40 L.H., Daily Totals		$1824.60		$2563.60	$45.62	$64.09

Crew B-21B	Hr.	Daily	Hr.	Daily	Bare Costs	Incl. O&P
1 Laborer Foreman	$28.70	$229.60	$44.70	$357.60	$28.94	$44.69
3 Laborers	26.70	640.80	41.55	997.20		
1 Equip. Oper. (crane)	35.90	287.20	54.10	432.80		
1 Hyd. Crane, 12 Ton		602.60		662.85	15.07	16.57
40 L.H., Daily Totals		$1760.20		$2450.45	$44.01	$61.26

Crew No.	Bare Costs		Incl. Subs O & P		Cost Per Labor-Hour	
Crew B-21C	**Hr.**	**Daily**	**Hr.**	**Daily**	**Bare Costs**	**Incl. O&P**
1 Laborer Foreman	$28.70	$229.60	$44.70	$357.60	$28.79	$44.34
4 Laborers	26.70	854.40	41.55	1329.60		
1 Equip. Oper. (crane)	35.90	287.20	54.10	432.80		
1 Equip. Oper. Oiler	30.10	240.80	45.35	362.80		
2 Cutting Torches		36.00		39.60		
2 Gases		129.60		142.55		
1 Crane, 90 Ton		1325.00		1457.50	26.62	29.28
56 L.H., Daily Totals		$3102.60		$4122.45	$55.41	$73.62
Crew B-22	**Hr.**	**Daily**	**Hr.**	**Daily**	**Bare Costs**	**Incl. O&P**
1 Labor Foreman (out)	$28.70	$229.60	$44.70	$357.60	$31.25	$48.27
1 Skilled Worker	34.85	278.80	54.20	433.60		
1 Laborer	26.70	213.60	41.55	332.40		
.75 Equip. Oper. (crane)	35.90	215.40	54.10	324.60		
.75 S.P. Crane, 5 Ton		237.75		261.50	7.93	8.72
30 L.H., Daily Totals		$1175.15		$1709.70	$39.18	$56.99
Crew B-22A	**Hr.**	**Daily**	**Hr.**	**Daily**	**Bare Costs**	**Incl. O&P**
1 Labor Foreman (out)	$28.70	$229.60	$44.70	$357.60	$30.29	$46.86
1 Skilled Worker	34.85	278.80	54.20	433.60		
2 Laborers	26.70	427.20	41.55	664.80		
.75 Equipment Oper. (crane)	35.90	215.40	54.10	324.60		
.75 Crane, 5 Ton		237.75		261.50		
1 Generator, 5 KW		32.20		35.40		
1 Butt Fusion Machine		432.80		476.10	18.49	20.34
38 L.H., Daily Totals		$1853.75		$2553.60	$48.78	$67.20
Crew B-22B	**Hr.**	**Daily**	**Hr.**	**Daily**	**Bare Costs**	**Incl. O&P**
1 Skilled Worker	$34.85	$278.80	$54.20	$433.60	$30.78	$47.88
1 Laborer	26.70	213.60	41.55	332.40		
1 Electro Fusion Machine		171.80		189.00	10.74	11.81
16 L.H., Daily Totals		$664.20		$955.00	$41.52	$59.69
Crew B-23	**Hr.**	**Daily**	**Hr.**	**Daily**	**Bare Costs**	**Incl. O&P**
1 Labor Foreman (outside)	$28.70	$229.60	$44.70	$357.60	$27.10	$42.18
4 Laborers	26.70	854.40	41.55	1329.60		
1 Drill Rig, Wells		3179.00		3496.90		
1 Light Truck, 3 Ton		176.20		193.80	83.88	92.27
40 L.H., Daily Totals		$4439.20		$5377.90	$110.98	$134.45
Crew B-23A	**Hr.**	**Daily**	**Hr.**	**Daily**	**Bare Costs**	**Incl. O&P**
1 Labor Foreman (outside)	$28.70	$229.60	$44.70	$357.60	$30.02	$46.15
1 Laborer	26.70	213.60	41.55	332.40		
1 Equip. Operator (medium)	34.65	277.20	52.20	417.60		
1 Drill Rig, Wells		3179.00		3496.90		
1 Pickup Truck, 3/4 Ton		78.00		85.80	135.71	149.28
24 L.H., Daily Totals		$3977.40		$4690.30	$165.73	$195.43
Crew B-23B	**Hr.**	**Daily**	**Hr.**	**Daily**	**Bare Costs**	**Incl. O&P**
1 Labor Foreman (outside)	$28.70	$229.60	$44.70	$357.60	$30.02	$46.15
1 Laborer	26.70	213.60	41.55	332.40		
1 Equip. Operator (medium)	34.65	277.20	52.20	417.60		
1 Drill Rig, Wells		3179.00		3496.90		
1 Pickup Truck, 3/4 Ton		78.00		85.80		
1 Pump, Cntfgl, 6"		217.00		238.70	144.75	159.23
24 L.H., Daily Totals		$4194.40		$4929.00	$174.77	$205.38

Crew No.	Bare Costs		Incl. Subs O & P		Cost Per Labor-Hour	
Crew B-24	**Hr.**	**Daily**	**Hr.**	**Daily**	**Bare Costs**	**Incl. O&P**
1 Cement Finisher	$32.85	$262.80	$48.35	$386.80	$31.27	$47.63
1 Laborer	26.70	213.60	41.55	332.40		
1 Carpenter	34.25	274.00	53.35	426.80		
24 L.H., Daily Totals		$750.40		$1146.00	$31.27	$47.63
Crew B-25	**Hr.**	**Daily**	**Hr.**	**Daily**	**Bare Costs**	**Incl. O&P**
1 Labor Foreman	$28.70	$229.60	$44.70	$357.60	$29.05	$44.74
7 Laborers	26.70	1495.20	41.55	2326.80		
3 Equip. Oper. (med.)	34.65	831.60	52.20	1252.80		
1 Asphalt Paver, 130 H.P		1590.00		1749.00		
1 Tandem Roller, 10 Ton		179.20		197.10		
1 Roller, Pneumatic Wheel		241.80		266.00	22.85	25.14
88 L.H., Daily Totals		$4567.40		$6149.30	$51.90	$69.88
Crew B-25B	**Hr.**	**Daily**	**Hr.**	**Daily**	**Bare Costs**	**Incl. O&P**
1 Labor Foreman	$28.70	$229.60	$44.70	$357.60	$29.52	$45.36
7 Laborers	26.70	1495.20	41.55	2326.80		
4 Equip. Oper. (medium)	34.65	1108.80	52.20	1670.40		
1 Asphalt Paver, 130 H.P.		1590.00		1749.00		
2 Rollers, Steel Wheel		358.40		394.25		
1 Roller, Pneumatic Wheel		241.80		266.00	22.81	25.10
96 L.H., Daily Totals		$5023.80		$6764.05	$52.33	$70.46
Crew B-25C	**Hr.**	**Daily**	**Hr.**	**Daily**	**Bare Costs**	**Incl. O&P**
1 Labor Foreman	$28.70	$229.60	$44.70	$357.60	$29.68	$45.63
3 Laborers	26.70	640.80	41.55	997.20		
2 Equip. Oper. (medium)	34.65	554.40	52.20	835.20		
1 Asphalt Paver, 130 H.P.		1590.00		1749.00		
1 Rollers, Steel Wheel		179.20		197.10	36.86	40.54
48 L.H., Daily Totals		$3194.00		$4136.10	$66.54	$86.17
Crew B-26	**Hr.**	**Daily**	**Hr.**	**Daily**	**Bare Costs**	**Incl. O&P**
1 Labor Foreman (outside)	$28.70	$229.60	$44.70	$357.60	$29.91	$46.27
6 Laborers	26.70	1281.60	41.55	1994.40		
2 Equip. Oper. (med.)	34.65	554.40	52.20	835.20		
1 Rodman (reinf.)	37.95	303.60	62.60	500.80		
1 Cement Finisher	32.85	262.80	48.35	386.80		
1 Grader, 30,000 Lbs.		456.60		502.25		
1 Paving Mach. & Equip.		1910.00		2101.00	26.89	29.58
88 L.H., Daily Totals		$4998.60		$6678.05	$56.80	$75.85
Crew B-27	**Hr.**	**Daily**	**Hr.**	**Daily**	**Bare Costs**	**Incl. O&P**
1 Labor Foreman (outside)	$28.70	$229.60	$44.70	$357.60	$27.20	$42.34
3 Laborers	26.70	640.80	41.55	997.20		
1 Berm Machine		216.20		237.80	6.76	7.43
32 L.H., Daily Totals		$1086.60		$1592.60	$33.96	$49.77
Crew B-28	**Hr.**	**Daily**	**Hr.**	**Daily**	**Bare Costs**	**Incl. O&P**
2 Carpenters	$34.25	$548.00	$53.35	$853.60	$31.73	$49.42
1 Laborer	26.70	213.60	41.55	332.40		
24 L.H., Daily Totals		$761.60		$1186.00	$31.73	$49.42
Crew B-29	**Hr.**	**Daily**	**Hr.**	**Daily**	**Bare Costs**	**Incl. O&P**
1 Labor Foreman (outside)	$28.70	$229.60	$44.70	$357.60	$28.79	$44.34
4 Laborers	26.70	854.40	41.55	1329.60		
1 Equip. Oper. (crane)	35.90	287.20	54.10	432.80		
1 Equip. Oper. Oiler	30.10	240.80	45.35	362.80		
1 Gradall, 3 Ton, 1/2 C.Y.		837.80		921.60	14.96	16.46
56 L.H., Daily Totals		$2449.80		$3404.40	$43.75	$60.80

Crew No.	Bare Costs		Incl. Subs O & P		Cost Per Labor-Hour	
Crew B-30	**Hr.**	**Daily**	**Hr.**	**Daily**	**Bare Costs**	**Incl. O&P**
1 Equip. Oper. (med.)	$34.65	$277.20	$52.20	$417.60	$29.92	$45.40
2 Truck Drivers (heavy)	27.55	440.80	42.00	672.00		
1 Hyd. Excavator, 1.5 C.Y.		720.40		792.45		
2 Dump Trucks, 16 Ton		953.20		1048.50	69.73	76.71
24 L.H., Daily Totals		$2391.60		$2930.55	$99.65	$122.11
Crew B-31	**Hr.**	**Daily**	**Hr.**	**Daily**	**Bare Costs**	**Incl. O&P**
1 Labor Foreman (outside)	$28.70	$229.60	$44.70	$357.60	$28.61	$44.54
3 Laborers	26.70	640.80	41.55	997.20		
1 Carpenter	34.25	274.00	53.35	426.80		
1 Air Compr., 250 C.F.M.		112.40		123.65		
1 Sheeting Driver		7.20		7.90		
2-50 Ft. Air Hoses, 1.5" Dia.		9.40		10.35	3.23	3.55
40 L.H., Daily Totals		$1273.40		$1923.50	$31.84	$48.09
Crew B-32	**Hr.**	**Daily**	**Hr.**	**Daily**	**Bare Costs**	**Incl. O&P**
1 Laborer	$26.70	$213.60	$41.55	$332.40	$32.66	$49.54
3 Equip. Oper. (med.)	34.65	831.60	52.20	1252.80		
1 Grader, 30,000 Lbs.		456.60		502.25		
1 Tandem Roller, 10 Ton		179.20		197.10		
1 Dozer, 200 H.P.		919.60		1011.55	48.61	53.47
32 L.H., Daily Totals		$2600.60		$3296.10	$81.27	$103.01
Crew B-32A	**Hr.**	**Daily**	**Hr.**	**Daily**	**Bare Costs**	**Incl. O&P**
1 Laborer	$26.70	$213.60	$41.55	$332.40	$32.00	$48.65
2 Equip. Oper. (medium)	34.65	554.40	52.20	835.20		
1 Grader, 30,000 Lbs.		456.60		502.25		
1 Roller, Vibratory, 29,000 Lbs.		433.20		476.50	37.08	40.78
24 L.H., Daily Totals		$1657.80		$2146.35	$69.08	$89.43
Crew B-32B	**Hr.**	**Daily**	**Hr.**	**Daily**	**Bare Costs**	**Incl. O&P**
1 Laborer	$26.70	$213.60	$41.55	$332.40	$32.00	$48.65
2 Equip. Oper. (medium)	34.65	554.40	52.20	835.20		
1 Dozer, 200 H.P.		919.60		1011.55		
1 Roller, Vibratory, 29,000 Lbs.		433.20		476.50	56.37	62.00
24 L.H., Daily Totals		$2120.80		$2655.65	$88.37	$110.65
Crew B-32C	**Hr.**	**Daily**	**Hr.**	**Daily**	**Bare Costs**	**Incl. O&P**
1 Labor Foreman	$28.70	$229.60	$44.70	$357.60	$31.01	$47.40
2 Laborers	26.70	427.20	41.55	664.80		
3 Equip. Oper. (medium)	34.65	831.60	52.20	1252.80		
1 Grader, 30,000 Lbs.		456.60		502.25		
1 Roller, Steel Wheel		179.20		197.10		
1 Dozer, 200 H.P.		919.60		1011.55	32.40	35.64
48 L.H., Daily Totals		$3043.80		$3986.10	$63.41	$83.04
Crew B-33A	**Hr.**	**Daily**	**Hr.**	**Daily**	**Bare Costs**	**Incl. O&P**
1 Equip. Oper. (med.)	$34.65	$277.20	$52.20	$417.60	$32.38	$49.16
.5 Laborer	26.70	106.80	41.55	166.20		
.25 Equip. Oper. (med.)	34.65	69.30	52.20	104.40		
1 Scraper, Towed, 7 C.Y.		150.75		165.80		
1.25 Dozer, 300 H.P.		1493.75		1643.15	117.47	129.21
14 L.H., Daily Totals		$2097.80		$2497.15	$149.85	$178.37
Crew B-33B	**Hr.**	**Daily**	**Hr.**	**Daily**	**Bare Costs**	**Incl. O&P**
1 Equip. Oper. (med.)	$34.65	$277.20	$52.20	$417.60	$32.38	$49.16
.5 Laborer	26.70	106.80	41.55	166.20		
.25 Equip. Oper. (med.)	34.65	69.30	52.20	104.40		
1 Scraper, Towed, 10 C.Y.		169.20		186.10		
1.25 Dozer, 300 H.P.		1493.75		1643.15	118.78	130.66
14 L.H., Daily Totals		$2116.25		$2517.45	$151.16	$179.82

Crew No.	Bare Costs		Incl. Subs O & P		Cost Per Labor-Hour	
Crew B-33C	**Hr.**	**Daily**	**Hr.**	**Daily**	**Bare Costs**	**Incl. O&P**
1 Equip. Oper. (med.)	$34.65	$277.20	$52.20	$417.60	$32.38	$49.16
.5 Laborer	26.70	106.80	41.55	166.20		
.25 Equip. Oper. (med.)	34.65	69.30	52.20	104.40		
1 Scraper, Towed, 12 C.Y.		169.20		186.10		
1.25 Dozer, 300 H.P.		1493.75		1643.15	118.78	130.66
14 L.H., Daily Totals		$2116.25		$2517.45	$151.16	$179.82
Crew B-33D	**Hr.**	**Daily**	**Hr.**	**Daily**	**Bare Costs**	**Incl. O&P**
1 Equip. Oper. (med.)	$34.65	$277.20	$52.20	$417.60	$32.38	$49.16
.5 Laborer	26.70	106.80	41.55	166.20		
.25 Equip. Oper. (med.)	34.65	69.30	52.20	104.40		
1 S.P. Scraper, 14 C.Y.		1552.00		1707.20		
.25 Dozer, 300 H.P.		298.75		328.65	132.20	145.42
14 L.H., Daily Totals		$2304.05		$2724.05	$164.58	$194.58
Crew B-33E	**Hr.**	**Daily**	**Hr.**	**Daily**	**Bare Costs**	**Incl. O&P**
1 Equip. Oper. (med.)	$34.65	$277.20	$52.20	$417.60	$32.38	$49.16
.5 Laborer	26.70	106.80	41.55	166.20		
.25 Equip. Oper. (med.)	34.65	69.30	52.20	104.40		
1 S.P. Scraper, 24 C.Y.		2317.00		2548.70		
.25 Dozer, 300 H.P.		298.75		328.65	186.84	205.52
14 L.H., Daily Totals		$3069.05		$3565.55	$219.22	$254.68
Crew B-33F	**Hr.**	**Daily**	**Hr.**	**Daily**	**Bare Costs**	**Incl. O&P**
1 Equip. Oper. (med.)	$34.65	$277.20	$52.20	$417.60	$32.38	$49.16
.5 Laborer	26.70	106.80	41.55	166.20		
.25 Equip. Oper. (med.)	34.65	69.30	52.20	104.40		
1 Elev. Scraper, 11 C.Y.		819.40		901.35		
.25 Dozer, 300 H.P.		298.75		328.65	79.87	87.85
14 L.H., Daily Totals		$1571.45		$1918.20	$112.25	$137.01
Crew B-33G	**Hr.**	**Daily**	**Hr.**	**Daily**	**Bare Costs**	**Incl. O&P**
1 Equip. Oper. (med.)	$34.65	$277.20	$52.20	$417.60	$32.38	$49.16
.5 Laborer	26.70	106.80	41.55	166.20		
.25 Equip. Oper. (med.)	34.65	69.30	52.20	104.40		
1 Elev. Scraper, 20 C.Y.		1648.00		1812.80		
.25 Dozer, 300 H.P.		298.75		328.65	139.05	152.96
14 L.H., Daily Totals		$2400.05		$2829.65	$171.43	$202.12
Crew B-33H	**Hr.**	**Daily**	**Hr.**	**Daily**	**Bare Costs**	**Incl. O&P**
.25 Laborer	$26.70	$53.40	$41.55	$83.10	$33.28	$50.36
1 Equipment Operator (med.)	34.65	277.20	52.20	417.60		
.2 Equipment Operator (med.)	34.65	55.44	52.20	83.52		
1 Scraper, 32-44 C.Y.		2728.00		3000.80		
.2 Dozer, 400 H.P.		304.80		335.30	261.45	287.59
11. L.H., Daily Totals		$3418.84		$3920.32	$294.73	$337.95
Crew B-33J	**Hr.**	**Daily**	**Hr.**	**Daily**	**Bare Costs**	**Incl. O&P**
1 Equipment Operator (med.)	$34.65	$277.20	$52.20	$417.60	$34.65	$52.20
1 Scraper 17 C.Y.		1552.00		1707.20	194.00	213.40
8 L.H., Daily Totals		$1829.20		$2124.80	$228.65	$265.60
Crew B-34A	**Hr.**	**Daily**	**Hr.**	**Daily**	**Bare Costs**	**Incl. O&P**
1 Truck Driver (heavy)	$27.55	$220.40	$42.00	$336.00	$27.55	$42.00
1 Dump Truck, 12 Ton		326.60		359.25	40.83	44.91
8 L.H., Daily Totals		$547.00		$695.25	$68.38	$86.91

Crew B-34B	Bare Costs Hr.	Bare Costs Daily	Incl. Subs O & P Hr.	Incl. Subs O & P Daily	Bare Costs	Incl. O&P
1 Truck Driver (heavy)	$27.55	$220.40	$42.00	$336.00	$27.55	$42.00
1 Dump Truck, 16 Ton		476.60		524.25	59.58	65.53
8 L.H., Daily Totals		$697.00		$860.25	$87.13	$107.53

Crew B-34C	Hr.	Daily	Hr.	Daily	Bare Costs	Incl. O&P
1 Truck Driver (heavy)	$27.55	$220.40	$42.00	$336.00	$27.55	$42.00
1 Truck Tractor, 40 Ton		301.60		331.75		
1 Dump Trailer, 16.5 C.Y.		103.00		113.30	50.58	55.63
8 L.H., Daily Totals		$625.00		$781.05	$78.13	$97.63

Crew B-34D	Hr.	Daily	Hr.	Daily	Bare Costs	Incl. O&P
1 Truck Driver (heavy)	$27.55	$220.40	$42.00	$336.00	$27.55	$42.00
1 Truck Tractor, 40 Ton		301.60		331.75		
1 Dump Trailer, 20 C.Y.		115.80		127.40	52.18	57.39
8 L.H., Daily Totals		$637.80		$795.15	$79.73	$99.39

Crew B-34E	Hr.	Daily	Hr.	Daily	Bare Costs	Incl. O&P
1 Truck Driver (heavy)	$27.55	$220.40	$42.00	$336.00	$27.55	$42.00
1 Truck, Off Hwy., 25 Ton		912.20		1003.40	114.03	125.43
8 L.H., Daily Totals		$1132.60		$1339.40	$141.58	$167.43

Crew B-34F	Hr.	Daily	Hr.	Daily	Bare Costs	Incl. O&P
1 Truck Driver (heavy)	$27.55	$220.40	$42.00	$336.00	$27.55	$42.00
1 Truck, Off Hwy., 22 C.Y.		933.40		1026.75	116.68	128.34
8 L.H., Daily Totals		$1153.80		$1362.75	$144.23	$170.34

Crew B-34G	Hr.	Daily	Hr.	Daily	Bare Costs	Incl. O&P
1 Truck Driver (heavy)	$27.55	$220.40	$42.00	$336.00	$27.55	$42.00
1 Truck, Off Hwy., 34 C.Y.		1181.00		1299.10	147.63	162.39
8 L.H., Daily Totals		$1401.40		$1635.10	$175.18	$204.39

Crew B-34H	Hr.	Daily	Hr.	Daily	Bare Costs	Incl. O&P
1 Truck Driver (heavy)	$27.55	$220.40	$42.00	$336.00	$27.55	$42.00
1 Truck, Off Hwy., 42 C.Y.		1276.00		1403.60	159.50	175.45
8 L.H., Daily Totals		$1496.40		$1739.60	$187.05	$217.45

Crew B-34J	Hr.	Daily	Hr.	Daily	Bare Costs	Incl. O&P
1 Truck Driver (heavy)	$27.55	$220.40	$42.00	$336.00	$27.55	$42.00
1 Truck, Off Hwy., 60 C.Y.		1649.00		1813.90	206.13	226.74
8 L.H., Daily Totals		$1869.40		$2149.90	$233.68	$268.74

Crew B-34K	Hr.	Daily	Hr.	Daily	Bare Costs	Incl. O&P
1 Truck Driver (heavy)	$27.55	$220.40	$42.00	$336.00	$27.55	$42.00
1 Truck Tractor, 240 H.P.		353.60		388.95		
1 Low Bed Trailer		170.20		187.20	65.48	72.02
8 L.H., Daily Totals		$744.20		$912.15	$93.03	$114.02

Crew B-34N	Hr.	Daily	Hr.	Daily	Bare Costs	Incl. O&P
1 Truck Driver (heavy)	$27.55	$220.40	$42.00	$336.00	$27.55	$42.00
1 Dump Truck, 12 Ton		326.60		359.25		
1 Flatbed Trailer, 40 Ton		119.40		131.35	55.75	61.33
8 L.H., Daily Totals		$666.40		$826.60	$83.30	$103.33

Crew B-34P	Hr.	Daily	Hr.	Daily	Bare Costs	Incl. O&P
1 Pipe Fitter	$40.95	$327.60	$61.55	$492.40	$34.13	$51.53
1 Truck Driver (light)	26.80	214.40	40.85	326.80		
1 Equip. Oper. (medium)	34.65	277.20	52.20	417.60		
1 Flatbed Truck, 3 Ton		176.20		193.80		
1 Backhoe Loader, 48 H.P.		215.80		237.40	16.33	17.97
24 L.H., Daily Totals		$1211.20		$1668.00	$50.46	$69.50

Crew B-34Q	Hr.	Daily	Hr.	Daily	Bare Costs	Incl. O&P
1 Pipe Fitter	$40.95	$327.60	$61.55	$492.40	$34.55	$52.17
1 Truck Driver (light)	26.80	214.40	40.85	326.80		
1 Eqip. Oper. (crane)	35.90	287.20	54.10	432.80		
1 Flatbed Trailer, 25 Ton		88.00		96.80		
1 Dump Truck, 12 Ton		326.60		359.25		
1 Hyd. Crane, 25 Ton		616.80		678.50	42.98	47.27
24 L.H., Daily Totals		$1860.60		$2386.55	$77.53	$99.44

Crew B-34R	Hr.	Daily	Hr.	Daily	Bare Costs	Incl. O&P
1 Pipe Fitter	$40.95	$327.60	$61.55	$492.40	$34.55	$52.17
1 Truck Driver (light)	26.80	214.40	40.85	326.80		
1 Eqip. Oper. (crane)	35.90	287.20	54.10	432.80		
1 Flatbed Trailer, 25 Ton		88.00		96.80		
1 Dump Truck, 12 Ton		326.60		359.25		
1 Hyd. Crane, 25 Ton		616.80		678.50		
1 Hyd. Excavator, 1 C.Y.		557.80		613.60	66.22	72.84
24 L.H., Daily Totals		$2418.40		$3000.15	$100.77	$125.01

Crew B-34S	Hr.	Daily	Hr.	Daily	Bare Costs	Incl. O&P
2 Pipe Fitters	$40.95	$655.20	$61.55	$984.80	$36.34	$54.80
1 Truck Driver (heavy)	27.55	220.40	42.00	336.00		
1 Eqip. Oper. (crane)	35.90	287.20	54.10	432.80		
1 Flatbed Trailer, 40 Ton		119.40		131.35		
1 Truck Tractor, 40 Ton		301.60		331.75		
1 Truck Crane, 80 Ton		995.00		1094.50		
1 Hyd. Excavator, 2 C.Y.		909.80		1000.80	72.68	79.95
32 L.H., Daily Totals		$3488.60		$4312.00	$109.02	$134.75

Crew B-34T	Hr.	Daily	Hr.	Daily	Bare Costs	Incl. O&P
2 Pipe Fitters	$40.95	$655.20	$61.55	$984.80	$36.34	$54.80
1 Truck Driver (heavy)	27.55	220.40	42.00	336.00		
1 Eqip. Oper. (crane)	35.90	287.20	54.10	432.80		
1 Flatbed Trailer, 40 Ton		119.40		131.35		
1 Truck Tractor, 40 Ton		301.60		331.75		
1 Truck Crane, 80 Ton		995.00		1094.50	44.25	48.68
32 L.H., Daily Totals		$2578.80		$3311.20	$80.59	$103.48

Crew B-35	Hr.	Daily	Hr.	Daily	Bare Costs	Incl. O&P
1 Laborer Foreman (out)	$28.70	$229.60	$44.70	$357.60	$32.85	$50.22
1 Skilled Worker	34.85	278.80	54.20	433.60		
1 Welder (plumber)	40.85	326.80	61.40	491.20		
1 Laborer	26.70	213.60	41.55	332.40		
1 Equip. Oper. (crane)	35.90	287.20	54.10	432.80		
1 Equip. Oper. Oiler	30.10	240.80	45.35	362.80		
1 Electric Welding Mach.		80.65		88.70		
1 Hyd. Excavator, .75 C.Y.		474.00		521.40	11.56	12.71
48 L.H., Daily Totals		$2131.45		$3020.50	$44.41	$62.93

Crew No.	Bare Costs		Incl. Subs O & P		Cost Per Labor-Hour	
Crew B-35A	**Hr.**	**Daily**	**Hr.**	**Daily**	**Bare Costs**	**Incl. O&P**
1 Laborer Foreman (out)	$28.70	$229.60	$44.70	$357.60	$31.97	$48.98
2 Laborers	26.70	427.20	41.55	664.80		
1 Skilled Worker	34.85	278.80	54.20	433.60		
1 Welder (plumber)	40.85	326.80	61.40	491.20		
1 Equip. Oper. (crane)	35.90	287.20	54.10	432.80		
1 Equip. Oper. Oiler	30.10	240.80	45.35	362.80		
1 Welder, 300 amp		81.20		89.30		
1 Crane, 75 Ton		1170.00		1287.00	22.34	24.58
56 L.H., Daily Totals		$3041.60		$4119.10	$54.31	$73.56
Crew B-36	**Hr.**	**Daily**	**Hr.**	**Daily**	**Bare Costs**	**Incl. O&P**
1 Labor Foreman (outside)	$28.70	$229.60	$44.70	$357.60	$30.28	$46.44
2 Laborers	26.70	427.20	41.55	664.80		
2 Equip. Oper. (med.)	34.65	554.40	52.20	835.20		
1 Dozer, 200 H.P.		919.60		1011.55		
1 Aggregate Spreader		35.00		38.50		
1 Tandem Roller, 10 Ton		179.20		197.10	28.35	31.18
40 L.H., Daily Totals		$2345.00		$3104.75	$58.63	$77.62
Crew B-36A	**Hr.**	**Daily**	**Hr.**	**Daily**	**Bare Costs**	**Incl. O&P**
1 Labor Foreman (outside)	$28.70	$229.60	$44.70	$357.60	$31.53	$48.09
2 Laborers	26.70	427.20	41.55	664.80		
4 Equip. Oper. (med.)	34.65	1108.80	52.20	1670.40		
1 Dozer, 200 H.P.		919.60		1011.55		
1 Aggregate Spreader		35.00		38.50		
1 Roller, Steel Wheel		179.20		197.10		
1 Roller, Pneumatic Wheel		241.80		266.00	24.56	27.02
56 L.H., Daily Totals		$3141.20		$4205.95	$56.09	$75.11
Crew B-36B	**Hr.**	**Daily**	**Hr.**	**Daily**	**Bare Costs**	**Incl. O&P**
1 Labor Foreman (outside)	$28.70	$229.60	$44.70	$357.60	$31.03	$47.33
2 Laborers	26.70	427.20	41.55	664.80		
4 Equip. Oper. (medium)	34.65	1108.80	52.20	1670.40		
1 Truck Driver, Heavy	27.55	220.40	42.00	336.00		
1 Grader, 30,000 Lbs.		456.60		502.25		
1 F.E. Loader, crl, 1.5 C.Y.		368.80		405.70		
1 Dozer, 300 H.P.		1195.00		1314.50		
1 Roller, Vibratory		433.20		476.50		
1 Truck, Tractor, 240 H.P.		353.60		388.95		
1 Water Tanker, 5000 Gal.		116.80		128.50	45.69	50.26
64 L.H., Daily Totals		$4910.00		$6245.20	$76.72	$97.59
Crew B-36C	**Hr.**	**Daily**	**Hr.**	**Daily**	**Bare Costs**	**Incl. O&P**
1 Labor Foreman (outside)	$28.70	$229.60	$44.70	$357.60	$32.04	$48.66
3 Equip. Oper. (medium)	34.65	831.60	52.20	1252.80		
1 Truck Driver, Heavy	27.55	220.40	42.00	336.00		
1 Grader, 30,000 Lbs.		456.60		502.25		
1 Dozer, 300 H.P.		1195.00		1314.50		
1 Roller, Vibratory		433.20		476.50		
1 Truck, Tractor, 240 H.P.		353.60		388.95		
1 Water Tanker, 5000 Gal.		116.80		128.50	63.88	70.27
40 L.H., Daily Totals		$3836.80		$4757.10	$95.92	$118.93
Crew B-37	**Hr.**	**Daily**	**Hr.**	**Daily**	**Bare Costs**	**Incl. O&P**
1 Labor Foreman (outside)	$28.70	$229.60	$44.70	$357.60	$28.09	$43.45
4 Laborers	26.70	854.40	41.55	1329.60		
1 Equip. Oper. (light)	33.05	264.40	49.80	398.40		
1 Tandem Roller, 5 Ton		107.40		118.15	2.24	2.46
48 L.H., Daily Totals		$1455.80		$2203.75	$30.33	$45.91

Crew No.	Bare Costs		Incl. Subs O & P		Cost Per Labor-Hour	
Crew B-38	**Hr.**	**Daily**	**Hr.**	**Daily**	**Bare Costs**	**Incl. O&P**
1 Labor Foreman (outside)	$28.70	$229.60	$44.70	$357.60	$29.96	$45.96
2 Laborers	26.70	427.20	41.55	664.80		
1 Equip. Oper. (light)	33.05	264.40	49.80	398.40		
1 Equip. Oper. (medium)	34.65	277.20	52.20	417.60		
1 Backhoe Loader, 48 H.P.		215.80		237.40		
1 Hyd.Hammer, (1200 lb)		112.00		123.20		
1 F.E. Loader (170 H.P.)		445.60		490.15		
1 Pavt. Rem. Bucket		51.80		57.00	20.63	22.69
40 L.H., Daily Totals		$2023.60		$2746.15	$50.59	$68.65
Crew B-39	**Hr.**	**Daily**	**Hr.**	**Daily**	**Bare Costs**	**Incl. O&P**
1 Labor Foreman (outside)	$28.70	$229.60	$44.70	$357.60	$28.09	$43.45
4 Laborers	26.70	854.40	41.55	1329.60		
1 Equip. Oper. (light)	33.05	264.40	49.80	398.40		
1 Air Compr., 250 C.F.M.		112.40		123.65		
2 Air Tools & Accessories		19.60		21.55		
2-50 Ft. Air Hoses, 1.5" Dia.		9.40		10.35	2.95	3.24
48 L.H., Daily Totals		$1489.80		$2241.15	$31.04	$46.69
Crew B-40	**Hr.**	**Daily**	**Hr.**	**Daily**	**Bare Costs**	**Incl. O&P**
1 Pile Driver Foreman (out)	$35.30	$282.40	$58.15	$465.20	$33.80	$53.91
4 Pile Drivers	33.30	1065.60	54.90	1756.80		
2 Equip. Oper. (crane)	35.90	574.40	54.10	865.60		
1 Equip. Oper. Oiler	30.10	240.80	45.35	362.80		
1 Crane, 40 Ton		888.20		977.00		
1 Vibratory Hammer & Gen.		1403.00		1543.30	35.80	39.38
64 L.H., Daily Totals		$4454.40		$5970.70	$69.60	$93.29
Crew B-40B	**Hr.**	**Daily**	**Hr.**	**Daily**	**Bare Costs**	**Incl. O&P**
1 Laborer Foreman	$28.70	$229.60	$44.70	$357.60	$29.13	$44.80
3 Laborers	26.70	640.80	41.55	997.20		
1 Equip. Oper. (crane)	35.90	287.20	54.10	432.80		
1 Equip. Oper. Oiler	30.10	240.80	45.35	362.80		
1 Crane, 40 Ton		951.90		1047.10	19.83	21.81
48 L.H., Daily Totals		$2350.30		$3197.50	$48.96	$66.61
Crew B-41	**Hr.**	**Daily**	**Hr.**	**Daily**	**Bare Costs**	**Incl. O&P**
1 Labor Foreman (outside)	$28.70	$229.60	$44.70	$357.60	$27.64	$42.87
4 Laborers	26.70	854.40	41.55	1329.60		
.25 Equip. Oper. (crane)	35.90	71.80	54.10	108.20		
.25 Equip. Oper. Oiler	30.10	60.20	45.35	90.70		
.25 Crawler Crane, 40 Ton		222.05		244.25	5.05	5.55
44 L.H., Daily Totals		$1438.05		$2130.35	$32.69	$48.42
Crew B-42	**Hr.**	**Daily**	**Hr.**	**Daily**	**Bare Costs**	**Incl. O&P**
1 Labor Foreman (outside)	$28.70	$229.60	$44.70	$357.60	$29.96	$47.33
4 Laborers	26.70	854.40	41.55	1329.60		
1 Equip. Oper. (crane)	35.90	287.20	54.10	432.80		
1 Equip. Oper. Oiler	30.10	240.80	45.35	362.80		
1 Welder	38.15	305.20	68.25	546.00		
1 Hyd. Crane, 25 Ton		616.80		678.50		
1 Gas Welding Machine		81.20		89.30		
1 Horz. Boring Csg. Mch.		372.40		409.65	16.73	18.40
64 L.H., Daily Totals		$2987.60		$4206.25	$46.69	$65.73

Crew No.	Bare Costs		Incl. Subs O & P		Cost Per Labor-Hour	
Crew B-43	**Hr.**	**Daily**	**Hr.**	**Daily**	**Bare Costs**	**Incl. O&P**
1 Labor Foreman (outside)	$28.70	$229.60	$44.70	$357.60	$29.13	$44.80
3 Laborers	26.70	640.80	41.55	997.20		
1 Equip. Oper. (crane)	35.90	287.20	54.10	432.80		
1 Equip. Oper. Oiler	30.10	240.80	45.35	362.80		
1 Drill Rig & Augers		3179.00		3496.90	66.23	72.85
48 L.H., Daily Totals		$4577.40		$5647.30	$95.36	$117.65
Crew B-44	**Hr.**	**Daily**	**Hr.**	**Daily**	**Bare Costs**	**Incl. O&P**
1 Pile Driver Foreman	$35.30	$282.40	$58.15	$465.20	$33.38	$53.44
4 Pile Drivers	33.30	1065.60	54.90	1756.80		
2 Equip. Oper. (crane)	35.90	574.40	54.10	865.60		
1 Laborer	26.70	213.60	41.55	332.40		
1 Crane, 40 Ton, & Access.		888.20		977.00		
45 L.F. Leads, 15K Ft. Lbs.		56.25		61.90	14.79	16.27
64 L.H., Daily Totals		$3080.45		$4458.90	$48.17	$69.71
Crew B-45	**Hr.**	**Daily**	**Hr.**	**Daily**	**Bare Costs**	**Incl. O&P**
1 Equip. Oper. (med.)	$34.65	$277.20	$52.20	$417.60	$31.10	$47.10
1 Truck Driver (heavy)	27.55	220.40	42.00	336.00		
1 Dist. Tank Truck, 3K Gal.		233.80		257.20		
1 Tractor, 4 x 2, 250 H.P.		291.60		320.75	32.84	36.12
16 L.H., Daily Totals		$1023.00		$1331.55	$63.94	$83.22
Crew B-46	**Hr.**	**Daily**	**Hr.**	**Daily**	**Bare Costs**	**Incl. O&P**
1 Pile Driver Foreman	$35.30	$282.40	$58.15	$465.20	$30.33	$48.77
2 Pile Drivers	33.30	532.80	54.90	878.40		
3 Laborers	26.70	640.80	41.55	997.20		
1 Chain Saw, 36" Long		33.40		36.75	.70	.77
48 L.H., Daily Totals		$1489.40		$2377.55	$31.03	$49.54
Crew B-47	**Hr.**	**Daily**	**Hr.**	**Daily**	**Bare Costs**	**Incl. O&P**
1 Blast Foreman	$28.70	$229.60	$44.70	$357.60	$29.48	$45.35
1 Driller	26.70	213.60	41.55	332.40		
1 Equip. Oper. (light)	33.05	264.40	49.80	398.40		
1 Crawler Type Drill, 4"		664.40		730.85		
1 Air Compr., 600 C.F.M.		297.80		327.60		
2-50 Ft. Air Hoses, 3" Dia.		35.40		38.95	41.57	45.72
24 L.H., Daily Totals		$1705.20		$2185.80	$71.05	$91.07
Crew B-47A	**Hr.**	**Daily**	**Hr.**	**Daily**	**Bare Costs**	**Incl. O&P**
1 Drilling Foreman	$28.70	$229.60	$44.70	$357.60	$31.57	$48.05
1 Equip. Oper. (heavy)	35.90	287.20	54.10	432.80		
1 Oiler	30.10	240.80	45.35	362.80		
1 Quarry Drill		822.60		904.85	34.28	37.70
24 L.H., Daily Totals		$1580.20		$2058.05	$65.85	$85.75
Crew B-47C	**Hr.**	**Daily**	**Hr.**	**Daily**	**Bare Costs**	**Incl. O&P**
1 Laborer	$26.70	$213.60	$41.55	$332.40	$29.88	$45.68
1 Equip. Oper. (light)	33.05	264.40	49.80	398.40		
1 Air Compressor, 750 CFM		317.40		349.15		
2-50' Air Hose, 3"		35.40		38.95		
1 Air Track Drill, 4"		664.40		730.85	63.58	69.93
16 L.H., Daily Totals		$1495.20		$1849.75	$93.46	$115.61
Crew B-47E	**Hr.**	**Daily**	**Hr.**	**Daily**	**Bare Costs**	**Incl. O&P**
1 Laborer Foreman	$28.70	$229.60	$44.70	$357.60	$27.20	$42.34
3 Laborers	26.70	640.80	41.55	997.20		
1 Truck, Flatbed, 3 Ton		176.20		193.80	5.51	6.06
32 L.H., Daily Totals		$1046.60		$1548.60	$32.71	$48.40

Crew No.	Bare Costs		Incl. Subs O & P		Cost Per Labor-Hour	
Crew B-47G	**Hr.**	**Daily**	**Hr.**	**Daily**	**Bare Costs**	**Incl. O&P**
1 Laborer Foreman	$28.70	$229.60	$44.70	$357.60	$28.79	$44.40
2 Laborers	26.70	427.20	41.55	664.80		
1 Equip. Oper. (light)	33.05	264.40	49.80	398.40		
1 Air Track Drill, 4"		664.40		730.85		
1 Air Compr., 600 C.F.M.		297.80		327.60		
2-50 Ft. Air Hoses, 3" Dia.		35.40		38.95		
1 Grout Pump		101.40		111.55	34.34	37.78
32 L.H., Daily Totals		$2020.20		$2629.75	$63.13	$82.18
Crew B-48	**Hr.**	**Daily**	**Hr.**	**Daily**	**Bare Costs**	**Incl. O&P**
1 Labor Foreman (outside)	$28.70	$229.60	$44.70	$357.60	$29.69	$45.51
3 Laborers	26.70	640.80	41.55	997.20		
1 Equip. Oper. (crane)	35.90	287.20	54.10	432.80		
1 Equip. Oper. Oiler	30.10	240.80	45.35	362.80		
1 Equip. Oper. (light)	33.05	264.40	49.80	398.40		
1 Centr. Water Pump, 6"		217.00		238.70		
1-20 Ft. Suction Hose, 6"		12.80		14.10		
1-50 Ft. Disch. Hose, 6"		8.90		9.80		
1 Drill Rig & Augers		3179.00		3496.90	61.03	67.13
56 L.H., Daily Totals		$5080.50		$6308.30	$90.72	$112.64
Crew B-49	**Hr.**	**Daily**	**Hr.**	**Daily**	**Bare Costs**	**Incl. O&P**
1 Labor Foreman (outside)	$28.70	$229.60	$44.70	$357.60	$30.95	$47.99
3 Laborers	26.70	640.80	41.55	997.20		
2 Equip. Oper. (crane)	35.90	574.40	54.10	865.60		
2 Equip. Oper. Oilers	30.10	481.60	45.35	725.60		
1 Equip. Oper. (light)	33.05	264.40	49.80	398.40		
2 Pile Drivers	33.30	532.80	54.90	878.40		
1 Hyd. Crane, 25 Ton		616.80		678.50		
1 Centr. Water Pump, 6"		217.00		238.70		
1-20 Ft. Suction Hose, 6"		12.80		14.10		
1-50 Ft. Disch. Hose, 6"		8.90		9.80		
1 Drill Rig & Augers		3179.00		3496.90	45.85	50.43
88 L.H., Daily Totals		$6758.10		$8660.80	$76.80	$98.42
Crew B-50	**Hr.**	**Daily**	**Hr.**	**Daily**	**Bare Costs**	**Incl. O&P**
2 Pile Driver Foremen	$35.30	$564.80	$58.15	$930.40	$32.31	$51.71
6 Pile Drivers	33.30	1598.40	54.90	2635.20		
2 Equip. Oper. (crane)	35.90	574.40	54.10	865.60		
1 Equip. Oper. Oiler	30.10	240.80	45.35	362.80		
3 Laborers	26.70	640.80	41.55	997.20		
1 Crane, 40 Ton		888.20		977.00		
60 L.F. Leads, 15K Ft. Lbs.		75.00		82.50		
1 Hammer, 15K Ft. Lbs.		359.80		395.80		
1 Air Compr., 600 C.F.M.		297.80		327.60		
2-50 Ft. Air Hoses, 3" Dia.		35.40		38.95		
1 Chain Saw, 36" Long		33.40		36.75	15.11	16.62
112 L.H., Daily Totals		$5308.80		$7649.80	$47.42	$68.33
Crew B-51	**Hr.**	**Daily**	**Hr.**	**Daily**	**Bare Costs**	**Incl. O&P**
1 Labor Foreman (outside)	$28.70	$229.60	$44.70	$357.60	$27.05	$41.96
4 Laborers	26.70	854.40	41.55	1329.60		
1 Truck Driver (light)	26.80	214.40	40.85	326.80		
1 Light Truck, 1.5 Ton		128.80		141.70	2.68	2.95
48 L.H., Daily Totals		$1427.20		$2155.70	$29.73	$44.91

Crew No.	Bare Costs		Incl. Subs O & P		Cost Per Labor-Hour	
Crew B-52	**Hr.**	**Daily**	**Hr.**	**Daily**	**Bare Costs**	**Incl. O&P**
1 Carpenter Foreman	$36.25	$290.00	$56.45	$451.60	$31.39	$48.55
1 Carpenter	34.25	274.00	53.35	426.80		
3 Laborers	26.70	640.80	41.55	997.20		
1 Cement Finisher	32.85	262.80	48.35	386.80		
.5 Rodman (reinf.)	37.95	151.80	62.60	250.40		
.5 Equip. Oper. (med.)	34.65	138.60	52.20	208.80		
.5 F.E. Ldr., T.M., 2.5 C.Y.		394.00		433.40	7.04	7.74
56 L.H., Daily Totals		$2152.00		$3155.00	$38.43	$56.29

Crew B-53	Hr.	Daily	Hr.	Daily	Bare Costs	Incl. O&P
1 Equip. Oper. (light)	$33.05	$264.40	$49.80	$398.40	$33.05	$49.80
1 Trencher, Chain, 12 H.P.		48.00		52.80	6.00	6.60
8 L.H., Daily Totals		$312.40		$451.20	$39.05	$56.40

Crew B-54	Hr.	Daily	Hr.	Daily	Bare Costs	Incl. O&P
1 Equip. Oper. (light)	$33.05	$264.40	$49.80	$398.40	$33.05	$49.80
1 Trencher, Chain, 40 H.P.		213.40		234.75	26.68	29.34
8 L.H., Daily Totals		$477.80		$633.15	$59.73	$79.14

Crew B-54A	Hr.	Daily	Hr.	Daily	Bare Costs	Incl. O&P
.17 Labor Foreman (outside)	$28.70	$39.03	$44.70	$60.79	$33.79	$51.11
1 Equipment Operator (med.)	34.65	277.20	52.20	417.60		
1 Wheel Trencher, 67 H.P.		816.20		897.80	87.20	95.92
9.36 L.H., Daily Totals		$1132.43		$1376.19	$120.99	$147.03

Crew B-54B	Hr.	Daily	Hr.	Daily	Bare Costs	Incl. O&P
.25 Labor Foreman (outside)	$28.70	$57.40	$44.70	$89.40	$33.46	$50.70
1 Equipment Operator (med.)	34.65	277.20	52.20	417.60		
1 Wheel Trencher, 150 H.P.		1460.00		1606.00	146.00	160.60
10 L.H., Daily Totals		$1794.60		$2113.00	$179.46	$211.30

Crew B-55	Hr.	Daily	Hr.	Daily	Bare Costs	Incl. O&P
2 Laborers	$26.70	$427.20	$41.55	$664.80	$26.73	$41.32
1 Truck Driver (light)	26.80	214.40	40.85	326.80		
1 Auger, 4" to 36" Dia		604.60		665.05		
1 Flatbed 3 Ton Truck		176.20		193.80	32.53	35.79
24 L.H., Daily Totals		$1422.40		$1850.45	$59.26	$77.11

Crew B-56	Hr.	Daily	Hr.	Daily	Bare Costs	Incl. O&P
1 Laborer	$26.70	$213.60	$41.55	$332.40	$29.88	$45.68
1 Equip. Oper. (light)	33.05	264.40	49.80	398.40		
1 Crawler Type Drill, 4"		664.40		730.85		
1 Air Compr., 600 C.F.M.		297.80		327.60		
1-50 Ft. Air Hose, 3" Dia.		17.70		19.45	61.24	67.37
16 L.H., Daily Totals		$1457.90		$1808.70	$91.12	$113.05

Crew B-57	Hr.	Daily	Hr.	Daily	Bare Costs	Incl. O&P
1 Labor Foreman (outside)	$28.70	$229.60	$44.70	$357.60	$30.19	$46.18
2 Laborers	26.70	427.20	41.55	664.80		
1 Equip. Oper. (crane)	35.90	287.20	54.10	432.80		
1 Equip. Oper. (light)	33.05	264.40	49.80	398.40		
1 Equip. Oper. Oiler	30.10	240.80	45.35	362.80		
1 Power Shovel, 1 C.Y.		833.20		916.50		
1 Clamshell Bucket, 1 C.Y.		43.40		47.75		
1 Centr. Water Pump, 6"		217.00		238.70		
1-20 Ft. Suction Hose, 6"		12.80		14.10		
20-50 Ft. Disch. Hoses, 6"		178.00		195.80	26.76	29.43
48 L.H., Daily Totals		$2733.60		$3629.25	$56.95	$75.61

Crew No.	Bare Costs		Incl. Subs O & P		Cost Per Labor-Hour	
Crew B-58	**Hr.**	**Daily**	**Hr.**	**Daily**	**Bare Costs**	**Incl. O&P**
2 Laborers	$26.70	$427.20	$41.55	$664.80	$28.82	$44.30
1 Equip. Oper. (light)	33.05	264.40	49.80	398.40		
1 Backhoe Loader, 48 H.P.		215.80		237.40		
1 Small Helicopter, w/pilot		2153.00		2368.30	98.70	108.57
24 L.H., Daily Totals		$3060.40		$3668.90	$127.52	$152.87

Crew B-59	Hr.	Daily	Hr.	Daily	Bare Costs	Incl. O&P
1 Truck Driver (heavy)	$27.55	$220.40	$42.00	$336.00	$27.55	$42.00
1 Truck, 30 Ton		208.20		229.00		
1 Water tank, 6000 Gal.		116.80		128.50	40.63	44.69
8 L.H., Daily Totals		$545.40		$693.50	$68.18	$86.69

Crew B-59A	Hr.	Daily	Hr.	Daily	Bare Costs	Incl. O&P
2 Laborers	$26.70	$427.20	$41.55	$664.80	$26.98	$41.70
1 Truck Driver (heavy)	27.55	220.40	42.00	336.00		
1 Water tank, 5K w/pum		116.80		128.50		
1 Truck, 30 Ton		208.20		229.00	13.54	14.90
24 L.H., Daily Totals		$972.60		$1358.30	$40.52	$56.60

Crew B-60	Hr.	Daily	Hr.	Daily	Bare Costs	Incl. O&P
1 Labor Foreman (outside)	$28.70	$229.60	$44.70	$357.60	$30.60	$46.69
2 Laborers	26.70	427.20	41.55	664.80		
1 Equip. Oper. (crane)	35.90	287.20	54.10	432.80		
2 Equip. Oper. (light)	33.05	528.80	49.80	796.80		
1 Equip. Oper. Oiler	30.10	240.80	45.35	362.80		
1 Crawler Crane, 40 Ton		888.20		977.00		
45 L.F. Leads, 15K Ft. Lbs.		56.25		61.90		
1 Backhoe Loader, 48 H.P.		215.80		237.40	20.76	22.83
56 L.H., Daily Totals		$2873.85		$3891.10	$51.36	$69.52

Crew B-61	Hr.	Daily	Hr.	Daily	Bare Costs	Incl. O&P
1 Labor Foreman (outside)	$28.70	$229.60	$44.70	$357.60	$28.37	$43.83
3 Laborers	26.70	640.80	41.55	997.20		
1 Equip. Oper. (light)	33.05	264.40	49.80	398.40		
1 Cement Mixer, 2 C.Y.		162.00		178.20		
1 Air Compr., 160 C.F.M.		87.40		96.15	6.24	6.86
40 L.H., Daily Totals		$1384.20		$2027.55	$34.61	$50.69

Crew B-62	Hr.	Daily	Hr.	Daily	Bare Costs	Incl. O&P
2 Laborers	$26.70	$427.20	$41.55	$664.80	$28.82	$44.30
1 Equip. Oper. (light)	33.05	264.40	49.80	398.40		
1 Loader, Skid Steer		153.40		168.75	6.39	7.03
24 L.H., Daily Totals		$845.00		$1231.95	$35.21	$51.33

Crew B-63	Hr.	Daily	Hr.	Daily	Bare Costs	Incl. O&P
4 Laborers	$26.70	$854.40	$41.55	$1329.60	$27.97	$43.20
1 Equip. Oper. (light)	33.05	264.40	49.80	398.40		
1 Loader, Skid Steer		153.40		168.75	3.84	4.22
40 L.H., Daily Totals		$1272.20		$1896.75	$31.81	$47.42

Crew B-64	Hr.	Daily	Hr.	Daily	Bare Costs	Incl. O&P
1 Laborer	$26.70	$213.60	$41.55	$332.40	$26.75	$41.20
1 Truck Driver (light)	26.80	214.40	40.85	326.80		
1 Power Mulcher (small)		103.20		113.50		
1 Light Truck, 1.5 Ton		128.80		141.70	14.50	15.95
16 L.H., Daily Totals		$660.00		$914.40	$41.25	$57.15

Crew No.	Bare Costs		Incl. Subs O & P		Cost Per Labor-Hour	
Crew B-65	**Hr.**	**Daily**	**Hr.**	**Daily**	**Bare Costs**	**Incl. O&P**
1 Laborer	$26.70	$213.60	$41.55	$332.40	$26.75	$41.20
1 Truck Driver (light)	26.80	214.40	40.85	326.80		
1 Power Mulcher (large)		184.00		202.40		
1 Light Truck, 1.5 Ton		128.80		141.70	19.55	21.51
16 L.H., Daily Totals		$740.80		$1003.30	$46.30	$62.71
Crew B-66	**Hr.**	**Daily**	**Hr.**	**Daily**	**Bare Costs**	**Incl. O&P**
1 Equip. Oper. (light)	$33.05	$264.40	$49.80	$398.40	$33.05	$49.80
1 Backhoe Ldr. w/Attchmt.		173.20		190.50	21.65	23.82
8 L.H., Daily Totals		$437.60		$588.90	$54.70	$73.62
Crew B-67	**Hr.**	**Daily**	**Hr.**	**Daily**	**Bare Costs**	**Incl. O&P**
1 Millwright	$35.50	$284.00	$52.45	$419.60	$34.28	$51.13
1 Equip. Oper. (light)	33.05	264.40	49.80	398.40		
1 Forklift		233.00		256.30	14.56	16.02
16 L.H., Daily Totals		$781.40		$1074.30	$48.84	$67.15
Crew B-68	**Hr.**	**Daily**	**Hr.**	**Daily**	**Bare Costs**	**Incl. O&P**
2 Millwrights	$35.50	$568.00	$52.45	$839.20	$34.68	$51.57
1 Equip. Oper. (light)	33.05	264.40	49.80	398.40		
1 Forklift		233.00		256.30	9.71	10.68
24 L.H., Daily Totals		$1065.40		$1493.90	$44.39	$62.25
Crew B-69	**Hr.**	**Daily**	**Hr.**	**Daily**	**Bare Costs**	**Incl. O&P**
1 Labor Foreman (outside)	$28.70	$229.60	$44.70	$357.60	$29.13	$44.80
3 Laborers	26.70	640.80	41.55	997.20		
1 Equip Oper. (crane)	35.90	287.20	54.10	432.80		
1 Equip Oper. Oiler	30.10	240.80	45.35	362.80		
1 Truck Crane, 80 Ton		995.00		1094.50	20.73	22.80
48 L.H., Daily Totals		$2393.40		$3244.90	$49.86	$67.60
Crew B-69A	**Hr.**	**Daily**	**Hr.**	**Daily**	**Bare Costs**	**Incl. O&P**
1 Labor Foreman	$28.70	$229.60	$44.70	$357.60	$29.38	$44.93
3 Laborers	26.70	640.80	41.55	997.20		
1 Equip. Oper. (medium)	34.65	277.20	52.20	417.60		
1 Concrete Finisher	32.85	262.80	48.35	386.80		
1 Curb Paver		582.80		641.10	12.14	13.36
48 L.H., Daily Totals		$1993.20		$2800.30	$41.52	$58.29
Crew B-69B	**Hr.**	**Daily**	**Hr.**	**Daily**	**Bare Costs**	**Incl. O&P**
1 Labor Foreman	$28.70	$229.60	$44.70	$357.60	$29.38	$44.93
3 Laborers	26.70	640.80	41.55	997.20		
1 Equip. Oper. (medium)	34.65	277.20	52.20	417.60		
1 Cement Finisher	32.85	262.80	48.35	386.80		
1 Curb/Gutter Paver		725.00		797.50	15.10	16.61
48 L.H., Daily Totals		$2135.40		$2956.70	$44.48	$61.54
Crew B-70	**Hr.**	**Daily**	**Hr.**	**Daily**	**Bare Costs**	**Incl. O&P**
1 Labor Foreman (outside)	$28.70	$229.60	$44.70	$357.60	$30.39	$46.56
3 Laborers	26.70	640.80	41.55	997.20		
3 Equip. Oper. (med.)	34.65	831.60	52.20	1252.80		
1 Motor Grader, 30,000 Lb.		456.60		502.25		
1 Grader Attach., Ripper		71.40		78.55		
1 Road Sweeper, S.P.		445.00		489.50		
1 F.E. Loader, 1-3/4 C.Y.		241.00		265.10	21.68	23.85
56 L.H., Daily Totals		$2916.00		$3943.00	$52.07	$70.41

Crew No.	Bare Costs		Incl. Subs O & P		Cost Per Labor-Hour	
Crew B-71	**Hr.**	**Daily**	**Hr.**	**Daily**	**Bare Costs**	**Incl. O&P**
1 Labor Foreman (outside)	$28.70	$229.60	$44.70	$357.60	$30.39	$46.56
3 Laborers	26.70	640.80	41.55	997.20		
3 Equip. Oper. (med.)	34.65	831.60	52.20	1252.80		
1 Pvmt. Profiler, 750 H.P.		4266.00		4692.60		
1 Road Sweeper, S.P.		445.00		489.50		
1 F.E. Loader, 1-3/4 C.Y.		241.00		265.10	88.43	97.27
56 L.H., Daily Totals		$6654.00		$8054.80	$118.82	$143.83
Crew B-72	**Hr.**	**Daily**	**Hr.**	**Daily**	**Bare Costs**	**Incl. O&P**
1 Labor Foreman (outside)	$28.70	$229.60	$44.70	$357.60	$30.93	$47.27
3 Laborers	26.70	640.80	41.55	997.20		
4 Equip. Oper. (med.)	34.65	1108.80	52.20	1670.40		
1 Pvmt. Profiler, 750 H.P.		4266.00		4692.60		
1 Hammermill, 250 H.P.		1344.00		1478.40		
1 Windrow Loader		791.20		870.30		
1 Mix Paver 165 H.P.		1656.00		1821.60		
1 Roller, Pneu. Tire, 12 T.		241.80		266.00	129.67	142.64
64 L.H., Daily Totals		$10278.20		$12154.10	$160.60	$189.91
Crew B-73	**Hr.**	**Daily**	**Hr.**	**Daily**	**Bare Costs**	**Incl. O&P**
1 Labor Foreman (outside)	$28.70	$229.60	$44.70	$357.60	$31.92	$48.60
2 Laborers	26.70	427.20	41.55	664.80		
5 Equip. Oper. (med.)	34.65	1386.00	52.20	2088.00		
1 Road Mixer, 310 H.P.		1546.00		1700.60		
1 Roller, Tandem, 12 Ton		179.20		197.10		
1 Hammermill, 250 H.P.		1344.00		1478.40		
1 Motor Grader, 30,000 Lb.		456.60		502.25		
.5 F.E. Loader, 1-3/4 C.Y.		120.50		132.55		
.5 Truck, 30 Ton		104.10		114.50		
.5 Water Tank 5000 Gal.		58.40		64.25	59.51	65.46
64 L.H., Daily Totals		$5851.60		$7300.05	$91.43	$114.06
Crew B-74	**Hr.**	**Daily**	**Hr.**	**Daily**	**Bare Costs**	**Incl. O&P**
1 Labor Foreman (outside)	$28.70	$229.60	$44.70	$357.60	$31.14	$47.38
1 Laborer	26.70	213.60	41.55	332.40		
4 Equip. Oper. (med.)	34.65	1108.80	52.20	1670.40		
2 Truck Drivers (heavy)	27.55	440.80	42.00	672.00		
1 Motor Grader, 30,000 Lb.		456.60		502.25		
1 Grader Attach., Ripper		71.40		78.55		
2 Stabilizers, 310 H.P.		2090.00		2299.00		
1 Flatbed Truck, 3 Ton		176.20		193.80		
1 Chem. Spreader, Towed		54.60		60.05		
1 Vibr. Roller, 29,000 Lb.		433.20		476.50		
1 Water Tank 5000 Gal.		116.80		128.50		
1 Truck, 30 Ton		208.20		229.00	56.36	62.00
64 L.H., Daily Totals		$5599.80		$7000.05	$87.50	$109.38
Crew B-75	**Hr.**	**Daily**	**Hr.**	**Daily**	**Bare Costs**	**Incl. O&P**
1 Labor Foreman (outside)	$28.70	$229.60	$44.70	$357.60	$31.65	$48.15
1 Laborer	26.70	213.60	41.55	332.40		
4 Equip. Oper. (med.)	34.65	1108.80	52.20	1670.40		
1 Truck Driver (heavy)	27.55	220.40	42.00	336.00		
1 Motor Grader, 30,000 Lb.		456.60		502.25		
1 Grader Attach., Ripper		71.40		78.55		
2 Stabilizers, 310 H.P.		2090.00		2299.00		
1 Dist. Truck, 3000 Gal.		233.80		257.20		
1 Vibr. Roller, 29,000 Lb.		433.20		476.50	58.66	64.53
56 L.H., Daily Totals		$5057.40		$6309.90	$90.31	$112.68

Crew No.	Bare Costs		Incl. Subs O & P		Cost Per Labor-Hour	

Crew B-76	Hr.	Daily	Hr.	Daily	Bare Costs	Incl. O&P
1 Dock Builder Foreman	$35.30	$282.40	$58.15	$465.20	$33.74	$54.02
5 Dock Builders	33.30	1332.00	54.90	2196.00		
2 Equip. Oper. (crane)	35.90	574.40	54.10	865.60		
1 Equip. Oper. Oiler	30.10	240.80	45.35	362.80		
1 Crawler Crane, 50 Ton		1206.00		1326.60		
1 Barge, 400 Ton		273.20		300.50		
1 Hammer, 15K Ft. Lbs.		359.80		395.80		
60 L.F. Leads, 15K Ft. Lbs.		75.00		82.50		
1 Air Compr., 600 C.F.M.		297.80		327.60		
2-50 Ft. Air Hoses, 3" Dia.		35.40		38.95	31.25	34.38
72 L.H., Daily Totals		$4676.80		$6361.55	$64.99	$88.40

Crew B-76A	Hr.	Daily	Hr.	Daily	Bare Costs	Incl. O&P
1 Laborer Foreman	$28.70	$229.60	$44.70	$357.60	$28.53	$43.99
5 Laborers	26.70	1068.00	41.55	1662.00		
1 Equip. Oper. (crane)	35.90	287.20	54.10	432.80		
1 Equip. Oper. Oiler	30.10	240.80	45.35	362.80		
1 Crawler Crane, 50 Ton		1206.00		1326.60		
1 Barge, 400 Ton		273.20		300.50	23.11	25.42
64 L.H., Daily Totals		$3304.80		$4442.30	$51.64	$69.41

Crew B-77	Hr.	Daily	Hr.	Daily	Bare Costs	Incl. O&P
1 Labor Foreman	$28.70	$229.60	$44.70	$357.60	$27.12	$42.04
3 Laborers	26.70	640.80	41.55	997.20		
1 Truck Driver (light)	26.80	214.40	40.85	326.80		
1 Crack Cleaner, 25 H.P.		44.40		48.85		
1 Crack Filler, Trailer Mtd.		157.00		172.70		
1 Flatbed Truck, 3 Ton		176.20		193.80	9.44	10.38
40 L.H., Daily Totals		$1462.40		$2096.95	$36.56	$52.42

Crew B-78	Hr.	Daily	Hr.	Daily	Bare Costs	Incl. O&P
1 Labor Foreman	$28.70	$229.60	$44.70	$357.60	$27.05	$41.96
4 Laborers	26.70	854.40	41.55	1329.60		
1 Truck Driver (light)	26.80	214.40	40.85	326.80		
1 Paint Striper, S.P.		134.20		147.60		
1 Flatbed Truck, 3 Ton		176.20		193.80		
1 Pickup Truck, 3/4 Ton		78.00		85.80	8.09	8.90
48 L.H., Daily Totals		$1686.80		$2441.20	$35.14	$50.86

Crew B-79	Hr.	Daily	Hr.	Daily	Bare Costs	Incl. O&P
1 Labor Foreman	$28.70	$229.60	$44.70	$357.60	$27.12	$42.04
3 Laborers	26.70	640.80	41.55	997.20		
1 Truck Driver (light)	26.80	214.40	40.85	326.80		
1 Thermo. Striper, T.M.		549.20		604.10		
1 Flatbed Truck, 3 Ton		176.20		193.80		
2 Pickup Truck, 3/4 Ton		156.00		171.60	22.04	24.24
40 L.H., Daily Totals		$1966.20		$2651.10	$49.16	$66.28

Crew B-80	Hr.	Daily	Hr.	Daily	Bare Costs	Incl. O&P
1 Labor Foreman	$28.70	$229.60	$44.70	$357.60	$28.81	$44.23
1 Laborer	26.70	213.60	41.55	332.40		
1 Truck Driver (light)	26.80	214.40	40.85	326.80		
1 Equip. Oper. (light)	33.05	264.40	49.80	398.40		
1 Flatbed Truck, 3 Ton		176.20		193.80		
1 Fence Post Auger, T.M.		357.40		393.15	16.68	18.34
32 L.H., Daily Totals		$1455.60		$2002.15	$45.49	$62.57

Crew B-80A	Hr.	Daily	Hr.	Daily	Bare Costs	Incl. O&P
3 Laborers	$26.70	$640.80	$41.55	$997.20	$26.70	$41.55
1 Flatbed Truck, 3 Ton		176.20		193.80	7.34	8.08
24 L.H., Daily Totals		$817.00		$1191.00	$34.04	$49.63

Crew B-80B	Hr.	Daily	Hr.	Daily	Bare Costs	Incl. O&P
3 Laborers	$26.70	$640.80	$41.55	$997.20	$28.29	$43.61
1 Equip. Oper. (light)	33.05	264.40	49.80	398.40		
1 Crane, Flatbed Mnt.		209.60		230.55	6.55	7.21
32 L.H., Daily Totals		$1114.80		$1626.15	$34.84	$50.82

Crew B-80C	Hr.	Daily	Hr.	Daily	Bare Costs	Incl. O&P
2 Laborers	$26.70	$427.20	$41.55	$664.80	$26.73	$41.32
1 Truck Driver (light)	26.80	214.40	40.85	326.80		
1 Light Truck, 1.5 Ton		128.80		141.70		
1 Manual fence post auger, gas		5.00		5.50	5.58	6.13
24 L.H., Daily Totals		$775.40		$1138.80	$32.31	$47.45

Crew B-81	Hr.	Daily	Hr.	Daily	Bare Costs	Incl. O&P
1 Laborer	$26.70	$213.60	$41.55	$332.40	$29.63	$45.25
1 Equip. Oper. (med.)	34.65	277.20	52.20	417.60		
1 Truck Driver (heavy)	27.55	220.40	42.00	336.00		
1 Hydromulcher, T.M.		196.00		215.60		
1 Tractor Truck, 4x2		208.20		229.00	16.84	18.53
24 L.H., Daily Totals		$1115.40		$1530.60	$46.47	$63.78

Crew B-82	Hr.	Daily	Hr.	Daily	Bare Costs	Incl. O&P
1 Laborer	$26.70	$213.60	$41.55	$332.40	$29.88	$45.68
1 Equip. Oper. (light)	33.05	264.40	49.80	398.40		
1 Horiz. Borer, 6 H.P.		62.60		68.85	3.91	4.30
16 L.H., Daily Totals		$540.60		$799.65	$33.79	$49.98

Crew B-83	Hr.	Daily	Hr.	Daily	Bare Costs	Incl. O&P
1 Tugboat Captain	$34.65	$277.20	$52.20	$417.60	$30.68	$46.88
1 Tugboat Hand	26.70	213.60	41.55	332.40		
1 Tugboat, 250 H.P.		440.60		484.65	27.54	30.29
16 L.H., Daily Totals		$931.40		$1234.65	$58.22	$77.17

Crew B-84	Hr.	Daily	Hr.	Daily	Bare Costs	Incl. O&P
1 Equip. Oper. (med.)	$34.65	$277.20	$52.20	$417.60	$34.65	$52.20
1 Rotary Mower/Tractor		228.80		251.70	28.60	31.46
8 L.H., Daily Totals		$506.00		$669.30	$63.25	$83.66

Crew B-85	Hr.	Daily	Hr.	Daily	Bare Costs	Incl. O&P
3 Laborers	$26.70	$640.80	$41.55	$997.20	$28.46	$43.77
1 Equip. Oper. (med.)	34.65	277.20	52.20	417.60		
1 Truck Driver (heavy)	27.55	220.40	42.00	336.00		
1 Aerial Lift Truck, 80'		496.80		546.50		
1 Brush Chipper, 130 H.P.		164.80		181.30		
1 Pruning Saw, Rotary		8.75		9.65	16.76	18.43
40 L.H., Daily Totals		$1808.75		$2488.25	$45.22	$62.20

Crew B-86	Hr.	Daily	Hr.	Daily	Bare Costs	Incl. O&P
1 Equip. Oper. (med.)	$34.65	$277.20	$52.20	$417.60	$34.65	$52.20
1 Stump Chipper, S.P.		70.70		77.75	8.84	9.72
8 L.H., Daily Totals		$347.90		$495.35	$43.49	$61.92

Crew B-86A	Hr.	Daily	Hr.	Daily	Bare Costs	Incl. O&P
1 Equip. Oper. (medium)	$34.65	$277.20	$52.20	$417.60	$34.65	$52.20
1 Grader, 30,000 Lbs.		456.60		502.25	57.08	62.78
8 L.H., Daily Totals		$733.80		$919.85	$91.73	$114.98

Crew B-86B	Hr.	Daily	Hr.	Daily	Bare Costs	Incl. O&P
1 Equip. Oper. (medium)	$34.65	$277.20	$52.20	$417.60	$34.65	$52.20
1 Dozer, 200 H.P.		919.60		1011.55	114.95	126.45
8 L.H., Daily Totals		$1196.80		$1429.15	$149.60	$178.65

Crew No.	Bare Costs		Incl. Subs O & P		Cost Per Labor-Hour	
Crew B-87	**Hr.**	**Daily**	**Hr.**	**Daily**	**Bare Costs**	**Incl. O&P**
1 Laborer	$26.70	$213.60	$41.55	$332.40	$33.06	$50.07
4 Equip. Oper. (med.)	34.65	1108.80	52.20	1670.40		
2 Feller Bunchers, 50 H.P.		864.80		951.30		
1 Log Chipper, 22″ Tree		1062.00		1168.20		
1 Dozer, 105 H.P.		453.80		499.20		
1 Chainsaw, Gas, 36″ Long		33.40		36.75	60.35	66.39
40 L.H., Daily Totals		$3736.40		$4658.25	$93.41	$116.46
Crew B-88	**Hr.**	**Daily**	**Hr.**	**Daily**	**Bare Costs**	**Incl. O&P**
1 Laborer	$26.70	$213.60	$41.55	$332.40	$33.51	$50.68
6 Equip. Oper. (med.)	34.65	1663.20	52.20	2505.60		
2 Feller Bunchers, 50 H.P.		864.80		951.30		
1 Log Chipper, 22″ Tree		1062.00		1168.20		
2 Log Skidders, 50 H.P.		1490.80		1639.90		
1 Dozer, 105 H.P.		453.80		499.20		
1 Chainsaw, Gas, 36″ Long		33.40		36.75	69.73	76.70
56 L.H., Daily Totals		$5781.60		$7133.35	$103.24	$127.38
Crew B-89	**Hr.**	**Daily**	**Hr.**	**Daily**	**Bare Costs**	**Incl. O&P**
1 Equip. Oper. (light)	$33.05	$264.40	$49.80	$398.40	$29.93	$45.33
1 Truck Driver (light)	26.80	214.40	40.85	326.80		
1 Truck, Stake Body, 3 Ton		176.20		193.80		
1 Concrete Saw		114.40		125.85		
1 Water Tank, 65 Gal.		13.60		14.95	19.01	20.91
16 L.H., Daily Totals		$783.00		$1059.80	$48.94	$66.24
Crew B-89A	**Hr.**	**Daily**	**Hr.**	**Daily**	**Bare Costs**	**Incl. O&P**
1 Skilled Worker	$34.85	$278.80	$54.20	$433.60	$30.78	$47.88
1 Laborer	26.70	213.60	41.55	332.40		
1 Core Drill (large)		103.05		113.35	6.44	7.08
16 L.H., Daily Totals		$595.45		$879.35	$37.22	$54.96
Crew B-89B	**Hr.**	**Daily**	**Hr.**	**Daily**	**Bare Costs**	**Incl. O&P**
1 Equip. Oper. (light)	$33.05	$264.40	$49.80	$398.40	$29.93	$45.33
1 Truck Driver, Light	26.80	214.40	40.85	326.80		
1 Wall Saw, Hydraulic, 10 H.P.		73.30		80.65		
1 Generator, Diesel, 100 KW		176.60		194.25		
1 Water Tank, 65 Gal.		13.60		14.95		
1 Flatbed Truck, 3 Ton		176.20		193.80	27.48	30.23
16 L.H., Daily Totals		$918.50		$1208.85	$57.41	$75.56
Crew B-90	**Hr.**	**Daily**	**Hr.**	**Daily**	**Bare Costs**	**Incl. O&P**
1 Labor Foreman (outside)	$28.70	$229.60	$44.70	$357.60	$28.75	$44.12
3 Laborers	26.70	640.80	41.55	997.20		
2 Equip. Oper. (light)	33.05	528.80	49.80	796.80		
2 Truck Drivers (heavy)	27.55	440.80	42.00	672.00		
1 Road Mixer, 310 H.P.		1546.00		1700.60		
1 Dist. Truck, 2000 Gal.		204.80		225.30	27.36	30.09
64 L.H., Daily Totals		$3590.80		$4749.50	$56.11	$74.21
Crew B-90A	**Hr.**	**Daily**	**Hr.**	**Daily**	**Bare Costs**	**Incl. O&P**
1 Labor Foreman	$28.70	$229.60	$44.70	$357.60	$31.53	$48.09
2 Laborers	26.70	427.20	41.55	664.80		
4 Equip. Oper. (medium)	34.65	1108.80	52.20	1670.40		
2 Graders, 30,000 Lbs.		913.20		1004.50		
1 Roller, Steel Wheel		179.20		197.10		
1 Roller, Pneumatic Wheel		241.80		266.00	23.83	26.21
56 L.H., Daily Totals		$3099.80		$4160.40	$55.36	$74.30

Crew No.	Bare Costs		Incl. Subs O & P		Cost Per Labor-Hour	
Crew B-90B	**Hr.**	**Daily**	**Hr.**	**Daily**	**Bare Costs**	**Incl. O&P**
1 Labor Foreman	$28.70	$229.60	$44.70	$357.60	$31.01	$47.40
2 Laborers	26.70	427.20	41.55	664.80		
3 Equip. Oper. (medium)	34.65	831.60	52.20	1252.80		
1 Roller, Steel Wheel		179.20		197.10		
1 Roller, Pneumatic Wheel		241.80		266.00		
1 Road Mixer, 310 H.P.		1546.00		1700.60	40.98	45.08
48 L.H., Daily Totals		$3455.40		$4438.90	$71.99	$92.48
Crew B-91	**Hr.**	**Daily**	**Hr.**	**Daily**	**Bare Costs**	**Incl. O&P**
1 Labor Foreman (outside)	$28.70	$229.60	$44.70	$357.60	$31.03	$47.33
2 Laborers	26.70	427.20	41.55	664.80		
4 Equip. Oper. (med.)	34.65	1108.80	52.20	1670.40		
1 Truck Driver (heavy)	27.55	220.40	42.00	336.00		
1 Dist. Truck, 3000 Gal.		233.80		257.20		
1 Aggreg. Spreader, S.P.		771.20		848.30		
1 Roller, Pneu. Tire, 12 Ton		241.80		266.00		
1 Roller, Steel, 10 Ton		179.20		197.10	22.28	24.51
64 L.H., Daily Totals		$3412.00		$4597.40	$53.31	$71.84
Crew B-92	**Hr.**	**Daily**	**Hr.**	**Daily**	**Bare Costs**	**Incl. O&P**
1 Labor Foreman (outside)	$28.70	$229.60	$44.70	$357.60	$27.20	$42.34
3 Laborers	26.70	640.80	41.55	997.20		
1 Crack Cleaner, 25 H.P.		44.40		48.85		
1 Air Compressor		55.40		60.95		
1 Tar Kettle, T.M.		40.95		45.05		
1 Flatbed Truck, 3 Ton		176.20		193.80	9.90	10.90
32 L.H., Daily Totals		$1187.35		$1703.45	$37.10	$53.24
Crew B-93	**Hr.**	**Daily**	**Hr.**	**Daily**	**Bare Costs**	**Incl. O&P**
1 Equip. Oper. (med.)	$34.65	$277.20	$52.20	$417.60	$34.65	$52.20
1 Feller Buncher, 50 H.P.		432.40		475.65	54.05	59.46
8 L.H., Daily Totals		$709.60		$893.25	$88.70	$111.66
Crew B-94A	**Hr.**	**Daily**	**Hr.**	**Daily**	**Bare Costs**	**Incl. O&P**
1 Laborer	$26.70	$213.60	$41.55	$332.40	$26.70	$41.55
1 Diaph. Water Pump, 2″		50.60		55.65		
1-20 Ft. Suction Hose, 2″		2.55		2.80		
2-50 Ft. Disch. Hoses, 2″		3.80		4.20	7.12	7.83
8 L.H., Daily Totals		$270.55		$395.05	$33.82	$49.38
Crew B-94B	**Hr.**	**Daily**	**Hr.**	**Daily**	**Bare Costs**	**Incl. O&P**
1 Laborer	$26.70	$213.60	$41.55	$332.40	$26.70	$41.55
1 Diaph. Water Pump, 4″		74.40		81.85		
1-20 Ft. Suction Hose, 4″		5.05		5.55		
2-50 Ft. Disch. Hoses, 4″		7.10		7.80	10.82	11.90
8 L.H., Daily Totals		$300.15		$427.60	$37.52	$53.45
Crew B-94C	**Hr.**	**Daily**	**Hr.**	**Daily**	**Bare Costs**	**Incl. O&P**
1 Laborer	$26.70	$213.60	$41.55	$332.40	$26.70	$41.55
1 Centr. Water Pump, 3″		53.60		58.95		
1-20 Ft. Suction Hose, 3″		4.25		4.70		
2-50 Ft. Disch. Hoses, 3″		5.00		5.50	7.86	8.64
8 L.H., Daily Totals		$276.45		$401.55	$34.56	$50.19
Crew B-94D	**Hr.**	**Daily**	**Hr.**	**Daily**	**Bare Costs**	**Incl. O&P**
1 Laborer	$26.70	$213.60	$41.55	$332.40	$26.70	$41.55
1 Centr. Water Pump, 6″		217.00		238.70		
1-20 Ft. Suction Hose, 6″		12.80		14.10		
2-50 Ft. Disch. Hoses, 6″		17.80		19.60	30.95	34.05
8 L.H., Daily Totals		$461.20		$604.80	$57.65	$75.60

Crews

Crew B-95A	Hr.	Daily	Hr.	Daily	Bare Costs	Incl. O&P
1 Equip. Oper. (crane)	$35.90	$287.20	$54.10	$432.80	$31.30	$47.83
1 Laborer	26.70	213.60	41.55	332.40		
1 Hyd. Excavator, 5/8 C.Y.		432.40		475.65	27.03	29.73
16 L.H., Daily Totals		$933.20		$1240.85	$58.33	$77.56

Crew B-95B	Hr.	Daily	Hr.	Daily	Bare Costs	Incl. O&P
1 Equip. Oper. (crane)	$35.90	$287.20	$54.10	$432.80	$31.30	$47.83
1 Laborer	26.70	213.60	41.55	332.40		
1 Hyd. Excavator, 1.5 C.Y.		720.40		792.45	45.03	49.53
16 L.H., Daily Totals		$1221.20		$1557.65	$76.33	$97.36

Crew B-95C	Hr.	Daily	Hr.	Daily	Bare Costs	Incl. O&P
1 Equip. Oper. (crane)	$35.90	$287.20	$54.10	$432.80	$31.30	$47.83
1 Laborer	26.70	213.60	41.55	332.40		
1 Hyd. Excavator, 2.5 C.Y.		1215.00		1336.50	75.94	83.53
16 L.H., Daily Totals		$1715.80		$2101.70	$107.24	$131.36

Crew C-1	Hr.	Daily	Hr.	Daily	Bare Costs	Incl. O&P
3 Carpenters	$34.25	$822.00	$53.35	$1280.40	$32.36	$50.40
1 Laborer	26.70	213.60	41.55	332.40		
32 L.H., Daily Totals		$1035.60		$1612.80	$32.36	$50.40

Crew C-2	Hr.	Daily	Hr.	Daily	Bare Costs	Incl. O&P
1 Carpenter Foreman (out)	$36.25	$290.00	$56.45	$451.60	$33.33	$51.90
4 Carpenters	34.25	1096.00	53.35	1707.20		
1 Laborer	26.70	213.60	41.55	332.40		
48 L.H., Daily Totals		$1599.60		$2491.20	$33.33	$51.90

Crew C-2A	Hr.	Daily	Hr.	Daily	Bare Costs	Incl. O&P
1 Carpenter Foreman (out)	$36.25	$290.00	$56.45	$451.60	$33.09	$51.01
3 Carpenters	34.25	822.00	53.35	1280.40		
1 Cement Finisher	32.85	262.80	48.35	386.80		
1 Laborer	26.70	213.60	41.55	332.40		
48 L.H., Daily Totals		$1588.40		$2451.20	$33.09	$51.01

Crew C-3	Hr.	Daily	Hr.	Daily	Bare Costs	Incl. O&P
1 Rodman Foreman	$39.95	$319.60	$65.90	$527.20	$34.78	$56.15
4 Rodmen (reinf.)	37.95	1214.40	62.60	2003.20		
1 Equip. Oper. (light)	33.05	264.40	49.80	398.40		
2 Laborers	26.70	427.20	41.55	664.80		
3 Stressing Equipment		40.80		44.90		
.5 Grouting Equipment		76.40		84.05	1.83	2.01
64 L.H., Daily Totals		$2342.80		$3722.55	$36.61	$58.16

Crew C-4	Hr.	Daily	Hr.	Daily	Bare Costs	Incl. O&P
1 Rodman Foreman	$39.95	$319.60	$65.90	$527.20	$38.45	$63.43
3 Rodmen (reinf.)	37.95	910.80	62.60	1502.40		
3 Stressing Equipment		40.80		44.90	1.28	1.40
32 L.H., Daily Totals		$1271.20		$2074.50	$39.73	$64.83

Crew C-5	Hr.	Daily	Hr.	Daily	Bare Costs	Incl. O&P
1 Rodman Foreman	$39.95	$319.60	$65.90	$527.20	$36.82	$59.39
4 Rodmen (reinf.)	37.95	1214.40	62.60	2003.20		
1 Equip. Oper. (crane)	35.90	287.20	54.10	432.80		
1 Equip. Oper. Oiler	30.10	240.80	45.35	362.80		
1 Hyd. Crane, 25 Ton		616.80		678.50	11.01	12.12
56 L.H., Daily Totals		$2678.80		$4004.50	$47.83	$71.51

Crew C-6	Hr.	Daily	Hr.	Daily	Bare Costs	Incl. O&P
1 Labor Foreman (outside)	$28.70	$229.60	$44.70	$357.60	$28.06	$43.15
4 Laborers	26.70	854.40	41.55	1329.60		
1 Cement Finisher	32.85	262.80	48.35	386.80		
2 Gas Engine Vibrators		48.00		52.80	1.00	1.10
48 L.H., Daily Totals		$1394.80		$2126.80	$29.06	$44.25

Crew C-7	Hr.	Daily	Hr.	Daily	Bare Costs	Incl. O&P
1 Labor Foreman (outside)	$28.70	$229.60	$44.70	$357.60	$28.87	$44.22
5 Laborers	26.70	1068.00	41.55	1662.00		
1 Cement Finisher	32.85	262.80	48.35	386.80		
1 Equip. Oper. (med.)	34.65	277.20	52.20	417.60		
1 Equip. Oper. (oiler)	30.10	240.80	45.35	362.80		
2 Gas Engine Vibrators		48.00		52.80		
1 Concrete Bucket, 1 C.Y.		16.00		17.60		
1 Hyd. Crane, 55 Ton		895.40		984.95	13.33	14.66
72 L.H., Daily Totals		$3037.80		$4242.15	$42.20	$58.88

Crew C-7A	Hr.	Daily	Hr.	Daily	Bare Costs	Incl. O&P
1 Labor Foreman (outside)	$28.70	$229.60	$44.70	$357.60	$27.16	$42.06
5 Laborers	26.70	1068.00	41.55	1662.00		
2 Truck Drivers(Heavy)	27.55	440.80	42.00	672.00		
2 Conc. Transit Mixers		1480.80		1628.90	23.14	25.45
64 L.H., Daily Totals		$3219.20		$4320.50	$50.30	$67.51

Crew C-7B	Hr.	Daily	Hr.	Daily	Bare Costs	Incl. O&P
1 Labor Foreman (outside)	$28.70	$229.60	$44.70	$357.60	$28.53	$43.99
5 Laborers	26.70	1068.00	41.55	1662.00		
1 Equipment Operator (heavy)	35.90	287.20	54.10	432.80		
1 Equipment Oiler	30.10	240.80	45.35	362.80		
1 Conc. Bucket, 2 C.Y.		24.60		27.05		
1 Truck Crane, 165 Ton		1912.00		2103.20	30.26	33.29
64 L.H., Daily Totals		$3762.20		$4945.45	$58.79	$77.28

Crew C-7C	Hr.	Daily	Hr.	Daily	Bare Costs	Incl. O&P
1 Labor Foreman (outside)	$28.70	$229.60	$44.70	$357.60	$28.94	$44.61
5 Laborers	26.70	1068.00	41.55	1662.00		
2 Equipment Operator (medium)	34.65	554.40	52.20	835.20		
2 Wheel Loader, 4 C.Y.		891.20		980.30	13.93	15.32
64 L.H., Daily Totals		$2743.20		$3835.10	$42.87	$59.93

Crew C-7D	Hr.	Daily	Hr.	Daily	Bare Costs	Incl. O&P
1 Labor Foreman (outside)	$28.70	$229.60	$44.70	$357.60	$28.12	$43.52
5 Laborers	26.70	1068.00	41.55	1662.00		
1 Equipment Operator (med.)	34.65	277.20	52.20	417.60		
1 Concrete Conveyer		152.80		168.10	2.73	3.00
56 L.H., Daily Totals		$1727.60		$2605.30	$30.85	$46.52

Crew C-8	Hr.	Daily	Hr.	Daily	Bare Costs	Incl. O&P
1 Labor Foreman (outside)	$28.70	$229.60	$44.70	$357.60	$29.88	$45.36
3 Laborers	26.70	640.80	41.55	997.20		
2 Cement Finishers	32.85	525.60	48.35	773.60		
1 Equip. Oper. (med.)	34.65	277.20	52.20	417.60		
1 Concrete Pump (small)		704.20		774.60	12.58	13.83
56 L.H., Daily Totals		$2377.40		$3320.60	$42.46	$59.19

Crew C-8A	Hr.	Daily	Hr.	Daily	Bare Costs	Incl. O&P
1 Labor Foreman (outside)	$28.70	$229.60	$44.70	$357.60	$29.08	$44.23
3 Laborers	26.70	640.80	41.55	997.20		
2 Cement Finishers	32.85	525.60	48.35	773.60		
48 L.H., Daily Totals		$1396.00		$2128.40	$29.08	$44.23

Crew C-8B	Bare Costs Hr.	Bare Costs Daily	Incl. Subs O & P Hr.	Incl. Subs O & P Daily	Bare Costs	Incl. O&P
1 Labor Foreman (outside)	$28.70	$229.60	$44.70	$357.60	$28.69	$44.31
3 Laborers	26.70	640.80	41.55	997.20		
1 Equipment Operator	34.65	277.20	52.20	417.60		
1 Vibrating Screed		43.05		47.35		
1 Vibratory Roller		433.20		476.50		
1 Dozer, 200 H.P.		919.60		1011.55	34.90	38.38
40 L.H., Daily Totals		$2543.45		$3307.80	$63.59	$82.69

Crew C-8C	Hr.	Daily	Hr.	Daily	Bare Costs	Incl. O&P
1 Labor Foreman (outside)	$28.70	$229.60	$44.70	$357.60	$29.38	$44.93
3 Laborers	26.70	640.80	41.55	997.20		
1 Cement Finisher	32.85	262.80	48.35	386.80		
1 Equipment Operator (med.)	34.65	277.20	52.20	417.60		
1 Shotcrete Rig, 12 CY/hr		223.80		246.20	4.66	5.13
48 L.H., Daily Totals		$1634.20		$2405.40	$34.04	$50.06

Crew C-8D	Hr.	Daily	Hr.	Daily	Bare Costs	Incl. O&P
1 Labor Foreman (outside)	$28.70	$229.60	$44.70	$357.60	$30.33	$46.01
1 Laborer	26.70	213.60	41.55	332.40		
1 Cement Finisher	32.85	262.80	48.35	386.80		
1 Equipment Operator (light)	33.05	264.40	49.80	398.40		
1 Compressor, 250 CFM		112.40		123.65		
2 Hoses, 1", 50'		7.30		8.05	3.74	4.11
32 L.H., Daily Totals		$1090.10		$1606.90	$34.07	$50.12

Crew C-8E	Hr.	Daily	Hr.	Daily	Bare Costs	Incl. O&P
1 Labor Foreman (outside)	$28.70	$229.60	$44.70	$357.60	$30.33	$46.01
1 Laborer	26.70	213.60	41.55	332.40		
1 Cement Finisher	32.85	262.80	48.35	386.80		
1 Equipment Operator (light)	33.05	264.40	49.80	398.40		
1 Compressor, 250 CFM		112.40		123.65		
2 Hoses, 1", 50'		7.30		8.05		
1 Concrete Pump (small)		704.20		774.60	25.75	28.32
32 L.H., Daily Totals		$1794.30		$2381.50	$56.08	$74.33

Crew C-10	Hr.	Daily	Hr.	Daily	Bare Costs	Incl. O&P
1 Laborer	$26.70	$213.60	$41.55	$332.40	$30.80	$45.85
2 Cement Finishers	32.85	525.60	48.35	773.60		
24 L.H., Daily Totals		$739.20		$1106.00	$30.80	$45.85

Crew C-10B	Hr.	Daily	Hr.	Daily	Bare Costs	Incl. O&P
3 Laborers	$26.70	$640.80	$41.55	$997.20	$29.16	$44.13
2 Cement Finishers	32.85	525.60	48.35	773.60		
1 Concrete mixer, 10 CF		123.20		135.50		
2 Concrete finisher, 48" dia		49.20		54.10	4.31	4.74
40 L.H., Daily Totals		$1338.80		$1960.40	$33.47	$48.87

Crew C-11	Hr.	Daily	Hr.	Daily	Bare Costs	Incl. O&P
1 Struc. Steel Foreman	$40.15	$321.20	$71.85	$574.80	$37.23	$64.53
6 Struc. Steel Workers	38.15	1831.20	68.25	3276.00		
1 Equip. Oper. (crane)	35.90	287.20	54.10	432.80		
1 Equip. Oper. Oiler	30.10	240.80	45.35	362.80		
1 Truck Crane, 150 Ton		1445.00		1589.50	20.07	22.08
72 L.H., Daily Totals		$4125.40		$6235.90	$57.30	$86.61

Crew C-12	Hr.	Daily	Hr.	Daily	Bare Costs	Incl. O&P
1 Carpenter Foreman (out)	$36.25	$290.00	$56.45	$451.60	$33.60	$52.03
3 Carpenters	34.25	822.00	53.35	1280.40		
1 Laborer	26.70	213.60	41.55	332.40		
1 Equip. Oper. (crane)	35.90	287.20	54.10	432.80		
1 Hyd. Crane, 12 Ton		602.60		662.85	12.55	13.81
48 L.H., Daily Totals		$2215.40		$3160.05	$46.15	$65.84

Crew C-13	Hr.	Daily	Hr.	Daily	Bare Costs	Incl. O&P
1 Struc. Steel Worker	$38.15	$305.20	$68.25	$546.00	$36.85	$63.28
1 Welder	38.15	305.20	68.25	546.00		
1 Carpenter	34.25	274.00	53.35	426.80		
1 Gas Welding Machine		81.20		89.30	3.38	3.72
24 L.H., Daily Totals		$965.60		$1608.10	$40.23	$67.00

Crew C-14	Hr.	Daily	Hr.	Daily	Bare Costs	Incl. O&P
1 Carpenter Foreman (out)	$36.25	$290.00	$56.45	$451.60	$33.21	$51.96
5 Carpenters	34.25	1370.00	53.35	2134.00		
4 Laborers	26.70	854.40	41.55	1329.60		
4 Rodmen (reinf.)	37.95	1214.40	62.60	2003.20		
2 Cement Finishers	32.85	525.60	48.35	773.60		
1 Equip. Oper. (crane)	35.90	287.20	54.10	432.80		
1 Equip. Oper. Oiler	30.10	240.80	45.35	362.80		
1 Crane, 80 Ton, & Tools		995.00		1094.50	6.91	7.60
144 L.H., Daily Totals		$5777.40		$8582.10	$40.12	$59.56

Crew C-14A	Hr.	Daily	Hr.	Daily	Bare Costs	Incl. O&P
1 Carpenter Foreman (out)	$36.25	$290.00	$56.45	$451.60	$34.28	$53.75
16 Carpenters	34.25	4384.00	53.35	6828.80		
4 Rodmen (reinf.)	37.95	1214.40	62.60	2003.20		
2 Laborers	26.70	427.20	41.55	664.80		
1 Cement Finisher	32.85	262.80	48.35	386.80		
1 Equip. Oper. (med.)	34.65	277.20	52.20	417.60		
1 Gas Engine Vibrator		24.00		26.40		
1 Concrete Pump (small)		704.20		774.60	3.64	4.01
200 L.H., Daily Totals		$7583.80		$11553.80	$37.92	$57.76

Crew C-14B	Hr.	Daily	Hr.	Daily	Bare Costs	Incl. O&P
1 Carpenter Foreman (out)	$36.25	$290.00	$56.45	$451.60	$34.22	$53.53
16 Carpenters	34.25	4384.00	53.35	6828.80		
4 Rodmen (reinf.)	37.95	1214.40	62.60	2003.20		
2 Laborers	26.70	427.20	41.55	664.80		
2 Cement Finishers	32.85	525.60	48.35	773.60		
1 Equip. Oper. (med.)	34.65	277.20	52.20	417.60		
1 Gas Engine Vibrator		24.00		26.40		
1 Concrete Pump (small)		704.20		774.60	3.50	3.85
208 L.H., Daily Totals		$7846.60		$11940.60	$37.72	$57.38

Crew C-14C	Hr.	Daily	Hr.	Daily	Bare Costs	Incl. O&P
1 Carpenter Foreman (out)	$36.25	$290.00	$56.45	$451.60	$32.66	$51.14
6 Carpenters	34.25	1644.00	53.35	2560.80		
2 Rodmen (reinf.)	37.95	607.20	62.60	1001.60		
4 Laborers	26.70	854.40	41.55	1329.60		
1 Cement Finisher	32.85	262.80	48.35	386.80		
1 Gas Engine Vibrator		24.00		26.40	.21	.24
112 L.H., Daily Totals		$3682.40		$5756.80	$32.87	$51.38

Crew C-14D	Bare Costs Hr.	Bare Costs Daily	Incl. Subs O & P Hr.	Incl. Subs O & P Daily	Cost Per Labor-Hour Bare Costs	Cost Per Labor-Hour Incl. O&P
1 Carpenter Foreman (out)	$36.25	$290.00	$56.45	$451.60	$33.98	$53.01
18 Carpenters	34.25	4932.00	53.35	7682.40		
2 Rodmen (reinf.)	37.95	607.20	62.60	1001.60		
2 Laborers	26.70	427.20	41.55	664.80		
1 Cement Finisher	32.85	262.80	48.35	386.80		
1 Equip. Oper. (med.)	34.65	277.20	52.20	417.60		
1 Gas Engine Vibrator		24.00		26.40		
1 Concrete Pump (small)		704.20		774.60	3.64	4.01
200 L.H., Daily Totals		$7524.60		$11405.80	$37.62	$57.02

Crew C-14E	Hr.	Daily	Hr.	Daily	Bare Costs	Incl. O&P
1 Carpenter Foreman (out)	$36.25	$290.00	$56.45	$451.60	$33.59	$53.29
2 Carpenters	34.25	548.00	53.35	853.60		
4 Rodmen (reinf.)	37.95	1214.40	62.60	2003.20		
3 Laborers	26.70	640.80	41.55	997.20		
1 Cement Finisher	32.85	262.80	48.35	386.80		
1 Gas Engine Vibrator		24.00		26.40	.27	.30
88 L.H., Daily Totals		$2980.00		$4718.80	$33.86	$53.59

Crew C-14F	Hr.	Daily	Hr.	Daily	Bare Costs	Incl. O&P
1 Laborer Foreman (out)	$28.70	$229.60	$44.70	$357.60	$31.02	$46.20
2 Laborers	26.70	427.20	41.55	664.80		
6 Cement Finishers	32.85	1576.80	48.35	2320.80		
1 Gas Engine Vibrator		24.00		26.40	.33	.37
72 L.H., Daily Totals		$2257.60		$3369.60	$31.35	$46.57

Crew C-14G	Hr.	Daily	Hr.	Daily	Bare Costs	Incl. O&P
1 Laborer Foreman (out)	$28.70	$229.60	$44.70	$357.60	$30.50	$45.69
2 Laborers	26.70	427.20	41.55	664.80		
4 Cement Finishers	32.85	1051.20	48.35	1547.20		
1 Gas Engine Vibrator		24.00		26.40	.43	.47
56 L.H., Daily Totals		$1732.00		$2596.00	$30.93	$46.16

Crew C-14H	Hr.	Daily	Hr.	Daily	Bare Costs	Incl. O&P
1 Carpenter Foreman (out)	$36.25	$290.00	$56.45	$451.60	$33.71	$52.55
2 Carpenters	34.25	548.00	53.35	853.60		
1 Rodman (reinf.)	37.95	303.60	62.60	500.80		
1 Laborer	26.70	213.60	41.55	332.40		
1 Cement Finisher	32.85	262.80	48.35	386.80		
1 Gas Engine Vibrator		24.00		26.40	.50	.55
48 L.H., Daily Totals		$1642.00		$2551.60	$34.21	$53.10

Crew C-15	Hr.	Daily	Hr.	Daily	Bare Costs	Incl. O&P
1 Carpenter Foreman (out)	$36.25	$290.00	$56.45	$451.60	$32.06	$49.60
2 Carpenters	34.25	548.00	53.35	853.60		
3 Laborers	26.70	640.80	41.55	997.20		
2 Cement Finishers	32.85	525.60	48.35	773.60		
1 Rodman (reinf.)	37.95	303.60	62.60	500.80		
72 L.H., Daily Totals		$2308.00		$3576.80	$32.06	$49.60

Crew C-16	Hr.	Daily	Hr.	Daily	Bare Costs	Incl. O&P
1 Labor Foreman (outside)	$28.70	$229.60	$44.70	$357.60	$31.67	$49.19
3 Laborers	26.70	640.80	41.55	997.20		
2 Cement Finishers	32.85	525.60	48.35	773.60		
1 Equip. Oper. (med.)	34.65	277.20	52.20	417.60		
2 Rodmen (reinf.)	37.95	607.20	62.60	1001.60		
1 Concrete Pump (small)		704.20		774.60	9.78	10.76
72 L.H., Daily Totals		$2984.60		$4322.20	$41.45	$59.95

Crew C-17	Hr.	Daily	Hr.	Daily	Bare Costs	Incl. O&P
2 Skilled Worker Foremen	$36.85	$589.60	$57.30	$916.80	$35.25	$54.82
8 Skilled Workers	34.85	2230.40	54.20	3468.80		
80 L.H., Daily Totals		$2820.00		$4385.60	$35.25	$54.82

Crew C-17A	Hr.	Daily	Hr.	Daily	Bare Costs	Incl. O&P
2 Skilled Worker Foremen	$36.85	$589.60	$57.30	$916.80	$35.26	$54.81
8 Skilled Workers	34.85	2230.40	54.20	3468.80		
.125 Equip. Oper. (crane)	35.90	35.90	54.10	54.10		
.125 Crane, 80 Ton, & Tools		124.38		136.80	1.54	1.69
81 L.H., Daily Totals		$2980.28		$4576.50	$36.80	$56.50

Crew C-17B	Hr.	Daily	Hr.	Daily	Bare Costs	Incl. O&P
2 Skilled Worker Foremen	$36.85	$589.60	$57.30	$916.80	$35.27	$54.80
8 Skilled Workers	34.85	2230.40	54.20	3468.80		
.25 Equip. Oper. (crane)	35.90	71.80	54.10	108.20		
.25 Crane, 80 Ton, & Tools		248.75		273.65		
.25 Walk Behind Power Tools		6.15		6.75	3.11	3.42
82 L.H., Daily Totals		$3146.70		$4774.20	$38.38	$58.22

Crew C-17C	Hr.	Daily	Hr.	Daily	Bare Costs	Incl. O&P
2 Skilled Worker Foremen	$36.85	$589.60	$57.30	$916.80	$35.27	$54.79
8 Skilled Workers	34.85	2230.40	54.20	3468.80		
.375 Equip. Oper. (crane)	35.90	107.70	54.10	162.30		
.375 Crane, 80 Ton & Tools		373.13		410.45	4.50	4.95
83 L.H., Daily Totals		$3300.83		$4958.35	$39.77	$59.74

Crew C-17D	Hr.	Daily	Hr.	Daily	Bare Costs	Incl. O&P
2 Skilled Worker Foremen	$36.85	$589.60	$57.30	$916.80	$35.28	$54.79
8 Skilled Workers	34.85	2230.40	54.20	3468.80		
.5 Equip. Oper. (crane)	35.90	143.60	54.10	216.40		
.5 Crane, 80 Ton & Tools		497.50		547.25	5.92	6.51
84 L.H., Daily Totals		$3461.10		$5149.25	$41.20	$61.30

Crew C-17E	Hr.	Daily	Hr.	Daily	Bare Costs	Incl. O&P
2 Skilled Worker Foremen	$36.85	$589.60	$57.30	$916.80	$35.25	$54.82
8 Skilled Workers	34.85	2230.40	54.20	3468.80		
1 Hyd. Jack with Rods		79.00		86.90	.99	1.09
80 L.H., Daily Totals		$2899.00		$4472.50	$36.24	$55.91

Crew C-18	Hr.	Daily	Hr.	Daily	Bare Costs	Incl. O&P
.125 Labor Foreman (out)	$28.70	$28.70	$44.70	$44.70	$26.92	$41.90
1 Laborer	26.70	213.60	41.55	332.40		
1 Concrete Cart, 10 C.F.		49.80		54.80	5.53	6.09
9 L.H., Daily Totals		$292.10		$431.90	$32.45	$47.99

Crew C-19	Hr.	Daily	Hr.	Daily	Bare Costs	Incl. O&P
.125 Labor Foreman (out)	$28.70	$28.70	$44.70	$44.70	$26.92	$41.90
1 Laborer	26.70	213.60	41.55	332.40		
1 Concrete Cart, 18 C.F.		76.80		84.50	8.53	9.39
9 L.H., Daily Totals		$319.10		$461.60	$35.45	$51.29

Crew C-20	Hr.	Daily	Hr.	Daily	Bare Costs	Incl. O&P
1 Labor Foreman (outside)	$28.70	$229.60	$44.70	$357.60	$28.71	$44.08
5 Laborers	26.70	1068.00	41.55	1662.00		
1 Cement Finisher	32.85	262.80	48.35	386.80		
1 Equip. Oper. (med.)	34.65	277.20	52.20	417.60		
2 Gas Engine Vibrators		48.00		52.80		
1 Concrete Pump (small)		704.20		774.60	11.75	12.93
64 L.H., Daily Totals		$2589.80		$3651.40	$40.46	$57.01

Crew No.	Bare Costs		Incl. Subs O & P		Cost Per Labor-Hour	
Crew C-21	**Hr.**	**Daily**	**Hr.**	**Daily**	**Bare Costs**	**Incl. O&P**
1 Labor Foreman (outside)	$28.70	$229.60	$44.70	$357.60	$28.71	$44.08
5 Laborers	26.70	1068.00	41.55	1662.00		
1 Cement Finisher	32.85	262.80	48.35	386.80		
1 Equip. Oper. (med.)	34.65	277.20	52.20	417.60		
2 Gas Engine Vibrators		48.00		52.80		
1 Concrete Conveyer		152.80		168.10	3.14	3.45
64 L.H., Daily Totals		$2038.40		$3044.90	$31.85	$47.53
Crew C-22	**Hr.**	**Daily**	**Hr.**	**Daily**	**Bare Costs**	**Incl. O&P**
1 Rodman Foreman	$39.95	$319.60	$65.90	$527.20	$38.10	$62.62
4 Rodmen (reinf.)	37.95	1214.40	62.60	2003.20		
.125 Equip. Oper. (crane)	35.90	35.90	54.10	54.10		
.125 Equip. Oper. Oiler	30.10	30.10	45.35	45.35		
.125 Hyd. Crane, 25 Ton		77.10		84.80	1.84	2.02
42 L.H., Daily Totals		$1677.10		$2714.65	$39.94	$64.64
Crew C-23	**Hr.**	**Daily**	**Hr.**	**Daily**	**Bare Costs**	**Incl. O&P**
2 Skilled Worker Foremen	$36.85	$589.60	$57.30	$916.80	$34.88	$53.93
6 Skilled Workers	34.85	1672.80	54.20	2601.60		
1 Equip. Oper. (crane)	35.90	287.20	54.10	432.80		
1 Equip. Oper. Oiler	30.10	240.80	45.35	362.80		
1 Crane, 90 Ton		1325.00		1457.50	16.56	18.22
80 L.H., Daily Totals		$4115.40		$5771.50	$51.44	$72.15
Crew C-23A	**Hr.**	**Daily**	**Hr.**	**Daily**	**Bare Costs**	**Incl. O&P**
1 Labor Foreman (outside)	$28.70	$229.60	$44.70	$357.60	$29.62	$45.45
2 Laborers	26.70	427.20	41.55	664.80		
1 Equip. Oper. (crane)	35.90	287.20	54.10	432.80		
1 Equip. Oper. Oiler	30.10	240.80	45.35	362.80		
1 Crane, 100 ton capacity		1519.00		1670.90		
3 Conc. bucket, 8 C.Y.		465.60		512.15	49.62	54.58
40 L.H., Daily Totals		$3169.40		$4001.05	$79.24	$100.03
Crew C-24	**Hr.**	**Daily**	**Hr.**	**Daily**	**Bare Costs**	**Incl. O&P**
2 Skilled Worker Foremen	$36.85	$589.60	$57.30	$916.80	$34.88	$53.93
6 Skilled Workers	34.85	1672.80	54.20	2601.60		
1 Equip. Oper. (crane)	35.90	287.20	54.10	432.80		
1 Equip. Oper. Oiler	30.10	240.80	45.35	362.80		
1 Truck Crane, 150 Ton		1445.00		1589.50	18.06	19.87
80 L.H., Daily Totals		$4235.40		$5903.50	$52.94	$73.80
Crew C-25	**Hr.**	**Daily**	**Hr.**	**Daily**	**Bare Costs**	**Incl. O&P**
2 Rodmen (reinf.)	$37.95	$607.20	$62.60	$1001.60	$29.78	$49.70
2 Rodmen Helpers	21.60	345.60	36.80	588.80		
32 L.H., Daily Totals		$952.80		$1590.40	$29.78	$49.70
Crew C-27	**Hr.**	**Daily**	**Hr.**	**Daily**	**Bare Costs**	**Incl. O&P**
2 Cement Finishers	$32.85	$525.60	$48.35	$773.60	$32.85	$48.00
1 Concrete Saw		114.40		125.85	7.15	7.87
16 L.H., Daily Totals		$640.00		$899.45	$40.00	$55.87
Crew C-28	**Hr.**	**Daily**	**Hr.**	**Daily**	**Bare Costs**	**Incl. O&P**
1 Cement Finisher	$32.85	$262.80	$48.35	$386.80	$32.85	$48.00
1 Portable Air Compressor		17.30		19.05	2.16	2.37
8 L.H., Daily Totals		$280.10		$405.85	$35.01	$50.37
Crew D-1	**Hr.**	**Daily**	**Hr.**	**Daily**	**Bare Costs**	**Incl. O&P**
1 Bricklayer	$35.25	$282.00	$53.70	$429.60	$31.08	$47.35
1 Bricklayer Helper	26.90	215.20	41.00	328.00		
16 L.H., Daily Totals		$497.20		$757.60	$31.08	$47.35

Crew No.	Bare Costs		Incl. Subs O & P		Cost Per Labor-Hour	
Crew D-2	**Hr.**	**Daily**	**Hr.**	**Daily**	**Bare Costs**	**Incl. O&P**
3 Bricklayers	$35.25	$846.00	$53.70	$1288.80	$32.12	$49.05
2 Bricklayer Helpers	26.90	430.40	41.00	656.00		
.5 Carpenter	34.25	137.00	53.35	213.40		
44 L.H., Daily Totals		$1413.40		$2158.20	$32.12	$49.05
Crew D-3	**Hr.**	**Daily**	**Hr.**	**Daily**	**Bare Costs**	**Incl. O&P**
3 Bricklayers	$35.25	$846.00	$53.70	$1288.80	$32.02	$48.85
2 Bricklayer Helpers	26.90	430.40	41.00	656.00		
.25 Carpenter	34.25	68.50	53.35	106.70		
42 L.H., Daily Totals		$1344.90		$2051.50	$32.02	$48.85
Crew D-4	**Hr.**	**Daily**	**Hr.**	**Daily**	**Bare Costs**	**Incl. O&P**
1 Bricklayer	$35.25	$282.00	$53.70	$429.60	$30.53	$46.38
2 Bricklayer Helpers	26.90	430.40	41.00	656.00		
1 Equip. Oper. (light)	33.05	264.40	49.80	398.40		
1 Grout Pump, 50 C.F./hr		107.75		118.55		
1 Hoses & Hopper		15.20		16.70		
1 Accessories		11.90		13.10	4.22	4.64
32 L.H., Daily Totals		$1111.65		$1632.35	$34.75	$51.02
Crew D-5	**Hr.**	**Daily**	**Hr.**	**Daily**	**Bare Costs**	**Incl. O&P**
1 Bricklayer	$35.25	$282.00	$53.70	$429.60	$35.25	$53.70
8 L.H., Daily Totals		$282.00		$429.60	$35.25	$53.70
Crew D-6	**Hr.**	**Daily**	**Hr.**	**Daily**	**Bare Costs**	**Incl. O&P**
3 Bricklayers	$35.25	$846.00	$53.70	$1288.80	$31.20	$47.59
3 Bricklayer Helpers	26.90	645.60	41.00	984.00		
.25 Carpenter	34.25	68.50	53.35	106.70		
50 L.H., Daily Totals		$1560.10		$2379.50	$31.20	$47.59
Crew D-7	**Hr.**	**Daily**	**Hr.**	**Daily**	**Bare Costs**	**Incl. O&P**
1 Tile Layer	$32.70	$261.60	$48.05	$384.40	$29.03	$42.65
1 Tile Layer Helper	25.35	202.80	37.25	298.00		
16 L.H., Daily Totals		$464.40		$682.40	$29.03	$42.65
Crew D-8	**Hr.**	**Daily**	**Hr.**	**Daily**	**Bare Costs**	**Incl. O&P**
3 Bricklayers	$35.25	$846.00	$53.70	$1288.80	$31.91	$48.62
2 Bricklayer Helpers	26.90	430.40	41.00	656.00		
40 L.H., Daily Totals		$1276.40		$1944.80	$31.91	$48.62
Crew D-9	**Hr.**	**Daily**	**Hr.**	**Daily**	**Bare Costs**	**Incl. O&P**
3 Bricklayers	$35.25	$846.00	$53.70	$1288.80	$31.08	$47.35
3 Bricklayer Helpers	26.90	645.60	41.00	984.00		
48 L.H., Daily Totals		$1491.60		$2272.80	$31.08	$47.35
Crew D-10	**Hr.**	**Daily**	**Hr.**	**Daily**	**Bare Costs**	**Incl. O&P**
1 Bricklayer Foreman	$37.25	$298.00	$56.75	$454.00	$32.44	$49.31
1 Bricklayer	35.25	282.00	53.70	429.60		
2 Bricklayer Helpers	26.90	430.40	41.00	656.00		
1 Equip. Oper. (crane)	35.90	287.20	54.10	432.80		
1 Truck Crane, 12.5 Ton		505.80		556.40	12.65	13.91
40 L.H., Daily Totals		$1803.40		$2528.80	$45.09	$63.22
Crew D-11	**Hr.**	**Daily**	**Hr.**	**Daily**	**Bare Costs**	**Incl. O&P**
1 Bricklayer Foreman	$37.25	$298.00	$56.75	$454.00	$33.13	$50.48
1 Bricklayer	35.25	282.00	53.70	429.60		
1 Bricklayer Helper	26.90	215.20	41.00	328.00		
24 L.H., Daily Totals		$795.20		$1211.60	$33.13	$50.48

Crew D-12	Hr.	Daily	Hr.	Daily	Bare Costs	Incl. O&P
1 Bricklayer Foreman	$37.25	$298.00	$56.75	$454.00	$31.58	$48.11
1 Bricklayer	35.25	282.00	53.70	429.60		
2 Bricklayer Helpers	26.90	430.40	41.00	656.00		
32 L.H., Daily Totals		$1010.40		$1539.60	$31.58	$48.11

Crew D-13	Hr.	Daily	Hr.	Daily	Bare Costs	Incl. O&P
1 Bricklayer Foreman	$37.25	$298.00	$56.75	$454.00	$32.74	$49.98
1 Bricklayer	35.25	282.00	53.70	429.60		
2 Bricklayer Helpers	26.90	430.40	41.00	656.00		
1 Carpenter	34.25	274.00	53.35	426.80		
1 Equip. Oper. (crane)	35.90	287.20	54.10	432.80		
1 Truck Crane, 12.5 Ton		505.80		556.40	10.54	11.59
48 L.H., Daily Totals		$2077.40		$2955.60	$43.28	$61.57

Crew E-1	Hr.	Daily	Hr.	Daily	Bare Costs	Incl. O&P
1 Welder Foreman	$40.15	$321.20	$71.85	$574.80	$37.12	$63.30
1 Welder	38.15	305.20	68.25	546.00		
1 Equip. Oper. (light)	33.05	264.40	49.80	398.40		
1 Gas Welding Machine		81.20		89.30	3.38	3.72
24 L.H., Daily Totals		$972.00		$1608.50	$40.50	$67.02

Crew E-2	Hr.	Daily	Hr.	Daily	Bare Costs	Incl. O&P
1 Struc. Steel Foreman	$40.15	$321.20	$71.85	$574.80	$36.96	$63.47
4 Struc. Steel Workers	38.15	1220.80	68.25	2184.00		
1 Equip. Oper. (crane)	35.90	287.20	54.10	432.80		
1 Equip. Oper. Oiler	30.10	240.80	45.35	362.80		
1 Crane, 90 Ton		1325.00		1457.50	23.66	26.03
56 L.H., Daily Totals		$3395.00		$5011.90	$60.62	$89.50

Crew E-3	Hr.	Daily	Hr.	Daily	Bare Costs	Incl. O&P
1 Struc. Steel Foreman	$40.15	$321.20	$71.85	$574.80	$38.82	$69.45
1 Struc. Steel Worker	38.15	305.20	68.25	546.00		
1 Welder	38.15	305.20	68.25	546.00		
1 Gas Welding Machine		81.20		89.30	3.38	3.72
24 L.H., Daily Totals		$1012.80		$1756.10	$42.20	$73.17

Crew E-4	Hr.	Daily	Hr.	Daily	Bare Costs	Incl. O&P
1 Struc. Steel Foreman	$40.15	$321.20	$71.85	$574.80	$38.65	$69.15
3 Struc. Steel Workers	38.15	915.60	68.25	1638.00		
1 Gas Welding Machine		81.20		89.30	2.54	2.79
32 L.H., Daily Totals		$1318.00		$2302.10	$41.19	$71.94

Crew E-5	Hr.	Daily	Hr.	Daily	Bare Costs	Incl. O&P
2 Struc. Steel Foremen	$40.15	$642.40	$71.85	$1149.60	$37.52	$65.27
5 Struc. Steel Workers	38.15	1526.00	68.25	2730.00		
1 Equip. Oper. (crane)	35.90	287.20	54.10	432.80		
1 Welder	38.15	305.20	68.25	546.00		
1 Equip. Oper. Oiler	30.10	240.80	45.35	362.80		
1 Crane, 90 Ton		1325.00		1457.50		
1 Gas Welding Machine		81.20		89.30	17.58	19.34
80 L.H., Daily Totals		$4407.80		$6768.00	$55.10	$84.61

Crew E-6	Hr.	Daily	Hr.	Daily	Bare Costs	Incl. O&P
3 Struc. Steel Foremen	$40.15	$963.60	$71.85	$1724.40	$37.56	$65.46
9 Struc. Steel Workers	38.15	2746.80	68.25	4914.00		
1 Equip. Oper. (crane)	35.90	287.20	54.10	432.80		
1 Welder	38.15	305.20	68.25	546.00		
1 Equip. Oper. Oiler	30.10	240.80	45.35	362.80		
1 Equip. Oper. (light)	33.05	264.40	49.80	398.40		
1 Crane, 90 Ton		1325.00		1457.50		
1 Gas Welding Machine		81.20		89.30		
1 Air Compr., 160 C.F.M.		87.40		96.15		
2 Impact Wrenches		24.00		26.40	11.86	13.04
128 L.H., Daily Totals		$6325.60		$10047.75	$49.42	$78.50

Crew E-7	Hr.	Daily	Hr.	Daily	Bare Costs	Incl. O&P
1 Struc. Steel Foreman	$40.15	$321.20	$71.85	$574.80	$37.52	$65.27
4 Struc. Steel Workers	38.15	1220.80	68.25	2184.00		
1 Equip. Oper. (crane)	35.90	287.20	54.10	432.80		
1 Equip. Oper. Oiler	30.10	240.80	45.35	362.80		
1 Welder Foreman	40.15	321.20	71.85	574.80		
2 Welders	38.15	610.40	68.25	1092.00		
1 Crane, 90 Ton		1325.00		1457.50		
2 Gas Welding Machines		162.40		178.65	18.59	20.45
80 L.H., Daily Totals		$4489.00		$6857.35	$56.11	$85.72

Crew E-8	Hr.	Daily	Hr.	Daily	Bare Costs	Incl. O&P
1 Struc. Steel Foreman	$40.15	$321.20	$71.85	$574.80	$37.27	$64.53
4 Struc. Steel Workers	38.15	1220.80	68.25	2184.00		
1 Welder Foreman	40.15	321.20	71.85	574.80		
4 Welders	38.15	1220.80	68.25	2184.00		
1 Equip. Oper. (crane)	35.90	287.20	54.10	432.80		
1 Equip. Oper. Oiler	30.10	240.80	45.35	362.80		
1 Equip. Oper. (light)	33.05	264.40	49.80	398.40		
1 Crane, 90 Ton		1325.00		1457.50		
4 Gas Welding Machines		324.80		357.30	15.86	17.45
104 L.H., Daily Totals		$5526.20		$8526.40	$53.13	$81.98

Crew E-9	Hr.	Daily	Hr.	Daily	Bare Costs	Incl. O&P
2 Struc. Steel Foremen	$40.15	$642.40	$71.85	$1149.60	$37.56	$65.46
5 Struc. Steel Workers	38.15	1526.00	68.25	2730.00		
1 Welder Foreman	40.15	321.20	71.85	574.80		
5 Welders	38.15	1526.00	68.25	2730.00		
1 Equip. Oper. (crane)	35.90	287.20	54.10	432.80		
1 Equip. Oper. Oiler	30.10	240.80	45.35	362.80		
1 Equip. Oper. (light)	33.05	264.40	49.80	398.40		
1 Crane, 90 Ton		1325.00		1457.50		
5 Gas Welding Machines		406.00		446.60	13.52	14.88
128 L.H., Daily Totals		$6539.00		$10282.50	$51.08	$80.34

Crew E-10	Hr.	Daily	Hr.	Daily	Bare Costs	Incl. O&P
1 Welder Foreman	$40.15	$321.20	$71.85	$574.80	$39.15	$70.05
1 Welder	38.15	305.20	68.25	546.00		
1 Gas Welding Machines		81.20		89.30		
1 Truck, 3 Ton		176.20		193.80	16.09	17.70
16 L.H., Daily Totals		$883.80		$1403.90	$55.24	$87.75

Crew E-11	Hr.	Daily	Hr.	Daily	Bare Costs	Incl. O&P
2 Painters, Struc. Steel	$31.00	$496.00	$57.05	$912.80	$30.44	$51.36
1 Building Laborer	26.70	213.60	41.55	332.40		
1 Equip. Oper. (light)	33.05	264.40	49.80	398.40		
1 Air Compressor 250 C.F.M.		112.40		123.65		
1 Sand Blaster		15.20		16.70		
1 Sand Blasting Accessories		11.90		13.10	4.36	4.80
32 L.H., Daily Totals		$1113.50		$1797.05	$34.80	$56.16

Crew No.	Bare Costs		Incl. Subs O & P		Cost Per Labor-Hour	
Crew E-12	**Hr.**	**Daily**	**Hr.**	**Daily**	**Bare Costs**	**Incl. O&P**
1 Welder Foreman	$40.15	$321.20	$71.85	$574.80	$36.60	$60.83
1 Equip. Oper. (light)	33.05	264.40	49.80	398.40		
1 Gas Welding Machine		81.20		89.30	5.08	5.58
16 L.H., Daily Totals		$666.80		$1062.50	$41.68	$66.41
Crew E-13	**Hr.**	**Daily**	**Hr.**	**Daily**	**Bare Costs**	**Incl. O&P**
1 Welder Foreman	$40.15	$321.20	$71.85	$574.80	$37.78	$64.50
.5 Equip. Oper. (light)	33.05	132.20	49.80	199.20		
1 Gas Welding Machine		81.20		89.30	6.77	7.44
12 L.H., Daily Totals		$534.60		$863.30	$44.55	$71.94
Crew E-14	**Hr.**	**Daily**	**Hr.**	**Daily**	**Bare Costs**	**Incl. O&P**
1 Welder Foreman	$40.15	$321.20	$71.85	$574.80	$40.15	$71.85
1 Gas Welding Machine		81.20		89.30	10.15	11.17
8 L.H., Daily Totals		$402.40		$664.10	$50.30	$83.02
Crew E-16	**Hr.**	**Daily**	**Hr.**	**Daily**	**Bare Costs**	**Incl. O&P**
1 Welder Foreman	$40.15	$321.20	$71.85	$574.80	$39.15	$70.05
1 Welder	38.15	305.20	68.25	546.00		
1 Gas Welding Machine		81.20		89.30	5.08	5.58
16 L.H., Daily Totals		$707.60		$1210.10	$44.23	$75.63
Crew E-17	**Hr.**	**Daily**	**Hr.**	**Daily**	**Bare Costs**	**Incl. O&P**
1 Structural Steel Foreman	$40.15	$321.20	$71.85	$574.80	$39.15	$70.05
1 Structural Steel Worker	38.15	305.20	68.25	546.00		
1 Power Tool		5.20		5.70	.33	.36
16 L.H., Daily Totals		$631.60		$1126.50	$39.48	$70.41
Crew E-18	**Hr.**	**Daily**	**Hr.**	**Daily**	**Bare Costs**	**Incl. O&P**
1 Structural Steel Foreman	$40.15	$321.20	$71.85	$574.80	$37.85	$65.76
3 Structural Steel Workers	38.15	915.60	68.25	1638.00		
1 Equipment Operator (med.)	34.65	277.20	52.20	417.60		
1 Crane, 20 Ton		797.10		876.80	19.93	21.92
40 L.H., Daily Totals		$2311.10		$3507.20	$57.78	$87.68
Crew E-19	**Hr.**	**Daily**	**Hr.**	**Daily**	**Bare Costs**	**Incl. O&P**
1 Structural Steel Worker	$38.15	$305.20	$68.25	$546.00	$37.12	$63.30
1 Structural Steel Foreman	40.15	321.20	71.85	574.80		
1 Equip. Oper. (light)	33.05	264.40	49.80	398.40		
1 Power Tool		5.20		5.70		
1 Crane, 20 Ton		797.10		876.80	33.43	36.77
24 L.H., Daily Totals		$1693.10		$2401.70	$70.55	$100.07
Crew E-20	**Hr.**	**Daily**	**Hr.**	**Daily**	**Bare Costs**	**Incl. O&P**
1 Structural Steel Foreman	$40.15	$321.20	$71.85	$574.80	$37.11	$64.07
5 Structural Steel Workers	38.15	1526.00	68.25	2730.00		
1 Equip. Oper. (crane)	35.90	287.20	54.10	432.80		
1 Oiler	30.10	240.80	45.35	362.80		
1 Power Tool		5.20		5.70		
1 Crane, 40 Ton		951.90		1047.10	14.95	16.45
64 L.H., Daily Totals		$3332.30		$5153.20	$52.06	$80.52
Crew E-22	**Hr.**	**Daily**	**Hr.**	**Daily**	**Bare Costs**	**Incl. O&P**
1 Skilled Worker Foreman	$36.85	$294.80	$57.30	$458.40	$35.52	$55.23
2 Skilled Workers	34.85	557.60	54.20	867.20		
24 L.H., Daily Totals		$852.40		$1325.60	$35.52	$55.23

Crew No.	Bare Costs		Incl. Subs O & P		Cost Per Labor-Hour	
Crew E-24	**Hr.**	**Daily**	**Hr.**	**Daily**	**Bare Costs**	**Incl. O&P**
3 Structural Steel Workers	$38.15	$915.60	$68.25	$1638.00	$37.28	$64.24
1 Equipment Operator (medium)	34.65	277.20	52.20	417.60		
1-25 Ton Crane		616.80		678.50	19.28	21.20
32 L.H., Daily Totals		$1809.60		$2734.10	$56.56	$85.44
Crew E-25	**Hr.**	**Daily**	**Hr.**	**Daily**	**Bare Costs**	**Incl. O&P**
1 Welder Foreman	$40.15	$321.20	$71.85	$574.80	$40.15	$71.85
1 Cutting Torch		18.00		19.80		
1 Gases		64.80		71.30	10.35	11.39
8 L.H., Daily Totals		$404.00		$665.90	$50.50	$83.24
Crew F-3	**Hr.**	**Daily**	**Hr.**	**Daily**	**Bare Costs**	**Incl. O&P**
4 Carpenters	$34.25	$1096.00	$53.35	$1707.20	$34.58	$53.50
1 Equip. Oper. (crane)	35.90	287.20	54.10	432.80		
1 Hyd. Crane, 12 Ton		602.60		662.85	15.07	16.57
40 L.H., Daily Totals		$1985.80		$2802.85	$49.65	$70.07
Crew F-4	**Hr.**	**Daily**	**Hr.**	**Daily**	**Bare Costs**	**Incl. O&P**
4 Carpenters	$34.25	$1096.00	$53.35	$1707.20	$33.83	$52.14
1 Equip. Oper. (crane)	35.90	287.20	54.10	432.80		
1 Equip. Oper. Oiler	30.10	240.80	45.35	362.80		
1 Hyd. Crane, 55 Ton		895.40		984.95	18.65	20.52
48 L.H., Daily Totals		$2519.40		$3487.75	$52.48	$72.66
Crew F-5	**Hr.**	**Daily**	**Hr.**	**Daily**	**Bare Costs**	**Incl. O&P**
1 Carpenter Foreman	$36.25	$290.00	$56.45	$451.60	$34.75	$54.13
3 Carpenters	34.25	822.00	53.35	1280.40		
32 L.H., Daily Totals		$1112.00		$1732.00	$34.75	$54.13
Crew F-6	**Hr.**	**Daily**	**Hr.**	**Daily**	**Bare Costs**	**Incl. O&P**
2 Carpenters	$34.25	$548.00	$53.35	$853.60	$31.56	$48.78
2 Building Laborers	26.70	427.20	41.55	664.80		
1 Equip. Oper. (crane)	35.90	287.20	54.10	432.80		
1 Hyd. Crane, 12 Ton		602.60		662.85	15.07	16.57
40 L.H., Daily Totals		$1865.00		$2614.05	$46.63	$65.35
Crew F-7	**Hr.**	**Daily**	**Hr.**	**Daily**	**Bare Costs**	**Incl. O&P**
2 Carpenters	$34.25	$548.00	$53.35	$853.60	$30.48	$47.45
2 Building Laborers	26.70	427.20	41.55	664.80		
32 L.H., Daily Totals		$975.20		$1518.40	$30.48	$47.45
Crew G-1	**Hr.**	**Daily**	**Hr.**	**Daily**	**Bare Costs**	**Incl. O&P**
1 Roofer Foreman	$31.45	$251.60	$53.60	$428.80	$27.49	$46.86
4 Roofers, Composition	29.45	942.40	50.20	1606.40		
2 Roofer Helpers	21.60	345.60	36.80	588.80		
1 Application Equipment		146.60		161.25		
1 Tar Kettle/Pot		52.50		57.75		
1 Crew Truck		98.80		108.70	5.32	5.85
56 L.H., Daily Totals		$1837.50		$2951.70	$32.81	$52.71
Crew G-2	**Hr.**	**Daily**	**Hr.**	**Daily**	**Bare Costs**	**Incl. O&P**
1 Plasterer	$31.35	$250.80	$47.70	$381.60	$28.38	$43.50
1 Plasterer Helper	27.10	216.80	41.25	330.00		
1 Building Laborer	26.70	213.60	41.55	332.40		
1 Grouting Equipment		107.75		118.55	4.49	4.94
24 L.H., Daily Totals		$788.95		$1162.55	$32.87	$48.44

Crew No.	Bare Costs		Incl. Subs O & P		Cost Per Labor-Hour	

Crew G-3	Hr.	Daily	Hr.	Daily	Bare Costs	Incl. O&P
2 Sheet Metal Workers	$40.20	$643.20	$61.85	$989.60	$33.45	$51.70
2 Building Laborers	26.70	427.20	41.55	664.80		
32 L.H., Daily Totals		$1070.40		$1654.40	$33.45	$51.70

Crew G-4	Hr.	Daily	Hr.	Daily	Bare Costs	Incl. O&P
1 Labor Foreman (outside)	$28.70	$229.60	$44.70	$357.60	$27.37	$42.60
2 Building Laborers	26.70	427.20	41.55	664.80		
1 Light Truck, 1.5 Ton		128.80		141.70		
1 Air Compr., 160 C.F.M.		87.40		96.15	9.01	9.91
24 L.H., Daily Totals		$873.00		$1260.25	$36.38	$52.51

Crew G-5	Hr.	Daily	Hr.	Daily	Bare Costs	Incl. O&P
1 Roofer Foreman	$31.45	$251.60	$53.60	$428.80	$26.71	$45.52
2 Roofers, Composition	29.45	471.20	50.20	803.20		
2 Roofer Helpers	21.60	345.60	36.80	588.80		
1 Application Equipment		146.60		161.25	3.67	4.03
40 L.H., Daily Totals		$1215.00		$1982.05	$30.38	$49.55

Crew G-6A	Hr.	Daily	Hr.	Daily	Bare Costs	Incl. O&P
2 Roofers Composition	$29.45	$471.20	$50.20	$803.20	$29.45	$50.20
1 Small Compressor		12.15		13.35		
2 Pneumatic Nailers		39.00		42.90	3.19	3.51
16 L.H., Daily Totals		$522.35		$859.45	$32.64	$53.71

Crew G-7	Hr.	Daily	Hr.	Daily	Bare Costs	Incl. O&P
1 Carpenter	$34.25	$274.00	$53.35	$426.80	$34.25	$53.35
1 Small Compressor		12.15		13.35		
1 Pneumatic Nailer		19.50		21.45	3.94	4.34
8 L.H., Daily Totals		$305.65		$461.60	$38.19	$57.69

Crew H-1	Hr.	Daily	Hr.	Daily	Bare Costs	Incl. O&P
2 Glaziers	$33.00	$528.00	$49.90	$798.40	$35.58	$59.08
2 Struc. Steel Workers	38.15	610.40	68.25	1092.00		
32 L.H., Daily Totals		$1138.40		$1890.40	$35.58	$59.08

Crew H-2	Hr.	Daily	Hr.	Daily	Bare Costs	Incl. O&P
2 Glaziers	$33.00	$528.00	$49.90	$798.40	$30.90	$47.12
1 Building Laborer	26.70	213.60	41.55	332.40		
24 L.H., Daily Totals		$741.60		$1130.80	$30.90	$47.12

Crew H-3	Hr.	Daily	Hr.	Daily	Bare Costs	Incl. O&P
1 Glazier	$33.00	$264.00	$49.90	$399.20	$29.28	$44.83
1 Helper	25.55	204.40	39.75	318.00		
16 L.H., Daily Totals		$468.40		$717.20	$29.28	$44.83

Crew J-1	Hr.	Daily	Hr.	Daily	Bare Costs	Incl. O&P
3 Plasterers	$31.35	$752.40	$47.70	$1144.80	$29.65	$45.12
2 Plasterer Helpers	27.10	433.60	41.25	660.00		
1 Mixing Machine, 6 C.F.		103.00		113.30	2.58	2.83
40 L.H., Daily Totals		$1289.00		$1918.10	$32.23	$47.95

Crew J-2	Hr.	Daily	Hr.	Daily	Bare Costs	Incl. O&P
3 Plasterers	$31.35	$752.40	$47.70	$1144.80	$29.95	$45.37
2 Plasterer Helpers	27.10	433.60	41.25	660.00		
1 Lather	31.45	251.60	46.60	372.80		
1 Mixing Machine, 6 C.F.		103.00		113.30	2.15	2.36
48 L.H., Daily Totals		$1540.60		$2290.90	$32.10	$47.73

Crew No.	Bare Costs		Incl. Subs O & P		Cost Per Labor-Hour	

Crew J-3	Hr.	Daily	Hr.	Daily	Bare Costs	Incl. O&P
1 Terrazzo Worker	$32.65	$261.20	$48.00	$384.00	$29.38	$43.18
1 Terrazzo Helper	26.10	208.80	38.35	306.80		
1 Terrazzo Grinder, Electric		70.00		77.00		
1 Terrazzo Mixer		139.60		153.55	13.10	14.41
16 L.H., Daily Totals		$679.60		$921.35	$42.48	$57.59

Crew J-4	Hr.	Daily	Hr.	Daily	Bare Costs	Incl. O&P
1 Tile Layer	$32.70	$261.60	$48.05	$384.40	$29.03	$42.65
1 Tile Layer Helper	25.35	202.80	37.25	298.00		
16 L.H., Daily Totals		$464.40		$682.40	$29.03	$42.65

Crew K-1	Hr.	Daily	Hr.	Daily	Bare Costs	Incl. O&P
1 Carpenter	$34.25	$274.00	$53.35	$426.80	$30.53	$47.10
1 Truck Driver (light)	26.80	214.40	40.85	326.80		
1 Truck w/Power Equip.		176.20		193.80	11.01	12.11
16 L.H., Daily Totals		$664.60		$947.40	$41.54	$59.21

Crew K-2	Hr.	Daily	Hr.	Daily	Bare Costs	Incl. O&P
1 Struc. Steel Foreman	$40.15	$321.20	$71.85	$574.80	$35.03	$60.32
1 Struc. Steel Worker	38.15	305.20	68.25	546.00		
1 Truck Driver (light)	26.80	214.40	40.85	326.80		
1 Truck w/Power Equip.		176.20		193.80	7.34	8.08
24 L.H., Daily Totals		$1017.00		$1641.40	$42.37	$68.40

Crew L-1	Hr.	Daily	Hr.	Daily	Bare Costs	Incl. O&P
1 Electrician	$40.75	$326.00	$60.65	$485.20	$40.80	$61.03
1 Plumber	40.85	326.80	61.40	491.20		
16 L.H., Daily Totals		$652.80		$976.40	$40.80	$61.03

Crew L-2	Hr.	Daily	Hr.	Daily	Bare Costs	Incl. O&P
1 Carpenter	$34.25	$274.00	$53.35	$426.80	$29.90	$46.55
1 Carpenter Helper	25.55	204.40	39.75	318.00		
16 L.H., Daily Totals		$478.40		$744.80	$29.90	$46.55

Crew L-3	Hr.	Daily	Hr.	Daily	Bare Costs	Incl. O&P
1 Carpenter	$34.25	$274.00	$53.35	$426.80	$37.36	$57.30
.5 Electrician	40.75	163.00	60.65	242.60		
.5 Sheet Metal Worker	40.20	160.80	61.85	247.40		
16 L.H., Daily Totals		$597.80		$916.80	$37.36	$57.30

Crew L-3A	Hr.	Daily	Hr.	Daily	Bare Costs	Incl. O&P
1 Carpenter Foreman (outside)	$36.25	$290.00	$56.45	$451.60	$37.57	$58.25
.5 Sheet Metal Worker	40.20	160.80	61.85	247.40		
12 L.H., Daily Totals		$450.80		$699.00	$37.57	$58.25

Crew L-4	Hr.	Daily	Hr.	Daily	Bare Costs	Incl. O&P
2 Skilled Workers	$34.85	$557.60	$54.20	$867.20	$31.75	$49.38
1 Helper	25.55	204.40	39.75	318.00		
24 L.H., Daily Totals		$762.00		$1185.20	$31.75	$49.38

Crew L-5	Hr.	Daily	Hr.	Daily	Bare Costs	Incl. O&P
1 Struc. Steel Foreman	$40.15	$321.20	$71.85	$574.80	$38.11	$66.74
5 Struc. Steel Workers	38.15	1526.00	68.25	2730.00		
1 Equip. Oper. (crane)	35.90	287.20	54.10	432.80		
1 Hyd. Crane, 25 Ton		616.80		678.50	11.01	12.12
56 L.H., Daily Totals		$2751.20		$4416.10	$49.12	$78.86

Crew No.	Bare Costs		Incl. Subs O & P		Cost Per Labor-Hour	
Crew L-5A	**Hr.**	**Daily**	**Hr.**	**Daily**	**Bare Costs**	**Incl. O&P**
1 Structural Steel Foreman	$40.15	$321.20	$71.85	$574.80	$38.09	$65.61
2 Structural Steel Workers	38.15	610.40	68.25	1092.00		
1 Equip. Oper. (crane)	35.90	287.20	54.10	432.80		
1 Crane,SP, 25 Ton		575.00		632.50	17.97	19.77
32 L.H., Daily Totals		$1793.80		$2732.10	$56.06	$85.38
Crew L-6	**Hr.**	**Daily**	**Hr.**	**Daily**	**Bare Costs**	**Incl. O&P**
1 Plumber	$40.85	$326.80	$61.40	$491.20	$40.82	$61.15
.5 Electrician	40.75	163.00	60.65	242.60		
12 L.H., Daily Totals		$489.80		$733.80	$40.82	$61.15
Crew L-7	**Hr.**	**Daily**	**Hr.**	**Daily**	**Bare Costs**	**Incl. O&P**
2 Carpenters	$34.25	$548.00	$53.35	$853.60	$33.02	$51.02
1 Building Laborer	26.70	213.60	41.55	332.40		
.5 Electrician	40.75	163.00	60.65	242.60		
28 L.H., Daily Totals		$924.60		$1428.60	$33.02	$51.02
Crew L-8	**Hr.**	**Daily**	**Hr.**	**Daily**	**Bare Costs**	**Incl. O&P**
2 Carpenters	$34.25	$548.00	$53.35	$853.60	$35.57	$54.96
.5 Plumber	40.85	163.40	61.40	245.60		
20 L.H., Daily Totals		$711.40		$1099.20	$35.57	$54.96
Crew L-9	**Hr.**	**Daily**	**Hr.**	**Daily**	**Bare Costs**	**Incl. O&P**
1 Labor Foreman (inside)	$27.20	$217.60	$42.35	$338.80	$30.92	$49.78
2 Building Laborers	26.70	427.20	41.55	664.80		
1 Struc. Steel Worker	38.15	305.20	68.25	546.00		
.5 Electrician	40.75	163.00	60.65	242.60		
36 L.H., Daily Totals		$1113.00		$1792.20	$30.92	$49.78
Crew L-10	**Hr.**	**Daily**	**Hr.**	**Daily**	**Bare Costs**	**Incl. O&P**
1 Structural Steel Foreman	$40.15	$321.20	$71.85	$574.80	$38.07	$64.73
1 Structural Steel Worker	38.15	305.20	68.25	546.00		
1 Equip. Oper. (crane)	35.90	287.20	54.10	432.80		
1 Hyd. Crane, 12 Ton		602.60		662.85	25.11	27.62
24 L.H., Daily Totals		$1516.20		$2216.45	$63.18	$92.35
Crew L-11	**Hr.**	**Daily**	**Hr.**	**Daily**	**Bare Costs**	**Incl. O&P**
2 Wreckers	$26.70	$427.20	$47.20	$755.20	$30.59	$49.58
1 Equip. Oper. (crane)	35.90	287.20	54.10	432.80		
1 Equip. Oper. (light)	33.05	264.40	49.80	398.40		
1 Hyd. Excavator, 2.5 C.Y.		1215.00		1336.50		
1 Skid steer loader		197.80		217.60	44.15	48.57
32 L.H., Daily Totals		$2391.60		$3140.50	$74.74	$98.15
Crew M-1	**Hr.**	**Daily**	**Hr.**	**Daily**	**Bare Costs**	**Incl. O&P**
3 Elevator Constructors	$45.05	$1081.20	$67.55	$1621.20	$42.80	$64.18
1 Elevator Apprentice	36.05	288.40	54.05	432.40		
5 Hand Tools		63.00		69.30	1.97	2.17
32 L.H., Daily Totals		$1432.60		$2122.90	$44.77	$66.35
Crew M-3	**Hr.**	**Daily**	**Hr.**	**Daily**	**Bare Costs**	**Incl. O&P**
1 Electrician Foreman (out)	$42.75	$342.00	$63.60	$508.80	$37.46	$56.42
1 Common Laborer	26.70	213.60	41.55	332.40		
.25 Equipment Operator, Medium	34.65	69.30	52.20	104.40		
1 Elevator Constructor	45.05	360.40	67.55	540.40		
1 Elevator Apprentice	36.05	288.40	54.05	432.40		
.25 Crane, SP, 4 x 4, 20 ton		145.35		159.90	4.28	4.70
34 L.H., Daily Totals		$1419.05		$2078.30	$41.74	$61.12

Crew No.	Bare Costs		Incl. Subs O & P		Cost Per Labor-Hour	
Crew M-4	**Hr.**	**Daily**	**Hr.**	**Daily**	**Bare Costs**	**Incl. O&P**
1 Electrician Foreman (out)	$42.75	$342.00	$63.60	$508.80	$37.12	$55.91
1 Common Laborer	26.70	213.60	41.55	332.40		
.25 Equipment Operator, Crane	35.90	71.80	54.10	108.20		
.25 Equipment Operator, Oiler	30.10	60.20	45.35	90.70		
1 Elevator Constructor	45.05	360.40	67.55	540.40		
1 Elevator Apprentice	36.05	288.40	54.05	432.40		
.25 Crane, Hyd, SP, 4WD, 40 Ton		217.40		239.15	6.04	6.64
36 L.H., Daily Totals		$1553.80		$2252.05	$43.16	$62.55
Crew Q-1	**Hr.**	**Daily**	**Hr.**	**Daily**	**Bare Costs**	**Incl. O&P**
1 Plumber	$40.85	$326.80	$61.40	$491.20	$36.78	$55.28
1 Plumber Apprentice	32.70	261.60	49.15	393.20		
16 L.H., Daily Totals		$588.40		$884.40	$36.78	$55.28
Crew Q-1C	**Hr.**	**Daily**	**Hr.**	**Daily**	**Bare Costs**	**Incl. O&P**
1 Plumber	$40.85	$326.80	$61.40	$491.20	$36.07	$54.25
1 Plumber Apprentice	32.70	261.60	49.15	393.20		
1 Equip. Oper. (medium)	34.65	277.20	52.20	417.60		
1 Trencher, Chain		1420.00		1562.00	59.17	65.08
24 L.H., Daily Totals		$2285.60		$2864.00	$95.24	$119.33
Crew Q-2	**Hr.**	**Daily**	**Hr.**	**Daily**	**Bare Costs**	**Incl. O&P**
2 Plumbers	$40.85	$653.60	$61.40	$982.40	$38.13	$57.32
1 Plumber Apprentice	32.70	261.60	49.15	393.20		
24 L.H., Daily Totals		$915.20		$1375.60	$38.13	$57.32
Crew Q-3	**Hr.**	**Daily**	**Hr.**	**Daily**	**Bare Costs**	**Incl. O&P**
1 Plumber Foreman (inside)	$41.35	$330.80	$62.15	$497.20	$38.94	$58.53
2 Plumbers	40.85	653.60	61.40	982.40		
1 Plumber Apprentice	32.70	261.60	49.15	393.20		
32 L.H., Daily Totals		$1246.00		$1872.80	$38.94	$58.53
Crew Q-4	**Hr.**	**Daily**	**Hr.**	**Daily**	**Bare Costs**	**Incl. O&P**
1 Plumber Foreman (inside)	$41.35	$330.80	$62.15	$497.20	$38.94	$58.53
1 Plumber	40.85	326.80	61.40	491.20		
1 Welder (plumber)	40.85	326.80	61.40	491.20		
1 Plumber Apprentice	32.70	261.60	49.15	393.20		
1 Electric Welding Mach.		80.65		88.70	2.52	2.77
32 L.H., Daily Totals		$1326.65		$1961.50	$41.46	$61.30
Crew Q-5	**Hr.**	**Daily**	**Hr.**	**Daily**	**Bare Costs**	**Incl. O&P**
1 Steamfitter	$40.95	$327.60	$61.55	$492.40	$36.85	$55.38
1 Steamfitter Apprentice	32.75	262.00	49.20	393.60		
16 L.H., Daily Totals		$589.60		$886.00	$36.85	$55.38
Crew Q-6	**Hr.**	**Daily**	**Hr.**	**Daily**	**Bare Costs**	**Incl. O&P**
2 Steamfitters	$40.95	$655.20	$61.55	$984.80	$38.22	$57.43
1 Steamfitter Apprentice	32.75	262.00	49.20	393.60		
24 L.H., Daily Totals		$917.20		$1378.40	$38.22	$57.43
Crew Q-7	**Hr.**	**Daily**	**Hr.**	**Daily**	**Bare Costs**	**Incl. O&P**
1 Steamfitter Foreman (inside)	$41.45	$331.60	$62.30	$498.40	$39.03	$58.65
2 Steamfitters	40.95	655.20	61.55	984.80		
1 Steamfitter Apprentice	32.75	262.00	49.20	393.60		
32 L.H., Daily Totals		$1248.80		$1876.80	$39.03	$58.65

Crew No.	Bare Costs		Incl. Subs O & P		Cost Per Labor-Hour	
Crew Q-8	**Hr.**	**Daily**	**Hr.**	**Daily**	**Bare Costs**	**Incl. O&P**
1 Steamfitter Foreman (inside)	$41.45	$331.60	$62.30	$498.40	$39.03	$58.65
1 Steamfitter	40.95	327.60	61.55	492.40		
1 Welder (steamfitter)	40.95	327.60	61.55	492.40		
1 Steamfitter Apprentice	32.75	262.00	49.20	393.60		
1 Electric Welding Mach.		80.65		88.70	2.52	2.77
32 L.H., Daily Totals		$1329.45		$1965.50	$41.55	$61.42
Crew Q-9	**Hr.**	**Daily**	**Hr.**	**Daily**	**Bare Costs**	**Incl. O&P**
1 Sheet Metal Worker	$40.20	$321.60	$61.85	$494.80	$36.18	$55.65
1 Sheet Metal Apprentice	32.15	257.20	49.45	395.60		
16 L.H., Daily Totals		$578.80		$890.40	$36.18	$55.65
Crew Q-10	**Hr.**	**Daily**	**Hr.**	**Daily**	**Bare Costs**	**Incl. O&P**
2 Sheet Metal Workers	$40.20	$643.20	$61.85	$989.60	$37.52	$57.72
1 Sheet Metal Apprentice	32.15	257.20	49.45	395.60		
24 L.H., Daily Totals		$900.40		$1385.20	$37.52	$57.72
Crew Q-11	**Hr.**	**Daily**	**Hr.**	**Daily**	**Bare Costs**	**Incl. O&P**
1 Sheet Metal Foreman (inside)	$40.70	$325.60	$62.60	$500.80	$38.31	$58.94
2 Sheet Metal Workers	40.20	643.20	61.85	989.60		
1 Sheet Metal Apprentice	32.15	257.20	49.45	395.60		
32 L.H., Daily Totals		$1226.00		$1886.00	$38.31	$58.94
Crew Q-12	**Hr.**	**Daily**	**Hr.**	**Daily**	**Bare Costs**	**Incl. O&P**
1 Sprinkler Installer	$40.75	$326.00	$61.55	$492.40	$36.68	$55.40
1 Sprinkler Apprentice	32.60	260.80	49.25	394.00		
16 L.H., Daily Totals		$586.80		$886.40	$36.68	$55.40
Crew Q-13	**Hr.**	**Daily**	**Hr.**	**Daily**	**Bare Costs**	**Incl. O&P**
1 Sprinkler Foreman (inside)	$41.25	$330.00	$62.35	$498.80	$38.84	$58.68
2 Sprinkler Installers	40.75	652.00	61.55	984.80		
1 Sprinkler Apprentice	32.60	260.80	49.25	394.00		
32 L.H., Daily Totals		$1242.80		$1877.60	$38.84	$58.68
Crew Q-14	**Hr.**	**Daily**	**Hr.**	**Daily**	**Bare Costs**	**Incl. O&P**
1 Asbestos Worker	$37.10	$296.80	$58.55	$468.40	$33.40	$52.70
1 Asbestos Apprentice	29.70	237.60	46.85	374.80		
16 L.H., Daily Totals		$534.40		$843.20	$33.40	$52.70
Crew Q-15	**Hr.**	**Daily**	**Hr.**	**Daily**	**Bare Costs**	**Incl. O&P**
1 Plumber	$40.85	$326.80	$61.40	$491.20	$36.78	$55.28
1 Plumber Apprentice	32.70	261.60	49.15	393.20		
1 Electric Welding Mach.		80.65		88.70	5.04	5.54
16 L.H., Daily Totals		$669.05		$973.10	$41.82	$60.82
Crew Q-16	**Hr.**	**Daily**	**Hr.**	**Daily**	**Bare Costs**	**Incl. O&P**
2 Plumbers	$40.85	$653.60	$61.40	$982.40	$38.13	$57.32
1 Plumber Apprentice	32.70	261.60	49.15	393.20		
1 Electric Welding Mach.		80.65		88.70	3.36	3.70
24 L.H., Daily Totals		$995.85		$1464.30	$41.49	$61.02
Crew Q-17	**Hr.**	**Daily**	**Hr.**	**Daily**	**Bare Costs**	**Incl. O&P**
1 Steamfitter	$40.95	$327.60	$61.55	$492.40	$36.85	$55.38
1 Steamfitter Apprentice	32.75	262.00	49.20	393.60		
1 Electric Welding Mach.		80.65		88.70	5.04	5.54
16 L.H., Daily Totals		$670.25		$974.70	$41.89	$60.92

Crew No.	Bare Costs		Incl. Subs O & P		Cost Per Labor-Hour	
Crew Q-17A	**Hr.**	**Daily**	**Hr.**	**Daily**	**Bare Costs**	**Incl. O&P**
1 Steamfitter	$40.95	$327.60	$61.55	$492.40	$36.53	$54.95
1 Steamfitter Apprentice	32.75	262.00	49.20	393.60		
1 Equip. Oper. (crane)	35.90	287.20	54.10	432.80		
1 Truck Crane, 12 Ton		602.60		662.85		
1 Electric Welding Mach.		80.65		88.70	28.47	31.32
24 L.H., Daily Totals		$1560.05		$2070.35	$65.00	$86.27
Crew Q-18	**Hr.**	**Daily**	**Hr.**	**Daily**	**Bare Costs**	**Incl. O&P**
2 Steamfitters	$40.95	$655.20	$61.55	$984.80	$38.22	$57.43
1 Steamfitter Apprentice	32.75	262.00	49.20	393.60		
1 Electric Welding Mach.		80.65		88.70	3.36	3.70
24 L.H., Daily Totals		$997.85		$1467.10	$41.58	$61.13
Crew Q-19	**Hr.**	**Daily**	**Hr.**	**Daily**	**Bare Costs**	**Incl. O&P**
1 Steamfitter	$40.95	$327.60	$61.55	$492.40	$38.15	$57.13
1 Steamfitter Apprentice	32.75	262.00	49.20	393.60		
1 Electrician	40.75	326.00	60.65	485.20		
24 L.H., Daily Totals		$915.60		$1371.20	$38.15	$57.13
Crew Q-20	**Hr.**	**Daily**	**Hr.**	**Daily**	**Bare Costs**	**Incl. O&P**
1 Sheet Metal Worker	$40.20	$321.60	$61.85	$494.80	$37.09	$56.65
1 Sheet Metal Apprentice	32.15	257.20	49.45	395.60		
.5 Electrician	40.75	163.00	60.65	242.60		
20 L.H., Daily Totals		$741.80		$1133.00	$37.09	$56.65
Crew Q-21	**Hr.**	**Daily**	**Hr.**	**Daily**	**Bare Costs**	**Incl. O&P**
2 Steamfitters	$40.95	$655.20	$61.55	$984.80	$38.85	$58.24
1 Steamfitter Apprentice	32.75	262.00	49.20	393.60		
1 Electrician	40.75	326.00	60.65	485.20		
32 L.H., Daily Totals		$1243.20		$1863.60	$38.85	$58.24
Crew Q-22	**Hr.**	**Daily**	**Hr.**	**Daily**	**Bare Costs**	**Incl. O&P**
1 Plumber	$40.85	$326.80	$61.40	$491.20	$36.78	$55.28
1 Plumber Apprentice	32.70	261.60	49.15	393.20		
1 Truck Crane, 12 Ton		602.60		662.85	37.66	41.43
16 L.H., Daily Totals		$1191.00		$1547.25	$74.44	$96.71
Crew Q-22A	**Hr.**	**Daily**	**Hr.**	**Daily**	**Bare Costs**	**Incl. O&P**
1 Plumber	$40.85	$326.80	$61.40	$491.20	$34.04	$51.55
1 Plumber Apprentice	32.70	261.60	49.15	393.20		
1 Laborer	26.70	213.60	41.55	332.40		
1 Equip. Oper. (crane)	35.90	287.20	54.10	432.80		
1 Truck Crane, 12 Ton		602.60		662.85	18.83	20.71
32 L.H., Daily Totals		$1691.80		$2312.45	$52.87	$72.26
Crew Q-23	**Hr.**	**Daily**	**Hr.**	**Daily**	**Bare Costs**	**Incl. O&P**
1 Plumber Foreman	$42.85	$342.80	$64.40	$515.20	$39.45	$59.33
1 Plumber	40.85	326.80	61.40	491.20		
1 Equip. Oper. (medium)	34.65	277.20	52.20	417.60		
1 Power Tools		5.20		5.70		
1 Crane, 20 Ton		797.10		876.80	33.43	36.77
24 L.H., Daily Totals		$1749.10		$2306.50	$72.88	$96.10
Crew R-1	**Hr.**	**Daily**	**Hr.**	**Daily**	**Bare Costs**	**Incl. O&P**
1 Electrician Foreman	$41.25	$330.00	$61.40	$491.20	$35.77	$53.81
3 Electricians	40.75	978.00	60.65	1455.60		
2 Helpers	25.55	408.80	39.75	636.00		
48 L.H., Daily Totals		$1716.80		$2582.80	$35.77	$53.81

Crews

Crew R-1A	Bare Costs Hr.	Bare Costs Daily	Incl. Subs O & P Hr.	Incl. Subs O & P Daily	Cost Per Labor-Hour Bare Costs	Cost Per Labor-Hour Incl. O&P
1 Electrician	$40.75	$326.00	$60.65	$485.20	$33.15	$50.20
1 Helper	25.55	204.40	39.75	318.00		
16 L.H., Daily Totals		$530.40		$803.20	$33.15	$50.20

Crew R-2	Hr.	Daily	Hr.	Daily	Bare Costs	Incl. O&P
1 Electrician Foreman	$41.25	$330.00	$61.40	$491.20	$35.79	$53.85
3 Electricians	40.75	978.00	60.65	1455.60		
2 Helpers	25.55	408.80	39.75	636.00		
1 Equip. Oper. (crane)	35.90	287.20	54.10	432.80		
1 S.P. Crane, 5 Ton		317.00		348.70	5.66	6.23
56 L.H., Daily Totals		$2321.00		$3364.30	$41.45	$60.08

Crew R-3	Hr.	Daily	Hr.	Daily	Bare Costs	Incl. O&P
1 Electrician Foreman	$41.25	$330.00	$61.40	$491.20	$39.98	$59.64
1 Electrician	40.75	326.00	60.65	485.20		
.5 Equip. Oper. (crane)	35.90	143.60	54.10	216.40		
.5 S.P. Crane, 5 Ton		158.50		174.35	7.93	8.72
20 L.H., Daily Totals		$958.10		$1367.15	$47.91	$68.36

Crew R-4	Hr.	Daily	Hr.	Daily	Bare Costs	Incl. O&P
1 Struc. Steel Foreman	$40.15	$321.20	$71.85	$574.80	$39.07	$67.45
3 Struc. Steel Workers	38.15	915.60	68.25	1638.00		
1 Electrician	40.75	326.00	60.65	485.20		
1 Gas Welding Machine		81.20		89.30	2.03	2.23
40 L.H., Daily Totals		$1644.00		$2787.30	$41.10	$69.68

Crew R-5	Hr.	Daily	Hr.	Daily	Bare Costs	Incl. O&P
1 Electrician Foreman	$41.25	$330.00	$61.40	$491.20	$35.27	$53.12
4 Electrician Linemen	40.75	1304.00	60.65	1940.80		
2 Electrician Operators	40.75	652.00	60.65	970.40		
4 Electrician Groundmen	25.55	817.60	39.75	1272.00		
1 Crew Truck		98.80		108.70		
1 Tool Van		122.15		134.35		
1 Pickup Truck, 3/4 Ton		78.00		85.80		
.2 Crane, 55 Ton		179.08		197.00		
.2 Crane, 12 Ton		120.52		132.55		
.2 Auger, Truck Mtd.		635.80		699.40		
1 Tractor w/Winch		260.00		286.00	16.98	18.68
88 L.H., Daily Totals		$4597.95		$6318.20	$52.25	$71.80

Crew R-6	Hr.	Daily	Hr.	Daily	Bare Costs	Incl. O&P
1 Electrician Foreman	$41.25	$330.00	$61.40	$491.20	$35.27	$53.12
4 Electrician Linemen	40.75	1304.00	60.65	1940.80		
2 Electrician Operators	40.75	652.00	60.65	970.40		
4 Electrician Groundmen	25.55	817.60	39.75	1272.00		
1 Crew Truck		98.80		108.70		
1 Tool Van		122.15		134.35		
1 Pickup Truck, 3/4 Ton		78.00		85.80		
.2 Crane, 55 Ton		179.08		197.00		
.2 Crane, 12 Ton		120.52		132.55		
.2 Auger, Truck Mtd.		635.80		699.40		
1 Tractor w/Winch		260.00		286.00		
3 Cable Trailers		490.65		539.70		
.5 Tensioning Rig		160.00		176.00		
.5 Cable Pulling Rig		934.00		1027.40	34.99	38.49
88 L.H., Daily Totals		$6182.60		$8061.30	$70.26	$91.61

Crew R-7	Hr.	Daily	Hr.	Daily	Bare Costs	Incl. O&P
1 Electrician Foreman	$41.25	$330.00	$61.40	$491.20	$28.17	$43.36
5 Electrician Groundmen	25.55	1022.00	39.75	1590.00		
1 Crew Truck		98.80		108.70	2.06	2.26
48 L.H., Daily Totals		$1450.80		$2189.90	$30.23	$45.62

Crew R-8	Hr.	Daily	Hr.	Daily	Bare Costs	Incl. O&P
1 Electrician Foreman	$41.25	$330.00	$61.40	$491.20	$35.77	$53.81
3 Electrician Linemen	40.75	978.00	60.65	1455.60		
2 Electrician Groundmen	25.55	408.80	39.75	636.00		
1 Pickup Truck, 3/4 Ton		78.00		85.80		
1 Crew Truck		98.80		108.70	3.68	4.05
48 L.H., Daily Totals		$1893.60		$2777.30	$39.45	$57.86

Crew R-9	Hr.	Daily	Hr.	Daily	Bare Costs	Incl. O&P
1 Electrician Foreman	$41.25	$330.00	$61.40	$491.20	$33.21	$50.29
1 Electrician Lineman	40.75	326.00	60.65	485.20		
2 Electrician Operators	40.75	652.00	60.65	970.40		
4 Electrician Groundmen	25.55	817.60	39.75	1272.00		
1 Pickup Truck, 3/4 Ton		78.00		85.80		
1 Crew Truck		98.80		108.70	2.76	3.04
64 L.H., Daily Totals		$2302.40		$3413.30	$35.97	$53.33

Crew R-10	Hr.	Daily	Hr.	Daily	Bare Costs	Incl. O&P
1 Electrician Foreman	$41.25	$330.00	$61.40	$491.20	$38.30	$57.29
4 Electrician Linemen	40.75	1304.00	60.65	1940.80		
1 Electrician Groundman	25.55	204.40	39.75	318.00		
1 Crew Truck		98.80		108.70		
3 Tram Cars		345.75		380.35	9.26	10.19
48 L.H., Daily Totals		$2282.95		$3239.05	$47.56	$67.48

Crew R-11	Hr.	Daily	Hr.	Daily	Bare Costs	Incl. O&P
1 Electrician Foreman	$41.25	$330.00	$61.40	$491.20	$38.12	$57.09
4 Electricians	40.75	1304.00	60.65	1940.80		
1 Equip. Oper. (crane)	35.90	287.20	54.10	432.80		
1 Common Laborer	26.70	213.60	41.55	332.40		
1 Crew Truck		98.80		108.70		
1 Crane, 12 Ton		602.60		662.85	12.53	13.78
56 L.H., Daily Totals		$2836.20		$3968.75	$50.65	$70.87

Crew R-12	Hr.	Daily	Hr.	Daily	Bare Costs	Incl. O&P
1 Carpenter Foreman	$34.75	$278.00	$54.10	$432.80	$31.94	$50.38
4 Carpenters	34.25	1096.00	53.35	1707.20		
4 Common Laborers	26.70	854.40	41.55	1329.60		
1 Equip. Oper. (med.)	34.65	277.20	52.20	417.60		
1 Steel Worker	38.15	305.20	68.25	546.00		
1 Dozer, 200 H.P.		919.60		1011.55		
1 Pickup Truck, 3/4 Ton		78.00		85.80	11.34	12.47
88 L.H., Daily Totals		$3808.40		$5530.55	$43.28	$62.85

Crew R-13	Hr.	Daily	Hr.	Daily	Bare Costs	Incl. O&P
1 Electrician Foreman	$41.25	$330.00	$61.40	$491.20	$38.59	$57.57
3 Electricians	40.75	978.00	60.65	1455.60		
.25 Equip. Oper. (crane)	35.90	71.80	54.10	108.20		
1 Equipment Oiler	30.10	240.80	45.35	362.80		
.25-1 Hyd. Crane, 33 Ton		160.40		176.45	3.82	4.20
42 L.H., Daily Totals		$1781.00		$2594.25	$42.41	$61.77

Crew No.	Bare Costs		Incl. Sub O & P		Cost Per Labor-Hour	
Crew R-15	**Hr.**	**Daily**	**Hr.**	**Daily**	**Bare Costs**	**Incl. O&P**
1 Electrician Foreman	$41.25	$330.00	$61.40	$491.20	$39.55	$58.97
4 Electricians	40.75	1304.00	60.65	1940.80		
1 Equipment Operator	33.05	264.40	49.80	398.40		
1 Aerial Lift Truck		252.60		277.85	5.26	5.79
48 L.H., Daily Totals		$2151.00		$3108.25	$44.81	$64.76
Crew R-18	**Hr.**	**Daily**	**Hr.**	**Daily**	**Bare Costs**	**Incl. O&P**
.25 Electrician Foreman	$41.25	$82.50	$61.40	$122.80	$31.43	$47.85
1 Electrician	40.75	326.00	60.65	485.20		
2 Helpers	25.55	408.80	39.75	636.00		
26 L.H., Daily Totals		$817.30		$1244.00	$31.43	$47.85
Crew R-19	**Hr.**	**Daily**	**Hr.**	**Daily**	**Bare Costs**	**Incl. O&P**
.5 Electrician Foreman	$41.25	$165.00	$61.40	$245.60	$40.85	$60.80
2 Electricians	40.75	652.00	60.65	970.40		
20 L.H., Daily Totals		$817.00		$1216.00	$40.85	$60.80
Crew R-21	**Hr.**	**Daily**	**Hr.**	**Daily**	**Bare Costs**	**Incl. O&P**
1 Electrician Foreman	$41.25	$330.00	$61.40	$491.20	$40.72	$60.63
3 Electricians	40.75	978.00	60.65	1455.60		
.1 Equip. Oper. (med.)	34.65	27.72	52.20	41.76		
.1 Hyd. Crane 25 Ton		57.50		63.25	1.75	1.93
32. L.H., Daily Totals		$1393.22		$2051.81	$42.47	$62.56
Crew R-22	**Hr.**	**Daily**	**Hr.**	**Daily**	**Bare Costs**	**Incl. O&P**
.66 Electrician Foreman	$41.25	$217.80	$61.40	$324.19	$34.30	$51.79
2 Helpers	25.55	408.80	39.75	636.00		
2 Electricians	40.75	652.00	60.65	970.40		
37.28 L.H., Daily Totals		$1278.60		$1930.59	$34.30	$51.79
Crew R-30	**Hr.**	**Daily**	**Hr.**	**Daily**	**Bare Costs**	**Incl. O&P**
.25 Electrician	$42.75	$85.50	$63.60	$127.20	$32.26	$49.12
1 Electrician	40.75	326.00	60.65	485.20		
2 Laborers, (Semi-Skilled)	26.70	427.20	41.55	664.80		
26 L.H., Daily Totals		$838.70		$1277.20	$32.26	$49.12
Crew R-31	**Hr.**	**Daily**	**Hr.**	**Daily**	**Bare Costs**	**Incl. O&P**
1 Electrician	$40.75	$326.00	$60.65	$485.20	$40.75	$60.65
1 Core Drill, Elec, 2.5 HP		50.30		55.35	6.28	6.91
8 L.H., Daily Totals		$376.30		$540.55	$47.03	$67.56
Crew W-41E	**Hr.**	**Daily**	**Hr.**	**Daily**	**Bare Costs**	**Incl. O&P**
1 Laborers, (Semi-Skilled)	$26.70	$213.60	$41.55	$332.40	$35.59	$54.06
1 Plumber	40.85	326.80	61.40	491.20		
.5 Plumber	42.85	171.40	64.40	257.60		
20 L.H., Daily Totals		$711.80		$1081.20	$35.59	$54.06

Historical Cost Indexes

The table below lists both the Means Historical Cost Index based on Jan. 1, 1993 = 100 as well as the computed value of an index based on Jan. 1, 2005 costs. Since the Jan. 1, 2005 figure is estimated, space is left to write in the actual index figures as they become available through either the quarterly "Means Construction Cost Indexes" or as printed in the "Engineering News-Record." To compute the actual index based on Jan. 1, 2005 = 100, divide the Historical Cost Index for a particular year by the actual Jan. 1, 2005 Construction Cost Index. Space has been left to advance the index figures as the year progresses.

Year	Historical Cost Index Jan. 1, 1993 = 100		Current Index Based on Jan. 1, 2005 = 100	
	Est.	Actual	Est.	Actual
Oct 2005				
July 2005				
April 2005				
Jan 2005	148.5		100.0	100.0
July 2004		143.7	96.8	
2003		132.0	88.9	
2002		128.7	86.7	
2001		125.1	84.2	
2000		120.9	81.4	
1999		117.6	79.2	
1998		115.1	77.5	
1997		112.8	76.0	
1996		110.2	74.2	
1995		107.6	72.5	
1994		104.4	70.3	
1993		101.7	68.5	
1992		99.4	67.0	
▼ 1991		96.8	65.2	

Year	Historical Cost Index Jan. 1, 1993 = 100	Current Index Based on Jan. 1, 2005 = 100	
	Actual	Est.	Actual
July 1990	94.3	63.5	
1989	92.1	62.0	
1988	89.9	60.5	
1987	87.7	59.0	
1986	84.2	56.7	
1985	82.6	55.6	
1984	82.0	55.2	
1983	80.2	54.0	
1982	76.1	51.3	
1981	70.0	47.1	
1980	62.9	42.4	
1979	57.8	38.9	
1978	53.5	36.0	
1977	49.5	33.3	
1976	46.9	31.6	
1975	44.8	30.2	
1974	41.4	27.9	
▼ 1973	37.7	25.4	

Year	Historical Cost Index Jan. 1, 1993 = 100	Current Index Based on Jan. 1, 2005 = 100	
	Actual	Est.	Actual
July 1972	34.8	23.4	
1971	32.1	21.6	
1970	28.7	19.3	
1969	26.9	18.1	
1968	24.9	16.8	
1967	23.5	15.8	
1966	22.7	15.3	
1965	21.7	14.6	
1964	21.2	14.3	
1963	20.7	13.9	
1962	20.2	13.6	
1961	19.8	13.3	
1960	19.7	13.3	
1959	19.3	13.0	
1958	18.8	12.7	
1957	18.4	12.4	
1956	17.6	11.9	
▼ 1955	16.6	11.2	

Adjustments to Costs

The Historical Cost Index can be used to convert National Average building costs at a particular time to the approximate building costs for some other time.

Example:

Estimate and compare construction costs for different years in the same city.

To estimate the National Average construction cost of a building in 1970, knowing that it cost $900,000 in 2005:

INDEX in 1970 = 28.7

INDEX in 2005 = 148.5

Note: The City Cost Indexes for Canada can be used to convert U.S. National averages to local costs in Canadian dollars.

Time Adjustment using the Historical Cost Indexes:

$$\frac{\text{Index for Year A}}{\text{Index for Year B}} \times \text{Cost in Year B} = \text{Cost in Year A}$$

$$\frac{\text{INDEX 1970}}{\text{INDEX 2005}} \times \text{Cost 2005} = \text{Cost 1970}$$

$$\frac{28.7}{148.5} \times \$900{,}000 = .193 \times \$900{,}000 = \$173{,}700$$

The construction cost of the building in 1970 is $173,700.

How to Use the City Cost Indexes

What you should know before you begin

Means City Cost Indexes (CCI) are an extremely useful tool to use when you want to compare costs from city to city and region to region.

This publication contains average construction cost indexes for 316 major U.S. and Canadian cities and Location Factors covering over 930 three-digit zip code locations.

Keep in mind that a City Cost Index number is a *percentage ratio* of a specific city's cost to the national average cost of the same item at a stated time period.

In other words, these index figures represent relative construction *factors* (or, if you prefer, multipliers) for Material and Installation costs, as well as the weighted average for Total In Place costs for each CSI MasterFormat division. Installation costs include both labor and equipment rental costs. When estimating equipment rental rates only, for a specific location, use 01590 EQUIPMENT RENTAL index.

The 30 City Average Index is the average of 30 major U.S. cities and serves as a National Average.

Index figures for both material and installation are based on the 30 major city average of 100 and represent the cost relationship as of July 1, 2003. The index for each division is computed from representative material and labor quantities for that division. The weighted average for each city is a weighted total of the components listed above it, but does not include relative productivity between trades or cities.

As changes occur in local material prices, labor rates and equipment rental rates, the impact of these changes should be accurately measured by the change in the City Cost Index for each particular city (as compared to the 30 City Average).

Therefore, if you know (or have estimated) building costs in one city today, you can easily convert those costs to expected building costs in another city.

In addition, by using the Historical Cost Index, you can easily convert National Average building costs at a particular time to the approximate building costs for some other time. The City Cost Indexes can then be applied to calculate the costs for a particular city.

Quick Calculations

Location Adjustment Using the City Cost Indexes:

$$\frac{\text{Index for City A}}{\text{Index for City B}} \times \text{Cost in City B} = \text{Cost in City A}$$

Time Adjustment for the National Average Using the Historical Cost Index:

$$\frac{\text{Index for Year A}}{\text{Index for Year B}} \times \text{Cost in Year B} = \text{Cost in Year A}$$

Adjustment from the National Average:

$$\frac{\text{Index for City A}}{100} \times \text{National Average Cost} = \text{Cost in City A}$$

Since each of the other RSMeans publications contains many different items, any *one* item multiplied by the particular city index may give incorrect results. However, the larger the number of items compiled, the closer the results should be to actual costs for that particular city.

The City Cost Indexes for Canadian cities are calculated using Canadian material and equipment prices and labor rates, in Canadian dollars. Therefore, indexes for Canadian cities can be used to convert U.S. National Average prices to local costs in Canadian dollars.

How to use this section

1. Compare costs from city to city.

In using the Means Indexes, remember that an index number is not a fixed number but a *ratio:* It's a percentage ratio of a building component's cost at any stated time to the National Average cost of that same component at the same time period. Put in the form of an equation:

$$\frac{\text{Specific City Cost}}{\text{National Average Cost}} \times 100 = \text{City Index Number}$$

Therefore, when making cost comparisons between cities, do not subtract one city's index number from the index number of another city and read the result as a percentage difference. Instead, divide one city's index number by that of the other city. The resulting number may then be used as a multiplier to calculate cost differences from city to city.

The formula used to find cost differences between cities for the purpose of comparison is as follows:

$$\frac{\text{City A Index}}{\text{City B Index}} \times \text{City B Cost (Known)} = \text{City A Cost (Unknown)}$$

In addition, you can use *Means CCI* to calculate and compare costs division by division between cities using the same basic formula. (Just be sure that you're comparing similar divisions.)

2. Compare a specific city's construction costs with the National Average.

When you're studying construction location feasibility, it's advisable to compare a prospective project's cost index with an index of the National Average cost.

For example, divide the weighted average index of construction costs of a specific city by that of the 30 City Average, which = 100.

$$\frac{\text{City Index}}{100} = \text{\% of National Average}$$

As a result, you get a ratio that indicates the relative cost of construction in that city in comparison with the National Average.

3. Convert U.S. National Average to actual costs in Canadian City.

$$\frac{\text{Index for Canadian City}}{100} \times \text{National Average Cost} = \text{Cost in Canadian City in \$ CAN}$$

4. Adjust construction cost data based on a National Average.

When you use a source of construction cost data which is based on a National Average (such as *Means cost data publications*), it is necessary to adjust those costs to a specific location.

$$\frac{\text{City Index}}{100} \times \text{"Book" Cost Based on National Average Costs} = \text{City Cost (Unknown)}$$

5. When applying the City Cost Indexes to demolition projects, use the appropriate division index. For example, for removal of existing doors and windows, use the Division 8 index.

What you might like to know about how we developed the Indexes

To create a reliable index, RSMeans researched the building type most often constructed in the United States and Canada. Because it was concluded that no one type of building completely represented the building construction industry, nine different types of buildings were combined to create a composite model.

The exact material, labor and equipment quantities are based on detailed analysis of these nine building types, then each quantity is weighted in proportion to expected usage. These various material items, labor hours, and equipment rental rates are thus combined to form a composite building representing as closely as possible the actual usage of materials, labor and equipment used in the North American Building Construction Industry.

The following structures were chosen to make up that composite model:

1. Factory, 1 story
2. Office, 2–4 story
3. Store, Retail
4. Town Hall, 2–3 story
5. High School, 2–3 story
6. Hospital, 4–8 story
7. Garage, Parking
8. Apartment, 1–3 story
9. Hotel/Motel, 2–3 story

For the purposes of ensuring the timeliness of the data, the components of the index for the composite model have been streamlined. They currently consist of:

- specific quantities of 66 commonly used construction materials;
- specific labor-hours for 21 building construction trades; and
- specific days of equipment rental for 6 types of construction equipment (normally used to install the 66 material items by the 21 trades.)

A sophisticated computer program handles the updating of all costs for each city on a quarterly basis. Material and equipment price quotations are gathered quarterly from over 316 cities in the United States and Canada. These prices and the latest negotiated labor wage rates for 21 different building trades are used to compile the quarterly update of the City Cost Index.

The 30 major U.S. cities used to calculate the National Average are:

Atlanta, GA	Memphis, TN
Baltimore, MD	Milwaukee, WI
Boston, MA	Minneapolis, MN
Buffalo, NY	Nashville, TN
Chicago, IL	New Orleans, LA
Cincinnati, OH	New York, NY
Cleveland, OH	Philadelphia, PA
Columbus, OH	Phoenix, AZ
Dallas, TX	Pittsburgh, PA
Denver, CO	St. Louis, MO
Detroit, MI	San Antonio, TX
Houston, TX	San Diego, CA
Indianapolis, IN	San Francisco, CA
Kansas City, MO	Seattle, WA
Los Angeles, CA	Washington, DC

F.Y.I.: The CSI MasterFormat Divisions

1. General Requirements
2. Site Construction
3. Concrete
4. Masonry
5. Metals
6. Wood & Plastics
7. Thermal & Moisture Protection
8. Doors & Windows
9. Finishes
10. Specialties
11. Equipment
12. Furnishings
13. Special Construction
14. Conveying Systems
15. Mechanical
16. Electrical

The information presented in the CCI is organized according to the Construction Specifications Institute (CSI) MasterFormat.

What the CCI does not *indicate*

The weighted average for each city is a total of the components listed above weighted to reflect typical usage, but it does *not* include the productivity variations between trades or cities.

In addition, the CCI does not take into consideration factors such as the following:

- managerial efficiency
- competitive conditions
- automation
- restrictive union practices
- unique local requirements
- regional variations due to specific building codes

	DIVISION	UNITED STATES			ALABAMA															
		30 CITY AVERAGE			BIRMINGHAM			HUNTSVILLE			MOBILE			MONTGOMERY			TUSCALOOSA			
		MAT.	INST.	TOTAL	MAT.	INST.	TOTAL	MAT.	INST.	TOTAL	MAT.	INST.	TOTAL	MAT.	INST.	TOTAL	MAT.	INST.	TOTAL	
01590	**EQUIPMENT RENTAL**	.0	100.0	100.0	.0	101.3	101.3	.0	101.2	101.2	.0	97.7	97.7	.0	97.7	97.7	.0	101.2	101.2	
02	**SITE CONSTRUCTION**	100.0	100.0	100.0	84.6	93.1	90.9	82.7	92.9	90.2	93.7	86.1	88.0	94.1	86.6	88.6	83.2	91.9	89.6	
03100	CONCRETE FORMS & ACCESSORIES	100.0	100.0	100.0	95.0	75.6	78.2	97.1	68.6	72.4	97.1	53.0	58.9	95.9	48.1	54.5	97.0	39.7	47.3	
03200	CONCRETE REINFORCEMENT	100.0	100.0	100.0	92.1	88.2	90.1	92.1	77.4	84.7	95.0	53.9	74.2	95.0	86.7	90.8	92.1	87.0	89.5	
03300	CAST-IN-PLACE CONCRETE	100.0	100.0	100.0	92.8	68.1	82.5	87.8	66.2	78.8	92.7	54.7	76.8	94.3	49.4	75.6	91.4	47.1	72.9	
03	**CONCRETE**	100.0	100.0	100.0	90.9	76.4	83.7	88.7	70.7	79.7	91.5	55.4	73.5	92.2	57.5	74.9	90.4	53.0	71.8	
04	**MASONRY**	100.0	100.0	100.0	85.9	76.0	79.7	85.8	65.2	72.8	86.3	52.3	65.0	86.9	36.3	55.2	86.0	38.6	56.3	
05	**METALS**	100.0	100.0	100.0	94.6	95.5	94.9	96.0	91.0	94.4	94.8	79.6	89.8	94.6	92.1	93.8	95.2	92.7	94.4	
06	**WOOD & PLASTICS**	100.0	100.0	100.0	96.7	76.0	85.9	96.7	68.4	81.9	96.7	52.7	73.7	95.3	48.3	70.7	96.7	38.3	66.2	
07	**THERMAL & MOISTURE PROTECTION**	100.0	100.0	100.0	95.7	80.9	88.4	95.4	75.1	85.4	95.4	69.5	82.7	95.1	63.0	79.3	95.3	61.8	78.8	
08	**DOORS & WINDOWS**	100.0	100.0	100.0	98.5	78.5	93.3	98.5	65.5	89.9	98.5	53.2	86.8	98.5	58.2	88.0	98.5	58.0	88.0	
09200	PLASTER & GYPSUM BOARD	100.0	100.0	100.0	103.1	75.9	85.9	100.5	68.1	80.0	100.5	52.0	69.7	100.5	47.4	66.9	100.5	37.2	60.4	
095,098	CEILINGS & ACOUSTICAL TREATMENT	100.0	100.0	100.0	102.7	75.9	86.6	102.7	68.1	81.9	102.7	52.0	72.3	102.7	47.4	69.6	102.7	37.2	63.4	
09600	FLOORING	100.0	100.0	100.0	104.4	50.5	90.1	104.4	54.2	91.1	114.7	54.7	98.8	112.9	28.3	90.5	104.4	41.5	87.7	
097,099	WALL FINISHES, PAINTS & COATINGS	100.0	100.0	100.0	95.3	70.8	80.4	95.3	64.2	76.4	100.1	56.9	73.9	95.3	56.0	71.4	95.3	48.2	66.7	
09	**FINISHES**	100.0	100.0	100.0	100.8	70.1	84.8	100.3	65.0	81.9	104.8	53.1	77.9	104.0	43.9	72.8	100.2	39.2	68.5	
10 - 14	**TOTAL DIV. 10000 - 14000**	100.0	100.0	100.0	100.0	86.6	97.1	100.0	84.5	96.7	100.0	69.0	93.4	100.0	77.1	95.1	100.0	74.5	94.6	
15	**MECHANICAL**	100.0	100.0	100.0	100.3	69.6	86.9	100.3	69.8	86.9	99.8	70.8	87.2	99.8	41.7	74.4	99.8	36.7	72.2	
16	**ELECTRICAL**	100.0	100.0	100.0	96.3	66.6	81.8	96.4	70.5	83.8	96.4	51.3	74.4	96.3	69.1	83.0	96.4	66.6	81.9	
01 - 16	**WEIGHTED AVERAGE**	100.0	100.0	100.0	96.4	76.5	87.4	96.3	73.3	85.8	97.1	63.0	81.6	97.0	58.4	79.5	96.3	56.4	78.2	

	DIVISION	ALASKA									ARIZONA								
		ANCHORAGE			FAIRBANKS			JUNEAU			FLAGSTAFF			MESA/TEMPE			PHOENIX		
		MAT.	INST.	TOTAL	MAT.	INST.	TOTAL	MAT.	INST.	TOTAL	MAT.	INST.	TOTAL	MAT.	INST.	TOTAL	MAT.	INST.	TOTAL
01590	**EQUIPMENT RENTAL**	.0	118.4	118.4	.0	118.4	118.4	.0	118.4	118.4	.0	94.9	94.9	.0	98.2	98.2	.0	98.8	98.8
02	**SITE CONSTRUCTION**	143.5	133.8	136.3	127.1	133.8	132.1	139.1	133.8	135.1	83.5	100.9	96.4	88.3	103.9	99.9	88.8	104.7	100.6
03100	CONCRETE FORMS & ACCESSORIES	132.4	113.9	116.4	134.1	119.7	121.6	133.8	113.9	116.6	102.8	61.5	67.1	97.3	63.1	67.7	98.4	69.8	73.6
03200	CONCRETE REINFORCEMENT	146.2	106.6	126.2	122.9	106.7	114.7	108.7	106.6	107.7	105.5	73.5	89.3	104.8	72.9	88.7	103.2	74.2	88.5
03300	CAST-IN-PLACE CONCRETE	195.7	114.6	161.8	163.1	115.1	143.0	196.5	114.6	162.3	91.8	78.3	86.2	98.3	71.4	87.1	98.4	78.9	90.3
03	**CONCRETE**	153.5	112.2	132.9	129.8	115.0	122.4	147.9	112.2	130.1	116.7	69.5	93.1	99.2	67.8	83.5	98.8	73.6	86.2
04	**MASONRY**	217.0	121.0	156.7	210.3	121.0	154.3	219.4	121.0	157.7	102.9	56.2	73.6	108.2	52.8	73.4	96.0	68.1	78.5
05	**METALS**	130.1	102.2	121.0	130.1	102.5	121.1	130.1	102.2	121.0	93.8	67.9	85.4	94.5	69.2	86.2	96.0	70.9	87.8
06	**WOOD & PLASTICS**	117.8	111.5	114.5	118.0	118.9	118.5	117.8	111.5	114.5	108.6	60.1	83.3	99.6	68.7	83.5	100.6	70.5	84.9
07	**THERMAL & MOISTURE PROTECTION**	198.5	115.8	157.8	194.6	118.6	157.2	195.2	115.8	156.2	107.3	66.9	87.4	105.4	63.8	85.0	105.3	70.8	88.4
08	**DOORS & WINDOWS**	126.2	108.8	121.7	123.3	112.9	120.6	123.3	108.8	119.6	102.2	64.2	92.3	98.9	65.2	90.2	100.0	69.9	92.2
09200	PLASTER & GYPSUM BOARD	132.6	111.6	119.3	132.6	119.3	124.1	132.6	111.6	119.3	90.8	59.0	70.6	91.6	67.7	76.4	93.9	69.6	78.5
095,098	CEILINGS & ACOUSTICAL TREATMENT	132.1	111.6	119.8	132.1	119.3	124.4	132.1	111.6	119.8	101.3	59.0	75.9	95.6	67.7	78.8	102.7	69.6	82.8
09600	FLOORING	164.9	127.0	154.9	164.9	127.0	154.9	164.9	127.0	154.9	95.2	51.8	83.8	97.3	64.6	88.7	97.6	67.5	89.7
097,099	WALL FINISHES, PAINTS & COATINGS	168.1	111.4	133.7	168.1	125.0	141.9	168.1	111.4	133.7	94.6	48.8	66.8	105.0	53.1	73.5	105.0	60.4	77.9
09	**FINISHES**	154.8	116.0	134.6	152.6	121.9	136.6	153.4	116.0	134.0	96.7	57.6	76.4	97.2	62.0	78.9	99.5	68.0	83.1
10 - 14	**TOTAL DIV. 10000 - 14000**	100.0	112.8	102.7	100.0	113.7	102.9	100.0	112.8	102.7	100.0	72.6	94.2	100.0	69.0	93.4	100.0	74.4	94.6
15	**MECHANICAL**	100.9	108.3	104.2	100.9	115.9	107.4	100.9	108.3	104.2	100.2	77.4	90.2	100.2	67.9	86.1	100.2	77.5	90.2
16	**ELECTRICAL**	149.1	112.3	131.2	150.9	112.3	132.1	150.9	112.3	132.1	96.9	48.0	73.1	90.6	60.3	75.8	99.6	65.7	83.1
01 - 16	**WEIGHTED AVERAGE**	134.5	113.8	125.1	130.7	116.9	124.4	133.6	113.8	124.6	100.5	67.6	85.5	97.8	67.7	84.1	98.8	74.3	87.7

	DIVISION	ARIZONA						ARKANSAS											
		PRESCOTT			TUCSON			FORT SMITH			JONESBORO			LITTLE ROCK			PINE BLUFF		
		MAT.	INST.	TOTAL	MAT.	INST.	TOTAL	MAT.	INST.	TOTAL	MAT.	INST.	TOTAL	MAT.	INST.	TOTAL	MAT.	INST.	TOTAL
01590	**EQUIPMENT RENTAL**	.0	94.9	94.9	.0	98.2	98.2	.0	86.0	86.0	.0	107.8	107.8	.0	86.0	86.0	.0	86.0	86.0
02	**SITE CONSTRUCTION**	71.2	100.3	92.7	85.1	104.4	99.4	77.6	83.8	82.2	100.5	99.2	99.6	77.4	83.8	82.1	79.6	83.8	82.7
03100	CONCRETE FORMS & ACCESSORIES	98.2	56.0	61.6	97.9	69.2	73.1	97.7	42.0	49.4	84.5	47.8	52.7	91.8	59.2	63.5	77.2	58.9	61.4
03200	CONCRETE REINFORCEMENT	105.5	72.5	88.8	86.6	73.5	80.0	96.9	74.6	85.6	92.7	49.6	70.9	97.1	72.2	84.5	97.1	72.1	84.5
03300	CAST-IN-PLACE CONCRETE	91.7	64.2	80.2	101.1	78.7	91.8	89.8	66.8	80.2	85.6	57.1	73.7	89.8	67.0	80.3	82.4	66.9	75.9
03	**CONCRETE**	102.6	62.0	82.3	97.4	73.1	85.3	87.4	57.4	72.4	84.0	53.0	68.6	87.0	64.7	75.9	85.0	64.5	74.8
04	**MASONRY**	103.3	58.8	75.4	97.8	56.2	71.7	97.5	53.3	69.8	92.8	46.9	64.0	95.9	53.3	69.2	118.2	53.3	77.5
05	**METALS**	93.8	66.7	85.0	95.3	68.7	86.6	95.7	71.3	87.7	90.0	75.9	85.4	91.6	70.8	84.8	94.3	70.6	86.6
06	**WOOD & PLASTICS**	103.8	54.6	78.1	99.8	70.5	84.5	100.4	40.6	69.2	86.4	48.8	66.8	97.4	63.3	79.6	78.9	63.3	70.7
07	**THERMAL & MOISTURE PROTECTION**	105.6	61.7	84.0	106.6	64.2	85.8	99.4	48.4	74.3	109.1	51.6	80.8	97.9	50.7	74.7	98.0	50.7	74.8
08	**DOORS & WINDOWS**	102.2	58.3	90.8	96.0	69.9	89.2	97.2	45.8	83.9	98.8	47.9	85.6	97.2	58.4	87.1	92.5	58.4	83.7
09200	PLASTER & GYPSUM BOARD	87.9	53.3	66.0	92.0	69.6	77.8	85.4	39.6	56.4	92.3	47.7	64.0	85.4	62.9	71.1	77.7	62.9	68.3
095,098	CEILINGS & ACOUSTICAL TREATMENT	99.6	53.3	71.8	95.6	69.6	80.0	91.2	39.6	60.3	89.0	47.7	64.2	91.2	62.9	74.2	86.8	62.9	72.4
09600	FLOORING	93.6	51.6	82.5	96.8	51.8	84.9	115.8	70.0	103.7	77.3	45.6	68.9	117.1	70.0	104.6	104.8	70.0	95.6
097,099	WALL FINISHES, PAINTS & COATINGS	94.6	48.8	66.8	102.6	48.8	70.0	96.4	58.3	73.3	84.9	52.4	65.2	96.4	59.9	74.2	96.4	59.9	74.2
09	**FINISHES**	94.3	53.8	73.2	96.9	63.8	79.6	96.4	48.4	71.4	87.7	47.9	67.0	96.7	61.9	78.6	91.1	61.9	75.9
10 - 14	**TOTAL DIV. 10000 - 14000**	100.0	71.1	93.9	100.0	74.4	94.6	100.0	63.4	92.2	100.0	54.9	90.4	100.0	66.4	92.8	100.0	66.4	92.8
15	**MECHANICAL**	100.2	73.6	88.6	100.1	69.2	86.6	100.1	43.2	75.2	100.2	43.1	75.2	100.1	58.4	81.8	100.1	45.6	76.3
16	**ELECTRICAL**	96.7	46.2	72.1	92.5	60.9	77.1	95.6	64.4	80.4	102.7	47.8	75.9	95.3	70.0	83.0	95.0	70.0	82.8
01 - 16	**WEIGHTED AVERAGE**	98.2	64.7	83.0	97.0	69.8	84.6	96.0	56.8	78.2	95.5	55.0	77.1	95.2	64.3	81.2	95.3	61.7	80.0

City Cost Indexes

	DIVISION	ARKANSAS			CALIFORNIA															
		TEXARKANA			ANAHEIM			BAKERSFIELD			FRESNO			LOS ANGELES			OAKLAND			
		MAT.	INST.	TOTAL	MAT.	INST.	TOTAL	MAT.	INST.	TOTAL	MAT.	INST.	TOTAL	MAT.	INST.	TOTAL	MAT.	INST.	TOTAL	
01590	EQUIPMENT RENTAL	.0	86.8	86.8	.0	102.1	102.1	.0	99.5	99.5	.0	99.5	99.5	.0	98.5	98.5	.0	103.3	103.3	
02	SITE CONSTRUCTION	95.3	84.6	87.4	103.2	109.0	107.5	108.1	106.0	106.6	109.6	106.0	106.9	98.1	107.8	105.3	155.8	104.6	117.9	
03100	CONCRETE FORMS & ACCESSORIES	85.0	39.6	45.7	105.0	121.5	119.3	97.5	121.1	117.9	101.2	127.2	123.7	105.5	121.4	119.3	107.4	133.5	130.0	
03200	CONCRETE REINFORCEMENT	96.6	48.1	72.1	95.6	112.5	104.1	108.9	112.5	110.7	92.2	112.8	102.6	113.0	112.7	112.9	103.1	113.4	108.3	
03300	CAST-IN-PLACE CONCRETE	89.8	44.6	70.9	107.2	117.2	111.4	102.3	116.1	108.1	112.0	113.5	112.6	87.3	115.4	99.1	139.5	118.9	130.9	
03	CONCRETE	84.3	43.9	64.1	107.8	117.4	112.6	107.1	116.8	112.0	109.3	118.7	114.0	103.1	116.7	109.9	126.4	123.6	125.0	
04	MASONRY	97.5	32.8	56.9	89.0	111.4	103.1	107.5	111.2	109.8	111.1	111.2	111.2	97.7	115.3	108.8	148.4	122.5	132.1	
05	METALS	86.9	59.4	77.9	110.9	99.8	107.2	105.4	99.0	103.3	111.1	100.2	107.6	112.8	98.4	108.1	105.1	104.9	105.0	
06	WOOD & PLASTICS	88.2	42.3	64.3	94.9	121.2	108.7	86.0	121.3	104.4	99.1	129.8	115.1	91.8	120.9	107.0	101.9	134.8	119.1	
07	THERMAL & MOISTURE PROTECTION	98.9	42.2	71.0	121.2	115.0	118.1	105.5	108.9	107.2	101.3	110.5	105.8	111.6	117.0	114.3	113.0	124.0	118.4	
08	DOORS & WINDOWS	97.6	41.1	83.0	102.8	116.2	106.2	101.6	113.3	104.6	101.9	118.6	106.2	96.7	116.0	101.7	106.4	125.7	111.4	
09200	PLASTER & GYPSUM BOARD	83.3	41.4	56.7	98.1	121.7	113.1	98.1	121.7	113.1	96.1	130.4	117.9	101.6	121.7	114.4	94.7	135.1	120.3	
095,098	CEILINGS & ACOUSTICAL TREATMENT	93.9	41.4	62.4	119.6	121.7	120.9	119.6	121.7	120.9	119.6	130.4	126.1	119.8	121.7	121.0	114.8	135.1	127.0	
09600	FLOORING	107.6	43.5	90.6	123.1	103.7	118.0	117.3	93.4	111.0	133.8	138.6	135.1	114.0	103.7	111.3	116.1	122.8	117.9	
097,099	WALL FINISHES, PAINTS & COATINGS	96.4	36.8	60.2	111.4	110.7	111.0	110.9	92.5	99.7	134.6	100.1	113.7	99.6	110.7	106.4	115.6	136.3	128.1	
09	FINISHES	95.0	40.1	66.4	114.3	117.4	115.9	114.5	113.6	114.0	120.9	127.5	124.3	110.8	117.1	114.1	115.9	132.6	124.6	
10 - 14	TOTAL DIV. 10000 - 14000	100.0	39.3	87.1	100.0	114.8	103.2	100.0	119.2	104.1	100.0	128.8	106.1	100.0	114.1	103.0	100.0	132.3	106.9	
15	MECHANICAL	100.1	38.1	73.0	100.2	108.4	103.8	100.2	99.6	99.9	100.2	111.6	105.2	100.1	108.4	103.7	100.2	127.0	111.9	
16	ELECTRICAL	96.8	39.8	69.0	88.6	107.9	98.0	89.5	93.6	91.5	87.5	97.1	92.2	96.5	112.9	104.5	102.6	134.4	118.1	
01 - 16	WEIGHTED AVERAGE	94.7	45.3	72.3	102.9	111.1	106.6	102.6	106.2	104.2	104.3	111.9	107.8	102.7	111.9	106.8	110.6	123.9	116.6	

	DIVISION	CALIFORNIA																	
		OXNARD			REDDING			RIVERSIDE			SACRAMENTO			SAN DIEGO			SAN FRANCISCO		
		MAT.	INST.	TOTAL	MAT.	INST.	TOTAL	MAT.	INST.	TOTAL	MAT.	INST.	TOTAL	MAT.	INST.	TOTAL	MAT.	INST.	TOTAL
01590	EQUIPMENT RENTAL	.0	98.1	98.1	.0	99.2	99.2	.0	100.6	100.6	.0	102.9	102.9	.0	97.1	97.1	.0	108.6	108.6
02	SITE CONSTRUCTION	109.7	103.8	105.3	114.4	105.2	107.6	100.8	106.7	105.1	119.5	110.9	113.2	102.2	100.6	101.0	158.2	111.2	123.4
03100	CONCRETE FORMS & ACCESSORIES	103.4	121.5	119.1	103.1	126.9	123.7	105.4	121.4	119.3	105.9	127.6	124.7	106.2	111.0	110.3	107.8	134.8	131.2
03200	CONCRETE REINFORCEMENT	108.9	112.2	110.6	105.6	112.6	109.1	107.9	112.5	110.2	95.3	112.9	104.2	105.6	112.4	109.0	117.5	113.9	115.7
03300	CAST-IN-PLACE CONCRETE	108.6	116.4	111.9	123.0	113.3	119.0	106.1	117.2	110.7	120.2	114.3	117.8	110.8	104.9	108.3	139.4	120.7	131.6
03	CONCRETE	110.5	117.0	113.8	120.2	118.4	119.3	109.3	117.4	113.3	115.8	119.1	117.5	113.0	108.4	110.7	128.7	124.9	126.8
04	MASONRY	112.3	105.9	108.2	114.4	108.2	110.5	85.6	111.0	101.5	119.0	112.2	114.7	98.1	106.4	103.3	148.8	129.0	136.4
05	METALS	105.2	99.0	103.1	110.3	99.4	106.7	111.1	99.6	107.4	100.4	99.8	100.2	108.8	98.8	105.6	111.2	107.1	109.9
06	WOOD & PLASTICS	94.0	121.3	108.3	94.2	129.8	112.8	94.9	121.2	108.7	96.7	130.0	114.1	100.3	109.2	104.9	101.9	135.0	119.2
07	THERMAL & MOISTURE PROTECTION	110.5	110.0	110.3	111.3	108.5	109.9	118.9	112.0	115.5	123.8	112.8	118.4	113.8	104.1	109.0	113.0	130.8	121.7
08	DOORS & WINDOWS	100.4	116.2	104.5	103.2	119.8	107.5	102.8	116.2	106.2	117.7	120.8	118.5	103.9	107.7	104.9	110.8	128.8	115.5
09200	PLASTER & GYPSUM BOARD	98.1	121.7	113.1	96.8	130.4	118.1	97.0	121.7	112.7	92.1	130.4	116.4	97.6	109.2	104.9	97.6	135.1	121.4
095,098	CEILINGS & ACOUSTICAL TREATMENT	119.6	121.7	120.9	126.7	130.4	128.9	115.1	121.7	119.1	114.8	130.4	124.2	103.7	109.2	107.0	123.5	135.1	130.5
09600	FLOORING	117.3	103.7	113.7	116.6	106.8	114.0	121.8	103.7	117.0	119.0	115.6	118.1	110.7	103.5	108.8	116.1	128.7	119.5
097,099	WALL FINISHES, PAINTS & COATINGS	110.3	102.7	105.7	110.3	112.3	111.5	107.8	110.7	109.6	112.9	116.2	114.9	109.4	105.0	106.7	115.6	151.7	137.5
09	FINISHES	114.3	116.5	115.5	116.1	123.1	119.8	112.1	117.4	114.8	114.5	125.5	120.2	107.8	109.1	108.5	118.6	136.0	127.7
10 - 14	TOTAL DIV. 10000 - 14000	100.0	115.1	103.2	100.0	128.9	106.1	100.0	114.8	103.1	100.0	129.6	106.3	100.0	112.5	102.7	100.0	133.1	107.0
15	MECHANICAL	100.2	108.4	103.8	100.2	105.7	102.6	100.1	108.4	103.7	100.2	111.2	105.0	100.3	106.8	103.1	100.2	151.6	122.7
16	ELECTRICAL	94.6	106.1	100.2	97.2	102.7	99.9	88.7	102.2	95.3	97.9	102.8	100.3	96.0	98.7	97.4	102.6	152.1	126.8
01 - 16	WEIGHTED AVERAGE	103.9	109.5	106.4	106.8	110.4	108.4	102.6	109.9	105.9	106.9	113.1	109.7	103.9	105.0	104.4	112.6	133.8	122.2

	DIVISION	CALIFORNIA												COLORADO					
		SAN JOSE			SANTA BARBARA			STOCKTON			VALLEJO			COLORADO SPRINGS			DENVER		
		MAT.	INST.	TOTAL	MAT.	INST.	TOTAL	MAT.	INST.	TOTAL	MAT.	INST.	TOTAL	MAT.	INST.	TOTAL	MAT.	INST.	TOTAL
01590	EQUIPMENT RENTAL	.0	100.0	100.0	.0	99.5	99.5	.0	99.2	99.2	.0	103.5	103.5	.0	95.2	95.2	.0	100.4	100.4
02	SITE CONSTRUCTION	148.5	99.7	112.4	109.6	106.1	107.0	107.6	105.3	105.9	116.2	111.1	112.4	95.0	97.1	96.5	93.6	106.4	103.0
03100	CONCRETE FORMS & ACCESSORIES	105.7	133.8	130.0	104.1	121.4	119.1	103.8	126.9	123.8	107.0	131.7	128.4	92.6	85.9	86.8	101.9	87.1	89.1
03200	CONCRETE REINFORCEMENT	96.2	113.6	105.0	108.9	112.5	110.7	109.5	112.6	111.1	102.7	113.5	108.2	113.0	84.6	98.6	113.0	84.7	98.7
03300	CAST-IN-PLACE CONCRETE	128.5	118.6	124.4	108.2	116.3	111.6	108.2	113.4	110.3	125.1	115.4	121.1	98.0	88.1	93.9	91.7	88.5	90.3
03	CONCRETE	118.1	123.8	120.9	110.4	117.0	113.7	110.4	118.5	114.5	119.4	121.4	120.4	108.6	86.5	97.5	102.6	87.2	94.9
04	MASONRY	147.2	125.5	133.6	107.9	111.1	109.9	111.9	108.2	109.5	84.4	121.6	107.7	108.7	82.6	92.3	108.4	85.8	94.2
05	METALS	104.5	106.9	105.3	105.7	99.3	103.6	107.1	99.8	104.7	103.8	101.4	103.0	97.9	88.1	94.7	100.2	88.4	96.3
06	WOOD & PLASTICS	101.0	134.5	118.5	94.0	121.3	108.3	95.9	129.8	113.6	95.1	134.5	115.6	93.9	88.3	91.0	103.3	88.8	95.7
07	THERMAL & MOISTURE PROTECTION	107.2	129.5	118.2	106.8	108.9	107.8	110.8	107.5	109.2	126.1	121.6	123.9	103.8	85.8	95.0	103.2	84.3	93.9
08	DOORS & WINDOWS	92.8	126.7	101.6	101.6	116.2	105.4	100.8	120.6	106.0	119.2	128.6	121.6	98.5	90.6	96.5	99.9	91.5	97.7
09200	PLASTER & GYPSUM BOARD	97.9	135.1	121.5	98.1	121.7	113.1	98.5	130.4	118.7	94.9	135.1	120.4	87.6	88.1	87.9	97.6	88.8	92.0
095,098	CEILINGS & ACOUSTICAL TREATMENT	106.2	135.1	123.5	119.6	121.7	120.9	119.6	130.4	126.1	123.7	135.1	130.5	99.2	88.1	92.5	98.3	88.8	92.6
09600	FLOORING	114.3	128.7	118.1	117.3	100.4	112.9	117.3	110.9	115.6	123.7	128.7	125.0	107.8	89.1	102.9	108.8	101.0	106.7
097,099	WALL FINISHES, PAINTS & COATINGS	112.0	139.4	128.6	110.3	102.7	105.7	110.3	101.4	104.9	112.1	136.0	126.6	106.7	53.6	74.5	106.7	77.7	89.1
09	FINISHES	111.8	134.0	123.3	114.6	116.0	115.3	114.4	122.6	118.7	116.0	132.8	124.8	98.2	83.0	90.3	99.6	88.8	94.0
10 - 14	TOTAL DIV. 10000 - 14000	100.0	131.7	106.7	100.0	115.1	103.2	100.0	128.9	106.1	100.0	129.9	106.4	100.0	91.3	98.1	100.0	91.5	98.2
15	MECHANICAL	100.2	141.3	118.2	100.2	108.4	103.8	100.2	101.3	100.7	100.2	118.4	108.2	100.2	81.3	91.9	100.2	86.4	94.1
16	ELECTRICAL	102.5	142.9	122.2	85.8	107.2	96.3	97.0	109.6	103.2	92.2	123.1	107.2	95.3	90.8	93.1	97.1	95.4	96.3
01 - 16	WEIGHTED AVERAGE	107.3	128.4	116.9	102.7	110.3	106.1	104.5	110.5	107.2	105.6	120.4	112.3	100.1	86.6	94.0	100.3	90.3	95.8

	DIVISION	COLORADO												CONNECTICUT					
		FORT COLLINS			GRAND JUNCTION			GREELEY			PUEBLO			BRIDGEPORT			BRISTOL		
		MAT.	INST.	TOTAL	MAT.	INST.	TOTAL	MAT.	INST.	TOTAL	MAT.	INST.	TOTAL	MAT.	INST.	TOTAL	MAT.	INST.	TOTAL
01590	**EQUIPMENT RENTAL**	.0	96.8	96.8	.0	100.0	100.0	.0	96.8	96.8	.0	96.9	96.9	.0	101.6	101.6	.0	101.6	101.6
02	**SITE CONSTRUCTION**	104.3	99.5	100.8	124.7	102.1	108.0	91.7	98.2	96.5	116.8	96.5	101.7	102.3	104.4	103.8	101.4	104.3	103.6
03100	CONCRETE FORMS & ACCESSORIES	101.3	80.4	83.2	106.7	81.0	84.4	98.7	49.5	56.1	103.3	86.1	88.4	101.0	110.7	109.4	101.0	110.3	109.1
03200	CONCRETE REINFORCEMENT	113.9	83.0	98.3	113.7	83.0	98.2	113.8	82.4	97.9	109.5	84.6	96.9	108.2	122.7	115.5	108.2	122.6	115.5
03300	CAST-IN-PLACE CONCRETE	106.4	83.5	96.9	113.2	84.6	101.2	89.0	60.8	77.2	101.5	89.3	96.4	105.2	117.7	110.4	98.6	117.6	106.5
03	**CONCRETE**	115.0	82.1	98.6	115.3	82.6	99.0	100.0	60.5	80.3	105.2	87.1	96.1	108.8	115.3	112.0	105.6	115.1	110.3
04	**MASONRY**	122.8	60.9	84.0	148.3	63.4	95.0	116.4	40.4	68.7	107.1	82.6	91.7	108.2	117.3	113.9	99.9	117.3	110.8
05	**METALS**	96.1	83.4	92.0	96.9	82.6	92.3	96.1	81.7	91.4	98.4	89.2	95.4	101.2	120.2	107.4	101.2	119.9	107.3
06	**WOOD & PLASTICS**	103.0	83.6	92.9	106.4	83.8	94.6	100.2	48.3	73.1	103.1	88.7	95.6	99.4	108.0	103.8	99.4	108.0	103.8
07	**THERMAL & MOISTURE PROTECTION**	103.8	72.5	88.4	104.9	68.6	87.0	102.9	60.7	82.2	104.2	85.5	95.0	101.6	117.5	109.4	101.8	113.6	107.6
08	**DOORS & WINDOWS**	95.9	87.9	93.8	100.8	88.0	97.5	95.9	68.7	88.8	95.2	90.8	94.1	107.1	121.0	110.7	107.1	113.0	108.6
09200	PLASTER & GYPSUM BOARD	96.0	83.4	88.0	106.7	83.4	91.9	94.7	47.1	64.5	85.2	88.1	87.0	104.4	107.2	106.2	104.4	107.2	106.2
095,098	CEILINGS & ACOUSTICAL TREATMENT	92.1	83.4	86.9	96.0	83.4	88.4	92.1	47.1	65.1	107.6	88.1	95.9	101.1	107.2	104.8	101.1	107.2	104.8
09600	FLOORING	108.6	71.5	98.8	116.1	71.5	104.4	107.2	71.5	97.8	112.5	101.0	109.5	99.8	115.1	103.9	99.8	115.1	103.9
097,099	WALL FINISHES, PAINTS & COATINGS	106.7	53.2	74.2	117.0	77.7	93.2	106.7	33.1	62.1	117.1	49.6	76.2	92.8	109.6	103.0	92.8	109.6	103.0
09	**FINISHES**	98.3	75.9	86.7	106.3	79.5	92.4	97.1	50.2	72.7	104.6	85.3	94.6	102.7	110.7	106.8	102.7	110.7	106.9
10 - 14	**TOTAL DIV. 10000 - 14000**	100.0	89.1	97.7	100.0	90.5	98.0	100.0	80.3	95.8	100.0	92.3	98.4	100.0	110.1	102.1	100.0	110.1	102.1
15	**MECHANICAL**	100.2	82.5	92.4	100.1	72.5	88.0	100.2	76.7	89.9	100.1	71.3	87.5	100.1	104.0	101.8	100.1	103.9	101.8
16	**ELECTRICAL**	92.3	93.0	92.6	88.5	65.4	77.3	92.3	93.0	92.6	89.3	82.6	86.0	100.3	105.8	103.0	100.3	108.4	104.2
01 - 16	**WEIGHTED AVERAGE**	101.0	82.7	92.7	103.8	77.6	91.9	98.5	71.1	86.0	100.0	83.9	92.6	102.7	110.8	106.4	101.9	110.6	105.9

	DIVISION	CONNECTICUT																	
		HARTFORD			NEW BRITAIN			NEW HAVEN			NORWALK			STAMFORD			WATERBURY		
		MAT.	INST.	TOTAL	MAT.	INST.	TOTAL	MAT.	INST.	TOTAL	MAT.	INST.	TOTAL	MAT.	INST.	TOTAL	MAT.	INST.	TOTAL
01590	**EQUIPMENT RENTAL**	.0	101.6	101.6	.0	101.6	101.6	.0	102.1	102.1	.0	101.6	101.6	.0	101.6	101.6	.0	101.6	101.6
02	**SITE CONSTRUCTION**	101.8	104.3	103.7	101.6	104.3	103.6	101.5	105.2	104.2	102.0	104.4	103.8	102.7	104.4	103.9	101.8	104.4	103.7
03100	CONCRETE FORMS & ACCESSORIES	99.9	110.3	108.9	101.2	110.4	109.1	100.8	110.6	109.3	101.0	110.7	109.4	101.0	110.9	109.6	101.0	110.6	109.3
03200	CONCRETE REINFORCEMENT	108.2	122.6	115.5	108.2	122.6	115.5	108.2	122.7	115.5	108.2	122.8	115.6	108.2	122.8	115.6	108.2	122.7	115.5
03300	CAST-IN-PLACE CONCRETE	99.7	117.6	107.2	100.1	117.6	107.4	101.9	117.7	108.5	103.5	119.3	110.1	105.2	119.4	111.1	105.2	117.7	110.4
03	**CONCRETE**	106.1	115.1	110.6	106.4	115.1	110.7	121.6	115.2	118.4	107.9	115.9	111.9	108.8	116.0	112.4	108.8	115.2	112.0
04	**MASONRY**	100.0	117.3	110.9	100.0	117.3	110.9	100.2	117.3	110.9	100.3	119.0	112.1	100.4	119.0	112.1	100.4	117.3	111.0
05	**METALS**	106.3	119.9	110.7	97.5	119.9	104.8	97.7	120.2	105.1	101.2	120.4	107.5	101.2	120.7	107.5	101.2	120.1	107.4
06	**WOOD & PLASTICS**	99.4	108.0	103.8	99.4	108.0	103.8	99.4	108.0	103.8	99.4	108.0	103.8	99.4	108.0	103.8	99.4	108.0	103.8
07	**THERMAL & MOISTURE PROTECTION**	100.3	113.6	106.9	101.8	115.2	108.4	101.9	115.2	108.5	101.8	118.2	109.9	101.8	118.2	109.9	101.8	115.2	108.4
08	**DOORS & WINDOWS**	107.1	113.0	108.6	107.1	113.0	108.6	107.1	121.0	110.7	107.1	121.0	110.7	107.1	121.0	110.7	107.1	121.0	110.7
09200	PLASTER & GYPSUM BOARD	104.4	107.2	106.2	104.4	107.2	106.2	104.4	107.2	106.2	104.4	107.2	106.2	104.4	107.2	106.2	104.4	107.2	106.2
095,098	CEILINGS & ACOUSTICAL TREATMENT	101.1	107.2	104.8	101.1	107.2	104.8	101.1	107.2	104.8	101.1	107.2	104.8	101.1	107.2	104.8	101.1	107.2	104.8
09600	FLOORING	99.8	115.1	103.9	99.8	115.1	103.9	99.8	115.1	103.9	99.8	115.1	103.9	99.8	115.1	103.9	99.8	115.1	103.9
097,099	WALL FINISHES, PAINTS & COATINGS	92.8	109.6	103.0	92.8	109.6	103.0	92.8	109.6	103.0	92.8	109.6	103.0	92.8	109.6	103.0	92.8	109.6	103.0
09	**FINISHES**	102.7	110.7	106.9	102.7	110.7	106.9	102.8	110.7	106.9	102.7	110.7	106.9	102.8	110.7	106.9	102.6	110.7	106.8
10 - 14	**TOTAL DIV. 10000 - 14000**	100.0	110.1	102.1	100.0	110.1	102.1	100.0	110.1	102.1	100.0	110.1	102.1	100.0	110.2	102.2	100.0	110.1	102.1
15	**MECHANICAL**	100.1	103.9	101.8	100.1	103.9	101.8	100.1	104.0	101.8	100.1	104.0	101.8	100.1	104.0	101.8	100.1	104.0	101.8
16	**ELECTRICAL**	99.9	103.1	101.5	100.3	108.4	104.3	100.2	108.4	104.2	100.3	108.4	104.2	100.3	137.7	118.5	99.8	105.8	102.7
01 - 16	**WEIGHTED AVERAGE**	102.7	109.9	106.0	101.4	110.7	105.6	103.2	111.2	106.8	102.2	111.5	106.4	102.3	115.6	108.4	102.2	110.7	106.1

	DIVISION	D.C.			DELAWARE			FLORIDA											
		WASHINGTON			WILMINGTON			DAYTONA BEACH			FORT LAUDERDALE			JACKSONVILLE			MELBOURNE		
		MAT.	INST.	TOTAL	MAT.	INST.	TOTAL	MAT.	INST.	TOTAL	MAT.	INST.	TOTAL	MAT.	INST.	TOTAL	MAT.	INST.	TOTAL
01590	**EQUIPMENT RENTAL**	.0	102.3	102.3	.0	118.4	118.4	.0	97.7	97.7	.0	89.3	89.3	.0	97.7	97.7	.0	97.7	97.7
02	**SITE CONSTRUCTION**	100.3	89.5	92.3	87.5	111.4	105.1	115.6	86.7	94.2	101.3	73.8	81.0	115.7	87.2	94.6	123.2	87.2	96.6
03100	CONCRETE FORMS & ACCESSORIES	98.0	80.7	83.0	96.7	101.4	100.7	98.2	69.8	73.6	95.6	68.4	72.0	98.0	53.3	59.3	92.5	73.7	76.2
03200	CONCRETE REINFORCEMENT	105.6	89.5	97.5	102.2	99.0	100.6	95.0	85.5	90.2	95.0	72.9	83.8	95.0	51.2	72.9	96.1	85.5	90.8
03300	CAST-IN-PLACE CONCRETE	114.7	88.1	103.6	78.8	94.4	85.3	89.9	70.5	81.8	94.3	69.4	83.9	90.8	58.7	77.4	104.5	75.8	92.5
03	**CONCRETE**	111.2	86.2	98.7	97.8	99.5	98.6	90.4	74.2	82.3	92.3	70.9	81.6	90.8	56.5	73.7	99.6	77.8	88.7
04	**MASONRY**	92.2	82.0	85.8	108.5	88.8	96.1	86.7	64.7	72.9	86.9	67.6	74.8	86.2	51.4	64.4	83.9	73.6	77.4
05	**METALS**	107.4	107.8	107.5	101.8	115.1	106.1	96.4	95.9	96.2	96.3	89.9	94.2	96.0	80.2	90.8	105.3	96.4	102.4
06	**WOOD & PLASTICS**	95.0	80.9	87.7	95.3	103.2	99.4	97.8	72.5	84.6	93.3	65.8	79.0	97.8	52.1	74.0	91.6	72.5	81.7
07	**THERMAL & MOISTURE PROTECTION**	98.3	81.7	90.2	100.6	104.3	102.4	95.4	69.7	82.7	95.4	74.7	85.2	95.7	58.0	77.2	95.6	76.3	86.1
08	**DOORS & WINDOWS**	102.6	90.7	99.5	94.4	106.2	97.4	100.9	69.9	92.8	98.5	63.3	89.4	100.9	49.7	87.6	100.0	75.4	93.7
09200	PLASTER & GYPSUM BOARD	101.1	80.5	88.0	112.3	103.1	106.5	100.5	72.3	82.6	100.1	65.5	78.1	100.5	51.3	69.3	96.2	72.3	81.1
095,098	CEILINGS & ACOUSTICAL TREATMENT	99.3	80.5	88.0	103.6	103.1	103.3	102.7	72.3	84.5	102.7	65.5	80.4	102.7	51.3	71.9	97.3	72.3	82.3
09600	FLOORING	113.7	102.8	110.8	83.7	98.3	87.5	118.6	73.5	106.7	118.6	61.0	103.4	118.6	50.1	100.5	115.3	73.5	104.2
097,099	WALL FINISHES, PAINTS & COATINGS	113.0	87.8	97.7	90.7	97.0	94.5	111.2	72.1	87.5	107.4	49.9	72.5	111.2	48.1	72.9	111.2	92.6	99.9
09	**FINISHES**	101.2	85.3	92.9	102.2	100.2	101.2	108.4	70.8	88.9	106.4	64.5	84.6	108.4	51.5	78.8	106.0	75.4	90.0
10 - 14	**TOTAL DIV. 10000 - 14000**	100.0	97.5	99.5	100.0	100.0	100.0	100.0	82.5	96.3	100.0	89.4	97.8	100.0	77.5	95.2	100.0	86.0	97.0
15	**MECHANICAL**	100.0	90.9	96.0	100.3	116.6	107.4	99.8	72.6	87.9	99.8	69.7	86.7	99.8	51.4	78.6	99.8	77.7	90.2
16	**ELECTRICAL**	97.7	97.6	97.6	97.2	104.8	100.9	96.3	67.0	82.0	96.3	75.2	86.0	96.1	67.2	82.0	96.2	73.9	85.3
01 - 16	**WEIGHTED AVERAGE**	102.0	90.8	96.9	99.5	105.7	102.3	98.4	74.5	87.5	97.7	72.3	86.2	98.3	61.0	81.4	100.5	79.0	90.7

City Cost Indexes

	DIVISION	FLORIDA																	
		MIAMI			ORLANDO			PANAMA CITY			PENSACOLA			ST. PETERSBURG			TALLAHASSEE		
		MAT.	INST.	TOTAL	MAT.	INST.	TOTAL	MAT.	INST.	TOTAL	MAT.	INST.	TOTAL	MAT.	INST.	TOTAL	MAT.	INST.	TOTAL
01590	EQUIPMENT RENTAL	.0	89.3	89.3	.0	97.7	97.7	.0	97.7	97.7	.0	97.7	97.7	.0	97.7	97.7	.0	97.7	97.7
02	SITE CONSTRUCTION	100.6	73.6	80.6	116.2	86.4	94.1	130.2	83.9	95.9	127.8	86.3	97.1	116.5	85.7	93.7	117.0	85.8	93.9
03100	CONCRETE FORMS & ACCESSORIES	95.8	68.4	72.0	98.0	71.3	74.9	97.0	27.7	37.0	87.4	51.1	55.9	95.5	47.2	53.7	97.9	39.0	46.9
03200	CONCRETE REINFORCEMENT	95.0	72.9	83.8	95.0	81.8	88.3	99.2	50.1	74.4	101.7	50.4	75.8	98.5	57.6	77.8	95.0	50.6	72.6
03300	CAST-IN-PLACE CONCRETE	91.9	69.6	82.5	97.7	70.4	86.3	95.4	34.6	70.0	95.4	54.9	78.5	101.7	55.3	82.3	94.1	48.4	75.0
03	CONCRETE	91.1	70.9	81.0	92.0	74.2	83.1	99.0	36.1	67.6	98.0	54.0	76.0	96.4	53.7	75.1	92.3	46.5	69.5
04	MASONRY	84.8	68.0	74.3	90.9	64.7	74.5	91.0	27.6	51.2	88.1	49.9	64.1	134.0	49.3	80.8	89.1	38.8	57.5
05	METALS	101.7	89.3	97.6	102.5	94.1	99.8	96.2	65.6	86.2	96.2	79.6	90.8	99.6	80.6	93.4	89.4	78.5	85.9
06	WOOD & PLASTICS	93.3	65.8	79.0	97.8	75.2	86.0	96.6	27.8	60.7	86.3	51.4	68.1	95.0	47.2	70.0	96.2	37.4	65.5
07	THERMAL & MOISTURE PROTECTION	99.3	71.4	85.6	95.7	69.9	83.1	96.0	31.4	64.2	95.7	52.8	74.6	95.3	49.6	72.8	95.7	46.5	71.5
08	DOORS & WINDOWS	98.5	63.9	89.5	100.9	69.6	92.8	98.4	26.7	79.9	98.4	50.2	85.9	99.5	46.5	85.8	99.5	41.0	84.3
09200	PLASTER & GYPSUM BOARD	100.1	65.5	78.1	103.1	75.0	85.3	98.4	26.4	52.7	94.0	50.6	66.5	97.6	46.3	65.1	100.5	36.2	59.7
095,098	CEILINGS & ACOUSTICAL TREATMENT	102.7	65.5	80.4	102.7	75.0	86.1	96.5	26.4	54.4	96.5	50.6	69.0	96.5	46.3	66.4	102.7	36.2	62.8
09600	FLOORING	126.0	61.7	109.0	118.6	73.5	106.7	118.0	19.2	91.9	112.2	53.2	96.6	116.9	53.0	100.0	118.6	38.9	97.5
097,099	WALL FINISHES, PAINTS & COATINGS	107.4	49.9	72.5	111.2	55.2	77.2	111.2	24.6	58.6	111.2	55.9	77.6	111.2	46.5	71.9	111.2	39.6	67.8
09	FINISHES	108.6	64.6	85.7	108.8	70.4	88.9	107.5	25.4	64.8	105.0	51.8	77.3	106.0	48.0	75.8	108.6	38.2	72.0
10 - 14	TOTAL DIV. 10000 - 14000	100.0	89.4	97.8	100.0	82.7	96.3	100.0	46.0	88.5	100.0	51.5	89.7	100.0	53.9	90.2	100.0	63.9	92.3
15	MECHANICAL	99.8	74.2	88.6	99.8	63.8	84.1	99.8	25.2	67.2	99.8	50.0	78.1	99.8	49.9	78.0	99.8	39.9	73.6
16	ELECTRICAL	96.7	73.3	85.3	96.7	44.6	71.3	95.3	33.4	65.1	98.4	58.8	79.1	96.6	48.1	73.0	96.7	41.3	69.7
01 - 16	WEIGHTED AVERAGE	98.7	72.9	87.0	99.8	69.3	86.0	99.5	37.9	71.5	99.2	58.1	80.5	101.6	55.8	80.8	97.6	49.2	75.6

	DIVISION	FLORIDA			GEORGIA														
		TAMPA			ALBANY			ATLANTA			AUGUSTA			COLUMBUS			MACON		
		MAT.	INST.	TOTAL	MAT.	INST.	TOTAL	MAT.	INST.	TOTAL	MAT.	INST.	TOTAL	MAT.	INST.	TOTAL	MAT.	INST.	TOTAL
01590	EQUIPMENT RENTAL	.0	97.7	97.7	.0	90.1	90.1	.0	92.8	92.8	.0	92.1	92.1	.0	90.1	90.1	.0	102.6	102.6
02	SITE CONSTRUCTION	116.7	86.4	94.3	101.6	76.2	82.8	101.2	95.6	97.0	97.8	92.6	94.0	101.5	76.6	83.1	102.3	93.8	96.0
03100	CONCRETE FORMS & ACCESSORIES	99.4	74.9	78.2	97.1	45.9	52.8	96.0	78.8	81.1	93.1	51.9	57.4	97.0	58.7	63.8	96.4	55.5	61.0
03200	CONCRETE REINFORCEMENT	95.0	90.6	92.8	94.6	91.7	93.1	98.7	93.5	96.1	99.8	77.3	88.4	95.0	92.7	93.9	96.3	91.9	94.1
03300	CAST-IN-PLACE CONCRETE	99.5	61.6	83.7	96.0	46.2	75.2	99.6	74.8	89.2	94.1	48.9	75.2	95.7	55.0	78.7	94.4	45.8	74.1
03	CONCRETE	95.0	74.3	84.7	93.0	56.1	74.6	98.5	80.0	89.3	94.2	56.1	75.2	92.9	65.1	79.0	92.5	60.4	76.4
04	MASONRY	87.7	75.7	80.2	88.9	39.9	58.1	90.9	70.6	78.1	91.2	40.2	59.2	88.9	57.2	69.0	102.8	38.4	62.4
05	METALS	99.7	95.6	98.4	95.6	88.8	93.4	93.3	80.4	89.1	92.2	69.6	84.8	95.2	92.3	94.3	90.9	90.8	90.9
06	WOOD & PLASTICS	99.3	77.4	87.9	96.7	42.5	68.4	96.4	80.9	88.3	93.4	53.3	72.5	96.7	58.6	76.8	104.4	57.5	79.9
07	THERMAL & MOISTURE PROTECTION	95.7	62.3	79.3	95.5	54.1	75.1	92.3	75.5	84.0	91.7	49.8	71.1	95.1	61.9	78.8	93.9	56.8	75.7
08	DOORS & WINDOWS	100.9	71.2	93.2	98.5	50.1	85.9	100.0	77.5	94.2	94.1	54.7	83.9	98.5	60.9	88.7	96.9	59.3	87.2
09200	PLASTER & GYPSUM BOARD	100.5	77.4	85.8	100.2	41.5	63.0	113.1	80.8	92.7	111.9	52.5	74.2	100.2	58.0	73.5	106.2	56.9	75.0
095,098	CEILINGS & ACOUSTICAL TREATMENT	102.7	77.4	87.5	101.8	41.5	65.6	107.9	80.8	91.7	108.9	52.5	75.1	101.8	58.0	75.6	97.7	56.9	73.2
09600	FLOORING	118.6	53.0	101.3	118.6	32.5	95.8	84.0	78.0	82.4	82.8	41.4	71.9	118.6	54.8	101.7	92.5	38.2	78.1
097,099	WALL FINISHES, PAINTS & COATINGS	111.2	46.5	71.9	107.4	41.4	67.3	90.3	84.5	86.8	90.3	39.3	59.4	107.4	49.1	72.0	109.1	48.4	72.3
09	FINISHES	108.5	68.3	87.6	106.2	41.5	72.6	96.2	79.3	87.4	95.7	48.2	71.0	106.1	56.2	80.2	94.1	51.1	71.7
10 - 14	TOTAL DIV. 10000 - 14000	100.0	85.3	96.9	100.0	77.5	95.2	100.0	84.6	96.7	100.0	67.6	93.1	100.0	79.6	95.7	100.0	77.6	95.2
15	MECHANICAL	99.8	73.9	88.5	99.8	49.7	77.9	100.0	80.4	91.5	100.0	66.3	85.3	99.8	50.1	78.1	99.8	41.9	74.5
16	ELECTRICAL	95.7	48.1	72.5	91.9	60.4	76.5	96.6	82.1	89.5	96.7	49.3	73.6	92.3	50.9	72.1	90.0	59.6	75.2
01 - 16	WEIGHTED AVERAGE	99.4	72.8	87.3	97.3	56.8	78.9	97.4	80.7	89.8	96.0	59.2	79.3	97.2	61.7	81.1	95.7	59.2	79.1

	DIVISION	GEORGIA						HAWAII			IDAHO								
		SAVANNAH			VALDOSTA			HONOLULU			BOISE			LEWISTON			POCATELLO		
		MAT.	INST.	TOTAL	MAT.	INST.	TOTAL	MAT.	INST.	TOTAL	MAT.	INST.	TOTAL	MAT.	INST.	TOTAL	MAT.	INST.	TOTAL
01590	EQUIPMENT RENTAL	.0	91.2	91.2	.0	90.1	90.1	.0	99.3	99.3	.0	101.7	101.7	.0	94.6	94.6	.0	101.7	101.7
02	SITE CONSTRUCTION	102.1	77.1	83.6	111.8	76.4	85.6	139.1	106.5	115.0	79.9	102.7	96.8	84.1	96.9	93.6	81.1	102.6	97.0
03100	CONCRETE FORMS & ACCESSORIES	97.4	52.8	58.8	84.2	45.7	50.8	106.8	144.6	139.5	97.7	82.3	84.4	111.5	67.5	73.4	97.8	81.8	84.0
03200	CONCRETE REINFORCEMENT	96.0	77.5	86.6	97.0	44.0	70.2	108.9	126.2	117.6	108.2	78.0	92.9	115.6	97.7	106.5	108.5	77.5	92.8
03300	CAST-IN-PLACE CONCRETE	92.7	51.6	75.5	94.0	52.2	76.5	199.7	129.5	170.3	95.9	91.0	93.9	104.0	86.7	96.8	95.2	90.8	93.3
03	CONCRETE	91.7	58.4	75.1	97.0	49.3	73.2	154.0	134.6	144.3	104.2	84.4	94.3	111.9	80.0	96.0	100.9	84.0	92.4
04	MASONRY	91.9	53.9	68.0	94.8	49.0	66.1	132.4	129.8	130.8	137.5	74.5	98.0	138.3	86.6	105.8	132.9	66.5	91.2
05	METALS	91.2	84.3	89.0	94.9	72.5	87.6	145.3	111.2	134.1	99.7	77.6	92.5	92.6	88.0	91.1	107.8	76.4	97.6
06	WOOD & PLASTICS	110.2	50.2	78.9	82.3	41.5	61.0	97.7	148.7	124.3	95.7	81.9	88.5	103.9	60.5	81.2	95.7	81.9	88.5
07	THERMAL & MOISTURE PROTECTION	95.5	53.0	74.6	95.2	57.4	76.7	113.0	130.0	121.4	96.9	80.0	88.6	167.0	80.7	124.6	97.1	70.5	84.1
08	DOORS & WINDOWS	99.6	50.9	87.0	93.8	38.7	79.5	105.8	140.1	114.7	94.7	77.8	90.3	114.8	67.5	102.6	94.7	71.0	88.6
09200	PLASTER & GYPSUM BOARD	100.5	49.4	68.1	91.2	40.4	59.0	116.7	149.8	137.7	83.2	81.2	81.9	134.0	59.3	86.7	83.2	81.2	81.9
095,098	CEILINGS & ACOUSTICAL TREATMENT	102.7	49.4	70.7	96.5	40.4	62.8	119.6	149.8	137.7	107.6	81.2	91.7	128.4	59.3	87.0	107.6	81.2	91.7
09600	FLOORING	118.6	48.8	100.1	110.4	39.0	91.6	168.4	134.9	159.5	99.8	56.3	88.3	137.7	93.5	126.1	100.1	56.3	88.5
097,099	WALL FINISHES, PAINTS & COATINGS	108.1	49.1	72.3	107.4	35.8	64.0	104.9	148.5	131.4	103.7	52.2	72.5	129.6	74.3	96.0	103.7	54.3	73.7
09	FINISHES	106.7	51.4	77.9	101.8	42.7	71.1	134.0	144.6	139.5	98.5	74.5	86.0	157.0	71.7	112.6	98.6	74.7	86.2
10 - 14	TOTAL DIV. 10000 - 14000	100.0	70.6	93.8	100.0	69.4	93.5	100.0	121.1	104.5	100.0	86.4	97.1	100.0	88.3	97.5	100.0	86.4	97.1
15	MECHANICAL	99.8	47.8	77.1	99.8	42.1	74.6	100.2	117.8	107.9	100.1	77.1	90.1	101.3	85.9	94.6	100.1	77.1	90.0
16	ELECTRICAL	93.8	58.9	76.8	90.6	30.2	61.1	108.0	124.1	115.9	93.5	78.3	86.1	79.1	85.1	82.0	88.7	73.9	81.5
01 - 16	WEIGHTED AVERAGE	97.1	58.8	79.7	97.0	49.1	75.2	121.0	125.7	123.1	100.2	80.5	91.2	108.1	83.3	96.8	100.3	78.3	90.3

City Cost Indexes

	DIVISION	IDAHO			ILLINOIS														
		TWIN FALLS			CHICAGO			DECATUR			EAST ST. LOUIS			JOLIET			PEORIA		
		MAT.	INST.	TOTAL	MAT.	INST.	TOTAL	MAT.	INST.	TOTAL	MAT.	INST.	TOTAL	MAT.	INST.	TOTAL	MAT.	INST.	TOTAL
01590	**EQUIPMENT RENTAL**	.0	101.7	101.7	.0	91.5	91.5	.0	102.4	102.4	.0	109.0	109.0	.0	89.3	89.3	.0	101.6	101.6
02	**SITE CONSTRUCTION**	87.8	100.6	97.2	96.9	90.9	92.5	87.2	96.0	93.7	103.5	96.3	98.2	96.8	89.9	91.7	95.9	95.0	95.3
03100	CONCRETE FORMS & ACCESSORIES	98.5	35.2	43.7	101.6	140.3	135.2	97.1	103.2	102.4	91.0	102.4	100.9	102.5	131.8	127.8	95.4	108.0	106.3
03200	CONCRETE REINFORCEMENT	110.4	49.1	79.4	100.7	139.3	120.2	97.0	99.8	98.4	98.9	102.5	100.7	100.7	128.0	114.5	97.0	99.8	98.5
03300	CAST-IN-PLACE CONCRETE	97.7	45.2	75.7	113.2	136.6	123.0	98.9	103.5	100.8	92.6	108.3	99.2	113.1	121.2	116.4	96.6	106.9	100.9
03	**CONCRETE**	108.4	42.3	75.4	111.4	137.9	124.6	97.7	103.1	100.4	88.6	105.5	97.1	111.4	126.6	118.9	96.6	106.4	101.5
04	**MASONRY**	136.3	39.0	75.2	93.4	134.3	119.1	68.6	91.8	83.2	73.9	107.9	95.2	94.9	121.5	111.6	112.3	108.7	110.1
05	**METALS**	107.8	62.2	92.9	96.2	122.1	104.7	97.9	106.4	100.7	95.6	118.6	103.1	94.1	114.3	100.7	97.9	107.4	101.0
06	**WOOD & PLASTICS**	96.4	35.1	64.4	103.3	139.5	122.2	98.8	102.1	100.5	94.9	99.9	97.5	104.9	132.2	119.2	98.8	104.3	101.7
07	**THERMAL & MOISTURE PROTECTION**	98.2	46.7	72.9	100.0	131.3	115.4	97.2	94.0	95.6	92.3	101.5	96.8	99.8	123.0	111.2	97.1	105.3	101.1
08	**DOORS & WINDOWS**	98.0	36.0	82.0	103.4	139.3	112.7	96.3	101.4	97.6	86.1	107.7	91.7	101.4	132.5	109.4	96.3	102.6	98.0
09200	PLASTER & GYPSUM BOARD	82.9	33.1	51.4	102.8	140.5	126.7	104.4	102.0	102.9	100.5	99.8	100.1	100.0	133.1	121.0	104.4	104.2	104.3
095,098	CEILINGS & ACOUSTICAL TREATMENT	100.4	33.1	60.1	106.3	140.5	126.8	97.1	102.0	100.1	90.0	99.8	95.9	106.3	133.1	122.3	97.1	104.2	101.4
09600	FLOORING	101.5	56.3	89.5	83.2	127.6	94.9	96.4	84.6	93.3	105.1	104.5	104.9	82.7	116.2	91.6	96.4	99.7	97.3
097,099	WALL FINISHES, PAINTS & COATINGS	103.7	32.0	60.2	80.5	129.5	110.2	91.2	97.6	95.1	99.3	90.4	93.9	78.3	115.5	100.9	91.2	96.3	94.3
09	**FINISHES**	97.7	38.4	66.9	93.1	137.4	116.1	95.8	99.4	97.7	94.5	100.2	97.5	92.5	127.9	110.9	95.9	105.4	100.8
10 - 14	**TOTAL DIV. 10000 - 14000**	100.0	48.6	89.1	100.0	126.9	105.7	100.0	93.5	98.6	100.0	94.8	98.9	100.0	123.2	104.9	100.0	95.3	99.0
15	**MECHANICAL**	100.1	36.4	72.2	100.0	123.2	110.1	100.0	101.4	100.6	100.0	95.8	98.2	100.1	117.7	107.8	100.0	102.1	101.0
16	**ELECTRICAL**	81.0	31.5	56.9	98.0	119.9	108.7	98.8	91.3	95.1	96.1	91.7	94.0	96.3	106.3	101.2	98.1	88.7	93.5
01 - 16	**WEIGHTED AVERAGE**	100.8	45.8	75.8	99.7	125.8	111.6	96.5	98.5	97.4	94.2	101.1	97.3	99.0	117.2	107.3	98.7	101.7	100.1

	DIVISION	ILLINOIS						INDIANA											
		ROCKFORD			SPRINGFIELD			ANDERSON			BLOOMINGTON			EVANSVILLE			FORT WAYNE		
		MAT.	INST.	TOTAL	MAT.	INST.	TOTAL	MAT.	INST.	TOTAL	MAT.	INST.	TOTAL	MAT.	INST.	TOTAL	MAT.	INST.	TOTAL
01590	**EQUIPMENT RENTAL**	.0	101.6	101.6	.0	102.4	102.4	.0	96.7	96.7	.0	85.9	85.9	.0	121.3	121.3	.0	96.7	96.7
02	**SITE CONSTRUCTION**	95.0	95.6	95.4	93.3	96.0	95.3	87.6	96.6	94.2	77.1	95.6	90.8	82.1	130.1	117.6	88.6	96.5	94.4
03100	CONCRETE FORMS & ACCESSORIES	99.7	113.7	111.9	99.0	103.8	103.2	96.1	75.6	78.3	99.9	76.1	79.3	92.6	75.8	78.0	94.7	75.0	77.6
03200	CONCRETE REINFORCEMENT	97.0	130.2	113.8	97.0	99.8	98.4	93.8	81.2	87.4	83.9	81.0	82.5	92.0	81.2	86.6	93.8	78.0	85.8
03300	CAST-IN-PLACE CONCRETE	98.9	98.8	98.9	91.9	104.2	97.0	102.4	80.8	93.4	101.6	77.3	91.5	97.0	90.4	94.2	108.8	76.2	95.2
03	**CONCRETE**	97.9	112.0	104.9	94.5	103.6	99.0	96.7	79.1	87.9	103.1	77.2	90.2	103.4	82.1	92.8	99.6	76.6	88.2
04	**MASONRY**	83.1	110.8	100.5	69.8	92.8	84.3	91.3	77.5	82.6	90.6	75.9	81.4	86.8	81.2	83.3	91.6	77.7	82.9
05	**METALS**	97.9	123.4	106.2	100.5	106.6	102.5	90.9	89.1	90.3	93.4	75.1	87.4	86.2	86.3	86.2	90.9	87.1	89.7
06	**WOOD & PLASTICS**	98.8	112.0	105.7	101.5	102.1	101.8	113.2	74.1	92.8	118.9	74.7	95.8	97.7	72.2	84.4	112.9	73.6	92.4
07	**THERMAL & MOISTURE PROTECTION**	97.1	111.2	104.1	96.7	98.5	97.6	101.0	76.1	88.7	91.5	80.6	86.1	94.8	84.1	89.5	100.7	82.4	91.7
08	**DOORS & WINDOWS**	96.3	114.8	101.1	96.3	101.4	97.6	99.3	78.1	93.8	103.2	78.4	96.8	95.6	75.8	90.5	99.3	74.7	92.9
09200	PLASTER & GYPSUM BOARD	104.4	112.1	109.3	104.4	102.0	102.9	96.7	74.0	82.3	98.2	74.7	83.3	91.8	70.6	78.4	95.9	73.5	81.7
095,098	CEILINGS & ACOUSTICAL TREATMENT	97.1	112.1	106.1	97.1	102.0	100.1	89.4	74.0	80.2	81.7	74.7	77.5	86.4	70.6	77.0	89.4	73.5	79.9
09600	FLOORING	96.4	93.3	95.6	96.6	84.7	93.4	87.3	88.6	87.6	103.9	88.6	99.9	97.0	83.6	93.5	87.3	79.9	85.3
097,099	WALL FINISHES, PAINTS & COATINGS	91.2	108.8	101.9	91.2	90.2	90.6	92.1	67.6	77.3	89.8	86.2	87.7	96.5	88.1	91.4	92.1	74.7	81.6
09	**FINISHES**	95.9	109.3	102.9	95.9	98.8	97.4	88.9	76.8	82.6	94.7	79.2	86.7	92.9	77.9	85.1	88.7	75.6	81.8
10 - 14	**TOTAL DIV. 10000 - 14000**	100.0	103.9	100.8	100.0	93.9	98.7	100.0	81.7	96.1	100.0	81.2	96.0	100.0	84.4	96.7	100.0	83.6	96.5
15	**MECHANICAL**	100.0	105.7	102.5	100.0	103.7	101.7	99.9	81.0	91.7	99.6	82.9	92.3	99.8	85.0	93.3	99.9	78.6	90.6
16	**ELECTRICAL**	98.1	110.0	103.9	98.7	89.1	94.0	85.2	92.8	88.9	102.4	80.5	91.7	95.5	83.7	89.7	86.3	80.8	83.6
01 - 16	**WEIGHTED AVERAGE**	97.4	109.4	102.8	96.8	98.9	97.8	94.6	83.3	89.5	98.2	80.7	90.3	95.2	86.8	91.4	95.1	80.6	88.5

	DIVISION	INDIANA															IOWA		
		GARY			INDIANAPOLIS			MUNCIE			SOUTH BEND			TERRE HAUTE			CEDAR RAPIDS		
		MAT.	INST.	TOTAL	MAT.	INST.	TOTAL	MAT.	INST.	TOTAL	MAT.	INST.	TOTAL	MAT.	INST.	TOTAL	MAT.	INST.	TOTAL
01590	**EQUIPMENT RENTAL**	.0	96.7	96.7	.0	91.5	91.5	.0	96.2	96.2	.0	106.8	106.8	.0	121.3	121.3	.0	95.5	95.5
02	**SITE CONSTRUCTION**	88.2	98.3	95.7	87.5	100.4	97.0	77.2	95.8	90.9	87.7	96.1	93.9	83.4	130.0	117.9	87.6	94.6	92.8
03100	CONCRETE FORMS & ACCESSORIES	96.2	100.6	100.0	96.6	86.8	88.1	90.5	75.3	77.3	100.3	77.6	80.6	94.1	77.7	79.9	103.6	78.2	81.6
03200	CONCRETE REINFORCEMENT	93.8	99.6	96.7	93.2	85.4	89.2	92.9	81.1	86.9	93.8	76.4	85.0	92.0	80.7	86.3	93.3	82.6	87.9
03300	CAST-IN-PLACE CONCRETE	107.0	101.5	104.7	98.5	87.6	93.9	106.8	80.1	95.7	99.6	82.4	92.4	94.0	90.4	92.5	107.8	80.1	96.2
03	**CONCRETE**	98.9	100.9	99.9	98.8	86.4	92.6	101.8	78.7	90.3	91.4	80.5	86.0	106.3	82.9	94.6	99.7	80.4	90.1
04	**MASONRY**	92.7	93.1	93.0	98.2	86.6	90.9	92.8	77.5	83.2	89.8	77.4	82.0	93.7	78.2	84.0	106.2	79.6	89.5
05	**METALS**	90.9	101.1	94.2	92.8	80.1	88.7	95.1	88.8	93.0	90.9	100.9	94.2	86.9	86.4	86.8	87.2	92.2	88.8
06	**WOOD & PLASTICS**	111.2	100.9	105.8	110.1	86.4	97.7	111.0	73.9	91.6	113.1	77.0	94.3	99.9	76.2	87.5	109.6	76.9	92.5
07	**THERMAL & MOISTURE PROTECTION**	100.0	98.0	99.0	99.9	85.9	93.0	93.1	76.4	84.9	99.7	82.3	91.1	94.9	81.5	88.3	100.0	80.3	90.3
08	**DOORS & WINDOWS**	99.3	102.5	100.1	106.9	86.3	101.6	98.2	78.0	93.0	92.7	77.0	88.6	96.2	79.0	91.7	99.3	80.2	94.4
09200	PLASTER & GYPSUM BOARD	91.4	101.5	97.8	94.8	86.5	89.5	91.5	74.0	80.4	96.3	77.0	84.0	91.8	74.7	81.0	104.4	76.6	86.8
095,098	CEILINGS & ACOUSTICAL TREATMENT	89.4	101.5	96.7	93.9	86.5	89.4	82.5	74.0	77.4	89.4	77.0	82.0	86.4	74.7	79.4	112.4	76.6	90.9
09600	FLOORING	87.3	99.4	90.5	87.0	93.4	88.7	96.2	88.6	94.2	87.3	77.4	84.7	97.0	87.3	94.5	123.0	58.2	105.9
097,099	WALL FINISHES, PAINTS & COATINGS	92.1	101.8	98.0	92.1	89.3	90.4	89.8	67.6	76.4	92.1	81.0	85.4	96.5	87.7	91.2	107.7	77.4	89.3
09	**FINISHES**	88.1	100.5	94.6	90.0	88.4	89.1	91.4	76.7	83.7	88.8	77.8	83.1	92.9	80.2	86.3	111.8	73.5	91.8
10 - 14	**TOTAL DIV. 10000 - 14000**	100.0	98.5	99.7	100.0	85.1	96.8	100.0	81.2	96.0	100.0	77.6	95.2	100.0	85.4	96.9	100.0	81.8	96.1
15	**MECHANICAL**	99.9	94.7	97.6	99.9	88.6	95.0	99.6	81.0	91.4	99.9	76.9	89.8	99.8	78.6	90.5	100.4	83.6	93.0
16	**ELECTRICAL**	93.4	102.6	97.9	105.1	92.8	99.1	90.1	76.4	83.4	99.8	83.4	91.8	94.1	80.6	87.5	97.0	82.1	89.7
01 - 16	**WEIGHTED AVERAGE**	95.8	98.7	97.1	98.8	88.7	94.2	96.1	80.8	89.2	95.0	82.6	89.4	96.0	85.3	91.1	98.8	82.7	91.5

DIVISION		IOWA																	
		COUNCIL BLUFFS			DAVENPORT			DES MOINES			DUBUQUE			SIOUX CITY			WATERLOO		
		MAT.	INST.	TOTAL	MAT.	INST.	TOTAL	MAT.	INST.	TOTAL	MAT.	INST.	TOTAL	MAT.	INST.	TOTAL	MAT.	INST.	TOTAL
01590	EQUIPMENT RENTAL	.0	95.2	95.2	.0	99.5	99.5	.0	101.5	101.5	.0	94.1	94.1	.0	99.5	99.5	.0	99.5	99.5
02	SITE CONSTRUCTION	92.8	91.2	91.6	85.8	98.4	95.1	79.3	99.9	94.6	85.7	91.5	90.0	94.7	95.4	95.2	86.3	94.5	92.4
03100	CONCRETE FORMS & ACCESSORIES	81.5	58.4	61.5	103.3	90.8	92.5	105.3	78.1	81.8	83.3	67.3	69.5	103.6	62.2	67.7	104.2	49.3	56.6
03200	CONCRETE REINFORCEMENT	95.2	78.5	86.8	93.3	95.7	94.5	93.3	79.3	86.2	92.0	82.3	87.1	93.3	67.3	80.2	93.3	82.1	87.6
03300	CAST-IN-PLACE CONCRETE	112.1	74.3	96.3	103.8	100.9	102.6	108.1	78.8	95.8	105.5	87.9	98.2	106.9	57.2	86.1	107.8	48.4	83.0
03	CONCRETE	101.5	68.8	85.2	97.8	95.7	96.7	98.8	79.4	89.1	96.1	78.2	87.1	99.3	62.5	80.9	99.7	56.6	78.2
04	MASONRY	105.8	73.9	85.8	103.8	87.6	93.7	100.0	80.3	87.6	107.2	72.0	85.1	99.1	59.5	74.3	100.7	59.5	74.8
05	METALS	92.4	89.5	91.5	87.2	101.6	91.9	87.3	92.3	89.0	85.8	91.4	87.6	87.2	83.7	86.0	87.2	90.2	88.2
06	WOOD & PLASTICS	85.2	54.5	69.2	109.6	89.1	98.9	111.1	77.0	93.3	87.2	64.7	75.5	109.6	60.7	84.1	110.3	49.3	78.5
07	THERMAL & MOISTURE PROTECTION	99.2	66.8	83.2	99.4	88.3	93.9	100.3	78.5	89.6	99.5	69.5	84.7	99.4	57.4	78.8	99.1	53.6	76.7
08	DOORS & WINDOWS	98.3	61.6	88.8	99.3	91.9	97.4	99.3	81.1	94.6	98.4	75.4	92.4	99.3	62.3	89.7	94.9	61.5	86.3
09200	PLASTER & GYPSUM BOARD	93.7	53.5	68.3	104.4	88.9	94.6	101.3	76.4	85.5	94.1	64.1	75.1	104.4	59.6	76.0	104.4	48.0	68.6
095,098	CEILINGS & ACOUSTICAL TREATMENT	107.0	53.5	74.9	112.4	88.9	98.3	110.6	76.4	90.1	107.0	64.1	81.3	112.4	59.6	80.7	112.4	48.0	73.7
09600	FLOORING	98.1	47.5	84.7	109.4	82.3	102.2	109.2	43.4	91.8	112.0	40.5	93.1	109.9	61.3	97.0	111.1	61.6	98.0
097,099	WALL FINISHES, PAINTS & COATINGS	99.2	72.1	82.8	103.5	94.9	98.3	103.5	78.5	88.3	106.5	77.4	88.9	104.7	63.0	79.4	104.7	36.8	63.5
09	FINISHES	100.8	56.4	77.7	107.4	89.4	98.0	105.6	71.0	87.6	105.5	62.2	83.0	108.6	61.7	84.2	108.0	49.5	77.6
10 - 14	TOTAL DIV. 10000 - 14000	100.0	74.3	94.5	100.0	86.3	97.1	100.0	82.5	96.3	100.0	78.7	95.5	100.0	77.7	95.2	100.0	72.1	94.1
15	MECHANICAL	100.4	76.9	90.1	100.4	94.9	98.0	100.4	80.7	91.7	100.4	78.9	91.0	100.4	82.0	92.3	100.4	46.9	77.0
16	ELECTRICAL	101.7	81.3	91.7	95.5	89.1	92.4	97.5	83.2	90.5	100.4	70.3	85.7	97.0	77.6	87.6	97.0	63.8	80.8
01 - 16	WEIGHTED AVERAGE	99.1	74.7	88.0	97.8	93.1	95.7	97.7	82.4	90.7	97.7	76.5	88.1	98.3	73.3	87.0	97.7	62.0	81.5

DIVISION		KANSAS															KENTUCKY		
		DODGE CITY			KANSAS CITY			SALINA			TOPEKA			WICHITA			BOWLING GREEN		
		MAT.	INST.	TOTAL	MAT.	INST.	TOTAL	MAT.	INST.	TOTAL	MAT.	INST.	TOTAL	MAT.	INST.	TOTAL	MAT.	INST.	TOTAL
01590	EQUIPMENT RENTAL	.0	103.5	103.5	.0	100.1	100.1	.0	103.5	103.5	.0	101.6	101.6	.0	103.5	103.5	.0	96.0	96.0
02	SITE CONSTRUCTION	109.9	92.7	97.1	91.0	90.9	90.9	99.9	93.0	94.8	92.9	90.4	91.1	94.1	93.4	93.6	69.5	100.3	92.3
03100	CONCRETE FORMS & ACCESSORIES	93.4	43.4	50.1	99.0	91.9	92.8	89.2	46.0	51.8	98.5	45.2	52.3	95.3	58.1	63.1	83.8	82.9	83.0
03200	CONCRETE REINFORCEMENT	105.1	58.5	81.5	99.8	82.4	91.0	104.5	80.0	92.1	97.0	94.5	95.8	97.0	81.1	89.0	83.3	74.5	78.9
03300	CAST-IN-PLACE CONCRETE	114.8	57.4	90.8	90.0	91.6	90.7	99.6	49.1	78.5	91.0	53.6	75.4	86.8	57.6	74.6	88.6	82.0	85.8
03	CONCRETE	114.8	52.6	83.8	96.2	90.5	93.3	102.0	55.1	78.6	94.1	59.0	76.6	91.9	63.5	77.7	94.9	81.1	88.0
04	MASONRY	104.2	42.2	65.3	103.5	87.5	93.5	118.5	40.6	69.7	98.1	64.0	76.7	91.8	65.8	75.5	93.9	79.6	84.9
05	METALS	95.3	78.2	89.7	100.1	94.5	98.3	95.1	88.2	92.8	100.5	96.9	99.3	100.5	90.3	97.2	91.2	80.0	87.5
06	WOOD & PLASTICS	91.7	43.4	66.5	98.5	92.3	95.2	87.9	46.7	66.4	95.7	41.3	67.3	93.7	56.8	74.4	93.3	82.3	87.5
07	THERMAL & MOISTURE PROTECTION	98.2	49.2	74.1	96.1	90.1	93.2	97.5	54.7	76.5	97.4	71.3	84.6	97.1	65.8	81.7	84.7	80.2	82.5
08	DOORS & WINDOWS	96.2	47.0	83.5	95.1	89.4	93.6	96.1	52.5	84.8	96.3	60.2	87.0	96.3	63.5	87.8	95.3	76.0	90.3
09200	PLASTER & GYPSUM BOARD	100.8	41.7	63.3	97.0	91.9	93.8	100.0	45.1	65.2	102.3	39.5	62.5	102.3	55.4	72.6	87.4	82.2	84.1
095,098	CEILINGS & ACOUSTICAL TREATMENT	88.2	41.7	60.3	89.1	91.9	90.8	88.2	45.1	62.3	89.1	39.5	59.4	89.1	55.4	68.9	86.4	82.2	83.9
09600	FLOORING	95.7	51.2	83.9	86.7	93.1	88.4	93.7	37.4	78.8	97.2	47.5	84.1	96.4	79.3	91.9	92.9	63.0	85.0
097,099	WALL FINISHES, PAINTS & COATINGS	91.2	49.9	66.2	99.3	89.5	93.4	91.2	39.3	59.7	91.2	69.5	78.0	91.2	63.4	74.3	96.5	75.6	83.8
09	FINISHES	93.9	45.1	68.5	92.2	91.2	91.7	92.6	42.9	66.7	93.8	46.0	68.9	93.6	62.2	77.3	91.1	78.5	84.5
10 - 14	TOTAL DIV. 10000 - 14000	100.0	50.2	89.4	100.0	84.4	96.7	100.0	66.1	92.8	100.0	70.0	93.6	100.0	70.7	93.8	100.0	64.2	92.4
15	MECHANICAL	100.0	47.9	77.2	99.9	88.2	94.8	100.0	36.3	72.1	100.0	70.6	87.1	100.0	67.9	86.0	99.8	88.6	94.9
16	ELECTRICAL	99.7	50.3	75.6	104.4	102.1	103.3	99.4	72.6	86.3	103.3	75.1	89.5	101.5	72.6	87.4	94.5	84.2	89.5
01 - 16	WEIGHTED AVERAGE	100.3	54.8	79.6	98.7	91.6	95.5	99.0	57.2	80.1	98.5	69.3	85.2	97.7	71.1	85.6	94.4	83.6	89.5

DIVISION		KENTUCKY									LOUISIANA								
		LEXINGTON			LOUISVILLE			OWENSBORO			ALEXANDRIA			BATON ROUGE			LAKE CHARLES		
		MAT.	INST.	TOTAL	MAT.	INST.	TOTAL	MAT.	INST.	TOTAL	MAT.	INST.	TOTAL	MAT.	INST.	TOTAL	MAT.	INST.	TOTAL
01590	EQUIPMENT RENTAL	.0	103.5	103.5	.0	96.0	96.0	.0	121.3	121.3	.0	86.8	86.8	.0	86.2	86.2	.0	86.2	86.2
02	SITE CONSTRUCTION	77.3	102.4	95.8	67.0	100.2	91.5	82.1	129.6	117.2	102.8	85.7	90.1	109.3	85.3	91.5	110.9	85.0	91.8
03100	CONCRETE FORMS & ACCESSORIES	97.6	70.3	74.0	92.5	82.6	83.9	88.7	64.7	67.9	80.0	44.3	49.1	96.9	48.4	54.9	97.4	49.2	55.7
03200	CONCRETE REINFORCEMENT	92.0	96.6	94.3	92.0	97.8	94.9	83.4	93.9	88.7	98.4	63.8	80.9	93.6	51.2	72.2	93.6	51.2	72.2
03300	CAST-IN-PLACE CONCRETE	95.7	80.1	89.2	92.7	80.5	87.6	91.4	76.7	85.3	93.9	44.2	73.2	85.9	52.3	71.9	91.0	53.8	75.5
03	CONCRETE	97.3	79.0	88.2	95.5	84.8	90.1	104.3	74.8	89.5	89.0	49.1	69.1	93.5	51.1	72.3	96.0	52.0	74.0
04	MASONRY	90.8	48.1	64.0	92.4	81.8	85.8	90.8	52.0	66.4	115.3	54.0	76.9	96.6	55.6	70.9	94.4	56.6	70.7
05	METALS	92.6	89.3	91.5	100.0	90.3	96.8	83.1	87.4	84.5	85.7	74.4	82.0	95.7	65.5	85.8	90.1	65.7	82.1
06	WOOD & PLASTICS	104.3	73.3	88.1	102.8	82.3	92.1	93.5	63.6	77.9	82.5	43.9	62.3	103.8	48.3	74.8	102.0	49.0	74.3
07	THERMAL & MOISTURE PROTECTION	95.3	78.7	87.2	85.0	82.0	83.5	94.8	67.5	81.4	99.4	54.2	77.2	97.4	55.2	76.7	100.7	55.3	78.4
08	DOORS & WINDOWS	96.2	80.1	92.0	96.2	87.1	93.8	93.6	75.3	88.8	99.3	50.0	86.6	104.3	48.6	89.9	104.3	50.6	90.4
09200	PLASTER & GYPSUM BOARD	94.8	71.8	80.2	95.9	82.2	87.2	86.5	61.8	70.9	79.1	43.0	56.2	99.1	47.5	66.4	99.1	48.2	66.8
095,098	CEILINGS & ACOUSTICAL TREATMENT	91.8	71.8	79.8	91.8	82.2	86.0	75.7	61.8	67.4	90.3	43.0	61.9	96.7	47.5	67.2	96.7	48.2	67.6
09600	FLOORING	97.9	43.8	83.6	96.4	74.8	90.7	95.3	63.0	86.8	105.0	68.9	95.5	109.4	67.6	98.4	109.7	52.8	94.6
097,099	WALL FINISHES, PAINTS & COATINGS	96.5	59.7	74.2	96.5	84.5	89.2	96.5	99.6	98.4	96.4	42.1	63.5	98.7	48.3	68.1	98.7	44.5	65.8
09	FINISHES	95.0	63.7	78.7	94.4	81.1	87.5	89.0	67.4	77.8	93.2	48.1	69.7	102.1	51.8	75.9	102.2	49.0	74.5
10 - 14	TOTAL DIV. 10000 - 14000	100.0	81.9	96.2	100.0	85.3	96.9	100.0	84.3	96.7	100.0	71.6	94.0	100.0	75.3	94.7	100.0	75.5	94.8
15	MECHANICAL	99.8	48.8	77.5	99.8	87.3	94.3	99.8	53.4	79.5	100.1	38.5	73.1	100.0	43.7	75.4	100.0	59.3	82.2
16	ELECTRICAL	94.9	50.8	73.4	94.9	84.2	89.7	94.0	82.3	88.3	92.4	56.4	74.8	92.8	50.4	72.1	92.8	63.1	78.4
01 - 16	WEIGHTED AVERAGE	95.9	66.9	82.7	96.3	86.4	91.8	94.2	74.3	85.2	95.6	54.7	77.0	98.4	55.1	78.7	97.8	60.0	80.7

	DIVISION	LOUISIANA									MAINE								
		MONROE			NEW ORLEANS			SHREVEPORT			AUGUSTA			BANGOR			LEWISTON		
		MAT.	INST.	TOTAL	MAT.	INST.	TOTAL	MAT.	INST.	TOTAL	MAT.	INST.	TOTAL	MAT.	INST.	TOTAL	MAT.	INST.	TOTAL
01590	**EQUIPMENT RENTAL**	.0	86.8	86.8	.0	87.4	87.4	.0	86.8	86.8	.0	101.6	101.6	.0	101.6	101.6	.0	101.6	101.6
02	**SITE CONSTRUCTION**	102.8	85.2	89.8	113.4	87.8	94.5	101.5	85.1	89.3	83.4	102.3	97.4	83.3	100.7	96.2	82.0	100.7	95.8
03100	CONCRETE FORMS & ACCESSORIES	79.3	44.9	49.5	97.2	69.5	73.2	98.6	47.6	54.4	99.6	64.9	69.6	93.9	80.3	82.1	99.7	80.3	82.9
03200	CONCRETE REINFORCEMENT	97.4	64.0	80.5	93.6	65.0	79.2	96.9	63.7	80.1	88.1	112.7	100.5	88.1	113.0	100.7	108.2	113.0	110.7
03300	CAST-IN-PLACE CONCRETE	93.9	48.1	74.8	90.1	60.0	77.5	92.6	51.8	75.5	81.5	63.2	73.9	81.5	65.1	74.7	92.8	65.1	81.2
03	**CONCRETE**	88.8	50.6	69.7	95.5	65.8	80.7	88.8	53.0	70.9	98.6	73.3	86.0	97.7	80.7	89.2	102.8	80.7	91.7
04	**MASONRY**	110.2	58.1	77.5	96.3	64.3	76.2	102.4	51.1	70.2	96.1	52.3	68.6	113.7	60.2	80.1	97.9	60.2	74.2
05	**METALS**	85.7	73.0	81.5	103.1	75.1	94.0	82.1	72.2	78.8	87.9	84.9	86.9	87.5	83.1	86.1	90.7	83.1	88.2
06	**WOOD & PLASTICS**	81.7	44.3	62.2	98.4	72.7	85.0	102.2	48.2	74.0	98.5	63.4	80.2	92.3	82.6	87.2	98.5	82.6	90.2
07	**THERMAL & MOISTURE PROTECTION**	99.4	57.0	78.5	101.6	65.2	83.7	98.3	55.6	77.3	102.1	52.5	77.7	102.0	59.5	81.1	101.8	59.5	81.0
08	**DOORS & WINDOWS**	99.3	55.0	87.9	104.9	69.4	95.7	97.2	52.5	85.6	103.9	62.1	93.1	103.8	74.3	96.2	107.1	74.3	98.6
09200	PLASTER & GYPSUM BOARD	78.7	43.4	56.3	97.9	72.5	81.8	85.4	47.4	61.3	102.8	61.4	76.6	98.9	81.2	87.6	105.2	81.2	89.9
095,098	CEILINGS & ACOUSTICAL TREATMENT	90.3	43.4	62.2	95.7	72.5	81.8	91.2	47.4	65.0	92.2	61.4	73.7	89.5	81.2	84.5	101.1	81.2	89.1
09600	FLOORING	104.6	43.6	88.5	110.1	51.8	94.7	115.6	64.6	102.1	99.8	43.9	85.0	97.3	54.7	86.1	99.8	54.7	87.9
097,099	WALL FINISHES, PAINTS & COATINGS	96.4	50.3	68.4	100.5	63.9	78.3	96.4	42.1	63.5	92.8	36.4	58.6	92.8	33.5	56.8	92.8	33.5	56.8
09	**FINISHES**	93.0	44.3	67.7	102.3	65.8	83.3	97.4	49.7	72.6	99.1	57.1	77.3	97.2	70.9	83.5	101.4	70.9	85.5
10 - 14	**TOTAL DIV. 10000 - 14000**	100.0	65.4	92.6	100.0	80.3	95.8	100.0	65.7	92.7	100.0	63.9	92.3	100.0	83.0	96.4	100.0	83.0	96.4
15	**MECHANICAL**	100.1	52.0	79.1	100.0	65.9	85.1	100.1	54.6	80.2	100.1	72.3	88.0	100.1	71.5	87.6	100.1	71.5	87.6
16	**ELECTRICAL**	94.1	55.9	75.4	94.0	70.5	82.6	92.9	67.0	80.3	100.2	80.9	90.8	98.5	80.9	89.9	100.3	80.9	90.8
01 - 16	**WEIGHTED AVERAGE**	95.5	57.5	78.2	100.2	69.8	86.4	94.7	59.7	78.8	97.7	72.3	86.2	98.0	76.8	88.3	99.2	76.8	89.0

	DIVISION	MAINE			MARYLAND						MASSACHUSETTS								
		PORTLAND			BALTIMORE			HAGERSTOWN			BOSTON			BROCKTON			FALL RIVER		
		MAT.	INST.	TOTAL	MAT.	INST.	TOTAL	MAT.	INST.	TOTAL	MAT.	INST.	TOTAL	MAT.	INST.	TOTAL	MAT.	INST.	TOTAL
01590	**EQUIPMENT RENTAL**	.0	101.6	101.6	.0	103.4	103.4	.0	99.2	99.2	.0	108.4	108.4	.0	103.8	103.8	.0	105.0	105.0
02	**SITE CONSTRUCTION**	80.8	100.7	95.5	97.6	93.3	94.4	87.4	88.9	88.5	88.4	108.9	103.6	86.1	104.8	100.0	85.2	105.0	99.8
03100	CONCRETE FORMS & ACCESSORIES	98.8	80.3	82.7	101.7	74.8	78.4	90.3	76.1	78.0	105.1	135.6	131.5	104.7	120.9	118.8	104.7	121.2	119.0
03200	CONCRETE REINFORCEMENT	108.2	113.0	110.7	95.9	87.8	91.8	84.8	76.3	80.5	107.1	143.7	125.6	108.2	142.8	125.7	108.2	125.6	117.0
03300	CAST-IN-PLACE CONCRETE	84.5	65.1	76.4	98.3	78.7	90.1	83.6	61.7	74.4	105.2	143.5	121.2	100.2	129.2	112.3	96.9	141.6	115.5
03	**CONCRETE**	98.8	80.7	89.7	99.5	79.7	89.6	85.3	72.2	78.8	111.7	138.9	125.3	109.0	127.2	118.1	107.4	128.4	117.9
04	**MASONRY**	95.4	60.2	73.3	92.8	74.6	81.4	97.0	75.9	83.8	116.7	144.8	134.3	112.6	127.3	121.9	112.4	140.0	129.7
05	**METALS**	92.0	83.1	89.1	100.3	99.3	99.9	96.5	91.4	94.8	103.3	125.9	110.7	100.0	121.6	107.1	100.0	115.7	105.2
06	**WOOD & PLASTICS**	98.5	82.6	90.2	97.9	75.7	86.3	85.9	75.6	80.5	102.3	136.3	120.0	101.3	120.4	111.3	101.3	120.8	111.4
07	**THERMAL & MOISTURE PROTECTION**	101.6	59.5	80.9	94.2	81.7	88.1	93.4	71.6	82.7	102.4	141.9	121.8	102.1	130.1	115.9	102.1	127.9	114.8
08	**DOORS & WINDOWS**	107.1	74.3	98.6	93.8	83.3	91.1	91.1	77.1	87.5	103.1	135.4	111.4	101.0	123.2	106.8	101.0	119.9	105.9
09200	PLASTER & GYPSUM BOARD	105.2	81.2	89.9	108.7	75.3	87.5	103.5	75.3	85.6	107.8	136.5	126.0	102.5	120.1	113.6	102.5	120.1	113.6
095,098	CEILINGS & ACOUSTICAL TREATMENT	101.1	81.2	89.1	104.1	75.3	86.8	105.1	75.3	87.2	100.1	136.5	121.9	102.1	120.1	112.9	102.1	120.1	112.9
09600	FLOORING	99.8	54.7	87.9	92.8	82.2	90.0	87.9	78.8	85.5	100.4	153.6	114.5	100.9	153.6	114.8	100.7	153.6	114.7
097,099	WALL FINISHES, PAINTS & COATINGS	92.8	33.5	56.8	96.0	71.1	80.9	96.0	41.1	62.7	95.9	150.7	129.1	95.3	119.2	109.8	95.3	119.2	109.8
09	**FINISHES**	101.4	70.9	85.6	97.3	75.2	85.8	94.7	72.9	83.4	101.7	141.0	122.1	101.5	127.0	114.8	101.5	127.2	114.9
10 - 14	**TOTAL DIV. 10000 - 14000**	100.0	83.0	96.4	100.0	87.2	97.3	100.0	89.7	97.8	100.0	122.5	104.8	100.0	118.7	104.0	100.0	119.6	104.2
15	**MECHANICAL**	100.1	71.5	87.6	100.0	85.7	93.7	100.0	85.7	93.7	100.1	125.8	111.4	100.1	105.5	102.5	100.1	105.4	102.4
16	**ELECTRICAL**	100.4	80.9	90.9	107.1	90.7	99.1	103.6	82.1	93.1	99.3	123.3	111.0	98.3	96.3	97.3	97.8	102.2	99.9
01 - 16	**WEIGHTED AVERAGE**	98.8	76.8	88.8	99.4	84.8	92.7	96.0	80.8	89.1	102.9	130.5	115.4	101.4	115.5	107.8	101.1	117.0	108.4

	DIVISION	MASSACHUSETTS																	
		HYANNIS			LAWRENCE			LOWELL			NEW BEDFORD			PITTSFIELD			SPRINGFIELD		
		MAT.	INST.	TOTAL	MAT.	INST.	TOTAL	MAT.	INST.	TOTAL	MAT.	INST.	TOTAL	MAT.	INST.	TOTAL	MAT.	INST.	TOTAL
01590	**EQUIPMENT RENTAL**	.0	103.8	103.8	.0	103.8	103.8	.0	101.6	101.6	.0	105.0	105.0	.0	101.6	101.6	.0	101.6	101.6
02	**SITE CONSTRUCTION**	82.7	104.8	99.1	86.7	104.8	100.1	85.8	104.8	99.9	84.8	105.0	99.7	86.9	103.6	99.2	86.3	103.7	99.2
03100	CONCRETE FORMS & ACCESSORIES	95.4	120.9	117.5	104.5	121.4	119.1	101.2	121.6	118.9	104.7	121.2	119.0	101.2	101.2	101.2	101.4	109.8	108.7
03200	CONCRETE REINFORCEMENT	86.8	125.6	106.4	107.4	130.6	119.1	108.2	130.7	119.6	108.2	125.6	117.0	90.2	105.9	98.1	108.2	124.8	116.6
03300	CAST-IN-PLACE CONCRETE	91.4	140.8	112.0	101.0	139.5	117.1	91.8	139.6	111.8	94.0	141.6	113.9	100.2	116.0	106.8	95.5	117.3	104.6
03	**CONCRETE**	98.6	127.9	113.3	109.2	128.7	118.9	100.0	128.6	114.3	106.1	128.4	117.2	101.0	106.7	103.9	101.8	114.5	108.2
04	**MASONRY**	111.5	140.1	129.4	111.9	138.4	128.5	97.8	138.5	123.3	112.2	140.0	129.6	98.4	112.1	107.0	98.1	114.1	108.2
05	**METALS**	96.3	115.2	102.5	97.3	117.6	103.9	97.3	114.9	103.0	100.0	115.7	105.2	97.1	99.4	97.8	99.9	107.5	102.4
06	**WOOD & PLASTICS**	91.3	120.4	106.5	101.3	120.4	111.3	100.4	120.4	110.9	101.3	120.8	111.4	100.4	100.4	100.4	100.4	110.6	105.7
07	**THERMAL & MOISTURE PROTECTION**	101.6	128.8	115.0	102.1	134.4	118.0	101.8	134.4	117.8	102.1	127.9	114.8	101.9	109.1	105.4	101.8	111.0	106.3
08	**DOORS & WINDOWS**	97.2	119.7	103.0	101.0	119.8	105.9	107.1	119.8	110.4	101.0	119.9	105.9	107.1	103.9	106.2	107.1	114.5	109.0
09200	PLASTER & GYPSUM BOARD	94.2	120.1	110.6	105.2	120.1	114.6	105.2	120.1	114.6	102.5	120.1	113.6	105.2	99.4	101.5	105.2	110.0	108.2
095,098	CEILINGS & ACOUSTICAL TREATMENT	90.5	120.1	108.2	101.1	120.1	112.5	101.1	120.1	112.5	102.1	120.1	112.9	101.1	99.4	100.1	101.1	110.0	106.4
09600	FLOORING	96.9	153.6	111.9	99.8	153.6	114.0	99.8	153.6	114.0	100.7	153.6	114.7	100.0	126.7	107.1	99.7	126.7	106.9
097,099	WALL FINISHES, PAINTS & COATINGS	95.3	119.2	109.8	92.9	119.2	108.8	92.8	119.2	108.8	95.3	119.2	109.8	92.8	99.0	96.6	94.6	99.0	97.3
09	**FINISHES**	96.1	127.0	112.1	101.2	127.0	114.6	101.2	127.0	114.6	101.4	127.2	114.8	101.2	105.6	103.5	101.3	112.1	106.9
10 - 14	**TOTAL DIV. 10000 - 14000**	100.0	118.7	104.0	100.0	118.8	104.0	100.0	118.8	104.0	100.0	119.6	104.2	100.0	102.7	100.6	100.0	104.8	101.0
15	**MECHANICAL**	100.1	105.3	102.4	100.1	109.6	104.3	100.1	122.8	110.0	100.1	105.4	102.4	100.1	95.3	98.0	100.1	98.4	99.4
16	**ELECTRICAL**	96.5	102.2	99.3	99.8	123.3	111.3	100.3	123.3	111.5	99.3	102.2	100.7	100.3	88.7	94.7	100.3	88.0	94.3
01 - 16	**WEIGHTED AVERAGE**	98.3	116.9	106.7	101.1	121.0	110.2	100.0	123.5	110.7	101.1	117.0	108.4	100.2	101.1	100.6	100.7	105.1	102.7

DIVISION	MASSACHUSETTS			MICHIGAN														
	WORCESTER			ANN ARBOR			DEARBORN			DETROIT			FLINT			GRAND RAPIDS		
	MAT.	INST.	TOTAL	MAT.	INST.	TOTAL	MAT.	INST.	TOTAL	MAT.	INST.	TOTAL	MAT.	INST.	TOTAL	MAT.	INST.	TOTAL
01590 EQUIPMENT RENTAL	.0	101.6	101.6	.0	112.2	112.2	.0	112.2	112.2	.0	97.7	97.7	.0	112.2	112.2	.0	105.1	105.1
02 SITE CONSTRUCTION	85.8	104.8	99.8	78.1	95.2	90.7	77.9	95.5	90.9	91.5	97.2	95.8	69.0	94.2	87.6	82.2	87.8	86.3
03100 CONCRETE FORMS & ACCESSORIES	101.8	121.0	118.5	97.6	110.4	108.7	97.4	123.0	119.6	99.0	123.1	119.9	101.1	93.5	94.5	97.2	79.1	81.5
03200 CONCRETE REINFORCEMENT	108.2	142.5	125.6	97.1	124.3	110.9	97.1	125.1	111.3	96.5	125.1	110.9	97.1	124.0	110.7	92.0	86.7	89.3
03300 CAST-IN-PLACE CONCRETE	91.8	139.0	111.5	94.3	111.2	101.4	92.1	122.0	104.6	101.2	122.0	109.9	94.9	93.6	94.4	95.9	99.9	97.6
03 CONCRETE	100.1	130.4	115.2	93.0	113.9	103.4	91.9	123.3	107.6	96.2	121.9	109.1	93.5	100.2	96.9	97.3	87.1	92.2
04 MASONRY	97.8	138.0	123.0	98.5	106.2	103.3	98.3	122.1	113.2	97.3	122.1	112.9	98.5	92.6	94.8	94.9	56.0	70.5
05 METALS	100.0	119.1	106.2	100.6	126.0	108.9	100.7	127.9	109.6	105.2	105.7	105.4	100.7	123.3	108.1	92.9	80.6	88.9
06 WOOD & PLASTICS	100.9	120.4	111.1	101.4	109.6	105.7	101.4	123.0	112.7	102.3	123.0	113.1	104.9	94.1	99.3	98.2	79.7	88.5
07 THERMAL & MOISTURE PROTECTION	101.8	129.0	115.2	97.7	109.2	103.4	96.6	124.2	110.2	95.2	124.2	109.5	95.9	92.5	94.2	91.7	62.4	77.3
08 DOORS & WINDOWS	107.1	126.5	112.1	96.5	111.1	100.3	96.5	118.4	102.2	98.0	120.3	103.8	96.5	100.2	97.4	94.2	71.0	88.2
09200 PLASTER & GYPSUM BOARD	105.2	120.1	114.6	102.4	108.6	106.3	102.4	122.3	115.1	102.4	122.3	115.1	104.0	92.6	96.8	93.2	74.2	81.2
095,098 CEILINGS & ACOUSTICAL TREATMENT	101.1	120.1	112.5	95.6	108.6	103.4	95.6	122.3	111.6	96.6	122.3	112.0	95.6	92.6	93.8	91.8	74.2	81.2
09600 FLOORING	99.8	153.6	114.0	87.7	109.8	93.5	87.2	123.8	96.9	87.3	123.8	97.0	87.4	74.9	84.1	96.6	44.2	82.8
097,099 WALL FINISHES, PAINTS & COATINGS	92.8	119.2	108.8	86.5	103.3	96.7	86.5	114.0	103.1	88.3	114.0	103.9	86.5	88.1	87.5	96.5	42.7	63.8
09 FINISHES	101.2	127.0	114.6	93.9	109.5	102.0	93.8	122.5	108.7	94.8	122.5	109.2	93.6	89.0	91.2	94.3	68.5	80.9
10 - 14 TOTAL DIV. 10000 - 14000	100.0	108.3	101.8	100.0	111.3	102.4	100.0	114.9	103.2	100.0	114.9	103.2	100.0	96.0	99.2	100.0	101.4	100.3
15 MECHANICAL	100.1	108.8	103.9	99.9	103.3	101.4	99.9	116.9	107.3	99.9	119.7	108.6	99.9	93.6	97.1	99.8	57.0	81.1
16 ELECTRICAL	100.3	104.5	102.3	94.1	87.9	91.1	94.1	116.4	105.0	95.9	116.3	105.8	94.1	88.9	91.5	94.6	61.3	78.4
01 - 16 WEIGHTED AVERAGE	100.5	118.4	108.6	96.9	105.7	100.9	96.7	118.3	106.5	98.7	116.9	107.0	96.7	96.2	96.5	95.8	70.2	84.2

DIVISION	MICHIGAN												MINNESOTA					
	KALAMAZOO			LANSING			MUSKEGON			SAGINAW			DULUTH			MINNEAPOLIS		
	MAT.	INST.	TOTAL	MAT.	INST.	TOTAL	MAT.	INST.	TOTAL	MAT.	INST.	TOTAL	MAT.	INST.	TOTAL	MAT.	INST.	TOTAL
01590 EQUIPMENT RENTAL	.0	105.1	105.1	.0	112.2	112.2	.0	105.1	105.1	.0	112.2	112.2	.0	101.8	101.8	.0	106.1	106.1
02 SITE CONSTRUCTION	82.7	88.1	86.7	84.1	94.1	91.5	80.5	88.0	86.1	71.2	93.9	88.0	82.7	102.5	97.4	82.6	108.8	102.0
03100 CONCRETE FORMS & ACCESSORIES	96.7	88.5	89.6	101.1	91.5	92.8	97.4	85.8	87.3	97.6	92.2	92.9	103.5	123.7	121.0	104.2	137.5	133.0
03200 CONCRETE REINFORCEMENT	92.0	88.3	90.1	97.1	123.8	110.6	92.5	88.1	90.3	97.1	123.4	110.4	95.3	109.0	102.2	95.4	130.0	112.9
03300 CAST-IN-PLACE CONCRETE	97.6	100.4	98.8	94.3	94.7	94.5	95.3	98.3	96.5	93.1	93.5	93.3	109.6	104.2	107.3	107.9	126.9	115.8
03 CONCRETE	100.5	91.8	96.2	93.2	99.7	96.4	95.4	89.8	92.6	92.4	99.5	95.9	100.2	114.7	107.4	101.0	132.8	116.9
04 MASONRY	97.9	84.3	89.4	91.9	94.6	93.6	95.1	80.1	85.7	100.0	86.1	91.2	107.9	113.8	111.6	107.9	136.5	125.9
05 METALS	91.5	84.5	89.2	99.1	122.9	106.8	89.2	84.1	87.5	100.7	121.9	107.7	93.9	125.5	104.2	96.7	139.9	110.8
06 WOOD & PLASTICS	99.9	88.2	93.8	104.3	89.6	96.6	96.7	85.2	90.7	97.4	92.4	94.8	114.7	127.0	121.2	115.2	136.5	126.3
07 THERMAL & MOISTURE PROTECTION	91.3	84.4	87.9	96.7	92.5	94.6	90.4	78.7	84.7	96.5	88.9	92.8	101.5	116.6	108.9	101.3	136.3	118.5
08 DOORS & WINDOWS	90.9	83.1	88.9	96.5	97.7	96.8	90.1	82.0	88.0	96.0	98.2	96.6	96.8	124.6	104.0	99.9	143.7	111.2
09200 PLASTER & GYPSUM BOARD	93.2	83.0	86.8	105.1	87.9	94.2	84.1	79.9	81.4	102.4	90.9	95.1	102.3	128.3	118.8	102.5	137.8	124.9
095,098 CEILINGS & ACOUSTICAL TREATMENT	91.8	83.0	86.5	95.6	87.9	91.0	93.6	79.9	85.3	95.6	90.9	92.8	93.4	128.3	114.3	94.2	137.8	120.4
09600 FLOORING	96.6	70.1	89.6	95.8	83.0	92.4	95.8	81.5	92.0	87.7	59.5	80.2	105.8	121.7	110.0	103.2	126.4	109.3
097,099 WALL FINISHES, PAINTS & COATINGS	96.5	86.6	90.5	100.4	94.4	96.8	95.7	61.9	75.2	86.5	81.1	83.2	93.3	103.0	99.2	99.6	127.1	116.3
09 FINISHES	94.3	84.7	89.3	98.2	89.7	93.8	92.7	82.1	87.2	93.7	84.4	88.9	100.6	122.4	111.9	100.4	134.8	118.3
10 - 14 TOTAL DIV. 10000 - 14000	100.0	103.7	100.8	100.0	97.1	99.4	100.0	103.0	100.6	100.0	96.5	99.2	100.0	99.3	99.8	100.0	109.4	102.0
15 MECHANICAL	99.8	86.3	93.9	99.9	93.2	97.0	99.7	89.7	95.4	99.9	91.0	96.0	100.2	94.8	97.8	100.1	122.2	109.8
16 ELECTRICAL	94.2	71.8	83.2	92.3	88.2	90.3	94.7	67.3	81.4	93.5	83.2	88.5	105.8	95.1	100.6	107.7	112.4	110.0
01 - 16 WEIGHTED AVERAGE	95.7	84.9	90.8	96.7	96.1	96.4	94.3	83.6	89.5	96.5	93.2	95.0	99.7	108.9	103.8	100.7	126.8	112.6

DIVISION	MINNESOTA									MISSISSIPPI								
	ROCHESTER			SAINT PAUL			ST. CLOUD			BILOXI			GREENVILLE			JACKSON		
	MAT.	INST.	TOTAL	MAT.	INST.	TOTAL	MAT.	INST.	TOTAL	MAT.	INST.	TOTAL	MAT.	INST.	TOTAL	MAT.	INST.	TOTAL
01590 EQUIPMENT RENTAL	.0	101.8	101.8	.0	101.8	101.8	.0	101.5	101.5	.0	98.3	98.3	.0	98.3	98.3	.0	98.3	98.3
02 SITE CONSTRUCTION	81.8	101.7	96.5	84.7	103.3	98.5	78.3	105.2	98.2	103.3	86.2	90.6	107.4	86.0	91.5	99.9	86.0	89.6
03100 CONCRETE FORMS & ACCESSORIES	104.0	104.5	104.4	94.5	130.9	126.0	84.5	126.8	121.1	95.5	42.5	49.6	80.3	35.1	41.1	95.3	40.0	47.4
03200 CONCRETE REINFORCEMENT	95.3	129.2	112.4	91.9	129.9	111.1	94.3	129.2	111.9	95.0	62.4	78.6	103.0	44.9	73.6	95.0	48.7	71.6
03300 CAST-IN-PLACE CONCRETE	105.8	98.2	102.6	108.9	126.0	116.0	97.8	124.0	108.7	104.5	46.3	80.2	104.9	40.5	78.0	102.5	43.0	77.7
03 CONCRETE	98.4	108.1	103.2	102.5	129.5	116.0	90.0	126.7	108.4	97.0	49.5	73.3	100.2	40.9	70.6	96.0	44.7	70.4
04 MASONRY	107.2	109.5	108.6	117.7	136.5	129.5	109.2	122.2	117.4	89.3	38.1	57.2	131.8	39.5	73.9	91.3	39.5	58.8
05 METALS	93.7	137.1	107.9	93.2	139.2	108.3	90.0	136.2	105.1	92.1	83.2	89.2	90.4	74.9	85.4	92.1	76.8	87.1
06 WOOD & PLASTICS	115.2	103.7	109.2	104.4	128.1	116.8	91.7	124.5	108.8	96.6	42.4	68.3	79.0	34.4	55.7	96.7	40.8	67.5
07 THERMAL & MOISTURE PROTECTION	101.3	103.7	102.5	101.2	133.9	117.3	98.6	117.3	107.8	95.2	47.3	71.6	95.3	42.4	69.3	95.0	43.6	69.8
08 DOORS & WINDOWS	96.8	125.9	104.3	94.1	139.1	105.8	91.6	137.2	103.4	98.5	48.4	85.5	97.8	39.2	82.7	98.9	43.8	84.6
09200 PLASTER & GYPSUM BOARD	102.0	104.3	103.5	96.3	129.4	117.3	87.0	125.8	111.6	103.7	41.3	64.2	91.2	33.1	54.4	103.7	39.8	63.2
095,098 CEILINGS & ACOUSTICAL TREATMENT	92.5	104.3	99.6	90.7	129.4	113.9	72.1	125.8	104.3	102.7	41.3	65.9	96.5	33.1	58.5	102.7	39.8	65.0
09600 FLOORING	105.6	82.1	99.4	98.6	126.4	106.0	97.4	126.4	105.1	118.6	41.4	98.2	109.1	35.4	89.6	118.6	40.9	98.1
097,099 WALL FINISHES, PAINTS & COATINGS	95.7	103.5	100.4	99.6	120.5	112.3	103.4	127.1	117.8	107.4	36.5	64.4	107.4	35.9	64.0	107.4	35.9	64.0
09 FINISHES	100.3	100.6	100.5	97.5	129.1	113.9	90.9	127.0	109.7	106.9	41.1	72.7	101.1	34.9	66.7	106.9	39.9	72.0
10 - 14 TOTAL DIV. 10000 - 14000	100.0	97.4	99.4	100.0	107.9	101.7	100.0	102.8	100.6	100.0	55.2	90.5	100.0	53.4	90.1	100.0	54.3	90.3
15 MECHANICAL	100.1	99.2	99.7	100.1	113.2	105.8	99.7	120.5	108.8	99.8	55.9	80.6	99.8	35.6	71.7	99.8	35.3	71.6
16 ELECTRICAL	105.7	89.4	97.8	106.5	106.1	106.3	102.5	106.1	104.3	96.7	61.1	79.3	96.1	38.1	67.8	97.5	38.1	68.5
01 - 16 WEIGHTED AVERAGE	99.4	105.1	102.0	99.8	122.0	109.8	95.6	120.5	106.9	97.9	56.6	79.1	99.4	45.9	75.1	97.9	47.5	75.0

DIVISION		MISSISSIPPI			MISSOURI														
		MERIDIAN			CAPE GIRARDEAU			COLUMBIA			JOPLIN			KANSAS CITY			SPRINGFIELD		
		MAT.	INST.	TOTAL	MAT.	INST.	TOTAL	MAT.	INST.	TOTAL	MAT.	INST.	TOTAL	MAT.	INST.	TOTAL	MAT.	INST.	TOTAL
01590	EQUIPMENT RENTAL	.0	98.3	98.3	.0	108.3	108.3	.0	109.0	109.0	.0	106.9	106.9	.0	104.8	104.8	.0	102.7	102.7
02	SITE CONSTRUCTION	98.7	86.0	89.3	93.6	92.5	92.8	101.5	94.5	96.3	100.7	98.6	99.1	94.3	98.0	97.0	95.3	93.2	93.7
03100	CONCRETE FORMS & ACCESSORIES	78.8	34.1	40.1	84.4	74.9	76.2	85.0	72.4	74.0	101.2	65.4	70.2	100.2	105.2	104.5	100.2	65.6	70.2
03200	CONCRETE REINFORCEMENT	101.7	48.1	74.6	93.9	84.7	89.2	100.7	112.5	106.7	111.3	73.6	92.2	105.8	113.4	109.6	97.0	111.7	104.4
03300	CAST-IN-PLACE CONCRETE	99.0	43.0	75.6	84.6	84.5	84.6	88.5	68.9	80.3	101.2	73.1	89.5	94.1	107.5	99.7	101.5	63.6	85.6
03	CONCRETE	93.0	41.9	67.5	85.5	81.8	83.7	85.9	80.5	83.2	99.9	70.8	85.4	95.9	107.9	101.9	99.2	74.7	87.0
04	MASONRY	89.0	29.4	51.6	114.0	75.8	90.0	131.1	83.4	101.2	98.6	60.4	74.6	103.0	105.7	104.7	87.4	65.4	73.6
05	METALS	90.4	75.8	85.6	92.1	109.3	97.7	92.7	121.1	102.0	99.8	89.8	96.5	107.5	115.0	109.9	96.6	105.6	99.5
06	WOOD & PLASTICS	77.7	33.0	54.4	80.2	72.3	76.1	88.7	68.3	78.1	100.7	63.6	81.3	100.1	104.4	102.4	99.1	66.6	82.1
07	THERMAL & MOISTURE PROTECTION	94.7	40.2	67.9	95.5	79.8	87.7	92.4	81.4	87.0	98.6	63.8	81.5	97.8	107.1	102.4	96.5	70.1	83.5
08	DOORS & WINDOWS	97.8	37.9	82.3	93.9	75.0	89.1	90.9	91.7	91.1	91.9	71.0	86.5	98.0	107.0	100.3	96.3	78.2	91.6
09200	PLASTER & GYPSUM BOARD	91.2	31.7	53.5	96.1	71.4	80.4	98.1	67.3	78.6	103.9	62.3	77.5	100.3	104.3	102.8	102.5	65.5	79.0
095,098	CEILINGS & ACOUSTICAL TREATMENT	96.5	31.7	57.6	85.5	71.4	77.0	90.0	67.3	76.4	87.6	62.3	72.4	93.9	104.3	100.1	90.0	65.5	75.3
09600	FLOORING	108.0	28.7	87.0	92.4	78.2	88.7	101.9	90.5	98.9	111.7	52.6	96.1	91.7	90.5	91.4	104.9	52.6	91.1
097,099	WALL FINISHES, PAINTS & COATINGS	107.4	35.4	63.7	100.9	78.5	87.3	99.3	74.8	84.5	92.3	46.4	64.4	97.2	118.8	110.3	93.6	66.4	77.1
09	FINISHES	100.2	32.6	65.0	90.0	74.9	82.1	93.1	73.3	82.8	99.0	60.6	79.0	95.3	103.6	99.6	96.6	62.6	78.9
10 - 14	TOTAL DIV. 10000 - 14000	100.0	53.4	90.1	100.0	70.2	93.7	100.0	91.0	98.1	100.0	79.8	95.7	100.0	93.8	98.7	100.0	80.5	95.9
15	MECHANICAL	99.8	35.3	71.6	100.1	93.8	97.4	100.0	92.2	96.6	100.1	57.9	81.6	100.0	107.1	103.1	100.0	67.7	85.9
16	ELECTRICAL	95.7	60.2	78.4	102.4	108.8	105.5	99.4	82.4	91.1	97.6	63.6	81.1	109.0	108.1	108.6	102.6	68.7	86.1
01 - 16	WEIGHTED AVERAGE	96.0	47.8	74.1	96.1	89.3	93.0	96.9	88.1	92.9	98.7	69.0	85.2	101.1	106.3	103.4	98.2	74.5	87.4

DIVISION		MISSOURI						MONTANA											
		ST. JOSEPH			ST. LOUIS			BILLINGS			BUTTE			GREAT FALLS			HELENA		
		MAT.	INST.	TOTAL	MAT.	INST.	TOTAL	MAT.	INST.	TOTAL	MAT.	INST.	TOTAL	MAT.	INST.	TOTAL	MAT.	INST.	TOTAL
01590	EQUIPMENT RENTAL	.0	103.3	103.3	.0	109.5	109.5	.0	98.5	98.5	.0	98.2	98.2	.0	98.2	98.2	.0	98.2	98.2
02	SITE CONSTRUCTION	95.8	93.5	94.1	93.3	96.5	95.7	85.2	96.1	93.3	92.2	94.7	94.0	95.7	95.5	95.5	97.5	95.3	95.9
03100	CONCRETE FORMS & ACCESSORIES	100.1	86.0	87.9	99.6	105.7	104.9	98.1	65.3	69.7	85.4	62.9	65.9	99.4	66.5	70.9	99.5	65.9	70.4
03200	CONCRETE REINFORCEMENT	104.5	97.3	100.9	85.8	115.6	100.9	93.2	75.9	84.5	101.2	76.0	88.4	93.3	75.9	84.5	96.6	65.9	81.0
03300	CAST-IN-PLACE CONCRETE	94.0	100.6	96.8	84.6	108.2	94.5	119.0	68.6	97.9	122.5	68.7	100.0	129.4	58.6	99.9	131.9	67.1	104.8
03	CONCRETE	95.6	93.9	94.8	85.3	109.4	97.3	104.6	69.4	87.0	104.6	68.4	86.5	109.7	66.5	88.1	111.4	67.2	89.3
04	MASONRY	102.4	81.0	89.0	95.2	111.2	105.2	120.1	66.4	86.4	120.1	67.4	87.0	123.7	71.5	90.9	119.2	65.7	85.6
05	METALS	103.3	104.0	103.6	96.7	126.0	106.3	105.7	86.8	99.5	98.5	86.5	94.6	102.2	86.6	97.1	101.3	81.2	94.8
06	WOOD & PLASTICS	100.7	85.9	93.0	96.2	103.6	100.0	101.8	65.1	82.6	88.9	63.6	75.7	104.5	65.3	84.0	104.6	65.3	84.1
07	THERMAL & MOISTURE PROTECTION	98.4	90.3	94.4	95.4	106.2	100.7	99.9	68.5	84.5	99.2	67.4	83.6	100.1	68.5	84.5	100.0	68.2	84.4
08	DOORS & WINDOWS	96.8	92.9	95.8	92.9	112.3	97.9	99.1	64.1	90.0	95.8	63.2	87.4	99.3	64.2	90.2	98.8	61.6	89.2
09200	PLASTER & GYPSUM BOARD	105.3	85.2	92.6	105.2	103.5	104.1	104.4	64.5	79.1	94.8	63.0	74.6	104.4	64.7	79.2	103.7	64.7	79.0
095,098	CEILINGS & ACOUSTICAL TREATMENT	93.0	85.2	88.3	90.9	103.5	98.4	112.4	64.5	83.6	109.7	63.0	81.7	112.4	64.7	83.8	109.7	64.7	82.7
09600	FLOORING	94.0	84.2	91.4	99.7	98.5	99.4	107.8	51.2	92.9	99.1	40.8	83.7	107.8	55.3	94.0	107.8	53.6	93.5
097,099	WALL FINISHES, PAINTS & COATINGS	92.9	80.2	85.2	100.9	108.7	105.6	99.2	61.9	76.6	99.2	42.1	64.6	99.2	47.4	67.8	99.2	52.6	71.0
09	FINISHES	96.3	84.5	90.2	94.9	104.1	99.7	106.6	61.9	83.4	101.8	56.0	78.0	106.7	61.9	83.4	106.0	62.0	83.1
10 - 14	TOTAL DIV. 10000 - 14000	100.0	87.3	97.3	100.0	101.1	100.2	100.0	72.0	94.0	100.0	70.8	93.8	100.0	73.3	94.3	100.0	59.2	91.3
15	MECHANICAL	100.1	90.2	95.8	100.1	104.4	102.0	100.4	77.8	90.5	100.4	70.5	87.3	100.4	79.6	91.3	100.4	74.7	89.1
16	ELECTRICAL	105.9	83.4	94.9	105.3	114.1	109.6	95.0	74.7	85.1	101.6	69.6	86.0	95.0	75.1	85.3	95.0	74.1	84.8
01 - 16	WEIGHTED AVERAGE	100.1	89.6	95.3	96.7	108.6	102.1	102.0	74.4	89.5	101.0	71.1	87.4	102.6	74.9	90.0	102.4	72.3	88.7

DIVISION		MONTANA			NEBRASKA												NEVADA		
		MISSOULA			GRAND ISLAND			LINCOLN			NORTH PLATTE			OMAHA			CARSON CITY		
		MAT.	INST.	TOTAL	MAT.	INST.	TOTAL	MAT.	INST.	TOTAL	MAT.	INST.	TOTAL	MAT.	INST.	TOTAL	MAT.	INST.	TOTAL
01590	EQUIPMENT RENTAL	.0	98.2	98.2	.0	101.6	101.6	.0	101.6	101.6	.0	101.6	101.6	.0	90.8	90.8	.0	101.7	101.7
02	SITE CONSTRUCTION	75.6	94.6	89.6	98.1	91.0	92.9	89.5	91.1	90.6	98.8	89.8	92.2	78.9	89.7	86.9	63.0	104.1	93.4
03100	CONCRETE FORMS & ACCESSORIES	89.1	59.0	63.1	96.6	50.7	56.9	101.3	46.8	54.1	96.5	47.3	53.8	96.1	72.4	75.5	94.3	103.0	101.8
03200	CONCRETE REINFORCEMENT	102.9	78.4	90.5	105.6	74.5	89.9	97.0	75.4	86.1	107.1	74.2	90.5	101.6	76.0	88.7	113.7	118.6	116.2
03300	CAST-IN-PLACE CONCRETE	90.2	65.1	79.7	115.5	58.7	91.8	104.3	61.2	86.3	115.5	52.5	89.2	108.3	75.9	94.8	111.6	85.8	100.8
03	CONCRETE	83.6	65.8	74.7	108.5	59.3	83.9	100.6	58.5	79.6	108.7	55.6	82.2	102.0	74.4	88.2	109.3	99.8	104.5
04	MASONRY	145.0	60.5	91.9	105.5	49.9	70.6	95.8	66.7	77.5	91.1	40.5	59.4	100.9	75.4	84.9	135.0	80.6	100.9
05	METALS	95.3	86.6	92.5	92.8	83.3	89.7	97.9	84.8	93.6	93.0	83.2	89.8	97.3	76.1	90.3	93.4	104.5	97.0
06	WOOD & PLASTICS	93.4	59.8	75.9	95.5	45.5	69.4	100.1	39.8	68.6	95.4	45.5	69.3	94.7	73.3	83.5	91.2	106.4	99.1
07	THERMAL & MOISTURE PROTECTION	98.3	70.9	84.8	96.9	57.5	77.5	97.3	60.3	79.1	96.8	47.4	72.5	92.5	70.4	81.6	102.6	89.4	96.1
08	DOORS & WINDOWS	95.9	62.5	87.2	90.0	50.1	79.7	95.5	49.7	83.6	89.3	53.2	80.0	98.8	67.0	90.6	94.4	112.7	99.1
09200	PLASTER & GYPSUM BOARD	96.9	59.1	72.9	100.0	43.8	64.4	104.4	37.9	62.3	100.5	43.8	64.5	106.9	73.0	85.4	81.3	106.4	97.2
095,098	CEILINGS & ACOUSTICAL TREATMENT	109.7	59.1	79.3	88.2	43.8	61.6	97.1	37.9	61.6	90.0	43.8	62.3	107.9	73.0	87.0	100.4	106.4	104.0
09600	FLOORING	101.3	63.7	91.3	94.6	36.3	79.2	96.4	43.7	82.5	94.5	37.8	79.5	122.7	48.4	103.1	101.2	66.7	92.1
097,099	WALL FINISHES, PAINTS & COATINGS	99.2	47.4	67.8	91.2	42.8	61.9	91.2	43.2	62.1	91.2	51.8	67.3	148.5	73.6	103.0	103.7	81.6	90.3
09	FINISHES	102.0	58.2	79.2	93.1	45.7	68.4	96.3	43.9	69.0	93.6	45.3	68.4	114.6	67.9	90.3	96.1	94.0	95.0
10 - 14	TOTAL DIV. 10000 - 14000	100.0	55.2	90.5	100.0	73.9	94.4	100.0	73.2	94.3	100.0	55.2	90.5	100.0	75.9	94.9	100.0	114.9	103.2
15	MECHANICAL	100.4	70.2	87.2	100.0	77.5	90.2	100.0	77.6	90.2	100.0	73.4	88.4	99.8	78.4	90.4	100.1	92.9	96.9
16	ELECTRICAL	99.7	72.3	86.3	91.2	66.6	79.2	96.6	66.6	81.9	93.0	44.0	69.1	90.7	82.1	86.5	88.6	88.0	88.3
01 - 16	WEIGHTED AVERAGE	98.7	70.3	85.8	97.2	66.2	83.1	98.0	67.7	84.2	96.8	59.9	80.0	99.1	76.7	89.0	98.4	95.6	97.2

	DIVISION	NEVADA						NEW HAMPSHIRE									NEW JERSEY		
		LAS VEGAS			RENO			MANCHESTER			NASHUA			PORTSMOUTH			CAMDEN		
		MAT.	INST.	TOTAL	MAT.	INST.	TOTAL	MAT.	INST.	TOTAL	MAT.	INST.	TOTAL	MAT.	INST.	TOTAL	MAT.	INST.	TOTAL
01590	EQUIPMENT RENTAL	.0	101.7	101.7	.0	101.7	101.7	.0	101.6	101.6	.0	101.6	101.6	.0	101.6	101.6	.0	99.4	99.4
02	SITE CONSTRUCTION	62.5	106.3	94.9	62.7	104.1	93.3	86.7	100.2	96.7	88.0	100.2	97.0	82.2	99.2	94.8	91.3	105.0	101.4
03100	CONCRETE FORMS & ACCESSORIES	93.2	107.9	105.9	94.3	102.9	101.8	100.9	63.9	68.9	101.4	63.9	68.9	88.8	59.4	63.3	101.4	123.2	120.3
03200	CONCRETE REINFORCEMENT	108.1	118.7	113.4	108.1	118.6	113.4	108.2	91.9	100.0	108.2	91.9	100.0	86.6	91.8	89.2	108.2	104.9	106.6
03300	CAST-IN-PLACE CONCRETE	105.3	108.6	106.7	113.7	85.8	102.1	101.0	97.4	99.5	92.8	97.4	94.7	88.0	90.9	89.2	80.7	128.7	100.7
03	CONCRETE	105.3	109.8	107.6	109.4	99.7	104.6	106.8	81.0	93.9	102.9	81.0	91.9	93.8	76.6	85.2	97.1	120.6	108.8
04	MASONRY	125.8	89.8	103.2	134.3	80.6	100.6	97.9	92.8	94.7	98.1	92.8	94.8	93.5	82.2	86.4	90.3	125.0	112.1
05	METALS	94.0	105.9	97.9	93.8	104.5	97.3	100.0	87.1	95.8	99.9	87.1	95.7	96.3	85.0	92.6	99.8	97.9	99.2
06	WOOD & PLASTICS	90.9	106.4	99.0	91.2	106.4	99.1	100.4	55.9	77.2	100.4	55.9	77.2	86.2	55.9	70.4	100.4	122.9	112.1
07	THERMAL & MOISTURE PROTECTION	105.1	98.7	102.0	102.6	89.4	96.1	101.6	89.0	95.4	102.1	89.0	95.6	101.5	92.1	96.9	101.4	122.1	111.6
08	DOORS & WINDOWS	94.7	112.7	99.4	94.7	111.1	99.0	107.1	65.4	96.3	107.1	65.4	96.3	108.0	59.4	95.4	107.1	113.8	108.8
09200	PLASTER & GYPSUM BOARD	81.9	106.4	97.5	83.2	106.4	97.9	105.2	53.8	72.6	105.2	53.8	72.6	94.0	53.8	68.5	105.2	122.6	116.2
095,098	CEILINGS & ACOUSTICAL TREATMENT	105.8	106.4	106.2	107.6	106.4	106.9	101.1	53.8	72.7	101.1	53.8	72.7	89.5	53.8	68.1	101.1	122.6	114.0
09600	FLOORING	101.2	79.5	95.4	101.2	66.7	92.1	100.0	103.7	101.0	99.8	103.7	100.8	94.4	103.7	96.8	99.8	132.6	108.5
097,099	WALL FINISHES, PAINTS & COATINGS	103.7	105.1	104.6	103.7	81.6	90.3	92.8	90.6	91.4	92.8	90.6	91.4	92.8	40.3	61.0	92.8	119.7	109.1
09	FINISHES	97.5	101.9	99.8	98.2	94.0	96.0	101.3	72.5	86.4	101.6	72.5	86.5	95.1	64.2	79.0	101.8	125.0	113.9
10 - 14	TOTAL DIV. 10000 - 14000	100.0	110.2	102.2	100.0	114.9	103.2	100.0	80.0	95.7	100.0	80.0	95.7	100.0	75.8	94.9	100.0	117.8	103.8
15	MECHANICAL	100.1	111.1	104.9	100.1	92.9	96.9	100.1	86.1	94.0	100.1	86.1	94.0	100.1	80.4	91.5	100.1	109.6	104.3
16	ELECTRICAL	90.5	105.5	97.8	88.6	88.0	88.3	100.4	74.5	87.8	100.2	74.5	87.7	98.5	74.5	86.8	100.6	117.6	108.9
01 - 16	WEIGHTED AVERAGE	98.1	105.5	101.5	98.7	95.5	97.3	101.2	82.8	92.9	100.9	82.8	92.7	98.0	78.4	89.1	100.0	115.2	106.9

	DIVISION	NEW JERSEY															NEW MEXICO		
		ELIZABETH			JERSEY CITY			NEWARK			PATERSON			TRENTON			ALBUQUERQUE		
		MAT.	INST.	TOTAL	MAT.	INST.	TOTAL	MAT.	INST.	TOTAL	MAT.	INST.	TOTAL	MAT.	INST.	TOTAL	MAT.	INST.	TOTAL
01590	EQUIPMENT RENTAL	.0	101.6	101.6	.0	99.4	99.4	.0	101.6	101.6	.0	101.6	101.6	.0	98.9	98.9	.0	116.3	116.3
02	SITE CONSTRUCTION	105.3	105.3	105.3	91.3	104.9	101.3	109.8	105.3	106.5	103.7	104.8	104.5	93.0	105.4	102.1	80.2	111.5	103.3
03100	CONCRETE FORMS & ACCESSORIES	111.5	122.7	121.2	101.4	123.7	120.7	99.8	122.7	119.7	100.4	122.6	119.6	100.3	123.4	120.3	96.1	69.0	72.6
03200	CONCRETE REINFORCEMENT	83.4	118.6	101.2	108.2	118.6	113.5	108.2	118.6	113.5	108.2	118.6	113.5	108.2	114.7	111.5	108.8	67.2	87.8
03300	CAST-IN-PLACE CONCRETE	88.9	128.5	105.4	80.7	130.0	101.3	94.2	128.6	108.5	102.8	128.5	113.6	95.5	119.9	105.7	104.5	76.4	92.8
03	CONCRETE	99.4	123.1	111.2	97.1	123.8	110.5	103.4	123.1	113.2	107.6	123.0	115.3	104.1	119.4	111.8	105.2	72.2	88.7
04	MASONRY	112.4	123.0	119.0	90.3	125.5	112.4	100.1	123.0	114.5	95.8	123.0	112.9	89.9	116.0	106.3	122.3	66.3	87.2
05	METALS	96.0	108.6	100.1	99.9	105.6	101.7	99.8	108.6	102.7	94.7	108.5	99.2	94.6	101.2	96.8	100.2	88.0	96.2
06	WOOD & PLASTICS	115.3	122.9	119.3	100.4	122.9	112.1	102.2	122.9	113.0	102.2	122.9	113.0	100.4	122.8	112.1	95.7	70.1	82.4
07	THERMAL & MOISTURE PROTECTION	102.4	126.0	114.0	101.4	127.1	114.1	101.7	126.0	113.7	102.2	119.7	110.8	100.0	116.0	107.9	101.6	72.6	87.3
08	DOORS & WINDOWS	108.0	118.2	110.7	107.1	118.2	109.9	113.1	118.2	114.4	113.1	118.2	114.4	107.1	116.4	109.5	94.7	72.5	89.0
09200	PLASTER & GYPSUM BOARD	108.9	122.6	117.6	105.2	122.6	116.2	105.2	122.6	116.2	105.2	122.6	116.2	105.2	122.6	116.2	85.4	68.8	74.8
095,098	CEILINGS & ACOUSTICAL TREATMENT	89.5	122.6	109.3	101.1	122.6	114.0	101.1	122.6	114.0	101.1	122.6	114.0	101.1	122.6	114.0	105.8	68.8	83.6
09600	FLOORING	104.5	132.6	111.9	99.8	132.6	108.5	100.0	132.6	108.6	99.8	132.6	108.5	100.0	117.0	104.5	101.2	65.0	91.6
097,099	WALL FINISHES, PAINTS & COATINGS	92.7	119.7	109.1	92.8	119.7	109.1	92.7	119.7	109.1	92.7	119.7	109.1	92.8	119.7	109.1	103.7	60.7	77.6
09	FINISHES	101.6	124.1	113.3	101.8	124.8	113.8	102.4	124.1	113.7	102.0	124.1	113.5	101.9	121.8	112.3	98.5	67.2	82.2
10 - 14	TOTAL DIV. 10000 - 14000	100.0	124.0	105.1	100.0	125.1	105.3	100.0	124.1	105.1	100.0	124.0	105.1	100.0	117.7	103.8	100.0	76.4	95.0
15	MECHANICAL	100.1	121.4	109.4	100.1	126.8	111.8	100.1	122.5	109.9	100.1	125.4	111.2	100.1	126.1	111.5	100.1	72.9	88.2
16	ELECTRICAL	98.5	128.9	113.3	102.2	132.5	117.0	102.1	132.5	116.9	102.2	128.9	115.2	100.8	129.8	114.9	81.8	76.3	79.1
01 - 16	WEIGHTED AVERAGE	101.1	120.7	110.0	100.2	122.5	110.3	102.5	121.4	111.1	101.8	121.2	110.6	99.9	119.1	108.6	98.3	76.8	88.6

	DIVISION	NEW MEXICO												NEW YORK					
		FARMINGTON			LAS CRUCES			ROSWELL			SANTA FE			ALBANY			BINGHAMTON		
		MAT.	INST.	TOTAL	MAT.	INST.	TOTAL	MAT.	INST.	TOTAL	MAT.	INST.	TOTAL	MAT.	INST.	TOTAL	MAT.	INST.	TOTAL
01590	EQUIPMENT RENTAL	.0	116.3	116.3	.0	86.7	86.7	.0	116.3	116.3	.0	116.3	116.3	.0	115.7	115.7	.0	115.8	115.8
02	SITE CONSTRUCTION	85.7	111.5	104.8	91.2	86.7	87.9	90.3	111.4	105.9	79.9	111.5	103.3	71.8	106.5	97.5	93.8	89.8	90.9
03100	CONCRETE FORMS & ACCESSORIES	96.1	69.0	72.6	93.5	67.2	70.7	96.1	68.9	72.5	96.1	69.0	72.6	98.7	92.1	93.0	103.9	76.5	80.2
03200	CONCRETE REINFORCEMENT	118.7	67.2	92.7	112.8	58.8	85.5	117.9	59.3	88.3	116.6	67.2	91.6	103.1	89.8	96.4	102.0	87.8	94.8
03300	CAST-IN-PLACE CONCRETE	105.1	76.4	93.1	91.4	66.7	81.1	96.8	76.4	88.3	98.8	76.4	89.4	88.7	100.5	93.6	105.9	86.7	97.9
03	CONCRETE	109.0	72.2	90.6	87.3	66.2	76.7	109.3	70.7	90.0	103.8	72.2	88.0	101.3	95.5	98.4	101.1	84.2	92.6
04	MASONRY	127.3	66.3	89.1	110.9	60.7	79.4	127.4	66.3	89.1	119.4	66.3	86.1	89.3	97.0	94.1	106.0	79.0	89.1
05	METALS	97.9	88.0	94.7	97.8	75.2	90.4	97.9	84.1	93.4	97.9	88.0	94.6	96.9	104.9	99.5	93.4	114.1	100.2
06	WOOD & PLASTICS	95.7	70.1	82.4	87.4	68.7	77.6	95.7	70.1	82.4	95.7	70.1	82.4	98.3	91.3	94.6	107.3	75.0	90.4
07	THERMAL & MOISTURE PROTECTION	102.5	72.6	87.8	88.7	65.5	77.3	102.8	72.6	87.9	101.9	72.6	87.5	90.8	92.8	91.8	101.5	80.3	91.1
08	DOORS & WINDOWS	97.6	72.5	91.1	87.1	69.2	82.5	93.6	70.0	87.5	93.7	72.5	88.2	95.7	85.0	92.9	90.7	73.4	86.2
09200	PLASTER & GYPSUM BOARD	80.4	68.8	73.0	82.7	68.8	73.9	80.4	68.8	73.0	80.4	68.8	73.0	111.0	90.8	98.2	116.3	73.7	89.3
095,098	CEILINGS & ACOUSTICAL TREATMENT	96.9	68.8	80.0	96.8	68.8	80.0	96.9	68.8	80.0	96.9	68.8	80.0	99.8	90.8	94.4	99.8	73.7	84.2
09600	FLOORING	101.2	65.0	91.6	133.4	65.2	115.4	101.2	65.0	91.6	101.2	65.0	91.6	86.8	94.1	88.7	97.6	80.2	93.0
097,099	WALL FINISHES, PAINTS & COATINGS	103.7	60.7	77.6	97.1	60.7	75.0	103.7	60.7	77.6	103.7	60.7	77.6	84.9	78.0	80.7	90.1	79.1	83.4
09	FINISHES	95.9	67.2	81.0	111.0	66.2	87.7	96.3	67.2	81.2	95.7	67.2	80.9	97.5	91.0	94.1	98.8	77.0	87.5
10 - 14	TOTAL DIV. 10000 - 14000	100.0	76.4	95.0	100.0	72.3	94.1	100.0	76.4	95.0	100.0	76.4	95.0	100.0	87.3	97.3	100.0	89.3	97.7
15	MECHANICAL	100.1	72.9	88.2	100.3	72.4	88.1	100.1	72.8	88.2	100.1	72.9	88.2	100.4	88.9	95.3	100.6	82.4	92.6
16	ELECTRICAL	80.2	76.3	78.3	82.5	60.4	71.7	80.9	76.3	78.7	81.8	76.3	79.1	103.1	89.4	96.4	101.1	80.2	90.9
01 - 16	WEIGHTED AVERAGE	98.6	76.8	88.7	95.8	69.1	83.7	98.5	76.1	88.4	97.3	76.8	88.0	97.9	93.9	96.1	98.6	84.5	92.2

	DIVISION	NEW YORK																	
		BUFFALO			HICKSVILLE			NEW YORK			RIVERHEAD			ROCHESTER			SCHENECTADY		
		MAT.	INST.	TOTAL	MAT.	INST.	TOTAL	MAT.	INST.	TOTAL	MAT.	INST.	TOTAL	MAT.	INST.	TOTAL	MAT.	INST.	TOTAL
01590	**EQUIPMENT RENTAL**	.0	93.1	93.1	.0	117.2	117.2	.0	117.6	117.6	.0	117.2	117.2	.0	116.5	116.5	.0	115.7	115.7
02	**SITE CONSTRUCTION**	96.1	93.4	94.1	112.1	130.3	125.6	134.2	128.5	129.9	112.3	130.4	125.7	73.9	107.2	98.6	71.5	106.3	97.3
03100	CONCRETE FORMS & ACCESSORIES	100.3	113.7	111.9	90.5	149.8	141.9	113.8	180.5	171.6	94.6	149.9	142.5	100.2	94.2	95.0	103.5	91.0	92.7
03200	CONCRETE REINFORCEMENT	104.0	105.9	105.0	103.3	184.3	144.2	111.2	187.4	149.7	105.3	184.3	145.2	105.0	86.5	95.7	101.8	89.8	95.7
03300	CAST-IN-PLACE CONCRETE	118.7	119.3	118.9	99.1	151.6	121.0	121.6	160.8	137.9	97.2	151.7	120.0	112.9	100.5	107.8	97.3	99.0	98.0
03	**CONCRETE**	108.2	113.3	110.7	104.6	155.5	130.0	120.2	172.8	146.5	103.7	155.5	129.6	113.6	96.0	104.8	105.5	94.5	100.0
04	**MASONRY**	104.9	119.8	114.2	109.5	152.4	136.4	105.3	161.1	140.3	116.0	152.4	138.9	103.0	98.0	99.8	90.8	94.1	92.9
05	**METALS**	100.4	94.8	98.5	103.5	137.5	114.6	108.1	140.9	118.8	103.6	137.9	114.8	96.8	107.3	100.2	97.0	104.9	99.6
06	**WOOD & PLASTICS**	104.3	112.9	108.8	88.5	151.2	121.2	111.4	187.9	151.4	92.4	151.2	123.1	101.1	93.8	97.3	103.9	91.3	97.3
07	**THERMAL & MOISTURE PROTECTION**	98.9	110.7	104.7	104.0	145.1	124.2	105.5	160.3	132.4	104.0	145.1	124.2	95.6	96.9	96.2	91.0	91.6	91.3
08	**DOORS & WINDOWS**	92.7	103.3	95.5	88.3	153.6	105.2	97.7	176.6	118.1	88.4	153.6	105.2	96.0	86.2	93.5	95.7	85.0	92.9
09200	PLASTER & GYPSUM BOARD	102.4	113.0	109.1	102.1	152.7	134.2	115.8	189.9	162.8	103.0	152.7	134.5	104.6	93.6	97.6	111.0	90.8	98.2
095,098	CEILINGS & ACOUSTICAL TREATMENT	94.6	113.0	105.6	79.3	152.7	123.3	109.4	189.9	157.7	79.3	152.7	123.3	99.1	93.6	95.8	99.8	90.8	94.4
09600	FLOORING	97.7	121.0	103.9	97.4	90.1	95.5	97.4	171.3	117.0	98.7	90.1	96.5	86.8	85.8	86.6	86.8	94.1	88.7
097,099	WALL FINISHES, PAINTS & COATINGS	94.8	114.8	106.9	112.6	146.7	133.3	97.2	150.2	129.3	112.6	146.7	133.3	92.5	98.4	96.1	84.9	78.0	80.7
09	**FINISHES**	95.7	115.9	106.2	104.0	137.5	121.4	108.9	176.8	144.2	104.4	137.5	121.6	96.8	93.1	94.9	97.3	90.3	93.6
10 - 14	**TOTAL DIV. 10000 - 14000**	100.0	108.0	101.7	100.0	120.4	104.3	100.0	151.7	111.0	100.0	120.4	104.3	100.0	91.5	98.2	100.0	86.2	97.1
15	**MECHANICAL**	99.8	97.4	98.8	99.8	144.9	119.6	100.3	163.2	127.8	99.8	144.9	119.6	99.8	88.7	95.0	100.4	87.4	94.7
16	**ELECTRICAL**	99.7	97.6	98.7	101.6	152.1	126.3	107.8	178.5	142.3	102.2	152.1	126.6	103.4	89.1	96.4	100.5	89.4	95.1
01 - 16	**WEIGHTED AVERAGE**	99.9	104.7	102.1	101.2	145.0	121.1	106.6	163.5	132.4	101.6	145.1	121.4	100.1	94.9	97.7	98.2	93.0	95.8

	DIVISION	NEW YORK															NORTH CAROLINA		
		SYRACUSE			UTICA			WATERTOWN			WHITE PLAINS			YONKERS			ASHEVILLE		
		MAT.	INST.	TOTAL	MAT.	INST.	TOTAL	MAT.	INST.	TOTAL	MAT.	INST.	TOTAL	MAT.	INST.	TOTAL	MAT.	INST.	TOTAL
01590	**EQUIPMENT RENTAL**	.0	115.7	115.7	.0	115.7	115.7	.0	115.7	115.7	.0	117.3	117.3	.0	117.3	117.3	.0	93.3	93.3
02	**SITE CONSTRUCTION**	92.6	106.6	103.0	69.6	105.6	96.2	77.5	107.6	99.8	123.8	125.2	124.9	133.5	125.1	127.3	101.6	72.3	79.9
03100	CONCRETE FORMS & ACCESSORIES	102.7	87.4	89.5	104.0	83.3	86.1	85.8	90.6	90.0	113.4	136.1	133.0	113.6	136.3	133.3	94.3	43.0	49.8
03200	CONCRETE REINFORCEMENT	103.1	88.4	95.7	103.1	86.9	94.9	103.7	88.4	96.0	104.0	183.5	144.2	108.1	183.5	146.2	95.1	39.5	67.0
03300	CAST-IN-PLACE CONCRETE	98.3	93.7	96.4	89.9	91.2	90.4	104.7	85.7	96.8	107.1	134.2	118.4	119.9	134.3	125.9	96.3	47.3	75.8
03	**CONCRETE**	104.4	90.8	97.6	102.3	87.8	95.1	115.2	89.5	102.4	108.1	143.1	125.6	119.0	143.3	131.1	99.5	45.8	72.7
04	**MASONRY**	96.9	92.4	94.1	89.4	89.1	89.2	90.6	94.0	92.7	100.0	134.1	121.4	104.8	134.1	123.2	79.1	37.1	52.7
05	**METALS**	96.8	104.2	99.3	94.9	103.2	97.6	94.9	103.9	97.9	94.8	134.3	107.7	104.0	134.6	114.0	93.0	74.5	87.0
06	**WOOD & PLASTICS**	103.9	86.8	95.0	103.9	82.8	92.9	84.1	91.8	88.2	112.3	135.7	124.5	112.2	135.7	124.5	93.1	44.6	67.8
07	**THERMAL & MOISTURE PROTECTION**	99.1	93.2	96.2	91.0	92.0	91.5	91.2	93.6	92.4	106.2	137.5	121.6	106.6	137.5	121.8	96.8	42.2	70.0
08	**DOORS & WINDOWS**	93.6	80.7	90.3	95.7	77.8	91.0	95.7	82.0	92.1	92.2	145.1	105.9	95.8	147.6	109.2	92.2	41.3	79.0
09200	PLASTER & GYPSUM BOARD	111.0	86.2	95.3	111.0	82.0	92.6	102.6	91.4	95.5	108.4	136.3	126.1	115.3	136.3	128.6	100.7	42.9	64.0
095,098	CEILINGS & ACOUSTICAL TREATMENT	99.8	86.2	91.6	99.8	82.0	89.1	99.8	91.4	94.7	79.9	136.3	113.7	107.6	136.3	124.8	91.9	42.9	62.5
09600	FLOORING	88.2	89.4	88.6	86.8	89.5	87.5	79.2	89.5	81.9	93.6	158.9	110.9	93.2	158.9	110.6	101.3	41.4	85.5
097,099	WALL FINISHES, PAINTS & COATINGS	90.3	86.8	88.2	84.9	84.6	84.8	84.9	73.4	78.0	94.8	146.7	126.3	94.8	146.7	126.3	108.5	39.1	66.4
09	**FINISHES**	98.6	87.6	92.9	97.3	84.3	90.5	94.3	88.3	91.2	98.7	141.6	121.0	107.0	141.6	125.0	96.4	42.3	68.3
10 - 14	**TOTAL DIV. 10000 - 14000**	100.0	97.7	99.5	100.0	84.1	96.6	100.0	83.7	96.5	100.0	117.6	103.7	100.0	141.6	108.9	100.0	62.5	92.0
15	**MECHANICAL**	100.4	84.3	93.3	100.4	80.8	91.8	100.4	77.0	90.1	100.6	127.0	112.1	100.6	127.0	112.1	100.0	43.1	75.1
16	**ELECTRICAL**	100.5	82.5	91.7	97.1	81.7	89.6	100.5	79.0	90.0	97.3	143.1	119.6	104.3	143.1	123.2	97.9	40.4	69.9
01 - 16	**WEIGHTED AVERAGE**	99.0	90.5	95.2	97.0	87.8	92.8	98.7	88.4	94.0	99.9	135.4	116.0	105.1	136.2	119.2	96.3	48.3	74.5

	DIVISION	NORTH CAROLINA																	
		CHARLOTTE			DURHAM			FAYETTEVILLE			GREENSBORO			RALEIGH			WILMINGTON		
		MAT.	INST.	TOTAL	MAT.	INST.	TOTAL	MAT.	INST.	TOTAL	MAT.	INST.	TOTAL	MAT.	INST.	TOTAL	MAT.	INST.	TOTAL
01590	**EQUIPMENT RENTAL**	.0	93.3	93.3	.0	99.7	99.7	.0	99.7	99.7	.0	99.7	99.7	.0	99.7	99.7	.0	93.3	93.3
02	**SITE CONSTRUCTION**	101.8	72.3	80.0	101.6	82.2	87.2	99.7	82.2	86.7	101.5	82.2	87.2	102.6	82.2	87.5	102.5	72.3	80.2
03100	CONCRETE FORMS & ACCESSORIES	101.7	43.0	50.9	96.2	43.3	50.4	92.0	43.3	49.8	96.3	43.3	50.4	99.0	43.3	50.7	95.8	43.3	50.3
03200	CONCRETE REINFORCEMENT	95.5	39.5	67.2	95.5	54.7	74.9	94.5	54.7	74.4	95.5	54.7	74.9	95.5	54.7	74.9	95.8	54.7	75.0
03300	CAST-IN-PLACE CONCRETE	98.7	47.4	77.3	96.5	47.5	76.0	93.2	47.5	74.1	95.7	47.5	75.6	102.2	47.5	79.3	95.9	47.5	75.6
03	**CONCRETE**	100.5	45.8	73.2	99.1	48.8	74.0	96.7	48.8	72.8	98.7	48.8	73.8	102.0	48.8	75.4	99.4	48.8	74.2
04	**MASONRY**	84.0	37.1	54.6	80.8	37.1	53.4	84.4	37.1	54.7	80.6	37.1	53.3	85.5	37.1	55.1	69.7	37.1	49.2
05	**METALS**	95.5	74.5	88.7	98.1	81.1	92.6	98.1	81.1	92.5	98.7	81.1	92.9	95.9	81.1	91.1	92.5	81.1	88.8
06	**WOOD & PLASTICS**	102.0	44.6	72.0	95.4	44.6	68.9	90.4	44.6	66.5	95.4	44.6	68.9	98.6	44.6	70.4	94.9	44.6	68.6
07	**THERMAL & MOISTURE PROTECTION**	96.9	43.2	70.5	97.5	42.8	70.6	97.2	42.8	70.4	97.5	42.8	70.6	97.3	42.8	70.6	96.9	42.8	70.3
08	**DOORS & WINDOWS**	96.3	41.3	82.1	96.3	47.1	83.6	92.3	47.1	80.6	96.3	47.1	83.6	93.0	47.1	81.1	92.3	47.1	80.6
09200	PLASTER & GYPSUM BOARD	107.2	42.9	66.4	107.2	42.9	66.4	101.0	42.9	64.1	107.2	42.9	66.4	107.2	42.9	66.4	101.8	42.9	64.4
095,098	CEILINGS & ACOUSTICAL TREATMENT	98.1	42.9	65.0	98.1	42.9	65.0	92.8	42.9	62.8	98.1	42.9	65.0	98.1	42.9	65.0	92.8	42.9	62.8
09600	FLOORING	105.1	41.4	88.3	105.3	41.4	88.4	101.4	41.4	85.6	105.3	41.4	88.4	105.3	41.4	88.4	102.3	41.4	86.2
097,099	WALL FINISHES, PAINTS & COATINGS	108.5	39.1	66.4	108.5	39.1	66.4	108.5	39.1	66.4	108.5	39.1	66.4	108.5	39.1	66.4	108.5	39.1	66.4
09	**FINISHES**	100.0	42.3	70.0	100.1	42.3	70.0	96.7	42.3	68.4	100.1	42.3	70.0	100.1	42.3	70.1	97.0	42.3	68.6
10 - 14	**TOTAL DIV. 10000 - 14000**	100.0	62.5	92.0	100.0	65.2	92.6	100.0	65.2	92.6	100.0	62.5	92.0	100.0	65.2	92.6	100.0	65.2	92.6
15	**MECHANICAL**	100.0	43.1	75.1	100.0	43.3	75.2	100.0	43.3	75.2	100.0	43.3	75.2	100.0	43.3	75.2	100.0	43.3	75.2
16	**ELECTRICAL**	99.7	36.3	68.8	97.4	38.6	68.8	94.8	38.6	67.4	98.0	38.6	69.1	98.0	38.6	69.1	98.5	38.6	69.3
01 - 16	**WEIGHTED AVERAGE**	98.2	47.8	75.3	97.9	50.3	76.3	96.7	50.3	75.6	98.0	50.3	76.3	97.9	50.3	76.3	96.0	49.4	74.8

DIVISION	NORTH CAROLINA			NORTH DAKOTA												OHIO		
	WINSTON-SALEM			BISMARCK			FARGO			GRAND FORKS			MINOT			AKRON		
	MAT.	INST.	TOTAL	MAT.	INST.	TOTAL	MAT.	INST.	TOTAL	MAT.	INST.	TOTAL	MAT.	INST.	TOTAL	MAT.	INST.	TOTAL
01590 EQUIPMENT RENTAL	.0	99.7	99.7	.0	98.2	98.2	.0	98.2	98.2	.0	98.2	98.2	.0	98.2	98.2	.0	97.4	97.4
02 SITE CONSTRUCTION	101.9	82.2	87.3	88.4	96.0	94.0	87.8	96.0	93.9	96.3	94.1	94.7	93.9	96.0	95.5	100.8	106.8	105.2
03100 CONCRETE FORMS & ACCESSORIES	97.6	43.1	50.4	96.4	51.0	57.0	97.4	51.4	57.5	93.1	44.8	51.3	88.1	57.8	61.9	98.1	99.0	98.9
03200 CONCRETE REINFORCEMENT	95.5	39.5	67.2	101.2	75.5	88.2	93.3	74.6	83.8	99.6	81.6	90.5	103.1	75.6	89.2	94.1	98.0	96.1
03300 CAST-IN-PLACE CONCRETE	98.7	47.4	77.3	104.9	56.3	84.6	109.9	58.6	88.5	107.4	54.4	85.2	107.4	54.9	85.5	99.1	106.2	102.0
03 CONCRETE	100.2	45.8	73.1	99.1	58.5	78.9	109.5	59.3	84.4	105.1	56.1	80.6	103.7	61.1	82.4	99.4	100.2	99.8
04 MASONRY	80.8	37.1	53.4	104.0	66.9	80.7	106.3	50.0	71.0	105.6	66.2	80.9	106.7	68.2	82.5	90.0	102.4	97.8
05 METALS	95.9	74.6	88.9	92.0	81.2	88.5	94.4	79.8	89.6	92.0	81.0	88.4	92.3	81.3	88.7	88.2	82.6	86.3
06 WOOD & PLASTICS	95.4	44.6	68.9	88.9	46.2	66.6	88.9	46.8	66.9	85.3	42.4	62.9	80.3	55.4	67.3	92.2	97.8	95.2
07 THERMAL & MOISTURE PROTECTION	97.5	42.8	70.6	100.5	56.9	79.1	101.2	54.2	78.1	101.2	57.1	79.5	100.9	58.1	79.9	104.0	100.4	102.2
08 DOORS & WINDOWS	96.3	43.1	82.5	99.5	50.5	86.8	99.4	50.9	86.9	99.5	47.1	85.9	99.6	55.6	88.2	104.2	98.9	102.8
09200 PLASTER & GYPSUM BOARD	107.2	42.9	66.4	115.0	45.0	70.6	115.0	45.7	71.1	113.0	41.2	67.5	111.4	54.5	75.4	95.2	97.3	96.6
095,098 CEILINGS & ACOUSTICAL TREATMENT	98.1	42.9	65.0	134.3	45.0	80.7	134.3	45.7	81.2	134.3	41.2	78.5	134.3	54.5	86.5	91.9	97.3	95.2
09600 FLOORING	105.3	41.4	88.4	108.4	78.4	100.4	108.2	46.9	91.9	106.6	46.9	90.8	103.9	81.0	97.8	103.2	91.7	100.2
097,099 WALL FINISHES, PAINTS & COATINGS	108.5	39.1	66.4	99.2	38.9	62.6	99.2	73.6	83.7	99.2	29.7	57.1	99.2	32.6	58.8	107.1	122.7	116.5
09 FINISHES	100.1	42.3	70.0	113.9	53.8	82.6	113.7	51.7	81.4	113.6	42.6	76.7	112.3	59.0	84.6	99.5	100.2	99.8
10 - 14 TOTAL DIV. 10000 - 14000	100.0	62.5	92.0	100.0	74.2	94.5	100.0	74.2	94.5	100.0	48.3	89.0	100.0	75.4	94.8	100.0	92.6	98.4
15 MECHANICAL	100.0	43.1	75.1	100.6	64.2	84.7	100.6	69.7	87.0	100.6	46.9	77.1	100.6	61.8	83.6	99.8	101.4	100.5
16 ELECTRICAL	98.0	38.6	69.1	94.4	70.1	82.5	94.2	69.8	82.3	96.8	71.4	84.4	99.7	71.4	85.9	97.9	90.5	94.3
01 - 16 WEIGHTED AVERAGE	97.8	49.1	75.7	99.2	66.9	84.5	100.8	65.9	85.0	100.4	60.7	82.4	100.5	68.2	85.8	97.7	98.0	97.8

DIVISION	OHIO																	
	CANTON			CINCINNATI			CLEVELAND			COLUMBUS			DAYTON			LORAIN		
	MAT.	INST.	TOTAL	MAT.	INST.	TOTAL	MAT.	INST.	TOTAL	MAT.	INST.	TOTAL	MAT.	INST.	TOTAL	MAT.	INST.	TOTAL
01590 EQUIPMENT RENTAL	.0	97.4	97.4	.0	102.5	102.5	.0	97.7	97.7	.0	96.4	96.4	.0	96.8	96.8	.0	97.4	97.4
02 SITE CONSTRUCTION	100.9	106.9	105.4	74.9	108.7	99.9	100.8	106.8	105.3	86.5	101.8	97.8	73.7	108.2	99.2	100.2	106.0	104.5
03100 CONCRETE FORMS & ACCESSORIES	98.1	86.1	87.7	95.9	90.9	91.6	98.3	105.6	104.6	98.6	86.7	88.3	95.8	80.2	82.3	98.2	106.2	105.1
03200 CONCRETE REINFORCEMENT	94.1	84.1	89.0	98.7	86.7	92.6	94.6	98.3	96.5	76.1	86.5	81.4	98.7	82.1	90.3	94.1	98.1	96.1
03300 CAST-IN-PLACE CONCRETE	100.1	103.4	101.5	84.8	91.5	87.6	97.2	114.1	104.3	89.6	93.7	91.3	77.3	89.2	82.3	94.3	108.9	100.4
03 CONCRETE	99.8	91.0	95.4	92.9	90.4	91.6	98.6	106.1	102.3	92.2	88.7	90.5	89.3	83.4	86.4	97.1	104.4	100.8
04 MASONRY	90.7	93.3	92.4	84.9	95.9	91.8	94.2	110.6	104.5	95.4	93.8	94.4	84.3	85.3	84.9	87.0	100.5	95.5
05 METALS	88.2	76.0	84.2	86.8	89.6	87.7	89.6	85.8	88.3	96.5	82.0	91.8	86.1	77.3	83.3	88.7	83.5	87.0
06 WOOD & PLASTICS	92.6	84.1	88.2	96.4	89.5	92.8	91.5	102.4	97.2	104.3	84.8	94.1	97.5	78.2	87.4	92.2	108.3	100.6
07 THERMAL & MOISTURE PROTECTION	104.7	95.9	100.3	88.9	97.8	93.3	102.9	114.0	108.3	99.4	95.6	97.5	94.5	87.6	91.1	104.6	108.1	106.3
08 DOORS & WINDOWS	98.6	77.9	93.3	95.9	87.6	93.8	95.0	101.3	96.7	99.5	84.1	95.5	96.2	78.1	91.5	98.6	104.6	100.2
09200 PLASTER & GYPSUM BOARD	96.0	83.2	87.9	101.2	89.5	93.8	94.3	102.0	99.2	99.3	84.2	89.8	101.2	77.9	86.5	95.2	108.1	103.4
095,098 CEILINGS & ACOUSTICAL TREATMENT	91.9	83.2	86.7	95.5	89.5	91.9	90.1	102.0	97.2	96.2	84.2	89.0	96.5	77.9	85.3	91.9	108.1	101.6
09600 FLOORING	103.4	83.5	98.1	107.2	101.8	105.8	103.0	112.2	105.4	97.6	91.0	95.9	110.0	84.0	103.1	103.4	109.5	105.0
097,099 WALL FINISHES, PAINTS & COATINGS	107.1	85.8	94.2	104.8	96.1	99.5	107.1	122.4	116.4	99.3	99.6	99.5	104.8	92.2	97.1	107.1	122.4	116.4
09 FINISHES	99.7	84.6	91.8	98.1	93.4	95.6	98.9	108.3	103.8	98.9	88.4	93.5	99.1	81.6	90.0	99.5	109.2	104.6
10 - 14 TOTAL DIV. 10000 - 14000	100.0	80.1	95.8	100.0	94.7	98.9	100.0	108.1	101.7	100.0	94.5	98.8	100.0	91.6	98.2	100.0	105.5	101.2
15 MECHANICAL	99.8	85.4	93.5	99.8	91.6	96.2	99.8	108.1	103.5	99.9	92.6	96.7	100.7	84.4	93.6	99.8	89.4	95.3
16 ELECTRICAL	97.1	90.1	93.7	99.5	83.0	91.4	98.0	109.7	103.7	99.7	87.2	93.6	97.4	80.7	89.3	97.2	90.8	94.1
01 - 16 WEIGHTED AVERAGE	97.2	88.5	93.2	94.7	92.3	93.6	97.1	106.0	101.1	97.8	90.5	94.5	94.4	84.9	90.1	96.8	98.1	97.4

DIVISION	OHIO									OKLAHOMA								
	SPRINGFIELD			TOLEDO			YOUNGSTOWN			ENID			LAWTON			MUSKOGEE		
	MAT.	INST.	TOTAL	MAT.	INST.	TOTAL	MAT.	INST.	TOTAL	MAT.	INST.	TOTAL	MAT.	INST.	TOTAL	MAT.	INST.	TOTAL
01590 EQUIPMENT RENTAL	.0	96.8	96.8	.0	99.4	99.4	.0	97.4	97.4	.0	77.1	77.1	.0	78.2	78.2	.0	86.8	86.8
02 SITE CONSTRUCTION	74.2	107.3	98.7	85.8	102.9	98.4	100.7	107.2	105.5	108.8	89.1	94.3	104.2	90.8	94.3	94.8	85.2	87.7
03100 CONCRETE FORMS & ACCESSORIES	95.8	85.1	86.6	98.6	98.5	98.5	98.1	92.4	93.2	93.7	40.4	47.5	97.5	55.4	61.0	99.0	36.6	44.9
03200 CONCRETE REINFORCEMENT	98.7	82.1	90.3	76.1	95.6	85.9	94.1	98.0	96.0	96.7	82.7	89.6	96.9	82.7	89.7	96.7	39.3	67.7
03300 CAST-IN-PLACE CONCRETE	81.0	89.8	84.7	89.6	108.4	97.5	98.2	105.1	101.1	92.6	51.6	75.5	89.6	51.7	73.7	83.3	39.7	65.1
03 CONCRETE	91.1	85.8	88.4	92.2	100.9	96.5	98.9	97.0	97.9	90.6	52.4	71.5	87.3	59.1	73.2	83.1	39.3	61.3
04 MASONRY	84.6	86.4	85.7	105.3	104.7	104.9	90.0	98.1	95.1	103.5	60.2	76.3	97.5	60.2	74.1	113.4	43.2	69.4
05 METALS	86.1	77.2	83.2	96.3	89.5	94.1	88.2	83.0	86.5	89.2	67.3	82.0	92.6	67.4	84.4	89.1	59.1	79.3
06 WOOD & PLASTICS	98.7	84.4	91.2	104.3	97.2	100.6	92.2	90.2	91.2	97.4	37.5	66.1	100.4	57.6	78.1	102.4	37.1	68.3
07 THERMAL & MOISTURE PROTECTION	94.4	88.7	91.6	101.0	109.3	105.1	104.8	100.9	102.9	99.8	60.3	80.4	99.5	62.4	81.2	99.3	45.5	72.9
08 DOORS & WINDOWS	94.2	81.4	90.9	97.5	93.6	96.5	98.6	96.3	98.0	95.6	51.5	84.2	97.2	62.5	88.2	95.6	36.6	80.4
09200 PLASTER & GYPSUM BOARD	101.2	84.2	90.5	99.3	97.0	97.8	95.2	89.5	91.6	82.7	36.6	53.5	85.4	57.3	67.6	86.6	36.1	54.5
095,098 CEILINGS & ACOUSTICAL TREATMENT	96.5	84.2	89.1	96.2	97.0	96.7	91.9	89.5	90.4	81.4	36.6	54.6	91.2	57.3	70.9	91.2	36.1	58.1
09600 FLOORING	110.0	84.0	103.1	96.8	100.8	97.8	103.4	97.9	102.0	112.9	53.7	97.2	115.8	53.7	99.4	117.1	43.3	97.5
097,099 WALL FINISHES, PAINTS & COATINGS	104.8	92.2	97.1	99.4	110.4	106.1	107.1	103.6	105.0	96.4	57.1	72.6	96.4	57.1	72.6	96.4	36.3	59.9
09 FINISHES	99.1	85.5	92.1	98.6	100.1	99.4	99.6	94.0	96.7	94.6	42.8	67.6	97.8	54.6	75.3	97.7	37.7	66.5
10 - 14 TOTAL DIV. 10000 - 14000	100.0	92.9	98.5	100.0	93.4	98.6	100.0	91.9	98.3	100.0	63.9	92.3	100.0	66.5	92.9	100.0	63.3	92.2
15 MECHANICAL	100.7	85.0	93.8	99.9	99.9	99.9	99.8	90.7	95.8	100.1	63.7	84.2	100.1	63.7	84.2	100.1	30.3	69.5
16 ELECTRICAL	97.4	84.0	90.9	99.5	99.4	99.4	97.2	95.4	96.3	95.3	68.5	82.3	96.8	68.5	83.0	95.0	35.1	65.8
01 - 16 WEIGHTED AVERAGE	94.4	86.7	90.9	98.0	99.6	98.7	97.1	94.8	96.0	96.1	61.6	80.4	96.5	65.0	82.2	95.7	43.7	72.1

City Cost Indexes

	DIVISION	OKLAHOMA						OREGON											
		OKLAHOMA CITY			TULSA			EUGENE			MEDFORD			PORTLAND			SALEM		
		MAT.	INST.	TOTAL	MAT.	INST.	TOTAL	MAT.	INST.	TOTAL	MAT.	INST.	TOTAL	MAT.	INST.	TOTAL	MAT.	INST.	TOTAL
01590	**EQUIPMENT RENTAL**	.0	78.5	78.5	.0	86.8	86.8	.0	99.5	99.5	.0	99.5	99.5	.0	99.5	99.5	.0	99.5	99.5
02	**SITE CONSTRUCTION**	105.5	91.4	95.0	101.2	86.2	90.1	108.8	104.8	105.8	118.4	104.8	108.3	109.9	104.8	106.1	108.4	104.8	105.7
03100	CONCRETE FORMS & ACCESSORIES	98.6	47.7	54.5	98.7	48.2	54.9	103.5	105.8	105.5	98.3	105.6	104.6	104.9	106.3	106.1	104.8	106.0	105.9
03200	CONCRETE REINFORCEMENT	96.9	82.7	89.7	96.9	82.6	89.7	106.0	99.5	102.7	103.4	99.4	101.4	106.9	99.9	103.3	107.0	99.9	103.4
03300	CAST-IN-PLACE CONCRETE	96.9	55.0	79.4	90.8	50.4	73.9	106.0	107.9	106.8	109.5	107.8	108.7	113.8	108.0	111.4	109.9	107.9	109.1
03	**CONCRETE**	90.9	56.8	73.8	87.9	56.4	72.2	108.8	105.0	106.9	115.6	104.8	110.2	112.8	105.3	109.0	110.9	105.1	108.0
04	**MASONRY**	99.3	62.4	76.1	96.8	62.4	75.2	112.0	101.4	105.3	108.9	101.4	104.2	113.2	107.8	109.8	116.7	107.8	111.1
05	**METALS**	94.5	67.3	85.6	92.0	81.9	88.7	94.4	97.0	95.3	94.0	96.8	94.9	95.6	98.0	96.4	94.8	97.7	95.8
06	**WOOD & PLASTICS**	102.2	46.2	73.0	101.1	47.6	73.1	94.1	105.7	100.1	87.8	105.7	97.1	95.3	105.7	100.7	95.3	105.7	100.7
07	**THERMAL & MOISTURE PROTECTION**	98.5	62.1	80.6	98.8	59.9	79.7	110.3	92.6	101.6	111.1	89.6	100.5	110.1	102.6	106.4	110.1	98.3	104.3
08	**DOORS & WINDOWS**	97.2	56.2	86.6	97.2	56.5	86.7	98.8	106.3	100.7	101.5	106.3	102.7	96.3	106.3	98.9	98.1	106.3	100.2
09200	PLASTER & GYPSUM BOARD	85.4	45.5	60.1	85.4	46.8	60.9	94.3	105.7	101.5	92.5	105.7	100.9	92.3	105.7	100.8	92.3	105.7	100.8
095,098	CEILINGS & ACOUSTICAL TREATMENT	91.2	45.5	63.8	91.2	46.8	64.6	105.3	105.7	105.5	115.1	105.7	109.5	105.3	105.7	105.5	105.3	105.7	105.5
09600	FLOORING	115.8	53.7	99.4	115.6	55.9	99.8	115.8	103.1	112.4	113.3	103.1	110.6	115.8	103.1	112.4	115.8	103.1	112.4
097,099	WALL FINISHES, PAINTS & COATINGS	96.4	57.1	72.6	96.4	53.6	70.4	115.8	74.6	90.8	115.8	62.8	83.6	115.8	74.6	90.8	115.8	74.6	90.8
09	**FINISHES**	97.9	48.4	72.1	97.3	49.3	72.3	110.2	102.1	106.0	112.3	100.7	106.3	110.0	102.1	105.9	109.8	102.1	105.8
10 - 14	**TOTAL DIV. 10000 - 14000**	100.0	65.9	92.7	100.0	66.8	92.9	100.0	108.2	101.7	100.0	108.1	101.7	100.0	108.2	101.7	100.0	108.2	101.7
15	**MECHANICAL**	100.1	64.9	84.7	100.1	62.6	83.7	100.2	103.9	101.8	100.2	103.9	101.8	100.2	109.5	104.2	100.2	104.0	101.8
16	**ELECTRICAL**	96.3	68.5	82.7	96.8	45.8	71.9	97.3	97.7	97.5	100.6	87.7	94.3	97.8	105.8	101.7	97.4	97.7	97.6
01 - 16	**WEIGHTED AVERAGE**	97.3	64.0	82.1	96.3	61.2	80.4	101.7	102.1	101.9	103.4	100.3	102.0	102.2	105.4	103.7	102.2	103.0	102.5

	DIVISION	PENNSYLVANIA																	
		ALLENTOWN			ALTOONA			ERIE			HARRISBURG			PHILADELPHIA			PITTSBURGH		
		MAT.	INST.	TOTAL	MAT.	INST.	TOTAL	MAT.	INST.	TOTAL	MAT.	INST.	TOTAL	MAT.	INST.	TOTAL	MAT.	INST.	TOTAL
01590	**EQUIPMENT RENTAL**	.0	115.7	115.7	.0	115.7	115.7	.0	115.7	115.7	.0	114.9	114.9	.0	94.7	94.7	.0	115.5	115.5
02	**SITE CONSTRUCTION**	91.4	106.4	102.5	95.3	106.6	103.7	92.0	107.1	103.2	82.5	104.9	99.1	101.1	94.8	96.4	98.5	109.1	106.3
03100	CONCRETE FORMS & ACCESSORIES	101.9	105.2	104.8	83.9	86.3	86.0	101.2	96.9	97.4	94.4	85.2	86.4	99.9	134.0	129.5	101.5	98.8	99.1
03200	CONCRETE REINFORCEMENT	103.1	102.0	102.5	100.0	97.1	98.5	102.0	94.3	98.1	103.1	96.9	100.0	104.0	135.7	120.0	99.3	106.2	102.8
03300	CAST-IN-PLACE CONCRETE	89.0	94.9	91.5	99.3	86.9	94.1	97.6	84.9	92.3	98.3	89.2	94.5	97.1	128.4	110.2	99.8	97.3	98.7
03	**CONCRETE**	98.8	102.1	100.5	94.0	90.0	92.0	93.1	93.5	93.3	100.9	90.3	95.6	107.0	131.9	119.4	95.9	100.9	98.4
04	**MASONRY**	94.0	90.7	91.9	96.6	80.7	86.6	86.4	92.3	90.1	94.6	84.3	88.1	97.0	129.3	117.2	89.6	97.4	94.5
05	**METALS**	97.3	120.2	104.8	91.0	117.0	99.5	91.2	116.7	99.5	99.3	117.4	105.2	104.5	127.7	112.1	95.6	123.0	104.6
06	**WOOD & PLASTICS**	103.2	108.5	105.9	80.4	88.2	84.4	99.6	98.1	98.8	96.8	84.4	90.4	97.9	133.8	116.6	100.1	98.0	99.0
07	**THERMAL & MOISTURE PROTECTION**	99.1	109.8	104.3	97.6	92.0	94.9	97.8	96.5	97.2	102.8	103.3	103.0	101.2	132.8	116.7	97.5	101.1	99.2
08	**DOORS & WINDOWS**	93.6	105.3	96.6	88.1	95.2	90.0	88.3	94.3	89.8	93.6	93.1	93.5	96.7	139.5	107.8	93.8	107.6	97.4
09200	PLASTER & GYPSUM BOARD	108.4	108.4	108.4	100.7	87.6	92.4	108.4	97.8	101.7	108.4	83.7	92.8	105.1	134.6	123.8	99.3	97.7	98.3
095,098	CEILINGS & ACOUSTICAL TREATMENT	90.0	108.4	101.1	94.5	87.6	90.3	90.0	97.8	94.7	90.0	83.7	86.2	101.9	134.6	121.5	89.8	97.7	94.5
09600	FLOORING	88.2	83.4	87.0	82.3	59.5	76.3	90.2	88.6	89.8	88.5	78.4	85.8	85.7	134.9	98.7	96.3	106.2	98.9
097,099	WALL FINISHES, PAINTS & COATINGS	90.3	93.3	92.1	85.4	111.7	101.4	96.9	88.6	91.8	90.3	87.0	88.3	92.5	141.7	122.3	96.3	117.6	109.2
09	**FINISHES**	95.8	100.0	98.0	94.2	83.3	88.5	97.2	94.5	95.8	94.8	83.4	88.9	101.2	135.1	118.9	96.0	101.6	98.9
10 - 14	**TOTAL DIV. 10000 - 14000**	100.0	96.8	99.3	100.0	99.7	99.9	100.0	104.0	100.9	100.0	98.0	99.6	100.0	126.5	105.6	100.0	104.8	101.0
15	**MECHANICAL**	100.4	99.8	100.1	99.8	84.7	93.2	99.8	92.9	96.8	100.4	90.3	95.9	100.1	132.3	114.2	100.1	100.6	100.3
16	**ELECTRICAL**	100.1	92.4	96.3	89.3	102.0	95.5	90.7	86.4	88.6	99.1	83.7	91.6	97.2	134.2	115.2	95.4	102.0	98.6
01 - 16	**WEIGHTED AVERAGE**	98.0	101.2	99.4	94.3	93.3	93.8	94.3	96.2	95.1	98.1	92.3	95.5	100.8	128.9	113.6	96.7	103.8	99.9

	DIVISION	PENNSYLVANIA									PUERTO RICO			RHODE ISLAND			SOUTH CAROLINA		
		READING			SCRANTON			YORK			SAN JUAN			PROVIDENCE			CHARLESTON		
		MAT.	INST.	TOTAL	MAT.	INST.	TOTAL	MAT.	INST.	TOTAL	MAT.	INST.	TOTAL	MAT.	INST.	TOTAL	MAT.	INST.	TOTAL
01590	**EQUIPMENT RENTAL**	.0	118.2	118.2	.0	115.7	115.7	.0	114.9	114.9	.0	89.1	89.1	.0	103.4	103.4	.0	99.3	99.3
02	**SITE CONSTRUCTION**	98.7	110.7	107.6	91.9	106.3	102.6	81.6	104.8	98.8	115.6	92.1	98.2	80.9	104.0	98.0	95.2	81.2	84.9
03100	CONCRETE FORMS & ACCESSORIES	99.4	86.2	88.0	102.0	84.5	86.9	82.4	79.7	80.1	94.1	20.1	30.0	103.5	115.4	113.8	96.3	38.4	46.1
03200	CONCRETE REINFORCEMENT	102.2	103.1	102.7	103.1	101.9	102.5	102.0	96.8	99.4	194.4	12.5	102.5	108.2	117.3	112.8	95.5	65.6	80.4
03300	CAST-IN-PLACE CONCRETE	71.0	91.3	79.5	93.0	88.7	91.2	87.6	83.9	86.1	105.2	31.4	74.4	86.1	113.2	97.4	85.1	49.3	70.2
03	**CONCRETE**	94.3	92.7	93.5	100.7	90.8	95.8	98.9	86.0	92.5	113.9	23.3	68.7	102.3	114.6	108.4	93.7	49.1	71.4
04	**MASONRY**	98.2	87.2	91.3	94.3	92.8	93.4	93.9	81.7	86.2	219.3	17.3	92.6	107.7	126.5	119.5	88.6	36.1	55.7
05	**METALS**	100.9	120.7	107.4	99.4	119.5	106.0	96.8	117.0	103.4	115.2	31.3	87.8	100.0	110.3	103.3	93.3	81.2	89.4
06	**WOOD & PLASTICS**	98.1	84.4	90.9	103.2	81.9	92.1	87.4	78.0	82.5	91.1	20.4	54.2	101.2	113.9	107.8	95.4	37.7	65.3
07	**THERMAL & MOISTURE PROTECTION**	101.2	105.4	103.3	98.9	99.1	99.0	97.2	98.7	97.9	167.7	23.8	97.0	100.6	114.4	107.4	97.1	43.8	70.9
08	**DOORS & WINDOWS**	94.4	92.5	93.9	93.6	93.2	93.5	90.5	89.6	90.3	153.8	17.1	118.5	101.0	115.1	104.7	96.3	42.6	82.4
09200	PLASTER & GYPSUM BOARD	108.7	83.8	92.9	111.0	81.2	92.1	104.4	77.1	87.1	252.1	18.0	103.7	102.0	113.4	109.2	107.2	35.7	61.9
095,098	CEILINGS & ACOUSTICAL TREATMENT	90.2	83.8	86.4	99.8	81.2	88.6	91.8	77.1	83.0	343.0	18.0	148.1	100.3	113.4	108.2	98.1	35.7	60.7
09600	FLOORING	83.9	87.6	84.9	88.2	82.1	86.6	84.2	83.8	84.1	200.8	17.1	152.3	100.7	133.9	109.5	105.3	41.0	88.3
097,099	WALL FINISHES, PAINTS & COATINGS	90.7	98.8	95.6	90.3	95.3	93.3	90.3	76.4	81.8	201.8	15.8	89.0	95.3	113.4	106.3	108.5	39.5	66.7
09	**FINISHES**	98.3	86.8	92.3	98.5	84.4	91.2	93.5	79.0	86.0	251.5	19.3	130.7	100.9	119.0	110.3	100.3	38.4	68.1
10 - 14	**TOTAL DIV. 10000 - 14000**	100.0	92.1	98.3	100.0	93.0	98.5	100.0	96.6	99.3	100.0	21.0	83.2	100.0	102.8	100.6	100.0	63.8	92.3
15	**MECHANICAL**	100.3	107.3	103.4	100.4	87.4	94.7	100.4	89.5	95.6	102.4	15.9	64.5	100.1	112.5	105.5	100.0	44.6	75.8
16	**ELECTRICAL**	98.9	90.9	95.0	100.1	84.5	92.5	94.0	71.7	83.1	133.6	15.7	76.1	99.0	101.4	100.1	97.5	33.2	66.1
01 - 16	**WEIGHTED AVERAGE**	98.6	98.5	98.6	98.8	92.9	96.1	96.2	88.6	92.7	136.6	26.3	86.5	100.2	112.4	105.8	96.8	48.9	75.1

City Cost Indexes

DIVISION		SOUTH CAROLINA												SOUTH DAKOTA					
		COLUMBIA			FLORENCE			GREENVILLE			SPARTANBURG			ABERDEEN			PIERRE		
		MAT.	INST.	TOTAL	MAT.	INST.	TOTAL	MAT.	INST.	TOTAL	MAT.	INST.	TOTAL	MAT.	INST.	TOTAL	MAT.	INST.	TOTAL
01590	EQUIPMENT RENTAL	.0	99.3	99.3	.0	99.3	99.3	.0	99.3	99.3	.0	99.3	99.3	.0	98.2	98.2	.0	98.2	98.2
02	SITE CONSTRUCTION	94.5	81.2	84.7	105.8	81.2	87.6	100.8	80.8	86.0	100.7	80.8	86.0	84.2	93.6	91.1	82.5	93.6	90.7
03100	CONCRETE FORMS & ACCESSORIES	100.9	40.7	48.7	82.0	40.6	46.1	96.1	40.3	47.8	99.8	40.3	48.3	97.0	40.1	47.7	95.3	41.6	48.8
03200	CONCRETE REINFORCEMENT	95.5	65.6	80.4	95.2	65.6	80.2	95.1	40.0	67.2	95.1	40.0	67.2	99.8	49.3	74.3	99.4	61.2	80.1
03300	CAST-IN-PLACE CONCRETE	80.3	49.0	67.2	72.5	49.2	62.8	72.5	49.1	62.8	72.5	49.1	62.8	102.3	51.2	81.0	99.4	46.9	77.5
03	CONCRETE	91.7	50.0	70.9	94.3	50.0	72.2	93.0	45.2	69.1	93.2	45.2	69.2	98.4	47.1	72.8	96.1	48.5	72.4
04	MASONRY	87.3	37.1	55.8	73.3	36.1	50.0	71.2	36.1	49.2	73.3	36.1	50.0	107.7	58.0	76.5	104.2	56.2	74.0
05	METALS	93.3	81.1	89.3	92.3	81.0	88.6	92.3	71.6	85.5	92.3	71.6	85.5	98.3	70.8	89.3	98.3	75.9	91.0
06	WOOD & PLASTICS	101.0	40.9	69.6	79.5	40.9	59.4	95.2	40.9	66.9	99.5	40.9	68.9	101.8	38.2	68.6	99.8	40.3	68.8
07	THERMAL & MOISTURE PROTECTION	97.0	43.7	70.8	97.4	44.2	71.2	97.4	44.2	71.2	97.4	44.2	71.3	99.0	52.2	76.0	99.2	49.9	75.0
08	DOORS & WINDOWS	96.3	44.4	82.8	92.3	44.4	79.9	92.2	38.4	78.3	92.2	38.4	78.3	95.0	41.1	81.1	98.3	45.7	84.7
09200	PLASTER & GYPSUM BOARD	107.2	39.1	64.1	95.8	39.1	59.8	102.4	39.1	62.3	104.4	39.1	63.0	106.3	36.8	62.3	104.4	39.0	63.0
095,098	CEILINGS & ACOUSTICAL TREATMENT	98.1	39.1	62.7	92.8	39.1	60.6	91.9	39.1	60.2	91.9	39.1	60.2	112.0	36.8	66.9	109.4	39.0	67.2
09600	FLOORING	105.1	41.0	88.2	95.4	41.0	81.0	102.7	41.8	86.6	104.4	41.8	87.9	108.7	58.7	95.5	107.8	41.8	90.4
097,099	WALL FINISHES, PAINTS & COATINGS	108.5	39.5	66.7	108.5	39.5	66.7	108.5	39.5	66.7	108.5	39.5	66.7	99.2	35.6	60.6	99.2	40.7	63.7
09	FINISHES	100.3	40.4	69.1	95.3	40.4	66.7	97.9	40.5	68.1	98.7	40.5	68.4	106.8	42.6	73.4	105.5	41.0	72.0
10 - 14	TOTAL DIV. 10000 - 14000	100.0	64.2	92.4	100.0	64.2	92.4	100.0	64.2	92.4	100.0	64.2	92.4	100.0	57.5	91.0	100.0	68.4	93.3
15	MECHANICAL	100.0	37.6	72.7	100.0	37.6	72.7	100.0	37.5	72.7	100.0	37.5	72.7	100.2	40.1	73.9	100.2	38.4	73.2
16	ELECTRICAL	98.0	35.0	67.3	95.7	20.0	58.8	97.6	31.8	65.5	97.6	31.8	65.5	97.3	60.8	79.5	93.7	51.9	73.3
01 - 16	WEIGHTED AVERAGE	96.6	48.3	74.7	95.0	46.1	72.8	95.2	46.0	72.9	95.5	46.0	73.0	99.4	54.6	79.0	98.7	53.7	78.2

DIVISION		SOUTH DAKOTA						TENNESSEE											
		RAPID CITY			SIOUX FALLS			CHATTANOOGA			JACKSON			JOHNSON CITY			KNOXVILLE		
		MAT.	INST.	TOTAL	MAT.	INST.	TOTAL	MAT.	INST.	TOTAL	MAT.	INST.	TOTAL	MAT.	INST.	TOTAL	MAT.	INST.	TOTAL
01590	EQUIPMENT RENTAL	.0	98.2	98.2	.0	99.4	99.4	.0	104.5	104.5	.0	104.9	104.9	.0	97.4	97.4	.0	97.4	97.4
02	SITE CONSTRUCTION	82.7	93.3	90.6	83.7	95.5	92.4	103.2	96.5	98.2	100.6	95.7	97.0	113.1	85.2	92.5	89.9	85.3	86.5
03100	CONCRETE FORMS & ACCESSORIES	104.4	37.0	46.0	95.4	42.6	49.6	98.7	47.1	54.0	88.7	38.4	45.1	83.5	47.9	52.7	97.8	47.9	54.6
03200	CONCRETE REINFORCEMENT	93.3	61.3	77.1	93.3	61.3	77.1	87.9	44.6	66.0	89.6	44.3	66.7	88.5	44.8	66.4	87.9	44.8	66.1
03300	CAST-IN-PLACE CONCRETE	98.7	44.3	76.0	102.8	49.8	80.7	100.8	49.4	79.3	100.5	42.7	76.4	81.0	54.1	69.8	94.6	49.6	75.8
03	CONCRETE	95.4	45.6	70.6	95.6	50.0	72.8	92.6	49.3	71.0	93.9	43.1	68.5	97.6	51.3	74.5	89.9	49.8	69.9
04	MASONRY	104.1	52.2	71.6	101.5	56.2	73.0	96.9	44.4	64.0	111.4	34.5	63.2	108.8	44.2	68.3	75.1	44.2	55.7
05	METALS	100.1	76.2	92.3	100.2	76.3	92.4	95.4	77.7	89.6	92.8	75.9	87.3	92.7	77.3	87.7	96.0	77.6	90.0
06	WOOD & PLASTICS	106.1	36.1	69.5	99.8	41.0	69.1	102.3	48.1	74.0	87.2	38.2	61.6	76.1	49.2	62.0	91.5	49.2	69.4
07	THERMAL & MOISTURE PROTECTION	99.6	48.1	74.3	99.1	49.5	74.7	100.3	51.7	76.4	99.3	40.6	70.5	96.1	49.3	73.1	93.7	49.5	72.0
08	DOORS & WINDOWS	99.3	43.4	84.8	99.3	46.0	85.5	101.4	48.2	87.6	102.1	40.6	86.2	97.2	49.3	84.8	93.7	49.3	82.2
09200	PLASTER & GYPSUM BOARD	106.3	34.6	60.9	106.3	39.7	64.1	84.2	47.2	60.8	89.3	37.1	56.2	90.1	48.4	63.6	98.0	48.4	66.6
095,098	CEILINGS & ACOUSTICAL TREATMENT	116.5	34.6	67.4	116.5	39.7	70.4	98.1	47.2	67.6	95.1	37.1	60.3	93.4	48.4	66.4	96.1	48.4	67.5
09600	FLOORING	107.8	73.6	98.8	107.8	72.1	98.4	104.0	51.6	90.1	92.8	25.0	74.8	98.0	53.1	86.1	103.6	53.1	90.2
097,099	WALL FINISHES, PAINTS & COATINGS	99.2	40.7	63.7	99.2	40.7	63.7	109.4	47.4	71.8	96.5	32.1	57.5	107.2	55.7	75.9	107.2	55.7	75.9
09	FINISHES	107.6	43.7	74.4	107.6	47.5	76.4	98.6	47.9	72.3	95.6	35.2	64.2	99.2	49.6	73.4	94.6	49.6	71.2
10 - 14	TOTAL DIV. 10000 - 14000	100.0	66.0	92.8	100.0	68.6	93.3	100.0	55.4	90.5	100.0	52.3	89.8	100.0	63.7	92.3	100.0	63.7	92.3
15	MECHANICAL	100.2	35.8	72.1	100.2	37.4	72.7	99.8	46.5	76.5	100.0	57.6	81.4	99.8	61.3	83.0	99.8	61.3	83.0
16	ELECTRICAL	94.2	51.9	73.6	93.6	73.7	83.9	103.3	54.1	79.3	100.7	52.3	77.1	91.6	52.4	72.4	96.8	61.2	79.4
01 - 16	WEIGHTED AVERAGE	99.3	52.5	78.0	99.1	57.8	80.3	98.8	55.8	79.3	98.5	53.3	78.0	97.6	58.2	79.7	94.9	59.2	78.7

DIVISION		TENNESSEE						TEXAS											
		MEMPHIS			NASHVILLE			ABILENE			AMARILLO			AUSTIN			BEAUMONT		
		MAT.	INST.	TOTAL	MAT.	INST.	TOTAL	MAT.	INST.	TOTAL	MAT.	INST.	TOTAL	MAT.	INST.	TOTAL	MAT.	INST.	TOTAL
01590	EQUIPMENT RENTAL	.0	102.2	102.2	.0	105.5	105.5	.0	86.8	86.8	.0	86.8	86.8	.0	85.9	85.9	.0	88.7	88.7
02	SITE CONSTRUCTION	95.7	91.8	92.8	95.0	99.2	98.1	101.5	85.4	89.6	101.9	86.3	90.3	89.8	85.3	86.5	96.5	87.4	89.7
03100	CONCRETE FORMS & ACCESSORIES	99.3	67.1	71.4	97.8	65.2	69.5	95.4	43.9	50.8	97.6	53.2	59.1	94.1	56.6	61.6	103.1	54.5	61.0
03200	CONCRETE REINFORCEMENT	102.2	70.5	86.2	87.1	67.0	76.9	94.1	53.6	73.6	94.1	56.1	74.9	98.7	54.0	76.1	92.5	45.7	68.9
03300	CAST-IN-PLACE CONCRETE	88.9	71.2	81.5	94.4	69.7	84.1	94.2	43.6	73.0	99.0	49.9	78.5	90.0	50.8	73.6	86.2	56.3	73.7
03	CONCRETE	89.6	70.7	80.2	90.0	68.6	79.3	88.9	46.7	67.8	91.3	53.4	72.4	81.8	54.7	68.3	85.6	54.3	70.0
04	MASONRY	84.4	78.0	80.4	85.7	65.9	73.3	100.4	54.4	71.5	102.4	52.9	71.3	94.7	56.4	70.7	104.9	59.1	76.1
05	METALS	97.0	96.1	96.7	96.7	91.9	95.1	95.7	67.7	86.5	95.7	68.9	86.9	94.0	66.7	85.1	97.1	65.7	86.8
06	WOOD & PLASTICS	97.4	67.7	81.9	98.4	66.2	81.6	97.1	43.4	69.0	98.4	55.7	76.1	94.6	59.8	76.4	110.9	55.1	81.7
07	THERMAL & MOISTURE PROTECTION	96.6	74.9	85.9	97.0	66.7	82.1	99.4	51.0	75.6	102.0	50.1	76.5	96.1	55.4	76.1	103.3	59.6	81.8
08	DOORS & WINDOWS	100.9	69.8	92.9	98.5	67.5	90.5	92.9	46.4	80.9	92.9	51.4	82.2	94.5	59.2	85.4	97.6	51.0	85.5
09200	PLASTER & GYPSUM BOARD	95.3	67.3	77.5	92.3	65.7	75.4	85.4	42.5	58.2	85.4	55.1	66.2	90.7	59.3	70.8	90.5	54.5	67.7
095,098	CEILINGS & ACOUSTICAL TREATMENT	106.3	67.3	82.9	96.5	65.7	78.0	91.2	42.5	62.0	91.2	55.1	69.6	88.0	59.3	70.8	93.0	54.5	69.9
09600	FLOORING	98.3	47.1	84.7	106.8	77.5	99.1	115.8	69.1	103.4	115.6	66.7	102.7	101.4	51.2	88.1	110.1	72.6	100.2
097,099	WALL FINISHES, PAINTS & COATINGS	99.0	60.6	75.7	112.7	70.1	86.9	95.2	56.6	71.8	95.2	39.2	61.3	92.2	44.0	63.0	91.4	52.9	68.0
09	FINISHES	97.3	62.1	79.0	104.0	68.0	85.3	97.3	49.6	72.5	97.3	54.3	74.9	92.1	54.2	72.4	91.8	57.5	74.0
10 - 14	TOTAL DIV. 10000 - 14000	100.0	79.1	95.6	100.0	70.8	93.8	100.0	73.9	94.5	100.0	66.0	92.8	100.0	66.2	92.8	100.0	78.2	95.4
15	MECHANICAL	99.8	76.6	89.7	99.8	73.6	88.3	100.1	46.8	76.8	100.1	52.9	79.5	99.9	56.9	81.1	100.0	64.4	84.4
16	ELECTRICAL	95.8	78.5	87.4	102.9	68.2	86.0	96.9	50.0	74.0	97.4	61.5	79.9	99.0	69.4	84.6	95.0	65.3	80.5
01 - 16	WEIGHTED AVERAGE	96.6	77.2	87.8	98.0	74.0	87.1	96.8	54.5	77.6	97.3	59.1	79.9	94.9	61.8	79.8	96.6	63.5	81.6

DIVISION		TEXAS																	
		CORPUS CHRISTI			DALLAS			EL PASO			FORT WORTH			HOUSTON			LAREDO		
		MAT.	INST.	TOTAL	MAT.	INST.	TOTAL	MAT.	INST.	TOTAL	MAT.	INST.	TOTAL	MAT.	INST.	TOTAL	MAT.	INST.	TOTAL
01590	**EQUIPMENT RENTAL**	.0	94.7	94.7	.0	98.1	98.1	.0	86.8	86.8	.0	86.8	86.8	.0	98.9	98.9	.0	85.9	85.9
02	**SITE CONSTRUCTION**	124.9	80.6	92.1	128.3	86.3	97.2	102.0	85.0	89.4	101.4	85.7	89.8	123.8	85.8	95.7	90.3	85.1	86.5
03100	CONCRETE FORMS & ACCESSORIES	97.7	39.9	47.6	93.6	59.3	63.9	94.8	48.5	54.7	96.3	58.6	63.6	91.4	66.7	70.0	91.7	40.2	47.1
03200	CONCRETE REINFORCEMENT	97.8	52.0	74.7	96.5	58.4	77.2	94.1	50.7	72.2	94.1	58.3	76.0	94.2	64.8	79.3	98.7	52.3	75.3
03300	CAST-IN-PLACE CONCRETE	106.7	47.0	81.8	97.1	57.4	80.5	96.5	38.2	72.1	92.6	52.4	75.8	88.8	69.5	80.8	82.1	63.3	74.3
03	**CONCRETE**	91.6	46.6	69.1	88.2	60.0	74.2	89.9	46.2	68.1	88.2	57.1	72.7	84.9	68.9	76.9	81.6	51.4	66.6
04	**MASONRY**	86.0	49.4	63.1	102.7	61.3	76.8	99.4	49.4	68.0	95.7	61.3	74.1	100.6	65.3	78.4	95.0	55.4	70.1
05	**METALS**	93.5	79.9	89.1	93.0	84.3	90.2	95.5	63.7	85.1	92.4	70.0	85.1	101.5	90.9	98.1	94.9	65.6	85.3
06	**WOOD & PLASTICS**	107.2	39.3	71.8	96.8	60.2	77.7	97.1	52.1	73.6	103.8	60.0	80.9	96.2	66.4	80.6	90.5	39.2	63.7
07	**THERMAL & MOISTURE PROTECTION**	99.1	47.8	73.9	94.3	60.9	77.9	98.9	52.1	75.9	99.8	55.5	78.0	94.7	67.6	81.3	95.1	52.6	74.2
08	**DOORS & WINDOWS**	101.6	41.4	86.0	105.6	56.7	93.0	92.9	47.5	81.2	87.5	56.6	79.5	107.1	66.0	96.5	95.1	42.0	81.4
09200	PLASTER & GYPSUM BOARD	90.7	38.2	57.4	91.7	59.5	71.3	85.4	51.4	63.9	85.4	59.5	69.0	88.6	66.0	74.3	87.4	38.2	56.2
095,098	CEILINGS & ACOUSTICAL TREATMENT	88.0	38.2	58.1	94.6	59.5	73.6	91.2	51.4	67.4	91.2	59.5	72.2	94.6	66.0	77.4	88.0	38.2	58.1
09600	FLOORING	114.7	50.6	97.7	105.1	62.0	93.7	115.8	66.4	102.7	150.1	49.6	123.5	98.2	68.2	90.3	97.9	50.6	85.4
097,099	WALL FINISHES, PAINTS & COATINGS	105.7	43.1	67.7	104.7	56.1	75.2	95.2	37.7	60.3	96.4	55.9	71.9	99.8	64.0	78.1	92.2	57.5	71.2
09	**FINISHES**	98.0	41.4	68.6	99.9	59.3	78.8	97.3	51.2	73.3	107.8	56.6	81.2	96.8	66.5	81.0	90.8	43.0	65.9
10 - 14	**TOTAL DIV. 10000 - 14000**	100.0	74.7	94.6	100.0	78.8	95.5	100.0	64.3	92.4	100.0	78.3	95.4	100.0	83.2	96.4	100.0	64.4	92.4
15	**MECHANICAL**	99.9	47.8	77.1	99.8	67.0	85.4	100.1	35.7	71.9	100.1	61.0	83.0	100.0	73.2	88.3	99.8	43.1	75.0
16	**ELECTRICAL**	97.2	53.6	75.9	98.7	64.7	82.1	96.2	56.9	77.0	96.1	62.4	79.6	97.7	68.6	83.5	98.9	63.9	81.8
01 - 16	**WEIGHTED AVERAGE**	97.7	54.1	77.8	98.7	67.1	84.3	96.7	52.4	76.6	96.4	63.3	81.4	99.1	72.8	87.2	94.9	55.0	76.8

DIVISION		TEXAS															UTAH		
		LUBBOCK			ODESSA			SAN ANTONIO			WACO			WICHITA FALLS			LOGAN		
		MAT.	INST.	TOTAL	MAT.	INST.	TOTAL	MAT.	INST.	TOTAL	MAT.	INST.	TOTAL	MAT.	INST.	TOTAL	MAT.	INST.	TOTAL
01590	**EQUIPMENT RENTAL**	.0	96.9	96.9	.0	86.8	86.8	.0	88.8	88.8	.0	86.8	86.8	.0	86.8	86.8	.0	100.7	100.7
02	**SITE CONSTRUCTION**	131.4	83.5	96.0	101.7	85.8	89.9	90.0	89.8	89.9	100.4	85.7	89.5	101.1	85.4	89.5	92.0	100.8	98.5
03100	CONCRETE FORMS & ACCESSORIES	94.9	43.2	50.1	95.3	41.1	48.3	91.8	57.9	62.4	96.0	41.9	49.2	96.0	44.2	51.1	101.0	61.5	66.8
03200	CONCRETE REINFORCEMENT	95.4	53.5	74.2	94.1	53.4	73.6	105.4	54.3	79.6	94.1	53.8	73.8	94.1	53.8	73.8	108.5	80.7	94.5
03300	CAST-IN-PLACE CONCRETE	94.3	50.0	75.8	94.2	45.8	73.9	80.6	70.6	76.4	85.0	54.0	72.1	90.8	49.5	73.5	88.3	72.8	81.8
03	**CONCRETE**	88.0	49.5	68.8	88.9	46.1	67.5	82.0	62.3	72.2	84.5	49.4	67.0	87.3	48.8	68.1	108.4	69.3	88.9
04	**MASONRY**	99.8	51.2	69.3	100.4	51.5	69.7	94.9	65.3	76.3	97.2	59.9	73.8	97.6	58.0	72.7	121.3	62.3	84.3
05	**METALS**	99.2	82.5	93.8	95.1	66.9	85.8	95.6	69.6	87.1	95.6	67.5	86.4	95.5	68.3	86.6	99.5	77.4	92.3
06	**WOOD & PLASTICS**	97.4	43.5	69.2	97.1	41.2	67.9	90.5	56.5	72.8	102.8	38.3	69.1	102.8	43.3	71.8	88.1	59.7	73.3
07	**THERMAL & MOISTURE PROTECTION**	90.6	52.3	71.8	99.4	47.6	73.9	95.1	66.0	80.8	100.1	51.4	76.1	100.1	55.0	77.9	102.2	68.7	85.8
08	**DOORS & WINDOWS**	104.2	44.6	88.8	92.9	42.9	80.0	97.0	56.7	86.6	87.5	40.2	75.3	87.5	46.3	76.9	89.3	61.8	82.2
09200	PLASTER & GYPSUM BOARD	86.1	42.4	58.4	85.4	40.2	56.8	87.4	55.9	67.5	85.4	37.3	54.9	85.4	42.4	58.1	80.6	58.4	66.5
095,098	CEILINGS & ACOUSTICAL TREATMENT	93.9	42.4	63.0	91.2	40.2	60.6	88.0	55.9	68.8	91.2	37.3	58.8	91.2	42.4	61.9	97.8	58.4	74.1
09600	FLOORING	107.2	42.1	90.0	115.8	41.5	96.2	97.9	57.8	87.3	150.3	38.6	120.7	150.7	75.3	130.8	101.2	62.3	90.9
097,099	WALL FINISHES, PAINTS & COATINGS	106.8	35.2	63.4	95.2	35.2	58.8	92.2	57.5	71.2	96.4	36.3	59.9	98.2	57.5	73.5	103.7	49.6	70.9
09	**FINISHES**	99.3	41.3	69.1	97.3	39.9	67.4	90.8	57.1	73.3	107.8	39.4	72.2	108.0	50.9	78.3	97.0	59.6	77.5
10 - 14	**TOTAL DIV. 10000 - 14000**	100.0	73.6	94.4	100.0	63.5	92.2	100.0	70.7	93.8	100.0	75.5	94.8	100.0	64.4	92.4	100.0	75.8	94.9
15	**MECHANICAL**	99.7	50.1	78.0	100.1	39.1	73.4	99.8	69.5	86.6	100.1	56.6	81.1	100.1	51.0	78.6	100.1	71.3	87.5
16	**ELECTRICAL**	95.5	47.0	71.9	97.0	43.8	71.0	99.0	63.9	81.9	97.0	63.2	80.5	100.1	57.4	79.3	88.6	76.4	82.7
01 - 16	**WEIGHTED AVERAGE**	98.8	54.9	78.9	96.7	49.9	75.4	95.2	66.8	82.3	96.6	57.8	79.0	97.4	57.2	79.1	98.9	72.1	86.7

DIVISION		UTAH									VERMONT						VIRGINIA		
		OGDEN			PROVO			SALT LAKE CITY			BURLINGTON			RUTLAND			ALEXANDRIA		
		MAT.	INST.	TOTAL	MAT.	INST.	TOTAL	MAT.	INST.	TOTAL	MAT.	INST.	TOTAL	MAT.	INST.	TOTAL	MAT.	INST.	TOTAL
01590	**EQUIPMENT RENTAL**	.0	100.7	100.7	.0	99.4	99.4	.0	100.7	100.7	.0	101.6	101.6	.0	101.6	101.6	.0	101.2	101.2
02	**SITE CONSTRUCTION**	80.1	100.8	95.4	88.4	98.8	96.1	79.8	100.7	95.3	79.6	99.1	94.0	79.6	99.1	94.0	112.3	86.1	92.9
03100	CONCRETE FORMS & ACCESSORIES	101.0	61.5	66.8	102.5	61.5	67.0	99.5	61.5	66.6	91.4	51.9	57.2	101.8	52.0	58.6	91.7	75.0	77.2
03200	CONCRETE REINFORCEMENT	113.4	80.7	96.9	117.4	80.7	98.9	110.7	80.7	95.6	108.2	56.0	81.8	108.2	56.0	81.8	84.8	81.8	83.3
03300	CAST-IN-PLACE CONCRETE	89.6	72.8	82.6	88.4	72.8	81.9	97.2	72.8	87.0	97.2	64.9	83.7	92.6	64.9	81.1	99.6	82.7	92.6
03	**CONCRETE**	99.2	69.3	84.3	108.4	69.4	88.9	116.9	69.4	93.2	104.3	57.7	81.0	102.8	57.8	80.3	101.8	80.0	90.9
04	**MASONRY**	114.6	62.3	81.8	127.3	62.3	86.5	132.9	62.3	88.6	98.7	62.0	75.7	86.5	62.0	71.1	85.8	74.6	78.8
05	**METALS**	99.9	77.4	92.6	97.8	77.5	91.1	103.1	77.5	94.7	101.7	67.6	90.5	99.9	67.7	89.4	102.4	96.0	100.3
06	**WOOD & PLASTICS**	88.1	59.7	73.3	89.7	59.7	74.0	88.3	59.7	73.4	88.6	50.1	68.5	100.9	50.1	74.4	94.5	75.4	84.5
07	**THERMAL & MOISTURE PROTECTION**	100.7	68.7	85.0	104.7	68.7	87.0	103.8	68.7	86.6	101.4	56.9	79.5	101.2	58.1	80.0	96.0	78.9	87.6
08	**DOORS & WINDOWS**	89.3	61.8	82.2	93.7	61.8	85.4	89.3	61.8	82.2	107.1	46.9	91.5	107.1	46.9	91.5	96.3	77.5	91.4
09200	PLASTER & GYPSUM BOARD	80.6	58.4	66.5	81.0	58.4	66.7	80.6	58.4	66.5	103.5	47.8	68.2	103.5	47.8	68.2	107.2	74.6	86.5
095,098	CEILINGS & ACOUSTICAL TREATMENT	97.8	58.4	74.1	97.8	58.4	74.1	97.8	58.4	74.1	94.9	47.8	66.6	94.9	47.8	66.6	98.1	74.6	84.0
09600	FLOORING	101.2	62.3	90.9	102.0	62.3	91.5	100.8	62.3	90.6	99.8	72.0	92.4	99.8	72.0	92.4	105.3	91.2	101.6
097,099	WALL FINISHES, PAINTS & COATINGS	103.7	49.6	70.9	103.7	63.3	79.2	103.7	63.3	79.2	92.8	41.1	61.4	92.8	41.1	61.4	119.6	86.4	99.4
09	**FINISHES**	95.9	59.6	77.0	97.6	61.1	78.6	96.2	61.1	77.9	98.8	53.8	75.4	98.7	53.8	75.4	100.7	79.1	89.5
10 - 14	**TOTAL DIV. 10000 - 14000**	100.0	75.8	94.9	100.0	75.8	94.9	100.0	75.8	94.9	100.0	84.8	96.8	100.0	84.8	96.8	100.0	88.0	97.4
15	**MECHANICAL**	100.1	71.3	87.5	100.1	71.3	87.5	100.2	71.3	87.6	100.1	68.5	86.3	100.1	68.5	86.3	100.0	85.8	93.8
16	**ELECTRICAL**	88.7	76.4	82.7	88.8	76.4	82.8	92.5	76.4	84.6	100.9	67.4	84.6	100.2	67.4	84.2	98.5	94.4	96.5
01 - 16	**WEIGHTED AVERAGE**	97.2	72.1	85.8	99.5	72.1	87.1	101.2	72.3	88.1	100.8	66.0	85.0	99.8	66.1	84.5	99.6	84.6	92.8

DIVISION		VIRGINIA																
		ARLINGTON			NEWPORT NEWS			NORFOLK			PORTSMOUTH			RICHMOND			ROANOKE		
		MAT.	INST.	TOTAL	MAT.	INST.	TOTAL	MAT.	INST.	TOTAL	MAT.	INST.	TOTAL	MAT.	INST.	TOTAL	MAT.	INST.	TOTAL
01590	EQUIPMENT RENTAL	.0	99.7	99.7	.0	104.9	104.9	.0	105.6	105.6	.0	104.8	104.8	.0	104.8	104.8	.0	99.7	99.7
02	SITE CONSTRUCTION	122.9	82.7	93.2	105.5	84.5	90.0	104.8	85.6	90.6	104.1	84.1	89.3	106.1	84.9	90.4	103.2	81.0	86.8
03100	CONCRETE FORMS & ACCESSORIES	90.5	71.9	74.4	96.4	58.0	63.2	100.7	58.3	64.0	84.5	58.2	61.7	97.3	57.3	62.7	96.2	36.9	44.8
03200	CONCRETE REINFORCEMENT	95.7	65.1	80.3	95.5	72.0	83.6	95.5	72.0	83.6	95.2	72.0	83.5	95.5	71.2	83.2	95.5	64.6	79.9
03300	CAST-IN-PLACE CONCRETE	96.9	74.3	87.5	97.6	56.2	80.3	100.5	56.3	82.0	96.6	56.0	79.7	103.8	56.0	83.8	109.6	47.7	83.8
03	CONCRETE	106.7	72.6	89.7	99.6	61.5	80.6	101.3	61.7	81.5	98.3	61.5	79.9	102.6	61.0	81.8	105.3	47.7	76.6
04	MASONRY	98.5	68.6	79.8	89.6	55.3	68.1	96.2	55.3	70.6	94.6	55.3	69.9	88.5	50.8	64.9	90.0	40.0	58.6
05	METALS	101.1	87.7	96.7	102.4	89.6	98.2	101.5	89.8	97.7	101.4	89.1	97.4	104.4	89.6	99.5	102.2	82.0	95.6
06	WOOD & PLASTICS	91.0	75.4	82.9	95.4	61.5	77.7	100.7	61.5	80.2	82.2	61.5	71.4	96.5	59.5	77.2	95.4	36.4	64.6
07	THERMAL & MOISTURE PROTECTION	98.1	72.7	85.6	97.1	52.3	75.1	96.8	52.4	75.0	97.0	52.4	75.1	96.6	54.3	75.8	97.0	45.9	71.9
08	DOORS & WINDOWS	94.2	72.6	88.6	96.3	58.8	86.6	96.3	61.5	87.3	96.3	61.5	87.3	96.3	56.6	86.0	96.3	43.6	82.6
09200	PLASTER & GYPSUM BOARD	103.4	74.6	85.1	107.2	59.4	76.9	107.2	59.4	76.9	98.8	59.4	73.8	107.2	57.4	75.6	107.2	34.4	61.1
095,098	CEILINGS & ACOUSTICAL TREATMENT	92.8	74.6	81.9	98.1	59.4	74.9	98.1	59.4	74.9	98.1	59.4	74.9	98.1	57.4	73.7	98.1	34.4	59.9
09600	FLOORING	103.5	70.5	94.8	105.3	41.1	88.4	105.1	41.1	88.2	96.5	41.1	81.9	105.1	65.4	94.6	105.3	36.9	87.3
097,099	WALL FINISHES, PAINTS & COATINGS	119.6	86.4	99.4	108.5	43.5	69.1	108.5	64.1	81.6	108.5	64.1	81.6	108.5	56.9	77.2	108.5	34.0	63.3
09	FINISHES	99.3	73.5	85.9	100.1	53.5	75.9	100.1	55.8	77.1	96.3	55.8	75.2	100.0	58.9	78.6	100.0	36.0	66.7
10 - 14	TOTAL DIV. 10000 - 14000	100.0	72.8	94.2	100.0	70.0	93.6	100.0	70.0	93.6	100.0	70.0	93.6	100.0	74.9	94.7	100.0	62.9	92.1
15	MECHANICAL	100.0	82.2	92.2	100.0	58.5	81.9	100.0	62.3	83.5	100.0	62.3	83.5	100.0	62.3	83.5	100.0	40.7	74.1
16	ELECTRICAL	96.0	92.1	94.1	97.9	60.6	79.7	98.0	62.5	80.7	96.3	62.5	79.8	98.5	69.6	84.4	97.9	35.0	67.2
01 - 16	WEIGHTED AVERAGE	100.2	79.4	90.7	99.2	63.6	83.0	99.6	65.2	84.0	98.4	65.0	83.2	99.9	65.9	84.5	99.8	48.4	76.5

DIVISION		WASHINGTON																
		EVERETT			RICHLAND			SEATTLE			SPOKANE			TACOMA			VANCOUVER		
		MAT.	INST.	TOTAL	MAT.	INST.	TOTAL	MAT.	INST.	TOTAL	MAT.	INST.	TOTAL	MAT.	INST.	TOTAL	MAT.	INST.	TOTAL
01590	EQUIPMENT RENTAL	.0	104.0	104.0	.0	90.0	90.0	.0	103.8	103.8	.0	90.0	90.0	.0	104.0	104.0	.0	97.2	97.2
02	SITE CONSTRUCTION	94.3	115.2	109.8	104.4	89.5	93.3	98.4	113.3	109.4	105.2	89.5	93.6	97.1	115.5	110.7	109.6	100.9	103.1
03100	CONCRETE FORMS & ACCESSORIES	106.4	101.3	102.0	112.9	81.2	85.4	97.1	103.0	102.2	119.8	81.4	86.5	97.1	102.5	101.8	97.9	97.0	97.1
03200	CONCRETE REINFORCEMENT	107.9	93.6	100.7	103.1	88.0	95.5	106.7	93.9	100.2	103.8	88.1	95.8	106.7	93.7	100.1	107.5	93.5	100.4
03300	CAST-IN-PLACE CONCRETE	97.8	103.4	100.2	117.4	85.4	104.1	102.7	109.9	105.7	121.6	85.5	106.5	100.6	109.7	104.4	112.3	102.6	108.3
03	CONCRETE	98.5	100.0	99.2	108.0	83.8	95.9	101.3	103.2	102.2	110.7	83.9	97.3	100.3	102.7	101.5	110.5	98.0	104.2
04	MASONRY	130.9	99.1	110.9	113.3	84.1	95.0	125.6	103.9	112.0	114.9	84.8	96.0	125.4	101.3	110.3	125.9	98.3	108.6
05	METALS	105.4	87.3	99.4	94.3	82.5	90.4	107.0	90.6	101.6	96.9	82.8	92.3	107.0	88.2	100.8	104.2	89.6	99.5
06	WOOD & PLASTICS	97.2	101.4	99.4	95.3	80.5	87.6	87.7	101.4	94.8	104.6	80.5	92.0	86.7	101.4	94.3	81.0	96.7	89.2
07	THERMAL & MOISTURE PROTECTION	107.2	98.5	102.9	172.6	80.1	127.1	107.2	102.1	104.7	169.8	81.7	126.5	107.0	100.0	103.6	109.6	91.1	100.5
08	DOORS & WINDOWS	98.6	98.2	98.5	112.0	75.9	102.6	100.8	99.1	100.3	111.7	77.3	102.8	99.2	99.1	99.2	96.2	94.5	95.8
09200	PLASTER & GYPSUM BOARD	104.4	101.3	102.5	119.1	79.8	94.2	97.8	101.3	100.0	123.9	79.8	95.9	100.0	101.3	100.8	98.3	96.9	97.4
095,098	CEILINGS & ACOUSTICAL TREATMENT	114.7	101.3	106.7	107.1	79.8	90.7	118.0	101.3	108.0	107.1	79.8	90.7	118.2	101.3	108.1	113.0	96.9	103.4
09600	FLOORING	120.1	101.6	115.2	110.8	45.2	93.5	112.0	105.8	110.4	114.4	76.1	104.3	112.6	101.6	109.7	118.5	83.1	109.2
097,099	WALL FINISHES, PAINTS & COATINGS	113.4	81.4	94.0	113.5	74.1	89.6	113.4	95.6	102.6	113.5	76.8	91.3	113.4	95.6	102.6	121.5	70.5	90.6
09	FINISHES	114.8	99.3	106.8	122.8	73.1	96.9	112.5	102.4	107.2	124.6	79.7	101.2	113.0	101.5	107.0	110.9	91.4	100.8
10 - 14	TOTAL DIV. 10000 - 14000	100.0	95.7	99.1	100.0	88.9	97.6	100.0	105.6	101.2	100.0	88.9	97.6	100.0	105.5	101.2	100.0	85.6	96.9
15	MECHANICAL	99.9	95.7	98.1	100.5	94.5	97.9	99.9	114.1	106.1	100.5	84.6	93.5	100.0	95.1	97.9	100.1	100.2	100.2
16	ELECTRICAL	104.4	95.3	100.0	94.9	86.7	90.9	104.2	109.1	106.6	93.2	83.3	88.4	104.2	97.1	100.8	115.3	103.5	109.5
01 - 16	WEIGHTED AVERAGE	104.0	98.3	101.4	105.5	85.1	96.2	104.3	105.8	105.0	106.2	83.7	96.0	104.1	99.7	102.1	106.0	97.2	102.0

DIVISION		WASHINGTON			WEST VIRGINIA												WISCONSIN		
		YAKIMA			CHARLESTON			HUNTINGTON			PARKERSBURG			WHEELING			EAU CLAIRE		
		MAT.	INST.	TOTAL	MAT.	INST.	TOTAL	MAT.	INST.	TOTAL	MAT.	INST.	TOTAL	MAT.	INST.	TOTAL	MAT.	INST.	TOTAL
01590	EQUIPMENT RENTAL	.0	104.0	104.0	.0	99.7	99.7	.0	99.7	99.7	.0	99.7	99.7	.0	99.7	99.7	.0	100.5	100.5
02	SITE CONSTRUCTION	100.5	114.0	110.5	101.0	85.9	89.8	102.4	86.6	90.7	107.1	85.8	91.3	107.6	85.6	91.3	84.7	102.4	97.8
03100	CONCRETE FORMS & ACCESSORIES	97.5	96.2	96.4	103.9	93.8	95.1	97.3	95.2	95.5	87.0	88.2	88.1	88.8	88.2	88.3	99.8	94.8	95.5
03200	CONCRETE REINFORCEMENT	107.2	92.5	99.8	95.5	88.0	91.7	95.5	88.8	92.1	94.1	90.2	92.1	93.5	92.7	93.1	91.8	100.2	96.1
03300	CAST-IN-PLACE CONCRETE	107.5	83.5	97.5	95.5	108.2	100.8	103.0	108.8	105.4	96.2	95.3	95.8	96.2	99.9	97.7	100.1	94.8	97.9
03	CONCRETE	105.2	90.6	97.9	99.1	98.2	98.7	102.2	99.2	100.7	102.3	91.8	97.1	102.3	93.8	98.1	96.4	96.2	96.3
04	MASONRY	117.4	73.2	89.7	87.2	93.7	91.3	89.5	94.6	92.7	76.5	85.5	82.1	99.6	90.5	93.9	92.3	96.2	94.7
05	METALS	105.3	82.9	98.0	102.4	100.2	101.7	102.4	100.5	101.8	101.0	100.3	100.8	101.2	101.9	101.4	91.3	102.5	94.9
06	WOOD & PLASTICS	87.2	101.4	94.6	103.6	93.4	98.3	95.4	94.7	95.0	84.6	86.7	85.7	86.4	86.2	86.3	112.8	93.9	102.9
07	THERMAL & MOISTURE PROTECTION	107.2	78.2	93.0	97.0	90.7	93.9	97.4	91.2	94.4	97.4	86.0	91.8	97.6	89.0	93.4	98.8	88.2	93.5
08	DOORS & WINDOWS	98.8	85.4	95.3	97.5	86.0	94.5	96.3	86.9	93.8	96.9	80.6	92.7	97.8	85.4	94.6	100.7	93.5	98.9
09200	PLASTER & GYPSUM BOARD	99.8	101.3	100.8	106.8	93.0	98.1	105.6	94.4	98.5	99.3	86.2	91.0	99.7	85.6	90.8	106.3	94.1	98.5
095,098	CEILINGS & ACOUSTICAL TREATMENT	112.9	101.3	106.0	96.3	93.0	94.3	91.9	94.4	93.4	91.0	86.2	88.1	91.0	85.6	87.8	97.2	94.1	95.3
09600	FLOORING	114.0	67.9	101.8	105.1	104.2	104.9	105.1	109.0	106.1	99.8	95.1	98.5	100.8	105.0	101.9	90.9	105.2	94.7
097,099	WALL FINISHES, PAINTS & COATINGS	113.4	76.8	91.2	108.5	96.6	101.3	108.5	93.2	99.2	108.5	100.2	103.4	108.5	89.0	96.7	89.5	84.8	86.7
09	FINISHES	112.2	89.5	100.4	99.6	96.3	97.9	98.1	97.9	98.0	95.9	90.7	93.2	96.3	91.1	93.6	97.3	95.8	96.5
10 - 14	TOTAL DIV. 10000 - 14000	100.0	100.7	100.2	100.0	101.4	100.3	100.0	102.0	100.4	100.0	100.3	100.1	100.0	103.6	100.8	100.0	92.8	98.5
15	MECHANICAL	100.0	93.6	97.2	100.0	85.4	93.6	100.0	82.2	92.3	100.0	90.0	95.6	100.0	94.9	97.8	100.2	86.7	94.3
16	ELECTRICAL	106.9	86.7	97.0	97.9	89.4	93.8	97.9	81.9	90.1	98.3	93.5	96.0	95.7	98.7	97.2	102.3	85.2	93.9
01 - 16	WEIGHTED AVERAGE	104.2	90.0	97.8	99.1	92.0	95.9	99.3	90.9	95.5	98.3	90.6	94.8	99.3	93.7	96.8	97.7	93.3	95.7

City Cost Indexes

DIVISION		WISCONSIN																
		GREEN BAY			KENOSHA			LA CROSSE			MADISON			MILWAUKEE			RACINE	
		MAT.	INST.	TOTAL	MAT.	INST.	TOTAL	MAT.	INST.	TOTAL	MAT.	INST.	TOTAL	MAT.	INST.	TOTAL	MAT.	INST.	TOTAL
01590	EQUIPMENT RENTAL	.0	98.2	98.2	.0	97.7	97.7	.0	100.5	100.5	.0	100.0	100.0	.0	86.1	86.1	.0	100.0	100.0
02	SITE CONSTRUCTION	87.2	98.3	95.4	92.8	101.6	99.3	78.9	102.4	96.3	87.5	104.6	100.1	88.8	93.3	92.1	87.9	105.5	100.9
03100	CONCRETE FORMS & ACCESSORIES	110.9	94.0	96.2	108.5	101.2	102.2	83.9	94.3	92.9	99.2	94.9	95.5	101.5	108.6	107.7	98.7	101.3	100.9
03200	CONCRETE REINFORCEMENT	90.1	89.5	89.8	89.3	97.9	93.7	91.5	88.2	89.9	89.5	88.5	89.0	89.5	98.3	93.9	89.5	97.9	93.7
03300	CAST-IN-PLACE CONCRETE	103.5	96.7	100.7	115.2	98.3	108.1	90.0	94.2	91.7	103.8	97.3	101.1	103.8	104.4	104.1	103.8	98.0	101.4
03	CONCRETE	97.6	94.4	96.0	103.3	99.7	101.5	87.9	93.5	90.7	97.3	94.7	96.0	97.7	104.5	101.1	97.2	99.6	98.4
04	MASONRY	123.1	95.2	105.6	98.3	106.4	103.4	91.5	96.2	94.4	102.4	99.1	100.4	101.9	115.1	110.2	101.9	106.4	104.7
05	METALS	93.8	97.5	95.0	93.2	100.8	95.7	91.2	96.8	93.0	94.3	95.2	94.6	96.0	92.0	94.7	94.3	100.8	96.4
06	WOOD & PLASTICS	118.8	93.9	105.8	114.7	99.9	107.0	95.1	93.9	94.5	109.5	93.8	101.3	112.1	108.0	110.0	109.7	99.9	104.6
07	THERMAL & MOISTURE PROTECTION	100.7	86.4	93.7	96.8	101.3	99.0	98.0	87.3	92.7	96.5	95.2	95.9	95.8	105.6	100.6	97.2	100.8	99.0
08	DOORS & WINDOWS	99.1	87.0	96.0	96.9	100.5	97.9	100.7	83.7	96.3	101.8	93.7	99.7	104.0	105.0	104.3	101.8	100.5	101.5
09200	PLASTER & GYPSUM BOARD	97.9	94.1	95.5	94.4	100.4	98.2	99.0	94.1	95.9	103.3	94.1	97.4	104.2	108.5	106.9	103.3	100.4	101.4
095,098	CEILINGS & ACOUSTICAL TREATMENT	91.2	94.1	92.9	84.3	100.4	93.9	95.4	94.1	94.6	90.2	94.1	92.5	93.8	108.5	102.6	90.2	100.4	96.3
09600	FLOORING	108.8	105.2	107.9	110.0	110.6	110.1	83.5	105.2	89.3	93.6	106.0	96.9	96.3	110.2	100.0	93.6	110.6	98.1
097,099	WALL FINISHES, PAINTS & COATINGS	99.2	75.8	85.0	103.6	91.9	96.5	89.5	69.8	77.6	93.3	91.7	92.4	95.7	102.6	99.9	93.3	88.0	90.1
09	FINISHES	100.9	94.5	97.6	98.5	102.2	100.4	93.2	94.1	93.7	95.6	97.0	96.3	97.6	108.8	103.4	95.6	101.7	98.8
10 - 14	TOTAL DIV. 10000 - 14000	100.0	92.4	98.4	100.0	91.7	98.2	100.0	92.8	98.5	100.0	92.6	98.4	100.0	94.4	98.8	100.0	91.8	98.2
15	MECHANICAL	100.6	83.2	93.0	100.1	94.2	97.5	100.2	86.2	94.1	99.8	89.7	95.4	99.8	95.7	98.0	99.8	94.2	97.4
16	ELECTRICAL	96.9	84.8	91.0	95.8	85.9	91.0	102.5	85.2	94.1	96.6	85.2	91.1	96.9	101.8	99.3	95.7	93.9	94.8
01 - 16	WEIGHTED AVERAGE	99.5	90.9	95.6	98.2	97.8	98.0	96.0	91.7	94.0	97.9	93.9	96.1	98.7	101.6	100.0	97.8	99.2	98.4

DIVISION		WYOMING									CANADA								
		CASPER			CHEYENNE			ROCK SPRINGS			CALGARY, ALBERTA			EDMONTON, ALBERTA			HALIFAX, NOVA SCOTIA	
		MAT.	INST.	TOTAL	MAT.	INST.	TOTAL	MAT.	INST.	TOTAL	MAT.	INST.	TOTAL	MAT.	INST.	TOTAL	MAT.	INST.	TOTAL
01590	EQUIPMENT RENTAL	.0	101.7	101.7	.0	101.7	101.7	.0	101.7	101.7	.0	107.0	107.0	.0	107.0	107.0	.0	98.6	98.6
02	SITE CONSTRUCTION	80.9	100.3	95.2	80.7	100.2	95.2	83.8	99.8	95.6	113.8	105.1	107.3	117.5	105.1	108.3	97.2	95.8	96.1
03100	CONCRETE FORMS & ACCESSORIES	97.8	43.6	50.8	98.5	43.4	50.8	99.3	30.6	39.8	122.1	76.4	82.5	119.8	76.4	82.2	91.0	66.3	69.6
03200	CONCRETE REINFORCEMENT	115.6	52.0	83.5	108.5	52.1	80.0	117.9	51.8	84.5	161.3	66.7	113.5	161.3	66.7	113.5	148.1	49.7	98.3
03300	CAST-IN-PLACE CONCRETE	99.4	58.0	82.1	99.4	57.9	82.1	99.9	43.6	76.4	191.6	87.1	147.9	206.9	87.1	156.9	178.4	68.3	132.4
03	CONCRETE	104.0	50.9	77.5	102.9	50.8	76.9	104.7	40.2	72.5	160.8	78.9	119.9	168.0	78.9	123.5	151.7	64.6	108.2
04	MASONRY	109.8	39.3	65.5	107.0	41.4	65.9	174.9	32.1	85.3	180.3	74.1	113.7	181.6	74.1	114.1	172.3	69.6	107.9
05	METALS	97.3	63.0	86.1	98.1	63.0	86.6	95.1	61.6	84.1	135.7	85.6	119.3	136.5	85.6	119.9	120.8	72.5	105.0
06	WOOD & PLASTICS	95.5	41.8	67.5	95.5	41.8	67.5	98.4	29.6	62.5	116.3	76.0	95.3	112.9	76.0	93.6	85.0	65.6	74.8
07	THERMAL & MOISTURE PROTECTION	102.4	54.3	78.8	102.4	54.9	79.1	103.0	41.3	72.7	125.9	79.9	103.3	126.2	79.9	103.4	105.4	67.7	86.9
08	DOORS & WINDOWS	93.5	44.4	80.8	94.3	44.4	81.4	100.6	37.7	84.4	91.8	73.2	87.0	91.8	73.2	87.0	82.5	61.1	77.0
09200	PLASTER & GYPSUM BOARD	80.4	40.0	54.8	80.4	40.0	54.8	86.4	27.5	49.1	151.0	74.8	102.7	153.8	74.8	103.7	166.0	64.7	101.7
095,098	CEILINGS & ACOUSTICAL TREATMENT	96.9	40.0	62.8	96.9	40.0	62.8	96.9	27.5	55.3	107.3	74.8	87.8	125.1	74.8	94.9	108.1	64.7	82.1
09600	FLOORING	101.2	43.1	85.8	101.2	69.2	92.7	103.5	42.2	87.3	132.6	86.2	120.4	132.6	86.2	120.4	110.0	62.6	97.4
097,099	WALL FINISHES, PAINTS & COATINGS	103.7	56.9	75.3	103.6	56.9	75.3	103.7	34.6	61.8	109.1	76.5	89.3	109.2	76.5	89.4	109.2	60.4	79.6
09	FINISHES	95.9	43.9	68.8	95.8	49.2	71.6	97.4	32.3	63.6	121.0	78.0	98.7	125.9	78.0	101.0	115.7	65.1	89.4
10 - 14	TOTAL DIV. 10000 - 14000	100.0	75.8	94.8	100.0	75.7	94.8	100.0	50.6	89.5	140.0	85.6	128.4	140.0	85.6	128.4	140.0	62.3	123.5
15	MECHANICAL	100.1	68.4	86.2	100.1	44.8	75.9	100.1	31.6	70.1	101.5	77.6	91.1	101.7	77.6	91.2	101.3	69.0	87.1
16	ELECTRICAL	88.2	62.8	75.8	88.2	52.6	70.8	87.6	45.3	67.0	117.3	86.3	102.2	118.2	86.3	102.6	127.3	70.0	99.3
01 - 16	WEIGHTED AVERAGE	97.6	60.1	80.5	97.5	54.7	78.0	101.4	44.8	75.7	124.6	82.0	105.2	126.2	82.0	106.2	119.3	70.3	97.0

DIVISION		CANADA																
		HAMILTON, ONTARIO			KITCHENER, ONTARIO			LAVAL, QUEBEC			LONDON, ONTARIO			MONTREAL, QUEBEC			OSHAWA, ONTARIO	
		MAT.	INST.	TOTAL	MAT.	INST.	TOTAL	MAT.	INST.	TOTAL	MAT.	INST.	TOTAL	MAT.	INST.	TOTAL	MAT.	INST.	TOTAL
01590	EQUIPMENT RENTAL	.0	103.1	103.1	.0	103.0	103.0	.0	99.2	99.2	.0	103.2	103.2	.0	101.0	101.0	.0	103.0	103.0
02	SITE CONSTRUCTION	110.0	103.5	105.2	99.2	102.9	101.9	92.2	96.8	95.6	110.0	103.2	105.0	91.9	97.1	95.7	111.5	103.1	105.3
03100	CONCRETE FORMS & ACCESSORIES	121.5	86.8	91.4	115.8	79.5	84.4	130.1	81.3	87.8	127.2	82.4	88.4	130.4	81.6	88.1	122.7	83.2	88.5
03200	CONCRETE REINFORCEMENT	174.6	89.1	131.4	110.3	89.0	99.5	156.8	83.7	119.9	123.3	87.5	105.2	159.4	83.8	121.2	174.6	89.7	131.7
03300	CAST-IN-PLACE CONCRETE	157.2	94.9	131.2	146.7	76.5	117.3	140.1	89.5	119.0	157.2	92.5	130.2	137.5	91.1	118.1	169.7	81.8	133.0
03	CONCRETE	146.6	90.1	118.4	124.4	80.5	102.5	136.2	84.8	110.5	138.6	87.0	112.9	135.4	85.5	110.5	152.6	84.1	118.4
04	MASONRY	167.2	91.2	119.5	162.8	87.1	115.3	160.7	81.5	111.0	167.3	88.4	117.8	163.9	81.5	112.2	165.8	90.5	118.5
05	METALS	139.4	89.2	123.0	116.4	88.7	107.3	105.6	85.4	99.0	124.5	88.8	112.8	125.1	85.9	112.3	107.2	89.0	101.2
06	WOOD & PLASTICS	117.7	85.6	100.9	111.2	77.9	93.8	131.7	81.2	105.3	117.7	80.6	98.3	131.7	81.5	105.5	118.6	81.4	99.2
07	THERMAL & MOISTURE PROTECTION	110.7	88.3	99.7	108.8	83.9	96.5	105.4	85.7	95.7	114.1	86.5	100.5	106.2	86.4	96.5	110.1	84.4	97.5
08	DOORS & WINDOWS	91.8	85.7	90.2	83.2	79.9	82.4	91.8	72.3	86.8	92.7	82.0	90.0	91.8	72.4	86.8	90.9	83.1	88.9
09200	PLASTER & GYPSUM BOARD	190.9	85.2	123.9	157.4	77.3	106.6	156.3	80.7	108.4	191.4	80.2	120.9	159.4	80.7	109.5	159.9	81.0	109.9
095,098	CEILINGS & ACOUSTICAL TREATMENT	112.6	85.2	96.2	107.3	77.3	89.3	95.7	80.7	86.7	114.4	80.2	93.9	107.3	80.7	91.3	100.1	81.0	88.6
09600	FLOORING	132.6	91.9	121.9	128.3	91.9	118.7	132.6	92.6	122.1	134.7	91.9	123.4	132.6	92.6	122.1	132.6	94.3	122.5
097,099	WALL FINISHES, PAINTS & COATINGS	109.2	99.2	103.2	109.2	90.9	98.1	109.2	86.3	95.3	109.2	99.2	103.2	109.2	86.3	95.3	109.2	104.3	106.2
09	FINISHES	127.7	89.2	107.7	119.7	83.0	100.6	118.0	84.3	100.4	128.9	85.9	106.5	121.3	84.5	102.1	120.2	87.6	103.2
10 - 14	TOTAL DIV. 10000 - 14000	140.0	91.5	129.7	140.0	89.3	129.2	140.0	79.2	127.1	140.0	90.2	129.4	140.0	80.0	127.2	140.0	90.7	129.5
15	MECHANICAL	101.7	85.5	94.7	101.3	81.9	92.8	101.3	74.6	89.6	101.8	82.1	93.2	101.8	74.7	89.9	101.3	83.5	93.5
16	ELECTRICAL	130.4	93.9	112.6	125.1	90.7	108.3	121.9	73.3	98.2	130.7	90.9	111.3	120.9	73.3	97.7	126.2	91.4	109.2
01 - 16	WEIGHTED AVERAGE	124.7	90.6	109.2	115.6	86.3	102.3	115.4	81.1	99.8	121.8	88.0	106.5	118.8	81.4	101.8	118.9	88.4	105.1

DIVISION		CANADA																	
		OTTAWA, ONTARIO			QUEBEC, QUEBEC			REGINA, SASKATCHEWAN			SASKATOON, SASKATCHEWAN			ST CATHARINES, ONTARIO			ST JOHNS, NEWFOUNDLAND		
		MAT.	INST.	TOTAL	MAT.	INST.	TOTAL	MAT.	INST.	TOTAL	MAT.	INST.	TOTAL	MAT.	INST.	TOTAL	MAT.	INST.	TOTAL
01590	**EQUIPMENT RENTAL**	.0	103.0	103.0	.0	101.4	101.4	.0	98.6	98.6	.0	98.6	98.6	.0	100.4	100.4	.0	98.6	98.6
02	**SITE CONSTRUCTION**	110.3	103.1	105.0	93.7	97.2	96.2	110.0	94.7	98.7	105.2	94.8	97.5	99.6	98.8	99.0	112.5	94.5	99.2
03100	CONCRETE FORMS & ACCESSORIES	120.2	83.2	88.2	130.2	81.8	88.3	102.4	56.0	62.2	102.4	55.9	62.1	113.3	81.2	85.5	100.6	56.9	62.8
03200	CONCRETE REINFORCEMENT	174.6	87.5	130.6	148.1	83.8	115.6	113.8	65.8	89.6	118.9	65.8	92.1	111.2	89.0	100.0	148.1	51.5	99.3
03300	CAST-IN-PLACE CONCRETE	159.6	92.4	131.5	151.5	91.6	126.5	162.8	65.7	122.2	148.0	65.6	113.6	140.1	80.9	115.4	179.6	64.7	131.6
03	**CONCRETE**	147.6	87.4	117.6	140.2	85.7	113.0	131.6	61.9	96.8	125.3	61.8	93.6	121.2	82.8	102.1	164.6	59.5	112.1
04	**MASONRY**	167.4	88.0	117.6	164.9	81.5	112.6	166.9	60.7	100.3	167.1	60.7	100.3	162.3	84.5	113.5	160.7	60.0	97.5
05	**METALS**	116.5	88.8	107.4	120.7	86.2	109.4	105.7	72.8	94.9	105.7	72.7	94.9	106.6	88.8	100.8	108.6	70.3	96.0
06	**WOOD & PLASTICS**	117.2	82.2	99.0	132.3	81.6	105.8	95.6	54.5	74.2	94.2	54.5	73.5	108.5	82.0	94.7	95.0	56.0	74.6
07	**THERMAL & MOISTURE PROTECTION**	110.9	85.1	98.2	105.7	86.6	96.3	106.5	61.8	84.5	104.9	60.8	83.2	108.8	85.1	97.1	109.9	59.4	85.1
08	**DOORS & WINDOWS**	91.8	82.8	89.5	91.8	79.6	88.7	86.7	53.7	78.2	85.7	53.7	77.5	82.7	82.5	82.7	98.1	53.8	86.6
09200	PLASTER & GYPSUM BOARD	231.0	81.8	136.4	198.1	80.7	123.7	166.3	53.3	94.7	147.7	53.3	87.9	143.3	81.6	104.2	172.3	54.8	97.8
095,098	CEILINGS & ACOUSTICAL TREATMENT	106.4	81.8	91.6	97.4	80.7	87.4	123.3	53.3	81.3	123.3	53.3	81.3	100.1	81.6	89.0	106.4	54.8	75.4
09600	FLOORING	132.6	90.6	121.5	132.6	92.6	122.1	120.0	58.9	103.9	120.0	58.9	103.9	126.5	91.9	117.4	114.9	53.3	98.6
097,099	WALL FINISHES, PAINTS & COATINGS	109.2	92.9	99.3	109.2	86.3	95.3	109.2	62.9	81.1	109.2	53.6	75.5	109.2	92.9	99.3	109.2	58.2	78.2
09	**FINISHES**	132.0	85.7	107.9	124.5	84.5	103.7	123.6	57.0	88.9	120.7	55.9	87.0	115.3	84.7	99.4	118.7	56.3	86.2
10 - 14	**TOTAL DIV. 10000 - 14000**	140.0	88.2	129.0	140.0	80.2	127.3	140.0	60.0	123.0	140.0	60.1	123.0	140.0	68.0	124.7	140.0	60.8	123.2
15	**MECHANICAL**	101.8	82.4	93.3	101.7	74.7	89.9	101.5	62.0	84.2	101.4	62.0	84.2	101.3	80.3	92.1	101.7	58.7	82.9
16	**ELECTRICAL**	121.9	91.9	107.3	125.7	73.3	100.2	126.3	63.4	95.6	126.5	63.4	95.7	128.2	91.9	110.5	120.7	58.9	90.6
01 - 16	**WEIGHTED AVERAGE**	120.7	88.1	105.9	119.6	81.8	102.5	116.0	64.9	92.8	114.7	64.7	92.0	113.6	85.6	100.9	120.0	62.8	94.0

DIVISION		CANADA																	
		THUNDER BAY, ONTARIO			TORONTO, ONTARIO			VANCOUVER, B C			WINDSOR, ONTARIO			WINNIPEG, MANITOBA					
		MAT.	INST.	TOTAL	MAT.	INST.	TOTAL	MAT.	INST.	TOTAL	MAT.	INST.	TOTAL	MAT.	INST.	TOTAL	MAT.	INST.	TOTAL
01590	**EQUIPMENT RENTAL**	.0	100.4	100.4	.0	103.2	103.2	.0	111.3	111.3	.0	100.4	100.4	.0	105.3	105.3	.0	.0	.0
02	**SITE CONSTRUCTION**	105.3	98.9	100.6	111.2	103.9	105.8	115.5	107.5	109.6	95.6	99.1	98.2	109.6	100.3	102.7	.0	.0	.0
03100	CONCRETE FORMS & ACCESSORIES	122.7	82.3	87.7	123.1	93.3	97.3	114.6	84.9	88.9	122.7	82.9	88.2	122.3	67.3	74.7	.0	.0	.0
03200	CONCRETE REINFORCEMENT	99.4	88.4	93.8	170.9	90.0	130.0	161.3	81.7	121.1	109.0	87.6	98.2	161.3	56.5	108.4	.0	.0	.0
03300	CAST-IN-PLACE CONCRETE	154.2	92.6	128.5	163.2	102.1	137.7	153.2	94.6	128.7	143.5	93.7	122.7	157.5	72.5	122.0	.0	.0	.0
03	**CONCRETE**	130.0	87.1	108.6	149.0	95.7	122.4	146.0	88.0	117.0	123.1	87.7	105.5	144.4	67.8	106.1	.0	.0	.0
04	**MASONRY**	163.0	85.4	114.3	186.3	99.1	131.6	175.3	88.8	121.0	162.4	91.7	118.1	177.7	63.4	106.0	.0	.0	.0
05	**METALS**	106.4	87.9	100.4	130.5	91.1	117.6	141.7	90.9	125.1	106.5	89.1	100.8	139.3	75.8	118.5	.0	.0	.0
06	**WOOD & PLASTICS**	118.6	82.4	99.7	118.6	91.7	104.6	114.3	82.7	97.8	118.6	81.0	99.0	116.3	68.0	91.1	.0	.0	.0
07	**THERMAL & MOISTURE PROTECTION**	109.1	84.3	96.9	111.0	95.0	103.1	114.4	89.5	102.2	108.8	88.1	98.6	106.1	70.2	88.5	.0	.0	.0
08	**DOORS & WINDOWS**	81.7	82.1	81.8	90.9	90.8	90.9	94.3	81.2	90.9	81.5	81.8	81.5	91.8	61.3	83.9	.0	.0	.0
09200	PLASTER & GYPSUM BOARD	169.3	82.0	113.9	173.3	91.5	121.5	147.7	81.7	105.9	164.0	80.5	111.1	153.3	66.6	98.4	.0	.0	.0
095,098	CEILINGS & ACOUSTICAL TREATMENT	95.7	82.0	87.5	120.6	91.5	103.2	109.0	81.7	92.7	95.7	80.5	86.6	107.3	66.6	82.9	.0	.0	.0
09600	FLOORING	132.6	52.9	111.6	132.6	97.5	123.3	132.6	89.2	121.2	132.6	92.6	122.1	132.6	65.3	114.8	.0	.0	.0
097,099	WALL FINISHES, PAINTS & COATINGS	109.2	94.9	100.5	112.2	104.3	107.4	109.1	98.7	102.8	109.2	94.5	100.3	109.2	55.3	76.5	.0	.0	.0
09	**FINISHES**	120.1	79.1	98.8	127.5	95.4	110.8	121.6	86.7	103.4	119.1	85.9	101.8	121.5	65.9	92.6	.0	.0	.0
10 - 14	**TOTAL DIV. 10000 - 14000**	140.0	69.0	124.9	140.0	93.7	130.2	140.0	89.3	129.2	140.0	70.3	125.2	140.0	63.9	123.8	.0	.0	.0
15	**MECHANICAL**	101.3	81.0	92.4	101.7	91.4	97.2	101.7	84.6	94.2	101.3	83.6	93.5	101.7	68.5	87.1	.0	.0	.0
16	**ELECTRICAL**	125.1	90.1	108.0	129.8	94.6	112.6	132.9	82.8	108.5	133.5	91.9	113.2	130.9	70.3	101.3	.0	.0	.0
01 - 16	**WEIGHTED AVERAGE**	114.8	85.4	101.5	124.4	95.0	111.0	125.6	88.2	108.6	114.7	88.0	102.6	124.3	71.0	100.1	.0	.0	.0

Location Factors

Costs shown in *Means cost data publications* are based on National Averages for materials and installation. To adjust these costs to a specific location, simply multiply the base cost by the factor and divide by 100 for that city. The data is arranged alphabetically by state and postal zip code numbers. For a city not listed, use the factor for a nearby city with similar economic characteristics.

STATE/ZIP	CITY	MAT.	INST.	TOTAL
ALABAMA				
350-352	Birmingham	96.4	76.5	87.4
354	Tuscaloosa	96.3	56.4	78.2
355	Jasper	96.7	53.5	77.1
356	Decatur	96.3	57.6	78.8
357-358	Huntsville	96.3	73.3	85.8
359	Gadsden	96.3	60.0	79.8
360-361	Montgomery	97.0	58.4	79.5
362	Anniston	96.0	47.7	74.1
363	Dothan	96.5	50.5	75.6
364	Evergreen	95.8	51.8	75.8
365-366	Mobile	97.1	63.0	81.6
367	Selma	96.1	53.0	76.5
368	Phenix City	96.8	56.4	78.5
369	Butler	96.2	50.5	75.5
ALASKA				
995-996	Anchorage	134.5	113.8	125.1
997	Fairbanks	130.7	116.9	124.4
998	Juneau	133.6	113.8	124.6
999	Ketchikan	143.2	113.4	129.6
ARIZONA				
850,853	Phoenix	98.8	74.3	87.7
852	Mesa/Tempe	97.8	67.7	84.1
855	Globe	98.2	62.9	82.2
856-857	Tucson	97.0	69.8	84.6
859	Show Low	98.3	66.9	84.1
860	Flagstaff	100.5	67.6	85.5
863	Prescott	98.2	64.7	83.0
864	Kingman	96.8	65.6	82.6
865	Chambers	96.8	62.5	81.2
ARKANSAS				
716	Pine Bluff	95.3	61.7	80.0
717	Camden	93.4	40.1	69.2
718	Texarkana	94.7	45.3	72.3
719	Hot Springs	92.6	39.6	68.5
720-722	Little Rock	95.2	64.3	81.2
723	West Memphis	95.1	55.0	76.9
724	Jonesboro	95.5	55.0	77.1
725	Batesville	93.5	49.9	73.7
726	Harrison	94.8	49.9	74.4
727	Fayetteville	92.1	48.1	72.1
728	Russellville	93.4	48.0	72.8
729	Fort Smith	96.0	56.8	78.2
CALIFORNIA				
900-902	Los Angeles	102.7	111.9	106.8
903-905	Inglewood	98.5	108.2	102.9
906-908	Long Beach	100.0	108.3	103.8
910-912	Pasadena	100.5	108.3	104.0
913-916	Van Nuys	103.8	108.3	105.8
917-918	Alhambra	102.8	108.2	105.3
919-921	San Diego	103.9	105.0	104.4
922	Palm Springs	100.5	105.8	102.9
923-924	San Bernardino	98.3	106.7	102.2
925	Riverside	102.6	109.9	105.9
926-927	Santa Ana	100.4	106.7	103.2
928	Anaheim	102.9	111.1	106.6
930	Oxnard	103.9	109.5	106.4
931	Santa Barbara	102.7	110.3	106.1
932-933	Bakersfield	102.6	106.2	104.2
934	San Luis Obispo	103.8	106.3	105.0
935	Mojave	100.8	103.7	102.1
936-938	Fresno	104.3	111.9	107.8
939	Salinas	104.5	116.1	109.8
940-941	San Francisco	112.6	133.8	122.2
942,956-958	Sacramento	106.9	113.1	109.7
943	Palo Alto	105.5	125.9	114.8
944	San Mateo	108.4	125.1	116.0
945	Vallejo	105.6	120.4	112.3
946	Oakland	110.6	123.9	116.6
947	Berkeley	110.1	122.5	115.7
948	Richmond	109.5	122.8	115.6
949	San Rafael	110.5	122.3	115.9
950	Santa Cruz	109.6	116.1	112.6
CALIFORNIA (CONT'D)				
951	San Jose	107.3	128.4	116.9
952	Stockton	104.5	110.5	107.2
953	Modesto	104.4	111.6	107.7
954	Santa Rosa	104.7	121.5	112.4
955	Eureka	106.3	103.7	105.1
959	Marysville	105.3	113.4	109.0
960	Redding	106.8	110.4	108.4
961	Susanville	106.1	111.1	108.4
COLORADO				
800-802	Denver	100.3	90.3	95.8
803	Boulder	97.6	86.7	92.7
804	Golden	99.9	86.2	93.7
805	Fort Collins	101.0	82.7	92.7
806	Greeley	98.5	71.1	86.0
807	Fort Morgan	98.4	85.9	92.7
808-809	Colorado Springs	100.1	86.6	94.0
810	Pueblo	100.0	83.9	92.6
811	Alamosa	101.6	81.3	92.4
812	Salida	101.6	81.1	92.3
813	Durango	102.0	79.2	91.6
814	Montrose	100.6	79.0	90.8
815	Grand Junction	103.8	77.6	91.9
816	Glenwood Springs	101.4	84.6	93.8
CONNECTICUT				
060	New Britain	101.4	110.7	105.6
061	Hartford	102.7	109.9	106.0
062	Willimantic	102.2	109.6	105.6
063	New London	98.6	110.7	104.1
064	Meriden	100.6	111.1	105.4
065	New Haven	103.2	111.2	106.8
066	Bridgeport	102.7	110.8	106.4
067	Waterbury	102.2	110.7	106.1
068	Norwalk	102.2	111.5	106.4
069	Stamford	102.3	115.6	108.4
D.C.				
200-205	Washington	102.0	90.8	96.9
DELAWARE				
197	Newark	100.5	105.7	102.8
198	Wilmington	99.5	105.7	102.3
199	Dover	100.6	105.7	102.9
FLORIDA				
320,322	Jacksonville	98.3	61.0	81.4
321	Daytona Beach	98.4	74.5	87.5
323	Tallahassee	97.6	49.2	75.6
324	Panama City	99.5	37.9	71.5
325	Pensacola	99.2	58.1	80.5
326,344	Gainesville	99.7	58.7	81.1
327-328,347	Orlando	99.8	69.3	86.0
329	Melbourne	100.5	79.0	90.7
330-332,340	Miami	98.7	72.9	87.0
333	Fort Lauderdale	97.7	72.3	86.2
334,349	West Palm Beach	96.6	68.5	83.8
335-336,346	Tampa	99.4	72.8	87.3
337	St. Petersburg	101.6	55.8	80.8
338	Lakeland	98.5	72.3	86.6
339,341	Fort Myers	97.9	64.3	82.6
342	Sarasota	99.6	66.7	84.6
GEORGIA				
300-303,399	Atlanta	97.4	80.7	89.8
304	Statesboro	97.1	45.0	73.4
305	Gainesville	95.8	57.4	78.4
306	Athens	95.2	62.1	80.1
307	Dalton	97.5	50.9	76.3
308-309	Augusta	96.0	59.2	79.3
310-312	Macon	95.7	59.2	79.1
313-314	Savannah	97.1	58.8	79.7
315	Waycross	97.2	51.6	76.5
316	Valdosta	97.0	49.1	75.2
317	Albany	97.3	56.8	78.9
318-319	Columbus	97.2	61.7	81.1

Location Factors

STATE/ZIP	CITY	MAT.	INST.	TOTAL
HAWAII				
967	Hilo	115.3	125.7	120.0
968	Honolulu	121.0	125.7	123.1
STATES & POSS.				
969	Guam	193.9	54.3	130.5
IDAHO				
832	Pocatello	100.3	78.3	90.3
833	Twin Falls	100.8	45.8	75.8
834	Idaho Falls	98.2	55.3	78.7
835	Lewiston	108.1	83.3	96.8
836-837	Boise	100.2	80.5	91.2
838	Coeur d'Alene	107.3	56.8	84.4
ILLINOIS				
600-603	North Suburban	99.1	116.3	106.9
604	Joliet	99.0	117.2	107.3
605	South Suburban	99.1	116.3	106.9
606-608	Chicago	99.7	125.8	111.6
609	Kankakee	95.4	105.3	99.9
610-611	Rockford	97.4	109.4	102.8
612	Rock Island	95.1	97.7	96.3
613	La Salle	96.5	102.7	99.3
614	Galesburg	96.2	100.8	98.3
615-616	Peoria	98.7	101.7	100.1
617	Bloomington	95.5	101.2	98.1
618-619	Champaign	99.0	100.0	99.5
620-622	East St. Louis	94.2	101.1	97.3
623	Quincy	95.6	95.4	95.5
624	Effingham	94.9	97.3	96.0
625	Decatur	96.5	98.5	97.4
626-627	Springfield	96.8	98.9	97.8
628	Centralia	93.1	98.0	95.3
629	Carbondale	92.8	97.8	95.1
INDIANA				
460	Anderson	94.6	83.3	89.5
461-462	Indianapolis	98.8	88.7	94.2
463-464	Gary	95.8	98.7	97.1
465-466	South Bend	95.0	82.6	89.4
467-468	Fort Wayne	95.1	80.6	88.5
469	Kokomo	92.9	82.2	88.0
470	Lawrenceburg	92.2	80.1	86.7
471	New Albany	93.7	74.8	85.1
472	Columbus	96.3	80.2	89.0
473	Muncie	96.1	80.8	89.2
474	Bloomington	98.2	80.7	90.3
475	Washington	94.1	85.2	90.1
476-477	Evansville	95.2	86.8	91.4
478	Terre Haute	96.0	85.3	91.1
479	Lafayette	95.9	79.5	88.4
IOWA				
500-503,509	Des Moines	97.7	82.4	90.7
504	Mason City	96.0	63.3	81.2
505	Fort Dodge	96.2	59.4	79.5
506-507	Waterloo	97.7	62.0	81.5
508	Creston	96.6	66.3	82.8
510-511	Sioux City	98.3	73.3	87.0
512	Sibley	97.0	52.9	77.0
513	Spencer	98.8	51.0	77.1
514	Carroll	95.8	57.0	78.2
515	Council Bluffs	99.1	74.7	88.0
516	Shenandoah	96.1	52.2	76.2
520	Dubuque	97.7	76.5	88.1
521	Decorah	96.8	54.5	77.6
522-524	Cedar Rapids	98.8	82.7	91.5
525	Ottumwa	96.9	70.7	85.0
526	Burlington	96.0	73.7	85.9
527-528	Davenport	97.8	93.1	95.7
KANSAS				
660-662	Kansas City	98.7	91.6	95.5
664-666	Topeka	98.5	69.3	85.2
667	Fort Scott	97.5	66.8	83.6
668	Emporia	97.4	59.4	80.1
669	Belleville	99.2	54.8	79.0
670-672	Wichita	97.7	71.1	85.6
673	Independence	99.0	53.0	78.1
674	Salina	99.0	57.2	80.1
675	Hutchinson	94.5	51.1	74.7
676	Hays	98.6	54.8	78.7
677	Colby	99.2	54.8	79.0

STATE/ZIP	CITY	MAT.	INST.	TOTAL
KANSAS (CONT'D)				
678	Dodge City	100.3	54.8	79.6
679	Liberal	98.4	45.6	74.4
KENTUCKY				
400-402	Louisville	96.3	86.4	91.8
403-405	Lexington	95.9	66.9	82.7
406	Frankfort	96.1	68.5	83.6
407-409	Corbin	93.3	44.6	71.2
410	Covington	94.0	93.6	93.8
411-412	Ashland	92.6	98.6	95.3
413-414	Campton	94.2	45.1	71.9
415-416	Pikeville	95.1	62.1	80.1
417-418	Hazard	93.5	44.8	71.4
420	Paducah	92.2	88.8	90.7
421-422	Bowling Green	94.4	83.6	89.5
423	Owensboro	94.2	74.3	85.2
424	Henderson	91.9	89.2	90.7
425-426	Somerset	91.5	46.5	71.0
427	Elizabethtown	91.1	82.8	87.3
LOUISIANA				
700-701	New Orleans	100.2	69.8	86.4
703	Thibodaux	98.1	64.4	82.8
704	Hammond	95.4	63.9	81.1
705	Lafayette	97.7	55.0	78.3
706	Lake Charles	97.8	60.0	80.7
707-708	Baton Rouge	98.4	55.1	78.7
710-711	Shreveport	94.7	59.7	78.8
712	Monroe	95.5	57.5	78.2
713-714	Alexandria	95.6	54.7	77.0
MAINE				
039	Kittery	95.6	72.4	85.1
040-041	Portland	98.8	76.8	88.8
042	Lewiston	99.2	76.8	89.0
043	Augusta	97.7	72.3	86.2
044	Bangor	98.0	76.8	88.3
045	Bath	96.7	72.4	85.7
046	Machias	96.3	71.8	85.2
047	Houlton	96.5	75.5	87.0
048	Rockland	95.5	70.6	84.2
049	Waterville	96.8	67.3	83.4
MARYLAND				
206	Waldorf	98.0	73.1	86.7
207-208	College Park	98.0	80.6	90.1
209	Silver Spring	97.3	78.2	88.6
210-212	Baltimore	99.4	84.8	92.7
214	Annapolis	98.8	79.9	90.2
215	Cumberland	95.5	80.8	88.8
216	Easton	96.9	45.4	73.5
217	Hagerstown	96.0	80.8	89.1
218	Salisbury	97.4	54.8	78.0
219	Elkton	94.6	67.2	82.2
MASSACHUSETTS				
010-011	Springfield	100.7	105.1	102.7
012	Pittsfield	100.2	101.1	100.6
013	Greenfield	98.3	103.3	100.5
014	Fitchburg	97.0	116.9	106.0
015-016	Worcester	100.5	118.4	108.6
017	Framingham	96.4	122.8	108.4
018	Lowell	100.0	123.5	110.7
019	Lawrence	101.1	121.0	110.2
020-022, 024	Boston	102.9	130.5	115.4
023	Brockton	101.4	115.5	107.8
025	Buzzards Bay	95.8	116.9	105.4
026	Hyannis	98.3	116.9	106.7
027	New Bedford	101.1	117.0	108.4
MICHIGAN				
480,483	Royal Oak	94.9	106.2	100.0
481	Ann Arbor	96.9	105.7	100.9
482	Detroit	98.7	116.9	107.0
484-485	Flint	96.7	96.2	96.5
486	Saginaw	96.5	93.2	95.0
487	Bay City	96.4	93.3	95.0
488-489	Lansing	96.7	96.1	96.4
490	Battle Creek	95.4	88.1	92.1
491	Kalamazoo	95.7	84.9	90.8
492	Jackson	93.9	94.7	94.3
493,495	Grand Rapids	95.8	70.2	84.2
494	Muskegon	94.3	83.6	89.5

STATE/ZIP	CITY	MAT.	INST.	TOTAL
MICHIGAN (CONT'D)				
496	Traverse City	93.4	73.9	84.6
497	Gaylord	94.6	77.9	87.0
498-499	Iron Mountain	96.5	87.1	92.2
MINNESOTA				
550-551	Saint Paul	99.8	122.0	109.8
553-555	Minneapolis	100.7	126.8	112.6
556-558	Duluth	99.7	108.9	103.8
559	Rochester	99.4	105.1	102.0
560	Mankato	96.3	102.1	99.0
561	Windom	95.1	81.4	88.9
562	Willmar	94.5	87.7	91.4
563	St. Cloud	95.6	120.5	106.9
564	Brainerd	96.1	101.0	98.3
565	Detroit Lakes	98.0	99.1	98.5
566	Bemidji	97.3	97.7	97.5
567	Thief River Falls	96.4	94.0	95.3
MISSISSIPPI				
386	Clarksdale	96.0	31.7	66.8
387	Greenville	99.4	45.9	75.1
388	Tupelo	97.4	38.4	70.6
389	Greenwood	97.3	34.3	68.7
390-392	Jackson	97.9	47.5	75.0
393	Meridian	96.0	47.8	74.1
394	Laurel	97.3	34.4	68.7
395	Biloxi	97.9	56.6	79.1
396	McComb	95.8	53.3	76.5
397	Columbus	97.2	37.8	70.2
MISSOURI				
630-631	St. Louis	96.7	108.6	102.1
633	Bowling Green	95.6	89.4	92.8
634	Hannibal	94.4	85.1	90.2
635	Kirksville	96.7	78.3	88.3
636	Flat River	96.6	93.7	95.3
637	Cape Girardeau	96.1	89.3	93.0
638	Sikeston	94.3	84.3	89.8
639	Poplar Bluff	93.9	85.0	89.8
640-641	Kansas City	101.1	106.3	103.4
644-645	St. Joseph	100.1	89.6	95.3
646	Chillicothe	97.0	70.7	85.1
647	Harrisonville	96.7	97.7	97.2
648	Joplin	98.7	69.0	85.2
650-651	Jefferson City	95.9	87.3	92.0
652	Columbia	96.9	88.1	92.9
653	Sedalia	96.1	84.3	90.7
654-655	Rolla	95.0	77.8	87.2
656-658	Springfield	98.2	74.5	87.4
MONTANA				
590-591	Billings	102.0	74.4	89.5
592	Wolf Point	100.8	70.5	87.0
593	Miles City	98.8	71.0	86.2
594	Great Falls	102.6	74.9	90.0
595	Havre	99.6	70.7	86.5
596	Helena	102.4	72.3	88.7
597	Butte	101.0	71.1	87.4
598	Missoula	98.7	70.3	85.8
599	Kalispell	97.8	69.8	85.1
NEBRASKA				
680-681	Omaha	99.1	76.7	89.0
683-685	Lincoln	98.0	67.7	84.2
686	Columbus	96.0	51.2	75.6
687	Norfolk	97.7	62.5	81.7
688	Grand Island	97.2	66.2	83.1
689	Hastings	97.0	58.5	79.5
690	Mccook	96.8	48.8	75.0
691	North Platte	96.8	59.9	80.0
692	Valentine	99.2	39.7	72.2
693	Alliance	98.9	37.0	70.8
NEVADA				
889-891	Las Vegas	98.1	105.5	101.5
893	Ely	98.7	81.2	90.7
894-895	Reno	98.7	95.5	97.3
897	Carson City	98.4	95.6	97.2
898	Elko	97.5	87.9	93.1
NEW HAMPSHIRE				
030	Nashua	100.9	82.8	92.7
031	Manchester	101.2	82.8	92.9

STATE/ZIP	CITY	MAT.	INST.	TOTAL
NEW HAMPSHIRE (CONT'D)				
032-033	Concord	98.6	82.8	91.4
034	Keene	97.6	52.5	77.1
035	Littleton	97.7	62.0	81.5
036	Charleston	97.0	49.3	75.4
037	Claremont	96.3	49.3	75.0
038	Portsmouth	98.0	78.4	89.1
NEW JERSEY				
070-071	Newark	102.5	121.4	111.1
072	Elizabeth	101.1	120.7	110.0
073	Jersey City	100.2	122.5	110.3
074-075	Paterson	101.8	121.2	110.6
076	Hackensack	99.7	121.7	109.7
077	Long Branch	99.4	120.0	108.7
078	Dover	100.0	121.5	109.8
079	Summit	100.1	120.7	109.4
080,083	Vineland	97.8	117.7	106.8
081	Camden	100.0	115.2	106.9
082,084	Atlantic City	98.6	115.7	106.4
085-086	Trenton	99.9	119.1	108.6
087	Point Pleasant	99.8	116.8	107.5
088-089	New Brunswick	100.3	119.4	108.9
NEW MEXICO				
870-872	Albuquerque	98.3	76.8	88.6
873	Gallup	98.3	76.8	88.6
874	Farmington	98.6	76.8	88.7
875	Santa Fe	97.3	76.8	88.0
877	Las Vegas	96.8	76.8	87.7
878	Socorro	96.4	76.8	87.5
879	Truth/Consequences	97.1	71.6	85.5
880	Las Cruces	95.8	69.1	83.7
881	Clovis	96.9	76.1	87.4
882	Roswell	98.5	76.1	88.4
883	Carrizozo	99.0	76.8	88.9
884	Tucumcari	97.8	76.1	87.9
NEW YORK				
100-102	New York	106.6	163.5	132.4
103	Staten Island	102.1	155.2	126.2
104	Bronx	100.2	155.2	125.2
105	Mount Vernon	100.2	135.4	116.2
106	White Plains	99.9	135.4	116.0
107	Yonkers	105.1	136.2	119.2
108	New Rochelle	100.7	135.4	116.5
109	Suffern	100.5	123.7	111.0
110	Queens	101.4	155.0	125.8
111	Long Island City	103.1	155.0	126.7
112	Brooklyn	103.4	156.5	127.5
113	Flushing	103.6	155.0	127.0
114	Jamaica	101.7	155.2	126.0
115,117,118	Hicksville	101.2	145.0	121.1
116	Far Rockaway	103.7	155.0	127.0
119	Riverhead	101.6	145.1	121.4
120-122	Albany	97.9	93.9	96.1
123	Schenectady	98.2	93.0	95.8
124	Kingston	101.1	114.3	107.1
125-126	Poughkeepsie	100.3	119.0	108.8
127	Monticello	99.6	113.8	106.1
128	Glens Falls	92.8	89.9	91.5
129	Plattsburgh	97.0	82.8	90.6
130-132	Syracuse	99.0	90.5	95.2
133-135	Utica	97.0	87.8	92.8
136	Watertown	98.7	88.4	94.0
137-139	Binghamton	98.6	84.5	92.2
140-142	Buffalo	99.9	104.7	102.1
143	Niagara Falls	97.6	102.8	100.0
144-146	Rochester	100.1	94.9	97.7
147	Jamestown	96.6	85.4	91.5
148-149	Elmira	96.5	84.8	91.2
NORTH CAROLINA				
270,272-274	Greensboro	98.0	50.3	76.3
271	Winston-Salem	97.8	49.1	75.7
275-276	Raleigh	97.9	50.3	76.3
277	Durham	97.9	50.3	76.3
278	Rocky Mount	95.4	36.2	68.5
279	Elizabeth City	96.1	38.6	70.0
280	Gastonia	97.2	47.5	74.6
281-282	Charlotte	98.2	47.8	75.3
283	Fayetteville	96.7	50.3	75.6
284	Wilmington	96.0	49.4	74.8
285	Kinston	94.2	35.8	67.6

STATE/ZIP	CITY	MAT.	INST.	TOTAL
NORTH CAROLINA (CONT'D)				
286	Hickory	94.5	34.9	67.5
287-288	Asheville	96.3	48.3	74.5
289	Murphy	95.5	34.0	67.6
NORTH DAKOTA				
580-581	Fargo	100.8	65.9	85.0
582	Grand Forks	100.4	60.7	82.4
583	Devils Lake	100.0	61.0	82.3
584	Jamestown	100.0	54.2	79.2
585	Bismarck	99.2	66.9	84.5
586	Dickinson	100.8	62.3	83.3
587	Minot	100.5	68.2	85.8
588	Williston	99.3	61.7	82.2
OHIO				
430-432	Columbus	97.8	90.5	94.5
433	Marion	94.4	88.3	91.6
434-436	Toledo	98.0	99.6	98.7
437-438	Zanesville	94.8	83.9	89.8
439	Steubenville	95.7	94.3	95.1
440	Lorain	96.8	98.1	97.4
441	Cleveland	97.1	106.0	101.1
442-443	Akron	97.7	98.0	97.8
444-445	Youngstown	97.1	94.8	96.0
446-447	Canton	97.2	88.5	93.2
448-449	Mansfield	94.5	93.4	94.0
450	Hamilton	94.3	90.4	92.5
451-452	Cincinnati	94.7	92.3	93.6
453-454	Dayton	94.4	84.9	90.1
455	Springfield	94.4	86.7	90.9
456	Chillicothe	93.5	92.4	93.0
457	Athens	96.5	82.3	90.1
458	Lima	97.1	87.9	92.9
OKLAHOMA				
730-731	Oklahoma City	97.3	64.0	82.1
734	Ardmore	94.2	63.3	80.2
735	Lawton	96.5	65.0	82.2
736	Clinton	95.7	61.6	80.2
737	Enid	96.1	61.6	80.4
738	Woodward	94.5	61.7	79.6
739	Guymon	95.6	33.9	67.6
740-741	Tulsa	96.3	61.2	80.4
743	Miami	93.3	67.1	81.4
744	Muskogee	95.7	43.7	72.1
745	Mcalester	92.9	56.1	76.2
746	Ponca City	93.4	61.5	78.9
747	Durant	93.4	61.7	79.0
748	Shawnee	94.9	60.1	79.1
749	Poteau	92.5	64.3	79.7
OREGON				
970-972	Portland	102.2	105.4	103.7
973	Salem	102.2	103.0	102.5
974	Eugene	101.7	102.1	101.9
975	Medford	103.4	100.3	102.0
976	Klamath Falls	103.7	100.1	102.1
977	Bend	102.5	102.2	102.4
978	Pendleton	96.6	100.7	98.5
979	Vale	94.4	93.7	94.1
PENNSYLVANIA				
150-152	Pittsburgh	96.7	103.8	99.9
153	Washington	93.8	102.1	97.6
154	Uniontown	94.0	99.6	96.5
155	Bedford	95.0	91.8	93.5
156	Greensburg	95.1	99.6	97.1
157	Indiana	93.9	97.3	95.4
158	Dubois	95.2	95.7	95.5
159	Johnstown	94.9	96.2	95.5
160	Butler	92.3	101.6	96.5
161	New Castle	92.3	100.2	95.9
162	Kittanning	92.8	103.2	97.5
163	Oil City	92.3	95.8	93.9
164-165	Erie	94.3	96.2	95.1
166	Altoona	94.3	93.3	93.8
167	Bradford	95.9	93.3	94.7
168	State College	95.5	93.7	94.7
169	Wellsboro	96.5	87.4	92.4
170-171	Harrisburg	98.1	92.3	95.5
172	Chambersburg	96.1	87.2	92.0
173-174	York	96.2	88.6	92.7
175-176	Lancaster	94.9	87.5	91.5
PENNSYLVANIA (CONT'D)				
177	Williamsport	93.4	81.8	88.2
178	Sunbury	95.5	87.9	92.0
179	Pottsville	94.6	89.3	92.2
180	Lehigh Valley	96.0	102.7	99.0
181	Allentown	98.0	101.2	99.4
182	Hazleton	95.3	92.4	94.0
183	Stroudsburg	95.3	95.8	95.5
184-185	Scranton	98.8	92.9	96.1
186-187	Wilkes-Barre	95.3	91.4	93.5
188	Montrose	94.9	92.0	93.6
189	Doylestown	94.8	116.5	104.6
190-191	Philadelphia	100.8	128.9	113.6
193	Westchester	97.6	115.6	105.7
194	Norristown	96.5	116.5	105.6
195-196	Reading	98.6	98.5	98.6
PUERTO RICO				
009	San Juan	136.6	26.3	86.5
RHODE ISLAND				
028	Newport	99.9	112.4	105.6
029	Providence	100.2	112.4	105.8
SOUTH CAROLINA				
290-292	Columbia	96.6	48.3	74.7
293	Spartanburg	95.5	46.0	73.0
294	Charleston	96.8	48.9	75.1
295	Florence	95.0	46.1	72.8
296	Greenville	95.2	46.0	72.9
297	Rock Hill	95.0	33.8	67.2
298	Aiken	95.9	70.4	84.3
299	Beaufort	96.7	37.4	69.8
SOUTH DAKOTA				
570-571	Sioux Falls	99.1	57.8	80.3
572	Watertown	97.9	52.0	77.0
573	Mitchell	96.9	51.6	76.3
574	Aberdeen	99.4	54.6	79.0
575	Pierre	98.7	53.7	78.2
576	Mobridge	97.5	51.9	76.8
577	Rapid City	99.3	52.5	78.0
TENNESSEE				
370-372	Nashville	98.0	74.0	87.1
373-374	Chattanooga	98.8	55.8	79.3
375,380-381	Memphis	96.6	77.2	87.8
376	Johnson City	97.6	58.2	79.7
377-379	Knoxville	94.9	59.2	78.7
382	Mckenzie	96.5	49.7	75.3
383	Jackson	98.5	53.3	78.0
384	Columbia	95.0	56.9	77.7
385	Cookeville	96.3	51.4	75.9
TEXAS				
750	Mckinney	98.2	57.1	79.5
751	Waxahackie	98.1	57.8	79.8
752-753	Dallas	98.7	67.1	84.3
754	Greenville	98.3	41.9	72.7
755	Texarkana	97.2	53.0	77.2
756	Longview	97.5	43.9	73.1
757	Tyler	98.1	56.4	79.2
758	Palestine	94.1	44.3	71.5
759	Lufkin	95.1	48.1	73.7
760-761	Fort Worth	96.4	63.3	81.4
762	Denton	96.9	53.2	77.0
763	Wichita Falls	97.4	57.2	79.1
764	Eastland	95.9	43.3	72.0
765	Temple	94.8	51.4	75.1
766-767	Waco	96.6	57.8	79.0
768	Brownwood	96.9	40.7	71.4
769	San Angelo	96.6	49.0	75.0
770-772	Houston	99.1	72.8	87.2
773	Huntsville	97.6	41.6	72.1
774	Wharton	98.8	46.5	75.1
775	Galveston	97.0	70.7	85.0
776-777	Beaumont	96.6	63.5	81.6
778	Bryan	94.5	65.2	81.2
779	Victoria	99.0	49.1	76.4
780	Laredo	94.9	55.0	76.8
781-782	San Antonio	95.2	66.8	82.3
783-784	Corpus Christi	97.7	54.1	77.8
785	Mc Allen	97.8	48.2	75.2
786-787	Austin	94.9	61.8	79.8

STATE/ZIP	CITY	MAT.	INST.	TOTAL
TEXAS (CONT'D)				
788	Del Rio	97.1	34.2	68.6
789	Giddings	94.5	43.4	71.3
790-791	Amarillo	97.3	59.1	79.9
792	Childress	96.5	53.5	76.9
793-794	Lubbock	98.8	54.9	78.9
795-796	Abilene	96.8	54.5	77.6
797	Midland	99.1	51.4	77.4
798-799,885	El Paso	96.7	52.4	76.6
UTAH				
840-841	Salt Lake City	101.2	72.3	88.1
842,844	Ogden	97.2	72.1	85.8
843	Logan	98.9	72.1	86.7
845	Price	99.6	51.1	77.6
846-847	Provo	99.5	72.1	87.1
VERMONT				
050	White River Jct.	99.0	47.9	75.8
051	Bellows Falls	97.5	50.3	76.0
052	Bennington	97.8	50.6	76.4
053	Brattleboro	98.2	50.8	76.7
054	Burlington	100.8	66.0	85.0
056	Montpelier	97.7	66.1	83.3
057	Rutland	99.8	66.1	84.5
058	St. Johnsbury	99.2	50.6	77.1
059	Guildhall	97.8	50.1	76.1
VIRGINIA				
220-221	Fairfax	99.2	81.2	91.0
222	Arlington	100.2	79.4	90.7
223	Alexandria	99.6	84.6	92.8
224-225	Fredericksburg	97.8	68.0	84.3
226	Winchester	98.5	55.4	78.9
227	Culpeper	98.3	56.6	79.4
228	Harrisonburg	98.6	48.1	75.7
229	Charlottesville	98.9	60.5	81.5
230-232	Richmond	99.9	65.9	84.5
233-235	Norfolk	99.6	65.2	84.0
236	Newport News	99.2	63.6	83.0
237	Portsmouth	98.4	65.0	83.2
238	Petersburg	98.6	65.9	83.8
239	Farmville	98.1	42.3	72.8
240-241	Roanoke	99.8	48.4	76.5
242	Bristol	97.8	48.7	75.5
243	Pulaski	97.5	43.7	73.1
244	Staunton	98.2	46.5	74.8
245	Lynchburg	98.4	51.0	76.8
246	Grundy	97.8	45.8	74.2
WASHINGTON				
980-981,987	Seattle	104.3	105.8	105.0
982	Everett	104.0	98.3	101.4
983-984	Tacoma	104.1	99.7	102.1
985	Olympia	102.8	99.7	101.4
986	Vancouver	106.0	97.2	102.0
988	Wenatchee	104.3	84.0	95.1
989	Yakima	104.2	90.0	97.8
990-992	Spokane	106.2	83.7	96.0
993	Richland	105.5	85.1	96.2
994	Clarkston	104.7	82.3	94.6
WEST VIRGINIA				
247-248	Bluefield	96.6	80.2	89.1
249	Lewisburg	98.1	83.9	91.6
250-253	Charleston	99.1	92.0	95.9
254	Martinsburg	98.0	80.0	89.8
255-257	Huntington	99.3	90.9	95.5
258-259	Beckley	96.4	89.5	93.3
260	Wheeling	99.3	93.7	96.8
261	Parkersburg	98.3	90.6	94.8
262	Buckhannon	97.9	93.2	95.8
263-264	Clarksburg	98.3	92.2	95.5
265	Morgantown	98.4	93.2	96.0
266	Gassaway	97.7	92.3	95.3
267	Romney	97.7	86.4	92.6
268	Petersburg	97.7	89.1	93.8
WISCONSIN				
530,532	Milwaukee	98.7	101.6	100.0
531	Kenosha	98.2	97.8	98.0
534	Racine	97.8	99.2	98.4
535	Beloit	97.7	96.4	97.1
537	Madison	97.9	93.9	96.1

STATE/ZIP	CITY	MAT.	INST.	TOTAL
(CONT'D)				
538	Lancaster	95.6	91.4	93.7
539	Portage	94.2	93.3	93.8
540	New Richmond	95.5	94.6	95.1
541-543	Green Bay	99.5	90.9	95.6
544	Wausau	94.9	90.9	93.1
545	Rhinelander	98.0	90.7	94.7
546	La Crosse	96.0	91.7	94.0
547	Eau Claire	97.7	93.3	95.7
548	Superior	95.4	94.8	95.1
549	Oshkosh	95.5	90.5	93.3
WYOMING				
820	Cheyenne	97.5	54.7	78.0
821	Yellowstone Nat'l Park	96.7	49.1	75.1
822	Wheatland	98.1	48.0	75.4
823	Rawlins	99.4	44.1	74.3
824	Worland	97.3	44.8	73.5
825	Riverton	98.4	47.3	75.2
826	Casper	97.6	60.1	80.5
827	Newcastle	97.2	44.1	73.1
828	Sheridan	98.1	52.8	77.5
829-831	Rock Springs	101.4	44.8	75.7
CANADIAN FACTORS (reflect Canadian currency)				
ALBERTA				
	Calgary	124.6	82.0	105.2
	Edmonton	126.2	82.0	106.2
	Fort McMurray	116.8	81.6	100.8
	Lethbridge	117.7	81.0	101.1
	Lloydminster	116.9	81.6	100.8
	Medicine Hat	117.0	81.0	100.6
	Red Deer	117.4	81.0	100.9
BRITISH COLUMBIA				
	Kamloops	117.8	84.3	102.6
	Prince George	119.1	84.3	103.3
	Vancouver	125.6	88.2	108.6
	Victoria	119.1	84.8	103.6
MANITOBA				
	Brandon	117.1	70.7	96.0
	Portage la Prairie	117.1	70.7	96.0
	Winnipeg	124.3	71.0	100.1
NEW BRUNSWICK				
	Bathurst	115.2	62.8	91.4
	Dalhousie	115.2	62.8	91.4
	Fredericton	117.1	66.9	94.3
	Moncton	115.5	62.8	91.6
	Newcastle	115.2	62.8	91.4
	Saint John	118.3	66.9	94.9
NEWFOUNDLAND				
	Corner Brook	120.1	62.8	94.1
	St. John's	120.0	62.8	94.0
NORTHWEST TERRITORIES				
	Yellowknife	114.3	79.8	98.6
NOVA SCOTIA				
	Dartmouth	117.7	69.2	95.7
	Halifax	119.3	70.3	97.0
	New Glasgow	115.8	69.2	94.6
	Sydney	113.3	69.2	93.3
	Yarmouth	115.6	69.2	94.5
ONTARIO				
	Barrie	120.0	86.7	104.9
	Brantford	119.2	90.3	106.1
	Cornwall	119.1	86.9	104.5
	Hamilton	124.7	90.6	109.2
	Kingston	120.1	87.3	105.2
	Kitchener	115.6	86.3	102.3
	London	121.8	88.0	106.5
	North Bay	119.2	85.0	103.6
	Oshawa	118.9	88.4	105.1
	Ottawa	120.7	88.1	105.9
	Owen Sound	120.3	85.0	104.2
	Peterborough	119.2	86.7	104.4
	Sarnia	119.4	90.7	106.3
	St. Catharines	113.6	85.6	100.9
	Sudbury	113.6	85.2	100.7

Location Factors

STATE/ZIP	CITY	MAT.	INST.	TOTAL
ONTARIO (CONT'D)				
	Thunder Bay	114.8	85.4	101.5
	Toronto	124.4	95.0	111.0
	Windsor	114.7	88.0	102.6
PRINCE EDWARD ISLAND				
	Charlottetown	117.8	58.7	91.0
	Summerside	117.5	58.7	90.8
QUEBEC				
	Cap-de-la-Madeleine	116.0	81.4	100.3
	Charlesbourg	116.0	81.4	100.3
	Chicoutimi	115.0	81.1	99.6
	Gatineau	115.4	81.1	99.9
	Laval	115.4	81.1	99.8
	Montreal	118.8	81.4	101.8
	Quebec	119.6	81.8	102.5
	Sherbrooke	115.8	81.1	100.1
	Trois Rivieres	116.3	81.4	100.4
SASKATCHEWAN				
	Moose Jaw	114.2	64.8	91.8
	Prince Albert	113.6	64.7	91.4
	Regina	116.0	64.9	92.8
	Saskatoon	114.7	64.7	92.0
YUKON				
	Whitehorse	114.0	63.8	91.2

Abbreviations

A	Area Square Feet; Ampere
ABS	Acrylonitrile Butadiene Stryrene; Asbestos Bonded Steel
A.C.	Alternating Current; Air-Conditioning; Asbestos Cement; Plywood Grade A & C
A.C.I.	American Concrete Institute
AD	Plywood, Grade A & D
Addit.	Additional
Adj.	Adjustable
af	Audio-frequency
A.G.A.	American Gas Association
Agg.	Aggregate
A.H.	Ampere Hours
A hr.	Ampere-hour
A.H.U.	Air Handling Unit
A.I.A.	American Institute of Architects
AIC	Ampere Interrupting Capacity
Allow.	Allowance
alt.	Altitude
Alum.	Aluminum
a.m.	Ante Meridiem
Amp.	Ampere
Anod.	Anodized
Approx.	Approximate
Apt.	Apartment
Asb.	Asbestos
A.S.B.C.	American Standard Building Code
Asbe.	Asbestos Worker
A.S.H.R.A.E.	American Society of Heating, Refrig. & AC Engineers
A.S.M.E.	American Society of Mechanical Engineers
A.S.T.M.	American Society for Testing and Materials
Attchmt.	Attachment
Avg.	Average
A.W.G.	American Wire Gauge
AWWA	American Water Works Assoc.
Bbl.	Barrel
B&B	Grade B and Better; Balled & Burlapped
B.&S.	Bell and Spigot
B.&W.	Black and White
b.c.c.	Body-centered Cubic
B.C.Y.	Bank Cubic Yards
BE	Bevel End
B.F.	Board Feet
Bg. cem.	Bag of Cement
BHP	Boiler Horsepower; Brake Horsepower
B.I.	Black Iron
Bit.; Bitum.	Bituminous
Bk.	Backed
Bkrs.	Breakers
Bldg.	Building
Blk.	Block
Bm.	Beam
Boil.	Boilermaker
B.P.M.	Blows per Minute
BR	Bedroom
Brg.	Bearing
Brhe.	Bricklayer Helper
Bric.	Bricklayer
Brk.	Brick
Brng.	Bearing
Brs.	Brass
Brz.	Bronze
Bsn.	Basin
Btr.	Better
BTU	British Thermal Unit
BTUH	BTU per Hour
B.U.R.	Built-up Roofing
BX	Interlocked Armored Cable
c	Conductivity, Copper Sweat
C	Hundred; Centigrade
C/C	Center to Center, Cedar on Cedar
Cab.	Cabinet
Cair.	Air Tool Laborer
Calc	Calculated
Cap.	Capacity
Carp.	Carpenter
C.B.	Circuit Breaker
C.C.A.	Chromate Copper Arsenate
C.C.F.	Hundred Cubic Feet
cd	Candela
cd/sf	Candela per Square Foot
CD	Grade of Plywood Face & Back
CDX	Plywood, Grade C & D, exterior glue
Cefi.	Cement Finisher
Cem.	Cement
CF	Hundred Feet
C.F.	Cubic Feet
CFM	Cubic Feet per Minute
c.g.	Center of Gravity
CHW	Chilled Water; Commercial Hot Water
C.I.	Cast Iron
C.I.P.	Cast in Place
Circ.	Circuit
C.L.	Carload Lot
Clab.	Common Laborer
Clam	Common maintenance laborer
C.L.F.	Hundred Linear Feet
CLF	Current Limiting Fuse
CLP	Cross Linked Polyethylene
cm	Centimeter
CMP	Corr. Metal Pipe
C.M.U.	Concrete Masonry Unit
CN	Change Notice
Col.	Column
CO_2	Carbon Dioxide
Comb.	Combination
Compr.	Compressor
Conc.	Concrete
Cont.	Continuous; Continued
Corr.	Corrugated
Cos	Cosine
Cot	Cotangent
Cov.	Cover
C/P	Cedar on Paneling
CPA	Control Point Adjustment
Cplg.	Coupling
C.P.M.	Critical Path Method
CPVC	Chlorinated Polyvinyl Chloride
C.Pr.	Hundred Pair
CRC	Cold Rolled Channel
Creos.	Creosote
Crpt.	Carpet & Linoleum Layer
CRT	Cathode-ray Tube
CS	Carbon Steel, Constant Shear Bar Joist
Csc	Cosecant
C.S.F.	Hundred Square Feet
CSI	Construction Specifications Institute
C.T.	Current Transformer
CTS	Copper Tube Size
Cu	Copper, Cubic
Cu. Ft.	Cubic Foot
cw	Continuous Wave
C.W.	Cool White; Cold Water
Cwt.	100 Pounds
C.W.X.	Cool White Deluxe
C.Y.	Cubic Yard (27 cubic feet)
C.Y./Hr.	Cubic Yard per Hour
Cyl.	Cylinder
d	Penny (nail size)
D	Deep; Depth; Discharge
Dis.;Disch.	Discharge
Db.	Decibel
Dbl.	Double
DC	Direct Current
DDC	Direct Digital Control
Demob.	Demobilization
d.f.u.	Drainage Fixture Units
D.H.	Double Hung
DHW	Domestic Hot Water
Diag.	Diagonal
Diam.	Diameter
Distrib.	Distribution
Dk.	Deck
D.L.	Dead Load; Diesel
DLH	Deep Long Span Bar Joist
Do.	Ditto
Dp.	Depth
D.P.S.T.	Double Pole, Single Throw
Dr.	Driver
Drink.	Drinking
D.S.	Double Strength
D.S.A.	Double Strength A Grade
D.S.B.	Double Strength B Grade
Dty.	Duty
DWV	Drain Waste Vent
DX	Deluxe White, Direct Expansion
dyn	Dyne
e	Eccentricity
E	Equipment Only; East
Ea.	Each
E.B.	Encased Burial
Econ.	Economy
E.C.Y	Embankment Cubic Yards
EDP	Electronic Data Processing
EIFS	Exterior Insulation Finish System
E.D.R.	Equiv. Direct Radiation
Eq.	Equation
Elec.	Electrician; Electrical
Elev.	Elevator; Elevating
EMT	Electrical Metallic Conduit; Thin Wall Conduit
Eng.	Engine, Engineered
EPDM	Ethylene Propylene Diene Monomer
EPS	Expanded Polystyrene
Eqhv.	Equip. Oper., Heavy
Eqlt.	Equip. Oper., Light
Eqmd.	Equip. Oper., Medium
Eqmm.	Equip. Oper., Master Mechanic
Eqol.	Equip. Oper., Oilers
Equip.	Equipment
ERW	Electric Resistance Welded
E.S.	Energy Saver
Est.	Estimated
esu	Electrostatic Units
E.W.	Each Way
EWT	Entering Water Temperature
Excav.	Excavation
Exp.	Expansion, Exposure
Ext.	Exterior
Extru.	Extrusion
f.	Fiber stress
F	Fahrenheit; Female; Fill
Fab.	Fabricated
FBGS	Fiberglass
F.C.	Footcandles
f.c.c.	Face-centered Cubic
f'c.	Compressive Stress in Concrete; Extreme Compressive Stress
F.E.	Front End
FEP	Fluorinated Ethylene Propylene (Teflon)
F.G.	Flat Grain
F.H.A.	Federal Housing Administration
Fig.	Figure
Fin.	Finished
Fixt.	Fixture
Fl. Oz.	Fluid Ounces
Flr.	Floor
F.M.	Frequency Modulation; Factory Mutual
Fmg.	Framing
Fndtn.	Foundation

Abbreviations

Fori.	Foreman, Inside
Foro.	Foreman, Outside
Fount.	Fountain
FPM	Feet per Minute
FPT	Female Pipe Thread
Fr.	Frame
F.R.	Fire Rating
FRK	Foil Reinforced Kraft
FRP	Fiberglass Reinforced Plastic
FS	Forged Steel
FSC	Cast Body; Cast Switch Box
Ft.	Foot; Feet
Ftng.	Fitting
Ftg.	Footing
Ft. Lb.	Foot Pound
Furn.	Furniture
FVNR	Full Voltage Non-Reversing
FXM	Female by Male
Fy.	Minimum Yield Stress of Steel
g	Gram
G	Gauss
Ga.	Gauge
Gal.	Gallon
Gal./Min.	Gallon per Minute
Galv.	Galvanized
Gen.	General
G.F.I.	Ground Fault Interrupter
Glaz.	Glazier
GPD	Gallons per Day
GPH	Gallons per Hour
GPM	Gallons per Minute
GR	Grade
Gran.	Granular
Grnd.	Ground
H	High; High Strength Bar Joist; Henry
H.C.	High Capacity
H.D.	Heavy Duty; High Density
H.D.O.	High Density Overlaid
Hdr.	Header
Hdwe.	Hardware
Help.	Helper Average
HEPA	High Efficiency Particulate Air Filter
Hg	Mercury
HIC	High Interrupting Capacity
HM	Hollow Metal
H.O.	High Output
Horiz.	Horizontal
H.P.	Horsepower; High Pressure
H.P.F.	High Power Factor
Hr.	Hour
Hrs./Day	Hours per Day
HSC	High Short Circuit
Ht.	Height
Htg.	Heating
Htrs.	Heaters
HVAC	Heating, Ventilation & Air-Conditioning
Hvy.	Heavy
HW	Hot Water
Hyd.;Hydr.	Hydraulic
Hz.	Hertz (cycles)
I.	Moment of Inertia
I.C.	Interrupting Capacity
ID	Inside Diameter
I.D.	Inside Dimension; Identification
I.F.	Inside Frosted
I.M.C.	Intermediate Metal Conduit
In.	Inch
Incan.	Incandescent
Incl.	Included; Including
Int.	Interior
Inst.	Installation
Insul.	Insulation/Insulated
I.P.	Iron Pipe
I.P.S.	Iron Pipe Size
I.P.T.	Iron Pipe Threaded
I.W.	Indirect Waste
J	Joule
J.I.C.	Joint Industrial Council
K	Thousand; Thousand Pounds; Heavy Wall Copper Tubing, Kelvin
K.A.H.	Thousand Amp. Hours
KCMIL	Thousand Circular Mils
KD	Knock Down
K.D.A.T.	Kiln Dried After Treatment
kg	Kilogram
kG	Kilogauss
kgf	Kilogram Force
kHz	Kilohertz
Kip.	1000 Pounds
KJ	Kiljoule
K.L.	Effective Length Factor
K.L.F.	Kips per Linear Foot
Km	Kilometer
K.S.F.	Kips per Square Foot
K.S.I.	Kips per Square Inch
kV	Kilovolt
kVA	Kilovolt Ampere
K.V.A.R.	Kilovar (Reactance)
KW	Kilowatt
KWh	Kilowatt-hour
L	Labor Only; Length; Long; Medium Wall Copper Tubing
Lab.	Labor
lat	Latitude
Lath.	Lather
Lav.	Lavatory
lb.; #	Pound
L.B.	Load Bearing; L Conduit Body
L. & E.	Labor & Equipment
lb./hr.	Pounds per Hour
lb./L.F.	Pounds per Linear Foot
lbf/sq.in.	Pound-force per Square Inch
L.C.L.	Less than Carload Lot
L.C.Y.	Loose Cubic Yard
Ld.	Load
LE	Lead Equivalent
LED	Light Emitting Diode
L.F.	Linear Foot
Lg.	Long; Length; Large
L & H	Light and Heat
LH	Long Span Bar Joist
L.H.	Labor Hours
L.L.	Live Load
L.L.D.	Lamp Lumen Depreciation
lm	Lumen
lm/sf	Lumen per Square Foot
lm/W	Lumen per Watt
L.O.A.	Length Over All
log	Logarithm
L-O-L	Lateralolet
L.P.	Liquefied Petroleum; Low Pressure
L.P.F.	Low Power Factor
LR	Long Radius
L.S.	Lump Sum
Lt.	Light
Lt. Ga.	Light Gauge
L.T.L.	Less than Truckload Lot
Lt. Wt.	Lightweight
L.V.	Low Voltage
M	Thousand; Material; Male; Light Wall Copper Tubing
M^2CA	Meters Squared Contact Area
m/hr; M.H.	Man-hour
mA	Milliampere
Mach.	Machine
Mag. Str.	Magnetic Starter
Maint.	Maintenance
Marb.	Marble Setter
Mat; Mat'l.	Material
Max.	Maximum
MBF	Thousand Board Feet
MBH	Thousand BTU's per hr.
MC	Metal Clad Cable
M.C.F.	Thousand Cubic Feet
M.C.F.M.	Thousand Cubic Feet per Minute
M.C.M.	Thousand Circular Mils
M.C.P.	Motor Circuit Protector
MD	Medium Duty
M.D.O.	Medium Density Overlaid
Med.	Medium
MF	Thousand Feet
M.F.B.M.	Thousand Feet Board Measure
Mfg.	Manufacturing
Mfrs.	Manufacturers
mg	Milligram
MGD	Million Gallons per Day
MGPH	Thousand Gallons per Hour
MH, M.H.	Manhole; Metal Halide; Man-Hour
MHz	Megahertz
Mi.	Mile
MI	Malleable Iron; Mineral Insulated
mm	Millimeter
Mill.	Millwright
Min., min.	Minimum, minute
Misc.	Miscellaneous
ml	Milliliter, Mainline
M.L.F.	Thousand Linear Feet
Mo.	Month
Mobil.	Mobilization
Mog.	Mogul Base
MPH	Miles per Hour
MPT	Male Pipe Thread
MRT	Mile Round Trip
ms	Millisecond
M.S.F.	Thousand Square Feet
Mstz.	Mosaic & Terrazzo Worker
M.S.Y.	Thousand Square Yards
Mtd.	Mounted
Mthe.	Mosaic & Terrazzo Helper
Mtng.	Mounting
Mult.	Multi; Multiply
M.V.A.	Million Volt Amperes
M.V.A.R.	Million Volt Amperes Reactance
MV	Megavolt
MW	Megawatt
MXM	Male by Male
MYD	Thousand Yards
N	Natural; North
nA	Nanoampere
NA	Not Available; Not Applicable
N.B.C.	National Building Code
NC	Normally Closed
N.E.M.A.	National Electrical Manufacturers Assoc.
NEHB	Bolted Circuit Breaker to 600V.
N.L.B.	Non-Load-Bearing
NM	Non-Metallic Cable
nm	Nanometer
No.	Number
NO	Normally Open
N.O.C.	Not Otherwise Classified
Nose.	Nosing
N.P.T.	National Pipe Thread
NQOD	Combination Plug-on/Bolt on Circuit Breaker to 240V.
N.R.C.	Noise Reduction Coefficient
N.R.S.	Non Rising Stem
ns	Nanosecond
nW	Nanowatt
OB	Opposing Blade
OC	On Center
OD	Outside Diameter
O.D.	Outside Dimension
ODS	Overhead Distribution System
O.G.	Ogee
O.H.	Overhead
O&P	Overhead and Profit
Oper.	Operator
Opng.	Opening
Orna.	Ornamental
OSB	Oriented Strand Board

O.S.&Y.	Outside Screw and Yoke
Ovhd.	Overhead
OWG	Oil, Water or Gas
Oz.	Ounce
P.	Pole; Applied Load; Projection
p.	Page
Pape.	Paperhanger
P.A.P.R.	Powered Air Purifying Respirator
PAR	Parabolic Reflector
Pc., Pcs.	Piece, Pieces
P.C.	Portland Cement; Power Connector
P.C.F.	Pounds per Cubic Foot
P.C.M.	Phase Contract Microscopy
P.E.	Professional Engineer; Porcelain Enamel; Polyethylene; Plain End
Perf.	Perforated
Ph.	Phase
P.I.	Pressure Injected
Pile.	Pile Driver
Pkg.	Package
Pl.	Plate
Plah.	Plasterer Helper
Plas.	Plasterer
Pluh.	Plumbers Helper
Plum.	Plumber
Ply.	Plywood
p.m.	Post Meridiem
Pntd.	Painted
Pord.	Painter, Ordinary
pp	Pages
PP; PPL	Polypropylene
P.P.M.	Parts per Million
Pr.	Pair
P.E.S.B.	Pre-engineered Steel Building
Prefab.	Prefabricated
Prefin.	Prefinished
Prop.	Propelled
PSF; psf	Pounds per Square Foot
PSI; psi	Pounds per Square Inch
PSIG	Pounds per Square Inch Gauge
PSP	Plastic Sewer Pipe
Pspr.	Painter, Spray
Psst.	Painter, Structural Steel
P.T.	Potential Transformer
P. & T.	Pressure & Temperature
Ptd.	Painted
Ptns.	Partitions
Pu	Ultimate Load
PVC	Polyvinyl Chloride
Pvmt.	Pavement
Pwr.	Power
Q	Quantity Heat Flow
Quan.;Qty.	Quantity
Q.C.	Quick Coupling
r	Radius of Gyration
R	Resistance
R.C.P.	Reinforced Concrete Pipe
Rect.	Rectangle
Reg.	Regular
Reinf.	Reinforced
Req'd.	Required
Res.	Resistant
Resi.	Residential
Rgh.	Rough
RGS	Rigid Galvanized Steel
R.H.W.	Rubber, Heat & Water Resistant; Residential Hot Water
rms	Root Mean Square
Rnd.	Round
Rodm.	Rodman
Rofc.	Roofer, Composition
Rofp.	Roofer, Precast
Rohe.	Roofer Helpers (Composition)
Rots.	Roofer, Tile & Slate
R.O.W.	Right of Way
RPM	Revolutions per Minute
R.S.	Rapid Start
Rsr	Riser
RT	Round Trip
S.	Suction; Single Entrance; South
SCFM	Standard Cubic Feet per Minute
Scaf.	Scaffold
Sch.; Sched.	Schedule
S.C.R.	Modular Brick
S.D.	Sound Deadening
S.D.R.	Standard Dimension Ratio
S.E.	Surfaced Edge
Sel.	Select
S.E.R.; S.E.U.	Service Entrance Cable
S.F.	Square Foot
S.F.C.A.	Square Foot Contact Area
S.F. Flr.	Square Foot of Floor
S.F.G.	Square Foot of Ground
S.F. Hor.	Square Foot Horizontal
S.F.R.	Square Feet of Radiation
S.F. Shlf.	Square Foot of Shelf
S4S	Surface 4 Sides
Shee.	Sheet Metal Worker
Sin.	Sine
Skwk.	Skilled Worker
SL	Saran Lined
S.L.	Slimline
Sldr.	Solder
SLH	Super Long Span Bar Joist
S.N.	Solid Neutral
S-O-L	Socketolet
sp	Standpipe
S.P.	Static Pressure; Single Pole; Self-Propelled
Spri.	Sprinkler Installer
spwg	Static Pressure Water Gauge
S.P.D.T.	Single Pole, Double Throw
SPF	Spruce Pine Fir
S.P.S.T.	Single Pole, Single Throw
SPT	Standard Pipe Thread
Sq.	Square; 100 Square Feet
Sq. Hd.	Square Head
Sq. In.	Square Inch
S.S.	Single Strength; Stainless Steel
S.S.B.	Single Strength B Grade
sst	Stainless Steel
Sswk.	Structural Steel Worker
Sswl.	Structural Steel Welder
St.;Stl.	Steel
S.T.C.	Sound Transmission Coefficient
Std.	Standard
STK	Select Tight Knot
STP	Standard Temperature & Pressure
Stpi.	Steamfitter, Pipefitter
Str.	Strength; Starter; Straight
Strd.	Stranded
Struct.	Structural
Sty.	Story
Subj.	Subject
Subs.	Subcontractors
Surf.	Surface
Sw.	Switch
Swbd.	Switchboard
S.Y.	Square Yard
Syn.	Synthetic
S.Y.P.	Southern Yellow Pine
Sys.	System
t.	Thickness
T	Temperature; Ton
Tan	Tangent
T.C.	Terra Cotta
T & C	Threaded and Coupled
T.D.	Temperature Difference
T.E.M.	Transmission Electron Microscopy
TFE	Tetrafluoroethylene (Teflon)
T. & G.	Tongue & Groove; Tar & Gravel
Th.;Thk.	Thick
Thn.	Thin
Thrded	Threaded
Tilf.	Tile Layer, Floor
Tilh.	Tile Layer, Helper
THHN	Nylon Jacketed Wire
THW.	Insulated Strand Wire
THWN;	Nylon Jacketed Wire
T.L.	Truckload
T.M.	Track Mounted
Tot.	Total
T-O-L	Threadolet
T.S.	Trigger Start
Tr.	Trade
Transf.	Transformer
Trhv.	Truck Driver, Heavy
Trlr	Trailer
Trlt.	Truck Driver, Light
TV	Television
T.W.	Thermoplastic Water Resistant Wire
UCI	Uniform Construction Index
UF	Underground Feeder
UGND	Underground Feeder
U.H.F.	Ultra High Frequency
U.L.	Underwriters Laboratory
Unfin.	Unfinished
URD	Underground Residential Distribution
US	United States
USP	United States Primed
UTP	Unshielded Twisted Pair
V	Volt
V.A.	Volt Amperes
V.C.T.	Vinyl Composition Tile
VAV	Variable Air Volume
VC	Veneer Core
Vent.	Ventilation
Vert.	Vertical
V.F.	Vinyl Faced
V.G.	Vertical Grain
V.H.F.	Very High Frequency
VHO	Very High Output
Vib.	Vibrating
V.L.F.	Vertical Linear Foot
Vol.	Volume
VRP	Vinyl Reinforced Polyester
W	Wire; Watt; Wide; West
w/	With
W.C.	Water Column; Water Closet
W.F.	Wide Flange
W.G.	Water Gauge
Wldg.	Welding
W. Mile	Wire Mile
W-O-L	Weldolet
W.R.	Water Resistant
Wrck.	Wrecker
W.S.P.	Water, Steam, Petroleum
WT., Wt.	Weight
WWF	Welded Wire Fabric
XFER	Transfer
XFMR	Transformer
XHD	Extra Heavy Duty
XHHW; XLPE	Cross-Linked Polyethylene Wire Insulation
XLP	Cross-linked Polyethylene
Y	Wye
yd	Yard
yr	Year
Δ	Delta
%	Percent
~	Approximately
Ø	Phase
@	At
#	Pound; Number
<	Less Than
>	Greater Than

A

B

C

D

E

F

INDEX

G

J

K

L

M

N

O

Q

R

S

U

V

W

X

Y

Z

Notes

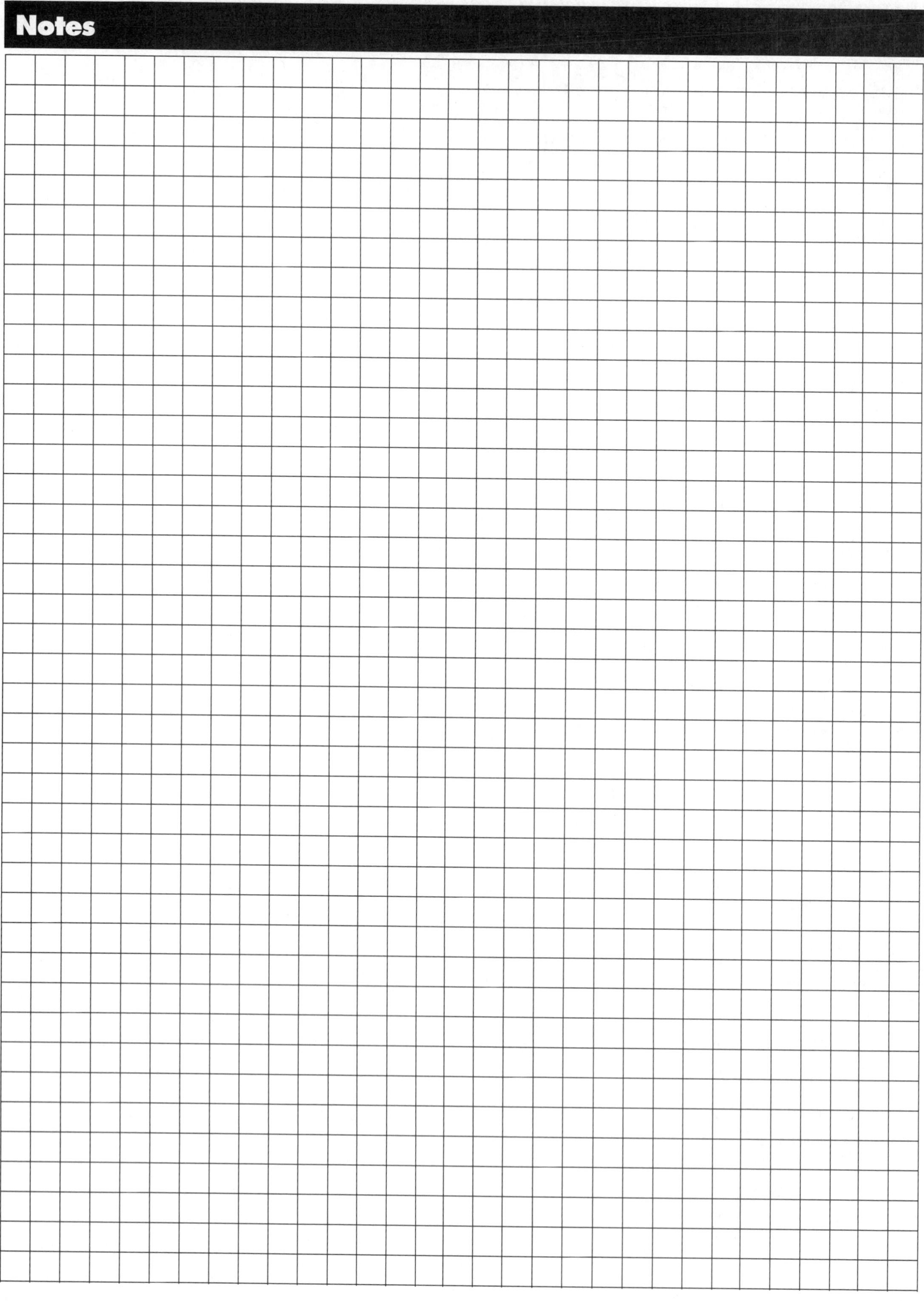

Notes

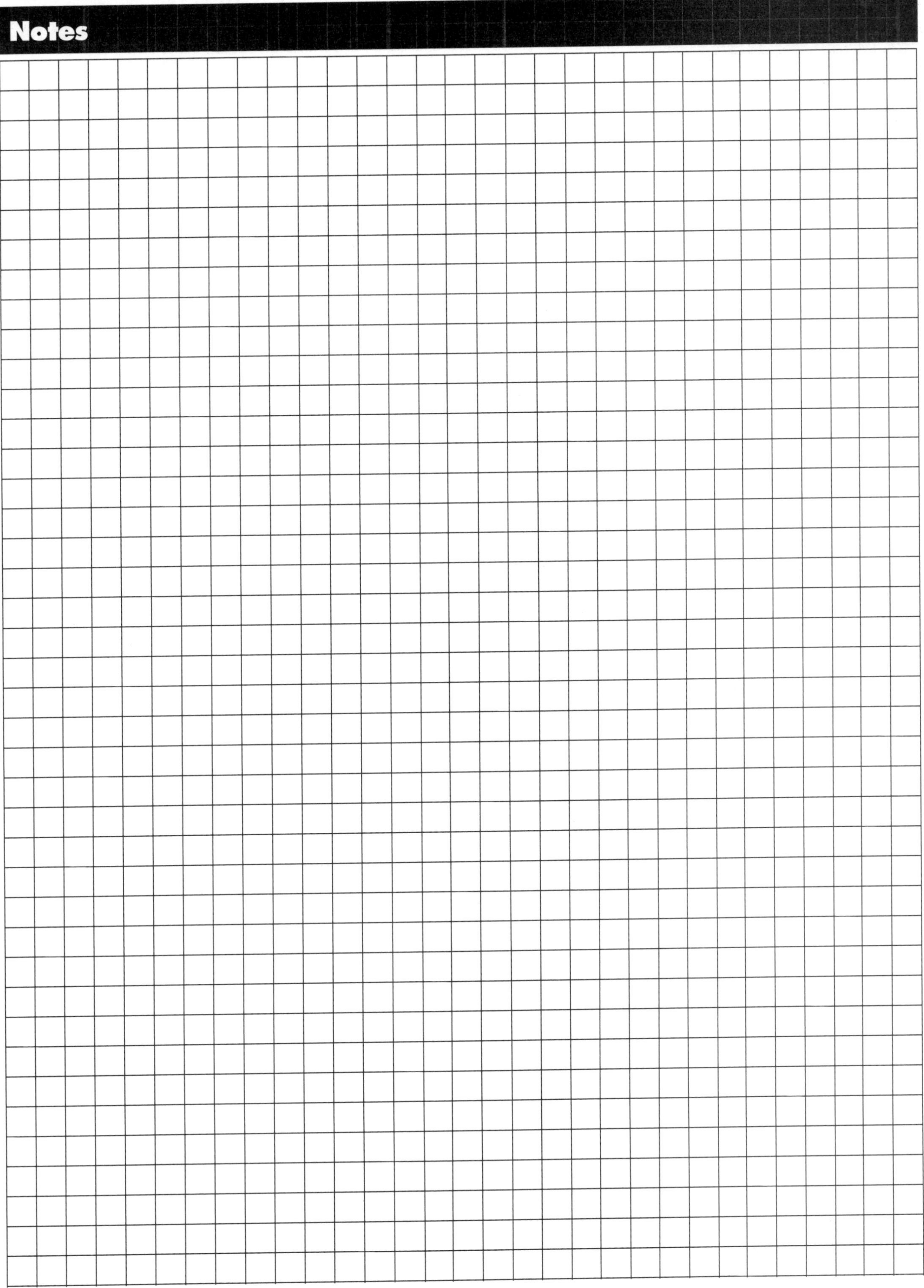

Notes

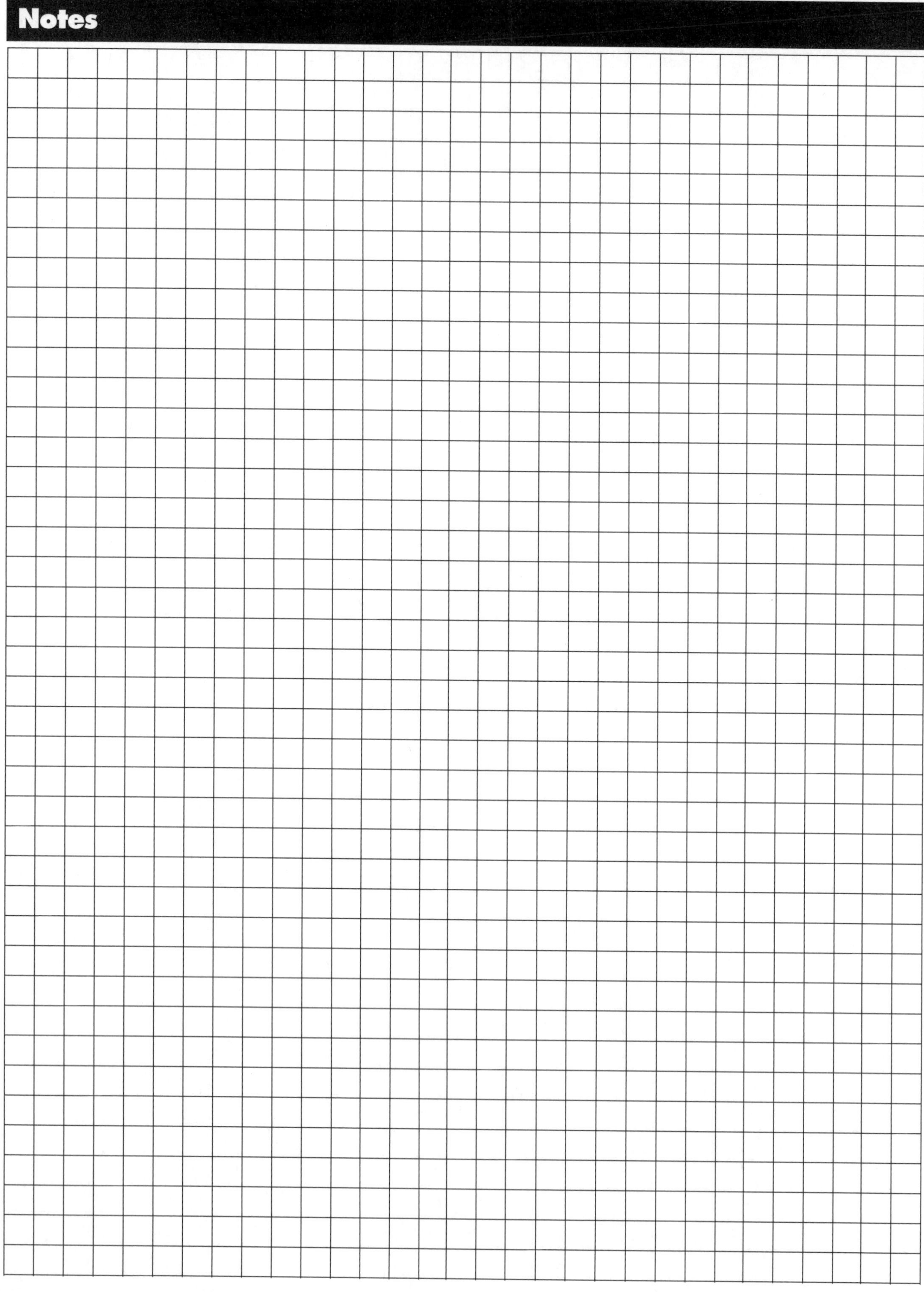

Notes

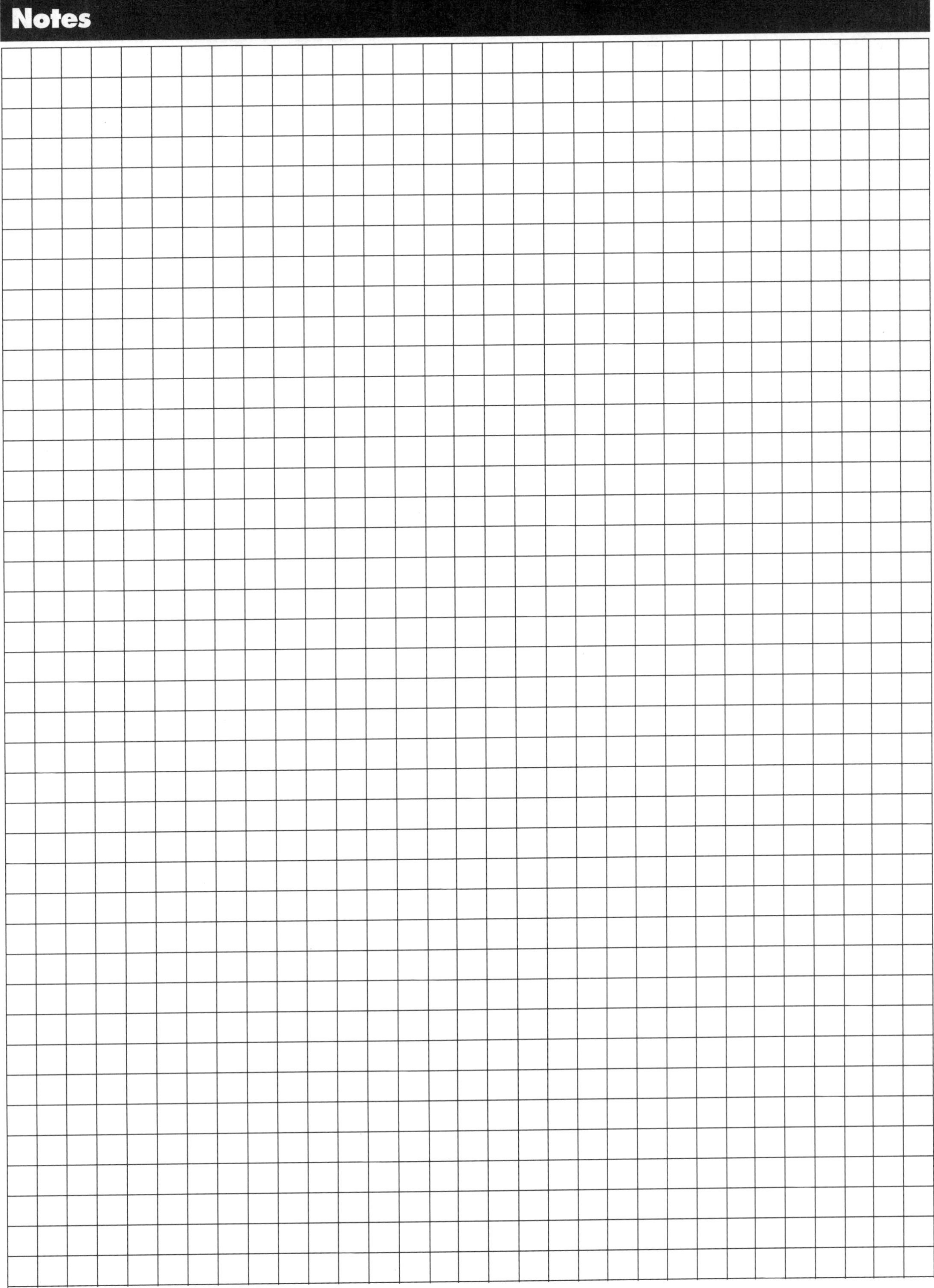

Notes

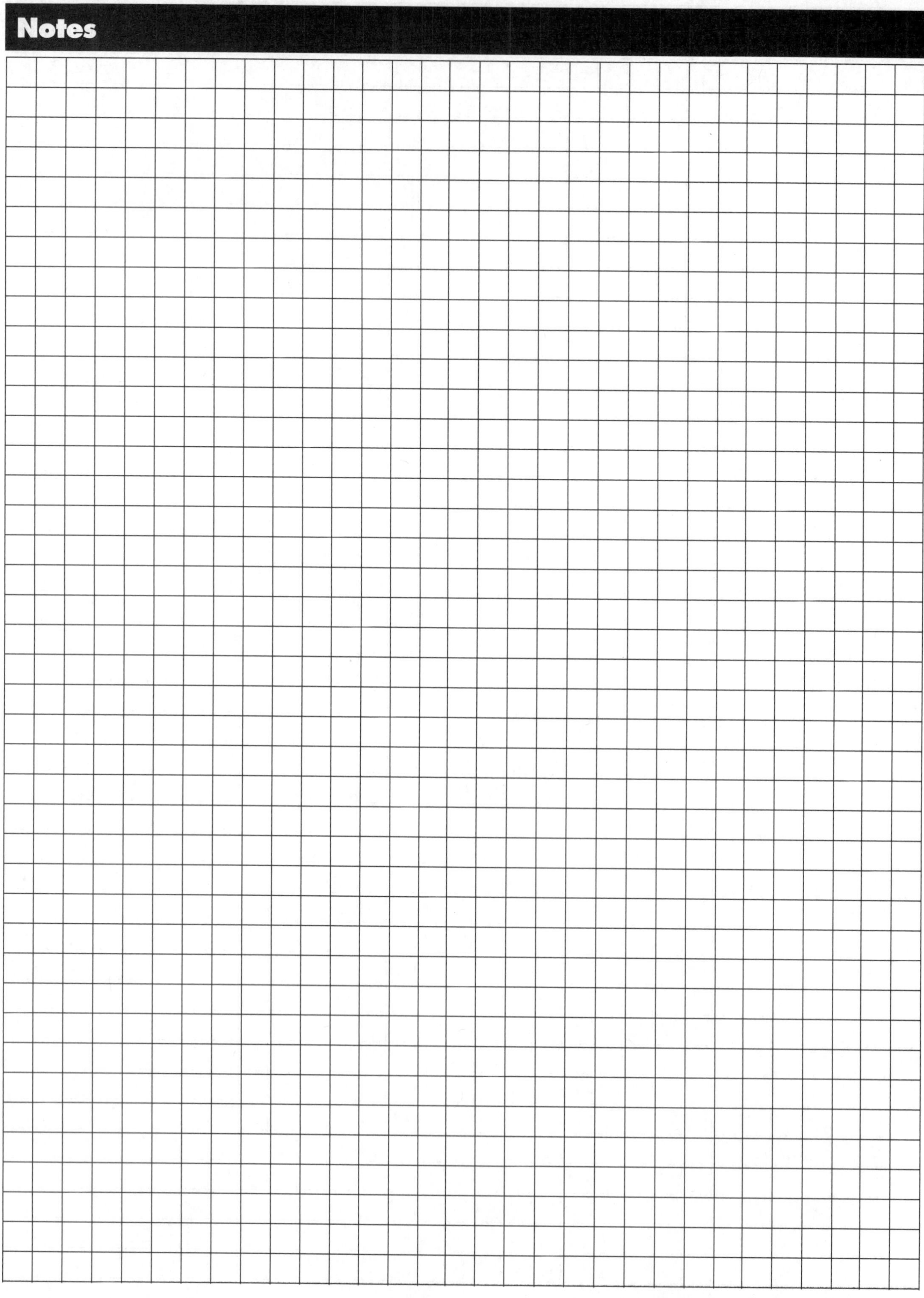

Notes

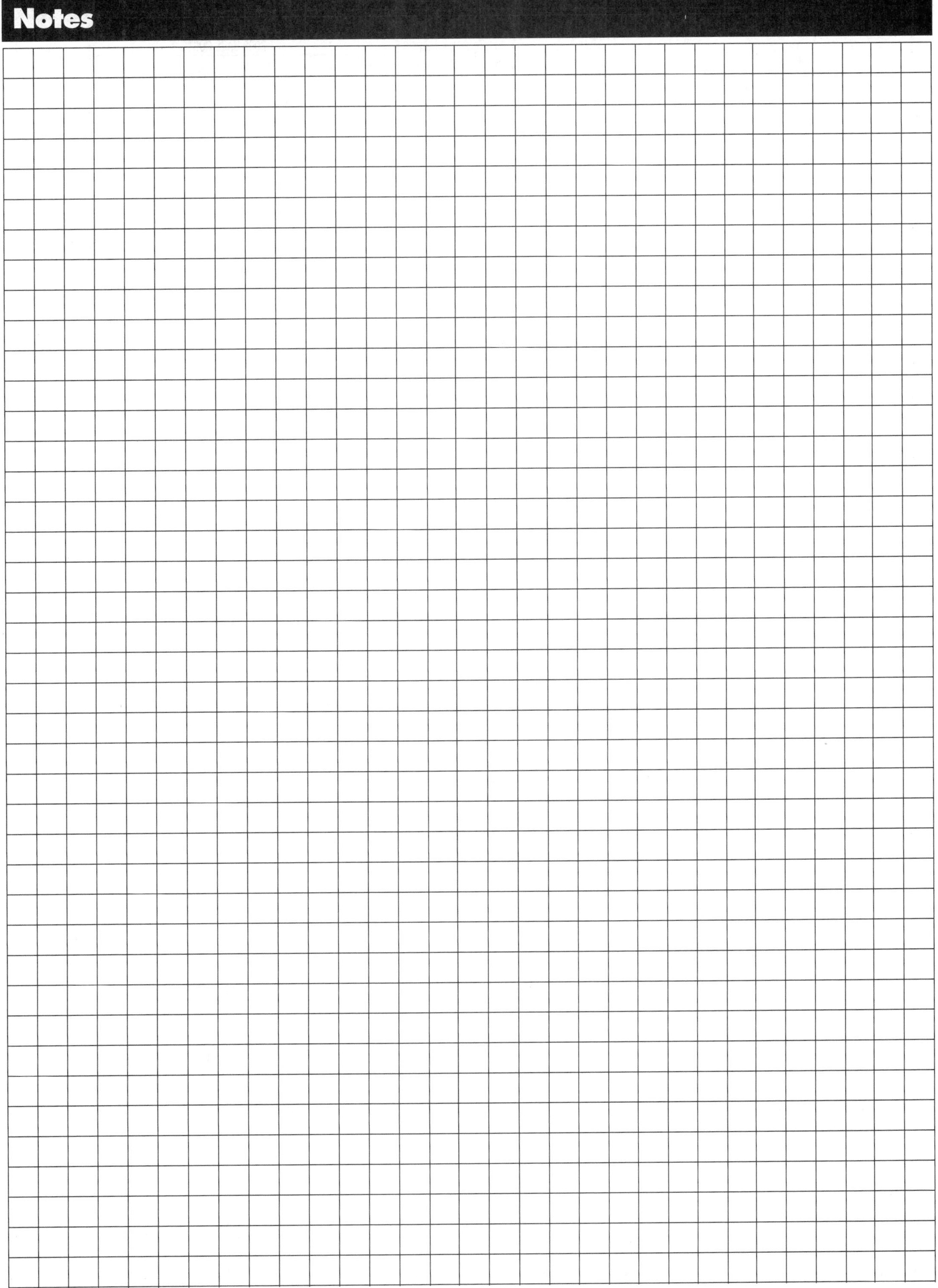

Division Notes

	CREW	DAILY OUTPUT	LABOR-HOURS	UNIT	2005 BARE COSTS MAT.	LABOR	EQUIP.	TOTAL	TOTAL INCL O&P

Division Notes

		CREW	DAILY OUTPUT	LABOR-HOURS	UNIT	2005 BARE COSTS MAT.	LABOR	EQUIP.	TOTAL	TOTAL INCL O&P

Division Notes

	CREW	DAILY OUTPUT	LABOR-HOURS	UNIT	2005 BARE COSTS MAT.	LABOR	EQUIP.	TOTAL	TOTAL INCL O&P

Division Notes

	CREW	DAILY OUTPUT	LABOR-HOURS	UNIT	2005 BARE COSTS MAT.	LABOR	EQUIP.	TOTAL	TOTAL INCL O&P

		CREW	DAILY OUTPUT	LABOR-HOURS	UNIT	2005 BARE COSTS MAT.	LABOR	EQUIP.	TOTAL	TOTAL INCL O&P

Division Notes

	CREW	DAILY OUTPUT	LABOR-HOURS	UNIT	2005 BARE COSTS				TOTAL INCL O&P
					MAT.	LABOR	EQUIP.	TOTAL	

Division Notes

		CREW	DAILY OUTPUT	LABOR-HOURS	UNIT	2005 BARE COSTS MAT.	LABOR	EQUIP.	TOTAL	TOTAL INCL O&P

Reed Construction Data, Inc.

Reed Construction Data, Inc., a leading worldwide provider of total construction information solutions, is comprised of three main product groups designed specifically to help construction professionals advance their businesses with timely, accurate and actionable project, product, and cost data. Reed Construction Data is a division of Reed Business Information, a member of the Reed Elsevier plc group of companies.

The *Project, Product, and Cost & Estimating* divisions offer a variety of innovative products and services designed for the full spectrum of design, construction, and manufacturing professionals. Through it's *International* companies, Reed Construction Data's reputation for quality construction market data is growing worldwide.

Cost Information

RSMeans, the undisputed market leader and authority on construction costs, publishes current cost and estimating information in annual cost books and on the CostWorks CD-ROM. RSMeans furnishes the construction industry with a rich library of complementary reference books and a series of professional seminars that are designed to sharpen professional skills and maximize the effective use of cost estimating and management tools. RSMeans also provides construction cost consulting for Owners, Manufacturers, Designers, and Contractors.

Project Data

Reed Construction Data provides complete, accurate and relevant project information through all stages of construction. Customers are supplied industry data through leads, project reports, contact lists, plans and specifications surveys, market penetration analyses and sales evaluation reports. Any of these products can pinpoint a county, look at a state, or cover the country. Data is delivered via paper, e-mail, CD-ROM or the Internet.

Building Product Information

The First Source suite of products is the only integrated building product information system offered to the commercial construction industry for comparing and specifying building products. These print and online resources include *First Source,* CSI's SPEC-DATA™, CSI's MANU-SPEC™, First Source CAD, and Manufacturer Catalogs. Written by industry professionals and organized using CSI's MasterFormat™, construction professionals use this information to make better design decisions.

FirstSourceONL.com combines Reed Construction Data's project, product and cost data with news and information from Reed Business Information's *Building Design & Construction* and *Consulting-Specifying Engineer,* this industry-focused site offers easy and unlimited access to vital information for all construction professionals.

International

BIMSA/Mexico provides construction project news, product information, cost-data, seminars and consulting services to construction professionals in Mexico. Its subsidiary, PRISMA, provides job costing software.

Byggfakta Scandinavia AB, founded in 1936, is the parent company for the leaders of customized construction market data for Denmark, Estonia, Finland, Norway and Sweden. Each company fully covers the local construction market and provides information across several platforms including subscription, ad-hoc basis, electronically and on paper.

Reed Construction Data Canada serves the Canadian construction market with reliable and comprehensive project and product information services that cover all facets of construction. Core services include: *BuildSource, BuildSpec, BuildSelect,* product selection and specification tools available in print and on the Internet; Building Reports, a national construction project lead service; CanaData, statistical and forecasting information; *Daily Commercial News,* a construction newspaper reporting on news and projects in Ontario; and *Journal of Commerce,* reporting news in British Columbia and Alberta.

Cordell Building Information Services, with its complete range of project and cost and estimating services, is Australia's specialist in the construction information industry. Cordell provides in-depth and historical information on all aspects of construction projects and estimation, including several customized reports, construction and sales leads, and detailed cost information among others.

For more information, please visit our Web site at www.reedconstructiondata.com.

Reed Construction Data, Inc., Corporate Office
30 Technology Parkway South
Norcross, GA 30092-2912
(800) 322-6996
(800) 895-8661 (fax)
info@reedbusiness.com
www.reedconstructiondata.com

Means Project Cost Report

By filling out and returning the Project Description, you can receive a discount of $20.00 off any one of the Means products advertised in the following pages. The cost information required includes all items marked (✔) except those where no costs occurred. The sum of all major items should equal the Total Project Cost.

$20.00 Discount per product for each report you submit.

DISCOUNT PRODUCTS AVAILABLE—FOR U.S. CUSTOMERS ONLY—STRICTLY CONFIDENTIAL

Project Description (No remodeling projects, please.)

✔ Type Building ______________________

✔ Location ______________________

Capacity ______________________

✔ Frame ______________________

✔ Exterior ______________________

✔ Basement: full ☐ partial ☐ none ☐ crawl ☐

✔ Height in Stories ______________________

✔ Total Floor Area ______________________

Ground Floor Area ______________________

✔ Volume in C.F. ______________________

% Air Conditioned ______________ Tons ________

Comments ______________________

Owner ______________________

Architect ______________________

General Contractor ______________________

✔ Bid Date ______________________

Typical Bay Size ______________________

✔ Labor Force: _____ % Union _____ % Non-Union

✔ Project Description (Circle one number in each line)

1. Economy 2. Average 3. Custom 4. Luxury

1. Square 2. Rectangular 3. Irregular 4. Very Irregular

	✔	**Total Project Cost**		$
A	✔	**General Conditions**		$
B	✔	**Site Work**		$
BS		**Site Clearing & Improvement**		
BE		*Excavation*	(C.Y.)	
BF		*Caissons & Piling*	(L.F.)	
BU		*Site Utilities*		
BP		*Roads & Walks Exterior Paving*	(S.Y.)	
C	✔	**Concrete**		$
C		*Cast in Place*	(C.Y.)	
CP		*Precast*	(S.F.)	
D	✔	**Masonry**		$
DB		*Brick*	(M)	
DC		*Block*	(M)	
DT		*Tile*	(S.F.)	
DS		*Stone*	(S.F.)	
E	✔	**Metals**		$
ES		*Structural Steel*	(Tons)	
EM		*Misc. & Ornamental Metals*		
F	✔	**Wood & Plastics**		$
FR		*Rough Carpentry*	(MBF)	
FF		*Finish Carpentry*		
FM		*Architectural Millwork*		
G	✔	**Thermal & Moisture Protection**		$
GW		*Waterproofing-Dampproofing*	(S.F.)	
GN		*Insulation*	(S.F.)	
GR		*Roofing & Flashing*	(S.F.)	
GM		*Metal Siding/Curtain Wall*	(S.F.)	
H	✔	**Doors and Windows**		$
HD		*Doors*	(Ea.)	
HW		*Windows*	(S.F.)	
HH		*Finish Hardware*		
HG		*Glass & Glazing*	(S.F.)	
HS		*Storefronts*	(S.F.)	

J	✔	**Finishes**		$
JL		*Lath & Plaster*	(S.Y.)	
JD		*Drywall*	(S.F.)	
JM		*Tile & Marble*	(S.F.)	
JT		*Terrazzo*	(S.F.)	
JA		*Acoustical Treatment*	(S.F.)	
JC		*Carpet*	(S.Y.)	
JF		*Hard Surface Flooring*	(S.F.)	
JP		*Painting & Wall Covering*	(S.F.)	
K	✔	**Specialties**		$
KB		*Bathroom Partitions & Access.*	(S.F.)	
KF		*Other Partitions*	(S.F.)	
KL		*Lockers*	(Ea.)	
L	✔	**Equipment**		$
LK		*Kitchen*		
LS		*School*		
LO		*Other*		
M	✔	**Furnishings**		$
MW		*Window Treatment*		
MS		*Seating*	(Ea.)	
N	✔	**Special Construction**		$
NA		*Acoustical*	(S.F.)	
NB		*Prefab. Bldgs.*	(S.F.)	
NO		*Other*		
P	✔	**Conveying Systems**		$
PE		*Elevators*	(Ea.)	
PS		*Escalators*	(Ea.)	
PM		*Material Handling*		
Q	✔	**Mechanical**		$
QP		*Plumbing*	*(No. of fixtures)*	
QS		*Fire Protection (Sprinklers)*		
QF		*Fire Protection (Hose Standpipes)*		
QB		*Heating, Ventilating & A.C.*		
QH		*Heating & Ventilating*	*(BTU Output)*	
QA		*Air Conditioning*	(Tons)	
R	✔	**Electrical**		$
RL		*Lighting*	(S.F.)	
RP		*Power Service*		
RD		*Power Distribution*		
RA		*Alarms*		
RG		*Special Systems*		
S	✔	**Mech./Elec. Combined**		$

Product Name ______________________

Product Number ______________________

Your Name ______________________

Title ______________________

Company ______________________

☐ Company

☐ Home Street Address ______________________

City, State, Zip ______________________

☐ Please send _______ forms.

Please specify the Means product you wish to receive. Complete the address information as requested and return this form with your check (product cost less $20.00) to address below.

RSMeans Company, Inc.,
Square Foot Costs Department
P.O. Box 800
Kingston, MA 02364-9988

For more information
visit Means Web Site
at www.rsmeans.com

Reed Construction Data/RSMeans . . . a tradition of excellence in Construction Cost Information and Services since 1942.

Table of Contents

Book Selection Guide

The following table provides definitive information on the content of each cost data publication. The number of lines of data provided in each unit price or assemblies division, as well as the number of reference tables and crews is listed for each book. The presence of other elements such as an historical cost index, city cost indexes, square foot models or cross-referenced index is also indicated. You can use the table to help select the Means' book that has the quantity and type of information you most need in your work.

Unit Cost Divisions	Building Construction Costs	Mechanical	Electrical	Repair & Remodel.	Square Foot	Site Work Landsc.	Assemblies	Interior	Concrete Masonry	Open Shop	Heavy Construc.	Light Commercial	Facil. Construc.	Plumbing	Western Construction Costs	Residential
1	1104	822	907	1013		1045		847	1014	1102	1051	752	1528	874	1102	710
2	2505	1468	460	1450		7794		547	1387	2467	4978	692	4305	1584	2490	772
3	1401	113	100	726		1259		201	1782	1396	1404	229	1312	82	1399	248
4	835	18	0	647		657		574	1056	811	594	407	1058	0	821	334
5	1820	239	193	957		768		927	691	1788	1023	805	1802	316	1802	752
6	1489	82	78	1475		452		1406	318	1481	608	1626	1596	47	1826	1757
7	1264	158	74	1241		468		489	408	1264	349	956	1318	168	1264	752
8	1839	28	0	1907		313		1673	645	1821	51	1227	2034	0	1841	1193
9	1595	47	0	1424		233		1688	360	1547	159	1337	1809	47	1586	1216
10	861	47	25	498		193		700	170	863	0	394	906	233	861	217
11	1018	322	169	502		137		813	44	925	107	218	1173	291	925	104
12	307	0	0	47		210		1416	27	298	0	62	1435	0	298	64
13	1153	1005	375	494		388		884	75	1136	279	484	1783	928	1119	193
14	345	36	0	258		36		292	0	344	30	12	343	35	343	6
15	1987	13138	636	1779		1569		1174	59	1996	1766	1217	10865	9695	2017	826
16	1277	470	10063	1002		739		1108	55	1265	760	1081	9834	415	1226	552
17	427	354	427	0		0		0	0	427	0	0	427	356	427	0
Totals	**21227**	**18347**	**13507**	**15420**		**16261**		**14739**	**8091**	**20961**	**13159**	**11499**	**43528**	**15071**	**21347**	**9696**

Assembly Divisions	Building Construction Costs	Mechanical	Elec-trical	Repair & Remodel.	Square Foot	Site Work Landsc.	Assemblies	Interior	Concrete Masonry	Open Shop	Heavy Construc.	Light Commercial	Facil. Construc.	Plumbing	Western Construction Costs	Asm Div	Residential
A		19	0	192	150	540	612	0	550		542	149	24	0		1	374
B		0	0	809	2480	0	5590	333	1914		0	2024	144	0		2	217
C		0	0	635	862	0	1220	1568	145		0	767	238	0		3	588
D		1031	780	693	1823	0	2439	753	0		0	1310	1027	896		4	871
E		0	0	85	255	0	292	5	0		0	255	5	0		5	393
F		0	0	0	123	0	126	0	0		0	123	3	0		6	358
G		465	172	332	111	1856	584	0	482		432	110	113	560		7	299
																8	760
																9	80
																10	0
																11	0
																12	0
Totals		**1515**	**952**	**2746**	**5804**	**2396**	**10863**	**2659**	**3091**		**974**	**4738**	**1554**	**1456**			**3940**

Reference Section	Building Construction Costs	Mechanical	Electrical	Repair & Remodel.	Square Foot	Site Work Landsc.	Assemblies	Interior	Concrete Masonry	Open Shop	Heavy Construc.	Light Commercial	Facil. Construc.	Plumbing	Western Construction Costs	Residential
Tables	130	45	85	69	4	80	219	46	71	130	61	57	104	51	131	42
Models					102							43				32
Crews	410	410	410	391		410		410	410	393	410	393	391	410	410	393
City Cost Indexes	yes	yes	yes	yes	yes	yes	yes	yes	yes	yes	yes	yes	yes	yes	yes	yes
Historical Cost Indexes	yes	yes	yes	yes	yes	yes	yes	yes	yes	yes	yes	yes	yes	yes	yes	no

Annual Cost Guides

For more information visit Means Web Site at www.rsmeans.com

Means Building Construction Cost Data 2005

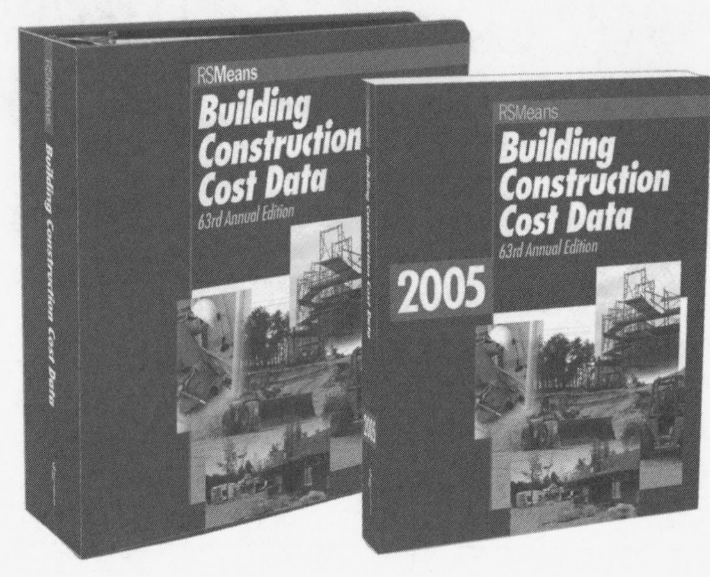

Available in Both Softbound and Looseleaf Editions

The "Bible" of the industry comes in the standard softcover edition or the looseleaf edition.

Many customers enjoy the convenience and flexibility of the looseleaf binder, which increases the usefulness of *Means Building Construction Cost Data 2005* by making it easy to add and remove pages. You can insert your own cost information pages, so everything is in one place. Copying pages for faxing is easier also. Whichever edition you prefer, softbound or the convenient looseleaf edition, you'll be eligible to receive *The Change Notice* FREE. Current subscribers receive *The Change Notice* via e-mail.

$115.95 per copy, Softbound
Catalog No. 60015

$144.95 per copy, Looseleaf
Catalog No. 61015

Means Building Construction Cost Data 2005

Offers you unchallenged unit price reliability in an easy-to-use arrangement. Whether used for complete, finished estimates or for periodic checks, it supplies more cost facts better and faster than any comparable source. Over 23,000 unit prices for 2005. The City Cost Indexes cover over 930 areas, for indexing to any project location in North America. Order and get *The Change Notice* FREE. You'll have year-long access to the Means Estimating **HOTLINE** FREE with your subscription. Expert assistance when using Means data is just a phone call away.

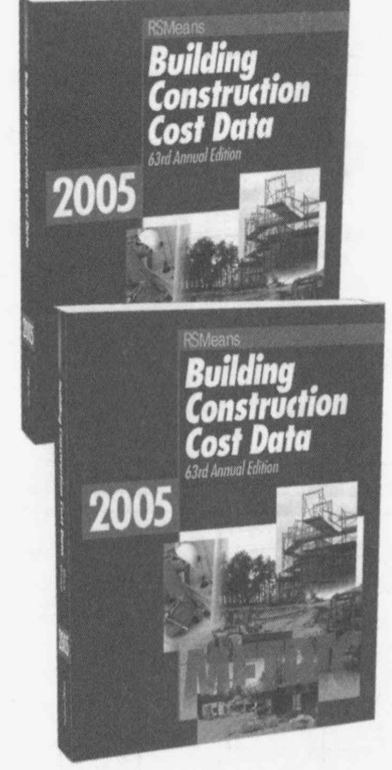

$115.95 per copy
Over 700 pages, illustrated, available Oct. 2004
Catalog No. 60015

Means Building Construction Cost Data 2005

Metric Version

The Federal Government has stated that all federal construction projects must now use metric documentation. The *Metric Version* of *Means Building Construction Cost Data 2005* is presented in metric measurements covering all construction areas. Don't miss out on these billion dollar opportunities. Make the switch to metric today.

$115.95 per copy
Over 700 pages, illus., available Dec. 2004
Catalog No. 63015

For more information
visit Means Web Site
at www.rsmeans.com

Annual Cost Guides

Means Mechanical Cost Data 2005

• HVAC • Controls

Total unit and systems price guidance for mechanical construction...materials, parts, fittings, and complete labor cost information. Includes prices for piping, heating, air conditioning, ventilation, and all related construction.

Plus new 2005 unit costs for:

- Over 2500 installed HVAC/controls assemblies
- "On Site" Location Factors for over 930 cities and towns in the U.S. and Canada
- Crews, labor and equipment

$115.95 per copy
Over 600 pages, illustrated, available Oct. 2004
Catalog No. 60025

RSMeans
Mechanical Cost Data
2005
RSMeans
Plumbing Cost Data
2005

Means Plumbing Cost Data 2005

Comprehensive unit prices and assemblies for plumbing, irrigation systems, commercial and residential fire protection, point-of-use water heaters, and the latest approved materials. This publication and its companion, *Means Mechanical Cost Data*, provide full-range cost estimating coverage for all the mechanical trades.

$115.95 per copy
Over 550 pages, illustrated, available Oct. 2004
Catalog No. 60215

Means Electrical Cost Data 2005

Pricing information for every part of electrical cost planning: More than 15,000 unit and systems costs with design tables; clear specifications and drawings; engineering guides and illustrated estimating procedures; complete labor-hour and materials costs for better scheduling and procurement; the latest electrical products and construction methods.

- A Variety of Special Electrical Systems including Cathodic Protection
- Costs for maintenance, demolition, HVAC/ mechanical, specialties, equipment, and more

$115.95 per copy
Over 450 pages, illustrated, available Oct. 2004
Catalog No. 60035

Means Electrical Change Order Cost Data 2005

You are provided with electrical unit prices exclusively for pricing change orders based on the recent, direct experience of contractors and suppliers. Analyze and check your own change order estimates against the experience others have had doing the same work. It also covers productivity analysis and change order cost justifications. With useful information for calculating the effects of change orders and dealing with their administration.

$115.95 per copy
Over 450 pages, available Oct. 2004
Catalog No. 60235

Means Facilities Maintenance & Repair Cost Data 2005

Published in a looseleaf format, *Means Facilities Maintenance & Repair Cost Data* gives you a complete system to manage and plan your facility repair and maintenance costs and budget efficiently. Guidelines for auditing a facility and developing an annual maintenance plan. Budgeting is included, along with reference tables on cost and management and information on frequency and productivity of maintenance operations.

The only nationally recognized source of maintenance and repair costs. Developed in cooperation with the Army Corps of Engineers.

$252.95 per copy
Over 600 pages, illustrated, available Dec. 2004
Catalog No. 60305

Means Square Foot Costs 2005

It's Accurate and Easy To Use!

- **Updated 2005 price information,** based on nationwide figures from suppliers, estimators, labor experts and contractors.
- "How-to-Use" Sections, with **clear examples** of commercial, residential, industrial, and institutional structures.
- Realistic graphics, offering true-to-life illustrations of building projects.
- Extensive information on using square foot cost data, including **sample estimates** and **alternate pricing methods.**

$126.95 per copy
Over 450 pages, illustrated, available Nov. 2004
Catalog No. 60055

Annual Cost Guides

Means Repair & Remodeling Cost Data 2005

Commercial/Residential

You can use this valuable tool to estimate commercial and residential renovation and remodeling.

Includes: New costs for hundreds of unique methods, materials and conditions that only come up in repair and remodeling. PLUS:

- Unit costs for over 16,000 construction components
- Installed costs for over 90 assemblies
- Costs for 300+ construction crews
- Over 930 "On Site" localization factors for the U.S. and Canada.

$99.95 per copy
Over 650 pages, illustrated, available Nov. 2004
Catalog No. 60045

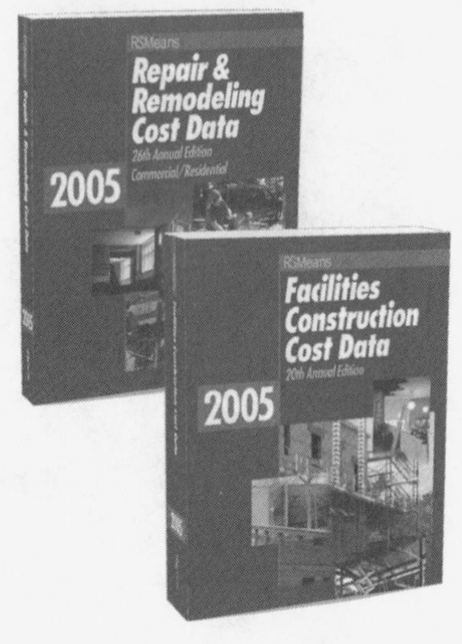

Means Facilities Construction Cost Data 2005

For the maintenance and construction of commercial, industrial, municipal, and institutional properties. Costs are shown for new and remodeling construction and are broken down into materials, labor, equipment, overhead, and profit. Special emphasis is given to sections on mechanical, electrical, furnishings, site work, building maintenance, finish work, and demolition. More than 45,000 unit costs, plus assemblies and reference sections are included.

$279.95 per copy
Over 1200 pages, illustrated, available Nov. 2004
Catalog No. 60205

Means Residential Cost Data 2005

Contains square foot costs for 30 basic home models with the look of today, plus hundreds of custom additions and modifications you can quote right off the page. With costs for the 100 residential systems you're most likely to use in the year ahead. Complete with blank estimating forms, sample estimates and step-by-step instructions.

$99.95 per copy
Over 600 pages, illustrated, available Oct. 2004
Catalog No. 60175

Means Light Commercial Cost Data 2005

Specifically addresses the light commercial market, which is an increasingly specialized niche in the industry. Aids you, the owner/designer/contractor, in preparing all types of estimates, from budgets to detailed bids. Includes new advances in methods and materials. Assemblies section allows you to evaluate alternatives in the early stages of design/planning.

Over 13,000 unit costs for 2005 ensure you have the prices you need...when you need them.

$99.95 per copy
Over 650 pages, illustrated, available Nov. 2004
Catalog No. 60185

Means Assemblies Cost Data 2005

Means Assemblies Cost Data 2005 takes the guesswork out of preliminary or conceptual estimates. Now you don't have to try to calculate the assembled cost by working up individual components costs. We've done all the work for you.

Presents detailed illustrations, descriptions, specifications and costs for every conceivable building assembly—240 types in all—arranged in the easy-to-use UNIFORMAT II system. Each illustrated "assembled" cost includes a complete grouping of materials and associated installation costs including the installing contractor's overhead and profit.

$189.95 per copy
Over 600 pages, illustrated, available Oct. 2004
Catalog No. 60065

Means Site Work & Landscape Cost Data 2005

Means Site Work & Landscape Cost Data 2005 is organized to assist you in all your estimating needs. Hundreds of fact-filled pages help you make accurate cost estimates efficiently.

Updated for 2005!

- Demolition features—including ceilings, doors, electrical, flooring, HVAC, millwork, plumbing, roofing, walls and windows
- State-of-the-art segmental retaining walls
- Flywheel trenching costs and details
- Updated Wells section
- Landscape materials, flowers, shrubs and trees

$115.95 per copy
Over 600 pages, illustrated, available Nov. 2004
Catalog No. 60285

For more information visit Means Web Site at www.rsmeans.com

Annual Cost Guides

Means Open Shop Building Construction Cost Data 2005

The latest costs for accurate budgeting and estimating of new commercial and residential construction... renovation work... change orders... cost engineering. *Means Open Shop BCCD* will assist you to...

- Develop benchmark prices for change orders
- Plug gaps in preliminary estimates, budgets
- Estimate complex projects
- Substantiate invoices on contracts
- Price ADA-related renovations

$115.95 per copy
Over 700 pages, illustrated, available Dec. 2004
Catalog No. 60155

Means Heavy Construction Cost Data 2005

A comprehensive guide to heavy construction costs. Includes costs for highly specialized projects such as tunnels, dams, highways, airports, and waterways. Information on different labor rates, equipment, and material costs is included. Has unit price costs, systems costs, and numerous reference tables for costs and design. Valuable not only to contractors and civil engineers, but also to government agencies and city/town engineers.

$115.95 per copy
Over 450 pages, illustrated, available Nov. 2004
Catalog No. 60165

Means Building Construction Cost Data 2005

Western Edition

This regional edition provides more precise cost information for western North America. Labor rates are based on union rates from 13 western states and western Canada. Included are western practices and materials not found in our national edition: tilt-up concrete walls, glu-lam structural systems, specialized timber construction, seismic restraints, landscape and irrigation systems.

$115.95 per copy
Over 650 pages, illustrated, available Dec. 2004
Catalog No. 60225

Means Heavy Construction Cost Data 2005

Metric Version

Make sure you have the Means industry standard metric costs for the federal, state, municipal and private marketplace. With thousands of up-to-date metric unit prices in tables by CSI standard divisions. Supplies you with assemblies costs using the metric standard for reliable cost projections in the design stage of your project. Helps you determine sizes, material amounts, and has tips for handling metric estimates.

$115.95 per copy
Over 450 pages, illustrated, available Dec. 2004
Catalog No. 63165

Means Construction Cost Indexes 2005

Who knows what 2005 holds? What materials and labor costs will change unexpectedly? By how much?

- Breakdowns for 316 major cities.
- National averages for 30 key cities.
- Expanded five major city indexes.
- Historical construction cost indexes.

$252.95 per year
$63.00 individual quarters
Catalog No. 60145 A,B,C,D

Means Interior Cost Data 2005

Provides you with prices and guidance needed to make accurate interior work estimates. Contains costs on materials, equipment, hardware, custom installations, furnishings, labor costs . . . every cost factor for new and remodel commercial and industrial interior construction, including updated information on office furnishings, plus more than 50 reference tables. For contractors, facility managers, owners.

$115.95 per copy
Over 600 pages, illustrated, available Oct. 2004
Catalog No. 60095

Means Concrete & Masonry Cost Data 2005

Provides you with cost facts for virtually all concrete/masonry estimating needs, from complicated formwork to various sizes and face finishes of brick and block, all in great detail. The comprehensive unit cost section contains more than 8,500 selected entries. Also contains an assemblies cost section, and a detailed reference section which supplements the cost data.

$105.95 per copy
Over 450 pages, illustrated, available Dec. 2004
Catalog No. 60115

Means Labor Rates for the Construction Industry 2005

Complete information for estimating labor costs, making comparisons and negotiating wage rates by trade for over 300 cities (United States and Canada). With 46 construction trades listed by local union number in each city, and historical wage rates included for comparison. No similar book is available through the trade.

Each city chart lists the county and is alphabetically arranged with handy visual flip tabs for quick reference.

$253.95 per copy
Over 300 pages, available Dec. 2004
Catalog No. 60125

Building Security: Strategies & Costs

By David Owen

Unauthorized systems access, terrorism, hurricanes and tornadoes, sabotage, vandalism, fire, explosions, and other threats... All are considerations as building owners, facility managers, and design and construction professionals seek ways to meet security needs.

This comprehensive resource will help you evaluate your facility's security needs — and design and budget for the materials and devices needed to fulfill them. The text and cost data will help you to:

- Identify threats, probability of occurrence, and the potential losses— to determine and address your real vulnerabilities.
- Perform a detailed risk assessment of an existing facility, and prioritize and budget for security enhancement.
- Evaluate and price security systems and construction solutions, so you can make cost-effective choices.

Includes over 130 pages of Means Cost Data for installation of security systems and materials, plus a review of more than 50 security devices and construction solutions—how they work, and how they compare.

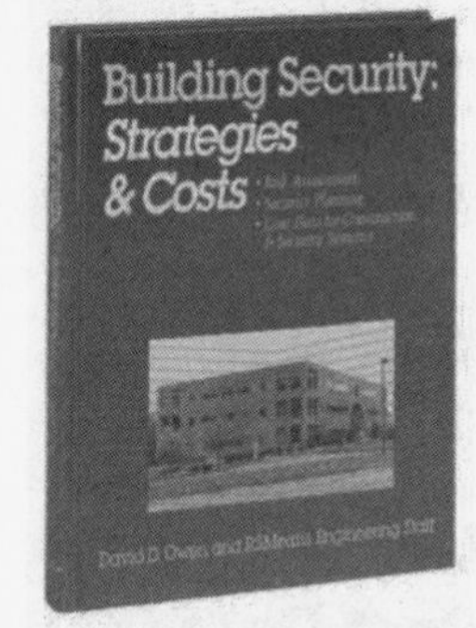

$89.95 per copy
Over 400 pages, Hardcover
Catalog No. 37339

Green Building: Project Planning & Cost Estimating

By RSMeans and Contributing Authors

Written by a team of leading experts in sustainable design, this new book is a complete guide to planning and estimating green building projects, a growing trend in building design and construction – commercial, industrial, institutional and residential. It explains:

- All the different criteria for "green-ness"
- What criteria your building needs to meet to get a LEED, Energy Star, or other recognized rating for green buildings
- How the project team works differently on a green versus a traditional building project
- How to select and specify green products
- How to evaluate the cost and value of green products versus conventional ones — not only for their first (installation) cost, but their cost over time (in maintenance and operation).

Features an extensive Green Building Cost Data section, which details the available products, how they are specified, and how much they cost.

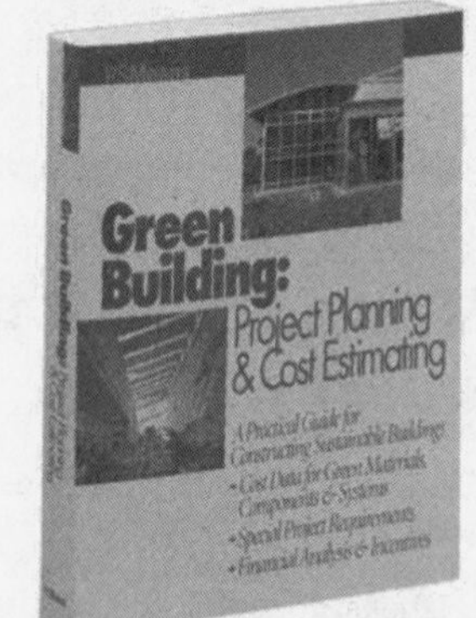

$89.95 per copy
350 pages, illustrated, Hardcover
Catalog No. 67338

Life Cycle Costing for Facilities

By Alphonse Dell'Isola and Dr. Steven Kirk

Guidance for achieving higher quality design and construction projects at lower costs!

Facility designers and owners are frustrated with cost-cutting efforts that yield the cheapest product, but sacrifice quality. Life Cycle Costing, properly done, enables them to achieve both—high quality, incorporating innovative design, and costs that meet their budgets.

The authors show how LCC can work for a broad variety of projects – from several types of buildings, to roads and bridges, to HVAC and electrical upgrades, to materials and equipment procurement. Case studies include:

- Health care and nursing facilities
- College campus and high schools
- Office buildings, courthouses, and banks
- Exterior walls, elevators, lighting, HVAC, and more

The book's extensive cost section provides maintenance and replacement costs for facility elements.

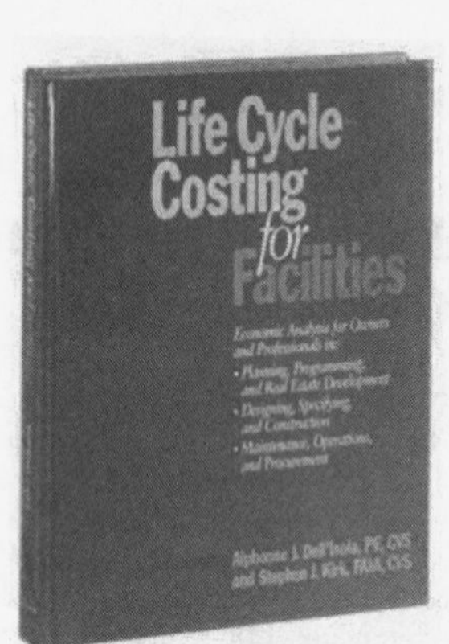

$99.95 per copy
450 pages, Hardcover
Catalog No. 67341

Reference Books

Value Engineering: Practical Applications

. . . For Design, Construction, Maintenance & Operations

By Alphonse Dell'Isola, PE

A tool for immediate application—for engineers, architects, facility managers, owners, and contractors. Includes: making the case for VE—the management briefing, integrating VE into planning and budgeting, conducting life cycle costing, integrating VE into the design process, using VE methodology in design review and consultant selection, case studies, VE workbook, and a life cycle costing program on disk.

$79.95 per copy
Over 450 pages, illustrated, Hardcover
Catalog No. 67319

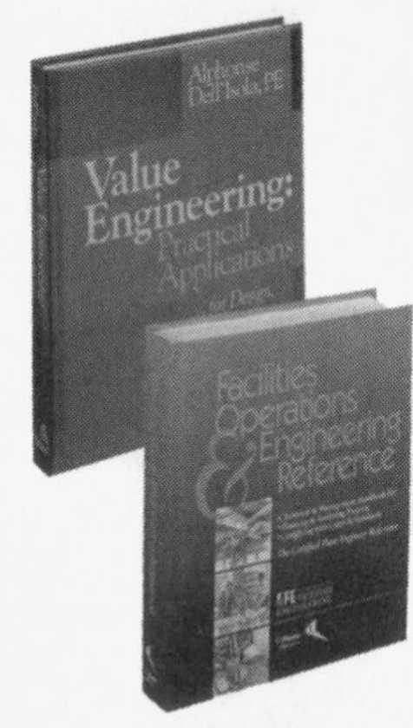

Facilities Operations & Engineering Reference

By the Association for Facilities Engineering and RSMeans

An all-in-one technical referance for planning and managing facility projects and solving day-to-day operations problems. Selected as the official Certified Plant Engineer reference, this handbook covers financial analysis, maintenance, HVAC and energy efficiency, and more.

$109.95 per copy
Over 700 pages, illustrated, Hardcover
Catalog No. 67318

The Building Professional's Guide to Contract Documents

3rd Edition

By Waller S. Poage, AIA, CSI, CVS

This comprehensive treatment of Contract Documents is an important reference for owners, design professionals, contractors, and students.

- Structure your Documents for Maximum Efficiency
- Effectively communicate construction requirements to all concerned
- Understand the Roles and Responsibilities of Construction Professionals
- Improve Methods of Project Delivery

$64.95 per copy, 400 pages
Diagrams and construction forms, Hardcover
Catalog No. 67261A

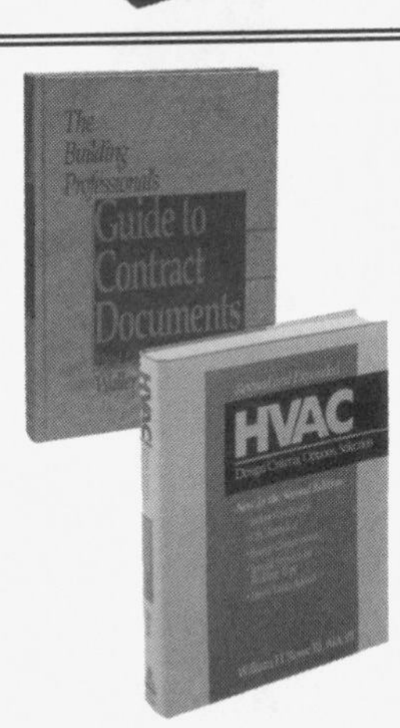

HVAC: Design Criteria, Options, Selection

Expanded 2nd Edition

By William H. Rowe III, AIA, PE

Includes Indoor Air Quality, CFC Removal, Energy Efficient Systems, and Special Systems by Building Type. Helps you solve a wide range of HVAC system design and selection problems effectively and economically. Gives you clear explanations of the latest ASHRAE standards.

$84.95 per copy
Over 600 pages, illustrated, Hardcover
Catalog No. 67306

Cost Planning & Estimating for Facilities Maintenance

In this unique book, a team of facilities management authorities shares their expertise at:

- Evaluating and budgeting maintenance operations
- Maintaining & repairing key building components
- Applying ***Means Facilities Maintenance & Repair Cost Data*** to your estimating

Covers special maintenance requirements of the 10 major building types.

$89.95 per copy
Over 475 pages, Hardcover
Catalog No. 67314

Facilities Maintenance Management

By Gregory H. Magee, PE

Now you can get successful management methods and techniques for all aspects of facilities maintenance. This comprehensive reference explains and demonstrates successful management techniques for all aspects of maintenance, repair, and improvements for buildings, machinery, equipment, and grounds. Plus, guidance for outsourcing and managing internal staffs.

$86.95 per copy
Over 280 pages with illustrations, Hardcover
Catalog No. 67249

Builder's Essentials: Advanced Framing Methods

By Scot Simpson

A highly illustrated, "framer-friendly" approach to advanced framing elements. Provides expert, but easy-to-interpret, instruction for laying out and framing complex walls, roofs, and stairs, and special requirements for earthquake and hurricane protection. Also helps bring framers up to date on the latest building code changes, and provides tips on the lead framer's role and responsibilities, how to prepare for a job, and how to get the crew started.

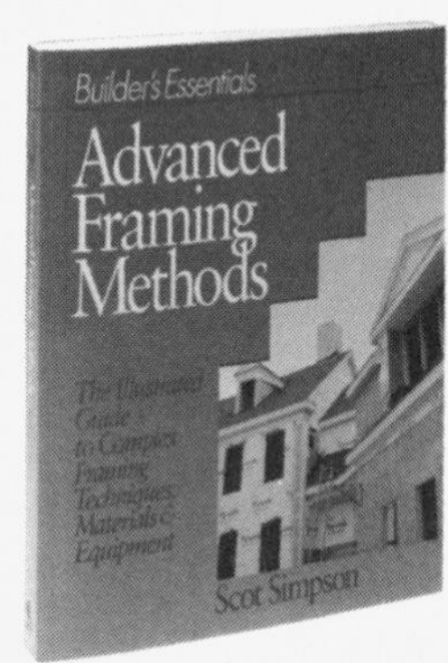

$24.95 per copy
250 pages, illustrated, Softcover
Catalog No. 67330

Reference Books

For more information visit Means Web Site at www.rsmeans.com

Interior Home Improvement Costs,

New 9th Edition

Estimates for the most popular remodeling and repair projects—from small, do-it-yourself jobs—to major renovations and new construction. Includes: Kitchens & Baths; New Living Space from Your Attic, Basement or Garage; New Floors, Paint & Wallpaper; Tearing Out or Building New Walls; Closets, Stairs & Fireplaces; New Energy-Saving Improvements, Home Theatres, and More!

$24.95 per copy
250 pages, illustrated, Softcover
Catalog No. 67308E

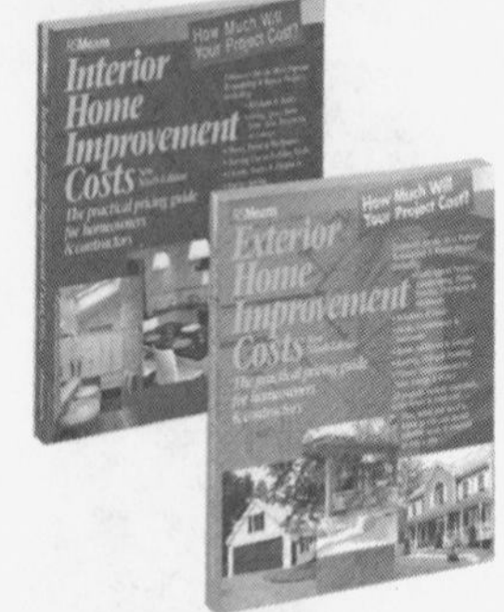

Exterior Home Improvement Costs

New 9th Edition

Estimates for the most popular remodeling and repair projects—from small, do-it-yourself jobs, to major renovations and new construction. Includes: Curb Appeal Projects—Landscaping, Patios, Porches, Driveways and Walkways; New Windows and Doors; Decks, Greenhouses, and Sunrooms; Room Additions and Garages; Roofing, Siding, and Painting; "Green" Improvements to Save Energy & Water

$24.95 per copy
Over 275 pages, illustrated, softcover
Catalog No. 67309E

Builder's Essentials: Plan Reading & Material Takeoff

By Wayne J. DelPico

For Residential and Light Commercial Construction

A valuable tool for understanding plans and specs, and accurately calculating material quantities.
Step-by-step instructions and takeoff procedures based on a full set of working drawings.

$35.95 per copy
Over 420 pages, Softcover
Catalog No. 67307

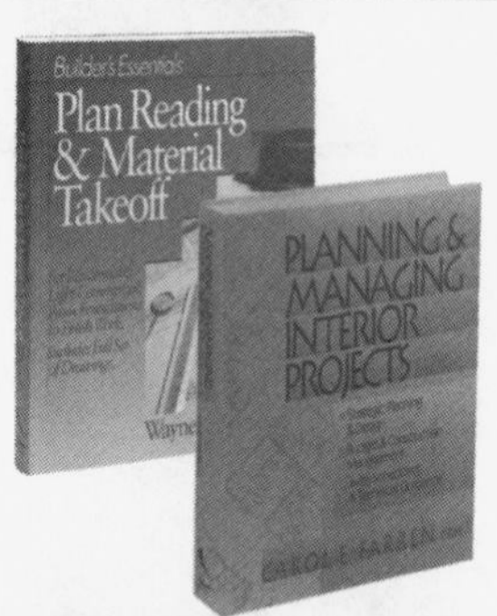

Planning & Managing Interior Projects

2nd Edition

By Carol E. Farren, CFM

Addresses changes in technology and business, guiding you through commercial design and construction from initial client meetings to post-project administration. Includes: evaluating space requirements, alternative work models, telecommunications and data management, and environmental issues.

$69.95 per copy
Over 400 pages, illustrated, Hardcover
Catalog No. 67245A

Builder's Essentials: Best Business Practices for Builders & Remodelers:

An Easy-to-Use Checklist System

By Thomas N. Frisby

A comprehensive guide covering all aspects of running a construction business, with more than 40 user-friendly checklists. This book provides expert guidance on: increasing your revenue and keeping more of your profit; planning for long-term growth; keeping good employees and managing subcontractors.

$29.95 per copy
Over 220 pages, Softcover
Catalog No. 67329

Builder's Essentials: Framing & Rough Carpentry 2nd Edition

By Scot Simpson

A complete training manual for apprentice and experienced carpenters. Develop and improve your skills with "framer-friendly," easy-to-follow instructions, and step-by-step illustrations. Learn proven techniques for framing walls, floors, roofs, stairs, doors, and windows. Updated guidance on standards, building codes, safety requirements, and more. Also available in Spanish!

$24.95 per copy
Over 150 pages, Softcover
Catalog No. 67298A Spanish Catalog No. 67298AS

How to Estimate with Means Data & CostWorks

By RSMeans and Saleh Mubarak

New 2nd Edition!

Learn estimating techniques using Means cost data. Includes an instructional version of Means CostWorks CD-ROM with Sample Building Plans.
The step-by-step guide takes you through all the major construction items. Over 300 sample estimating problems are included.

$59.95 per copy
Over 190 pages, Softcover
Catalog No. 67324A

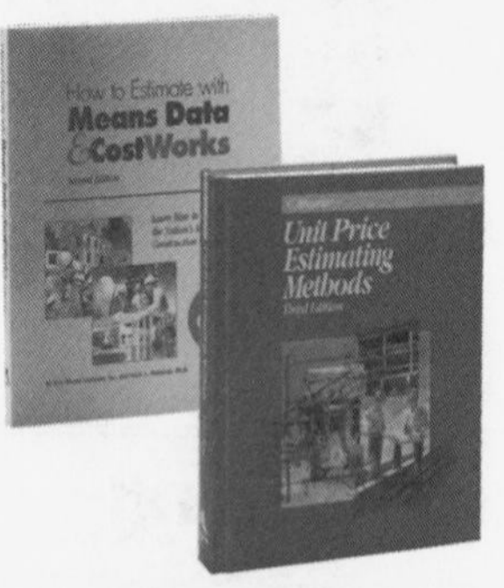

Unit Price Estimating Methods

New 3rd Edition

This new edition includes up-to-date cost data and estimating examples, updated to reflect changes to the CSI numbering system and new features of Means cost data. It describes the most productive, universally accepted ways to estimate, and uses checklists and forms to illustrate shortcuts and timesavers. A model estimate demonstrates procedures. A new chapter explores computer estimating alternatives.

$59.95 per copy
Over 350 pages, illustrated, Hardcover
Catalog No. 67303A

Means Landscape Estimating Methods

4th Edition

By Sylvia H. Fee

This revised edition offers expert guidance for preparing accurate estimates for new landscape construction and grounds maintenance. Includes a complete project estimate featuring the latest equipment and methods, and **two chapters on Life Cycle Costing, and Landscape Maintenance Estimating.**

$62.95 per copy
Over 300 pages, illustrated, Hardcover
Catalog No. 67295B

Means Environmental Remediation Estimating Methods, 2nd Edition

By Richard R. Rast

Guidelines for estimating 50 standard remediation technologies. Use it to prepare preliminary budgets, develop estimates, compare costs and solutions, estimate liability, review quotes, negotiate settlements.

A valuable support tool for ***Means Environmental Remediation Unit Price*** and ***Assemblies*** books.

$99.95 per copy
Over 750 pages, illustrated, Hardcover
Catalog No. 64777A

Means Illustrated Construction Dictionary, Condensed Edition, 2nd Edition

By RSMeans

Recognized in the industry as the best resource of its kind, this has been further enhanced with updates to existing terms and the addition of hundreds of new terms and illustrations . . . in keeping with the most recent developments in the industry.
The best portable dictionary for office or field use—an essential tool for contractors, architects, insurance and real estate personnel, homeowners, and anyone who needs quick, clear definitions for construction terms.

$59.95 per copy
Over 500 pages
Catalog No. 67282A

Means Repair and Remodeling Estimating

New 4th Edition
By Edward B. Wetherill & RSMeans

Focuses on the unique problems of estimating renovations of existing structures. It helps you determine the true costs of remodeling through careful evaluation of architectural details and a site visit.
New section on disaster restoration costs.

$69.95 per copy
Over 450 pages, illustrated, Hardcover
Catalog No. 67265B

Facilities Planning & Relocation

New, lower price and user-friendly format.
By David D. Owen

A complete system for planning space needs and managing relocations. Includes step-by-step manual, over 50 forms, and extensive reference section on materials and furnishings.

$89.95 per copy
Over 450 pages, Softcover
Catalog No. 67301

Means Square Foot & Assemblies Estimating Methods

3rd Edition!

Develop realistic Square Foot and Assemblies Costs for budgeting and construction funding. The new edition features updated guidance on square foot and assemblies estimating using UNIFORMAT II. An essential reference for anyone who performs conceptual estimates.

$69.95 per copy
Over 300 pages, illustrated, Hardcover
Catalog No. 67145B

Means Electrical Estimating Methods 3rd Edition

Expanded new edition includes sample estimates and cost information in keeping with the latest version of the CSI MasterFormat and UNIFORMAT II. Complete coverage of Fiber Optic and Uninterruptible Power Supply electrical systems, broken down by components and explained in detail. Includes a new chapter on computerized estimating methods. A practical companion to *Means Electrical Cost Data.*

$64.95 per copy
Over 325 pages, Hardcover
Catalog No. 67230A

Means Mechanical Estimating Methods 3rd Edition

This guide assists you in making a review of plans, specs, and bid packages with suggestions for takeoff procedures, listings, substitutions and pre-bid scheduling. Includes suggestions for budgeting labor and equipment usage. Compares materials and construction methods to allow you to select the best option.

$64.95 per copy
Over 350 pages, illustrated, Hardcover
Catalog No. 67294A

Means Spanish/English Construction Dictionary

By RSMeans, The International Conference of Building Officials (ICBO), and Rolf Jensen & Associates (RJA)

Designed to facilitate communication among Spanish- and English-speaking construction personnel—improving performance and job-site safety. Features the most common words and phrases used in the construction industry, with easy-to-follow pronunciations. Includes extensive building systems and tools illustrations.

$22.95 per copy
250 pages, illustrated, Softcover
Catalog No. 67327

Project Scheduling & Management for Construction

New 3rd Edition
By David R. Pierce, Jr.

A comprehensive yet easy-to-follow guide to construction project scheduling and control—from vital project management principles through the latest scheduling, tracking, and controlling techniques. The author is a leading authority on scheduling with years of field and teaching experience at leading academic institutions. Spend a few hours with this book and come away with a solid understanding of this essential management topic.

$64.95 per copy
Over 300 pages, illustrated, Hardcover
Catalog No. 67247B

Reference Books

For more information visit Means Web Site at www.rsmeans.com

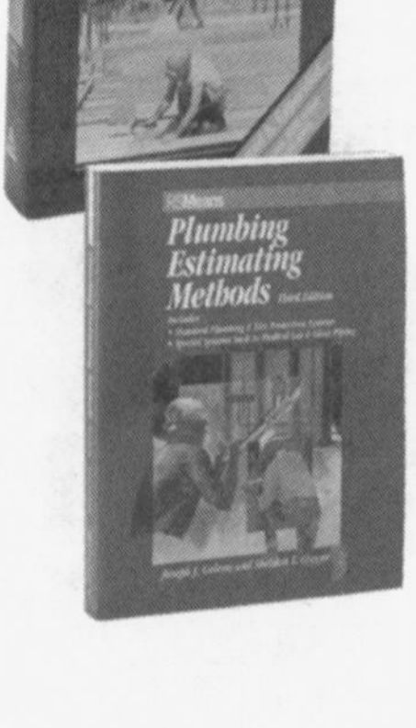

Concrete Repair and Maintenance Illustrated
By Peter H. Emmons
$69.95 per copy
Catalog No. 67146

Superintending for Contractors:
How to Bring Jobs in On-Time, On-Budget
By Paul J. Cook
$35.95 per copy
Catalog No. 67233

HVAC Systems Evaluation
By Harold R. Colen, PE
$84.95 per copy
Catalog No. 67281

Basics for Builders: How to Survive and Prosper in Construction
By Thomas N. Frisby
$34.95 per copy
Catalog No. 67273

Successful Interior Projects Through Effective Contract Documents
By Joel Downey & Patricia K. Gilbert
Now $24.98 per copy, limited quantity
Catalog No. 67313

Building Spec Homes Profitably
By Kenneth V. Johnson
$29.95 per copy
Catalog No. 67312

Estimating for Contractors
How to Make Estimates that Win Jobs
By Paul J. Cook
$35.95 per copy
Catalog No. 67160

Successful Estimating Methods:
From Concept to Bid
By John D. Bledsoe, PhD, PE
$32.48 per copy
Catalog No. 67287

Total Productive Facilities Management
By Richard W. Sievert, Jr.
$29.98 per copy
Over 270 pages, illustrated, Hardcover
Catalog No. 67321

Means Productivity Standards for Construction
Expanded Edition *(Formerly Man-Hour Standards)*
$49.98 per copy
Over 800 pages, Hardcover
Catalog No. 67236A

Means Plumbing Estimating Methods, 3rd Edition
By Joseph Galeno and Sheldon Greene
Now $29.98 per copy
Catalog No. 67283B

Reference Books

Preventive Maintenance Guidelines for School Facilities

By John C. Maciha

A complete PM program for K-12 schools that ensures sustained security, safety, property integrity, user satisfaction, and reasonable ongoing expenditures.

Includes schedules for weekly, monthly, semiannual, and annual maintenance with hard copy and electronic forms.

$149.95 per copy
Over 225 pages, Hardcover
Catalog No. 67326

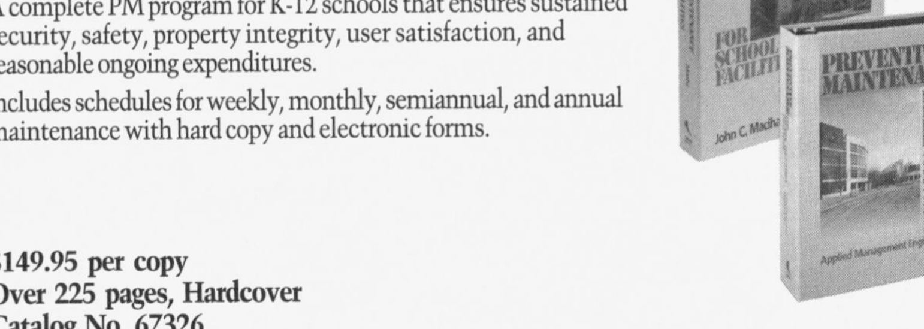

Preventive Maintenance for Higher Education Facilities

By Applied Management Engineering, Inc.

An easy-to-use system to help facilities professionals establish the value of PM, and to develop and budget for an appropriate PM program for their college or university. Features interactive campus building models typical of those found in different-sized higher education facilities, and PM checklists linked to each piece of equipment or system in hard copy and electronic format.

$149.95 per copy
150 pages, Hardcover
Catalog No. 67337

Historic Preservation: Project Planning & Estimating

By Swanke Hayden Connell Architects

Expert guidance on managing historic restoration, rehabilitation, and preservation building projects and determining and controlling their costs. Includes:

- How to determine whether a structure qualifies as historic
- Where to obtain funding and other assistance
- How to evaluate and repair more than 75 historic building materials

$99.95 per copy
Over 675 pages, Hardcover
Catalog No. 67323

Means Illustrated Construction Dictionary, 3rd Edition

Long regarded as the Industry's finest, the Means *Illustrated Construction Dictionary* is now even better.
With the addition of over 1,000 new terms and hundreds of new illustrations, it is the clear choice for the most comprehensive and current information. **The companion CD-ROM that comes with this new edition adds many extra features: larger graphics, expanded definitions, and links to both CSI MasterFormat numbers and product information.**

$99.95 per copy
Over 790 pages, Illustrated, Hardcover
Catalog No. 67292A

Designing & Building with the IBC, 2nd Edition

By Rolf Jensen & Associates, Inc.

This updated comprehensive guide helps building professionals make the transition to the 2003 *International Building Code®*. Includes a side-by-side code comparison of the IBC 2003 to the IBC 2000 and the three primary model codes, a quick-find index, and professional code commentary. With illustrations, abbreviations key, and an extensive Resource section.

$99.95 per copy
Over 875 pages
Catalog No. 67328A

Residential & Light Commercial Construction Standards, 2nd Edition

By RSMeans and Contributing Authors

New, updated second edition of this unique collection of industry standards that define quality construction. For contractors, subcontractors, owners, developers, architects, engineers, attorneys, and insurance personnel, this book provides authoritative requirements and recommendations compiled from the nation's leading professional associations, industry publications, and building code organizations.

$59.95 per copy
600 pages, illustrated, Softcover
Catalog No. 67322A

Means Estimating Handbook, 2nd Edition

By RSMeans

Updated new Second Edition answers virtually any estimating technical question - all organized by CSI MasterFormat. This comprehensive reference covers the full spectrum of technical data required to estimate construction costs. The book includes information on sizing, productivity, equipment requirements, code-mandated specifications, design standards and engineering factors.

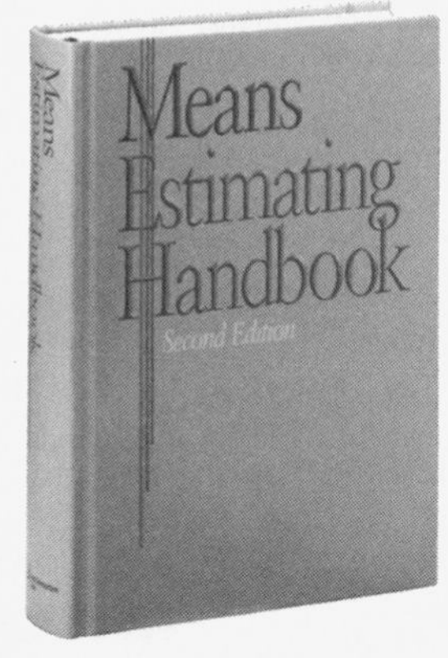

$99.95 per copy
Over 900 pages, Hardcover
Catalog No. 67276A

For more information visit Means Web Site at www.rsmeans.com

Seminars

Means CostWorks Training

This one-day seminar course has been designed with the intention of assisting both new and existing users to become more familiar with CostWorks program. The class is broken into two unique sections: (1) A one-half day presentation on the function of each icon; and each student will be shown how to use the software to develop a cost estimate. (2) Hands-on estimating exercises that will ensure that each student thoroughly understands how to use CostWorks. You must bring your own laptop computer to this course.

CostWorks Benefits/Features:
- Estimate in your own spreadsheet format
- Power of Means National Database
- Database automatically regionalized
- Save time with keyword searches
- Save time by establishing common estimate items in "Bookmark" files
- Customize your spreadsheet template
- Hot key to Product Manufacturers' listings and specs
- Merge capability for networking environments
- View crews and assembly components
- AutoSave capability
- Enhanced sorting capability

Unit Price Estimating

This interactive two-day seminar teaches attendees how to interpret project information and process it into final, detailed estimates with the greatest accuracy level.

The single most important credential an estimator can take to the job is the ability to visualize construction in the mind's eye, and thereby estimate accurately.

Some Of What You'll Learn:
- Interpreting the design in terms of cost
- The most detailed, time tested methodology for accurate "pricing"
- Key cost drivers—material, labor, equipment, staging and subcontracts
- Understanding direct and indirect costs for accurate job cost accounting and change order management

Who Should Attend: Corporate and government estimators and purchasers, architects, engineers...and others needing to produce accurate project estimates.

Square Foot and Assemblies Cost Estimating

This two-day course teaches attendees how to quickly deliver accurate square foot estimates using limited budget and design information.

Some Of What You'll Learn:
- How square foot costing gets the estimate done faster
- Taking advantage of a "systems" or "assemblies" format
- The Means "building assemblies/square foot cost approach"
- How to create a very reliable preliminary and systems estimate using bare-bones design information

Who Should Attend: Facilities managers, facilities engineers, estimators, planners, developers, construction finance professionals...and others needing to make quick, accurate construction cost estimates at commercial, government, educational and medical facilities.

Repair and Remodeling Estimating

This two-day seminar emphasizes all the underlying considerations unique to repair/remodeling estimating and presents the correct methods for generating accurate, reliable R&R project costs using the unit price and assemblies methods.

Some Of What You'll Learn:
- Estimating considerations—like labor-hours, building code compliance, working within existing structures, purchasing materials in smaller quantities, unforeseen deficiencies
- Identifies problems and provides solutions to estimating building alterations
- Rules for factoring in minimum labor costs, accurate productivity estimates and allowances for project contingencies
- R&R estimating examples are calculated using unite prices and assemblies data

Who Should Attend: Facilities managers, plant engineers, architects, contractors, estimators, builders...and others who are concerned with the proper preparation and/or evaluation of repair and remodeling estimates.

Mechanical and Electrical Estimating

This two-day course teaches attendees how to prepare more accurate and complete mechanical/electrical estimates, avoiding the pitfalls of omission and double-counting, while understanding the composition and rationale within the Means Mechanical/Electrical database.

Some Of What You'll Learn:
- The unique way mechanical and electrical systems are interrelated
- M&E estimates, conceptual, planning, budgeting and bidding stages
- Order of magnitude, square foot, assemblies and unit price estimating
- Comparative cost analysis of equipment and design alternatives

Who Should Attend: Architects, engineers, facilities managers, mechanical and electrical contractors...and others needing a highly reliable method for developing, understanding and evaluating mechanical and electrical contracts.

Plan Reading and Material Takeoff

This two-day program teaches attendees to read and understand construction documents and to use them in the preparation of material takeoffs.

Some of What You'll Learn:
- Skills necessary to read and understand typical contract documents—blueprints and specifications
- Details and symbols used by architects and engineers
- Construction specifications' importance in conjunction with blueprints
- Accurate takeoff of construction materials and industry-accepted takeoff methods

Who Should Attend: Facilities managers, construction supervisors, office managers...and other responsible for the execution and administration of a construction project including government, medical, commercial, educational or retail facilities.

Facilities Maintenance and Repair Estimating

This two-day course teaches attendees how to plan, budget, and estimate the cost of ongoing and preventive maintenance and repair for existing buildings and grounds.

Some Of What You'll Learn:
- The most financially favorable maintenance, repair and replacement scheduling and estimating
- Auditing and value engineering facilities
- Preventive planning and facilities upgrading
- Determining both in-house and contract-out service costs; annual, asset-protecting M&R plan

Who Should Attend: Facility managers, maintenance supervisors, buildings and grounds superintendents, plant managers, planners, estimators...and others involved in facilities planning and budgeting.

Scheduling and Project Management

This two-day course teaches attendees the most current and proven scheduling and management techniques needed to bring projects in on time and on budget.

Some Of What You'll Learn:
- Crucial phases of planning and scheduling
- How to establish project priorities, develop realistic schedules and management techniques
- Critical Path and Precedence Methods
- Special emphasis on cost control

Who Should Attend: Construction project managers, supervisors, engineers, estimators, contractors...and others who want to improve their project planning, scheduling and management skills.

Advanced Project Management

This two-day seminar will teach you how to effectively manage and control the entire design-build process and allow you to take home tangible skills that will be immediately applicable on existing projects.

Some Of What You'll Learn:
- Value engineering, bonding, fast-tracking and bid package creation
- How estimates and schedules can be integrated to provide advanced project management tools
- Cost engineering, quality control, productivity measurement and improvement
- Front loading a project and predicting its cash flow

Who Should Attend: Owners, project managers, architectural and engineering managers, construction managers, contractors...and anyone else who is responsible for the timely design and completion of construction projects.

For more information visit Means Web Site at www.rsmeans.com

Seminars

2005 Means Seminar Schedule

Location	Dates
Las Vegas, NV	March
Washington, DC	April
Phoenix, AZ	April
Denver, CO	May
San Francisco, CA	June
Philadelphia, PA	June
Washington, DC	September
Dallas, TX	September
Las Vegas, NV	October
Orlando, FL	November
Atlantic City, NJ	November
San Diego, CA	December

Note: Call for exact dates and details.

Registration Information

Register Early... Save up to $100! Register 30 days before the start date of a seminar and save $100 off your total fee. *Note: This discount can be applied only once per order. It cannot be applied to team discount registrations or any other special offer.*

How to Register Register by phone today! Means toll-free number for making reservations is: **1-800-334-3509.**

Individual Seminar Registration Fee $895. Individual CostWorks Training Registration Fee $349. To register by mail, complete the registration form and return with your full fee to: Seminar Division, Reed Construction Data, RSMeans Seminars, 63 Smiths Lane, Kingston, MA 02364.

Federal Government Pricing All Federal Government employees save 25% off regular seminar price. Other promotional discounts cannot be combined with Federal Government discount.

Team Discount Program Two to four seminar registrations: Call for pricing.

Multiple Course Discounts When signing up for two or more courses, call for pricing.

Refund Policy Cancellations will be accepted up to ten days prior to the seminar start. There are no refunds for cancellations received later than ten working days prior to the first day of the seminar. A $150 processing fee will be applied for all cancellations. Written notice of cancellation is required . Substitutions can be made at anytime before the session starts. **No-shows are subject to the full seminar fee.**

AACE Approved Courses The RSMeans Construction Estimating and Management Seminars described and offered to you here have each been approved for 14 hours (1.4 recertification credits) of credit by the AACE International Certification Board toward meeting the continuing education requirements for re-certification as a Certified Cost Engineer/Certified Cost Consultant.

AIA Continuing Education We are registered with the AIA Continuing Education System (AIA/CES) and are committed to developing quality learning activities in accordance with the CES criteria. RSMeans seminars meet the AIA/CES criteria for Quality Level 2. AIA members will receive (14) learning units (LUs) for each two-day RSMeans Course.

NASBA CPE Sponsor Credits We are part of the National Registry of CPE Sponsors. Attendees may be eligable for (16) CPE credits.

Daily Course Schedule The first day of each seminar session begins at 8:30 A.M. and ends at 4:30 P.M. The second day is 8:00 A.M.–4:00 P.M. Participants are urged to bring a hand-held calculator since many actual problems will be worked out in each session.

Continental Breakfast Your registration includes the cost of a continental breakfast, a morning coffee break, and an afternoon break. These informal segments will allow you to discuss topics of mutual interest with other members of the seminar. (You are free to make your own lunch and dinner arrangements.)

Hotel/Transportation Arrangements RSMeans has arranged to hold a block of rooms at each hotel hosting a seminar. To take advantage of special group rates when making your reservation, be sure to mention that you are attending the Means Seminar. You are, of course, free to stay at the lodging place of your choice. **(Hotel reservations and transportation arrangements should be made directly by seminar attendees.)**

Important Class sizes are limited, so please register as soon as possible.

Note: *Pricing subject to change.*

Registration Form

Call 1-800-334-3509 to register or FAX 1-800-632-6732. Visit our Web site www.rsmeans.com

Please register the following people for the Means Construction Seminars as shown here. Full payment or deposit is enclosed, and we understand that we must make our own hotel reservations if overnight stays are necessary.

☐ Full payment of $ ______________________ enclosed.

☐ Bill me

Name of Registrant(s)
(To appear on certificate of completion)

____

____

____

P.O. #: ______________________________

GOVERNMENT AGENCIES MUST SUPPLY PURCHASE ORDER NUMBER

Firm Name ______________________________

Address ______________________________

City/State/Zip ______________________________

Telephone No. fax No. ______________________________

E-mail Address ______________________________

Charge our registration(s) to: ☐ MasterCard ☐ VISA ☐ American Express ☐ Discover

Account No. ______________________ Exp. Date __________

Cardholder's Signature ______________________________

Seminar Name City Dates

____

____

Please mail check to: Seminar Division, Reed Construction Data, RSMeans Seminars, 63 Smiths Lane, P.O.Box 800, Kingston, MA 02364 USA

Builder's Essentials: Estimating Building Costs for the Residential & Light Commercial Contractor

By Wayne J. DelPico

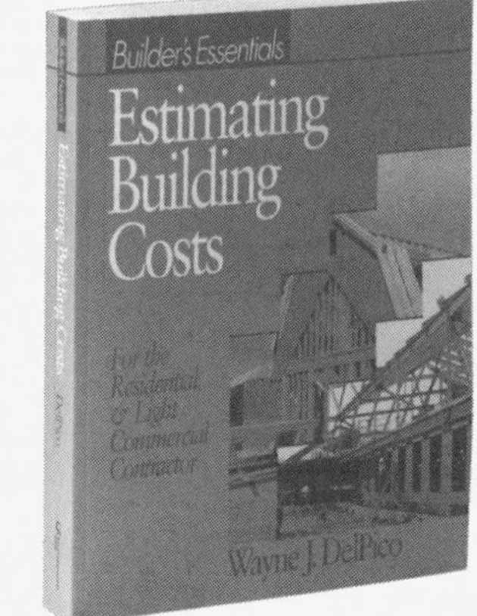

Step-by-step estimating methods for residential and light commercial contractors. Includes a detailed look at every construction specialty—explaining all the components, takeoff units, and labor needed for well-organized, complete estimates covers:

Correctly interpreting plans and specifications.
Developing accurate and complete labor and material costs.
Understanding direct and indirect overhead costs... and accounting for time-sensitive costs.
Using historical cost data to generate new project budgets.

Plus hard-to-find, professional guidance on what to consider so you can allocate the right amount for profit and contingencies.

$29.95 per copy
Over 400 pages, illustrated, softcover
Catalog No. 67343

Building & Renovating Schools

This all-inclusive guide covers every step of the school construction process—from initial planning, needs assessment, and design, right through moving into the new facility. A must-have resource for anyone concerned with new school construction or renovation, including architects and engineers, contractors and project managers, facility managers, school administrators and school board members, building committees, community leaders, and anyone else who wants to ensure that the project meets the schools' needs in a cost-effective, timely manner. With square foot cost models for elementary, middle, and high school facilities and real-life case studies of recently completed school projects.

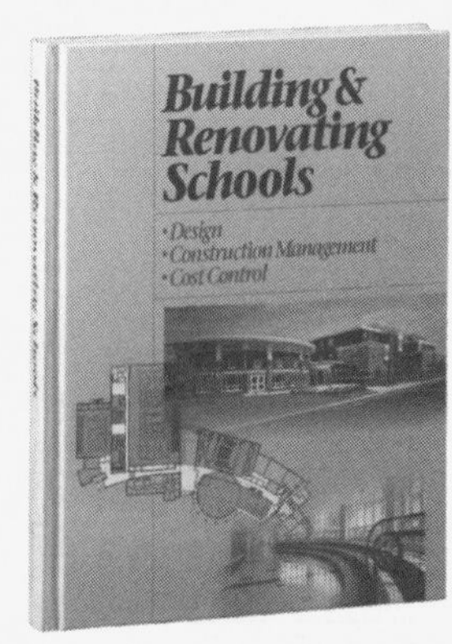

The contributors to this book – architects, construction project managers, contractors, and estimators who specialize in school construction – provide start-to-finish, expert guidance on the process.

$99.95 per copy
Over 425 pages
Catalog No. 67342

Means ADA Compliance Pricing Guide, New Second Edition

By Adaptive Environments and RSMeans

Completely updated and revised to the new 2004 Americans with Disabilities Act Accessible Guidelines, this book features more than 70 of the most commonly needed modifications for ADA compliance—their design requirements, suggestions, and final cost. Projects range from installing ramps and walkways, widening doorways and entryways, and installing and refitting elevators, to relocating light switches and signage, and remodeling bathrooms and kitchens. Also provided are:

Detailed cost estimates for budgeting modification projects, including estimates for each of 260 alternates.
An assembly estimate for every project, with detailed cost breakdown including materials, labor hours, and contractor's overhead.
3,000 Additional ADA compliance-related unit cost line items.
Costs that are easily adjusted to over 900 cities and towns.

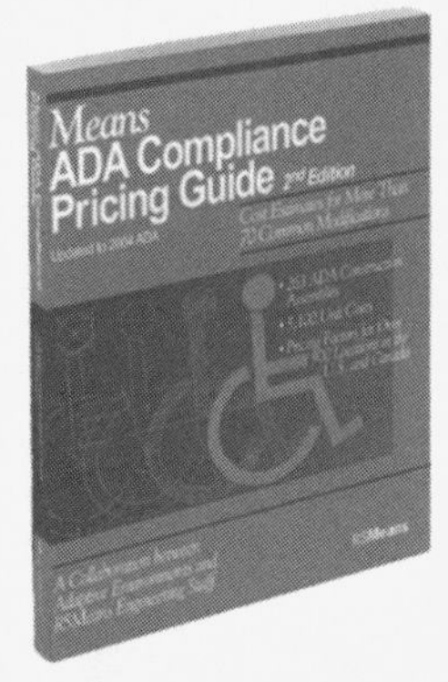

Over 350 pages
$79.99 per copy
Catalog No. 67310A

ORDER TOLL FREE 1-800-334-3509
OR FAX 1-800-632-6732.

2005 Order Form

Qty.	Book No.	COST ESTIMATING BOOKS	Unit Price	Total
	60065	Assemblies Cost Data 2005	$189.95	
	60015	Building Construction Cost Data 2005	115.95	
	61015	Building Const. Cost Data–Looseleaf Ed. 2005	144.95	
	63015	Building Const. Cost Data–Metric Version 2005	115.95	
	60225	Building Const. Cost Data–Western Ed. 2005	115.95	
	60115	Concrete & Masonry Cost Data 2005	105.95	
	60145	Construction Cost Indexes 2005	251.95	
	60145A	Construction Cost Index–January 2005	63.00	
	60145B	Construction Cost Index–April 2005	63.00	
	60145C	Construction Cost Index–July 2005	63.00	
	60145D	Construction Cost Index–October 2005	63.00	
	60345	Contr. Pricing Guide: Resid. R & R Costs 2005	39.95	
	60335	Contr. Pricing Guide: Resid. Detailed 2005	39.95	
	60325	Contr. Pricing Guide: Resid. Sq. Ft. 2005	39.95	
	64025	ECHOS Assemblies Cost Book 2005	189.95	
	64015	ECHOS Unit Cost Book 2005	126.95	
	54005	ECHOS (Combo set of both books)	251.95	
	60235	Electrical Change Order Cost Data 2005	115.95	
	60035	Electrical Cost Data 2005	115.95	
	60205	Facilities Construction Cost Data 2005	279.95	
	60305	Facilities Maintenance & Repair Cost Data 2005	252.95	
	60165	Heavy Construction Cost Data 2005	115.95	
	63165	Heavy Const. Cost Data–Metric Version 2005	115.95	
	60095	Interior Cost Data 2005	115.95	
	60125	Labor Rates for the Const. Industry 2005	253.95	
	60185	Light Commercial Cost Data 2005	99.95	
	60025	Mechanical Cost Data 2005	115.95	
	60155	Open Shop Building Const. Cost Data 2005	115.95	
	60215	Plumbing Cost Data 2005	115.95	
	60045	Repair and Remodeling Cost Data 2005	99.95	
	60175	Residential Cost Data 2005	99.95	
	60285	Site Work & Landscape Cost Data 2005	115.95	
	60055	Square Foot Costs 2005	126.95	
		REFERENCE BOOKS		
	67147A	ADA in Practice	59.98	
	67310A	ADA Compliance Pricing Guide, 2nd Ed.	79.95	
	67273	Basics for Builders: How to Survive and Prosper	34.95	
	67330	Bldrs Essentials: Adv. Framing Methods	24.95	
	67329	Bldrs Essentials: Best Bus. Practices for Bldrs	29.95	
	67298A	Bldrs Essentials: Framing/Carpentry 2nd Ed.	24.95	
	67298AS	Bldrs Essentials: Framing/Carpentry Spanish	24.95	
	67307	Bldrs Essentials: Plan Reading & Takeoff	35.95	
	67261A	Bldg. Prof. Guide to Contract Documents 3rd Ed.	64.95	
	67342	Building & Renovating Schools	99.95	
	67339	Building Security: Strategies & Costs	89.95	
	67312	Building Spec Homes Profitably	29.95	
	67146	Concrete Repair & Maintenance Illustrated	69.95	
	67314	Cost Planning & Est. for Facil. Maint.	89.95	
	67317A	Cyberplaces: The Internet Guide 2nd Ed.	59.95	
	67328A	Designing & Building with the IBC, 2nd Ed.	99.95	
	67230B	Electrical Estimating Methods 3rd Ed.	64.95	

Qty.	Book No.	REFERENCE BOOKS (Cont.)	Unit Price	Total
	64777A	Environmental Remediation Est. Methods 2nd Ed.	$ 99.95	
	67160	Estimating for Contractors	35.95	
	67276A	Estimating Handbook 2nd Ed.	99.95	
	67249	Facilities Maintenance Management	86.95	
	67246	Facilities Maintenance Standards	79.95	
	67318	Facilities Operations & Engineering Reference	109.95	
	67301	Facilities Planning & Relocation	89.95	
	67231	Forms for Building Const. Professional	47.48	
	67260	Fundamentals of the Construction Process	34.98	
	67323	Historic Preservation: Proj. Planning & Est.	99.95	
	67308E	Home Improvement Costs–Int. Projects 9th Ed.	29.95	
	67309E	Home Improvement Costs–Ext. Projects 9th Ed.	29.95	
	67324A	How to Est. w/Means Data & CostWorks 2nd Ed.	59.95	
	67304	How to Estimate with Metric Units	9.98	
	67306	HVAC: Design Criteria, Options, Select. 2nd Ed.	84.95	
	67281	HVAC Systems Evaluation	84.95	
	67282A	Illustrated Construction Dictionary, Condensed	59.95	
	67292A	Illustrated Construction Dictionary, w/CD-ROM	99.95	
	67295B	Landscape Estimating 4th Ed.	62.95	
	67341	Life Cycle Costing	99.95	
	67302	Managing Construction Purchasing	19.98	
	67294A	Mechanical Estimating 3rd Ed.	64.95	
	67245A	Planning and Managing Interior Projects 2nd Ed.	69.95	
	67283A	Plumbing Estimating Methods 2nd Ed.	59.95	
	67326	Preventive Maint. Guidelines for School Facil.	149.95	
	67337	Preventive Maint. for Higher Education Facilities	149.95	
	67236A	Productivity Standards for Constr.–3rd Ed.	49.98	
	67247B	Project Scheduling & Management for Constr. 3rd Ed.	64.95	
	67265B	Repair & Remodeling Estimating 4th Ed.	69.95	
	67322A	Resi. & Light Commercial Const. Stds. 2nd Ed.	59.95	
	67327	Spanish/English Construction Dictionary	22.95	
	67145A	Sq. Ft. & Assem. Estimating Methods 3rd Ed.	69.95	
	67287	Successful Estimating Methods	32.48	
	67313	Successful Interior Projects	24.98	
	67233	Superintending for Contractors	35.95	
	67321	Total Productive Facilities Management	29.98	
	67284	Understanding Building Automation Systems	29.98	
	67303A	Unit Price Estimating Methods 3rd Ed.	59.95	
	67319	Value Engineering: Practical Applications	79.95	
		MA residents add 5% state sales tax		
		Shipping & Handling**		
		Total (U.S. Funds)*		

Prices are subject to change and are for U.S. delivery only. *Canadian customers may call for current prices. **Shipping & handling charges: Add 7% of total order for check and credit card payments. Add 9% of total order for invoiced orders.

Send Order To: **ADDV-1001**

Name (Please Print) ____________________

Company ____________________

☐ **Company**
☐ **Home** Address ____________________

____

City/State/Zip ____________________

Phone # ____________ P.O. # ____________

(Must accompany all orders being billed)

Mail To: **RSMeans** P.O. Box 800, Kingston, MA 02364-0800